DNA
Damage
Recognition

DNA
Damage
Recognition

edited by

Wolfram Siede
University of North Texas Health Science Center
Fort Worth, Texas, U.S.A.

Yoke Wah Kow
Emory University School of Medicine
Atlanta, Georgia, U.S.A.

Paul W. Doetsch
Emory University School of Medicine
Atlanta, Georgia, U.S.A.

Taylor & Francis
Taylor & Francis Group
New York   London

Published in 2006 by
Taylor & Francis Group
270 Madison Avenue
New York, NY 10016

© 2006 by Taylor & Francis Group, LLC

No claim to original U.S. Government works
Printed in the United States of America on acid-free paper
10 9 8 7 6 5 4 3 2 1

International Standard Book Number-10: 0-8247-5961-3 (Hardcover)
International Standard Book Number-13: 978-0-8247-5961-2 (Hardcover)

Library of Congress Cataloging-in-Publication Data

Catalog record is available from the Library of Congress

Taylor & Francis Group
is the Academic Division of T&F Informa plc.

Visit the Taylor & Francis Web site at
http://www.taylorandfrancis.com

Preface

The topics of this book are the various molecular mechanisms that are involved in the process of DNA damage recognition as the initial step of DNA damage repair and of other related responses, such as damage tolerance and cell cycle checkpoint regulation. The authors were asked to provide review-type articles designed with the researcher in the field in mind. But sufficient introductory comments were requested so that a non-expert or an interested advanced student with some background knowledge can follow. While we did not have a narrow definition of damage recognition in mind, a complete description of cellular DNA repair mechanisms was certainly not our goal. However, looking at the scope of this impressive collection of in-depth reviews, it almost happened...

In the beginning of the book, certain theoretical aspects of damage recognition that are common themes throughout the book have been addressed. How can one imagine that recognition proteins find their rare targets? How can we envisage protein movement—by random diffusion or "patrolling" along DNA? Another equally important topic addresses protein cooperation that enhances recognition specificity.

There is clearly an emphasis on structural aspects of DNA damage recognition throughout the book. Wherever such information is available, it is explained in detail how protein/DNA damage contacts are being accomplished and which types of structural features or consequences of DNA damage are being probed by the recognition apparatus in order to permit a distinction from undamaged DNA. In this fashion, damage recognition is addressed within the major pathways of DNA repair, i.e., simple damage reversal, nucleotide excision repair, base excision repair, mismatch repair, recombinational repair, and DNA endjoining. We interpret the pathways of damage tolerance as a consequence of a combined recognition/accommodation process and thus we have also included chapters on translesion synthesis.

DNA recognition steps occur at several levels within a single repair pathway and this complexity has been considered. How are repair intermediates being recognized? How is a repair intermediate handed over to the next player? If there is a competition between different mechanisms: how is a pathway choice accomplished? While we do not emphasize downstream reactions, the strategy of originating a transmissible downstream signal will be addressed wherever appropriate, especially in the context of regulatory responses.

We were equally interested in putting specific aspects of DNA damage recognition in the cellular context. Certain chapters provide the necessary backdrop by giving up-to-date reviews of certain pathways. However, also chromosome structure,

DNA structure, and sequence context, which may affect DNA damage recognition positively or negatively, are being addressed. The interference with cellular processes such as replication and transcription is discussed for several examples since such interference as a consequence of DNA damage may influence, aide, or even initiate a recognition process.

All that is left is to thank our authors for their hard work and superb contributions. We would like to acknowledge the help of Stacy Harman Holloway. We are also indebted to Anita Lekhwani, Moraima Suarez, Joseph Stubenrauch, and their colleagues at Taylor & Francis Books for pursuing the idea and providing such excellent editorial support. We also thank the National Institute of Environmental Health Sciences for supporting a research collaboration amongst the three editors as well as two other authors of individual chapters that made this book possible ("*Cellular Responses to Genotoxic Stress*" Program Project ES11163).

Wolfram Siede
Yoke Wah Kow
Paul W. Doetsch

Contents

Contributors

Samir Acharya Department of Molecular Virology, Immunology and Medical Genetics, Ohio State University, Columbus, Ohio, U.S.A.

Karen H. Almeida Hillman Cancer Center, University of Pittsburgh Cancer Institute, Pittsburgh, Pennsylvania, U.S.A.

Jean-Christophe Amé Unité 9003 du CNRS, Ecole Supérieure de Biotechnologie de Strasbourg, Illkirch, France

Gali Arad Department of Biological Chemistry, Weizmann Institute of Science, Rehovot, Israel

Nikolaos A. A. Balatsos Institute of Molecular Biology and Genetics, B.S.R.C. Alexander Fleming, and Department of Biochemistry and Biotechnology, School of Health Sciences, University of Thessaly, Varkiza, Greece

Vladimir Beljanski Department of Biochemistry, Emory University School of Medicine, Atlanta, Georgia, U.S.A.

Ashok S. Bhagwat Department of Chemistry, Wayne State University, Detroit, Michigan, U.S.A.

Suse Broyde Department of Biology, New York University, New York, New York, U.S.A.

James M. Bugni Department of Biological Engineering, Massachusetts Institute of Technology, Cambridge, Massachusetts, U.S.A.

Bernard Connolly School of Cell and Molecular Biosciences, University of Newcastle, Newcastle Upon Tyne, U.K.

Deborah L. Croteau Laboratory of Molecular Genetics, National Institute of Environmental Health Sciences, National Institutes of Health, Research Triangle Park, North Carolina, U.S.A.

Françoise Dantzer   Unité 9003 du CNRS, Ecole Supérieure de Biotechnologie de Strasbourg, Illkirch, France

Gilbert de Murcia   Unité 9003 du CNRS, Ecole Supérieure de Biotechnologie de Strasbourg, Illkirch, France

Matthew J. DellaVecchia   Laboratory of Molecular Genetics, National Institute of Environmental Health Sciences, National Institutes of Health, Research Triangle Park, North Carolina, U.S.A.

Julie Della-Maria Goetz   Radiation Oncology Research Laboratory, Department of Radiation Oncology and Greenebaum Cancer Center, University of Maryland School of Medicine, Baltimore, Maryland, U.S.A.

M. L. Dodson   Sealy Center for Molecular Science, University of Texas Medical Branch, Galveston, Texas, U.S.A.

Paul W. Doetsch   Department of Biochemistry, Emory University School of Medicine, Atlanta, Georgia, U.S.A.

Zhiwan Dong   Radiation Oncology Research Laboratory, Department of Radiation Oncology and Greenebaum Cancer Center, University of Maryland School of Medicine, Baltimore, Maryland, U.S.A.

Hong Dou   Sealy Center for Molecular Science and Department of Human Biological Chemistry and Genetics, University of Texas Medical Branch, Galveston, Texas, U.S.A.

Richard Fishel   Department of Molecular Virology, Immunology and Medical Genetics, Ohio State University, Columbus, Ohio, U.S.A.

Marco Foiani   Istituto FIRC di Oncologia Molecolare, Dipartimento di Scienze Biomolecolari e Biotecnologie, Università di Milano, Milan, Italy

Nicholas E. Geacintov   Department of Chemistry, New York University, New York, New York, U.S.A.

Moshe Goldsmith   Department of Biological Chemistry, Weizmann Institute of Science, Rehovot, Israel

Gérard Gradwohl   Unité INSERM U682, Strasbourg, France

Tapas K. Hazra   Sealy Center for Molecular Science and Department of Human Biological Chemistry and Genetics, University of Texas Medical Branch, Galveston, Texas, U.S.A.

Ayal Hendel   Department of Biological Chemistry, Weizmann Institute of Science, Rehovot, Israel

Eric A. Hendrickson   University of Minnesota Medical School, Minneapolis, Minnesota, U.S.A.

Karl-Peter Hopfner   Gene Center, University of Munich, Munich, Germany

Joy L. Huffman   Department of Molecular Biology—MB4, Skaggs Institute for Chemical Biology and The Scripps Research Institute, La Jolla, California, U.S.A.

Lior Izhar   Department of Biological Chemistry, Weizmann Institute of Science, Rehovot, Israel

Tadahide Izumi   Sealy Center for Molecular Science and Department of Human Biological Chemistry and Genetics, University of Texas Medical Branch, Galveston, Texas, U.S.A.

Roland Kanaar   Departments of Radiation Oncology and Cell Biology and Genetics, Erasmus Medical Center, Rotterdam, The Netherlands

Caroline Kisker   Department of Pharmacological Sciences, Center for Structural Biology, State University of New York at Stony Brook, Stony Brook, New York, U.S.A.

Yoke Wah Kow   Department of Radiation Oncology, Division of Cancer Biology, Emory University School of Medicine, Atlanta, Georgia, U.S.A.

John B. Leppard   Department of Molecular Medicine, Institute of Biotechnology, The University of Texas Health Science Center, San Antonio, Texas, U.S.A.

Michael R. Lieber   USC Norris Comprehensive Cancer Center, University of Southern California Keck School of Medicine, Los Angeles, California, U.S.A.

Stephen J. Lippard   Department of Chemistry, Massachusetts Institute of Technology, Cambridge, Massachusetts, U.S.A.

Zvi Livneh   Department of Biological Chemistry, Weizmann Institute of Science, Rehovot, Israel

R. Stephen Lloyd   Center for Research on Occupational and Environmental Toxicology, Oregon Health and Science University, Portland, Oregon, U.S.A.

Maria Pia Longhese   Dipartimento di Biotecnologie e Bioscienze, Università di Milano-Bicocca, Milan, Italy

David F. Lowry   Macromolecular Structure and Dynamics, Pacific Northwest National Laboratory, Richland, Washington, U.S.A.

Yunmei Ma   USC Norris Comprehensive Cancer Center, University of Southern California Keck School of Medicine, Los Angeles, California, U.S.A.

Ayelet Maor-Shoshani   Department of Biological Chemistry, Weizmann Institute of Science, Rehovot, Israel

A. K. McCullough   Center for Research on Occupational and Environmental Toxicology, Oregon Health and Science University, Portland, Oregon, U.S.A.

Isabel Mellon   Department of Pathology and Laboratory Medicine, Markey Cancer Center, Graduate Center for Toxicology, University of Kentucky, Lexington, Kentucky, U.S.A.

Sankar Mitra   Sealy Center for Molecular Science and Department of Human Biological Chemistry and Genetics, University of Texas Medical Branch, Galveston, Texas, U.S.A.

Josiane Ménissier-de Murcia   Unité 9003 du CNRS, Ecole Supérieure de Biotechnologie de Strasbourg, Illkirch, France

Teresa A. Motycka   Radiation Oncology Research Laboratory, Department of Radiation Oncology and Greenebaum Cancer Center, University of Maryland School of Medicine, Baltimore, Maryland, U.S.A.

Hanspeter Naegeli   Institute of Pharmacology and Toxicology, University of Zürich-Tierspital, Zürich, Switzerland

Timothy R. O'Connor   Biology Department, Beckman Research Institute, City of Hope National Medical Center, Duarte, California, U.S.A.

Uta-Maria Ohndorf   Department of Chemistry, Massachusetts Institute of Technology, Cambridge, Massachusetts, U.S.A.

Roman Osman   Department of Physiology and Biophysics, Mount Sinai School of Medicine, New York, New York, U.S.A.

Ulrich Pannicke   Department of Transfusion Medicine, Institute for Clinical Transfusion Medicine and Immunogenetics, University of Ulm, Ulm, Germany

Dinshaw J. Patel   Cellular Biochemistry and Biophysics Program, Memorial Sloan-Kettering Cancer Center, New York, New York, U.S.A.

Emmanuelle Pion   Unité 7034 du CNRS, Laboratoire de Pharmacologie et Physico-chimie des Interactions cellulaires et Moléculaires, Faculté de Pharmacie, Université Louis Pasteur, Illkirch, France

Emmy P. Rogakou   Institute of Molecular Biology and Genetics, B.S.R.C. Alexander Fleming, Varkiza, Greece

J. B. Alexander Ross   Department of Chemistry, University of Montana, Missoula, Montana, U.S.A.

Leona D. Samson   Department of Biological Engineering, Massachusetts Institute of Technology, Cambridge, Massachusetts, U.S.A.

Gwendolyn B. Sancar   Department of Biochemistry and Biophysics, School of Medicine, University of North Carolina at Chapel Hill, Chapel Hill, North Carolina, U.S.A.

Valérie Schreiber, Unité 9003 du CNRS, Ecole Supérieure de Biotechnologie de Strasbourg, Illkirch, France

Klaus Schwarz Department of Transfusion Medicine, Institute for Clinical Transfusion Medicine and Immunogenetics, University of Ulm, Ulm, Germany

Eleanore Seibert Department of Physiology and Biophysics, Mount Sinai School of Medicine, New York, New York, U.S.A.

Gerald S. Shadel Department of Pathology, Yale University School of Medicine, New Haven, Connecticut, U.S.A.

Wolfram Siede Department of Cell Biology and Genetics, University of North Texas Health Science Center, Fort Worth, Texas, U.S.A.

Milan Skorvaga Department of Molecular Genetics, Cancer Research Institute, Slovak Academy of Sciences, Bratislava, Slovakia

Robert W. Sobol Hillman Cancer Center, University of Pittsburgh Cancer Institute, Pittsburgh, Pennsylvania, U.S.A.

Binwei Song Department of Biochemistry, Emory University School of Medicine, Atlanta, Georgia, U.S.A.

Wei Song Radiation Oncology Research Laboratory, Department of Radiation Oncology and Greenebaum Cancer Center, University of Maryland School of Medicine, Baltimore, Maryland, U.S.A.

Catherine Spenlehauer Unité 9003 du CNRS, Ecole Supérieure de Biotechnologie de Strasbourg, Illkirch, France

Ottar Sundheim Department of Molecular Biology—MB4, Skaggs Institute for Chemical Biology and The Scripps Research Institute, La Jolla, California, U.S.A.

Mark D. Sutton Department of Biochemistry, School of Medicine and Biomedical Sciences, University at Buffalo, State University of New York, Buffalo, New York, U.S.A.

John A. Tainer Department of Molecular Biology—MB4, Skaggs Institute for Chemical Biology and The Scripps Research Institute, La Jolla, California, U.S.A.

Moon-shong Tang Departments of Environmental Medicine, Pathology, and Medicine, New York University School of Medicine, Tuxedo, New York, U.S.A.

Laurence Tartier Unité 9003 du CNRS, Ecole Supérieure de Biotechnologie de Strasbourg, Illkirch, France

John-Stephen Taylor Department of Chemistry, Washington University, St. Louis, Missouri, U.S.A.

Fritz Thoma   Institut für Zellbiologie, ETH-Hönggerberg, Zürich, Switzerland

Alan E. Tomkinson   Radiation Oncology Research Laboratory, Department of Radiation Oncology and Greenebaum Cancer Center, University of Maryland School of Medicine, Baltimore, Maryland, U.S.A.

Hui-Min Tseng   Radiation Oncology Research Laboratory, Department of Radiation Oncology and Greenebaum Cancer Center, University of Maryland School of Medicine, Baltimore, Maryland, U.S.A.

Helle D. Ulrich   Cancer Research UK, Clare Hall Laboratories, South Mimms, U.K.

Bennett Van Houten   Laboratory of Molecular Genetics, National Institute of Environmental Health Sciences, National Institutes of Health, Research Triangle Park, North Carolina, U.S.A.

Lieneke van Veelen   Department of Radiation Oncology, Erasmus Medical Center-Daniel, Rotterdam, The Netherlands

Sangeetha Vijayakumar   Radiation Oncology Research Laboratory, Department of Radiation Oncology and Greenebaum Cancer Center, University of Maryland School of Medicine, Baltimore, Maryland, U.S.A.

Zhigang Wang   Graduate Center for Toxicology, University of Kentucky, Lexington, Kentucky, U.S.A.

Bernard Weiss   Department of Pathology, Emory University School of Medicine, Atlanta, Georgia, U.S.A.

Joanna Wesoly   Department of Cell Biology and Genetics, Erasmus Medical Center, Rotterdam, The Netherlands

Lee R. Wiederhold   Sealy Center for Molecular Science and Department of Human Biological Chemistry and Genetics, University of Texas Medical Branch, Galveston, Texas, U.S.A.

David M. Wilson III   Laboratory of Molecular Gerontology, GRC, National Institute on Aging, Baltimore, Maryland, U.S.A.

Kefei Yu   USC Norris Comprehensive Cancer Center, University of Southern California Keck School of Medicine, Los Angeles, California, U.S.A.

Part I

Mechanisms of Damage Recognition: Theoretical Considerations

This section deals with the mechanism(s) by which DNA glycosylases and AP endonucleases search and recognize DNA damage. DNA base lesions and AP sites in genomic DNA are generated spontaneously in cells via oxidative respiration and also as a result of exposure to exogenous agents, such as ultraviolet, ionizing radiation, and various oxidation, deamination, and alkylation agents. Cells are equipped with a multitude of DNA glycosylases and AP endonucleases that recognize various base lesions and AP sites, respectively. These are key enzymes for the initiation of a ubiquitous base excision repair process by which DNA damage is repaired.

The recognition of DNA damage by repair enzymes is a dynamic process that involves the initial non-targeted binding to DNA repair enzymes, then the relocation of these proteins to the damaged site. Chapter 1 treats this dynamic process from a thermodynamic perspective, providing important insights into the roles of protein, DNA, and lesion structure in this process.

Chapter 2 provides an overview of this process from an experimental standpoint, in particular the mechanism by which DNA glycosylases search for the damaged base. Chapter 3 provides an overview of the complexity of the base excision repair pathway and underscore the importance of accessory proteins in enhancing the rate and efficiency of DNA glycosylases and AP endonucleases. It also addresses their role in the formation of base excision protein complexes for increasing the specificity and efficiency of the overall base excision repair pathway.

1
Dynamics of DNA Damage Recognition

Eleanore Seibert and Roman Osman
Department of Physiology and Biophysics, Mount Sinai School of Medicine, New York, New York, U.S.A.

J. B. Alexander Ross
Department of Chemistry, University of Montana, Missoula, Montana, U.S.A.

1. INTRODUCTION

Maintenance of the native DNA sequence and structure is essential for normal function and ultimately to survival. DNA damage can result from numerous external agents and from normal cellular processes. For example, ionizing radiation can induce single- and double-strand breaks, exposure to UV light can produce pyrimidine dimers, spontaneous depurination or enzyme-mediated base removal can produce abasic sites, and cytosine deamination can yield a non-native DNA base—uracil. These lesions produce structural and dynamic changes in DNA, which impair its function. In fact, failure to correctly repair DNA damage can result in mutations, cancer, and death. To minimize the detrimental effects of DNA damage, evolution has provided cells with DNA repair systems, each eliminating a different kind of DNA damage and restoring its normal function.

Critical features of any DNA repair pathway are the ability to specifically recognize the damage and efficiently remove the lesion. The complexity of damage recognition is highlighted by the fact that many of the damaged bases do not differ substantially from their native forms, e.g., the potentially mutagenic uracil differs from thymine only by the absence of a methyl at C_5. In some instances, damage consists of a mismatch between two normal DNA bases, generating only a small perturbation in DNA structure and properties. The efficient removal of the damage is also an important contributor to the fidelity of the repair process. As in other enzymatic systems, repair enzymes optimally stabilize transition states of the reactions they catalyze, while taking advantage of the special properties of damaged DNA. Thus, the fidelity of DNA repair that depends on specific recognition of the damage and its efficient removal, is intimately linked with the structural, energetic, and dynamic properties of damaged DNA.

In an effort to understand the mechanisms involved in specific recognition of DNA damage by repair enzymes, this chapter focuses on the base excision repair (BER) pathway. This highly conserved and coordinated pathway is responsible for initiating repair of chemically altered bases, such as alkylated adenine, oxidized bases, and deoxyuridine (1,2). The BER proceeds via a multi-enzyme pathway, which can be divided into damage-specific and damage-general phases (3). In the damage-specific phase, a glycosylase recognizes the particular damaged base and cleaves the $N-C_1'$ bond between the damaged base and deoxyribose, producing an apurinic/apyrimidinic (AP) site and the free damaged base. Subsequently, in the damage-general phase the 5'-phosphodiester bond is hydrolyzed by an apurinic/apyrimidinic endonuclease, leaving a free 3'-hydroxyl group on the ribose and a 5'-phosphodeoxyribose. Next, polymerase β, acting as a 5'-deoxyribose phosphodiesterase, cleaves the 3'-phosphodiester bond subsequently reverting to a polymerase and inserting the appropriate nucleotide complementary to the intact strand. Finally, DNA ligase joins the 3'-hydroxyl and 5'-phosphate in an energy-dependent manner (3).

At least seven members of the damage-specific glycosylase family have been reported to date (2). Crystal structures of glycosylase–DNA complexes have been determined for uracil–DNA glycosylase (UDG) (4–6), alkyladenine glycosylase (AlkA) (7–9), 8-oxoguanine DNA glycosylase (hOGG1) (10), and formamidopyrimidine DNA glycosylase (Fpg) (11). So far as we know, all of the crystal structures reveal that the target base for the glycosylase activity is everted into an extrahelical position and inserted into the "specificity" pocket in the enzyme. Furthermore, the DNA in the complexes is distorted in the vicinity of the damaged base. Interestingly, DNA distortion as reflected in local DNA flexibility is sequence dependent (12–14). Finally, several of these enzymes exhibit different efficiencies for their substrate depending upon the sequence context in which the damage occurs (15–18). The origin of the sequence-dependent efficiency is presently not clear (15–22). Taken together, these factors lead to an interesting hypothesis regarding the mechanism of specific damage recognition by BER glycosylases. Flexible DNA sequences are better substrates because less energy is required to distort them into the conformation of the productive enzyme–substrate complex. Thus, local DNA dynamics plays a crucial role in specific recognition by DNA glycosylases.

UDG, which removes uracil from DNA, serves as an excellent model enzyme to test this hypothesis because its structure in complex with damaged DNA has been determined (4–6,23,24) and its catalytic mechanism has been largely elucidated (25–27). We describe here the results of a combined experimental and theoretical approach to explore the hypothesis that sequences requiring less energy for distortion will be better substrates for UDG. We emphasize DNA bending and base-pair opening because they are the essential molecular elements that contribute to the formation of a productive enzyme–DNA complex.

We first describe combined experimental and theoretical studies of the sequence-dependent nature of UDG activity towards substrates containing A•U mismatches. We extend this approach to the potentially mutagenic G•U mismatches, adding to the investigation mismatches with 6-methylisoxanthopterin (6MI), a fluorescent analog of guanine. We present results from molecular dynamics simulations that demonstrate the sequence-dependent spontaneous base-pair opening as a sensitive model of local dynamics of G•U mismatches.

2. ROLE OF DNA FLEXIBILITY IN SEQUENCE-DEPENDENT ACTIVITY OF UDG

Uracil arises in DNA from misincorporation during DNA replication (A•U) or more frequently as a result of the spontaneous hydrolytic deamination of cytosine

(G•U). While the A•U base pair is not directly mutagenic, some transcription factors have significantly reduced binding affinity for A•U compared to A•T base pairs (28). Thus, although the eukaryotic replicative polymerases are high fidelity enzymes, with misincorporation occurring at a rate of approximately 10^{-7} per base pair per generation (29), this non-negligible finite probability leads to undesired biological consequences. In humans, spontaneous deamination of cytosine occurs at a rate of approximately 100–500 events per cell per day (30). Failure to repair the G•U base pair leads to a G•C transition mutation to A•T during DNA replication. Consequently, the presence of uracil in DNA is detrimental, and its removal by repair enzymes is essential to normal cell function. This repair function is initiated by UDG, an enzyme of the BER pathway (3,31), which is found in all prokaryotes and eukaryotes. The UDG recognizes and removes uracil in U•A and U•G base pairs from both double- and single-stranded DNA (32,33).

The UDG discriminates uracil from the closely related thymine through an active-site tyrosine (Tyr 147 in human UDG), which acts as a steric exclusion gate (34–36). Hydrogen bonding between the uracil and an active-site asparagine (Asn 204 in human UDG) selects against cytosine (4). Mutation of Tyr 147 to Ala, Cys or Ser, or of Asn 204 to Asp, produces a mutant UDG that catalyzes the removal of either thymine or cytosine, respectively. However, these mutant enzymes are still 1–2 orders of magnitude more selective for uracil than either cytosine or thymine (37), suggesting that other factors contribute to the specificity of UDG.

Crystal structures have shown that DNA in complex with UDG is bent by about 40° and that the uracil is extrahelical. DNA bending is thought to occur via a "Ser-Pro pinch" mechanism, in which the DNA backbone is compressed by serine–proline-rich loops of UDG (5). Parikh and coworkers suggest that damaged sites along the DNA may allow greater compression than undamaged sites and that the combination of backbone compression and insertion of the Leu 272 side chain into the minor groove facilitates flipping of the uracil through the major groove. Base flipping requires the loss of favorable stacking interactions between adjacent bases and the breakage of hydrogen bonds with the complementary base, which can amount to approximately 3–7 kcal/mol depending on the particular dinucleotide step (38,39). Most importantly, base flipping in DNA is coupled to bending (40,41) suggesting that flexibility should play an important role in UDG activity. Since DNA flexibility is a sequence-dependent property (14), UDG activity should depend on the nucleotide sequence surrounding the uracil. Slupphaug et al. (17) have reported sequence-dependent variations of up to 20-fold in UDG efficiency for different DNA sequences for human (16–18). Attempts to correlate changes in UDG efficiency with DNA stability did not lead to a clear relationship. A recent kinetic analysis of HSV1 UDG activity indicated that the sequence dependence results from differences in binding energy rather than catalysis (42).

It is reasonable to propose at this stage that flexible sequences should be better substrates of DNA repair enzymes than rigid ones since distorting a flexible DNA would require less energy. This hypothesis was tested by Seibert et al. (43) on two sequences with high and low UDG efficiency (17). We used a combined theoretical/experimental approach to determine the effect of sequence on local DNA flexibility and on the enzymatic constants of uracil removal by UDG. Using molecular dynamics simulations we have developed a sequence-dependent model of DNA flexibility. The flexibility model was corroborated by the analysis of lifetime data and quantum yields of the fluorescence spectroscopy of the DNA sequences with an adenine analog, 2-aminopurine (2AP). Finally, the role of DNA flexibility in enzyme–DNA interaction was linked to the flexibility model through kinetic assays on each sequence.

The sequences that were used for the experimental part were taken from the work of Slupphaug et al. (17) and are shown below. They are named AUA and GUG to represent the surrounding bases in the sequence. **P** represents 2-aminopurine, which does not perturb the DNA structure and properties (44). For the computational simulations the sequence has been shortened by 4 base pairs on each side to be able to conduct nearly converged simulations.

$$5'\text{-CCGGAAGCAUAAAGTGCGC-}3'$$
$$3'\text{-GGCCTTCGTPTTTCACGCG-}5' \quad \text{AUA}$$

$$5'\text{-CCGGCAGGGUGGTTTGCGC-}3' \quad \text{GUG}$$
$$3'\text{-GGCCGTCCCPCCAAACGCG-}5'$$

UDG efficiency on the AUA sequence was approximately 20-fold greater than on the GUG. Since the crystal structures of UDG in complex with DNA show that the DNA is bent in the region of the uracil, we focused on this region for the analysis of the MD simulations. Bending of the DNA at the uracil towards the grooves and towards the backbone was characterized by axis curvature parameters A_{tip} (bending towards the grooves) and A_{inc} (bending towards the backbone). The probability distributions from the simulations were converted to a two-dimensional potential of mean force described by the following relationship:

$$\Delta\omega(x_0y_0 \rightarrow x_1y_1) = -RT\ln(N_{x1y1}/N_{x0y0}) + \omega_0$$

$\Delta\omega$ is the work (free energy) required to rearrange the system from its minimum energy position (x_0y_0) to any other position (x_1y_1) and N is the value of the probability distribution function at a given point. If the distributions are normalized to the most likely positions, then ω_0 can be set to 0. To quantify flexibility we represent it by the effective force constant, which can be obtained as the second derivative of the energy with respect to the two variables. The force constants for bending towards the grooves or towards the backbone are smaller for the AUA than for the GUG context by 13–17%. Hence, the AUA sequence is locally more flexible, or more easily bent, in the region of the uracil than the GUG sequence. Consequently, if the necessary condition for the formation of a productive enzyme–substrate complex of UDG with the damaged DNA is DNA bending, then U in a context of a flexible sequence will be excised more efficiently.

To test the conclusions from the computational simulations and to generalize them, we have conducted spectroscopic measurements of the *local* flexibility in the region of the uracil using the base analog 2AP (45,46). The observed fluorescence lifetime of 2AP is reduced due to collisions with a quencher molecule during the excited-state lifetime. We have demonstrated previously that the average fluorescence lifetime of 2AP is shorter in flexible sequences as compared to rigid sequences (45,46). Furthermore, the greater reduction in quantum yield of 2AP fluorescence in DNA than in the lifetime indicates the presence of static quenching. Reduction in lifetime indicates flexibility, whereas changes in quantum yield suggest the presence of stacking interactions between 2AP and the adjacent bases.

Time-resolved fluorescence decay curves of 2AP differ dramatically between the two sequences. Approximately 93% of the fluorescence intensity of 2AP in the

AUA context is accounted for by two short lifetimes below 1 ns, indicating that 2AP in this sequence context undergoes efficient collisional quenching. In the GUG context, the individual lifetimes are longer than in the AUA context, and the contribution of the short lifetime component to the total fluorescence intensity is only 5%. The average lifetime, τ_{num}, is 0.32 ns in the AUA sequence and 2.48 ns in the GUG sequence, reaffirming that the frequency of collision of 2AP with its neighbors in the GUG context is considerably less than in the AUA context. Thus, the conclusions from the spectroscopic experiments are that the AUA sequence is locally more flexible than GUG, clearly in agreement with the results of our MD simulations.

To characterize the origin of the rigidity in the GUG sequence we evaluated the degree of stacking of the DNA in the region of the uracil by measuring the relative quantum yields for each oligo (45,46). The quantum yield of 2AP in the AUA context is approximately 12.5 times larger than in the GUG context. In the more flexible AUA context about 47% of the 2AP is statically quenched, whereas in the GUG context the intensity is reduced by 99%, presumably due to efficient stacking interactions. Thus, it appears that stacking contributes to increased rigidity. Further analysis of the interbase parameters (*Shift, Slide, Rise, Tilt, Roll,* and *Twist*) that describe the translations and rotations of adjacent bases, and can be directly obtained from the simulation results, show that most of the dynamic quenching arises from collisions with the bases on the 5' side, in good agreement with quantum chemical calculations of Jean and Hall (47). Similarly, static quenching, as inferred from the most probable positions of the interbase parameters, is most efficiently induced by the base on the 5' side. Thus, the GC base pair stacks better on the UA base pair, reducing the fluorescence quantum yield by static quenching and contributing to the rigidity of the GUG sequence. On the other hand, the AT base pair through its dynamic collisions with the UA base pair induces an efficient dynamic quenching which reflects the larger flexibility of the AUA sequence. The greater flexibility of the AUA suggests a molecular mechanism for the differential sequence-dependent UDG

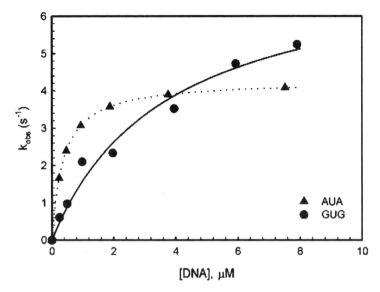

Figure 1 Characterization of catalytic activity of UDG on two different sequences: A-U-A/T-2AP-T; ℓ: G-U-G/C-2AP-C.

efficiency. The smaller energy required to distort the AUA oligo should be reflected in the binding step, rather than in the catalytic step.

A full kinetic analysis on the 19-mer oligos using a fluorescence-based assay was used to determine K_M and k_{cat}. The rate (k_{obs}) as a function of DNA concentration for both sequences is shown in Fig. 1. The K_M of UDG for the AUA sequence is more than 11-fold smaller than the K_M for GUG. Since UDG is inhibited by the product (5,48), K_M approximates the apparent affinity constant. In contrast, k_{cat} for the GUG sequence is 1.7 times greater than for AUA, which may be due to faster product release of the more rigid sequence.

Thus, a detailed analysis of *local* differences in the DNA dynamical properties accounts for the variations in UDG activity as a function of DNA sequence. A combination of MD simulations and fluorescence spectroscopy illustrates key differences in the structure and dynamics of the two DNA sequences. The AUA sequence is more flexible and less stacked than the GUG sequence. The direct correlation between DNA flexibility and UDG activity demonstrates that flexibility has an important contribution to the efficiency of DNA repair.

3. OPENING AND BENDING DYNAMICS OF G•U MISMATCHES IN DNA

While an A•U base pair can be removed by UDG, the natural substrate is actually a G•U mismatch, which occurs as a consequence of spontaneous hydrolytic deamination of cytosine in the cell. Such events occur at a rate of approximately 100–500 per cell per day in humans (33). While G•U wobble base pairs are an essential element of RNA structure (49), their presence in DNA is harmful to normal cell function. If these lesions are not repaired prior to a subsequent round of DNA replication, a transition mutation from G•C to an A•T base pair may result. DNA repair enzymes that initiate this repair, such as UDG (17), and thymine DNA glycosylase (TDG) (50), specifically recognize the mismatch, and cleave the attachment of U or T to the deoxyribose at the N_1–C_1' bond. It is well known that these enzymes need to flip the damaged base from inside the double helix in order to carry out their function. The energetics of base flipping is controlled by several factors. On the base pair level hydrogen bonding is probably the most important property, but locally the stacking interaction with adjacent bases has an important contribution. Since opening and bending are coupled (41), base flipping will also depend on flexibility properties of the DNA sequence. Recent work from our laboratory demonstrated that opening is influenced by small-angle base-pair opening (40). Thus, investigation of these properties can provide an important link between DNA flexibility and DNA repair efficiency.

DNA base-pair opening is energetically unfavorable because it requires breaking the hydrogen bonds in the base pair as well as the loss of stacking interactions with adjacent bases. In a G•U base pair, two hydrogen bonds replace the three hydrogen bonds formed in a Watson–Crick G•C base pair. The O_6 and H_1–N_1 of guanine form hydrogen bonds with the N_3–H_3 and O_2 of uracil, respectively (39). Such an arrangement in DNA leads to a wobble base pair, which is expected to increase the probability of base-pair opening. In RNA, G•U base pairs are very stable, with melting temperatures significantly greater than those of other mismatched base pairs (49). High-resolution crystal structures of RNA oligomers containing G•U mismatches indicate that water-mediated hydrogen bonds partially compensate for the loss of the third hydrogen bond (51). A water molecule in the minor groove bridges guanine N_2, uracil O_2, and O_2' of the uracil ribose (52), an arrangement that is obviously impossible in DNA.

While there is a large body of literature regarding the structural properties of G•U base pairs in an RNA context, relatively little has been reported about such mismatches in DNA. G•T mismatches have been studied in DNA, and their similarity to G•U suggests similar thermodynamic properties. NMR (53) and X-ray crystallography (54) of DNA containing G•T mismatches show local structural deformations of the sugar phosphate backbone and small changes in base stacking. However, the overall conformation is minimally perturbed. The G•T mismatch in DNA is also stabilized by a water molecule in the minor groove, which bridges the guanine N_2 and thymine O_2 atoms, and two additional major groove waters that bridge guanine N_7 and O_6 with thymine O_4 (54).

NMR proton exchange experiments in DNA containing a G•T mismatch show that the opening equilibrium constant at a G•T site is substantially larger than that of a G•C base pair (55). This has been attributed in part to the need to break one less hydrogen bond. NMR experiments provide also kinetic information about the lifetime of the base pair. The G•C base-pair lifetime is on the order of milliseconds, whereas the lifetime of the G•T mismatch was estimated at microseconds, which is at the lower limit of time constants accessible by NMR. Interestingly, the opening equilibrium constants for the neighboring base pairs and the lifetime are altered by the presence of the G•T mismatch, suggesting that sequence context may play a role in determining the degree of opening experienced by a particular base pair. Specifically, the dynamics of opening are more rapid in sequence contexts containing G•C tracts as compared to A•T tracts (56). However, interpretation of these measurements is complex because of the effects of buffer as well as due to the estimation of rate constants from the extrapolation to infinite buffer concentration (56).

The properties of base-pair opening have also been investigated by computational methods. In several reports, the base was opened incrementally by applying a biasing potential during a molecular dynamics simulation, and the free energy was determined as a function of the opening angle (57–60). These reports indicate that the energy for opening into the major groove is substantially lower than that for opening into the minor groove. Furthermore, opening of one member of a base pair is symmetrically coupled to the opening of its complementary base. Such simulations have been conducted in the free DNA as well as in the DNA in a complex with a protein. Estimation of the free energy of the base-pair opening derived from equilibrium molecular dynamics simulations showed that the barrier to base flipping is reduced by 11.6 kcal/mol for a G•U base pair as compared to a G•C base pair in the same sequence context (40). This reduction was partly attributed to the increased sampling of small-angle opening in G•U vs. G•C base pairs. The reduced barrier and the sequence dependence of the open lifetime imply that spontaneous base opening of G•U base pairs occurs on a very rapid time scale.

Fluorescence spectroscopy of base analogs incorporated into DNA is a particularly suitable method for assessing local dynamics and for providing a basis to compare theoretical results with experiment. Suitable probes minimally perturb the DNA structure, report on local properties of DNA, and have spectroscopic properties that enable their selective excitation. For example, using 2AP as a fluorescent reporter, it has been shown that divalent cations induce opening of the base opposite to an abasic site and, interestingly, that the observed fraction of molecules in the open conformation depends on the sequence surrounding the opening base (46). The 2-AP was also used to probe the relationship between flexibility and UDG efficiency in excising uracil from DNA.(43) 6-Methylisoxanthopterin (6MI) is a recently reported fluorescent analog of guanine (61) that can form three hydrogen bonds with the complementary cytosine. Figure 2 shows a comparison of the structures of guanine and 6MI. Melting temperature experiments indicate that the substitution of guanine by 6MI in G•C base pairs only slightly reduces the stability of the double helix (62). 6MI and related probes have shown significant promise in studying DNA structure and dynamics, as well as protein–DNA interactions (63–65). Potentially, 6MI could be employed as a fluorescent probe in the study of local structural and dynamical properties of G•U wobble base pairs. We have analyzed the sequence dependence of spontaneous opening of G•U wobble base pairs, and compared the properties of guanine-containing DNA oligomers with those of 6MI-containing DNA oligomers (66). This study provided an understanding of the local structural fluctuations during G•U base opening and the origin of stabilization of the open state.

To probe the dynamic properties of G•U mismatches molecular dynamics (MD) simulations were performed for 5 ns on the following sequences.

$$5'-\text{AAG CAU AAA GT}-3'$$
$$3'-\text{TTC GTX TTT CA}-5'$$ TXT/AUA

$$5'-\text{CAG GGU GGT TT}-3'$$
$$3'-\text{GTC CCX CCA AA}-5'$$ CXC/GUG

X stands here either for G or for 6MI (M). These sequences were chosen because A•U mismatches in these sequence contexts exhibited large differences in their local structural and dynamical properties (43) and Slupphaug et al. (17) have shown that the differences in UDG activity for G•U and A•U mismatches in these sequence

Guanine   6-Methylisoxanthopterin (6MI)

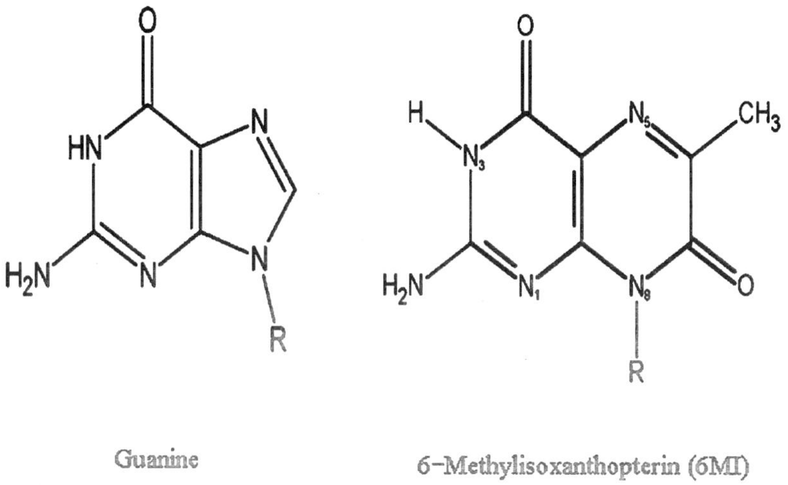

Figure 2 Molecular diagrams of guanine and 6-methylisoxanthopterin (6MI).

contexts are similar. The simulations showed that root-mean-square (RMS) deviations were similar for all sequences, indicating that substitution of guanine by 6MI does not significantly disrupt global DNA dynamics.

G•U mismatches form a wobble base pair whose major characteristic is displacement of the U towards the major groove and the G slightly towards the minor groove (54). This restores some of the hydrogen bonding (two out of the original three in G•C) but generates a fault whose dynamic stability is not well understood. An important consequence of such a fault is that all sequences open to some extent, with the proportion of structures in the open state being sequence dependent. The CGC/GUG sequence has a very small population of the open form ca. 4%, which indicates that a G•U mismatch in the context of adjacent C•G base pairs opens infrequently. In the TGT/AUA sequence, the open state of the G•U base pair in the context of adjacent T•A base pairs is considerably more populated, ~22%. Thus, the stability of the wobble base pair is sequence dependent. The 6MI-containing sequences mirror those with the G•U mismatch in their sequence dependence but show a larger population of the open state. In the T•A context, the fraction of open states is more than 57% whereas in the C•G context it is only 36%. The larger proportion of open states in the 6MI•U base pair indicates a reduced stability of the mismatched base pair, which may be due to the methyl group and the carbonyl on 6MI that faces the major groove. Thus, it appears that spontaneous base-pair opening occurs in each sequence to varying degrees. As a result, the sequences in which the U is surrounded by adenines exhibit more opening than those in which it is surrounded by guanines.

The structural dynamic nature of the spontaneous opening of G•U (M•U) mismatches can be characterized by their localization to the mismatch base pair, since the opening motion does influence the distribution of the opening angles of the adjacent base pairs. The opening motion is composed of a symmetric motion of both bases in the base pair towards the major groove. This is in agreement with previous work (57,58), and because a symmetric motion requires a larger energetic cost it suggests that the open states are stabilized by other factors. The main cost of opening a base pair comes from breaking the hydrogen bonds and reducing the stacking interactions. In order to maximize the hydrogen bonding in the mismatch, the wobble base pair displaces U towards the major groove and G(M) towards the minor groove. Such a rearrangement reduces the strength of the hydrogen bonding and provides an explanation for the relatively high fraction of the open states. However, in the open states the bases also lose the stabilization due to stacking for which there must be a compensatory stabilizing interaction.

One important source of stabilization is the formation of a bifurcated hydrogen bond between O_2 of U as an acceptor and the N_3–H and N_2–H of G(M) as donors. This is illustrated in Fig. 3A. The wobble base pair is clearly seen in the upper panel with hydrogen bonds between N_1–H(G)$\cdots O_2$(U) and O_6(G)$\cdots$H–N_3(U). As a consequence of the opening motion, the length of the $O_6 \cdots$H–N_3 bond is extended to 5 Å rendering it effectively broken. The change in the length of the hydrogen bond and the angle between the acceptor and the donor have been used to cluster the trajectory into the closed and open population. The distribution of the two populations is shown in Fig. 3B. The consequence of the opening motion is the exposure of the N_3–H in the major groove presenting an opportunity for effective solvation by water. Indeed, a proximity analysis (67) of the solvation shows that the open state has on the average 0.8 waters near N_3–H, providing additional

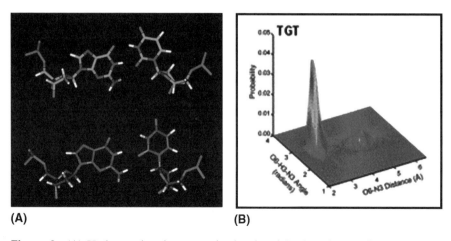

Figure 3 (A) Hydrogen bond patterns in the closed (top) and open (bottom) states of the G•U wobble base pair. Similar structures are observed in 6MI•U mismatch base pairs. (B) Probability distribution of the closed and open states in the TGT sequence as a function of the hydrogen bonding distance and angle.

stabilization of the open state. In contrast, the closed state has no water hydrogen bonded to N_3–H. This is a unique situation only available in G•U mismatches.

An examination of the possible contribution of the bifurcated hydrogen bonding to the stabilization of the open state proves not to be the reason for the sequence dependence. Both in the TGT/AUA and the CGC/GUG sequences, the geometrical properties of the bifurcated bond and the presence of the water molecule next to N_3–H of U are the same. The differential stabilization comes from the sequence-dependent loss of favorable stacking interactions. The energy of stacking interactions depends on the nature of the bases involved in the interaction as well as on the sequence context in which the interaction occurs (68). The work of SantaLucia et al. (69) suggests that neighboring G•C base pairs provide greater thermodynamic stability than do A•T base pairs, in agreement with our results. A recent empirical approach allows the prediction of base pair stacking from temperature-dependent changes in overlap between adjacent DNA bases as determined by NMR (70). The "stacking sum" is such a measure and can be obtained as the sum of the absolute values of *Shift* and *Slide*, which describe the relative lateral displacement of one base with respect to the other. Since in unperturbed DNA this sum is zero, deviations indicate a decrease in stacking. Application of this approach shows that bases on the 3′ side, but not on the 5′ side, stabilize the sequences containing a G•U mismatch by stacking interactions. Furthermore, the deviations from zero in the TGT/AUA sequence are larger than in the CGC/GUG sequence. A similar behavior is observed in the 6MI-containing sequences, except the deviations are larger, implying that the 6MI sequences stack less well. Thus, it is clear that the origin of the sequence dependence flexibility associated with the opening motion is due to differences in stacking and not hydrogen bonding.

Several previous reports have indicated that DNA base opening and local bending are coupled processes (40,41). A PMF analysis of the bending angle distributions provides the force constants for bending in the open and closed structures. In general, the structures in the open state are more bent than those in the closed state and more rigid for bending into the grooves. Interestingly, the characteristic anisotropy of bending into the grooves vs. into the backbone is substantially reduced in the open state because the force constants for bending towards the grooves are increased. In the open state, flexibility of bending towards the grooves, particularly the minor groove is reduced due to the opening of the base pair. Consequently, the opening/closing motion generates nearly discrete bent states: "closed/straight" and "open/bent." The open/bent states convert into closed/straight states in a concerted motion; the existence of intermediate states is energetically not accessible.

As was demonstrated in the AUA and GUG sequences (43), bending flexibility correlated with UDG efficiency of uracil removal from damaged DNA. A parallel behavior is observed in the sequences containing a G•U mismatch. The harmonic force constants for bending the TGT/AUA sequence are substantially lower than those of the CGC/GUG sequences. Thus, the dynamic flexibility properties of DNA sequences reflected in opening and in bending both contribute to the efficiency of UDG activity as a repair enzyme.

Recent discovery of the fluorescent guanine analog 6MI (61,62) could possibly test the predictions from the computational simulations. However, the spectroscopy of 6MI in DNA is not easy to interpret due to several complicating factors. First, the N_3 proton of 6MI is ionizable with a pK_a of approximately 8.3 (71). This can lead to spectroscopic heterogeneity whose interpretation may be difficult. It is not known whether in normal double-stranded DNA such spectroscopic heterogeneity exists. Second, the efficiency of both dynamic and static quenching of 6MI depends on the nature of the nucleosides and nucleotides (71). This is in distinct contrast from 2AP, whose quenching is independent of the specific nucleotide (45). To obtain reliable spectroscopic estimates of DNA flexibility, the variation in efficiency must be accounted for when using 6MI as a probe of DNA dynamics. At the present time no reliable representation of this property is available in DNA. Third, our unpublished work shows that cytidine forms a weak ground-state complex with 6MI and is a very inefficient quencher. A small shift in the emission spectrum and an apparent increase in the fluorescence lifetime of 6MI in the presence of cytidine suggest that this complex is fluorescent. Further experiments and theory are required to understand the changes in fluorescence lifetime observed when 6MI is present in a cytidine context.

4. CONCLUSIONS

A combined experimental/theoretical approach to the elucidation of the role of local DNA dynamics in repair enzyme efficiency shows the strength of this synergistic approach. Fluorescence spectroscopic measurements of DNA oligos containing a fluorescent base analog at a specific sequence position provide information regarding the collision of the analog with its neighbors. With the aid of MD simulations, the translation of the spectroscopic results into a measure of local DNA allows dissection of the experimental flexibility parameter into individual base motions. Such an approach not only provides an explanation of sequence-dependent UDG efficiency in molecular terms but also defines the framework for the formulation of a comprehensive description of the role of DNA dynamics in repair enzyme efficiency.

The studies of the sequences containing A•U mismatches demonstrated that the sequence with greater flexibility had a smaller K_M, implying that UDG binding to the substrate is controlled in part by the energy required to distort the substrate in the course of forming the enzyme–substrate complex. It is interesting to note that the relative efficiency of UDG as determined from the ratio of k_{cat}/K_M values for the two sequences is 6.8. The relative flexibility of the two sequences as determined from the ratio of τ_{num} values is 7.7. Thus, it appears that enzyme efficiency for different sequences is strongly correlated with their flexibility. Solidification of this conclusion will require a characterization of the local dynamics of a variety of DNA sequences for which the enzyme has different efficiencies.

Sequence-dependent partial base-pair opening was observed in G•U and 6MI•U mismatches. The sequence-dependent differences in opening are primarily due to differences in stacking interactions. These studies demonstrate that both local DNA flexibility and the relative stabilities of the open and closed states play a role in UDG efficiency. Sequences requiring a smaller investment of energy for bending and base-pair opening will be better UDG substrates and that the effect of sequence on UDG efficiency will be manifested primarily in K_M. Clearly, motions other than bending and opening may contribute to the energetics of UDG binding as well.

The role of DNA flexibility has been observed in other systems. The binding of TATA box-binding protein (TBP) to the TATA transcription regulatory element represents a biological system in which local DNA structure and flexibility correlate with binding efficiency. Crystallographic studies showed that binding of TATA box-binding protein induces a severe distortion in the DNA (72). Further, bend angles of DNA bound to TBP estimated from solution experiments correlated with transcriptional activity (73). These results are supported by a molecular dynamics simulations and potential of mean force studies, which indicated that sequences requiring less energy to kink into a particular conformation are bound more efficiently by TBP (74). Other molecular dynamics studies demonstrated a correlation between increased DNA flexibility and transcriptional activity (75). Recent molecular dynamics simulations on a series of TATA elements indicated a correlation between DNA flexibility and transcriptional activity (76). In addition, they also identified other local structural properties that correlate with transcriptional activity. Taken together, these results demonstrate the importance of local structure, dynamics, and flexibility in determining transcriptional activity for different promoter sequences. Further support for the importance of these factors in DNA–protein interactions is demonstrated by a combination of simulations and bioinformatics to construct a Hidden Markov Model based on static and dynamic bending properties of DNA. The HMM predicts with very high fidelity binding sites of catabolite activator protein (CAP) (77). Thus, it is clear that DNA flexibility is important not only in the repair of DNA damage but also in numerous other functions regulated by DNA–protein complexes.

Although further work is required to definitively relate local DNA flexibility to recognition and repair of damaged DNA, a model of dynamic recognition of damaged DNA by UDG may be constructed and is presented in Fig. 4. Non-specific binding of UDG to undamaged DNA does not induce DNA bending because the force constants for bending are too high. Thus, the energy required to induce a bent structure that will lead to a stable complex is too high, and consequently a productive enzyme substrate complex is not formed. As in other protein–DNA non-specific complexes, the thermal fluctuations drive a unidimensional diffusion of the protein on the DNA (78–80) and UDG translocates along the DNA until it encounters the damaged site. The increased DNA flexibility at the damaged site allows the enzyme to bend the DNA at a cost that is now within the appropriate range. However, the energetics of DNA bending depends on the sequence in which the damage occurs and therefore more flexible sequences will bind to UDG with higher affinity.

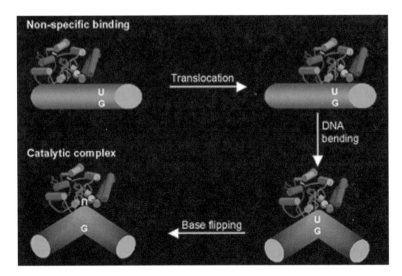

Figure 4 Model of dynamic recognition of damage by UDG. Cylinders indicate DNA. Uracil and guanine are indicated by U and G, respectively.

Once the DNA is bent, the damaged base is now able to flip out of the DNA helix to form the catalytic complex. Sequences requiring more distortion energy will be less good UDG substrates, and therefore will be repaired less frequently. This schematic representation integrates DNA flexibility with the process of damage recognition and its efficient repair.

ACKNOWLEDGMENTS

This work was supported by PHS Grant CA 63317 and by training grants GM08553 and CA78207.

REFERENCES

1. Krokan HE, Standal R, Slupphaug G. DNA glycosylases in the base excision repair of DNA. Biochem J 1997; 325(Pt 1):1–16.
2. Schärer OD, Jiricny J. Recent progress in the biology, chemistry and structural biology of DNA glycosylases. Bioessays 2001; 23(3):270–281.
3. Mol CD, Parikh SS, Putnam CD, Lo TP, Tainer JA. DNA repair mechanisms for the recognition and removal of damaged DNA bases. Annu Rev Biophys Biomol Struct 1999; 28:101–128.
4. Slupphaug G, Mol CD, Kavli B, Arvai AS, Krokan HE, Tainer JA. A nucleotide-flipping mechanism from the structure of human uracil–DNA glycosylase bound to DNA [see comments]. Nature 1996; 384(6604):87–92.
5. Parikh SS, Mol CD, Slupphaug G, Bharati S, Krokan HE, Tainer JA. Base excision repair initiation revealed by crystal structures and binding kinetics of human uracil–DNA glycosylase with DNA. EMBO J 1998; 17(17):5214–5226.
6. Parikh SS, Walcher G, Jones GD, Slupphaug G, Krokan HE, Blackburn GM, Tainer JA. Uracil–DNA glycosylase–DNA substrate and product structures: conformational

strain promotes catalytic efficiency by coupled stereoelectronic effects [In Process Citation]. Proc Natl Acad Sci USA 2000; 97(10):5083–5088.

7. Lau AY, Schärer OD, Samson L, Verdine GL, Ellenberger T. Crystal structure of a human alkylbase–DNA repair enzyme complexed to DNA: mechanisms for nucleotide flipping and base excision. Cell 1998; 95(2):249–258.

8. Lau AY, Wyatt MD, Glassner BJ, Samson LD, Ellenberger T. Molecular basis for discriminating between normal and damaged bases by the human alkyladenine glycosylase, AAG. Proc Natl Acad Sci USA 2000; 97(25):13573–13578.

9. Hollis T, Ichikawa Y, Ellenberger T. DNA bending and a flip-out mechanism for base excision by the helix-hairpin-helix DNA glycosylase, *Escherichia coli* AlkA. EMBO J 2000; 19(4):758–766.

10. Bruner SD, Norman DP, Verdine GL. Structural basis for recognition and repair of the endogenous mutagen 8-oxoguanine in DNA. Nature 2000; 403(6772):859–866.

11. Gilboa R, Zharkov DO, Golan G, Fernandes AS, Gerchman SE, Matz E, Kycia JH, Grollman AP, Shoham G. Structure of formamidopyrimidine–DNA glycosylase covalently complexed to DNA. J Biol Chem 2002; 277(22):19811–19816.

12. Beveridge DL, McConnell KJ. Nucleic acids: theory and computer simulation, Y2K. Curr Opin Struct Biol 2000; 10(2):182–196.

13. Lafontaine I, Lavery R. Collective variable modeling of nucleic acids. Curr Opin Struct Biol 1999; 9(2):170–176.

14. Rachofsky EL, Ross JB, Osman R. Conformation and dynamics of normal and damaged DNA. Comb Chem High Throughput Screen 2001; 4(8):675–706.

15. Eftedal I, Guddal PH, Slupphaug G, Volden G, Krokan HE. Consensus sequences for good and poor removal of uracil from double stranded DNA by uracil–DNA glycosylase. Nucleic Acids Res 1993; 21(9):2095–2101.

16. Eftedal I, Volden G, Krokan HE. Excision of uracil from double-stranded DNA by uracil–DNA glycosylase is sequence specific. Ann N Y Acad Sci 1994; 726: 312–314.

17. Slupphaug G, Eftedal I, Kavli B, Bharati S, Helle NM, Haug T, Levine DW, Krokan HE. Properties of a recombinant human uracil–DNA glycosylase from the UNG gene and evidence that UNG encodes the major uracil–DNA glycosylase. Biochemistry 1995; 34(1):128–138.

18. Nilsen H, Yazdankhah SP, Eftedal I, Krokan HE. Sequence specificity for removal of uracil from U.A pairs and U.G mismatches by uracil–DNA glycosylase from *Escherichia coli*, and correlation with mutational hotspots. FEBS Lett 1995; 362(2):205–209.

19. Lari SU, Al-Khodairy F, Paterson MC. Substrate specificity and sequence preference of G:T mismatch repair: incision at G:T, O6-methylguanine:T, and G:U mispairs in DNA by human cell extracts. Biochemistry 2002; 41(29):9248–9255.

20. Hang B, Sági J, Singer B. Correlation between sequence-dependent glycosylase repair and the thermal stability of oligonucleotide duplexes containing 1, N6-ethenoadenine. J Biol Chem 1998; 273(50):33406–33413.

21. Asagoshi K, Yamada T, Terato H, Ohyama Y, Monden Y, Arai T, Nishimura S, Aburatani H, Lindahl T, Ide H. Distinct repair activities of human 7,8-dihydro-8-oxoguanine DNA glycosylase and formamidopyrimidine DNA glycosylase for formamidopyrimidine and 7,8-dihydro-8-oxoguanine. J Biol Chem 2000; 275(7):4956–4964.

22. Ye N, Holmquist GP, O'Connor TR. Heterogeneous repair of *N*-methylpurines at the nucleotide level in normal human cells. J Mol Biol 1998; 284(2):269–285.

23. Parikh SS, Putnam CD, Tainer JA. Lessons learned from structural results on uracil–DNA glycosylase. Mutat Res 2000; 460(3–4):183–199.

24. Drohat AC, Xiao G, Tordova M, Jagadeesh J, Pankiewicz KW, Watanabe KA, Gilliland GL, Stivers JT. Heteronuclear NMR and crystallographic studies of wild-type and H187Q *Escherichia coli* uracil DNA glycosylase: electrophilic catalysis of uracil expulsion by a neutral histidine 187. Biochemistry 1999; 38(37):11876–11886.

25. Luo N, Mehler E, Osman R. Specificity and catalysis of uracil DNA glycosylase. A molecular dynamics study of reactant and product complexes with DNA. Biochemistry 1999; 38(29):9209–9220.

26. Stivers JT, Pankiewicz KW, Watanabe KA. Kinetic mechanism of damage site recognition and uracil flipping by *Escherichia coli* uracil DNA glycosylase. Biochemistry 1999; 38(3):952–963.

27. Drohat AC, Jagadeesh J, Ferguson E, Stivers JT. Role of electrophilic and general base catalysis in the mechanism of *Escherichia coli* uracil DNA glycosylase. Biochemistry 1999; 38(37):11866–11875.

28. Verri A, Mazzarello P, Biamonti G, Spadari S, Focher F. The specific binding of nuclear protein(s) to the cAMP responsive element (CRE) sequence (TGACGTCA) is reduced by the misincorporation of U and increased by the deamination of C. Nucleic Acids Res 1990; 18(19):5775–5780.

29. Kunkel TA, Bebenek K. DNA replication fidelity. Annu Rev Biochem 2000; 69:497–529.

30. Mosbaugh DW, Bennett SE. Uracil-excision DNA repair. Prog Nucleic Acid Res Mol Biol 1994; 48:315–370.

31. Srivastava DK, Berg BJ, Prasad R, Molina JT, Beard WA, Tomkinson AE, Wilson SH. Mammalian abasic site base excision repair. Identification of the reaction sequence and rate-determining steps. J Biol Chem 1998; 273(33):21203–21209.

32. Krokan H, Wittwer CU. Uracil DNA–glycosylase from HeLa cells: general properties, substrate specificity and effect of uracil analogs. Nucleic Acids Res 1981; 9(11): 2599–2613.

33. Delort AM, Duplaa AM, Molko D, Teoule R, Leblanc JP, Laval J. Excision of uracil residues in DNA: mechanism of action of *Escherichia coli* and *Micrococcus luteus* uracil–DNA glycosylases. Nucleic Acids Res 1985; 13(2):319–335.

34. Savva R, McAuley-Hecht K, Brown T, Pearl L. The structural basis of specific base-excision repair by uracil–DNA glycosylase. Nature 1995; 373(6514):487–493.

35. Xiao G, Tordova M, Jagadeesh J, Drohat AC, Stivers JT, Gilliland GL. Crystal structure of *Escherichia coli* uracil DNA glycosylase and its complexes with uracil and glycerol: structure and glycosylase mechanism revisited. Proteins 1999; 35(1):13–24.

36. Pearl LH. Structure and function in the uracil–DNA glycosylase superfamily. Mutat Res 2000; 460(3–4):165–181.

37. Kavli B, Slupphaug G, Mol CD, Arvai AS, Peterson SB, Tainer JA, Krokan HE. Excision of cytosine and thymine from DNA by mutants of human uracil–DNA glycosylase. EMBO J 1996; 15(13):3442–3447.

38. Friedman RA, Honig B. A free energy analysis of nucleic acid base stacking in aqueous solution. Biophys J 1995; 69(4):1528–1535.

39. Bloomfield VA, Crothers DM, Tinoco I. Nucleic Acids: Structures, Properties and Functions. Sausalito, CA: University Science Books, 2000.

40. Fuxreiter M, Luo N, Jedlovszky P, Simon I, Osman R. Role of base flipping in specific recognition of damaged DNA by repair enzymes. J Mol Biol 2002; 323(5):823–834.

41. Ramstein J, Lavery R. Energetic coupling between DNA bending and base pair opening. Proc Natl Acad Sci USA 1988; 85(19):7231–7235.

42. Bellamy SR, Baldwin GS. A kinetic analysis of substrate recognition by uracil–DNA glycosylase from herpes simplex virus type 1. Nucleic Acids Res 2001; 29(18):3857–3863.

43. Seibert E, Ross JBA, Osman R. Role of DNA flexibility in sequence-dependent activity of uracil DNA glycosylase. Biochemistry 2002; 41(36):10976–10984.

44. Verma S, Eckstein F. Modified oligonucleotides: synthesis and strategy for users. Annu Rev Biochem 1998; 67:99–134.

45. Rachofsky EL, Osman R, Ross JBA. Probing structure and dynamics of DNA with 2-aminopurine: effects of local environment on fluorescence. Biochemistry 2001; 40(4):946–956.

46. Rachofsky EL, Seibert E, Stivers JT, Osman R, Ross JBA. Conformation and dynamics of abasic sites in DNA investigated by time-resolved fluorescence of 2-aminopurine. Biochemistry 2001; 40(4):957–967.

47. Jean JM, Hall KB. 2-Aminopurine fluorescence quenching and lifetimes: role of base stacking. Proc Natl Acad Sci USA 2001; 98(1):37–41.

48. Stivers JT. 2-Aminopurine fluorescence studies of base stacking interactions at abasic sites in DNA: metal-ion and base sequence effects. Nucleic Acids Res 1998; 26(16): 3837–3844.

49. Varani G, McClain WH. The G•U wobble base pair. A fundamental building block of RNA structure crucial to RNA function in diverse biological systems. EMBO Rep 2000; 1(1):18–23.

50. Saparbaev M, Laval J. 3,N4-ethenocytosine, a highly mutagenic adduct, is a primary substrate for *Escherichia coli* double-stranded uracil–DNA glycosylase and human mismatch-specific thymine–DNA glycosylase. Proc Natl Acad Sci USA 1998; 95(15): 8508–8513.

51. Sundaralingam M, Pan B. Hydrogen and hydration of DNA and RNA oligonucleotides. Biophys Chem 2002; 95(3):273–282.

52. Shi K, Wahl M, Sundaralingam M. Crystal structure of an RNA duplex r(G GCGC CC)2 with non-adjacent G∗U base pairs. Nucleic Acids Res 1999; 27(10):2196–2201.

53. Hare D, Shapiro L, Patel DJ. Wobble dG•dT pairing in right-handed DNA: solution conformation of the d(C-G-T-G-A-A-T-T-C-G-C-G) duplex deduced from distance geometry analysis of nuclear Overhauser effect spectra. Biochemistry 1986; 25(23): 7445–7456.

54. Hunter WN, Brown T, Kneale G, Anand NN, Rabinovich D, Kennard O. The structure of guanosine–thymidine mismatches in B-DNA at 2.5-A resolution. J Biol Chem 1987; 262(21):9962–9970.

55. Moe JG, Russu IM. Kinetics and energetics of base-pair opening in 5′-d(CGCGAATTCGCG)-3′ and a substituted dodecamer containing G.T mismatches. Biochemistry 1992; 31(36):8421–8428.

56. Dornberger U, Leijon M, Fritzsche H. High base pair opening rates in tracts of GC base pairs. J Biol Chem 1999; 274(11):6957–6962.

57. Giudice E, Várnai P, Lavery R. Energetic and conformational aspects of A:T base-pair opening within the DNA double helix. CHEMPHYSCHEM 2001; 11:673–677.

58. Várnai P, Lavery R. Base Flipping in DNA: pathways and energetics studied with molecular dynamic simulations. J Am Chem Soc 2002; 124(25):7272–7273.

59. Huang N, Banavali NK, MacKerell AD Jr. Protein-facilitated base flipping in DNA by cytosine-5-methyltransferase. Proc Natl Acad Sci USA 2003; 100(1):68–73.

60. Banavali NK, MacKerell AD Jr. Free energy and structural pathways of base flipping in a DNA GCGC containing sequence. J Mol Biol 2002; 319(1):141–160.

61. Hawkins ME. Fluorescent pteridine nucleoside analogs: a window on DNA interactions. Cell Biochem Biophys 2001; 34(2):257–281.

62. Hawkins ME, Pfleiderer W, Balis FM, Porter D, Knutson JR. Fluorescence properties of pteridine nucleoside analogs as monomers and incorporated into oligonucleotides. Anal Biochem 1997; 244(1):86–95.

63. Moser AM, Patel M, Yoo H, Balis FM, Hawkins ME. Real-time fluorescence assay for O6-alkylguanine-DNA alkyltransferase. Anal Biochem 2000; 281(2):216–222.

64. Wojtuszewski K, Hawkins ME, Cole JL, Mukerji I. HU binding to DNA: evidence for multiple complex formation and DNA bending. Biochemistry 2001; 40(8):2588–2598.

65. Deprez E, Tauc P, Leh H, Mouscadet JF, Auclair C, Hawkins ME, Brochon JC. DNA binding induces dissociation of the multimeric form of HIV-1 integrase: a time-resolved fluorescence anisotropy study. Proc Natl Acad Sci USA 2001; 98(18): 10090–10095.

66. Seibert E, Ross JB, Osman R. Contribution of opening and bending dynamics to specific recognition of DNA damage. J Mol Biol 2003; 330(4):687–703.

67. Mezei M. Modified proximity criteria for the analysis of the solvation of a polyfunctional solvent. Mol Simul 1988; 1:327–332.

68. Sponer J, Leszczynski J, Hobza P. Nature of nucleic acid-base stacking: nonempirical ab initio and empirical potential characterization of 10 stacked base dimers. Comparison of stacked and H-bonded base pairs. J Phys Chem 1996; 100(13):5590–5596.

69. SantaLucia J, Allawi HT, Seneviratne A. Improved nearest-neighbor parameters for predicting DNA duplex stability. Biochemistry 1996; 35(11):3555–3562.

70. Lam SL, Ip LN. Low temperature solution structures and base pair stacking of double helical d(CGTACG)(2). J Biomol Struct Dyn 2002; 19(5):907–917.

71. Seibert E, Chin AS, Pfleiderer W, Hawkins ME, Laws WR, Osman R, Ross JBA. Spectroscopy and electronic structure of 6,8-dimethylisoxanthopterin, a guanine analogue. J Phys Chem A 2003; 107:178–185.

72. Nikolov DB, Chen H, Halay ED, Hoffman A, Roeder RG, Burley SK. Crystal structure of a human TATA box-binding protein/TATA element complex. Proc Natl Acad Sci USA 1996; 93(10):4862–4867.

73. Wu J, Parkhurst KM, Powell RM, Brenowitz M, Parkhurst LJ. DNA bends in TATA-binding protein-TATA complexes in solution are DNA sequence-dependent. J Biol Chem 2001; 276(18):14614–14622.

74. Pardo L, Pastor N, Weinstein H. Selective binding of the TATA box-binding protein to the TATA box-containing promoter: analysis of structural and energetic factors. Biophys J 1998; 75(5):2411–2421.

75. de Souza ON, Ornstein RL. Inherent DNA curvature and flexibility correlate with TATA box functionality. Biopolymers 1998; 46(6):403–415.

76. Qian X, Strahs D, Schlick T. Dynamic simulations of 13 TATA variants refine kinetic hypotheses of sequence/activity relationships. J Mol Biol 2001; 308(4):681–703.

77. Thayer KM, Beveridge DL. Hidden Markov models from molecular dynamics simulations on DNA. Proc Natl Acad Sci USA 2002; 99(13):8642–8647.

78. von Hippel PH, Berg OG. On the specificity of DNA–protein interactions. Proc Natl Acad Sci USA 1986; 83(6):1608–1612.

79. Berg OG. Diffusion-controlled protein–DNA association: influence of segmental diffusion of the DNA. Biopolymers 1984; 23(10):1869–1889.

80. von Hippel PH, Berg OG. Facilitated target location in biological systems. J Biol Chem 1989; 264(2):675–678.

2

In Search of Damaged Bases

R. Stephen Lloyd and A. K. McCullough
Center for Research on Occupational and Environmental Toxicology, Oregon Health and Science University, Portland, Oregon, U.S.A.

M. L. Dodson
Sealy Center for Molecular Science, University of Texas Medical Branch, Galveston, Texas, U.S.A.

1. INTRODUCTION

Cells continuously face a daunting challenge to find, remove and repair damaged DNA bases within vast excesses of undamaged DNA. Failure to correct these lesions, prior to subsequent rounds of DNA replication, can lead to mutagenic or lethal consequences. In this review, we will attempt to summarize the sequential events by which DNA glycosylases reduce the complexities associated with locating altered bases. For this process to be successful, these enzymes must discriminate between bases that will ultimately be a substrate for catalytic excision and those that escape the repair process. The intent of this review is not to summarize each individual DNA glycosylase or family of related glycosylases, but rather to discuss the topics of damaged DNA target site location, DNA bending, and nucleotide flipping. We will draw upon key examples from the literature to illustrate both the diversity and commonality of the biophysical and biochemical processes that constitute the initiation of base excision repair (BER). Concerning the topics of substrate recognition and the structures of DNA glycosylases and their associated DNA complexes, the reader is referred to Chapter 14. Further, the subject of DNA glycosylase-mediated catalysis will not be reviewed within this chapter but the reader is referred to a comprehensive review (1), which is an outstanding compilation and analysis of the literature on DNA glycosylases from a mechanistic chemical perspective.

2. MECHANISM FOR AN INCREASED RATE OF TARGET SITE LOCATION

Imagine that you enter a darkened room ($\sim 10^8$ cm^3), in which a continuous piece of string (occupying a total volume of $\sim 10^2$ cm^3) has been randomly attached to points on the walls, ceiling and floor, such that the string is distributed equally throughout the room. A piece of tape has been attached to one site on the string—your assignment is to find the tape. Several strategies could be employed to find the tape and given sufficient time, any active search mechanism will find the correct site. One could search the entire volume of the room in a random manner, but using this approach, the majority of the search time would be investigating the void space within the room. Alternatively, one could use such a random walk to initially locate the string and with the knowledge that the tape is located somewhere on the string, simply investigate the volume occupied by the string (a walk along the string); thus in this example, reducing the volume to be analyzed by a factor of 10^6.

In the context of a cell, many of the interactions between proteins and DNA occur at very specific or specialized sites on the DNA. These include, but are not limited to, the initiation, regulation and termination of replication and transcription, restriction/modification, recombination, and various DNA repair process. The enzymes and proteins that carry out these transactions face similar challenges to those described above, in that they must locate their cognate DNA sequence(s) or structure (target) in a timely manner within vast excesses of nonspecific (nontarget) DNA. With the discovery of sequence-specific DNA binding, it has been recognized that proteins must utilize specialized strategies to accelerate target site location (2). In this analysis, a random-collision, 3-dimensional diffusion-based model could not account for the kinetics and specificity of bacteriophage λ repressor binding. Kinetic analyses of the lac repressor–operator binding specificity, in which rate accelerations were reported to be ~1000-fold faster than diffusion-controlled processes support similar conclusions (3). Recently, Stivers and Jiang (1) considered whether physiologically relevant amounts of uracil DNA glycosylase (UDG) could be expected to remove, in a reasonable amount of time, the ~100 uracil residues that form spontaneously within every human cell per day without using a form of accelerated target site location and came to similar interpretations. In their model, they assumed that nonspecific DNA did not accelerate rates at which uracil residues could be found. Using known rate and equilibrium constants for UDG, a ratio of specific to nonspecific DNA of $10^2/10^9$, and a UDG concentration of ~7 μM, they calculated that <0.001% of the 100 uracils could be excised per day. These calculations emphasize the necessity to accelerate specific target site location.

An eloquent articulation of mechanisms that could promote the apparent rate acceleration by proteins in target site location came from the laboratories of Peter von Hippel and Otto Berg (4), who described several possible scenarios including sliding, hopping (or jumping), and looping (including intersegment transfer). Sliding is a term that denotes the following: after an initial random encounter with DNA, a protein remains closely associated with that DNA molecule, such that following the initial release of cations from the DNA, the protein is energetically free to slide on DNA, displacing and replacing an equal number of counterions as the protein translocates in one dimension along the DNA by Brownian motion. This walk on DNA could not have a directionality component unless there was a utilization of an additional energy source. To date, there are no definitive examples of protein sliding, even though this is a term commonly associated with processive enzymatic activities on defined genomes. Hopping (or jumping) is a term that implies multiple associative/dissociative encounters that all occur within the same "domain" of DNA. In this review, we will use the term microscopic association/dissociation to describe the random diffusion process in which a protein continuously binds to and dissociates from a DNA molecule through many cycles. Eventually, the enzyme dissociates from the original DNA domain and macroscopically diffuses until it encounters another DNA domain. Experimentally, sliding and hopping are difficult to distinguish, since in both cases multiple binding (and potentially catalytic events) would occur within one DNA domain. The net effect would be experimentally observed as a clustering of product sites. Examples of these types of observations will be discussed below. Looping is a term that describes the process by which a protein or enzyme with two different binding sites within one polypeptide or a protein dimer (or higher order) can bind to two DNA sites simultaneously, thus promoting multiple DNA encounters within the same DNA domain. Examples of looping will also be discussed below. Intersegment transfer is a specialized form of looping in which the two DNAs that are bound by the same protein, represent two distinct functional DNA domains. Enzymes that utilize any of these strategies should experience a significant rate enhancement of target site location.

3. IN VITRO EVIDENCE FOR PROCESSIVE NICKING ACTIVITY OF DNA GLYCOSYLASES

3.1. Bacteriophage T4 Pyrimidine Dimer Glycosylase

In the following section, we will summarize the literature that is germane to the question of whether DNA glycosylases can catalyze multiple incision events within discrete domains of DNA, prior to macroscopic diffusion from that DNA and reinitiation additional catalytic cycles on other DNA molecules. The discovery of processive (or clustered) nicking activity of the T4 pyrimidine dimer glycosylase/abasic site (AP) lyase (T4-pdg, formerly named T4 endonuclease V) was made by incubating limiting concentrations of T4-pdg with form I covalently closed circular supercoiled DNA that had been heavily irradiated with short wave ultraviolet (UV) light. For these initial observations, the influences of UV were sufficient that nearly one third of all the plasmid molecules contained at least one location where two *cis, syn* cyclobutane pyrimidine dimers were produced in close proximity (within ~12 base pairs) (bp) on complementary strands. When following the kinetics of the glycosylase/AP lyase activity on pyrimidine dimers, it was observed that while large percentages of unincised form I DNA were still present in the reaction, there was a very significant and linear accumulation of form III (full length linear) DNA molecules. After ruling out any contaminating activities that could account for these double-strand breaks, it was established that at low ionic strengths, T4-pdg incises, on average, most of the pyrimidine dimers on the plasmid DNA to which it was originally bound, prior to dissociation from that DNA molecule (5). In that study, the average number of dimers was varied from 2.5 to 35 dimers per molecule. At all dimer substrate concentrations above 5 dimers per plasmid, form III DNA accumulated linearly and proportionately with UV dose. It was also shown that varying the concentration of the plasmid DNA molecules by more than 8-fold did not alter the linearity of the accumulation of form III DNAs. Additionally, velocity sedimentation analyses of the DNA products through alkaline sucrose gradients, revealed that there was a bimodal distribution in the number average molecular weights of the DNAs, such that the plasmid DNAs were either fully incised or contained no single-strand breaks. This result not only validated the use of the kinetics of the accumulation of form III DNA as a measure of the enzyme's processivity, but also demonstrated that for in vitro reactions, T4-pdg can incise all dimers within a plasmid with the average distance between dimers varying from 0.27 to 2.7 kb.

The greatest limitation of these data was that they did not address the molecular mechanism by which the clustering of the incision events occurs. Interestingly, for the *Eco*RV restriction enzyme, it has been possible to experimentally distinguish sliding from hopping and these data clearly favor the hopping mechanism (6) and reviewed in Ref. 7. Thus, it is reasonable to hypothesize that T4-pdg uses a microscopic hopping mechanism in which on average, the majority of association/dissociation events do not allow the enzyme to macroscopically diffuse from the original DNA site until all dimers had been incised. This speculation is further reinforced by the fact that it is known that T4-pdg has one active site based on both X-ray crystallography (8,9) and active site mutational analyses (10–13). Additionally, it exists as a monomer in free solution (14). These data suggest that in order for T4-pdg to incise at dimers in close proximity on complementary strands, if originally bound to a dimer in the Watson strand, it must dissociate from the DNA, rebind, and incise the same DNA molecule but at a site containing a dimer in the Crick strand. Sliding on DNA could not account for such reactions in the absence of enzyme dimerization. The only data that might suggest that T4-pdg can dimerize on UV irradiated DNA comes from an electron microscopy study of plasmid DNA molecules that had been reacted with T4-pdg, where it was observed that a high percentage of the DNA molecules in the Kleinschmidt spread, contained discrete loops that were measured to be on average, the size of the interdimer distance (15). Further analyses revealed that changing the interdimer distance proportionately changed the size of the loops observed.

Additional data that would favor a model of repetitive association/dissociation events vs. sliding come from experiments in which variation in the salt concentration of the reaction ($\sim$5–200 mM) dramatically changed the apparent clustering of DNA incisions by T4-pdg (16,17). Collectively, those data demonstrated that clustering of incisions within plasmid DNA molecules was highly dependent on the salt concentration of the reaction. For salt concentrations below 40–50 mM, complete dimer incision was observed in UV irradiated plasmids. In fact, it was experimentally possible to decrease the salt concentration sufficiently, such that macroscopic dissociation of the T4-pdg was not observed from the plasmid DNA to which it is was originally bound. In these experiments, there was an initial conversion of form I DNAs to forms II and III in which the percent conversion of form I DNA was directly proportional to the amount of enzyme added. Following the initial burst of incision, the percent of form I DNAs remained constant upon further incubation, suggesting that the T4-pdg molecules could not diffuse a sufficient distance from the DNA to which they were bound to initiate further catalytic cycles. However, if in these reactions, the salt concentration was rapidly raised to 100 mM, further catalysis was immediately observed (16). Further analyses of DNA reaction products observed at salt concentration >100 mM revealed that the accumulation of form III DNA molecules was no longer linear. Additionally, analyses of the number average molecular weights of the products of the DNA incision reactions revealed that at 100 mM salt, there was a global progressive shortening of the lengths of all DNAs, in contrast to the bimodal distribution observed at low salt. Collectively, these data demonstrated that at salt concentrations well below what is physiologically relevant, it is possible to observe a strong clustering effect of incisions by T4-pdg. However, at salt concentrations within a physiological range (120–140 mM), there was no evidence for processivity.

3.2. Other DNA Glycosylases

Although T4-pdg is the most extensively studied glycosylase relative to its mechanism of target site location, data have been reported for other glycosylases including the *Micrococcus luteus* pyrimidine dimer glycosylase (18), the *Chlorella* virus pyrimidine dimer glycosylase (19), UDG (20–22) and *E. coli* MutY and Fpg (23). Relative to T4-pdg, the other two cyclobutane dimer specific glycosylases mentioned above, exhibited similar processive, clustered nicking activities on heavily irradiated plasmids, except that the salt concentration at which the transition from a processive scarch to a distributive search process was significantly higher.

The data concerning the processivity of UDG has been somewhat controversial with both a limited processivity (20,22) and a distributive process (21) being reported. Nicking assays using plasmids that contained a high density of uracil residues and limiting amounts of UDG, revealed an accumulation of form III DNA molecules, while there were still significant quantities of form I DNA present. However, these data differed significantly from those obtained using T4-pdg because form III DNAs continued to increase well after all form I DNA had been converted to form II. This limited processivity was strongly inhibited by salt concentrations greater than 50 mM. Using a different strategy, Bennett et al. (22) ligated 25-mer oligodeoxynucleotides containing either U:G or U:A mismatches into long DNAs, reacted them with UDG and analyzed the size distribution of the incised DNA products. A processive mechanism was predicted to release primarily 25-mer products without significant accumulation of intermediate oligodeoxynucleotides, while a distributive mechanism of target site location would yield oligomeric multiples of 25-mers. At low salt concentrations, their data clearly demonstrated a strong processive nicking activity such that almost exclusively, 25-mer product DNAs accumulated with little evidence of intermediates. However, raising the salt concentration to as little as 50 mM began to shift the search mechanism to a more distributive pattern (22). In contrast, Purmal et al. (21) using a DNA ligation strategy similar to the one described above, concluded that the mechanism by which uracil was released from these oligomeric DNAs was random. Their data clearly showed a laddering of DNAs with a broad distribution of size fragments, a result consistent with a random incision mechanism. Although the apparent discrepancies in these data have not been experimentally resolved, it is likely that these differences reflect subtle changes in the final salt concentrations of the reactions and the sequence contexts of the uracil residues. All of these investigations would indicate that the UDG target site location mechanism is very sensitive to the salt concentration of the reaction.

Recently, Francis and David (23) have also adopted the oligomeric substrate approaches described above to investigate the mechanism of target site location for *E. coli* MutY and Fpg. DNAs were ligated such that A:G and A:8-oxoG or 8-oxoG:C mispairs were spaced 25 base pairs apart and subsequently reacted with MutY or Fpg, respectively. Fpg showed a highly processive incision activity that was not significantly affected until salt concentrations exceeded 100 mM. Analyses of the MutY reactions were more complex, with the 26 kDa catalytic domain showing a salt-dependent processive release of A from both the DNAs containing A:G and A:8-oxoG mismatches, while the full length MutY functioned processively only on DNAs containing an A:G, but not an A:8-oxoG mispair.

In addition to DNA glycosylases, the second major enzyme in the BER pathway, AP endonuclease, has been investigated for processive nicking activity (24). Using strategies similar in concept to those described above, Carey and Strauss

(24) determined that human AP endonuclease clustered multiple incisions in DNA prior to macroscopic diffusion from the DNA to which it was originally bound. These data suggest that multiple enzymes in the BER pathway may function processively.

4. DISCOVERY AND SIGNIFICANCE OF IN VIVO PROCESSIVE NICKING ACTIVITY BY T4-PDG

4.1. In Vivo Evidence for Processive Incision and Repair of UV Irradiated DNAs

The conclusions drawn from the analyses of the in vitro processivity reactions of T4-pdg would suggest that clustering of incisions at dimer sites in purified plasmid DNAs is likely to be an in vitro biochemical "artifact," since observation of the clustering of incisions required salt concentrations <50 mM. However, in order to test whether there might be physiological mechanisms within *E. coli* to compensate for these elevated salt concentrations, the following experimental strategy was developed by Dr. Elliott Gruskin (25,26). Nucleotide excision repair deficient (*uvr*A) and homologous recombination deficient (*rec*A) *E. coli* were transformed with pBR322-based plasmids that either expressed low quantities of T4-pdg or served as negative controls. Exponentially growing cultures were UV irradiated to introduce 5 or 10 pyrimidine dimers per plasmid molecule. At various time intervals, plasmid DNAs were extracted and analyzed for the persistence of dimers in plasmid DNAs. The kinetics of the accumulation of fully repaired (dimer-free) plasmids was measured by their lack of ability to be cleaved by T4-pdg in subsequent in vitro reactions. An assumption in the design and interpretation of these data was that the rate-limiting step in the completion of repair was the recognition and incision of the dimers and that the subsequent steps of BER would be relatively fast. Therefore, if limiting amounts of T4-pdg initiates repair of all dimers within a subset of plasmids, then these DNAs would be free of lesions, while other plasmids would retain the same number of dimers as were present following the original irradiation. Further, it was predicted that the percent of dimer-free form I DNA would accumulate linearly and that the rate of accumulation of fully repaired plasmids would be inversely proportional to the UV dose.

The experimental data characterizing the kinetics of the accumulation of fully repaired form I DNA showed all the expected parameters of a processive nicking activity. Plasmid DNAs that contained no dimers accumulated linearly and with no time lag throughout the time course of repair (25,26). Further, DNAs containing an average of 5 dimers per plasmid were fully repaired at twice the rate as that observed for the plasmids containing 10 dimers per plasmid. Additionally, in vivo kinetic analyses of the number average molecular weights of plasmid DNAs revealed a bimodal distribution, strongly suggesting that at physiological salt concentrations, T4-pdg initiated repair at all dimer sites within a subset of plasmids prior to reinitiating repair on additional plasmids.

Similar strategies were also used to study whether the UvrABC complex or DNA photolyase processively removed or reversed, respectively, pyrimidine dimers from plasmid DNAs (25,26). These data revealed that UvrABC exhibited a limited processivity, while photolyase reversed dimers by a completely random process throughout the plasmid population. Thus, this experimental design provided a robust assay that was capable of distinguishing highly clustered repair from distributive repair.

4.2. Biological Significance of the Processive Incision Activity of T4-pdg

It is assumed that the interactions between glycosylases and DNA is dominated by electrostatics. Examination of the predicted surface charge distribution on glycosylases reveals that the DNA binding interface is heavily weighted toward basic residues, while often the opposite side of the enzyme has clusters of acidic residues (reviewed in Ref. 27 and Tainer, this volume). However, prior to the determination of any glycosylase crystal structures, we hypothesized that T4-pdg would have a positively charged face that was necessary for the enzyme's ability to cluster incisions in discrete DNA domains. Dr. Diane Dowd was the first to investigate whether site-directed mutagenesis could be used to neutralize amino acids in T4-pdg that were critical for the electrostatic interactions leading to processivity (28–30). Although multiple mutations in T4-pdg were created, only a subset of these was shown to dramatically affect the processive nicking activity of the enzyme. Mutations at two residues (Arg 3 and Lys 26) did not compromise the catalytic activities of T4-pdg in the assays used, but the ability of these mutant enzymes to processively incise plasmid DNAs was abolished even at low salt concentrations. Additionally, in vivo processive repair of UV irradiated plasmids was also lost, and the plasmid repair kinetics resembled that of photolyase or UvrABC. Most significantly, the inability of these mutant enzymes to carry out processive nicking activities resulted in a total loss of the ability of T4-pdg to enhance survival in $uvrA^- recA^- E. coli$ following UV irradiation.

Further evidence that electrostatic interactions dominate the processivity of T4-pdg was obtained by engineering additional basic residues into the DNA binding face (31–33). In these experiments, neutral amino acid residues were converted to basic residues and many of these were successful in increasing nonspecific DNA binding and expanded the salt concentration over which the clustering of dimer incision could be observed. However, in vivo UV survival assays showed that genetically enhancing the density of basic residues on the binding face of T4-pdg did not enhance survival beyond that of the wild type enzyme, and even in some cases, decreased survival was measured.

Overall, these data clearly reveal not only that T4-pdg utilizes a mechanism to cluster its DNA incision activity within defined DNA domains, but also that this in vivo processivity can be as essential for promoting survival as would be observed for catalytic amino acids. Finally, to the best of our knowledge, no other glycosylases have been examined for their abilities to function processively within cells, and thus, the question still remains as to whether this is a generalizable function of DNA glycosylases or whether this is a specialized feature of the T4-pdg.

5. DNA BENDING AS A POTENTIAL PREREQUISITE FOR NUCLEOTIDE FLIPPING

As will be discussed later in the review, DNA glycosylases and glycosylase/AP lyases are not able to completely discriminate substrate from nonsubstrate bases while the nucleotides are buried within duplex B-form DNA. Rather, all of these enzymes utilize a form of nucleotide flipping to establish specific binding of the enzyme to target DNA. Most often, the damaged nucleotide is rotated out of the duplex into a binding/catalytic site on the enzyme. However, other enzymes may flip the nucleotide opposite the damaged base, or in some cases, both the damaged nucleotide and the opposite base may be flipped.

A mechanism by which glycosylases could sample extrahelical bases is to monitor spontaneous base pair openings; however, this is likely to be very inefficient since the life-times of the extrahelical bases will vary significantly depending on both the strength of the hydrogen bonding and base stacking interactions (34–36). Another mechanism to prepay a portion of the binding energy is for the enzyme to bend nontarget as well as target DNA. The structures of DNAs in cocrystal complexes with various glycosylases that flip the damaged nucleotide are germane to this point (37 42). These structures reveal that the DNA is severely bent away from the enzyme as much as 60°–75°. However, it should be noted that these structures represent intermediates that are well along the reaction pathway, and since to date, there are no cocrystal structures of glycosylases complexed to nonspecific DNA, one could infer that bent DNA is a likely intermediate that precedes the formation of a catalytically competent complex.

Chen et al. (43) used atomic force microscopy to examine nonspecific DNA bending. Control linear DNAs or DNAs containing a site-specific 8-oxoguanine: cytosine pair was reacted with human OGG1 (hOGG1). In both cases, hOGG1 bent both the damaged and undamaged DNA to the same extent (∼72°), a value which is in excellent agreement with the bend angle measured in the cocrystal structure of OGG1 bound to substrate DNA (39). In addition, they also observed that nonspecifically bound DNAs could also be unbent, and this bimodal distribution raised the possibility that nucleotide flipping could be the parameter distinguishing the two bound forms.

In addition to hOGG1, T4-pdg has also been investigated relative to its nontarget DNA binding (44). In this investigation, NMR analyses revealed a limited number of amino acid side chains that were perturbed upon the addition of undamaged DNA, while addition of lesion-containing DNA revealed a different set of contacts. This approach will permit a rigorous investigation of the interactions between glycosylases and undamaged DNAs.

T4-pdg is the only enzyme known to date to flip the nucleotide opposite the damaged base (45,46). The X-ray cocrystal structure of a catalytically inactive mutant of T4-pdg (E23Q) with pyrimidine dimer-containing DNA revealed that the adenine opposite the 5′ pyrimidine of the dimer was flipped into a cleft in the side of the enzyme with the extrahelical adenine primarily stabilized between Tyr21 and Pro25 (45). Biochemical evidence for nucleotide flipping was demonstrated using 2-aminopurine (2-AP) in place of the adenine in the T4-pdg substrate, in which the 2-AP was positioned opposite either a pyrimidine dimer or substrate analogs (pyrrolidine or reduced abasic site). Either wild type T4-pdg or the catalytically inactive E23Q mutant was added to these substrates, and the fluorescence enhancement of the 2-AP measured. Depending on the specific type of DNA, 2-AP fluorescence could be enhanced up to ∼7-fold (46). However, it was also shown that specific binding could be achieved without fluorescence enhancement, suggesting that binding and flipping can be uncoupled under certain scenarios.

Very recently our laboratory has utilized fluorescence spectroscopy to examine the temporal relationship between T4-pdg-induced DNA bending and nucleotide flipping (R. K. Walker, in preparation). The experimental design utilized not only the relative fluorescence changes associated with the flipping of the 2-AP as described above, but also added fluorescent probes to the ends of the DNAs to monitor DNA bending using analyses of fluorescence resonance energy transfer. By monitoring the kinetics of both the T4-pdg induced bending and flipping, it has been possible to conclude that these reactions are uncoupled and occur at significantly different rates, with bending preceding the extrahelical movement of the 2-aminopurine. These data are in good agreement with the conclusions drawn from the hOGG1 data suggesting that bending can precede specific base binding.

6. MECHANISMS OF NUCLEOTIDE FLIPPING

6.1. Flipping of a Single Nucleotide

As was mentioned above, the chemistry for the removal of damaged bases generally occurs within a specific binding pocket with the base completely extrahelical. Although DNA bending is likely to be a necessary prerequisite for nucleotide flipping, simply bending the DNA is probably insufficient to achieve specific binding of the damaged base. Insight into the mechanism by which glycosylases promote full extrahelical extrusion of bases comes from analyses of numerous cocrystal structures in which a bulky hydrophobic wedge inserts into the space previously occupied by the base (37–42). In most cases, it is likely that the exact identity of the amino acid side chain is not the most critical characteristic, but rather, the bulk of the side chain to prevent reinsertion of the extrahelical base. The following are the residues that have been implicated in the wedge function of various DNA glycosylases: AAG–Y162; AlkA–L125; FPG–R108; hOGG1–N149 and UDG–L191. Experimental data have been obtained to establish the role of L191 of UDG as a wedge extrusion group (47,48). Analyses of a site-directed mutant of UDG (L191G) revealed that it was significantly compromised in its ability to bind uracil, with binding reduced 60-fold. Fluorescence spectroscopy revealed that this mutant was able to only partially flip the uracil. Thus, in this case, an amino acid side chain without sufficient bulk created an intermediate structure that did not support efficient catalysis. The Stivers' group has carried out a series of elegant experiments to chemically reverse the deleterious effects caused by mutations at L191 (48,49). They reasoned that if the UDG L191G mutant was unable to achieve a full extrahelical flipping of the target uracil, then insertion of a pyrene residue opposite the uracil in duplex DNA could possibly restore activity. Pyrene is sufficiently bulky that it fully occupies the space normally taken up by a complete base pair. When the activities of the UDG L191 mutants were examined using DNAs containing a pyrene:uracil pair, it was determined that wild type levels of activity had been restored. These data directly implicate the role of L191 in UDG base flipping.

Further mutational analyses of residues within UDG have tested hypotheses concerning roles for S88 and S189 in nucleotide flipping due to their phosphodiester backbone interactions at +1 and –1 relative to uracil (50). Although individual mutations only modestly affected the ultimate formation of a bound complex, the double mutant S88A, S189A was determined to proceed only through a metastable, partially flipped stage. These data suggest that for UDG, stabilizing phosphodiester backbone interactions surrounding the target uracil are critical for the overall catalytic efficiency of the enzyme. Again, the demonstration of multiple discrete steps along the reaction pathway reveals the complexity of the processes by which glycosylases progressively discriminate between target and nontarget DNAs.

6.2. Double Flipping

Replication of DNA containing 8-oxoG lesions can result in the misincorporation of adenine opposite the lesion. Once replicated, the 8-oxoG:A mispair poses a significant recognition challenge to cells because the undamaged adenine is the incorrect base, while the oxidized guanine represents the best linkage back to the original G:C base pair. To combat this potentially mutagenic lesion, cells possess a glycosylase, MutY, that removes the adenine from the 8-oxoG:A mismatch. In *E. coli*, MutY is composed of two distinct folding domains of 26 and 13 kDa that can be revealed by limited proteolysis (51). The 26 kDa domain contains the catalytic residues necessary for the release of adenine (52,53), while the 13 kDa domain shares a very limited sequence homology with MutT, a pyrophosphohydrolase that cleaves 8-oxo-dGTP to 8-oxodGMP (54). Subsequent NMR analyses have revealed that the fold of the 13 kDa domain and MutT are very similar (55,56).

Collectively, these data suggested that MutY could carry out a dual flipping mechanism for the recognition of the A:8-oxoG mismatch (56). This hypothesis has been supported from a recent study by Bernards et al. (57), in which stopped flow fluorescent studies revealed kinetically distinct steps in flipping the 8-oxoG and A. This study demonstrates that MutY flips the 8-oxoG at a rate of $\sim$110/sec while the adenine flip occurs much slower, at $\sim$15/sec. They also reported that following the double flip, there was an even slower protein isomerization step, occurring at $\sim$2/ sec. Thus, the preponderance of evidence suggests that MutY carries out a multistepped process to discriminate its complex target.

7. SPECIFICITY OF GLYCOSYLASE BINDING SITES AND CATALYTIC ACTIVITIES

In the previous sections, we have provided a synopsis of the events that precede the specific binding of the damaged base and its catalytic removal. In these concluding remarks, we will only mention the series of steps that ultimately identify the damaged base and remove it.

The "purpose" of nucleotide flipping is to bring bases into a region of the glycosylase which contains amino acid residues that analyze the properties of the incoming base and either sterically exclude it or make a series of interactions that collectively determine the K_m for that base. These specific interactions may include hydrogen bonding, hydrophobic interactions, stabilization of electron orbitals and salt bridges, and are completely analogous to the kind of enzyme–substrate interactions operative in all other enzyme classes (Tainer, this volume).

Once a base is bound within the active site pocket of a glycosylase, the final discrimination step between target and nontarget DNA is the catalytic removal (or not) of the damaged base (i.e., k_{cat} for the substrate base). This step operationally initiates the BER pathway. Depending on the enzyme, the glycosylase reaction may result in either the breakage of the glycosyl bond alone, or a combined glycosyl bond scission with additional β- or β/β-elimination reactions. The DNA products of these reactions are an abasic site with the phosphodiester backbone intact, or an abasic site with the backbone cleaved, accompanied by either a 3′-α,β-unsaturated aldehyde, 5′-phosphate or 3′- and 5′-phosphates, respectively.

The determinant of the spectrum of the final DNA reaction products has been shown to be the identity of the nucleophile that collapses onto the oxocarbenium ion product of the glycosylase step (at C1′ of the deoxyribose) (reviewed in Refs. 1, 27, 58). For DNA glycosylases that catalyze only cleavage of the glycosylase bond, it has been suggested that this nucleophile is an activated water molecule or a direct SN1 attack (59,60). For the enzymes that cleave the glycosyl bond, and additionally catalyze β- or β/β-elimination reactions, the identity of the nucleophile has been established to be either the α-amino group of the N-terminus or an ε-amino group of lysine (10,11,27,61). The secondary amine of an N-terminal proline residue leads to a β/δ-elimination chemistry almost exclusively (41,42,62,63) The identity of the residues that catalyze β- or β/δ-elimination reactions has been determined by reduction of the covalent imine intermediate by NaBH₄ or NaCNBH₃. Stivers and Jiang (1) provide an outstanding review of the chemistry of these reactions. Further, the reader is referred to the contribution by Roman Osman in this volume () for an excellent review of the molecular dynamics of DNA flexibility, bending, flipping, and base catalysis

REFERENCES

1. Stivers JT, Jiang YL. Chem Rev 2003; 103:2729–2759.
2. Ptashne M. Nature 1967; 214:232–234.
3. Riggs AD, Bourgeois S, Cohn M. J Mol Biol 1970; 53:401–417.
4. von Hippel PH, Berg OG. J Biol Chem 1989; 264:675–678.
5. Lloyd RS, Hanawalt PC, Dodson ML. Nucleic Acids Res 1980; 8:5113–5127.
6. Stanford NP, Szczelkun MD, Marko JF, Halford SE. Embo J 2000; 19:6546–6557.
7. Halford SE. Biochem Soc Trans 2001; 29:363–374.
8. Morikawa K, Matsumoto O, Tsujimoto M, Katayanagi K, Ariyoshi M, Doi T, Ikehara M, Inaoka T, Ohtsuka E. Science 1992; 256:523–526.
9. Morikawa K, Ariyoshi M, Vassylyev DG, Matsumoto O, Katayanagi K, Ohtsuka E. J Mol Biol 1995; 249:360–375.
10. Schrock RDd, Lloyd RS. J Biol Chem 1991; 266:17631–17639.
11. Schrock RDd, Lloyd RS. J Biol Chem 1993; 268:880–886.
12. Manuel RC, Latham KA, Dodson ML, Lloyd RS. J Biol Chem 1995; 270:2652–2661.
13. Latham KA, Manuel RC, Lloyd RS. J Bacteriol 1995; 177:5166–5168.
14. Latham KA, Rajendran S, Carmical JR, Lee JC, Lloyd RS. Biochim Biophys Acta 1996; 1292:324–334.
15. Lloyd RS, Dodson ML, Gruskin EA, Robberson DL. Mutat Res 1987; 183:109–115.
16. Gruskin EA, Lloyd RS. J Biol Chem 1986; 261:9607–9613.
17. Ganesan AK, Seawell PC, Lewis RJ, Hanawalt PC. Biochemistry 1986; 25:5751–5755.
18. Hamilton RW, Lloyd RS. J Biol Chem 1989; 264:17422–17427.
19. McCullough AK, Romberg MT, Nyaga S, Wei Y, Wood TG, Taylor JS, Van Etten JL, Dodson ML, Lloyd RS. J Biol Chem 1998; 273:13136–131342.
20. Higley M, Lloyd RS. Mutat Res 1993; 294:109–116.
21. Purmal AA, Lampman GW, Pourmal EI, Melamede RJ, Wallace SS, Kow YW. J Biol Chem 1994; 269:22046–22053.
22. Bennett SE, Sanderson RJ, Mosbaugh DW. Biochemistry 1995; 34:6109–6119.
23. Francis AW, David SS. Biochemistry 2003; 42:801–810.
24. Carey DC, Strauss PR. Biochemistry 1999; 38:16553–16560.
25. Gruskin EA, Lloyd RS. J Biol Chem 1988; 263:12728–12737.
26. Gruskin EA, Lloyd RS. J Biol Chem 1988; 263:12738–12743.
27. McCullough AK, Dodson ML, Lloyd RS. Annu Rev Biochem 1999; 68:255–285.
28. Dowd DR, Lloyd RS. Biochemistry 1989; 28:8699–8705.
29. Dowd DR, Lloyd RS. J Mol Biol 1989; 208:701–707.

30. Dowd DR, Lloyd RS. J Biol Chem 1990; 265:3424–3431.
31. Nickell C, Lloyd RS. Biochemistry 1991; 30:8638–8648.
32. Nickell C, Anderson WF, Lloyd RS. J Biol Chem 1991; 266:5634–5642.
33. Nickell C, Prince MA, Lloyd RS. Biochemistry 1992; 31:4189–4198.
34. Gueron M, Leroy JL. Methods Enzymol 1995; 261:383–413.
35. Moe JG, Russu IM. Nucleic Acids Res 1990; 18:821–827.
36. Dornberger U, Leijon M, Fritzsche H. J Biol Chem 1999; 274:6957–6962.
37. Barrett TE, Scharer OD, Savva R, Brown T, Jiricny J, Verdine GL, Pearl LH. Embo J 1999; 18:6599–6609.
38. Parikh SS, Walcher G, Jones GD, Slupphaug G, Krokan HE, Blackburn GM, Tainer JA. Proc Natl Acad Sci USA 2000; 97:5083–5088.
39. Bruner SD, Norman DP, Verdine GL. Nature 2000; 403:859–866.
40. Fromme JC, Verdine GL. Nat Struct Biol 2002; 9:544–552.
41. Zharkov DO, Golan G, Gilboa R, Fernandes AS, Gerchman SE, Kycia JH, Rieger RA, Grollman AP, Shoham G. Embo J 2002; 21:789–800.
42. Gilboa R, Zharkov DO, Golan G, Fernandes AS, Gerchman SE, Matz E, Kycia JH, Grollman AP, Shoham G. J Biol Chem 2002; 277:19811–19816.
43. Chen L, Haushalter KA, Lieber CM, Verdine GL. Chem Biol 2002; 9:345–350.
44. Ahn HC, Ohkubo T, Iwai S, Morikawa K, Lee BJ. J Biol Chem 2003; 278:30985–30992.
45. Vassylyev DG, Kashiwagi T, Mikami Y, Ariyoshi M, Iwai S, Ohtsuka E, Morikawa K. Cell 1995; 83:773–782.
46. McCullough AK, Dodson ML, Scharer OD, Lloyd RS. J Biol Chem 1997; 272: 27210–27217.
47. Handa P, Roy S, Varshney U. J Biol Chem 2001; 276:17324–17331.
48. Jiang YL, Kwon K, Stivers JT. J Biol Chem 2001; 276:42347–42354.
49. Jiang YL, Stivers JT, Song F. Biochemistry 2002; 41:11248–11254.
50. Jiang YL, Stivers JT. Biochemistry 2002; 41:11236–11247.
51. Manuel RC, Czerwinski EW, Lloyd RS. J Biol Chem 1996; 271:16218–16226.
52. Guan Y, Manuel RC, Arvai AS, Parikh SS, Mol CD, Miller JH, Lloyd S, Tainer JA. Nat Struct Biol 1998; 5:1058–1064.
53. Manuel RC, Lloyd RS. Biochemistry 1997; 36:11140–11152.
54. Noll DM, Gogos A, Granek JA, Clarke ND. Biochemistry 1999; 38:6374–6379.
55. Volk DE, House PG, Thiviyanathan V, Luxon BA, Zhang S, Lloyd RS, Gorenstein DG. Biochemistry 2000; 39:7331–7336.
56. House PG, Volk DE, Thiviyanathan V, Manuel RC, Luxon BA, Gorenstein DG, Lloyd RS. Prog Nucleic Acid Res Mol Biol 2001; 68:349–364.
57. Bernards AS, Miller JK, Bao KK, Wong I. J Biol Chem 2002; 277:20960–20964.
58. Dodson ML, Michaels ML, Lloyd RS. J Biol Chem 1994; 269:32709–32712.
59. Hollis T, Lau A, Ellenberger T. Mutat Res 2000; 461:201–210.
60. Hollis T, Ichikawa Y, Ellenberger T. Embo J 2000; 19:758–766.
61. Sun B, Latham KA, Dodson ML, Lloyd RS. J Biol Chem 1995; 270:19501–19508.
62. Zharkov DO, Rieger RA, Iden CR, Grollman AP. J Biol Chem 1997; 272:5335–5341.
63. Rieger RA, McTigue MM, Kycia JH, Gerchman SE, Grollman AP, Iden CR. J Am Soc Mass Spectrom 2000; 11:505–515.

3

Increased Specificity and Efficiency of Base Excision Repair Through Complex Formation

Karen H. Almeida and Robert W. Sobol
Hillman Cancer Center, University of Pittsburgh Cancer Institute,
Pittsburgh, Pennsylvania, U.S.A.

1. INTRODUCTION

The genome is continuously damaged by metabolites from cellular processes and by environmental agents. The inability to correct DNA damage brings about unwanted genetic changes in critical genes and eventually can lead to aberrant cellular growth or cancer. Protecting the genome from changes leading to disease and/or death is therefore vital to the survival of a species. Consequently, it is important to understand the diverse cellular systems for DNA damage repair available for protection from an enormous array of DNA lesions. Base excision repair (BER) is considered the predominant pathway for repair of small DNA lesions resulting from exposure to either environmental agents or cellular metabolic processes that produce alkylating agents, reactive oxygen species, and/or other reactive metabolites able to modify DNA (1–4).

BER is a highly versatile DNA repair mechanism, employing at least 10 known mammalian DNA glycosylases for the recognition and removal of a variety of different base modifications or lesions (Table 1) (5,6). Repair is initiated by a lesion-specific DNA glycosylase (mono- or bi-functional) and can be completed by either of two sub-pathways: short-patch BER; a mechanism whereby only one nucleotide is replaced or long-patch BER; a mechanism whereby 2–13 nucleotides are replaced. The majority of repair is currently thought to occur via the short-patch pathway. The paradigm for the short-patch BER pathway is as follows: the DNA base lesion is removed by a damage-specific glycosylase; the resulting apurinic/apyrimidinic (AP) site is recognized by an AP endonuclease (Ape1/APEX/HAP1/Ref-1 in mammals) (5), catalyzing the incision of the damaged strand, leaving a 3′ OH and a 5′ deoxyribose-phosphate moiety (5′ dRP) at the margins. Polymerase β (pol-β) hydrolyzes the 5′ dRP moiety and fills the single nucleotide gap, preparing the strand for ligation by either DNA ligase I (LigI) or a complex of DNA ligase III (LigIII) and XRCC1 (Fig. 1).

Table 1   Summary of Mammalian BER Proteins

Function	Gene (3,4,6,148)	Known substrate[a] (3,6,148)	Reference
DNA glycosylase			
Uracil DNA glycosylase	UNG/UDG	ssU, U:G, U:A, 5-FU	
	UNG2	ssU, U:G, U:A, 5-FU	
	UDG2	U:A	
	DUG	U:G, T:G, EthenoC	
	SMUG	ssU, U:G, U:A, 5-HmU	(149)
Thymine/Uracil DNA glycosylase	TDG	U:G, ethenoC:G, T:G, Tg:G	(150)
	MBD4	U or T in U/TpG:5-meCpG, Tg:G	(150)
3MeA DNA glycosylase	Aag/MPG	3-meA, 7-meA, 3-meG, 7-meG, hypoxanthine, ethenoA, ethenoG	
8-OxoG DNA glycosylase/AP lyase	OGG1	8-oxoG:C, 8-oxoG:T, 8-oxoG:G, me-FapyG:C, FapyG:C	
	OGG2	8-oxoG:G, 8-oxoG:A	(151)
MutY G:A mismatch glycosylase/AP lyase	MYH	A:G, A:8-oxoG, C:A, 2-OH-A	
	MTH	8-oxo-dGTP	
Thymine glycol DNA glycosylase/AP lyase	NTH1	T/C-glycol, dihydrouracil, fapy	
	NEIL1	TgG, 5-OH and 6-OH dihydrothymine, 5-OH-C, 5-OH-U	
	NEIL2	5-OH-U, 5,6-dihydrouracil, 5-OH-C	(152)
	NEIL3		(153)
AP endonuclease			
	Ape1/ HAP1/ REF1	AP sites, 3′ phosphate, 3′ phosphoglycolate	
	Ape2	AP sites, 3′ phosphate, 3′ phosphoglycolate	
DNA polymerases			
Short or long patch	pol-β	5′ dRP, gapped DNA	
Long patch	pol-δ	gapped DNA	
	pol-ε	gapped DNA	
DNA ligase			
	LigI	Nicked DNA	
	LigIII	Nicked DNA	
Additional factors			
Flap endonuclease	Fen1	Displaced flap structures	
Proliferating cell nuclear antigen	PCNA		
Replication factor C	RFC		
Scaffold protein	XRCC1	ssBreak	
DNA kinase-phosphatase	PNKP	gapped DNA with 3′ PO_4 and 5′ OH	(87)
Poly(ADP-ribose)polymerase	PARP1	ssBreak	(154,155)
	PARP2	ssBreak	(156)

[a]Target base on left in mismatches

Almeida and Sobol

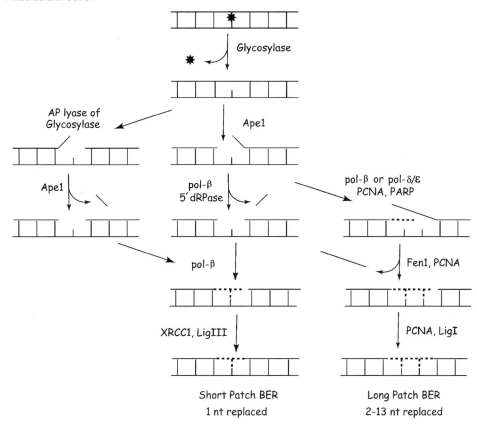

Figure 1 Schematic diagram of the BER pathway.

Bi-functional glycosylases contain an additional 3′ AP lyase activity. Upon recognition of a base lesion by a bi-functional DNA glycosylase, the lesion (e.g., 8-oxoG) is excised from the DNA strand in a mechanism similar to mono-functional glycosylases. However, the DNA backbone is then incised 3′ to the damage site, leaving a 3′ unsaturated aldehyde (after β-elimination) and a 5′ phosphate at the termini. A 3′ phosphodiesterase activity, supplied by Ape1 (7), cleaves this terminus in preparation for polymerase extension by pol-β and ligation by LigI or LigIII/XRCC1. Alternatively, the processing of the strand break induced by oxidative damage may be accomplished by polynucleotide kinase 3′ phosphatase (PNKP), a bi-functional enzyme containing both a polynucleotide kinase activity and an associated 3′ phosphatase activity (8).

Long-patch BER is initiated in a similar fashion: glycosylase-initiated lesion removal and subsequent strand excision by Ape1. However, in cases where the 5′ dRP moiety is refractory to pol-β 5′ dRP lyase activity (9), polymerase ε, δ, or β, coupled with proliferating cell nuclear antigen (PCNA) and replication factor C (RFC) synthesizes the DNA to fill the gap, resulting in a displaced DNA flap of 2–13 bases in length. DNA synthesis and strand displacement by pol-β is stimulated by a combination of the structure-specific endonuclease Fen1 and poly(ADP-ribose) polymerase (PARP1) (10,11). Fen1 then catalyzes the removal of this flap, leaving a nick that has been transferred 2–13 nucleotides downstream of the original damage site. The intact DNA strand is restored with LigI (Fig. 1). The components described here are essential to in vitro BER; however, there are numerous accessory factors and/or post-translational modifications shown to assist in processing the DNA in vivo, as will be described (Table 1).

An estimated rate of 10^4 damaging events/mammalian cell/day underscores the importance of the BER pathway (6,12,13). Accumulation of base lesions has severe consequences to the cell, including replication blocks, increased recombination, increased mutations and cytotoxicity (14–20). Furthermore, the DNA intermediates generated during BER can themselves be considered lesions and as such have dire consequences: AP sites, generated either spontaneously (e.g., spontaneous depurination) or as products of DNA glycosylase activity, are chemically labile and degrade into DNA single-strand breaks. If left intact, AP sites cause base mis-incorporation (21) and are highly mutagenic at the transcriptional level (22). An in-depth review of the consequences and repair of AP sites can be found in a subsequent chapter. The BER intermediate 5′ dRP, if left un-repaired, produces replication blocks, increased recombination, chromosomal aberrations and cytotoxicity (1,23–26). And finally, DNA strand breaks are also highly cytotoxic and are implicated in the cellular sensitivity to many DNA damaging agents (27).

Attempts to establish in vivo animal model systems exploring BER have proven difficult. Whereas mice carrying null mutations in most, if not all, DNA glycosylase genes are viable and develop normally, all BER proteins acting subsequent to gylcosylase initiation (with the exception of PARP1) are embryonic or perinatal lethal, implying that interrupted BER has more serious consequences than base damage (28). Considering the severe consequences of un-repaired BER intermediates, it seems unlikely that a biological system would evolve to leave these reactive repair intermediates exposed to the cellular milieu. Therefore, for BER to be beneficial to the cell, intermediates must be transferred down the pathway in a protected state, which requires rigid coordination or orchestration of the proteins involved. Based on structural and biochemical studies involving uracil DNA glycosylase (UNG), Ape1 and pol-β, Wilson and Kunkel (29) proposed a molecular relay-style mechanism driven by sequential protein–protein interactions (30,31). This and subsequent studies have suggested that BER functions in a coordinated fashion, likely involving either pre-formed repair complexes, such as the so-called "repairosome," or repair complexes that assemble at the site of damage. This review presents evidence to date in support of this latter hypothesis that complexes of BER proteins assemble at the site of the DNA lesion and mediate repair in a coordinated fashion involving protein–protein interactions that dictate subsequent steps or sub-pathway choice. This mechanism or process of complex formation appears to provide an increase in specificity and efficiency to the BER pathway, thereby facilitating the maintenance of genome integrity by preventing the accumulation of highly toxic repair intermediates.

2. DNA LESION RECOGNITION AND REMOVAL

The initial step in any DNA repair reaction is the recognition of the DNA lesion. A critical component of this first step in BER is the specificity of recognition and the capacity to recognize subtle DNA lesions such as uracil, for example, resulting from the deamination of cytosine. The various mechanisms and active site residues responsible for lesion recognition and removal have attracted considerable study.

As will be presented below, the specificity and activity of DNA glycosylases are further refined through the formation of protein complexes to direct or target the start of repair and facilitate the identification of subtle lesions in DNA and associated alterations in chromatin, etc. This is particularly important since DNA damage in nucleosome core particles, for example, is more resistant to removal by glycosylases than free DNA (32,33). In one of the earliest reports of increased specificity of a glycosylase (34), it was demonstrated that the alkyl-adenine DNA glycosylase (Aag/MPG) exists in a complex with the Rad23 proteins hHR23-A and -B (35). Such an interaction was found to elevate the binding affinity for damaged DNA and to mediate an increased rate of lesion removal. Hence, lesion recognition by DNA glycosylases may be enhanced through protein complex formation and therefore protein partners may provide another layer of DNA repair specificity and regulation. Aag was more recently found to exist in a complex with methylated DNA binding domain 1 (MBD1), suggesting that such a hetero-dimer can promote recognition of subtle base damage in chromatinized DNA (36). Increased lesion recognition and lesion removal following protein complex formation is, however, not unique to Aag/MPG. Thymine DNA glycosylase (TDG) interacts physically with Ape1 (37), the subsequent enzyme in the BER pathway, an interaction that enhances the ability of TDG to excise damaged bases. TDG activity may also be enhanced through other protein partnerships. For example, functional interactions have been detected between TDG and xeroderma pigmentosum group C protein (XPC-HR23B), where XPC-HR23B was shown to promote dissociation from AP sites, stimulating TDG turn over (38). Lesion removal via bi-functional glycosylase activity may also be stimulated through cofactor interactions. Xeroderma pigmentosum G protein, XPG, a 3′ endonuclease more commonly associated with nucleotide excision repair, promotes the binding of human endonuclease III (hNTH1) to its substrate DNA and increases glycosylase activity approximately 2-fold through direct protein–protein interactions (39,40). Furthermore, co-existence of the human 8-oxoguanine-DNA N-glycosylase homolog (hOGG1) and Cockayne syndrome B group protein (CSB) was demonstrated in vivo under conditions of cellular stress, although no physical or functional interaction could be confirmed using purified proteins (41). More recently, hOGG1 activity has been demonstrated to depend on the presence of XRCC1 and will be discussed later.

PCNA appears to play a central role in many steps of the BER pathway, not the least of which is lesion recognition. There are several examples of glycosylase/ PCNA interactions and it may be that PCNA facilitates an enhanced capacity for lesion recognition and removal. UNG was the first of the glycosylases proposed to interact with PCNA, an interaction identified in HeLa cell extracts (42). In subsequent studies, it was suggested that the UNG/PCNA complex formation directs BER to replication foci (43). SMUG1 or the 5-meC DNA glycosylase was purified in a complex with several nuclear proteins, PCNA among them (44). It is not known if this SMUG1/PCNA partnership affects specificity. Finally, the hMYH protein (specific for the removal of adenine or 2-hydroxyadenine when mis-paired with guanine or 8-oxo-guanine) was observed to co-localize with PCNA (45,46) and again suggests that glycosylase/PCNA interactions facilitate the sub-cellular targeting of BER initiation at replication foci, allowing for a high degree of specificity to mediate a "post-replication BER" mechanism (Table 1).

2.1. Glycosylase Stimulation by AP Endonuclease

Proteins generally evolve to optimize the rate of turnover for a particular biological reaction. Incorporated into this rate is the ability of the enzyme in question to release its product, allowing catalysis of another substrate and imparting a level of regulation to the amount of product generated. Many glycosylases, however, are inhibited by their product AP sites (34,47–52) and addition of Ape1, the successive protein in the BER pathway, stimulates turnover through release of this product inhibition (Table 2). For example, kinetic studies of TDG exhibited a one-to-one stoichiometry for thymine removal from G:T mismatched DNA (48), presumably due to the protein remaining bound to the AP site (53). No formation of a ternary complex between TDG, AP:DNA, and Ape1 was established although recently a direct interaction between TDG and Ape1 was reported (37). Furthermore, Ape1 could not incise DNA in the presence of TDG. Instead, molar equivalents of Ape1 increased the rate of dissociation of TDG from the AP:DNA 2–3-fold but a significant increase in turnover was observed only after addition of 100-fold excess of Ape1. Facilitation of this dissociation did not require a catalytically active Ape1 molecule, similar to reports for the uracil DNA glycosylases UNG2 and SMUG1. A 125-fold molar excess of Ape1 approximately doubled the catalytic activity of UNG2 (54), while a 15-fold molar excess of Ape1 stimulated uracil release by SMUG1 approximately 3-fold even under conditions designed to impair the catalytic activity of Ape1 without affecting binding capacity (49). To date there is no report of Ape1 stimulating the activity of Aag/MPG even though this glycosylase has similarly low turnover capabilities, characteristic of product inhibition kinetics (see Table 2).

2.2. Post-Translational Modification of DNA Glycosylases

Post-translational modifications (PTM) have also been shown to affect lesion removal to provide tighter regulation of the BER pathway. Recently, two-hybrid analysis revealed a specific interaction between TDG and the human ubiquitin-like modifiers SUMO-1 and SUMO-3 (55). TDG was covalently modified by SUMO proteins in a reversible manner and this SUMO modification effectively abrogated the binding capacity of TDG for its AP:DNA product. As a result, SUMO-modified TDG continued to process G:U mismatched DNA (previously determined to be a better substrate than G:T mismatches) (56), at a steady-state rate of 0.027/min, unlike its unmodified counterpart that leveled off in a plateau corresponding to the characteristic product inhibition kinetics.

Addition of Ape1 stimulated the turnover kinetics of the sumoylated TDG while having little effect on the unmodified enzyme (55). Equimolar concentrations of Ape1 produced an additional 3-fold increase in the enzymatic turnover of sumoylated TDG (0.079/min), measured through thymine release, whereas unmodified TDG remained undetectable. Non-sumoylated TDG was proposed to be a high-affinity state of the glycosylase, able to bind and process damaged DNA. Once bound, TDG hydrolyzes the damaged base, retaining its position on the resulting AP site. Possibly, it is this complex that undergoes SUMO conjugation, reducing the DNA binding affinity of TDG for its product DNA and providing an opportunity for Ape1 displacement which ultimately results in the transfer of the AP site from the glycosylase to the AP endonuclease for further processing.

TDG has also been demonstrated to associate with transcriptional co-activators CBP and p300 and is a substrate for CBP/p300 acetylation (37).

Table 2 Summary of Kinetics and Binding of BER Proteins[a]

Protein	Substrate[b]	Product[b]	K_d(nM)	k_{on}(M^{-1}s^{-1})	k_{off}(s^{-1})	k_{cat}(min^{-1})	K_m(nM)	Turnover stimulated	Reference
hTDG	G:T					0.91		hAPE1	(48,53)
	G:U					11		SUMO, hAPE1	(53,55)
	C:U					0.021	12	hAPE1	(157)
	sMeG:T					4.7			(53)
						0.026			(53)
		G:AP			2.8×10^{-5}				(48)
		sMeG:AP			7.3×10^{-4}				(48)
		C:AP			1.0×10^{-3}				(48)
		G:PYR	0.016						(47)
		G:THF	0.023						(47)
hMBD4	G:T			$> 200 \times 10^6$	8×10^{-6}	0.72[c]			(50)
	G:U			$> 200 \times 10^6$	8×10^{-6}	3[c]			(50)
hUNG	A:4'-S-dU		310	0.16×10^6	0.05				(54)
	G:4'-S-dU		200	0.20×10^6	0.04				(54)
	U					[d]			(54)
	G:U					276	1.0×10^7	hAPE1	(158)
	ssU					546	400		(158)
hSMUG1		AP	6	30.3×10^6	0.2		1100[e]		(54,159)
	G:U					0.084[f]	35[f]	hAPE1	(49,160)
	ssU					150[f]	1090[f]	hAPE1	(49,160)
hAag	3-MeA					0.78[g]	130[g]		(161)
						10.8	8		(162)
	7-Me-G					0.462[g]	500[g]		(161)
						0.348	25		(162)
	Hx		8.6			0.096	11		(34)

(*Continued*)

Table 2 Summary of Kinetics and Binding of BER Proteins[a] (Continued)

Protein	Substrate[b]	Product[b]	K_d(nM)	k_{on}(M^{-1}s^{-1})	k_{off}(s^{-1})	k_{cat}(min^{-1})	K_m(nM)	Turnover stimulated	Reference
εA						0.096[g]	40–100[g]		(163)
							24		(164)
		AP	1.6						(34)
		T:PYR	0.023						(47)
		T:THF	0.16						(47)
mMyh	GO:A					[d]		hAPE1	(51)
hOGG1	C:8-oxoG		23.4			0.108	9.9	hAPE1	(59,60)
						0.5[c]		hAPE1	(60)
	C:8-OH-G					0.034	23		(165)
	C:β-elim					0.042	513		(166)
		C:AP	223		5.3×10^{-5}				(59,60)
						0.0532	7.2		(60)
		C:F	2.8		5.00×10^{-3}				(59)
hNTH1	A:Tg		0.8			3.7		hAPE1	(57,167)
hApe1	G:AP		1.7	5×10^7	0.04	600	200		(62)
	C:AP		72.7			21.67	13.7		(60)
	C:β-elim		3						(60)
	G:F							pol-β	(70)
		3'OH	4.3						(168)
		G:Incised-F	4						(60)
Average						184.62	32.5		(70)

[a]Values are apparent values, i.e., dependent on conditions unless otherwise stated.
[b]Target base is on the right.
[c]Non-steady-state conditions.
[d]Characteristic kinetics of product inhibition were demonstrated, but no k_{cat} value was calculated.
[e]Mitochodrial rat homolog. Reported K_i for AP site of 1200 nM.
[f]Xenopus homolog.
[g]Murine homolog.

Interestingly, TDG/CBP complex formation does not affect the binding or activity of TDG nor does it affect the histone acetylation capabilities of CBP (37). Acetylated TDG leads to the release of CBP/p300 from the DNA bound complex and effectively abrogates TDG interaction with Ape1. It is currently unclear why a cell might modify TDG during the repair process; however, it is possible that acetylation of TDG by CBP modulates the interaction between TDG and Ape1 in an effort to regulate the recruitment of BER enzymes.

It remains to be seen whether other DNA glycosylases undergo similar PTMs that affect glycosylase efficiency and/or complex formation with downstream BER proteins. For TDG, at least, PTMs may provide an additional level of complexity and specificity to the regulation of BER.

2.3. Bi-functional Glycosylases

Bi-functional glycosylases add an additional level of complexity due to their 3' AP lyase activity, yielding a replication-blocking 3' β-eliminated unsaturated aldehyde and a 5' phosphate at the margins of the repair gap. Until recently, glycosylase action and catalysis to yield the β-elimination product were thought to occur concomitantly. However, studies with bi-functional glycosylases such as human endonuclease III (hNTH1) and mammalian 8-oxoguanine-DNA N-glycosylase homolog (OGG1) suggest a dissociation of the two activities. Studies of hNTH1 activity and binding affinity to thymine glycol (Tg) containing DNA revealed that the 3' AP lyase activity of hNTH1 is dependent on the binding affinity of the enzyme to its product AP site (57). Initially, the Tg:A substrate exhibited little product inhibition to its corresponding AP:A product. Upon addition of a 4-fold excess of Ape1, whose catalytic activity was abrogated in the absence of $MgCl_2$, glycosylase activity of hNTH1 increased 2–3-fold, indicating that in the absence of Ape1 activity, hNTH1 was inhibited by its AP site product. Presumably, this increase was due to Ape1 stimulation of hNTH1 dissociation from its AP:A product, enabling the enzyme to bind additional substrate and effectively circumventing its 3' lyase activity. Indeed, Ape1-assisted dissociation of hNTH1 was so effective in abrogating the 3' lyase activity that upon activation of Ape1, only 5' dRP (Ape1 product) was detectable, not the products of β-elimination. Therefore, under physiological conditions (i.e., abundant Ape1), hNTH1 may act as a mono-functional glycosylase on Tg:A damaged DNA, shuttling damage towards cleavage by Ape1, which has a greater capacity to cleave 5' to the AP site than 3' to the β-elimination product (58). Similarly, hOGG1 and mMyh showed an enhancement of glycosylase activity and suppression of the glycosylase 3' lyase function upon addition of Ape1, suggesting that circumventing the 3' lyase activity is common to bi-functional glycosylase activity in vivo (51,59,60). It should be noted that there remains a discrepancy as to whether hOGG1 is actively displaced by Ape1 or if displacement occurs spontaneously and immediately prior to Ape1 association (59,60).

Interestingly, in the case of the Tg:G substrate, hNTH1 was significantly inhibited by its product AP:G site and the rate of glycosylase activity was not stimulated by addition of Ape1 (57). This implies that Ape1 does not effect the dissociation of hNTH1 from its AP:G product. Furthermore, Ape1 stimulated the rate of 3' lyase activity under single-turnover conditions only, suggesting that the 3' lyase activity of hNTH1 is inhibited by its β-elimination product and that Ape1 does not alleviate that inhibition (57). The disparity in the processing of Tg:A and Tg:G substrates is attributed to differences in the binding affinities of hNTH1 to their respective AP site products (57). A relatively low affinity for AP:A enables Ape1 to effectively replace hNTH1 at the site of damage, while a higher affinity of hNTH1 to its AP:G product prevents Ape1 from performing this displacement. hNTH1 remains bound to the AP:G site, the 3′ lyase activity cleaves the backbone and hNTH1 then remains bound to the 3′-elimination product. It has been reported previously that the 3′ lyase activity of hNTH1 is dependent on the base opposite the AP site, being 100-fold greater for AP:G than AP:A (61). Interestingly, it is the nature of the orphan base pairing with the AP site that determines whether the DNA is cleaved by Ape1 or hNTH1 and therefore bestowing a sequence dependence to the choice of BER initiation. It has yet to be determined if other bi-functional glycosylases exhibit a similar sequence-dependent pathway selection.

Processing of the α-β unsaturated aldehyde, resulting from 3′ lyase activity, is still unclear. For hOGG1, the binding affinity for the product of 3′ lyase activity (the β-elimination product; $K_d = 223$ nM) was 100-fold weaker than for the product of glycosylase activity (AP site; $K_d = 2.8$ nM), but Ape1 affinity for the hOGG1 mediated β-elimination product ($K_d = 72.7$ nM) was ~3-fold stronger than hOGG1's affinity, implying that Ape1 could displace hOGG1 off the β-eliminated DNA (60). However, it has also been reported that hOGG1 remains tightly bound to the β-elimination product ($k_{off} = 5.3 \times 10^{-5}$ s^{-1}) and that Ape1 has no effect on that affinity ($k_{off} = 4.8 \times 10^{-5}$ s^{-1}) (59), similar to hNTH1, as discussed above. In the absence of glycosylase, Ape1 strongly bound the β-elimination product, preferring it to the intact AP site, although no K_d or k_{off} values were reported (62). Taken together, this data suggest that, in vitro, the processing of a 3′ α-β unsaturated aldehyde (β-elimination product) through the BER pathway is limited. Clearly, the interactions of Ape1 and bi-functional glycosylases with the β-elimination DNA product need to be further elucidated in order to determine the in vivo processing of these dangerous intermediates.

As mediators of the initial step of BER, glycosylases are critical to the recognition and removal of DNA base lesions. Their mechanism of action has been extensively studied on a number of different levels. Kinetic data with purified protein suggest that glycosylases may be inhibited by their product AP sites. Evaluation of protein partnerships, such as with Ape1, demonstrates a stimulation of overall catalytic efficiency through glycosylase displacement resulting in an alleviation of product inhibition. Specificity of BER initiation is emerging through evidence of PTMs that may contribute to the overall regulation of the pathway. Importantly, regulation of glycosylase activity via protein–protein interactions has been shown to facilitate improved lesion recognition and to ensure that initiated repair may be completed via the formation of lesion-specific repair complexes.

3. STRAND INCISION

AP sites are detrimental cellular DNA lesions that give rise to an increase in genetic mutations and other genetic rearrangements (63). In addition to the AP sites generated through glycosylase action, it has been estimated that at least 10,000 AP sites are generated spontaneously per mammalian cell per day (13). The sheer quantity of this lesion suggests that a tightly controlled system is necessary for accurate repair. Ape1 recognizes these AP sites and incises the DNA 5′ to the lesion. In the mouse, Ape1 is an essential gene. Mice with a targeted homozygous null mutation in the Ape1 (Ref-1) gene die during early embryonic development (64). Detailed analysis of Ape1 null embryos (pre-implantation Ape1 null embryos) following ionizing radiation indicates a role for Ape1 in the repair of ionizing-induced DNA damage (65). No other Ape1-deficient mammalian cellular models are available, making in vivo studies in mammalian systems challenging. The recent demonstration that siRNA may be used to down-regulate Ape1 does however show promise for future studies to define the role of Ape1 in mammalian cells in vivo (66). Although AP-site specific 5′ endonucleolytic activity is the major function of Ape1, there are a number of minor functions associated with Ape1 as well, including 3′ phosphodiesterase, 3′ phosphatase and 3′–5′exonuclease capabilities. All of these alternate functions participate in the processing of 3′ blocked termini within a repair patch. Ape1 also encodes a redox function, reported to mediate redox activation of stress-inducible transcription factors (67). Physical and functional partnerships with Ape1 include a number of DNA repair proteins such as LigI, Fen1, PCNA, XRCC1 and most importantly for short-patch repair, pol-β (see Table 3). Additionally, partnership with p53 (identified via Ape1 over-expression) enhanced the ability of p53 to *trans*-activate various target promoters and increased the ability of p53 to stimulate endogenous p21 expression (68). Moreover, the extent of apoptosis induced by p53 correlated with endogenous Ape1 levels (68). The multi-functional role of Ape1 and its participation in many protein–protein and multi-protein complexes support the hypothesis that the specificity of BER is facilitated by complex formation (see Fig. 2). The following section will focus on Ape1 interactions with pol-β, reflecting a critical role in both short-patch repair and essential partnerships with long-patch repair proteins. The influence of XRCC1 on Ape1 will be discussed in a subsequent section.

Specific in vivo protein interactions between Ape1 and pol-β were initially identified by two-hybrid analysis (69). Ternary complex formation between Ape1/pol-β/AP:DNA did not require a functional Ape1 molecule and was only seen in the absence of $MgCl_2$. Indeed, addition of $MgCl_2$, activating 5′ endonuclease activity of Ape1, catalyzed DNA strand incision and possibly complex dissociation, as only pol-β was found to be associated with the incised DNA under these conditions. However, a weak association of Ape1 to the pol-β/5′ dRP:DNA complex was also possible (69). Thus, Ape1 was proposed to act as a loading factor for pol-β onto AP sites.

The physical interactions of Ape1 and pol-β confer a potential cooperativity to each enzymatic function, suggesting that Ape1 incision activity may be affected by the presence of pol-β. In vitro Ape1 processing of an intact AP site seems to function in a similar fashion to the glycosylases; i.e., an affinity for both substrate and product and a role in promoting recruitment of the next enzyme in the pathway, pol-β (29). Ape1 binds both the intact AP site substrate and the incised AP site (5′ dRP), product with high and relatively equal affinity, with calculated K_m values in the range of $3–4 \times 10^{-9}$ M (70). Steady-state kinetic studies reflect the classic characteristics of product inhibition kinetics: the product of Ape1 incision competes for binding of free Ape1 molecules thereby limiting turnover. Addition of a 40-fold excess of pol-β protein marginally alleviated this inhibition, providing less than a 2-fold increase in incised DNA, and this enhancement appears to be due to pol-β gap-filling activity as activation was most effective in the presence of nucleotide triphosphates, allowing for DNA synthesis (70). Although these conditions are unlikely to occur in vivo, evidence of Ape1 enhancement in vitro highlights the possibility that cooperativity occurs within the cell.

Ape1 also possesses 3′–5′exonuclease activities that process 3′ replication-blocking termini of nicked and gapped DNA. The exonuclease activity increased

Table 3 Known Interactions with BER Proteins[a]

	Ape1[b]	pol-β	LigIII	PCNA	PARP1	RPA	XRCC1	XPG	p300	CSB	XPC-HR23B	H-R23A	WRN	BLM	MBD1
TDG	(37,48)														
UNG2				(42,43)[c]		(43)[c]			(37)		(38)				
Aag												(35)			(36)
SMUG1															
MYH	(45,51)			(44)[c] (45,46)[c]		(45)[c]									
NTH1								(39,40)							
OGG1							(100) (106)			(41)[d]					
Ape1		(69,169)		(75)[c]											
pol-β	(69,169)				(11,92)		(93)		(95)						
LigI	(76)[e]	(86)[c]		(119,122)									(94)		
LIGIII					(113)			(93,171)[c]							
Fen1	(75,76)	(11)[e]		(120,121)	(11)[e] (127)				(134) (132,133)				(129)	(130)	
PCNA	(75)[c]	(118,170)[c] (11,92)						(128)[c]							
PARP1		(104)[c]					(103) (104)		(172)						
PARP2		(104)[c]	(104)[c]		(104)								(131)		
PNKP		(87)[c]	(87)[c]				(87)			(173)[e]					
p53	(68)	(174)[c]					(141)		(175)						
Tdp1															
p21				(126)	(127)[c]										

[a]Denotes both physical and functional interactions unless indicated otherwise.
[b]Functional interactions between Ape1 and gylcosylases are shown in Table 2.
[c]Denotes physical interaction only.
[d]Functional cross-talk reported, but no physical or functional interactions established.
[e]Denotes functional interaction only.

Almeida and Sobol

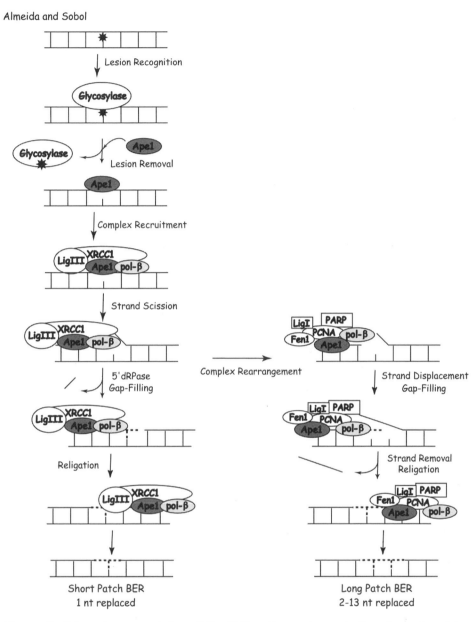

Figure 2 Schematic representation of the BER pathway, demonstrating observed protein–protein interactions for each of the steps in the reaction mechanism. Shown are the major protein complexes observed in lesion recognition and removal, strand scission, 5′ dRP lesion removal, gap-filling DNA synthesis, and variations in complex formation in the short-patch and long-patch sub-pathways. Note that only pol-β is indicated in the long-patch sub-pathway in this diagram; however, pol-δ and pol-ε may also fulfill the gap-filling reaction. *Source*: From Refs. 91,115–117.

in the presence of 3′ mismatched bases, implying that Ape1 exonuclease activity may proofread DNA synthesized in repair patches by low-fidelity polymerases such as pol-β (71). Recently, it has been reported that the 3′–5′ exonuclease activity of Ape1 is attenuated ten-fold by the 5′dRP lesion (the product of Ape1-mediated

5′endonuclease activity), suggesting that pol-β lyase activity precedes Ape1 exonuclease activity (72,73). Furthermore, removal of 5′ dRP by pol-β alleviated this inhibition, implying an orchestration of the Ape1 exonuclease function after excision of the 5′dRP residue. The rate of lyase activity was increased 2-fold under DNA synthesis conditions, although the lyase step still remained rate limiting (74), implying that synthesis occurs prior to lyase activity. This would be consistent with Ape1 3′–5′exonuclease functioning in a proofreading capacity for pol-β (73). Activation of the proofreading function of Ape1 by removal of the 5′ dRP moiety would ensure that no degradation of the DNA occurs prior to repair synthesis. This coordination suggests a role for Ape1 exonuclease activity after the completion of both pol-β-mediated repair synthesis and excision of 5′ dRP steps, a previously unexplored possibility. Exactly when Ape1 dissociates is not yet known. Ape1 may remain bound to the pol-β/DNA complex until after excision of the 5′ dRP and repair synthesis has been completed, at which time any mis-incorporation of nucleotides by pol-β would be recognized and excised by the Ape1 3′–5′exonuclease activity. Alternatively, Ape1 may dissociate from the pol-β/DNA complex until repair synthesis is complete. If a mismatch is then detected, Ape1 could re-associate to execute its 3′–5′exonuclease proofreading function.

Ape1 is also a necessary component of the long-patch BER pathway. Interactions have been established between Ape1 and LigI, Fen1 and PCNA (see Table 3). Specifically, addition of Ape1 slightly enhanced the removal of the displaced flap by Fen1 (75). Moreover, Ape1 stimulated the endonucleolytic activity of Fen1 on flaps of increasing length, suggesting that the Ape1 influence on Fen1 catalytic capacity increases as more strand displacement occurs (76). Ape1 also stimulated the resealing capability of LigI (76) and has been demonstrated to interact physically with PCNA (75). All these proteins are essential for long-patch repair, implying a coordination of the long-patch pathway by Ape1.

Optimal processing of AP sites through the short-patch BER pathway appears to be dependent on the formation of a complex between the DNA, Ape1, and pol-β (Fig. 2). The absence or imbalance of this complex may result in the accumulation of 5′ dRP ends, rendering the cell refractory to repair through short-patch BER and potentially triggering repair through strand displacement (long-patch BER). Additionally, maintaining the stability of this complex until completion of short-patch repair synthesis or initiation of the long-patch pathway may protect the integrity of the undamaged DNA upstream of the AP site by attenuating the exonucleolytic capacity of Ape1 and providing an opportunity for the verification of accurate repair.

4. GAP FILLING AND RELIGATION

The gap-filling step in BER is predominantly conducted by pol-β, the central participant of both short- and long-patch BER (Fig. 2). However, pol-β encodes two functionally independent domains. A 5′dRP lyase activity is encoded in the 8 kDa N-terminal domain and is necessary for the preparation of DNA termini within the gap for ligation (77–79). In addition, the 31 kDa C-terminal domain of pol-β encodes the nucleotidyltransferase activity, essential for DNA synthesis in both short-patch and long-patch BER (2). Unexpectedly, the 5′dRP lyase activity was found to be necessary and sufficient to confer cellular survival following alkylation damage (23), as deletion of this domain rendered cells extremely sensitive to DNA alkylating agents. These studies suggested that in vivo recovery from alkylation-induced toxicity is dependent on active removal of the 5'dRP moiety (23). It has been reported that pol-λ and pol-ι can substitute for pol-β in BER assays in vitro by facilitating both gap filling and 5' dRP lesion recognition and removal reactions via their polymerase domains and an intrinsic 5' dRP lyase activity similar to that found for pol-β (80–82). However, a role for pol-λ and pol-ι in DNA repair has not yet been established in vivo and therefore will not be discussed further in this text. Resealing of the newly repaired DNA strand is accomplished by either LigI or the XRCC1/LigIII complex in the short-patch pathway and by LigI in the long-patch pathway (83). Recent studies demonstrate an increased resistance to DNA alkylating agents as a function of LigIII over-expression, whereas no increase was detected for LigI over-expression (27). Thus, LigIII may mediate the predominant repair pathway in response to alkylation damage. There are currently many known interactions with pol-β, including the repair proteins Ape1, LigI, Fen1, PCNA, PARP1, PARP2, PNKP, XRCC1, p300, p53, and WRN (see Table 3).

4.1. Protein Interactions with pol-β in Short-Patch BER

There is an abundance of evidence in support of both physical and functional interactions between Ape1 and pol-β. (Many of the interactions between pol-β and Ape1 have been discussed in the context of Ape1 activity; see Sec. 2). Mechanistically, Ape1 is an essential enzyme in the BER pathway and pol-β is the prevailing participant in the subsequent steps in the BER mechanism (2). It has been proposed that pol-β may be recruited to the site of damage by Ape1, as both Ape1 and pol-β exhibit measurable protein–protein interactions in vivo (69). Interestingly, the functional interaction between Ape1 and pol-β is reciprocal. Ape1 was found to stimulate pol-β 5'dRP lyase activity (69). A 3–4-fold enhancement of the 5' dRP lyase activity of pol-β was demonstrated upon addition of Ape1 in vitro, although a potential enhancement of the gap-filling activity was not assessed (69). Conversely, Ape1 was found to exhibit product (5' dRP) inhibition and pol-β stimulated Ape1 activity by eliminating the inhibitory (5' dRP) lesion (70). This co-dependence of pol-β and Ape1 enzymatic activities and affinity for respective substrate and product implies that, at least in vitro, a ternary complex is required during Ape1 incision and pol-β lyase activities in order to achieve optimal processing. This has also been supported by recent in vivo observations. Utilizing an Ape1 haplo-insufficient mouse model, it was suggested that even a 50% deficiency in Ape1 may render a cell more susceptible to cellular stress due to de-regulation of the pol-β-dependent short-patch BER pathway (84).

Although there may be a role for PARP1 in short-patch BER (85), it is thought to play a more important role in long-patch BER (11) and therefore will be discussed below (see Sect. 3.2). The final step in short-patch BER is re-ligation of the nicked DNA, a step conducted by LigI or by the LigIII/XRCC1 complex (83). There is no direct interaction between pol-β and LigIII currently known. Restoration of the newly repaired DNA strand via a LigIII mechanism is mediated by XRCC1, which can associate with LigIII and pol-β simultaneously and will be discussed later with respect to XRCC1 coordination of the short-patch pathway (see Sec. 4). It has long been recognized that pol-β forms a direct protein–protein interaction with LigI (86). Affinity chromatography was used to isolate a 180 kDa multi-protein complex from crude nuclear extracts of bovine testis. Interestingly, this complex had the ability to complete repair of a uracil-containing double-stranded DNA oligimer, suggesting that the complex contained all the proteins needed for the complete BER reaction (86). It is possible that the ~180 kDa complex is comprised of stoichiometric quantities of UDG (32 kDa), Ape1 (34 kDa), pol-β (39 kDa), and DNA ligase I (~80 kDa), i.e., reconstitution of the short-patch pathway, thereby accommodating its BER activity and supporting the idea of short-patch BER complex formation. A significant portion of this complex, however, dissociated into the eluant, leaving only the pol-β/LigI interaction intact (86). Thus, it is likely that the components of this complex are only weakly associated. DNA ligases are not the only short-patch repair proteins that interact with pol-β. PNKP, responsible for the restoration of termini that are suitable for DNA polymerase function in response to oxidative stress (8), associates directly with pol-β (87), although any potential functional interactions have not yet been evaluated.

4.2. Protein Interactions with pol-β in Long-Patch BER

The short-patch sub-pathway is the predominant mechanism utilized by mammalian cells for BER (1). However, the long-patch sub-pathway is observed to contribute approximately 10% of the BER capacity in wild-type cells (88) and is the predominant sub-pathway when the 5′ dRP lesion is refractory to repair (88), under ATP-limiting conditions (89) or as a function of the initiating lesion (90). Similar to that observed for short-patch repair, pol-β appears to be the major polymerase in the long-patch sub-pathway (2), although pol-δ and pol-ε have both been shown to be involved in this sub-pathway (9,91). Other proteins required for complete repair include PARP1, Fen1, and PCNA, with possible involvement of PARP2, RFC, and RPA, among others (see Table 3). PARP1 is more commonly thought to recognize single-strand breaks and initiate repair but is also involved, in a capacity that remains unclear, with long-patch repair. Interestingly, PARP1 recognizes the 5′ dRP lesion formed subsequent to Ape1-mediated incision as part of a pol-β/ Fen1/PARP1 multi-protein complex, identified in DNA/protein cross-linking studies (92). Fen1 cleaves the displaced DNA strand following synthesis and PCNA facilitates long-patch repair as an accessory in the DNA synthesis step (9). It has now been demonstrated that pol-β associates with each of these proteins either physically or functionally (see Table 3). Specifically, a combination of PARP1 and Fen1 stimulated long-patch repair by enhancing pol-β-mediated DNA synthesis (11), suggesting that these proteins act as accessory factors for pol-β strand displacement activity. Interestingly, the association of pol-β with XRCC1, a protein thought to provide a scaffold for the processing of short-patch repair, has been demonstrated to suppress stand displacement activity of pol-β (93). This implies that there is a cellular preference for repair to proceed through short-patch, but if the long-patch machinery is engaged, pol-β plays a central role. Additionally, pol-β enhanced the excision activity of Fen1 4-fold and addition of PARP1 to the Fen1/pol-β reaction mixture raised the excision activity to 10 times the activity of Fen1 alone (11). No physical interaction has been established for pol-β and Fen1 although an association was verified for pol-β and PARP1 (92). Photoaffinity labeling of the DNA substrate and cross-linking the substrate DNA to associating proteins routinely isolated three proteins linked to the DNA: pol-β, Fen1, and PARP1 (92). Recently, Werner Syndrome Protein (WRN) was reported to interact directly with pol-β, stimulating pol-β strand displacement activities (94). This stimulation was dependent on the helicase activity of WRN, implying that WRN plays a role in pol-β-mediated long-patch repair (94). Finally, it was demonstrated that pol-β interacts directly with the tumor suppressor protein p53. This interaction stabilized the association of pol-β and AP site DNA but only in the presence of Ape1 (174). Furthermore, p53 stimulated a reconstituted BER system up to 9-fold, a significant increase in overall repair. Conversely, p53 did not effect overall repair in response to alkylation damage in vivo, as deletion of p53 caused no change in the cellular sensitivity to these DNA damaging agents, regardless of the p53 status of the cell (26). Thus, p53 may enhance specificity and functionality of BER but is not required for efficient processing of DNA damage.

4.3. Post-Translational Modifications

Recently, it was reported that pol-β directly interacts in vivo with the transcriptional coactivator p300, suggesting that this protein may regulate BER via a post-translational mechanism (95). p300 was found to acetylate pol-β at an amino acid critical for the 5′ dRP lyase function: lysine 72. Acetylation at this site blocks formation of the Schiff-base intermediate and results in abrogation of the 5′ dRP lyase activity of pol-β but leaves its gap-filling and DNA binding functions unaffected (95). Whether acetylated pol-β interacts with other BER proteins in the same manner as un-acetylated pol-β remains to be seen. Nevertheless, p300 has a novel regulatory role in the critical pol-β 5′ dRP lyase activity required for cellular survival (23). Given the dire consequences of abrogation of the pol-β 5′ dRP lyase activity, it seems likely that a de-acetylase, designed to reactivate pol-β 5′dRP lyase function, is present in vivo but to date has not been identified. It is tempting to speculate that the reason for 5′ dRP lyase abolition is to force repair through long-patch BER, however, the cellular conditions necessary and the advantage gained from a preference for long-patch repair remains to be determined.

Pol-β is critical to cellular survival following stress (96). More specifically, pol-β is essential to short-patch DNA repair by removing the toxic intermediate, 5′dRP (23). Accumulation of this intermediate in cells resulted in a dramatic sensitivity to DNA alkylating agents (23). Pol-β has been demonstrated to interact either physically or functionally with proteins acting both upstream (Ape1) and downstream (LigI and LigIII/XRCC1) of this pathway, implying that pol-β coordinates lesion processing through short-patch repair. Additionally, pol-β is involved in the orchestration of long-patch repair. Interactions with essential proteins of the long-patch pathway have been clearly established and include: Ape1, Fen1, PARP1, and PCNA (see Fig. 2 and Table 3). Finally, pol-β is linked with p53-mediated cell death. p53 stimulates repair but is not absolutely required for repair to occur (26). To our knowledge pol-β stimulation of p53-induced apoptosis has not been assessed. It is possible that under conditions of severe cellular stress (DNA damage), p53 stimulates repair but that if damage is too great, the same interaction affects the induction of the apoptotic response.

5. XRCC1 COORDINATION

XRCC1 interacts with most, if not all components of the BER short-patch pathway yet has no known catalytic function (97). However, XRCC1 is required for effective and efficient DNA repair in mammalian cells. For example, XRCC1-deficient cell lines exhibit a reduced repair capacity and are highly sensitive to DNA alkylating agents. These deficient cells accumulate DNA strand breaks in response to damage and exhibit increased levels of both SCE and chromosomal aberrations (98). The human XRCC1 protein is 633 amino acids in length, containing two unique BRCT domains that mediate protein–protein interactions. The BRCT domains (BRCT-I, 315–404 aa; BRCT-II, 538–633 aa) and the N-terminal domain (1–183aa) are tethered together by flexible hinged regions. Only the BRCT-I domain is conserved across all known XRCC1 homologs and is critical for cellular survival, suggesting that this domain, hence protein–protein interactions, is fundamental to overall XRCC1 function (99). Discrete binding domains within XRCC1 for AP site repair proteins as well as a newly identified region binding the bi-functional glycosylase hOGG1 support the idea of XRCC1 functioning as a platform for the coordination of short-patch BER, via protein–protein interactions and complex formation.

5.1. DNA Damage Recognition

Recently, XRCC1 has been shown to interact directly with hOGG1 (100). This interaction stimulated the glycosylase activity but not the lyase activity of the enzyme. XRCC1 and Ape1 had an additive affect on the glycosylase activity of hOGG1, resulting in a final XRCC1/Ape1/hOGG1 heterotrimeric complex that mediated the repair of 8-oxoG to yield the $5'$ dRP-containing cleaved abasic site, a substrate for pol-β. This supports a model of protein complex formation facilitating efficient lesion repair, with XRCC1 acting as a scaffold to orchestrate the movement of DNA repair intermediates from the glycosylase to Ape1 and then to pol-β and LigIII (100).

An interaction between XRCC1 and PNKP, a protein necessary for the initial processing of DNA single-strand break termini into ends that are amenable to religation, was shown to be involved in BER following oxidative stress (8,87). Interestingly, XRCC1 stimulated both the kinase and phosphatase activity of PNKP at damaged termini, resulting in an overall acceleration of single-strand break repair (87). A partnership between PNKP and XRCC1 suggests that single-strand break repair may be considered a sub-pathway of base excision repair, where BER proteins are utilized or recruited for the replacement of damaged nucleotides and restoration of the intact strands (see Table 3).

PARP1 is expressed abundantly in the nucleus where it senses DNA nicks and breaks in response to ionizing radiation and alkylating DNA damage (101). At the site of a DNA strand break, the enzyme synthesizes poly(ADP-ribose) using NAD+ as a substrate and catalyzes the transfer of the ADP-ribose moiety to itself, as well as a limited number of other proteins including XRCC1. The enzyme is modified by long chains of negatively charged ribose polymers, whose cumulative charge repels the DNA phosphate backbone, resulting in a loss of affinity for the DNA and thus inactivation (102). PARP1 interacts exclusively with the BRCT-I region of XRCC1 and preferentially binds when it is in an active poly(ADP-ribosyl)ated state. This apparent conflict could be explained by the fact that XRCC1 negatively regulates PARP activity, suggesting that XRCC1 may limit the auto-modification of PARP1 thus regulating the accumulation of repellent charge and indirectly the dissociation of PARP1 from the damage site (103). Indeed, XRCC1 has increased affinity for modified PARP1 and PARP2, an enzyme with similar characteristics as PARP1 (104). Furthermore, PARP1 has recently been shown to actively recruit XRCC1 at sites of oxidative stress (105), suggesting that PARP1 senses damage and recruits XRCC1 to act as the scaffold upon which a repair complex can be constructed.

5.2. XRCC1 Scaffold Coordination

XRCC1 associates physically with Ape1 and promotes AP endonuclease activity by greater than 5-fold (106). The $3'–5'$ exonuclease activity of Ape1 was also increased with addition of XRCC1 which would allow for enhanced proof-reading capabilities for the error-prone pol-β as well as more efficient processing of any $3'$ β-elimination products that are formed (106). Both hOGG1 and Ape1 occupy the same region on XRCC1, encompassing the first hinge and BRCT1 domains, from amino acids 183 to 404. This is consistent with biochemical data demonstrating that Ape1 displaces the glycosylase from the damaged DNA (59). In such a scenario, XRCC1 could remain bound to the DNA providing a scaffold for the replacement of hOGG1 by Ape1.

In 1999, the NMR solution structure of the XRCC1 N-terminal domain provided the first evidence of XRCC1 interacting directly with DNA, specifically gapped and nicked residues (107). A ternary complex of XRCC1/gapped DNA/pol-β was mapped, providing more detailed interactions consistent with previous reports (93). This complex significantly reduced the strand displacement and excessive gap-filling activities of pol-β during DNA repair (93). Suppression of these activities would facilitate processing of damage through the short-patch BER pathway, since strand displacement and gap filling would be necessary to replace a 2–13 nucleotide gap. Interestingly, the NMR solution structure of the XRCC1/pol-β/gapped DNA complex places the DNA securely inside the surrounding proteins, rendering the BER intermediate inaccessible and thus protected from the cellular milieu (107).

It has long been established that XRCC1 interacts physically with DNA Ligase III (108). The gene for DNA Ligase III has two known nuclear isoforms, LigIIIα and LigIIIβ, whose sequence differ only in the C-termini (109). As a result, LigIIIα interacts with XRCC1 while LigIIIβ does not (109). LigIIIα polypeptide levels and activity are both effected by the XRCC1 protein. Indeed, XRCC1-deficient cell lines exhibit a marked reduction in LigIIIα protein, in some cases reducing the level to only 25% of WT protein levels and concomitant ligation activity (110). XRCC1 may facilitate the targeting of LigIIIα to strand breaks, as deletion of the C-terminal BRCT-II binding domain of XRCC1 results in a decreased ability to rejoin DNA ends (111). However, the majority of nuclear LigIIIα is thought to bind XRCC1 prior to any DNA damaging event, as cellular levels of ligase are attenuated 3–6-fold in XRCC1 mutants (112).

There is strong evidence of a complex comprised of XRCC1, pol-β, and LigIIIα that can orchestrate three of the five short-patch BER steps (87). Additional proteins could assemble onto this complex. There is support for a functional interaction between Ape1 and pol-β (69) as well as interactions between both of these proteins and XRCC1 (93,106). Furthermore, a multi-protein complex of XRCC1, pol-β, LigIIIα, and PNKP co-associated in human cell extracts and together repaired damage caused by reactive oxygen species and ionizing radiation (87), highlighting the formation of repair protein complexes to effect optimal restoration of damaged DNA. Likewise, it is reasonable to speculate that the DNA damage sensing protein PARP1 may form a similar repair complex with XRCC1, pol-β, and LigIIIα. Along these lines, it has now been shown that PARP1 forms a protein–protein interaction with LigIIIα, supporting the hypothesis that once a strand break is recognized by PARP1, it may actively recruit the XRCC1/LigIIIα complex to the damaged site to facilitate repair (113). In addition, both PARP and Ape1 could bind to the XRCC1/pol-β/LigIII complex. Ape1 affinity for the first hinge region of XRCC1 is comparable with or without the BRCT-I domain, whereas binding decreased substantially with removal of the first hinge region (100), indicating that the interaction of Ape1 is strongest at the first hinge region but also encompasses the BRCT-I domain. PARP1 on the other hand, binds exclusively to the BRCT-I domain. Which complex actually forms may be dependent on the nature of the DNA termini at the gap.

Functionally, the XRCC1/pol-β interaction was observed to suppress excessive gap filling and strand displacement functions of pol-β, whereas the association of pol-β with PARP1 stimulated these same functions (11,92), suggesting that these three proteins also may play a role in the regulation of pathway selection. Recently, inhibition of the activity of the XRCC1/LigIII complex was shown to be responsible for the initiation of long-patch repair, following ionizing radiation damage (114), further highlighting the involvement of these proteins in pathway selection.

6. LONG-PATCH REPAIR

In the event that pol-β is unable to excise a 5′ blocking lesion (5′dRP), thus preventing the formation of a terminus amenable to ligation, a PCNA-dependent alternative pathway (long-patch BER) is available for repair. Cellular extracts of pol-β-deficient lines showed that pol-β exerted less influence on long-patch repair than on short-patch (91), although dependence of pol-β in the long-patch pathway was independently substantiated (115). Further, compelling evidence established that pol-δ and pol-ε can substitute for pol-β during the repair of AP sites (91,115–117). Unlike pol-β, these polymerases do not possess a 5′ lyase activity. Therefore, necessary processing of the gap is completed by flap endonuclease (Fen1), an enzyme that catalyzes the removal of displaced stands of DNA (9) following multi-nucleotide gap filling by pol-β, pol-δ, or pol-ε. Restoration of the intact DNA strand is accomplished by Lig I. In vitro pathway reconstitution using purified proteins verified the necessary components of long-patch BER repair as follows: Ape1, PCNA, Fen1, pol-δ/pol-β, and Lig I (115).

6.1. Protein Interactions

Evidence in support of complex formation during long-patch repair is substantial and PCNA appears to anchor this complex (see Fig. 2 and Table 3). Physical interactions of PCNA with Ape1, Fen1, pol-β and LigI, i.e., all essential components of the long-patch pathway, are well established (75,118–120). Several of these interactions promote long-patch repair, implying that PCNA may be functioning as a molecular adaptor for the recruitment and facilitation of repair proteins at DNA damage sites. Specifically, PCNA enhances both Fen1 and LigI activities (115,121,122) and disruption of the PCNA-binding site of either Fen1 or LigI significantly reduced the efficiency of AP site repair although repair patch size was not affected (123). Interestingly, two human DNA-N-glycosylases, UNG2 and MYH, have been confirmed to interact with PCNA (124). Both these glycosylases excise damage generated during DNA replication, suggesting that long-patch BER is coupled to the replication machinery through PCNA. Alternatively, the p21 regulatory protein can bind PCNA and prohibit PCNA-directed stimulation of Fen1, LigI, and pol-δ (125,126), suggesting that the link between replication and repair is tightly controlled. PCNA also associates with proteins known to effect long-patch BER but are not essential. Recently, PCNA co-immunoprecipitated with PARP1 from S-phase synchronized HeLa cells, the effect of this interaction is a marked inhibition of both functions (127). Moreover, a direct physical interaction of PCNA with XPG, a structure-specific repair endonuclease that is homologous to Fen1 but closely associated with NER repair, was reported (128).

Fen1 is also important for the formation and efficient functioning of the long-patch BER machinery. Interaction and activity enhancement of Fen1 and LigI by Ape1 has previously been discussed (see Sec. 2). Fen1 physically interacts with PARP1 and assists in the enhancement of gap-filling and strand displacement activities of pol-β (11). Werner Syndrome protein (WRN) dramatically stimulated the rate of Fen1 cleavage of a 5' flap DNA substrate (129). Similarly, Bloom Syndrome (BLM) protein stimulated both the endonucleolytic and exonucleolytic cleavage activity of Fen1 (130). Both the BLM and WRN helicases co-immunoprecipitated with Fen1 from HeLa nuclear extracts and the stimulation of Fen1 was independent of helicase catalytic activity (129,130). PARP1 has also been confirmed to interact with the WRN protein (131). In response to DNA damaging agents, cells defective in the WRN protein were deficient in the poly(ADP-ribosyl)ation-mediated repair pathway. Although PARP1 was activated, the subsequent poly(ADP-ribosyl)ation of other cellular proteins was severely impaired (131). Given the number of functional and physical interactions currently established for proteins known to be involved in long-patch BER (Table 3), it is likely that a coordinated complex is functioning to facilitate repair.

6.2. Post-Translational Modifications

As with short-patch BER, PTMs add an additional layer to the regulation of DNA repair, not only through altered specificity of an individual protein but also by modifying the potential for complex formation. It is currently unclear if and how the PTM state of BER proteins affect overall DNA repair. The transcriptional cofactor p300 may have a role in DNA repair regulation through association with long-patch repair proteins. Complex formation in vivo between PCNA and p300 was independent of S-phase DNA synthesis, but p300 was associated with freshly synthesized DNA following UV irradiation, suggesting that p300 may participate in chromatin remodeling at sites of DNA damage in order to facilitate synthesis mediated by PCNA or long-patch BER (132). Association significantly attenuated the acetyltransferase activity and transcriptional activation properties of p300 particularly when targeted to chromatin (133). Further, p300 complexed with and modified Fen1 in vivo (134). Fen1 acetylation was enhanced in human cells upon UV exposure, resulting in a surprising repression of both DNA binding and nuclease activity. PCNA, however, was able to stimulate Fen1 activity regardless of the modification state (134). This data would be consistent with p300 modulating pathway selection for optimal repair. For example, p300 acetylation of pol-β abrogates the lyase activity required for short-patch repair (95), implying that repair in the presence of p300 may be accomplished through long-patch BER. Meanwhile, p300 complex formation with PCNA stimulates long-patch repair synthesis and PCNA can enhance strand cleavage regardless of the acetylation state of Fen1 (134). It is tempting to speculate that p300 plays a regulatory role in DNA repair through both individual protein modifications as well as through altered complex formation. However a more thorough understanding of the PTM state of BER proteins may be critical to understanding global repair.

Repair through long-patch BER is orchestrated by a protein complex that coordinates the specificity and facilitates the overall processing of DNA damage. Most, if not all, of the necessary repair proteins are associated in vivo and suggests that complex formation in long-patch BER is critical to facilitate efficient and accurate repair.

7. EMERGING SUBPATHWAYS

A new branch of base excision repair is currently emerging. In this case, stalled DNA topoisomerase 1 (Top1) forms the DNA lesion. Top1 is essential for the relaxation of DNA supercoiling ahead of the replication machinery by cleaving the DNA and resulting in the reversible transfer of a DNA phosphodiester onto a tyrosine residue (Try723) of the protein (135,136). Once the supercoiling is alleviated, Top1 religates the DNA, reversing the DNA–protein covalent bond and restoring the intact DNA strand (135,136). If this re-ligation step is chemically inhibited (for example, through the chemotherapeutic drug camptothecin), the protein remains bound to the DNA creating a toxic single-strand break in actively replicating cells (137). Tyrosyl DNA phosphodiesterase (Tdp1) catalyzes the hydrolysis of Top1 from the 3′ terminus of DNA, resulting in a gap with a 3′ phosphate and a 5′ hydroxyl group at the margins (138), although there are alternative mechanisms of repair for this lesion (139). After lesion removal by Tdp1, PNKP can then transfer the phosphate group from the 3′ to the 5′ terminus, thus preparing the DNA ends for repair (8). If any gap-filling activity is required, pol-β is considered a likely candidate (136,140,141). Interestingly, both PNKP and pol-β are known to interact with XRCC1 (see Sec. 5) and it has recently been reported that Tdp1, critical for the repair of these Top1-mediated lesions, co-immunoprecipitated with XRCC1 (141). Further, XRCC1 complementation enhanced Tdp1 activity in EM9 cell lines (deficient in XRCC1 activity) (141). The PNKP-mediated transfer of a phosphate group was also stimulated by XRCC1 association (87). Although the actual site of the association between Tdp1 and XRCC1 was not clarified, it is tempting to speculate that Tdp1 may interact with the BRCT-I domain as do the other proteins known to sense or recognize damage; such as Ape1, hOGG1, and PARP1. This would allow XRCC1 to coordinate the repair of Top1-mediated lesions similar to oxidation and alkylation lesions.

8. CONCLUSIONS

It is becoming clear that the BER pathway is a tightly controlled repair mechanism, with many levels of regulation and specificity. The studies described above suggest that each protein in the BER pathway is interactive with either the previous or subsequent enzyme in the pathway. It is therefore not surprising that an imbalance in BER protein expression can lead to an overall defect in the pathway. For example, an abundance of a BER protein could produce increased levels of toxic intermediates as well as interfere with the balance of protein–protein interactions. Indeed, it has recently been reported that Aag and/or Ape1 over-expression in human cells was associated with frameshift mutations and microsatellite instability (142). Such genetic problems may result from imbalanced BER processing. Ulcerative colitis epithelium cells that were undergoing inflammation and microsatellite instability exhibited increased levels of activity for both these BER proteins, implying that an imbalance of BER enzymatic activity and potentially altered complex formation could contribute to this disease (142). Over-expression of pol-β can be equally deleterious. Indeed, many human tumors exhibit elevated levels of pol-β (143). Cells over-expressing pol-β acquire a spontaneous mutator phenotype and display an attenuated sensitivity to cancer chemotherapeutics (144). An elevated level of pol-β could occlude other higher fidelity polymerases from performing gap-filling functions necessary for repair, replication, and recombination, resulting in genetic instability (145). De-regulated expression of pol-β induces aneuploidy, promotes tumerogenesis in nude immunodeficient mice (146), and increases apoptosis (147). Furthermore, an engineered abundance of pol-β in lens epithelium of mice resulted in early onset of severe cortical cataracts (26). Even haplo-insufficiency in the Ape1 gene leads to an overall defect in the BER pathway (84). In summary, the normal, accurate and efficient processing of base lesions by the BER pathway depends on productive protein–protein interactions and specific multi-protein complexes and therefore de-regulated expression of BER proteins may perturb normal repair processing and could represent predisposition factors for a variety of human health issues.

ACKNOWLEDGMENTS

We would like to thank Dr. B. Köberle (University of Pittsburgh) for critically reading the manuscript.

REFERENCES

1. Sobol RW, Wilson SH. Mammalian DNA β-polymerase in base excision repair of alkylation damage. Progr Nucleic Acid Res Mol Biol 2001; 68:57–74.
2. Wilson SH, Sobol RW, Beard WA, Horton JK, Prasad R, Vande Berg BJ. DNA Polymerase β and Mammalian Base Excision Repair. Cold Spring Harbor Symp Quant Biol 2001; 65:143–155.
3. Nilsen H, Krokan HE. Base excision repair in a network of defence and tolerance. Carcinogenesis 2001; 22:987–998.
4. Memisoglu A, Samson L. Base excision repair in yeast and mammals. Mutat Res 2000; 451:39–51.
5. Wood RD, Mitchell M, Sgouros J, Lindahl T. Human DNA repair genes. Science 2001; 291:1284–1289.
6. Lindahl T, Wood RD. Quality control by DNA repair. Science 1999; 286:1897–1905.
7. Wallace SS. Enzymatic processing of radiation-induced free radical damage in DNA. Radiat Res 1998; 150:S60–S79.
8. Jilani A, Ramotar D, Slack C, Ong C, Yang XM, Scherer SW, Lasko DD. Molecular cloning of the human gene, PNKP, encoding a polynucleotide kinase 3′-phosphatase and evidence for its role in repair of DNA strand breaks caused by oxidative damage. J Biol Chem 1999; 274:24176–24186.
9. Gary R, Kim K, Cornelius HL, Park MS, Matsumoto Y. Proliferating cell nuclear antigen facilitates excision in long-patch base excision repair. J Biol Chem 1999; 274: 4354–4363.
10. Prasad R, Dianov GL, Bohr VA, Wilson SH. FEN1 stimulation of DNA polymerase β mediates an excision step in mammalian long patch base excision repair. J Biol Chem 2000; 275:4460–4466.

11. Prasad R, Lavrik OI, Kim SJ, Kedar P, Yang XP, Vande Berg BJ, Wilson SH. DNA polymerase β-mediated long patch base excision repair: Poly(ADP-ribose)polymerase-1 stimulates strand displacement DNA synthesis. J Biol Chem 2001; 276:32411–32414.

12. Nakamura J, Walker VE, Upton PB, Chiang SY, Kow YW, Swenberg JA. Highly sensitive apurinic/apyrimidinic site assay can detect spontaneous and chemically induced depurination under physiological conditions. Cancer Res 1998; 58:222–225.

13. Lindahl T. Instability and decay of the primary structure of DNA. Nature 1993; 362:709–715.

14. Engelward B, Dreslin A, Christensen J, Huszar D, Kurahara C, Samson L. Repair deficient 3-methyladenine DNA glycosylase homozygous mutant mouse cells have increased sensitivity to alkylation induced chromosome damage and cell killing. EMBO J 1996; 15:945–952.

15. Engelward BP, Allan JM, Dreslin AJ, Kelly JD, Gold B, Samson LD. A chemical and genetic approach together define the biological consequences of 3-methyladenine lesions in the mammalian genome. J Biol Chem 1998; 273:5412–5418.

16. Alanazi M, Leadon SA, Mellon I. Global genome removal of thymine glycol in *Escherichia coli* requires endonuclease III but the persistence of processed repair intermediates rather than thymine glycol correlates with cellular sensitivity to high doses of hydrogen peroxide. Nucleic Acids Res 2002; 30:4583–4591.

17. Cerda SR, Chu SS, Garcia P, Chung J, Grumet JD, Thimmapaya B, Weitzman SA. Regulation of expression of N-methylpurine DNA glycosylase in human mammary epithelial cells: role of transcription factor AP-2. Chem Res Toxicol 1999; 12:1098–1109.

18. Asaeda A, Ide H, Tano K, Takamori Y, Kubo K. Repair kinetics of abasic sites in mammalian cells selectively monitored by the aldehyde reactive probe (ARP). Nucleosides Nucleotides 1998; 17:503–513.

19. Monti P, Campomenosi P, Ciribilli Y, Iannone R, Inga A, Shah D, Scott G, Burns PA, Menichini P, Abbondandolo A, Gold B, Fronza G. Influences of base excision repair defects on the lethality and mutagenicity induced by Me-lex, a sequence-selective N_3-adenine methylating agent. J Biol Chem 2002; 277:28663–28668.

20. Friedberg EC, Walker GC, Siede W. DNA Repair. Washington, DC: ASM Press, 1995.

21. Xiao W, Samson L. *In vivo* evidence for endogenous DNA alkylation damage as a source of spontaneous mutation in eukaryotic cells. Proc Natl Acad Sci USA 1993; 90:2117–2121.

22. Zhou W, Doetsch PW. Effects of abasic sites and DNA single-strand breaks on prokaryotic RNA polymerases. Proc Natl Acad Sci USA 1993; 90:6601–6605.

23. Sobol RW, Prasad R, Evenski A, Baker A, Yang XP, Horton JK, Wilson SH. The lyase activity of the DNA repair protein β-polymerase protects from DNA-damage-induced cytotoxicity. Nature 2000; 405:807–810.

24. Ochs K, Sobol RW, Wilson SH, Kaina B. Cells deficient in DNA polymerase β are hypersensitive to alkylating agent-induced apoptosis and chromosomal breakage. Cancer Res 1999; 59:1544–1551.

25. Sobol RW, Watson DE, Nakamura J, Yakes FM, Hou E, Horton JK, Ladapo J, Van Houten B, Swenberg JA, Tindall KR, Samson LD, Wilson SH. Mutations associated with base excision repair deficiency and methylation-induced genotoxic stress. Proc Natl Acad Sci USA 2002; 99:6860–6865.

26. Sobol RW, Kartalou M, Almeida KH, Joyce DF, Engelward BP, Horton JK, Prasad R, Samson LD, Wilson SH. Base excision repair intermediates induce p53-independent cytotoxic and genotoxic responses. J Biol Chem 2003; 278:39951–39959.

27. Ho EL, Satoh MS. Repair of single-strand DNA interruptions by redundant pathways and its implication in cellular sensitivity to DNA-damaging agents. Nucleic Acids Res 2003; 31:7032–7040.

28. Friedberg EC, Meira LB. Database of mouse strains carrying targeted mutations in genes affecting biological responses to DNA damage: Version 5. DNA Repair 2003; 2:501–530.

29. Wilson SH, Kunkel TA. Passing the baton in base excision repair. Nat Struct Biol 2000; 7:176–178.

30. Mol CD, Hosfield DJ, Tainer JA. Abasic site recognition by two apurinic/apyrimidinic endonuclease families in DNA base excision repair: the 3′ ends justify the means. Mutat Res 2000; 460:211–229.

31. Mol CD, Izumi T, Mitra S, Tainer JA. DNA-bound structures and mutants reveal abasic DNA binding by APE1 and DNA repair coordination. Nature 2000; 403: 451–456.

32. Nilsen H, Lindahl T, Verreault A. DNA base excision repair of uracil residues in reconstituted nucleosome core particles. EMBO J 2002; 21:5943–5952.

33. Beard BC, Wilson SH, Smerdon MJ. Suppressed catalytic activity of base excision repair enzymes on rotationally positioned uracil in nucleosomes. Proc Natl Acad Sci USA 2003; 100:7465–7470.

34. Miao F, Bouziane M, O'Connor TR. Interaction of the recombinant human methyl-purine-DNA glycosylase (MPG protein) with oligodeoxyribonucleotides containing either hypoxanthine or abasic sites. Nucleic Acids Res 1998; 26:4034–4041.

35. Miao F, Bouziane M, Dammann R, Masutani C, Hanaoka F, Pfeifer G, O'Connor TR. 3-Methyladenine-DNA glycosylase (MPG protein) interacts with human RAD23 proteins. J Biol Chem 2000; 275:28433–28438.

36. Watanabe S, Ichimura T, Fujita N, Tsuruzoe S, Ohki I, Shirakawa M, Kawasuji M, Nakao M. Methylated DNA-binding domain 1 and methylpurine-DNA glycosylase link transcriptional repression and DNA repair in chromatin. Proc Natl Acad Sci USA 2003; 100:12859–12864.

37. Tini M, Benecke A, Um SJ, Torchia J, Evans RM, Chambon P. Association of CBP/p300 acetylase and thymine DNA glycosylase links DNA repair and transcription. Mol Cell 2002; 9:265–277.

38. Shimizu Y, Iwai S, Hanaoka F, Sugasawa K. Xeroderma pigmentosum group C protein interacts physically and functionally with thymine DNA glycosylase. EMBO J 2003; 22:164–173.

39. Klungland A, Hoss M, Gunz D, Constantinou A, Clarkson SG, Doetsch PW, Bolton PH, Wood RD, Lindahl T. Base excision repair of oxidative DNA damage activated by XPG protein. Mol Cell 1999; 3:33–42.

40. Bessho T. Nucleotide excision repair 3′ endonuclease XPG stimulates the activity of base excision repairenzyme thymine glycol DNA glycosylase. Nucleic Acids Res 1999; 27:979–983.

41. Tuo J, Chen C, Zeng X, Christiansen M, Bohr VA. Functional crosstalk between hOgg1 and the helicase domain of Cockayne syndrome group B protein. DNA Repair 2002; 1:913–927.

42. Muller-Weeks SJ, Caradonna S. Specific association of cyclin-like uracil-DNA glycosylase with the proliferating cell nuclear antigen. Exp Cell Res 1996; 226:346–355.

43. Otterlei M, Warbrick E, Nagelhus TA, Haug T, Slupphaug G, Akbari M, Aas PA, Steinsbekk K, Bakke O, Krokan HE. Post-replicative base excision repair in replication foci. EMBO J 1999; 18:3834–3844.

44. Vairapandi M, Liebermann DA, Hoffman B, Duker NJ. Human DNA-demethylating activity: a glycosylase associated with RNA and PCNA. J Cell Biochem 2000; 79: 249–260.

45. Parker A, Gu Y, Mahoney W, Lee SH, Singh KK, Lu AL. Human homolog of the MutY repair protein (hMYH) physically interacts with proteins involved in long patch DNA base excision repair. J Biol Chem 2001; 276:5547–5555.

46. Boldogh I, Milligan D, Lee MS, Bassett H, Lloyd RS, McCullough AK. hMYH cell cycle-dependent expression, subcellular localization and association with replication foci: evidence suggesting replication-coupled repair of adenine: 8-oxoguanine mispairs. Nucleic Acids Res 2001; 29:2802–2809.

47. Scharer OD, Nash HM, Jiricny J, Laval J, Verdine GL. Specific binding of a designed pyrrolidine abasic site analog to multiple DNA glycosylases. J Biol Chem 1998; 273: 8592–8597.

48. Waters TR, Gallinari P, Jiricny J, Swann PF. Human thymine DNA glycosylase binds to apurinic sites in DNA but is displaced by human apurinic endonuclease 1. J Biol Chem 1999; 274:67–74.

49. Nilsen H, Haushalter KA, Robins P, DE Barnes, Verdine GL, Lindahl T. Excision of deaminated cytosine from the vertebrate genome: role of the SMUG1 uracil-DNA glycosylase. EMBO J 2001; 20:4278–4286.

50. Petronzelli F, Riccio A, Markham GD, Seeholzer SH, Stoerker J, Genuardi M, Yeung AT, Matsumoto Y, Bellacosa A. Biphasic kinetics of the human DNA repair protein MED1 (MBD4), a mismatch-specific DNA N-glycosylase. J Biol Chem 2000; 275: 32422–32429.

51. Yang H, Clendenin WM, Wong D, Demple B, Slupska MM, Chiang JH, Miller JH. Enhanced activity of adenine-DNA glycosylase (Myh) by apurinic/apyrimidinic endo-nuclease (Ape1) in mammalian base excision repair of an A/GO mismatch. Nucleic Acids Res 2001; 29:743–752.

52. Barrett TE, Savva R, Panayotou G, Barlow T, Brown T, Jiricny J, Pearl LH. Crystal structure of a G:T/U mismatch-specific DNA glycosylase: mismatch recognition by complementary-strand interactions. Cell 1998; 92:117–129.

53. Waters TR, Swann PF. Kinetics of the action of thymine DNA glycosylase. J Biol Chem 1998; 273:20007–20014.

54. Parikh SS, Mol CD, Slupphaug G, Bharati S, Krokan HE, Tainer JA. Base excision repair initiation revealed by crystal structures and binding kinetics of human uracil-DNA glycosylase with DNA. EMBO J 1998; 17:5214–5226.

55. Hardeland U, Steinacher R, Jiricny J, Schar P. Modification of the human thymine-DNA glycosylase by ubiquitin-like proteins facilitates enzymatic turnover. EMBO J 2002; 21:1456–1464.

56. Hardeland U, Bentele M, Jiricny J, Schar P. Separating substrate recognition from base hydrolysis in human thymine DNA glycosylase by mutational analysis. J Biol Chem 2000; 275:33449–33456.

57. Marenstein DR, Chan MK, Altamirano A, Basu AK, Boorstein RJ, Cunningham RP, Teebor GW. Substrate specificity of human endonuclease III (hNTH1): Effect of human APE1 on hNTH1 activity. J Biol Chem 2003; 278:9005–9012.

58. Demple B, Harrison L. Repair of oxidative damage to DNA: enzymology and biology. Annu Rev Biochem 1994; 63:915–948.

59. Vidal AE, Hickson ID, Boiteux S, Radicella JP. Mechanism of stimulation of the DNA glycosylase activity hOGG1 by the major human endonuclease: bypass of the AP lyase activity step. Nucleic Acids Res 2001; 29:1285–1292.

60. Hill JW, Hazra TK, Izumi T, Mitra S. Stimulation of human 8-oxoguanine-DNA gly-cosylase by AP-endonuclease: potential coordination of the initial steps in base excision repair. Nucleic Acids Res 2001; 29:430–438.

61. Eide L, Luna L, Gustad EC, Henderson PT, Essigmann JM, Demple B, Seeberg E. Human endonuclease III acts perferentially on DNA damage opposite guanine residues in DNA. Biochemistry 2001; 40:6653–6659.

62. Strauss PR, Beard WA, Patterson TA, Wilson SH. Substrate binding by human apurinic/apyrimidinic endonuclease indicates a Briggs-Haldane mechanism. J Biol Chem 1997; 272:1302–1307.

63. Loeb LA, Preston BD. Mutagenesis by apurinic/apyrimidinic sites. Annu Rev Genet 1986; 20:201–230.

64. Xanthoudakis S, Smeyne RJ, Wallace JD, Curran T. The redox/DNA repair protein, Ref-1, is essential for early embryonic development in mice. Proc Natl Acad Sci USA 1996; 93:8919–8923.

65. Ludwig DL, MacInnes MA, Takiguchi Y, Purtymun PE, Henrie M, Flannery M, Meneses J, Pedersen RA, Chen DJ. A murine AP-endonuclease gene-targeted deficiency with post-implantation embryonic progression and ionizing radiation sensitivity. Mutat Res 1998; 409:17–29.

66. Fan Z, Beresford PJ, Zhang D, Xu Z, Novina CD, Yoshida A, Pommier Y, Lieberman J. Cleaving the oxidative repair protein Ape1 enhances cell death mediated by granzyme A. Nat Immunol 2003; 4:145–153.

67. Fritz G, Grosch S, Tomicic M, Kaina B. APE/Ref-1 and the mammalian response to genotoxic stress. Toxicology 2003; 193:67–78.

68. Gaiddon C, Moorthy NC, Prives C. Ref-1 regulates the transactivation and pro-apoptotic functions of p53 in vivo. EMBO J 1999; 18:5609–5621.

69. Bennett RAO, Wilson DM III, Wong D, Demple B. Interaction of human apurinic endonuclease and DNA polymerase β in the base excision repair pathway. Proc Natl Acad Sci USA 1997; 94:7166–7169.

70. Masuda Y, Bennett RA, Demple B. Dynamics of the interaction of human apurinic endonuclease (Ape1) with its substrate and product. J Biol Chem 1998; 273: 30352–30359.

71. Chou KM, Cheng YC. An exonucleolytic activity of human apurinic/apyrimidinic endonuclease on 3′ mispaired DNA. Nature 2002; 415:655–659.

72. Wilson DM III. Properties of and substrate determinants for the exonuclease activity of human apurinic endonuclease Ape1. J Mol Biol 2003; 330:1027–1037.

73. Wong D, DeMott MS, Demple B. Modulation of the 3′- > 5′-Exonuclease Activity of Human Apurinic Endonuclease (Ape1) by Its 5′-incised Abasic DNA Product. J Biol Chem 2003; 278:36242–36249.

74. Srivastava DK, Berg BJ, Prasad R, Molina JT, Beard WA, Tomkinson AE, Wilson SH. Mammalian abasic site base excision repair. Identification of the reaction sequence and rate-determining steps. J Biol Chem 1998; 273:21203–21209.

75. Dianova II, Bohr VA, Dianov GL. Interaction of human AP endonuclease 1 with flap endonuclease 1 and proliferating cell nuclear antigen involved in long-patch base excision repair. Biochemistry 2001; 40:12639–12644.

76. Ranalli TA, Tom S, Bambara RA. AP endonuclease 1 coordinates flap endonuclease 1 and DNA ligase I activity in long patch base excision repair. J Biol Chem 2002; 277:41715–41724.

77. Matsumoto Y, Kim K. Excision of deoxyribose phosphate residues by DNA polymerase β during DNA repair. Science 1995; 269:699–702.

78. Prasad R, Beard WA, Chyan JY, Maciejewski MW, Mullen GP, Wilson SH. Functional analysis of the amino-terminal 8-kDa domain of DNA polymerase β as revealed by site-directed mutagenesis. DNA binding and 5′-deoxyribose phosphate lyase activities. J Biol Chem 1998; 273:11121–11126.

79. Prasad R, Beard WA, Strauss PR, Wilson SH. Human DNA polymerase β deoxyribose phosphate lyase. Substrate specificity and catalytic mechanism. J Biol Chem 1998; 273:15263–15270.

80. Prasad R, Bebenek K, Hou E, Shock DD, Beard WA, Woodgate R, Kunkel TA, Wilson SH. Localization of the deoxyribose phosphate lyase active site in human DNA polymerase iota by controlled proteolysis. J Biol Chem 2003; 278:29649–29654.

81. Bebenek K, Tissier A, Frank EG, McDonald JP, Prasad R, Wilson SH, Woodgate R, Kunkel TA. 5′-Deoxyribose phosphate lyase activity of human DNA polymerase iota in vitro. Science 2001; 291:2156–2159.

82. Garcia-Diaz M, Bebenek K, Kunkel TA, Blanco L. Identification of an intrinsic 5′-deoxyribose-5-phosphate lyase activity in human DNA polymerase lambda: a possible role in base excision repair. J Biol Chem 2001; 276:34659–34663.

83. Tomkinson AE, Chen L, Dong Z, Leppard JB, Levin DS, Mackey ZB, Motycka TA. Completion of base excision repair by mammalian DNA ligases. Progr Nucleic Acid Res Mol Biol 2001; 68:151–164.

84. Raffoul JJ, Cabelof DC, Nakamura J, Meira LB, Friedberg EC, Heydari AR. Apurinic/ apyrimidinic endonuclease (APE/ref-1) haploinsufficient mice display tissue-specific differences in DNA polymerase β-dependent base excision repair. J Biol Chem In press, 2004.

85. Dantzer F, de La Rubia G, Menissier-De Murcia J, Hostomsky Z, de Murcia G, Schreiber V. Base excision repair is impaired in mammalian cells lacking Poly(ADP-ribose) polymerase-1. Biochemistry 2000; 39:7559–7569.

86. Prasad R, Singhal RK, Srivastava DK, Molina JT, Tomkinson AE, Wilson SH. Specific interaction of DNA polymerase β and DNA ligase I in a multiprotein base excision repair complex from bovine testis. J Biol Chem 1996; 271:16000–16007.

87. Whitehouse CJ, Taylor RM, Thistlethwaite A, Zhang H, Karimi-Busheri F, Lasko DD, Weinfeld M, Caldecott KW. XRCC1 stimulates human polynucleotide kinase activity at damaged DNA termini and accelerates DNA single-strand break repair. Cell 2001; 104:107–117.

88. Horton JK, Prasad R, Hou E, Wilson SH. Protection against methylation-induced cytotoxicity by DNA polymerase β-dependent long patch base excision repair. J Biol Chem 2000; 275:2211–2218.

89. Petermann E, Ziegler M, Oei SL. ATP-dependent selection between single nucleotide and long patch base excision repair. DNA Repair 2003; 2:1101–1114.

90. Fortini P, Parlanti E, Sidorkina OM, Laval J, Dogliotti E. The type of DNA glycosylase determines the base excision repair pathway in mammalian cells. J Biol Chem 1999; 274:15230–15236.

91. Fortini P, Pascucci B, Parlanti E, Sobol RW, Wilson SH, Dogliotti E. Different DNA polymerases are involved in the short- and long-patch base excision repair in mammalian cells. Biochemistry 1998; 37:3575–3580.

92. Lavrik OI, Prasad R, Sobol RW, Horton JK, Ackerman EJ, Wilson SH. Photoaffinity labeling of mouse fibroblast enzymes by a base excision repair intermediate: Evidence for the role of poly (ADP-ribose) polymerase-1 in DNA repair. J Biol Chem 2001; 276:25541–25548.

93. Kubota Y, Nash RA, Klungland A, Schar P, Barnes DE, Lindahl T. Reconstitution of DNA base excision-repair with purified human proteins: interaction between DNA polymerase β and the XRCC1 protein. EMBO J 1996; 15:6662–6670.

94. Harrigan JA, Opresko PL, von Kobbe C, Kedar PS, Prasad R, Wilson SH, Bohr VA. The Werner syndrome protein stimulates DNA polymerase β strand displacement synthesis via its helicase activity. J Biol Chem 2003; 278:22686–22695.

95. Hasan S, El-Andaloussi N, Hardeland U, Hassa PO, Burki C, Imhof R, Schar P, Hottiger MO. Acetylation regulates the DNA end-trimming activity of DNA polymerase β. Mol Cell 2002; 10:1213–1222.

96. Sobol RW, Horton JK, Kuhn R, Gu H, Singhal RK, Prasad R, Rajewsky K, Wilson SH. Requirement of mammalian DNA polymerase-β in base-excision repair. Nature 1996; 379:183–186.

97. Caldecott KW. Protein-protein interactions during mammalian DNA single-strand break repair. Biochem Soc Transact 2003; 31:247–251.

98. Thompson LH, Brookman KW, Jones NJ, Allen SA, Carrano AV. Molecular cloning of the human XRCC1 gene, which corrects defective DNA strand break repair and sister chromatid exchange. Mol Cell Biol 1990; 10:6160–6171.

99. Taylor RM, Thistlethwaite A, Caldecott KW. Central role for the XRCC1 BRCT I domain in mammalian DNA single-strand break repair. Mol Cell Biol 2002; 22: 2556–2563.

100. Marsin S, Vidal AE, Sossou M, Menissier-De Murcia J, Le Page F, Boiteux S, De Murcia G, Radicella JP. Role of XRCC1 in the coordination and stimulation of oxidative DNA damage repair initiated by the DNA glycosylase hOGG1. J Biol Chem 2003; 278:44068–44074.

101. Bouchard VJ, Rouleau M, Poirier GG. PARP-1, a determinant of cell survival in response to DNA damage. Exp Hematol 2003; 31:446–454.

102. Zahradka P, Ebisuzaki K. A shuttle mechanism for DNA-protein interactions. The regulation of poly (ADP-ribose) polymerase. Eur J Biochem 1982; 127:579–585.

103. Masson M, Niedergang C, Schreiber V, Muller S, Menissier-de Murcia J, de Murcia G. XRCC1 is specifically associated with poly(ADP-ribose) polymerase and negatively regulates its activity following DNA damage. Mol Cell Biol 1998; 18:3563–3571.

104. Schreiber V, Ame JC, Dolle P, Schultz I, Rinaldi B, Fraulob V, Menissier-de Murcia J, de Murcia G. Poly(ADP-ribose) polymerase-2 (PARP-2) is required for efficient base excision DNA repair in association with PARP-1 and XRCC1. J Biol Chem 2002; 277:23028–23036.

105. El-Khamisy SF, Masutani M, Suzuki H, Caldecott KW. A requirement for PARP-1 for the assembly or stability of XRCC1 nuclear foci at sites of oxidative DNA damage. Nucleic Acids Res 2003; 31:5526–5533.

106. Vidal AE, Boiteux S, Hickson ID, Radicella JP. XRCC1 coordinates the initial and late stages of DNA abasic site repair through protein-protein interactions. EMBO J 2001; 20:6530–6539.

107. Marintchev A, Mullen MA, Maciejewski MW, Pan B, Gryk MR, Mullen GP. Solution structure of the single-strand break repair protein XRCC1 N-terminal domain. Nat Struct Biol 1999; 6:884–893.

108. Caldecott KW, McKeown CK, Tucker JD, Ljungquist S, Thompson LH. An interaction between the mammalian DNA repair protein XRCC1 and DNA ligase III. Mol Cell Biol 1994; 14:68–76.

109. Martin IV, MacNeill SA. ATP-dependent DNA ligases. Genome Biol 2002; 3:3005.

110. Ljungquist S, Kenne K, Olsson L, Sandstrom M. Altered DNA ligase III activity in the CHO EM9 mutant. Mutat Res 1994; 314:177–186.

111. Moore DJ, Taylor RM, Clements P, Caldecott KW. Mutation of a BRCT domain selectively disrupts DNA single-strand break repair in noncycling Chinese hamster ovary cells. Proc Natl Acad Sci USA 2000; 97:13649–13654.

112. Caldecott KW, Tucker JD, Stanker LH, Thompson LH. Characterization of the XRCC1-DNA ligase III complex in vitro and its absence from mutant hamster cells. Nucleic Acids Res 1995; 23:4836–4843.

113. Leppard JB, Dong Z, Mackey ZB, Tomkinson AE. Physical and functional interaction between DNA ligase IIIalpha and poly(ADP-Ribose) polymerase 1 in DNA single-strand break repair. Mol Cell Biol 2003; 23:5919–5927.

114. Lomax ME, Cunniffe S, O'Neill P. 8-OxoG retards the acitivty of the ligase III/ XRCC1 complex during repair of a single-strand break, when present within a clustered DNA damage site. DNA Repair 2004; 3:289–299.

115. Klungland A, Lindahl T. Second pathway for completion of human DNA base excision-repair: reconstitution with purified proteins and requirement for DNase IV (FEN1). EMBO J 1997; 16:3341–3348.

116. Matsumoto Y, Kim K, Bogenhagen DF. Proliferating cell nuclear antigen-dependent abasic site repair in *Xenopus laevis* oocytes: an alternative pathway of base excision DNA repair. Mol Cell Biol 1994; 14:6187–6197.

117. Stucki M, Pascucci B, Parlanti E, Fortini P, Wilson SH, Hubscher U, Dogliotti E. Mammalian base excision repair by DNA polymerases delta and epsilon. Oncogene 1998; 17:835–843.

118. Kedar PS, Kim SJ, Robertson A, Hou E, Prasad R, Horton JK, Wilson SH. Direct interaction between mammalian DNA polymerase β and proliferating cell nuclear antigen. J Biol Chem 2002; 277:31115–31123.

119. Levin DS, McKenna AE, Motycka TA, Matsumoto Y, Tomkinson AE. Interaction between PCNA and DNA ligase I is critical for joining of Okazaki fragments and long-patch base-excision repair. Curr Biol 2000; 10:919–922.

120. Wu X, Li J, Li X, Hsieh CL, Burgers PM, Lieber MR. Processing of branched DNA intermediates by a complex of human FEN-1 and PCNA. Nucleic Acids Res 1996; 24:2036–2043.

121. Tom S, Henricksen LA, Bambara RA. Mechanism whereby proliferating cell nuclear antigen stimulates flap endonuclease 1. J Biol Chem 2000; 275:10498–10505.

122. Tom S, Henricksen LA, Park MS, Bambara RA. DNA ligase I and proliferating cell nuclear antigen form a functional complex. Biol J Chem 2001; 276:24817–24825.

123. Matsumoto Y, Kim K, Hurwitz J, Gary R, Levin DS, Tomkinson AE, Park MS. Reconstitution of proliferating cell nuclear antigen-dependent repair of apurinic/apyrimidinic sites with purified human proteins. J Biol Chem 1999; 274:33703–33708.

124. Matsumoto Y. Molecular mechanism of PCNA-dependent base excision repair. Progr Nucl Acid Res Mol Biol 2001; 68:129–138.

125. Tom S, Ranalli TA, Podust VN, Bambara RA. Regulatory roles of p21 and apurinic/apyrimidinic endonuclease 1 in base excision repair. J Biol Chem 2001; 276: 48781–48789.

126. Dotto GP. p21 (WAF1/Cip1): more than a break to the cell cycle? Biochimica et Biophysica Acta 2000; 1471:M43–M56.

127. Frouin I, Maga G, Denegri M, Riva F, Savio M, Spadari S, Prosperi E, Scovassi AI. Human proliferating cell nuclear antigen, poly(ADP-ribose) polymerase-1, and p21 waf1/cip: A Dynamic exchange of partners. J Biol Chem 2003; 278:39265–39268.

128. Gary R, Ludwig DL, Cornelius HL, MacInnes MA, Park MS. The DNA repair endonuclease XPG binds to proliferating cell nuclear antigen (PCNA) and shares sequence elements with the PCNA-binding regions of FEN-1 and cyclin-dependent kinase inhibitor p21. J Biol Chem 1997; 272:24522–24529.

129. Brosh RM Jr, von Kobbe C, Sommers JA, Karmakar P, Opresko PL, Piotrowski J, Dianova I, Dianov GL, Bohr VA. Werner syndrome protein interacts with human flap endonuclease 1 and stimulates its cleavage activity. EMBO J 2001; 20:5791–5801.

130. Sharma S, Sommers JA, Wu L, Bohr VA, Hickson ID, Brosh RM Jr. Stimulation of flap endonuclease-1 by the Bloom's syndrome protein. J Biol Chem In Press, 2004.

131. von Kobbe C, Harrigan JA, May A, Opresko PL, Dawut L, Cheng WH, Bohr VA. Central role for the Werner syndrome protein/poly(ADP-ribose) polymerase 1 complex in the poly (ADP-ribosyl)ation pathway after DNA damage. Mol Cell Biol 2003; 23: 8601–8613.

132. Hasan S, Hassa PO, Imhof R, Hottiger MO. Transcription coactivator p300 binds PCNA and may have a role in DNA repair synthesis. Nature 2001; 410:387–391.

133. Hong R, Chakravarti D. The human proliferating cell nuclear antigen regulates transcriptional coactivator p300 activity and promotes transcriptional repression. J Biol Chem 2003; 278:44505–44513.

134. Hasan S, Stucki M, Hassa PO, Imhof R, Gehrig P, Hunziker P, Hubscher U, Hottiger MO. Regulation of human flap endonuclease-1 activity by acetylation through the transcriptional coactivator p300. Mol Cell 2001; 7:1221–1231.

135. Bjornsti MA, Osheroff N. Introduction to DNA topoisomerases. Meth Mol Biol 1999; 94:1–8.

136. Pommier Y, Redon C, Rao VA, Seiler JA, Sordet O, Takemura H, Antony S, Meng L, Liao Z, Kohlhagen G, Zhang H, Kohn KW. Repair of and checkpoint response to topoisomerase I-mediated DNA damage. Mutat Res 2003; 532:173–203.

137. Liu LF, Desai SD, Li TK, Mao Y, Sun M, Sim SP. Mechanism of action of camptothecin. Ann New York Acad Sci 2000; 922:1–10.

138. Interthal H, Pouliot JJ, Champoux JJ. The tyrosyl-DNA phosphodiesterase Tdp1 is a member of the phospholipase D superfamily. Proc Natl Acad Sci USA 2001; 98: 12009–12014.

139. Liu C, Pouliot JJ, Nash HA. Repair of topoisomerase I covalent complexes in the absence of the tyrosyl-DNA phosphodiesterase Tdp1. Proc Natl Acad Sci USA 2002; 99:14970–14975.

140. Trivedi R, Sobol RW. Unpublished. 2004.
141. Plo I, Liao ZY, Barcelo JM, Kohlhagen G, Caldecott KW, Weinfeld M, Pommier Y. Association of XRCC1 and tyrosyl DNA phosphodiesterase (Tdp1) for the repair of topoisomerase I-mediated DNA lesions. DNA Repair 2003; 2:1087–1100.
142. Hofseth LJ, Khan MA, Ambrose M, Nikolayeva O, Xu-Welliver M, Kartalou M, Hussain SP, Roth RB, Zhou X, Mechanic LE, Zurer I, Rotter V, Samson LD, Harris CC. The adaptive imbalance in base excision-repair enzymes generates microsatellite instability in chronic inflammation. J Clin Invest 2003; 112:1887–1894.
143. Srivastava DK, Husain I, Arteaga CL, Wilson SH. DNA polymerase β expression differences in selected human tumors and cell lines. Carcinogenesis 1999; 20:1049–1054.
144. Canitrot Y, Cazaux C, Frechet M, Bouayadi K, Lesca C, Salles B, Hoffmann JS. Overexpression of DNA polymerase β in cell results in a mutator phenotype and a decreased sensitivity to anticancer drugs. Proc Natl Acad Sci USA 1998; 95:12586–12590.
145. Canitrot Y, Frechet M, Servant L, Cazaux C, Hoffmann JS. Overexpression of DNA polymerase β: a genomic instability enhancer process. FASEB J 1999; 13:1107–1111.
146. Bergoglio V, Pillaire MJ, Lacroix-Triki M, Raynaud-Messina B, Canitrot Y, Bieth A, Gares M, Wright M, Delsol G, Loeb LA, Cazaux C, Hoffmann JS. Deregulated DNA polymerase β induces chromosome instability and tumorigenesis. Cancer Res 2002; 62:3511–3514.
147. Frechet M, Canitrot Y, Cazaux C, Hoffmann JS. DNA polymerase β imbalance increases apoptosis and mutagenesis induced by oxidative stress. FEBS Lett 2001; 505:229–232.
148. Kelley MR, Kow YW, Wilson DM III. Disparity between DNA base excision repair in yeast and mammals: translational implications. Cancer Res 2003; 63:549–554.
149. Wibley JE, Waters TR, Haushalter K, Verdine GL, Pearl LH. Structure and specificity of the vertebrate anti-mutator uracil-DNA glycosylase SMUG1. Mol Cell 2003; 11:1647–1659.
150. Yoon JH, Iwai S, O'Connor TR, Pfeifer GP. Human thymine DNA glycosylase (TDG) and methyl-CpG-binding protein 4 (MBD4) excise thymine glycol (Tg) from a Tg:G mispair. Nucleic Acids Res 2003; 31:5399–5404.
151. Hazra TK, Izumi T, Maidt L, Floyd RA, Mitra S. The presence of two distinct 8-oxoguanine repair enzymes in human cells: their potential complementary roles in preventing mutation. Nucleic Acids Res 1998; 26:5116–5122.
152. Hazra TK, Izumi T, Boldogh I, Imhoff B, Kow YW, Jaruga P, Dizdaroglu M, Mitra S. Identification and characterization of a human DNA glycosylase for repair of modified bases in oxidatively damaged DNA. Proc Natl Acad Sci USA 2002; 99:3523–3528.
153. Rosenquist TA, Zaika E, Fernandes AS, Zharkov DO, Miller H, Grollman AP. The novel DNA glycosylase, NEILI, protects mammalian cells from radiation-mediated cell death. DNA Repair 2003; 2:581–591.
154. Shall S, de Murcia G. Poly(ADP-ribose) polymerase-1: what have we learned from the deficient mouse model? Mutat Res 2000; 460:1–15.
155. Herceg Z, Wang ZQ. Functions of poly(ADP-ribose) polymerase (PARP) in DNA repair, genomic integrity and cell death. Mutat Res 2001; 477:97–110.
156. Ame JC, Rolli V, Schreiber V, Niedergang C, Apiou F, Decker P, Muller S, Hoger T, Menissier-de Murcia J, de Murcia G. PARP-2, A novel mammalian DNA damage-dependent poly(ADP-ribose) polymerase. J Biol Chem 1999; 274:17860–17868.
157. Saparbaev M, Laval J. 3,N4-ethenocytosine, a highly mutagenic adduct, is a primary substrate for *Escherichia coli* double-stranded uracil-DNA glycosylase and human mismatch-specific thymine-DNA glycosylase. Proc Natl Acad Sci USA 1998; 95:8508–8513.
158. Slupphaug G, Eftedal I, Kavli B, Bharati S, Helle NM, Haug T, Levine DW, Krokan HE. Properties of a recombinant human uracil-DNA glycosylase from the UNG gene and evidence that UNG encodes the major uracil-DNA glycosylase. Biochemistry 1995; 34:128–138.

159. Domena JD, Timmer RT, Dicharry SA, Mosbaugh DW. Purification and properties of mitochondrial uracil-DNA glycosylase from rat liver. Biochemistry 1988; 27: 6742–6751.

160. Haushalter KA, Todd Stukenberg MW, Kirschner MW, Verdine GL. Identification of a new uracil-DNA glycosylase family by expression cloning using synthetic inhibitors. Curr Biol 1999; 9:174–185.

161. Roy R, Brooks C, Mitra S. Purification and biochemical characterization of recombinant N-methylpurine-DNA glycosylase of the mouse. Biochemistry 1994; 33: 15131–15140.

162. O'Connor TR. Purification and characterization of human 3-methyladenine-DNA glycosylase. Nucleic Acids Res 1993; 21:5561–5569.

163. Roy R, Biswas T, Hazra TK, Roy G, Grabowski DT, Izumi T, Srinivasan G, Mitra S. Specific interaction of wild-type and truncated mouse N-methylpurine-DNA glycosylase with ethenoadenine-containing DNA. Biochemistry 1998; 37:580–589.

164. Saparbaev M, Kleibl K, Laval J. *Escherichia coli, Saccharomyces cerevisiae,* rat and human 3-methyladenine DNA glycosylases repair 1,N_6-ethenoadenine when present in DNA. Nucleic Acids Res 1995; 23:3750–3755.

165. Asagoshi K, Yamada T, Terato H, Ohyama Y, Monden Y, Arai T, Nishimura S, Aburatani H, Lindahl T, Ide H. Distinct repair activities of human 7,8-dihydro-8oxoguanine DNA glycosylase and formamidopyrimidine DNA glycosylase for formamidopyrimidine and 7,8-dihydro-8-oxoguanine. J Biol Chem 2000; 275:4956–4964.

166. Audebert M, Radicella JP, Dizdaroglu M. Effect of single mutations in the OGG1 gene found in human tumors on the substrate specificity of the Ogg1 protein. Nucleic Acids Res 2000; 28:2672–2678.

167. Marenstein DR, Ocampo MT, Chan MK, Altamirano A, Basu AK, Boorstein RJ, Cunningham RP, Teebor GW. Stimulation of human endonuclease III by Y box-binding protein 1 (DNA-binding protein B). J Biol Chem 2001; 276:21242–21249.

168. Sokhansanj BA, Rodrigue GR, Fitch JP, Wilson DM III. A quantitative model of human DNA base excision repair I: Mechanistic insights. Nucleic Acids Res 2002; 30:1817–1825.

169. Masuda Y, Bennett RA, Demple B. Rapid dissociation of human apurinic endonuclease (Ape1) from incised DNA induced by magnesium. J Biol Chem 1998; 273:30360–30365.

170. Naryzhny SN, Lee H. The post-translational modifications of proliferating cell nuclear antigen (PCNA): acetylation, not phosphorylation, plays an important role in the regulation of its function. J Biol Chem. 2004; 279:20194–20199.

171. Caldecott KW, Aoufouchi S, Johnson P, Shall S. XRCC1 polypeptide interacts with DNA polymerase β and possibly poly(ADP-ribose) polymerase, and DNA ligase III is a novel molecular 'nick-sensor' *in vitro*. Nucleic Acids Res 1996; 24:4387–4394.

172. Hassa PO, Buerki C, Lombardi C, Imhof R, Hottiger MO. Transcriptional coactivation of nuclear factor-kappaB-dependent gene expression by p300 is regulated by poly-(ADP)-ribose polymerase-1. J Biol Chem 2003; 278:45145–45153.

173. Flohr C, Burkle A, Radicella JP, Epe B. Poly(ADP)-ribosylation accelerates DNA repair in a pathway dependent on Cockayne syndrome B protein. Nucleic Acids Res 2003; 31:5332–5337.

174. Zhou J, Ahn J, Wilson SH, Prives C. A role for p53 in base excision repair. EMBO J 2001; 20:914–923.

175. Wadgaonkar R, Collins T. Murine double minute (MDM2) blocks p53-coactivator interaction, a new mechanism for inhibition of p53-dependent gene expression. J Biol Chem 1999; 274:13760–13767.

Part II

UV Damage and Other Bulky DNA-Adducts

The largest section of this book, UV Damage and Other Bulky DNA-adducts includes ten chapters that span a diverse range of topics but are united by the consideration of events that respond to the presence of DNA modifications that cause substantial structural distortions of B-form, duplex DNA, or act as structural barriers to RNA polymerases during the elongation phase of transcription.

The section begins with a comprehensive review of UV light-induced DNA photoproducts in Chapter 4. UV light-induced DNA photoproducts represent an important class of damages that have long served as important models for studying DNA damage recognition and repair pathways as well as for understanding the biological endpoints of DNA damage. As discussed in Chapter 3, much is now known about the structural similarities and differences among the major classes of DNA photoproducts.

Recognition of one particularly important class of UV light-induced photoproducts, namely the cyclobutane pyrimidine dimer (CPD) and its specific recognition and reversal by DNA photolyases is the subject of Chapter 5. These direct reversal DNA repair enzymes are among the most extensively characterized with respect to the molecular mechanisms utilized in substrate recognition and discrimination.

The topic of damage recognition is then expanded considerably in Chapter 6 in a discussion of how the prokaryotic (bacterial) nucleotide excision repair (NER) machinery, comprised of multiple proteins, operates on a variety of structurally diverse substrates.

This topic is then extended to eukaryotic systems in Chapter 7, where the numbers of protein participants in NER expand considerably into a more complex system of damage recognition and processing. However, we see that despite this increase in complexity, the basic biochemical steps in prokaryotic and eukaryotic NER are quite similar.

Attention is next turned to encounters of the transcription machinery with DNA damage and its role in mediating the prioritization of the repair of DNA damage via transcription coupled repair (TCR). Chapter 8 reviews this situation in prokaryotes for a variety of damages and discusses the relationships between TCR and NER.

The DNA repair processes taking place in actively transcribed genes in eukaryotes is then addressed in Chapter 9 where the situation again becomes more complicated and varies among different organisms. It should be emphasized that much of our knowledge of NER and TCR is based on studies examining these processes on CPDs in a variety of in vitro and in vivo model systems.

It is also important to realize that, in eukaryotic cells, for example, these processes must take place not on naked DNA containing just a bulky lesion, but within the context of chromatin structure. Here, the dynamics of nucleosome-based DNA packaging and its modulation by nucleic acid transactions exert effects on damage deposition into the genome as well as on repair processes. This is the focus of Chapter 10 where once again, the model bulky lesions are UV light-induced DNA photoproducts.

From this topic, we then move to the realm of an unusual example of DNA damage recognition and repair. Chapter 11 reviews the remarkable range of DNA damage recognition exhibited by the ultraviolet damage endonuclease (UVDE) protein from fission yeast. The UVDE protein is capable of initiating a repair process termed alternative excision repair (AER) and operates within a limited range of prokaryotes and simple eukaryotes.

Next, Chapter 12 moves us from the area of UV photoproduct recognition to another important class of bulky DNA lesions caused by crosslinking agents. The focus is on recognition of DNA crosslinks caused by the antitumor agent cisplatin and its ability to induce structural distortions that provoke interactions with a host of diverse cellular proteins ranging from repair proteins to transcription factors containing HMG domains.

Finally, Chapter 13 explores and discusses the types of distortions caused by another class of DNA damages, polycyclic aromatic carcinogens. The emphasis here is on the molecular nature of the structural distortions studied by various methodologies, including computational analysis as well as some discussion on their repair by NER.

4

Structure and Properties of DNA Photoproducts

John-Stephen Taylor
Department of Chemistry, Washington University, St. Louis, Missouri, U.S.A.

1. INTRODUCTION

Exposure of DNA to light leads to a variety of types of DNA damage, some that result from direct absorption of light by the DNA, and others that result from indirect (photosensitized) pathways involving absorption of light by other molecules that then cause damage. While many photoproducts are stable on repair and replication time-scales, others are converted to other products, either spontaneously, or following absorption of light. The photochemistry of bases, dinucleotides, oligonucleotides, and DNA has been extensively reviewed in a number monographs (1–4) and review articles (5–7). Because of the ubiquitous nature of DNA photoproducts, many organisms have evolved specific enzymes for their repair, and most, if not all, photoproducts are also subject to repair by general excision repair systems. These include the base excision repair (BER) *cis–syn* dimer-specific glycosylases (8), the *cis–syn* and photoproduct-specific photolyases (9,10), the spore photoproduct-specific lyase (11), and the nucleotide excision repair (NER) systems such as the *Escherichia coli* uvrABC and human excinuclease (12,13), as well as the UVDE-initiated alternate excision repair (AER) system found in yeast (14,15). This chapter will focus on the structure and properties of photoproducts that result from direct absorption of light by DNA and their secondary products, and how some of these features may contribute to their recognition by repair enzymes. Discussions of photosensitized reactions of DNA can be found elsewhere (5,7).

1.1. DNA Photochemistry

Primary photoproducts arise from absorption of a photon of light by DNA, whereas secondary photoproducts arise from primary photoproducts following absorption of a second photon of light. The quantum yield for formation of a photoproduct is a measure of the efficiency of the reaction and is calculated as the fraction of the excited molecules that are converted to that photoproduct. The absorption maximum of the bases in DNA is about 260 nm with an absorption tail that reaches about 300 nm, corresponding to the UVC (240–280 nm) and half of the UVB (280–320) regions. Unmodified DNA does not absorb significantly in the UVA (320–400) region. Whereas the absorption maximum of thymidine is 267 nm under neutral conditions, it is 271 nm for cytidine, and 278 nm for 5-methylcytidine (m^{5}C) (16). The shift in the absorption maximum of C upon enzymatic methylation of the 5 position of C's at CG sites in vivo results in about a 5–10-fold increase in the absorption of sunlight which increases the relative frequency of photoproduct formation at Pym^{5}CG sites compared to PyCG sites (17,18).

1.2. Primary Photoproducts

The primary photoproducts of DNA occur from either the singlet or triplet excited states of pyrimidines, and primarily involve formation of a four-membered ring intermediate or product (Fig. 1). The four-membered ring results from a (2+2) cycloaddition reaction between the C5,C6-double bond of a pyrimidine and a double

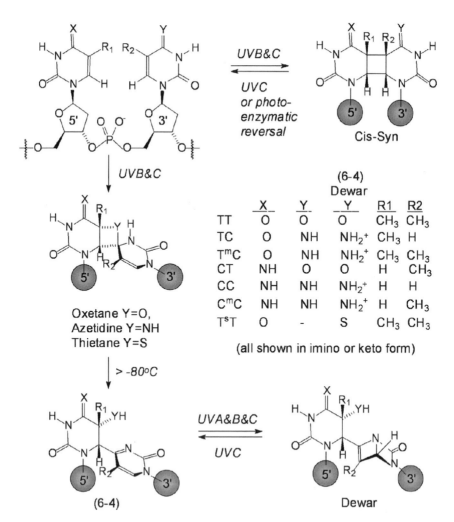

	X	Y	Y	R1	R2
TT	O	O	O	CH$_3$	CH$_3$
TC	O	NH	NH$_2$$^+$	CH$_3$	H
T^{m}C	O	NH	NH$_2$$^+$	CH$_3$	CH$_3$
CT	NH	O	O	H	CH$_3$
CC	NH	NH	NH$_2$$^+$	H	H
C^{m}C	NH	NH	NH$_2$$^+$	H	CH$_3$
T^{s}T	O	–	S	CH$_3$	CH$_3$

(all shown in imino or keto form)

Figure 1 Photochemistry of a dipyrimidine sites in B form DNA. For simplicity, cytosine bases are shown in the imino tautomeric form that must precede azetidine formation. See Fig. 4 for other tautomeric forms of C. mC refers to 5-methylcytosine.

bond of another base upon which the pyrimidine is stacked. Depending on the nature of the four-membered ring formed, this initial photoproduct may be stable, or may rearrange to another more stable product. The principal class of primary photoproducts is the cyclobutane pyrimidine dimer (CPD) in which a cyclobutane ring forms between the C5,C6-double bond of one pyrimidine with the C5,C6-double bond of another pyrimidine. The second major class of primary photoproducts is the (6–4) product. This class of photoproducts results from a (2+2) reaction between the C5,C6-double bond of one pyrimidine and the C4,X4-double bond of another pyrimidine to give an unstable four-membered ring intermediate when X=O or N, and a relatively stable product when X–S. The four-membered ring intermediate is known as an oxetane (X=O) if thymine or uracil is involved, an azetidine (X=N) if cytosine or 5-methyl cytosine is involved, or a thietane (X=S) if 4-thiothymine or 4-thiouracil is involved. These initial products spontaneously convert to pyrimidine-(6–4)-pyrimidone products, otherwise known as (6–4) products by opening of the four-membered ring. Pyrimidines also form unstable (2+2) cycloadducts with adenine which rearrange to give TA∗ photoproducts. Adenine has also been found to form dimeric products with another A, though at a very low frequency. Other types of photoproducts are also formed that do not involve an initial (2+2) cycloaddition, such as the spore photoproduct.

2. CYCLOBUTANE PYRIMIDINE DIMERS

The major direct photoproduct of native duplex DNA is the CPD, and is produced with a quantum yield of about 1%. Site-specific CPDs can be conveniently prepared by irradiating short oligodeoxynucleotides (19–22) or through the use of DNA synthesis building blocks which have been developed for the TT CPD (23–26), and one for the TU CPD (27). Because these photoproducts lose their 5,6-double bonds, they no longer have an absorption maximum at 260 nm, and instead only have an absorption tail in the UVC region (240–280 nm). Because of their low absorptivity in the UVB region (280–320 nm), CPDs are very stable in sunlight, but can be reversed by shorter wavelength UVC light, such as 254 nm light from a germicidal lamp, or by visible light and photolyases or photolyase mimics (28). In principle, eight different stereoisomers of CPDs can be produced from a (2+2) cycloaddition reaction in which the bases are in a head-to-head (syn) or head-to-tail (anti) alignment, and in a base stacked (cis) or unstacked (trans) orientation (Fig. 2). The stereochemistry of the CPD that ultimately forms is controlled by the structure of the DNA in which it is produced. In duplex B form DNA, the bases are sequentially stacked upon each other in the same antiglycosyl bond orientation. Consequently, a CPD can only form between adjacent pyrimidines on the same strand (intrastrand, adjacent photoproduct) in a head-to-head alignment (syn) with the C5 and C6 substituents all on the same side of the four-membered ring (cis), which corresponds to the *cis–syn* dimer. When one of the two pyrimidines is in a syn glycosyl conformation and the other is in an antiglycosyl conformation, as can occur in single-stranded DNA, a *trans-syn* CPD can form (Fig. 3). Photodimerization with the 5′-pyrimidine in the syn conformation results in a *trans-syn*-I dimer, whereas photodimerization with the 3′-pyrimidine in the syn conformation results in the *trans-syn*-II dimer (29,30). There have been no reports of the isolation of a *cis–syn*-II CPD, which would result from photodimerization with both pyrimidines in the syn glycosyl conformation. The CPDs can also form between non-adjacent

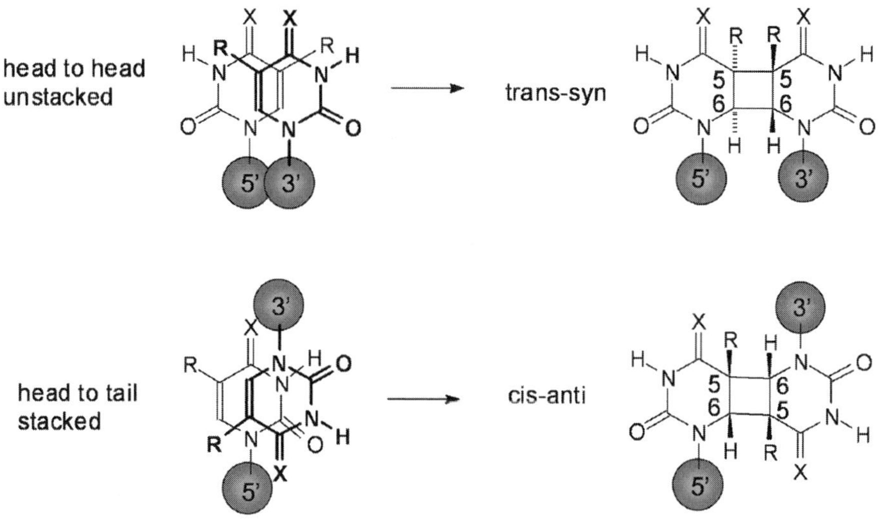

Figure 2 Stereochemistry of CPDs. Syn products result from a head-to-head arrangement of the bases as would occur within a strand, and antiproducts result from a head-to-tail arrangement as might occur between strands. Cis and trans refer to the relative orientation of the C5 and C6 substituents on the four-membered ring. *Trans-syn* dimers are described in more detail in Fig. 3.

pyrimidines, if the pyrimidines become stacked upon each other in a bulge loop structure (31–34). It is also possible to imagine antipyrimidine dimers that form from photodimerization of two pyrimidines in a head-to-tail arrangement as might occur between strands, and there is some evidence that such antiphotoproducts form in denatured DNA (35).

2.1. Secondary Reactions of CPDs

Whereas a thymine in a CPD is stable at pH 7 and 37°C, a C or 5-methyl C is not, and can isomerize (tautomerize) to one of several different hydrogen-bonding structures, or can hydrolyze (deaminate) to give the corresponding U or T containing CPDs (Fig. 4). Tautomerization and deamination change the base pairing properties of the photoproducts, which will affect the tertiary structure and properties of the DNA, and consequently their rates of recognition and repair. Some of these processes occur with half-lives of the order of hours which further complicates the study of their recognition and repair (Table 1).

2.2. Tautomerization

When the C5–C6 double bond of cytosine becomes saturated and loses the pi bond, as happens upon CPD formation, the cytosine ring also loses aromatic stabilization. As a result, the amino tautomer is no longer preferentially stabilized, and there is an increase in the proportion of the otherwise minor imino tautomers which have different base pairing properties. The first evidence for an increase in the proportion of these minor imino tautomers came from molecular orbital calculations (36) and

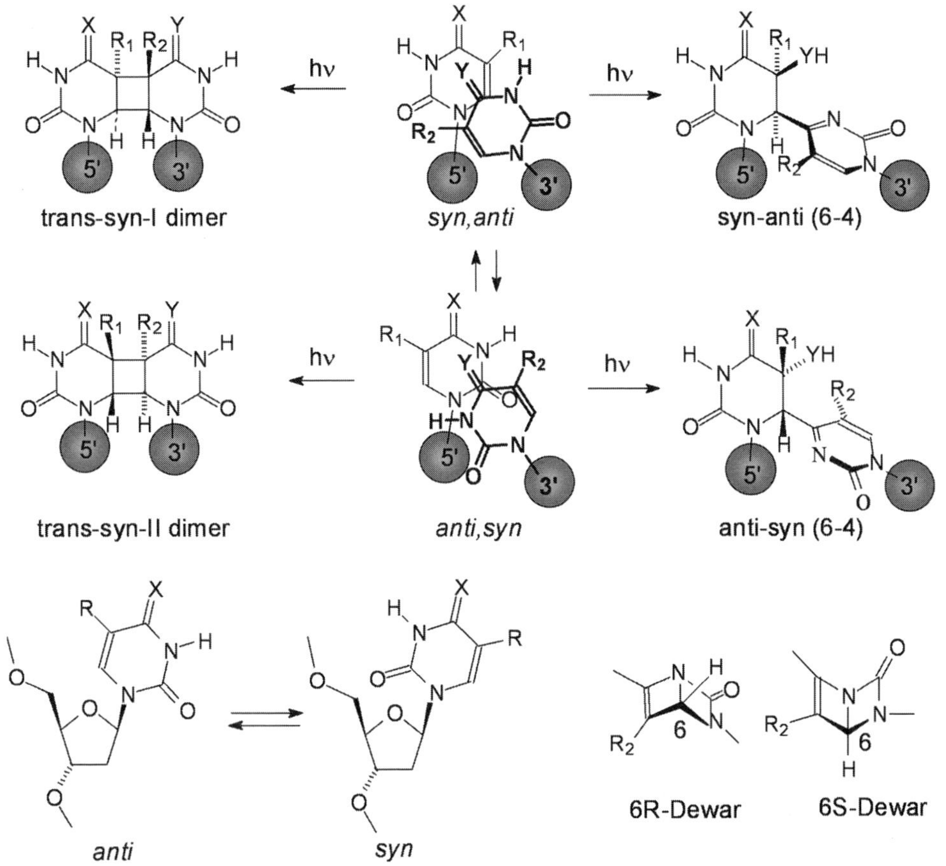

Figure 3 Stereochemistry of dipyrimidine photoproducts arising from syn glycosyl bond conformers. *Trans-syn*-I and -II CPDs are minor photoproducts and arise from a conformation of DNA in which one pyrimidine is in the minor syn glycosyl conformation (the base on top is shown in bold). (6–4) products with alternate stereochemistry could also arise from such minor glycosyl conformations, but none have thus far been isolated or characterized. The Dewar product of TpT has been assigned the 6R stereochemistry by NMR, but in principle could also form with the 6S stereochemistry.

spectrophotometric studies (37) on 5,6-dihydrocytosine. In the latter study, it was found that whereas only 4% of the imino tautomer was present in water, it was the exclusive tautomer in the less polar organic solvent chloroform. An equilibrium between the amino and E-imino tautomer would explain how 5,6-dihydrocytidine triphosphate can substitute for either CTP or UTP during transcription by *M. lysodeikticus* RNA polymerase (38). The E-imino tautomer would also explain why A is incorporated exclusively opposite a C6 hydroxylamine adduct of C by RNA polymerase (39). On the other hand, saturation of the 5,6-double bond of thymine does not change its tautomeric state as revealed by the exclusive insertion of 5,6-dihydrothymidine triphosphate opposite As by DNA polymerase I of *E. coli* (40). In a study in which care was taken to minimize deamination, the *cis–syn* cyclobutane dimer of TpC was found to be non-mutagenic within the limits of detection in *E. coli* under SOS conditions, indicating that the C was predominantly in the amino tautomeric state during *trans*-dimer synthesis (22). Evidence has also been

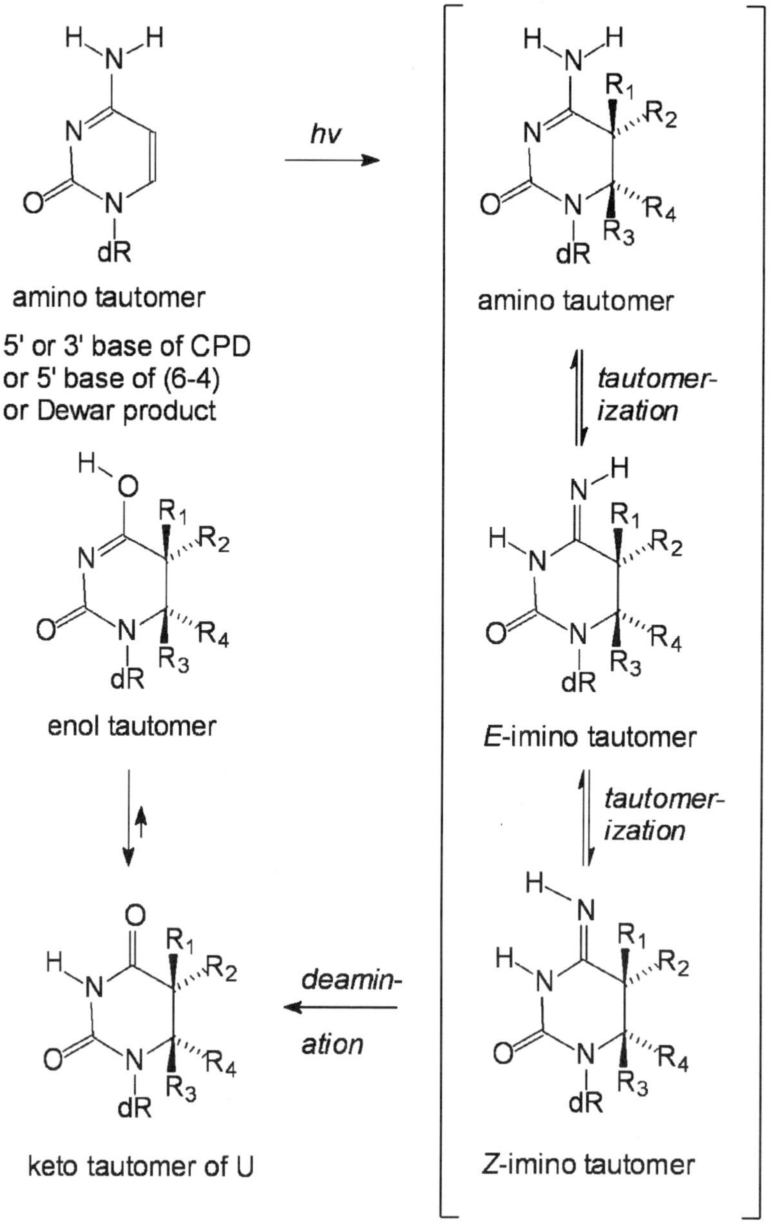

Figure 4 Deamination and tautomerization pathways for dipyrimidine photoproducts. Both the 5'- and 3'-Cs of a CPD can tautomerize and deaminate, but only the 5'-C of a (6–4) or Dewar product can deaminate or tautomerize. Thymine exists exclusively in the keto form.

obtained to suggest that *trans*-lesion synthesis past C-containing CPDs by pol η is likewise non-mutagenic (41). These results contrast with recent ab initio quantum mechanical calculations of C-containing dimers which conclude that the *E*-imino tautomer is the most stable tautomer in both the gas phase and in water (42). Because of the ready deamination of C in a dimer (see next section), no NMR or

Table 1 Deamination Half-lives of C-Containing Photoproducts

Substrate	pH	T	$t_{\frac{1}{2}}$ (min)	Reference
d(T[c,s]C)[a]	7.0	37	178	(43)
d(C[c,s]T)[a]	7.0	37	193	(43)
d(C[c,s]T)	7.0	25	401	(160)
d(mC[c,s]T)	7.0	25	48 d	(53)
d(C(6–4)T)	7.0	25	152 hr	(160)
Dimers in poly(dC•dG)	7.0	37	120	(45)
Dimers in ss-phage	7.5	37	29	(50)
Dimers in ds-phage	7.5	37	55	(50)
Dimers in ds-plasmid[b]	7.5	37	288	(51)
Dimers at CCCC in ds-phage	7.4	37	185 hr	(52)
TC̲A dimmer in *E. coli*		42	28 hr	(46,47)
TC̲T dimmer in *E. coli*		42	231	(46,47)
C̲ in ss DNA	7.4	37	219 y	(44)
C in duplex DNA	7.4	37	31,400 y	(44)

[a]Calculated from published parameters (43) assuming that the temperature dependence for deamination of the C[c,s]T is the same as for T[c,s]T.
[b]Calculated from the data.

crystal structure has yet been reported for a duplex with a C-containing CPD, and so it is not known which tautomer of C in a dimer will predominate in duplex DNA and how it will affect the stability of base pairs with G.

2.3. Deamination of C-Containing CPDs

In addition to increasing the relative proportion of the otherwise minor tautomers of C, saturation of the 5,6-double bond greatly accelerates deamination by lowering the transition state energy for the addition of water to the C4 position. The kinetics and mechanism of deamination of C-containing CPDs of dinucleotides have been studied in detail, leading to the conclusion that the rate determining step is the attack of hydroxide on the protonated base (43). Whereas the half-life for deamination of cytosine in DNA is on the order of thousands of years (44), deamination of cytosine in CPDs takes place in hours to days. Since C-containing dimers were first shown to deaminate in poly(dI•dC) with a half-life of about 2 hr (45) there have been numerous studies of deamination in vitro and in vivo with wide ranging results (Table 1). In one set of studies, deamination in *E. coli* was found to follow first order kinetics with a half-life that ranges between 4 and >24 hr depending on the photoproduct site (46–49). In another study, deamination of dimers in duplex phage DNA was reported not to follow first order kinetics and to occur within 14 min following a lag time of about 40 min (50). In most reports, however, deamination appears to follow simple first order kinetics with half-lives ranging from 5 hr to 8 days (51,52). There have been very limited data concerning the rates of deamination of 5-methyl C-containing CPDs. In an in vitro study with dinucleotides, the deamination rate of the *cis–syn* dimer of m^5CpT was reported to be about 10^{-5} min^{-1} (corresponding to a half-life of 48 days) (53), which is about 1000-fold slower than that reported for the corresponding unmethylated dimer (43). In contrast, the double deamination of the CPD of Cm^{5}C was estimated to occur at about the same rate as the unmethylated dimer in human cells (48). Deamination of a C or m^5C-containing CPD in duplex

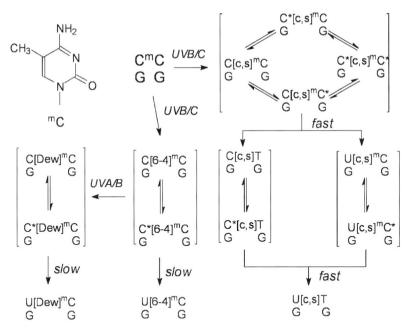

Figure 5 Photochemical, tautomerization and deamination pathways for a Cm^{5}C site. An asterisk denotes either the Z or E imino tautomeric form of the C. Deamination of the 3′-base in a (6–4) or Dewar product is not possible, and deamination of the amino group in the 5′- or 3′-C of a CPD is faster than the 5′-C in a (6–4) or Dewar product. See Fig. 4 for the structures of the tautomers.

DNA results in the formation of a U•G or T•G mispaired dimer. Recent studies of deamination in *E. coli* suggest that the second deamination is almost seven times faster than the first, possibly due to destabilization of the duplex that results from the first deamination event (49). Overall, the photochemistry of CC and Cm^{5}C is the most complicated of all the dipyrimidine sites in duplex DNA because of the multiple tautomerization and deamination pathways as illustrated in Fig. 5.

2.4. Tertiary Structure of *Cis–Syn* Cyclobutane Dimer-Containing DNA

The structure of DNA containing a *cis–syn* dimer has been the focus of many studies over the past two decades with conflicting results regarding its effect on DNA bending (Table 2). Before DNA-containing CPDs could be prepared in the amounts needed for experimental 3D structural studies, they were studied by theoretical molecular mechanics calculations. Depending on the calculation, the *cis–syn* thymine dimer was predicted to either not bend DNA to any great extent (<15°) (54–56), or to bend it by 27° (57). These early calculations were carried out by molecular mechanics methods that did not enable the structures to vary greatly from the initial geometry. Later studies included molecular dynamics and were better able to sample alternate conformation and find the lowest energy structures (58–60). These later, more sophisticated modeling studies, were carried out on both the dimmer containing and parental duplexes, and concluded that a thymine dimer bends DNA by less than 15°. In one study, a DNA duplex containing a thymine dimer was calculated to have a 39° bend, but since the parent duplex was calculated to have a 28° bend, the

Table 2 Bending and Unwinding Angles (°) for *Cis–Syn* Thymine Dimer-Containing DNA Duplexes and Their Parent Duplexes[a]

Structure	Method[a]	Bending	Unwinding	Reference
CGCGT[c,s]TCGCG GCGCA AGCGC	Modeling	27	20	(57)
...AAGT[c,s]TGAA... ...TTCA ACTT...	Circulariza-tion	30	nd	(61)
CCGTTT[c,s]TTTG GGCAAA AAAC	Electrophor-esis	9	nd	(62,63)
CGCAT[c,s]TACGC GCGTA ATGCG	NMR	9	15	(64)
CGCGAAT[c,s]TCGCG GCGCTTA AGCGC	Modeling	39	nd	(58)
CGCGAATTCGCG GCGCTTAAGCGC	Modeling	28	nd	(58)
CGCAT[c,s]TACGC GCGTA ATGCG	Modeling	22	nd	(59)
CGCATTACGC GCGTAATGCG	Modeling	8	nd	(59)
GCACGAAT[c,s]TAAG CGTGCTTA ATTC	NMR (1ttd.pdb)	23	−26	(67)
GCACGAATTAAG CGTGCTTAATTC	NMR (1coc.pdb)	31	19	(67)
CGCAT[c,s]TACGC GCGTA TTGCG	NMR (1ql5.pdb)	12 ± 7	< 15° −7, 6	(161)
CGTAT[c,s]TATGC [b] GCATA ATACG	x-ray (1n4e.pdb)	23, 33		(72)
CGT AT[c,s]T ATGC[b] GCA^{Br}UA A^{Br}UACG	x-ray (1mv7.pdb)	21, 32	4, 8	(72)

[a]The protein data bank structure file is given in parentheses.
[b]Data are given for the two duplexes present in the unit cell.

bend induced by the dimer was only 11° (58). Likewise, in another study, a DNA duplex containing a thymine dimer was calculated to have a bend of 22°, but since the parent duplex was calculated to have an 8° bend, the dimer only induced a bend of about 14° (59). These studies underscore the importance of determining the bend of both the parent and photodamaged duplexes when trying to determine the effect of photoproduct formation on DNA structure.

Experimental support for the early molecular modeling study that predicted a bend of 27° (57) came shortly thereafter from circularization assays of dimer-containing multimers which concluded that a thymine dimer bends DNA by about 30° (Table 2) (61). To complicate matters, subsequent studies of the electrophoretic mobility of phased *cis–syn* dimer-containing multimers in a different sequence context concluded that a *cis–syn* dimer only bends DNA by 9° (62,63). The large difference in the experimentally determined bending angles may either be due to differences in the methods being used, their interpretation, or to the effect of sequence context on the bending angle. With regard to the latter possibility, a *cis–syn* thymine dimer was found to affect the bending of the intrinsically bent T_6-tract sequence to different extents, depending on its position within the T_6-tract (62). It is also possible that the circularization assay is more sensitive to the bendability of the DNA than is the electrophoretic mobility assay.

The first experimental 3D structure of a *cis–syn* thymine dimer-containing duplex in solution was calculated from proton NMR data and found that the dimer bends DNA by 9° and unwinds it by 15° in comparison to an idealized duplex DNA (64). The bending angle of the NMR structure is very close to that estimated for a thymine dimer from the electrophoretic mobility studies (62,63), while the unwinding angle is very close to that of about 10° previously determined by an electrophoretic study of irradiated plasmids (65,66). A subsequent NMR study of a different dimer-containing DNA dodecamer duplex also concluded that there was little bending in comparison to the NMR structure for the parent duplex (67). Analysis of these latter two structures with the CURVES program (68,69) indicates that while the dimer-containing duplex has an overall bend of 23° and is over wound by 26°, the parent duplex is bent by 31° and is unwound by 20°. In this case, the *cis–syn* dimer appears to reduce the bending instead of increase it and over wind the DNA rather than unwind it as expected, which may be the result of how the NMR data were processed. In this regard, using a limited set of interproton NOE-derived distances of about 5 Å or less to constrain molecular dynamics calculations of DNA duplexes with approximate force fields has been shown to lead to significant errors over the length of a single turn of a DNA duplex (35 Å) (70,71).

When the first crystal structure of a *cis–syn* thymine dimer-containing duplex was finally solved in 2002 it was found that the DNA of one of the two duplexes in the unit cell was bent by 33° into the major groove in remarkable agreement with one of the first modeling studies (57) and with the circularization experiments (72) (Fig. 6). Because the crystal structure of the undamaged parent duplex could not be obtained for comparison purposes, it is not known how much of the bend is due to the native DNA sequence, and how much is due to crystal packing forces. Comparison of the crystal structures of the two duplexes in the unit cell, however, suggests that crystal packing forces do influence bending, as the bending angle of the second duplex is 23°. A similar set of bending angles was observed for the bromo-U derivatives used to help solve the structure (Table 2). The bend induced by the dimer can be attributed in part to a positive roll angle of 22° at the dimer that is caused by the wedge-like structure of the thymine dimer. The bend is also in the same direction as the bend observed in the crystal structure of T4 *denV* endonuclease V complex with a thymine dimer (73,74), suggesting that this enzyme may recognize the dimer in part by the bend that it induces (72).

The average twist angles of 33.7° and 35.0° in the crystal structures correspond to unwinding angles of 6° and –7° based on 10.5 bp/turn for B DNA. The average of these values is somewhat less than the 10° found in solution (64–66), while the 8° and 4° unwinding angles of the corresponding bromo-U derivatives are much closer. The unwinding of the helix can be attributed in part to a low twist angle between the thymine dimer base pairs seen in all the structures that presumably results from formation of the cyclobutane ring which forces the two T's to become more aligned. It is important to note that the cyclobutane ring in the dimer is not flat, which would have caused the two thymines to line up exactly. Instead, the cyclobutane ring adopts a puckered form that maintains a right-handed twist between the two T's as the bases normally do in B DNA (see Fig. 6D). This particular conformation was first deduced from early NMR experiments on a *cis–syn* dimer containing duplex (75) and is referred to as a CB+ conformation (56) and is opposite to that found in the cis–*syn* dimer of the dinucleotide TpT (76,77). This observation indicates that the 3D structure of photoproducts of dinucleotides cannot be used to reliably predict the structure of photoproducts in duplex DNA.

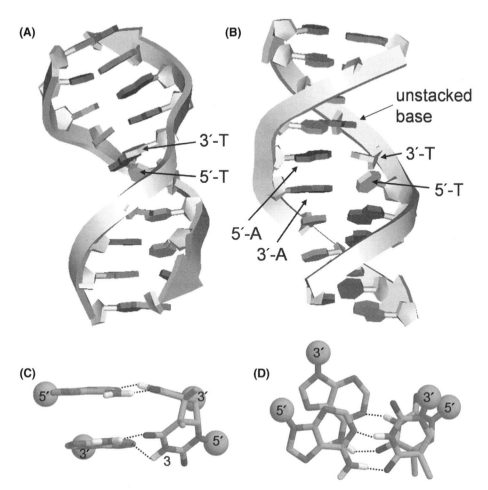

Figure 6 Tertiary structure and H-bonding in *cis–syn* thymine dimer-containing decamer duplex crystal structure. The structures were generated from Protein Data Bank crystal structure 1n4e, Table 1. (A) View showing bending towards major groove, and distortions in the backbone of both strands. (B) View showing distortion from planarity of 5'-T and absence of base stacking of the base flanking the 3'-T. (C) Major groove and, (D) Top view of base pairing interactions of the thymine dimer showing the poor alignment of the N3H of the 5'-T with the N1 of A. Views A and B were created with Accelrys Viewer Lite version 4.2 and views C and D were created with Rasmol version 2.6.

The crystal structure also reveals that the intradimer torsion angle β (P(*n*)–O5'(*n*+1)-C5'-C4') drops from the canonical value of 175–148° presumably to accommodate the structure of the thymine dimer. The decrease in this torsion angle causes a pinching effect in the minor groove that was also observed in one of the NMR-derived structures as a zigzag conformation of the phosphodiester backbone (67) (Fig. 6). The interchain phosphate distances in both the minor and major grooves of the crystal structure are 2–3 Å greater at the site of the CPD than for idealized B-DNA, and wider than expected for adenine tracts (78). The wider minor groove may be an important recognition element for T4 *denV* endo V which interacts with the minor groove of DNA (74). Another unusual feature of the crystal structure is the presence of the BII type phosphodiester conformation of the phosphodiester to the 3′-side of the dimer that was also deduced from an NMR study (67). The BII type conformation is consistent with an unusual upfield shifted ^{31}P NMR signal that was observed at low temperatures in this NMR study and in an earlier one (75). This BII conformation has been associated with destacking interactions a TpA steps (79,80) and may also be an important feature in DNA damage recognition.

2.5. Base Pairing and Thermodynamic Properties

Most of the modeling, NMR, and X-ray structures are in general agreement on the local structure and base pairing of the thymine dimer in a DNA duplex. All structures show that pairing of the bases flanking the dimer are unaffected, though base stacking is disrupted between the 3′-T of the dimer and the flanking base (Fig. 6B). They also show that the 3′-T forms a fairly coplanar Watson–Crick base pair with an A, whereas the 5′-T is out of the plane of the A, and does not form a very good H-bond between the N3H of T and the N1 of A (Figs. 6C and D and 7). The hydrogen bond between the imino proton of the 5′-T of the dimer and the A is about 2.5 Å in the crystal structure, and much longer than the 1.89 Å observed for the 3′-T (72). The H-N3-N1 angle of about 40° for the H-bond of the 5′-T in the crystal structure is also outside the normal range of 2–17° observed for normal base pairs (81). Numerous NMR studies have found that the chemical shift of the imino proton of the 5′-T is shifted upfield to a greater extent than that of the 3′-T as would be expected if it were not as H-bonded (64,67,82). It has also been suggested, however, that some of the upfield shift is due to ring current effects of the flanking base (67,75). The distorted nature of the base pair formed with the 5′-T would explain the 1.5 kcal/mol decrease in the stability of the DNA duplex upon thymine dimer formation (Fig. 8).

The diminished base pairing ability of the 5′-T of the thymine dimer has been suggested to play a role in the mechanism of T4 *denV* endo V recognition of the

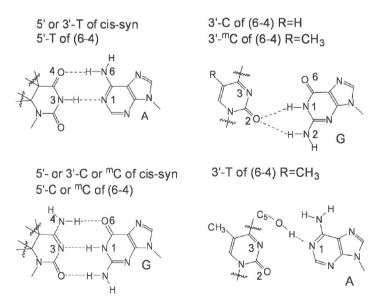

Figure 7 Schematized view of hydrogen bonding in the *cis–syn* and (6–4) photoproduct containing duplexes. In a duplex with A opposite the 3′-T of a (6–4) product, the N1 of A may be H-bonded to the C5 OH group of the 5′-T.

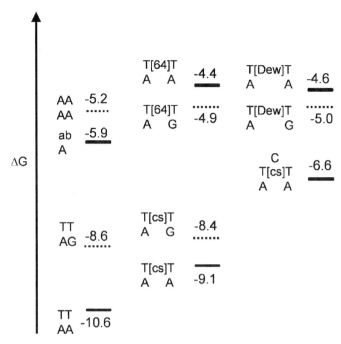

Figure 8   Free energies ($\triangle$G) for duplex formation for dithymidine photoproduct containing duplexes. The data are for d(GAGTAxyATGAG)•d(CTCATzATACTC) where xy = TT, T[c,s]T, T(6–4)T, T[Dew]T, and the non-adjacent dimer T^C[c,s]T, and z = A or G. Dashed lines refer to the mismatched sequence. For comparison a doubly mismatched TT•TT sequence is shown along with an abasic site (ab). *Source*: Data from Ref. 104.

thymine dimer by weakening base pairing to the A which then flips out of the helix into a binding pocket in the enzyme (74,83). Other repair enzymes also appear to make use of base flipping mechanisms that would be facilitated by weakened H-bonding such as *E. coli* CPD photolyase (84,85) or the damage binding subunit of *E. coli* uvrABC excinuclease (86,87). The *cis–syn* thymine dimer is repaired about nine times slower than the (6–4) photoproduct of TT (88), and has an approximately 10-fold lower binding constant for the uvrA DNA damage recognition subunit (89), which correlates well with its lesser perturbation of DNA duplex formation. A duplex containing a G opposite the 5′-T of a TT CPD, which corresponds to the product of deamination of a Tm^{5}C CPD, has been found to be 0.7 kcal/mol less stable than with an A opposite the 5′-T (Fig. 8). Double deamination at a CC or Cm^{5}C site would result in a doubly mismatched duplex containing GG opposite a U[*cis–syn*]U or U[*cis–syn*]T dimer, which would presumably be even less stable. In this regard, replacement of two A's opposite a *cis–syn* thymine dimer with two G's have been found to result in a four-fold enhancement in repair by the human excinuclease (90).

3. OTHER DIMER-RELATED PRODUCTS

3.1. Dimers with a Cleaved Intradimer Phosphodiester Backbone

In addition to chemical and photochemical reactions that can modify the structure of the primary photoproducts, there is evidence that enzymes can do the same. A number of years ago, Paterson and coworkers (91) reported that thymine dimers

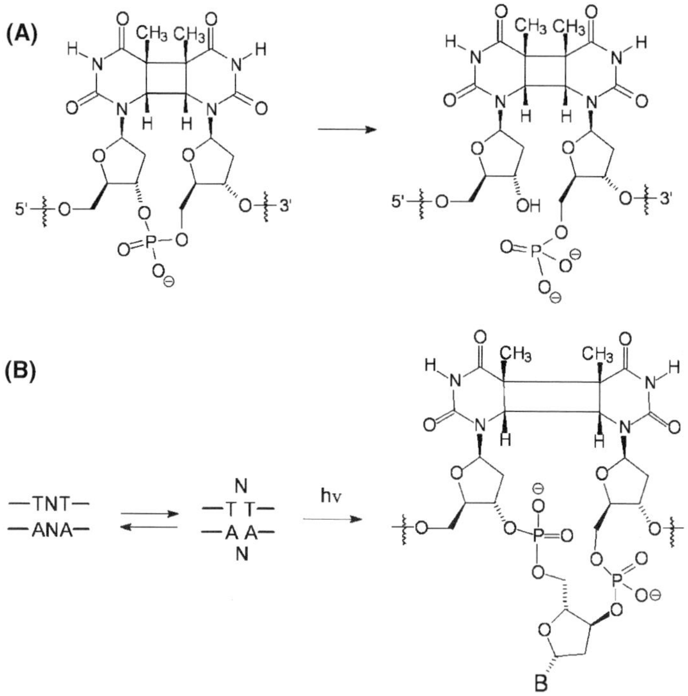

Figure 9 Other *cis–syn*-related photoproducts. (A) *Cis–syn* thymine dimer with a cleaved intradimer phosphodiester bond that results from enzymatic processing. (B) Non-adjacent *cis–syn* thymine dimer that could form via interior loop structures, or slipped intermediates.

isolated from human cell extracts contained a cleaved intradimer phosphodiester bond (Fig. 9A). Based on this and subsequent work, it was proposed that enzymatic cleavage of the intradimer phosphodiester bond makes the thymine dimer more detectable by the human excision repair system (92–97). In support of this proposal it was found that a thymine dimer with a cleaved intradimer phosphodiester (CPD*) was cleaved faster by the uvrABC excinuclease than a normal CPD (98). In spite of the numerous studies describing the existence and repair of CPD*, nothing is known about its structure or physical properties. DNA containing a site-specific CPD* can be prepared by photoligation of oligonucleotides (98). Alternatively and automated DNA synthesis building block for a *cis–syn* thymine dimer lacking an intradimer linkage and phosphates has been described that could be adapted for the large-scale synthesis of CPD* (99).

3.2. Non-Adjacent Dimers

Cyclobutane pyrimidine dimer formation is not restricted to sequential pyrimidines in DNA but can also occur between non-sequential pyrimidines through bulge loop or interior loop structures (Fig. 9B). Irradiation of single-strand, alternating poly[d(GT)] or poly[d(CT)] produces non-adjacent *cis–syn* dimers as the major product in about 1% yield when irradiated with 254 nm (31). The yield of non-adjacent dimers increases to 40% when wavelengths >280 nm and a triplet sensitizer such as acetone is used (32). Non-adjacent CC dimers, which contain an extrahelical thymine, are also produced in low yield upon irradiation of poly[d(TC)]•poly[d(GGA)] with 254 nm light (100).

More recently, a site-specific non-adjacent T^C[c,s]T dimer was prepared by irradiation of a dodecamer duplex which contains a C bulge between the two central T's (34). The NMR spectra of the non-adjacent dimer-containing duplex had many features in common with that of the corresponding dimer-containing duplex. Most notably was the upfield shifted 5'-methyl proton signal of the dimer, and NOE crosspeaks indicative of a right hand twisted cyclobutane ring conformation (CB$^+$) (56,64,75). Whereas a thymine dimer only decreases the stability of duplex formation by 1.5 kcal/mol, the non-adjacent dimer decreases the stability by 4.0 kcal/mol and is more similar in effect to the parental bulge loop-containing duplex. The greater destabilization of the non-adjacent thymine dimer compared to a thymine dimer may be due to differences in the conformation of the non-adjacent dimer that more greatly affect base stacking and base pairing interactions. Nothing is known about the repair of this lesion.

3.3. *Trans-Syn* Dimers

Trans-syn dimers are very minor photoproducts of duplex DNA and are formed with a frequency that is about 12% that of *cis–syn* dimers in denatured DNA and 2% in native DNA (101). They can be produced in low yield by irradiating oligodeoxynucleotides (102), or through the use of a *trans-syn*-I or -II DNA synthesis building block (30,103). *Trans-syn* dimers arise from a photochemical (2+2) cycloaddition reaction between two pyrimidines, one of which becomes locked in a syn glycosyl conformation and can no longer base pair with the complementary base (Fig. 3). Both the *trans-syn*-I and -II thymine dimers decrease duplex stability by about 5 kcal/mole (Ren and Taylor, unpublished results), which is slightly less than 6 kcal/mol for the (6–4) product, and much greater than 1.5 kcal/mol for the *cis–syn* dimer (Fig. 8) (104). This correlates well with the repair rate of the *trans-syn*-I thymine dimer by the *E. coli* uvrABC excinuclease which is about 6-fold greater than a *cis–syn* dimer and 0.66-fold slower than the (6–4) product (88). The *trans-syn*-I thymine dimer has about the same binding constant as the (6–4) product for the uvrA subunit (89). One of the *trans-syn* thymine dimers, originally identified as *trans-syn*-I, but possibly *trans-syn*-II, was found to bend DNA by 22° and unwind it by about 15° based on the analysis of the gel electrophoretic mobility of phased multimers (63). Authentic *trans-syn*-II thymine dimer, but not *trans-syn*-I dimer, was found to be a good substrate together with the *cis–syn* thymine dimer, for the *Paramecium bursaria chorella* virus pyrimidine dimer glycosylase (105). The *trans-syn*-II dimer is also a good substrate for this enzyme, presumably because the enzyme operates on the 5'-T, and both the *trans-syn*-II dimer and the *cis–syn* dimer have the 5'-T in the same anti glycosyl bond orientation.

4. (6–4) PRODUCTS

Pyrimidine-(6,4)-pyrimidone or (6–4) photoproducts (Fig. 1) are produced with a quantum yield of about 0.1% in DNA, and form more frequently at TpC sites. They can be prepared by direct irradiation of oligodeoxynucleotides with 254 nm light followed by HPLC purification (106–108), or through the use of T(6–4) T and T(6–4) C DNA synthesis building blocks (109,110). The (6–4) products are produced via a (2+2) cycloaddition reaction between the 5,6-double bond of the 5'-pyrimidine and a hetero double bond of the 3'-pyrimidine. If the 3'-pyrimidine is a T, an unstable oxetane intermediate is formed which spontaneously opens to yield the

(6–4) product at temperatures above –80°C (111). If the 3′-pyrimidine is a C or m⁵C, then cycloaddition presumably takes place with a photochemically induced imino tautomer to give a four-membered ring azetidine, which spontaneously opens to give the (6–4) product. The stereochemistry of the intermediate oxetane or azetidine and the resulting (6–4) product are determined by the orientation of the two pyrimidines prior to cycloaddition. The only (6–4) products that have been reported to date have the stereochemistry that arises from both pyrimidines having been in an antiglycosyl bond orientation, though it is conceivable that stereoisomers arising from syn glycosyl conformations can also be produced (Fig. 3).

The (6–4) products formed at TT, TC, CT, or CC sites all share the same 3′-pyrimidone ring system which is roughly perpendicular to the 5′-ring, and only differs by the substituent at the C5 position, which is a methyl group for TT, CT, Tm⁵C, and Cm⁵C sequences, and a hydrogen at TC and CC sequences. Unlike the *cis–syn* dimers, the pyrimidone ring of the (6–4) products has a UV absorption maximum at approximately 325 nm which leads to a secondary photoreaction in sunlight (wavelengths >290 nm) that produces the Dewar photoproduct (see next section). While the pyrimidone ring is incapable of deamination, and is stable under neutral conditions, it will degrade under alkaline conditions to give a characteristic strand break (112).

The 5′-pyrimidine of the (6–4) product is otherwise identical in structure to the parental pyrimidine except that the 5,6-double bond has become saturated by the addition of the pyrimidone ring to the C6 position and either a hydroxyl or an ammonium group to the C5 position (Fig. 1). The C5 substituent is transferred from the original 3′-pyrimidine and is either a hydroxyl group (–OH) when the 3′-base is a T, or an ammonium group (–NH₃⁺) when the 3′-base is a C or m⁵C. Because the 5,6-double bond of the 5′-pyrimidine of a (6–4) product is saturated, the 5′-C in a (6–4) product derived from CT, CC, and Cm⁵C is prone to tautomerization and deamination, just as it is for a *cis–syn* dimer (Fig. 4). The rate of deamination, however, is much slower than for *cis–syn* dimers, and has been estimated to occur with a half-life of a week for CpT (Table 1) (113), and to be immeasurably long for m⁵CpT (53). It is not known at this time which tautomer of C predominates in the (6–4) products of CpT, CpC, and CpᵐC, or how this might affect base pairing and the stability of the duplex.

4.1. Tertiary Structure of (6–4)-Containing DNA

As with CPDs, the first structural study of (6–4) product-containing duplexes was by molecular modeling (114). Unfortunately, the molecular modeling calculations were carried out with the wrong stereochemistry at the C5 position of the 5′-pyrimidine of the (6–4) products which resulted in structures that were stabilized by interactions that are not present in the actual structures. It was not until 10 years later that a solution structure of a duplex decamer containing a T(6–4) T product opposite AA was obtained by NMR. The DNA was bent by 44° towards the major groove, and unwound by 32° compared to idealized B DNA (Table 3) (64,115). In contrast, molecular dynamics calculations on the same (6–4) product-containing duplex decamer concluded that the (6–4) product bends DNA by only about 5° (59). In support of the NMR structure, analysis of the electrophoretic mobility of phased multimers of the 5′-C5 thio analog of the T(6–4) T product opposite AA indicates that the (6–4) photoproduct bends DNA by 47° (116). A more recent study that used FRET to measure the end-to-end distance of a (6–4) product containing duplex could not

Table 3 Bending and Unwinding Angles (°) for (6–4) and Dewar Photoproduct-Containing DNA

Substrate	Method[a]	Bending	Unwinding	Reference
CGCAT(6–4)TACGC GCGTA ATGCG	NMR	44	32	(64,115)
CGCAT(6–4)TACGC GCGTA ATGCG	Modeling	5	nd	(59)
CGCAT(6–4)TACGC GCGTA GTGCG	NMR (1cfl.pdb)	27	2	(119)
CGCAT[Dew]TACGC[b] GCGTA ATGCG	NMR	21	16	(131)
CGCAT[Dew]TACGC[b] GCGTA GTGCG	NMR (1qkg.pdb)	43	39	(132,133)
...ACC^{s}T(6–4)TCGCT... ...TGGA AGCGA...	Electrophoresis	47	nd	(137)
...ACC^{s}T[Dew]TCGCT... ...TGGA AGCGA...	Electrophoresis	28	nd	(137)
...ATCGT(6–4)TCTCA... ...TAGCA AGAGT...	FRET	0	nd	(117)

[a]The protein data bank structure file is given in parentheses.
[b]Structures in which the C6 carbon of the 3′;-T of the Dewar product is inverted (6S, see Fig. 3).

detect any significant bending induced by the T(6–4) T product opposite AA (117). The 32° unwinding angle of the NMR decamer structure is about one-third of the 97° estimated from a study with supercoiled DNA which found that a (6–4) product unwinds DNA 6.5 times more than a *cis–syn* dimer (118) which unwinds DNA by 15° (63,66). Because of the uncertainty of the bending and unwinding angles, it is difficult to conclude anything about the role of these features in the recognition and repair of (6–4) products by any repair system.

The structure of the same decamer duplex with GA instead of AA opposite the T(6–4) T product has also been solved by NMR (Fig. 10) (119). This structure is of interest because it corresponds to the major product of DNA synthesis past the T(6–4) T product by *E. coli* pol V which preferentially inserts a G opposite the 3′-T (106,120). It also corresponds to the structure of a (6–4) product of a Tm^{5}C site, except that an OH group is attached to the C5 position of the 5′-T instead of an NH$_3^+$ group. Comparison of the two NMR structures indicates that replacing the A opposite the 3′-T of T(6–4) T with a G causes the DNA bend to decrease from 44° to 27° and the unwinding angle to decrease from 32° to only 2° (Table 3).

4.2. Base Pairing and Thermodynamic Properties of (6–4)-Containing DNA

Because of the structural rearrangement that occurs in forming the (6–4) product, the 3′-pyrimidone ring becomes oriented roughly parallel to the helix axis. Reorientation of the pyrimidone ring leads to the formation of a hole in the DNA at this position, which causes disruption in the base stacking with the base to the 3′-side and between the two As opposite the (6–4) product (Fig. 10). In the NMR study of the T(6–4) T•AA decamer duplex, a weak NOE was detected between the imino proton of the 5′-T and the opposed A that is indicative of Watson–Crick base pairing (Fig. 7) (64). Because the 3′-pyrimidone ring lacks any H-bonding protons,

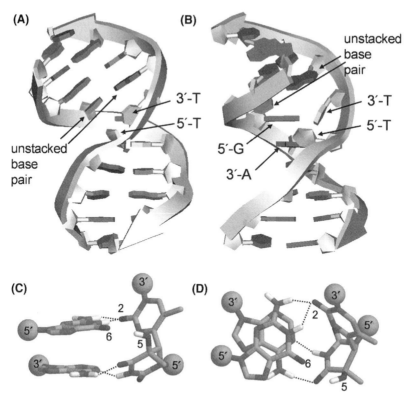

Figure 10 Tertiary structure and H-bonding in the T(6–4)T•GA duplex decamer NMR structure. The pictures were generated from Protein Data Bank structure 1cfl, Table 3. (A) View showing bending toward major groove and distortions in the backbone of both strands. (B) View showing distortion from planarity of 5'- and 3'-Ts, and absence of base stacking of the base pair flanking the 3'-T. (C) Major groove and (D) Top view of base pairing interactions of the (6–4) product showing H-bonding of the N1 imino and N2 amino groups of G H-bonding with of the O_2 carbonyl of the 3'-T. (Views A and B were created with Accelrys Viewer Lite version 4.2, and views C and D were created with Rasmol version 2.6.)

H-bonding interactions with the opposed A cannot be directly monitored by proton NMR. The calculated NMR structure, however, suggests that N1 of the A opposite the 3'-T of the (6–4) product may be H-bonded to the OH attached to the C5 position of the 5'-T. A molecular dynamics study of the T(6–4) T•AA decamer duplex did not detect this H-bond, but did detect a very weak interaction between the N1 of the pyrimidone ring and the N6 amino group of the A (average of 3 Å) (59).

The NMR structure of the T(6–4) T•GA duplex decamer is similar to that of the T(6–4) T•AA duplex, except that stacking is interrupted between the G opposite the 3'-T and the base to its 5'-side (Fig. 10 A and B), rather than between the two bases opposite the (6–4) product (119). H-bonding was detected between the 5'-T and the A, and there was evidence that the N1 imino proton of G is H-bonded. In the NMR structure, both the imino and amino protons of the G were found to H-bond to the O_2 carbonyl of the 3'-T of the (6–4) product (Figs. 10C and D and 7). In an earlier study of a T(6–4) T•GA dodecamer duplex, the imino proton NMR signal of the G was used to monitor the extent of H-bonding to the 3'-T (104). In sharp contrast to the behavior of imino proton signal of the G in the parental and CPD duplexes, the imino proton signal in the (6–4) duplex appears to be in rapid exchange with solvent. The rapid exchange is highly indicative of little or no H-bonding interactions between the G and the pyrimidone ring of the (6–4) product. The low value of 10.5 ppm for the proton NMR chemical shift of the imino proton of the G is almost identical to that observed in G•A mismatches in which the imino proton is thought to be exposed to water and not involved in base pairing (121,122). It is also similar to chemical shifts of 10.3–10.4 ppm for the imino proton of G that is involved in base pairing with G or an N6-benzopyrene adduct of A (123,124).

An unusual set of NOEs were observed in the proton NMR of the T(6–4) T•AA dodecamer duplex structure between the methyl group of the 5′-T of the (6–4) product and all the sugar protons of the 5′-flanking A (64,115). A similar set of NOEs were observed in the T(6–4) T•GA dodecamer duplex that also included NOEs to the H2 and H8 protons of the 5′-flanking A (104). These NOEs are hard to explain by a single structure, but could be explained by a dynamic structure in which the (6–4) product is flipping between an intrahelical H-bonded conformation, and an extrahelical conformation, which would also explain the rapid exchangeability of the G opposite the 3′-T of the (6–4) product (104). Though the molecular dynamics study did not show such a flipping process, it is likely that this process would occur on a timescale that is much greater than the 1 ns period of time that the simulations were allowed to run (59).

The 6 kcal/mol drop in thermal stability in going from TT•AA dodecamer duplex to the corresponding T(6–4) T product-containing duplex is consistent with a structure in which base pairing and pi-stacking at the site of the photoproducts are greatly disrupted. This drop in stability is much greater than that of 2.0 kcal/mol for forming the T•G mismatch in parent duplex, and comparable to that of 4.3 and 7.6 kcal/mol determined experimentally for the replacement of the T with an abasic site analog in GTG•CAC and CTC•GAG sequence contexts (125). Calculations based on thermodynamic parameters for predicting nucleic acid duplex stability further suggest that the central four nucleotides of the duplex behave more like a non-base paired interior loop structure than a base paired structure. Using optimized nearest neighbor parameters for DNA duplex stability and mismatches (126,127), the free energy of duplex formation for the dodecamer in which the T(6–4) T•AA is replaced by AA•AA is –5.2 kcal/mol which is very close to that of –4.4 determined experimentally for the (6–4) containing duplex (Fig. 8). An interior loop structure in which there is poor H-bonding and pi-stacking would also account for the rapid exchange of the imino proton of G opposite the T(6–4) T product in the same dodecamer duplex.

Additional evidence that the T(6–4) T•AA product behaves more like a doubly mismatched structure comes from the estimated unwinding angle for (6–4) products of 97° which corresponds to the unwinding of 2.5 base pairs (118). Evidence for a loss of base stacking in addition to base pairing comes from the observation that the hypochromicity of the T(6–4) T-containing duplex dodecamers in 1 M salt is less than that of the corresponding *cis–syn* dimer-containing duplexes (104). A further indication of a substantial disruption of base pairing and base stacking at the site of the T(6–4) T product comes from an additional loss of about half of the hypochromicity when the salt concentration was reduced to 250 mM. This loss of hypochromicity is indicative of the presence of only half a duplex and an inability to propagate a duplex past the (6–4) product.

Based on the thermodynamic and H-bonding data, it is likely that DNA containing a (6–4) product is best described by an unstable fluxional structure in which the (6–4) product equilibrates between base paired and non-based paired structures, unlike the more stable *cis–syn* dimer. It is also possible that non-specific, and (6–4)-specific repair enzymes make use of this fluxional behavior to recognize this type of photodamaged DNA. In this regard, the (6–4) product of TT is repaired 9-times faster than the *cis–syn* dimer by the uvrABC system (88), and binds about 10-fold more tightly to the DNA damage recognition subunit (89). A photoproduct flipping mechanism for (6–4) photolyase has been proposed that is supported by the observation that introducing a double T mismatch opposite the (6–4) further enhances the rate of photorepair four-fold, presumably by further reducing base pairing with the photoproduct (128).

5. DEWAR PHOTOPRODUCT

The (6–4) product is not stable to prolonged irradiation with UV light and is isomerized to the Dewar product with a quantum yield of about 1% (Fig. 1). The photoisomerization reaction leads to the formation of a bond between N3 and C6 of the 3′-pyrimidone ring of the (6–4) product which results in a unique structure containing two four-membered rings fused together. The Dewar photoproduct was named after the unique structure that the pyrimidone ring adopts following isomerization, which is known as a Dewar valence bond isomer (129). The Dewar product can best be prepared synthetically in essentially quantitative yield from the (6–4) photoproduct by irradiation with UVB light or Pyrex-filtered medium pressure mercury arc light to excite the 3′-pyrimidone ring which absorbs at about 325 nm (106,108,130).

5.1. Tertiary Structure of Dewar-Containing DNA

Unlike the case for the *cis–syn* dimers and the (6–4) photoproducts, the tertiary structure of the Dewar photoproduct-containing duplexes has never been studied by molecular modeling alone, but only in conjunction with NMR. NMR-derived structures have been proposed for the Dewar product of TT in duplex DNA opposite both complementary and mismatched sequences (131–133). Unfortunately, the structures used to model the NMR data have the stereochemistry of the C6 carbon of the 3′-ring inverted (6S) despite being shown in the figures as the 6R stereochemistry (Fig. 3). The fact that models could be made with the 6S stereochemistry calls into question either the stereochemistry of the Dewar photoproduct or the validity of the duplex structures proposed. The structure of the Dewar product with C6 inverted is not consistent with the NMR data reported for the dinucleotide product (129,134). The distance between the methyl group of the 3′-T in the inverted structure is about 7 Å away from the H3′ proton and outside the limit of 5 Å for producing an NOE, whereas an NOE between these two groups is clearly seen in the dinucleotide photoproduct. It is possible, however, that the stereochemistry of the Dewar product might be different when produced in single strand or duplex DNA. Evidence that the stereochemistry of the Dewar product formed in a dinucleotide is the same as that formed in duplex DNA, however, comes from enzymatic degradation/ HPLC/MS assays (135,136). It is not known to what extent the bending of the calculated structures of the T[Dewar]T•AA and T[Dewar]T•GA duplex decamers would be affected by the incorrect use of a Dewar structure with an inverted center. That being said, the T[Dewar]T•AA photoproduct DNA was calculated to have smaller bending angle than the corresponding T(6–4) T•AA duplex (21° vs. 44°), and a smaller unwinding angle (16° vs. 32°) (131). In contrast, the T[Dewar]T•GA structure was calculated to have a larger bending angle than the corresponding (6–4) product (43° vs. 27°) and a much larger unwinding angle (132,133). The bending angle of 21° for the T[Dewar]T•AA structure is consistent with the value of 28° from an electrophoresis study of multimers of the thio analog, i.e., s^5T[Dewar]T•AA (137).

5.2. Base Pairing and Thermodynamic Stability of Dewar-Containing Duplexes

Certain features of the Dewar-containing duplexes can still be ascertained from the NMR data without resorting to the computed structures. The first is that, unlike the corresponding (6–4) product, there is no evidence of H-bonding in the T[Dewar]T•AA dodecamer duplex between N3 of the 5′-T and a directly opposed A in the complementary strand (131). Likewise for the T[Dewar]T•GA decamer duplex in which a G is opposite the 3′-T of the Dewar product, no signal could be assigned to the N3H of the 5′-T (132,133). A signal for the imino proton of G was observed that is indicative of weak hydrogen bonding to the Dewar product, but hard to assign without a valid model. The lack of detectable H-bonding of the 5′-T of the Dewar product, and evidence for only weak H-bonding between the 3′-T and an opposed G are reflected in the 6 and 5.5 kcal/mol destabilizations of the T[Dewar]T•AA and T[Dewar]T•GA duplexes compared to TT•AA (104). The amount of destabilization caused by the Dewar is almost identical to that caused by the (6–4) product, and is similar to that of a doubly mismatched interior loop structure (Fig. 8). The Dewar photoproduct of TT is repaired at about the same rate as a (6–4) product by the uvrABC system (88), though it is bound about 2.5-fold less tightly by the uvrA DNA damage recognition subunit (89).

5.3. Thio Analogs of (6–4) and Dewar Products

Thiocarbonyl analogs of the bases have been used extensively as photocrosslinking agents between nucleic acids, and as model systems for studying the repair and mutagenesis by DNA photoproducts. They are useful because they form photoproducts at 360 nm, compared to 260 nm for the normal nucleic acid bases, which allows them to be produced selectively in DNA (138). The photochemistry of a dipyrimidine containing a 4-thiothymidine in the 3′-position is similar to that of the normal base (Fig. 1), but with some notable differences. Unlike TT and TC sequences which form an unstable four-membered ring oxetane and azetidine intermediates, respectively, a Ts^{4}T sequence forms a stable four-membered ring thietane intermediate that is in a 3:1 equilibrium with the (6–4) product at room temperature (139). This equilibrium can be driven completely to the s^5T(6–4) T product by alkylation or thiolation of the C5 thiol group (139). Unexpectedly, the s^5T(6–4) T products reverse to the parent duplex upon irradiation with 254 nm light, which mimics the photoenzymatic reversal of (6–4) products by (6–4) photolyase (140,141). The unique photochemistry of s^4T has made it useful for the preparation of site-specific (6–4) products in oligonucleotides that might otherwise contain multiple dipyrimidine sites

(116,140). A building block is now available for the site-specific introduction of the s⁵T(6–4) T product into DNA (142,143).

6. SPORE PHOTOPRODUCT

The ability of proteins to affect DNA photoproduct formation has been known for some time and is most striking in bacterial spores that are very resistant to UV light (144). In *Bacillus subtilis* spores, an alpha/beta-type small, acid-soluble protein (SASP) causes the DNA to adopt an A conformation which inhibits formation of the *cis–syn* and (6–4) products and enhances formation of the unusual spore photoproduct (Fig. 11) (145). It also appears that spore photoproduct formation is further enhanced by dipicolinic acid that is present in the spores (146). The structure of the spore product has not been rigorously proven, but is proposed to be the one shown in Fig. 11 based on synthetic studies with a dinucleotide substrate and upon a consideration of the mechanism of the reaction and the structure of A DNA (147). The spore photoproduct is repaired by enzymatic reversal to the parent nucleotides by a spore product-specific lyase (148,149). Oligodeoxynucleotides containing a site-specific spore photoproduct can be prepared in about 5% yield by irradiation of the DNA at 10% relative humidity (150). There is also recent evidence that an interstrand spore product can form in denatured DNA (35). There have been no studies on the effects of this photoproduct on the structure or properties of duplex DNA, and there is no building block for its site-specific incorporation into DNA.

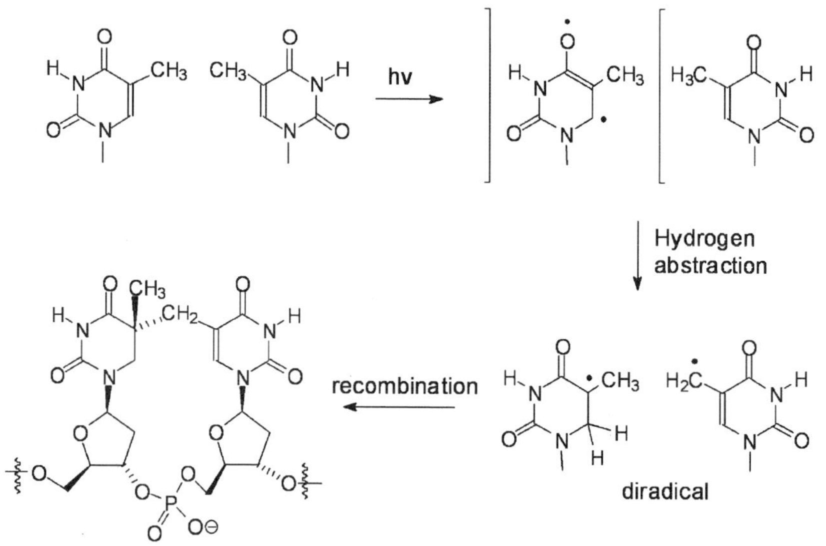

Figure 11   Formation of the spore photoproduct. In the first step an electronically excited T abstracts a hydrogen atom from an adjacent T resulting in a diradical that then recombines to form the spore product. The structure and stereochemistry of the product are based on the structure of the dinucleotide product and modeling studies.

7. TA* PRODUCT

The TA* photoproduct (Fig. 12A) is a very minor product and is produced with a quantum yield of about 0.01% in duplex poly(dA-dT) and calf thymus DNA (151), but became of immediate interest because of the ubiquitous nature of TATA sequences in promoters. The structure of the TA* product was first proposed to be a cyclobutane adduct resulting from the (2+2) cycloaddition between the 5,6 double bond of T and the 5,6 double bond of A (152). The stereochemistry of this adduct was originally assigned as *trans-syn* cyclobutane based on 2D proton NMR data and molecular dynamics calculations (153). Later it was shown that the carbon NMR spectrum was inconsistent with a cyclobutane adduct but consistent with the product of a subsequent electrocyclic rearrangement of a *cis–syn* adduct resulting in an eight-membered ring product (154). Substrates containing a site-specific TA* product can be prepared in good yield upon UV irradiation of single strand oligo-deoxynucleotides (155). Nothing is known about the effect of this product on the

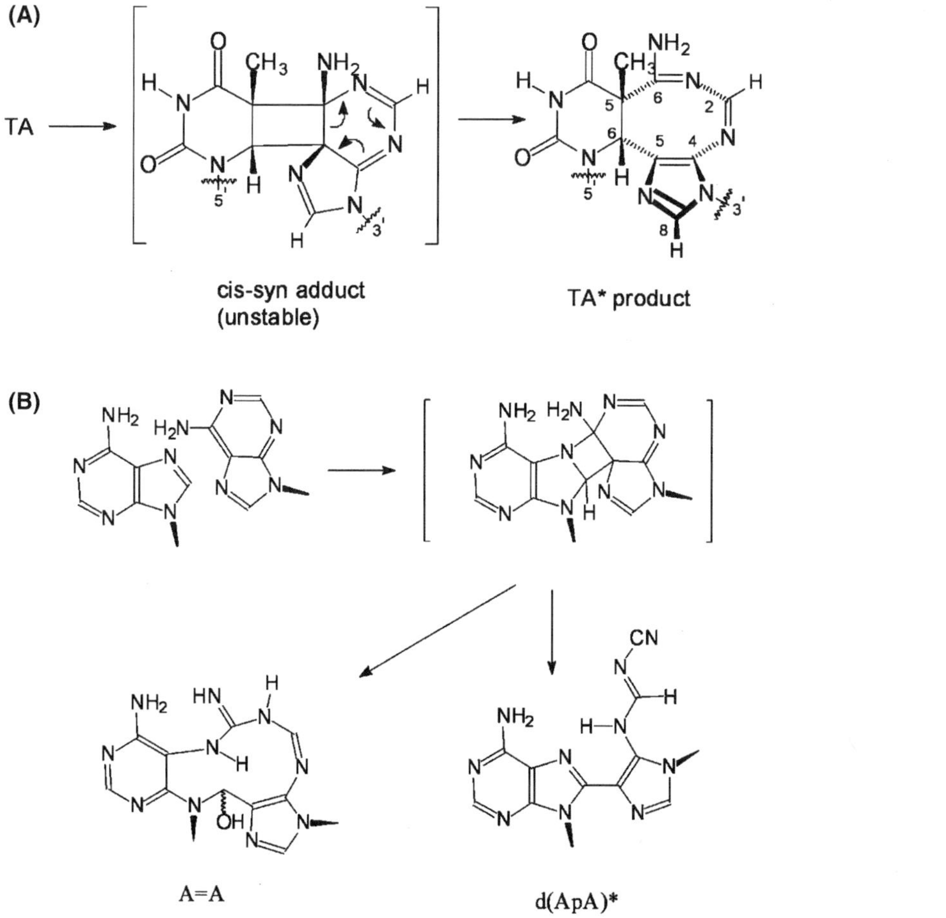

Figure 12 Other minor direct photoproducts of DNA. (A) The TA* photoproduct results from a (2+2) cycloaddition reaction followed by an electrocyclic ring opening to yield an eight-membered ring. (B) The A = A and AA* photoproducts result from a (2+2) cycloaddition followed by two alternative ring opening pathways.

structure or properties of duplex DNA or its repair, and there is no building block for this product.

7.1. AA* and Porschke Photoproducts

Another class of very minor photoproducts of DNA is produced at AA sequences. Irradiation of d(ApA) with 254 nm light produces two major products A = A and (AA)* from what appears to be a common (2+2) azetidine photoadduct (Fig. 12B) (156–158). The A = A product is formed with a very low quantum yield of about 0.2%, 0.006%, and 0.001% in single-stranded poly(dA), single- and double-stranded *E. coli* DNA, respectively (157). The low quantum yield for formation of these products in duplex DNA indicates that they may have limited biological significance, and have not been extensively studied. Oligodeoxynucleotides containing site-specific A = A and (AA)* products have been isolated by HPLC and characterized by mass spectrometry (159), but have not been further studied.

8. CONCLUSIONS

Despite many years of effort by many independent groups, there still is no consensus on the tertiary structure of DNA-containing TT CPDs, (6–4) products, and their Dewar valence isomers, though there seems to be some agreement on their local structure and base pairing properties that relate to their recognition and repair. There is very little known about the effect of sequence context on the structure and properties of DNA containing these photoproducts and how this might affect their recognition and repair. There is almost nothing known about the structure and properties of DNA containing dipyrimidine photoproducts of TC, TmC, CT, CC, and CmC sites and their deamination products, which are most relevant to mutagenesis, not to mention minor DNA photoproducts. It is clear that much more work is needed before one can hope to understand how repair enzymes detect and repair DNA photodamage. Especially needed are data on the structure and dynamics of photodamaged DNA on biologically relevant timescales, and pre-steady state kinetic studies of their interactions with repair enzymes and systems. Photoproduct analogs in which the backbone and bases are isotopically labeled and chemically modified may turn out to be very useful for such studies.

REFERENCES

1. Wang SY. Vol. 1. USA: Academic Press, Inc., 1976.
2. Wang SY. Vol. 1. USA: Academic Press, Inc., 1976.
3. Cadet J, Vigny P. Morrison H, ed. Bioorganic Photochemistry. Vol. 1. USA: John Wiley & Sons, 1990:1–272.
4. Ruzsicska BP, Lemaire DGE. In: Horspool WM, Song P-S, eds. CRC Handbook of Organic Photochemistry and Photobiology. Boca Raton: CRC Press, 1995:1289–1317.
5. Cadet J, Berger M, Douki T, Morin B, Raoul S, Ravanat JL, Spinelli S. Biol Chem 1997; 378:1275–1286.
6. Begley TP. Compr. Nat Prod Chem 1999; 5:371–399.
7. Ravanat JL, Douki T, Cadet J. J Photochem Photobiol B 2001; 63:88–102.
8. Lloyd RS. Prog. Nucleic Acid Res Mol Biol 1999; 62:155–175.

9. Sancar A. Chem Rev 2003; 103:2203–2237.
10. Deisenhofer J. Mutat Res 2000; 460:143–149.
11. Setlow P. Environ Mol Mutagen 2001; 38:97–104.
12. Petit C, Sancar A. Biochimie 1999; 81:15–25.
13. Reardon JT, Sancar A. Cell Cycle 2004; 3:141–4.
14. Avery AM, Kaur B, Taylor JS, Mello JA, Essigmann JM, Doetsch PW. Nucleic Acids Res 1999; 27:2256–2264.
15. Alleva JL, Zuo S, Hurwitz J, Doetsch PW. Biochemistry 2000; 39:2659–2666.
16. Fasman GD, ed., Handbook of Biochemistry and Molecular Biology. Nucleic Acids. Vol. I. 3rd ed. Cleveland: CRC Press, 1975.
17. Tommasi S, Denissenko MF, Pfeifer GP. Cancer Res 1997; 57:4727–4730.
18. You YH, Li C, Pfeifer GP. J Mol Biol 1999; 293:493–503.
19. Banerjee SK, Christensen RB, Lawrence CW, LeClerc JE. Proc. Natl Acad Sci USA 1988; 85:8141–8145.
20. Gibbs PEM, Lawrence CW. Nucleic Acids Res 1993; 21:4059–4065.
21. Jiang N, Taylor J-S. Biochemistry 1993; 32:472–481.
22. Horsfall MJ, Borden A, Lawrence CW. J Bacteriol 1997; 179:2835–2839.
23. Taylor J-S, Brockie IR, O'Day CL. J Am Chem Soc 1987; 109:6735–6742.
24. Murata T, Iwai S, Ohtsuka E. Nucleic Acids Res 1990; 18:7279–7286.
25. Ordoukhanian P, Taylor JS. Nucleic Acids Res 1997; 25:3783–3786.
26. Kosmoski JV, Smerdon MJ. Biochemistry 1999; 38:9485–9494.
27. Taylor J-S, Nadji S. Tetrahedron 1991; 47:2579–2590.
28. Carell T, Burgdorf LT, Kundu LM, Cichon M. Curr Opin Chem Biol 2001; 5:491–498.
29. Liu F-T, Yang NC. Biochemistry 1978; 17:4865–4876.
30. Kao JL-F, Nadji S, Taylor J-S. Chem Res Toxicol 1993; 6:561–567.
31. Nguyen HT, Minton KW. J Mol Biol 1988; 200:681–693.
32. Nguyen HT, Minton KW. J Mol Biol 1989; 210:869–874.
33. Love JD, Minton KW. J Biol Chem 1992; 267:24953–24959.
34. Lingbeck JM, Taylor JS. Biochemistry 1999; 38:13717–13724.
35. Douki T, Laporte G, Cadet J. Nucleic Acids Res 2003; 31:3134–3142.
36. Dupuy-Mamelle N, Pullman B. J Chim Phys Phys Chim Biol 1967; 64:708–712.
37. Brown DM, Hewlins MJE. J Chem Soc C 1968:2050–2055.
38. Grossman L, Kato K, Orce L. Fed. Proc 1966; 25:276.
39. Phillips JH, Brown DM. J Mol Biol 1966; 21:405–419.
40. Ide H, Wallace SS. Nucleic Acids Res 1988; 16:11339–11353.
41. Yu SL, Johnson RE, Prakash S, Prakash L. Mol Cell Biol 2001; 21:185–188.
42. Danilov VI, Les A, Alderfer JL. J Biomol Struct Dyn 2001; 19:179–191.
43. Lemaire DGE, Ruzsciska BP. Biochemistry 1993; 32:2525–2533.
44. Frederico LA, Kunkel TA, Shaw BR. Biochemistry 1990; 29:2532–2537.
45. Setlow RB, Carrier WL, Bollum FJ. Proc Natl Acad Sci USA 1965; 53:1111–1118.
46. Fix D, Bockrath R. Mol Gen Genet 1981; 182:7–11.
47. Fix D. Mol. Gen. Genet 1986; 204:452–456.
48. Tu Y, Dammann R, Pfeifer GP. J Mol Biol 1998; 284:297–311.
49. Burger A, Fix D, Liu H, Hays J, Bockrath R. Mutat Res 2003; 522:145–156.
50. Tessman I, Kennedy MA, Liu SK. J Mol Biol 1994; 235:807–812.
51. Barak Y, Cohen-Fix O, Livneh Z. J Biol Chem 1995; 270:24174–24179.
52. Peng W, Shaw BR. Biochemistry 1996; 35:10172–10181.
53. Douki T, Cadet J. Biochemistry 1994; 33:11942–11950.
54. Rao SN, Keepers JW, Kollman P. Nucleic Acids Res 1984; 12:4789–4807.
55. Raghunathan G, Kieber-Emmons T, Rein R, Alderfer JL. J Biomol Struct Dyn 1990; 7:899–913.
56. Kim J-K, Alderfer JL. J. Biomol Struct Dyn 1992; 9:705–718.
57. Pearlman DA, Holbrook SR, Pirkle DH, Kim S-H. Science 1985; 227:1304–1308.

58. Miaskiewicz K, Miller J, Cooney M, Osman R. J Am Chem Soc 1996; 118:9156–9163.
59. Spector TI, Cheatham TE III, Kollman PA. J. Am Chem Soc 1997; 119:7095–7104.
60. Cooney M, Miller JH. Nucleic Acids Res 1997; 25:1432–1436.
61. Husain I, Griffith J, Sancar A. Proc. Natl Acad Sci USA 1988; 85:2558–2562.
62. Wang C-I, Taylor J-S. Proc. Natl Acad Sci USA 1991; 88:9072–9076.
63. Wang C-I, Taylor J-S. Proc Res Toxicol 1993; 6:519–523.
64. Kim J-K, Patel D, Choi B-S. Photochem. Photobiol 1995; 62:44–50.
65. Spadari S, Sutherland BM, Pedrali-Noy G, Focher F, Chiesa MT, Ciarrocchi G. Toxicol Pathol 1987; 15:82–87.
66. Ciarrocchi G, Pedrini AM. J Mol Biol 1982; 155:177–183.
67. McAteer K, Jing J, Kao J, Taylor JS, Kennedy MA. J Mol Biol 1998; 282:1013–32.
68. Lavery R, Sklenar H. J Biomol Struct Dyn 1989; 6:655–667.
69. Lavery R, Sklenar H. J Biomol Struct Dyn 1988; 6:63–91.
70. Kuszewski J, Schwieters C, Clore GM. J Am Chem Soc 2001; 123:3903–3918.
71. McAteer K, Kennedy MA. J Biomol Struct Dyn 2003; 20:487–506.
72. Park H, Zhang K, Ren Y, Nadji S, Sinha N, Taylor JS, Kang C. Proc Natl Acad Sci USA 2002; 99:15965–15970.
73. Morikawa K, Matsumoto O, Tsujimoto M, Katayanagi K, Ariyoshi M, Doi T, Ikehara M, Inaoka T, Ohtsuka E. Science 1992; 256:523–525.
74. Morikawa K, Shirakawa M. Mutat Res 2000; 460:257–275.
75. Taylor J-S, Garrett DS, Brockie IR, Svoboda DL, Telser J. Biochemistry 1990; 29: 8858–8866.
76. Cadet J, Voituriez FE, Hruska FE, Grand A. Biopolymers 1985; 24:897–903.
77. Hruska FE, Voituriez L, Grand A, Cadet J. Biopolymers 1986; 25:1399–1417.
78. Hizver J, Rozenberg H, Frolow F, Rabinovich D, Shakked Z. Proc Natl Acad Sci USA 2001; 98:8490–8495.
79. Lefevre JF, Lane AN, Jardetzky O. FEBS Lett 1985; 190:37–40.
80. Kennedy MA, Nuutero ST, Davis JT, Drobny GP, Reid BR. Biochemistry 1993; 32:8022–8035.
81. Saenger W. In: Cantor CR, ed. Springer Advanced Texts in Chemistry. New York: Springer-Verlag, 1984.
82. Kemmink J, Boelens R, Koning T, van der Marel GA, van Boom JH, Kaptein R. Nucleic Acids Res 1987; 15:4645–4653.
83. Vassylyev DG, Kashiwagi T, Mikami Y, Ariyoshi M, Iwai S, Ohtsuka E, Morikawa K. Cell 1995; 83:773–782.
84. Park H-W, Kim S-T, Sancar A, Deisenhofer J. Science 1995; 268:1866.
85. Vande Berg BJ, Sancar GB. J Biol Chem 1998; 273:20276–20284.
86. Hsu DS, Kim ST, Sun Q, Sancar A J Biol Chem 1995; 270:8319–8327.
87. Moolenaar GF, Hoglund L, Goosen N. EMBO J 2001; 20:6140–6149.
88. Svoboda DL, Smith CA, Taylor J-S, Sancar A. J Biol Chem 1993; 268:10694–10700.
89. Reardon JT, Nichols AF, Keeney S, Smith CA, Taylor J-S, Linn S, Sancar A. J. Biol. Chem 1993; 268:21301–21308.
90. Mu D, Tursun M, Duckett DR, Drummond JT, Modrich P, Sancar A. Mol. Cell Biol 1997; 17:760–769.
91. Weinfeld M, Gentner NE, Johnson LD, Paterson MC. Biochemistry 1986; 25: 2656–2664.
92. Paterson MC, Middlestadt MV, MacFarlane SJ, Gentner NE, Weinfeld M, Eker AP. J Cell Sci 1987; (suppl) 6:161–176.
93. Weinfeld M, Paterson MC. Nucleic Acids Res 1988; 16:5693.
94. Liuzzi M, Weinfeld M, Paterson MC. J Biol Chem 1989; 264:6355–6363.
95. Liuzzi M, Paterson MC. J Biol Chem 1992; 267:22421–22427.
96. Galloway AM, Liuzzi M, Paterson MC. J Biol Chem 1994; 269:974–980.
97. Famulski KS, Liuzzi M, Bashir S, Mirzayans R, Paterson MC. Biochem. J 2000; 345(pt 3): 583–593.

98. Zheng Y, Hunting D, Tang M. Biochemistry 1998; 37:3243–3249.
99. Nadji S, Wang C-I, Taylor J-S. J Am Chem Soc 1992; 114:9266–9269.
100. Evans DH, Morgan AR. J Mol Biol 1982; 160:117–122.
101. Patrick MH, Rahn RO. In: Wang SY, ed. Photochemistry and Photobiology of Nucleic Acids. Vol. II. New York: Academic Press, 1976:35–95.
102. Banerjee SK, Borden A, Christensen RB, LeClerc JE, Lawrence CW. J Bacteriol 1990; 172:2105–2112.
103. Taylor J-S, Brockie IR. Nucleic Acids Res 1988; 16:5123–5136.
104. Jing Y, Kao JF-L, Taylor J-S. Nucleic Acids Res 1998; 26:3845–3853.
105. McCullough AK, Romberg MT, Nyaga S, Wei Y, Wood TG, Taylor JS, Van Etten JL, Dodson ML, Lloyd RS. J Biol Chem 1998; 273:13136–13142.
106. LeClerc JE, Borden A, Lawrence CW. Proc Natl Acad Sci USA 1991; 88:9685–9689.
107. Smith CA, Taylor J-S. Biochemistry 1992; 31:2208–2209.
108. Horsfall MJ, Lawrence CW. J Mol Biol 1994; 235:465–471.
109. Iwai S, Shimizu M, Kamiya H, Ohtsuka E. J Am Chem Soc 1996; 118:7642–7643.
110. Mizukoshi T, Hitomi K, Todo T, Iwai S. J Am Chem Soc 1998; 120:10634–10642.
111. Rahn RO, Hosszu JL. Photochem Photobiol 1969; 10:131–137.
112. Franklin WA, Lo KL, Haseltine WA. J Biol Chem 1982; 257:13535–13543.
113. Douki T, Voituriez L, Cadet J. Photochem Photobiol 1991; 53:293–297.
114. Rao SN, Kollman PA. Photochem Photobiol 1985; 42:465–475.
115. Kim J-K, Choi B-S. Eur. J Biochem 1995; 228:849–854.
116. Warren MA, Murray JB, Connolly BA. J Mol Biol 1998; 279:89–100.
117. Mizukoshi T, Kodama TS, Fujiwara Y, Furuno T, Nakanishi M, Iwai S. Nucleic Acids Res 2001; 29:4948–4954.
118. Rosenberg M, Echols H. J Biol Chem 1990; 265:20641–20645.
119. Lee JH, Hwang GS, Choi BS. Proc Natl Acad Sci USA 1999; 96:6632–6636.
120. Tang M, Pham P, Shen X, Taylor J-S, O'Donnell M, Woodgate R, Goodman MF. Nature 2000; 404:1014–1018.
121. Li Y, Agrawal S. Biochemistry 1995; 34:10056–10062.
122. Maskos K, Gunn BM, LeBlanc DA, Morden KM. Biochemistry 1993; 32:3583–3595.
123. Yeh HJC, Sayer JMa, Liu X, Altieri AS, Byrd RA, Lakshman MK, Yagi H, Schurter EF, Gorenstein DG, Jerina DM. Biochemistry 1995; 34:13570–13581.
124. Faibis V, Gognet JAH, Boulard Y, Sowers L, Fazakerley GV. Biochemistry 1996; 35:14452–14464.
125. Gelfand CA, Plum GE, Grollman AP, Johnson F, Breslauer KJ. Biochemistry 1998; 37:7321–7327.
126. Peyret N, Seneviratne PA, Allawi HT, SantaLucia J Jr. Biochemistry 1999; 38: 3468–3477.
127. SantaLucia J Jr. Proc Natl Acad Sci USA 1998; 95:1460–1465.
128. Zhao X, Liu J, Hus DS, Zhao S, Taylor J-S, Sancar A. J Biol Chem 1997; 272: 32580–32590.
129. Taylor J-S, Cohrs MP. J Am Chem Soc 1987; 109:2834–2835.
130. Smith CA, Taylor J-S. J Biol Chem 1993; 268:11143–11151.
131. Lee JH, Hwang GS, Kim JK, Choi BS. FEBS Lett 1998; 428:269–274.
132. Lee J-H, Choi B-S. J Biochem Mol Biol 2000; 33:268–275.
133. Lee JH, Bae SH, Choi BS. Proc Natl Acad Sci USA 2000; 97:4591–4596.
134. Taylor J-S, Garrett DS, Cohrs MP. Biochemistry 1988; 27:7206–7215.
135. Douki T, Court M, Cadet J. J Photochem Photobiol B 2000; 54:145–154.
136. Douki T, Court M, Sauvaigo S, Odin F, Cadet J. J Biol Chem 2000; 275:11678–11685.
137. Connolly BA, Newman PC. Nucleic Acids Res 1989; 17:4957–4974.
138. Favre A. In: Morrison H, ed. Bioorganic Photochemistry. Vol. 1. New York: Wiley and Sons, 1990:379–425.
139. Clivio P, Fourrey J-L, Gasche J, Favre A. J Am Chem Soc 1991; 113:5481–5483.
140. Liu J, Taylor J-S. J Am Chem Soc 1996; 118:3287–3288.

141. Clivio P, Fourrey J-L. JCSSC 1996:2203–2204.
142. Guerineau V, Matus SK, Halgand F, Laprevote O, Clivio P. Org Biomol Chem 2004; 2:899–907.
143. Matus SK, Fourrey JL, Clivio P. Org Biomol Chem 2003; 1:3316–3320.
144. Donnellan JE Jr, Setlow RB. Science 1965; 14:308–310.
145. Nicholson WL, Setlow B, Setlow P. Proc Natl Acad Sci USA 1991; 88:8288–8292.
146. Setlow B, Setlow P. Appl Environ Microbiol 1993; 59:3418–3423.
147. Kim SJ, Lester C, Begley TP. J Org Chem 1995; 60:6256–6257.
148. Rebeil R, Sun Y, Chooback L, Pedraza-Reyes M, Kinsland C, Begley TP, Nicholson WL. J Bacteriol 1998; 180:4879–4885.
149. Rebeil R, Nicholson WL. Proc Natl Acad Sci USA 2001; 98:9038–9043.
150. Slieman TA, Rebeil R, Nicholson WL. J Bacteriol 2000; 182:6412–6417.
151. Bose SN, Davies RJH. Nucleic Acids Res 1984; 12:7903–7914.
152. Bose SN, Davies RJH, Sethi SK, McCloskey JA. Science 1983; 220:723 725.
153. Koning TMG, Davies RJH, Kaptein R. Nucleic Acids Res 1990; 18:277–284.
154. Zhao X, Nadji S, Kao JL-F, Taylor J-S. Nucleic Acids Res 1996; 24:1554–1560.
155. Zhao X, Kao JLF, Taylor J-S. Biochemistry 1995; 34:1386–1392.
156. Kumar S, Sharma ND, Davies RJH, Phillipson DW, McCloskey JA. Nucleic Acids Res 1987; 15:1199–1216.
157. Sharma ND, Davies RJ. J. Photochem Photobiol B 1989; 3:247–258.
158. Kumar S, Joshi PC, Sharma ND, Bose SN, Davies RJH, Takeda N, McCloskey JA. Nucleic Acids Res 1991; 19:2841–2847.
159. Wang Y, Taylor JS, Gross ML. Chem Res Toxicol 2001; 14:738–745.
160. Douki T, Cadet J. J Photochem Photobiol B, Biol 1992; 15:199–213.
161. Lee JH, Choi YJ, Choi BS. Nucleic Acids Res 2000; 28:1794–1801.

5

Damage Recognition by DNA Photolyases

Gwendolyn B. Sancar
Department of Biochemistry and Biophysics, School of Medicine, University of North Carolina at Chapel Hill, Chapel Hill, North Carolina, U.S.A.

1. OVERVIEW OF PHOTOLYASES

Photolyases are enzymes that catalyze repair of damaged nucleic acid bases in a light-driven cyclic electron transfer reaction (1–4), a process known as photoreactivation. The primary substrates of these enzymes are cyclobutane-type pyrimidine dimmers (5–7) and pyrimidine–pyrimidone (4–6) photoproducts (4,8), each of which forms efficiently when adjacent pyrimidine bases in nucleic acids absorb far UV radiation (250–300 nm). Photolyases bind to these lesions, absorb a photon of near UV or visible light (350–450 nm), and utilize this energy to repair the damaged bases with a quantum yield of 0.1–1.0 (see Ref. 9 for a recent review). Given the large UV flux that prevailed during the early evolution of life on earth, it is likely that photolyases appeared quite early. Indeed, both their relatively simple composition (see in what follows) and the fact that photolyases have been identified in all phylogenetic lineages including viruses, Archae, Eubacteria, and Eukaryotes, suggest an early origin. However, among extant organisms, the occurrence of these enzymes is by no means universal [the most notable absence being placental mammals (10,11)], nor, surprisingly, does it correspond precisely to potential exposure to sunlight (12).

Although it is theoretically possible for photolyases to repair dimers in DNA or RNA, and RNA photolyases have been reported in some plant species (13), only the photolyases that act on DNA bases preferentially have been studied in detail. For the purpose of this chapter, the term photolyase will be used to indicate the DNA photolyases. On the basis of their substrate specificities, photolyases are classified as either CPD photolyases (repairs the *cis–syn* pyrimidine dimer) or 6–4 photolyases (repair the 6–4 pyrimidine–pyrimidone). In no case has a single photolyase been found that can act on both types of substrate (5,8). As a group, the photolyases share a number of features: the enzymes are monomers with molecular weights of 55–65 kDa (14); the amino acid sequences of the enzymes are moderately conserved, with homology ranging from 15% to 75% depending upon the sources of the enzyme (14); each enzyme contains a reduced flavin chromophore that is the electron donor in the reaction that initiates repair (2,4,15–17); each enzyme contains a second, "antennae" chromophore that is the primary absorber of light for the photolysis reaction (4,18,19). These similarities suggest that photolyases are derived from a single ancestral gene that was later duplicated and modified to produce the extant enzymes with their substrate specificities. This argument is bolstered by the relatively recent discovery of the cryptochromes, circadian photoreceptors that are structurally related to the photolyases but have lost the ability to repair DNA (11).

2. THE NATURE OF THE SUBSTRATES

The preferred substrates for photolyases are *cis–syn* cyclobutane dimers (CPDs; Py <> Py) (5–7) or 6–4 pyrimidine–pyrimidone photoproducts (6–4 photoproducts) (8). The structures of these substrates for two thymidine nucleotides are shown in Fig. 1. It should be noted that cytidines at one or both positions in the lesion are also possible only in CPDs TT, TC, and CC form 6–4 photoproducts. Because the substrate structures are dissimilar, CPDs and 6–4 photoproducts pose different challenges for the photolyases that repair them. Addition of an electron to the cyclobutane ring linking the two pyrimidines in CPDs initiates a spontaneous rearrangement of the bonding electrons, which restores the double bonds between C5 and C6 [2 + 2 cycloreversion (9)]. In contrast, repair of 6–4 photoproducts requires both cleavage of a bond linking the two pyrimidines and a group transfer to restore the bases to their undamaged state. Despite these differences, repair by both enzymes is effected by light-driven electron donation. While the structures of CPDs and 6–4 photoproducts suggest significant

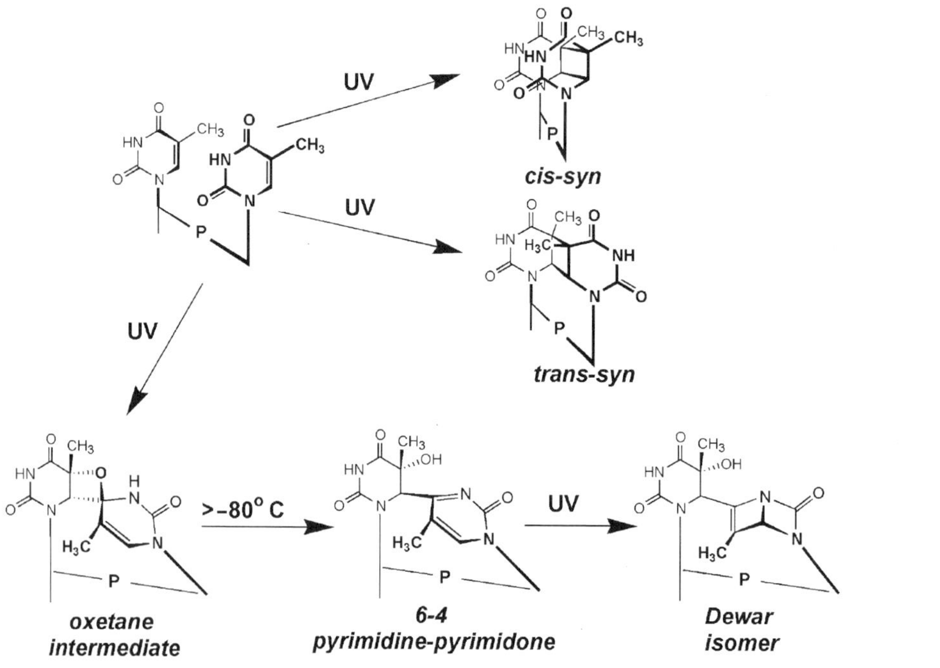

Figure 1   Substrates for the DNA photolyases. Pyrimidine dimers and 6–4 photoproducts are formed between adjacent pyrimidine bases in DNA upon absorption of UV light (λ_{max} = 254 nm). *Cis–syn* and *trans–syn* dimers are direct photoproducts, whereas the 6–4 pyrimidine– pyrimidone photoproduct is formed from an unstable oxetane (or azetidine for cytosine nucleo- tides) intermediate. Continued exposure of the 6–4 photoproduct to UV converts it to the Dewar isomer not detectably repaired by the known photolyases.

differences in the substrate binding sites for the two classes of photolyases, as is discussed later, the amino acids surrounding the active sites of the enzymes are highly homologous.

3. CHARACTERIZATION OF SUBSTRATE BINDING AND DISCRIMINATION BY PHOTOLYASES

3.1. CPD Photolyases

The CPD Photolyases bind pyrimidine dimers efficiently ($K_A \sim 10^8$–5×10^9 M^{-1}) in high molecular weight double-stranded or single-stranded DNA, regardless of whether the substrate is supercoiled, nicked, or linear (20–23). Pyrimidine dimers in substrates as small as T<>T dinucleotides and T<>T base dimers are repaired; however, the affinity of the enzyme for these substrates is 10^4–10^5 lower than the affinity for dimers in double-stranded high molecular weight DNA. Thus, while a CPD in a dinucleotide has all of the structural determinants necessary to align the substrate correctly in the enzyme active site, structural features of the DNA surrounding the dimer are important determinants of binding affinity. Results of studies using oligonucleotides of defined length and DNA footprinting analyses indicate that four to six nucleotides is the minimum length substrate that can be bound with high affinity (24–28). Binding of *Escherichia coli* photolyase to U<>U in RNA has also been observed, albeit with $\sim 10^5$-fold lower affinity than for U<>U or T<>T in DNA (29). As the ratio of RNA to DNA in most cells is about 3–10:1, this level of substrate discrimination is sufficient to prevent sequestering of DNA photolyases by dimers in RNA.

In addition to discriminating between DNA and RNA, CPD photolyases also discriminate between geometric isomers of pyrimidine dimers. Although there are eight potential isomers of pyrimidine dimers, only two form in DNA: the *cis–syn* dimer, which forms in both double-stranded and single-stranded DNA, and the *trans–syn* dimer, which forms in single-stranded DNA (Fig. 1) (30). In vitro, *E. coli* CPD photolyase binds to and repairs both of these stereoisomers; however, the *cys–syn* dimer is bound preferentially by a factor of 10^4–10^5 (31). As photolyases are generally present in very low numbers in cells (32–34), it seems unlikely that the low level of binding to *trans–syn* dimers contributes substantially to repair in vivo unless another mechanism augments recruitment of the photolyase to this lesion. Nevertheless, this binding preference is of interest as it establishes that the orientation of the two pyrimidine bases around the cyclobutane ring is a crucial-binding determinant. Photolyases have also been shown to repair pyrimidine dimers formed between bases separated by a single intervening purine base (35), which is consistent with the result of phosphate modification studies (24,25) suggesting that the enzyme has little interaction with the sugar-phosphate backbone between the dimerized bases.

Cyclobutane dimer photolyases exhibit a marked preference for dimers of specific base composition. Compared with T<>T dimers, in vitro the *E. coli* enzyme exhibits a 10-fold lower affinity for C<>C dimers, a 3-fold lower affinity for U<>U dimers, and about equal affinity for U<>T dimers (29). Once bound, U<>U dimers are repaired about 60% as efficiently as T<>T dimers; however, C<>C dimers are repaired with only about 5% efficiency. This pattern of substrate preference has the consequence that photolyase can either increase or decrease the frequency of transition mutations at C-containing dimers. Cytosines in potentially lethal dimers are deaminated to U at a substantial rate (36–38) and, in this form, are efficiently bound and photoreactivated, ultimately producing C → T transition mutations (39,40). However, binding of photolyase to T < > C dimers reduces the rate of deamination of C (41) and increases the efficiency of lesion recognition by the nucleotide excision repair (NER) apparatus (42,43); so, at the moment, it is unclear to what extent these two effects counterbalance one another in vivo.

3.2. 6–4 Photolyases

The 6–4 photolyases were discovered only in the last decade (8) and thus have been less exhaustively characterized than the CPD photolyases. In fact, initial reports of a photolyase that could repair the 6–4 photoproduct were met with some skepticism because the structure of the 6–4 photoproduct appears to preclude a simple cycloreversion reaction of the type catalyzed by the CPD photolyases. As can be seen in Fig. 1, the 6–4 photoproduct is formed by transfer of the group at C-4 of the 3′ base in the dinucleotide to the C-5 position of the 5′ base and the formation of a single bond between C-6 of the 5′ base and C-4 of the 3′ base. Breaking either of these bonds would yield only damaged bases. Nevertheless, identification of the final reaction products has definitively shown that the 6–4 photolyases restore the two pyrimidines to their original structures (8,44) in a light-dependent reaction. The proposed solution to this apparent paradox is that the binding of 6–4 photolyase leads to the formation of a four-membered oxetane or azetidine ring in a light-independent reaction, that this normally unstable intermediate is stabilized by amino acids at the active site, and that the light-dependent reversal of the lesion proceeds upon electron donation from flavin to this intermediate (4,44). This reaction sequence seems plausible considering that the four-membered oxetane or azetidine ring is the presumed intermediate in the formation of the 6–4 photoproduct (Fig. 1); however, definitive proof for this reaction scheme is lacking.

The 6–4 pyrimidine–pyrimidone is itself a reaction intermediate; upon prolonged UV irradiation, this lesion is converted to the Dewar isomer in which a single bond is formed between N3 and C6 of the 3′ base (45). Thus, an important question has been the identity of the true substrate of the 6–4 photolyases. Using defined substrates imbedded in oligonucleotides, it has been demonstrated that these enzymes bind the 6–4 pyrimidine–pyrimidone photoproduct with high affinity ($K_A \sim$ 2–$5 \times 10^8 M^{-1}$) and display a 4–10-fold lower affinity for the Dewar isomer. Importantly, however, photolysis of the Dewar isomer is <1% as efficient as that of the 6–4 photoproduct (3,4), indicating that the physiologically relevant substrate of these photolyases is the 6–4 photoproduct. The enzymes bind TT and TC 6–4 photoproducts with similar affinity (4). Binding to CPDs is undetectable (4,8,46). Detailed studies testing the ability of these enzymes to bind lesions in RNA or in small nucleotides have not been reported; however, it is clear that the enzymes bind 6–4 photoproducts in single-stranded DNA with a similar (3) or slightly higher (4) affinity than in double-stranded DNA.

4. INTERACTIONS AT THE PHOTOLYASE–PHOTOPRODUCT INTERFACE: THE MOLECULAR BASIS FOR SUBSTRATE BINDING AND DISCRIMINATION

The two prototypes for the CPD photolyases, the *E. coli* and *Saccharomyces cerevisiae* enzymes, are present in approximately 20 and 150 molecules per cell, respectively

(32–34), and are yet capable of repairing hundreds of dimers in the genome in a 5-min photoreactivation period. Although the number of 6–4 photolyase molecules per cell has not been measured, it is clear from attempts to purify the enzymes from native sources that the same must be true for these enzymes. Thus the photolyases must be highly effective at discriminating between pyrimidine dimers and nondamaged DNA, as well as potentially competing lesions. Comparison of the binding constants of the CPD photolyases to dimers vs. nondamaged DNA indicates that the discrimination ratio is $\sim 10^5$ (23). This is the same order of magnitude seen for many sequence-specific DNA-binding proteins (47); yet to be effective, the photolyases must recognize dimers and 6–4 photoproducts in a variety of different sequence contexts (48). The structural basis for high-affinity binding and efficient substrate discrimination has been revealed through a combination of DNA footprinting, mutational, and structural studies.

4.1. Contacts on DNA

Chemical modification of residues within the photolyase–DNA-binding interface can interfere with substrate binding while exposure of preformed photolyase–DNA complexes to chemical and enzymatic agents can identify regions on the DNA that are in sufficiently close contact with the enzyme to protect it from attack by these agents. These "footprinting" techniques have revealed that the CPD photolyases interact with groups on 3–4 nucleotides on each side of the dimer (Fig. 2) (24,25,49). The majority of the contacts are with groups on the dimer-containing strand, where the most prominent interactions are with the sugar-phosphate backbone; ethylation of the phosphate immediately 5′ to the dimer, as well as phosphates 3–4 residues 3′ to the dimer, inhibit photolyase binding. The small footprint size and skewed pattern of binding are consistent with the ability of the enzymes to repair closely spaced dimers on opposite strands (50). Modification of the phosphate between the two bases in the dimer does not affect binding; however, it is clear from studies cited earlier that the enzyme must interact with the bases in the dimer. Weaker and more variable interactions with bases 5′ and 3′ to the dimer indicate that the enzymes interact primarily with the major groove of the DNA. Only a single phosphate on the nondimer strand interacts strongly, consistent with the similar binding constants for single and double-stranded DNA. The roles of these interactions have been probed by site-directed mutagenesis coupled with footprinting analysis (51,52), which have revealed that the network of interactions along the phosphodiester backbone contributes both to binding affinity and to the ability of the enzyme to discriminate between dimer and nondimer DNA. Thus, the CPD photolyases must be recognizing the unique distorted structure of the phosphodiester backbone surrounding the dimer and the dimerized bases. A structural basis for backbone recognition was recently revealed by the crystal structure of a pyrimidine dimer in a DNA decamer (53). As had been previously predicted from theoretical (54) and experimental circularization studies (55), the dimer bends DNA by about 30° toward the major groove and produces substantial distortion of the phosphodiester backbone 5′ and 3′ to the dimer (Fig. 2) (53). On the basis of measurements of the affinity of E. coli photolyase for dimers in oligonucleotides of varying length, about 50% of the binding energy for the photolyase–DNA interaction comes from interactions with the phosphodiester backbone (27,29,26).

The interactions between the 6–4 photolyases and their substrates have not been studied in the same detail. Compared with the E. coli and S. cerevisiae CPD photolyases, the 6–4 photolyases appear to protect a slightly longer stretch of DNA from attack by DNAseI, particularly 5′ to the lesion, and, as indicated by enhanced base methylation, distort the DNA more 3′ to the lesion (Fig. 2) (3,4,24). It remains to be determined to what extent these differences reflect differences in the sequences used for footprinting or authentic differences in the binding interface.

(A) CPD Photolyase

5′-pNpNpNpNpNpNpNpNpNpNp**N***p*Py<>Py*p*N*p***N***p*N*p***N***p*NpNpNpNp-3′
3′- NpNpNpNpNpN**p**NpNpN**p**NpNpPu p **Pu**pNpNpNpNpNpNpNpNp-5′

(B) 6-4 Photolyase

5′-pNpNpNpNpNpNpNpNpNpNpNpPy<>Py**p**NpNpNpNpNpNpNpN-3′
3′- NpNpNpNpNpNpNpNpNp**N**pNpPu p PupN**p**NpNpN**p**NpNpNpNp-5′

(C) Pyrimidine Dimer in DNA

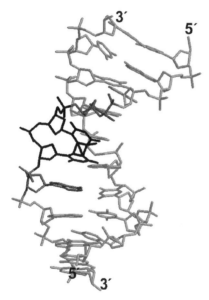

Figure 2 Comparison of the contacts made on DNA by CPD A and 6–4 photolyases B, and structure of a pyrimidine dimer in DNA C. (A and B) represent a compilation of data from different sources in which the sequence surrounding the lesion is not conserved (3,4,24,25). Nucleotides surrounding the lesions are indicated by pN, the pyrimidines in the lesions are indicated as Py <> Py. The bold lines above and below the sequences show the region protected from DNAseI digestion when the photolyase is bound to the lesion. Bold "p"s indicate sites at which ethylation of a phosphate inhibits photolyase binding (CPD photolyases) or which become hypersensitive to DNAseI upon photolyase binding (6–4 photoproduct). Bold "N"s indicate positions at which methylation interferes with binding or which become hypersensitive to methylation upon photolyase binding. In (C) the crystal structure of a decamer containing a pyrimidine dimer is shown (co-ordinates for the structure from Ref. 53). The nucleotides in the dimer are shown in black, and the nucleotide 5′ (above) and 3′ (below) to the dimer that display the greatest distortion are shown in gray. Note the bend in the helix at the site of the dimer and the widening of both major and minor grooves.

4.2. Structure of CPD Photolyases and a Model for DNA Binding

Many of the sites involved in substrate binding and discrimination were initially identified through mutation studies on the *E. coli* and *S. cerevisiae* CPD enzymes (51,52). However, the key to understanding the results of these studies has been the crystal structure of the *E. coli* enzyme (56). Overall, the structure is composed of two domains (Fig. 3): an N-terminal α/β domain, which is the binding site for the folate chromophore of the enzyme, and a C-terminal α-helical domain, which binds the flavin chromophore. These two domains are connected by a long interdomain loop that wraps around the N-terminal domain. The most striking feature of the enzyme structure is that the FAD chromophore is largely buried in the C-terminal domain, with the only access to solvent (and therefore substrate) via a hole leading from the surface of the α-helical domain to the isoalloxazine ring of FADH⁻ (Fig. 3). This hole is of a size and polarity to fit the bases of the pyrimidine dimer. The hole is flanked on two sides by a band of positive electrostatic potential. Remarkably, many of the amino acids identified by mutational studies as being involved in photolyase binding to dimer-containing DNA lie along this band or at the lip of the hole (52). It should be noted that, in the crystal structure of dimer-containing DNA, the bases and cyclobutane ring of the dimer are still within the central axis of the double helix (Fig. 2 and (53). These features have led to the proposal that the enzyme binds the DNA backbone through the positively charged residues flanking the hole and that the dimer is flipped out of the double helix and into the hole (56). Although the cocrystal structure of the photolyase–DNA complex has not been solved, this "dimer flipping" model is supported by a number of observations and is now widely accepted. *Thermus thermophilus* photolyase has been crystallized with a thymine monomer in the hole, and many of the interactions predicted for the flipped dimer are seen in this structure (57). The amino acid

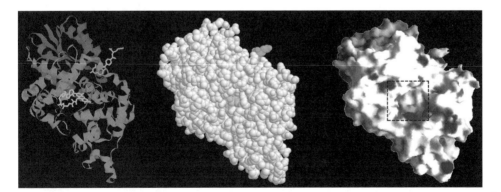

Figure 3 Molecular structure of *E. coli* CPD photolyase as determined by x-ray crystallography (56). Left: Ribbon diagram of the backbone showing the three structural domains (N-terminal α/β domain in red, C-terminal α-helical domain in green, and the interdomain loop in orange). The folate (cyan) and flavin (yellow) chromophores are shown in "sticks" format and lie, respectively, on the right edge and center of the enzyme in this view. Center: Space-filling model of photolyase shown from the same view as the left figure. Note that the flavin chromophore is almost completely buried in the structure with the exception of a "hole" leading from the surface to the alloxazine ring. Right: Electrostatic potential of the photolyase molecule shown from the same view. Positive potential is indicated by blue, and negative potential is indicated by red. The dashed box outlines the hole leading to the flavin chromophore. *Source*: Adapted from Ref. 56. (*See color insert*.)

sequences of the *E. coli* and *S. cerevisiae* photolyases are 50% identical over the region comprising the proposed DNA-binding domain (58), permitting the structure of the yeast enzyme to be modeled on that of the bacterial enzyme (Fig. 4). Alanine substitution of amino acids lying deep in the hole reduces the affinity of *S. cerevisiae* photolyase for dimer-containing DNA (59), a result expected only if the dimer enters the hole. In addition, upon formation of the ES complex, the environment around

```
                              ▼    *           ▼▼▼▼▼
E.c. CPD   (200) FPVEEKAAIAQLRQFCQNGAGEYEQQRD----FPAVEGTSRLSASLATGGLSPR
S.c. CPD   (305) PDVSEEAALSRLKDFLGTKSSKYNNEKD----MLYLGGTSGLSVYITTGRISTR
A.n. CPD   (208) P--GETAAIARLQEFCDRAIADYDPQRN----FPAEAGTSGLSPALKFGAIGIR
T.t. CPD   (177) P--GEEAALAGLRAFLEAKLPRYAEERDR----LDGEGGSRLSPYFALGVLSPR
D.m. 6-4   (220) FPGGETEALRRMEESLKDEIWVARFEKPNTAPNSLEPSTTVLSPYLKFGCLSAR
X.l. 6-4   (210) YPGGESEALSRLDLHMKRTSWVCNFKKPETEPNSLTPSTTVLSPYVKFGCLSAR

                          ▼ +*   *▼
E.c. CPD   (250) QCLHRLLAEQPQALD---G---GAGSVWLNELIWREFYRHLITYHPSLCKHRPFI
S.c. CPD   (354) LIVNQAFQSCNGQIMSKALKDNSSTQNFIKEVAWRDFYRHCMCNWPYTSMGMPYR
A.n. CPD   (256) QAWQAASAAHALSRS---DEARNSIRVWQQELAWREFYQHALYHFPSLADGP-YR
T.t. CPD   (225) LAAWEAERRGGEG----------ARKWVAELLWRDFSYHLLYHPWMAERP-LD
D.m. 6-4   (274) LFNQKLKEIIKRQP-----KHSQPPVSLIGQLMWREFYYTVAAAEPNFDRML-GN
X.l. 6-4   (264) TFWWKIADIYQGK------KHSDPPVSLHGQLLWREFYYTTGAGIPNFNKME-GN

                                              ▼   ▼+ ▼*
E.c. CPD   (299) AWTDRVQWQSNPAHLQAWQEGKTGYPIVDAAMRQLNSTGWMHNRLRMITASFLVK
S.c. CPD   (408) LDTLDIKWENNPVAFEKWCTGNTGIPIVDAIMRKLLYTGYINNRSRMITASFLSK
A.n. CPD   (306) SLWQQFPWENREALFTAWTQAQTGYPIVDAAMRQLTETGWMHNRCRMIVASFLTK
T.t. CPD   (267) PRFQAFPWQEDEALFQAWYEGKTGVPLVDAAMRELHATGFLSNRARMNAAQFAVK
D.m. 6-4   (322) VYCMQIPWQEHPDHLEAWTHGRTGYPFIDAIMRQLRQECWIHHLARHAVACFLTR
X.l. 6-4   (311) PVCVQVDWDNNKEHLEAWSEGRTGYPFIDAIMTQLRTECWIHHLARHAVACFLTR

                              ▼  ▼  ▼▼   *      ▼       +       +  +
E.c. CPD   (354) -DLLIDWREGERYFMSQLIDGDLAANNGGWQWAASTGTDAAPYFRIFNPTTQGEK
S.c. CPD   (464) -NLLIDWRWGERWFMKHLIDGDSSSNVGGWGFCSSTGIDGQPYFRVFNMDIQAKK
A.n. CPD   (362) -DLIIDWRRGEQFFMQHLVDGDLAANNGGWQWSASSGMDPK-PLRIFNPASQAKK
T.t. CPD   (323) -HLLLPWKRCEEAFRHLLLDGDRAVNLQGWQWAGGLGVDAAPYFRVFNPVLQGER
D.m. 6-4   (378) GDLWISWEEGQRVFEQLLLDQDWAINAGNWMWLSASAFFHQ-YFRVYSPVAFGKK
X.l. 6-4   (367) GDLWISWEEGQKVFEELLLDADWSLNAGNWLWLSASAFFHQ-FFRVYSPVAFGKK

E.c. CPD   (408) FDHEGEFIRQWLPELRDVPGKVVHEPWKWAQKAG------VTLDYPQPIVEHKEA
S.c. CPD   (518) YDPQMIFVKQWVPELISSENKRPEN-------------------YPKPLVDLKHS
A.n. CPD   (415) FDATATYIKRWLPELRHVHPKDLISGEITP---------IERRGYPAPIVNHNLR
T.t. CPD   (377) HDPEGRWLKRWAPEYPSYAPKDPVVDLEEAR-----------R
D.m. 6-4   (432) TDPQGHYIRKYVPELSKYPATCIYEPWKASLVDQRAYGCVLGTDYPHRIVKHEVV
X.l. 6-4   (421) TDKNGDYIKKYLPILKKFPAEYIYEPWKSPRSLQERAGCIIGKDYPKPIVEHNVV

E.c. CPD   (457) RVQTLAAYEAARKGK
S.c. CPD   (554) RALKVYKDAM
A.n. CPD   (461) QKQ----FKALYNQLKAAIAEPEAEPDS
T.t. CPD
D.m. 6-4   (487) HKENIKRMGAAYKVNREVRTGKEEESSFEEKSETSTSGKRKVRRATGSAPKRKR
X.l. 6-4   (476) SKQNIQRMKAAYARRSGSTEGVDKDSGQNNKK----GGKRKVAAGTSVAELFKKK
```

Figure 4 Alignment of the amino acid sequences of the C-terminal α-helical domains of CPD (E.c., S.c., A.n., and T.t.) and 6–4 photolyases (D.m. and X.l.). Residues shown as white on black are identical in all the enzymes, whereas residues shown as black on gray are similar. Stars are shown over residues that line the "hole" leading to FADH⁻, "+" indicates residues that form the region of electronegativity on each side of the hole, and the filled triangles indicate residues that interact with FADH⁻. Among the CPD photolyases, E.c. and S.c. enzymes utilize a folate chromophore in addition to FADH⁻, whereas A.n. and T.t. utilize deazaflavin instead of folate. *Note*: E.c., *Escherichia coli*; S.c., *Saccharomyces cerevisiae*; A.n., *Anacystis nidulans*; T.t., *Thermus thermophilus*; D.m., *Drosophila melanogaster*; X.l., *Xenopus laevis*.

the isoalloxazine ring of flavin becomes more nonpolar (60) and bases across the helix from the dimer become markedly more solvent accessible, consistent with "severe distortion of local helical structure" (61).

4.3. Substrate Discrimination

The role of specific amino acids in substrate binding and discrimination has been delineated via site-directed mutagenesis of the *S. cerevisiae* photolyase while comparison of ethylation interference patterns of mutant and wild-type enzymes has yielded a correlation between changes in specific contacts on the DNA and changes in specific amino acids at the DNA–photolyase interface (59). On the basis of these results, a model for interactions at the interface has been proposed and is shown in Fig. 5. In this model, the 5′ base in the dimer is involved in π–π stacking interactions with Trp^{387} (*S. cerevisiae* numbering), N-3 of both bases are within hydrogen bonding distance of O–ε of Glu^{384}, and the phosphate 5′ to the dimer is within hydrogen bonding distance of the side chains of Lys^{336} and Lys^{383}. Other potential interactions involving the dimer include a hydrogen bond between the exocyclic amine of the adenine base in FAD and O-4' of the 5′ dimer and a "hydrogen bond" between the sulfur in Met^{455} and the methyl group of the 3′ base. Among the residues lining the hole, alanine substitution of Trp^{387} has the largest effect on substrate binding and discrimination, reducing discrimination by 6-fold and decreasing the quantum yield for photoreactivation by 80%. This profound effect suggests that interaction between the Trp^{387} side chain and the 5′ base orients the dimer in the

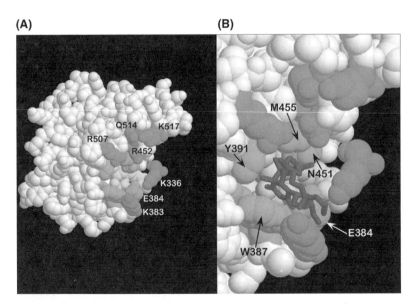

Figure 5 Space filling model of the *C*-terminal domain of *S. cerevisiae* photolyase. (A) The FADH$^-$ chromophore is shown in yellow at the bottom of the hole leading from the surface of the molecule to the interior. Amino acids lining the hole and thought to interact with the bases of the dimer are shown in cyan. Amino acids thought to interact with the phosphodiester backbone of the DNA are shown in green and are labeled. (B) Close-up view of the dimer-binding site, with a pyrimidine dimer (shown in stick format) manually docked in the proposed binding orientation. Amino acids proposed to interact with the dimer are labeled. (*See color insert.*)

cavity. Additional interactions with Trp[387] and Phe[494] are predicted for T-containing dimers specifically and may explain the preference of the enzyme for these dimers over C-containing dimers. Arg[452], Gln[514], Arg[507], and Lys[517] lie on the surface of the enzyme immediately outside of the active site cavity and are positioned to interact with DNA residues 3′ to the dimer. Mutations at Lys[383], Arg[452], Arg[507], Gln[514], and Lys[517] decrease both the affinity of the enzyme for DNA and the ability of the enzyme to discriminate between dimer-containing and nondamaged DNA. In general, the decrease in affinity for Ala substitution mutants of these sites is greater than for single amino acid substitutions within the hole, once again confirming the critical role of DNA backbone–photolyase interactions in binding. The loss of discrimination seen with these mutants is further support for the conclusion that photolyase recognizes the distortion of the sugar-phosphate backbone (24,25,52). It is particularly intriguing that mutation of Arg[452] *increases* nonspecific binding of the enzyme to DNA and decreases specific binding and the quantum yield for photoreactivation at least as much as the Trp[387] substitution. Furthermore, substitution of this residue decreases interaction of the enzyme with the phosphodiester bonds immediately flanking the dimer. These observations suggest that Arg[452] recognizes a key distortion of the phosphodiester backbone and plays an important role in orienting the dimer in the hole, perhaps by facilitating the flipping of the dimer into the hole or stabilizing the flipped dimer. Insertion of an amino acid side chain into the helix is a common method for obtaining base flipping (62–66).

The amino acid sequence of the 6–4 photolyases is highly similar to that of the CPD photolyases (Fig. 4). Using the *E. coli* photolyase structure as a starting point, molecular modeling studies similar to those described for *S. cerevisiae* photolyase have been used to model the DNA-binding domain of the *Xenopus laevis* 6–4 photolyase (67). Once again, the flavin chromophore is deeply buried in the center of the C-terminal domain and accessible only from a "hole" on the enzyme surface, and the general geometry of the binding site is similar to that of the CPD photolyases. Residues equivalent to Arg[507] and Lys[517] of *S. cerevisiae* photolyase are conserved and, together with Glu[384] → Gln, define a region of positive electrostatic potential surrounding and extending out from the hole. However, residues equivalent to Arg[452] and Gln[514] have changed to Leu (Leu[355] in the *X. laevis* enzyme) and Phe, respectively, thereby defining a new hydrophobic region on the surface of the enzyme. Lining the cavity leading to FAD, hydrophobic residues equivalent to *S. cerevisiae* Trp[387] and Tyr[391] are retained while Asn[451] and Met[455] have changed to His (*X. laevis* His[354] and His[358], respectively). Thus, Hitomi et al. (3) have suggested that changes in these four residues may account in large part for the differences in binding specificities of the CPD and 6–4 photolyases. This proposal is supported by the results of alanine substitution mutagenesis and molecular docking studies. Substitution of *X. laevis* Leu[355] strongly inhibits substrate binding (67), just as the substitution of the equivalent residue in the yeast enzyme strongly inhibits binding, and Leu[355] appears to be positioned to interact with the 3′-pyrimidine ring of the substrate. Substitution of *X. laevis* His[354] abolishes catalytic activity and substitution of His[358] reduces the rate of repair by >90%. In addition, the photolysis reaction is significantly inhibited below pH 8.5 and a significant deuterium isotope effect is apparent, both of which are consistent with the involvement of protonated histidines in catalysis. A model consistent with these observations is that His[354] and His[358] are involved in hydrogen bonding to the 3′ pyrimidone and 5′-pyrimidine of the substrate, respectively, and, via acid–base catalysis, in light-independent formation of the oxetane or azetidine intermediate. However, due to the proximity of these residues to the flavin chromophore, additional studies are needed to firmly establish this reaction mechanism.

5. SUBSTRATE BINDING IN VIVO

While we have a rather good understanding of how photolyases recognize their substrates in vitro, additional levels of complexity are apparent when considering photolyases function in vivo. Here, the photolyases must recognize not only different types of lesions, but also lesions imbedded in different structural contexts within the complex milieu of chromatin. In addition, photolyases are just one group of repair enzymes that act on pyrimidine dimers and 6–4 photoproducts, and thus the potential exists for competition or synergy between repair pathways. Recent and not-so-recent studies indicate that the substrate specificity of photolyases has been exploited to enhance repair at important and specific regions of chromosomes and to enhance the efficiency of other repair pathways. It should be noted that at present, this work has utilized solely the CPD photolyases; although it is assumed that the 6–4 photolyases behave similarly, the validity of that assumption remains to be tested.

5.1. Substrate Binding in Chromatin

A significant fraction of CPDs are subject to photoreactivation, regardless of their packaging in chromatin (8,69). However, the kinetics of repair indicate that the susceptibility of CPDs to repair changes with time (70). The basic subunit of chromatin, the nucleosome is composed of a central histone octamer with two turns of the DNA wrapped around it and an adjacent linker region connecting nucleosome cores. Detailed studies of the effect of chromatin structure on repair by CPD photolyases have been carried out in *S. cerevisiae*. In this organism, minichromosomes have been constructed in which the precise position of each nucleosome has been mapped. In such plasmids, photoreactivation of CPDs in linker DNA and in micrococcal nuclease sensitive regions of promoters and replication origins requires 15–30 min to complete, whereas about 2 hr are required to remove dimers from positioned nucleosomes (71). Thus, in living cells, repair by CPD photolyase is strongly modulated by chromatin structure. The rate of repair is slowest in the center of nucleosomes and increases towards the periphery (72). This pattern is most consistent with a model in which dimers introduced into nucleosomal DNA move out of the nucleosome and into the adjacent linker regions. This does not preclude repair of some dimers on the nucleosome surface, but rather suggests that dimers are repaired most efficiently by photolyase when they are in or near linker regions.

5.2. Photoreactivation in Actively Transcribed Genes

Photoreactivation is also strongly modulated by the transcriptional state of the DNA. In *S. cerevisiae*, actively transcribed genes are repaired by CPD photolyase more rapidly than inactive genes, regardless of the RNA polymerase involved in transcription (73–75). While this may, to some extent, be explicable by differences in nucleosome density and dynamics, the situation is clearly more complex. For genes transcribed by RNA polymerases II and III, dimers in the actively transcribed strand are photoreactivated much more slowly than are dimers in the nontranscribed strand (74,75). This has been attributed to inhibition of CPD photolyase binding by stalled RNA polymerase molecules on the transcribed strand, rather than stimulation of repair of the nontranscribed strand, a view that is supported by the observation that active transcription is required for the strand preference to be manifest (71,74,76). In RNAP II-transcribed genes, the strand preference for photoreactivation nicely complements the preferential repair of dimers in the transcribed strand by the NER pathway, a phenomenon that has been termed transcription-coupled repair (77–79). Interestingly, in RNAP I-transcribed genes, there is no pronounced inhibition of photoreactivation on the transcribed strand (73), whereas in RNAP III-transcribed genes, the nontranscribed strand is preferentially repaired by both photolyase and NER (75). This suggests that interactions between photolyase and the various RNA polymerases or their transcription factors, as well as the stability of stalled transcription complexes, play an important role in determining whether and where dimers are efficiently repaired in vivo.

5.3. Photoreactivation at Replication Origins, Promoters, and Telomeres

The eukaryotic chromosome is composed of diverse functional and structural regions. Regions near telomeres and centromeres are organized into compact heterochromatin while replication origins and promoters have a less condensed chromatin structure but are decorated with DNA-binding proteins specific to the functions of these regions. Several studies have probed whether and how CPD photolyase recognizes lesions in these various milieus and the extent to which sequence vs. chromatin structure affects recruitment of photolyase to dimers. At both promoters and replication origins, repair of CPDs by photolyase is substantially faster than by the NER pathway (71,80). Of course, this is only true with sufficient light flux; nevertheless, these results argue that photoreactivation is a predominant mechanism of repair at these important regulatory regions under certain conditions. The high AT content of these regions may help to target photolyase to these regions in the absence of damage. Proteins bound to the regulatory regions modulate efficient repair. TATA-binding protein inhibits repair of the *SNR6* promoter in vivo (81), and heterogeneity of repair sites in the replication origin *ARS1* is consistent with the ORC and cell cycle-dependent binding of pre-replication and replication-initiation complexes (80,82). Perhaps, the most convincing evidence for the role of chromatin structure is a recent study that examined repair in a *URA3* gene placed 2 kbp from a telomere (83). In cells in which the *URA3* gene was silenced by overexpression of the chromatin-binding protein Sir3, photoreactivation and NER were inhibited in the *URA3* promoter and silenced regions adjacent to the gene. In contrast, cells with a deletion of *SIR3* displayed rapid repair of the promoter and flanking regions by photolyase. This rules out sequence effects and argues that chromatin structure plays an important role in the efficiency of photoreactivation throughout the genome. Remodeling of chromatin may also play a role in the efficiency of repair; a recently published study has demonstrated that in vitro the nucleosome remodeling complexes SWI/SNF and yISW2 alter the conformation of nucleosomal DNA containing dimers and cause nucleosome migration (84). Thus, the dynamic properties of nucleosomes and chromatin remodeling activities are also likely to contribute to differences in binding of photolyases throughout the genome.

6. SUMMARY AND FUTURE DIRECTIONS

Photolyases, particularly the CPD photolyases, are among the best characterized of the DNA repair enzymes in terms of our understanding of the molecular mechanisms employed in substrate recognition and discrimination. The discovery that the enzyme employs interactions with both bases of the dimer and the surrounding phosphodiester backbone to achieve binding specificity and discrimination has served as a paradigm for understanding the binding by other enzymes that recognize specific structures in DNA. The "dimer flip" mechanism utilized by these enzymes is unique; in that, it is the only known situation among repair enzymes in which two bases are extruded from the helix. While structural and mutagenesis studies have provided strong evidence for this binding mechanism, the details, such as the specific interactions that lead to flipping of the dimer, as well as the ultimate test of the proposed mechanism of photolysis by the 6–4 photolyases, await the "holy grail" of the field, the solution of the cocrystal structure of photolyase with its substrate in DNA. For most organisms living on earth, photolyases are a substantial part of the armature used to protect the genome from the deleterious effects of UV damage, and thus understanding the mechanisms used by these enzymes to deal with damage in DNA is intrinsically important. In addition, by exploiting the light dependence of the photoreactivation reaction, it is possible to manipulate repair to define the roles of specific lesions or pathways. Studies on the role of chromatin structure and transcription on photoreactivation may well provide a basis for identifying factors and interactions important for more ubiquitous repair pathways.

REFERENCES

1. Rupert CSJ. J Gen Phys 1960; 43:573–595.
2. Sancar GB, Jorns MS, Payne G, Fluke DJ, Rupert CS, Sancar AJ. J Biol Chem 1987; 262:492–498.
3. Hitomi K, Kim S-T, Iwai S, Harima N, Otoshi E, Ikenaga M, Todo TJ. J Biol Chem 1997; 272:32591–32596.
4. Zhao X, Liu J, Hsu DS, Zhao S, Taylor J-S, Sancar A. J Biol Chem 1997; 272: 32580–32590.
5. Brash DE, Franklin WA, Sancar GB, Sancar A, Haseltine WA. J Biol Chem 1985; 260:11438–11441.
6. Wulff DL, Rupert CS. Biochem Biophys Res Commun 1962; 7:237–240.
7. Setlow JK, Setlow RB. Nature 1963; 197:560–561.
8. Todo T, Takemori H, Ryo H, Ihara M, Matsunaga T, Nikaido O, Sato K, Nomura T. Nature 1993; 361:371–374.
9. Sancar A. Chem Rev 2003; 103:2203–2237.
10. Li YF, Kim S-T, Sancar A. Proc Natl Acad Sci USA 1993; 90:4389–4393.
11. Hsu DS, Zhao X, Zhao S, Kazantsev A, Wang RP, Todo T, Sancar A. Biochemistry 1996; 35:13871–13877.
12. Harm H. Wand SY, ed. Photochemistry and Photobiology of Nucleic Acids Vol. 2. 2nd ed. New York: Academic Press 1976:192–263.
13. Gordon MP, Huang CW, Hurtler J. In: Wand SY, ed. Photochemistry and Photobiology of Nucleic Acids, Vol. 2. 2nd ed. New York: Academic Press 1976:265–308.
14. Sancar GB. J Bacteriol 1985; 161:769–771.
15. Todo T, Kim S-T, Hitomi K, Otoshi E, Inui T, Morioka H, Kobayashi H, Ohtsuka E, Toh H, Ikenaga M. Nucleic Acids Res 1997; 25:764–768.

16. Nakajima S, Sugiyama M, Iwai S, Hitomi K, Otoshi E, Kim S-T, Jiang C-Z, Todo T, Britt AB, Yamamoto K. Nucleic Acids Res 1998; 26:638–644.

17. Sancar GB, Smith FW, Heelis PF. J Biol Chem 1987; 262:15457–15465.

18. Johnson JL, Hamm-Alvarez S, Payne G, Sancar GB, Rajagopalan KV, Sancar A. Proc Natl Acad Sci, USA 1988; 85:2046–2050.

19. Kim S-T, Sancar A. Photochem Photobiol 1993; 57:895–904.

20. Setlow JK, Boling ME, Bollum FJ. Proc Natl Acad Sci USA 1965; 53:1430–1436.

21. Sancar GB, Smith FW, Sancar A. Biochemistry 1985; 24:1849–1855.

22. Rupert CS, Goodgal SH, Herriott RM. J Gen Phys 1958; 41:451–471.

23. Sancar GB. Mutat Res 1990; 236:147–160.

24. Baer M, Sancar GB. Mol. Cell Biol 1989; 9:4777–4788.

25. Husain I, Sancar GB, Holbrook SR, Sancar A. J Biol Chem 1987; 262:13188–13197.

26. Jordan SP, Alderfer JL, Chanderkar LP, Jorns MS. Biochemistry 1989; 28:8149–8153.

27. Jorns MS, Sancar GB, Sancar A. Biochemistry 1985; 24:1856–1861.

28. Setlow JK, Bollum FJ. Biochim Biophys Acta 1968; 157:233–237.

29. Kim S-T, Sancar A. Biochemistry 1991; 30:8623–8630.

30. Wang SY. In: Wand SY, ed. Photochemistry and Photobiology of Nucleic Acids, Vol. 1. 2nd ed. New York: Academic Press 1976:295–356.

31. Kim S-T, Malhotra K, Smith CA, Taylor J-S, Sancar A. Biochemistry 1993; 32: 7065–7068.

32. Harm W, Harm H, Rupert CS. Mutat Res 1968; 6:371–385.

33. Yasui A, Laskowski W. Int J Radiat Biol 1975; 28:511–518.

34. Fukui A, Hieda K, Matsudaira Y. Mutat Res 1978; 51:435–439.

35. Kim S-T, Sancar A. Photochem Photobiol 1995; 61:171–174.

36. Setlow RB, Carrier WL, Bollum FJ. Biochemistry 1965; 53:1111–1118.

37. Tessman I, Liu SK, Kennedy MA. Proc Natl Acad Sci USA 1992; 89:1159–1163.

38. Burger A, Fix D, Liu H, Hays J, Bockrath R. Mutat Res 2003; 522:145–156.

39. Tessman I, Kennedy MA. Mol Gen Genet 1991; 227:144–148.

40. Fix, D, Bockrath R. Mol Gen Genet 1981; 182:7–11.

41. Ruiz-Rubio M, Yamamoto K, Bockrath RJ. Bacteriol 1988; 170:5371–5374.

42. Sancar A, Franklin KA, Sancar GB. Proc Natl Acad Sci USA 1984; 81:7397–7401.

43. Sancar GB, Smith FW. Mol Cell. Biol 1989; 9:4767–4776.

44. Kim S-T, Malhotra K, Smith CA, Taylor J-S, Sancar A. J Biol Chem 1994; 269: 8535–8540.

45. Taylor JS, Cohrs MP. J Am Chem Soc 1987; 109:2834–2835.

46. Todo T, Ryo H, Yamamoto K, Toh H, Inui T, Ayaki H, Nomura T, Ikenaga M. Science 1996; 272:109–112.

47. von Hippel PH, Berg OG. Proc Natl Acad Sci USA 1986; 83:1608–1612.

48. Elledge SJ, Davis RW. Molec Cell Biol 1987; 7:2783–2793.

49. Kiener A, Husain I, Sancar A, Walsh CT. J Biol Chem 1989; 264:13880–13887.

50. Sung P, Prakash S, Prakash L. Genes Dev 1988; 2:1476–1485.

51. Li YF, Sancar A. Biochemistry 1990; 29:5698–5706.

52. Baer ME, Sancar GB. J Biol Chem 1993; 268:16717–16724.

53. Park HJ, Zhang K, Ren Y, Nadji S, Sinha N, Taylor J-S, Kang CH. Proc Natl Acad Sci USA 2002; 99:15965–15970.

54. Pearlman DA, Holbrook SR, Pirkle DH, Kim SH. Science 1985; 227:1304–1308.

55. Husain I, Griffith J, Sancar A. Proc Natl Acad Sci USA 1988; 85:2558–2562.

56. Park H-W, Kim S-T, Sancar A, Deisenhofer J. J Science 1995; 268:1866–1872.

57. Komori H, Masui R, Kuramitsu S, Yokoyama S, Shibata T, Inoue Y, Miki K. Proc Natl Acad Sci USA 2001; 98:13560–13565.

58. Tran H, Brunet A, Grenier JM, Datta SR, Fornace AJ Jr, DiStefano PS, Chiang LW, Greenberg ME. Science 2002; 296:530–534.

59. Vande Berg BJ, Sancar GB. J Biol Chem 1998; 273:20276–20284.

60. Weber S, Richter G, Schleicher E, Bacher A, Möbius K, Kay CWM. Biophys J 2001; 81:1195–1204.
61. Christine KS, MacFarlane AW, Yang K, Stanley RJ. J Biol Chem 2002; 277:38339–38344.
62. Slupphaug G, Mol CD, Kavli B, Arvai AS, Krokan HE, Trainer JA. Nature 1996; 384:87–92.
63. Reinisch KM, Chen L, Verdine GL, Lipscomb WN. Cell 1995; 82:143–153.
64. Nelson HCM, Bestor TH. Chem Biol 1996; 3:419–423.
65. Roberts RJ. Cell 1995; 82:9–12.
66. Klimasaukas S, Kumar S, Roberts RJ, Cheng X. Cell 1994; 76:357–369.
67. Hitomi K, Nakamura H, Kim ST, Mizukoshi T, Ishikawa T, Iwai S, Todo TJ. Bio Chem 2001; 276:10103–10109.
68. van de Merwe W, Bronk BV. Mutat Res 1981; 84:429–441.
69. Zwetsloot JCM, Vermeulen W, Hoeijmakers JHJ, Yasui A, Eker APM, Bootsma D. Mutat Res 1985; 146:71–77.
70. Pendrys JP. Mutat Res 1983; 122:129–133.
71. Suter B, Livingstone-Zatchej M, Thoma F. EMBO J 1997; 16:2150–2160.
72. Suter B, Thoma FJ. J Mol Biol 2002; 319:395–406.
73. Meier A, Livingstone-Zatchej M, Thoma F. J Biol Chem 2002; 277:11845–11852.
74. Livingstone-Zatchej M, Meier A, Suter B, Thoma F. Nucleic Acids Res 1997; 25:3795–3800.
75. Aboussekhra A, Thoma F. Genes Dev 1998; 12:411–421.
76. Selby CP, Drapkin R, Reinberg D, Sancar A. Nucleic Acids Res 1997; 25:787–793.
77. Selby CP, Sancar A. Science 1993; 260:53–57.
78. Mellon I, Spivak G, Hanawalt PC. Cell 1987; 51:241–249.
79. Bohr VA, Smith CA, Okumoto DS, Hanawalt PC. Cell 1985; 40:359–369.
80. Suter B, Wellinger RE, Thoma F. Nucleic Acids Res 2000; 28:2060–2068.
81. Aboussekhra A, Thoma F. EMBO J 1999; 18:433–443.
82. Fujita M, Hori Y, Shirahige K, Tsurimoto T, Yoshikawa H, Obuse C. Genes Cells 1998; 3:737–749.
83. Livingstone-Zatchej M, Marcionelli R, Möller K, de Pril R, Thoma F. J Biol Chem 2003; 278:37471–3779.
84. Gaillard H, Fitzgerald DJ, Smith CL, Peterson CL, Richmond TJ, Thoma F. J Biol Chem 2003; 278:17655–17663.

6

Damage Recognition by the Bacterial Nucleotide Excision Repair Machinery

Deborah L. Croteau and Matthew J. DellaVecchia
Laboratory of Molecular Genetics, National Institute of Environmental Health Sciences, National Institutes of Health, Research Triangle Park, North Carolina, U.S.A.

Milan Skorvaga
Department of Molecular Genetics, Cancer Research Institute, Slovak Academy of Sciences, Bratislava, Slovakia

Bennett Van Houten
Laboratory of Molecular Genetics, National Institute of Environmental Health Sciences, National Institutes of Health, Research Triangle Park, North Carolina, U.S.A.

1. INTRODUCTION

Nucleotide excision repair (NER) in prokaryotes was first discovered in 1964 (1,2) and is one of the most extensively studied DNA repair systems. One key question concerning NER, nicely outlined by Hanawalt and Haynes (3), is: How can one repair complex work on such a wide range of DNA lesions with different chemical properties and conformational properties? While the past 30 years has shed much light on this damage recognition problem, a clear picture of how this protein machine measures the DNA helix to find lesions has remained elusively in the shadows. Over the past 15 years several outstanding reviews have appeared and the reader is encouraged to revisit them (4–7). This chapter will primarily discuss the overall structures of the three principal bacterial proteins involved, UvrA, UvrB, and UvrC, which mediate damage recognition and processing during NER.

1.1. Overview of Nucleotide Excision Repair

The current overall mechanism of NER in prokaryotes is described in Figure 1. In solution, UvrA dimerizes in an ATP-dependent fashion. Although $UvrA_2$ can independently bind to and recognize altered nucleotides within the DNA helix under physiological conditions, the $UvrA_2B$ complex is probably the actual DNA damage recognition unit. We wish to acknowledge at the onset that, while we have chosen to use the $UvrA_2B$ complex nomenclature, we recognize that the stoichiometry of this complex is still controversial. The $UvrA_2B$ complex possesses ATPase activity and a

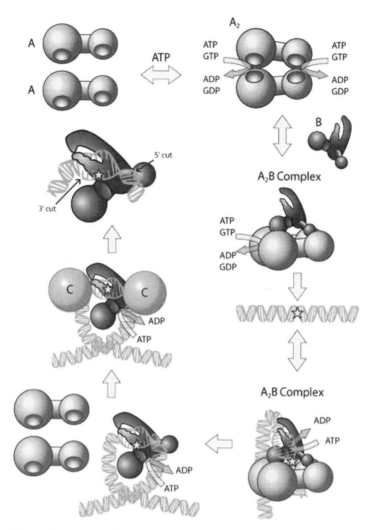

Figure 1 Damage detection, verification, and incision by the prokaryotic NER system. A hypothetical scheme for the key steps in the mechanism is shown; see the text for references and a more complete description. Each monomer of UvrA contains two ABC ATPase domains. In solution, two molecules of UvrA form a dimer presumably with the ABC ATPase modules and ATP binding drives dimer formation. As expected, the $UvrA_2$ complex possesses ATP/GTPase activity. UvrB interacts with this $UvrA_2$ dimer in solution, creating the $UvrA_2B$ complex. This complex has altered ATPase activity and directly recognizes damaged DNA. Upon binding to DNA, the $UvrA_2B$–DNA complex undergoes conformational changes. The DNA lesion remains in close contact with UvrA and then it is transferred to UvrB. UvrB is endowed with a cryptic ATPase activity (the red nodule on UvrB) that is activated in the context of the $UvrA_2B$–DNA. In this complex, the DNA is unwound around the site of the lesion because UvrB has inserted its β-hairpin structure between the two strands of the DNA to facilitate damage verification. The DNA is also wrapped around UvrB. The UvrA molecules hydrolyze ATP and dissociate from the complex. A stable UvrB–DNA complex is generated. Before UvrC can make the 3′ incision, UvrB must bind ATP, but not hydrolyze it. After the 3′ incision is generated a second incision event on the 5′ side of the DNA lesion is produced, thus UvrC forms a dual incision approximately 12 nucleotides apart. After the incision events, the DNA remains stably bound to UvrB until UvrD, DNA polI and ligase can come in and perform the repair synthesis reaction (not shown). (*See color insert.*)

limited strand-destabilizing activity, presumably to facilitate damage recognition. The UvrA$_2$ component of the UvrA$_2$B complex conducts the first level of damage recognition for DNA damage. If no lesion is encountered the UvrA$_2$B complex dissociates. If a DNA perturbation is discovered the UvrA$_2$B–DNA complex binds tightly to the site of the lesion. Upon forming this tight complex, there are conformational changes in both the DNA and proteins: UvrB verifies that the damage is present and binds tightly to the DNA; UvrA dissociates leaving a securely bound UvrB molecule loaded onto the site of the damage. UvrC binds to the UvrB–DNA complex and makes a dual incision, first on the 3′ side of the lesion then subsequently on the 5′ side of the DNA damage. The first incision event occurs 4–5 phosphodiester bonds 3′ to the lesion while the subsequent incision is 8 phosphodiester bonds 5′ to the lesion. In Figure 1, we have chosen to depict two independent molecules of UvrC to perform the individual incision reactions, but acknowledge that the actual stoichiometry is not definitively known.

Following the dual incisions, DNA helicase II (UvrD) is required for the release of UvrC and the incised oligonucleotide, while UvrB is removed from the nondamaged DNA strand during the repair synthesis reaction by DNA polymerase I. DNA ligase I ligates the newly synthesized DNA to the parent DNA.

In prokaryotes, UvrA, UvrB, and UvrC are the three key proteins that carry out the damage recognition and incision process. In higher eukaryotes, many more proteins are required for the same biological function. The prokaryotic system may seem simple by comparison to NER in higher eukaryotes, but the exact determinants as to how DNA damage is recognized awaits more structural information. No crystal structures are yet available for UvrA. However, the structures of several other proteins with similar folds have been structurally characterized. Therefore, the common features that UvrA possesses with reference to these other recently solved structures will be discussed. Of the UvrABC proteins, only UvrB's overall crystal structure has been solved (8–10). Analysis of these crystal structures has revealed many new and exciting features of UvrB and has provided clues as to how this protein might recognize DNA damage. Like UvrA, there are no crystal structures of full length UvrC. However, recently the N-terminal domain of UvrC that mediates the 3′ incision has been solved (10b). While UvrC does not recognize DNA damage per se, it does contribute to the overall spectrum of lesions that the UvrABC system can incise. The proteins' role in the overall NER reaction will be discussed.

2. DIVERSITY OF DNA LESIONS RECOGNIZED

The list of DNA lesions that can be recognized and repaired by the bacterial NER system is extensive (reviewed in Refs. 4, 7). Base damage, sugar damage, phosphate backbone modifications, chemicals which bind noncovalently to the DNA disrupting base-stacking interactions (reviewed in Ref. 4), protein–DNA cross-links (11), one base gaps, and even strand breaks (12) are just a few of the many categories of lesions recognized. During the 1990s, NMR spectroscopy led to the solution structure of a wide variety of DNA adducts including a series of benzo[a]pyrene diol epoxide N$_2$-guanine diasteromers (13), pyrmidine dimers (14,15), cisplatin-GG intrastrand adducts (16), and thymine glycol (17). While these structures have given us great insight into some of the structural and conformational determinants that provide efficient damage recognition and subsequent repair, currently, *there is no consensus regarding the physical characteristics of a particular DNA damage that*

can be used to predict how readily the UvrABC repair system will utilize a DNA
fragment containing a damaged nucleotide as a substrate.

When one looks at B-form DNA, one is struck by three important structural features that a protein machine could monitor to find DNA lesions: hydrogen bonding between complementary bases, sugar-phosphate backbone, and base stacking interactions. Since mismatched bases and extrahelical bases are very poorly incised by this system, hydrogen bonding, while it may contribute to recognition, is not a key structural feature. The sugar-phosphate backbone can be described by a number of key dihedral angles that can accommodate fairly large changes within B-DNA, and it has been tempting to suggest that the orientation of the negatively charged phosphates could provide a key recognition feature (18). However, energetic analysis of UvrA binding to benzo[a]pyrene diol epoxide adducts would suggest that relatively few phosphate interactions are important for binding. Additionally, a large driving force for UvrA recognition might be through a hydrophobic effect generated by a large van der Waal contact within the major or minor groove of the DNA and the surface of DNA (18). Finally, as we will discuss in more detail below, base stacking interactions between the Pi electron clouds of adjacent base pairs is one of the key driving forces that create the 36° offset between adjacent base pairs and thus causes the 10 bp rise per helical turn B-form DNA molecule (4).

To ensure that the nucleotide excision repair proteins maintain their high specificity, it has been proposed that damage recognition is achieved in at least two dynamic steps: initial damage recognition followed by closer inspection through damage verification. In this review, we will develop a model in which UvrA identifies the site of damage, and UvrB, through a series of dynamic interactions alters the structure of the DNA lesion and surrounding bases to inspect the damaged site possibly by probing the stacking interactions. Furthermore, it is known that bending and base pair opening are energetically linked: bending DNA promotes base pair opening and helical unwinding, and similarly, unwinding and opening the DNA bases leads to DNA bending. Thus, one key feature in damage recognition by the NER proteins is that they participate in a dynamic dance with DNA: as the recognition proteins embrace the DNA, the DNA helical structure is altered.

3. THE PROTEINS AND THEIR STRUCTURAL DOMAINS

In this review we will describe in detail the known functional domains within the UvrABC proteins. We will compare the functional motifs of these proteins to other proteins whose crystal structures have been solved. This will provide insight into how the UvrABC proteins may function in vivo to identify and incise DNA damage. After discussing some of the key structural features and motifs of the individual proteins, we will return to the question of how damage recognition is achieved by these three remarkable proteins.

3.1. UvrA Protein Motifs

Sequence analysis and mutagenesis have revealed conserved functional domains within each of the UvrABC proteins. Bacterial UvrA (Mw 102–110 kD) belongs to the ATP-Binding Cassette superfamily; it contains two independent ABC-type ATPase domains, each containing several conserved subdomains. In addition, the protein contains two zinc finger domains. Figure 2B shows a ClustalWProf alignment of UvrA's N-terminal and C-terminal ABC ATPase domains from

Bacillus caldotenax, *E. coli* and *Bacillis subtilis* along with several other known ABC ATPases whose crystal structures are known.

UvrA's two ABC domains are not contiguous. Instead, they are interrupted by the presence of intervening sequences. As depicted in Fig. 2, both of UvrA's ABC-type ATPases are broken up as follows: the Walker A sequence, Q-loop and the ENI-equivalent PRS motif which is followed by an insertion of variable length, containing a zinc finger motif. After the intervening sequence is the signature sequence, the LSGG, which is then immediately followed by the highly conserved Walker B sequence. The final subdomain of the ABC ATPases is the conserved Histidine-loop. UvrA's amino-terminal ABC domain is connected to the carboxy-terminal ABC domain through a flexible protease-sensitive hinge region. The carboxy-terminal ABC domain has the same overall schematics as the amino-terminal ABC domain.

The duplication of the ABC-type ATPase domain within one protein is not unique. It occurs in several other well-known disease causing proteins such as cystic fibrosis transmembrane conductance regulator (CFTR) and the multidrug resistance locus (MDR1). The ABC-type ATPase motif is also present in other DNA repair proteins such as the eukaryotic MutS protein, which is involved in the DNA mismatch repair pathway, and the double-strand break repair enzyme, Rad50. Both Rad50 and UvrA have discontinuous ABC-type ATPase subdomains and both are disrupted after the Q-loop and before the signature sequence.

3.1.1. The ABC ATPase Motif

Several X-ray crystal structures for ABC-type ATPases have been solved and an overall global conformation for this cassette has been established (19–22). The Walker A sequence (denoted A on UvrA schematic, Fig. 2) is thought to bind to the phosphate residues of a bound nucleotide during NER. The Q-loop (Q) and Histidine loop (H) make contacts with the gamma phosphate of the bound nucleotide and participate in the activation of the water molecule that will eventually cleave the β–γ phosphate bond. The signature sequence, LSGG, interacts with the gamma phosphate as well as the ribose ring of the nucleotide. The Walker B sequence (B) is the magnesium-binding site. The PRS motifs in UvrA are similar to the ENI motif, which was proposed to communicate the activity of the ATPase between the ABC motifs and the transmembrane spanning domains of ABC transporters (22). For more information on the ABC-type ATPases see Refs. 23, 24.

The amino acids adjacent to the Q-loop in many membrane-bound ABC ATPase transporters have the consensus sequence of ENI (23). This motif is generally written as follows: S/T φ X D/E N φ, where X is any residue and φ is a hydrophobic residue. UvrA matches the consensus in the carboxy-terminal ABC ATPase domain with the exception that the D/E residue is replaced with a conserved serine; the UvrA sequence is TPRSNP. Alternatively, the amino-terminal ABC ATPase fits the consensus motif less well with the sequence NPRSTV.

In ABC transporter proteins there is a conserved spatial relationship between the LSGG and the ENI motif (23). The importance of the amino acids adjacent to the Q-loop is illustrated by the fact that in ABC transporters this domain binds to the transmembrane domain. The importance of this domain is further exemplified by the fact that 70% of all cystic fibrosis cases results in mutations within this region (25). The amino acids adjacent to the Q-loop are not conserved between the different

(A)

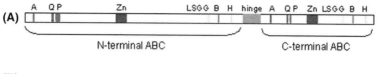

(B)

```
                   A
NtEcA   24 DKL IVVT GLSGSGKSSL AFDTL YAEGQRRYVESLSA----YARQFL SLMEKPDVDH IEGL
NtBcA   24 GKLVVLT GLSGSGKSSL AFDT IYAEGQRRYVESLSA----YARQFL GQMEKPDVDA IEGL
CtEcA  633 GLFTC IT GVSGSGKSTL INDTL FP IAQRQLNG---------AT IAEP APYRD IQGLEHF
CtBcA  630 GT FVAVT GVSGSGKSTLVNEVL YKAL AQKLH-----------RAKAKP GEHRD IRGLEHL
HisP    32 GDVIS I IGSSGSGKST FLRC INFLEKP SEGAI IVNGQN INLVRDKDGQLKVADKNQLRLL
MJ0796  31 GEFVS IHGP SGSGKSTMLN I IGCLDKPTEGEVYIDN------IKTN--DLDDDELTK IR

              Q      PRS/ENI
NtEcA   81 SP-AIS IEQKSTSHNPRSTVGT ITE IHDYLRLL FAR--Zn---RLKFLVNVGLNYLTLSR
NtBcA   81 SP-AIS IDQKTTSRNPRSTVGTVTE IYDYLRLL FAR--Zn---RLGFL QNVGLDYLTLSR
CtEcA  684 DK-VID IDQSP IGRTPRSNP ATYTGVFTPVREL FAG--Zn---KLQTLMDVGLTYIRLGQ
CtBcA  680 DK-VID IDQSP IGRTPRSNP ATYTGVFDD IRDVFAS--Zn---KLETLYDVGLGYMKLGQ
HisP    92 RTRLTMVFQHFNLWSHMTVLENVMEAP IQVLGLSKHDA--RERALKYLAKVG IDERAQGK
MJ0796  82 RDKIGFVFQQFNLIPLLT ALENVELPL IFKYRGAMSGEERRKRALECLKMAELEERFANH

             LSGG                 B
NtEcA  483 SAETLSGGEAQRIRLASQIGAGLVG-VMYVLDEPS IGLHQRDNERLLGTL IHLRDL-GNT
NtBcA  481 SAGTLSGGEAQRIRLATQ IGSRLTG-VLYVLDEPS IGLHQRDNDRL IATLKSMRDL-GNT
CtEcA  826 SATTLSGGEAQRVKL AREL SKRGTGQTLYILDEPTTGLHFAD IQQLLDVLHKLRDQ-GNT
CtBcA  822 PATTLSGGEAQRVKL AAELHRRSNGRTLY ILDEPTTGLHVDD IARLLDVLHRLVDN-GDT
HisP   150 YPVHLSGGQQQRVS IARAL AMEPDV---LF-DEPTS ALDPELVGEVLR IMQQLAEE-GKT
MJ0796 142 KPNQLSGGQQQRVAIARAL ANNPP I----ILADEPTGALDSKTGEK IMQLLKKLNEEDGKT

              H
NtEcA  167 VIVVEHDEDA IRAADHV ID IGP GAGVHGG------------------
NtBcA  167 L IVVEHDEDTML AADYL ID IGP GAG IHGG------------------
CtEcA  163 IVVIEHNLDVIKTADW IVDLGPEGGSGGE ILVSGTPETVAECEASHTA
CtBcA  161 VLVIEHNLDV IKTADYI IDLGPEGGDRGG------------------
HisP   206 MVVVTHEMGFARHVSSHV IFLHQGK IEEEG------------------
MJ0796 199 VVVVTHMD INVARFGER I IYLKDG-----------------------
```

(C)

```
Tth_N   82 AIS IDQKTTSHNPRSTVGTVTE IHDYLRLL FAR 114
Drd_N   82 AIS IDQKTTSHNPRSTVGTVTE IHDYLRLL YAR 114
Bsu_N   84 AIS IDQKTTSRNPRSTVGTVTE IYDYLRLL YAR 116
Eco_N   82 AIS IDQKTTSRNPRSTVGTVTE IYDYLRLL FAR 114
Bpe_N   82 AIS IEQKSAGHNPRSTVGT ITE IHDYLRLL YAR 114
Tth_C  687 VIE IDQSP IGRTPRSNP ATYTGVFDE IRDLFAK 719
Drd_C  723 VIE IDQSP IGRTPRSNP ATYTGVFTE IRDLFTR 755
Bsu_C  683 VID IDQAP IGRTPRSNP ATYTGVFDD IRDVFAQ 715
Eco_C  681 VID IDQSP IGRTPRSNP ATYTGVFDD IRDVFAS 713
Bpe_C  691 T ISVDQSP IGRTPRSNP ATYTGLFTP IRELFAG 723
```

Figure 2 UvrA. (A) Linear model of UvrA showing major motifs. The conserved motifs of the ABC ATPase are noted above the appropriate sequence: Walker A (A, red); Q-loop (Q, green); PRS/ENI motif (P, magenta); signature sequence (LSGG, yellow); Walker B (B, orange); Histidine-loop (H, tangerine). (B) Sequence alignment of known ABC ATPase proteins with UvrA's N- and C-terminal ABC ATPase domains. *S. typhymurium* HisP (20); *M. jannaschii* MJ0796 (21); NTC, N-terminal *Bacillus caldotenax* UvrA ATPase with the insertion domain deleted which includes amino acids 115–465 (Genbank AAK29748); NTE, N-terminal *E. coli* UvrA ATPase with the insertion domain deleted which includes amino acids 115–465 (Genbank P07671); CTC, C-terminal *Bacillus caldotenax* UvrA ATPase with the insertion domain deleted which includes amino acids 714–804 (Genbank AAK29748); CTE, C-terminal *E. coli* UvrA ATPase with the insertion domain deleted which includes amino acids 718–807 (Genbank P07671). (C) Sequence alignment of UvrA's Q-loop and adjacent sequences. Tht_N or _C, *Thermus thermophilus* UvrA N- or C-terminal Q-loop, respectively (GenBank Q56242); Drd_N or _C, *Deinococcus radiodurans* UvrA N- or C-terminal Q-loop, respectively (GenBank Q46577); Bsu_N or _C, *Bacillus subtilis* UvrA N- or C-terminal Q-loop, respectively (GenBank O34863); Eco_N or _C, *E. coli* UvrA N- or C-terminal Q-loop, respectively (GenBank P07671); Bpe_N or _C *Bordetella pertussis* UvrA N- or C-terminal Q-loop, respectively (GenBank NP88216). Coloring based on conservation (80%) and amino acid property. (*See color insert.*)

ABC ATPase family members but are highly divergent and consequently thought to be specific to the function of the particular protein.

The global alignment of UvrA's amino- and carboxy-terminal ABC ATPase domains reveals that the amino acids adjacent to the Q-loop, the PRS motif, show a high level of sequence conservation, see Fig. 2C. Thus, this domain of UvrA should be investigated for its potential role in transmitting ATPase activity either between the two ABC ATPase domains of UvrA or alternatively to UvrA's partner, UvrB.

3.1.2. Zinc Finger Motif

Proteins utilize zinc finger domains for several functions including DNA binding, RNA binding, and protein–protein interactions (26). UvrA has the consensus sequence of $CXXCX_{18-20}CXXC$ in which the four cysteine residues are used to coordinate one zinc molecule. Extended X-ray absorption studies were employed to confirm the existence of two zinc atoms within the protein, one in each zinc finger (27). The amino-terminal zinc finger is less well conserved than the C-terminal zinc finger motif and is not essential for NER *in vitro* and *in vivo* (28). Amino acid substitutions that disrupted the C-terminal zinc finger lead to insoluble proteins and in vivo made the bacterium profoundly sensitive to cell killing by UV (28). Visse et al. concluded that disruption of the C-terminal zinc finger is detrimental for the structural integrity of the protein.

In an independent study, Wang et al. (29) created random mutations in the first cysteine of the C-terminal zinc finger. Of the eight mutants generated, only one was purified and analyzed, the C763F UvrA. This mutant protein retained no in vivo repair activity, failed to bind to DNA, but retained vigorous ATPase activity. They concluded that C-terminal zinc finger is primarily responsible for UvrA's DNA binding capacity. However, the inability to purify the remaining seven out of eight mutants also suggests that the C-terminal zinc finger may be required to enforce the correct spatial orientation of the C-terminal ABC ATPase subdomains, which are important for dimerization.

3.2. UvrB Protein Structure

Analysis of the primary amino acid sequence of UvrB protein (75 kD) reveals that it is composed of six motifs as shown in Fig. 3A: 1a, 1b, 2, 3, β-hairpin and UVR. Three groups have independently solved four crystal structures of UvrB. Two structures were obtained from *Thermus thermophilus* UvrB (PDB codes 1C4O (8) and 1D2M (9)). Two structures solved with and without ATP were obtained from *Bacillus caldotenax* UvrB (PDB codes 1D9Z and 1D9X) (10). The structure of the *Bacillus caldotenax* UvrB is displayed in Fig. 3B with conserved color-coding between Figs. 3A and 3B.

3.2.1. Helicase Motifs

UvrB shares sequence homology to the type II superfamily of helicases. The six helicase motifs (I–VI) are annotated above the sequence in Fig. 3A. The ATP binding site is located at the interface between domains 1a and 3 as can be seen in Fig. 2B. A search of structurally similar proteins identified the helicases NS3 (Protein Data Bank code 1HEI) and PcrA (Protein Data Code 1PJR) as UvrB's nearest neighbors for domains 1 and 3 (10). Domain motions within helicases are coupled to the

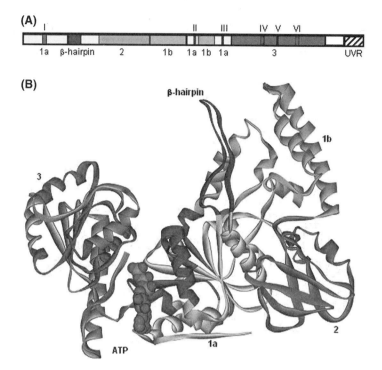

Figure 3 UvrB. (A) Linear model of UvrB showing major motifs. Helicase motifs are noted above the graphic (red). UvrB's structural features are noted below the graphic: domain 1a, yellow; β-hairpin, blue; domain 1b, green; domain 2, orange; domain 3, grey; UVR, white striped. (B) Crystal structure of *Bacillus caldotenax* UvrB (PDB 1D9Z, (10)). Coloration is conserved between panels A and B. The UVR domain of UvrB is not represented in the crystal structure. (*See color insert.*)

hydrolysis of ATP in order to drive DNA unwinding. The use of UvrB's helicase fold to recognize DNA damage will be discussed below.

3.2.2. The β-Hairpin Motif

The β-hairpin motif of UvrB plays a direct role in damage recognition by UvrB and in the handoff of DNA from UvrA to UvrB (12,30,31). It is a highly conserved element within the UvrB proteins and contains an unusually high degree of hydrophobic amino acid residues (12,30,31). The β-hairpin motif within other DNA helicases and RNA polymerase II is thought to be essential for separating the two DNA strands (31).

In an attempt to locate the DNA binding site of UvrB, Theis et al. (10) superimposed the structure of UvrB onto the solved structure of NS3 in complex with DNA [Protein Data Bank code 1A1V, (32)]. From this analysis, they suggested that the flexible β-hairpin might open and close around one strand of the DNA (10). Figure 4 depicts the putative trajectory of DNA through the gap created between the β-hairpin structure and domains 1a/1b of *Bacillus caldotenax* UvrB. A closer analysis of this domain's role in damage recognition will follow below.

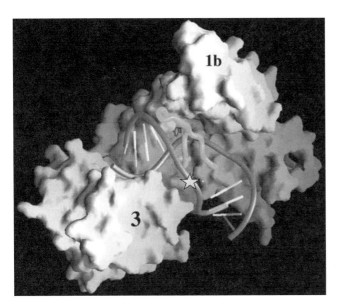

Figure 4 Hypothetical space filling model of UvrB bound to DNA. UvrB's β-hairpin element is depicted in blue and the yellow star represents the location of the DNA lesion. *Source*: Adapted from Ref. 10. (*See color insert.*)

3.2.3. Domain 2-UvrA Interaction Domain

Domain 2, within the crystal structures of UvrB, was poorly ordered and therefore segments were modeled as polyalanines (10). However, the function of this segment of the protein has been inferred by Selby and Sancar (33) who observed that this domain shares similarity to a fragment within the transcription repair-coupling factor, TRCF or Mfd. Therefore, this domain was proposed to be the domain of interaction with UvrA since both Mfd and UvrB interact with UvrA. Hsu et al. (34) provided experimental support for this hypothesis by fusing fragments of UvrB with the maltose binding protein. In their studies, they were able to identify two domains that interacted with UvrA, domain 2, as was proposed, but also the C-terminal domain of UvrB. The determination of the specific amino acids responsible for the interaction of UvrB with UvrA were recently defined using a new crystal structure of a *B. caldotenax* Y96A UvrB mutant that gave a high resolution structure of domain 2 (33b). This study showed that mutating highly conserved arginine residues (R183 or R194/R196 on the surface to glutamic acid (E) leads to decreased binding of UvrB to UvrA, a decrease in UvrB loading at the damaged site, and subsequent reduction in incision efficiency. Surprisingly, mutating the strictly conserved R215 to alanine had little effect.

3.2.4. C-terminal Coiled–Coiled (UVR) Domain

The UVR domain of UvrB was disordered in the X-ray structures of the proteins analyzed (8,9,35). Therefore, its position is not known with respect to the full-length protein and is omitted from Fig. 3B. Both UvrB and UvrC share this domain. Fortunately the C-terminal end of this domain of *E. coli* UvrB has been analyzed as an independently folded unit using NMR and X-ray crystallographic techniques (35,36). This domain adopted a coiled–coiled structure and formed a dimer when it was crystallized (Protein Data Bank 1QOJ) (35,36). The contacts between the two monomers were through hydrophobic interactions, as was shown by mutagenesis of residue F652L in UvrB and residue F223L in UvrC. Both mutations lead to a disrupted interaction between the two proteins (37). Based on the observation that UvrB's UVR element could dimerize, it was suggested that UvrB and UvrC would dimerize through this domain forming an UvrBC heterodimer (35,36). In support of this, hydrodynamic and crosslinking studies suggested that this domain was important for the multimerization of UvrB (38).

3.3. UvrC Protein Structure

Analysis of UvrC's primary amino acid sequence reveals that it contains four conserved motifs including two endonuclease elements, the UVR motif and a tandem helix hairpin helix motif, see Fig. 5. The N-terminal end of UvrC, which is responsible for the 3′ incision event (39), shares sequence homology to the Uri family of

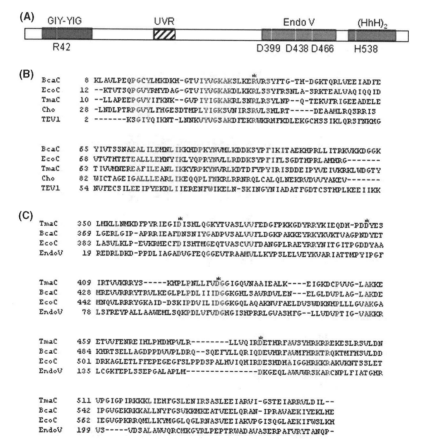

Figure 5 UvrC (A) Linear model of UvrC showing major motifs. The domains are as follows: GIY-YIG endonuclease motif, blue; UVR, white striped; Endo V endonuclease motif, orange; (HhH)$_2$, tandem helix–hairpin–helix motif, green. Catalytic amino acids within the endonuclease motifs are noted below the sequences. (B) Sequence alignment of UvrC's 3′ endonuclease, the GIY-YIG homologous region with the R42 amino acid denoted by the asterisk. (C) Sequence alignment of UvrC's 5′ endonuclease, the Endo V homologous region with the catalytic amino acids denoted by the asterisks. (*See color insert.*)

endonucleases (named after UvRC and Intron-encoded endonucleases) also known as GIY-YIG endonucleases (40). The C-terminal nuclease shares homology with DNA endonuclease V. As discussed above, the UVR motif is important in mediating the UvrB–UvrC interaction because, based on homology modeling it is proposed to be a domain of interaction between UvrB and UvrC (35). This domain was discussed above and will not be developed further here. The final element in UvrC that we will discuss is the helix–hairpin–helix motif that is a well-recognized DNA binding motif. Currently, how these elements come together in three-dimensional space to facilitate the dual DNA incision events is not known. However, a comparison of crystal structures of the UVR motif and other proteins with homologous domains can shed light on how UvrC probably utilizes these elements for its role in the NER pathway.

3.3.1. N-Terminal Nuclease Domain

The GIY-YIG family of endonucleases shares four invariant residues, Gly, Arg, Glu and Asn, and two tyrosines, $Y-X_{10}-Y$, (40,41). As for UvrA, no crystal structure has been solved for the UvrC protein. However, sequence alignments have shown various similarities between UvrC and the protein Tev1, an intron-encoded endonuclease, whose crystal structure has been solved (41). The conserved amino acids are shown in red in the alignment of UvrC with Tev1 in Fig. 5B. The GIY-YIG family of endonucleases uses the conserved polar residue to affect incision (42).

Tev1 consists of two functional domains: an N-terminal catalytic and a C-terminal DNA binding domain, separated by a flexible linker. The DNA binding domain of Tev1 is a sequence-specific DNA binding motif and therefore shares no homology to UvrC. Although Tev1-mediated cleavage occurs in a sequence-specific manner, the N-terminal catalytic domain contributes little to the DNA-binding affinity of Tev1 (42). This raises several questions: what domain(s) of UvrC confer DNA binding ability to UvrC; will UvrC be like Tev1; will the stronger DNA binding motifs be localized to the C-terminal nuclease and helix–hairpin–helix domain of UvrC? Alternatively, does the protein–protein interaction between UvrC and UvrB within the UvrBC–DNA complex provide sufficient stability that neither nuclease or helix–hairpin–helix motif need to have a high DNA binding affinity? A systematic analysis of the individual nuclease motifs should be adequate to resolve the issue. Recently crystal structures of the N-terminal catalytic domain of *B. caldotenax* and *T.maritima* have been solved to 2.0 and 1.5 angstroms, respectively (42b).

3.3.2. C-terminal Nuclease Domain

The C-terminal half of UvrC contains an endonuclease domain plus a tandem helix–hairpin–helix (HhH_2) motif and is responsible for the 5′ incision (39). UvrC's C-terminal nuclease sequence shares homology to Endonuclease V (nfi gene in *E. coli*) and ERCC1 (43), which in complex with XPF is responsible for the 5′ incision event in the human NER pathway (44,45). An alignment of the C-terminal catalytic domain of UvrC with Endonuclease V and ERCC1 is shown in Fig. 5C. UvrC and ERCC1 share homology throughout the nuclease and HhH motif while Endonuclease V does not share the HhH element. However, the HhH motif of UvrC shares homology with the HhH of Mus81, a mammalian structure-specific endonuclease.

3.3.3. The Helix–Hairpin–Helix Motif

The helix–hairpin–helix motif is the most evolutionarily conserved domain that DNA repair systems rely upon for interacting with DNA (40). Fourteen homologous families of helix–hairpin–helix proteins have been identified and each contain the α-helix–loop–α-helix signature structure (46). Numerous other DNA repair proteins share this motif, including glycosylases, ligases, and endo- and exonucleases (40,46).

The tandem HhH domain of UvrC has been subjected to biochemical and NMR analysis. The structure was shown to be similar to HhH motifs found in RuvA and DNA ligase (47). NMR chemical shift mapping after DNA binding revealed that hydrogen bonds between the backbone of the protein, amide protons, and the phosphate oxygens of DNA were created. This is similar to what has been found for DNA pol β. The helix–hairpin–helix motif in the crystal structure of pol β was shown to interact with DNA via the formation of hydrogen bonds between the protein backbone nitrogens and the DNA phosphate groups (48). This type of inter-action with DNA obviates the need to interact with the bases and is probably the structural basis for the sequence-independent DNA binding mode of action.

UvrC's HhH motifs can bind to a bubble DNA structure containing at least six unpaired nucleotides but not to ssDNA or dsDNA (47). DNA binding was coopera-tive suggesting that at least two UvrC molecules were involved, although the real number of molecules participating could not be calculated due to the low affinity. In an independent study, the role of the tandem HhH domain was probed by dele-tion analysis (49,50). These studies showed that, depending on the type of DNA damage and sequence context surrounding the lesion, deletion of the tandem HhH motif could result in defective 3′ and/or 5′ incisions.

3.3.4. UvrC Stoichiometry

The stoichiometry of the UvrC protein within the nucleotide excision reaction is under debate. Based on the self-association of the homologous UVR domain of UvrB in the NMR and crystallographic studies, it was proposed that UvrC might dimerize through interactions within this domain (35,51). The protein has been purified as a monomer, UvrCI, or tetramer, UvrCII (52). From the same study, it was demonstrated that the two species of UvrC possessed distinct DNA binding properties; UvrCII bound dsDNA while UvrCI did not. While both species of UvrC were competent to participate in the NER reaction (52), the tetramer species suppressed damage recognition by UvrA and UvrB by binding to the double-stranded substrate.

Of the proteins whose crystal structures have been solved, and those that share some similarity to UvrC, the Tev1 and the Hef proteins may provide insight to the multimerization of UvrC. Tev1 was found to be a monomer in the crystal structure. However, Tev1 induces a double-stranded break in the DNA. Therefore, a proposal arose stating that a transient dimer was formed creating two active sites (41). A clear dimer interface within Tev1 has not been identified; likewise no dimer interface for the corresponding homologous region in UvrC has been identified either.

The C-terminal nuclease-tandem HhH may play a role in the multimerization of UvrC. In the crystal structure of another nuclease containing the nuclease-tandem HhH domain organization, *Pyrococcus furiosus*, Hef, the protein was a dimer (53). This domain organization is shared by other structure-specific eukaryotic nucleases including XPF, ERCC1 (an inactive nuclease), Rad1 and Mus81. The crystal struc-ture and mutagenesis of *Pfu* Hef demonstrated that there are two domains within the protein that contributed to dimerization. This was revealed when individual mutations within the HhH or the nuclease domain, which only disrupted one of the dimerization interfaces but not the nuclease activity, still generated proteins that form dimers and thus function in the nuclease assay. However, when a protein was generated that disrupted both dimerization interfaces the protein was no longer functional. This study clearly demonstrated that the nuclease and tandem HhH domains each contributed, independently of the other element, to the dimerization of the proteins (53). It remains to be determined whether the C-terminal nuclease domain of UvrC participates in the multimerization of the protein.

Historically, UvrC was difficult to work with due to it's instability (54). Recently, analysis of *Bacillus caldotenax* UvrC revealed that extended storage of the protein caused inactivation due to oxidation of sulfhydryls and the activity of the inactivated protein could be restored by pretreatment with a reducing agent (55). UvrC contains four closely positioned cysteine residues. From these observations, the conclusion that inappropriate disulfide bond formation leads to defective UvrB–DNA and UvrC interactions was made. In addition, it should be noted that all of the nuclease-HhH-type nucleases mentioned above function as dimers and if their binding partners are not present, then the proteins tend to fall apart and thus aggregate. Therefore, UvrC may also share this property lending support to the notion that UvrC is probably a stable dimer or tetramer in vivo.

4. REACTION PATHWAY FOR DAMAGE DETECTION AND PROCESSING

Using Fig. 1 as an outline, the next section explores the steps leading to efficient DNA damage detection and processing, culminating in the dual incision pathway.

4.1. Dimerization of UvrA

UvrA can exist as a monomer or dimer. Conditions that favor dimer formation are high protein concentration and ATP binding, but not hydrolysis (56,57). Incubation of UvrA with a nonhydrolyzable ATP analog, ATPγS, also favors dimer formation (58). Furthermore, mutations within the protein that promote binding of ATP, but not hydrolysis, also shift the monomer–dimer equilibrium towards the dimer. The association constant for *E. coli* UvrA has been reported to be $K_A \sim 10^8 \, M^{-1}$(59) and the stoichiometry of ATP binding to UvrA is one molecule of ATP per UvrA dimer (60).

ABC-type ATPase proteins always function as either homodimers or heterodimers (61). The crystal structures of other ABC ATPases, such as Rad50, MutS, MJ0796, and BtuD have been solved and all share a common dimer architecture in that the ABC ATPase module is the dimer interface (19,22,24,62,63,21). The ABC ATPase subdomains from each monomer are oriented such that two nucleotides are bound across the dimer interface. We know UvrA functions as a homodimer and self-associates in a head-to-head fashion. This was based on the ability of the N-terminal fragment of UvrA to dimerize independently of the C-terminus, but not vice versa (64).

Mutagenesis of the Walker A sequences of UvrA's two ABC ATPase domains revealed that the two ATPase modules are not identical, but rather showed cooperativity (64). The K_m values for the individual sites have been reported to be $60 \, \mu M$ for the N-terminal site and 312 µM for the C-terminal site (65). The suggestion was put forth that ATP binding by the N-terminal ABC domain promotes dimerization while the second ABC ATPase domain in the C-terminal end of the protein has another functional role.

All of the data above agree with the new information gleaned from the crystal structures of other ABC ATPases in that the ABC ATPase domains create the dimer interface for UvrA. Furthermore, because UvrA functions as a dimer, those factors that promote dimer formation favor DNA binding (57).

4.2. DNA Damage Detection by UvrA

Several models have been put forth to suggest how UvrA detects DNA damage. We do not know what structural elements of the DNA the protein is monitoring at the molecular level, and thus in lieu of a crystal structure of the UvrA$_2$–DNA complex, we can only speculate on how the protein interacts with DNA.

UvrA binds to single- and double-stranded DNA. UvrA has a higher affinity for damaged DNA than nondamaged DNA; the binding constant for damaged DNA is approximately 10^3-fold above undamaged DNA (57). UvrA does not require ATP to bind to DNA but ATP binding and hydrolysis modulates its DNA binding activity (56,57). The initial DNA binding event by UvrA is most likely a nonspecific interaction. Hydrolysis of ATP via the C-terminal ABC domain leads to dissociation of the protein enabling it to sample other DNA sites until it finds DNA damage (66). Consequently, DNA binding to nonspecific sites is enhanced by either the use of UvrA mutants that cannot hydrolyze ATP (65) or by the addition of nonhydrolyz-able ATP analogs to the reaction mixture, such as ATPγS (56,66). It is believed that ATP hydrolysis is required for UvrA dissociation and is thus an important part of the mechanism that UvrA employs for damage recognition.

In an attempt to map the DNA binding domain to one of UvrA's known motifs, UvrA was physically separated into two fragments, the N-terminal ABC domain and the C-terminal ABC domain (64). From these studies it was discovered that both domains possessed nonspecific DNA binding properties.

In another study, site-specific mutagenesis of *E. coli* UvrA was conducted to examine a putative helix–turn–helix motif (67), a well-recognized structural element used by many enzymes to bind to DNA (68). This motif was identified in UvrA; the amino acids of the putative helix–turn–helix are underlined in Fig. 2B. Several mutants within this motif were identified by a screening approach and two were selected for further analysis, G502D and V508D. Upon purification, these mutant proteins were able to bind to DNA as strongly as wild type UvrA and made UvrA$_2$B complexes; however, both mutant proteins failed to show enhanced binding to UV-irradiated DNA and failed to load UvrB onto damaged DNA (67). When analyzed for ATPase activity, the mutants displayed an elevated level of ATPase activity rela-tive to wild type. At the time of this study, it was not yet recognized that the helix–turn–helix motif was actually an integral part of UvrA's N-terminal ABC ATPase subdomain, see (Fig. 2B.)

Mutageneses of similar residues within the ABC ATPase of CFTR are disease-causing mutations suggesting that these amino acids play an important role in the ABC ATPase architecture. In addition, from the crystal structures of the solved ATPases, these amino acids are not on the surface; rather they are buried within the structure. For example, in the dimeric crystal structure of MJ0796 (69) depicted in Fig. 6 the amino acids in question are highlighted in yellow. The catalytically

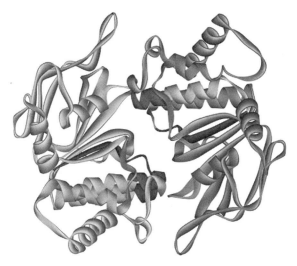

Figure 6 Helix–turn–helix region of the ABC ATPase MJ0796 dimer (PDB 1L2T). One monomer is depicted in pale green while the other monomer is shown in grey. The position of the signature sequences, red, and the Walker B sequences, blue, are shown. The amino acids homologous to those of UvrA's putative helix–turn–helix are depicted in yellow. *Source*: From Ref. 69. (*See color insert.*)

important ABC subdomains, the signature sequence and Walker B motifs, are shown in red and blue, respectively. Due to the buried nature of this region, it is unlikely that these amino acids of UvrA would participate in the DNA binding reaction. Instead, the experimental evidence supports the notion that both the N-terminal ABC ATPase as well as the C-terminal ABC ATPase is indispensable for damage recognition.

DNA binding by wild type UvrA is stabilized by the presence of UV DNA lesions. The glycine-rich residues located at the very C-terminus of the UvrA protein have been proposed to be involved in the damage-specific recognition by UvrA (70). When the C-terminal 40 amino acids of UvrA were deleted, the protein was no longer able to stably bind to UV irradiated DNA. Thus, the investigators suggested that this domain represents a damage-stabilizing functional domain. In the same study, the authors noted that deletions of UvrA beyond the C-terminal 40 amino acids rendered the mutant proteins unable to bind DNA. This observation can be reconciled by noting that a conserved ABC ATPase subdomain, the Histidine-loop, would be deleted if larger C-terminal deletions were constructed, as can be seen in Fig. 2B.

Other groups have speculated that the C-terminal zinc finger of UvrA facilitates DNA binding. However, direct crosslinking or selected mutagenesis, which do not disrupt the overall fold, have not been conducted to confirm this hypothesis. Whether or not this motif is engaged in nonspecific or damage-specific DNA binding remains to be determined.

The analysis of UvrA was historically complicated by the fact that the *E. coli* UvrA protein is rapidly inactivated (71) and that many of the mutant proteins created were insoluble. This was probably due to the distribution of the ABC ATPase subdomains throughout UvrA's sequence, and when folding of these elements was impaired, insoluble proteins were obtained. Now that the crystal structures of several ABC ATPase proteins have been solved, it is clear that UvrA possesses all of the conserved motifs and these domains are essential in the proper functioning of the UvrA$_2$ dimer.

It has been well documented that the UvrA protein is a functional dimer and that ATP binding is not required for DNA binding. However, ATP binding and hydrolysis play a role in the damage recognition mechanism employed by the UvrA$_2$ dimer. It is also clear that the two ATPase domains must be intact for UvrA$_2$ to bind to damaged DNA specifically. However, at this time, the structural elements that the UvrA$_2$ dimer exploits for damage recognition are not thoroughly understood. A re-examination of UvrA in the context of what is known about the ABC ATPase motifs now implies that these domains create the UvrA$_2$ dimer interface and that conserved amino acids outside of this region will likely be required for DNA binding. An obvious candidate region includes the C-terminal zinc finger domain.

4.3. Damage Verification by the UvrA$_2$B Complex

Under physiological conditions, it is believed that UvrA would entirely be present in the UvrA$_2$B complex (72). The dimerization of UvrA is required for UvrB binding to the UvrA$_2$ complex; therefore the binding site for UvrB is probably created as a consequence of dimerization. The domain of UvrA responsible for this interaction was mapped by deletion analysis to the first 230 amino acids (70). The domain of UvrB responsible for interaction with UvrA is proposed to be domain 2, based on the finding that Mfd and UvrB share this domain and both interact with UvrA (33). In addition, the C-terminal UvrB fragment containing amino acids 547–673 has been shown to interact with UvrA (34).

While studied extensively, the stoichiometry of proteins within the UvrA–UvrB complex remains controversial. From gel filtration and sedimentation studies, it was concluded that UvrB was a monomer in the UvrA$_2$B complex (73). In crystal structures, UvrB was also visualized as a monomer (8–10). However, by using a crosslinking strategy, it was shown that UvrB could dimerize in an ATP-independent manner (38). Further support for an UvrB$_2$ dimer comes from the NMR and X-ray crystallography analysis of the C-terminal domain of UvrB (35,36). It was shown that this domain of UvrB adopts an unusual helix–loop–helix conformation and dimerizes in a head-to-head fashion. Yet another study investigated the complex by atomic force microscopy (74). Volume analysis of the visualized complexes suggested that UvrB existed as both a monomer and a dimer bound to DNA. Definitive proof of the stoichiometry of the UvrA–UvrB complex awaits further analysis.

Functional ATPase domains in UvrA are required for the association of UvrB with UvrA. The nucleotide co-factor requirements for the UvrA$_2$B complex formation in solution were analyzed by gel filtration and velocity sedimentation (73). ATP binding and hydrolysis was shown to be necessary because neither ADP nor ATPγS could promote this interaction. In support of the necessity for ATP hydrolysis, size exclusion chromatography showed that amino acid substitutions in either Walker A motif rendered the ABC ATPases unable to hydrolyze ATP and failed to load UvrB onto DNA (60). It remains to be determined why UvrA must hydrolyze ATP in order to bind to UvrB productively. Alternatively a functional ATPase in UvrB is not required to generate the UvrA$_2$B complex in solution, however ATP hydrolysis by UvrB is essential for damage recognition by the UvrA$_2$B complex on DNA (75).

4.3.1. Role of UvrB in Damage Detection

As discussed above, UvrA is a high affinity DNA binding protein with the ability to bind to both nonspecific and damaged DNA sites. Independently, UvrB is a low affinity damage specific DNA binding protein on ss DNA (34). UvrB's K_d is $\sim 5 \times 10^{-6}$ M when binding to a psoralen or cisplatin-modified oligonucleotide verses $K_d > 10^{-5}$ M when binding to undamaged or an AP site (34). The association constant for UvrA$_2$B binding to a psoralen adduct in duplexed DNA is 1×10^9 M^{-1} while that of UvrA$_2$ alone is $\sim 5 \times 10^7$ M^{-1} (72). Thus, neither protein alone is sufficient for damage recognition, but rather it is the combination of the two that confers upon the system the remarkable spectrum of lesions that can be repaired.

One hypothesis for the role of UvrB is to modulate the stability of the UvrA$_2$ dimer. Footprinting experiments with UvrA$_2$ and UvrA$_2$B indicated that the rate of appearance of the specific UvrA$_2$ footprint increased when UvrB was present, suggesting that UvrB promotes the association of UvrA and consequently DNA binding (reviewed in Ref. 5).

UvrB is proposed to increase the stability and specificity of the UvrA$_2$B–DNA complex relative to that of the UvrA$_2$–DNA complex (5). A benzoxyamine-modified AP site containing oligonucleotide was used in combination with gel shift analysis to investigate the stability of the various protein–DNA reaction intermediates (5). When UvrB is present, the stability of the UvrA$_2$–DNA complex is very short ($t_{1/2} \sim 15$ sec) whereas the UvrA$_2$B–DNA and UvrB–DNA complexes are long lived ($t_{1/2} \sim 2$ hr) and appear to be in equilibrium (5). Thus, the presence of UvrB decreases binding to nondamaged DNA sites, thereby increasing the specificity for lesion-containing sites (76).

Upon UvrA$_2$B binding to a DNA lesion, the DNA undergoes conformational changes. In the complex, the DNA is bent by approximately 130° (77) and the area around the lesion is unwound by about five base pairs (75,78). Analysis of the DNA in the UvrA$_2$B–DNA and UvrB–DNA complexes by atomic force microscopy revealed that the DNA is wrapped around one UvrB molecule (79). The consequences of wrapping may be important for damage recognition and may facilitate damage verification by UvrB.

4.3.2. Conformational Changes in the UvrA$_2$B–DNA Complex

There are conformational changes within the DNA and proteins in the UvrA$_2$B–DNA complex. Upon UvrB binding to the UvrA$_2$–DNA complex, there is a DNase I-hypersensitive site at the 11th phosphodiester bond 5′ to the lesion and the DNA footprint is reduced from 33 to a 19 bp footprint pattern (80). Obviously, UvrA undergoes considerable conformational changes because it becomes destabilized and dissociates from the UvrB–DNA complex. As mentioned above, UvrB possesses a helicase fold and it is endowed with a limited amount of strand-destabilizing activity. This activity, when stimulated in the UvrA$_2$B–DNA complex, may facilitate damage recognition by UvrB (58,81,82). After UvrB's cryptic ATPase activity is turned on, the UvrA$_2$B–DNA interaction is distinctly different (83). The nature of this activity is now proposed to be the result of UvrB inserting its β-hairpin structure between the two strands of the DNA in order to specifically bind to the DNA (7,12,31).

4.3.3. Beta-Hairpin Acts as a Padlock for Binding to DNA

Theis et al. (7) suggested that the stable UvrB–DNA complex is formed when one DNA strand is locked between the β-hairpin and domain 1b of UvrB. A more detailed discussion of how the β-hairpin structure may participate in damage recognition is as follows. Allosteric interactions between $UvrA_2$ and ATP binding by UvrB induce a conformational change in UvrB such that an "open" channel between the flexible β-hairpin structure and domain 1b is created for DNA to pass through. Upon UvrA departure and ATP hydrolysis by UvrB, the flexible β-hairpin would return to its closed conformation. In this scenario, the energy harvested from ATP binding by UvrB is used to drive DNA into an unfavorable conformation, i.e., the generation of ssDNA. Thus, the stability of the UvrB–DNA complex is explained in part because the structure is physically constrained by the double-stranded DNA outside of the limited single-stranded region surrounding the damaged site and the β-hairpin structure clamped around one DNA strand (7,12).

In support of the model described above, gel shift assays reveal that a mutant *Bacillus caldotenax* UvrB protein, which lacks the tip of the β-hairpin forms the $UvrA_2B$-DNA complex but fails to generate a stable UvrB–DNA complex, and thus is inactive in the overall reaction scheme (31). In addition, site-directed mutagenesis of the aromatic residues at the tip of the β-hairpin structure, Y101 and Y108, render the mutant UvrB protein able to bind the DNA, albeit with low affinity, but dysfunctional when it comes to the UvrA-mediated loading of UvrB. This suggests that these amino acids are important for the strand-separating phase of the reaction (12). Ultimately a co-crystal of UvrB with DNA will resolve where the DNA interacts with the protein.

The direct role of several other hydrophobic amino acids at the base of the β-hairpin has also been investigated. Mutagenesis of the highly conserved tyrosine residues, Y92, Y93, Y95, and Y96, are important for the discrimination of damaged vs. undamaged DNA (12). When the double mutant Y92A and Y93A was made, the mutant UvrB protein bound to and incised nondamaged DNA. The phenotype of the double mutant Y95A and Y96A UvrB protein was somewhat different. This mutant lost damage-specific binding but was able to promote incision if a nicked or gapped DNA substrate was used. Therefore, these amino acids participate in damage recognition mediated by UvrB. From these observations, Moolenaar et al. suggested that within the UvrB–DNA complex, the DNA lesion is flipped out of the helix and that the hydrophobic residues are inserted to maintain the base stacking interactions at the site of the lesion. They further speculate that these residues inhibit binding to nondamaged DNA because of steric interference that would be encountered upon trying to disrupt a normal base's stacking interactions, while any real lesion might have disrupted stacking interactions already and be more susceptible to flipping." More detailed analysis of several mutations in or around the beta-hairpin domain of UvrB have revealed that Y96, E99 and R123 are essential for DNA damage processing by UvrB, where as Y93 and Y92 play a less important role (83b). Further analysis using a unique photoaffinity cross-linking reagent that mimics a DNA lesion has revealed that the beta-hairpin deletion, Y96A, E99A, and R123A mutants of UvrB are completely defective in the DNA hand-off from UvrA to UvrB. (83c).

There is other support for this "flipped out base" hypothesis. Previously, it was suggested that UvrB monitors the base-stacking interactions based on the differential incision efficiencies of certain DNA adducts (4,5). In addition, it has been demonstrated that protein–DNA complexes and crosslinks are incised by the

UvrABC system (84,85), and in some cases actually stimulate incision rates (84). In such substrates the presence of a protein–DNA crosslink would, more than likely, cause the base to be flipped out.

The fact that protein–DNA crosslinks are repaired also has other implications for the model. It is highly unlikely that the protein-crosslinked damage-containing strand could be accommodated behind the β-hairpin. Therefore, the nondamaged strand probably is the DNA strand sandwiched between the β-hairpin and domain 1b of UvrB. This observation is consistent with the observed DNA footprint on the nondamaged strand (58,80). An analysis of DNA substrates, which are incised by the UvrABC system, demonstrated that sequences on the 5′ side of the lesion are more important than those on the 3′ side, thus the notion that UvrB approaches the lesion from the 5′ side (86). If the model is correct, UvrB has to be defined as a 3′ to 5′ helicase because UvrB binds the nondamaged strand (7) and not a 5′ to 3′ helicase that has been reported previously (81).

4.3.4. Role of ATP in the Reaction Mechanism of the UvrABC System

In the reaction scheme shown in Fig. 7, the binding and hydrolysis of ATP by UvrB leads to a propreincision complex. This step in the reaction was revealed when streptavidin-coated magnetic beads and biotinylated DNA substrates were exploited to pull-down stable protein–DNA reaction intermediates (6). After incubation with UvrC, it was clearly shown that the UvrB–ADP–DNA ternary complexes did not support incision while the UvrB–ATP or ATPγS–DNA complexes did. While binding of UvrC to the UvrB–DNA complex is co-factor independent, generation of a productive UvrBC–DNA complex and progression of the reaction cycle requires that UvrB is in the ATP-bound configuration (6).

4.3.5. UvrC Mediated Incisions

The stoichiometry of UvrC in the UvrBC–DNA complex is yet another point of controversy in the NER pathway. It is generally recognized that UvrC possesses

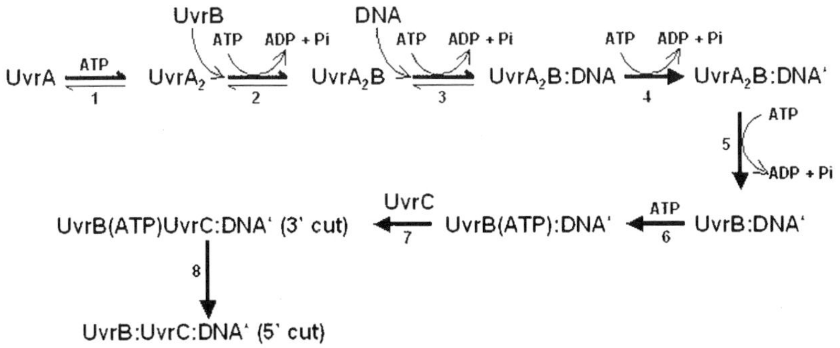

Figure 7 Role of ATP in the mechanism of the UvrABC nuclease system. (1) ATP binding by UvrA promotes UvrA$_2$ dimer formation; (2,3) ATP binding and hydrolysis is required in order to generate the UvrA$_2$B and UvrA$_2$B–DNA complexes; (4) ATP hydrolysis is required to minimize binding to nondamaged DNA sites and promote complex formation at sites containing a lesion. Asterisk denotes specific DNA binding events at sites of damage; (5) ATP hydrolysis is required for UvrB to form a stable UvrB–DNA∗ complex; (6) UvrB must bind ATP prior to UvrC making the 3′ incision and (7) UvrC makes the 3′ incision on damaged DNA when ATP is bound to UvrB; (8) ATP is not necessary for the 5′ incision.

the nuclease centers for both the 3′ and 5′ incision events (39,87); however, it is less clear how many molecules of UvrC are needed to accomplish the incision events. UvrC's nuclease centers were identified by site-directed mutagenesis of the respective catalytic amino acids (39,87). Incision is bimodal and occurs on either side of the lesion (88–90). The two nucleases are separable but allosterically connected because the 3′ incision event must precede the 5′ cut (91).

The precise role of UvrC in the damage recognition phase of the reaction is not fully explored and merits further attention. However, the notion that UvrC detects the DNA lesion can perhaps be best appreciated from studies using a truncated version of UvrC that lacks the tandem HhH motif (50). These studies indicated that different DNA lesions in the same DNA sequence or the same lesion in slightly altered DNA sequence context produced dramatically different incision results. In some cases, both the 3′ and 5′ incisions events were inhibited while in other examples only one incision was defective. Clearly, UvrC contributes to the spectrum of lesions incised, and thus to the overall recognition of the system, but exactly how UvrC manages this is unknown. UvrC participates in lesion recognition and its mechanism of recognition involves the tandem HhH motif (50).

4.3.6. The UvrBC Nuclease

The *E. coli* UvrBC complex alone is sufficient for incision and this activity requires both UVR motifs. When a DNA lesion is placed close to the end of a substrate, the UvrBC complex can incise this substrate at the same position where the 3′ incision would normally occur (86). Additionally, if a DNA lesion is placed in a bubble or Y structure, the UvrBC complex can cut this substrate, suggesting that the UvrBC complex is a damage-dependent structure-specific endonuclease (91).

4.4. Local Conformation of DNA Influences Efficiency of Incision

An analysis of several DNA lesions within three different sequence contexts demonstrated that the DNA sequence context influences the ability of the UvrABC system to recognize and incise the DNA lesions (49,92) At first it was thought that the stability of the UvrB–DNA complex was inversely proportional to the overall incision efficiency (92), however, now it appears to be more complicated than that. Therefore, the reader is cautioned on the interpretation of data that were obtained by the use of one DNA lesion in a single DNA sequence context.

5. DNA DAMAGE RECOGNITION WITHIN THE BIOLOGICAL CONTEXT OF THE CELL

While much has been learned about the reaction mechanism of UvrABC system using purified proteins in biochemical assays, it is essential that nucleotide excision repair be placed in the biological context of the entire cell. Two complementary approaches suggest that the UvrABC system does not work in isolation but is part of a complex dynamic network that responds to an onslaught of DNA damage. Using UvrA fused to the green-fluorescence protein, Walker and coworkers were able to visualize the location of UvrA in living cells of *Bacillus subtilis* (93). They found that UvrA was uniformly localized to chromatin during normal growth, but underwent a dramatic redistribution to specific sites during DNA damage. This redistribution was reversible. Using polyclonal antibodies to UvrA, UvrB, and

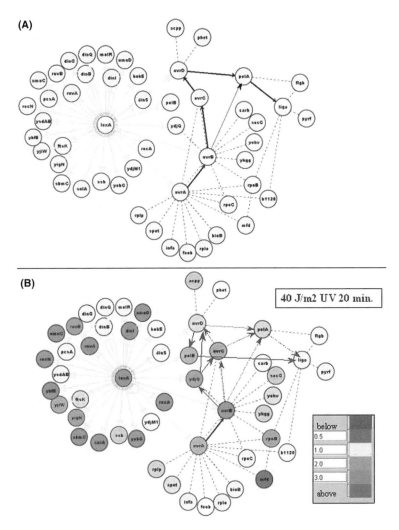

Figure 8 Dynamic network of bacterial NER. (A) Protein–protein and protein–DNA interactions of the UvrABC system using Cytoscape (96b). Protein–protein interactions are shown in dashed blue lines; protein–DNA interactions are shown as orange arrows. Nodes are proteins or in the case of LexA interactions promoter sequences in genes. Reaction pathway is shown as black arrow. Nucletide excision repair interacting proteins: *Acpp* = Acyl carrier protein (ACP); B1120 = hypothetical protein; *BioB* = Biotin synthase (EC 2.8.1.6) *Carb* = Carbamoyl-phosphate synthase large chain; *YdjQ* Cho = UvrC homolog, b1741; *Feob* = IRON(II) transport protein; *Flgb* = Flagellar basal-body ROD protein (FLGB) (Proximal ROD protein); *Infa* = Translation initiation factor IF-1; *Phet* = Phenylalanyl-tRNA synthetase beta chain; *Rplo* = 50S ribosomal protein L15; *Rplp* = 50 S ribosomal protein L16; *RpoB* = DNA-directed RNA polymerase, beta subunit (RPOB); *RpoC* = DNA-directed RNA polymerase, beta subunit (RPOB'); *SecG* = Protein-export membrane protein; *Spot* = Penta-phosphate guanosine-3'-pyrophosphohydrolase (Spot); *Yehv* = HspR; *Ykgg* = Hypothetical protein HP0137. See text for references. (B) Alterations in the NER network under UV-stress. Layered onto the network in panel A are gene expression changes that occur in *E. coli* 20 min. after 40 J/m^2 of UV light. Red, genes that are repressed; Green, genes that are induced; Yellow, genes that showed no change; White indicate no data. Green lines indicate possible remodeling of the NER system in response to UV damage (solid lines indicate new interactions predicted after UV light, dashed, preexisting interactions). Note the induction of *ydjQ* CHO, the UvrC homolog, *polB* and the repression of polA, UvrC, and Mfd. (*See color insert.*)

UvrC, and subcellular fractionation methods, Grossman and coworkers found that following UV irradiation, UvrA and UvrC joined an ensemble of 15 other proteins, including three subunits of RNA polymerase, Topo I, and DNA gyrase, to relocate near the inner membrane of *E. coli*, and located at DNA-membrane junctions (94). Using immuno-gold labeling they also showed that antibodies to 6–4 photoproducts co-localized with UvrA and UvrC to the inner membrane fraction. Strangely, UvrB could not be identified in this fraction in their study.

Using high density DNA microarrays, Hanawalt and coworkers performed a global genome analysis of genes induced by UV light in *E. coli* (95). They found a number of new genes which were induced with putative LexA binding sites, and many more, which did not apparently have LexA SOS boxes. They also observed a number of repressed genes. Using these data, and interaction maps in bacteria (96), we have assembled a bacterial nucleotide excision repair interactome, see Fig. 8. This dynamic network undergoes significant change following UV irradiation and suggests that bacteria employ alternative repair proteins, and may follow a significantly different reaction pathway in response to DNA damage (Fig. 8B).

UvrB

XPD

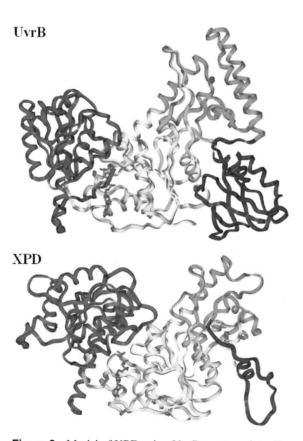

Figure 9 Model of XPD using UvrB as a template. Domain 1a, in yellow, shows the helicase motifs 1–3. Domain 3, in red, shows helicase motifs 4–6; ATP is in gray. Nonhomologous regions are shown in dark blue and green. The beta-hairpin of UvrB and the possibly similar structure of XPD are shown in light blue. (*See color insert.*)

6. SIMILARITIES IN DAMAGE RECOGNITION AND VERIFICATION BETWEEN BACTERIAL AND EUKARYOTIC NUCLEOTIDE EXCISION REPAIR SYSTEMS

While it is beyond the scope of this review to compare and contrast the differences between bacterial and eukaryotic repair systems, it is interesting to note that one of the key steps in damage recognition in bacterial repair is damage verification by UvrB. Nature has selected a protein with a helicase fold to fulfill this task. However, UvrB does not act as a true helicase, but uses the energy of ATP binding/hydrolysis to probe sites identified through the help of UvrA. This strand opening and bending of the DNA is believed to allow damage verification on the damaged strand, and act as a landing site for UvrC to mediate the dual incisions. In this regard, eukaryotes use a very similar process in which the two helicases, as part of TFIIH, work to open up the helix at the site of a damaged base in preparation for the action of the two nucleases, XPG, and a heterodimer of ERCC-1-XP-F. Using UvrB as a model we have developed a plausible structure for XPD (97), which suggests that most of the XPD causing mutations lie within the helicase motifs at the interface between domains 1 and 3, (Fig. 9). Using reconstituted NER and transcription assays, Egly and coworkers have shown that many of the XPD mutations leading to trichothiodystrophy lie outside the helicase motifs, and lead to a basal change in transcription with little effect on repair (98).

ACKNOWLEDGMENTS

We would like to thank Caroline Kisker and Rachelle Bienstock for critically reading the manuscript. In addition, we would like to thank graphic artist Tom Buhrman.

REFERENCES

1. Setlow RB, Carrier WL. The Disappearance of Thymine Dimers from DNA: An Error-Correcting Mechanism. Proc Natl Acad Sci USA 1964; 51:226–231.
2. Boyce RP, Howard-Flanders P. Release of Ultraviolet Ligh-Induced Thymine Dimers from DNA in E. Coli K-12. Proc Natl Acad Sci USA 1964; 51:293–300.
3. Hanawalt PC, Haynes RH. Repair Replication of DNA in Bacteria: Irrelevance of Chemical Nature of Base Defect. Biochem Biophys Res Commun 1965; 19:462–467.
4. Van Houten B. Nucleotide excision repair in Escherichia coli. Microbiol Rev 1990; 54:18–51.
5. Van Houten B, Snowden A. Mechanism of action of the Escherichia coli UvrABC nuclease: clues to the damage recognition problem. Bioessays 1993; 15:51–59.
6. Moolenaar GF, Herron MF, Monaco V, van der Marel GA, van Boom JH, Visse R, Goosen N. The role of ATP binding and hydrolysis by UvrB during nucleotide excision repair. J Biol Chem 2000; 275:8044–8050.
7. Theis K, Skorvaga M, Machius M, Nakagawa N, Van Houten B, Kisker C. The nucleotide excision repair protein UvrB, a helicase-like enzyme with a catch. Mutat Res 2000; 460:277–300.
8. Machius M, Henry L, Palnitkar M, Deisenhofer J. Crystal structure of the DNA nucleotide excision repair enzyme UvrB from Thermus thermophilus. Proc Natl Acad Sci USA 1999; 96:11717–11722.

9. Nakagawa N, Sugahara M, Masui R, Kato R, Fukuyama K, Kuramitsu S. Crystal structure of Thermus thermophilus HB8 UvrB protein, a key enzyme of nucleotide excision repair. J Biochem (Tokyo) 1999; 126:986–990.

10a. Theis K, Chen PJ, Skorvaga M, Van Houten B, Kisker C. Crystal structure of UvrB, a DNA helicase adapted for nucleotide excision repair. Embo J 1999; 18:6899–6907.

10b. Truglio JJ, Rhau B, Croteau DL, DellaVecchia MJ, Wang H, Theis K, Wang ML, Karakas E, Skorvaga M, Van Houten B, Kisker C. Crystal Structure of the N-terminal GIY-YIG Endonuclease Domain of UVrC. EMBO J 2005; 24(5):885–894.

11. Minko IG, Zou Y, Lloyd RS. Incision of DNA-protein crosslinks by UvrABC nuclease suggests a potential repair pathway involving nucleotide excision repair. Proc Natl Acad Sci USA 2002; 99:1905–1909.

12. Moolenaar GF, Hoglund L, Goosen N. Clue to damage recognition by UvrB: residues in the beta-hairpin structure prevent binding to non-damaged DNA. Embo J 2001; 20:6140–6149.

13. Geacintov NE, Cosman M, Hingerty BE, Amin S, Broyde S, Patel DJ. NMR solution structures of stereoisometric covalent polycyclic aromatic carcinogen-DNA adduct: principles, patterns, and diversity. Chem Res Toxicol 1997; 10:111–146.

14. Lee JH, Choi YJ, Choi BS. Solution structure of the DNA decamer duplex containing a $3'$-T $\times$T basepair of the cis-syn cyclobutane pyrimidine dimer: implication for the mutagenic property of the cis-syn dimer. Nucleic Acids Res 2000; 28:1794–1801.

15. Lee JH, Hwang GS, Choi BS. Solution structure of a DNA decamer duplex containing the stable $3'$ T.G base pair of the pyrimidine (6-4)pyrimidone photoproduct [(6-4) adduct]: implications for the highly specific $3'$ T→C transition of the (6-4) adduct. Proc Natl Acad Sci USA 1999; 96:6632–6636.

16. Gelasco A, Lippard SJ. NMR solution structure of a DNA dodecamer duplex containing a cis-diammineplatinum (II) d(GpG) intrastrand cross-link, the major adduct of the anticancer drug cisplatin. Biochemistry 1998; 37:9230–9239.

17. Kung HC, Bolton PH. Structure of a duplex DNA containing a thymine glycol residue in solution. J Biol Chem 1997; 272:9227–9236.

18. Zou Y, Bassett H, Walker R, Bishop A, Amin S, Geacintov NE, Van Houten B. Hydrophobic forces dominate the thermodynamic characteristics of UvrA-DNA damage intractions. J Mol Biol 1998; 281:107–119.

19. Hopfner KP, Karcher A, Shin DS, Craig L, Arthur LM, Carney JP, Tainer JA. Structural biology of Rad50 ATPase: ATP-driven conformation control in DNA double-strand break repair and the ABC-ATPase superfamily. Cell 2000; 101:789–800.

20. Hung LW, Wang IX, Nikaido K, Liu PQ, Ames GF, Kim SH. Crystal structure of the ATP-binding subunit of an ABC transporter. Nature 1998; 396:703–707.

21. Yuan YR, Blecker S, Martsinkevich O, Millen L, Thomas PJ, Hunt JF. The crystal structure of the MJ0796 ATP-binding cassette. Implications for the structural consequences of ATP hydrolysis in the active site of an ABC transporter. J Biol Chem 2001; 276:32313–32321.

22. Locher KP, Lee AT, Rees DC. The E. coli BtuCD structure: a framework for ABC transporter architecture and mechanism. Science 2002; 296:1091–1098.

23. Jones PM, George AM. Mechanism of ABC transporters: a molecular dynamics simulation of a well characterized nucleotide-binding subunit. Proc Natl Acad Sci USA 2002; 99:12639–12644.

24. Hopfner KP, Tainer JA. RAD50/SMC proteins and ABC trasporters: unifying concepts from high-resolution structures. Curr Opin Struct Biol 2003; 13:249–255.

25. http://www.ornl.gov/sci/techresources/human_genome/posters/chromosome/cftr.shtml.

26. Matthews JM, Sunde M. Zinc fingers–folds for many occasions. IUBMB Life 2002; 54:351–355.

27. Navaratnam S, Myles GM, Strange RW, Sancar A. Evidence from extended X-ray absorption fine structure and site-specific mutagenesis for zinc fingers in UvrA protein of Escherichia coli. J Biol Chem 1989; 264:16067–16071.

28. Visse R, de Ruijter M, Ubbink M, Brandsma JA, van de Putte P. The first zinc-binding domain of UvrA is not essential for UvrABC-mediated DNA excision repair. Mutat Res 1993; 294:263–274.

29. Wang J, Mueller KL, Grossman L. A mutational study of the C-terminal zinc-finger motif of the Escherichia coli UvrA protein. J Biol Chem 1994; 269:10771–10775.

30. Goosen N, Moolenaar GF. Role of ATP hydrolysis by UvrA and UvrB during nucleotide excision repair. Res Microbiol 2001; 152:401–409.

31. Skorvaga M, Theis K, Mandavilli BS, Kisker C, Van Houten B. The beta-hairpin motif of UvrB is essential for DNA binding, damage processing, and UvrC-mediated incisions. J Biol Chem 2002; 277:1553–1559.

32. Kim JL, Morgenstern KA, Griffith JP, Dwyer MD, Thomson JA, Murcko MA, Lin C, Caron PR. Hepatitis C virus NS3 RNA helicase domain with a bound oligonucleotide: the crystal structure provides insights into the mode of unwinding. Structure 1998; 6: 89–100.

33a. Selby CP, Sancar A. Molecular mechanism of transcription-repair coupling. Science 1993; 260:53–58.

33b. Truglio JJ, Croteau DL, DellaVecchia MJ, Theis K, Skorvaga M, Mandivilli BS, Van Houten B, Kisker C. Interactions between UvrA and UvrB: the role of UvrB's domain 2 in nucleotide excision repair. EMBO J 2004; 23(13):2498–2509.

34. Hsu DS, Kim ST, Sun Q, Sancar A. Structure and function of the UvrB protein. J Biol Chem 1995; 270:8319–8327.

35. Sohi M, Alexandrovich A, Moolenaar G, Visse R, Goosen N, Vernede X, Fontecilla-Camps JC, Champness J, Sanderson MR. Crystal structure of Escherichia coli UvrB C-teriminal domain, and a model for UvrB-uvrC interaction. FEBS Lett 2000; 465:161–164.

36. Alexandrovich A, Sanderson MR, Moolenaar GF, Goosen N, Lane AN. NMR assignments and secondary structure of the UvrC binding domain of UvrB. FEBS Lett 1999; 451:181–185.

37. Moolenaar GF, Bazuine M, van Knippenberg IC, Visse R, Goosen N. Characterization of the Escherichia coli damage-independent UvrBC endonuclease activity. J Biol Chem 1998; 273:34896–34903.

38. Hildebrand EL, Grossman L. Oligomerization of the UvrB nucleotide excision repair protein of Escherichia coli. J Biol Chem 1999; 274:27885–27890.

39. Verhoeven EE, van Kesteren M, Moolenaar GF, Visse R, Goosen N. Catalytic sites for 3′ and 5′ incision of Escherichia coli nucleotide excision repair are both located in UvrC. J Biol Chem 2000; 275:5120–5123.

40. Aravind L, Walker DR, Koonin EV. Conserved domains in DNA repair proteins and evolution of repair systems. Nucleic Acids Res 1999; 27:1223–1242.

41. Van Roey P, Meehan L, Kowalski JC, Belfort M, Derbyshire V. Catalytic domain structure and hypothesis for function of GIY-YIG intron endonuclease I-TevI. Nat Struct Biol 2002; 9:806–811.

42a. Derbyshire V, Kowalski JC, Dansereau JT, Hauer CR, Belfort M. Two-domain structure of the td intron-encoded endonuclease I-TevI correlates with the two-domain configuration of the homing site. J Mol Biol 1997; 265:494–506.

42b. Truglio J, Rhau B, Croteau DL, DellaVecchia MJ, Wang H, Theis K, Wang ML, Karakas E, Skorvaga M, Van Houten B, Kisker C. Crystal Structure of the N-terminal GIY-YIG Endonuclease Domain of UvrC. EMBO J 2005; 24(5):885–94.

43. Westerveld A, Hoeijmakers JH, van Duin M, de Wit J, Odijk H, Pastink A, Wood RD, Bootsma D. Molecular cloning of a human DNA repair gene. Nature 1984; 310: 425–429.

44. Matsunaga T, Mu D, Park CH, Reardon JT, Sancar A. Human DNA repair excision nuclease. Analysis of the roles of the subunits involved in dual incisions by using anti-XPG and anti-ERCC1 antibodies. J Biol Chem 1995; 270:20862–20869.

45. Davies AA, Friedberg EC, Tomkinson AE, Wood RD, West SC. Role of the Rad1 and Rad10 proteins in nucleotide excision repair and recombination. J Biol Chem 1995; 270:24638–24641.

46. Doherty AJ, Serpell LC, Ponting CP. The helix-hairpin-helix DNA-binding motif: a structural basis for non-sequence-specific recognition of DNA. Nucleic Acids Res 1996; 24:2488–2497.

47. Singh S, Folkers GE, Bonvin AM, Boelens R, Wechselberger R, Niztayev A, Kaptein R. Solution structure and DNA-binding properties of the C-terminal domain of UvrC from E.coli. Embo J 2002; 21:6257–6266.

48. Pelletier H, Sawaya MR, Wolfle W, Wilson SH, Kraut J. Crystal structures of human DNA polymerase beta complexed with DNA: implications for catalytic mechanism, processivity, and fidelity. Biochemistry 1996; 35:12742–12761.

49. Moolenaar GF, Uiterkamp RS, Zwijnenburg DA, Goosen N. The C-terminal region of the Escherichia coli UvrC protein, which is homologous to the C-terminal region of the human ERCC1 protein, is involved in DNA binding and 5′-incision. Nucleic Acids Res 1998; 26:462–468.

50. Verhoeven EE, van Kesteren M, Turner JJ, van der Marel GA, van Boom JH, Moolenaar GF, Goosen N. The C-terminal region of Escherichia coli UvrC contributes to the flexibility of the UvrABC nucleotide excision repair system. Nucleic Acids Res 2002; 30:2492–2500.

51. Alexandrovich A, Czisch M, Frenkiel TA, Kelly GP, Goosen N, Moolenaar GF, Chowdhry BZ, Sanderson MR, Lane AN. Solution structure, hydrodynamics and thermodynamics of the UvrB C-terminal domain. J Biomol Struct Dyn 2001; 19:219–236.

52. Tang M, Nazimiec M, Ye X, Iyer GH, Eveleigh J, Zheng Y, Zhou W, Tang YY. Two forms of UvrC protein with different double-stranded DNA binding affinites. J Biol Chem 2001; 276:3904–3910.

53. Nishino T, Komori K, Ishino Y, Morikawa K. X-Ray and Biochemical Anatomy of an Archaeal XPF/Rad1/Mus81 Family Nuclease. Similarity between Its Endonuclease Domain and Restriction Enzymes. Structure (Camb) 2003; 11:445–457.

54. Thomas DC, Kunkel TA, Casna NJ, Ford JP, Sancar A. Activites and incision patterns of ABC excinuclease on modified DNA containing single-base mismatches and extrahelical bases. J Biol Chem 1986; 261:14496–14505.

55. Jiang G, Skorvaga M, Van Houten B, States JC. Reduced sulfhydryls maintain specific incision of BPDE-DNA adducts by recombinant thermoresistant Bacillus caldotenax UvrABC endonuclease. Protein Expr Purif 2003; 31:88–98.

56. Seeberg E, Steinum AL. Purification and properties of the uvrA protein from Escherichia coli. Proc Natl Acad Sci USA 1982; 79:988–992.

57. Mazur S, Grossman L. Dimerization of Echerichia colli UvrA and Its Binding to Undamaged and Ultraviolet Light Damaged DNA. Biochemistry 1991; 30:4432–4443.

58. Oh EY, Claassen L, Thiagalingam S, Mazur S, Grossman L. ATPase activity of the UvrA and UvrAB protein complexes of the Escherichia coli UvrABC endonuclease. Nucleic Acids Res 1989; 17:4145–4159.

59. Orren DK, Sancar A. Subnits of ABC excinuclease interact in solution in the abscence of DNA. UCLA Symp Mol Cell Biol 1988; 83:87–94.

60. Myles GM, Hearst JE, Sancar A. Site-specific mutagenesis of conserved residues within Walker A and B sequences of Escherichia coli UvrA protein. Biochemistry 1991; 30:3824–3834.

61. Jones PM, George AM. Subunit interactions in ABC transporters: towards a functional architecture. FEMS Microbiol Lett 1999; 179:187–202.

62. Obmolova G, Ban C, Hsieh P, Yang W. Crystal structures of mismatch repair protein MutS and its complex with a substrate DNA. Nature 2000; 407:703–710.

63. Lamers MH, Perrakis A, Enzlin JH, Winterwerp HH, de Wind N, Sixma TK. The crystal structure of DNA mismatch repair protein MutS binding to a G×T mismatch. Nature 2000; 407:711–717.

64. Myles GM, Sancar A. Isolation and characterization of functional domains of UvrA. Biochemistry 1991; 30:3834–3840.

65. Thiagalingam S, Grossman L. Both ATPase sites of Escherichia coli UvrA have functional roles in nucleotide excision repair. J Biol Chem 1991; 266:11395–11403.

66. Thiagalingam S, Grossman L. The multiple roles for ATP in the Escherichia coli UvrABC endonuclease-catalyzed incision reaction. J Biol Chem 1993; 268:18382–18389.

67. Wang J, Grossman L. Mutations in the helix-turn-helix motif of the Escherichia coli UvrA protein eliminate its specificity for UV-damaged DNA. J Biol Chem 1993; 268:5323–5331.

68. Rosinski JA, Atchley WR. Molecular evolution of helix-turn-helix proteins. J Mol Evol 1999; 49:301–309.

69. Smith PC, Karpowich N, Millen L, Moody JE, Rosen J, Thomas PJ, Hunt JF. ATP binding to the motor domain from an ABC transporter drives formation of a nucleotide sandwich dimer. Mol Cell 2002; 10:139–149.

70. Claassen LA, Grossman L. Deletion mutagenesis of the Escherichia coli UvrA protein localizes domains for DNA binding, damage recognition, and protein-protein interactions. J Biol Chem 1991; 266:11388–11394.

71. Zou Y, Crowley DJ, Van Houten B. Involvement of molecular chaperonins in nucleotide excision repair. Dnak leads to increased thermal stability of UvrA, catalytic UvrB loading, enhanced repair, and increased UV resistance. J Biol Chem 1998; 273:12887–12892.

72. Sancar A, Sancar GB. DNA repair enzymes. Annu Rev Biochem 1988; 57:29–67.

73. Orren DK, Sancar A. The (A)BC excinuclease of Escherichia coli has only the UvrB and UvrC subunits in the incision complex. Proc Natl Acad Sci USA 1989; 86:5237–5241.

74. Verhoeven EE, Wyman C, Moolenaar GF, Goosen N. The presence of two UvrB subunits in the UvrAB complex ensures damage detection in both DNA strands. EMBO J 2002; 21:4196–4205.

75. Lin JJ, Phillips AM, Hearst JE, Sancar A. Active site of (A)BC excinuclease. II. Binding, bending, and catalysis mutants of UvrB reveal a direct role in 3′ and an indirect role in 5′ incision. J Biol Chem 1992; 267:17693–17700.

76. Hoare S, Zou Y, Purohit V, Krishnasamy R, Skorvaga M, Van Houten B, Geacintov NE, Basu AK. Differential incision of bulky carcinogen-DNA adducts by the UvrABC nuclease: comparison of incision rates and the interactions of Uvr subunits with lesions of different structures. Biochemistry 2000; 39:12252–12261.

77. Shi Q, Thresher R, Sancar A, Griffith J. Electron microscopic study of (A)BC excinuclease. DNA is sharply bent in the UvrB-DNA complex. J Mol Biol 1992; 226:425–432.

78. Visse R, King A, Moolenaar GF, Goosen N, van de Putte P. Protein-DNA interactions and alterations in the DNA structure upon UvrB-DNA preincision complex formation during nucleotide excision repair in Escherichia coli. Biochemistry 1994; 33:9881–9888.

79. Verhoeven EE, Wyman C, Moolenaar GF, Hoeijmakers JH, Goosen N. Architecture of nucleotide excision repair complexes: DNA is wrapped by UvrB before and after damage recognition. Embo J 2001; 20:601–611.

80. Van Houten B, Gamper H, Sancar A, Hearst JE. DNase I footprint of ABC excinuclease. J Biol Chem 1987; 262:13180–13187.

81. Oh EY, Grossman L. Helicase properties of the Escherichia coli UvrAB protein complex. Proc Natl Acad Sci USA 1987; 84:3638–3642.

82. Koo HS, Claassen L, Grossman L, Liu LF. ATP-dependent partitioning of the DNA template into supercoiled domains by Escherichia coli UvrAB. Proc Natl Acad Sci USA 1991; 88:1212–1216.

83a. Caron PR, Grossman L. Involvement of a cryptic ATPase activity of UvrB and its proteolysis product, UvrB* in DNA repair. Nucleic Acids Res 1988; 16:10891–10902.

83b. Skorvaga M, DellaVecchia MJ, Croteau DL, Theis K, Truglio JJ, Mandavilli BS, Kisker C, Van Houten B. Identification of residues within UvrB that are important for efficient DNA binding and damage processing. J Biol Chem 2004; 279(49):51574–51580.

83c. DellaVecchia MJ, Croteau DL, Skorvaga M, Dezhurov SV, Lavrik OI, Van Houten B. Analyzing the handoff of DNA from UvrA to UvrB utilizing DNA-protein photoaffinity labeling. J. Biol. Chem 2004; 279(43):45245–45256.

84. Sancar A, Franklin KA, Sancar GB. Escherichia coli DNA photolyase stimulates uvrABC excision nuclease in vitro. Proc Natl Acad Sci USA 1984; 81:7397–7401.

85. Kurtz AJ, Dodson ML, Lloyd RS. Evidence for multiple imino intermediates and identification of reactive nucleophiles in peptide-catalyzed beta-elimination at abasic sites. Biochemistry 2002; 41:7054–7064.

86. Moolenaar GF, Monaco V, van der Marel GA, van Boom JH, Visse R, Goosen N. The effect of the DNA flanking the lesion on formation of the UvrB-DNA preincision complex. J Biol Chem 2000; 275:8038–8043.

87. Lin JJ, Sancar A. Active site of (A)BC excinuclease. I. Evidence for 5′ incision by UvrC through a catalytic site involving Asp399, Asp438, Asp466, and His538 residues. J Biol Chem 1992; 267:17688–17692.

88. Rupp WD, Sancar A, Sancar GB. Properties and regulation of the UVRABC endonuclease. Biochimie 1982; 64:595–598.

89. Sancar A, Rupp WD. A novel repair enzyme: UVRABC excision nuclease of Escherichia coli cuts a DNA strand on both sides of the damaged region. Cell 1983; 33:249–260.

90. Yeung AT, Mattes WB, Oh EY, Grossman L. Enzymatic properties of purified Escherichia coli uvrABC proteins. Proc Natl Acad Sci USA 1983; 80:6157–6161.

91. Zou Y, Walker R, Bassett H, Geacintov NE, Van Houten B. Formation of DNA repair intermediates and incision by the ATP-dependent UvrB-UvrC endonuclease. J Biol Chem 1997; 272:4820–4827.

92. Delagoutte E, Bertrand-Burggraf E, Dunand J, Fuchs RP. Sequence-dependent modulation of nucleotide excision repair: the efficiency of the incision reaction is inversely correlated with the stability of the pre-incision UvrB-DNA complex. J Mol Biol 1997; 266:703–710.

93. Smith BT, Grossman AD, Walker GC. Localization of UvrA and effect of DNA damage on the chromosome of Bacillus subtilis. J Bacteriol 2002; 184:488–493.

94. Lin CG, Kovalsky O, Grossman L. DNA damage-dependent recruitment of nucleotide excision repair and transcription proteins to Escherichia coli inner membranes. Nucleic Acids Res 1997; 25:3151–3158.

95. Courcelle J, Khodursky A, Peter B, Brown PO, Hanawalt PC. Comparative gene expression profiles following UV exposure in wild-type and SOS-deficient Escherichia coli. Genetics 2001; 158:41–64.

96a. Wojcik J, Boneca IG, Legrain P. Prediction, assessment and validation of protein interaction maps in bacteria. J Mol Biol 2002; 323:763–770.

96b. Ideker T, Ozier O, Schwikowski B, Siegel AF. Discovering regulatory and signalling circuits in molecular interaction networks. Bioinformatics 2002; 18(Suppl 1):S233–S240.

97. Bienstock RJ, Skorvaga M, Mandavilli BS, Van Houten B. Structural and functional characterization of the human DNA repair helicase XPD by comparative molecular modeling and site-directed mutagenesis of the bacterial repair protein UvrB. J Biol Chem 2003; 278:5309–5316.

98. Dubaele S, Proietti De Santis L, Bienstock RJ, Keriel A, Stefanini M, Van Houten B, Egly JM. Basal transcription defect discriminates between xeroderma pigmentosum and trichothiodystrophy in XPD patients. Mol Cell 2003; 11:1635–1646.

7

Recognition of DNA Damage During Eukaryotic Nucleotide Excision Repair

Hanspeter Naegeli

Institute of Pharmacology and Toxicology, University of Zürich-Vetsuisse, Zürich, Switzerland

1. INTRODUCTION

The genome of all living organisms is under permanent attack from endogenous metabolic byproducts and environmental factors that alter the chemical structure of DNA and corrupt its nucleotide sequence. A network of DNA repair systems has evolved to cope with these genotoxic insults by eliminating DNA damage from the genome. Nucleotide excision repair (NER) is the only pathway in mammalian cells that removes bulky DNA adducts induced by ultraviolet (UV) light and electrophilic chemicals. It consists of a ubiquitous "cut and patch" mechanism that operates by excision of a short single-stranded DNA fragment, followed by restoration of the duplex DNA structure through repair synthesis. This particular DNA repair mode was first discovered in *Escherichia coli* as an enzymatic system that excises UV-radiation products from DNA (1,2). Subsequently, a functionally similar excision process was identified in humans (3,4). It was found that patients with the rare disease xeroderma pigmentosum (XP), characterized by extreme photosensitivity and a 2000-fold increased incidence of sunlight-induced skin cancer, are defective in NER of UV damage. Individuals who suffer from this autosomal recessive disorder are classified into seven repair-deficient complementation groups designated XPA through XPG. These patients also have an increased incidence of internal tumors and, in some cases, neurological abnormalities, probably reflecting the importance of NER in the repair of endogenous DNA damage (5). Cells derived from XP individuals show elevated mutation rates because of misincorporation of bases opposite to the unexcised lesions during replication of the damaged DNA template.

2. NUCLEOTIDE EXCISION REPAIR SUBSTRATES

Nucleotide excision repair represents the most important pathway for the removal of DNA damage inflicted by UV light and a variety of other DNA damaging agents that distort the DNA helix. Ultraviolet irradiation of DNA generates cross-links between adjacent pyrimidines, either cyclobutane pyrimidine dimers or pyrimidine–pyrimidone (6–4) photoproducts. Both UV cross-links are repaired through the NER pathway, but much confusion has been generated in the past by the fact that the two lesions impose different structural deformations on the DNA double helix. As a consequence, the human NER system recognizes and removes cyclobutane dimers only at approximately 10–20% the rate of repair of (6–4) photoproducts (6,7). Many organisms, including bacteria, yeast, plants, *Drosophila, Xenopus*, and even marsupials, have specialized enzymes that reverse UV-radiation damage using the energy of sunlight (reviewed in Ref. 8). Such DNA photolyases, specific for either cyclobutane dimers or (6–4) photoproducts, provide very useful tools to analyze the contribution of each individual lesion to the overall UV-radiation response. In fact, the heterologous expression of DNA photolyases has been employed to remove one particular lesion from the genome of UV-irradiated human cells or transgenic mice. All these studies suggest that the major mutagenic, apoptotic, and cytotoxic effects of UV light are attributable to the formation of cyclobutane dimers (9–11), which are more refractory than (6–4) photoproducts to excision by the NER system.

With the advent of targeted gene replacement in embryonic stem cells, several mouse models have been generated to analyze the consequences of defective NER activity following exposure to various genotoxic agents (12–17). Such knock-out experiments lend support to the notion that NER activity is an essential part of the cellular defense system that protects the genome against UV-radiation products and bulky DNA adducts generated by other carcinogens. In fact, the lack of NER activity in mice not only recapitulates the predisposition to UV-induced skin cancer, but also results in increased tumorigenesis following exposure to electrophilic chemicals. For example, the homozygous $XPA^{-/-}$ mutant mouse, completely deficient in NER, is prone to tumors of the skin, lymphoid system, and liver after treatment with polycyclic aromatic hydrocarbons.

3. EUKARYOTIC NER REACTION

The multistep NER mechanism involves (i) damage recognition and assembly of the incision complex, (ii) dual DNA incision and damage excision, (iii) DNA repair synthesis and ligation (Fig. 1). The initial molecular recognition step in the NER system should be highly specific for damaged DNA to avoid futile repair cycles among the 3 billion base pairs in the human genome and, at the same time, versatile in order to detect a broad spectrum of chemically unrelated lesions. In fact, the NER system is able to detect a nearly infinite variety of DNA adducts despite its very limited repertoire of damage recognition subunits. This extraordinary substrate versatility has generally been ascribed to an indirect readout mechanism, whereby particular distortions of the double helix, induced by a damaged nucleotide, provide the molecular determinants not only for lesion recognition, but also for subsequent demarcation and verification processes. In this chapter, we will discuss the evidence in support of an alternative mechanism of substrate discrimination that is initiated by the detection of thermodynamically unstable base pairs followed by direct localization of the lesion through an enzymatic proofreading activity.

Most enzymatic steps that follow the successful recognition of a lesion in the eukaryotic NER pathway have been characterized in detail (see, for example, Refs. 18–20).

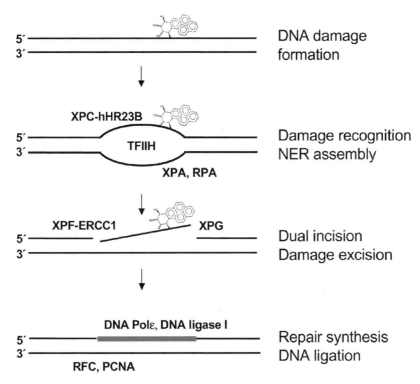

Figure 1 Scheme of the human NER pathway. This DNA repair process is highly conserved among eukaryotes. In this example, excision of the DNA adduct is shown following damage by benzo(a)pyrene diol epoxide.

Figure 1 illustrates that a key NER intermediate is generated by local unwinding of the duplex substrate, resulting in a partially open DNA conformation with "Y-shaped" double-stranded to single-stranded transitions around the lesion site (21). In this open intermediate, the two strands of duplex DNA are separated, allowing for discrimination between the damaged residue and the undamaged complementary sequence. Structure-specific endonucleases cleave the damaged DNA strand at the borders of this "bubble" intermediate, thereby releasing the adducted residues as part of oligomeric segments of 24–32 nucleotides in length (22). The duplex DNA structure is then restored through the synthesis of small repair patches that fill in the gap generated by oligonucleotide excision (Fig. 1).

The geometry of the NER reaction is asymmetric in two respects. First, the incisions are asymmetric relative to the long axis of DNA because the cleavages occur 15–25 nucleotides away from the damaged base on the 5′ side but only 3–9 nucleotides away on the 3′ side. Second, the incisions must be restricted to the damaged strand, as the complementary undamaged sequence, later in the reaction, has to serve as a template for repair patch synthesis by DNA polymerases. Obviously, the localization of the target lesion defines the DNA strand on which both incisions are made, while the undamaged strand is protected from endonucleolytic attack. However, it is not clear what molecular constraints determine the sites of cleavage and what exactly directs the nucleases selectively to the damaged strand.

4. SUBUNITS OF THE EUKARYOTIC NER MACHINERY

Distinct sets of proteins, necessary and sufficient to carry out the NER reaction (referred to as core NER proteins), have been identified in prokaryotic and eukaryotes. Only six polypeptides (UvrA, UvrB, UvrC, UvrD, DNA polymerase I, and DNA ligase) are required for the NER process in prokaryotes (23). In contrast, the eukaryotic NER system displays a considerably higher degree of genetic complexity and its core subunits can be divided in two groups: (i) the factors necessary for damage recognition and double DNA incision and (ii) others required for DNA repair synthesis. These factors assemble into two distinct multienzyme machines (24–27). The term "excinuclease" has been used (28) to denote an initial multiprotein complex that carries out damage recognition and DNA incision. Subsequently, repair patch synthesis is carried out by a second large machine with DNA polymerase and DNA ligase activity.

Although the order of arrival and departure of each factor is still intensively debated, the favored model involves the sequential assembly of an "excinuclease" complex that includes XPC–hHR23B (a dimer composed of XPC and a human homolog of RAD23), transcription factor IIH (TFIIH, ten subunits), XPA (a possible homodimer), replication protein A (RPA, three subunits), XPG, which is responsible for the 3' incision, and XPF–ERCC1 (a dimer composed of XPF and excision repair cross-complementing 1 protein), which makes the 5' incision (29,30). The co-ordinated action of these six core factors is sufficient to carry out oligonucleotide excision on naked substrates in vitro without the aid of any other accessory protein. Multiple interaction domains, which promote assembly of the active "excinuclease," have been identified among these NER subunits (reviewed by Ref. 31). For example, XPC and XPG are both stably associated with the TFIIH complex. Transcription factor IIH has also been shown to interact with XPA; XPG and XPA in turn interact with RPA, and XPA protein associates with the ERCC1–XPF endonuclease (Fig. 2). However, there is still limited information about the location of these interacting proteins relative to one another and relative to the lesion in the NER complex.

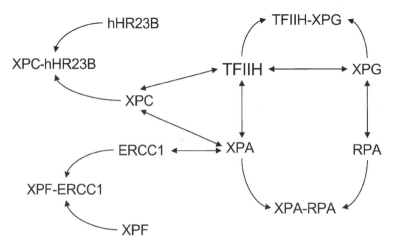

Figure 2 Interactions between core NER factors.

After DNA incision by XPG and XPF–ERCC1, a DNA polymerase complex is recruited to carry out repair patch synthesis and DNA ligation. This second category of NER factors involves a series of DNA replication enzymes and accessory factors, among which RPA may adopt a crucial function in coupling DNA incision to the subsequent DNA synthesis step. Following its participation in the "excinuclease" complex, RPA is already prebound to the gapped excision intermediate and, as a consequence, is in the position to co-ordinate the dissociation of early incision factors with the assembly of the DNA polymerase complex (18,30). The synthesis of repair patches is further dependent on replication factor C (RFC), a pentameric matchmaker that binds to the excision gap and mediates the entry of proliferating cell nuclear antigen (PCNA), which in turn acts as a sliding clamp for DNA polymerases δ and ε (32). Finally, the newly synthesized repair patches are ligated to the pre-existing DNA through the action of DNA ligase I.

5. STEPWISE ASSEMBLY OF THE MAMMALIAN NER RECOGNITION COMPLEX

It is becoming increasingly clear that, in mammalian cells, NER is executed by the sequential recruitment of repair proteins to the site of the DNA lesion, rather than by the action of a preassembled "repairosome" complex. It can be estimated that a fully functional "repairosome" may achieve a mass of ≥2.7 MDa (33), but in vivo studies monitoring the movement of NER factors to sites of DNA repair are not compatible with such large machines and instead favor the stepwise assembly of individual core subunits (34).

A fundamental process in the assembly of the NER complex, its initiation by recognition of DNA damage, remains poorly understood. Three general models have been proposed for this early lesion recognition step: "XPC first," "XPA first," or "RPA first" (7). In the "XPC first" model, the XPC–hHR23B complex represents the primary molecular recognition component that binds to damaged sites and initiates the NER pathway by recruiting TFIIH and other factors (35–37). This model is mainly supported by competition experiments aimed at determining the order in which NER proteins interact with lesion sites. Sugasawa et al. (36) found that damaged plasmids preincubated with XPC–hHR23B are more rapidly repaired than those preincubated with the XPA–RPA complex. These results supported a previous model proposed by Evans et al. (38), who suggested that XPA–RPA might not be the initial DNA damage recognition factor but rather serves during subsequent steps to assist TFIIH in opening the duplex in the preincision complex. Further support for the "XPC first" model came from reconstitution experiments demonstrating that ATP is not required for the recruitment of purified XPC and TFIIH to damaged DNA fragments, In contrast, ATP is necessary for the recruitment of XPA (and other core subunits) to the lesion, suggesting that the local ATP-dependent unwinding mediated by TFIIH is a prerequisite for the subsequent inclusion of XPA into the growing NER complex (30). In apparent conflict with these reports, Wakasugi and Sancar (39) observed that preincubation of damaged DNA with XPA–RPA promotes the repair more rapidly than that of damaged substrate with XPC–hHR23B. These authors concluded that, at least with some lesions, XPA–RPA is the initial DNA damage recognition subunit (see in what follows).

To study in detail the order of assembly of the human NER complex, the nuclear trafficking of each core subunit was analyzed in intact living cells. For that purpose, cell monolayers were exposed to UV light through filters with defined pore sizes to obtain localized foci of DNA damage and repair. The movement of XPC–hHR23B and XPA from unirradiated regions of the nuclei to these damaged foci was monitored by staining the factors with fluorescently tagged antibodies (29). Interestingly, XPC–hHR23B has been shown to accumulate in DNA repair foci in both wild-type and XPA cells, whereas XPA did not accumulate in the damaged foci of XPC cells. These results are consistent with XPC being the first factor that recognizes the lesions. In contrast, XPA is not recruited to DNA lesions unless XPC–hHR23B is already present at the site of damage. Similarly, using XP cells of other complementation groups, these focal UV-irradiation studies were extended to demonstrate that XPG and RPA are recruited to the complex prior to and independently of XPA (40). As the XPA subunit is apparently not required for inclusion of RPA into the preincision complex, it may be concluded that the association between these two subunits only occurs at their site of action in the repair process. An attractive advantage of the "XPC first" model is that it accommodates the much higher affinity of XPC for damaged DNA duplexes in comparison with XPA or RPA (37,41). Conversely, a possible problem associated with this model is that XPC has no detectable affinity for cyclobutane pyrimidine dimers in vitro (37,42). Furthermore, XPC is apparently not recruited to nuclear foci of UV damage containing exclusively cyclobutane dimers (43); yet, it is one of the six core factors that are absolutely required to excise such dimers in reconstituted systems (7).

In the "XPA first" model, it is proposed that XPA (or the XPA–RPA complex) recognizes the damage and then recruits TFIIH followed by the other repair factors (39,44). The "RPA first" model was prompted more recently to explain the results of a psoralen cross-linking experiment. A furan-side psoralen adduct was constructed to be used both as a substrate for in vitro NER reactions and as a cross-linker that immobilizes repair subunits located in close proximity to the lesion (45). When the furan-side psoralen adduct was incubated with a reconstituted NER complex, two different subunits of RPA (RPA70 and RPA32) and a single TFIIH subunit (XPD) were cross-linked to the DNA substrate. However, neither XPC nor XPA was immobilized to the psoralen substituent, indicating that these factors do not make intimate contacts with the adduct in the ultimate incision complex, even though they bind preferentially to damaged DNA. These results suggest a prominent role of RPA in damage recognition but could also be taken as evidence for a function of RPA in binding to the undamaged strand directly opposite to the lesion (46), from where it may become cross-linked to the psoralen adduct. Another interpretation is that there may not be a rigid order of assembly, nor a universal recognition factor. Any of the subunits with affinity for damaged DNA may mark the site to be repaired and subsequently recruit the remaining components of the NER system. To account for the apparent failure of purified XPC–hHR23B to discriminate cyclobutane pyrimidine dimers from undamaged DNA, a more recent version of this model proposes that three factors (XPA, RPA, and XPC) may act in a co-operative manner to locate the lesions and subsequently recruit the TFIIH complex (7).

6. A PREASSEMBLED REPAIROSOME IN YEAST?

Homologs of all the perviously mentioned mammalian NER factors have been identified in yeast (20), indicating a similar mechanism of damage recognition and "excinuclease" assembly throughout eukaryotes (see Table 1 for the homology between NER genes of humans and yeast). However, the absence of detectable preassembled NER holocomplexes in intact mammalian cells contrasts with the results of earlier studies that identified a high molecular weight NER complex in extracts of *Saccharomyces cerevisiae* (47). Gentle purification of tagged TFIIH led to the isolation of a protein fraction that besides the TFIIH subunits included at least Rad4–Rad23, Rad2, Radl4, and Radl–Radl0. The affinity purification of tagged Radl4 from *S. cerevisiae* extracts yielded a large complex that includes all the previously mentioned proteins, as well as Rad7, Radl6, and RPA (48). The fraction containing all these subunits was functional, as it could carry out double DNA incisions on damaged substrates. These data suggest that, unlike the human NER system, the proteins required for NER in yeast might assemble into a so-called "repairosome" independently of DNA damage. As an alternative to the "repairosome" model, a sequential assembly has also been observed upon incubation of yeast NER proteins with damaged DNA substrates (49). Several explanations may account for these different findings. For example, the experimental protocols for cell lysis, extract preparation, and chromatographic fractionation are not identical in the different studies. In general, the conditions used during these isolation steps differ from those encountered by NER proteins in living cells as competing factors may be absent, or protein and substrate concentrations may change. However, the different models are not mutually exclusive and it remains possible that part of the NER factors in yeast are preassembled in "repairosomes," which are dedicated to continuous surveillance of the genome and thus facilitate the fast repair of DNA lesions. In support of this idea, it has been proposed that the assembly of NER complexes in yeast may diverge from the situation in mammalian cells due to differences in genome size and chromatin structure (40).

Table 1 Nomenclature of Homologous NER Subunits in Eukaryotes

Homo sapiens	*Saccharomyces cerevisiae*	*Schizosaccharomyces pombe*[a]
DDB1	Unknown	Ddbl
DDB2	Unknown	Unknown
XPC	Rad4	Rhp41,Rhp42
hHR23A, hHR23B	Rad23	Rhp23
XPB (ERCC3)	Rad25	Ercc3sp
XPD (ERCC2)	Rad3	Radl5,Rhp3
XPG	Rad2	Radl3
XPA	Radl4	Rhpl4
RPA	Rfa	RPA
XPF	Radl	Radl6
ERCC1	Radl0	Swil0
Unknown	Rad7	Rhp7
Unknown	Radl6	Rhpl6
CSA	Rad28	SPBC577.09
CSB	Rad26	Rhp26

[a]See http ://www. genedb. org/genedb/pombe/index.jsp for further details.

7. ROLE OF DAMAGED DNA BINDING IN DAMAGE RECOGNITION

Another factor with an affinity for damaged DNA (damaged DNA binding-DDB) stimulates excision of bulky UV lesions in humans and, therefore, has also been implicated in the damage recognition process. It is a heterodimer of p1 27 and p48 polypeptides (50) whose small subunit is encoded by the *XPE* gene (51). Mutations in the 48-kDa subunit (DDB2) are found in all cases of XP complementation group E (52,53).

Damaged DNA binding appeared to be a candidate for the initial damage-recognition function for several reasons. First, DDB has an extraordinarily high preference for damaged DNA duplexes, including UV-irradiated substrates (54–56). Unlike XPC–hHR23B, DDB binds tightly to DNA fragments containing cyclobutane pyrimidine dimers. The strong affinity of DDB for irradiated DNA is illustrated by the fact that it is the only such factor that can be detected in crude mammalian cell extracts by electrophoretic mobility shift or filter-binding assays (57). Second, the binding of DDB to damaged substrates leads to bending of the DNA by an angle of 55° (58). This observation prompted the hypothesis that DDB binds to cyclobutane dimers and distorts the DNA helix around the lesion so that XPC (or XPA) can now bind to it. Third, XPE cells are severely compromised in the repair of cyclobutane dimers but show normal excision of (6–4) photoproducts (59). Next, Chinese hamster ovary cells, which lack p48 expression because of transcriptional silencing, are inefficient in cyclobutane dimer repair compared with human cells, but this deficiency can be corrected by reactivating *DDB2* gene expression or by exogenous expression of p48 (60,61). Finally, p48 localizes to sites of UV-induced lesions within minutes following UV irradiation, and the binding of XPC to these DNA repair foci is accelerated when p48 is present (43). In addition, microinjection of the DDB heterodimer into nuclei of XPE cells restores their NER activity (62).

In vitro reconstitution experiments demonstrated that human NER could be performed on naked DNA in the absence of DDB protein even when cyclobutane pyrimidine dimers are used as the substrate (7,25,63). In electrophoretic mobility shift assays, purified DDB and XPC–hHR23B form distinct complexes with a damaged DNA probe (37). A mixture of DDB and XPC–hHR23B yields two independently shifted complexes of the same mobility as seen for the individual proteins. As no new shifts are formed that would be diagnostic of a ternary intermediate of the two factors with a single DNA molecule, there is no biochemical proof that XPC–hHR23B is recruited to DNA lesions by DDB or that XPC–hHR23B may be capable of displacing DDB from the lesion site. In combination, these findings led to a re-evaluation of the role of DDB in repair and to the speculation that this factor may facilitate repair through acetylation of histones (64–66). In fact, DDB becomes tightly bound to chromatin soon after UV irradiation and can only be extracted from the chromatin fraction by nuclease treatment (67). DDB1 and DDB2 (p48) are also found in complexes together with cullin 4A and Roc1, which display ubiquitin E3 ligase activity, consistent with another possible role of DDB in post-translational chromatin modification (68–70). A mechanism has been suggested whereby p48 mediates ubiquitin ligation of protein substrates around cyclobutane dimers, possibly including histones, which may cause nucleosome unfolding and thereby allow access of XPC–hHR23B and the remaining components of the NER machinery to cyclobutane dimers (43). This auxiliary activity is not required for the recognition of (6–4) photoproducts. DDB2 (p48) is itself rapidly degraded after UV irradiation via the ubiquitin-mediated proteasome pathway (71,72), suggesting that the interaction with ubiquitin ligase may also serve to regulate the cellular DDB2 pool while it participates in DNA repair.

8. RECOGNITION OF BULKY LESIONS DURING TRANSCRIPTION-COUPLED DNA REPAIR

DNA repair is highly nonuniform in the chromosomal context. While "global genome" repair (GGR) is active in the entire genome, regardless of whether any specific sequence is transcribed or not, living organisms have set up a more efficient "transcription-coupled" repair (TCR) pathway that eliminates DNA lesions only from DNA that is undergoing active transcription. For example, cyclobutane dimers are removed more rapidly in transcribed genes than from transcriptionally silent regions or the genome as a whole (73) and, in particular, are removed from the template strand of RNA polymerase II-transcribed genes more rapidly than from the nontranscribed coding strand (74). Humans and mice that are genetically defective in XPC retain the capacity for transcription-coupled NER, indicating that this particular factor is not essential for the preferential repair of template strands in transcribed genes (75). Thus, the TCR process is dependent on RNA polymerase II but requires neither XPC–hHR23B nor DDB, whereas GGR is independent of transcription but needs the XPC and, for some lesions, the DDB complexes. Transcription-coupled repair is important for the rapid recovery of transcription activity and thus protects cells from apoptosis induced by transcription blocking lesions (76).

   A widely accepted mechanistic model assumes that XPC is not needed in TCR because the damaged base is recognized when it physically blocks the transcription machinery. Several bulky lesions, including cyclobutane pyrimidine dimers, polycyclic aromatic hydrocarbon adducts, or cisplatin intrastrand cross-links, have been shown to block the RNA polymerase II holoenzyme (77). Mammalian RNA polymerase I and III do not elicit TCR (78,79), but the RNA polymerase I of S. cerevisiae mediates the TCR of UV damage in ribosomal genes (80,81). On undamaged DNA, the RNA polymerase II complex proceeds unhindered until transcription of a gene is completed. However, in the presence of a transcription blocking base lesion, progression of RNA polymerase II is arrested, eventually leading to the recruitment of a large complex that includes CSA and CSB, several NER proteins (XPB, XPD, and XPG), one or more proteins involved in mismatch repair, as well as the gene product mutated in UV-sensitive syndrome (see, for example, Refs. 82–84). Xeroderma pigmentosum G is unique among the factors required for TCR in also stimulating the removal of oxidative lesions from the genome overall (85). Inactivation of CSA or CSB results in the genetic disease Cockayne syndrome (CS), characterized by sun sensitivity, mental retardation, short stature, and an early-aging phenotype, while a defective mismatch repair system predisposes to cancer. By analogy to the ordered recruitment of the GGR complex, it is generally assumed that the TCR machinery is sequentially assembled rather than being recruited as part of the RNA polymerase II holoenzyme (86). For example, TFIIH dissociates from the transcription machinery soon after promoter clearance (87) and, therefore, needs to be recruited again by the stalled RNA polymerase II enzyme to take part in the TCR process. Although the function of each individual subunit remains to be established, the multiprotein complex assembled during TCR is thought to dislocate the stalled transcription machinery from the site of damage in the transcribed strand, allowing access to this site to proteins required for NER or other DNA repair processes. The recruitment of repair factors to such damaged sites is thus facilitated, and this mechanism accounts for the accelerated removal of DNA lesions from transcriptionally active loci.

9. BIPARTITE SUBSTRATE DISCRIMINATION IN THE GGR PATHWAY

The molecular mechanism of damage recognition has been a matter of much debate, in part raised by the fact that none of the individual core factors display a high enough specificity and versatility to function as a unique sensor of damaged substrates in the GGR pathway (7). In addition, with exception of DDB, none of these factors have a sufficiently high affinity for cyclobutane dimers to be detected in standard DNA-binding assays. How does the NER complex, then, recognize, base lesions that produce little distortion of DNA? In principle, a higher degree of selectivity for damaged substrates may be achieved by a composite mechanism that employs more than one protein subunit to recognize different features of damaged DNA.

Hanawalt and Haynes (88) first postulated a model of DNA damage recognition termed the "close-fitting sleeve" model. They proposed that the need for DNA repair is determined by comparing the secondary structure of the DNA surrounding the lesion to that of the normal Watson–Crick double helix. Elaborating on this concept, Gunz et al. (89) showed that the efficiency of bulky lesion recognition can vary over several orders of magnitude and that there is a general correlation between the efficiency with which a particular DNA lesion is recognized by the human NER system and the amount of helical destabilization it causes.

It is, therefore, expected that at least some individual NER factors responsible for the initial damage recognition would show high affinity and specificity for the helical distortions that result from the presence of DNA lesions. Simple mismatches or bubbles are, however, not processed by the NER machinery, indicating that the local thermodynamic destabilization is not sufficient to qualify as a NER substrate (42,90). Similarly, the human NER complex remains inactive on DNA substrates in which only the backbone of the duplex had been modified, leaving the hydrogen bonding between complementary bases intact (90). However, the presence of covalent DNA modifications in conjunction with disruption of the canonical base pairing resulted in a robust response by the NER machinery (Fig. 3). These experiments performed with artificially manipulated DNA substrates indicate that the molecular signal leading to recognition of base adducts during GGR consists of two fundamentally distinct elements, i.e., disruption of Watson–Crick base pairing and altered chemistry of the damaged deoxyribonucleotide residue (90,91). Neither defective base pairing in duplex DNA nor defective chemistry of its deoxyribonucleotide components induces NER activity, but the combination of these two substrate alterations results in the assembly of active incision complexes. The term of "bipartite recognition" has therefore been proposed to indicate that human NER factors utilize two principal levels of discrimination by recognizing distinct characteristic features of DNA carrying a bulky adduct.

Further evidence supporting a bipartite model of substrate discrimination was sought by studying the excision of DNA adducts caused by benzo[a]pyrene (B[a]P) and the related fjord region benzo[c]phenanthrene (B[c]Ph) diol epoxides (92). A fjord region (+)-trans-anti-B[c]Ph-N^6-dA adduct, which maintains the normal

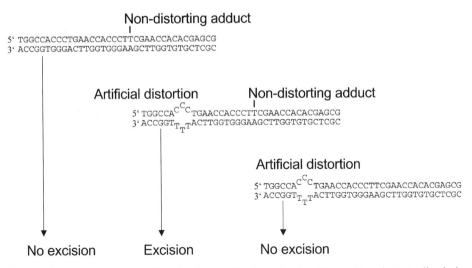

Figure 3 Experimental findings leading to the hypothesis of bipartite substrate discrimination in the human NER pathway.

Watson–Crick base pairing throughout the modified duplex, is not excised during incubation in human cell extracts when situated in a fully complementary duplex (Fig. 4, lane 4). Similarly, an artificial distortion consisting of three consecutive mismatches in the unmodified control duplex is not processed by the human NER system (Fig. 4, lane 5). However, the same (+)-*trans*-anti-B[*c*]Ph-N^6-dA adduct in combination with three mismatched bases stimulates excision activity in human cell extracts, resulting in oligonucleotide excision products that are characteristic of NER activity (Fig. 4, lane 6). Besides, in accord with the hypothesis of bipartite recognition is the more efficient excision of *cis*-anti-B[*a*]P-N^2-dG adducts compared with the stereoisomeric *trans*-anti-B[*a*]P-dG isomers, as the Watson–Crick alignment between base pairs is disrupted in the *cis* conformation but retained in the *trans* conformation (93). In fact, the *cis*-anti-B[*a*]P-N^2-dG adducts result in displacement of both the adducted guanine and the complementary cytosine from their normal position in the double helix. An intriguing observation is that excision of these bay region B[*a*]P-*N*^{2}-dG adducts (as monitored by the appearance of excision fragments 24–32 nucleotides long) was abolished when the whole dCMP residue, which is normally located across the lesion in the fully complementary duplex, was removed (94). The complete lack of excision from such deletion duplexes raises the possibility that the undamaged base in the complementary strand, once it is dislocated into a "flip-out" position by the presence of an adduct, may provide a molecular docking station for the loading of GGR subunits onto DNA. Apparently, the absence of this loading site in deletion duplexes generates defective substrates that escape processing by the NER system.

10. XPC–hHR23B AS A SENSOR OF DEFECTIVE BASE PAIRING

XPC–hHR23B consists of the 125-kDa *XPC* gene product associated with hHR23B, a 58-kDa homolog of the yeast NER protein RAD23 (95). Additionally, the XPC–hHR23B complex contains centrin 2, an 18-kDa centrosome component (96). The

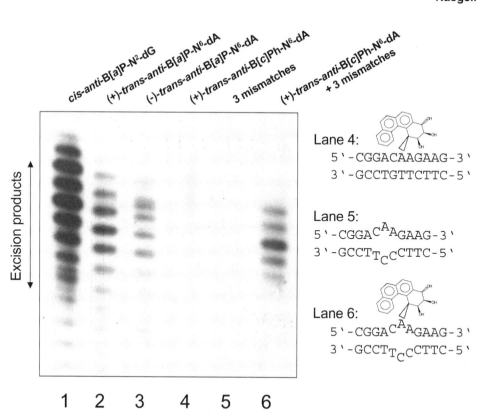

Figure 4 The hypothesis of bipartite substrate discrimination is confirmed by the differential excision by the human NER system of bay (B[*a*]P) and fjord region (B[*c*]Ph) polycyclic aromatic hydrocarbon adducts.

stability of the *XPC* gene product is entirely dependent on the presence of its binding cofactor hHR23B (97). In cell extracts, virtually all XPC is complexed with hHR23B, whereas a trace amount copurifies with the mammalian ortholog hHR23A (98). The domain participating in binding to hHR23B has been mapped within the evolutionarily conserved carboxy-terminal half of XPC (Fig. 5). The NER activity of purified XPC protein is stimulated in vitro by the addition of hHR23B, but the 56-amino acid XPC binding domain of hHR23B is sufficient to mediate this effect (99,100).

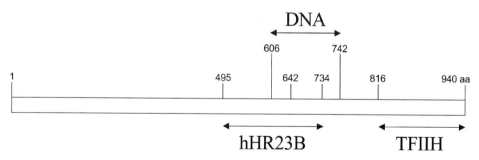

Figure 5 Domain structure of human XPC protein.

XPC-hHR23B binds preferentially to a number of damaged DNA substrates containing, for example, (6-4) photoproducts or AAF adducts (36,37,101). Recombinant XPC protein itself possesses this DNA-binding property, whereas hHR23B has no affinity for DNA. The structural determinants underlying the interaction with damaged DNA have been probed using a series of artificial DNA substrates and these experiments indicate that XPC is attracted to sites of damage by the helical distortion that is typically induced by bulky base adducts (42). It appears that XPC protein detects helix distortions regardless of the presence or absence of damaged bases and, therefore, it binds to small regions of mismatched or unpaired bases and also displays a significant affinity for single-stranded DNA (41). Additionally, it binds preferentially to junction regions at the transition between double-stranded and single-stranded DNA (102), which occur in the open intermediate during the NER reaction. Scanning force microscopy studies showed that binding of XPC to DNA induces a bending of the DNA substrate that may enhance the recruitment of other factors to the NER complex (103). Although these biochemical findings provide a general mechanism of XPC–hHR23B as the initiator of GGR, they do not adequately explain its involvement in the repair of cyclobutane pyrimidine dimers. As already mentioned earlier, XPC displays no detectable preference for DNA fragments containing cyclobutane dimers (7) despite its stringent requirement for the removal of all UV lesions.

By interaction with the XPB and p62 subunits, XPC is responsible for the recruitment of TFIIH to sites of helical distortion (104). Separate regions of the human XPC protein are involved in the interactions with DNA, hHR23B, and TFIIH (Fig. 5). In fact, the carboxy-terminal 125 amino acids are dispensable for both DNA and hHR23B binding while interactions with TFIIH are strongly reduced by truncation of this domain (105). Another recent study suggests that XPC–hHR23B is also able to interact with XPA during the transition from an initial recognition intermediate (involving XPC and TFIIH) and the ultimate incision complex (106). In reconstituted NER assays, XPC does not persist in the NER complex, as it is released from the DNA substrate with the arrival of XPG and XPA (30,107).

To summarize, the following functions can be assigned to XPC. First, XPC recognizes DNA distortions through interactions with bases that cannot form normal Watson–Crick hydrogen bonds in duplex DNA. Second, XPC–hHR23B associates with TFIIH and recruits this large DNA helicase complex to the sites of helical distortion.

11. TRANSCRIPTION FACTOR IIH AS A SENSOR OF DEFECTIVE DEOXYRIBONUCLEOTIDE CHEMISTRY

The entry of TFIIH increases the molecular complexity of the eukaryotic NER system, In fact, TFIIH is made up of ten distinct polypeptides: XPB, XPD, p62, p52, p44, p34, cdk7, cyclin H, MAT1, and TTDA/TFB5/GTF2H5 (108). The multifunctional TFIIH complex is able to shuttle rapidly between transcription initiation and the NER process (109), but competition studies suggest that TFIIH may exhibit a higher affinity for NER than for transcription (110). Indeed, this hierarchy could be important to ensure the immediate removal of RNA polymerase blocking lesions and the rapid resumption of RNA synthesis. In transcription, DNA unwinding by TFIIH allows the nascent mRNA molecules to progress from initiation to the elongation phase. In NER, local DNA unwinding by TFIIH produces double-stranded to single-stranded transitions at the edges of a central bubble of 20–25 nucleotides (21), thus providing an adequate substrate for incision by the two structure-specific endonucleases XPG and XPF–ERCC1 (Fig. 1).

Central to the local unwinding process at the site of the lesion are the two DNA helicases, XPB and XPD, which translocate on single-stranded DNA in the $3' \rightarrow 5'$ and $5' \rightarrow 3'$ polarity, respectively. More recent studies on the prokaryotic recombination factor RecBCD (111) suggests that such bipolar DNA helicase complexes may unwind the double helix by a co-ordinated mechanism whereby the two subunits translocate with opposite polarities, but in the same direction, on each strand of the antiparallel DNA duplex (Fig. 6). A strand-specific block of one of the two DNA helicases, as well as continued translocation of the partner enzyme along the opposing undamaged strand, is expected to produce a severe distortion of the double helix. Thus, the concomitant unwinding and bending of the DNA substrate through the action of TFIIH (Fig. 6) is likely to constitute an essential prerequisite for the continued recruitment of NER proteins to the growing incision complex. Because unwinding activity occurs at the site of damage, TFIIH may be the first factor that comes in direct contact with the offending lesion. Low resolution analyses have shown that TFIIH is organized in a ring-like structure with a 2.5–3.0 nm wide central hole, which is likely to surround the DNA in such a way that the helicase subunits would be located in close proximity to the damaged base (112,113). This hypothesis is supported by a site-directed cross-linking study, indicating that the XPD subunit of TFIIH (together with RPA) is situated near the lesion in the ultimate (excinuclease) complex (45). It, therefore, seems intuitive to make the short step to proposing that inhibition of the TFIIH-associated helicases may serve as a damage detector in the eukaryotic NER process (114–116). If TFIIH does perform such a recognition function in NER, then it may participate as a proofreading enzyme at the level of damage verification, as it would provide a possible mechanism to discriminate between simple DNA distortions and those sites that carry an actual damage. Such a proofreading activity is required not only in GGR, but also in TCR to confirm that

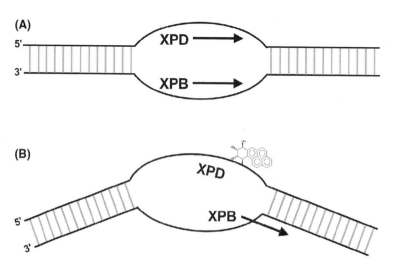

Figure 6 Model of substrate scanning by TFIIH. (A) Mode of DNA unwinding by helicase complexes composed of enzymatic subunits with opposite polarity. (B) The unilateral inhibition of one DNA helicase generates a site-specific DNA distortion.

transcription is arrested by the presence of a bulky lesion before recruiting the remaining NER factors.

To summarize, the following NER functions can be assigned to TFIIH. First, the ATP-dependent DNA unwinding activity leads to formation of an open intermediate that constitutes the target for structure-specific endonucleases. Second, the DNA helicase activity of TFIIH may have been adopted to scan conformationally distorted sites in the DNA duplex for the presence of bulky lesions. Third, TFIIH is directly responsible for the recruitment of XPG and XPA to the nascent "excinuclease" complex (discussed in the following).

12. ROLE OF XPA–RPA IN INTEGRATING DIFFERENT RECOGNITION SIGNALS

The critical requirement of XPA and its interaction partner RPA in both GGR and TCR placed this complex at the forefront of DNA damage recognition. Numerous studies have provided evidence that both XPA and RPA display a binding preference for damaged DNA (117–120; reviewed in Ref. 33). In addition, RPA stabilizes the association of XPA with DNA (121). Compared with XPC, however, the affinity and selectivity of the XPA–RPA complex for damaged duplexes are of lower orders of magnitude (see, for example, Ref. 122). Nevertheless, a DNA damage "verification" function for XPA–RPA has been proposed by several authors (29,36,38).

Xeroderma pigmentosum A is a 36-kDa zinc metalloprotein that can form homodimers (123) and also associates with many other core NER subunits (124–127). The N-terminal portion (residues 1–97) contains regions for binding to RPA34 and ERCC1. The C-terminal domain (residues 226–273) has been shown to bind to TFIIH. The central domain (residues 98–219) contains the zinc finger, is required for binding to RPA70, and has been identified as the minimal polypeptide necessary for binding to DNA (128). The NMR solution structure analysis of this fragment, comprising residues 98–219, revealed a positively charged cleft that exhibits the appropriate curvature and size to accommodate the DNA double helix (129). Further NMR studies conducted in the presence of either a DNA fragment or a short RPA peptide sequence led to the surprising conclusion that the zinc-finger domain of XPA (residues 105–129) is not involved in DNA binding but, instead, is required for the interaction with RPA. The domains of XPA mediating associations with other core NER factors are illustrated in Fig. 7.

Replication protein A represents the most abundant single-stranded DNA-binding factor in human cells (130). The interaction of RPA with single DNA filaments occurs through the 70-kDa subunit, and each RPA monomer occupies ~30 nucleotides, which corresponds roughly to the length of the gapped DNA intermediate generated in the NER process. In DNA replication, RPA is required for the initiation of DNA synthesis by the DNA polymerase α/primase, as well as for the elongation of nascent DNA (120,130). In the NER system, RPA is necessary together with XPA to assist TFIIH in the opening of the DNA double helix around sites of damage (38,44). Additionally, the XPA–RPA complex interacts with XPG on its 3′-oriented side and with XPF-ERCC1 on its 5′ facing side and, therefore, plays a crucial role in the positioning of these repair endonucleases for incision of the damaged strand (131–133). Finally, RPA is also an integral component of the DNA resynthesis machinery, as it remains associated with the DNA substrate after incision, whereas all the other core factors are progressively released. Accordingly,

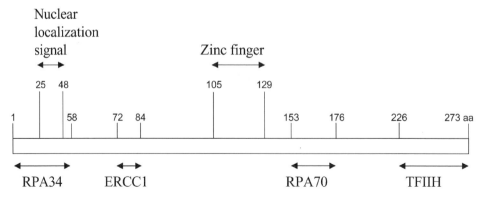

Figure 7 Domain structure of human XPA protein.

RPA may regulate the transition from dual DNA incision to DNA synthesis and "hand over" the gapped DNA intermediate to the DNA polymerase complex. Replication protein A promotes the recruitment of RFC (134) and PCNA (135) to initiate DNA synthesis, although an adjuvant role in recruiting PCNA has also been assigned to XPG (136,137).

Biochemical studies argue against a direct participation of XPA–RPA as a sensor of DNA damage. Both subunits display an affinity for distorted DNA structures carrying mismatches, loops, or bubbles, even if no actual DNA lesion has been introduced into the substrate (138). Xeroderma pigmentosum A has a strong preference for binding to artificially distorted DNA molecules that share the architectural feature of presenting two double strands emerging from a central bend. Thus, it is possible that the function of XPA is to recognize a DNA kink that is introduced as an obligatory intermediate in both GGR and TCR. Similarly, RPA binding is mediated by the appearance of single-stranded DNA, which is again a structural intermediate in the NER reaction. On the basis of these findings, Missura et al. (138) proposed that the XPA–RPA complex recognizes specific architectural features of DNA that are generated only when TFIIH encounters a bulky lesion. In this scenario, XPA and RPA are not damage recognition or "verification" subunits but rather act as regulatory components that control the correct three-dimensional arrangement of the NER complex before activating the nuclease components. This view is supported by Riedl et al. (30), who noted that XPA protein might act as a wedge to keep the DNA structure ready for the arrival of XPF–ERCC1. Xeroderma pigmentosum A also contributes to stabilization of the incision intermediate by inhibiting the strand-separation activity of RPA (138,139). The single-stranded DNA-binding properties of its interaction partner, RPA, are compatible with a role in protecting the undamaged strand from inadvertent nuclease attack (46,140). In DNA-binding experiments using purified factors and short DNA fragments, RPA interacts preferentially with the undamaged strand, while rejecting the damaged strand, and this strand-specific bias is further increased by the addition of XPA (141).

In summary, there is no evidence for a role of XPA and RPA in recognizing or "verifying" DNA damage. Instead, the characterization of these factors suggest an alternative hypothesis, whereby XPA–RPA first replaces the XPC–hHR23B heterodimer in the incision complex and then monitors the characteristic changes of DNA architecture (unwinding and kinking of the duplex) induced following recognition of damage by XPC and TFIIH. Thus, it may be concluded that the primary function of

XPA and RPA is to integrate the different recognition activities of XPC (the sensor of defective base pairing) and TFIIH (the sensor of defective DNA chemistry), such that the nucleases are selectively targeted to the lesion on the damaged strand.

13. DAMAGE-SPECIFIC RECRUITMENT OF XPG AND XPF–ERCC1

Because XPG (but not XPF) associates with foci of DNA damage in XPA cells (29), the 3′ endonuclease appears to be incorporated in the incision complex ahead of the 5′ endonuclease. Xeroderma pigmentosum G is recruited to the NER complex before the XPA subunit, presumably due to its strong interaction with TFIIH (142). In reconstituted systems, however, XPA, RPA, and XPG mutually stabilize their association within the NER complex, suggesting that these three factors bind to DNA in a synergistic manner (30). An interaction between RPA and XPG may be needed to change the spatial relation of XPG to the other subunits and place the 3′ endonuclease in close contact with the lesion.

Xeroderma pigmentosum G is a member of the FEN-1 family of structure-specific nucleases and shows activity against substrates containing double-stranded to single-stranded transitions with a 5′ single-stranded overhang (143). Similar to the complex abnormalities of some XP-B and XP-D patients, individuals in the XP-G group frequently have features in common with CS (144,145). Patients with large truncations in the XPG protein are frequently affected by XP/CS while missense mutations generally give rise to XP only. In agreement with these clinical findings, complete inactivation of the *XPG* gene in mice leads to severe developmental defects (146). This complexity could be explained by multiple functions of XPG protein in GGR, TCR, transcription and base excision repair of oxidative DNA damage (147). This view is supported by the finding that XPG protein binds most avidly to DNA containing bubbles, preferring those that resemble in size (i.e., 20 nucleotides) the region of helix opening in transcription (148).

XPF–ERCC1 is the last factor that binds to the NER incision complex in biochemically reconstituted systems (149). This structure-specific endonuclease has a polarity opposite to that of XPG and cleaves DNA 5′ of the lesion in the NER pathway (132,133,150). XPF–ERCC1 is also involved in recombinational repair processes where it is required to remove nonhomologous 3′ DNA tails projecting from heteroduplex intermediates (151,152). XPF- or ERCC1-deficient cells are characteristically hypersensitive to DNA-interstrand cross-linking agents, presumably because the XPF–ERCC1 complex has the ability to cut adjacent to such cross-links (153). The large XPF subunit of the heterodimer contains a conserved nuclease motif that catalyzes the incision (154). The function of the ERCC1 subunit is to stabilize XPF and provide a link to the NER machinery through its interaction with XPA (131,155).

14. REGULATION OF THE DAMAGE RECOGNITION PROCESS

The p53 tumor suppressor protein is stabilized, in response to a variety of genotoxic stimuli, to regulate downstream target genes involved in cell-cycle control and apoptosis. A direct molecular link between p53 function and NER activity was suggested by the observation that homozygous p53 mutant cells are deficient in GGR but proficient for TCR of bulky lesions (156,157). Although several different

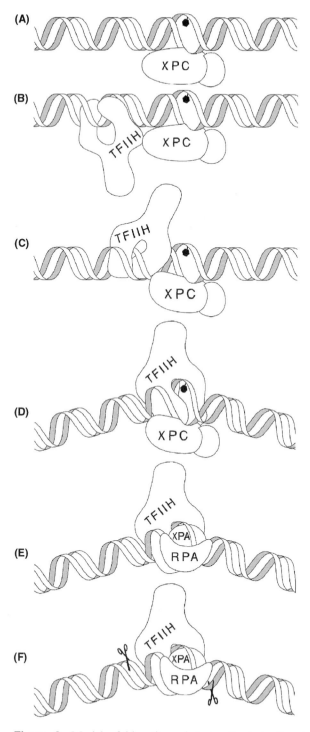

Figure 8 Model of bipartite substrate discrimination by a dynamic reaction cycle. (A) Recognition of disrupted base pairs by XPC protein. (B) Recruitment of TFIIH. (C) Local scanning of DNA through DNA helicase activity. (D) Inhibition of TFIIH translocation; formation of a kinked and locally unwound intermediate. (E) Recruitment of XPA and RPA. (F) Recruitment and activation of structure-specific endonucleases.

mechanisms have been proposed, the evidence is now accumulating that p53 protein is responsible for the concomitant induction of two early factors participating in the GGR pathway.

Recent studies showed that expression of XPC, the initiator of GGR, is induced by p53 protein after exposure to DNA-damaging agents (158). Analysis of the human *XPC* gene revealed a sequence element in the promoter region that mediates DNA binding by p53 protein, indicating that p53 acts as a crucial transcription factor not only in delaying cell cycle progression and triggering apoptosis, but also by increasing GGR efficiency. Upregulation by p53 has also been reported for the *DDB2* gene (60). Overexpression of the *DDB2* gene product, p48, has been shown to enhance GGR even in the background of p53 deficiency, demonstrating that wild-type p53 protein itself is not required for efficient repair (159,160). As in the case of XPC, the identification of a region in the human *DDB2* gene that binds and responds transcriptionally to p53 protein further confirmed the direct link between p53 function and GGR efficiency. In summary, these reports indicate that p53 protein controls the earliest step of GGR through the co-ordinated regulation of genes involved in the detection of bulky lesions. Interestingly, p53 appears to act synergistically with another tumor suppressor protein. In fact, overexpression of BRCA1 increases the expression of both XPC and p48, again resulting in higher GGR of UV photoproducts (161).

The intracellular level of XPC protein is additionally regulated by its interaction partner hHR23B. Knock-out mice revealed that the two mammalian orthologs hHR23A and hHR23B have a fully redundant function in NER. As a consequence, hHR23A is able to compensate for the role of hHR23B in NER, such that single mutants display normal repair activities (162). In the absence of both HR23 proteins, however, the *XPC* gene product on its own is intrinsically unstable. Both HR23 proteins contain a ubiquitin-like N-terminus and two ubiquitin-associated domains that provide a direct link to the 26S proteasome degradation machinery. Under normal conditions, hHR23B protects XPC from proteolysis and, consequently, results in increased steady-state levels of this recognition subunit. Interestingly, following treatment with genotoxic agents that induce bulky lesions, hHR23B further suppresses the proteasomal destruction of XPC (97). It appears that the interaction with hHR23B provides a flexible mechanism to adapt the level of XPC protein to the changing requirements after exposure to genotoxic insults.

It has become clear that NER activity in living cells is modulated by the nucleosomal organization of chromatin (163,164, reviewed in Ref. 165). The core of a nucleosome is composed of an octamer of histone proteins and 145 base pairs of DNA wrapped around this octamer. Different proteins, including DDB (discussed earlier) and the p53 tumor suppressor (157), have been discussed as possible factors that mediate the accessibility of DNA repair subunits to DNA in the chromatin context, but it is likely that extraregulatory systems are required to overcome the structural barriers that chromatin poses to the removal of DNA damage. Using synthetic oligonucleotides with a site-directed lesion, it is possible to introduce DNA damage at a specific position within reconstituted chromatin for in vitro repair studies. Such an approach demonstrated a severe reduction of UV photoproduct excision when the target lesion is located in the center of reconstituted nucleosome cores (166,167). Unexpectedly, strong inhibition of NER activity in physiologically spaced dinucleosome templates was also observed when the lesion was placed in the linker DNA (168). Chromatin remodeling complexes containing an ATPase subunit of the SWI2/SNF2 superfamily have been implicated as possible candidates for assisting

NER through transient disruption or movement of nucleosomes in vivo. These complexes have recently been shown to facilitate excision of bulky lesions situated within reconstituted nucleosomes (167,168). Interestingly, the TCR component CSB displays homology to the SWI/SNF family and is indeed able to promote ATP-dependent nucleosome remodeling in vitro (169). Another NER factor with homology to the SWI/SNF family is Radl6, which together with Rad7 is required for the GGR of nontranscribed sequences, including repressed loci, in yeast (170). This finding suggests that Rad7 and Radl6 may be subunits of a complex that promotes relaxation of the chromatin structure in order to facilitate NER of transcriptionally silent sequences. Interestingly, no mammalian homologs of these yeast *RAD7* and *RAD16* genes have been identified.

15. CONCLUSIONS

How the human NER machinery recognizes many kinds of bulky DNA base adducts, as well as discriminates between these lesions and undamaged DNA (including the undamaged strand directly opposite the adduct), poses an experimental challenge that has not yet been fully resolved. Figure 8 depicts a simplified model for DNA-damage recognition in the human GGR pathway. On the basis of the evidence that has been reviewed here, we propose that XPC–hHR23B is the initial sensor for disrupted base pairs in the GGR pathway, whereas TFIIH functions as a proofreading enzyme that verifies the presence of bulky lesions. This bipartite model for the assembly of the human NER system provides a mechanistic basis to explain both selectivity and versatility of the NER system.

Numerous structure-activity studies show that the efficiency of bulky lesion excision depends on the base-pairing properties in the immediate vicinity to the damaged nucleotide. Thus, DNA-damage recognition begins when XPC–hHR23B, the initiator of GGR, probes the thermodynamic stability of the double helix and discriminates between normal duplex DNA and DNA that has departed from the canonical Watson–Crick base pairing conformation (Fig. 8A). Xeroderma pigmentosum C then attracts TFIIH to the distorted site and loads the ring-like helicase domain of this large multifunctional factor onto the damaged strand (Fig. 8B). This step serves to probe the chemical composition of the target strand in order to identify the precise location of the adducted nucleotide (Fig. 8C). Damage recognition is completed when TFIIH encounters the bulky adduct and one of the two helicases becomes sequestered on the damaged strand, generating a kinked and unwound DNA structure (Fig. 8D). Transcription factor IIH intervenes in the reaction pathway in a way that nonspecific or erroneous intermediates can be aborted before generating spurious incision events. If the assembly occurs at undamaged sites, ATP hydrolysis by TFIIH leads to dissociation of nascent NER intermediates. In the presence of a bulky adduct, however, TFIIH is frozen in a stable nucleoprotein complex at the damaged site, and the kinked and unwound DNA constitutes a high-affinity binding substrate for XPA and RPA, respectively.

Although local unwinding by TFIIH generates the substrate for structure-specific endonucleases, incision is not carried out until the inclusion of XPA and RPA into the NER complex. Thus, the involvement of XPA and RPA may serve to ensure that the endonucleolytic scissions at the damage site occur in a precise and co-ordinated manner. A "licensing" concept has been introduced previously to describe the finding that a particular replication factor is used in higher eukaryotes to prevent the semiconservative DNA synthesis at inappropriate sites or during inappropriate stages of the cell cycle (171). A similar concept may be used for the role of the XPA–RPA complex during assembly of the incision machinery. We propose that RPA and XPA acquire an essential function in conveying the bipartite recognition of DNA damage to the nuclease subunits. Replication protein A monitors local unwinding in the NER intermediate, whereas XPA controls the degree of DNA bending, thereby double-checking the three-dimensional architecture of the incision complex. If this nucleoprotein intermediate is correctly assembled and the DNA strands are properly located in the complex (Fig. 8E), XPA and RPA bring the two structure-specific endonucleases in a position that leads to DNA incision (Fig. 8F). This regulatory function of XPA–RPA excludes the risk that DNA may be processed by endonucleases at an inappropriate (undamaged) site or at an improper (premature) step during the assembly of the incision complex.

ACKNOWLEDGMENT

Research in the authors' laboratory is supported by the Swiss National Science Foundation (grant 3100A0-101747).

REFERENCES

1. Setlow RB, Carrier WL. Proc Natl Acad Sci USA 1964; 51:226–231.
2. Boyce RP, Howard-Flanders P. Proc Natl Acad Sci USA 1964; 51:293–300.
3. Cleaver JE. Nature 1968; 218:652–656.
4. Setlow RB, Regan JD, German J, Carrier WL. Proc Natl Acad Sci USA 1969; 64: 1035–1041.
5. Kraemer KH, Lee M-M, Andrews AD, Lambert WC. Arch Dermatol 1994; 130: 1018–1021.
6. Ford JM, Hanawalt PC. J Biol Chem 1997; 272:28073–28080.
7. Reardon JT, Sancar A. Genes Dev 2003; 17:2539–2551.
8. Sancar A. Science 1996; 272:48–49.
9. Chigancas V, Miyaji EN, Muotri AR, de Fatima JJ, Amarante-Mendes GP, Yasui A, Menck CF. Cancer Res 2000; 60:2458–2463.
10. You YH, Lee DH, Yoon JH, Nakajiama S, Yasui A, Pfeifer GP. J Biol Chem 2001; 276:44688–44694.
11. Schul W, Jans J, Rijksen YM, Klemann KH, Eker AP, de Wit J, Nikaido O, Nakajima S, Yasui A, Hoeijmakers JH, van der Horst GT. EMBO J 2002; 21:4719–4729.
12. McWhir J, Selfridge J, Harrison DJ, Squires S, Melton DW. Nat Genet 1993; 5: 217–224.
13. de Vries A, van Oostrom CT, Hofhuis FM, Dortant PM, Berg RJ, de Gruijl FR, Wester PW, van Kreijl CF, Capel PJ, van Steeg H, Verbeek SJ. Nature 1995; 377:169–173.
14. Nakane H, Takeuchi S, Yuba S, Saijo M, Nakatsu Y, Murai H, Nakatsuru Y, Ishikawa T, Hirota S, Kitamura Y, Kato Y, Tsunoda Y, Miyauchi H, Horio T, Tokunaga T, Matsunaga T, Nikaido O, Nishimune Y, Okada Y, Tanaka K. Nature 1995; 377: 165–168.
15. Sands AT, Abuin A, Sanchez A, Conti CJ, Bradley A. Nature 1995; 377:162–165.
16. de Vries A, van Oostrom CT, Dortant PM, Beems RB, van Kreijl CF, Capel PJ, van Steeg H. Mol Carcinog 1997; 19:46–53.
17. Ide F, Iida N, Nakatsuru Y, Oda H, Tanaka K, Ishika T. Carcinogenesis 2000; 21: 1263–1265.

18. Sancar A. Annu Rev Biochem 1996; 65:43–81.
19. Wood RD. J Biol Chem 1997; 272:23465–23468.
20. Prakash S, Prakash L. Mutat Res 2000; 451:13–24.
21. Evans E, Fellows J, Coffer A, Wood RD. EMBO J 1997; 16:625–638.
22. Huang JC, Svoboda D, Reardon JT, Sancar A. Proc Natl Acad Sci USA 1992; 89: 3664–3668.
23. Lin JJ, Sancar A. Mol Microbiol 1992; 6:2219–2224.
24. Aboussekhra M, Biggerstaff M, Shivji MK, Vilpo JA, Moncollin V, Podust VN, Protic M, Hübscher U, Egly JM, Wood RD. Cell 1995; 80:859–868.
25. Mu D, Park CH, Matsunaga T, Hsu DS, Reardon JT, Sancar A. J Biol Chem 1995; 270:2415–2418.
26. Guzder SN, Habraken Y, Sung P, Prakash L, Prakash S. J Biol Chem 1995; 270: 12973–12976.
27. Araujo SJ, Tirode F, Coin F, Pospiech H, Syvaoja JE, Stucki M, Hiibscher U, Egly JM, Wood RD. Genes Dev 2000; 14:349–359.
28. Huang JC, Hsu DS, Kazantsev A, Sancar A. Proc Natl Acad Sci USA 1994; 91: 12213–12217.
29. Volker M, Mone MJ, Karmakar P, van Hoffen A, Schul W, Vermeulen W, Hoeijmakers JH, van Driel R, van Zeeland AA, Mullenders LH. Mol Cell 2001; 8:213–224.
30. Riedl T, Hanaoka F, Egly JM. EMBO J 2003; 22:5293–5303.
31. Araujo SJ, Wood RD. Mutat Res 1999; 435:23–33.
32. Shivji MK, Podust VN, Hübscher U, Wood RD. Biochemistry 1995; 34:5011–5017.
33. Thoma BS, Vasquez KM. Mol Carcinog 2003; 38:1–13.
34. Houtsmuller AB, Rademakers S, Nigg AL, Hoogstraten D, Hoeijmakers JH, Vermeulen W. Science 1999; 284:958–961.
35. Naegeli H. FASEB J 1995; 9:1043–1050.
36. Sugasawa K, Ng JM, Masutani C, Iwai S, van der Spek PJ, Eker AP, Hanaoka F, Bootsma D, Hoeijmakers JH. Mol Cell 1998; 2:223–232.
37. Batty D, V Rapic-Otrin, Levine AS, Wood RD. J Mol Biol 2000; 300:275–290.
38. Evans E, Moggs JG, Hwang JR, Egly J-M, Wood RD. EMBO J 1997; 16:6559–6573.
39. Wakasugi M, Sancar A. J Biol Chem 1999; 274:18759–18768.
40. Rademakers S, Volker M, Hoogstraten D, Nigg AL, Mone MJ, AA van Zeeland, Hoeijmakers JH, Houtsmuller AB, Vermeulen W. Mol Cell Biol 2003; 23: 5755–5767.
41. Reardon JT, Mu D, Sancar A. J Biol Chem 1996; 271:19451–19456.
42. Sugasawa K, Okamoto T, Shimizu Y, Masutani C, Iwai S, Hanaoka F. Genes Dev 2001; 15:507–521.
43. Fitch ME, Nakajima S, Yasui S, Ford J. J Biol Chem 2003; 278:46906–46910.
44. Mu D, Wakasugi M, Hsu DS, Sancar A. J Biol Chem 1997; 272:28971–28979.
45. Reardon JT, Sancar A. Mol Cell Biol 2002; 22:5938–5945.
46. de Laat WL, Appeldoorn E, Sugasawa K, Weterings E, Jaspers NG, Hoeijmakers JH. Genes Dev 1998; 12:2598–2609.
47. Svejstrup JQ, Wang ZG, Feaver WJ, Wu XH, Bushnell DA, Donahue TF, Friedberg EC, Kornberg RD. Cell 1995; 80:21–28.
48. Rodriguez K, Talamantez J, Huang W, Reed SH, Wang Z, Chen L, Feaver WJ, Friedberg EC, Tomkinson AE. J Biol Chem 1998; 273:34180–34189.
49. Guzder SN, Sung P, Prakash L, Prakash S. J Biol Chem 1996; 271:8903–8910.
50. Dualan R, Brody T, Keeney S, Nichols AF, Admon A, Linn S. Genomics 1995; 29: 62–69.
51. Nichols AF, Ong P, Linn S. J Biol Chem 1996; 271:24317–24320.
52. Itoh T, Linn S, Ono T, Ymaizumi M. J Invest Dermatol 2000; 114:1022–1029.
53. Rapic-Otrinv, Navazza V, Nardo T, Botta E, McLenigan MP, Bisi DC, Levine AS, Stefanini M. Hum Mol Genet 2003; 12:1507–1522.
54. Hwang BJ, Chu G. Biochemistry 1993; 32:1657–1666.

55. Reardon JT, Nichols AF, Keeney S, Smith CA, Taylor JS, Linn S, Sancar A. J Biol Chem 1993; 268:21301–21308.
56. Payne A, Chu G. Mutat Res 1994; 310:89–102.
57. Feldberg RS, Grossman L. Biochemistry 1976; 15:2402–2408.
58. Fujiwara Y, Masutani C, Mizukoshi T, Kondo J, Hanaoka F, Iwai S. J Biol Chem 1999; 274:20027–20033.
59. Hwang BJ, Ford JM, Hanawalt PC, Chu G. Proc Natl Acad Sci USA 1999; 96: 424–428.
60. Hwang BJ, Toering S, Francke U, Chu G. Mol Cell Biol 1998; 18:4391–4399.
61. Tang JY, Hwang BJ, Ford JM, Hanawalt PC, Chu G. Mol Cell 2000; 5:737–744.
62. Keeney S, Eker AP, Brody T, Vermeulen W, Bootsma D, Hoeijmakers JH, Linn S. Proc Natl Acad Sci USA 1994; 91:4053–4056.
63. Wakasugi M, Kawashima A, Morioka H, Linn S, Sancar A, Mori T, Nikaido O, Matsunaga T. J Biol Chem 2002; 277:1637–1640.
64. Brand M, Moggs JG, Oulad-Abdelghani M, Lejeune F, Dilworth FJ, Stevenin J, Almouzni G, Tora L. EMBO J 2001; 20:3187–3196.
65. Datta A, Bagchi S, Nag A, Shiyanov P, Adami GR, Yoon T, Raychaudhuri P. Mutat Res 2001; 486:89–97.
66. Martinez E, Palhan VB, Tjernberg A, Lymar ES, Gamper AM, Kundu TK, Chait BT, Roeder RG. Mol Cell Biol 2001; 21:6782–6795.
67. Otrin V, McLenigan M, Takao M, Levine A, Protic M. J Cell Sci 1997; 110:1159–1168.
68. Chen X, Zhang Y, Douglas L, Zhou P. J Biol Chem 2001; 22:48175–48182.
69. Nag A, Bondar T, Shiv S, Raychaudhuri P. Mol Cell Biol 2001; 21:6738–6747.
70. Groisman R, Polanowska J, Kuraoka I, Sawada J, Saijo M, Drapkin R, Kisselev AF, Tanaka K, Ynakatani K. Cell 2003; 113:357–367.
71. Rapic-Otrin V, McLenigan MP, Bisi DC, Gonzalez M, Levine AS. Nucleic Acids Res 2002; 30:2588–2598.
72. Fitch ME, Cross IV, Turner SJ, Adimoolam S, Lin CX, Williams KG, Ford JM. DNA Repair 2003; 2:819–826.
73. Bohr VA, Smith CA, Okumoto DS, Hanawalt PC. Cell 1985; 40:359–369.
74. I Mellon, Spivak G, Hanawalt PC. Cell 1987; 51:241–249.
75. Venema J, van Hoffen A, Karcagi V, Natarajan AT, van Zeeland AA, Mullenders LH. Mol Cell Biol 1991; 11:4128–4134.
76. Proietti De Santis L, Lorenti Garcia C, Balajee AS, Latini P, Pichierri P, Nikaido O, Stefanini M, Palitti F. DNA Repair 2002; 1:209–223.
77. Tornaletti S, Hanawalt PC. Biochimie 1999; 81:139–146.
78. Christians FC, Hanawalt PC. Biochemistry 1993; 32:10512–10518.
79. Dammann R, Pfeifer GP. Mol Cell Biol 1997; 17:219–229.
80. Conconi A, Bespalov VA, Smerdon MJ. Proc Natl Acad USA 2002; 99:649–654.
81. Meier A, M Livingstone-Zatchej, Thoma F. J Biol Chem 2002; 277:11845–11852.
82. Mellon I, Fajpal DK, Koi M. Science 1996; 272:557–560.
83. Le Page F, Kwoh EE, Avrutskaya A, Gentil A, Leadon SA, Sarasin A, PK Cooper PK. Cell 2000; 101:159–171.
84. Spivak G, Itoh T, Matsunaga T, Nikaido O, Hanawalt P, Yamaizumi M. DNA Repair 2002; 1:629–643.
85. Klungland A, Hoss M, Gunz D, Constantinou A, Clarkson SG, Doetsch PW, Bolton PH, Wood RD, Lindahl T. Mol Cell 1999; 3:33–42.
86. Maldonado E, Shiekhattar R, Sheldon M, Cho H, Drapkin R, Rickert P, Lees E, Anderson CW, Linn S, Reinberg D. Nature 1996; 381:86–89.
87. Zawel L, Kumar KP, Reinberg D. Genes Dev 1995; 9:1479–1490.
88. Hanawalt PC, Haynes RH. Biochem Biophys Res Commun 1965; 19:462–467.
89. Gunz D, Hess MT, Naegeli H. J Biol Chem 1996; 271:25089–25098.
90. Hess MT, Schwitter U, Petretta M, Giese B, Naegeli H. Proc Natl Acad Sci USA 1997; 94:6664–6669.

91. Buschta-Hedayat N, Buterin T, Hess MT, Missura M, H Naegeli H. Proc Natl Acad Sci USA 1999; 96:6090–6095.

92. Buterin T, Hess MT, Luneva N, Geacintov NE, Amin S, Kroth H, Seidel A, Naegeli H. Cancer Res 2000; 60:1849–1856.

93. Hess MT, Gunz D, Luneva N, Geacintov NE, Naegeli H. Mol Cell Biol 1997; 17: 7069–7076.

94. Buterin T, Hess MT, Gunz D, Geacintov NE, Mullenders LH, Naegeli H. Cancer Res 2002; 62:4229–4235.

95. Masutani C, Sugasawa K, Yanagisawa J, Sonoyama T, Ui M, Enomoto T, Takio K, Tanaka K, van der Spek PJ, Bootsma D, Hoeijmakers JH, Hanaoka F. EMBO J 1994; 13:1831–1843.

96. Araki M, Masutani C, Takemura M, Uchida A, Sugasawa K, Kondoh J, Ohkuma Y, Hanaoka F. J Biol Chem 2001; 276:18665–18672.

97. Ng JM, Vermeulen W, van der Horst GT, Bergink S, Sugasawa K, Vrieling H, Hoeijmakers JH. Genes Dev 2003; 17:1630–1645.

98. van der Spek PJ, Eker A, Rademakers S, Visser C, Sugasawa K, Masutani C, Hanaoka F, Bootsma D, Hoeijmakers JH. Nucleic Acids Res 1996; 24:2551–2559.

99. Sugasawa K, Masutani C, Uchida A, Maekawa T, van der Spek PJ, Bootsma D, Hoeijmakers JH, Hanaoka F. Mol Cell Biol 1996; 16:4852–4861.

100. Masutani C, Araki M, Sugasawa K, van der Spek PJ, Yamada A, Uchida A, Maekawa T, Bootsma D, Hoeijmakers JH, Hanaoka F. Mol Cell Biol 1997; 17:6915–6923.

101. Kusumoto R, Masutani C, Sugasawa K, Iwai S, Araki M, Uchida A, Mizukoshi T, Hanaoka F. Mutat Res 2001; 485:219–227.

102. Sugasawa K, Shimizu Y, Iwai S, Hanaoka F. DNA Repair 2002; 1:95–107.

103. Janicijevic A, Sugasawa K, Shimizu Y, Hanaoka F, Wijgers N, Djurica M, Hoeijmakers JH, Wyman C. DNA Repair 2003; 2:325–336.

104. Yokoi M, Masutani C, Maekawa T, Sugasawa K, Ohkuma Y, Hanaoka F. J Biol Chem 2000; 275:9870–9875.

105. Uchida A, Sugasawa K, Masutani C, Dohmae N, Araki M, Yokoi M, Ohkuma Y, Hanaoka F. DNA Repair 2002; 1:449–461.

106. You JS, Wang M, Lee SH. J Biol Chem 2003; 278:7476–7485.

107. Wakasugi M, Sancar A. Proc Natl Acad Sci USA 1998; 95:6669–6674.

108. Drapkin R, Reinberg D. Trends Biochem Sci 1994; 19:504–508.

109. Hoogstraten D, Nigg AL, Heath H, Mullenders LH, van Driel R, Hoeijmakers JH, Vermeulen W, Houtsmuller AB. Mol Cell 2002; 10:1163–1174.

110. You Z, Feaver WJ, Friedberg EC. Mol Cell Biol 1998; 18:2668–2676.

111. Dillingham MS, Spies M, Kowalczykowski SC. Nature 2003; 423:893–897.

112. Schultz P, Fribourg S, Poterszman A, Mallouh V, Moras D, Egly JM. Cell 2000; 102:599–607.

113. Chang W-H, Kornberg RD. Cell 2000; 102:609–613.

114. Naegeli H, Bardwell L, Friedberg EC. J Biol Chem 1992; 267:392–398.

115. Wood RD. Biochimie 1999; 81:39–44.

116. Villani G, LeGac NT. J Biol Chem 2000; 275:33185–33188.

117. Robins P, Jones CJ, Biggerstaff M, Lindahl T, Wood RD. EMBO J 1991; 10: 3913–3921.

118. Clugston CK, McLaughlin K, Kenny MK, Brown R. Cancer Res 1992; 52: 6375–6379.

119. Burns JL, Guzder SN, Sung P, Prakash S, Prakash L. J Biol Chem 1996; 271: 11607–11610.

120. Lao Y, Lee CG, Wold MS. Biochemistry 1999; 38:3974–3984.

121. He Z, Henricksen LA, Wold MS, Ingles CJ. Nature 1995; 374:566–569.

122. Lao Y, Gomes XV, Ren Y, Taylor JS, Wold MS. Biochemistry 2000; 39:850–859.

123. Yang-Z G, Liu Y, Mao LY, Zhang J-T, Zou Y. Biochemistry 2002; 41:13012–13020.

124. Park C-H, Sancar A. Proc Natl Acad Sci USA 1994; 91:5017–5021.

125. Li L, Elledge SJ, Peterson CA, Bates ES, Legerski RJ. Proc Natl Acad Sci USA 1994; 91:5012–5016.

126. Li L, Peterson CA, Lu X, Legerskil RJ. Mol Cell Biol 1995; 15:1993–1998.

127. Li L, Lu X, Peterson CA, Legerski RJ. Mol Cell Biol 1995; 15:5396–5402.

128. Kuraoka I, Morita EH, Saijo M, Matsuda T, Morikawa K, Shirakawa M, Tanaka K. Mutat Res 1996; 362:87–95.

129. Ikegami T, Kuraoka I, Saijo M, Kodo N, Kyogoku Y, Morikawa K, Tanaka K, Shirakawa M. Nat Struct Biol 1998; 5:701–706.

130. Wold MS. Annu Rev Biochem 1997; 66:61–62.

131. Matsunaga T, Park C-H, Bessho T, Mu D, Sancar A. J Biol Chem 1996; 271:11047–11050.

132. Bessho T, Sancar A, Thompson LH, Thelen MP. J Biol Chem 1997; 272:3833–3837.

133. de Laat WL, Appeldorn E, Jaspers NG, Hoeijmakers JH. J Biol Chem 1998; 273:7835–7842.

134. Yuzhakov A, Kelman Z, Hurwitz J, O'Donnell M. EMBO J 1999; 18:6189–6199.

135. Gomes XV, Burgers PM. J Biol Chem 2001; 276:34768–34775.

136. Miura M, Nakamura S, Sasaki T, Takasaki Y, Shiomi T, Yamaizumi M. Exp Cell Res 1996; 226:126–132.

137. Gary R, Ludwig DL, Cornelius HL, MacInnes MA, Park MS. J Biol Chem 1997; 272:24522–24529.

138. Missura M, Buterin T, Hindges R, Hübscher U, Kaspárková J, Brabec V, Naegeli H. EMBO J 2001; 20:3554–3564.

139. Patrick SM, Turchi JJ. J Biol Chem 2002; 277:16096–16101.

140. Hermanson-Miller IL, Turchi JJ. Biochemistry 2002; 41:2402–2408.

141. Lee J-H, Park C-J, Arunkumar AI, Chazin WJ, Choi BS. Nucleic Acids Res 2003; 31:4747–4754.

142. Araujo SJ, EANigg, Wood RD. Mol Cell Biol 2001; 21:2281–2291.

143. O'Donovan A, Davies AA, Moggs JG, West SC, Wood RD. Nature 1994; 371:432–435.

144. Nouspikel T, Lalle P, Leadon SA, Cooper PK, Clarkson SG. Proc Natl Acad Sci USA 1997; 94:3116–3121.

145. Emmert S, Slor H, Busch DB, Batko S, Albert RB, Coleman D, Khan SG, B Abu-Libdeh, DiGiovanna JJ, Cunningham BB, Lee MM, Crollick J, Inui H, Ueda T, Hedayati M, Grossman L, Shahlavi T, Cleaver JE, Kraemer KH. J Invest Dermatol 2002; 118:972–982.

146. Harada YN, Shiomi N, Koike M, Ikawa M, Okabe M, Hirota S, Kitamura Y, Kitagawa M, Matsunaga T, Nikaido O, Shiomi T. Mol Cell Biol 1999; 19:2366–2372.

147. Lee SK, Yu SL, Prakash L, Prakash S. Cell 2002; 109:823–834.

148. Spivak G, Lloyd RS, Sweder KS. DNA Repair 2003; 2:235–242.

149. Mu D, Hsu DS, Sancar A. J Biol Chem 1996; 271:8285–8294.

150. Sijbers AM, de Laat WL, Ariza RR, Biggerstaff M, Wei YF, Moggs JG, Carter KC, Shell BK, Evans E, MC de Jong, Rademakers S, de Rooij J, Jaspers NG, Hoeijmakers JH, Wood RD. Cell 1996; 86:811–822.

151. Adair GM, Rolig RL, Moore-Faver D, Zabelshansky M, Wilson JH, Nairn RS. EMBO J 2000; 19:3771–3778.

152. Sargent RG, Meservy JL, Perkins BD, Kilburn AE, Intody Z, Adair GM, Nairn RS, Wilson JH. Nucleic Acids Res 2000; 28:3771–3778.

153. Kuraoka I, Kobertz WR, Ariza RR, Biggerstaff M, Essigmann JM, Wood RD. J Biol Chem 2000; 275:26632–26636.

154. Enzlin JH, Schärer OD. EMBO J 2002; 21:2045–2053.

155. Wakasugi M, Reardon JT, Sancar A. J Biol Chem 1997; 272:16030–16034.

156. Ford JM, Hanawalt PC. Proc Natl Acad Sci USA 1995; 92:8876–8880.

157. Wang XW, Yeh H, Schaeffer L, Roy R, Moncollin V, Egly JM, Wang Z, Friedberg EC, Evans MK, Taffe BG, Bohr VA, Weeda G, Hoeijmakers JH, Forrester K, Harris CC. Nat Genet 1995; 10:188–195.
158. Adimoolam S, Ford JM. Proc Natl Acad Sci USA 2002; 99:12985–12990.
159. Tan T, Chu G. Mol Cell Biol 2002; 22:3247–3254.
160. Fitch ME, Cross IV, Ford JM. Carcinogenesis 2003; 24:843–850.
161. Hartman AR, Ford JM. Nat Genet 2003; 32:180–184.
162. Ng JM, Vrieling H, Sugasawa K, Ooms MP, Grootegoed JA, Vreeburg JT, Visser P, Beems RB, Gorgels TG, Hanaoka F, Hoeijmakers JH, van der Horst GT. Mol Cell Biol 2002; 22:1233–1245.
163. Wellinger RE, Thoma F. EMBO J 1997; 16:5046–5056.
164. Liu X, Smerdon MJ. J Biol Chem 2000; 275:23729–23735.
165. Thoma F. EMBO J 1999; 18:6585–6598.
166. Wang ZG, Wu XH, Friedberg EC. J Biol Chem 1991; 266:22472–22478.
167. Hara R, Sancar A. Mol Cell Biol 2001; 22:6779–6787.
168. Ura K, Araki M, Saeki H, Masutani C, Ito T, Iwai S, Mizukoshi T, Kaneda Y, Hanaoka F. EMBO J 2001; 20:2004–2014.
169. Citterio E, Van Den Boom V, Schnitzler G, Kanaar R, Bonte E, Kingston RE, Hoeijmakers JH, Vermeulen W. Mol Cell Biol 2000; 20:7643–7653.
170. Verhage R, Zeeman A, de Groot N, Gleig F, Bang D, van der Putte P, Brouwer J. Mol Cell Biol 1994; 14:6135–6142.
171. Blow JJ. J Cell Biol 1993; 122:993–1002.

8
Interactions of the Transcription Machinery with DNA Damage in Prokaryotes

Isabel Mellon

Department of Pathology and Laboratory Medicine, Markey Cancer Center, Graduate Center for Toxicology, University of Kentucky, Lexington, Kentucky, U.S.A.

1. GENERAL OVERVIEW

DNA damage can influence interactions between the transcription machinery and the DNA template (reviewed in Refs. 1 and 2) This has been investigated in phage, prokaryotic, and eukaryotic systems. Many studies have focused on how DNA damage can influence the ability of RNA polymerase to either bypass a specific lesion or become arrested at the damaged site (1). The behavior of RNA polymerase elongation complexes at damaged sites is generally thought to play an important role in mechanisms underlying the DNA repair pathway termed transcription-coupled repair (TCR) (2–4). In this chapter, the influence of different types of DNA damage on RNA polymerase progression and how this may impact mechanisms of TCR are described with an emphasis on studies performed in prokaryotic systems.

1.1. Transcription

Transcription is the copying of the template strand of a gene into a complementary RNA and is the initial step in gene expression. The synthesis of the nascent transcript is catalyzed by RNA polymerase. Transcription is generally divided into several steps. Initiation includes activation, promoter binding, and RNA chain initiation. Elongation can include translocation, pausing, editing of the transcript by endonucleolytic cleavage near the 3′ end and resynthesis. Termination occurs when the transcript and RNA polymerase complex are released. Each step can be subject to regulation. While in theory, DNA damage has the potential to impact any stage of transcription: initiation, elongation, or termination, this chapter will focus on how different types of DNA lesions can impact transcription elongation and termination.

The "core" RNA polymerase elongation complex in *E. coli* is comprised of the $\alpha_2\beta\beta'$ complex that is formed after the release of sigma and additional initiation accessory proteins. Once in the elongation mode, it is highly stable and processive (5). When elongation complexes are artificially stalled in studies performed in vitro by omission of NTPs, they remain bound to the DNA templates and transcripts for extremely long periods of time. Footprint analyses indicate that RNA polymerase complex protects ~30 bp of DNA from nuclease digestion that includes the melted transcription bubble (12–14 bps in length) and duplex DNA upstream (4-5 bps) and downstream (8 bp) of the bubble (6–16). There is evidence that it contains protein binding sites that bind and maintain the melted transcription bubble, that bind the RNA transcript, and that bind the incoming nucleotide triphosphate. There has been controversy regarding the presence and size of an RNA:DNA hybrid within the transcription bubble but many studies indicate that during translocation there is a transient formation of an RNA:DNA hybrid of 9–12 bps in length (7,8,12,15). Structural studies of elongation complexes isolated from either *E. coli* (6,13,14,16) or *Thermus aquaticus* (17) indicate that they are shaped like a partially open crab claw with an internal groove that runs between the claws. In addition to the internal groove, there is a secondary channel where it is thought that the NTPs enter into the active site and a tunnel where the nascent RNA is thought to exit. One arm of the claw is primarily the β subunit while the other arm is primarily the β' subunit. The DNA within the elongation complex is kinked between the downstream double-stranded DNA and the transcription bubble, with a ~90° bend angle. The architecture and the specific interactions between RNA polymerase and the transcription bubble, the transcript and DNA downstream and upstream of the bubble are thought to be the basis of the extreme stability of the elongation complex.

RNA polymerase can also move backward or "backtrack" along the template (5,9–11,16,18). This can occur at a pause site or when there has been a misincorporation event at the 3' terminus of the nascent RNA. Some models propose that forward elongation and backtracking are static processes. As the polymerase moves forward, the downstream DNA is unwound and the transcription bubble moves forward as the newly synthesized RNA base pairs with the template strand. As this occurs, the 5' end of the transcript within the transcription bubble is displaced from the template to become free in solution and the upstream DNA reanneals to form double-stranded DNA. Thus, during RNA synthesis and translocation, the size of the transcription bubble and the RNA:DNA hybrid remains constant. Backtracking can be viewed as a reversal of this process. As the polymerase moves backwards, the upstream DNA unwinds and the 5' region of the transcript reanneals with the template strand. The 3' terminus of the transcript becomes displaced from the template strand and the DNA reanneals. As in forward elongation, the size of the transcription bubble and the RNA:DNA hybrid remains largely unchanged when the polymerase backtracks. In addition, there is evidence that RNA polymerase can edit the transcript by spontaneous endonucleolytic cleavage of a variable length of the 3' end (19) or cleavage can occur with the participation of the accessory factors GreA and GreB (10). RNA synthesis and elongation can be resumed from the newly formed 3' terminus.

Others have proposed that elongation and backtracking are not static processes and that RNA polymerase complexes can adopt multiple conformational states (reviewed in Ref. 18). There is evidence that RNA polymerase can exist in at least two states: one in which the rate of elongation occurs rapidly and the other in which elongation occurs slowly. In addition, an allosteric (noncatalytic) binding site that is specific for the templated NTP has been discovered. The elongation rate of RNA polymerase switches from the slow rate to the rapid rate when the templated

NTP binds to the allosteric site. During the slow rate of elongation, the RNA polymerase complex is proposed to be in a more "open" state while during rapid elongation it may be more tightly clamped onto the DNA. In addition, structural and modeling studies (16) have indicated that there is flexibility in the RNA polymerase "jaws." Crosslinking experiments indicate that the DNA downstream of the complex is more exposed when RNA polymerase is in the "open" conformation than when it is the "closed" or "clamped down" conformation (16). In addition to elongation and backtracking, there is evidence that RNA polymerase can be in a hypertranslocated state and skip forward along the template (18).

Transcription termination generally occurs when the RNA polymerase complex encounters a termination sequence at the end of a gene or operon (5,20,21). There are two classes of termination sequences. Rho-dependent terminators require the participation of additional proteins including the hexameric RNA/DNA helicase known as Rho to bring about the release of the transcript and the RNA polymerase complex (22). Rho binds to a region of the nascent transcript exiting the RNA polymerase complex and displaces it from the template DNA strand and the polymerase complex. In contrast, intrinsic terminators do not require the participation of additional proteins to bring about transcription termination (21). Intrinsic termination is mediated by the formation of hairpin structures within the nascent transcript that destabilize interactions between the RNA and the RNA polymerase complex and the dissociation of a weak rU/dA enriched RNA/DNA hybrid. Termination at both classes of terminators can be regulated by other proteins. In addition, termination can occur prematurely within a gene. Rho can promote premature termination by binding regions of the nascent transcript that can become exposed when the RNA is not properly translated.

1.2. Transcription-Coupled Repair

Transcription-coupled repair is generally measured as more rapid or more efficient removal of certain types of DNA damage from the transcribed strands of expressed genes compared with the nontranscribed strands (23–25). Many types of DNA damage including cyclobutane pyrimidine dimers (CPDs) formed by exposure to UV light pose as blocks to transcription elongation. The CPDs in the transcribed strands of expressed genes pose blocks to RNA polymerase elongation, while those in the nontranscribed strand are generally bypassed (26). It is generally believed that the blockage of the RNA polymerase complex at the damaged site is an early event that initiates TCR. In general, RNA synthesis is globally reduced when cells are exposed to UV light. One of the earliest indications of the existence of TCR of UV damage was the key observation that when mammalian cells are exposed to UV light, RNA synthesis resumes before any significant amount of UV-induced damage is removed from the bulk of the genome (27). This key early observation is likely explained by the subsequent observations of more rapid removal of UV-induced damage from active genes (28,29) and from the transcribed strands of expressed genes (23) compared with the nontranscribed strands and unexpressed regions of the genomes of mammalian cells. The selective repair of damage from the transcribed strands of expressed genes, which only comprises a small percentage of the mammalian genome, allows transcription to resume in the absence of significant repair in the bulk of the DNA. This strand-selective repair that was first described for mammalian cells (23) was then subsequently documented in *E. coli* (24) and in yeast (25).

The TCR has been clearly demonstrated to be a subpathway of nucleotide excision repair (NER) in *E. coli*, yeast, and mammalian cells. Substrates include CPDs and (6–4) photoproducts produced by UV light and certain bulky lesions induced by chemical agents (30,31). Hence, this subpathway of NER has been conserved from bacteria to humans and operates on many different lesions. In addition, there is evidence that the repair of oxidative damage can be coupled to transcription (32,33). Since many lesions formed by oxidative damage are substrates for base excision repair (BER) pathways, it has been suggested that BER pathways can also be coupled to transcription. However, there has been no direct genetic or biochemical demonstration of a role of BER in TCR. In addition, this area has been confounded by recent retractions of several papers in this area (34–36).

The TCR in *E. coli* was first alluded to by studies of mutation frequency decline (MFD) in the glutamine tRNA gene (37) before it was directly measured in the lactose operon (23) and the *tyrT* tRNA gene (38). The frequency of mutations generated by UV damage in the transcribed strand of certain tRNA operons was found to decline more rapidly than the frequency of mutations generated by damage in the nontranscribed strand (37). In addition, a gene was identified (39), *mfd*, and mutations in *mfd* were later found to abolish or reduce TCR (40,41). In addition, biochemical studies demonstrated that the Mfd protein promotes the release of RNA polymerase complexes stalled at lesions or artificially stalled by ribonucleotide depletion (40). It is generally thought that the stalled polymerase complex or perhaps the transcription bubble is an important signal for TCR. Hence the observation that Mfd displaces lesion-blocked RNA polymerases from the template provides a conundrum, in that the presumed signal for TCR should be lost when the polymerase complex becomes displaced from the lesion. Attempts to reconcile this have proposed that Mfd also recruits repair proteins to the damaged site as it displaces the RNA polymerase complex (40). However, more recent studies (described in sect. 2) have provided additional insights into possible mechanisms.

2. THE BEHAVIOR OF RNA POLYMERASE COMPLEXES WITH DIFFERENT TYPES OF DNA DAMAGE

A number of studies have examined the influence of different lesions on transcription elongation by introducing them into specific sites in either the transcribed or nontranscribed strand of an expressed gene (reviewed in Ref. 1). Studies have been carried out with phage, *E. coli* and mammalian RNA polymerases. A summary of results obtained studying phage and *E. coli* RNA polymerase is presented (Table 1). Several "bulky" adducts have been found to pose strong blocks to elongation by *E. coli* RNA polymerase. These include CPDs (40), or psoralen monoadducts and interstrand crosslinks produced by psoralen diadducts (42). In contrast, the nonbulky or smaller, less distorting base modifications, 8-oxoguanine and O^6-methylguanine, are efficiently bypassed by phage and *E. coli* RNA polymerase (43). In addition, uracil is efficiently bypassed when it is present in template DNA (43–45). Some base modifications exhibit more of an intermediate effect on elongation. Abasic sites and dihydrouracil can cause RNA polymerase to pause but it then is able to bypass the site (43,45,46). Similar results have been obtained studying T7 RNA polymerase (Table 1). A psoralen monoadduct or psoralen interstrand crosslink are potent blocks to elongation (47). The bulky lesion, guanine C-8 amino-fluorene, exhibits an intermediate level of bypass by T7 RNA polymerase (48). The

Table 1 Efficiency of Elongation Blockage by RNA Polymerases at Sites of DNA Damage

DNA damage type	Relative efficiency of blockage	References
E. coli		
Psoralen monoadduct	High	(42)
Psoralen interstrand crosslink	High	(42)
CPD	High	(40)
Abasic site	Low	(44,57)
8-Oxoguanine	Low	(43)
Dihydrouracil	Low	(46)
O^6-methylG	Low	(43)
Uracil	Low	(43,44,57)
1 bp gap	Low	(52)
single strand breaks	High	(44)
T7 Phage		
Psoralen monoadduct	High	(47)
Psoralen interstrand crosslink	High	(47)
AAF-guanine	Intermediate	(48)
AF-guanine	Low	(48)
BPDE- guanine	High	(49)
BPDE-adenine (-)	Intermediate	(50)
BPDE-adenine (+)	High	(50)
Abasic site	Low	(44,58)
8-Oxoguanine	Low	(48)
Dihydrouracil	Low	(57)
Gaps	Low	(52)

N-acetyl-2-aminofluorene modified guanine is a more efficient block to elongation by T7 RNA polymerase but it is not a complete block (48). Interesting observations have also been obtained comparing the influence of different isomers of certain chemical adducts on elongation. The (+)-*trans* form of N7-benzo[a]pyrene diol epoxide (BPDE)-adducted guanine is an efficient block to the progression of T7 RNA polymerase (49). The (−) isomer of BPDE-adducted guanine also inhibits elongation but to a lesser extent than the (+) isomer (49). Interestingly, the (+) isomer of BPDE-adducted adenine is more efficiently bypassed by T7 RNA polymerase than the (+) isomer of guanine (50). In general, when it has been studied, base modifications that block or reduce RNA polymerase elongation do so only when they are present in the transcribed strand of a gene. When the same lesions are placed in the nontranscribed strand of the same gene they do not have a significant or measurable impact on RNA polymerase elongation.

The influence of strand breaks and gaps on phage and *E. coli* RNA polymerase elongation has also been examined (44,51–53). Quite surprisingly, T7 RNA polymerase is able to bypass gaps of between 1 and 24 nucleotides (52) and SP6 RNA polymerase can bypass gaps of between 1 and 19 nucleotides (53). The efficiency of bypass decreases as the size of the gap increases. *E. coli* RNA polymerase can bypass a 1 nucleotide gap but with greatly reduced efficiency compared with phage RNA polymerases (53). In contrast, strand breaks formed by cleavage of abasic sites are efficient blocks to phage and *E. coli* RNA polymerase elongation (44). This includes a break with a normal 3′ hydroxyl and a 5′ phosphoryl group at the abasic site, a break with a 3′ unsaturated aldehyde and a 5′ phosphoryl group, and a break with a 3′ phosphoryl group and a 5′ phosphoryl group at the abasic site. These types of strand breaks appear to be complete blocks to RNA polymerase elongation.

It is not currently clear why some lesions are potent blocks to elongation while others are either readily bypassed or produce a transient pausing of the RNA polymerase complex. Certain lesions may alter the structure of the template such that it may not be accommodated within the internal groove of the RNA polymerase complex that associates with the transcription bubble and upstream and downstream regions of duplex DNA. Alternatively certain lesions may not be accommodated in the active site or may interfere with the secondary channel of the RNA polymerase complex through which the incoming nucleotide travels to reach the active site. Molecular modeling of the behavior of human RNA polymerase II at sites of different isomers of benzo[c]phenanthrene diol epoxide adducts proposes a model whereby bypass of these lesions is achieved when the orientation and conformation of the adduct in the active site permit nucleotide incorporation while arrest occurs when the orientation and conformation of the adduct blocks the nucleotide entry channel (54). It is also not clear what conformational changes to the RNA polymerase complex occur when it becomes either blocked or paused at a lesion. It is possible that blockage or pausing may result in a conformational change in which the tightly clamped down "jaws" of the polymerase complex become more open. It is also unclear whether the blockage of E. coli RNA polymerase complex by different types of lesions or breaks formed in vivo results in prolonged arrest, dissociation and/or backtracking.

Transcription bypass of lesions or abnormal bases can have important implications for mutagenesis (1,55,56). Bypass of abasic sites generally results in the incorporation of adenine opposite the abasic site (44,57). Uracil is incorporated opposite O^6-methylguanine (43). Either adenine or cytosine can be incorporated opposite 8-oxoguanine (44,48) and either adenine or guanine can be incorporated opposite dihydrouracil (46,58). It is clear that incorporation of the incorrect base results in mutagenic events in vitro and in vivo and can result in what is termed "transcription mutagenesis" (55,56). There has been a wealth of information on determinants that contribute to miscoding events for DNA polymerases. However, much less is known for RNA polymerases and it is unclear if the mechanisms defined for DNA polymerases are applicable to RNA polymerases.

3. THE BEHAVIOR OF RNA POLYMERASE COMPLEXES AT LESIONS AND NER

There are two subpathways of NER. One is TCR. The other is generally referred to as global genome repair (GGR) which represents repair of the nontranscribed strands of expressed genes and repair of unexpressed regions of the genome. Many of the same proteins are required for TCR and GGR (2,4). However, the two pathways likely differ at the damage recognition step. For TCR, damage recognition is initiated by the stalling of RNA polymerase complexes at lesions in the transcribed strands of expressed genes. For GGR, damage recognition is initiated by other proteins (59).

3.1. NER in *E. coli*

NER removes an assortment of different types of DNA damage. It removes chemical adducts introduced by exposure to chemical carcinogens and CPDs and (6–4) photo-products produced by UV light. Given that this pathway removes many different types of lesions that can be structurally dissimilar, it is likely that it recognizes the distortion of the DNA helix rather than the lesion itself. There are several steps in the process (60): damage recognition and lesion verification, unwinding of the DNA at the lesion, two incisions, one on each side of the lesion, removal (excision) of a stretch of DNA containing the lesion, DNA synthesis to replace the excised DNA, and ligation of the newly synthesized DNA to the parental strand. While the repertoire of proteins that orchestrate the process differ, the general strategy has been conserved in *E. coli*, yeast, and mammalian systems.

NER in *E. coli* is understood in detail and has served as a paradigm for the investigation of other organisms (61). Damage recognition and processing is carried out by the UvrABC system. UvrA dimerizes and binds UvrB to form $UvrA_2B$ complex and the $UvrA_2B$ complex binds DNA. The helicase activity of the complex may enable scanning for damage by translocating along the DNA and unwinding the DNA at the site of the lesion. It has been generally believed that damage recognition is achieved by UvrA since UvrA alone has specificity for damaged substrates (62). In the $UvrA_2B$ complex, it is thought that UvrA recognizes the damage and brings or loads UvrB. It is also thought that one of the functions of UvrB is in lesion verification (63,64). A region of UvrB is inserted into the DNA helix to verify that the distortion represents bona fide DNA damage and to determine which strand contains the damage. $UvrA_2$ then dissociates leaving an unwound preincision complex containing UvrB that is recognized and bound by UvrC. UvrBC produces an incision on each side of the lesion: the first incision is made at the 4th or 5th phophodiester bond 3′ to the lesion and the second incision is made at the 8th phosphodiester bond 5′ to the lesion. The catalytic sites for the 3′ and 5′ incisions are located in separate regions of the N-terminal and C-terminal regions of UvrC (65). UvrD unwinds and displaces the damaged oligonucleotide produced by the incisions. The resulting gap is filled in by DNA polymerase I and the repair patch is sealed by DNA ligase. While it is generally held that UvrA is responsible for damage recognition, more recent studies have found that UvrB can recognize damage in the absence of UvrA when the damage is present in a bubble substrate (63,66) or when it is close to the end of a double-stranded fragment. The ability of UvrB to recognize damage in the absence of UvrA may have implications for TCR.

3.2. TCR and NER

Genetic and biochemical studies have interpreted that both TCR and GGR require UvrA, B, C, and D (40,41). TCR also requires Mfd and active transcription (24). Initially Sancar and Selby (40) found that Mfd promotes the release of RNA polymerases stalled at lesions in the transcribed strand of an expressed gene in a cell-free system. However, more recent novel observations by Roberts and colleagues (67) that Mfd has the ability to reverse "backtracked" RNA polymerase complexes may be more relevant to the mechanism of TCR (67). In addition, they found that Mfd binds to DNA upstream of the RNA polymerase complex and to the β subunit of the RNA polymerase complex. While it has not been demonstrated for *E. coli* RNA polymerase, mammalian RNA polymerase II complexes arrested at CPDs can backtrack for up to 25 nucleotides with the assistance of the transcription elongation factor, SII (26). A model is proposed for TCR of UV damage (Fig. 1) that incorporates backtracking, the reversal of backtracking and loading of NER factors onto the transcription bubble.

The model is as follows. After UV-irradiation, RNA polymerase complex elongates until it encounters a CPD on the transcribed strand. The polymerase then translocates backwards. This may be an intrinsic property of the RNA polymerase complex or it may be assisted by the stimulatory factors GreA and GreB. Mfd recognizes the backtracked complex and binds to DNA upstream of the RNA polymerase complex and to the RNA polymerase complex (67). Mfd then induces forward translocation of the polymerase until it re-encounters and perhaps

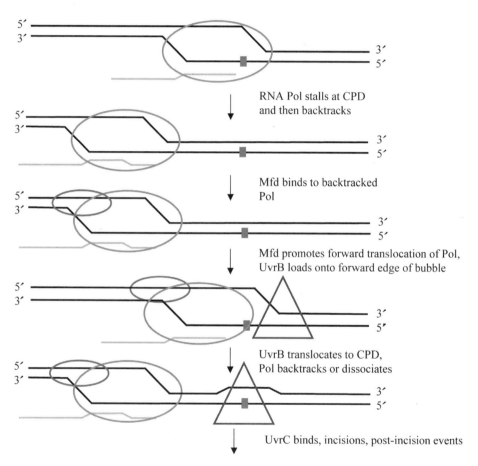

Figure 1 A model for transcription-coupled repair in *E. coli*. Elongating RNA polymerase complex (RNA pol; large oval) stalls at damage (small red square) in the transcribed strand. The polymerase complex, transcription bubble and nascent RNA (blue line) translocate backwards. The Mfd protein (green oval) binds backtracked polymerase and DNA upstream of the bubble (after Park et al. (67)). Mfd promotes the forward translocation of the polymerase complex (after Park et al. (67)). UvrB protein (blue triangle) binds 5′ (relative to the damaged strand), loads onto the forward edge of the bubble, and translocates to the lesion. The polymerase complex backtracks or is dissociated by Mfd. Subsequent NER processing events continue as they would in nontranscribed DNA.

even by-passes the lesion for a short distance (perhaps achieving a hypertranslocated conformation). When in this mode, the "jaws" of the RNA polymerase complex may be in a more open state that opens up the accessibility of downstream DNA. UvrA$_2$B or perhaps UvrB alone loads 5′ to the lesion (relative to the damaged strand). The loading of UvrB is facilitated by features of the transcription bubble brought about by the forward translocation induced by MFD. At this point the polymerase may backtrack again or may be completely released by Mfd. UvrC then binds to the lesion-bound UvrB complex resulting in a stable preincision complex and this and subsequent downstream NER events continue as they would in nontranscribed DNA. The salient point of this model is that the "coupling" of NER to transcription is mediated by the correct positioning of the transcription bubble at the lesion rather than by direct physical interactions between NER proteins and transcription factors. During NER in nontranscribed DNA, the loading of UvrB is an asymmetric process that initiates on the 5′ side of the damage and involves denaturing the DNA near the site of the lesion (68). In a transcribed DNA, RNA polymerase denatures DNA at the lesion when it forms a bubble around it. The damage containing bubble substrate may be recognized by UvrA$_2$B or UvrB alone when it encounters the stalled polymerase complex from the 5′ side (relative to the damaged strand). Mfd may serve two functions. One is to maintain the transcription bubble at the site of the lesion by reversing backtracked complexes. The other may be to ultimately displace the complex from the damaged site to allow incision and DNA synthesis.

Genetic studies have suggested a requirement of UvrA in TCR based on the observation that TCR is abolished in *uvrA* mutants (41). However, given that UvrA and UvrB form partners in vivo, it is unclear if UvrB is stable in *uvrA* mutants. Hence, the absence of TCR in *uvrA* mutants may be a consequence of the absence or instability of UvrB when UvrA is absent. In addition, while biochemical studies have also suggested a requirement of UvrA in TCR (40), these assays may be more representative of the detection of shielding of lesions from GGR by stalled RNA polymerase complexes rather than the detection of bona fide TCR (69). Future studies are necessary to test these and other models.

4. THE BEHAVIOR OF RNA POLYMERASE COMPLEXES AT LESIONS AND BER

BER represents a collection of repair pathways that operate on a variety of different lesions induced by oxidative damage, alkylation damage, and other types of damage (70). The broad substrate specificity is accomplished by a large number of different damage-specific glycosylases. Hence, this differs from NER where the broad substrate specificity is accomplished by assembling a multiprotein complex.

4.1. Alkylation Damage

Alkylating agents represent a broad class of DNA damaging agents that are present in the environment, are used as chemotherapeutic agents and can be formed endogenously during cellular metabolism (70). *N*-methylpurines (NMPs) are the most abundant lesions produced by simple alkylating agents such as methyl methanesulfonate and dimethyl sulfate. 7-Methylguaine and 3-methyladenine are the most abundant NMPs formed by these agents (71). The NMPs are removed by BER in *E. coli*, yeast, and mammalian cells and repair is initiated by specific glycosylases (72,73). The removal of NMPs has been compared in the transcribed and nontranscribed strands of the DHFR gene in mammalian cells (74), the GAL1 gene in *S. cerevisiae* (75), and the lactose operon of *E. coli* (76). No significant difference was found in the repair of the transcribed and nontranscribed strands of these genes. Hence, TCR does not appear to be a subpathway of methylation damage-specific BER.

4.2. Oxidative Damage

Oxidative damage is formed as a consequence of exposure to ionizing radiation and a variety of chemical agents and as by-products of normal cellular metabolism. These agents introduce a large number of modifications to DNA including alterations of bases, the deoxyribose sugar and cleavage of the phosphodiester backbone (70). Several studies have found more rapid removal of oxidative damage from the transcribed strands of expressed genes in yeast and mammalian cells (32,33,77,78). However, several of these papers have been retracted or "corrected" (34–36,79). The retractions have focused on the validity of the immunological-based assays that were used to measure TCR of oxidative damage. It is important to note that there are reports of TCR of oxidative damage that are independent of the retracted data based on the immunological assays. Le Page et al. (33) have investigated the repair of 8-oxoguanine in human cells using site-specific lesions in an episomal system and have found TCR of 8-oxoguanine. In addition, recent evidence for TCR of 8-oxoguanine in *E. coli* has been obtained by Doetsch and colleagues (56) studying transcriptional mutagenesis.

It has been suggested that TCR of oxidative damage reflects a coupling of BER to transcription. An alternative is that components of NER are directly involved in TCR of oxidative damage. As in the model for TCR of UV damage (Fig. 1), the coupling of repair of oxidative damage to transcription is mediated by the correct positioning of the transcription bubble at the lesion. After the introduction of oxidative damage, RNA polymerase elongates until it either encounters an oxidative lesion or a repair intermediate formed by the processing of the oxidative lesion which then blocks RNA polymerase progression. The polymerase then translocates backwards. Mfd recognizes the backtracked complex and binds to DNA upstream of the RNA polymerase complex. Mfd then induces forward translocation of the polymerase until it re-encounters and perhaps by-passes the lesion for a short distance. $UvrA_2B$ or perhaps UvrB alone loads 5' to the lesion (relative to the damaged strand). The loading of UvrB is facilitated by features of the transcription bubble brought about by forward translocation induced by Mfd. At this point the polymerase may backtrack again or be completely released by Mfd. UvrC then binds and makes a 5' incision (relative to the oxidative lesion) or both 3' and 5' incisions and dissociates leaving a gap that is filled in by repair synthesis.

Doetsch and Viswanathan (43) have found that 8-oxoguanine is not a block to transcription elongation by *E. coli* RNA polymerase. However, as described above (Sect. 2 and Table 1) they have found that a substrate that reflects a repair intermediate formed by the action of the lyase activity of the glycosylase, endo III, is an efficient block (44). Hence the contribution of BER to TCR may be to generate the substrates that arrest the polymerase complexes as has been suggested by Doetsch and colleagues (56). This could explain the absence of TCR of alkylation damage (74–76). The glycosylases that mediate repair of alkylation damage are simple glycosylases that produce abasic sites (72) and abasic sites do not block elongation of

E. coli RNA polymerase at least when examined in a cell-free system (44). Future studies are needed to test these and other models.

5. SUMMARY AND FUTURE DIRECTIONS

A wealth of knowledge has been obtained about interactions between the transcription machinery and DNA damage in both prokaryotes and eukaryotes. The strategy of using site-specific lesions in studies performed in vitro to study stalling or bypass of the transcription complex has yielded information that has important implications for mutagenesis and for mechanisms underlying TCR. However, a great deal remains to be learned. Structural studies are needed to define in detail how lesions are accommodated within an RNA polymerase complex and how this influences arrest, pausing, or bypass. What conformational changes occur to RNA polymerase when it encounters different types of DNA damage? Do RNA polymerases backtrack at DNA damage roadblocks? If so, can backtracking occur spontaneously or does it require accessory proteins? What is the mechanism by which Mfd or other proteins bring about termination and RNA polymerase release at DNA damage roadblocks? Based on what is already known, it is unlikely that Mfd employs the same mechanism as the transcription termination protein Rho. Much remains to be learned about the specific mechanisms that "couple" transcription and DNA repair.

Lastly, an interesting question from a teleologic viewpoint relates to why cells possess mechanisms that couple DNA repair and transcription. One reason may be that TCR serves to repair transcription-blocking lesions and hence, it facilitates a rapid recovery of transcription. However, transcription complexes are extremely stable when they are stalled at endogenous pause sites or at sites of damage. In the absence of a mechanism to specifically find transcription-blocking lesions, lesions would be shielded from the repair machinery by the RNA polymerase complex and hence, refractory to repair (69). Furthermore, stable arrested complexes would inhibit gene expression and perhaps interfere with or block the DNA replication machinery. Recent studies have found that eukaryotic RNA pol II complexes are degraded in response to DNA damage (80,81). Svejstrup has suggested that degradation of damage-stalled pol II complexes might be an alternative to TCR (2). Hence, the importance of removing stalled RNA polymerase complexes may be indicated by the development of specific repair mechanisms that remove transcription-blocking damage and if TCR fails to occur, then the RNA polymerase complex stalled at the damaged site may be actually degraded. It will be important to determine whether degradation of RNA polymerases arrested at damage sites occurs in *E. coli*.

REFERENCES

1. Doetsch PW. Translesion synthesis by RNA polymerases: occurrence and biological implications for transcriptional mutagenesis. Mutat Res 2002; 510:131–140.
2. Svejstrup JQ. Mechanisms of transcription-coupled DNA repair. Nat Rev Mol Cell Biol 2002; 3:21–29.
3. Scicchitano DA, Mellon I. Transcription and DNA damage: a link to a kink. Environ Health Perspect 1997; 105(suppl 1):145–153.

4. Mellon I. Transcription-coupled repair. Encyclopedia of the Human Genome. Nature Publishing Group, 2003.

5. von Hippel PH. An integrated model of the transcription complex in elongation, termination, and editing. Science 1998; 281:660–665.

6. Darst SA, Kubalek EW, Kornberg RD. Three-dimensional structure of *Escherichia coli* RNA polymerase holoenzyme determined by electron crystallography. Nature 1989; 340:730–732.

7. Nudler E, Avetissova E, Markovtsov V, Goldfarb A. Transcription processivity: protein-DNA interactions holding together the elongation complex. Science 1996; 273:211–217.

8. Nudler E, Mustaev A, Lukhtanov E, Goldfarb A. The RNA-DNA hybrid maintains the register of transcription by preventing backtracking of RNA polymerase. Cell 1997; 89:33–41.

9. Komissarova N, Kashlev M. RNA polymerase switches between inactivated and activated states By translocating back and forth along the DNA and the RNA. J Biol Chem 1997; 272:15329–15338.

10. Komissarova N, Kashlev M. Transcriptional arrest: *Escherichia coli* RNA polymerase translocates backward, leaving the 3' end of the RNA intact and extruded. Proc Natl Acad Sci USA 1997; 94:1755–1760.

11. Korzheva N, Mustaev A, Nudler E, Nikiforov V, Goldfarb A. Mechanistic model of the elongation complex of *Escherichia coli* RNA polymerase. Cold Spring Harb Symp Quant Biol 1998; 63:337–345.

12. Nudler E, Gusarov I, Avetissova E, Kozlov M, Goldfarb A. Spatial organization of transcription elongation complex in *Escherichia coli*. Science 1998; 281:424–428.

13. Darst SA, Polyakov A, Richter C, Zhang G. Insights into *Escherichia coli* RNA polymerase structure from a combination of x-ray and electron crystallography. J Struct Biol 1998; 124:115–122.

14. Darst SA, Polyakov A, Richter C, Zhang G. Structural studies of *Escherichia coli* RNA polymerase. Cold Spring Harb Symp Quant Biol 1998; 63:269–276.

15. Nudler E. Transcription elongation: structural basis and mechanisms. J Mol Biol 1999; 288:1–12.

16. Korzheva N, Mustaev A, Kozlov M, Malhotra A, Nikiforov V, Goldfarb A, Darst SA. A structural model of transcription elongation. Science 2000; 289:619–625.

17. Zhang G, Campbell EA, Minakhin L, Richter C, Severinov K, Darst SA. Crystal structure of Thermus aquaticus core RNA polymerase at 3.3 A resolution. Cell 1999; 98:811–824.

18. Erie DA. The many conformational states of RNA polymerase elongation complexes and their roles in the regulation of transcription. Biochem Biophys Acta 2002; 1577:224–239.

19. Surratt CK, Milan SC, Chamberlin MJ. Spontaneous cleavage of RNA in ternary complexes of *Escherichia coli* RNA polymerase and its significance for the mechanism of transcription. Proc Natl Acad Sci USA 1997; 88:7983–7987.

20. Uptain SM, Kane CM, Chamberlin MJ. Basic mechanisms of transcript elongation and its regulation. Annu Rev Biochem 1997; 66:117–172.

21. Yarnell WS, Roberts JW. Mechanism of intrinsic transcription termination and antitermination. Science 1999; 284:611–615.

22. Richardson JP. Loading Rho to terminate transcription. Cell 2003; 114:157–159.

23. Mellon I, Spivak G, Hanawalt PC. Selective removal of transcription-blocking DNA damage from the transcribed strand of the mammalian DHFR gene. Cell 1987; 51:241–249.

24. Mellon I, Hanawalt PC. Induction of the *Escherichia coli* lactose operon selectively increases repair of its transcribed DNA strand. Nature 1989; 342:95–98.

25. Sweder KS, Hanawalt PC. Preferential repair of cyclobutane pyrimidine dimers in the transcribed strand of a gene in yeast chromosomes and plasmids is dependent on transcription. Proc Natl Acad Sci USA 1992; 89:10696–10700.

26. Tornaletti S, Reines D, Hanawalt PC. Structural characterization of RNA polymerase II complexes arrested by a cyclobutane pyrimidine dimer in the transcribed strand of template DNA. J Biol Chem 1999; 274:24124–24130.

27. Mayne LV, Lehmann AR. Failure of RNA synthesis to recover after UV irradiation: an early defect in cells from individuals with Cockayne's syndrome and xeroderma pigmentosum. Cancer Res 1982; 42:1473–1478.

28. Bohr VA, Smith CA, Okumoto DS, Hanawalt PC. DNA repair in an active gene: removal of pyrimidine dimers from the DHFR gene of CHO cells is much more efficient than in the genome overall. Cell 1985; 40:359–369.

29. Mellon I, Bohr VA, Smith CA, Hanawalt PC. Preferential DNA repair of an active gene in human cells. Proc Natl Acad Sci USA 1986; 83:8878–8882.

30. Chen RH, Maher VM, Brouwer J, van de Putte P, McCormick JJ. Preferential repair and strand-specific repair of benzo[a]pyrene diol epoxide adducts in the HPRT gene of diploid human fibroblasts. Proc Natl Acad Sci USA 1992; 89:5413–5417.

31. van Hoffen A, Venema J, Meschini R, van Zeeland AA, Mullenders LH. Transcription-coupled repair removes both cyclobutane pyrimidine dimers and 6-4 photoproducts with equal efficiency and in a sequential way from transcribed DNA in xeroderma pigmentosum group C fibroblasts. EMBO J 1995; 14:360–367.

32. Cooper PK, Nouspikel T, Clarkson SG, Leadon SA. Defective transcription-coupled repair of oxidative base damage in Cockayne syndrome patients from XP group G. Science 1997; 275:990–993.

33. Le Page F, Kwoh EE, Avrutskaya A, Gentil A, Leadon SA, Sarasin A, Cooper PK. Transcription-coupled repair of 8-oxoguanine: requirement for XPG, TFIIH, and CSB and implications for Cockayne syndrome. Cell 2000; 101:159–171.

34. Leadon SA. Retraction. DNA Repair (Amst) 2003; 2:361.

35. Leadon SA, Avrutskaya AV. Retraction: Differential involvement of the human mismatch repair proteins, hMLH1 and hMSH2, in transcription-coupled repair. Cancer Res 2003; 63:3846.

36. Gowen LC, Avrutskaya AV, Latour AM, Koller BH, Leadon SA. Retraction. Science 2003; 300:1657.

37. Engstrom J, Larsen S, Rogers S, Bockrath R. UV-mutagenesis at a cloned target sequence: converted suppressor mutation is insensitive to mutation frequency decline regardless of the gene orientation. Mutat Res 1984; 132:143–152.

38. Li S, Waters R. Induction and repair of cyclobutane pyrimidine dimers in the *Escherichia coli* tRNA gene tyrT: Fis protein affects dimer induction in the control region and suppresses preferential repair in the coding region of the transcribed strand, except in a short region near the transcription start site. J Mol Biol 1997; 271:31–46.

39. Witkin EM. Mutation frequency decline revisited. Bioessays 1994; 16:437–444.

40. Selby CP, Sancar A. Molecular mechanism of transcription-repair coupling. Science 1993; 260:53–58.

41. Mellon I, Champe GN. Products of DNA mismatch repair genes mutS and mutL are required for transcription-coupled nucleotide-excision repair of the lactose operon in *Escherichia coli*. Proc Natl Acad Sci USA 1996; 93:1292–1297.

42. Shi YB, Gamper H, Hearst JE. The effects of covalent additions of a psoralen on transcription by *E. coli* RNA polymerase. Nucleic Acids Res 1987; 15:6843–6854.

43. Viswanathan A, Doetsch PW. Effects of nonbulky DNA base damages on *Escherichia coli* RNA polymerase-mediated elongation and promoter clearance. J Biol Chem 1998; 273:21276–21281.

44. Zhou W, Doetsch PW. Effects of abasic sites and DNA single-strand breaks on prokaryotic RNA polymerases. Proc Natl Acad Sci USA 1993; 90:6601–6605.

45. Viswanathan A, Liu J, Doetsch PW. *E. coli* RNA polymerase bypass of DNA base damage. Mutagenesis at the level of transcription. Ann N Y Acad Sci 1999; 870:386–388.

46. Liu J, Doetsch PW. *Escherichia coli* RNA and DNA polymerase bypass of dihydrouracil: mutagenic potential via transcription and replication. Nucleic Acids Res 1998; 26: 1707–1712.
47. Shi YB, Gamper H, Hearst JE. Interaction of T7 RNA polymerase with DNA in an elongation complex arrested at a specific psoralen adduct site. J Biol Chem 1988; 263: 527–534.
48. Chen YH, Bogenhagen DF. Effects of DNA lesions on transcription elongation by T7 RNA polymerase. J Biol Chem 1993; 268:5849–5855.
49. Choi DJ, Marino-Alessandri DJ, Geacintov NE, Scicchitano DA. Site-specific benzo[a]pyrene diol epoxide-DNA adducts inhibit transcription elongation by bacteriophage T7 RNA polymerase. Biochemistry 1994; 33:780–787.
50. Remington KM, Bennett SE, Harris CM, Harris TM, Bebenek K. Highly mutagenic bypass synthesis by T7 RNA polymerase of site-specific benzo[a]pyrene diol epoxide-adducted template DNA. J Biol Chem 1998; 273:13170–13176.
51. Zhou W, Doetsch PW. Transcription bypass or blockage at single-strand breaks on the DNA template strand: effect of different 3′ and 5′ flanking groups on the T7 RNA polymerase elongation complex. Biochemistry 1994; 33:14926–14934.
52. Zhou W, Reines D, Doetsch PW. T7 RNA polymerase bypass of large gaps on the template strand reveals a critical role of the nontemplate strand in elongation. Cell 1995; 82:577–585.
53. Liu J, Doetsch PW. Template strand gap bypass is a general property of prokaryotic RNA polymerases: implications for elongation mechanisms. Biochemistry 1996; 35: 14999–15008.
54. Schinecker TM, Perlow RA, Broyde S, Geacintov NE, Scicchitano DA. Human RNA polymerase II is partially blocked by DNA adducts derived from tumorigenic benzo[c]phenanthrene diol epoxides: relating biological consequences to conformational preferences. Nucleic Acids Res 2003; 31:6004–6015.
55. Viswanathan A, You HJ, Doetsch PW. Phenotypic change caused by transcriptional bypass of uracil in nondividing cells. Science 1999; 284:159–162.
56. Bregeon D, Doddridge ZA, You HJ, Weiss B, Doetsch PW. Transcriptional mutagenesis induced by uracil and 8-oxoguanine in *Escherichia coli*. Mol Cell 2003; 12:959–970.
57. Zhou W, Doetsch PW. Efficient bypass and base misinsertions at abasic sites by prokaryotic RNA polymerases. Ann NY Acad Sci 1994; 726:351–354.
58. Liu J, Zhou W, Doetsch PW. RNA polymerase bypass at sites of dihydrouracil: implications for transcriptional mutagenesis. Mol Cell Biol 1995; 15:6729–6735.
59. Sugasawa K, Ng JM, Masutani C, Iwai S, van der Spek PJ, Eker AP, Hanaoka F, Bootsma D, Hoeijmakers JH. Xeroderma pigmentosum group C protein complex is the initiator of global genome nucleotide excision repair. Mol Cell 1998; 2:223–232.
60. Batty DP, Wood RD. Damage recognition in nucleotide excision repair of DNA. Gene 2000; 241:193–204.
61. Petit C, Sancar A. Nucleotide excision repair: from *E. coli* to man. Biochimie 1999; 81:15–25.
62. Mazur SJ, Grossman L. Dimerization of *Escherichia coli* UvrA and its binding to undamaged and ultraviolet light damaged DNA. Biochemistry 1991; 30:4432–4443.
63. Zou Y, Van Houten B. Strand opening by the UvrA(2)B complex allows dynamic recognition of DNA damage. EMBO J 1999; 18:4889–4901.
64. Theis K, Skorvaga M, Machius M, Nakagawa N, Van Houten B, Kisker C. The nucleotide excision repair protein UvrB, a helicase-like enzyme with a catch. Mutat Res 2000; 460:277–300.
65. Verhoeven EE, van Kesteren M, Moolenaar GF, Visse R, Goosen N. Catalytic sites for 3′ and 5′ incision of *Escherichia coli* nucleotide excision repair are both located in UvrC. J Biol Chem 2000; 275:5120–5123.
66. Zou Y, Luo C, Geacintov NE. Hierarchy of DNA damage recognition in *Escherichia coli* nucleotide excision repair. Biochemistry 2001; 40:2923–2931.

67. Park JS, Marr MT, Roberts JW. *E. coli* Transcription repair coupling factor (Mfd protein) rescues arrested complexes by promoting forward translocation. Cell 2002; 109:757–767.

68. Moolenaar GF, Monaco V, van der Marel GA, van Boom JH, Visse R, Goosen N. The effect of the DNA flanking the lesion on formation of the UvrB-DNA preincision complex. Mechanism for the UvrA-mediated loading of UvrB onto a DNA damaged site. J Biol Chem 2000; 275:8038–8043.

69. Selby CP, Sancar A. Transcription preferentially inhibits nucleotide excision repair of the template DNA strand in vitro. J Biol Chem 1990; 265:21330–21336.

70. Friedberg EC, Walker GC, Siede W. DNA Repair and Mutagenesis. Washington, DC: ASM Press, 1995.

71. Beranek DT. Distribution of methyl and ethyl adducts following alkylation with monofunctional alkylating agents. Mutat Res 1990; 231:11–30.

72. Seeberg E, Eide L, Bjoras M. The base excision repair pathway. Trends Biochem Sci 1995; 20:391–397.

73. Wyatt MD, Allan JM, Lau AY, Ellenberger TE, Samson LD. 3-methyladenine DNA glycosylases: structure, function, and biological importance. Bioessays 1999; 21: 668–676.

74. Plosky B, Samson L, Engelward BP, Gold B, Schlaen B, Millas T, Magnotti M, Schor J, Scicchitano DA. Base excision repair and nucleotide excision repair contribute to the removal of N-methylpurines from active genes. DNA Repair (Amst) 2002; 1:683–696.

75. Li S, Smerdon MJ. Base excision repair of *N*-methylpurines in a yeast minichromosome. Effects of transcription, dna sequence, and nucleosome positioning. J Biol Chem 1999; 274:12201–12204.

76. Mellon I, Alanazi M. unpublished data.

77. Leadon SA, Cooper PK. Preferential repair of ionizing radiation-induced damage in the transcribed strand of an active human gene is defective in Cockayne syndrome. Proc Natl Acad Sci USA 1993; 90:10499–10503.

78. Gowen LC, Avrutskaya AV, Latour AM, Koller BH, Leadon SA. BRCA1 required for transcription-coupled repair of oxidative DNA damage. Science 1998; 281:1009–1012.

79. Cozzarelli NR. Editorial Expression of Concern: Preferential repair of ionizing radiation-induced damage in the transcribed strand of an active human gene is defective in Cockayne syndrome. Proc Natl Acad Sci USA 2003; 100:11816.

80. Bregman DB, Halaban R, van Gool AJ, Henning KA, Friedberg EC, Warren SL. UV-induced ubiquitination of RNA polymerase II: a novel modification deficient in Cockayne syndrome cells. Proc Natl Acad Sci USA 1996; 93:11586–11590.

81. Ratner JN, Balasubramanian B, Corden J, Warren SL, Bregman DB. Ultraviolet radiation-induced ubiquitination and proteasomal degradation of the large subunit of RNA polymerase II. Implications for transcription-coupled DNA repair. J Biol Chem 1998; 273:5184–5189.

9

DNA Repair in Actively Transcribed Genes in Eukaryotic Cells

Moon-shong Tang

Departments of Environmental Medicine, Pathology, and Medicine, New York University School of Medicine, Tuxedo, New York, U.S.A.

1. INTRODUCTION

Genomic DNA, because of its reactivity, is constantly reacting with environmental chemicals and physical agents, as well as endogenously generated metabolites. The consequences of these reactions can jeopardize the integrity of the genetic information. As cells from most organisms contain only one or a few copies of the genetic information in their chromosome, cells have evolved many defense mechanisms to repair the modified DNA and to maintain the integrity of the genetic information. The importance and the complexity of DNA repair systems is manifested in the fact that more than 100 genes in most organisms code for proteins that are involved in DNA repair mechanisms (1,2).

Although ample evidence has shown that the formation of DNA damage may have specific preferences, be it due to properties of damaging agents, DNA base composition, primary sequence, epigenetic modification, or protein association and chromatin structure, overall, due to the enormous size of genomic DNA (about 4×10^6 bp in *Escherichia coli* cells to 6×10^9 bp in human cells), the distribution of DNA damage in genomic DNA is rather even (3). Although a substantial proportion of genes codes for repair functions, given the amount of DNA damage occurring under normal physiological conditions, the available concentrations of repair proteins remain a limiting factor for repair efficiency. Finding the damaged bases in genomic DNA is like finding a needle in a haystack. Cells must have many important mechanisms to allow the limited number of repair proteins to find this needle in order to guarantee survival with genetic integrity. What are these mechanisms?

Chromatin structure, DNA sequence, and secondary DNA structure may play important roles in determining the efficiency of DNA repair. However, the dynamic status of genomic DNA may also play an important role in this repair. Many activities, such as replication and transcription, can and do take place at different regions of the genome. These activities may affect chromatin structure and they may interact with DNA repair machinery. How do these activities affect DNA repair?

It should be noted that the most studied type of, and the model for, DNA damage, ultraviolet (UV) light-induced cyclobutane pyrimidine dimers (CPDs), may have unique features in terms of repair. As sunlight is ubiquitous, it is conceivable that even single-celled organism mechanisms would have evolved to cope with this type of DNA damage. Although CPD repair is important and most DNA repair mechanisms are tailored to effectively deal with this type of DNA damage, the mechanism of CPD repair is by no means identical to the repair of other types of DNA damage, even under the same pathway, such as nucleotide excision repair (NER).

It is understandable that answers for these questions have been very much dependent on the available technology at a given period of time. During the past two-and-a-half decades, many ingenious studies have been performed to answer these questions at different levels and have contributed greatly to the understanding of the regulation of DNA repair. This chapter will briefly discuss the results obtained without using DNA hybridization and DNA sequencing techniques. The results from these studies demonstrate different aspects of the heterogeneity of DNA repair. One significant development in the DNA repair field is the discovery by Hanawalt and colleagues (4–6) (quickly confirmed by numerous laboratories) that DNA repair, particularly NER, is tightly associated with transcription in numerous genes. This finding was made possible by the advent of modern molecular biology tools, especially Southern DNA hybridization, DNA sequencing, and ligation-mediated polymerase chain reaction (LMPCR), which enabled researchers to determine DNA damage at the gene-fragment and sequence level make (7–11). These techniques also allowed for the identification of gene products that are involved in DNA repair, and their development represents a totally new era for studying DNA repair. This chapter will place more emphasis on discussing the results from these studies.

2. HETEROGENEITY OF DNA REPAIR

2.1. Effects of DNA Structure on DNA Repair

2.1.1. DNA Sequence

The notion that DNA composition and sequence not only affect the efficiency of DNA adduct formation, but also may affect their repair that has been investigated at three levels: in AT-rich regions vs. general genomic DNA, in repetitive sequences vs. general genomic regions, and at the sequence level in a gene (12–16). It is understandable that before DNA sequencing techniques and the availability of DNA restriction enzymes, investigators could only explore whether DNA repair is conducted with different efficiency at different regions of genomic DNA based on the repetitiveness of the regions and by their differences in density and initial concentration $\times$ hybridization time (C_0t) value. Using both CsCl gradient density centrifugation and C_0t analysis, Lieberman and Poirier (12,14) found no differences in CPD and bulky chemical-DNA adduct repair in different genomic regions. These investigators chose to determine repair at 24 hr after initial damage. It is now clear that most mammalian cells are able to repair DNA damage within 24 hr of incubation at 37°C under most laboratory experimental conditions.

Whether DNA sequence plays a role in determining repair efficiency is a difficult question to address in vivo because it is difficult to differentiate between the roles that chromatin structure and sequence may play in repair. This question is best to be answered by in vitro experiments that determine the sequence effect on the repair.

2.1.2. Chromatin Structure

The nucleosome is the basic unit of the chromatin fiber in the eukaryotic cell. Each nucleosome consists of an octamer of four core histone proteins—H2A, H2B, H3, and H4—and 168 bp of DNA wrapped around the octamer surface. These nucleosomes are connected by linker DNA of different lengths (0–60 bp) (17). Because of its histone protein–DNA interactions, the nucleosomal structure may restrict the access of protein–DNA binding; therefore, it is conceivable that chromatin structure may affect many processes of DNA metabolism, including DNA repair. The first indication that chromatin structure may play a role in determining the preference of DNA repair induced by UV and chemical carcinogens was demonstrated more than two decades ago by Smerdon and colleagues (18–20). These researchers found that newly synthesized DNA induced by DNA damaging agents such as UV and N-acetoxy-2-acetylaminofluorene was initially sensitive to staphylococcal, micrococcal nuclease, or DNAse 1 digestion and then gradually becomes resistant (18–20). These nucleases are known to preferentially digest linker DNA regions over nucleosome core regions. Two possible explanations for their results are (i) the DNA repair event is able to open chromatin structure to allow repair to take place and (ii) DNA repair occurs preferentially at linker regions and/or at open chromatin structures. Using synthetic nucleosome constructs as substrates, it has been demonstrated that the nucleosome structure inhibits photoproduct repair (21). Results from fine mapping of repair at the sequence level using cell extracts demonstrate that CPD at linker regions is indeed repaired more quickly than DNA damage at nucleosome core regions (21). Interestingly, it has been found that the extent of inhibition for <6–4> pyrimidine–pyrimidone photoproduct (<6–4> photoproducts) repair is the same in both linker and core regions using purified NER factors (TFIIH, XPC-hHR-23B, XPA, RPA, ERCC1-XPF) (22). However, addition of ATP-dependent chromatin remodeling factors can facilitate repair of <–6–4> photoproducts in the linker region but not in the core regions (23).

These results together strongly suggest that DNA "free" of nucleosomal proteins is likely a prerequisite for NER to take place. These results also suggest that, depending on the location of damage in the linker, nucleosomal core, and chromatin region, different remodeling factors are probably required for efficient repair.

Higher order chromatin structures beyond the nucleosome also play an important role in determining CPD repair. For example, CPD repair in the transcriptionally inactive gene 754 and coagulation factor IX in the inactive X chromosome is much lower than CPD repair in global genomic DNA regions (24). The inactive X chromosome is known to be highly methylated and its chromatin structure is denser than that found in autosomes (17). There are several factors known that can affect the tightness of chromatin structure, such as acetylation, the phosphorylation and ubiquitination state of the histone, and binding of transcription factors (17). There is indirect evidence suggesting that histone acetylation may enhance CPD repair. For example, the presence of sodium butyrate enhances CPD repair (25). Binding of transcription factors greatly affects the extent of CPD repair (26). A compelling example to support this notion is that a wide range of CPD repair efficiencies have been found in the promoter and immediate adjacent regions in several genes (26–30).

However, concrete evidence to demonstrate the causative relationship between nucleosome acetylation, phosphorylation, ubiquitination, methylation, and transcription factor binding remains to be established.

2.1.3. *Non-randomness of Repair Synthesis*

Data from genome sequencing projects indicates that only a small fraction of genomic DNA in mammalian cell codes for genes; more than 90% of genomic DNA is not coding for genes (31). In contrast, in *E. coli* cells, most genomic DNA is presented in genes (31). It is therefore not surprising that the first evidence of heterogeneity in DNA repair was found in mammalian cells. Using 5-bromo-2'-deoxyuridine (BrdUrd) to label repair-patched DNA and an antibody against BrdUrd repair-synthesized patches, Cohen and Lieberman (32,33) found that UV light-induced repair synthesis in both rodent and human cells is nonrandom at early repair times. However, at completion, they found that repair patches are evenly distributed in the genomic DNA (32). These results demonstrate not only that mammalian cells have the capacity to selectively repair DNA damage, but also that the overall distribution of DNA damage is random. These results are worth noting not only from a historical point of view, but also because the technique of using antibodies to pull down the repair synthesis patches later proved to be very useful in assessing the heterogeneity of DNA repair mediated by different repair pathways, such as NER and base excision repair, induced by different DNA damaging agents (34,35).

2.2. **Effect of Transcription on DNA Repair**

Hanawalt and colleagues (36–38) found that cells, ranging from *E. coli* to rodent and human, have the capacity to preferentially repair CPDs formed in the transcribed (T) strand of transcriptionally active genes. This phenomenon has been termed transcription-coupled repair (TCR) to distinguish it from the less-efficient repair in the nontranscribed (NT) strand of transcriptionally active genes and the rest of genomic DNA, including both strands in transcriptionally inactive genes; repair of DNA damage in the latter regions has been termed global genomic repair (GGR) (39–42). These findings implied that separate repair pathways might have evolved in cells to cope with DNA damage in the T strand of transcriptionally active genes and to ensure quick recovery of gene expression. These concepts were quickly proven to be correct. Cells from individuals suffering from two different genetic diseases— Cockayne syndrome (CS) and xeroderma pigmentosum complementation group C (XPC)—show discrete defects in these two types of repair. While CS cells are proficient in GGR repair but defective in TCR, XPC cells are, in contrast, proficient in TCR but are defective in GGR (39–44).

Although TCR and GGR pathways have been found in different organisms, it appears that the commonality stops there. The biochemical details, the efficiency of TCR and GGR, and the phenotypes of genes that regulate TCR and GGR are by no means universal across the eukaryotic kingdom. There are both subtle and significant differences in TCR and GGR between organisms, reflecting different aspects of the complexity of DNA repair in different organisms (see in what follows). Therefore, it is best to study each organism as an individual entity rather than trying to generalize. With this in mind, the phenomenon, the potential mechanisms, and the genes involved in these two pathways will be described separately in four representative organisms: humans, rodents, yeast, and *E. coli*.

3.　METHODS FOR DETECTING TCR AND GGR

3.1.　Quantification at the Defined DNA Fragment Level

The TCR and GGR in mammalian cells is mainly measured by quantifying DNA damage and repair in the T strand vs. the NT strand of a transcriptionally active gene or in the coding region of a gene domain vs. the noncoding region adjacent to an active gene. The method developed by Mellon and colleagues (36–38) for CPD detection that was later modified by others to detect a variety of DNA damage has five crucial steps (Fig. 1). The first step is to label the newly synthesized DNA with BrdUrd. The second step is to digest the genomic DNA isolated from treated cells with restriction enzymes to trim the target sequence to the proper size and to separate the parental DNA from replicated DNA by CsCl gradient density centrifugation. The third step is to either react the trimmed DNA with repair enzymes, such as CPD lyase (36–38) and UvrABC nuclease (10,45–49), or subject the trimmed DNA to chemical reactions, such as heat and alkaline conditions, to produce single-strand breaks (SSBs) at damaged DNA sites specifically and quantitatively (10,45–49). The fourth step is to manifest the SSBs by denaturing the DNA and then electrophoretically separating the denatured DNA by size. The fifth step is to quantify the full-length DNA fragment by DNA hybridization and calculate the DNA damage per DNA fragment on the basis of the Poisson distribution equation, $P(0) = e^{-n}$, where $P(0)$ represents the fraction of DNA fragment in full-length DNA after repair enzyme treatment or chemical reactions to induced damage specific SSB and n represents the average number of DNA damage per DNA fragment.

　　For studying the repair of the DNA adduct, the n formed at time 0 can be compared with the remaining n at any time point. By the nature of this equation, the sensitivity of this method of calculation $\Delta n/\Delta p(0)$ is n dependent; when n is within 0.25–2.5, a reasonably reliable $\Delta n/\Delta p(0)$ value can be obtained by the current experimental method (10). One important factor for proper application of this method is that the SSB induced either by repair enzyme treatment or by chemical reactions has to be equal or at least proportional to the amount of DNA damage.

　　A method using BrdUrd to label repair synthesis and then using a BrdUrd antibody to pull down repaired DNA fragments has been successfully used for quantifying DNA repair of damage that is not readily converted to strand breaks either enzymatically or by chemical reactions such as thymine glycol and 8-oxo-deoxyguanine (34,35).

3.2.　Quantification at the Sequence Level

A more sensitive method using LMPCR to detect DNA damage at sequence level has been developed and is widely used (9,11,27,28,45–51) (Fig. 2). This method uses repair enzymes, such as T4 endo V, UvrABC nuclease treatment, or chemical reactions, to induce DNA-damage specific SSB (9,10,45–48). Primer extension is then used to produce double-stranded DNA fragments with blunt ends specifically at target sequences and followed by LMPCR to amplify different-sized DNA fragments (9,10,27,28,45–51). The selection of the template strand for primer extension determines whether lesions on the T or NT strand will be detected. The different sizes of DNA fragments are then separated by sequencing gel electrophoresis and identified by using ^{32}P-labeled probes and DNA hybridization techniques. Details of this method are described by Pfeifer and colleagues (9,11). This method allows for the mapping of the relative extent of DNA damage formation at different sequences

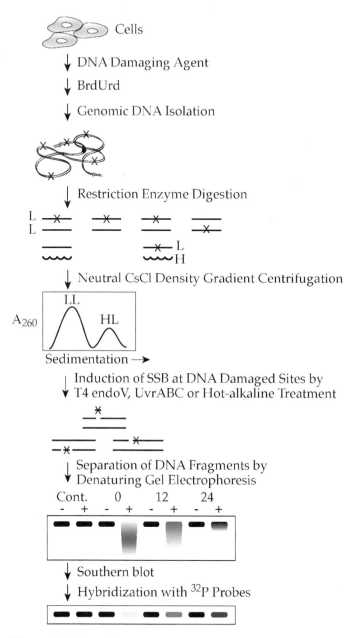

Figure 1 Schematic representations of the method for determining strand-specific DNA repairs. L and H designate nonreplicated light and newly replicated heavy (BrdUrd)-labeled DNA chains, respectively. Southern-blot lanes correspond to undamaged control DNA or DNA samples isolated at 0, 12, and 24 hr following exposure of cells to DNA damaging agent and subsequently treated (+) or left untreated (–) with reagent to induce SSB of unrepaired damage.

within a small DNA fragment. Comparing the relative extent of DNA damage formation at different sequences at different time points lends insights into the effect of transcription, DNA sequence, epigenetic modifications, and chromatin structure on DNA repair and DNA damage formation. The calculation of DNA damage by this

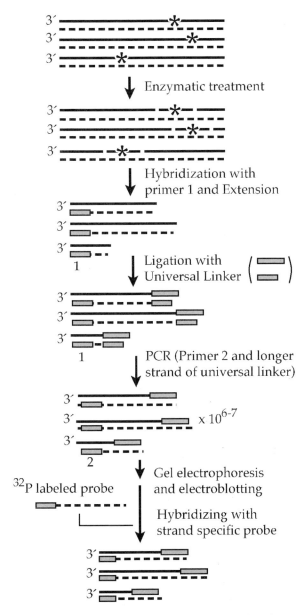

Figure 2 Schematic representation of the method for mapping DNA damage at the sequence level by ligation-mediated PCR. Asterisks indicate sites of lesions that result in DNA SSB following appropriate enzymatic or chemical treatment.

method is no longer dependent on the Poisson distribution equation and therefore allows detection of the nonrandomness of DNA damage and repair. In addition, because PCR is used to amplify the DNA damage-specific fragments produced by DNA damage-specific repair enzyme treatment or chemical reactions, the sensitivity of this method in detecting DNA damage is at least two orders of magnitude better than Hanawalt's method. This method, however, does not allow for the determination of the amount of DNA damage in a DNA fragment.

4. DNA REPAIR IN TRANSCRIPTIONALLY ACTIVE GENES IN DIFFERENT ORGANISMS

4.1. Rodent Cells

It has been known for decades that cultured rodent cells are unable to repair the majority of CPDs in their genome, but yet rodent cells have similar UV sensitivity as human cells in which CPDs are proficiently repaired; this puzzling phenomenon has been dubbed as the rodent cell repair paradox (52–55). In search of an answer for this paradox, Bohr et al. (36) developed a method combining T4 endo V, alkaline electrophoresis, and Southern DNA hybridization techniques to detect CPD repair in the actively expressed DHFR gene region vs. its 3′ downstream noncoding region in cultured Chinese hamster ovary (CHO) cells. They discovered that CPDs are more efficiently repaired in the actively expressed DHFR gene region than in the 3′ downstream noncoding region (36). They termed the more efficient CPD repair in the actively expressed gene vs. noncoding region as gene-specific preferential repair (36). This observation was interpreted as meaning that rodent cells have the capacity to repair CPDs in the regions, such as actively transcribed genes, to ensure their survival and are believed to be a plausible explanation for why rodent cells have similar UV sensitivity to humans even though these cells are unable to remove most of the CPDs in their genome (36,55). It was quickly found by Mellon et al. (37) in Hanawalt's laboratory that the preferential repair in the actively expressed gene is due to preferential CPD repair in the T strand of the active gene, and that CPD repair in the NT strand of the active gene is the same as in the rest of the genomic DNA region. The term TCR was thus coined (37). This claim was supported by the observation that TCR in the DHFR gene was greatly suppressed in CHO cells treated with the transcription inhibitor α-amanitin and most recently by the observation that there is a lack of CPD repair in the T strand, as well as in the NT strand, in the DHFR gene domain of a cell line with a deletion encompassing the promoter and the first four exons (27,56,57). It is, however, worth noting that the term "coupled" is an operational rather than a mechanistic description. Two decades after the first observation was reported, the mechanistic relationship between transcription and CPD repair has yet to be established.

Because of the deficiency of CPD repair in the majority of genomic regions, the rodent cells are a superior system to demonstrate TCR. In rodent cells, numerous mutants related to DNA repair have been generated in the laboratory. Six classes of mutants (ERCC1–ERCC6), which are deficient in NER and can complement each other for the NER deficiency, have been established (for a review see Ref. 61). Five (ERCC2–ERCC6) of the six classes of mutants have NER deficiencies that can be complemented by human cells (58–61). Cells from five complementation groups (ERCC1–ERCC5) are totally deficient in CPD repair while cells from ERCC6 have residual CPD repair capacity (58,59,61). When TCR was determined in these six groups of cells, it was found that ERCC6 cells have residual GGR capacity but lack TCR (62,63). In contrast, ERCC1 and ERCC2 cells lack both GGR and TCR (Tang, unpublished results). These results suggest that the ERCC6 gene codes for a factor that is essential for TCR. In fact, because of excision repair cross-complementation between rodent and human cells, it was found that CSB is the human counterpart of rodent ERCC6, whereas XPD and XPB are the human counterparts of ERCC2 and ERCC3, respectively. These human NER XP and CSB genes were eventually cloned (59–61,63).

It has been found that the deficiency in GGR in cultured rodent cells, CHO cells in particular, is due to these rodent cells having a dysfunctional *p53* gene (64). The *p53* gene regulates the expression of the *p48* gene, whose gene product is a component of the DNA damage binding (DDB) protein that, apparently, is important for the recognition of CPD in genomic DNA (65–67). Introduction of a functional *p53* gene or *p48* gene into rodent cells enhances CPD repair in the genomic DNA in these cells (67). Intriguingly, functional *p48* or *p53* does not enhance the UV resistance of these cells (67). Recently, it has been found that in vitro DDB does not affect CPD excision, and furthermore, certain rodent cell lines, which have DDB, showed the same deficiency in GGR as CHO cells (68,69).

Making things even more complicated is the recent observation that efficient CPD repair occurs in the NT strand of the *APRT* gene and exon 1 of the *DHFR* gene when CPD repair was determined at the sequence level (27,28,50,51). However, when these genes are translocated to different genomic positions, CPD repair in these translocated genes is different from that in the endogenous positions (51). These results strongly suggest that CPD repair in mammalian cells is probably a very complex process and highly dependent on genomic context and chromatin structures; it is under multiple levels of regulation, of which transcription is just one component. Classifying rodent CPD repair in genomic DNA, as occurring through two repair pathways although convenient, is probably too simplistic.

Furthermore, it seems that TCR in rodent cells is limited to CPD only. It has been found that in the *DHFR* gene, *N*-(deoxyguanosine-8-yl) -2-aminoflourene (dG-C8-AF), N-(deoxyguanosine-8-yl) -2-acetylaminofluorene (dG-C8-AAF), <6–4> photoproducts, 4-nitroquinoline 1-oxide-1-DNA adducts, and benzo(a) pyrene diol epoxide (BPDE)–DNA adducts are almost equally, if not identically, repaired in both T and NT strands (70–73). These findings raise the possibility that CPD repair in rodent cells may be unique, and further investigation of the regulation of CPD repair in rodent cells through genetic and biochemical approaches is warranted.

4.2. Human Cells

Transcription-coupled repair for CPD was found to occur in transcriptionally active genes, such as the *DHFR* gene in cultured human fibroblasts at the same time as in CHO cells (37). However, unlike rodent cells that are deficient in CPD repair, human cells are proficient in CPD repair (37,74). Therefore, CPDs in both T and NT strands are eventually repaired in these cells and the TCR is manifested by CPD repair kinetics that is much faster in the T strand than in the NT strand (24,37,74). Indeed, TCR for CPD in human cells was best demonstrated in XPC cells in which the GGR was defective in a manner similar to CHO cells; therefore, large differences in CPD repair between the T and NT strands were more easily observed (43,44). It was expected that, with the exception of XPC, cells with defective genes for NER, such as *XPA, XPG, XPB, XPD*, and *XPF*, should also be defective in TCR (61). In contrast, cells from CS groups A and B (CSA and CSB) are defective in TCR but are still proficient in GGR (39–42). It is worth noting that, because of the lack of sensitivity in differentiating CPD repair in the coding vs. the noncoding region in the *DHFR* gene domain, XPC was first reported to be defective in gene-specific preferential repair for CPD (53). Indeed, the rate and the kinetics of CPD repair in XPC cells are not as robust as in normal human cells (43,44). Although these findings are more than a decade old and despite the fact that proteins coded by these genes have been purified, TCR has not been demonstrated in vitro and its mechanisms remain unknown.

All six crucial factors (TFIIH, XPC-hHR-23B, XPA, RPA, ERCC1-XPF, XPG) involved in CPD excision repair have been purified and demonstrated to be necessary in the in vitro reconstituted NER system (for review, see Refs. 75–77). The XPC protein has been shown to be involved in one of the three initial steps, which are crucial for CPD recognition, which eventually leads to excision repair of CPD (78,79). In contrast, the exact function of CSA and CSB in TCR remains unclear. Cockayne syndrome A is a five-WD-40-repeat-containing protein, and it has potential to interact with other proteins (80–82). Indeed, purified CSA shows affinity toward CSB (80,81). Intriguingly, this CSA and CSB association has not been shown in cell extracts (83). Cockayne syndrome B is a nuclear protein of 168 kDa. It contains helicase motif but does not function as a helicase (84). Cockayne syndrome B is homologous to nucleosome remodeling factors in the SWI/SNF2 family and has double-stranded DNA-dependent ATPase (40,84,85). It has been found that the ATPase activity of CSB is crucial for its function in changing DNA conformation and rearranging nucleosome (86). Cockayne syndrome B interacts with RNA polymerase II (RNAPII) and TFIIH (87,88). It has been suggested that the nucleosome remodeling activity may allow NER proteins to have better access to damaged bases (86). Besides the lack of TCR, one other distinct feature of CS cells is the failure to resume RNA synthesis after UV or DNA damaging agent treatment (40,89–91). It has been demonstrated that the latter, rather than the former, is responsible for CS cell sensitivity towards DNA damaging agents (90–93). Cockayne syndrome cells also show deficiency in ubiquitination of RNAPII after UV irradiation (94–96). It has been suggested that ubiquitination is important for releasing stalled RNAPII from a damaged site either through dissociation from T strand or through retraction (97). As both CS proteins do not affect NER directly (Sancar and colleagues, personal communication), their roles in TCR may be through three possible mechanisms: (i) release or retraction of stalled RNAPII, allowing NER proteins to have better access to damaged bases; (ii) reinitiation of RNA synthesis, which may result in efficient CPD repair similar to the proficient CPD repair in both strands at transcription initiation site; and (iii) remodeling of the chromatin structure to allow NER proteins to conduct repair.

Unlike rodent cells in which TCR is mainly manifested in CPD repair, in human cells, TCR has been found for repair not only of CPD, but also of other bulky DNA damage, such as <6–4> photoproducts, and BPDE-DNA adducts (44,47,98).

It has been demonstrated that *p53*-mutant cells are proficient in TCR for CPD but deficient in GGR for CPD (99–101). *p53*-mutant cells have also been found to be proficient in TCR but deficient in GGR for BPDE-DNA adducts (100). As previously described, *p53* regulates the expression of *p48*, and the *p48* gene product DDB2 is one of the two components of DDB (65–67). DNA damage binding protein has been shown to have a higher affinity toward UV-induced DNA damage than XPA and XPC and, therefore, has been proposed to participate in NER, particularly in the NT strand and bulk genomic DNA region, because in these regions the recognition of DNA damage is not assisted by its effect on stalling transcription (78,102). However, recently, it has been reported that *p53*-mutant cells are defective in TCR for UVB-induced photoproducts (103). The role of *p53* in TCR remains unclear. Although in the in vitro reconstituted NER systems, the *p53* protein does not affect NER; it is, however, able to interact with many NER factors including TFIIH, XPD,

XPB, and RPA (104,105). The question of how these interactions affect TCR and GGR remains to be answered. If CSB/A directly participates in TCR, then it is possible that the TCR pathway is involved in the repair of a particular type of DNA damage and is dependent on its effect on the induction of the expression of the *CSB* and *CSA* genes. If this is the case, perhaps, UVB irradiation may be able to induce these two genes via its effect on the *p53* gene. This possibility remains to be tested.

In *E. coli* cells, it has been found that cells with mutations in the mismatch repair genes *mutS* and *mutL*, but not *mutH*, are deficient in TCR (106). Similar results were found in human cells; it was found that lymphoblastoid cell lines derived from hereditary nonpolyposis colorectal cancer (HNPCC), which are defective in genes homologous to yeast mismatch repair genes (hMSH1, hMSH2), are also deficient in TCR in the DHFR gene (107). However, several laboratories have recently reported that human adenocarcinoma strains, deficient in hMSH1 and hMSH2, are proficient in TCR in c-jun and *p53* genes (108–110). The role of mismatch repair genes in repair of DNA damage in the transcriptionally active genes remains controversial. It has been found that mutations in mismatch repair genes in yeast do not affect TCR (111). Thus, it appears that the regulation of TCR in yeasts is different than in human cells.

4.3. *Escherichia coli* cells

Although TCR was originally found in mammalian cells, it was quickly found that *E. coli* and baker's yeast such as *S. cerevisiae* also have a similar capacity to preferentially repair CPDs in the T strand (38,112). In the *E. coli* system, preferential repair in the T strand of the *lacZ* gene was found to occur in association with transcription (38,113). Transcription-coupled repair was also demonstrated in an in vitro reconstituted system; Selby and Sancar (114–116) elegantly reconstituted an in vitro system that is functional in both transcription and NER. These workers found that CPD efficiently blocks transcription, and the stalled RNA polymerase complex hampers NER. However, in the presence of purified *mfd* gene product transcription-coupled repair factor, which binds to the stalled RNAP–DNA template complex, two consequences were observed: (i) RNAP and nascent transcript are released from the stalled RNAP–DNA template complex and (ii) binding of UvrA to the damaged site eventually leads to UvrB and UvrC binding and results in excision of the CPD-containing oligonucleotide fragment (114–116).

Cells with an *mfd* mutation are not more UV sensitive than cells without an *mfd* mutation, indicating that TCR may not be a major repair mechanism in *E. coli* cells (115). However, in cells that are deficient in recombination repair mechanism such as in *recA* deficient cells, an additional *mfd* mutation significantly increases cell UV sensitivity (115). These results suggest that GGR and/or recombination repair can eliminate CPD efficiently in *mfd*-mutant cells, and that TCR becomes a significant repair pathway only when GGR or recombination repair is missing. Interestingly, it has been shown in the *lacZ* system that *mfd* is required for TCR when the expression of this gene is at basal levels (113). When the *lacZ* gene is induced by isopropyl-β-D-thiogalactosidase, it results in a high level of expression, and the CPDs in the T strand are efficiently repaired regardless of *mfd* status (113). It should be noted that even though *mfd*-mutant cells have the same UV sensitivity as wild-type cells, *mfd* cells are more susceptible to UV-induced mutagenesis (115,117).

Transcription-coupled repair was greatly reduced in *mutS* and *mutL* mutants but not in *mutH* mutants (106). As *mutS* and *mutL* are involved in recognition of mismatch bases, while *mutH* is an endonuclease rather than a DNA-damage recognition protein, it has been suggested that the mechanism that recognizes mismatches also contributes to repair of DNA damage in the T strand by facilitating the recognition of DNA damage (106). Intriguingly, in the in vitro reconstituted NER systems, *MutS* and *MutL* proteins do not affect NER (118,119). Perhaps, in order to have an effect on TCR, the mismatch repair system requires DNA replication. The interplay of the mismatch repair pathway with the NER pathway, particularly TCR, remains to be explored.

4.4. Yeast

In yeast, NER is controlled by genes that belong to the RAD3 epitasis group (for review, see Refs. 61,120). Transcription-coupled repair appears to be partially regulated by the *RAD26* and *RAD28* genes, whereas GGR is regulated by the *RAD7*, *RAD16*, and *RAD23* genes (30,112,121–126). The *RAD26* and *RAD28* genes are *CSA* and *CSB* homologs, respectively (122,123). Unlike *CSA* or *CSB*-mutant cells that are modestly more UV sensitive than normal human cells, *RAD26*- and *RAD28*-mutant cells have the same UV sensitivity as their parental cells (122). However, in the *RAD7*- or *RAD16*-mutant background, an additional *RAD26* or *RAD28* mutation sensitizes the cells to UV-induced killing (122). In these double mutants, while CPD repair in the NT strand is completely deficient, as is the case in *rad7* and *rad16* single mutant cells, CPD repair in the T strand remains at a significant level (122). While CPD repair in the T strand in the *rad26* mutant cells is lower than RAD^+ cells, it is significantly higher than in the $rad26^- rad7^-$ double mutants (122,127). These results suggest that GGR overlaps with TCR in yeast. Indeed, it has been demonstrated that both TCR and GGR are partially defective in *rad23* mutants. Cells that are completely deficient in NER, such as *rad14*- and *rad*-mutant cells, are much more UV sensitive than $rad7^- rad26^-$ double mutants (122).

Since to date an in vitro reconstituted system to demonstrate TCR has not been successfully established, yeast has become one of the most convenient eukaryotic systems to investigate genes that are involved in TCR and molecular analysis of TCR. For example, deletion of the *SPT4* gene in a $rad26^-$ mutant background results in the partial defect in TCR being restored (128). The *SPT4* gene is involved in the repression of transcription elongation, and its repression is modulated by the phosphorylation of the C-terminal domain (CTD) of RNA polymerase (128,129). Mutation at another locus, *Kin28*, which codes a subunit of TFIIH that is required for phosphorylation of the CTD region of RNAPII, also impairs TCR but not GGR (130). These results indicate that transcription elongation is a prerequisite for TCR and that transcription initiation may not be sufficient for TCR.

Results from mapping CPD repair at single nucleotide resolution have shown that Rad26p is dispensable for CPD repair in the promoter and transcription termination regions but is required for CPD repair at the initiation and elongation regions of an active gene (29,127).

It is widely accepted that TCR occurs in the genes transcribed by RNAPII and does not occur in genes transcribed by RNAPI, such as rDNA and genes transcribed by RNAPIII (29,131–133). Recently, by separating actively transcribed rDNA genes from transcriptionally inactive rDNA genes, it has been found that TCR occurs in actively transcribed rDNA genes (134). As there are multiple copies of tRNA genes and the majority of tRNA genes are transcriptionally inactive, it is possible that TCR occurs in transcriptionally active tRNA genes at the same rate as found in rDNA genes.

5. MODELS OF TCR IN EUKARYOTIC CELLS

Although the exact mechanisms and the regulation of TCR have not been worked out, several conditions that are essential for TCR to take place have been discovered. These conditions are active transcription, functional CSA/CSB proteins, RNAP stalled by DNA lesions, and proficient NER function. Transcription-coupled repair mainly occurs at 3' downstream of the transcription site; it appears that due to the binding of transcription initiation factors, DNA damage at the transcription initiation site is consequently repaired efficiently and independently from TCR. Using naked DNA as template, it has been demonstrated that a stalled RNAP per se does not hinder NER. Nonetheless, based on the TCR mechanism in the *E. coli* model, it is reasonable to assume that in order to allow NER to take place, the stalled RNAP must be cleared far enough away from the DNA damage site so that the NER complex has adequate space to function. One possible mechanism to achieve this is through ubiquitination of RNAP, followed by release of RNAP from DNA template. However, no evidence has demonstrated that this process facilitates NER, and therefore, it seems that this process is an abortive action rather than a signal for NER. The RNAP backtracking process that takes place when the transcription machinery meets a hairpin or other structural hurdle during normal transcription is likely a mechanism that occurs during TCR. To resume transcription, the backtracked RNAP probably needs to be dephosphorylated at its CTD and, therefore, recruiting transcription factors such as TFIIH and other transcription factors can restore its activity. TFIIH is a crucial factor involved in DNA damage recognition and unwinding of damaged stretches of the DNA to allow NER-mediated dual incision to occur. It is possible that because of the recruitment of TFIIH for reinitiating stalled RNAP, this process fortuitously facilitates NER, and factors involved in these processes are thus crucial for TCR. Many elegant models to account for TCR have been proposed, but the foregoing factors must be taken into consideration for models explaining the sequential events and regulation of TCR in eukaryotic cells. The following model summarizes our current understanding of TCR in eukaryotic systems using human cells as a model (Fig. 3):

1. recruitment of RNAPII, TFIIH, and other transcription factors to the promoter of an active gene to form a ternary complex;
2. initiation of transcription by RNAPII–TFIIH–others complex;
3. release of TFIIH and other transcription factors from the complex;
4. phosphorylation of the CTD of RNAPII;
5. transcription elongation;
6. blockage of RNAPII movement at damaged bases site;
7. backtracking of stalled RNAPII;
8. dephosphorylation of RNAPII;
9. recruitment of TFIIH and other transcription factors to RNAPII to facilitate repair of CPD and allow resumption of transcription;
10. completion of transcription.

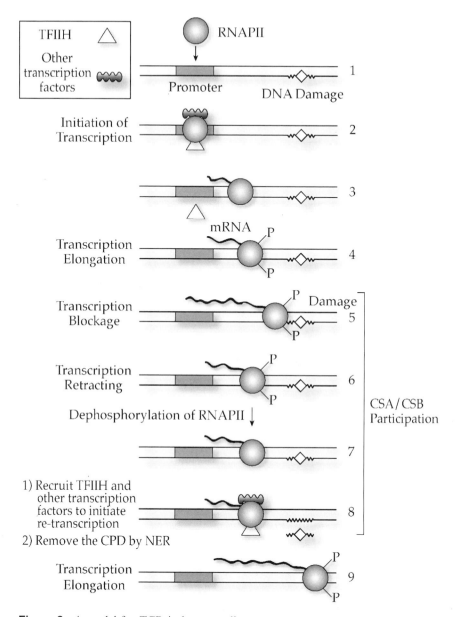

Figure 3 A model for TCR in human cells.

6. EFFECT OF DIFFERENT KINDS OF DNA DAMAGE ON TCR

On the basis of our current understanding, it seems logical to envision that TCR should take place in response to any bulky DNA damage that is both able to stall transcription and is also a substrate for NER. Transcription-coupled repair, however, has been mainly observed for the repair of CPDs. Several types of bulky DNA damage are repaired efficiently in bulk genomic DNA and in the T strand of transcriptionally active genes. For example, <6–4> photoproducts and dG-C8-AF adducts, which are able to block transcription, show little, if any, TCR (70–73,135). Although BPDE-dG adducts are preferentially repaired in the T strand of the *p53* and HPRT genes in human cells, these adducts are proficiently repaired in both strands of the *DHFR* gene in CHO cells (47,70,98). Furthermore, although CPD is preferentially repaired in the T strand of most genes tested in both rodent and human cells, it is repaired equally efficiently in both strands of the endogenous *APRT* gene in rodent cells (50,51). Intriguingly, CPD repair in translocated *APRT* genes is different from CPD repair in the endogenous APRT gene (51). These results suggest that while TCR is an important mechanism to facilitate DNA removal, many other factors, such as genomic context and chromatin structure, may determine how the CPD will be repaired and whether TCR will occur. It is reasonable to envision that the major factor determining the efficiency of DNA repair is the accessibility of repair enzymes to the damaged bases, and this is determined by the nature of DNA damage and chromatin structure. Active genes, which have very different chromatin structures from inactive genes or noncoding regions, may be much more accessible to or have a greater affinity for DNA repair proteins. Transcription may enhance the accessibility of a localized region or enhance a region's affinity for damage-recognition/repair proteins. However, if the chromatin structure surrounding a gene has already rendered the region maximally accessible to repair proteins (as may be the case for the endogenous CHO *APRT* gene locus), then transcription would not be necessary to facilitate repair, and both strands might be repaired with equal efficiency. On the other hand, if the chromatin structure of an active gene is not already optimal for the repair process, then transcription-induced chromatin changes may enhance the repair or promote TCR in the T strand. Similar reasoning can also be applied to different kinds of DNA damage in regard to the effect of transcription on their repair. In the case where the DNA damage is a <6–4> photoproduct, which greatly affects chromatin structure and consequently would have greater accessibility for repair enzymes, the transcription may not affect its repair. However, DNA damage such as the CPD, which does not have a high affinity toward repair enzymes, may enhance its recognition by repair enzymes by stalling transcription.

ACKNOWLEDGMENTS

The author would like to thank Drs. Aziz Sancar, Darel Hunting, Gerry Adair, and Yen-Yee Tang for their critical review of the manuscript. This work was supported by NIH grants ES03124, ES08389, ES00260, and ES10344.

REFERENCES

1. Aravind L, Walker DR, Koonin EV. Nucleic Acids Res 1999; 27:1223–1289.
2. Wood RD, Mitchell M, Sgouros J, Lindahl T. Science 2001; 291:1284–1289.
3. Kornberg A, Baker TA. DNA Replication. New York: W.H. Freeman and Co., 1992:19–20.
4. Hanawalt PC. Oncogene 2002; 21:8949–8956.
5. Balajee AS, Bohr VA. Gene 2000; 250:15–30.
6. Mullenders LHF. Mutat Res 1998; 409:59–64.
7. Smith CA, Hanawalt PC. In: Pfeifer GP, ed. Technologies for Detection of DNA Damage and Mutations. New York: Plenum Press, 1996:117–128.

8. Bohr VA. In: Pfeifer GP, ed. Technologies for Detection of DNA Damage and Muta-
    tions. New York: Plenum Press, 1996:131–136.
9. Tornaletti S, Pfeifer GP. In: Pfeifer GP, ed. Technologies for Detection of DNA
    Damage and Mutations. New York: Plenum Press, 1996:199–208.
10. Tang M-s. In: Pfeifer GP, ed. Technologies for Detection of DNA Damage. New York:
    Plenum Press, 1996:139–153.
11. Drouin R, Rodriguez H, Hoslmquist GP, Akman SA. In: Pfeifer GP, ed. Technologies
    for detection of DNA Damage and Mutations. New York: Plenum Press, 1996:211–223.
12. Lieberman MW, Poirier MC. Proc Natl Acad Sci USA 1974; 71:2461–2465.
13. Zolan ME, Cortopassi GA, Smith CA, Hanawalt PC. Cell 1982; 71:613–619.
14. Lieberman MW, Poirier MC. Biochemistry 1974; 13:3018–3023.
15. Tornaletti S, Pfeifer GP. Science 1994; 263:1436–1438.
16. Drouin GS, Holmquist GP. Science 1994; 263:1438–1440.
17. Wolffe A. Chromatin: Structure and Function. 3rd ed. San Diego: Acdemic Press,
    1998:16–34.
18. Smerdon MJ, Tlsty TD, Lieberman MW. Biochemistry 1978; 17:2377–2386.
19. Smerdon MJ, Lieberman MW. Proc Natl Acad Sci USA 1978; 75:4238–4241.
20. Tlsty TD, Lieberman MW. Nucleic Acids Res 1978; 5:3261–3273.
21. Liu X, Smerdon MJ. J Biol Chem 2000; 275:23729–23735.
22. Hara R, Mo J, Sancar A. Mol Cell Biol 2000; 20:9173–9181.
23. Ura K, Araki M, Saeki H, Masutani C, Ito T, Iwai S, Mizukoshi T, Kaneda Y,
    Hanaoka F. EMBO J 2000; 20:2004–2014.
24. Venema J, Bartosova Z, Natarajan AT, van Zeeland AA, Mullenders LHF. J Biol
    Chem 1992; 267:8852–8856.
25. Smerdon MJ, Conconi A. Prog Nucleic Acid Res Mol Biol 1999; 62:227–255.
26. Tu Y, Tornaletti S, Pfeifer GP. EMBO J 1996; 15:675–683.
27. Hu W, Feng Z, Chasin LA, Tang M-s. J Biol Chem 2000; 277:38305–38310.
28. Feng Z, Hu W, Chasin LA, Tang M-s. Nucleic Acids Res 2003; 31:5897–5906.
29. Teng Y, Waters R. Nucleic Acids Res 2000; 28:1114–1119.
30. Yu S, Teng Y, Lowndes NF, Waters R. Mutat Res 2001; 485:229–236.
31. Brown TA. Genomes 2. Oxford, UK: Bios Scientific Publishers Ltd., 2002:22–30.
32. Cohn SM, Lieberman MW. J Biol Chem 1984; 259:12463–12469.
33. Cohn SM, Lieberman MW. J Biol Chem 1984; 259:12456–12462.
34. Leadon SA. Nucleic Acids Res 1986; 14:8979–8995.
35. Avrutskaya AV, Leadon SA. Methods 2000; 22:127–134.
36. Bohr VA, Smith CA, Okumoto DS, Hanawalt PC. Cell 1985; 40:359–369.
37. Mellon I, Spivak G, Hanawalt PC. Cell 1987; 51:241–249.
38. Mellon I, Hanawalt PC. Nature 1989; 342:95–98.
39. Venema J, Mullenders LHF, Natarajan AT, van Zeeland AA, Mayne LV. Proc Natl
    Acad Sci USA 1990; 87:4707–4711.
40. Troelstra C, van Gool A, de Wit J, Vermeulen W, Bootsma D, Hoeijmakers JHJ. Cell
    1992; 71:939–953.
41. van Hoffen A, Natarajan AT, Mayne LV, van Zeeland AA, Mullenders LHF, Venema
    J. Nucleic Acids Res 1993; 21:5890–5895.
42. Tu Y, Bates S, Pfeifer GP. Mutat Res 1998; 400:143–151.
43. Venema J, van Hoffen A, Natarajan AT, van Zeeland AA, Mullenders LHF. Nucleic
    Acids Res 1990; 18:443–448.
44. van Hoffen A, Venema J, Meschini R, van Zeeland AA, Mullenders LHF. EMBO J
    1995; 14:360–367.
45. Tang M-s, Qian M, Pao A. Biochemistry 1994; 33:2726–2732.
46. Denissenko MF, Pao A, Tang M-s, Pfeifer GP. Science 1996; 274:430–432.
47. Denissenko MF, Pao A, Pfeifer GP, Tang M-s. Oncogene 1998; 16:1241–1247.
48. Feng Z, Hu W, Chen JX, Pao A, Li H, Rom W, Hung M-C, Tang M-s. Natl Cancer
    Inst 94; 2002:1527–1536.

49. Thomas DC, Morton AG, Bohr VA, Sancar A. Proc Natl Acad Sci USA 1988; 85: 3723–3727.
50. Zheng Y, Pao A, Adair GM, Tang M-s. J Biol Chem 2001; 276:16786–16796.
51. Feng Z, Hu W, Komissarova E, Pao A, Hung M-C, Adair GM, Tang M-s. J Biol Chem 2002; 277:12777–12783.
52. van Zeeland AA, Smith CA, Hanawalt PC. Mutat Res 1981; 82:173–189.
53. Bohr VA, Okumoto DS, Hanawalt PC. Proc Natl Acad Sci USA 1986; 83:3830–3833.
54. Hanawalt PC. Environ Mol Mutagen 2001; 38:89–96.
55. Zheng Y. Nucleotide Excision Repair in Mammalian Cells. Ph.D. dissertation, University of Texas, Austin, TX, 1999.
56. Carreau M, Hunting D. Mutat Res 1992; 274:57–64.
57. Christians FC, Hanawalt PC. Mutat Res 1992; 274:93–101.
58. Busch D, Greiner C, Lewis K, Ford R, Adair G, Thompson L. Mutagenesis 1989; 4:349–354.
59. Thompson LH, Salazar EP, Brookman KW, Collins CC, Stewart SA, Busch DB, Weber CA. J Cell Sci Suppl 1987; 6:97–110.
60. Thompson LH. Bioessays 1998; 20:589–597.
61. Friedberg EC, Walker GC, Siede W. DNA repair and mutagenesis. Washington, DC: ASM Press, 1995:317–366.
62. Vreeswijk MPG, Overkamp MWJI, Westland BE, van Hees-Stuivenberg S, Vrieling H, Zdzienicka MZ, van Zeeland AA, Mullenders LHF. Mutat Res 1998; 409:49–56.
63. Orren DK, Dianov GL, Bohr VA. Nucleic Acids Res 1996; 24:3317–3323.
64. Lee H, Larner JM, Hamlin JL. Gene 1997; 184:177–183.
65. Hwang BJ, Toering S, Francke U, Chu G. Mol Cell Biol 1998; 18:4391–4399.
66. Hwang BJ, Ford JM, Hanawalt PC, Chu G. Proc Natl Acad Sci USA 1999; 96:424–428.
67. Tang JY, Hwang BJ, Ford JM, Hanawalt PC, Chu G. Mol Cell 2000; 5:737–744.
68. Zolezzi F, Linn S. Gene 2000; 245:151–159.
69. Wittschieben BO, Wood RD. DNA Repair 2003; 2:1065–1069.
70. Tang M-s, Pao A, Zhang X-s. J Biol Chem 1994; 269:12749–12754.
71. Tang M-s, Bohr VA, Zhang X-s, Pierce J, Hanawalt PC. J Biol Chem 1989; 264: 14455–14462.
72. Vreeswijk MPG, van Hoffen A, Westland BE, Vrieling H, van Zeeland AA, Mullenders LHF. J Biol Chem 1994; 269:31858–31863.
73. Snyderwine EG, Bohr VA. Cancer Res 1992; 52:4183–4189.
74. Mellon I, Bohr VA, Smith AC, Hanawalt PC. Proc Natl Acad Sci USA 1986; 83: 8878–8882.
75. Sancar A. Annu Rev Biochem 1996; 65:43–81.
76. Wood RD. J Biol Chem 1997; 272:23465–23468.
77. de Laat WL, Jasper NGJ, Hoeijmakers JHJ. Gene Dev 1999; 13:768–785.
78. Reardon JT, Sancar A. Genes Dev 2003; 17:2539–2551.
79. Sugasawa K, Ng JMY, Masutani C, Iwai S, van der Spek PJ, Eker APM, Hanaoka F, Bootsma D, Hoeijmakers JHJ. Mol Cell 1998; 2:223–232.
80. Henning KA, Li L, Iyer N, McDaniel LD, Reagan MS, Legerski R, Schultz RA, Stefanini M, Lehmann AR, Mayne LV, Friedberg EC. Cell 1995; 82:555–564.
81. Neer EJ, Schmidt CJ, Nambudripad R, Smith TF. Nature 1994; 371:297–300.
82. Tantin D, Kansal A, Carey M. Mol Cell Biol 1997; 17:6803–6814.
83. van der Horst GTJ, Meira L, Gorgels TGMF, de Wit J, Velasco-Miguel S, Richardson JA, Kamp Y, Vreeswijk MPG, Smit B, Bootsma D, Hoeijmakers JHJ, Friedberg EC. DNA Repair 2002; 1:143–157.
84. Selby CP, Sancar A. J Biol Chem 1997; 272:1885–1890.
85. Eisen JA, Sweder KS, Hanawalt PC. Nucleic Acid Res 1995; 23:2715–2723.
86. Citterio E, van den Boom V, Schnitzler G, Kanaar R, Bonte E, Kingston RE, Hoeijmakers JHJ, Vermeulen W. Mol Cell Biol 2000; 20:7643–7653.
87. Tantin D. J Boil Chem 1998; 273:27794–27799.

88. van Gool AJ, Citterio E, Rademakers S, van Os R, Vermeulen W, Constantinou A, Egly J-M, Bootsma D, Hoeijmakers JHJ. EMBO J 1997; 16:5955–5965.

89. Mayne LV, Lehmann AR. Cancer Res 1982; 42:1473–1478.

90. van Oosterwijk MF, Versteeg A, Filon R, van Zeeland AA, Mullenders LHF. Mol Cell Biol 1996; 16:4436–4444.

91. van Oosterwijk MF, Filon R, de Groot AJL, van Zeeland AA, Mullenders LHF. J Biol Chem 1998; 273:13599–13604.

92. Rockx DAP, Mason R, van Hoffen A, Barton MC, Citterio E, Bregman DB, van Zeeland AA, Vrieling H, Mullenders LHF. Proc Natl Acad Sci 2000; 97:10503–10508.

93. van Oosterwijk MF, Filon R, Kalle WHJ, Mullenders LHF, van Zeeland AA. Nucleic Acids Res 1996; 24:4653–4659.

94. Bregman DB, Halaban R, van Gool AJ, Henning KA, Friedberg EC, Warren SL. Proc Natl Acad Sci USA 1996; 93: 1996; 93:11586–11590.

95. Ratner JN, Balasubramanian B, Corden J, Warren SL, Bregman DB. J Biol Chem 1998; 273:5184–5189.

96. McKay BC, Chen F, Clarke ST, Wiggin HE, Harley LM, Ljungman M. Mutat Res 2001; 485:93–105.

97. Tornaletti S, Hanawalt PC. Biochimie 1999; 81:139–146.

98. Chen R-H, Maher VM, Bouwer J, van de Putte P, McCormick JJ. Proc Natl Acad Sci USA 1992; 89:5413–5417.

99. Ford JM, Hanawalt PC. Proc Natl Acad Sci USA 1995; 92:8876–8880.

100. Wani MA, Zhu Q, El-Mahdy M, Venkatachalam S, Wani AA. Cancer Res 2000; 60:2273–2280.

101. Ford JM, Hanawalt PC. J Biol Chem 1997; 272:28073–28080.

102. Wakasugi M, Kawashima A, Morioka H, Linn S, Sancar A, Mori T, Nikaido O, Matsunaga T. J Biol Chem 2002; 277:1637–1640.

103. Mathonnet G, Leger C, Desnoyers J, Drouin R, Therrien J-P, Drobetsky EA. Proc Natl Acad Sci 2003; 100:7219–7224.

104. Wang XW, Yeh H, Schaeffer L, Roy R, Moncollin V, Egly J-M, Wang Z, Friedberg EC, Evans MK, Taffe BG, Bohr VA, Hoeijmakers JHJ, Forrester K, Harris CC. Nat Genet 1995; 10:188–195.

105. Leveillard T, Andera L, Bissonnette N, Schaeffer L, Bracco L, Egly J-M, Wasylyk B. EMBO J 1996; 15:1615–1624.

106. Mellon I, Champe GN. Proc Natl Acad Sci USA 1996; 93:1292–1297.

107. Mellon I, Rajpal DK, Koi M, Boland CR, Champe GN. Science 1996; 272:557–560.

108. Rochette PJ, Bastien N, McKay BC, Therrien J-P, Drobetsky EA, Drouin R. Oncogene 2002; 21:5743–5752.

109. Adimoolam S, Lin CX, Ford JM. J Biol Chem 2001; 276:25813–25822.

110. Therrien J-P, Loignon M, Drouin R, Drobetsky EA. Cancer Res 2001; 61:3781–3786.

111. Sweder KS, Verhage RA, Crowley DJ, Crouse GF, Brouwer J, Hanawalt PC. Genetics 1996; 143:1127–1135.

112. van Gool AJ, Verhage R, Swagemakers SMA, van de Putte P, Brouwer J, Troelstra C, Bootsma D, Hoeijmakers JHJ. EMBO J 1994; 13:5361–5369.

113. Kunala S, Brash DE. J Mol Biol 1995; 246:264–272.

114. Selby CP, Sancar A. Science 1993; 260:53–57.

115. Selby CP, Sancar A. Microbiol Rev 1994; 58:317–329.

116. Selby CP, Sancar A. Methods Enzymol 2003; 371:300–324.

117. Oller AR, Fijalkowska IJ, Dunn RL, Schaaper RM. Proc Natl Acad Sci USA 1992; 89:11036–11040.

118. Selby CP, Sancar A. J Biol Chem 1995; 270:4890–4895.

119. Mu D, Tursun M, Duckett DR, Drummond JT, Modrich P, Sancar A. Mol Cell Biol 1997; 17:760–769.

120. Friedberg EC. Microbiol Rev 1988; 52:536–553.

121. Verhage RA, Zeeman A-M, Lombaerts M, van de Putte P, Brouwer J. Mutat Res 1996; 362:155–165.
122. Verhage RA, van Gool AJ, de Groot N, Hoeijmakers JHJ, van de Putte P, Brouwer J. Mol Cell Biol 1996; 16:496–502.
123. Tijsterman M, Brouwer J. J Biol Chem 1999; 274:1199–1202.
124. Tijsterman M, Verhage RA, van de Putte P, Tasseron-De Jong JG, Brouwer J. Proc Natl Acad Sci USA 1997; 94:8027–8032.
125. Mueller JP, Smerdon MJ. Mol Cell Biol 1996; 16:2361–2368.
126. Svejstrup JQ. Nat Rev 2002; 3:21–29.
127. Gregory SM, Sweder KS. Nucleic Acids Res 2001; 29:3080–3086.
128. Jansen LET, den Dulk H, Brouns RM, de Ruijter M, Brandsma JA, Brouwer J. EMBO J 2000; 19:6498–6507.
129. Jansen LET, Belo AI, Hulsker R, Brouwer J. Nucleic Acids Res 2002; 30:3532–3539.
130. Tijsterman M, Tasseron-de Jong JG, Verhage RA, Brouwer J. Mutat Res 1998; 409:181–188.
131. Vos J-MH, Wauthier EL. Mol Cell Biol 1991; 11:2245–2252.
132. Christians FC, Hanawalt PC. Biochemistry 1993; 32:10512–10518.
133. Dammann R, Pfeifer GP. Mol Cell Biol 1997; 17:219–229.
134. Conconi A, Bespalov VA, Smerdon MJ. Proc Natl Acad Sci 2002; 99:649–654.
135. McGregor WG, Mah MC-M, Chen R-H, Maher VM, McCormick JJ. J Biol Chem 1995; 270:27222–27227.

10

Chromatin Structure and the Repair of UV Light-Induced DNA Damage

Fritz Thoma
Institut für Zellbiologie, ETH-Hönggerberg, Zürich, Switzerland

1. INTRODUCTION

This chapter is focused on repair of UV lesions in nucleosomes by photolyase and nucleotide excision repair (NER). Eukaryotic genomes are folded by histone proteins into an array of nucleosomes, which is further condensed into chromatin fibers and higher order structures. Nucleosomes perform complex roles in DNA function. As they present the DNA in two supercoils on the outside, they modulate the accessibility of DNA and they are obstacles for proteins that translocate along the DNA. On the other hand, nucleosomes are dynamic and undergo various structural transitions that affect the accessibility of DNA and can be induced or suppressed by chromatin remodeling activities or by polymerases moving along the DNA during transcription and replication.

In the first portion of the chapter, structural and dynamic properties of nucleosomes are reviewed. Recent crystal data of nucleosome core particles provide information on how nucleosomes may accommodate DNA lesions and modulate damage accessibility. Since photolyases, discussed in the second part, are monomeric enzymes that are regulated by light, the photolyase experiments provide most direct insight on DNA damage accessibility in chromatin. These data suggest that damage recognition is regulated by intrinsic properties of nucleosomes. Nucleosome mobility or disruption and unfolding may expose DNA lesions and make them accessible to repair enzymes. The initial step of damage recognition by NER, discussed in the third section, may occur in a similar way as damage recognition by photolyase, yet the subsequent steps require additional remodeling of nucleosomes to provide space for excision and DNA synthesis and to regenerate chromatin after repair. In the final section, it is discussed how chromatin-remodeling activities may contribute to damage recognition. Since recent work showed that chromatin remodeling activities can act on UV-damaged nucleosomes and facilitate repair, such remodeling factors might act randomly to keep chromatin in a fluid state and facilitate DNA accessibility.

2. NUCLEOSOMES: HETEROGENEITY IN A CONSERVED STRUCTURE

Nucleosomes consist of a core and linker DNA, which makes the connection to the next core. The nucleosome core itself contains of about 147 bp DNA wrapped around an octamer of core histones, two each of H2A, H2B, H3, and H4. In most eukaryotes, but not in the budding yeast *Saccharomyces cerevisiae*, the linker DNA is associated with an additional histone of the H1 class, which stabilizes nucleosomes and chromatin fibers. The core-to-core linker DNA varies in length from about 15 to 90 bp in different organisms, tissues and between individual nucleosomes (1).

During the last few years, several crystal structures of nucleosome core particles were published containing different DNA sequences, histones, and histone variants, and DNA-ligands bound to minor grooves (2–7). While the overall architecture of nucleosomes was unchanged, these structures revealed subtle differences in histone–histone contacts as well as an unexpected flexibility of nucleosomal DNA that makes it possible to understand, how DNA lesions are tolerated and processed on the nucleosome surface.

The first crystals were obtained from chicken erythrocyte nucleosome cores and contained mixed sequence DNA and an unknown fraction of modified histones. They showed that B-DNA forms a left-handed superhelix on the outside of the histone octamer, contains several sharp bends and makes numerous interactions with the histone octamer within (8).

The resolution of crystals was improved to 2.8 Å by using a defined sequence of 146 bp palindromic DNA and recombinant *Xenopus* histones. These crystals revealed a detailed picture on the interactions of histones with DNA and the restricted accessibility of the DNA (Fig. 1A) (2). Each histone consists of a structured, three helix domain called histone fold, and two unstructured tails. The histone fold domains pack to form the heterodimers H2A–H2B and H3–H4. These oligomerize to form the histone octamer, which binds the DNA with major contacts to the phosphates in intervals of 10 bp. The [H3–H4]$_2$ tetramer, binds to the central 60 bp of the

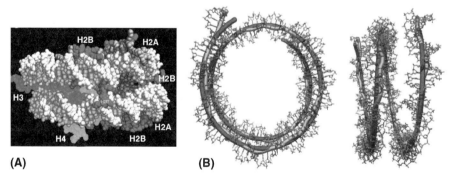

(A)   **(B)**

Figure 1 Structural features of nucleosomes. (A) Space filling view of a nucleosome core particle. Indicated are the DNA (white) and histone proteins H2A (yellow), H2B (red), H3 (blue), and H4 (green). N-terminal tails of histones protrude outside of the particle and are available for modifications and interactions with proteins and other nucleosomes. *Source*: From Ref. 2. (B) The DNA structure in the 147 bp nucleosome core particle. Indicated are the superhelix path of DNA (gold) and best-fit, ideal superhelix (red). *Source*: From Ref. 6. (*See color insert.*)

nucleosomal DNA, while the H2A–H2B dimer organizes 30 bp towards either end of the DNA. The preultimate 10 bp of nucleosomal DNA are bound by a region of H3, which is not an integral part of the [H3–H4]$_2$ tetramer. The N-terminal tails of the histones protrude outside of the disc-shaped particle, where they are accessible to interact with adjacent nucleosomes, enzymes, that modify histone tails, and proteins that interact with modified histones. The N-terminal tails of H3 and H2B pass through channels in the DNA superhelix created by two juxtaposed minor grooves (2).

A very similar structure was reported for palindromic DNA reconstituted with native chicken erythrocyte histones (7). However, crystals made with recombinant yeast histones showed a lack of stabilizing interactions between the two H2A–H2B dimers at the region that is likely involved in holding together the two gyres of the DNA superhelix (3). Similarly, crystals made with H2A.Z, which is a variant of histone H2A, revealed a subtle destabilization of the interactions between the H2A.Z–H2B dimer and the [H3–H4]$_2$-tetramer (4).

With respect to the DNA, the crystal structures revealed a much higher degree of structural deformations and flexibility than originally expected for a conserved particle. In the 146 bp *Xenopus* nucleosome core particle, the two DNA arms extended 72 and 73 bp, respectively, from the pseudo-dyad axis demonstrating that 1 bp difference was accommodated without disruption of the particle (2). Recently, a nucleosome core structure was published at 1.9 Å that contained 147 bp DNA and elucidated the structural properties of DNA with unprecedented accuracy (Fig. 1B) (6). The superhelix path substantially deviated from the ideal superhelix. Overall, the DNA base-pair–step geometry had twice the curvature necessary to accommodate the DNA superhelical path in the nucleosome. This means that the DNA was not only bent towards the histones, but also in other directions. Bending of DNA towards the major groove was smooth. Bending into the minor groove was smooth in the H3–H4 contact domains, but kinked in the H2A–H2B region. A comparison of the 147 bp and 146 bp core structures revealed that different length of DNA is accommodated in a nucleosome core by alterations in the DNA twist (DNA stretching) (6). The crystal structures of nucleosome cores in complex with minor groove DNA-binding ligands further demonstrated that the structure of the histone octamer and its interaction with the DNA remained unaffected by ligand binding, but the DNA structure changed at the binding sites and in adjacent regions (5).

In summary, the structural data established that nucleosomal DNA is highly flexible. It tolerates distortions, bending, kinking as well as ligand binding. Differential stretching and unwinding of DNA likely have consequences on the formation and accommodation of DNA lesions as well as on the mechanisms of nucleosome positioning and nucleosome remodeling. In view of the many histone variants and histone modifications, the large number of nucleosomes per cell, and the variability already observed in the few crystallized examples, we can now accept the heterogeneity of nucleosomes as a regulatory principle of eukaryotic genomes.

3. DYNAMIC PROPERTIES OF NUCLEOSOMES REGULATE DNA ACCESSIBILITY

Nucleosomes regulate the accessibility of DNA to proteins by steric hindrance from histones, the histone tails and from the adjacent DNA supercoil. However, the structure and composition of nucleosomes can change and different conformations can coexist in a dynamic equilibrium (Fig. 2A).

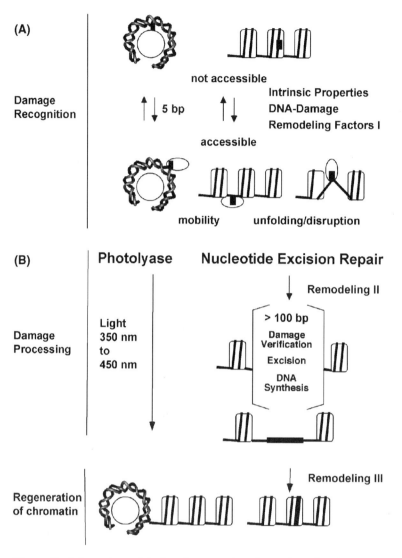

Figure 2 Schematic model of nucleosome dyamics and repair. (A) Structural transitions regulate the accessibility of DNA lesions. A change of the histone octamer position on the DNA
sequence (nucleosome position) by mobility may expose a damage in linker DNA or on the
nucleosome surface. Alternatively, the damage becomes accessible by (partial) disruption or
unfolding of nucleosomes. The structural transitions are affected by intrinsic properties of
nucleosomes (composition, DNA-sequence, and histone modifications), structural distortions
generated by DNA damage (square), nontargeted chromatin remodeling activities (remodeling
factors I). Indicated are histone octamers (white circles, ovals), DNA (black lines), one superhelical turn of DNA wrapped around the histone octamer (left side) and (B) Damage processing. Photolyase binds DNA as a monomeric protein, flips the pyrimidine dimer out into the
active site and reverts it to the native bases with the energy of light. NER is a multistep pathway. Damage recognition (A) is followed by damage verification, excision of an oligonucleotide containing the damage, and DNA-repair synthesis to generate a repair patch (thick line).
NER requires >100 bp space which implies disruption or displacement of nucleosomes (remodeling II). After repair, the repair patches get incorporated in nucleosomes (regeneration of
chromatin; remodeling III). *Source*: From Ref. 88.

In vitro experiments established that nucleosomes undergo salt and temperature dependent unfolding as well as dissociation and reassembly. H2A and H2B are less firmly bound to the ends of nucleosomal DNA, while the H3- and H4- interactions are tighter and less frequently disrupted. The ends of nucleosomal DNA are therefore, preferentially accessible to restriction endonucleases and transcription factors or readily invaded by exonucleases and RNA polymerases (1,10). Histone octamers may occupy a dominant position surrounded by minor positions when assembled on DNA in vitro. The distribution of positions can change with different temperature and salt conditions in a process referred to as nucleosome mobility (Fig. 2A). Sequences that accommodate bending and kinking as it occurs in nucleosome cores act as preferred positioning elements (10–13). Consequently, deformation of DNA as it occurs by damage formation may alter those properties and affect positioning and mobility.

Nucleosome positioning was also investigated in vivo using yeast as a model organism. By insertion and deletion of DNA sequences into specific chromatin regions of minichromosomes, it was demonstrated, that nucleosome positions could be changed in vivo. Positioning is influenced by several parameters including the DNA sequence, chromatin folding, and flanking structures or proteins which act as boundaries and restrict mobility (14–17). Similar to the observations made in vitro, a few high-resolution studies indicated that individual nucleosomes can occupy multiple and overlapping positioning frames in vivo (18–20). One example is the yeast *URA3* gene (Fig. 3C) (15,21). These data support a model that nucleosome positions are in a dynamic equilibrium and that nucleosomes are mobile in living cells.

Several mechanisms of nucleosome translocation have been discussed (e.g., in Refs. 10, 13, 22). One possibility is a rotation of the whole nucleosomal DNA about its long axis with respect to the surface of the histone octamer or the rotation of the histone octamer as a unit within the superhelix, or a complete dissociation of the histone octamer and reassembly close by. These three mechanisms are unlikely, since they require breakage of many histone–DNA-contacts. Another possibility is overtwisting or undertwisting of DNA at one end and step-by-step propagation of the twist through the nucleosome. Finally, local uncoiling of DNA from the histone surface could provide a mechanism for translocation. DNA might move on the histone surface by formation of a DNA bulge at one end and propagation of the bulge through the nucleosome. Both, movement of a bulge or twist require only local disruption and reformation of histone–DNA contacts and appear therefore to be favored mechanisms to explain nucleosome mobility. These mechanisms are supported by the local variations of DNA structure and twist observed in the crystals.

So far, the structural and dynamic properties of nucleosomes were discussed as intrinsic properties provided by the DNA sequence and histone composition. However, chromatin-remodeling complexes have been identified which interact with nucleosomes and alleviate or enforce their repressive nature. Some are targeted to specific promoters via interactions with DNA binding transcription factors or they interact and comigrate with elongating RNA polymerases. Finally, abundant remodeling activities may act globally to promote structural adjustments in constrained chromatin thereby performing a chromatin surveillance function (Fig. 2A) (22–25).

One group of remodeling activities consists of ATP-dependent complexes. The energy of ATP hydrolysis is used to disrupt the chromatin structure which can be scored by enhanced factor binding, disruption of the DNase I cleavage pattern of mononucleosomes, formation of dinucleosomes, movements of histone octamers, and by generation of nuclease hypersensitive sites. The other group of complexes, referred to as modifying complexes, acts by chemical modification of the histones mostly at their N-terminal tails that protrude outside of the particle. Those modifications

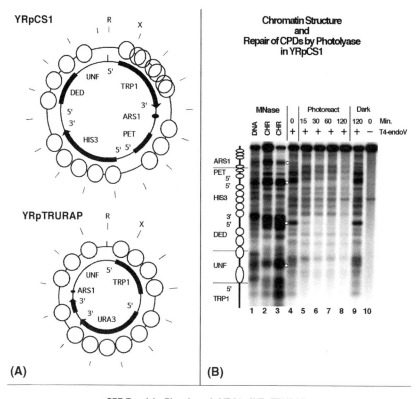

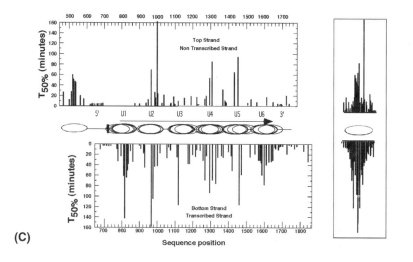

Figure 3 (*Caption on facing page*)

include the acetylation of lysines, the methylation of lysines and arginines, the phosphorylation of serines and threonines, the ubiquitylation of lysines, the sumoylation of lysines, and the poly-ADP-ribosylation of glutamic acids (26).

The consequences of histone modifications are two fold. As long as the modifications persist they may alter the intrinsic properties of nucleosomes or the interactions of histones with other nucleosomes and chromatin fibers to stabilize or (de)stabilize the structures. For example, loss of positive charges by acetylation may compromise DNA binding and nucleosome stability. However, hyperacetylation and removal of histone tails (the sites of acetylation) only moderately increased the accessibility of nucleosomal DNA for restriction endonucleases and repair enzymes in vitro (27,28). Thus, hyperacetylation might act above the nucleosome level and destabilize higher order structures thereby making DNA accessible for repair. Indeed, inhibition of histone deacetylation by sodium butyrate stimulated NER in human cells (29). Whether an opening of chromatin structure was the cause for enhanced repair remains to be demonstrated.

Alternatively, histone modifications act by recruiting modification specific proteins which themselves influence the stability and accessibility of nucleosomes and chromatin domains. A well-known example is formation of heterochromatin

---

Figure 3 (*Facing page*) Repair of CPDs by photolyase is modulated by chromatin structure. (A) Yeast minichromosomes (YRpCS1 and YRpTRURAP) serve as model substrates with defined chromatin structures and transcriptional properties. ARS1 is an origin of replication; arrows indicate genes (TRP1, URA3, and HIS3); DED, PET are truncated genes; $5'$ and $3'$ are promoter regions and $3'$ ends, respectively; circles represent positioned nucleosomes. R, X indicate restriction sites for EcoRI and XbaI. Radial dashes indicate intervals of 0.2 kb. (B) Comparison of chromatin structure and CPD repair by photolyase. For chromatin analysis (lanes 1–3), chromatin (CHR) and DNA (DNA) were partially digested with micrococcal nuclease (MNase), and the cutting sites were displayed by indirect endlabeling. A schematic interpretation of chromatin structure is shown (left side). Chromatin regions of 140–200 bp that are protected against MNase cleavage represent footprints of positioned nucleosomes (ovals), cutting sites between nucleosomes represent linker DNA, long regions with multiple cutting sites represent nuclease sensitive regions (ARS1; $5'$PET-$5'$HIS3; $3'$HIS3-$5'$DED). Repair analysis is shown for FTY117 (YRpCS1 $rad1\delta$), which is deficient in NER (lanes 4–10). Cells were UV irradiated with 100 J/m^2 and exposed to photoreactivating light for 15–120 min (lanes 5–8). DNA was extracted and cut at CPDs with T4-endonuclease V (+ T4-endoV, lanes 4–9). Lane 10 shows irradiated DNA (same as lane 4) without T4-endoV cleavage. An aliquot of cells was kept in the dark for 120 min (lane 9). Dots indicate fast repair in nuclease sensitive regions and linker DNA. CPD bands, which correspond to the footprints of positioned nucleosomes, are slowly repaired indirect endlabeling: DNA digested with XbaI hybridized to a radioactively labelled XbaI-EcoRI, fractionated on an alkaline agarose gel, blotted to a membrane and hybridized to a radio actively labelled XbaI-EcoRI fragment. (A) and (C) Correlation of photorepair in the *URA3* gene of YRpTRURAP with chromatin structure at high resolution using primer extension (left). The superposition of all nucleosomal regions (right) shows slow repair in the center of nucleosomes and increased repair rates towards the end suggesting that nucleosome mobility may expose DNA lesions in linker DNA and facilitate repair. Bars show the time (minutes) used to remove 50% of the lesions (T$_{50\%}$). Indicated are: the major nucleosome positions of *URA3* (ovals U1–U6) and the flanking *TRP1* region, the *URA3* promoter region ($5'$), TATA-box (black square), the direction of transcription (arrow). Overlapping ovals indicate that each nucleosome can adopt multiple positioning frames. *Source*: Adapted from Refs. 51, 54.

domains by a coordinated deacetylation and methylation of histones combined with a recruitment of silencing proteins that bind to the modified histones and generate a repressive chromatin structure (30). Although heterochromatin in yeast silences transcription, DNA repair of UV lesions was still possible which illustrates that the DNA remains accessible and dynamic despite a more compact chromatin structure (31).

How all these covalent modifications and the combination of modifications impact the structural and dynamic properties of nucleosomes is unknown, but it adds an additional dimension with respect to structural and functional heterogeneity of chromatin.

4. DAMAGE TOLERANCE OF NUCLEOSOMES

Cyclobutane pyrimidine dimers (CPDs) and pyrimidine–pyrimidone (6–4) photoproducts (6–4PPs) are the two major classes of DNA lesions generated by UV light. The solution structure of DNA duplex-decamer containing a (6–4)PP revealed a bending of the DNA helix by 44° (32). The crystal structure of a DNA decamer containing a *cis, syn* thymine dimer showed that the CPD bends DNA approximately 30° toward the major groove and unwinds approximately 9° (33). The severe distortions in the DNA imply that the formation of lesions depends on the conformational flexibility of the DNA in protein/DNA complexes (34).

Analysis of the UV-damage distribution in nucleosomes (UV-photofootprinting) should provide insight into structural constraints of DNA and sequence dependent heterogeneity of nucleosomes. No UV-footprint was detected in a nucleosome reconstituted on 5S-rDNA (35). However, UV-irradiation of nucleosomes containing poly(dA)·(dT) DNA (T-tracts) generated a CPD pattern much different from that obtained in naked DNA indicating that the DNA structure was altered in these nucleosomes (28,36,37). While these examples illustrate the sequence dependent heterogeneity of damage formation in individual nucleosomes, more general information was obtained by analysis of UV-lesions in a population of nucleosome cores isolated from irradiated human cell cultures. The CPD distribution within the nucleosomal DNA was modulated with a periodicity of 10.3 bp closely reflecting the average periodicity of nucleosomal DNA. The sites of maximum CPD formation in core DNA mapped to positions where the phosphate backbone was farthest from the core histone surface (38). In contrast to the CPD pattern, the distribution of 6–4PPs was much more random (39). The reasons for the different modulation of CPD and 6–4PP formation remain to be elucidated. It must be emphasized that not one site was identified so far that was completely resistant to damage formation. Thus, damage distribution is another piece of information that illustrates the astonishing conformational flexibility of nucleosomal DNA.

In view of the structural and dynamic properties of nucleosomes discussed above, it is important to know whether DNA lesions by themselves may change the structure and stability of nucleosomes (Fig. 2A). This seems not the case for UV lesions, since UV damaged nucleosomes could be purified (38,39) and reconstituted nucleosome cores could be irradiated without remarkable disruption, destabilization or change in the rotational setting (36,37). Under physiologically relevant conditions, damaged nucleosomes contain not more than one DNA lesion. In view of the numerous histone DNA contact sites and the conformational flexibility of nucleosomal DNA, the accommodation of DNA lesions, here referred to as damage tolerance, might therefore be the rule rather than exception.

However, when nucleosomes were assembled on damaged DNA in vitro, CPDs reduced the efficiency of core histone binding (40) and influenced the rotational setting of the DNA on the histone octamer (36,41). Chromatin assembly on damaged DNA in vivo is probably a rare event and its biological significance is not known. It may occur in the context of DNA replication, if the DNA lesions are bypassed by translesion synthesis and chromatin is reassembled behind the replication fork. It may also happen during transcription, if the damage is located on the nontranscribed strand and does not block the elongating RNA polymerase. Transcription dependent chromatin transitions involve disruption and reassembly of nucleosomes around the RNA polymerase (42).

In summary, nucleosomes tolerate UV lesions. Disruption or destabilization of nucleosomes by damage formation per se is not the predominant mechanism that allows damage recognition and repair in nucleosomes.

5. REPAIR OF NUCLEOSOMES BY PHOTOLYASE

DNA repair by photolyase is a specialized system for removing the major UV-induced DNA lesions, CPDs and 6–4PPs. The structure and reaction mechanisms have been reviewed in detail (43). Briefly, photolyases are monomeric proteins of about 450–550 amino acids. They contain FAD as an essential cofactor for catalysis and binding to damaged DNA. Methenyltetrahydrofolate or 8-hydroxy-7,8-dimethyl-5-deazariboflavin acts as a second cofactor and chromophore to enhance repair under limiting light conditions. Photolyases are structure specific enzymes, that bind to DNA, bend DNA by about 36° and most likely flip out the pyrimidine dimer from the double helix into the active site cavity. Following repair, the dinucleotide moves out of the cavity and the enzyme dissociates from the DNA (43).

To address how DNA lesions are recognized and processed in nucleosomes, it was important to investigate repair in vitro with defined sequence nucleosomes as substrates. Initial studies were done with *E. coli* photolyase and a nucleosome reconstituted on a 134 bp long DNA fragment. The nucleosome was precisely positioned with a defined rotational setting and contained polypyrimidine tracts, which allowed one to monitor repair of CPDs over three helical turns. Moreover, the DNA fragment was short and did not provide space for nucleosome mobility and repositioning. In this nucleosome, repair was severely inhibited compared with efficient repair in naked DNA (28). A strong inhibition of *E.coli* photolyase was also observed when nucleosomes were reconstituted on a 5S-rDNA sequence containing a CPD at a specific site (44). In another construct, a nucleosome was reconstituted at one end of a 226-bp-long DNA fragment to allow space for damage or repair-induced rearrangement. Repair was inhibited in the nucleosome, but efficient outside, and no rearrangement of the positioned nucleosome was observed (37).

T4 endonuclease V is a base excision repair enzyme from bacteriophage T4 that generates single strand cuts at CPDs (45). It kinks the DNA with a 60° inclination at the central thymine dimer. The adenine base complementary to the 5' side of the thymine dimer is completely flipped out of the DNA duplex and trapped in a cavity on the protein surface (46). When T4-endonuclease V was tested on the same nucleosome substrates, the reaction was as severely inhibited as with photolyase (28,44).

Taken together, those experiments demonstrated that reconstituting DNA into nucleosomes caused a strong inhibition. Only little repair was observed at some specific sites (see below). Moreover, exposure of nucleosomes to photoreactivating enzymes and T4-endonuclease V did not remarkably destabilize nucleosomes or alter nucleosome positions. Thus, in vitro, nucleosomes are severe obstacles for processing of CPDs, but lesions immediately outside of nucleosomes are efficiently processed.

In living cells DNA lesions are completely repaired, despite packaging of DNA into chromatin. Removal of CPDs by photolyase has been observed in many eukaryotic cells (references in 47), including mice and human cells carrying photolyase as a transgene (48,49). The extent of repair observed in these studies implies that mechanisms exist in vivo to make CPDs and 6–4PPs accessible in chromatin.

Studies of photolyase in yeast turned out to be most rewarding with respect to understanding damage recognition in chromatin for several reasons. Damage formation by UV light and repair by photolyase are regulated by light. This allows doing precise time course experiments to assess dynamic properties of chromatin. Moreover, no data support a requirement of additional proteins for photorepair, although photolyase may stimulate NER in the absence of light (50). Thus, the photorepair results can be interpreted as an interaction of a monomeric enzyme with chromatin and serve as a model for damage recognition by other pathways.

A typical example of a repair experiment with yeast is presented in Fig. 3. Yeast cells containing minichromosomes were irradiated in suspension with UV light and further exposed to photoreactivating light. The DNA was extracted, cut at CPDs with T4-endonuclease V and the cutting sites were displayed by indirect end-labeling. For comparison, chromatin was isolated, cut with micrococcal nuclease (MNase) and the cutting sites are displayed on the same gel side by side with cuts at CPDs. Preferentially MNase cuts in the linker DNA between nucleosomes and in nonnucleosomal regions, while nucleosomal DNA is resistant. Positioned nucleosomes are identified as regions of about 140–200 bp that are protected against MNase digestion (Fig. 3).

Under light saturating conditions, CPDs that map in open, nonnucleosomal regions were repaired by photolyase in less than 15 min. These sites include lesions in linker DNA between nucleosome cores and the nuclease sensitive promoter regions, 3'ends of genes, and the open origin of replication (ARS1) (51–53). In contrast to nonnucleosomal DNA, DNA lesions that correlate with nucleosome footprints were repaired in about 2 hr. This means, that the presence of nucleosomes slowed down repair compared with open regions, but the inhibition was not complete. Given that 15 min were required to repair nucleosome free DNA, the data suggest a dynamic process by which nucleosomal DNA is transiently exposed for 15 min within two hours.

Detailed insight on nucleosome repair was obtained by nucleotide resolution analysis of the URA3 gene using a primer extension technique for mapping CPDs (54). The URA3 gene is transcribed, but maintains six positioned nucleosomes, each of them represents multiple positioning frames that vary within a few base pairs (15,21). Repair of CPDs by photolyase was heterogeneous along the gene with rapid repair in the nucleosome free promoter and in linker DNA. Moreover, repair was slow in the center and increased gradually towards the periphery. This modulation of repair was most obvious when the repair patterns were superimposed (54) (Fig. 3C).

Taken together, the strong inhibition of repair observed in nucleosomes in vitro and complete repair observed in vivo imply that nucleosomes in vivo are less stable, undergo structural transitions that allow the damage to be recognized and repaired. Assuming that photolyase does not recruit remodeling activities to the DNA lesion, the photorepair data provide direct information on the intrinsic structural and dynamic properties of chromatin in living cells.

The combination of multiple nucleosome positions observed by nuclease footprinting with a gradual increase of repair rates towards the end of nucleosomes can be explained in different ways (54). The simplest model is based on nucleosome mobility. Nucleosomes may change their positions in a dynamic equilibrium thereby releasing DNA lesions into linker DNA where they are rapidly recognized and repaired (Fig. 2A). DNA lesions generated in the center of nucleosomes will appear less frequently in linker DNA than lesions generated towards the ends of nucleosomes. Alternatively, one might consider that partial dissociation or uncoiling of nucleosomes regulates damage recognition. An increase of repair rates towards the periphery is consistent with such an uncoiling model, but it is less consistent with the multiple positions observed by footprinting.

Although nucleosome disruption and mobility were discussed separately it is not unlikely that both mechanisms are involved. Nucleosome disruption might predominate, if nucleosome mobility is restricted. These models of course imply that nucleosomes can move despite the damage and that the damage does not inhibit dynamic properties of nucleosomes. Moreover, we need to consider that nucleosomes in yeast and in particular in the *URA3* gene are closely spaced and might frequently be in face-to-face contact. Hence, nucleosome movement for more than a linker length can only occur together with displacement of the flanking nucleosomes. Such a coordinated move may occur by the DNA-bulging mechanism. A bulge that moves along the nucleosome surface may be propagated to the next nucleosome as soon as the nucleosomes happen to be in face-to-face contact (54).

6. REPAIR OF NUCLEOSOMES BY NER

In contrast to photoreactivation, NER is a multistep pathway, which removes a wide variety of bulky adducts as well as CPDs and 6-4PPs. Nucleotide excision repair is divided into two subpathways. Transcription coupled repair (TCR) removes lesions from the transcribed strand of genes transcribed by RNAP2 and also from yeast ribosomal genes transcribed by RNAP1. In TCR, elongating RNA polymerases are blocked at lesions and thereby act in damage recognition. Global genome repair (GGR), on the other hand, removes DNA lesions from nontranscribed chromatin, including the nontranscribed strand of transcribed genes (55–59). Thus, it is GGR that must access DNA lesions in nucleosomes.

The excision reaction has been reconstituted with human and yeast components on naked DNA substrates containing different lesions. According to a recent model, the mammalian NER steps include damage recognition by XPC-hHR23B, melting of about 25 bp DNA around the lesion by the helicase activities of TFIIH (XPB, XPD), DNA damage verification by XPA and RPA, excision of an oligonucleotide containing the DNA lesion by XPG and ERCC1/XPF. Finally, the gap is filled by DNA synthesis using the replication machinery (56,60). Data obtained by local UV irradiation combined with fluorescent antibody labeling in normal and repair-deficient human cells support a sequential assembly of repair proteins at the site of the damage. The earliest known NER factor in the reaction mechanism to be identified was XPC. The damage recognition complex XPC–hHR23B appeared to be essential for the recruitment of all subsequent NER factors (61).

In the yeast *S. cerevisiae*, the basic reaction requires the damage binding factors Rad14 (the yeast homolog of XPA), RPA, and the Rad4–Rad23 complex (the yeast homolog of XPC–hHR23B), the transcription factor TFIIH with the two DNA helicases Rad3 and Rad25, and the two endonucleases, the Rad1–Rad10 complex and Rad2. In addition, the Rad7–Rad16 complex plays a role in damage recognition, stimulates the incision reaction (57) and is required for excision or the damaged oligonucleotide (62).

In vitro, NER of cell extracts was severely inhibited in reconstituted nucleosome arrays (63) and in *Simian virus 40* minichromosomes (64). Studies with defined sequence nucleosomes confirmed the repressive role of nucleosomes on NER. *Xenopus* nuclear extracts repaired a single UV photoproduct located five bases from the dyad center of a positioned nucleosome at about half the rate at which the naked DNA fragment was repaired (44). Purified human NER factors were unable to excise dinucleosome substrates that contained 6–4PPs at the center of the nucleosome or in the linker DNA (65). Other studies reported that nucleosomes inhibited the repair of a CPD more severely than repair of the acetylaminofluorene–guanine adduct or a (6–4) photoproduct (66). Similarly, repair of a platinum–DNA adduct was inhibited by nucleosomes (67). Thus, all NER reactions were severely inhibited in nucleosomes. Some variation of repair inhibition observed in different experiments might reflect different affinities of damage recognition proteins for different lesions and/or differential stabilities of damaged nucleosomes.

As with photoreactivation, detailed insight in NER of chromatin was obtained by direct comparison of chromatin structure and NER in yeast. A modulation of NER by chromatin and transcription was observed in the YRpTRURAP minichromosome that contains the *URA3* gene (68). High-resolution analysis of CPD removal from the *URA3* gene showed that repair was fast in linker DNA and slow in positioned nucleosome. This modulation was observed on the nontranscribed strand, while the transcribed strand was repaired homogenously by TCR (69). Similar results were reported for the genomic copy of *URA3* and for removal of 6–4PPs indicating that modulation by chromatin structure is not damage dependent (70). A modulation of NER by positioned nucleosomes was also reported for the inactive yeast *MET17* gene (71). Taken together, the chromatin modulation of GGR observed in yeast and the strong inhibition reported in vitro resembled the repair properties observed by photolyase. Thus, in analogy to photoreactivation, exposure of DNA lesions in linker DNA or a disruption or unfolding of nucleosomes by natural dynamic properties would be sufficient to sense the damage by NER proteins and to start the reaction (Fig. 2A).

Much in contrast to photoreactivation, however, the basic NER reactions require more space. About 25 bp of DNA are unwound in the open complex (72) and the human excision complex needs about 100 bp of DNA to excise the lesion in vitro (73). Since the linker DNA between nucleosomes is too short to accommodate a repair reaction, nucleosomes must be disrupted or rearranged (indicated as remodeling II in Fig. 2B). Probably sufficient space could be provided by disruption of one nucleosome. Alternatively, the ends of nucleosomal DNA could be unwrapped from the two nucleosomes flanking the DNA lesion, e.g., by disruption of H2A–H2B interactions with DNA. Finally, nucleosomes flanking the lesion might be forced to slide away by a few base pairs. There is no obvious need for opening up a chromatin domain of several nucleosomes. How this remodeling occurs and how large the remodeled regions might be are unknown.

In yeast, GGR depends on Rad7 and Rad16 (74–77). The Rad7p–Rad16p complex has several properties that suggest a chromatin remodeling activity combined with damage recognition. First, Rad16 has homology to Snf2, a protein of the SWI/SNF nucleosome-remodeling complex (78). Second, the Rad7–Rad16 complex recognizes UV lesion in an ATP-dependent way suggesting that ATP-hydrolysis promotes translocation of the complex on the DNA in search of DNA lesions (79,80). Third, Rad7p–Rad16p together with Abf1p can generate superhelical torsion in DNA in vitro (62). These properties put Rad7p–Rad16p at the beginning of the NER reaction. Both, the energy dependent search for DNA lesions and the potential to generate torsional stress may help to remove or to destabilize nucleosomes. Since Rad7p–Rad16p also stimulate the incision and excision reaction on naked DNA (62,79,80), Rad7p–Rad16p may contribute to those NER steps in vivo, but probably independent of nucleosome remodeling.

No Rad7–Rad16 homologues were found in mammalian cells (81). However, recent work pointed out a central role for the DNA damage binding protein complex (DDB) in GGR (82). The damage binding protein complex is composed of two subunits, DDB1 (p127) and DDB2 (p48).Global genome repair is apparently controlled through the activated product of the p53 tumor suppressor gene in human cells and appears to be mediated through DDB2 (83). In vitro, DDB binds to UV-irradiated DNA and stimulates a reconstituted excision reaction (84). Since binding of DDB to damaged DNA bends DNA by 54°–57° (85), it was hypothesized that DDB binds to the CPD and distorts the DNA helix so that XPC–HR23B can be recruited (86). In vivo, DDB rapidly translocates to UV-damaged DNA sites (84). The binding characteristics of DDB2 and XPC to either CPDs or 6–4PPs were tested in repair-deficient xeroderma pigmentosum (XP)-A cells that stably expressed photoproduct-specific photolyases. DDB2 localized to UV-irradiated sites that contained either CPDs or 6–4PPs. However, XPC localized only to UV-irradiated sites that contained 6–4PPs, suggesting that XPC does not efficiently recognize CPDs in vivo. Only XPC did localize to CPDs when DDB2 was overexpressed in the same cell, signifying that DDB2 activates the recruitment of XPC to CPDs and may be the initial recognition factor in the NER pathway (48). In addition, it was reported that DDB interacts with the human STAGA complex, a chromatin acetylating transcription coactivator. Thus, DDB has the potential to bind to DNA lesions and to recruit a remodeling complex to the site of DNA lesions (87).

DNA helicases are another class of enzymes that can disrupt nucleosomes. It seems conceivable that the TFIIH helicases of the NER process could disrupt nucleosomes and open up the repair site. Once space is available, excision and DNA-repair synthesis may occur generating a repair patch of newly synthesized DNA (47).

In a last step of NER, the repair patch gets incorporated into nuclease resistant nucleosomes by a process originally referred to as nucleosome rearrangements (88). The mechanism of this chromatin regeneration step (indicated as remodeling III, Fig. 2B) is unknown, but there is evidence that chromatin assembly factor CAF-1 is involved. First, in *S. cerevisiae*, the *CAC1*, *CAC2*, and *CAC3* genes encode the three CAF-I subunits. Deletion of any of the three *CAC* genes conferred an increase in sensitivity to killing by UV radiation. Epistatic analysis suggested that CAF-I contributes to error-free postreplicative damage repair, but may have an auxiliary role in NER (89). Second, it was demonstrated that the NER pathway was required for recruitment of CAF-1 to sites of UV damage. Thus, these observations support the hypothesis that CAF-1 contributes to regeneration of chromatin after repair synthesis and ensures the maintenance of epigenetic information by acting locally at repair sites (90).

7. SITE-SPECIFIC REPAIR IN NUCLEOSOME AND DAMAGE RECOGNITION

So far, the repair mechanisms have been discussed starting from damage sensing outside of nucleosomes or in disrupted nucleosomes. However, the questions remain whether and how DNA lesions can be recognized and repaired on the nucleosome surface. An obvious problem is that the accessibility of nucleosomal DNA to proteins is sterically restricted by the histones in the inside of the particle, the histone tails protruding outside and by the proximity of the two DNA gyres (Fig. 1A). Thus, small and monomeric enzymes like photolyase have a better chance to access DNA lesions than multiprotein complexes. In addition, binding of repair enzymes could be influenced by surface properties, e.g., by the negative charges of the proximal DNA turn or the charged residues of the histones. Moreover, modifications of histone tails alter both, the steric and charge contributions, and may recruit proteins that further reduce the accessibility of DNA. Finally, binding and processing of the lesions involve substantial distortions in the DNA that have to be tolerated in the nucleosome or, alternatively, may lead to a partial disruption or dissociation of the DNA.

The accessibility of CPDs in nucleosomes was tested for *E. coli* photolyase and by T4-endonuclease V in a nucleosome containing long stretches of pyrimidines (28). Both proteins are monomeric enzymes, but different with respect of the extent of DNA bending, flipping out of the pyrimidine dimer or the base opposite to the dimer, and the reaction mechanisms leading to reversal of the damage or a single strand cut at the CPD site (see above). Despite the general inhibition, both enzymes did recognize and process CPDs at some sites of nucleosomal DNA. Surprisingly, these repair sites, did not coincide with sites where the sugar-phosphate backbone faces outside and could be cut by DNaseI. Thus, DNaseI and the repair enzymes recognized different features of nucleosomal DNA. Repair was found at sites where the DNA backbone was exposed on the top of the nucleosome, but other sites were slowly repaired, although the DNA was exposed. Hence, exposure of DNA on the top turn is not a sufficient criterion for efficient repair. In view of the recent crystal data (Fig. 1), it seems possible that the direction of bending of the DNA on the nucleosome surface may play a crucial role for damage recognition.

Information from in vivo experiments with respect to site-specific repair in nucleosomes is scarce. When nucleosomes were isolated from mammalian cells after different repair times neither preferential removal of CPDs from the ends nor from the outside, nor from any particular site was obvious (91). This result suggested that the location of the lesion in the nucleosome was not critical for damage removal by NER. On the other hand, the yeast experiments showed enhanced repair towards the end of positioned nucleosomes for NER and photolyase. This could be interpreted in terms of preferential accessibility and repair of nucleosomes, if we assume that nucleosomes do not change positions. The same experiments, however, revealed substantial heterogeneity in repair for sites mapping within nucleosomes, but being only a few base pairs apart (Fig. 3) (54,69). This observation is consistent with a modulation of CPD accessibility on the nucleosome surface.

8. CHROMATIN REMODELING AND DNA REPAIR

The mechanisms described above for damage recognition are based on intrinsic properties of nucleosomes. From many examples in transcription regulation, we have learned that accessibility of DNA could be facilitated by chromatin remodeling complexes. Some activities are recruited by transcription factors and act locally to modify or disrupt individual nucleosomes in promoter regions, while others act more broadly and modify entire chromatin domains. Thus, it is conceptually appealing that the same or similar complexes are used to relieve the nucleosomal constraints in DNA repair and to facilitate access of repair proteins.

Several reports indicate that remodeling might contribute to DNA repair. Yeast cells mutated in *INO80*, a gene of an ATPase-dependent remodeling complex in yeast, were found to be sensitive to agents that cause DNA damage (92). Similarly, mutations in *NPS1/STH1*, the catalytic subunit of the RSC complex (remodels structure of chromatin), or its reduced expression enhanced sensitivity to DNA damaging treatments including UV light (93). Histone acetyltransferase Gcn5, which is part of the SAGA and ADA complexes, influences photorepair and NER at the *MFA2* locus of yeast, yet it has little effect on these processes for most of the genome (94). Moreover, it was reported that p53 acts as a chromatin accessibility factor, mediating UV-induced global chromatin relaxation (95).

There are two different ways how remodeling activities could facilitate repair. A recruitment model implies, that the damage is recognized first by a damage recognition protein which then recruits a remodeling activity that "opens up" chromatin and provides the space for the repair reactions. Such a mechanism is conceivable for the early steps of NER as discussed above. Alternatively, remodeling might occur by random interactions of remodeling complexes with chromatin (Fig. 2A). This model, however, implies that damaged chromatin can be remodeled and that DNA lesions do not inhibit the mechanisms of chromatin remodeling.

To address this topic, several ATP-dependent remodeling activities have been tested on various nucleosome substrates and DNA lesions in vitro. ACF (ATP-utilizing chromatin assembly and remodeling factor) facilitated dual incision by NER at a 6–4PP placed in linker DNA, but not in the nucleosomes (65).

SWI/SNF (switch/sucrose nonfermenting) was first discovered in yeast and later in higher eukaryotes and is probably the best-characterized remodeling complex. It increases the accessibility of nucleosomal DNA to transcription factors, DNaseI and restriction endonucleases, and induces octamer sliding (22,24,25). Recent UV-photofootprint experiments showed that the DNA structure changed when nucleosomes were complexed with ySWI/SNF in the presence of ATP. Moreover, the inhibition by nucleosomes was relieved and allowed almost uniform repair of CPDs by photolyase along the DNA. Thus, SWI/SNF can remodel UV-damaged nucleosomes and destabilized the structure of the complex to such an extent that photolyase can access and process the lesion (37). ySWI/SNF also stimulated the excision by human excision nuclease of acetylaminofluorene–guanine and (6–4) PPs, but CPDs were not removed from nucleosomal DNA (66,96).

Yeast ISW2 is a two-subunit complex belonging to the ISWI group of ATP-dependent remodeling factors which influence nucleosome positioning in vivo (97) as well as spacing of nucleosomes in vitro (98). When yISW2 was tested on UV-damaged nucleosomes, it moved the nucleosome from the end to a more central position in an ATP-dependent manner. The CPDs inside the nucleosome remained refractory to repair by photolyase, while the DNA lesions outside of the nucleosome were efficiently repaired. These data showed that yISW2 could alter positions of nucleosomes on UV-damaged DNA, thereby releasing CPDs towards linker DNA (37).

The above experiments provided the proof of principle that UV-damaged nucleosomes could be remodeled. Several mechanisms have been discussed how nucleosomes are remodeled by ATP-dependent complexes, from disruption and reformation of all histone–DNA contacts to the formation and propagation of a DNA-bulge or local twist (see above). Irrespective of the underlying mechanism, UV lesions do not inhibit these structural transitions. Based on these observations it seems reasonable to strengthen the random remodeling hypothesis. Since ATP-dependent remodeling complexes change the structure or positions of nucleosomes, these complexes might act randomly on chromatin and keep chromatin in a fluid state. This would facilitate DNA recognition in transcriptional regulation as well as in repair. Thus, remodeling complexes might perform a rather general role in maintenance of chromosome structure.

9. CONCLUSIONS

Recent years have provided deep insight into the locus and site-specific heterogeneity of chromatin with respect to sequence, composition, and chromatin dynamics. Simultaneously, it became obvious that this heterogeneity projects into DNA-damage formation and damage recognition with consequences for mutagenesis and genome stability. While the basic components and reactions of several repair pathways are known, we are just at the beginning to learn how UV lesions are removed from nucleosomes. Thus, more studies will be required to elucidate the molecular mechanisms of DNA repair in living cells and to verify the mechanism in a reconstituted chromatin repair system in vitro. Moreover, the experience with UV lesions, photolyase, and NER may promote chromatin studies on as yet unexplored DNA lesions and repair pathways.

ACKNOWLEDGMENTS

We thank U. Suter for continuous support and A. Bucceri, M. Fink, and M. Lopes for comments on the manuscript. This work was supported by the Swiss National Science Foundation and the Swiss Federal Institute of Technology (ETH).

REFERENCES

1. Van Holde KE. Chromatin. Berlin: Springer Verlag, 1989.
2. Luger K, Mader AW, Richmond RK, Sargent DF, Richmond TJ. Crystal structure of the nucleosome core particle at 2.8 A resolution. Nature 1997; 389(6648):251–260.
3. White CL, Suto RK, Luger K. Structure of the yeast nucleosome core particle reveals fundamental changes in internucleosome interactions. Embo J 2001; 20(18):5207–5218.
4. Suto RK, Clarkson MJ, Tremethick DJ, Luger K. Crystal structure of a nucleosome core particle containing the variant histone H2A.Z. Nat Struct Biol 2000; 7(12):1121–1124.
5. Suto RK, Edayathumangalam RS, White CL, Melander C, Gottesfeld JM, Dervan PB, Luger K. Crystal structures of nucleosome core particles in complex with minor groove DNA-binding ligands. J Mol Biol 2003; 326(2):371–380.
6. Richmond TJ, Davey CA. The structure of DNA in the nucleosome core. Nature 2003; 423(6936):145–150.

7. Harp JM, Hanson BL, Timm DE, Bunick GJ. Asymmetries in the nucleosome core particle at 2.5 A resolution. Acta Crystallogr D Biol Crystallogr 56 Pt 2000; 12: 1513–1534.

8. Richmond TJ, Finch JT, Rushton B, Rhodes D, Klug A. Structure of the nucleosome core particle at 7 A resolution. Nature 1984; 311(5986):532–537.

9. Luger K, Rechsteiner TJ, Flaus AJ, Waye M, Richmond TJ. Characterization of nucleosome core particles containing histone proteins made in bacteria. J Mol Biol 1997; 272(3): 301–311.

10. Widom J. Structure, dynamics, and function of chromatin in vitro. Annu Rev Biophys Biomol Struct 1998; 27:285–327.

11. Simpson RT, Stafford DW. Structural features of a phased nucleosome core particle. Proc Natl Acad Sci USA 1983; 80(1):51–55.

12. Pennings S, Meersseman G, Bradbury EM. Mobility of positioned nucleosomes on 5 S rDNA. J Mol Biol 1991; 220(1):101–110.

13. Flaus A, Richmond TJ. Positioning and stability of nucleosomes on MMTV 3'LTR sequences. J Mol Biol 1998; 275(3):427–441.

14. Thoma F, Simpson RT. Local protein-DNA interactions may determine nucleosome positions on yeast plasmids. Nature 1985; 315(6016):250–252.

15. Thoma F. Protein-DNA interactions and nuclease-sensitive regions determine nucleosome positions on yeast plasmid chromatin. J Mol Biol 1986; 190(2):177–190.

16. Thoma F, Zatchej M. Chromatin folding modulates nucleosome positioning in yeast minichromosomes. Cell 1988; 55(6):945–953.

17. Tanaka S, Zatchej M, Thoma F. Artificial nucleosome positioning sequences tested in yeast minichromosomes: a strong rotational setting is not sufficient to position nucleosomes in vivo. Embo J 1992; 11(3):1187–1193.

18. Zhang X-Y, Hörz W. Nucleosomes are positioned on mouse satellite DNA in multiple highly specific frames that are correlated with a diverged subrepeat of nine base-pairs. J Mol Biol 1984; 176:105–129.

19. Buttinelli M, Di Mauro E, Negri R. Multiple nucleosome positioning with unique rotational setting for the Saccharomyces cerevisiae 5S rRNA gene in vitro and in vivo. Proc Natl Acad Sci USA 1993; 90(20):9315–9319.

20. Fragoso G, John S, Roberts MS, Hager GL. Nucleosome positioning on the MMTV LTR results from the frequency- biased occupancy of multiple frames. Genes Dev 1995; 9(15):1933–1947.

21. Tanaka S, Livingstone-Zatchej M, Thoma F. Chromatin structure of the yeast URA3 gene at high resolution provides insight into structure and positioning of nucleosomes in the chromosomal context. J Mol Biol 1996; 257(5):919–934.

22. Narlikar GJ, Fan HY, Kingston RE. Cooperation between complexes that regulate chromatin structure and transcription. Cell 2002; 108(4):475–487.

23. Svejstrup JQ. Transcription. Histones face the FACT. Science 2003; 301(5636):1053–1055.

24. Peterson CL. ATP-dependent chromatin remodeling: going mobile. FEBS Lett 2000; 476(1–2):68–72.

25. Becker PB, Horz W. ATP-dependent nucleosome remodeling. Annu Rev Biochem 2002; 71:247–273.

26. Felsenfeld G, Groudine M. Controlling the double helix. Nature 2003; 421(6921): 448–453.

27. Anderson JD, Lowary PT, Widom J. Effects of histone acetylation on the equilibrium accessibility of nucleosomal DNA target sites. J Mol Biol 2001; 307(4):977–985.

28. Schieferstein U, Thoma F. Site-specific repair of cyclobutane pyrimidine dimers in a positioned nucleosome by photolyase and T4 endonuclease V in vitro. Embo J 1998; 17(1):306–316.

29. Smerdon MJ, Lan SY, Calza RE, Reeves R. Sodium butyrate stimulates DNA repair in UV-irradiated normal and xeroderma pigmentosum human fibroblasts. J Biol Chem 1982; 257(22):13441–13447.

30. Kurdistani SK, Grunstein M. Histone acetylation and deacetylation in yeast. Nat Rev Mol Cell Biol 2003; 4(4):276–284.
31. Livingstone-Zatchej M, Marcionelli R, Moller K, De Pril R, Thoma F. Repair of UV lesions in silenced chromatin provides in vivo evidence for a compact chromatin structure. J Biol Chem 2003; 278(39):37471–37479.
32. Kim JK, Choi BS. The solution structure of DNA duplex-decamer containing the (6–4) photoproduct of thymidylyl (3′->5′) thymidine by NMR and relaxation matrix refinement. Eur J Biochem 1995; 228(3):849–854.
33. Park H, Zhang K, Ren Y, Nadji S, Sinha N, Taylor JS, Kang C. Crystal structure of a DNA decamer containing a *cis-syn* thymine dimer. Proc Natl Acad Sci USA 2002; 99(25):15965–15970.
34. Becker MM, Wang JC. Use of light for footprinting DNA in vivo. Nature 1984; 309: 682–687.
35. Wang Z, Becker MM. Selective visualization of gene structure with ultraviolet light. Proc Natl Acad Sci USA 1988; 85(3):654–658.
36. Schieferstein U, Thoma F. Modulation of cyclobutane pyrimidine dimer formation in a positioned nucleosome containing poly(dA.dT) tracts. Biochemistry 1996; 35(24): 7705–7714.
37. Gaillard H, Fitzgerald DJ, Smith CL, Peterson CL, Richmond TJ, Thoma F. Chromatin remodeling activities act on UV-damaged nucleosomes and modulate DNA damage accessibility to photolyase. J Biol Chem 2003; 278(20):17655–17663.
38. Gale JM, Nissen KA, Smerdon MJ. UV-induced formation of pyrimidine dimers in nucleosome core DNA is strongly modulated with a period of 10.3 bases. Proc Natl Acad Sci USA 1987; 84(19):6644–6648.
39. Gale JM, Smerdon MJ. UV induced (6–4) photoproducts are distributed differently than cyclobutane dimers in nucleosomes. Photochem Photobiol 1990; 51(4):411–417.
40. Mann DB, Springer DL, Smerdon MJ. DNA damage can alter the stability of nucleosomes: effects are dependent on damage type. Proc Natl Acad Sci USA 1997; 94(6): 2215–2220.
41. Suquet C, Smerdon MJ. UV damage to DNA strongly influences its rotational setting on the histone surface of reconstituted nucleosomes. J Biol Chem 1993; 268(32): 23755–23757.
42. Cavalli G, Thoma F. Chromatin transitions during activation and repression of galactose- regulated genes in yeast. Embo J 1993; 12(12):4603–4613.
43. Sancar A. Structure and function of DNA photolyase and cryptochrome blue-light photoreceptors. Chem Rev 2003; 103(6):2203–2238.
44. Kosmoski JV, Ackerman EJ, Smerdon MJ. DNA repair of a single UV photoproduct in a designed nucleosome. Proc Natl Acad Sci USA 2001; 98(18):10113–10118.
45. Gordon LK, Haseltine WA. Comparison of the cleavage of pyrimidine dimers by the bacteriophage T4 and Micrococcus luteus UV-specific endonucleases. J Biol Chem 1980; 255(24):12047–12050.
46. Vassylyev DG, Kashiwagi T, Mikami Y, Ariyoshi M, Iwai S, Ohtsuka E, Morikawa K. Atomic model of a pyrimidine dimer excision repair enzyme complexed with a DNA substrate: structural basis for damaged DNA recognition. Cell 1995; 83(5):773–782.
47. Thoma F. Light and dark in chromatin repair: repair of UV-induced DNA lesions by photolyase and nucleotide excision repair. Embo J 1999; 18(23):6585–6598.
48. Fitch ME, Nakajima S, Yasui A, Ford JM. In vivo recruitment of XPC to UV-induced cyclobutane pyrimidine dimers by the DDB2 gene product. J Biol Chem 2003; 278(47):46906–46910.
49. Schul W, Jans J, Rijksen YM, Klemann KH, Eker AP, de Wit J, Nikaido O, Nakajima S, Yasui A, Hoeijmakers JH, van der Horst GT. Enhanced repair of cyclobutane pyrimidine dimers and improved UV resistance in photolyase transgenic mice. Embo J 2002; 21(17):4719–4729.

50. Sancar GB, Smith FW. Interactions between yeast photolyase and nucleotide excision repair proteins in *Saccharomyces cerevisiae* and *Escherichia coli*. Mol Cell Biol 1989; 9(11):4767–4776.

51. Suter B, Livingstone-Zatchej M, Thoma F. Chromatin structure modulates DNA repair by photolyase in vivo. EMBO J 1997; 16(8):2150–2160.

52. Suter B, Wellinger RE, Thoma F. DNA repair in a yeast origin of replication: contributions of photolyase and nucleotide excision repair. Nucleic Acids Res 2000; 28(10):2060–2068.

53. Suter B, Schnappauf G, Thoma F. Poly(dA.dT) sequences exist as rigid DNA structures in nucleosome-free yeast promoters in vivo. Nucleic Acids Res 2000; 28(21):4083–4089.

54. Suter B, Thoma F. DNA-repair by photolyase reveals dynamic properties of nucleosome positioning in vivo. J Mol Biol 2002; 319(2):395–406.

55. Hanawalt PC. Transcription-coupled repair and human disease. Science 1994; 266(5193):1957–1958.

56. Hoeijmakers JH. Genome maintenance mechanisms for preventing cancer. Nature 2001; 411(6835):366–374.

57. Prakash S, Prakash L. Nucleotide excision repair in yeast. Mutat Res 2000; 451(1–2): 13–24.

58. Conconi A, Bespalov VA, Smerdon MJ. Transcription-coupled repair in RNA polymerase I-transcribed genes of yeast. Proc Natl Acad Sci USA 2002; 99(2):649–654.

59. Meier A, Livingstone-Zatchej M, Thoma F. Repair of active and silenced rDNA in yeast: the contributions of photolyase and transcription-coupled nucleotide excision repair. J Biol Chem 2002; 277(14):11845–11852.

60. Wood RD. DNA damage recognition during nucleotide excision repair in mammalian cells. Biochimie 1999; 81(1–2):39–44.

61. Volker M, Mone MJ, Karmakar P, van Hoffen A, Schul W, Vermeulen W, Hoeijmakers JH, van Driel R, van Zeeland AA, Mullenders LH. Sequential assembly of the nucleotide excision repair factors in vivo. Mol Cell 2001; 8(1):213–224.

62. Yu S, Owen-Hughes T, Friedberg EC, Waters R, Reed SH. The yeast Rad7/Rad16/ ABf1 complex generates superhelical torsion in DNA that is required for nucleotide excision repair. DNA Repair 2004; 3:277–287.

63. Wang ZG, Wu XH, Friedberg EC. Nucleotide excision repair of DNA by human cell extracts is suppressed in reconstituted nucleosomes. J Biol Chem 1991; 266:22472–22478.

64. Sugasawa K, Masutani C, Hanaoka F. Cell-free repair of UV-damaged simian virus-40 chromosomes in human cell extracts .1. Development of a Cell-Free System Detecting Excision Repair of UV-Irradiated SV40 Chromosomes. J Biol Chem 1993; 268(12): 9098–9104.

65. Ura K, Araki M, Saeki H, Masutani C, Ito T, Iwai S, Mizukoshi T, Kaneda Y, Hanaoka F. ATP-dependent chromatin remodeling facilitates nucleotide excision repair of UV-induced DNA lesions in synthetic dinucleosomes. Embo J 2001; 20(8):2004–2014.

66. Hara R, Sancar A. Effect of damage type on stimulation of human excision nuclease by SWI/SNF chromatin remodeling factor. Mol Cell Biol 2003; 23(12):4121–4125.

67. Wang D, Hara R, Singh G, Sancar A, Lippard SJ. Nucleotide excision repair from site-specifically platinum-modified nucleosomes. Biochemistry 2003; 42(22):6747–6753.

68. Smerdon MJ, Thoma F. Site-specific DNA repair at the nucleosome level in a yeast minichromosome. Cell 1990; 61:675–684.

69. Wellinger RE, Thoma F. Nucleosome structure and positioning modulate nucleotide excision repair in the non-transcribed strand of an active gene. Embo J 1997; 16(16): 5046–5056.

70. Tijsterman M, de Pril R, Tasseron-de Jong JG, Brouwer J. RNA polymerase II transcription suppresses nucleosomal modulation of UV- induced (6–4) photoproduct and cyclobutane pyrimidine dimer repair in yeast. Mol Cell Biol 1999; 19(1):934–940.

71. Powell NG, Ferreiro J, Karabetsou N, Mellor J, Waters R. Transcription, nucleosome positioning and protein binding modulate nucleotide excision repair of the Saccharomyces cerevisiae MET17 promoter. DNA Repair (Amst) 2003; 2(4):375–386.

72. Evans E, Moggs JG, Hwang JR, Egly JM, Wood RD. Mechanism of open complex and dual incision formation by human nucleotide excision repair factors. Embo J 1997; 16(21):6559–6573.

73. Huang JC, Sancar A. Determination of minimum substrate size for human excinuclease. J Biol Chem 1994; 269(29):19034–19040.

74. Bang DD, Verhage R, Goosen N, Brouwer J, Vandeputte P. Molecular cloning of RAD16, a gene involved in differential repair in saccharomyces-cerevisiae. Nucleic Acids Res 1992; 20(15):3925–3931.

75. Verhage R, Zeeman AM, de Groot N, Gleig F, Bang DD, van de Putte P, Brouwer J. The RAD7 and RAD16 genes, which are essential for pyrimidine dimer removal from the silent mating type loci, are also required for repair of the nontranscribed strand of an active gene in Saccharomyces cerevisiae. Mol Cell Biol 1994; 14(9):6135–6142.

76. Waters R, Rong Z, Jones NJ. Inducible removal of UV-induced pyrimidine dimers from transcriptionally active and inactive genes of saccharomyces-cerevisiae. Mol Gen Genet 1993; 239(1–2):28–32.

77. Mueller JP, Smerdon MJ. Repair of plasmid and genomic DNA in a rad7 delta mutant of yeast. Nucleic Acids Res 1995; 23(17):3457–3464.

78. Schild D, Glassner BJ, Mortimer RK, Carlson M, Laurent BC. Identification of RAD16, a yeast excision repair gene homologous to the recombinational repair gene RAD54 and to the SNF2 gene involved in transcriptional activation. Yeast 1992; 8(5):385–395.

79. Guzder SN, Sung P, Prakash L, Prakash S. The DNA-dependent ATPase activity of yeast nucleotide excision repair factor 4 and its role in DNA damage recognition. J Biol Chem 1998; 273(11):6292–6296.

80. Guzder SN, Sung P, Prakash L, Prakash S. Synergistic interaction between yeast nucleotide excision repair factors NEF2 and NEF4 in the binding of ultraviolet-damaged DNA. J Biol Chem 1999; 274(34):24257–24262.

81. Wood RD, Mitchell M, Sgouros J, Lindahl T. Human DNA repair genes. Science 2001; 291(5507):1284–1289.

82. Tang JY, Hwang BJ, Ford JM, Hanawalt PC, Chu G. Xeroderma pigmentosum p48 gene enhances global genomic repair and suppresses UV-induced mutagenesis. Mol Cell 2000; 5(4):737–744.

83. Hwang BJ, Ford JM, Hanawalt PC, Chu G. Expression of the p48 xeroderma pigmentosum gene is p53-dependent and is involved in global genomic repair. Proc Natl Acad Sci USA 1999; 96(2):424–428.

84. Wakasugi M, Kawashima A, Morioka H, Linn S, Sancar A, Mori T, Nikaido O, Matsunaga T. DDB accumulates at DNA damage sites immediately after UV irradiation and directly stimulates nucleotide excision repair. J Biol Chem 2002; 277(3):1637–1640.

85. Fujiwara Y, Masutani C, Mizukoshi T, Kondo J, Hanaoka F, Iwai S. Characterization of DNA recognition by the human UV-damaged DNA-binding protein. J Biol Chem 1999; 274(28):20027–20033.

86. Tang J, Chu G. Xeroderma pigmentosum complementation group E and UV-damaged DNA-binding protein. DNA Repair (Amst) 2002; 1(8):601–616.

87. Martinez E, Palhan VB, Tjernberg A, Lymar ES, Gamper AM, Kundu TK, Chait BT, Roeder RG. Human STAGA complex is a chromatin-acetylating transcription coactivator that interacts with pre-mRNA splicing and DNA damage-binding factors in vivo. Mol Cell Biol 2001; 21(20):6782–6795.

88. Smerdon MJ, Lieberman MW. Nucleosome rearrangement in human chromatin during UV-induced DNA-reapir synthesis. Proc Natl Acad Sci USA 1978; 75(9):4238–4241.

89. Game JC, Kaufman PD. Role of Saccharomyces cerevisiae chromatin assembly factor-I in repair of ultraviolet radiation damage in vivo. Genetics 1999; 151(2):485–497.

90. Green CM, Almouzni G. Local action of the chromatin assembly factor CAF-1 at sites of nucleotide excision repair in vivo. Embo J 2003; 22(19):5163–5174.

91. Jensen KA, Smerdon MJ. DNA repair within nucleosome cores of UV-irradiated human cells. Biochemistry 1990; 29(20):4773–4782.

92. Shen X, Mizuguchi G, Hamiche A, Wu C. A chromatin remodelling complex involved in transcription and DNA processing. Nature 2000; 406(6795):541–544.

93. Koyama H, Itoh M, Miyahara K, Tsuchiya E. Abundance of the RSC nucleosome-remodeling complex is important for the cells to tolerate DNA damage in Saccharomyces cerevisiae. FEBS Lett 2002; 531(2):215–221.

94. Teng Y, Yu Y, Waters R. The saccharomyces cerevisiae histone acetyltransferase Gcn5 has a role in the photoreactivation and nucleotide excision repair of UV-induced cyclobutane pyrimidine dimers in the MFA2 gene. J Mol Biol 2002; 316(3):489–499.

95. Rubbi CP, Milner J. p53 is a chromatin accessibility factor for nucleotide excision repair of DNA damage. Embo J 2003; 22(4):975–986.

96. Hara R, Sancar A. The SWI/SNF chromatin-remodeling factor stimulates repair by human excision nuclease in the mononucleosome core particle. Mol Cell Biol 2002; 22(19):6779–6787.

97. Kent NA, Karabetsou N, Politis PK, Mellor J. In vivo chromatin remodeling by yeast ISWI homologs Isw1p and Isw2p. Genes Dev 2001; 15(5):619–626.

98. Tsukiyama T, Palmer J, Landel CC, Shiloach J, Wu C. Characterization of the imitation switch subfamily of ATP-dependent chromatin-remodeling factors in Saccharomyces cerevisiae. Genes Dev 1999; 13(6):686–697.

11

The Ultraviolet Damage Endonuclease (UVDE) Protein and Alternative Excision Repair: A Highly Diverse System for Damage Recognition and Processing

Paul W. Doetsch, Vladimir Beljanski, and Binwei Song
Department of Biochemistry, Emory University School of Medicine, Atlanta, Georgia, U.S.A.

1. INTRODUCTION

The major cytotoxic and mutagenic UV light-induced DNA photoproducts, cyclobutane pyrimidine dimers (CPDs) and (6–4) pyrimidine–pyrimidone photoproducts (6–4 PPs) have been and continue to be widely studied for characterizing DNA repair pathways and for elucidating the biological endpoints of unrepaired DNA damage. Early models for the recognition and initial enzymatic processing of UV light-induced DNA damage included a direct endonucleolytic incision event occurring 5′ to the site of the lesion followed by exonucleolytic removal of a DNA segment containing the damage, repair synthesis, and ligation (1). The notion of a single, direct acting 5′-endonuclease capable of damage recognition and initiating DNA repair was abandoned following the discovery and characterization of the ATP-dependent, 5′ and 3′-acting, dual incision nucleotide excision repair (NER) machinery in *Escherichia coli, Saccharomyces cerevisiae*, and humans (2,3). In addition, the discovery and characterization of a second major DNA excision repair pathway, the base excision repair (BER) pathway in numerous organisms which is initiated by various DNA *N*-glycosylases (4) further discouraged searches for additional excision repair proteins in prokaryotes or eukaryotes that might initiate the repair of a wide range of DNA damages. The discovery of a NER-independent alternative excision repair (AER) pathway for UV photoproducts in the fission yeast *Schizosaccharomyces pombe* led to the isolation and characterization of a remarkably diverse single protein, UV damage endonuclease (UVDE or Uvelp) that functions as a broad specificity, ATP-independent 5′ endonuclease for initiating the repair of CPDs and 6–4 photoproducts (6–4 PPs) as well as a range of structurally heterogeneous DNA lesions. This chapter will emphasize the initial discovery, properties, substrate specificity, and cellular function of *S. pombe* UVDE in the AER pathway with its known components. A comparison of UVDE homologs from other organisms will be made in order to gain insight into the regions of UVDE required for recognition and enzymatic processing of damaged DNA. Several earlier reviews have discussed various aspects of the *S. pombe* AER pathway (5–7). The AER pathway is also sometimes referred to as the UV damage excision repair (UVER) pathway (6).

2. DISCOVERY AND INITIAL CHARACTERIZATION OF *S. POMBE* UVDE

2.1. A NER-Independent Excision Repair Pathway for UV Photoproducts

Analysis of radiation-sensitive mutants in *S. pombe* (8,9) led to the identification of a novel group of mutants that could not be classified as possessing defects in NER, recombination or damage tolerance/checkpoint functions (10). The NER mutants of *S. cerevisiae* and *E. coli* were highly UV sensitive and completely deficient in the removal of both CPDs and 6–4 PPs (11). In contrast, it was determined that *S. pombe rad* mutants deficient in NER were only moderately UV sensitive and were still capable of efficient removal of both CPDs and 6–4 PPs via a light-independent pathway (12). It was proposed that fission yeast possessed a novel excision repair pathway, for CPDs and 6–4 PPs. The first biochemical evidence that this novel pathway existed was the identification of an ATP-independent, pyrimidine dimer incision activity present in cell-free *S. pombe* extracts (13). Subsequently, an ATP-independent damage-dependent endonuclease was found in extracts and partially purified preparations of *S. pombe* cells (14,15). The enzyme was initially named *S. pombe* DNA endonuclease (SPDE) but was renamed UV damage endonuclease (UVDE). Initial biochemical analysis revealed that UVDE (also designated Uvelp) cleaves the phosphodiester backbone of substrates containing both CPDs and 6–4 PPs immediately 5′ to the site of the lesion. Analyzing strains defective in Uvelp activity in conjunction with NER mutants revealed the double mutant to be more sensitive to UV light than either single mutant, indicating this endonuclease did in fact represent a component of an excision repair pathway in vivo that is distinct from NER (15,16). *uvel* (which encodes UVDE) homologs have also been identified in *Neurospora crassa* and *Bacillus subtilis* (17,18). The isolation of the *uvel* gene by Yasui's group (18) and the subsequent biochemical characterization of recombinant Uvelp by Kaur et al. (19) led to the determination that the substrate specificity of Uvelp is much broader than originally suspected.

2.2. Biochemical Characterization of *S. pombe* UVDE

Initial biochemical characterizations of UVDE were carried out using substrates containing CPDs or 6–4 PPs. The full length *uvel* gene encodes a 599 amino acid protein (Fig. 1) of approximately 68.8 kDa (18) which, when overexpressed in either *E. coli* or *S. cerevisiae* cells, yields a relatively unstable protein that rapidly loses activity (18,19). Three putative initiation methionine codons, Met-1, Met-56, and Met-64 reside at the *N*-terminus. Features of a mitochondrial protein between the first and second methionine as well as motifs for nuclear localization after the third methionine were identified by Yasuhira and Yasui (20) from a scan of full-length UVDE with the PSORTII program (21). Various affinity-tagged, truncated versions of UVDE are considerably more stable and have been used in enzymological studies

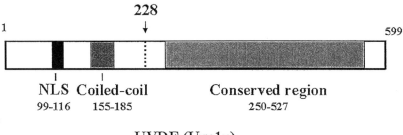

UVDE (Uve1p)

Figure 1 *Schizosaccharomyces pombe* UVDE protein. The *uve1* gene encodes a 599 amino acid protein containing putative nuclear localization signal (NLS) region (amino acids 99–116), a coiled coil region (amino acids 155–185), and a conserved region (amino acids 250–527) similar to regions found in the *N. crassa* and *B. subtilis* UVDE functional homologs that is thought to be required for enzymatic activity. Numerous studies have been conducted with a N-terminal 228 amino acid truncated version (arrow) of UVDE.

with success (18,19). Such modified UVDE proteins contain N-terminal deletions of various lengths with a corresponding sharp decline in activity against UV photoproducts for truncations larger that 232 amino acids. In contrast even short deletions of 35 residues at the C-terminus abrogate activity (18). Based on these findings together with the sequence alignments against the UVDE homologs from *N. crassa* and *B. subtilis* that indicate a conserved region among the three proteins (Fig. 2), a critical region for activity has been proposed. This region includes the C-terminal two-thirds of the *S. pombe* UVDE protein and are thought to be essential for mediating enzymatic activity (Ref. 18 and Fig. 1). Kaur et al. (19) conducted a detailed enzymological characterization utilizing a highly stable, 228 residue N-terminal truncated version of UVDE. The enzyme exhibits high activity over a broad range of salt concentrations and optimal activity in an environment which provides an overall net positive charge (pH 6.0–6.5). The apparent K_m against CPD-containing oligonucleotide substrates is approximately 50 nM.

Figure 2 *Schizosaccharomyces pombe* UVDE conserved region sequence alignments with *N. crassa* and *B. subtilis* functional homologs. Alignments are colored using the ClustalX scheme in Jalview: orange: glycine (G); yellow: proline (P); blue: small and hydrophobic amino acids (A, V, L, I, M, F, W); green: hydroxyl and amine amino acids (S, T, N, Q); red: charged amino acids (D, E, R, K); cyan: histidine (H) and tyrosine (Y). *Source*: Adapted from Michele Clamp, Sanger Institute, Cambridge, U.K. (*See color insert.*)

3. RECOGNITION AND PROCESSING OF UV PHOTOPRODUCTS

Initial studies had demonstrated that UVDE was capable of recognizing both *cis–syn* CPDs and 6–4 PPs (14). It was unique in this respect as no other single protein endonuclease could recognize both types of these UV photoproducts. The CPDs and 6–4PPs are the most frequently occurring forms of UV-induced damage, but there are significant differences in the structural distortions associated with these lesions. Incorporation of a *cis–syn* CPD into double-stranded DNA causes a 7–30° distortion in the DNA helix (22–24) and destabilizes the duplex by ~1.5 kcal/mol (25). It has been demonstrated that this relatively small structural distortion allows CPDs to retain their ability to form Watson–Crick hydrogen bonds in much the same way as a mismatched dimer. On the other hand, when a 6–4 PP is incorporated into DNA the plane of the 3' base moiety is shifted 90° relative to that of the 5' thymine (26). This destabilizes the duplex by ~6 kcal/mol and is believed to cause 6–4 PPs to lose their ability to form hydrogen bonds with the opposite strand. The ability of Uvelp to recognize such different structural distortions suggested that it might also recognize other, less frequently occurring UV photoproducts. The CPDs can occur significantly in DNA in four different isoforms (*cis–syn* I, *cis–syn* II, *trans–syn* I, and *trans–syn* II) (27). The CPDs exist predominately in the *cis–syn* I form in double-stranded DNA whereas *trans–syn* dimers are found mainly in single-stranded regions of DNA (28). 6–4 PPs are alkali labile lesions at positions of cytosine (and much less frequently thymine) located 3' to pyrimidine nucleosides (29). 6–4 PPs are not stable in sunlight and are converted to their Dewar valence isomers upon exposure to 313 nm light (26,30). The specificity of UVDE was determined for a series of UV photodamages: *cis–syn* CPD, *trans–syn* I CPD, *trans–syn* II CPD, 6–4 PP and the Dewar 6–4 PP isomer (Ref. 31 and Fig. 3). This study also established that UVDE cleaves photoproduct-containing duplex DNA substrates at two sites; the primary site is immediately 5' to the damage and the secondary site is one nucleotide 5' to the site of damage. Each of the five UV photoproducts shown in Figure 3 causes different structural distortions when incorporated into DNA (24). The UVDE recognizes and processes bipyrimidine photoproduct substrates in a similar manner with

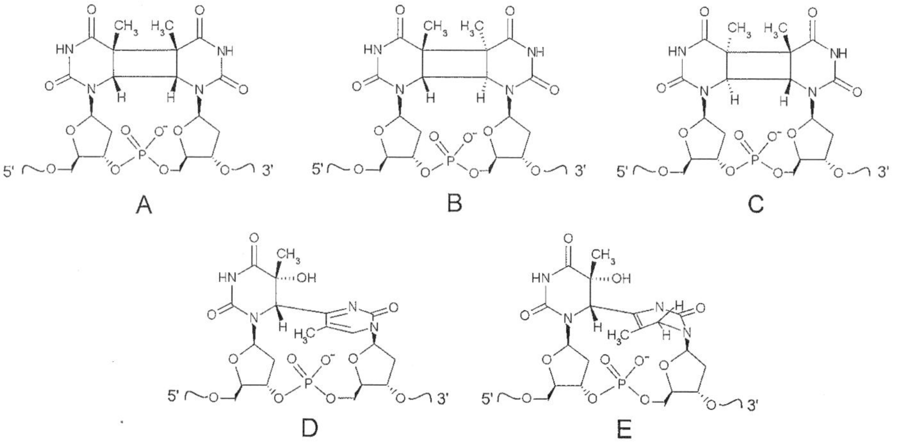

Figure 3 UV photoproducts that are strong UVDE substrates. (A) *cis–syn* cyclobutane dimer, (B) *trans–syn* I cyclobutane dimer, (C) *trans–syn* II cyclobutane dimer, (D) (6–4) photoproducts, (E) Dewar valence isomer (Dewar).

respect to both the site and extent of duplex DNA cleavage (31). This property of UVDE is remarkable and it is worth noting that this is the first example of a single protein endonuclease capable of recognizing such a broad range of UV photoproducts. These findings raised the issue of the nature of the structural distortion recognized by the UVDE protein and prompted subsequent investigations examining other potential UVDE substrates that produce a range of structural distortions in duplex, B-form DNA.

4. RECOGNITION AND PROCESSING OF PLATINUM G-G DIADDUCTS

Cis-diamminedichloroplatinum (II) (cisplatin) is a widely used antitumor drug that induces several types of mono- and diadducts in DNA. The major adduct formed results from the coordination of two adjacent Gs to platinum to form the intrastrand crosslink *cis*-[Pt(NH3) 2{d(GpG)-N7(l),-N7(2)}] (*cis*-Pt-GG) (32). The formation of the *cis*-Pt-GG crosslink unwinds DNA by 13° and bends it by 34–55° in the direction of the major grove (33,34). The UVDE was shown to be capable of cleaving a duplex DNA oligonucleotide (32-mer) containing a centrally located platinum G-G diadduct at sites located two and three nucleotides 5′ to the lesion site (31). The extent of cleavage on platinum adducts was weak (about 40-fold less) relative to that observed for UV photoproducts and may also reflect a requirement for interaction with specific proteins to enhance substrate recognition and cleavage. It is possible that in vivo other accessory proteins may contribute to the efficiency of UVDE, thereby enhancing its ability to initiate the repair of a wide variety of DNA damages. Despite this substantial decrease in substrate recognition and cleavage it can be concluded that UVDE is capable of recognizing and processing (cleaving duplex DNA) a non-UV dimer lesion, in this case, a platinum–DNA intrastrand crosslink (Fig. 4A).

5. RECOGNITION AND PROCESSING OF ABASIC SITES

Apurinic/apyrimidinic (AP) or abasic sites in DNA are frequently formed by the spontaneous hydrolysis of *N*-glycosyl bonds, can be induced by radiation and chemicals and are formed as intermediates during BER following DNA *N*-glycosylase removal of damaged bases (Ref. 35 and Fig. 4B). The AP endonucleases, which are thought to be the major enzymes responsible for initiation of repair of abasic sites, cleave DNA phosphodiester bonds hydrolytically 5′ to the site to yield a 3′-hydroxyl terminus (35). AP lyases associated with some DNA repair *N*-glycosylases cleave DNA phosphodiester bonds by a beta-elimination mechanism to yield a 3′ alpha, beta-unsaturated aldehyde (35). The significant differences in duplex DNA structural distortion caused by CPDs and 6–4PPs, prompted the search by several groups to identify other UVDE substrates, including those that are less bulky and induce less severe structural perturbations such as abasic sites (36). The nearly simultaneous reports by two different groups that abasic sites were recognized and cleaved by UVDE immediately expanded the known range of types of duplex DNA distortions that could be processed by this enzyme (31,37). This finding also further complicated the concept of a common structural feature among UV photoproducts, platinum–DNA diadducts, and abasic sites that might be recognized by UVDE (31). The DNA structural distortions caused by abasic sites and platinum G-G diadducts are compared to normal B-form duplex DNA in Figure 5. UVDE

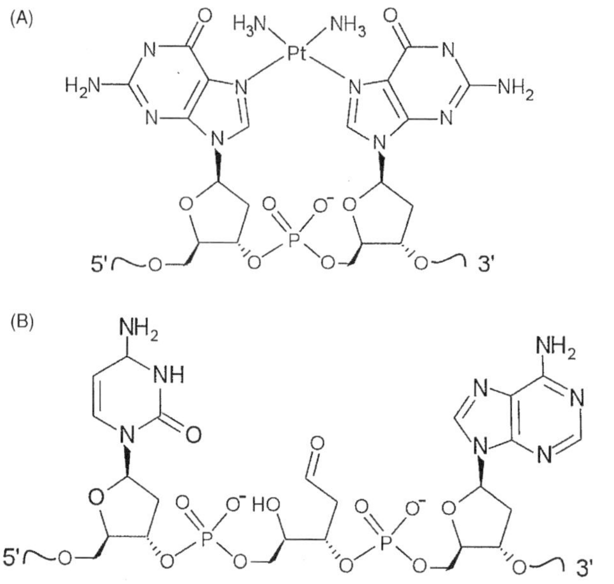

(A)

(B)

Figure 4 DNA lesions that are weak-to-moderate UVDE substrates. (A) Platinum G-G diadduct, *cis*-[Pt(NH$_3$)$_2$\{d(GpG)-N7(1),-N7(2)\}] (weak substrate), (B) abasic site (moderate substrate).

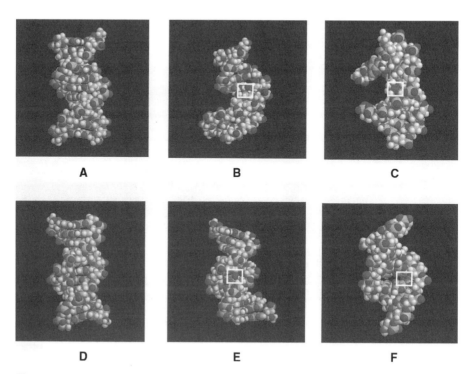

A   B   C

D   E   F

Figure 5 Duplex DNA structural distortion comparisons of oligomers containing (A,D) no damage, (B,E) an abasic site (moderate UVDE substrate) and (C,F) platinum G-G diadduct (weak UVDE substrate). White squares indicate location of DNA lesion. Oligomers shown are space-filling representations of duplex DNA oligomers with major (A–C) and minor (D–F) groove views of molecules. *Source*: Adapted from Refs. 60,61. (*See color insert.*)

processes abasic sites in a manner similar to hydrolytic AP endonucleases cleaving immediately 5′ to the lesion site and producing 3′-hydroxyl and 5′-deoxyribose phosphate terminal as well as a secondary cleavage event (unlike hydrolytic AP endonucleases) one nucleotide 5′ to the lesion (31). In addition, abasic site substrates containing either A or G opposite to the lesions are processed with equal efficiency (31,37), whereas lesions with C or T opposite are processed less efficiently (37). We can conclude that UVDE is capable of recognizing abasic sites and in so doing can repair two different classes of major species of DNA damage (UV photoproducts and abasic sites) that are induced in abundance in the genomes of most terrestrial organisms. Yasui and colleagues (37) have demonstrated that introduction of *uve1* into *E. coli* mutants lacking both of its major AP endonucleases (exonuclease III and endonuclease IV) conferred resistance to methyl methane sulfonate (MMS) and t-butylhydroperoxide. An important additional finding by this group was the discovery that Uve1p could trim 3′ ends of AP sites that have been cleaved by either an AP lyase or bleomycin cleavage sites containing 3′-phosphoglycolate groups. This result suggests that an additional repair function of Uve1p might be under certain circumstances to trim 3′-blocking termini and convert them to 3′ hydroxyl groups for DNA repair synthesis to occur.

6. MODIFIED BASES NOT RECOGNIZED BY UVDE

Despite the seemingly diverse range of structural distortions recognized by UVDE in duplex DNA, a number of base modifications are not substrates for this enzyme. DNA damages that are not recognized by UVDE include DNA deamination and oxidation products such as xanthine, inosine, and 8-oxoguanine (Fig. 6) (31). Curiously, uracil and dihydrouracil are recognized as poor substrates for UVDE with highest activity when the opposite base is G (normal hydrogen bonding disrupted) as opposed to A (base-paired). The observed higher activity on such mispaired bases provides a clue that prompted studies using unmodified base–base mispair combinations (discussed in Sect.7). These findings suggest that perhaps UVDE recognizes a distortion caused by the formation of a wobble base pair between uracil (or dihydrouracil) and the opposite G rather than the modified base itself.

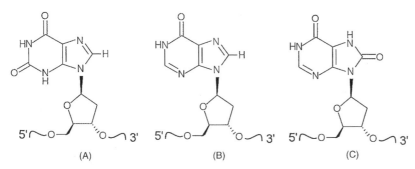

Figure 6 DNA lesions that are not recognized by UVDE: (A) xanthine, (B) inosine, (C) 8-oxoguanine.

7. RECOGNITION AND PROCESSING OF BASE–BASE MISMATCHES

There are multiple causes of single base mismatches in cells. The most common of these is nucleotide misincorporation by DNA polymerase during replication. Mismatches can also arise following deamination of cytosine to uracil forming U/G mispairs or upon recombination between homologous sequences (38). To correct these lesions, cells have developed several mechanisms for mismatch repair (MMR) that are essential for maintaining the integrity of the genome. The MMR also functions in maintaining the stability of simple DNA repeat tracts during replication including insertions caused by slippage loops in the primer strand and deletions caused by failure to repair loops in the template strand.

In contrast to *S. cerevisiae* and humans, less is known about MMR in *S. pombe*. The MutL homologue *pms1* has been identified (39,40). Disruption of the *S. pombe pms1* gene confers a spontaneous mutator phenotype, reduction of spore viability and an increase in postmeiotic segregation (PMS) indicating that it plays a role in mismatch correction (40). Two other genes, *swi4* and *swi8* are homologues of *S. cerevisiae MSH3* and *MSH2*, respectively, and it has been proposed that they may mediate roles in loop repair and in the case of *swi8*, correction of single base mismatches (41). *Schizosaccharomyces pombe* Exo1p (encoded by the *exo1* gene) is a meiotically induced 5' to 3' double-stranded DNA exonuclease, is a homolog of *S. cerevisiae EXO1*, and has been proposed to play a role in mutation avoidance and mismatch correction (42–44). Genetic analysis of meiotic recombination events has indicated the existence of at least two pathways responsible for MMR in *S. pombe*; a major, long patch MMR system (mediated by *msh1* and *pms1*) which recognizes all mismatch combinations except C/C, and a minor, short-patch MMR system which recognizes all combinations, including C/C mismatches (45,46). Further support for these observations was provided by the discovery of two distinct mismatch-binding activities in *S. pombe* crude cell extracts (47). In addition, the *S. pombe* NER genes *rhp14*, *swi10*, and *rad16* (homologs of the *S. cerevisiae RAD 14*, *RAD10*, and *RAD1* genes, respectively) have been identified as components of the short-patch MMR system and function independently of *msh2 pms1* (48). A more recent genetic study of MMR in *S. pombe rad13*, *rad2*, and *uve1* mutants showed only a weak dependence for UVDE on both meiotic mismatch repair and mutation avoidance (49). Thus, the available genetic and more limited biochemical evidence suggests that *S. pombe* possesses multiple pathways for conducting MMR.

Because of significant structural differences between CPDs, 6–4 PPs, platinum G-G diadducts, abasic sites, and other UVDE substrates, it is not obvious what distortions present in damaged DNA are recognized by UVDE. One possibility is that Watson–Crick base pairing is disrupted for the 3' pyrimidines in both CPDs and 6–4 PPs. This suggests that UVDE might recognize mispaired bases in duplex DNA. In vitro studies by Kaur et al. (50) established that to various extents, UVDE recognizes and cleaves all 12 possible base mispair combinations 5' to the mismatched base in a strand-specific manner within the same flanking sequence context (i.e., cleaves only one of the two strands and shows a preference for the base closest to the 3' terminus of a duplex). Furthermore, the UVDE mismatch endonuclease activity and UVDE UV endonuclease activity exhibit similar properties and compete for the same substrates (50). These biochemical results suggest that UVDE might be involved in mismatch repair in *S. pombe*. This notion was confirmed by the observation that *uve1* null mutants exhibit a mutator phenotype in experiments comparing wild type, *uve1* mutants and a known mismatch repair-deficient strain (*pms1*

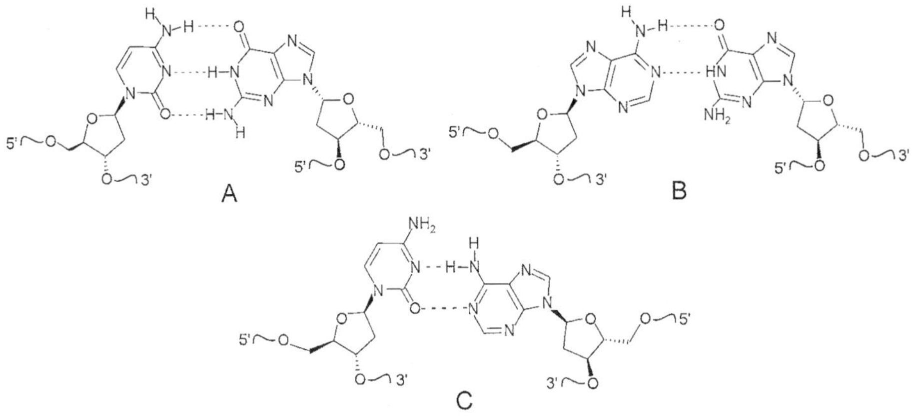

Figure 7 Comparisons of (A) Watson/Crick C/G base pair (non-UVDE substrate) with corresponding hydrogen bonds indicated; (B) A/G mismatch (moderate UVDE substrate) with non-standard hydrogen bonds indicated; (C) C/A mismatch (strong UVDE substrate) with non-standard hydrogen bonds indicated. *Source*: Adapted from Refs. 58,59.

mutants) to form colonies resistant to the toxic arginine analog L-canavanine (50). The cleavage efficiency of UVDE is variable depending on the nature of the base mispair with low efficiency observed at certain base/base combinations (50). The site of cleavage, similar to that seen on other UVDE substrates, is immediately 5′ as well as one and two nucleotides 5′ to the site of the mismatch. The extent of cleavage at these three locations depends on the exact nature of the base–base mispair. In general, robust cleavage is observed at ∗C/A, ∗C/C, and ∗G/G sites, moderate cleavage at ∗G/A, ∗A/G, and ∗T/G sites, and weak cleavage at ∗G/T, ∗A/A, ∗A/C, ∗C/T, ∗T/T, and ∗T/C (asterisks indicate base located on cleaved strand) sites (Fig. 7). This low efficiency of cleavage for certain base–base mismatch combinations is similar to *E. coli* MutH cleavage in the absence of MutL (51). Likewise, the human XPG protein greatly enhances the efficiency of the human BER *N*-glycosylase hNthl-mediated cleavage of substrates containing oxidative base damage (52). It may turn out to be the case that UVDE interacting proteins modulate its activity and increase the extent to which substrates are processed.

8. RECOGNITION AND PROCESSING OF INSERTION–DELETION LOOPS

Studies have been carried out comparing the ability of UVDE to incise duplex DNA containing insertion deletion loops (IDLs) and hairpins (53). Primer/template slippage can occur at repetitive DNA sequences during replication, resulting in single-stranded IDLs of one or more unpaired nucleotides that can be mutagenic. The UVDE recognizes and cleaves heteroduplex DNA with small unpaired loops (IDLs) 2–4 nucleotides in length but not larger loops 6–8 nucleotides in length (Fig. 8). In addition, the enzyme does not recognize palindromic insertions that can form hairpin structures. These results further support a role for Uvelp in mismatch repair in *S. pombe*. In addition, these findings add another element to the complex picture of the nature of the structural distortion(s) recognized by UVDE.

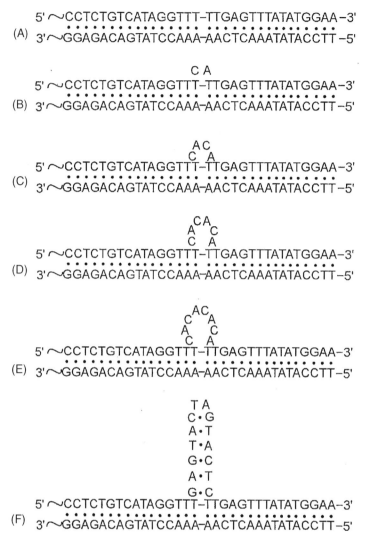

Figure 8 Comparison of insertion-deletion loop (IDL)-containing oligomers that correspond to IDL sizes of: (A) 0 nucleotides (non-IDL, non-UVDE substrate); (B) 2 nucleotides (UVDE substrate); (C) four nucleotides (UVDE substrate; (D) 6 nucleotides (non-UVDE substrate); (E) 8 nucleotides (non-UVDE substate); and (F) 14 nucleotide hairpin (non-UVDE substrate). *Source*: Adapted from Ref. 53.

The available evidence suggests that UVDE may be an example of a DNA repair enzyme with a wide substrate specificity that may be involved in interactions between excision repair and mismatch repair pathways.

9. SUBSEQUENT STEPS FOLLOWING UVDE-INITIATED ALTERNATIVE EXCISION REPAIR

Although the emphasis of this chapter is on the recognition of different types of DNA lesions by UVDE, a brief discussion of the post-UVDE steps in the AER pathway is necessary. Of significant interest has been to determine the nature of the steps following UVDE-mediated DNA incision events. It had been initially proposed that these events include lesion removal by an endo-or exonucleolytic activity to generate a gap that is filled in and ligated by DNA polymerase and DNA ligase, respectively (5). Epistasis analysis has implicated several gene products in UVDE-initiated DNA repair, including the *S. pombe* FEN-1 (flap endonuclease) homolog, Rad2p. Two other genes implicated in this pathway are *rad18* and *rhp51,* which are thought to function in recombination events (54). *uve1* mutants are more UV sensitive than *uve1 rad2* double mutants, suggesting that the presence of the unrepaired cleavage product near the site of a particular lesion is more toxic to the cell than the uncleaved, original lesion itself (16). Further genetic and immunochemical analyses have suggested that DNA incised by UVDE is processed by two separate mechanisms, one dependent and one independent of Rad2p (16). The biochemical reconstitution of the AER pathway has been reported and consists of (as a minimum of proteins) UVDE, Rad2p (FEN-1), DNA polymerase delta, the accessory proteins PCNA and RFC, and DNA ligase (55). It should also be noted that the steps in the AER pathway subsequent to the action of Uve1p are similar to those employed in long-patch BER (4). The interconnections of AER with other DNA repair pathways are currently unknown. The UVDE is induced by UV light both at the level of transcription and enzyme activity (56) and unlike NER and BER, there does not appear to be any coupling of AER to transcription (57). The AER pathway appears to operate in both the nucleus and mitochondria in *S. pombe,* as subcellular localization studies have identified the presence of UVDE in these locations (20).

10. *SCHIZOSACCHAROMYCES POMBE* UVDE HOMOLOGS

Several studies indicate that structural and functional (biochemical) UVDE homologs exist in *S. pombe. N. crassa,* and *B. subtilis.* (17,18). In addition, sequence database searches have revealed UVDE homologs in a number of other organisms. Using Blast or FastA on the NCBI (www.nlm.nih.gov) or TIGR (www.tigr.org) databases, putative UVDE homologs have been identified in a variety of organisms including (but not limited to) *Deinococcus radiodurans, Bacillus anthracus, Clostidium perfingens, Methanococcus jannaschii, Thermotoga maritima,* and *Halobacterium* sp. A UVDE consensus sequence has been identified by using the Vector NTI AlignX program, which spans amino acids 308–465 in the C-terminal region of *S. pombe* UVDE. This region shows considerable sequence similarity with specific regions of the *B. subtilis* and *N. crassa* UVDE proteins. The alignments for this region as well as flanking regions are shown in Fig. 2. If one assumes that the sequence homologies reflect functional UVDE homologs in the above species and taking into account the known *S. pombe. N. crassa,* and *B. subtilis* enzymes, the evolutionary distance between these organisms suggests that the Uve1p-mediated AER system is a DNA repair pathway that arose early in the evolution of life and is maintained in species that live in a diverse range of environments. An important goal of future work in this area will be to obtain a better picture of the species distribution of UVDE homologs and AER. It is anticipated that such studies will provide important insights into the extent that the AER pathway is utilized in nature as well as its relative standing in comparison with other pathways such as NER, BER, and MMR as a general and frequently utilized system for the reversal of genetic damage.

11. CONCLUSIONS

The remarkable versatility of the UVDE protein with respect to the spectrum of DNA lesions recognized poses a problem for understanding the structural basis for substrate recognition. Obviously, the availability of X-ray crystallographic information or NMR structural information of the UVDE protein in solution would provide valuable insights into the molecular basis for DNA damage recognition. Unfortunately, such information is not available as attempts to crystallize UVDE have been, to date, unsuccessful.

ACKNOWLEDGMENT

Work from our laboratory on the UVDE protein was supported by NIH grant CA73041 from the National Cancer Institute.

REFERENCES

1. Hanawalt PC. Repair of genetic material in living cells. Endeavour 1972; 31(113):83–87.
2. Sancar A. Mechanisms of DNA excision repair. Science 1994; 266(5193):1954–1956.
3. Sancar A, Tang M-S. Nucleotide excision repair. Photochem Photobiol 1993; 57(5): 905–921.
4. Nilsen H, Krokan HE. Base excision repair in a network of defence and tolerance. Carcinogenesis 2001; 22(7):987–998.
5. Doetsch PW. What's old is new: an alternative DNA excision repair pathway. Trends Biochem Sci 1995; 20(10):384–386.
6. Yasui A, McCready SJ. Alternative repair pathways for UV-induced DNA damage. BioEssays 1998; 20(4):291–297.
7. McCready SJ, Osman F, Yasui A. Repair of UV damage in the fission yeast Schizosaccharomyces pombe. Mutat Res 2000; 451(1–2):197–210.
8. Schupbach M. The isolation and genetic classification of UV-sensitive mutants of Schizosaccharomyces pombe. Mutat Res 1971; 11(4):361–371.
9. Nasim A, Smith BP. Genetic control of radiation sensitivity in Schizosaccharomyces pombe. Genetics 1975; 79(4):573–582.
10. Phipps J, Nasim A, Miller DR. Recovery, repair, and mutagenesis in Schizosaccharomyces pombe. Adv Genet 1985; 23:1–72.
11. McCready S, Box B. Repair of 6-4 photoproducts in Saccharomyces cerevisiae. Mutat Res 1993; 293(3):233–240.
12. McCready S, Carr AM, Lehmann AR. Repair of cyclobutane pyrimidine dimers and 6-4 photoproducts in the fission yeast Schizosaccharomyces pombe. Mol Microbiol 1993; 10(4):885–890.
13. Sidik K, Lieberman HB, Freyer GA. Repair of DNA damaged by UV light and ionizing radiation by cell-free extracts prepared from Schizosaccharomyces pombe. Proceedings of the National Academy of Sciences of the United States of America 1992; 89(24):12112–12116.
14. Bowman KK, Sidik K, Smith CA, Taylor JS, Doetsch PW, Freyer GA. A new ATP-independent DNA endonuclease from Schizosaccharomyces pombe that recognizes cyclobutane pyrimidine dimers and 6-4 photoproducts. Nucleic Acids Res 1994; 22(15):3026–3032.
15. Freyer GA, Davey S, Ferrer JV, Martin AM, Beach D, Doetsch PW. An alternative eukaryotic DNA excision repair pathway. Mol Cell Biol 1995; 15(8):4572–4577.

16. Yonemasu R, McCready SJ, Murray JM, Osman F, Takao M, Yamamoto K, Lehmann AR, Yasui A. Characterization of the alternative excision repair pathway of UV-damaged DNA in Schizosaccharomyces pombe. Nucleic Acids Res 1997; 25(8):1553–1558.

17. Yajima H, Takao M, Yasuhira S, Zhao JH, Ishii C, Inoue H, Yasui A. A eukaryotic gene encoding an endonuclease that specifically repairs DNA damaged by ultraviolet light. EMBO J 1995; 14(10):2393–2399.

18. Takao M, Yonemasu R, Yamamoto K, Yasui A. Characterization of a UV endonuclease gene from the fission yeast Schizosaccharomyces pombe and its bacterial homolog. Nucleic Acids Res 1996; 24(7):1267–1271.

19. Kaur B, Avery AM, Doetsch PW. Expression, purification, and characterization of ultraviolet DNA endonuclease from Schizosaccharomyces pombe. Biochemistry 1998; 37(33): 11599–11604.

20. Yasuhira S, Yasui A. Alternative excision repair pathway of UV-damaged DNA in Schizosaccharomyces pombe operates both in nucleus and in mitochondria. J Biol Chem 2000; 275(16):11824–11828.

21. Nakai K, Kanehisa M. A knowledge base for predicting protein localization sites in eukaryotic cells. Genomics 1992; 14(4):897–911.

22. Pearlman DA, Holbrook SR, Pirkle DH, Kim SH. Molecular models for DNA damaged by photoreaction. Science 1985; 227(4692):1304–1308.

23. Husain I, Griffith J, Sancar A. Thymine dimers bend DNA. Proceedings of the National Academy of Sciences of the United States of America 1988; 85(8):2558–2562.

24. Taylor JS. Structure and Properties of DNA photoproducts. In: Siede W, Kow YW, Doetsch PW, eds. DNA Damage Recognition. New York: Marcel Dekker, 2005: 63–90.

25. Jing Y, Kao JF, Taylor JS. Thermodynamic and base-pairing studies of matched and mismatched DNA dodecamer duplexes containing *cis-syn*, (6-4) and Dewar photoproducts of TT. Nucleic Acids Res 1998; 26(16):3845–3853.

26. Taylor JS, Garrett DS, Cohrs MP. Solution-state structure of the Dewar pyrimidinone photoproduct of thymidylyl-(3′–5′)-thymidine. Biochemistry 1988; 27(19):7206–7215.

27. Khattack MN, Wang SY. The photochemical mechanism of pyrimidine cyclobutyl dimerization. Tetrahedron 1992; 28(4):945–957.

28. Taylor JS, Brockie IR. Synthesis of a *trans-syn* thymine dimer building block. Solid phase synthesis of CGTAT [t,s]TATGC. Nucleic Acids Res 1988; 16(11):5123–5136.

29. Lippke JA, Gordon LK, Brash DE, Haseltine WA. Distribution of UV light-induced damage in a defined sequence of human DNA: detection of alkaline-sensitive lesions at pyrimidine nucleoside-cytidine sequences. Proceedings of the National Academy of Sciences of the United States of America 1981; 78(6):3388–3392.

30. Taylor JS, Cohrs MP. DNA, Light, and Dewar Pyrimidinones - the Structure and Biological Significance of Tpt3. J Am Chem Soc 1987; 109(9):2834–2835.

31. Avery AM, Kaur B, Taylor JS, Mello JA, Essigmann JM, Doetsch PW. Substrate specificity of ultraviolet DNA endonuclease (UVDE/Uve1p) from Schizosaccharomyces pombe. Nucleic Acids Res 1999; 27(11):2256–2264.

32. Sherman SE, Lippard SJ. Structural Aspects of Platinum Anticancer Drug-Interactions with DNA. Chem Rev 1987; 87(5):1153–1181.

33. Takahara PM, Rosenzweig AC, Frederick CA, Lippard SJ. Crystal structure of double-stranded DNA containing the major adduct of the anticancer drug cisplatin. [see comment]. Nature 1995; 377(6550):649–652.

34. Ohndorf UM, Lippard SJ. Structural aspects of Pt-DNA adduct. In: Seide W, Kow YW, Doetsch PW, eds. DNA Damage Recognition. New York: Marcel Dekker, 2005: 225–244.

35. Doetsch PW, Cunningham RP. The enzymology of apurinic/apyrimidinic endonucleases. Mutat Res 1990; 236(2–3):173–201.

36. Berger RD, Bolton PH. Structures of apurinic and apyrimidinic sites in duplex DNAs. J Biol Chem 1998; 273(25):15565–15573.

37. Kanno S, Iwai S, Takao M, Yasui A. Repair of apurinic/apyrimidinic sites by UV damage endonuclease; a repair protein for UV and oxidative damage. Nucleic Acids Res 1999; 27(15):3096–3103.

38. Modrich P. Strand-specific mismatch repair in mammalian cells. J Biol Chem 1997; 272(40):24727–24730.

39. Fleck O, Michael H, Heim L. The swi4+ gene of Schizosaccharomyces pombe encodes a homologue of mismatch repair enzymes. Nucleic Acids Res 1992; 20(9):2271–2278.

40. Schar P, Baur M, Schneider C, Kohli J. Mismatch repair in Schizosaccharomyces pombe requires the mutL homologous gene pms1: molecular cloning and functional analysis. Genetics 1997; 146(4):1275–1286.

41. Crouse G. Mismatch repair systems in Saccharomyces cerevisiae. In: Nickoloff JA, Hoekstra MF, eds. DNA Damage and Repair, Volume 1: DNA Repair in Prokaryotes and Lower Eukaryotes. Totowa: Humana Press, 1998:411–448.

42. Szankasi P, Smith GR. A DNA exonuclease induced during meiosis of Schizosaccharomyces pombe. J Biol Chem 1992; 267(5):3014–3023.

43. Szankasi P, Smith GR. A single-stranded DNA exonuclease from Schizosaccharomyces pombe. Biochemistry 1992; 31(29):6769–6773.

44. Szankasi P, Smith GR. A role for exonuclease I from S. pombe in mutation avoidance and mismatch correction. Science 1995; 267(5201):1166–1169.

45. Schar P, Kohli J. Marker effects of G to C transversions on intragenic recombination and mismatch repair in Schizosaccharomyces pombe. Genetics 1993; 133(4):825–835.

46. Schar P, Munz P, Kohli J. Meiotic mismatch repair quantified on the basis of segregation patterns in Schizosaccharomyces pombe. Genetics 1993; 133(4):815–824.

47. Fleck O, Schar P, Kohli J. Identification of two mismatch-binding activities in protein extracts of Schizosaccharomyces pombe. Nucleic Acids Res 1994; 22(24):5289–5295.

48. Fleck O, Lehmann E, Schar P, Kohli J. Involvement of nucleotide-excision repair in msh2 pms1-independent mismatch repair. [see comment]. Nat Genet 1999; 21(3):314–317.

49. Kunz C, Fleck O. Role of the DNA repair nucleases Rad13, Rad2 and Uve1 of Schizosaccharomyces pombe in mismatch correction. J Mol Biol 2001; 313(2):241–253.

50. Kaur B, Fraser JL, Freyer GA, Davey S, Doetsch PW. A Uve1p-mediated mismatch repair pathway in Schizosaccharomyces pombe. Mol Cell Biol 1999; 19(7): 4703–4710.

51. Hall MC, Matson SW. The *Escherichia coli* MutL protein physically interacts with MutH and stimulates the MutH-associated endonuclease activity. J Biol Chem 1999; 274(3):1306–1312.

52. Klungland A, Hoss M, Gunz D, Constantinou A, Clarkson SG, Doetsch PW, Bolton PH, Wood RD, Lindahl T. Base excision repair of oxidative DNA damage activated by XPG protein. Mol Cell 1999; 3(1):33–42.

53. Kaur B, Doetsch PW. Ultraviolet damage endonuclease (Uve1p): a structure and strand-specific DNA endonuclease. Biochemistry 2000; 39(19):5788–5796.

54. Lehmann AR, Walicka M, Griffiths DJ, Murray JM, Watts FZ, McCready S. The rad18 gene of Schizosaccharomyces pombe defines a new subgroup of the SMC superfamily involved in DNA repair. Mol Cell Biol 1995; 15(12):7067–7080.

55. Alleva JL, Hurwitz J, Zuo S, Doetsch PW. In vitro reconstitution of the Schizosaccharomyces pombe alternative excision repair pathway. Biochemistry 2000; 39(10):2659–2666.

56. Davey S, Nass ML, Ferrer JV, Sidik K, Eisenberger A, Mitchell DL, Freyer GA. The fission yeast UVDR DNA repair pathway is inducible. Nucleic Acids Res 1997; 25(5):1002–1008.

57. Yasuhira S, Morimyo M, Yasui A. Transcription dependence and the roles of two excision repair pathways for UV damage in fission yeast Schizosaccharomyces pombe. J Biol Chem 1999; 274(38):26822–26827.

58. Gao YG, Robinson H, Sanishvili R, Joachimiak A, Wang AH. Structure and recognition of sheared tandem G x A base pairs associated with human centromere DNA sequence at atomic resolution. Biochemistry 1999; 38(50):16452–16460.

59. Natarajan G, Lamers MH, Enzlin JH, Winterwerp HH, Perrakis A, Sixma TK. Structures of *Escherichia coli* DNA mismatch repair enzyme MutS in complex with different mismatches: a common recognition mode for diverse substrates. Nucleic Acids Res 2003; 31(16):4814–4821.

60. Feng B, Stone MP. Solution structure of an oligodeoxynucleotide containing the human n-ras codon 61 sequence refined from 1H NMR using molecular dynamics restrained by nuclear Overhauser effects. Chem Res Toxicol 1995; 8(6):821–832.

61. Marzilli LG, Sadd JS, Kuklenyik Z, Keating KA, Xu Y. Relationship of solution and protein-bound structures of DNA duplexes with the major intrastrand cross-link lesions formed on cisplatin binding to DNA. J Am Chem Soc 2001; 123(12):2764–2770.

12

Structural Aspects of Pt-DNA Adduct Recognition by Proteins

Uta-Maria Ohndorf and Stephen J. Lippard
Department of Chemistry, Massachusetts Institute of Technology, Cambridge, Massachusetts, U.S.A.

1. BACKGROUND

The ability to introduce DNA damage selectively into a specific cell type eventually to confer cell death is a highly desirable strategy in cancer chemotherapy. One of the most successfully employed agents is the inorganic drug *cis*-diamminedichloro platinum(II) (*cis*-DDP), or cisplatin (1). The numerous advances made in understanding the mechanism of action of the drug are summarized in several recent review articles covering a broad range of topics including DNA as the drug target (2), the recognition and processing of cisplatin–DNA adducts (3), proteins recognizing Pt–DNA damage (4,5), cellular consequences of protein binding (6), repair of cisplatin-damage (7) and new platinum complexes as potential therapeutics (8). The present chapter focuses on the structural aspects of cisplatin–DNA–protein interactions with an emphasis on recognition of the major adduct, a 1,2-intrastrand cross-link. A wealth of structural information recently made available, including knowledge of a ternary cisplatin–protein–DNA complex and DNA complexes of proteins with the ability to recognize *cis*-DDP–DNA adducts bound to their natural DNA substrates, reveal striking similarities in the DNA recognition mechanisms.

2. INTRODUCTION

The high mortality rate of patients with cancer led to the establishment of extensive and more vigorous national cancer programs in several industrialized countries in the 1970s such as the "National Cancer Act of 1971" in the United States, which is also referred to as a declaration of "war on cancer". Curiously, the introduction of this program coincided with the start of clinical trials for a new "weapon" in this war, the inorganic compound cisplatin (*cis*-diamminedichloroplatinum(II), or *cis*-DDP). The cell growth inhibitory properties of this and related platinum compounds had been discovered serendipitously by Barnett Rosenberg et al. (9). Today, over 90% of patients with testicular cancer are cured due to chemotherapy with cisplatin, in combination with the advent of improved diagnostic and surgical techniques (10).

Perhaps the most famous case is that of cyclist Lance Armstrong, whose testicular cancer had already metastasized to the brain prior to his treatment with cisplatin, although he reported of being cured with plutonium (11).

Cisplatin is a square-planar, neutral compound having two chloride and two ammonia ligands in *cis* positions. The stereoisomer *trans*-DDP is inactive. This geometric structure–activity relationship led to the development of several second generation platinum drugs with different amine substituents (Fig. 1). The specific ligand-exchange kinetics of platinum compounds is important for determining the pharmacological reactivity of the drug. Following aquation through substitution of the chloride ions by water molecules, the cytotoxic effect results from cross-linking DNA through the N_7 nitrogen heteroatoms of the DNA purine bases (12,13). Cisplatin-DNA adducts inhibit DNA replication, block transcription and repair, and ultimately trigger apoptosis (3,7). Cells deficient in repair are hypersensitive to *cis*-DDP (7, and references therein).

Having a platinum(II) atom at the focal point for the cross-link imposes stringent requirements for a well-defined, square-planar coordination sphere in which DNA serves as a bidentate ligand (14). The X-ray structure of a cisplatin-cross-linked dinucleotide d(pGpG), the smallest, most flexible DNA ligand, revealed that the maximum possible distance spanned by the cross-linked atoms is 2.9 Å (15,16). Cross-linking induces the guanine base planes to rotate towards each other in a head-to-head conformation with an interbase dihedral angle of 76–87° to minimize steric crowding (Fig. 2A). By comparison, the ligands in the clinically inactive *trans*-DDP bridge a distance of 3.9 Å.

3. STRUCTURAL CONSEQUENCES OF PLATINUM-BINDING TO DOUBLE-STRANDED DNA

Kinetic and stereochemical constraints and the binding of platinum almost exclusively to N_7 nitrogen atoms of purine bases lead to a unique adduct distribution in double-stranded DNA. Over 90% are 1,2-intrastrand cross-links, either between two adjacent guanine bases (> 65%) or between an adenine base and a guanine base (20%–25%); minor adducts include the 1,3-d(GpNpG) intrastrand adduct and inter-strand adducts (4%) (17). In contrast, the related drug carboplatin (Fig. 1) forms a substantially greater number of 1,3-intrastrand cross-links (18).

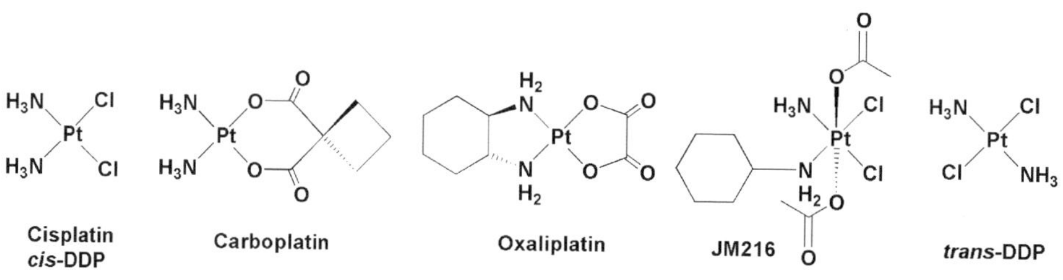

Figure 1 Structures of platinum compounds. Cisplatin and carboplatin are the most extensively used anticancer drugs and cisplatin is curative for testicular cancer. Oxaliplatin has been approved in Europe and the USA for the treatment of colorectal cancer. JM216, an orally active platinum(IV) compound, which functions much like cisplatin upon reduction by intracellular agents and loss of the axial ligands, is currently in phase III clinical trials for prostate cancer. The isomer *trans*-DDP is clinically inactive.

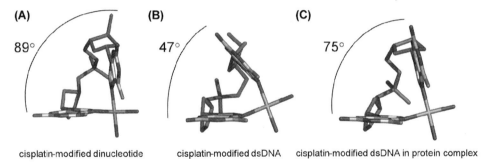

Figure 2 Comparison of the structures of a single-stranded cis-$\{[Pt(NH_3)_2]\}^{2+}$ platinated dinucleotide (A), with the site in a cisplatin-modified DNA dodecamer duplex (B), and in a protein complex with cisplatin-modified DNA (C). The interbase dihedral angles are indicated. *Source*: From Refs. 15,26,66. (*See color insert*.)

The N_7 atoms of adjacent purine bases in B-type DNA are separated by $\sim 4.2\,\text{Å}$; thus the formation of a platinum cross-link must significantly perturb the DNA structure. Initial studies by gel electrophoresis indicated that cisplatin adducts bend and unwind duplex DNA (19,20). Thermodynamic studies revealed a helix destabilization of 6.3 kcal per mol for the cis-DDP-G^*G^* 1,2-intrastrand cross-link, where the asterisks denote the sites of platinum modification, modulated by the nature of the bases flanking the adduct (21,22). The advent of X-ray and NMR solution structures helped to place these parameters in a geometric context (Fig. 2) (reviewed in Ref. 23). The following sections discuss the major structural differences among the cisplatin–DNA adducts.

3.1. The 1,2-d(GpG) Intrastrand Cross-Link

Several structure determinations of duplex DNA containing a 1,2-intrastrand platinum cross-link with amine ligands as in cisplatin (24–27), with a 1,2-diaminocyclohexane ligand as in oxaliplatin (Fig. 1) (28), or with a cyclohexylamine ligand as in JM216 (Fig. 1) (29), reveal that the DNA duplex is substantially bent towards the major groove (Fig. 3). The minor groove is widened and shallow, due to helix unwinding centered around the platination site, exposing previously inaccessible hydrophobic regions of the minor groove floor and the sugar-phosphate backbone. DNA bending results from canting of the cross-linked bases out of their natural positions to satisfy a planar platinum coordination sphere. The bend of 39°–55° in the crystal structures is a result of an interbase angle of $\sim 26°$ at the cross-linked guanines with base pairing maintained around the platinum site (24). The DNA is B-form on the 3′-side to the adduct and A-form on the 5′-side, probably due to crystal packing forces, which most likely also influence the degree of bending observed.

In contrast, a larger bend of $\sim 79°$ is observed in solution, accompanied by a significantly larger interbase angle of $\sim 47°$ (Figs. 2 and 3) (26,30). It has been argued, however, that the choice of the DNA starting model significantly influences the outcome of the refined minimized structure (31). NMR/modeling methods may not impose restraint sets sufficient to delineate novel structural features, such as the cross-linked base pair step, present in cisplatin-modified DNA. Only when the conformation of the sugar-phosphate backbone from the cisplatin–DNA–HMG domain structure (vide infra), the so called Lippard base pair step (32), was used did the

(A)   **(B)**   **(C)**

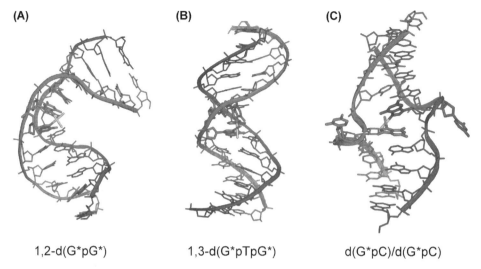

1,2-d(G*pG*)   1,3-d(G*pTpG*)   d(G*pC)/d(G*pC)

Figure 3 Structures of double-stranded DNA modified with cisplatin. (A) The 1,2-d(G*pG*) intrastrand cross-link (PDB entry 1A84); (B) the 1,3-d(G*pTpG*) intrastrand adduct (PDB entry 1DA5); and (C) the d(G*pC)/d(G*pC) interstrand adduct (PDB entry 1A2E), where G* denotes the location of the platinated nucleotides. *Source*: From Refs. 26,33,37. (*See color insert.*)

NMR data refine to a satisfactory result. Accordingly, the following descriptions of the NMR determinations of the 1,3-intrastrand and the interstrand adducts may not be accurate at the sites of platination.

In summary, the NMR and crystal structures show that a range of bending values for DNA containing the 1,2-intrastrand cross-link is possible. The bend angle increases with increasing canting of the guanine bases. The interbase dihedral angle between cross-linked guanines in duplex DNA is in all cases substantially lower than in the [Pt(NH$_3$)$_2$d(pGpG)] complex, however (Fig. 2, Table 1) (14). It is thus likely that a continuum of conformations with different bend angles exist, balanced by a reduced strain on the platinum coordination sphere.

3.2. The 1,3-d(GpNpG) Cross-Link

For the 1,3-d(GpNpG) adduct two structures, an NMR-based model and an NMR solution structure were reported (33,34). The NMR solution structure is well determined except for the central thymine in the platinum lesion due to a lack of NOE data restraining it to a fixed position in space. It is clear, however, that it is extruded from the helix unlike predictions of the NMR model, where it retained its stacking interactions with the 3'-platinated guanine (Fig. 3). The central base pairing around the cross-link is significantly distorted. The DNA helix is unwound at the platination site and kinked by 30° towards the major groove, although the exact curvature is difficult to determine by NMR due to a lack of long distance restraints.

3.3. The Interstrand d(G*pC)/d(G*pC) Cross-Link

The NMR and crystal structures of the cisplatin–DNA interstrand adduct reveal significant structural distortions in the DNA duplex (Fig. 3C) (35–37). The helix is curved, with different bending directions observed in the NMR and crystal

Table 1 Selected Proteins Recognizing the 1,2-Intrastrand Cisplatin-DNA Adduct

Protein	Function	Natural DNA substrate	K_d/Specificity for *cis*-DDP DNA (Ref.)	DNA binding domain	Recognition site
1. Chromatin remodeling/ Architectural proteins					
rHMGB1	Stabilizing bent DNA	n.d.[a]	120 nM/100 (58,59)	2 HMG	Minor groove
yNHP6A	Stabilizing bent DNA	n.d.[a]	0.1 nM/40 (70)	HMG	Minor groove
hHistone H1	Linker histone	Linker DNA between nucleosomes	n.d.[a] (71)	Winged helix	Major groove?
dHMG-D	Architectural protein	n.d.[a]	200 nM/2–3 (120)	HMG	Minor groove
hSSRP1/FACT	Enabling transcription past nucleosomes	n.d.[a]	n.d.[a] (53,54)	HMG	Minor groove
T4Endonuclease VII	Cleavage of junction DNA	Junction DNA	n.d.[a] (121)	n.d.[b]	Minor groove
2. Transcription factors					
hLEF-1	Transcription enhancer	DNA sequence	100 nM/n.d.[a] (57)	HMG	Minor groove
hSRY/mSRY	Sex-determining factor	DNA sequence	120 nM/20 (72)	HMG	Minor groove
tsHMG	Chromatin remodeling during spermatogenesis	n.d.[a]	24 nM/230 (122)	2 HMG	Minor groove
mtTFA	Mitochondrial DNA transcription	DNA sequence	100 nM/n.d.[a] (57)	2 HMG	Minor groove
hUBF	rRNA transcription	DNA sequence	60 pM/n.d.[a] (123)	6 HMG	Minor groove
yIxr1	Cox5b promoter recognition	DNA sequence	250 nM/10 (75,77)	2 HMG	Minor groove
hYB-1	Y-Box recognition	DNA sequence	n.d.[a] (124)	Coldshock domain	ssDNA
hTBP	Part of TFIID transcription factor	DNA sequence	0.3 nM/3000 (81)	n.d.[b]	Minor groove
p53	Tumor suppressor	DNA sequence	n.d.[b] (125)	n.d.[b]	Major and minor groove

(*Continued*)

Table 1 Selected Proteins Recognizing the 1,2-Intrastrand Cisplatin-DNA Adduct (*Continued*)

Protein	Function	Natural DNA substrate	K_d/Specificity for *cis*-DDP DNA (Ref.)	DNA binding domain	Recognition site
3. DNA repair proteins					
Mismatch repair					
scCmb1	Mismatch recognition?	Cytosine-containing mismatch DNA	n.d.[a] (126)	HMG	Minor groove
hMutSα	Mismatch recognition	dsDNA with mispaired bases	363 nM/1.5 (115)	n.a.[b]	Major and minor groove
hAAG	Base excision repair	3-methyladenine containing dsDNA	115 nM/n.d. (86)	n.a.[b]	Minor groove
Nucleotide excision repair					
hXPA	Damage recognition	Bent DNA	n.d.[a] (98)	α/β fold	n.d.[a]
hRPA	Stabilizing opened DNA complex, ssDNA recognition	Single-stranded DNA	n.d.[a] (105,106)	OB-fold	ssDNA
hXPE	Damage recognition	Pyrimidine dimers	n.d.[a] (109)	n.a.[b]	n.d.[a]
hXPC-HR23B	Damage recognition	DNA damage	n.d.[a] (127)	n.a.[b]	n.d.[a]
Direct reversal					
E.coli/yeast photolyase	UV-damage repair	Pyrimidine dimers	50 nM/n.d.[a] (90)	n.a.[b]	n.d.[a]

[a]Values not determined are denoted as n.d.

[b]Proteins with unclassified structures are denoted as not applicable, n.a.

structures, and unwound by $70°–87°$. The cis-$\{Pt(NH_3)_2\}^{2+}$ moiety lies in the minor rather the major groove due to a $180°$ rotation of the deoxyguanosines. Both guanine bases remain in the *anti* conformation, with their O_6 atoms positioned in a head-to-tail orientation. The $O_{4'}$ oxygen atoms of the sugar rings point towards the $3'$ direction instead of the $5'$ direction. The complementary cytosine bases are extruded extrahelically. Curiously, the mismatch repair protein DNA glycosylase generates a base excision product that is remarkably similar to that of the cisplatin-interstrand adduct (38).

4. RECOGNITION OF *cis*-DDP-1,2 INTRASTRAND CROSS-LINK BY CELLULAR PROTEINS

As outlined in the previous sections, each cisplatin adduct distorts the double-stranded DNA architecture in a unique manner. These structural differences are critically important for recognition of the adducts by proteins with resulting downstream consequences. Even though the exact signal transduction pathways leading to cell death are unknown, it is evident that timely removal of cisplatin-adducts from the genome is imperative for cell survival. Cisplatin adducts block DNA transcription and replication (39,40).

Nucleotide excision repair (NER) (see Chapter 16 on NER), comprising both global genomic repair and transcription coupled repair pathways, is a major cellular defense system against cisplatin (41,42). Its efficiency is highly dependent on the accessibility of the damaged site, modulated by protein binding and the chromatin structure. Several proteins involved in DNA transcription, packaging, rearrangement, replication, and repair recognize the major cisplatin–DNA adduct. The affinity of these proteins for platinated DNA may contribute to the therapeutic effect of the drug. By binding to the cisplatin–DNA lesions they obstruct damage recognition and processing, ultimately leading to cell death. Consistent with this model are in vitro experiments showing that the presence of HMG-domain proteins can block NER (vide infra) (41,42).

In contrast to the platinated DNA substrates employed for studies in vitro, the genome of eukaryotes is densely packaged into chromatin. The fundamental building block of chromatin is the nucleosome core particle, in which 147 bp of DNA are wrapped around a histone octamer connected by linker DNA (43). In a global genomic repair assay the nucleosome significantly retarded excision repair of cisplatin adducts. Repair of a site-specific GG-Pt DNA nucleosome substrate was $\sim$30% of the level observed with free DNA, whereas with a GTG-Pt DNA substrate excision levels of $\sim$10% of that with free DNA were measured (44). In order to allow for chromatin accessibility during transcription, the nucleosome positioning and structure undergo dynamic fluctuations with the aid of several enzymes. Histone–DNA interactions are modulated through covalent modifications of their component amino acids (45, and references therein). These histone modifications also affect repair levels of cisplatin adducts. Excision from native nucleosomal DNA is $\sim$2-fold higher than the level observed with recombinant, unmodified protein (44).

A second determinant for cisplatin toxicity is interference of the DNA adducts with cellular pathways. A correlation between arrest of transcription through ubiquitylation of RNA polymerase II (pol II) and the presence of cisplatin-damage has been demonstrated in an in vitro system (46). Pol II senses the DNA damage and stalls upon reaching the cisplatin-adduct. This transcription arrest apparently signals for ubiquitylation and degradation of pol II.

An important set of protein modifications observed upon cisplatin treatment are phosphorylation and acetylation of the core histones, which indirectly influences chromatin fluidity and transcription (68). Cellular pathways can further be manipulated if DNA-binding proteins are diverted from their natural sites due to their high affinity for *cis*-DDP adducts. Modulation of the effective cellular concentration of essential proteins could have fatal consequences for the cell. For all these reasons, it is crucial to understand the recognition mechanism between proteins and *cis*-DDP-modified DNA as the first step in the cellular response to cisplatin damage.

The following sections focus on selected members of three classes of proteins that recognize *cis*-DDP–DNA, transcription factors, repair proteins, and proteins involved in chromatin remodeling. Proteins having a high mobility group (or HMG) domain interacting with DNA appear in all of these classes, and many critical cellular functions may be interrupted as a consequence of HMG-domain recognition of cisplatin-modified DNA (47) (Table 1).

4.1. Chromatin Reorganization

4.1.1. hSSRP1/FACT

A novel chromatin remodeling factor, FACT (facilitates chromatin transcription), that activates transcription elongation was identified in HeLa nuclear extracts (48). It may function to unravel H2A/H2B histone dimers from the nucleosome cores (49). In yeast, FACT has been implicated in the regulation of transcription and replication (50). Drosophila FACT interacts physically with the GAGA transcription factor thus directing chromatin remodeling to its binding site (51).

Human FACT purified as a heterodimer of SSRP1 and SPT16. The 81-kDa protein SSRP1 (structure-specific recognition protein 1) component was the first protein directly identified to bind to cisplatin-modified DNA (52,53). The DNA-binding activity is conferred by the small 80 amino acid HMG-domain in SSRP1 (54). FACT and the isolated HMG domain of SSRP1 bind specifically to the 1,2-d(GpG) intrastrand cisplatin–DNA cross-link, whereas the full-length SSRP1 fails to form a complex with cisplatin-modified DNA. The observed differences in binding affinity can be reconciled by the fact that Spt16 is necessary to prime SSRP1 for DNA recognition. The HMG domain is most likely inaccessible in the native protein and unveiled only through a conformational change in the Spt16/SSRP1 dimer (54). The specific interaction of FACT with cisplatin-modified DNA could modulate the cytotoxicity of the drug by diverting the protein complex from its natural binding site or by shielding the DNA cross-link from repair, both having potentially lethal consequences for the cell. Moreover, the RNA pol II transcription machinery might stall upon encounter with the stable FACT–cisplatin–DNA complex leading to ubiquitination and proteolysis of the polymerase and, ultimately, to cell death (46).

4.1.2. HMGB1

HMGB1 is a nuclear protein with two tandem HMG box domains, HMGB1a and HMGB1b, and a C-terminal acidic tail. It binds with high affinity to unusual DNA structures like 4-way junctions and DNA bulges, suggesting that HMGB1 functions by distorting linear DNA into a bent conformation (55). By increasing chromatin flexibility HMGB1 fulfills two key tasks. It helps the formation of enhanceosomes by promoting protein interactions with their respective DNA binding sites and it modulates chromatin remodeling by binding to the entry and exit points of DNA at the nucleosomes (56).

HMGB1 has been linked to cisplatin activity in a number of studies. The protein binds specifically to the major adduct of cisplatin-modified DNA and further bends the double helix (57). Binding occurs mainly through the first HMG domain with a K_d of 120 nM, as determined by gel shift experiments (58–60). The nature of the platinum adduct flanking sequence is an important factor in determining binding specificity of the full-length protein and the individual domains. The affinity of HMGB1a is greatest with a neighboring A:T base pair, probably as a consequence of increased inherent flexibility in purine-rich sequences (60,61). Kinetic data obtained by stopped flow measurements indicate an on-rate for cisplatin-modified DNA near the diffusion limit for HMGB1 and its isolated domains (57,62).

Gel mobility shift assays show that the nature of the auxiliary ligands also influences HMGB1 affinity. The recognition of the sequence TG*G*A by HMGB1a decreases in the order of cisplatin $>$ cis-$\{Pt(NH_3)(NH_2Cba)\}^{2+}$ $>$ cis-$\{Pt(NH_3)$-$(2\text{-}pic)\}^{2+}$ $\sim$ cis-$\{Pt(NH_3)(NH_2Cy)\}^{2+}$ $>$ $\{Pt(en)\}^{2+}$ $>$ $>$ $\{Pt(dach)\}^{2+}$ (63).

The presence of HMG-domain proteins can block NER, as revealed by in vitro experiments, consistent with their postulated role in protecting platinated DNA from repair in the cell (42). Given the abundance of HMGB1 in nuclei it was hypothesized that the cisplatin sensitivity of cancer cells could be modulated by cellular HMGB1 levels. Consistent with this hypothesis, upregulation of HMGB1 production with hormones such as estrogen increases the cisplatin sensitivity of MCF-7 breast cancer cells (64). In contrast, a knock out of HMGB1 in an embryonic mouse cell line had no influence on cisplatin-sensitivity compared to the Hmgb1$^{+/+}$ cells (65). These results indicate that the mechanism of cisplatin sensitization by cellular proteins is complex and multifactorial, stressing the importance of cell type in determining the modulation of the efficacy of the drug by cellular factors.

4.1.3. The HMGB1a–Cisplatin–DNA Structure

The crystal structure of a 16-base-pair double stranded deoxyoligonucelotide containing a single cisplatin intrastrand cross-link in complex with HMGB1a reveals the molecular interactions responsible for the high specificity of HMG domains for cisplatin-modified DNA (66). The HMG domain binds in the minor groove of the DNA. The binding site extends solely to the 3' side of the double helix with respect to the cisplatin adduct (Fig. 4). The molecular interface between protein and DNA is largely hydrophobic and complex stability is aided by intermolecular stacking interactions. The combination of cisplatin-modification and HMG-domain binding results in a DNA conformation that is severely underwound with a helical twist angle of only 9°. The DNA is bent even more towards the major groove and adopts a widened minor groove, 12 Å across, even larger than in cisplatin-modified DNA alone. The two guanine bases involved in the drug-DNA cross-link are rolled open with an interbase dihedral angle of 75° rendering the platinum coordination sphere less strained than in the structures of cisplatin-modified DNA alone (Figs. 2 and 3). The resulting hydrophobic notch in the minor groove at the d(G*pG*) junction serves as an intercalation site for a phenylalanine side chain, which interacts in an edge-to-face manner with the guanine base on the 5'-side of the platinum adduct and stacks onto the second guanine heterocycle (Fig. 4). The π-π interactions

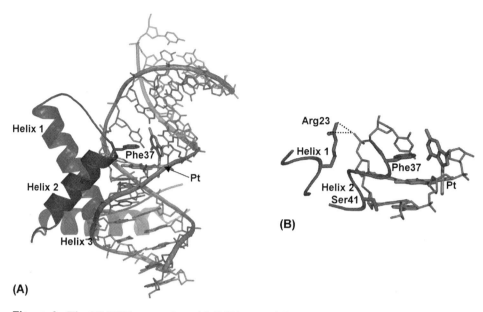

Figure 4 The HMGB1a complex with DNA containing a cisplatin-1,2-d(GpG) intrastrand adduct. (A) Overall structure (PDB entry 1CKT). The protein and DNA backbones are shown as a ribbon, the intercalating Phe 37 residue as a stick representation and (B) close-up of the intercalation site. (*See color insert.*)

between the intercalating residue and the cisplatin-modified bases may also be responsible for determining the binding orientation on the DNA. The significant overlap of the side chain phenyl ring with the purine base would be drastically reduced in the inverse orientation. Mutations of phenylalanine to the larger heterocycle tryptophan lowers binding affinity 5-fold, whereas replacement by alanine, where the side chain is too short for intercalation, nearly abrogates complex formation (66,67).

The HMGB1a–cisplatin–DNA structure presents an alternative protein binding mode compared to that of other HMG-domains such as those present in the transcription factors LEF-1 (Fig. 5) and SRY (vide infra), or HMGB1b. DNA footprinting analysis revealed that HMGB1b binds symmetrically to cisplatin-modified DNA, congruent with the cisplatin-induced bend (67). This binding mode is dictated by an intercalating residue at position 16 at the beginning of the first α-helix, just like in the transcription factors. A comparison of DNA substrates prior to protein binding could help rationalize the need for two different binding modes in the HMG-domain motif. HMG-domains recognize linear DNA with inherent flexibility by an induced-fit mechanism. The HMG domain serves as a clamp, binding into the minor groove and thus bending the DNA and compressing the major groove. It is conceivable that such a mechanism works best if the bending wedge is situated at the center of the protein, i.e., position 16. In contrast, HMG-domain binding to a prebent substrate could work according to a lock-and-key-type mechanism where the domain docks onto the DNA. An intercalating residue at one end of the protein, i.e., position 37, is sufficient to anchor it to the prebent DNA. Indeed, this design is well reflected in the properties of the two HMG domains of HMGB1. HMGB1a with intercalating residue at position 37 has a higher binding affinity for

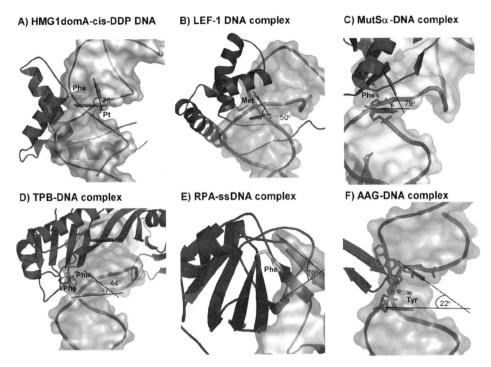

Figure 5 Comparison of intercalation sites and interbase dihedral angles in complexes of minor groove DNA-binding proteins as indicated. The DNA is shown as a surface representation with the bases forming the intercalation site and the intercalating residue in stick representation. (A) HMGB1a complex with *cis*-DDP DNA (pdb entry 1CKT); (B) LEF-1–DNA complex (pdb entry 2LEF); (C) MutSα-DNA complex (pdb entry 1EWQ); (D) TBP-DNA complex (pdb entry 1YTB), intercalation site shown for two of four intercalating Phe residues; (E) RPA complex with single-stranded DNA (pdb entry 1JMC), a total of four aromatic residues intercalate along the DNA single strand; and (F) AAG-DNA complex (pdb entry 1EWN). (*See color insert.*)

cisplatin-modified DNA, whereas HMGB1b with intercalating residue a position 16 can bend DNA more effectively (60). It is interesting to note that the yeast analogue of HMGB1, Nhp6A, containing only one HMG domain, utilizes hydrophobic wedges at both positions (69). In the structure of Nhp6A bound to a purine rich DNA substrate, a methionine residue at position 16 intercalates into the deformable base pair step, whereas Nhp6A binding to cisplatin-modified DNA is guided by a phenylalanine at position 37 (70). DNA footprinting revealed no preferred binding orientation in the latter complex.

4.1.4. Histone H1

Histones significantly inhibit the repair of cisplatin-modified DNA in a manner that can be modulated by post-translational modifications (44). Histone H1 binds to the linker DNA at the entry and exit points of the nucleosomes. It serves as the antagonist of HMGB1 (4) by stiffening the chromatin and locking the nucleosome into place. H1 is replaced by HMGB1 during chromatin remodeling. It is thus not surprising that H1 also binds to DNA globally modified with cisplatin, but not *trans*-DDP–DNA (71). The biological consequences of its interaction with *cis*-DDP–DNA have not yet been determined.

4.2. Transcription Factors

4.2.1. LEF-1 and SRY

Two sequence-specific transcription factors containing an HMG domain bind specifically to the major cisplatin–DNA adduct. The protein encoded by the sex determining region on the Y chromosome (SRY) is responsible for testis formation, and the lymphoid enhancer factor 1 (LEF-1) facilitates T-cell enhancer assembly (57,72). Upon binding in the minor groove both proteins bend DNA towards the major groove as determined by NMR spectral studies (73,74). The ability to interact with bent DNA structures and to bend linear DNA stems from an intercalating residues at a position equivalent to residue 16 in the HMG domain. In an in vitro assay, the testis-specific protein SRY inhibited repair of the major cisplatin–DNA adduct (72).

4.2.2. Ixr1

Ixr1 (cross-link recognition protein 1) was identified in a screen for cisplatin-DNA binding proteins in a yeast cDNA library (75). It is an HMG-domain protein that inhibits transcription of cytochrome c oxidase subunit V by binding to the Cox5b promoter (76). The protein binds at least an order of magnitude more tightly to DNA containing a single 1,2-intrastrand cisplatin cross-link than to unmodified DNA (77). Consistent with the shielding of platinated DNA, yeast deficient in Ixr1 are 2–6 fold more resistant to cisplatin and accumulated fewer cisplatin adducts than wild-type strains (75,78). This differential sensitivity to cisplatin is abolished in excision repair-deficient cells (78). Experiments failed to provide any evidence suggesting that Ixr1 could be removed from its natural binding site sufficiently to affect transcription from the Cox5b promoter (77).

4.2.3. TBP

The TATA-binding protein (TBP) is essential for transcription initiation in eukaryotes. It binds to consensus sequences TATa/tAa/t typically located 25–30 base pairs upstream of a transcription start site. The crystal structures of TBP in complex with the TATA box recognition sequence reveal several similarities with the structure of HMGB1a bound to cisplatin-modified DNA, even though the protein folds are unrelated (79,80). TBP binds to a widened and flattened minor groove. Phenylalanine residues are employed as hydrophobic wedges, thereby bending the DNA towards the major groove (Fig. 5D). The resulting DNA helix conformation closely resembles that found in the complex of HMGB1a with cisplatin-modified DNA. It is therefore not surprising that human TBP binds with a very low K_d of 0.3 nM to cisplatin-modified DNA and exhibits an exceptionally high specificity ratio of 3000 for cisplatin-modified vs. unmodified DNA (81). TBP displays sequence context selectivity for platinated DNA similar to that to HMGB1. In a competition assay, *cis*-DDP DNA extensively sequestered TBP from its natural binding site (81). Transcription could be restored upon addition of more TBP. Similar to its binding to the TATA box, TBP associates and dissociates from the cisplatin 1,2-d(GpG) intrastrand cross-links 1000-fold more slowly than HMGB1 (81). The presence of a platinum adduct in the vicinity of the TATA box even further increases the affinity of yeast

TBP as a consequence of a >30-fold slower dissociation rate (82). Platination of the promoter region at the TATA box in vivo could result in upregulated transcription from damaged promoters due to a decreased ability to execute promoter clearance (82). TBP is able to shield the 1,2-(GpG) intrastrand adduct when allowed to form the protein–DNA complex prior to exposure to the repair machinery (81). Like HMGB1, TBP can also discriminate between different spectator ligands in the platinum complex (63).

4.3. Repair Proteins

The anticancer efficacy of cisplatin is highly dependent on the efficiency of cisplatin–DNA adduct removal. Overexpression of repair factors confers cisplatin resistance (83,84). Because cisplatin resistance is a major reason for treatment failure, it is of great interest to understand how repair mechanisms participate in the modulation of the cellular sensitivity to the drug. Even though the NER pathway is the major detoxification process for cisplatin adducts, several other repair proteins also recognize the 1,2-intrastrand cross-link.

4.3.1. Base Excision Repair–3-Methyladenine DNA Glycosylase (Aalkyl Adenine Glycosylase, AAG)

The DNA N-glycosylases are base excision-repair proteins that locate and cleave damaged bases from DNA. Alkyl adenine glycosylase (AAG) removes 3-methyladenine and a wide variety of other damaged bases from DNA in vivo (See, Chapter on nonbulky base damage). A comparison of glycosylase activities in cell extracts from wild type and AAG knockout mice established that AAG is the principal enzyme acting on 3-methyladenine, $1,N^6$-ethenoadenine and hypoxanthine (85). AAG also recognizes cisplatin-modified DNA with dissociation constants of 71 nM, 115 nM, and 144 nM for 1,2-d(ApG), 1,2-d(GpG) and 1,3-d(GpTpG) intrastrand cross-links, respectively (86).

 The crystal structure of human AAG in complex with double-stranded DNA containing a transition state mimic of the glycosylase reaction, a pyrrolidine abasic nucleotide, lends insight into the recognition process of AAG for its broad range of natural substrates. The glycosylase binds into the minor groove. The pyrrolidine is flipped out of the DNA duplex into the active site of AAG (Fig. 5F) (87). A tyrosine inserts into the minor groove of the DNA duplex at the space left by the flipped-out abasic nucleotide, serving as a substitute base that stabilizes the DNA distortion. The DNA is kinked where the tyrosine intercalates, contributing to a widening of the minor groove and an overall bending angle of 22° across the central 8 base pairs. The amount of DNA bending is probably diminished by crystal packing contacts that realign the ends of neighboring DNAs. It is likely that, by analogy to the HMGB1a-co-crystal structure, the 1,2-d(GpG) cisplatin cross-link provides a prestructured intercalation site for Tyr-162. Even though human DNA glycosylases excise adducts formed by nitrogen mustards, AAG could not repair cisplatin-damaged DNA (86,88). Furthermore, the presence of cisplatin adducts inhibited the excision of $1,N^6$-ethenoadenine by AAG in vitro. Cisplatin adducts might divert human AAG from repairing its natural binding sites, leading to enhanced toxicity because of the persistence of AAG substrates in DNA (86), but there is yet no evidence to support this hypothesis.

4.3.2. Direct Reversal — DNA Photolyase

DNA photolyase reverses cyclobutane pyrimidine dimers upon excitation with blue light and restores the DNA to its native form (see Chapter 5 on photolyases). The structure of *E. coli* photolyase suggests substrate recognition through a base-flipping mechanism with the pyrimidine dimer rotated out of the DNA helix into the enzyme active site (89). Atomic force microscopy determined that photolyase can accommodate DNA substrates with an average bend of 36°. *E. coli* photolyase binds to duplex DNA containing a single 1,2-d(GpG) cisplatin adduct with a K_d of 50 nM and stimulates repair by the Uvr(A)BC excinuclease (90). Moreover, *E. coli* cells expressing photolyase stimulate excision of cisplatin by 1.3-fold and are more resistant to cisplatin than photolyase-deficient cells (90). In contrast, even though *S. cerevisiae* photolyase also recognizes cisplatin adducts, yeast strains deficient in photolyase are more resistant to cisplatin than wild type cells (91). This differential processing of cisplatin adducts in prokaryotic and eukaryotic cells stresses again that great care has to be taken in choosing a model system for determining the origins of cisplatin toxicity.

4.3.3. Nucleotide Excision Repair—XPA, RPA, XPE

Nucleotide excision repair is a multienzyme system whereby the individual factors assemble sequentially at the sites of DNA damage (92). Biological consequences of defective global genome repair are apparent from the inherited multisystem disorder known as xeroderma pigmentosum (XP) (93). Patients with XP are extremely sensitive to UV radiation and have a predisposition toward skin cancer. The NER pathway is the major detoxification process for cisplatin–DNA adducts. Several components of the NER systems specifically recognize cisplatin-damaged DNA.

Preincision complex formation is triggered by binding of XPC-HR23B to the cisplatin adduct (94). Both XP complementation group A (XPA) and replication protein A (RPA) have also been implicated in damage recognition (95). Also, XPA binds most efficiently to rigidly bent duplexes but not to single-stranded DNA. Conversely, RPA recognizes single-stranded sites but not backbone bending (92). Each protein alone binds weakly to cisplatin-modified DNA, but the association of XPA with RPA generates a sensor that detects, simultaneously, backbone and base-pair modifications of DNA (92).

XPA is a 273-amino acid zinc-finger protein. The central DNA recognition domain comprising residues 98–219 and the full-length protein have a higher affinity for cisplatin-damaged DNA than unmodified DNA (96–98). The NMR structure revealed that the central DNA-binding domain comprises a zinc-containing subdomain, presumed to serve as the binding surface for RPA, and a carboxy-terminal subdomain (99,100). XPA contacts both the damaged and undamaged strand and can be photocross-linked to DNA containing a single 1,3-d(GpTpG) adduct (101).

XPA-deficient cells are more sensitive to cisplatin than wild-type cells (102). Moreover, testicular tumor cells have low levels of XPA protein and the ERCC1–XPF endonuclease complex (103).

A heterotrimeric protein, RPA, is also involved in DNA replication and homologous recombination. The largest subunit, RPA70, is composed of two structurally similar domains, each of which binds single-stranded DNA (104). The interaction of RPA with XPA is essential for increased binding affinity to damaged DNA (92). The relative binding affinities of RPA for cisplatin adducts correlate with the degree of single-strand character induced, with a greater interaction for the 1,3-d(GpTpG) compared to the 1,2-d(GpG) cross-link and only poor recognition of the interstrand adduct (105,106). Binding occurs preferentially to the undamaged strand (101).

The crystal structure of RPA70 in complex with a single-stranded DNA substrate reveals that recognition arises through stacking of four aromatic residues with the DNA bases along the single strand (104) (Fig. 5E). Even though the DNA adopts no standard configuration, the interphosphate distance of 5.4 Å–6.3 Å is more consistent with a 3′-endo sugar pucker observed in A-form DNA. The fact that a platinum-1,2-cross-link, with low single-strand character, is recognized at all could be a result of the presentation of a prestructured intercalation site (Fig. 5).

XP complementation group E (XP-E) is the mildest form of the autosomal recessive disease XP. Cells generally show about 50% of normal repair levels (107). One protein factor implicated in the disease, detected by a UV-damaged DNA probe, also binds cisplatin-damaged DNA (108,109). It was designated XPE or UV-damaged DNA-binding activity (UV-DDB). Purified XPE recognizes cisplatin adducts, but has no affinity for *trans*-DDP adducts (110).

XPE was identified to be the human homologue of *S. cerevisiae* photolyase (111) and, like XPA and XPC-HR23B, it is believed to be a damage-specific DNA-binding protein (112). Due to the lack of a crystal structure, the exact recognition mechanism has not yet been determined. When human tumor cells were selected for resistance to cisplatin, they exhibited increased expression of XPE and, concomitantly, more efficient DNA repair in a transfection assay (109). Because NER is an extremely complex multienzyme process, the precise role of XPE in conveying cisplatin resistance remains to be established, however.

4.3.4. *Mismatch Repair — MSH2/MutSα*

Mismatch repair corrects mispaired and unpaired bases in duplex DNA that arise during replication (this volume, chapter on mismatch repair). In humans mismatch recognition is accomplished by two heterodimers of *E. coli* MutS homologues. MutSα is a heterodimer of hMSH2 and hMSH6. It binds to mismatches and small insertion/deletion loops, whereas the MutSβ heterodimer of hMSH2 and hMSH3 binds to larger insertion/deletion loops. Both the isolated MSH2 and the heterodimer MutSα recognize specifically the 1,2 d(GpG) cisplatin adduct but not the 1,3-adduct or *trans*-DDP-modified DNA (113,114). The purified hMSH2 protein binds to globally platinated DNA containing an average of six cisplatin adducts with a K_d of 67 nM. Gel mobility shift assays and surface plasmon resonance measurements of MutS showed that the major cisplatin intrastrand cross-link was recognized with only a 1.5-fold specificity. In contrast, MutS recognized with a specificity of 129 a G:T mismatch incorporated at the 3′-guanine of the cisplatin cross-link (114,115).

The crystal structure of *Thermus aquaticus* MutS with DNA containing an unpaired thymidine reveals an induced-fit recognition mechanism between four domains of MutS dimer and DNA duplex kinked at the mismatch (116). Again, the general architecture has striking similarity with the HMG-domain–cisplatin–DNA co-crystal structure (Fig. 5). The DNA is bent towards the major groove. MutS binds asymmetrically around the mismatch, with Phe39 intercalating from the minor groove side, stacking onto unpaired thymine with an edge-to-face interaction with a second thymine. Mutation of the intercalating phenylalanine to alanine abolishes binding (116). It seems reasonable to postulate that recognition of cisplatin-modified DNA is based on the presence of a preformed intercalation site.

Mismatch repair deficient cells are approximately two-fold more resistant to cisplatin than the corresponding wild-type cells (117). An abortive repair model has been proposed to link the role of mismatch repair proteins to cisplatin toxicity (113). Mismatch repair proteins recognize the cisplatin adducts in the template strand but then attempt to repair the damage in the newly synthesized unmodified strand. No correct base can be incorporated opposite the cisplatin adduct as long as it persists and a futile cycle of repair is initiated. Recurrent misdirected repair could eventually lead to cell cycle arrest and signal for apoptosis (113).

In summary, because proteins from several DNA repair systems are able to recognize cisplatin binding, the locus of platination may influence the pathway of damage removal. Depending upon which damage-recognition factor first gains access to — or, conversely, is obstructed from accessing—the cisplatin–DNA damage, the various repair systems might differentially affect cell survival. For example, it is conceivable that impeding the more versatile NER pathway may pose a larger threat to genomic stability than hampering the more specialized mismatch repair pathway.

5. SUMMARY AND OUTLOOK

A wide range of proteins with the ability to recognize specifically the 1,2-intrastrand cisplatin–DNA cross-link have been identified to date. These proteins bind to a variety of natural DNA substrates and discriminate for them based on structure. Even the sequence-specific transcription factors LEF-1 or TBP pose no exception because they identify their cognate DNA targets primarily by sequence-inherent flexibility rather than discrimination for specific nucleotides (118). Although their tertiary structures are unrelated, all of these proteins bind and bend DNA by a shared mechanism. Hydrophobic, often aromatic, side chains are employed as an intercalative wedge to pry open one or several base pair steps, thereby inducing a positive roll and bending the DNA away from the protein towards the major groove (Fig. 5). Based on the structural information available, it is unlikely that erroneous recognition of cisplatin-modified DNA arises due to the provision of a prebent DNA target (119), but rather by presenting a "hydrophobic notch" for intercalation at a much lower energetic cost.

The wide range of proteins recognizing cisplatin-modified DNA suggests that the mechanism of cytotoxicity is multifactorial, involving transcription, chromatin remodeling, and repair. Cisplatin sensitivity can be increased by rendering the genome less accessible for repair proteins, either through repair shielding or by modulating chromatin structure. Furthermore, cisplatin modification has the potential to alter cellular pathways by diverting proteins from carrying out their normal functions or by interfering with signaling cascades through modulation of post-translational modifications.

In summary, although protein binding to cisplatin-modified DNA may suggest a role in mediating its cytotoxicity, it has to be considered that the majority of such studies were carried out in vitro using naked DNA. In vivo systems are significantly more complex, rendering the outcome of overexpression or knockout studies deduced from in vitro results difficult if not impossible to predict. Nevertheless, protein binding to cisplatin damage is certain to modulate the cellular response to the drug and a strong body of evidence points to the involvement of HMG-domain proteins. Exactly how remains to be determined, but powerful new tools, such as RNAi, continue to emerge that one day should enable the quest to be accomplished. The ultimate goal, of course, is to marry Pt-DNA coordination chemistry with the modulation of cellular pathways and targets to provide improved drugs for cancer therapy.

ACKNOWLEDGMENT

Work from our laboratory has benefited from generous support from the National Cancer Institute under grant CA34992.

ABBREVIATIONS

NH_2Cba	cyclobutylamine
2-pic	2-picoline
NH_2Cy	cyclohexylamine
en	ethylenediamine
dach	diaminocyclohexane

REFERENCES

1. Hellerstedt BA, Pienta KJ. Testicular cancer. Curr Opin Oncol 2002; 14:260–264.
2. Reedijk J. The mechanism of action of platinum antitumor drugs. Pure Appl Chem 1987; 59:181–192.
3. Jamieson ER, Lippard SJ. Structure, recognition, and processing of cisplatin-DNA adducts. Chem Rev 1999; 99:2467–2498.
4. Zlatanova J, Yaneva J, Leuba SH. Proteins that specifically recognize cisplatin-damaged DNA: a clue to anticancer activity of cisplatin. FASEB J 1998; 12:791–799.
5. Kartalou M, Essigmann JM. Recognition of cisplatin adducts by cellular proteins. Mutat Res 2001; 478:1–21.
6. Cohen SM, Lippard SJ. Prog Nucleic Acid Res Mol Biol. San Diego: Academic Press, 2001:93–130.
7. Zamble DB, Lippard SJ. 30 years of cisplatin – chemistry and biochemistry of a leading anticancer drug. Basel: VHCA-Wiley, 1999:73–110.
8. Zhang CX, Lippard SJ. New metal complexes as potential therapeutics. Curr Opin Chem Biol 2003; 7:481–489.
9. Rosenberg B, Van Camp L, Krigas T. Inhibition of cell division in *Escherichia coli* by electrolysis products from a platinum electrode. Nature 1965; 205:698–699.
10. Bosl GJ, Motzer RJ. Testicular germ-cell cancer. N Engl J Med 1997; 337:242–253.
11. Hambley TW. Platinum binding to DNA: structural controls and consequences. J Chem Soc, Dalton Trans 2001:2711–2718.
12. Caradonna JP, Lippard SJ, Gait MJ, Singh M. The antitumor drug *cis*-[Pt(NH3)2Cl2] forms an intrastrand d(GpG) cross-link upon reaction with [d(ApGpGpCpCpT)]2. J Am Chem Soc 1982; 104:5793–5795.
13. Cohen GL, Ledner JA, Bauer WR, Ushay HM, Caravana C, Lippard SJ. Sequence dependent binding of *cis*-dichlorodiammineplatinum(II) to DNA. J Am Chem Soc 1980; 102:2487–2488.
14. Sherman SE, Lippard SJ. Structural aspects of platinum anticancer drug interactions with DNA. Chem Rev 1987; 87:1153–1181.

15. Sherman SE, Gibson D, Wang AHJ, Lippard SJ. X-ray structure of the major adduct of the anticancer drug cisplatin with DNA-cis-[Pt(NH$_3$)$_2$(d(pGpG))]. Science 1985; 230: 412–417.

16. Sherman SE, Gibson D, Wang AHJ, Lippard SJ. Crystal and molecular structure of cis-[Pt(NH$_3$)$_2${d(pGpG)}], the principal adduct formed by cis-diamminedichloroplatinum(II) with DNA. J Am Chem Soc 1988; 110:7368–7381.

17. Fichtinger-Schepman AMJ, van der Veer JL, den Hartog JHJ, Lohman PHM, Reedijk J. Adducts of the antitumor drug cis-diamminedichloroplatinum(II) with DNA: formation, identification, and quantitation. Biochemistry 1985; 24:707–713.

18. Blommaert FA, van Dijk-Knijnenburg HCM, Dijt FJ, den Engelse L, Baan RA, Berends F, Fichtinger-Schepman AMJ. Formation of DNA adducts by the anticancer drug carboplatin: different nucleotide sequence preferences in vitro and in cells. Biochemistry 1995; 34:8474–8480.

19. Rice JA, Crothers DM, Pinto AL, Lippard SJ. The major adduct of the antitumor drug cis-diamminedichloroplatinum(II) with DNA bends the duplex by approximately equal to 40 degrees toward the major groove. Proc Natl Acad Sci USA 1988; 85: 4158–4161.

20. Bellon SF, Coleman JH, Lippard SJ. DNA unwinding produced by site-specific intrastrand cross-links of the antitumor drug cis-diamminedichloroplatinum(II). Biochemistry 1991; 30:8026–8035.

21. Poklar N, Pilch DS, Lippard SJ, Redding EA, Dunham SU, Breslauer KJ. Influence of cisplatin intrastrand crosslinking on the conformation, thermal stability, and energetics of a 20-mer DNA duplex. Proc Natl Acad Sci USA 1996; 93:7606–7611.

22. Pilch DS, Dunham SU, Jamieson ER, Lippard SJ, Breslauer KJ. DNA sequence context modulates the impact of a cisplatin 1,2-d(GpG) intrastrand cross-link on the conformational and thermodynamic properties of duplex DNA. J Mol Biol 2000; 296:803–812.

23. Gelasco A, Lippard SJ. Topics in Biological Inorganic Chemistry. Heidelberg: Springer-Verlag, 1999:1–43.

24. Takahara PM, Rosenzweig AC, Frederick CA, Lippard SJ. Crystal structure of a double-stranded DNA dodecamer containing the major adduct of the anticancer drug cisplatin. Nature 1995; 377:649–652.

25. Dunham SU, Lippard SJ. Long-range distance constraints in platinated nucleotides: structure determination of the 5′ orientational isomer of cis-[Pt(NH$_3$)(4-aminoTEMPO)-{d(GpG)}]$^+$ from combined paramagnetic and diamagnetic NMR constraints with molecular modeling. J Am Chem Soc 1995; 117:10702–10712.

26. Gelasco A, Lippard SJ. NMR solution structure of a DNA dodecamer duplex containing a cis-diammineplatinum(II) d(GpG) intrastrand cross-link, the major adduct of the anticancer drug cisplatin. Biochemistry 1998; 37:9230–9239.

27. Yang D, van Boom SSGE, Reedijk J, van Boom JH, Wang AHJ. Structure and isomerization of an intrastrand cisplatin-cross-linked octamer DNA duplex by NMR analysis. Biochemistry 1995; 34:12912–12920.

28. Spingler B, Whittington DA, Lippard SJ. 2.4 angstrom crystal structure of an oxaliplatin 1,2-d(GpG) intrastrand cross-link in a DNA dodecamer duplex. Inorg Chem 2001; 40:5596–5602.

29. Silverman AP, Bu WM, Cohen SM, Lippard SJ. 2.4 Å crystal structure of the asymmetric platinum complex {Pt(ammine)(cyclohexylamine)}(2+) bound to a dodecamer DNA complex. J Biol Chem 2002; 277:49743–49749.

30. Dunham SU, Turner CJ, Lippard SJ. Solution structure of a DNA duplex containing a nitroxide spin-labeled platinum d(GpG) intrastrand cross-link refined with NMR-derived long-range electron-proton distance restraints. J Am Chem Soc 1998; 120: 5395–5406.

31. Marzilli LG, Saad JS, Kuklenyik Z, Keating KA, Xu Y. Relationship of solution and protein-bound structures of DNA duplexes with the major intrastrand cross-link lesions formed on cisplatin binding to DNA. J Am Chem Soc 2001; 123:2764–2770.

32. Sullivan ST, Ciccarese A, Fanizzi FP, Marzilli LG. A rare example of three abundant conformers in one retro model of the cisplatin-DNA d(GpG) intrastrand cross link. Unambiguous evidence that guanine O6 to carrier amine ligand hydrogen bonding is not important. Possible effect of the lippard base pair step adjacent to the lesion on carrier ligand hydrogen bonding in DNA adducts. J Am Chem Soc 2001; 123: 9345–9355.

33. van Garderen CJ, van Houte LPA. The solution structure of a DNA duplex containing the *cis*-Pt(NH$_3$)$_2$[d(-GTG-)-N7(G),N7(G)] adduct, as determined with high-field NMR and molecular mechanics/dynamics. Eur J Biochem 1994; 225:1169–1179.

34. Teuben JM, Bauer C, Wang AHJ, Reedijk J. Solution structure of a DNA duplex containing a *cis*-diammineplatinum(II) 1,3-d(GTG) intrastrand cross-link, a major adduct in cells treated with the anticancer drug carboplatin. Biochemistry 1999; 38:12305–12312.

35. Huang HF, Zhu LM, Reid BR, Drobny GP, Hopkins PB. Solution structure of a cis-platin-induced DNA interstrand cross-link. Science 1995; 270:1842–1845.

36. Paquet F, Perez C, Leng M, Lancelot G, Malinge JM. NMR solution structure of a DNA decamer containing an interstrand cross-link of the antitumor drug *cis*-diammine-dichloroplatinum(II). J Biomol Struct Dynam 1996; 14:67–77.

37. Coste F, Malinge J-M, Serre L, Shepard W, Roth M, Leng M, Zelwer C. Crystal structure of a double-stranded DNA containing a cisplatin interstrand cross-link at 1.63 Å resolution: hydration at the platinated site. Nucleic Acids Res 1999; 27:1837–1846.

38. Barrett TE, Savva R, Barlow T, Brown T, Jiricny J, Pearl LH. Structure of a DNA base-excision product resembling a cisplatin interstrand adduct. Nature Struct Biol 1998; 5:697–701.

39. Köberle B, Grimaldi KA, Sunters A, Hartley JA, Kelland LR, Masters JRW. DNA repair capacity and cisplatin sensitivity of human testis tumour cells. Int J Cancer 1997; 70:551–555.

40. Sark MWJ, Timmer-Bosscha H, Meijer C, Uges DRA, Sluiter WJ, Peters WHM, Mulder NH, De Vries EGE. Cellular basis for differential sensitivity to cisplatin in human germ-cell tumour and colon-carcinoma cell-lines. Brit J Cancer 1995; 71: 684–690.

41. Huang J-C, Zamble DB, Reardon JT, Lippard SJ, Sancar A. HMG-domain proteins specifically inhibit the repair of the major DNA adduct of the anticancer drug cisplatin by the human excision nuclease. Proc Natl Acad Sci USA 1994; 91:10394–10398.

42. Zamble DB, Mu D, Reardon JT, Sancar A, Lippard SJ. Repair of cisplatin-DNA adducts by the mammalian excision nuclease. Biochemistry 1996; 35:10004–10013.

43. Luger K, Mäder AW, Richmond RK, Sargent DF, Richmond TJ. Crystal structure of the nucleosome core particle at 2.8 Å resolution. Nature 1997; 389:251–260.

44. Wang D, Hara R, Singh G, Sancar A, Lippard SJ. Nucleotide excision repair from site-specifically platinum-modified nucleosomes. Biochemistry 2003; 42:6747–6753.

45. Fischle W, Wang Y, Allis CD. Binary switches and modification cassettes in histone biology and beyond. Nature 2003; 425:475–479.

46. Lee KB, Wang D, Lippard SJ, Sharp PA. Transcription-coupled and DNA damage dependent ubiquitination of RNA polymerase II in vitro. Proc Natl Acad Sci USA 2002; 99:4239–4244.

47. Thomas JO, Travers AA. HMG1 and 2, and related 'architectural' DNA-binding proteins. Trends Biochem Sci 2001; 26:167–174.

48. Orphanides G, LeRoy G, Chang C-H, Luse DS, Reinberg D. FACT, a factor that facilitates transcript elongation through nucleosomes. Cell 1998; 92:105–116.

49. Orphanides G, Wu W-H, Lane WS, Hampsey M, Reinberg D. The chromatin-specific transcription elongation factor FACT comprises human SPT16 and SSRP1 proteins. Nature 1999; 400:284–288.

50. Formosa T, Eriksson P, Wittmeyer J, Ginn J, Yu Y, Stillman DJ. Spt16-Pob3 and the HMG protein Nhp6 combine to form the nucleosome-binding factor SPN. EMBO J 2001; 20:3506–3517.

51. Shimojima T, Okada M, Nakayama T, Ueda H, Okawa K, Iwamatsu A, Handa H, Hirose S. Drosophila FACT contributes to Hox gene expression through physical and functional interactions with GAGA factor. Genes Dev 2003; 17:1605–1616.

52. Toney JH, Donahue BA, Kellett PJ, Bruhn SL, Essigmann JM, Lippard SJ. Isolation of cDNAs encoding a human protein that binds selectively to DNA modified by the anticancer drug cis-diamminedichloroplatinum(II). Proc Natl Acad Sci USA 1989; 86:8328–8332.

53. Bruhn SL, Pil PM, Essigman JM, Housman DE, Lippard SJ. Isolation and characterization of human cDNA clones encoding a high mobility group box protein that recognizes structural distortions to DNA caused by binding of the anticancer agent cisplatin. Proc Natl Acad Sci USA 1992; 89:2307–2311.

54. Yarnell AT, Oh S, Reinberg D, Lippard SJ. Interaction of FACT, SSRP1, and the high mobility group (HMG) domain of SSRP1 with DNA damaged by the anticancer drug cisplatin. J Biol Chem 2001; 276:25736–25741.

55. Bustin M, Reeves R. Prog Nucl Acid Res Mol Biol. San Diego: Academic Press, Inc, 1996:35–100.

56. Agresti A, Bianchi ME. HMGB proteins and gene expression. Curr Opin Genet Dev 2003; 13:170–178.

57. Chow CS, Whitehead JP, Lippard SJ. HMG domain proteins induce sharp bends in cisplatin-modified DNA. Biochemistry 1994; 33:15124–15130.

58. Pil PM, Lippard SJ. Specific binding of chromosomal protein HMG1 to DNA damaged by the anticancer drug cisplatin. Science 1992; 256:234–237.

59. Jung Y, Lippard SJ. Nature of full-length HMGB1 binding to cisplatin-modified DNA. Biochemistry 2003; 42:2664–2671.

60. Dunham SU, Lippard SJ. DNA sequence context and protein composition modulate HMG-domain protein recognition of cisplatin-modified DNA. Biochemistry 1997; 36: 11428–11436.

61. Cohen SM, Mikata Y, He Q, Lippard SJ. HMG-Domain protein recognition of cisplatin 1,2-intrastrand d(GpG) cross-links in purine-rich sequence contexts. Biochemistry 2000; 39:11771–11776.

62. Jamieson ER, Lippard SJ. Stopped-flow fluorescence studies of HMG-domain protein binding to cisplatin-modified DNA. Biochemistry 2000; 39:8426–8438.

63. Wei M, Cohen SM, Silverman AP, Lippard SJ. Effects of spectator ligands on the specific recognition of intrastrand platinum-DNA cross-links by high mobility group box and TATA-binding proteins. J Biol Chem 2001; 276:38774–38780.

64. He Q, Liang CH, Lippard SJ. Steroid hormones induce HMG1 overexpression and sensitize breast cancer cells to cisplatin and carboplatin. Proc Natl Acad Sci USA 2000; 97:5768–5772.

65. Wei M, Burenkova O, Lippard SJ. Cisplatin sensitivity in Hmgb1−/− and Hmgb1+/+ mouse cells. J Biol Chem 2003; 278:1769–1773.

66. Ohndorf U-M, Rould MA, He Q, Pabo CO, Lippard SJ. Basis for recognition of cisplatin-modified DNA by high-mobility group proteins. Nature 1999; 399:708–712.

67. He Q, Ohndorf UM, Lippard SJ. Intercalating residues determine the mode of HMG1 domains A and B binding to cisplatin-modified DNA. Biochemistry 2000; 39:14426–14435.

68. Wang D, Lippard SJ. Cisplatin-induced post-translational modification of histones H3 and H4. J Biol Chem 2004; 279:20622–20625.

69. Masse JE, Wong B, Yen YM, Allain FHT, Johnson RC, Feigon J. The S. cerevisiae architectural HMGB protein NHP6A complexed with DNA: DNA and protein conformational changes upon binding. J Mol Biol 2002; 323:263–284.

70. Wong B, Masse JE, Yen YM, Giannikoupoulos P, Feigon J, Johnson RC. Binding to cisplatin-modified DNA by the Saccharomyces cerevisiae HMGB protein Nhp6a. Biochemistry 2002; 41:5404–5414.

71. Yaneva J, Leuba SH, van Holde K, Zlatanova J. The major chromatin protein histone H1 binds preferentially to cis-platinum-damaged DNA. Proc Natl Acad Sci USA 1997; 94:10394–10398.

72. Trimmer EE, Zamble DB, Lippard SJ, Essigmann JM. Human testis-determining factor SRY binds to the major DNA adduct of cisplatin and a putative target sequence with comparable affinities. Biochemistry 1998; 37:352–362.

73. Love JJ, Li X, Case DA, Giese K, Grosschedl R, Wright PE. Structural basis for DNA bending by the architectural transcription factor LEF-1. Nature 1995; 376:791–795.

74. Werner MH, Huth JR, Gronenborn AM, Clore GM. Molecular basis of human 46X,Y sex reversal revealed from the three-dimensional solution structure of the human SRY-DNA complex. Cell 1995; 81:705–714.

75. Brown SJ, Kellett PJ, Lippard SJ. Ixr1, a yeast protein that binds to platinated DNA and confers sensitivity to cisplatin. Science 1993; 261:603–605.

76. Lambert JR, Bilanchone VW, Cumsky MG. The ORD1 gene encodes a transcription factor involved in oxygen regulation and is identical to IXR1, a gene that confers cisplatin sensitivity to *Saccharomyces cerevisiae*. Proc Natl Acad Sci USA 1994; 91: 7345–7349.

77. McANulty MM, Whitehead JP, Lippard SJ. Binding of Ixr1, a yeast HMG-domain protein, to cisplatin-DNA adducts in vitro and in vivo. Biochemistry 1996; 35:6089–6099.

78. McANulty MM, Lippard SJ. The HMG-domain protein Ixr1 blocks excision repair of cisplatin-DNA adducts in yeast. Mutat Res 1996; 362:75–86.

79. Kim JL, Nikolov DB, Burley SK. Co-crystal structure of TBP recognizing the minor-groove of a TATA element. Nature 1993; 365:520–527.

80. Kim YC, Geiger JH, Hahn S, Sigler PB. Crystal structure of a yeast TBP/TATA-box complex. Nature 1993; 365:512–520.

81. Jung Y, Mikata Y, Lippard SJ. Kinetic studies of the TATA-binding protein interaction with cisplatin-modified DNA. J Biol Chem 2001; 276:43589–43596.

82. Cohen SM, Jamieson ER, Lippard SJ. Enhanced binding of the TATA-binding protein to TATA boxes containing flanking cisplatin 1,2-cross-links. Biochemistry 2000; 39: 8259–8265.

83. Dabholkar M, Vionnet J, Bostick-Bruton F, Yu JJ, Reed E. Messenger RNA levels of XPAC and ERCC1 in ovarian cancer tissue correlate with response to platinum-based chemotherapy. J Clin Invest 1994; 94:703–708.

84. Ferry KV, Ozols RF, Hamilton TC, Johnson SW. Expression of nucleotide excision repair genes in CDDP-sensitive and resistant human ovarian cancer cell lines. Proc Am Assoc Cancer Res 1996; 37:365.

85. Hang B, Singer B, Margison GP, Elder RH. Targeted deletion of alkylpurine-DNA-N-glycosylase in mice eliminates repair of 1,N6-ethenoadenine and hypoxanthine but not of 3,N4-ethenocytosine or 8-oxoguanine. Proc Natl Acad Sci USA 1997; 94: 12869–12874.

86. Kartalou M, Samson LD, Essigmann JM. Cisplatin adducts inhibit 1,N^6-Ethenoadenine repair by interacting with the human 3-methyladenine DNA glycosylase. Biochemistry 2000; 39:8032–8038.

87. Lau AY, Scharer OD, Samson L, Verdine GL, Ellenberger T. Crystal structure of a human alkylbase-DNA repair enzyme complexed to DNA: mechanism for nucleotide flipping and base excision. Cell 1998; 95:249–258.

88. Mattes WB, Lee CS, Laval J, O'Connor TR. Excision of DNA adducts of nitrogen mustards by bacterial and mammalian 3-methyladenine-DNA glycosylases. Carcinogenesis 1996; 17:643–648.

89. Park HW, Kim ST, Sancar A, Deisenhofer J. Crystal structure of DNA photolyase from *Escherichia coli*. Science 1995; 268:1866–1872.

90. Oezer Z, Reardon JT, Hsu DS, Malhotra K, Sancar A. The other function of DNA photolyase: stimulation of excision repair of chemical damage to DNA. Biochemistry 1995; 34:15886–15889.

91. Fox ME, Feldman BJ, Chu G. A novel role for DNA photolyase: binding to DNA damaged by drugs is associated with enhanced cytotoxicity in *Saccharomyces cerevisiae*. Mol Cell Biol 1994; 14:8071–8077.

92. Riedl T, Hanaoka F, Egly J-M. The comings and goings of nucelotide excision repair factors on damaged DNA. EMBO J 2003; 22:5293–5303.
93. Berneburg M, Lehmann AR. Xeroderma pigmentosum and related disorders: defects in DNA repair and transcription. Adv Genet 2001; 43:71–102.
94. Volker M, Moné MJ, Karmakar P, van Hoffen A, Schul W, Vermeulen W, Hoeijmakers JHJ, van Driel R, van Zeeland AA, Mullenders LHF. Sequential assembly of the nucleotide excision repair factors in vivo. Mol Cell 2001; 8:213–224.
95. Reardon JT, Sancar A. Molecular anatomy of the human excision nuclease assembled at sites of DNA damage. Mol Cell Biol 2002; 22:5938–5945.
96. Missura M, Buterin T, Hindges R, Huebscher U, Kasparkova J, Brabec V, Naegeli H. Double-check probing of DNA bending and unwinding by XPA-RPA: an architectural function in DNA repair. EMBO J 2001; 20:3554–3564.
97. Jones CJ, Wood RD. Preferential binding of the xeroderma pigmantosum group A complementing protein to damaged DNA. Biochemistry 1993; 32:12096–12104.
98. Asahina H, Kuraoka I, Shirakawa M, Morita EH, Miura N, Miyamoto E, Ohtsuka E, Okada Y, Tanaka K. The XPA protein is a zinc metalloprotein with an ability to recognize various kinds of DNA damage. Mutat Res 1994; 315:229–237.
99. Ikegami T, Kuraoka I, Saijo M, Kodo N, Kyogoku Y, Morikawa K, Tanaka K, Shirakawa M. Solution structure of the DNA- and RPA-binding domain of the human repair factor XPA. Nature Struct Biol 1998; 5:701–706.
100. Buchko GW, Daughdrill GW, de Lorimier R, Rao BK, Isern NG, Lingbeck JM, Taylor JS, Wold MS, Gochin M, Spicer LD, Lowry DF, Kennedy MA. Interactions of human nucleotide excision repair protein XPA with DNA and RPA70 Delta C327: chemical shift mapping and 15N NMR relaxation studies. Biochemistry 1999; 38: 15116–15128.
101. Hermanson-Miller IL, Turchi JJ. Strand-specific binding of RPA and XPA to damaged duplex DNA. Biochemistry 2002; 41:2402–2408.
102. Dijt FJ, Fichtinger-Schepman AMJ, Berends F, Reedijk J. Formation and repair of cisplatin-induced adducts to DNA in cultured normal and repair-deficient human fibroblasts. Cancer Res 1988; 48:6058–6062.
103. Köberle B, Masters J, Hartley FR, Wood DP. Defective repair of cisplatin-induced DNA damage caused by reduced XPA protein in testicular germ cell tumours. Curr Biol 1999; 9:273–276.
104. Bochkarev A, Pfuetzner RA, Edwards AM, Frappier L. Structure of the single-stranded-DNA-binding domain of replication protein A bound to DNA. Nature 1997; 385:176–181.
105. Patrick SM, Turchi JJ. Replication protein A (RPA) binding to duplex cisplatin-damaged DNA is mediated through the generation of single-stranded DNA. J Biol Chem 1999; 274:14972–14978.
106. Patrick SM, Turchi JJ. Human replication protein A preferentially binds cisplatin-damaged duplex DNA in vitro. Biochemistry 1998; 37:8808–8815.
107. Bootsma D, Kraemer K, Cleaver J, Hoeijmakers JHJ. The genetic basis of human cancer. New York: McGraw-Hill, 1998:245–274.
108. Chu G, Chang E. Xeroderma pigmentosum group E cells lack a nuclear factor that binds to damaged DNA. Science 1988; 242:564.
109. Chu G, Chang E. Cisplatin-resistant cells express increased levels of a factor that recognizes damaged DNA. Proc Natl Acad Sci USA 1990; 87:3324–3327.
110. Payne A, Chu G. Xeroderma pigmentosum group E binding factor recognizes a broad spectrum of DNA damage. Mutat Res 1994:310.
111. Patterson M, Chu G. Evidence that xeroderma pigmentosum cells from complementation group E are deficient in a homolog of yeast photolyase. Mol Cell Biol 1989:5105–5112.
112. Tang JY, Hwang BJ, Ford JM, Hanawalt PC, Chu G. Xeroderma pigmentosum p48 gene enhances global genomic repair and suppresses UV-induced mutagenesis. Mol Cell 2000; 5:737–744.

113. Mello JA, Acharya S, Fishel R, Essigmann JM. The mismatch-repair protein hMSH2 binds selectively to DNA adducts of the anticancer drug cisplatin. Chem Biol 1996; 3:579–589.

114. Yamada M, O'Regan E, Brown P, Karran P. Selective recognition of a cisplatin-DNA adduct by human mismatch repair proteins. Nucleic Acids Res 1997; 25:491–496.

115. Fourrier L, Brooks P, Malinge J-M. Binding discrimination of MutS to a set of lesions and compound lesions (base damage and mismatch) reveals its potential role as a cisplatin-damaged DNA sensing protein. J Biol Chem 2003; 278:21267–21275.

116. Obmolova G, Ban C, Hsieh P, Yang W. Crystal structures of mismatch repair protein MutS and its complex with a substrate DNA. Nature 2000; 407:703–710.

117. Aebi S, Kurdi-Haidar B, Gordon R, Cenni B, Zheng H, Fink D, Christen RD, Boland CR, Koi M, Fishel R, Howell SB. Loss of DNA mismatch repair in acquired resistance to cisplatin. Cancer Res 1996; 56:3087–3090.

118. Juo ZS, Chiu TK, Leiberman PM, Baikalov I, Berk AJ, Dickerson RE. How proteins recognize the TATA box. J Mol Biol 1996; 261:239–254.

119. Vichi P, Coin F, Renaud JP, Vermeulen W, Hoeijmakers JHJ, Moras D, Egly JM. Cisplatin- and UV-damaged DNA lure the basal transcription factor TFIID/TBP. EMBO J 1997; 16:7444–7456.

120. Churchill MEA, Jones DNM, Glaser T, Hefner MA, Searles MA, Travers AA. HMG-D is an architecture-specific protein that preferentially binds to DNA containing the dinucleotide TG. EMBO J 1995; 14:1264–1275.

121. Murchie AIH, Lilley DMJ. T4 endonuclease VII cleaves DNA containing a cisplatin adduct. J Mol Biol 1993; 233:77–85.

122. Ohndorf U-M, Whitehead JP, Raju NL, Lippard SJ. Binding of tsHMG, a mouse testis-specific HMG-domain protein, to cisplatin-DNA adducts. Biochemistry 1997; 36: 14807–14815.

123. Treiber DK, Zhai X, Jantzen H-M, Essigmann JM. Cisplatin-DNA adducts are molecular decoys for the ribosomal RNA transcription factor hUBF (human upstream binding factor). Proc Natl Acad Sci USA 1994; 91:5672–5676.

124. Ise T, Nagatani G, Imamura T, Kato K, Takano H, Nomoto M, Izumi H, Ohmori H, Okamoto T, Ohga T, Uchiumi T, Kuwano M, Kohno K. Transcription factor Y-box binding protein 1 binds preferentially to cisplatin-modified DNA and interacts with pro-liferating cell nuclear antigen. Cancer Res 1999; 59:342–346.

125. Fojta M, Pivonkova H, Brazdova M, Kovarova L, Palecek E, Pospisilova S, Vojtesek B, Kasparkova J, Brabec V. Recognition of DNA modified by antitumor cis-platin by "latent" and "active" protein p53. Biochem Pharmacol 2003; 65:1305–1316.

126. Fleck O, Kunz C, Rudolph C, Kohli J. The high mobility group domain protein Cmb1 of *Schizosaccharomyces pombe* binds to cytosines in base mismatches and opposite chemically altered guanines. J Biol Chem 1998; 273:30398–30405.

127. Sugasawa K, Ng JMY, Masutani C, Iwai S, van der Spek PJ, Eker APM, Hanaoka F, Bootsma D, Hoeijmakers JHJ. Xeroderma pigmentosum group C protein complex is the initiator of global genome nucleotide excision repair. Mol Cell 1998; 2:223–232.

13

Structural Aspects of Polycyclic Aromatic Carcinogen-Damaged DNA and Its Recognition by NER Proteins

Nicholas E. Geacintov
Department of Chemistry, New York University, New York, New York, U.S.A.

Hanspeter Naegeli
Institute of Pharmacology and Toxicology, University of Zürich-Tierspital, Zürich, Switzerland

Dinshaw J. Patel
Cellular Biochemistry and Biophysics Program, Memorial Sloan-Kettering Cancer Center, New York, New York, U.S.A.

Suse Broyde
Department of Biology, New York University, New York, New York, U.S.A.

1. INTRODUCTION

Nucleotide excision repair (NER) is the major repair pathway in eukaryotes that removes bulky DNA adducts (1–4). A complex set of factors sequentially assemble on the damaged DNA region (5,6) and cause the incision of the DNA on both sides of the lesion, and the excision of oligonucleotides 24–32 nucleotides in length that contain the lesion. The mechanisms of this complex process have been extensively studied (7–13), and it is now accepted that the first and rate-determining step in NER (reviewed in Ref. 14) is the recognition of helix-distorting lesions by the XPC/HR23B protein complex. As described recently (5), once the XPC/HR23B complex has been recruited to the damaged site, the multiprotein transcription factor TFIIH binds in turn to this complex. The helicases XPB and XPD within TFIIH cause the unwinding of a 20–25 nucleotide patch around the site of the lesion in an ATP-dependent manner. The arrival of RPA, a single-strand binding protein, further stabilizes this bubble-like structure. Also, XPA then binds to this nucleo-protein complex, presumably to promote the correct positioning of the subsequently acting endonucleases, and XPC/HR23B is released. The structure-specific endo-nuclease XPG, followed by the endonuclease XPF-ERCC1 bind to the bubble-like intermediate and incise the damaged strand on the 3'- and 5'-sides of the lesion, respectively, thus releasing a 24–32 nucleotide long fragment containing the damage (5).

The single-stranded gap in the DNA molecule is restored by repair synthesis catalyzed by pol δ or pol ε using the undamaged strand as the template. Finally, the nick is closed by DNA ligase I, thus regenerating the DNA molecule in its undamaged form. This NER pathway is termed *global genomic repair* (GGR). A second mechanism, referred to as *transcription-coupled repair* (TCR), is selective for the transcribed DNA strand in genes that are being expressed (15). In contrast to GGR which requires the XPC/HR23B complex for initiation of NER, TCR does not depend on XPC/HR23B and repair is initiated when RNA polymerase II becomes stalled upon encountering a DNA adduct.

One of the key and vital features of the mammalian NER machinery is its ability to recognize and excise an astounding variety of bulky DNA lesions. Indeed, as was noted by Wood (4), "the rate of repair of various lesions by NER apparently varies over several orders of magnitude, and it is of interest to determine the precise structural features that define the efficiency of recognition during NER." This question has been the focus of our recent research using families of DNA adducts derived from the reactions of polycyclic aromatic hydrocarbon (PAH) metabolite model compounds with oligonucleotides of defined base composition and sequence. The conformational characteristics of such DNA adducts have been extensively investigated by us (16–21) and by other research groups (22–31). These different families of PAH–DNA adducts are ideal for examining the effects of adduct chemical structure, adduct stereochemistry-dependent conformations, and the distortions in the local DNA structure caused by these lesions. A recent review article summarizes the extensive information that has been garnered on the relationships between adduct structure, adduct conformation, and in vitro NER activity in response to different PAH–DNA adducts in cell extracts prepared from HeLa cells and other cell lines (32).

Our approach toward studying these relationships is multidisciplinary in nature and can be summarized in terms of distinct steps:

1. The site-specific, PAH-modified oligonucleotides are synthesized by either direct synthesis (33–35), or automated DNA synthesis approaches that involve the incorporation of PAH-nucleotide phosphoramidite derivatives into oligonucleotides of defined composition and sequence (36).
2. The solution structures of these PAH-modified oligonucleotides in the duplex form are then studied. Based on the sets of interproton distances and other NMR data, and using structural refinement-molecular mechanics/molecular dynamics techniques, three-dimensional models of the PAH-modified duplexes are generated that are fully consistent with the experimental data (16–18).
3. The thermal stabilities of these duplexes are independently measured to crudely assess the extent of DNA destabilization caused by the lesions (25,35–38).
4. Molecular dynamics (MD) simulations are employed to generate an ensemble of structures in explicit solvent and with counterions. Detailed molecular views of the time-dependent fluctuations of the conformations of the DNA adducts, including the time course of the different structural parameters, as well as thermodynamic properties, are then evaluated. Molecular mechanics Poisson–Boltzmann surface area (MM–PBSA) free energy calculations (39–41) are performed on the entire ensemble of structures to delineate thermodynamic properties. The structural and thermodynamic data provide

insights into the relative stabilities and structural distortions of duplexes associated with the presence of the bulky PAH-DNA adducts (42–45).

5. The detailed information thus gained is correlated with in vitro NER assays to assess the structural and associated energetic factors that account for the relative susceptibilities of the different adducts to DNA repair in vitro (38).

In this chapter, we highlight some of the recent successes in deducing the characteristics of different stereoisomeric PAH diol epoxide–DNA adducts that are efficiently recognized or not by human NER enzymes in vitro, stressing insights obtained from the structural data and the analysis of the adduct conformations utilizing computational techniques.

A bipartite model of NER was advanced by Naegeli and co-workers in the late 1990s (46–48). In this model, NER enzymes use a dual level of discrimination by recognizing (1) chemical alterations of the DNA nucleotides, and (2) disruption of Watson-Crick base pairing. The presence of the lesion is required because disruption of hydrogen bonding alone is not recognized by NER enzymes (13,48). Our structural analysis is focused on evaluation of the distortions in DNA structural parameters caused by bulky stereoisomeric PAH diol epoxide residues, including perturbation of hydrogen bonding. Our conclusions are that the perturbation of hydrogen bonding by bulky adducts is, in part, a consequence of certain structural alterations, but that other distortions which need not disrupt hydrogen bonding also can play a role. We thus provide an enhancement to the bipartite model, a multipartite model, in which other parameters, besides Watson–Crick hydrogen bonding, are affected by the bulky lesions (38). The distortions include base displacement, helix unwinding, helix bending, increased rise, base pair buckling, and changes in the helical backbone parameters as discussed later in this chapter (42–44). Striking differences in relative repair susceptibilities of stereoisomeric pairs of adducts can be explained on the basis of changes in these structural factors caused by the presence of the bulky adducts. Moreover, the thermodynamic calculations in these cases substantiate the relative impact of the distortions on the thermal stabilities of duplexes that contain a single PAH adduct.

2. METABOLISM OF PAH TO DIOL EPOXIDES AND FORMATION OF STEREOISOMERIC DNA ADDUCTS

The "bay" region benzo[*a*]pyrene (Fig. 1) is the best-known and most widely studied PAH compound. The "fjord" benzo[*c*]phenanthrene is a well-known example of a

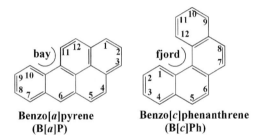

Benzo[*a*]pyrene
(B[*a*]P)

Benzo[*c*]phenanthrene
(B[*c*]Ph)

Figure 1 Structures of the bay region benzo[*a*]pyrene and the fjord region benzo[*c*]phenanthrene.

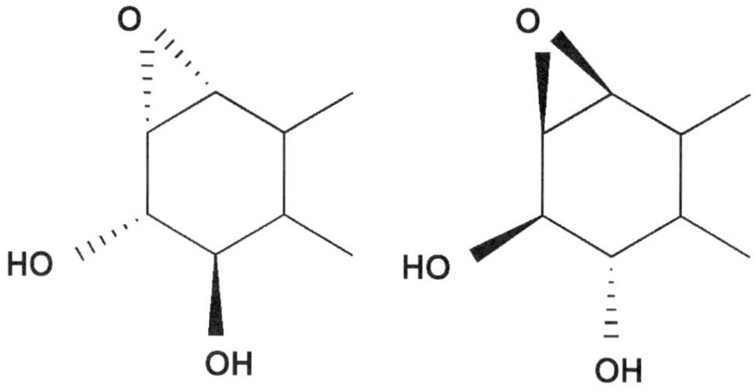

(+)-7R,8S,9S,10R anti-B[a]PDE   (−)-7S,8R,9R,10S anti-B[a]PDE
or   or
(−)-1R,2S,3S,4R anti-B[c]PhDE   (+)-1S,2R,3R,4S anti-B[c]PhDE

Figure 2 Absolute configurations of substituents in *r7,t8*-dihydroxy-*t9,10*-epoxy-7,8,9,10-tetrahydrobenzo[a]pyrene, (+)-*anti*-B[a]PDE, and *r4,t3*-dihydroxy-*t1,2*-epoxy-1,2,3,4-tetrahydrobenzo[c]phenanthrene, (−)-*anti*-B[c]PhDE, after metabolic activation of the parent compounds (Fig. 1) to the bay and fjord region diol epoxides.

differently shaped PAH. Human exposure occurs through ingestion of PAH-contaminated food and inhalation of PAH as air pollutants and tobacco smoke constituents. In living cells, PAH compounds are metabolized to numerous oxygenated derivatives, including the highly reactive and genotoxic PAH diol epoxides (49) that can bind covalently to cellular DNA (50). Some of these reactive metabolites are mutagenic in human and bacterial cells, are carcinogenic in experimental animals, and are therefore suspected to play a role in the etiology of many human cancers (51), especially lung-associated cancers (52,53). Positive correlations between DNA adduct levels and susceptibility to cancer have been documented (54), and the etiological relevance of stable DNA adducts in human carcinogenesis has been described (55–57).

The metabolic activation of the parent PAH benzo[a]pyrene (B[a]P) and benzo[c]phenanthrene (B[c]Ph) gives rise to the diol epoxides *anti*-B[a]PDE and *anti*-B[c]PhDE with the oxide groups positioned within the bay or fjord regions, and with absolute configurations relative to the distal 7-OH or 4-OH groups, shown in Fig. 2. These diol epoxides can exist as a pair of (+)- and (−) enantiomers that react with the exocyclic amino groups of adenine and guanine in DNA to form stable adducts involving bond formation between the C10-position of B[a]PDE, or C1 position of B[c]PhDE and $-N^2$-dG or $-N^6$-dA in DNA (50). Addition of the exocyclic amino groups of dA or dG can occur either *cis* or *trans* to the ring-opened (+)- or (−)-*anti* diol epoxides (Fig. 3), thus resulting in four stereoisomeric adducts, (+)-*trans*-, (−)-*trans*, (+)-*cis*, and (−)-*cis* with dG, and four with dA. Some of these stereoisomeric adducts adopt remarkably different conformations in double-stranded DNA. If these DNA adducts are not excised by normal cellular repair mechanisms, they can persist until DNA replication occurs, and cause mutations if the replication is error-prone. Insights into the molecular bases of the recognition and excision of the stereoisomeric PAH-N^2-dG and N^6-dA adducts in DNA by human NER enzymes adducts can be obtained by studying lesions with the same chemical structures but different absolute configurations and conformations.

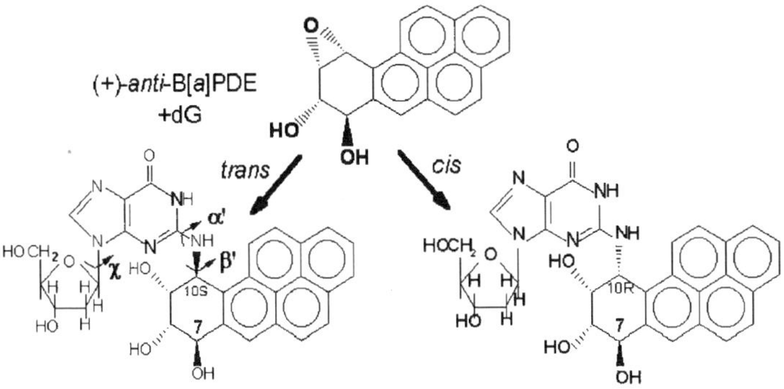

Figure 3 Formation of covalent adducts by either *cis* or *trans* addition of N^2-dG to the C10 position of (+)-*anti*-B[*a*]PDE. The glycosidic torsion angle χ, and the torsion angles α' and β' are designated.

3. METHODS

3.1. NER Assays In Vitro

A PAH-modified 11-mer is first ^{32}P radioactively end-labeled at its 5′-end, and then ligated into a ~140-mer oligonucleotide with the lesion positioned in its center. The purified 140-mers are then annealed with a fully complementary 140-mer unmodified strand to form a double-stranded DNA molecule containing a single modified guanine or adenine residue (58,59). An NER assay is employed to compare the relative excision efficiencies of different lesions in the modified oligonucleotide duplexes. Briefly, cell extracts are prepared from human HeLa cells, and aliquots containing about 50 μg protein equivalents are stored at –80°C until needed. The excision reactions containing one aliquot of the cell extracts in 25 μL of a 40 mM HEPES-KOH (pH 7.8), 5 mM MgCl$_2$, 0.5 mM DTT, 2 mM ATP, 20 μM each of dGTP, dATP, and dTTP, dATP, 22 mM phosphocreatine, 50 μg/mL creatine phosphokinase, and 5 fmol of the centrally ^{32}P-labeled ~140-mer duplex with the single PAH-modified residue (the radioactively labeled phosphodiester bond is positioned on the 5′-side and five nucleotides from the lesion). After incubation for 1.5 hr at 30°C, the reactions are stopped by the addition of EDTA, the solution is extracted with a phenol/chloroform solution, and the DNA in the aqueous phase is desalted by precipitation with 80% methanol. The desalted oligonucleotides are then dried and subjected to denaturing 20% acrylamide gel electrophoresis. If NER is successful, a series of radioactively labeled oligonucleotides about 24–32 bases in length is observed (4,58–60), which is the hallmark of NER. The extent of repair is quantitatively estimated (radioactivity associated with the excised 24–32-mers divided by the total radioactivity signal in a given lane in the gel). As a positive control of successful NER, an *N*-acetyl-2-aminofluorene-C8-dG adduct embedded in similar ~140 mer duplexes is employed (58).

3.2. Computational Analysis of Adduct Structure by Molecular Dynamic (MD) Simulations and Molecular Mechanics Poisson–Boltzmann Surface Area (MM–PBSA) Methods

Considerable progress has been made in recent years in improving MD simulation methods for studying the properties of biological macromolecules in aqueous solution with explicit counterions (39–41). The more recently developed MM–PBSA methodology has been employed to calculate the relative free energies of macromolecules and complex molecular systems. The results of this method have been benchmarked against various experimental systems and it has been shown to be reasonably robust for the kind of systems we are investigating here (39,44). In our research, the structural features of the PAH-modified (or unmodified) oligonucleotide duplexes are determined by MD simulations using available solution NMR conformations as starting structures. If not available, reasonable structures are modeled from closely analogous experimental structures. The latest available AMBER package (61) with the Cornell et al. force field (62,63) and improved parameter sets (41), with explicit water molecules and counterions in a periodic box to represent the environment around the macromolecule, is employed. Typically, an ensemble of several thousand structures is obtained that represents the evolution of the conformations of the PAH-modified duplexes at about 1 ps time intervals on a time scale of up to 5 ns in current work. The evolution of structural DNA parameters at different sites of the duplex, e.g., the integrity of Watson–Crick hydrogen bonding, glycosidic torsion angles, buckle, roll, twist, tilt, rise and the backbone torsion, and sugar pucker parameters, can be monitored as a function of time following the collection of the trajectory coordinates. The free energy, G_{tot}, is computed from the molecular mechanical energy (E_{MM}), solvation free energy ($G_{solvation}$), and solute entropic contributions to free energy, using the MM–PBSA methods (39,64–66). This approach combines an explicit molecular mechanical model for the solute with a continuum method for the solvation free energy. The enthalpies, entropies, and free energies are computed for each of the structures in the ensemble, and then these thermodynamic parameters are averaged over the entire ensemble. The interested reader may consult the references cited here for further details concerning these methods.

A few relevant examples illustrate the quality of these computational methods. Molecular dynamics simulations have been successfully applied to a number of damaged DNA structures, such as the cisplatin–DNA adduct (67), the *cis-syn* cyclobutane dimer (68), and the 6–4 photoproduct (68). The MD simulations not only were able to accurately reproduce the NMR structural features of the damaged DNA, but also provided new insights that could not be obtained from the NMR experiments. In the case of the cisplatin–DNA adduct (67), the MD simulation reproduced the inter-proton distances and the characteristic chemical shifts in the NMR data, and also revealed increased conformational flexibility at the platinum binding site that could not be determined by the NMR experiment. Our own benchmarks against experiment (see below) have also provided excellent agreement with experiment on both the structural and thermodynamic levels (44). We have applied the MM–PBSA methodology to study the thermodynamics of PAH diol epoxide-modified DNA duplexes, and were successful in reproducing measured thermodynamic properties and accounting for the differences in the thermal stabilities of pairs of stereoisomeric B[*a*]PDE-N^6-dA and B[*c*]PhDE-N^6-dA adducts in double-stranded oligonucleotides (42,45).

We are using the molecular mechanics/dynamics methods in several ways: (1) to construct three-dimensional models from the set of inter-proton distances with upper and lower bounds derived from the NMR studies by restrained energy minimization or MD calculations; (2) to further analyze experimental NMR structural data to assess the possible time-dependent fluctuations in the adduct conformations by MD methods; thereby, we gain further insights into the DNA structural parameters that are most affected by the presence of the bulky lesions that might explain their functional characteristics; (3) to predict possible adduct conformations in double-stranded DNA when no experimental data is available, and (4) to use MM–PBSA free energy calculations to gain insights into differences in thermal stabilities (T_m values) of pairs of stereoisomeric adducts, and to determine how the lesions perturb the relative free energies of the PAH-modified DNA duplexes. Finally, the structural data is correlated with the relative susceptibilities of the different adducts to excision by human NER enzymes in vitro.

The results of NMR structural studies of DNA adducts provide vital experimental insights that allow for studies of structure-NER relationships and that serve as benchmarks for validating the computational methods. The NMR structures of some lesions, e.g., the four stereoisomeric B[a]P-N^2-dG adducts (18), can provide good rationalizations of the observed NER patterns (58). In the case of some other adducts, however, we found that the NMR structures alone do not necessarily distinguish between intercalated adducts that are excised by NER enzymes from those that are not (42,59). However, in conjunction with NMR structural data, computational methods and thermal stability studies can provide critical additional information that distinguishes NER-resistant from NER-susceptible PAH–DNA adduct conformations (38,42,45). This measured approach has been our strategy for studying the structure–function relationships of a variety of PAH–DNA adducts, and a number of successes have been achieved in relating the conformations of different PAH-DNA adducts in different sequence contexts to their thermal stabilities (42,44,45) and to DNA repair (32,38,42,43,45). The MD method can provide insights into potentially critical interactions, e.g., quality of hydrogen bonding and stacking interactions between PAH residues and DNA bases, that can influence the thermodynamic stabilities of adducts and their susceptibilities to DNA repair (43). Free energy MM–PBSA calculations can, within the method limitations (45), reasonably account for trends in thermal stability differences of stereoisomeric adduct pairs that are identical in all respects except for adduct orientations, and provide information about the origins of these differences at the molecular level (42,44,45). We note that computation of free energy differences by these methods is applicable only to stereoisomeric lesions and is not feasible for comparing adducts that differ in chemical structures from one another, or in comparing adducts with unmodified DNA.

4. PAH–DNA ADDUCTS: CONFORMATIONAL MOTIFS

The diversity of absolute configurations about the chiral carbon atoms of the PAH diol epoxides, the shapes of the PAH molecules (fjord vs. bay), and the possibility of reaction with either N^2-dG or N^6-dA leads to different families of covalent adduct conformations in double-stranded DNA (18). The different kinds of structural motifs discovered so far are illustrated for guanine adducts in Fig. 4, and adenine adducts in Fig. 5. Earlier work, dealing mainly though not exclusively with B[a]PDE-N^2-dG adducts, was reviewed by us in 1997 (18). Since then, we have published the NMR solution structures of a number of other DNA adducts (19–21). The basic structural motifs, based on our work and that of others (22–27,29,30), are summarized in Figs. 4 and 5. Detailed computational analyses have provided insights into the origins and stereochemistry dependence of these remarkably different adduct conformations (69–72).

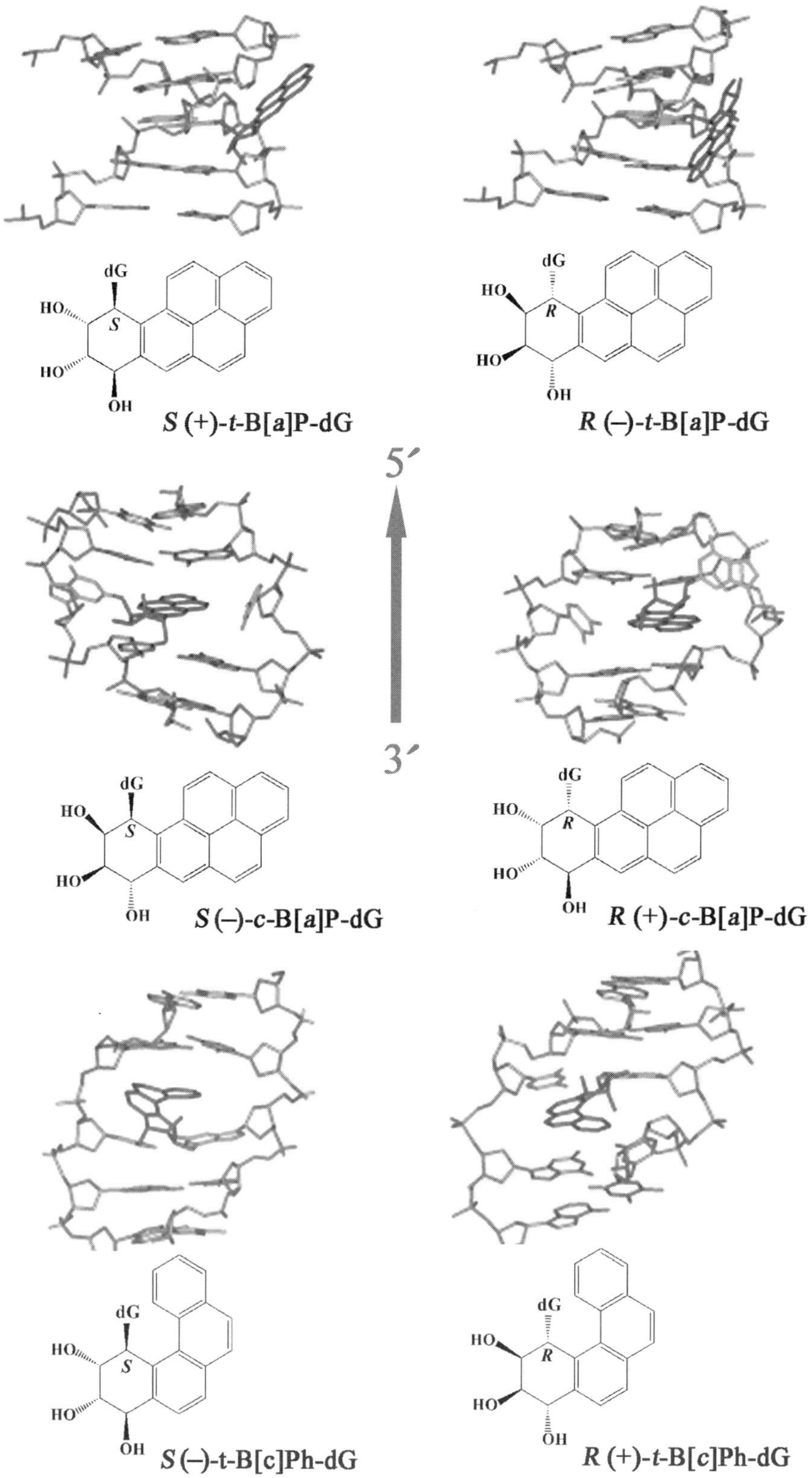

Figure 4 (*Caption on facing page*)

4.1. Guanine Adducts: Minor Groove Conformations

The B[a]P residues are bound to the exocyclic amino group of guanine. In the (+) - *trans*-B[a]P-N^2-dG adduct (*S* absolute configuration at the C10 linkage site), the pyrenyl residue (in red, Fig. 4) points towards the 5′-end of the modified strand, while in the stereoisomeric (−)-*trans*-adduct (*R*) it points towards the 3′-direction. All Watson–Crick base pairs are intact (16,17).

4.2. Guanine Adducts: Base-Displaced Intercalation

In the stereoisomeric *R* (+)-*cis*- and *S* (−)-*cis*-B[a]P-N^2-dG adducts (see Fig. 3 for definition), the B[a]P residue is intercalated with the benzylic ring in the minor or major grooves, respectively. The modified guanine (green) is displaced into the minor groove ((+)-*cis*) or the major groove ((−)-*cis*), while the partner C (green) residue is in the major groove in both cases (73,74).

4.3. Guanine Adducts: Intercalation from the Minor Groove Without Base Displacement

These adducts are derived from the *anti* diol epoxides of the fjord PAH B[c]Ph (Fig. 1). All Watson–Crick base pairs are intact in the duplexes with the *R* (+)-*trans*- and *S* (−)-*trans*-B[c]Ph-N^2-dG adducts. Note the remarkable differences in conformations of the bay *trans*-B[a]P-N^2-dG adducts vs. the fjord B[c]Ph-N^2-dG adducts with the identical stereochemical characteristics (Fig. 4). These are startlingly different adduct conformations that result from the different topologies of the aromatic ring systems of the fjord B[c]Ph and bay B[a]P residues (19).

4.4. Adenine Adducts: Intercalation from the Major Groove

All Watson–Crick base pairs, though perturbed, are maintained in the fjord 1*R* (+)-*trans*- and 1*S* (−)-*trans*-B[c]Ph-N^6-dA adducts (75,76) and a structurally related fjord *R* (+) -*trans-anti*-B[g]C-N^6-dA adduct (20). Some stretching and unwinding of the duplexes is, of course, necessary to accommodate intercalation of the bulky PAH residues. The key finding is that in the *R* adducts, the B[g]C and B[c]Ph residues are intercalated on the 5′-side of the modified adenine (20,75), while in the *S* adduct the B[c]Ph residue is intercalated on the 3′-side (76). Note the approximately parallel orientation of the B[c]Ph residue with respect to the neighboring base pairs, and the twist of the nonplanar aromatic ring system which is opposite in *S* and *R* isomers to optimize stacking.

Figure 4 (*Facing page*) Conformational motifs of adducts derived from the binding of the enantiomeric *anti*-B[a]PDE or *anti*-B[c]PhDE to N^2-guanine elucidated by NMR methods to form adducts with either *S* (left) or *R* (right) absolute configurations at the linkage site. Top: minor groove adducts derived from the reactions of either (+)- or (−)-*anti*-B[a]PDE (left and right, respectively) with N^2-dG (16,17) by *trans*-addition (*t*). Middle: adducts with base-displaced intercalative conformations derived from the reactions of either (−)- or (+)-*anti*-B[a]PDE (left and right, respectively) with N^2-dG (73,74) by *cis*-addition (*c*). Bottom: intercalation from the minor groove; adducts derived from the reactions of either (−)- or (+)-*anti*-B[c]PhDE (left and right, respectively) with N^2-dG by *trans*-addition (19). The arrow denotes the orientation of the modified strand.

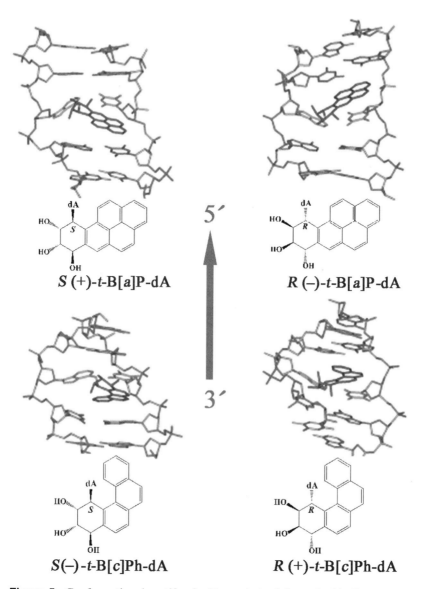

Figure 5 Conformational motifs of adducts derived from the binding of the enantiomeric bay region *anti*-B[*a*]PDE or fjord region *anti*-B[*c*]PhDE to N^6-adenine to form adducts with either *S* (left) or *R* (right) absolute configurations at the linkage site. Top: Intercalated adducts derived from the binding of (+)-*anti*-B[*a*]PDE or (−)-*anti*-B[*a*]PDE by *trans*-addition (*t*) to N^6-dA. The structures shown are taken from (45); NMR structures of the *R* adducts (21,23,24,26,28–30) and of *S* adducts (22,27) have been published (see the text). Bottom: Intercalative adduct conformations derived from the binding of (−)-*anti*-B[*c*]PhDE or (+)-*anti*-B[*c*]PhDE by *trans*-addition (*t*) to N^6-dA (75,76). The modified strands are oriented as in Fig. 4 (5′ on top, 3′ on the bottom).

4.5. Adenine Adducts: Distorting Intercalation from the Major Groove

The NMR structures of the bay *R* (−)-*trans*-B[*a*]P-N^6-dA adduct in different sequence contexts have been studied (21–30), and the B[*a*]P residues are also positioned on the

5′-side of the modified adenine as in the case of the two R fjord adducts (20,75). All Watson–Crick base pairs are intact in these R configuration bay and fjord N^6-dA adducts. However, the conformations of the S (+)-*trans*-B[*a*]P-N^6-dA adduct could not be determined accurately with a thymidine as the partner base to the modified adenine (23,26). On the other hand, with a guanine mismatch opposite the S (+)-*trans*-B[*a*]P-N^6-dA adduct (22,25), or with an adduct derived from the binding of a *syn* B[*a*]PDE stereoisomer to N^6-dA with a thymidine complementary base in the duplex (27), it turned out to be feasible to determine the conformations of the S B[*a*]P-N^6-dA adducts. In these duplexes with the adducts with an S absolute configuration at the adduct linkage sites, the B[*a*]P residues were found to be intercalated on the 3′-side of the modified adenine residues. Consequently, in all PAH-N^6-dA adducts studied so far, the paradigm of 5′-intercalated orientation for R adducts, and 3′-intercalative orientation for the S adducts, is maintained. Furthermore, the results of calculations indicate that in the case of the S (+)-*trans*-B[*a*]P-N^6-dA adducts, the B[*a*]P residue should be oriented on the 3′-side of the modified adenine (72), even though multiple conformations are evident in the fully complementary duplexes (26). The S (+)-*trans*-B[*a*]P-N^6-dA adduct structures shown in Fig. 5 are an average obtained by Yan et al. (45) from MD simulations of the time-dependent trajectory of these adducts in sequence *II* (see below). Note that the B[*a*]P residue is not parallel to neighboring base pairs, and that flanking base pairs appear to be more distorted than in the case of the fjord *trans*-B[*c*]Ph-dA adducts (Fig. 5).

4.6. External Major Groove Adducts

Such external, nonintercalated conformations for N^6-adenine adducts have been found only in the case of styrene oxide-N^6-adenine adducts in double-stranded DNA (77,78). The R and S α-(N^6-adenyl) styrene oxide adducts, with a single aromatic ring, were found to be positioned in the major groove, pointing in opposite directions of the modified strand as in the R and S B[*a*]P- and B[*c*]Ph-N^6-dA adducts. These are the only known examples of nonintercalated aromatic hydrocarbon -N^6-dA adducts. The processing of these adducts by NER enzymes has not been investigated.

5. INSIGHTS INTO THE STRUCTURAL MOTIFS AT THE NUCLEOSIDE ADDUCT LEVEL DERIVED FROM COMPUTATIONAL APPROACHES

The remarkable differences in the conformations of the four stereoisomeric B[*a*]P-N^2-dG lesions in double-stranded DNA prompted a detailed examination of the reasons for these diverse conformations, and the structural factors that govern these fascinating differences. We first noted that the torsion angles (Fig. 3) α′ and β′ (and to a much smaller extent χ) exhibit remarkably similar patterns in these four stereoisomeric, conformationally different *anti*-B[*a*]P-N^2-dG adducts in double-stranded DNA (18). From simple steric hindrance considerations, we realized that the sterically permitted values of α′ and β′ are limited in range for each stereoisomeric adduct because of "primary" steric hindrance effects between benzylic ring atoms on the B[*a*]P and the covalently linked purine residues (18,70–73). Detailed determination of the energetically most favorable values of α′ and β′ for different B[*a*]P-dG (69,70), B[*a*]P-dA (71,72), and B[*c*]Ph-dA (79) mononucleoside adducts, utilizing the AMBER suite of programs, permitted us to fully map the entire potential energy surfaces of these adducts. It was shown that only limited ranges of the torsion angles α' and β' are possible, and that the opposite orientations of adducts derived from two enantiomeric diol epoxides is rooted in the near mirror relationships of substituents about the chiral carbon atoms in the benzylic rings of B[a]PDE or B[c]PhDE. Further analysis suggests why the *trans*-B[a]P-dG adducts are characterized by minor groove conformations, while the *cis*-adducts disfavor the minor groove and hence are intercalated with base-displacement (69). The limited ranges of the torsion angles α' and β' are particularly useful in modeling viable adduct conformations when the structures of the adducts are not known.

6. PAH–DNA ADDUCT CONFORMATIONAL MOTIFS AND NER

The susceptibilities of different PAH–DNA lesions to NER depend markedly on: (1) PAH–DNA adduct stereochemistry and the purine modified (G or A), (2) adduct conformation, and (3) extent of the apparent local distortion of the intrinsic DNA structure caused by the adducts. Particularly interesting are the observations that chemically identical, stereoisomeric *anti*-B[a]P-N^2-dG adducts exhibit differences in susceptibilities to human NER enzymes in cell-free extracts by factors of ~100 or more, depending on the absolute configurations of substituents about the four chiral carbon atoms in the 7–10-ring, and the presence or absence of a cytidine or adenine residue opposite the lesion in the partner strand (58).

The efficiencies of excision of PAH–DNA adducts with different conformational motifs, relative to the excision of the control *N*-acetyl-2-aminofluorene-C8-dG adduct, is summarized in Fig. 6. All NER experiments were performed in the same sequence context as the NMR structural studies, except in the case of the *S* adducts where the role of sequence context on NER only was investigated.

6.1. Stereoisomeric Bay Region B[a]P-N^2-dG Adducts

6.1.1. Base-Displaced Intercalative cis-B[a]P-dG Adducts

> 5'-CCATC[G*]CTACC
> 3'-GGTAG C GATGG   *I*

These *cis*-adducts in the sequence context of the 11-mer duplex *I* embedded in 139-mer long double-stranded DNA molecules, are excised almost as easily as the *N*-acetyl-2-aminofluorene-C8-dG reference adduct that also has a base-displaced adduct conformation (80). The excision efficiency of the *R* (+)-*cis*-B[a]P-dG adducts is somewhat higher than that of the stereoisomeric (−)-*cis*-adduct (58). These two *cis* adducts are among the most easily excised PAH–DNA adducts studied up until now. In the base-displaced structural motif, the modified guanine residue is displaced into the minor groove in the case of the *R* (+)-*cis*-B[a]P-dG adduct, and into the major groove in the case of the *S* (−)-*cis*-B[a]P-dG adducts. In both cases, the cytidine base in the complementary strand that would have normally been paired with the modified guanine, is displaced into the major and minor grooves, respectively. In both *cis*-adducts, the pyrenyl residue is intercalatively inserted into the helix, thus taking the space normally occupied by the G:C Watson–Crick base pair. These are obviously major structural distortions at the lesion site that are readily recognized by the NER machinery, most likely the XPC–HR23B protein complex.

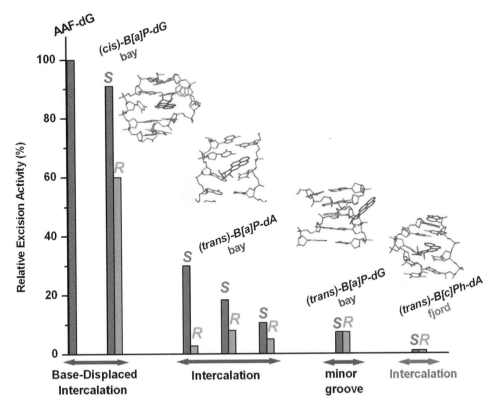

Figure 6 Summary of NER efficiencies in human (HeLa) cell extracts in vitro for different PAH–DNA adduct conformations relative to a 2-AAF-C8-dG adduct (see text). Examples of adduct structures are shown (from left to right): the base-displaced intercalated R (+)-cis-B[a]P-N^2-dG, the intercalated R (−)-trans-B[a]P-N^6-dA, the minor groove S (+)-trans-B[a]P-N^2-dG, and the intercalated S (−)-trans-B[c]Ph-N^2-dG adduct. (The structures are from Figs. 4 and 5.)

6.1.2. Minor Groove Adducts

The two minor groove R and S trans-B[a]P-N^2-dG adducts are excised with efficiencies ∼7–12 times lower than the stereoisomeric, base-displaced intercalated cis-B[a]P-N^2-dG adducts (Fig. 6). This difference is attributed to the lower degree of distortion of the DNA helices in the case of the minor groove trans-B[a]P-N^2-dG adducts as compared to the base-displaced intercalated cis-B[a]P-N^2-dG adducts (58).

6.2. Differences in the Processing of Bay Region *Trans*-B[*a*]P-N^6-dA and Fjord B[*c*]Ph-N^6-dA Adducts by NER Enzymes

5'-d(CGGAC[A*]AGAAG) d(GCCT G T TC TTC)	*II*
5'-d(GGTC[A*]CGAG) d(CCAG T GCTC)	*III*
5'-d(CTCTC [A*]CTTCC) d(GAGAG T GAAGG)	*IV*

Intercalated $B[a]P$-N^6-dA adducts resemble classical intercalation complexes since the bulky PAH residues are inserted into the helix without base displacement. These intercalated adenine adducts therefore appear to be less distorted than the base-displaced, intercalated cis-$B[a]P$-N^2-dG adducts. Indeed, the two $trans$-$B[a]P$-N^6-dA adducts are excised with lower efficiencies than the two base-displaced intercalated cis-$B[a]P$-N^2-dG adducts (Fig. 6). In all three sequences studied, *II–IV*, the adduct with *S* absolute configuration at the $B[a]P$/C10-N^6-dA linkage is excised with efficiencies ∼2 (in *III*) to 10 (in *IV*) times greater than the stereoisomeric adducts with *R* absolute configuration at the same site. In *R* adducts in double stranded DNA, the bulky $B[a]P$ residue is intercalated on the 5'-side of the modified adenine base in the *N-ras* CA*A and CAA* sequence context *II* (29,30), as well as in sequence *III* (23,24), and sequence *IV* (21), as shown in Fig. 5. However, as discussed above, the stereoisomeric *S* (+)-*trans*-$B[a]P$-N^6-dA adduct paired with a thymidine in the complementary strand, is most likely oriented on the 3'-side, but exhibits multiple conformations (26). Surprisingly, both the *R* or *S* fjord *trans*-$B[c]Ph$-N^6-dA adducts are resistant to NER with the adducts either in sequence II or IV (59), or in the *H-ras* CA*G sequence context (59).

7. STRUCTURAL DIFFERENCES BETWEEN BAY AND FJORD STEREOISOMERIC PAH-N^6–ADENINE ADDUCTS AND CORRELATIONS WITH NER SUSCEPTIBILITIES

7.1. Overview

The $B[a]P$-N^6-dA and $B[c]Ph$-N^6-dA adducts are all intercalated without base displacement. The differences in resistance to NER of the *R* adducts relative to the *S* $B[a]P$-N^6-dA adducts, though observed with a limited number of samples, is particularly intriguing (Fig. 8). The NMR properties of the *S* and *R* *trans*-$B[a]P$-N^6-dA adducts provide a clue to the uniformly higher efficiency of excision of the adducts with *S* configuration (Fig. 8). In the case of the *R* adducts, the $B[a]P$ residues are intercalated on the 5'-side of the modified adenine-thymidine base pair. There appears to be some conformational heterogeneity however, since minor proportions of other types of conformations, though still intercalated on the 5'-side, have been observed (25,26,30). However, in the case of the *S* adducts, multiple, inter-converting conformations were observed that precluded a detailed study of these adducts with a thymidine in the complementary strand opposite the $B[a]P$-modified adenine (23,26). The NMR structures suggest that the more facile excision of the *S* adducts is linked to their conformational heterogeneity and thus to a greater overall structural disorder. The lack of NER in the case of the fjord *S* and *R* $B[c]Ph$-N^6-dA adducts is surprising, even though they are both, like the stereochemically analogous bay *S* and *R* $B[a]P$-N^6-dA adducts, intercalated on the 3'- and 5'-sides of the modified adenine-thymidine base pairs, respectively. While the bay region *S* (+)-*trans*-$B[a]P$-N^6-dA adducts are defined by multiple conformers and definitive structures could not be established by NMR (26), the stereochemically analogous *S* (−)-*trans*-$B[c]Ph$-N^6-dA adducts (76) are well defined by an intercalative conformation on the 3'-side of the modified adenine-thymidine base pair (Fig. 5). We note in passing that the optical rotatory dispersion of the fjord $B[c]Ph$ adducts has an opposite sign to that of the bay $B[a]P$ adducts with the identical absolute configuration (50). Inspection of the conformations of the bay *R* (+)-*trans*-$B[a]P$-N^6-dA and fjord *R* (−) -*trans*-$B[c]Ph$-N^6-dA adducts (Fig. 5) does not provide a clear answer to the question of why the B[a]P adduct is excised while the B[c]Ph adduct is resistant to NER.

Insights into the reasons for these differences were sought on three different levels: (a) degree of structural distortion of the DNA as reflected in thermal destabilization measurements, (b) NMR solution structures, and (c) molecular dynamic simulations with structural and thermodynamic analyses of the resultant ensembles.

7.2. Thermal Dissociation of PAH-Modified Duplexes: Correlation of T_m with DNA Repair

In thermal melting experiments, the UV absorbance at 260 nm of the DNA duplexes is followed as a function of increasing temperature. As the double strands dissociate into single strands, the absorbance increases (81). The midpoint of the rising portion of these curves defines the "melting temperature" T_m, an indicator of the thermal stabilities of oligonucleotide duplexes. The T_m values represent a crude index of the overall degree of structural destabilization and distortions associated with the PAH lesions. More sophisticated and detailed characterizations of the thermodynamic profiles of these B[a]PDE-modified oligonucleotide duplexes require extensive microcalorimetric studies (82).

In their initial studies of the NMR solution structures of the R and S trans-B[a]P-N^6-dA studies, Schurter et al. (23) reported that the T_m values of duplexes containing S (+)-trans-B[a]P-N^6-dA adducts were significantly lower than those containing the stereoisomeric R (−)-trans-adducts in sequence III. The same effects on T_m are also observed in sequences II and IV (36). Typical UV melting profiles are shown in Fig. 7 for duplex II, either unmodified (UM), or with R (+)-trans- or S (−)-trans-B[a]P-N^6-dA lesions. Both the R and the S adducts destabilize duplexes II, and the duplex with the S adduct has a lower T_m than the one with the R adduct. The T_m values for duplexes II, III (from Ref. 23), and IV, are summarized in the form of a bar graph in Fig. 8 (upper panel). Similar patterns of melting profiles shown in Fig. 7 for duplex II are also observed in the other two sequences. The T_m values of duplexes with $10S$ (+)-trans-adducts are consistently lower than those with the $10R$ (−)-trans-adducts. The relative efficiencies of NER for the same sequences are compared in the bar graph shown in the lower panel of Fig. 8. Comparing the upper and lower panels in this Figure, it is evident that the higher efficiencies of NER are inversely correlated with lower T_m values for each sequence. *Our hypothesis is that the extent of structural disorder and destabilization caused by each adduct is crudely reflected in the T_m value associated with that duplex, in intercalated structural families, and that these distortions/destabilizations are also recognized by NER proteins, most likely the XPC/HR23B damage recognition complex* (5). Such an effect is consistent with a "thermodynamic probing" mechanism of DNA repair suggested by Gunz et al. (83).

This hypothesis is further supported by our results with the R and S fjord PAH-N^6-dA adducts. The T_m values of duplexes II (35) and IV (34) with the R and S adenine adducts derived from the fjord B[c]Ph diol epoxides are unchanged relative to the unmodified duplex IV. In contrast to the B[a]P-dA adducts, neither the S (−)-trans- nor the R (+) -trans-B[c]Ph-dA adducts destabilize double stranded DNA. A number of other fjord PAH-N^6-dA adducts (PAH: benzo[g]chrysene, dibenzo[a,l]pyrene) exhibit T_m values that are either higher (R adducts) or somewhat lower (S adducts) than the T_m of the unmodified duplexes (35). These enhancements in the T_m values are somewhat unusual and are attributed to the optimized

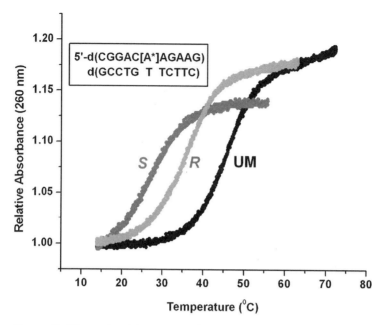

Figure 7 Thermal melting curves of unmodified (UM) and modified duplex *II* with either *R* (−)-*trans*- or *S* (+)-*trans*-B[*a*]P-N^6-dA adducts. *Source*: Adapted from Ref. 38.

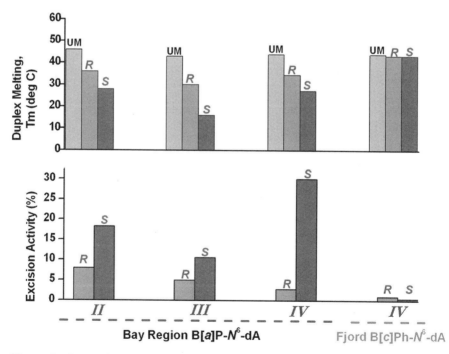

Figure 8 Comparison of thermal melting points, T_m, of duplexes, and NER efficiencies of *S* and *R* bay region B[*a*]P-N^6-dA in duplexes *II*, *III*, and *IV*, and fjord B[*c*]Ph-N^6-dA adducts in duplex IV. *Source*: The T_m values for sequence *II* are from Ref. 35; for sequence *III* from Ref. 23; for sequence *IV* (B[*a*]P-N^6-dA adduct) from Ref. 36; and for sequence *IV* (B[*c*]Ph-N^6-dA adduct) from Ref. 34. The NER data are from Ref. 59.

PAH–DNA base stacking interactions. These effects are particularly favored in the fjord compounds (18,20,35,74) because the sterically hindered fjord region has the aromatic ring system twisted in opposite directions in the S and R stereoisomeric adducts, which optimizes stacking with the neighboring base pairs. Neither of the two fjord B[c]Ph-N^6-dA adducts in duplex IV (Fig. 8), nor in duplex II (60), are excised by human NER enzymes. On the other hand, the more distorting R and S bay region B[a]P-N^6-dA adducts are excised with efficiencies that are correlated with the lowering in the T_m values (Fig. 8). The preliminary results summarized in Fig. 8 and in Ref. (59), suggest that certain types of bulky adduct-induced distortions in DNA conformation lead to efficient recognition and excision of the bulky adducts by NER enzymes, and also to a significant lowering in the T_m values. Clearly, more effort will be needed to substantiate the generality of these correlations and the validity of the hypothesis. This correlation has been observed up until now only in one type of conformational motif, namely intercalation without base displacement associated with the PAH-N^6-dA adducts studied.

7.3. B[a]P-N^2-Guanine Adducts

In the case of the stereoisomeric B[a]P-N^2-dG adducts, the T_m values of the modified duplexes with *cis*-adducts are lowered by 4–5°C, while the T_m of the two *trans*-adducts are lowered by 8–10°C (18,37). The base-displaced *cis*-adducts are probably stabilized by the PAH-base stacking interactions which compensate for the loss of one Watson–Crick base pair. On the other hand, the two minor groove *trans*-adducts are destabilized by a weakening of the Watson–Crick base pairing, the partial exposure of the hydrophobic B[a]P residues to the aqueous solvent, and some widening of the minor groove to accommodate the adduct (16,44). The relative NER efficiencies of these two different structural families are not correlated with the extent of lowering of the T_m values of the duplexes, since the minor groove *trans* adducts are more resistant to NER than the two *cis* adducts (Fig. 6). This highlights the likelihood that there is an interplay between distortion/destabilization, and stabilizing factors such as stacking which, together, govern the susceptibility to NER.

8. COMPUTATIONAL ANALYSIS

Molecular dynamics simulations provide molecular views of time-dependent structural fluctuations. These ensembles can be employed to derive thermodynamic parameters, including free energies. Together, the time-dependent structural perturbations induced by the bulky lesions and the relative energetic quantities helped to provide a rationale for: (1) the greater destabilization (lower T_m) of duplexes associated with the S (+)-*trans*-adducts relative to the $R(-)$-*trans*-B[a]P-N^6-dA adducts (45), and (2) the lack of thermal destabilization of duplexes with S $(-)$-*trans*- and R (+)-*trans*-B[c]Ph-N^6-dA adducts (34).

While NMR studies of the S B[a]P-N^6-dA adduct with a complementary T opposite the modified adenine adduct clearly indicate the co-existence of multiple conformations, all indications are that the B[a]P residue is intercalated on the 3′-side of the modified base (26,27). It was not clear, however, why such conformational heterogeneity is not observed in the case of the S-$(-)$-*trans*-B[c]Ph-N^6-dA adduct, nor why the S and R-*trans*-B[c]Ph-N^6-dA fjord region adducts are thermally more stable than the analogous bay B[a]P-N^6-dA adducts. The computational investigations have helped to elucidate the origins of the structural, dynamic, and thermodynamic properties of these S and R adducts in double-stranded DNA, and the results have provided insights into the relationships between adduct structure and the relative susceptibilities of the adducts to DNA repair (42).

8.1. The Bay Region *R* and *S Trans*-B[*a*]P-*N*⁶-dA Adducts

The dynamic structural features of the modified (or unmodified) oligonucleotide duplexes are determined by MD simulations using solution NMR conformations as starting structures. If not available (as in the case of the $10S$ (+)-*trans*-B[*a*]P-dA adduct, for example), reasonable structures are modeled from closely related NMR structures, and subjected to energy minimization and molecular dynamics equilibration. The key feature that distinguishes the R and S adducts is the opposite orientation of the B[*a*]P residues attached to the exocyclic amino groups of the modified adenine, A^*, relative to the plane of this base. Opposite orientations are a general paradigm that characterizes PAH-N^2-dG or N^6-dA adducts with opposite R or S absolute configurations at the linkage site irrespective of specific conformation (18,72,84). These differences in orientation have a profound effect on their conformational properties.

It has been suggested (25,26) that, because of the right-handed helical twist of B-DNA, the S B[*a*]P residues intercalated on the 3′-side are sterically more hindered than the R adducts intercalated on the 5′-side of the B[*a*]P-modified A^*:T base pair. The results of MD simulations of the trajectories of R and S B[*a*]P-N^6-dA adducts in the CA*A context of duplex *II* are consistent with this hypothesis and provide valuable insights into the structural and thermodynamic differences of these two stereoisomeric adducts (45). Furthermore, similar analysis of the same stereoisomeric R and S adducts in sequence *IV*, indicate that the conformational trajectories of the S *trans*-B[*a*]P-N^6-dA adducts are base sequence-dependent (43). Finally, the steric crowding effects that manifest themselves as multiple conformations of poorly resolved NMR spectra in the bay S *trans*-B[*a*]P-N^6-dA adducts in duplexes with complementary A^*:T base pairs (23,26), do not manifest themselves in the NMR solution structures of the stereochemically analogous fjord S (−)-*trans*-B[*c*]Ph-N^6-dA adducts (76). These results have been accounted for in terms of the topological differences between the bay region *trans*-B[*a*]P-N^6-dA and the fjord region *trans*-B[*c*]Ph-N^6-dA adducts (42).

8.1.1. Starting Models in the MD Simulations

An MD simulation was performed with R (−)-*trans*- and S (+)-*trans*-B[*a*]P-N^6-dA adducts in the fully complementary *N-ras* double-stranded sequence *II* with the lesions at the second position of codon 61, CA*A (45). As the starting structure for the R adduct, the NMR solution structure of this adduct in the same sequence determined by Zegar et al. (29) was employed. On the other hand, no structure was available for the stereoisomeric S (+)-*trans*-B[*a*]P-N^6-dA adduct with a T in the complementary strand opposite A^*. However, an experimental solution structure of an S (+)-*trans*-adduct with a mismatched G opposite A^* has been published (25), and was used as a reasonable starting structure in modeling sequence *II* (45). The MD trajectories for both adducts and the unmodified sequence were compared in the time interval from 500–2000 ps.

8.1.2. General Consequences of Intercalative Adduct Conformations

Due to the intercalative adduct conformations, there are significant distortions in the DNA backbone parameters at and near the site of the lesion. In both cases, there is significant local unwinding, about 41° and 29° in the case of the S and R adducts, respectively. In the R adduct, because of the presence of the B[a]P residue, there is an increase in the distance between the modified A*-T base pair and the G:C base pair on its 5′-side (this distance is called the "rise," Fig. 9). In the S adduct, the B[a]P residue is inserted on the 3′-side of A* and thus the rise between A*:T and the 3′-side A:T base pair is more than double the normal value of 3.4 Å. The 3′ or 5′-base pairs adjacent to A*:T in the S and R adducts, respectively, are significantly buckled to make room for the extended B[a]P ring systems on one side of the A*:T base pair or the other (Fig. 5).

8.1.3. Consequences of Steric Crowding in the S Adduct

A significant consequence of the 5′- or 3′-orientation of the intercalated aromatic B[a]P ring systems is the protrusion of the benzylic ring on the 5′-side of the plane of A* in the S (+)-*trans*-B[a]P-N^6-dA adducts, and on the 3′-side in the R (−)-*trans* adduct. Furthermore, because of the Watson–Crick alignment of the A*:T base pair, as well as the limited range of favorable values of the torsion angles α' and β' at the linkage site (Fig. 3), the range of energetically favorable adduct conformations is limited. Because of the right-handed helical twist, the C residue in the modified strand on the 5′-side of A* is twisted towards the benzylic ring in the case of the S (+)-*trans*-B[a]P-N^6-dA adduct, while in the R (−)-*trans*-adduct the A residue on the 3′-side of A* is twisted away from the benzylic ring. Therefore, while steric crowding is avoided in the case of the R adduct, there would be severe crowding between the benzylic ring and the adjacent C residue in the case of the S adduct, if no conformational rearrangement occurred. This crowding is relieved by a rotation of the glycosidic angle χ at the A* residue from the normal *anti* domain towards the *syn* domain (Fig. 10). This figure shows the distribution of the values of the glycosidic angles among the 1500 structures in the dynamics trajectory in the 500–2000 ps time range. In the case of the R adduct the mean value of χ is $-87° \pm 15°$, which is close to the mean value of $-98° \pm 18°$ in the *anti* domain characteristic of

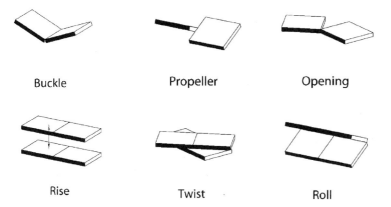

Buckle	Propeller	Opening
Rise	Twist	Roll

Figure 9 Typical DNA base pair parameters (buckle, propeller twist, opening, rise, twist, and roll) that can be affected by the presence of bulky PAH residues.

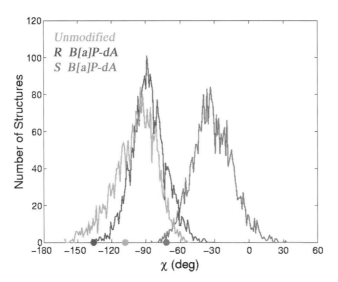

Figure 10 Comparisons of the distribution of the glycosidic torsion angle χ of the B[a]P-modified adenine (A*) for the unmodified (green) and modified duplexes *II* with *R* (−)-*trans*- (blue), and *S* (+)-*trans*-B[a]P-N^6-dA (red) adducts. The green, blue, and red dots denote the χ values of the starting structures in the molecular dynamic simulation (0.5–2.0 ns) for the unmodified duplex, and the *R*, and *S* adducts, respectively. *Source*: Adapted from Ref. 45. (*See color insert*.)

the unmodified duplex *II*. In contrast, in the *S* adduct, the mean value of χ is $-31° \pm 17°$ (the "high-*anti*" domain) which is significantly shifted towards the *syn* domain of glycosydic torsion angles. The range of observed χ values also includes structures that are fully in the *syn* domain.

8.1.4. Distorted Hydrogen Bonding at the Site of the Lesion

The progressive shift of the glycosidic angle from *anti* to *syn* entails a loss in the quality of Watson–Crick hydrogen bonding in the A*:T base pair. The quality of hydrogen bonding can be expressed in terms of the function I_H, a hydrogen bond quality index (85):

$$I_H = \Sigma(d - d_0)^2 + (1 + \cos\gamma)^2$$

where $d_0 = 2.95$ Å is an ideal donor–acceptor hydrogen_bond distance (86), d is the actual hydrogen bond distance in a given conformer within the MD simulation trajectory, while γ is the hydrogen bond angle (donor–H-acceptor) with the ideal value being 180°. The sum is over the two A*:T hydrogen bonds.

The fluctuations of the values of I_H within the MD trajectory are depicted in Fig. 11 for the *S* adduct (top) and the *R* adduct (bottom). It is evident that the I_H values are consistently larger and fluctuate more frequently towards higher magnitudes in the case of the *S* (+)-*trans*-B[a]P-N^6-dA than in the *R* adduct, indicating that the quality of hydrogen bonding is considerably worse in the former. The overall hydrogen bonding quality can be expressed in terms of the sum of I_H values, ΣI_H, over the entire MD trajectory. This sum would be equal to zero for perfect hydrogen bonding without any fluctuations and deviations of d and γ from their ideal values. The values of ΣI_H are ~600, 360, and 200 for the *S* (+)-*trans*-,

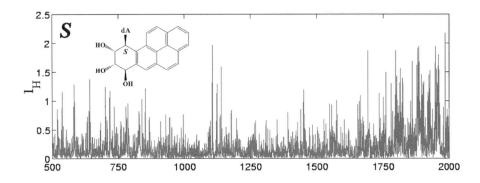

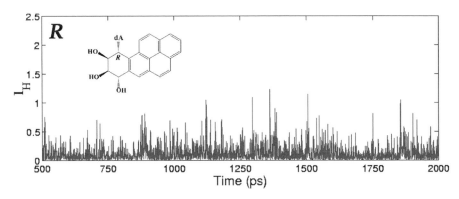

Figure 11 Depiction of the time-dependent fluctuations of the hydrogen-bond quality index, I_H, in duplex *II* with either S (+)-*trans*- (top) or R (−)-*trans*-B[a]P-N^6-dA (bottom) adducts as a function of time. The results of a MD simulation study in the time range of 0.5–2.0 ns are shown. The greater the value of I_H, the greater the deviation of the A*:T hydrogen bonding from ideal values for Watson–Crick base pairing. *Source*: Adapted from Ref. 45.

R (−)-*trans*-B[a]P-N^6-dA:dT, and the unmodified dA:dT base pairs, respectively in duplex *II*.

8.1.5. Free Energy Calculations for Duplexes with S and R Trans-B[a]
P-N^6-dA Adducts

DNA duplexes with S (+)-*trans*-B[a]P-N^6-dA lesions are thermally less stable than those with the stereoisomeric R (−)-*trans* adducts as indicated by the T_m values of 26° and 33°C in duplex *II*, respectively, measured under standard conditions (20 mM sodium phosphate buffer, 100 mM NaCl at pH 7 and ~10 µM oligonucleotide concentration, Fig. 7). Based on these differences in T_m values, the transition enthalpy of the duplex with the R adduct is lower by ~6 ± 2 kcal/mol than that of the duplex with the isomeric S adduct (45). Free energy calculations reveal that the free energy of the duplex with the R adduct is some 13 kcal/mol lower, and the enthalpy 10 kcal/mol lower, than the analogous values estimated for the S adduct. The experimental and calculated enthalpy differences are thus reasonably close to one another (45). The calculated relative free energies and entropy differences between the two duplexes with R and S adducts are consistent with the lower T_m of the S adducts. However, this comparison can only be approximate, since it rests on the assumption that the modified single strands of S and R adducts resulting from the dissociation of the duplexes have similar thermodynamic properties (45).

8.1.6. Differences in B[a]P-Base Stacking, Unwinding, and Backbone Parameters

A detailed analysis of the structural results of these MD simulations indicates that (1) the aromatic B[a]P residue is better stacked with adjacent base pairs on both sides of the R adduct, while only one face of the aromatic B[a]P residue in the S adduct is stacked with a base pair at the intercalation pocket. These differences arise because of the right-handed helical twist and the intercalation of the R adduct on the 5'-side and the S adduct on the 3'-side of the modified A*:T base pair, and the relative orientations of this base pair with respect to either the base pair on the 3' or 5'-side of A*:T. Furthermore, the relief of steric crowding between the benzylic ring of the S adduct and the 5'-neighboring cytidine base causes not only an *anti* to *syn* glycosidic angle rotation that weakens the overall quality of the A*:T base pairing, but also to a greater extent of unwinding than in the case of the R adduct. In turn these changes cause greater distortions in the backbone parameters of the S than in the R adduct, thus accounting for the lower thermal stability of duplexes with S (+)-*trans*- than R (−)-*trans*-B[a]P-N^6-dA adducts.

8.1.7. Effects of Base Sequence Context

It is shown in Fig. 8 that the R (−)-*trans*-B[a]P-N^6-dA adducts are excised more efficiently in the *N-ras* CA*A sequence context of duplex *II* than in the CA*C context of duplex *IV*. In contrast, the stereoisomeric S adducts are excised less efficiently in the CA*A than in the CA*C sequence context. The T_m values are similar for each of the two stereoisomeric adducts in both sequences, and thus do not provide any clues to these differences, and neither do the available NMR structures of the R B[a]P-N^6-dA adducts in these two sequence contexts (21,29). However, MD simulations of the trajectories of the S and R *trans*-B[a]P-N^6-dA adducts in the CA*C sequence context of duplex *IV* (43), reveal significant differences with those obtained in sequence *II* (45).

8.1.8. Effects of 3'-Bases Flanking the Lesion: Possible Sequence-Specific Hydrogen Bonds

Whereas in the CA*A sequence context the glycosidic angle χ of the modified A* residue of the S adduct freely samples both *anti* and *syn* domains (with maximal population in the ~−20–40° high *anti* region), separate maxima are found in the CA*C sequence in the *anti*- and the *syn* regions (maximum χ values at −85° and +30°, respectively). In duplex *IV*, the *anti* conformation appears to be stabilized by a hydrogen bond between the O-atom of the 9-OH group of the B[a]P residue and the amino group of C on the 3'-side of A* in the CA*C sequence context. Furthermore, in the *syn*-conformation, the 9-OH can form a hydrogen bond with O6 of the guanine residue complementary to the C flanking A* on the 3'-side. This results in a greater roll angle, and thus bending at the site of the lesion. The lack of overlap in the distributions of χ glycosidic angles between the *anti*- and *syn*-conformers suggests that these two conformers are positioned in different energy wells, and that a barrier hinders interconversion between the two. It was proposed by Yan et al. (43) that the higher susceptibility of the S (+)-*trans*-B[a]P-N^6-dA adduct in the CA*C than in the CA*A sequence context is due to the greater roll-derived bend, stabilized by a hydrogen bond between the B[a]P and 3'-flanking C residues in the *anti* conformation.

Another potential reason is a higher proportion of adducts in the *syn* conformation lacking normal Watson–Crick pairing in the CA*C than in the CA*A sequence. As in the CA*A sequence, the R adducts are considerably less distorted than the S adducts in the CA*C sequence, thus accounting for their lower overall free energies and greater resistance to NER. While these interpretations based on computational results are intriguing, a greater body of experimental and computational results will need to be acquired in order to test the interpretative and predictive usefulness of the computational methods utilized (43).

8.2. The Fjord Region *R* and *S Trans*-B[*c*]Ph-N^6-dA Adducts

The resistance to NER of both the S (−)-*trans*- and the R (+)-*trans*-B[*c*]Ph-N^6-dA adducts seems unusual for such bulky PAH–DNA adducts. The unusual lack of a thermal destabilization of the duplexes associated with these two adducts in sequence context *IV* (34), was also observed in the sequence context of duplex *II* since the T_m values are similar in the modified and unmodified duplexes (35). The structural characteristics of the S and R *trans*-B[*c*]Ph-N^6-dA adducts in the duplex sequence *II* were analyzed in detail by MD simulation and calculations of the free energies of this pair of adducts by Wu et al. (42). The analysis of the structures of the stereochemically analogous R and S *trans*-B[*a*]P-N^6-dA adducts were carried out in the same sequence (45) using identical computational methods. The results of these calculations show that the difference in free energies of the R and S *trans*-B[*c*]Ph-N^6-dA adducts is negligible ($\sim$0.2 kcal/mol), a result that is consistent with the same T_m values. This finding is in striking contrast to the results obtained with the R and S bay region *trans*-B[*a*]P-N^6-dA adducts.

8.2.1. The Distortion Free Energy

As discussed earlier, the free energies calculated for the B[*c*]Ph-modified sequences and the unmodified sequences cannot be compared with one another because the modified and unmodified duplexes are chemically not identical. Thus, the free energy calculations are not adapted to comparing the T_m values of the modified and unmodified duplexes. However, the free energies of the distorted duplexes relative to the unmodified DNA duplex can be estimated by calculating a *distortion free energy*. In this calculation, the B[*c*]Ph residues are removed from the modified duplexes and replaced by an H atom, leaving all distortions associated with the presence of the B[*c*]Ph (or B[*a*]P) residues intact. In this way, the distorted DNA duplexes are chemically identical to the unmodified duplexes and the calculated free energies are comparable. In these calculations, only the interactions of the PAH residues with the DNA bases are neglected. The distortion free energies relative to the unmodified duplexes *II* are found to be $\sim$0.2 and $\sim$0 kcal/mol for the R and S B[*c*]Ph-N^6-dA adducts, respectively, while in the case of the R and S *trans*-B[*a*]P-N^6-dA adducts the distortion free energies are 3.8 and 8.3 kcal/mol, respectively. The difference in the distortion free energies of the two fjord adducts is negligible, while the duplex with the bay region S (+)-*trans*-B[*a*]P-N^6-dA adduct is clearly significantly more distorted than the R adduct (42). These results are consistent with the differences in T_m values between the unmodified and the four modified duplexes studied.

8.2.2. Comparable Distortions in the Structural Parameters of Duplexes with Fjord R and S B[c]Ph-N^6-dA Adducts

The similar T_m values of these two adducts are due to similar distortions in the backbone parameters, twist or unwinding, stretching or rise (Fig. 9), and concomitant similar Van der Waals interaction energies between the aromatic B[c]Ph residues and neighboring base pairs in their respective intercalation pockets. The base pairs adjacent to the intercalated B[c]Ph residues on either the 3'-side (S adduct) or the 5'-side (R adduct) are buckled but not ruptured. In the case of the S adduct, the steric crowding between the benzylic ring and the 5'-flanking C next to A* is less severe than in the case of the S B[a]P-N^6-dA adduct. Thus, the mean value of the glycosidic angle χ is similar in the R B[c]Ph-N^6-dA adduct and the unmodified duplex II. However, in the S adduct, the mean value is $-61 \pm 14°$ and is thus shifted towards the *syn* domain, although much less so than in the case of the S B[a]P-N^6-dA adduct (Fig. 10). As a consequence, the summed hydrogen bonding quality index, ΣI_H is ~400 in the S B[c]Ph-N^6-dA (vs. ~600 in the case of the S (+)-*trans*-B[a]P-N^6-dA adduct), while in the R B[c]Ph adduct it is only ~250 (~360 in the R B[a]P adduct). Since the ΣI_H value is 206 in the unmodified duplex II, it is evident that the hydrogen bonding in both B[c]Ph-N^6-dA adducts is significantly less distorted than in the case of the bay region, stereochemically analogous B[a]P-N^6-dA adducts.

8.2.3. Origins of the Greater Thermal Stabilities of the Fjord B[c]Ph-N^6-dA than of the Bay B[a]P-N^6-dA Adducts

The distortion free energies reflect the structural distortions in the double-stranded DNA molecules caused by the PAH adducts. As outlined in Sec. 8.2.1, the distortion caused by the two B[c]Ph-N^6-dA adducts is significantly less than the distortion caused by the two B[a]P-N^6-dA adducts, and these distortion free energies are in line with the differences in T_m between the modified and unmodified duplexes (Fig. 8). A detailed comparison of these structures shows that in the B[a]P-N^6-dA adducts, (1) the backbone torsional parameters are more distorted, (2) the rise values are greater, (3) the unwinding at the site of the lesion is more extensive, and (4) the concomitant stabilizing stacking interactions between the B[a]P residues and neighboring base pairs are less extensive than in the case of the two B[c]Ph-N^6-dA adducts (42). What are the origins of these differences for the bay and fjord PAH-N^6-dA adducts? These issues were addressed by considering the arrangements of the aromatic rings, or topologies, of the bay and fjord adducts. The linkage sites to the exocyclic amino group of adenine occurs at the C10 position of the diol epoxide B[a]PDE, and the 1-position of the fjord B[c]PhDE diol epoxides (38,42). The most distant aromatic rings (the 1–5 positions, Fig. 1) in the B[a]P residue are rigid and extend a larger distance from the C10 linkage site than the most distant (positions 5–10) aromatic rings in the B[c]Ph residue from the C1 linkage site. Furthermore, because of steric hindrance effects between the C1-H atom with the C12-H atom in the B[c]Ph residue (Fig. 1), the distal aromatic 9–12-ring in B[c]Ph is twisted out of plane (87) as in other fjord region compounds (88). The nonplanarity of the aromatic ring system in B[c]Ph is such that the distal aromatic ring can twist out of plane to enhance stacking interactions with the neighboring base pairs (75,76); since this twist is in opposite directions in the S and R adducts, this effect contributes to an enhancement in the overall stabilities of the duplexes. In contrast, the B[a]P aromatic ring system is rigid and planar (88) and extends a greater distance from the C10 linkage site than the twisted B[c]Ph aromatic ring system from its C1 linkage site. At the intercalation sites, the planar B[*a*]P ring systems are not parallel to the adjacent base pairs, which causes the neighboring base pairs to buckle (Fig. 5), and the helix has to stretch more to accommodate the rigid aromatic B[*a*]P ring system. The rise is thus greater than in the case of the twisted and more adaptable fjord aromatic ring system. Because, the B[*a*]P aromatic moiety is extended further from its linkage site than the aromatic moiety of the aromatic B[*c*]Ph ring system, the B[*a*]P-N^6-dA adducts are more unwound, and the local helix untwisting is greater, than in the B[*c*]Ph-N^6-dA adducts (42). The adaptability of the twist of the distal ring in opposite directions in the case of these fjord *S* and *R* B[*c*]Ph adducts enhances the stacking interactions with the neighboring bases on the 3′- or 5′-side, respectively, of the A*:T base pair (75,76). This effect tends to increase the thermal stability of the duplexes, and thus their T_m values. Finally, in the *S* (−)-*trans*-B[*c*]Ph-N^6-dA adduct, a small movement of the benzylic ring towards the 3′-side and a rotation of the glycosidic angle χ from the *anti* to the high *anti* domain alleviates the steric crowding with the 5′-neighboring C base. In contrast, a greater shift in χ is necessary in the case of the stereochemically similar *S* (+) -*trans*-B[*a*]P-N^6-dA to alleviate steric crowding; this leads to a diminished quality of hydrogen bonding in this *S* bay region adduct than in the *S* fjord region adduct. In summary, the structural differences in ring shape and size account for the higher T_m values of the *S* and *R* fjord adducts than those of the *S* and *R* bay region adducts.

8.2.4. Introducing Structural Disorder into the Fjord B[c]Ph-N^6-dA by Mispairing of Bases

> 5′-d(CGGAC[A*]AGAAG) *II*
> d(GCCT G T   TC TTC)
>
> 5′-d(CGGAC[A*]AGAAG) *II*_{MM}
> d(GCCT **T C C**CTTC)

When the normal, complementary three bases 3′-….GTT…-5′ opposite the 5′-….CA*A…-3′ in sequence *II* are replaced by the noncomplementary 3′-….TCC…-5′ to yield the partially mismatched duplex sequence *II*_{MM}, the *R* (+) -*trans*-B[*c*]Ph-N^6-dA (A*) is readily excised by NER enzymes (32). In contrast, the fully complementary duplex with the same fjord lesion is refractory to excision repair (Fig. 8). This result not only highlights the important role of Watson–Crick hydrogen bonding in NER, but also points to local structural disorder as a signal of recognition of the lesion by the NER machinery. Furthermore, this experiment indicates that the nature of the bulky lesion itself is not as relevant to the resistance of the fjord *trans*-B[*c*]Ph-N^6-dA adduct to NER, as the insufficient extent of distortion/destabilization around the lesion. The nature of the distortions around the site of the adduct appear to be the key to the recognition and removal of DNA damage by the NER apparatus (4).

8.3. NER of Other Bulky Fjord PAH-N⁶-dA Adducts

The resistance to repair of the *R* and *S* B[*c*]Ph-N^6-dA adducts in fully complementary duplexes, as well as the unusual lack of an effect of the adducts on the thermal stabilities of the duplexes, is not solely due to the relatively small size of the aromatic B[*c*]Ph ring system. In fact, the T_m values of the more bulky *R* adducts of the (+)-*trans*-B[*g*]C-N^6-dA and (+)-*trans*-DB[*a,l*]P-N^6-dA adducts (Fig. 12) in duplex *IV* are 9° and 7° *higher*, respectively, than the T_m of the unmodified duplexes (35).

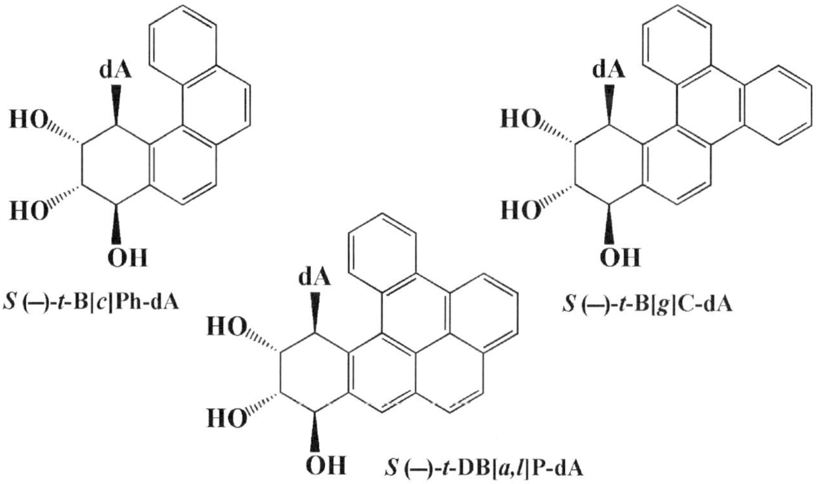

Figure 12 Comparison of fjord PAH-N^6-dA adducts studied.

The NMR solution structure of the R B[g]C adduct has been established (20) and resembles the intercalative conformation of the R (+)-*trans*-B[c]Ph-N^6-dA adduct in the same sequence context (75). The structures of the R (+)-*trans*- and S (−)-*trans*-DB[a,l]P-N^6-dA adducts have not yet been studied by NMR, but are likely to be similar, according to computational studies (89). The NMR solution structures of the S (−)-*trans*-B[g]C-N^6-dA adducts have not yet been investigated. Interestingly, the T_m values of the S B[g]C-N^6-dA and S DB[a,l]P-N^6-dA adducts are only 3° and 5° lower than the T_m of the unmodified duplex (35), thus indicating a much smaller extent of overall de-stabilization than observed in the case of the bay region S (+)-*trans*-B[a]P-N^6-dA adducts (the T_m is lowered by ~19°, Fig. 8). Consistent with the lack of excision observed in the case of the fjord *trans*-B[c]Ph-N^6-dA adducts, the fjord R and the S *trans*-B[g]C- and the *trans*-DB[a,l]P-N^6-dA adducts in sequence *IV* are also resistant to NER (59). These observations are remarkable because the bulky fjord adducts with as many as five aromatic rings are not recognized by the NER machinery.

8.4. Resistance of Fjord PAH-N^6-dA Adducts to NER and the Unusually High Tumorigenic Activities of Fjord PAH Compounds

The B[c]Ph, B[g]C, and DB[a,l]P are some of the fjord PAH compounds that have received the most attention from researchers in the cancer research community. The major DNA adducts formed result from the reactions of their fjord region diol epoxides with adenines in DNA (50). The fjord PAH (90), and especially their fjord diol epoxides (91–94) are significantly more tumorigenic in animal model systems than their bay region PAH counterparts. The unusually high reactivities of the fjord PAH diol epoxides with adenines in DNA has been correlated with their unusually high tumorigenic activities (50). A number of structural reasons have been advanced to explain these exceptional tumorigenicities (50). *Our results summarized here suggest that the exceptional tumorigenic properties of the fjord PAH may arise from their structural properties that lead to N^6-adenine adducts that are resistant to mammalian NER.*

9. CONCLUSIONS

9.1. Conformationally Related PAH-DNA Adducts for Studies of Structure–Function Relationships

Among the structurally diverse adducts derived from the reactions of stereoisomeric and structurally related PAH diol epoxides with guanine and adenine in DNA, the relative efficiencies of NER vary by factors of ~100, or more (32). In the series of six bay and fjord region PAH–adenine adducts studied up until now, four in sequence *II*, and two in sequence *IV* (Sec. 8.1.7–8.1.8), there is a correlation between the extent of distortion in double-stranded DNA caused by the bulky lesions, the associated changes in thermal stabilities (T_m) of the duplexes, and the efficiencies of NER. The NMR structural studies have shown that this class of PAH diol epoxide–DNA adducts assume intercalative adduct conformations without base displacement (Fig. 5), and thus form a family of structurally similar bulky DNA adducts with different degrees of structural distortions. These DNA adducts form an ideal set of lesions to assess the structural DNA distortions that are recognized by the human NER machinery, most likely by the XPC–HR23B protein complex in a rate-determining step (14).

9.2. Relationships Between Adduct Properties and NER

Among the intercalated PAH-N^6-dA lesions studied, the topologically more compact fjord PAH adducts are not recognized or excised by the human NER machinery in vitro in spite of the bulk of the PAH residues. On the other hand, the physically more extended and thus more bulky bay region B[*a*]P-N^6-dA adducts are excised under the same conditions, but with different efficiencies that depend on the adduct stereochemistry at the carcinogen-N^6-adenine linkage site. The T_m values of the modified duplexes are lower, and the efficiencies of NER higher, the greater the structural distortions/destabilizations in the DNA caused by these bulky lesions. Detailed structural analysis by MD simulation techniques and free energy calculations, indicate that these structural distortions include: (1) abnormal values of the glycosidic angle χ at the modified adenine residue, (2) an increased stretching (rise) at the intercalation site to make room for the bulky adduct and thus unwinding at the modified A*:T base pair and in some cases at the neighboring base pairs as well, (3) a distortion of the DNA backbone parameters, and (4) a lowering in the quality of hydrogen bonding, including loss of Watson–Crick base pairing. These cumulative distortions serve as a signal for recruiting the NER damage recognition proteins. The deviations of these structural parameters from their normal values in B-form unmodified DNA reduce the stabilities of the modified duplexes, while stacking interactions between the aromatic PAH residues and adjacent DNA bases at the intercalation sites tend to stabilize the carcinogen-modified duplexes. In the case of the topologically more compact fjord PAH-N^6-dA adducts, the thermal stabilities, as characterized by the T_m values, are often unchanged, and sometimes even higher than the T_m values of the unmodified duplexes (35); these effects are most likely due to enhanced stacking interactions that compensate for the distortions in the B-DNA conformation at or near the lesion sites. The adaptable fjord region twist and the compact topology are key to the enhanced stacking. Computational analysis of these structures indicate that the extent of distortion is lower in the fjord B[*c*]Ph-N^6-dA than in the sterically similar bay region B[*a*]P-N^6-dA adducts. Taken together, these structural perturbations, mitigated in part by base-stacking interactions, provide a good rationalization for the experimental observations that the fjord adducts are not susceptible to excision by human repair enzymes, while the bay region B[a]P-N^6-dA adducts are.

9.3. What Is Important in Recognition of Bulky Lesions by NER Enzymes?

The conclusion from this body of work is that the human NER enzymes recognize multiple structural features, including adduct stereochemistry and conformation-dependent unwinding, buckle, rise, roll, base-displacement (58), and deviations in Watson–Crick base pairing quality, including rupture. We term this a multipartite model of NER. This model is an extension of the bipartite model previously proposed by one of us which emphasized the importance of the disruption of hydrogen bonding (47), as demonstrated experimentally (46,48). This chapter describes how detailed studies of NMR solution structures of a family of intercalated bulky PAH–DNA adducts, coupled with MD simulations and free energy calculations, can provide deeper insights into the complex factors that are involved in the processing of such lesions by the human NER machinery.

ACKNOWLEDGMENTS

This work was supported by NIH grants CA 20851 and CA 99194 (N.E.G.), CA 46533 (D.J.P.), CA 28031 and CA 75449 (S.B.), and the Swiss National Science Foundation, 3100A0-101747 (H.N.).

REFERENCES

1. Braithwaite E, Wu X, Wang Z. Repair of DNA lesions: mechanisms and relative repair efficiencies. Mutat Res 1999; 424:207–219.
2. Mitchell JR, Hoeijmakers JH, Niedernhofer LJ. Divid and conquer: nucleotide excision repair battles cancer and ageing. Curr Opin Cell Biol 2003; 15:232–240.
3. Petit C, Sancar A. Nucleotide excision repair from *E. coil* to man. Biochimie 1999; 81: 15–25.
4. Wood RD. DNA damage recognition during necleotide excision repair in mammalian cells. Biochimie 1999; 81:39–44.
5. Riedl T, Hanaoka F, Egly JM. The comings and goings of nucleotide excision repair factors on damged DNA. Embo J 2003; 22:5293–5303.
6. Guzder SN, Sung P, Prakash L, Prakash S. Nucleotide excision repair in yeast is mediated by sequential assembly of repair factors and not by a pre-assembled repairosome. J Biol Chem 1996; 271:8903–8910.
7. Ng JM, Vermeulen W, van der Horst GT, Bergink S, Sugasawa K, Vrieling H, Hoeijmakers JH. A novel regulation mechanism of DNA repair by damage-induced and RAD23-dependent stabilization of xeroderma pigmentosum group C protein. Genes Dev 2003; 17:1630–1645.
8. Chen Z, Xu XS, Yang J, Wang G. Defining the function of XPC protein in psoralen and cisplatin-mediated DNA repair and mutagenesis. Carcinogenesis 2003; 24:1111–1121.
9. Kusumoto R, Masutani C, Sugasawa K, Iwai S, Araki M, Uchida A, Mizukoshi T, Hanaoka F. Diversity of the damage recognition step in the global genomic necleotide excision repair in vitro. Mutat Res 2001; 485:219–227.
10. Sugasawa K, Ng JM, Masutani C, Maekawa T, Uchida A, van der Spek PJ, Eker AP, Rademakers S, Visser C, Aboussekhra A, Wood RD, Hanaoka F, Bootsma D,

Hoeijmakers JH. Two human homologs of Rad23 are functionally interchangeable in complex formation and stimulation of XPC repair activity. Mol Cell Biol 1997; 17:6924–6931.

11. Sugasawa K, Ng JM, Masutani C, Iwai S, van der Spek PJ, Eker AP, Hanaoka F, Bootsma D, Hoeijmakers JH. Xeroderma pigmentosum group C protein complex is the initiator of global genome nucleotide excision repair. Mol Cell 1998; 2:223–232.

12. Sugasawa K, Okamoto T, Shimizu Y, Masutani C, Iwai S, Hanaoka F. A multistep damage recognition mechanism for global genomic necleotide excision repair Genes Dev 2001; 15:507–521.

13. Sugasawa K, Shimizu Y, Iwai S, Hanaoka F. A molecular mechanism for DNA damage recognition by the xeroderma pigmentosum group C protein complex. DNA Repair (Amst) 2002; 1:95–107.

14. Thoma BS, Vasquez KM. Critical DNA damage recognition functions of XPC-hHR23B and XPA-RPA in nucleotide excision repair. Mol Carcinog 2003; 38:1–13.

15. Hanawalt PC. Subpathways of nucleotide excision repair and their regulation. Oncogene 2002; 21:8949–8956.

16. Cosman M, de los Santos C, Fiala R, Hingerty BE, Singh SB, Ibanez V, Margulis LA, Live D, Geacintov NE, Broyde S. Solution confirmation of the major adduct between the carcinogen (+)-*anti*-benzo[a]pyrene diol epoxide and DNA. Proc Natl Acad Sci USA 1992; 89:1914–1918.

17. de los Santos C, Cosman M, Hingerty BE, Ibanez V, Margulis LA, Geacintov NE, Broyde S, Patel DJ. Influence of benzo[a]pyrene diol epoxide chirality on solution confirmations of DNA covalent adducts: the (−)-*trans-anti*-[BP]G.C adduct structure and comparision with the (+)-*trans-anti*-[BP]G.C enantiomer. Biochemistry 1992; 31:5245–5252.

18. Geacintov NE, Cosman M, Hingerty BE, Amin S, Broyde S, Patel DJ. NMR solution structures of stereoisometric covalent polycyclic aromatic carcinogen-DNA adduct: principles, patterns, and diversity. Chem Res Toxicol 1997; 10:111–146.

19. Lin CH, Huang X, Kolbanovsky A, Hingerty BE, Amin S, Broyde S, Geacintov NE, Patel DJ. Molecular topology of polycyclic aromatic carcinogens determines DNA adduct conformation: a link to tumorigenic activity. J Mol Biol 2001; 306:1059–1080.

20. Suri AK, Mao B, Amin S, Geacintov NE, Patel DJ. Solution cofirmation of the (+)-*trans-anti*-benzo[g]chrysene-dA opposite dT in a DNA duplex. J Mol Biol 1999; 292:289–307.

21. Mao B, Gu Z, Gorin A, Chen J, Hingerty BE, Amin S, Broyde S, Geacintov NE, Patel DJ. Solution structure of the (+)-*cis-anti*-benzo[a]pyrene-dA ([BP]dA) adduct opposite dT in a DNA duplex. Biochemistry 1999; 38:10831–10842.

22. Yeh HJ, Sayer JM, Liu X, Altieri AS, Byrd RA, Lakshman MK, Yagi H, Schurter EJ, Gorenstein DG, Jerina DM. NMR solution structure of a nonanucleotide duplex with a dG mismatch opposite a 10S adduct derived from *trans* addition of a deoxyadenosine N6-amino group to (+)-(7R,8S,9S,10R)-7,8-dihydroxy-9,10-epoxy-7,8,9,10- tetrahydro-benzo[a]pyrene: an unusual *syn* glycosidic torsion angle at the modified dA. Biochemistry 1995; 34:13570–13581.

23. Schurter EJ, Yeh HJ, Sayer JM, Lakshman MK, Yagi H, Jerina DM, Gorenstein DG. NMR solution structure of a nonanucleotide duplex with a dG mismatch opposite a 10R adduct derived from *trans* addition of a deoxyadenosine N6-amino group to (−)-(7S,8R,9R,10S)-7,8-dihydroxy-9,10-epoxy- 7,8,9,10- tetrahydrobenzo[a]pyrene. Biochemistry 1995; 34:1364–1375.

24. Schurter EJ, Sayer JM, Oh-hara T, Yeh HJ, Yagi H, Luxon BA, Jerina DM, Gorenstein DG. Nuclear magnetic resonance solution structure of an undecanucleotide duplex with a complementary thymidine base opposite a 10R adduct derived from trans addition of a deoxyadenosine N6-amino group to (−)-(7R,8S,9R,10S)-7,8-dihydroxy-9,10-expoxy-7,8,9,10-tetrahydrobenzo[a]pyrene. Biochemistry 1995; 34:9009–9020.

25. Schwartz JL, Rice JS, Luxon BA, Sayer JM, Xie G, Yeh HJ, Liu X, Jerina DM, Gorenstein DG. Solution structure of the minor conformer of a DNA duplex containing

a dG mismatch opposite a benzo[a]pyrene diol epoxide/dA adduct: glycosidic rotation from *syn* to *anti* at the modified deoxyadenosine. Biochemistry 1997; 36:11069–11076.

26. Volk DE, Rice JS, Luxon BA, Yeh HJ, Liang C, Xie G, Sayer JM, Jerina DM, Gorenstein DG. NMR evidence for *syn-anti* interconversion of a *trans* opened (10R)-dA adduct of benzo[a]pyrene (7S,8R)-diol (9R,10S)-epoxide in a DNA duplex. Biochemistry 2000; 39:14040–14053.

27. Pradhan P, Tirumala S, Liu X, Sayer JM, Jerina DM, Yeh HJ. Solution structure of a *trans*-opened (10S)-dA adduct of (+)-(7S,8R,9S,10R,)-7,8-dihydroxy-9,10-epoxy-7,8,9,10-tetrahydrobenzo[a]pyrene in a fully complementary DNA duplex:evidence for a major *syn* confirmation. Biochemistry 2001; 40:5870–5881.

28. Volk DE, Thiviyanathan V, Rice JS, Luxon BA, Shah JH, Yagi H, Sayer JM, Yeh HJ, Jerina DM, Gorenstein DG. Solution structure of a *cis*-opened (10R)-N6-deoxyadenosine adduct of (9S,10R)-9,10-epoxy-7,8,9,10-tetrahydrobenzo[a]pyrene in a DNA duplex. Biochemistry 2003; 42:1410–1420.

29. Zegar IS, Kim SJ, Johansen TN, Horton PJ, Harris CM, Harris TM, Stone MP. Adduction of the human N-ras codon 61 sequence with (−)-(7S,8R,9R,10S)-7,8-dihydroxy-9,10-epoxy-7,8,9,10-tetrahydrobenzo[a]pyrene: structural refinement of the intercalated SRSR(61,2) (−)-(7S,8R,9S,10R)-N6[10-(7,8,9,10- terahydrobenzo[a]pyrenyl)]-2′-deoxyadenosyl adduct from 1H NMR. Biochemistry 1996; 35:6212–6224.

30. Zegar IS, Chary P, Jabil RJ, Tamura PJ, Johansen TN, Lloyd RS, Harris CM, Harris TM, Stone MP. Multiple conformations of an intercalated (−)-(7S,8R,9S, 10R)-N6-[10-(7,8,9,10-tetrahydrobenzo[a]pyrenyl)]-2′-deoxyadenosyl adduct in the N-ras codon 61 sequence. Biochemistry 1998; 37:16516–16528.

31. Kim HY, Wilkinson AS, Harris CM, Harris TM, Stone MP. Minor groove orientation for the (1S,2R,3S,4R)-N2-[1-(1,2,3,4-tetrahydro-2,3,4-trihydroxybenz[a]anthracenyl)]-2′-deoxyguanosyl adduct in the N-ras codon 12 sequence. Biochemistry 2003; 42: 2328–2338.

32. Naegeli H, Geacintov NE. Mechanisms of repair of polycyclic aromatic hydrocarbon-induced DNA damage. In: Luch A, ed. The Carcinogenic Effects of Polycyclic Aromatic Hydrocarbons. London: Imperial College, 2005.

33. Mao B, Xu J, Li B, Margulis LA, Smirnov S, Ya NQ, Courtney SH, Geacintov NE. Synthesis and characterization of covalent adducts derived from the binding of benzo[a]-pyrene diol epoxide to a -GGG- sequence in a deoxyoligonucleotide. Carcinogenesis 1995; 16:357–365.

34. Laryea A, Cosman M, Lin JM, Liu T, Agarwal R, Smirnov S, Amin S, Harvey RG, Dipple A, Geacintov NE. Direct synthesis and characterization of site-specific adenosyl adducts derived from the binding of a 3,4-dihydroxy-1,2-epoxybenzo[c]phenanthrene stereoisomer to an 11-mer oligodeoxyribonucleotide. Chem Res Toxicol 1995; 8:444–454.

35. Ruan Q, Kolbanovskiy A, Zhuang P, Chen J, Krzeminski J, Amin S, Geacintov NE. Synthesis and characterization of site-specific and stereoisomeric fjord dibenzo[a,1]pyrene diol epoxide-N(6)-adenine adducts: unusual thermal stabilization of modified DNA duplexes. Chem Res Toxicol 2002; 15:249–261.

36. Krzeminski J, Ni JS, Zhuang P, Luneva N, Amin S, Geacintov NE. Total synthesis, mass spectrometric sequencing, and stabilities of oligonucleotide duplexes with single *trans-anti*-BDPE-N6-dA lesions in N-ras codon 61 and other sequence contexts. Polycycl Arom Compd 1999; 17:1–10.

37. Ya N-Q, Smirnov S, Cosman M, Bhanot S, Ibanez V, Geacintov NE. Thermal stabilities of benzo[a]pyrene diol epoxide-modified oligonucleotide duplexes. Effects of mismatched complementary strands and bulges. Struct Biol State of the Art, Proc on Conversation Discip Biomol Stereodyn, 8th, 1994; 2:349–366.

38. Geacintov NE, Broyde S, Buterin T, Naegeli H, Wu M, Yan SX, Patel DJ. Thermodynamic and structural factors in the removal of bulky DNA adducts by the nucleotide excision repair machinery. Biopolymers 2002; 65:202–210.

39. Kollman PA, Massova I, Reyes C, Kuhn B, Huo S, Chong L, Lee M, Lee T, Duan Y, Wang W, Donini O, Cieplak P, Srinivasan J, Case DA, Cheatham TE. Calculating structures and free energies of complex molecules: combining molecular mechanics and continuum models. Acc Chem Res 2000; 33:889–897.
40. Auffinger P, Westnof E. Simulations of the molecular dynamics of nucleic acids. Curr Opin Struct Biol 1998; 8:27–36.
41. Cheatham TE, Young MA. Molecular dynamics simulation of nucleic acids: successes, limitations, and promise. Biopolymers 2001; 56:232–256.
42. Wu M, Yan S, Patel DJ, Geacintov NE, Broyde S. Relating repair susceptibility of carcinogen-damaged DNA with structural distortion and thermodynamic stability. Nucleic Acids Res 2002; 30:3422–3432.
43. Yan S, Wu M, Buterin T, Naegeli H, Geacintov NE, Broyde S. Role of Base Sequence Context in Conformational Equilibria and Nucleotide Excision Repair of Benzo[a]pyrene Diol Epoxide-Adenine Adducts. Biochemistry 2003; 42:2339–2354.
44. Yan S, Wu M, Patel DJ, Geacintov NE, Broyde S. Simulating structural and thermodynamic properties of carcinogen-damaged DNA. Biophys J 2003; 84:2137–2148.
45. Yan S, Shapiro R, Geacintov NE, Broyde S. Stereochemical, structural and thermodynamic origins of stability differences between stereoisomeric benzo[a]pyrene diol epoxide deoxyadenosine adducts in a DNA mutational hot spot sequence. J Am Chem Soc 2001; 123:7054–7066.
46. Hess MT, Naegeli H, Capobianco M. Stereoselectivity of human nuecleotide excision repair promoted by defective hybridization. J Biol Chem 1998; 273:27867–27872.
47. Hess MT, Schwitter U, Petretta M, Giese B, Naegeli H. Bipartite substrate discrimination by human nucleotide excision repair. Proc Natl Acad Sci USA 1997; 94:6664–6669.
48. Buschta-Hedayat N, Buterin T, Hess MT, Missura M, Naegeli H. Recognition of non-hybridizing base pairs during nucleotide excision repair of DNA. Proc Natl Acad Sci USA 1999; 96:6090–6095.
49. Conney AH. Induction of microsomal enzymes by foreign chemicals and carcinogenesis by polycyclic aromatic hydrocarbons: G.H.A. Clowes Memorial Lecture. Cancer Res 1982; 42:4875–4917.
50. Szeliga J, Dipple A. DNA adduct formation by polycyclic aromatic hydrocarbon dihydrodiol epoxides. Chem Res Toxicol 1998; 11:1–11.
51. Greenblatt MS, Bennett WP, Hollstein M, Harris CC. Mutations in the p53 tumor suppressor gene: clues to cancer etiology and molecular pathogenesis. Cancer Res 1994; 54:4855–4878.
52. Yoon JH, Smith LE, Feng Z, Tang M, Lee CS, Pfeifer GP. Methylated CpG dinucleotides are the preferential targets for G-to-T transversion mutations induced by benzo[a]pyrene diol epoxide in mammalian cells: similarities with the p53 mutation spectrum in smoking-associated lung cancers. Cancer Res 2001; 61:7110–7117.
53. Pfeifer GP, Denissenko MF, Olivier M, Tretyakova N, Hecht SS, Hainaut P. Tobacco smoke carcinogens, DNA damage and p53 mutations in smoking-associated cancers. Oncogene 2002; 21:7435–7451.
54. Melendez-Colon VJ, Luch A, Seidel A, Baird WM. Cancer initiation by polycyclic aromatic hydrocarbons results from formation of stable DNA adducts rather than apurinic sites. Carcinogenesis 1999; 20:1885–1891.
55. Kriek E, Rojas M, Alexandrov K, Bartsch H. Polycyclic aromatic hydrocarbon-DNA adducts in humans: relevance as biomarkers for exposure and cancer risk. Mutat Res 1998; 400:215–231.
56. Smith LE, Denissenko MF, Bennett WP, Li H, Amin S, Tang M, Pfeifer GP. Targeting of lung cancer mutational hotspots by polycyclic aromatic hydrocarbons. J Natl Cancer Inst 2000; 92:803–811.
57. Ross JA, Nesnow S. Polycyclic aromatic hydrocarbons: correlations between DNA adducts and ras oncogene mutations. Mutat Res 1999; 424:155–166.

58. Hess MT, Gunz D, Luneva N, Geacintov NE, Naegeli H. Base pair conformation-dependent excision of benzo[a]pyrene diol epoxide-guanine adducts by human nucleotide excision repair enzymes. Mol Cell Biol 1997; 17:7069–7076.

59. Buterin T, Hess MT, Luneva NP, Geacintov NE, Amin S, Kroth H, Seidel A, Naegeli H. Unrepaired fjord region polycyclic aromatic hydrocarbon-DNA adducts in ras codon 61 mutational hot spots. Cancer Res 2000; 60:1849–1856.

60. Hess MT, Schwitter U, Petretta M, Giese B, Naegeli H. Site-specific DNA substrates for human excision repair: comparision between deoxyribose and base adducts. Chem Biol 1996; 3:121–128.

61. Case DA, Darden TA, Cheatham TE, Simmerling CL, Wang J, Duke RE, Luo R, Merz KM, Wang B, Pearlman DA, Crowley M, Brozell S, Tsui V, Gohlke H, Mongan J, Cui G, Beroza P, Schafmeister C, Caldwell JW, Ross WS, Kollman P. AMBER San Fransisco, CA: University of Calfornia, 2004.

62. Cornell W, Cieplak P, Bayly CI, Gould IR, Merz KM Jr, Ferguson DM, Spellmeyer DC, Fox T, Caldwell JW and Kollman PA. A second generation force field for the simulation of proteins, nucleic acids and organic molecules. J Am Chem Soc 1995; 117:5179–5197.

63. Cheatham TE, Cieplak P, Kollman PA. A modified version of the Cornell et al. force field with improved sugar pucker phases and helical repeat. J Biomol Struct Dynam 1999:16:845–862.

64. Srinivasan J, Cheatham TE, Cieplak P, Kollman PA, Case DA. Continuum solvent studies of the stability of DNA, RNA, and phosphoramidate-DNA helices. J Am Chem Soc 1998; 120:9401–9409.

65. Jayaram B, Sprous D, Young MA, Bereridge DL. Free energy analysis of the conformational preferences of A and B forms of DNA in solution. J Am Chem Soc 1998; 120: 10629–10633.

66. Cheatham TE, Srinivasan J, Case DA, Kollman PA. Molecular dynamics and continumm solvent studies of the stability of poly-G-polyC and polyA-polyT DNA duplexes in solution. J Biomol Struct Dyn 1998; 16:265–280.

67. Elizondo-Riojas M-A, Kozelka J. Unrestrained 5 ns molecular dynamics simulation of a cisplatin-DNA 1,2-GG adduct provides a rationale for the NMR features and reveals increased conformational flexibility at the platinum binding site. J Mol Biol 2001; 314:1227–1243.

68. Spector TI, Cheatham TE, Kollman PA. Unrestrained molecular dynamics of photodamaged DNA in aqueous solution. J Am Chem Soc 1997; 119:7095–7104.

69. Xie XM, Geacintov NE, Broyde S. Origins of conformational differences between *cis* and *trans* DNA adducts derived from enantiomeric *anti*-benzo[a]pyrene diol epoxides. Chem Res Toxicol 1999; 12:597–609.

70. Xie XM, Geacintov NE, Broyde S. Stereochemical origin of opposite orientations in DNA adducts derived from enantiomeric anti-benzo[a]pyrene diol epoxides with different tumorigenic potentials. Biochemistry 1999; 38:2956–2968.

71. Tan J, Geacintov NE, Broyde S. Conformational determines of structures in stereoisomeric *cis*-opened *anti*-benzo[a]pyrene diol epoxide adducts to adenine in DNA. Chem Res Toxicol 2000; 13:811–822.

72. Tan J, Geacintov NE, Broyde S. Principles governing conformations in stereoisomeric adducts of bay region benzo[a]pyrene diol epoxides to adenine in DNA: Steric and hydrophobic effects are dominant. J Am Chem Soc 2000; 122:3021–3032.

73. Cosman M, de los Santos C, Fiala R, Hingerty BE, Ibanez V, Luna E, Harvey R, Geacintov NE, Broyde S, Patel DJ. Solution conformation of the (+)-*cis-anti*-[BP]dG adduct in a DNA duplex: intercalation of the covalently attached benzo[a]pyrenyl ring into the helix and displacement of the modified deoxyguanosine. Biochemistry 1993; 32:4145–4155.

74. Cosman M, Hingerty BE, Luneva NP, Amin S, Geacintov NE, Broyde S, Patel DJ. Solution conformation of the (−)-*cis-anti*-benzo[a]pyrenyl-dG adduct opposite dC in a DNA duplex:

intercalation of the covalently attached BP ring in to the helix with base displacement of the modified deoxyguanosine into the major groove. Biochemistry 1996; 35:9850–9863.

75. Cosman M, Fiala R, Hingerty BE, Laryea A, Lee H, Harvey RG, Amin S, Geacintov NE, Broyde S, Patel DJ. Solution conformation of the (+)-*trans-anti*-[BPh]dA adduct opposite dT in a DNA duplex: intercalation of the covalently attached benzo[c]phenan-threne to the 5′-side of the adduct site without disruption of the modified base pair. Biochemistry 1993; 32:12488–12497.

76. Cosman M, Laryea A, Fiala R, Hingerty BE, Amin S, Geacintov NE, Broyde S, Patel DJ. Solution conformation of the (−)-*trans-anti*-benzo[c]phenanthrene-dA ([BPh]dA) adduct opposite dT in a DNA duplex: intercalation of the covalently attached benzo[c]-phenanthrenyl ring to the 3′-side of the adduct site and comparison with the (+)-*trans-anti*-[BPh]dA opposite dT stereoisomer. Biochemistry 1995; 34:1295–1307.

77. Feng B, Zhou L, Passarelli M, Harris CM, Harris TM, Stone MP. Major groove (R)-alpha-(N6-adenyl)styrene oxide adducts in an oligodeoxynucleotide containing the human N-ras codon 61 sequence: conformations of the R(61,2) and R (61,3) sequence isomers from 1H NMR. Biochemistry 1995; 34:14021–14036.

78. Feng B, Voehler M, Zhou L, Passarelli M, Harris CM, Harris TM, Stone MP. Major groove (S)-alpha-(N6-adenyl)styrene oxide adducts in an oligodeoxynucleotide contain-ing the human N-ras codon 61 sequences: conformations of the S(61,2) and S(61,3) sequence isomers from I H NMR. Biochemistry 1996; 35:7316–7329.

79. Wu M, Yan X, Tan J, Patel DJ, Geacintov NE, Broyde S. Coformational searches elucidate effects of stereochemistry on structures of deoxyadenosine covalently bound to tumorigenic metabolites of benzo[c]phenanthrene. Frontier in Bioscience 2004; 9: 2807–2818.

80. O'Handley SF, Sanford DG, Xu R, Lester CC, Hingerty BE, Broyde S, Krugh TR. Structural characterization of an N-acetyl-2 aminofluorene (AAF) modified DNA oligo-mer by NMR, energy minimization, and molecular dynamics. Biochemistry 1993; 32: 2481–2497.

81. Marky LA, Breslauer KJ. Calculating thermodynamic data for transitions of any mole-cularity from equilibrium melting curves. Biopolymers 1987; 26:1601–1620.

82. Marky LA, Rentzeperis D, Luneva NP, Cosman M, Geacintov NE, Kupke DW. Differ-ential hydration thermodynamics of stereoisomeric DNA-Benzo[a]pyrene adducts derived from diol epoxide enantiomers with different tumorigenic potentials. J Am Chem Soc 1996; 118:3804–3810.

83. Gunz D, Hess MT, Naegeli H. Recognition of DNA adducts by human nucleotide exci-sion repair. Evidence for a thermodynamic probing mechanism. J Biol Chem 1996; 271:25089–25098.

84. Wu M, Yan S, Patel DJ, Geacintov NE, Broyde S. Cyclohexene ring and Fjord region twist inversion in stereoisomeric DNA adducts of enantiomeric benzo[c]phenanthrene diol epoxides. Chem Res Toxicol 2001; 14:1629–1642.

85. Hingerty BE, Figueroa S, Hayden TL, Broyde S. Prediction of DNA structure from sequence: a build-up technique. Biopolymers 1989; 28:1195–1222.

86. Saenger W. Principles of Nucleic Acid Structure. New York: Springer-Verlag, 1984: 17.

87. Hirschfeld FL. The structure of overcrowded compounds. Part VII. Out-of-plane defor-mation in benzo[c]phenanthrene and 1,12-dimethyl-benzo[c]phenanthrene. J Chem Soc 1963; 11:2126–2135.

88. Kaufman-Katz A, Carrell HL, Glusker JP. Dibenzo[a,l]pyrene (dibenzo[def,p]chrysene): fjord-region distortions. Carcinogenesis 1998; 19:1641–1648.

89. Guo J, Geacintov NE, Broyde S. Structural and thermodynamic studies of dibenso[a,l]-pyrene diol epoxide DNA adducts. American Association for Cancer Research, Proceedings of the 94th Annual Meeting, Washington, D.C., 2003; 44:450.

90. Cavalieri EL, Higginbotham S, RamaKrishna NVS, Devanesan PD, Todorovic R, Rogan EG, Salmasi S. Comparative dose-response tumorigenicity studies of dibenz[a,l]-pyrene versus 7,12-dimethylbenzanthracene, benzo[a]pyrene and two dibenzo [a,l]pyrene dihydrodiols in mouse skin and rat mammary gland. Carcinogenesis 1991; 12:1939–1944.

91. Amin S, Krzeminski J, Rivenson A, Kurtzke C, Hecht SS, El-Bayoumi K. Mammary carcinogenicity in female CD rate of fjord region diol epoxides of benzo[c]phenanthrene, benzo[g]chrysene and dibenzo[a]pyrene. Carcinogenesis 1995; 16:1971–1974.

92. Amin S, Desai D, Hecht SS. Tumor initiating activity on mouse skin of bay region diol epoxides of 5,6-dimethylchrysene and benzo[c]phenanthrene. Carcinogenesis 1993; 14: 2033–2037.

93. Amin S, Desai D, Dai W, Harvey RG, Hecht SS. Tumorigenicity in newborn mice of fjord region and other sterically hindered diol epoxides of benzo[g]chrysene, dibenzo[a,1]-pyrene(dibenzo[def,p]chrysene), 4H-cycloppenta[def]chrysene, and fluoranthere. Carcinogenesis 1995; 16:2813–2817.

94. Hecht SS, El-Bayoumi K, Rivenson A, Amin S. Potent mammary carcinogenicity in female CD rats of a fjord region diol epoxide to benzo[c]phenanthrene compared to a bay region diol-epoxide of benzo[a]pyrene. Cancer Res 1994; 54:21–24.

Part III

Non-Bulky Base Damage

This section deals with the recognition and repair of non-bulky base damage in both prokaryotes and eukaryotes. Non-bulky base damage is generally the result of oxidation by reactive oxygen species, deamination of exocyclic amino groups by deaminating agents, and the addition of small alkyl groups to DNA bases by alkylating agents. In general, this DNA damage does not generate substantial changes to the overall DNA structure; however, significant structural perturbations in the vicinity of the damage site are observed. These non-bulky base damages are generally recognized by DNA glycosylases.

Chapter 14 provides a comparative treatment of the structures and functions of DNA glycosylases and AP endonucleases.

Chapter 15 provides an overview of the repair of oxidized DNA bases and the pathways involved in the repair of oxidized bases.

The repair of alkylation damage is exhaustively treated in Chapter 16.

Deamination of DNA bases can occur spontaneously and is increased substantially by deaminating agents. The formation and repair of deaminated bases in DNA are considered in Chapter 17.

Chapter 18 provides an exciting paradigm, suggesting the possibility of increasing the efficiency of DNA repair via a replication-coupled repair process.

The AP site is the most common spontaneous DNA damage location. The lesion is formed from the loss of DNA bases. It is also a repair intermediate of DNA glycosylases. Chapter 19 provides an overview on the recognition and repair of this most common DNA damage site.

Chapter 20 reviews the current understanding of the enzymes and pathways involved in the repair of mitochondrial DNA damage.

14

Structural Features of DNA Glycosylases and AP Endonucleases

Joy L. Huffman, Ottar Sundheim, and John A. Tainer
Department of Molecular Biology—MB4, Skaggs Institute for Chemical Biology and The Scripps Research Institute, La Jolla, California, U.S.A.

1. THE BASE EXCISION REPAIR PATHWAY

The most common and perhaps the most important type of DNA damage is base damage, which occurs at the rate of several thousand base pairs per cell per day in humans (1). This damage is primarily caused by endogenous metabolic and immune processes rather than environmental toxins, with the exception of UV damage to the skin from sunlight and oxidative damage to lung and blood from cigarette smoke (2,3). While the repair of DNA base damage can be complex, most such damage is prevented from becoming mutagenic or toxic by the base excision repair (BER) pathway. All base damage repair is initiated and, in special cases, completed by proteins that specifically recognize the damaged bases and either remove them to begin BER or repair them by direct damage reversal. Three-dimensional crystal structures of representatives of all major components of the BER pathway have been determined. Here, we review the enzymes that act in the initiation of BER and the proteins that complete the direct damage reversal process. The BER involves the remarkably specific detection and removal of damaged bases in the context of an enormous background of normal DNA, followed by DNA backbone cleavage. In the damage-general steps of BER that follow, DNA polymerases and ligases complete repair by synthesizing new DNA and rejoining the deoxyribonucleotide phosphate backbone. The elucidation of the initial steps of damaged base repair have provided critical insights into protein–DNA interactions and chemistry with broad and profound impacts on our understanding of biochemistry, cell biology, and life itself. For example, the DNA glycosylases that initiate BER use unique DNA-binding motifs, flip damaged nucleotides 180° into damage-specific pockets, and initiate a choreographed and coordinated handoff of damaged DNA intermediates to downstream pathway components. As will be discussed in the following sections, experimental characterizations of DNA base repair processes are providing an integrated understanding of structural cell biology at escalating levels of complexity, from DNA base damage to protein–DNA complexes to dynamically assembled macromolecular machines.

2. DNA GLYCOSYLASE STRUCTURAL FAMILIES

Crystal structures have been determined for a number of DNA glycosylases, allowing for classification into structural families by architectural folds (Fig. 1). A full
understanding of individual family members requires detailed structural and biochemical analysis, particularly of enzyme:DNA complexes, as overall folds do not

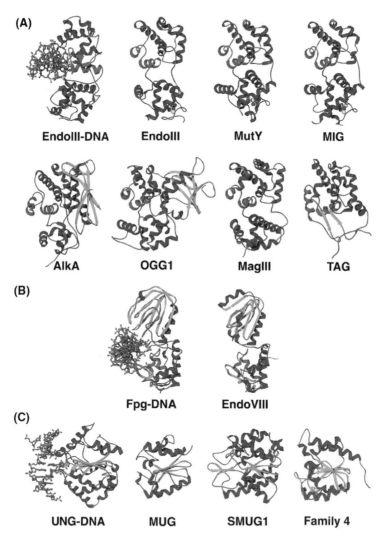

Figure 1   (A) Structures of selected HhH superfamily members. HhH motifs are colored red,
α helices are blue, and β strands are green. The EndoIII–DNA complex is rotated 90° about
the Z-axis relative to the other structures to illustrate DNA binding in the cleft between the
N- and C-terminal domains. Note that EndoIII, MutY, and MIG each contain an iron–sulfur
cluster, whereas AlkA and OGG1 have additional β sheet-containing domains. TAG displays
only limited structural homology with the other HhH family members outside the HhH motif.
Not pictured due to space limitations are MBD4 and T4 Endonuclease V. (B) H2TH family
members Fpg, bound to DNA, and EndoVIII. The H2TH motif is colored red and the
N-terminal six amino acids are colored purple. (C) Representative structures for the four
UDG family members. (*See color insert.*)

provide the mechanistic details of the varied specificity of these enzymes. In addition, each enzyme is regulated and targeted differentially, and the structural basis for these differences must lie in subtle changes in protein surfaces and protein–DNA conformations. While an exhaustive list of known glycosylase structures is not provided here, specific examples of the major folds (helix–hairpin–helix, helix–two turn–helix, and uracil DNA glycosylase, UDG) will be followed by discussion of specific lesion recognition (resulting from oxidation, deamination, or alkylation). Those folds that are not widely represented or found in only one or two enzymes will be examined according to the damage recognized. Finally, structural aspects of two classes of AP (apurinic/apyrimidinic) endonucleases are outlined.

A mechanistic distinction can be made between monofunctional DNA glycosylases and bifunctional DNA glycosylases/AP lyases. Bifunctional DNA glycosylases process the abasic site with an AP lyase activity inherent to the glycosylase itself. Monofunctional DNA glycosylases protect the abasic site until acted upon by an AP endonuclease. In both cases, the resulting strand break requires further processing by other proteins (lyases and/or nucleases) to remove the sugar–phosphate residue remaining at the 3′ or 5′ end, respectively. Repair is completed by the concerted actions of a DNA polymerase to fill the gap and a DNA ligase to seal the strand (reviewed in Refs. 4–6). Monofunctional glycosylases typically use an activated water as a nucleophile in attacking the C1′ of the target nucleotide, whereas bifunctional glycosylases/AP lyases often use a lysine side chain or an N-terminal proline (7). An intermediate step in the mechanism for AP lyase activity in the bifunctional enzymes is formation of a Schiff base between the nucleophilic lysine or proline and C1′ of the sugar. This Schiff base can be chemically reduced to form a covalently "trapped" complex resembling the Schiff base intermediate (8). This trapping reaction has been exploited for determination of several enzyme–DNA complex crystal structures, which have provided much insight into the mechanisms of DNA recognition and AP lyase activity for the bifunctional enzymes (9–13). This covalent enzyme–DNA trapping reaction is impossible with monofunctional DNA glycosylases that use an activated water as the attacking nucleophile, but has been observed in select cases where protein residues act as nucleophiles (14).

Despite differences in the folds and specific residues used to recognize damaged bases, several common themes for BER initiation have emerged. Among these, extrahelical flipping of the damaged base into a lesion-specific recognition pocket is particularly intriguing, as it must rely on an intrinsic property of the damaged DNA. Because all DNA glycosylases studied to date kink DNA at the site of damage and flip the lesion base out of the helix, an initial step in recognition must exploit the deformability of the DNA at a base pair destabilized by the presence of the lesion. Because each glycosylase is necessarily damage-specific, only bases that can be accommodated in a defined binding pocket upon flipping provide the necessary contacts and orientation for base excision. The focus of this chapter will be on overall folds and specific mechanisms for damage recognition employed by DNA glycosylases.

2.1. The Helix–Hairpin–Helix Motif

The HhH motif was first discovered in EndoIII (EndonucleaseIII or Nth) (15) as a sequence-independent DNA-binding motif. The HhH motif is a major motif for sequence-independent DNA-binding proteins that is present in a superfamily of glycosylases, including EndoIII, AlkA (3-methyladenine DNA glycosylase II), and

Ogg1 (8-oxoguanine glycosylase), which remove a broad spectrum of oxidized and alkylated base lesions (15–18). Structures of at least eight different HhH-containing DNA glycosylases have been determined (Fig. 1A) (16,18–24). Structural studies on bacterial EndoIII, AlkA, and human Ogg1 (hOGG1) in complex with DNA have shown that the HhH motifs participate in DNA recognition through interaction with phosphate and oxygen atoms of the DNA backbone (10,18,25,26). The HhH motif has also been found in a number of other proteins that bind DNA in a sequence-independent manner, such as DNA Polymerase β and NAD^+-dependent DNA ligase (27). The core fold of these enzymes is comprised of two α-helical N- and C-terminal domains. The N-terminal domain typically has four α helices, and the C-terminal domain has 6–7 α helices. A number of these point their helical N-termini and thus positively charged dipoles toward the DNA-binding site, located at the cleft between the two domains. Cocrystal structures have revealed that the DNA is bent 60–70° at the lesion site by HhH-containing enzymes.

The HhH motif, also found near the cleft, is composed of two α helices that cross at a conserved angle and are linked by a Type II β hairpin. Specificity of base removal strongly correlates with the amino acid sequence within this motif. In the HhH-containing glycosylases of known structure, the hairpin loop shows strong sequence conservation, with consensus sequence L/F-P/K/H-G-V/I-G-K/R/T (27). A conserved aspartate is also found ~20 residues C-terminally that is proposed to activate the nucleophile for attack of the scissile glycosylic bond. TAG (3-methyl-adenine glycosylase) is an HhH protein (24), yet this was an unexpected feature due to low sequence homology with other HhH enzymes of known structure.

The HhH DNA glycosylase structures show a variety of small additions, such as an $[Fe_4–S_4]$ iron sulfur cluster in EndoIII, MutY (adenine DNA glycosylase, ADG), and MIG (thymine DNA glycosylase, TDG), a β sheet in AlkA and hOGG1, a zinc-binding domain in TAG, and a methyl-CpG-binding domain in MBD4 (methyl-CpG-binding domain protein 4). The iron–sulfur cluster found in EndoIII-like enzymes is involved in DNA binding (10). The loop extending from the iron sulfur cluster protrudes into the minor groove of DNA and interacts with the HhH motif in DNA binding and damage recognition. Similar types of interactions involving different loop structures have been reported in other DNA glycosylases; i.e., the asparagine loop of hOGG1 or the leucine wedge loop of AlkA, which intercalate into the minor groove of DNA (18,26).

2.2. The Helix–Two Turn–Helix Motif

The helix–multi-turn–helix motif was first discovered in the flap endonuclease FEN-1 structure (28), but also occurs in several DNA glycosylases as a prevalent helix–two turn–helix (H2TH) motif. Family members identified to date include bacterial EndoVIII (Endonuclease VIII or Nei), MutM, Fpg, and mammalian Nei-like proteins (NEIL1, NEIL2, and NEIL3), and representative structures have been determined for all but the mammalian NEIL proteins (Fig. 1B) (9,12,13,29–33). Interestingly, these enzymes catalyze similar mechanisms of base removal and backbone cleavage as the HhH enzymes, though they use a completely different molecular scaffold. The overall topology of these enzymes is conserved across the family. Similar to the HhH proteins, N- and C-terminal domains create a cleft where DNA is bound, but the H2TH proteins contain α/β structures rather than all α helices. The N-terminal domain has conserved amino acids at positions 1–6, followed by two four-stranded β sheets that form an antiparallel β sandwich flanked by two helices. The C-terminal domain contains the H2TH motif and is helix-rich, with the zinc finger contributing the only two β strands. The β-hairpin loop of the zinc finger motif intercalates into the minor groove of DNA. Positively charged residues line the surface of the cleft to create an electrostatically positive surface for DNA binding, rather than the helix dipoles used by the HhH enzymes. The H2TH motifs are used in a manner similar to the HhH, namely recognizing DNA through interactions with the backbone.

2.3. Uracil DNA Glycosylases

The UDGs comprise a prominent and highly important glycosylase structural super-family, as UDGs are the major repair enzymes that recognize and repair uracil resulting from both misincorporation of dUTP and cytosine deamination in DNA. Four distinct families have been identified to date. Although the families share limited sequence similarity, structures have revealed that they possess a common core fold (Fig. 1C). Family 1, composed of UNG (uracil DNA *N*-glycosylase) and its close orthologs, are highly conserved DNA glycosylases present in most living organisms examined. Family 2 enzymes, the bacterial MUG (mismatch-specific UDG) and the eukaryotic homolog TDG, initiate BER of G:U/T mismatches. Single-strand-selective monofunctional UDG (SMUG1) comprises the third enzyme class in the UDG superfamily. The fourth family was first discovered in the thermophilic bacterium *Thermatoga maritima* and is the only UDG that possesses an iron sulfur cluster.

The topology of the common core of the UDG superfamily consists of a central four-stranded parallel twisted β sheet flanked by α helices. The β sheet in MUG is extended, with one extra strand oriented in an *anti*-parallel direction at the edge of the sheet. A positive DNA-binding groove traverses one face of the molecule, where the C-terminal ends of the sheet form the base of the cleft. The uracil-binding pocket penetrates back from the groove into the core of the enzyme. Unlike MUG, UNG and SMUG1 have an additional small β sheet made up of two and three strands, respectively, located on the larger lobe of the DNA-binding cleft. MUG also lacks the coil of helices at the N-terminal end present in both UNG and SMUG1. A short helix immediately follows strand β2 in MUG and SMUG1, and a ~40 residue segment leading to helix 5 is unique to the SMUG1 fold.

3. SPECIFIC MECHANISMS FOR RECOGNITION OF DAMAGE

3.1. Oxidation Damage

Oxidation of cellular macromolecules occurs at significant frequencies in aerobic organisms due to byproducts of normal metabolism and the immune response. Furthermore, the oxidation of both mitochondrial and nuclear DNA has been implicated in human disease and aging. Strand breaks, abasic sites, and oxidized base residues, with 7,8-dihydro-8-oxoguanine (8-oxoG) and 5-hydroxycytosine (5-OHC) representing the most frequent mutagenic base lesions, can all be caused by oxidative damage.

DNA glycosylases that remove oxidized base residues can be divided into two functional subgroups: those that repair oxidized purines [e.g. *E. coli* Fpg (formamidopyrimidine DNA glycosylase)] and those that repair oxidized pyrimidines [e.g. *E. coli* EndoIII (Endonuclease III or Nth) and Endo VIII (Endonuclease VIII or

Nei)]. In human cells, five DNA glycosylases for removal of oxidized bases have been cloned and characterized: hNth1 (human EndoIII), hOGG1 (human Ogg1), NEIL1, NEIL2, and NEIL3 (Nei-like) (32,34–38). hNth1, NEIL1, and NEIL2 catalyze excision of oxidized pyrimidines, such as 5-OHC, whereas hOGG1 repairs oxidized purines, such as 8-oxoG. However, NEIL1 also appears to be an alternative glyco-sylase for the removal of 8-oxoG. All five enzymes remove imidazole ring-fragmen-ted faPy (formamidopyrimidine) residues, which block replication and are cytotoxic.

Glycosylases that repair oxidative damage fall into two structural families, determined by the presence of either an HhH or an H2TH motif. Representative structures have been determined for both families. *E. coli* EndoIII was one of the first DNA repair protein structures elucidated (17), although cocrystal structures with DNA were not determined until very recently (10). Other HhH structures for oxidative damage-sensing glycosylases have included hOGG1 and MutY (18,23,39,40), and those carrying the H2TH motif include bacterial MutM, Fpg, and EndoVIII (9,12,13,29–31). Structures determined in the presence of damaged bases have provided invaluable details regarding specific lesion recognition, and these will be discussed briefly in the following sections.

3.1.1. hOGG1

Several distinct enzymes recognize 8-oxoG in different contexts because it is evi-dently a major mutagenic base lesion, pairing preferentially with A rather than C and thereby causing transversion G:C to T:A mutations upon replication (1). The structure of a catalytically inactive hOGG1 enzyme core bound to 8-oxoG-containing duplex DNA revealed that hOGG1 is a HhH family member and that it flips the 8-oxoG base out of the double helix into a specific recognition pocket (18). hOGG1 discriminates 8-oxoG from G using a single hydrogen bond between the Gly42 carbonyl and the purine N7, which is protonated only in 8-oxoG, and no direct contacts are made to the 8-oxo moiety (Fig. 2A). hOGG1 is a bifunctional glycosylase/AP lyase, and Lys249 has been proposed to attack C1′ and promote β-elimination. The base opposite 8-oxoG is preferentially $C > T > G >> A$, and the preference for C was shown to result from specific hydrogen bonds donated by Arg154 and Arg204 to N_3 and O_2 of the cytosine base. A structure of apo hOGG1 showed that, in the absence of DNA, the overall enzyme conformation is conserved but that key catalytic residues, such as Lys249, are positioned improperly for catalysis (39). Binding of the correct substrate is proposed to be coupled to reorientation of these side chains and subsequent catalysis.

3.1.2. MutY and MutT

As protection against 8-oxoG lesions in *E. coli*, three proteins work to avoid muta-tion by this deleterious base. MutM recognizes 8-oxoG:C pairs and excises the 8-oxoG base, MutY recognizes 8-oxoG:A pairs and excises the A, and MutT recog-nizes free 8-oxoG deoxyribonuclceotide triphosphates and catalyzes the removal of pyrophosphate, thus preventing misincorporation of this base into DNA (41). After MutY-initiated repair of the adenine, which likely results from misincorporation during replication, cytosine can be inserted opposite 8-oxoG, and MutM can initiate repair of the 8-oxoG, thus avoiding propagation of a GC to TA mutation upon further replication. MutY is a monofunctional HhH family member with an iron sul-fur cluster, and structures have been determined for: (i) a catalytically active MutY core domain alone (23), (ii) an inactive mutant alone and in complex with free

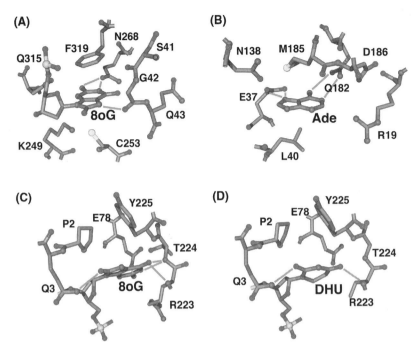

Figure 2 Specific recognition of oxidized bases. (A) Recognition of 8oxoG by hOGG1. (B) The adenine-binding pocket in MutY. (C and D) MutM specifically recognizes both 8oxoG (C) and DHU (D).

adenine, which is an inhibitory product (23), and (iii) other regulatory and catalytic mutants of the core domain (40). In the adenine-bound structure, the base is bound in a deep pocket surrounded by a large positively charged groove, ideal for DNA binding (Fig. 2B). The position of the adenine is such that it is likely to be flipped out of the double helix. MutY makes specific hydrogen bonds similar to Watson–Crick base pairing using Gln182 and Glu37 that discriminate against other bases in the adenine-binding pocket.

In each of the MutY structures determined to date, the C-terminal domain of MutY is missing. Interestingly, the C-terminal domain shows primary sequence homology to MutT, which binds free 8-oxo-dGTP and whose structure is known (42–44). This similarity has led to speculation that the C-terminal domain specifically recognizes the 8-oxoG base, and biochemical evidence supports a double base-flipping mechanism for recognition of the 8-oxoG:A pair (45).

3.1.3. MutM

MutM is a functional homolog of Ogg1 in *E. coli* and other prokaryotes in that it catalyzes the removal of 8-oxoG opposite C, but MutM binds DNA with a H2TH motif rather than an HhH motif. In structures of the H2TH BER enzymes, which have thus far all been bifunctional DNA glycosylases/AP lyases that recognize oxidized base lesions, an N-terminal proline has been identified as catalyzing base excision and β-elimination, while an internal lysine and arginine have been proposed to promote an additional activity resulting in a final β,δ-elimination product (12,13). Structures have been determined for several members of this family with and without

DNA, and these have revealed the mechanisms for recognition of both the damaged base and its opposite paired base (9,12,13,29–31). For example, recent structures of an inactive MutM mutant with lesion-containing DNA have shed light on recognition of oxidized bases (31), particularly in comparison to the hOGG1:8-oxoG structure. MutM initiates repair of 8-oxoG, DHU (5,6-dihydroxyuracil), and a number of other lesions. As in hOGG1, MutM makes its discriminating contacts with N7 of 8-oxoG, which is contacted by the main chain carbonyl of Ser220 in MutM (Fig. 2C). The structure of MutM with DNA-containing DHU, a non-aromatic pyrimidine ring that is also a substrate for MutM, revealed that the protonated N3 overlays with N7 of 8-oxoG and that the DHU carbonyls O2 and O4 are in identical positions to those of 8-oxoG (O8 and O6, respectively) (compare Fig. 2C and D). Structures of MutM with reduced abasic DNA and C, G, or T opposite have shown that a single residue, Arg112, specifies the preference for the opposing base by enforcing a hydrogen bonding pattern in which C is accommodated in the normal *anti* position, G takes the less favorable *syn* position, presenting its Hoogstein face, and T is pushed away from the Arg112 side chain but is still found in the *anti* position (9).

3.2. Deamination Products

Uracils occur in DNA at a frequency of 100–500 per cell per day (1), either by misincorporation or by spontaneous deamination of cytosine (46,47). Uracil can induce CG to TA transition mutations, as it pairs preferentially with adenine during replication. The UDG superfamily of glycosylases is comprised of four enzyme families, classified by primary sequence homology. Family 1 UNG enzymes are ubiquitous in most organisms, including bacteria, budding yeast, plants, and many eukaryotes (and their viruses), but not in fission yeast or insects. Family 2 enzymes are widespread in mammals (TDG or TDG), archaea, and some eubacterial species (MUG or mismatch-specific UDG). The SMUG1 (single-strand-selective monofunctional UDG 1) enzymes of family 3 are limited to insects, amphibians, and mammals. UDG family 4 homologs are present in a range of thermophilic and mesophilic eubacteria and archaea. The human glycosylases that remove uracil are uracil–DNA glycosylase (UNG/UDG), TDG, single-strand-selective monofunctional uracil–DNA glycosylase (SMUG1), and methyl-CpG-binding endonuclease 1 (MED1/MBD4).

3.2.1. UNG

The UNG initiates base excision repair by hydrolyzing the N-C1′ glycosylic bond of uridine in DNA to yield free uracil and an abasic site. UNG enzymes are able to excise uracil from single-stranded DNA and from double-stranded DNA, regardless of whether the opposite base is a G or an A. UNG enzymes are highly specific for uracil in DNA, with negligible activity against T or C or naturally occurring uracil in RNA. Other substrates reported for UNG are largely restricted to uracil analogs with minor modifications at the 5 position. In humans, UNG activity is targeted to both nuclei and mitochondria through alternative splicing of the same gene (48). Unique N-terminal regions determine the subcellular localization, but the catalytic domain, whose structure is known, is the same for both nuclear and mitochondrial forms.

Five conserved motifs are present in all crystal structures determined of UNG (49–52): (1) the water-activating loop (144-GQDPYH-148; human); (2) the 5′ side backbone compression loop (165-PPPPS-169); (3) the uracil recognition region (199-GVLLLN-204); (4) the 3' side backbone compression loop (246-GS-247); and (5) the minor groove-intercalation loop (268-HPSPLS-273) (53). UNG binds to DNA using three rigid loops made up of motifs 2, 4, and 5. These three loops largely consist of serine, proline, and glycine residues, which permit a close approach of the polypeptide chain to the DNA backbone. The loops compress the backbone (pinch) and slightly bend the DNA, which becomes fully bent (~45°) and kinked (~2°) when a push from the minor groove intercalation loop and a pull from the complementary uracil-specific recognition pocket flip the uridine into an extrahelical position (53,54).

The highly conserved substrate-binding pocket provides the shape and electrostatic complementarity to fit uracil in an extrahelical conformation but is too narrow to accommodate purines (Fig. 3A). Selection against thymine and 5-methylated pyrimidines is provided by the side chain of a tyrosine (Tyr147 in human UNG). Specific hydrogen bonds provide discrimination against cytosine. The O2 carbonyl of uracil makes a hydrogen bond to the UDG main chain NH that joins a conserved Gly-Gln sequence (Gly143-Gln144). The amide side chain of a conserved asparagine (Asn204) makes specific hydrogen bonds to N3 and O4 of uracil, whereas cytosine is excluded by unfavorable interactions with its exocyclic amine N4. A water cluster at the base of the uracil-binding pocket provides interactions that fix the proper orientation of the amide group (55).

3.2.2. TDG/MUG

Human TDG was first discovered for its ability to hydrolyze the *N*-glycosidic bond in T:G mismatches (56). The TDG was later shown to remove thymine from C:T and T:T mismatches, but much less efficiently. More importantly, it removes uracil mispaired with guanine with ~10-fold higher activity than thymine (reviewed in Ref. 57). MUG (mismatch UDG), the bacterial homolog of human TDG, only removes uracil and not thymine mismatched with guanine (58). *E.coli* MUG is closely related to TDG (37% sequence identity), and they are thought to possess the same fold. The cyclic alkylation product ethenocytosine (εC) and 5-fluorouracil (5-FU) are both efficiently removed by TDG/MUG enzymes. Removal of εC might be a major function of TDG in human cells (57,59,60). Remarkably, TDG can also remove 5-FU from single-stranded DNA, which was unexpected, as TDG/MUG are double-strand specific for all other substrates tested. The crystal structures of MUG (55,60,61) reveal a similar overall fold to the family 1 UDGs (Fig. 1C). Two highly conserved motifs in

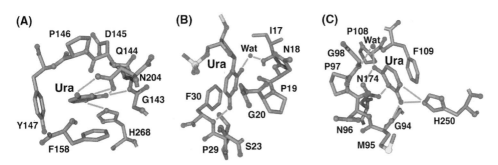

Figure 3 Uracil binds deep in a hydrophobic pocket in (A) UNG, (B) MUG, and (C) SMUG1.

UNG have topological and conformational equivalents in MUG: the water activating loop (GQDPY) and the minor groove-intercalating loop (HPSPLS). The corresponding motifs in MUG are 16-GINPGL-20 (identical in human TDG) and 140-NPSGLS-145 (MPSSS in human TDG). The latter motif forms specific hydrogen bonds with the orphan guanine and constitutes the basis of the mismatch specificity (61). The catalytic residues (underlined) of UNG are in both cases replaced by asparagines. The aspartate in the first motif in UNG activates a water molecule for nucleophilic attack on the C1′ of the deoxyribose. The asparagine in MUG cannot activate the nucleophilic water, but a water molecule is found in almost the same position as seen in UNG. The tyrosine residue that provides the barrier against thymine in UNG is replaced by a glycine (Gly20) in MUG (Fig. 3B). The preference for U:G over T:G in MUG is probably explained by the position of Ser23 hydroxyl group, which would clash with the 5′ methyl-group of thymine. In TDG, the residue corresponding to Ser23 is an alanine. The smaller alanine side chain allows better accommodation of thymine, explaining the expanded specificity of TDG for that base. The TDG/MUG specificity for G:U/T mispairs over G:C base pairs results not from the recognition of the scissile base itself, as in the family 1 enzymes, but rather from a combination of the ease of flipping out a base from an unstable pair over flipping from a Watson–Crick C:G pair and from the identity of the "widowed" base that remains.

3.2.3. SMUG1

When first discovered in *Xenopus laevis*, the xSMUG1 enzyme was characterized as single-strand-selective monofunctional UDG (62), but in the initial characterization of xSMUG1, the strong feedback inhibition of the abasic site was not taken into account. In fact, human SMUG1 removes uracil efficiently from both U:G mismatches and U:A base pairs (63). SMUG1, but not UNG, initiates BER of 5-hydroxymethyl-uracil lesions (64) generated in vivo by oxidation damage to thymine or oxidation and deamination of 5-methylcytosine.

In the crystal structure of xSMUG1 in complex with uracil containing double-stranded DNA, the enzyme had detached from the abasic product and rebound to the DNA end prior to crystallization (65). End binding was also observed with the substrate analog βFU (1-[2′-deoxy-2′-fluoro-β-D-arabinofuranosyl]-uracil). At the 5′ end of the damage-containing strand, a cytosine has an extrahelical conformation and points towards the pyrimidine specificity pocket of the xSMUG1. Upon replacing the 5′ end cytosine base with βFU, a mixed population of extrahelical cleaved abasic sites and βFU in a productive orientation in the active site was observed. Two motifs, the minor groove intercalation loop (251-PSPRN-255) and the short α helix unique to the SMUG1 family (256-PQANK-260) are inserted as a wedge into the DNA duplex, flipping the scissile nucleotide through the major groove. Penetration of both motifs into the base stack creates a more extensive disruption of the double-stranded DNA than seen for the other glycosylases in the UDG family. A conserved arginine residue, Arg254, occupies the gap left from the flipped-out base, whereas a proline from the unique α helix pushes into the base stack on the distal strand.

In the crystal structures, a second SMUG1 enzyme is bound to the 3′ end of the damage-containing strand. At this end, the base pairing remains intact and the active site is solvent accessible. Structures of the xSMUG1-dsDNA complex with free uracil (C5–H) and 5-hydroxymethyl-uracil (HmU) revealed the rather remarkable mechanism for achieving pyrimidine specificity in SMUG1 (Fig. 3C). The uracil

N_3 imino and O_4 carbonyl moieties hydrogen bond to the Asn174 side chain, and O_2 accepts hydrogen bonds from the Met95 main chain NH group and the imidazole ring of His250. This hydrogen bonding pattern implies that cytosine is rejected by SMUG1 in a manner analogous to that for UNG.

Rejection of thymine, however, is quite different in SMUG1 from the family 1 UDGs. The tyrosine that acts as the thymine barrier in UNG is replaced by a glycine (Gly98) in SMUG1. A well-ordered water molecule is found in place of the tyrosine side chain, which upon uracil binding retains a van der Waals contact with C_5 and a hydrogen bond to the O_4 carbonyl of the pyrimidine. Both lone-pairs of the water molecule accept hydrogen bonds from the NH groups of Gly98 and Met102, such that the water molecule specifically donates a hydrogen bond to the pyrimidine 4 position. This provides additional discrimination against cytosine, which has an amino group at this position. This tightly held water makes three hydrogen bonds in the absence of a base and would have to be displaced to accommodate a thymine in the binding pocket. HmU, however, is able to compensate for the energetic penalty of displacing the water molecule by binding in the same orientation and with the same hydrogen bonding patterns as uracil, with its hydroxyl group at the 5 position reinstating the hydrogen bonds for the water molecule.

3.2.4. Family 4 UDG

The first crystal structure of a family 4 UDG from *Thermus thermophilus* HB8 (TthUDG) was recently determined, confirming that this family possesses a very similar fold to the family 1 UDG enzymes (66). TthUDG contains an iron–sulfur cluster located distant from the putative DNA-binding site, which is more likely related to stability than to DNA binding or catalysis. The structure of TthUDG in complex with uracil reveals that family 4 enzymes recognize uracil in both double- and single-stranded DNA and discriminate against cytosine in the same manner as UNG family 1 enzymes. The thymine barrier contributed by the side chain of Tyr147 in human UNG is replaced by a conserved glutamate (Glu47) in the same position, which is thought to play an analogous role.

3.2.5. Mismatch Repair Glycosylases—MIG and MBD4

Deamination of 5-methylcytosine to thymine is more rapid than deamination of cytosine to uracil (1), explaining the existence of specific G:T mismatch repair glycosylases, such as MIG and MBD4. Both MIG and MBD4 are HhH-containing glycosylases with closer structural homology to EndoIII than to MUG, despite sharing similar substrates. MIG recognizes G:T mispairs and removes the thymine base (21), whereas MBD4 recognizes G:T preferentially but will also excise uracil from G:U mispairs (67). A mechanism for G:T mismatch recognition and glycosylic bond cleavage has been proposed that is consistent with structural analysis, complementary biochemistry, and characterization of key site-directed mutants in which MIG bond cleavage is enhanced by a physical distortion of the nucleotide that imparts a ~90° twist to the thymine base, away from its normal *anti* position in DNA (21), similar to a model proposed for UDG (68).

MBD4 is comprised of two DNA-binding domains: a G:T mismatch-specific DNA glycosylase and a methyl-CpG-binding domain. This apparent fusion of functions results in an enzyme with a preference for G:T mispairs in a CpG context. As methyl-CpG steps often occur in clusters, this may lead to a local increase in repair enzyme concentration for damage sensing. NMR structures have been elucidated for domains homologous to the methyl-CpG domain (69–71) and for the MBD4 HhH-containing glcosylase domain (72), raising questions regarding how DNA might be bound by both domains simultaneously, particularly if the HhH-containing glycosylase domain bends DNA as significantly as the other HhH family members.

3.2.6. dUTPases

As an alternative to BER, nucleotide pool "sanitizing" enzymes have been discovered that remove improper bases to prevent their misincorporation into DNA. Among these, dUTPases are of particular importance, as dUTP can be misincorporated opposite an A during replication but is often repaired to C, introducing TA to CG mutations. Structures of several of these ubiquitous enzymes have been determined, revealing a highly conserved fold and dUTP-binding pocket across kingdoms, as has been observed in the UDG family, reflecting the ancient and essential nature of deamination damage repair (73–78).

3.3. Alkylation Damage

The most common form of nonenzymatic methylation of DNA likely results from physiological exposure to endogenous S-adenosyl methionine (SAM), which is found in the nucleus and participates in targeted enzymatic DNA methylation (1). The primary substrates for nonenzymatic methylation are ring nitrogens of purine residues, with 3-methyladenine (3mA) and 7-methylguanine (7mG) being the predominant lesions formed. 7mG does not alter base pairing with C, but 3mA blocks replication and is cytotoxic. Each of the alkylated bases bears a formal positive charge likely to be important for recognition. Four classes of enzymes initiate BER of alkylated bases, typified by the following enzymes: (1) *E. coli* TAG (3mA DNA glycosylase I), (2) *E. coli* AlkA (3mA DNA glycosylase II), (3) *Heliobacter pylori* MagIII (3mA DNA glycosylase III), and (4) human AAG (alkyladenine DNA glycosylase). TAG, AlkA, and MagIII are HhH-containing enzymes, whereas AAG has an unusual fold not seen in other BER enzymes. Because 3mA is so deleterious, each of the alkyl base glycosylases efficiently removes this lesion. AlkA and AAG have wide specificity for a broad spectrum of substrates, including deamination and cyclic etheno adduct products. TAG removes 3mA preferentially and 3mG with lower affinity, but not 7mG, and MagIII has highest specificity for 3mA.

     At least two mechanisms for repair of alkylation damage have been described that do not rely on BER. Alkylation damage can be removed directly, without further modification of the nucleotide or DNA, by proteins using either suicidal (AGT or Ada) or non-suicidal (AlkB) reactions. These proteins have no homology to the BER enzymes or to each other. The AGT is covalently modified in the process of repairing 06-alkylguanine lesions, and structures of the modified enzyme are discussed below (79–82). AlkB removes alkyl groups from 1-methyladenine and 3-methylcytosine (83,84). Based on sequence homology, the AlkB enzymes belong to a structural superfamily of 2-oxoglutarate and iron-dependent oxygenases (85), but their molecular structures have not yet been determined. Interestingly, AlkB also repairs methylation damage to A and C bases within single-stranded DNA and RNA (86).

3.3.1. *AlkA*

The apo structure of AlkA was among the first determined for a DNA glycoyslase and revealed an HhH-containing fold very similar to that of EndoIII (16,22). AlkA contains an additional α/β domain of unknown function with structural similarity to TBP, the TATA-binding protein, consisting of a five-stranded antiparallel β sheet with two flanking helices. The later structure of AlkA bound to DNA was the first such complex structure for an HhH-containing DNA glycosylase and revealed that contacts to the DNA are predominantly made to the lesion-containing strand (25). The DNA used in co-crystallization contained an abasic site analog, 1-azaribose, designed to mimic a transition state, which was flipped out of the DNA helix and into the active-site cleft. Leu125 lies at the tip of a loop between two helices that is positioned in the minor groove, such that Leu125 intercalates into the DNA helix like a wedge, distorting the base stacking near the flipped out abasic site and causing the DNA to bend ~66° (Fig. 4A). A sodium ion in the AlkA–DNA structure bridges the DNA and the HhH motif, strengthening the interaction similarly to the sodium/ potassium ion bound near HhH motifs in Pol β. Therefore, the HhH motif assists in anchoring the distorted DNA to the protein rather than directly flipping the substrate out of the helix.

AlkA has broad specificity for alkylated and oxidized bases with a delocalized positive charge on the base. It has been suggested that, in the absence of an activated water molecule, which was not found in the AlkA–DNA structure, Asp238—an aspartate conserved across the HhH family—might stabilize a carbocation intermediate. The positively charged base may not require protonation and would have an already weakened glycosylic bond.

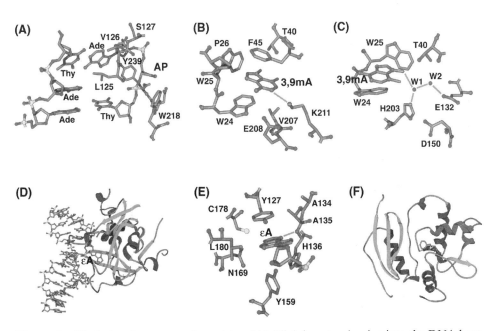

Figure 4 Alkylation damage repair proteins. (A) AlkA inserts a leucine into the DNA base stack to flip the AP site out of the helix. (B and C) MagIII specifically binds 3,9mA. (D) AAG flips εA out of the DNA helix into a specific recognition pocket. (E) Close-up view of AAG recognition of εA. (F) Ribbon representation of AGT with benzylated Cys145 shown as sticks. The HTH motif is colored red. (*See color insert.*)

3.3.2. MagIII

H. pylori MagIII has a similar topology to the EndoIII-like HhH enzymes but does not contain an iron–sulfur cluster or any additional domains (19). The major topological difference in MagIII is a longer helix in the C-terminal domain, which contributes to both a lysine carbamylation site (at Lys205) and the substrate-binding site. MagIII excises 3mA but not 7mG, and structures were determined for the protein alone and with 3,9mA (3,9-dimethyladenine) or εA (1,*N*-ethenoadenine) bound. 3,9mA has a positive charge and is therefore a good mimic of alkylated DNA base lesion substrates. The alkylbase-binding pocket is lined by hydrophobic residues (Trp24, Trp25, Pro26, and Phe45) and Lys211 (Fig. 4B). 3,9mA specifically donates a hydrogen bond from its 9-methyl group to a water in a network involving His203, Asp150, and Glu132, but direct contacts are not made to the 3-methyl moiety or the adenine N6 (Fig. 4C). 3mA is instead recognized by π–π stacking interactions between the base and nearby aromatic side chains. Modeling of 7mG based on the position of 3mA in the binding pocket revealed that steric hindrance likely precludes binding of that base.

3.3.3. AAG

The AAG is the only known human alkylbase DNA glycosylase, although other human enzymes exist that perform different types of alkylation damage repair, such as AGT, hABH2, and hABH3 (human AlkB homologs 2 and 3). The AAG is a structural outlier, with a topology unlike any of the other known BER glycosylases (26,87), consisting of a single α/β domain in which an antiparallel β sheet is surrounded by α helices (Fig. 4D). A β hairpin protrudes into the minor groove of DNA in co-crystal structures. A structure of AAG in complex with substrate εA-containing DNA revealed that the base is flipped out and inserted into a deep pocket, as occurs in the other structural families of DNA glycosylases. Alkylbases are specifically recognized using planar stacking and cation–π interactions by Tyr127, His136, and Tyr159, and the chemical instability of the glycosylic bond in alkylated nucleobases likely contributes to the catalytic specificity of AAG (Fig. 4E).

3.3.4. AGT and Ada

The AGT and Ada are homologous proteins that directly remove alkyl groups from the O6 position of guanine in a stoichiometric suicide reaction. Structures have been determined for bacterial, archaeal, and human AGT/Ada proteins (79–82). Human AGT is of particular interest because it repairs damage induced by some anticancer chemotherapeutics. The crystal structure of unreacted human AGT, as well as structures of the methylated and benzylated product complexes (79), revealed a two domain α/β fold and support a model for extrahelical alkylguanine nucleotide binding by AGT (Fig. 4F). The N-terminal domain consists of an antiparallel β sheet followed by two α helices. The C-terminal domain is comprised of a β hairpin, 4 α helices, and a 3_{10} helix, which harbors a conserved Pro-Cys-His-Arg motif. Human AGT also contains a novel zinc-binding site not seen in the bacterial or archaeal homologs that is likely to play a structural role. The C-terminal domain also contains a helix–turn–helix (HTH) motif, often used by DNA-binding proteins for sequence-specific recognition (88). The alkylated product structures, in which Cys145 has a covalently attached benzyl or methyl group, establish the active site as being near the recognition helix of the HTH motif. Alkylation of the active-site

AGT cysteine residue creates steric clash with the proposed DNA-binding region and is proposed to couple release of the repaired DNA with a destabilization of the AGT protein fold, thereby promoting the biological turnover of the alkylated protein.

4. AP ENDONUCLEASES

AP sites arise spontaneously due to the intrinsic instability of glycosyl bonds, with purines being lost roughly 20 times as fast as pyrimidines (1). AP endonucleases recognize and promote repair of AP sites generated either spontaneously or as the first step in BER. These enzymes process the products of both the monofunctional DNA glycosylases, which produce abasic sites, and the glycosylase/AP lyases, which cleave both the base—to generate an abasic site—and the phosphodiester DNA backbone—to produce a 5' phosphate and a 3' α β—unsaturated aldehyde group. AP sites and 3' aldehydes, like many of their damaged base predecessors, are often blocks to replication and may be mutagenic and cytotoxic if left unrepaired (89). Therefore, AP sites must be processed and handed off directly to the next enzyme in the BER repair pathway. There are two classes of AP endonucleases, typified by the *E. coli* enzymes ExoIII (Exonuclease III) and EndoIV (Endonuclease IV) (90–92). In bacteria, ExoIII is the major AP endonuclease, accounting for approximately 90% of the activity measured in crude cell extracts (93,94). EndoIV was discovered as a residual AP activity present in ExoIII-deficient *E. coli* cells (93–95). In humans, the major abasic endonuclease is AP endonuclease 1 (APE1), an ExoIII homolog. Mouse knock-outs of APE1 are embryonic lethal (96), underscoring the importance of AP endonucleases in development.

4.1. APE1

Both ExoIII and APE1 possess 5' AP endonuclease, 3'–5' double-stranded exonuclease, 3'-phosphodiesterase, and 3'-phosphatase activities, but with significantly different efficiencies. In addition to its catalytic roles, APE1 also coordinates both BER and the regulation of gene expression. For example, APE1 interacts directly with a number of DNA repair proteins, including Pol β, XRCC1, PCNA, FEN1, and DNA Ligase I (97–100). Since APE1 links the damage-specific DNA glycosylases to the damage-general BER enzymes, it is probable that conformational controls act in pathway coordination by aiding in the assembly of key DNA base repair components. APE1 also has multiple functions in the cell that are distinct from its AP endonuclease activity. It is also called Ref-1 and acts as a redox activator of many DNA-binding proteins, including c-jun, p53, Pax 5, and Pax 8 (101–104).

The crystal structures of *E. coli* ExoIII and an ExoIII-Mn^{2+}-dCMP ternary complex revealed a positively charged active-site groove formed between the ends of two six-stranded β-sheets (105). Structures of human APE1 with abasic site-containing DNA revealed that the enzyme inserts loops into both the major and minor grooves of DNA, severely kinking the DNA to bind a flipped-out AP site in a pocket that excludes DNA bases (Fig. 5A) (106). The APE1:DNA interface is centered around the extrahelical abasic deoxyribose and its flanking phosphates, which together contribute nearly one-third of the buried surface area of the complex. Both the amount of helical displacement and the extent of the APE1:DNA interface are larger than that seen in DNA complexes with damage-specific DNA glycosylases,

(A)   **(B)**

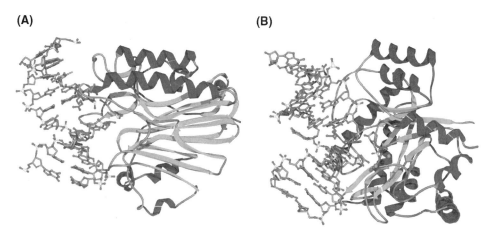

Figure 5 AP endonucleases. (A) The APE1–DNA crystal structure. R177, which intercalates into the DNA base stack, is shown as pink sticks. (B) The EndoIV–DNA structure, with intercalating residues R37, Y72, and L73 shown in pink. (*See color insert.*)

suggesting that APE1 may competitively displace these enzymes from their inhibitory AP site products, consistent with the observed APE1 enhancement of DNA glycosylase activities (106). Furthermore, site-directed mutagenesis of residues that penetrate the DNA minor groove (Met270 and Met271) and the DNA major groove (Arg177) surprisingly revealed that none of these residues is required to flip AP sites out of the DNA helix, indicating that APE1 functions as a structure-specific nuclease targeted to DNA that can adopt a kinked conformation. Arg177 intercalation into the DNA base stack and interaction with DNA phosphates together must slow APE1 dissociation from the cleaved product. Thus, human APE1 is structurally optimized to retain the cleaved product and therefore acts in vivo to coordinate the orderly transfer of potentially toxic DNA damage intermediates between the excision and synthesis steps of DNA repair.

4.2. Endo IV

EndoIV contains both zinc and manganese ions, and unlike magnesium-dependent ExoIII and APE1, it resists inactivation by EDTA (92,107). Detailed biochemical and structural analyses have revealed that both zinc and manganese are required for catalysis (108,109). The enzyme exhibits an additional 3′ diesterase repair activity, catalyzing the removal of DNA-blocking groups such as 3′ phosphates and 3′ phosphoglycolates, which are generated at single-strand breaks in DNA exposed to reactive oxygen species as well as the α,β-unsaturated aldehydes produced by AP lyases. Hence, AP endonucleases are required for both the processing of AP sites and the generation of free 3′ hydroxyl groups for subsequent DNA repair synthesis. EndoIV also plays a significant role in the repair of oxidative DNA lesions. EndoIV can cleave the DNA backbone 5′ to α-deoxyadenosine, 5,6-dihydrothymine, 5,6-dihydrouracil, and 5-hydroxyuracil residues, generating a free 3′ hydroxyl group and a 5′-dangling damaged nucleotide, which is a good substrate for human flap endonuclease (FEN1) and *E. coli* DNA pol I (110,111). However, EndoIV does not seem to be able to process thymine glycol residues (112). This incision repair pathway may circumvent the requirement for highly mutagenic AP sites in the repair of oxidative lesions, thus providing an alternative pathway to BER, as suggested by the pronounced radioresistance of the triple *nei, nth, fpg* mutant in *E. coli* (113).

EndoIV adopts an eight-stranded α/β barrel fold (TIM barrel) containing three zinc ions bound in a deep depression near the center of the barrel, partly accessible to the solvent (Fig. 5B) (108). In the structure of EndoIV bound to an abasic site-containing DNA oligonucleotide, EndoIV bends the DNA about 90°, flipping both the target abasic site and the nucleotide opposite out of the axis of the DNA helix by inserting three residues (Arg37, Tyr72, Leu73) into the DNA base stack. The DNA within the complex was hydrolyzed at the abasic site, with the expected 3′ hydroxyl and 5′-deoxyribose phosphate being produced, supporting a three metal ion mechanism for phosphodiester bond cleavage (108).

5. EMERGING QUESTIONS

As we begin to define common themes for recognition of damaged bases, such as sequence-independent DNA recognition motifs, minor groove intercalation, and major groove extrahelical flipping into lesion-specific active-site pockets, the known structural biology of base repair raises several important new questions. What aspects of repair enzyme structural chemistry may provide self-regulation and coordination of repair steps? What is the nature and role of conformational change in proteins and DNA in the major base repair pathways? We know that repair intermediates are as harmful as the initial damage itself, and that these intermediates are protected from one repair step to the next by the enzymes involved, such that pathway-specific handoffs must be efficiently coordinated. Do structures of repair protein:DNA complexes provide clues to pathway selection, the coordination of steps, and the avoidance of destructive interference? In general, direct visualizations of complexes both in solution and in crystal diffraction experiments have provided a molecular-level understanding based upon the discovery of testable general themes and principles for DNA base damage recognition, processing, and coordination. Such a detailed understanding of the molecular basis for DNA base integrity is fundamental to resolving many scientific, medical, and public health issues, including evaluation of the risks from inherited repair protein mutations, environmental toxins, and medical procedures.

ACKNOWLEDGMENTS

The work on DNA base repair in the Tainer lab is supported by NIH grants (GM4632, CA97209, and CA92584 to J.A.T.), and a grant from the Human Frontiers in Science Program to J.A.T., J.L.H. by NIH training grant (HL07781), and O.S. by a fellowship from the Skaggs Institute of Chemical Biology.

REFERENCES

1. Lindahl T. Instability and decay of the primary structure of DNA. Nature 1993; 362:709–715.
2. Das SK. Harmful health effects of cigarette smoking. Mol Cell Biochem 2003; 253: 159–165.

3. Ames BN, Gold LS, Willett WC. The causes and prevention of cancer. Proc Natl Acad Sci USA 1995; 92:5258–5265.

4. Memisoglu A, Samson L. Base excision repair in yeast and mammals. Mutat Res 2000; 451:39–51.

5. Dogliotti E, Fortini P, Pascucci B, Parlanti E. The mechanism of switching among multiple BER pathways. Prog Nucleic Acid Res Mol Biol 2001; 68:3–27.

6. Seeberg E, Eide L, Bjoras M. The base excision repair pathway. Trends Biochem Sci 1995; 20:391–397.

7. Dodson ML, Michaels ML, Lloyd RS. Unified catalytic mechanism for DNA glycosylases. J Biol Chem 1994; 269:32709–32712.

8. Tchou J, Grollman AP. The catalytic mechanism of Fpg protein. Evidence for a Schiff base intermediate and amino terminus localization of the catalytic site. J Biol Chem 1995; 270:11671–11677.

9. Fromme JC, Verdine GL. Structural insights into lesion recognition and repair by the bacterial 8-oxoguanine DNA glycosylase MutM. Nat Struct Biol 2002; 9:544–552.

10. Fromme JC, Verdine GL. Structure of a trapped endonuclease III-DNA covalent intermediate. EMBO J 2003; 22:3461–3471.

11. Fromme JC, Bruner SD, Yang W, Karplus M, Verdine GL. Product-assisted catalysis in base-excision DNA repair. Nat Struct Biol 2003; 10:204–211.

12. Gilboa R, Zharkov DO, Golan G, Fernandes AS, Gerchman SE, Matz E, Kycia JH, Grollman AP, Shoham G. Structure of formamidopyrimidine-DNA glycosylase covalently complexed to DNA. J Biol Chem 2002; 277:19811–19816.

13. Zharkov DO, Golan G, Gilboa R, Fernandes AS, Gerchman SE, Kycia JH, Rieger RA, Grollman AP, Shoham G. Structural analysis of an *Escherichia coli* endonuclease VIII covalent reaction intermediate. EMBO J 2002; 21:789–800.

14. Williams SD, David SS. Evidence that MutY is a monofunctional glycosylase capable of forming a covalent Schiff base intermediate with substrate DNA. Nucleic Acids Res 1998; 26:5123–5133.

15. Thayer MM, Ahern H, Xing D, Cunningham RP, Tainer JA. Novel DNA binding motifs in the DNA repair enzyme endonuclease III crystal structure. EMBO J 1995; 14:4108–4120.

16. Yamagata Y, Kato M, Odawara K, Tokuno Y, Nakashima Y, Matsushima N, Yasumura K, Tomita K, Ihara K, Fujii Y, Nakabeppu Y, Sekiguchi M, Fujii S. Three-dimensional structure of a DNA repair enzyme, 3-methyladenine DNA glycosylase II, from *Escherichia coli*. Cell 1996; 86:311–319.

17. Kuo CF, McRee DE, Fisher CL, O'Handley SF, Cunningham RP, Tainer JA. Atomic strucute of the DNA repair [4Fe-4S] enzyme endonuclease III. Science 1992; 258:434–440.

18. Bruner SD, Norman DP, Verdine GL. Structural basis for recognition and repair of the endogenous mutagen 8-oxoguanine in DNA. Nature 2000; 403:859–866.

19. Eichman BF, O'Rourke EJ, Radicella JP, Ellenberger T. Crystal structures of 3-mehtyladenine DNA glycoslyase MagIII and the recognition of alkylated bases. EMBO J 2003; 22:4898–4909.

20. Vassylyev DG, Kashiwagi T, Mikami Y, Ariyoshi M, Iwai S, Ohtsuka E, Morikawa K. Atomic model of a pyrimidine dimer excision repair enzyme complexed with a DNA substrate: structural basis for damaged DNA recognition. Cell 1995; 83:773–782.

21. Mol CD, Arvai AS, Begley TJ, Cunningham RP, Tainer JA. Structure and activity of a thermostable thymine-DNA glycosylase: evidence for base twisting to remove mismatched normal DNA bases. J Mol Biol 2002; 315:373–384.

22. Labahn J, Scharer OD, Long A, Ezaz-Nikpay K, Verdine GL, Ellenberger TE. Structural basis for the excision repair of alkylation-damaged DNA. Cell 1996; 86:321–329.

23. Guan Y, Manuel RC, Arvai AS, Parikh SS, Mol CD, Miller JH, Lloyd S, Tainer JA. MutY catalytic core, mutant and bound adenine structures define specificity for DNA repair enzyme superfamily. Nat Struct Biol 1998; 5:1058–1064.

24. Drohat AC, Kwon K, Krosky DJ, Stivers JT. 3-Methyladenine DNA glycosylase I is an unexpected helix-hairpin-helix superfamily member. Nat Struct Biol 2002; 9:659–664.

25. Hollis T, Ichikawa Y, Ellenberger T. DNA bending and a flip-out mechanism for base excision by the helix-hairpin-helix DNA glycosylase, *Escherichia coli* AlkA. EMBO J 2000; 19:758–766.

26. Lau AY, Scharer OD, Samson L, Verdine GL, Ellenberger T. Crystal structure of a human alkylbase-DNA repair enzyme complexed to DNA: mechanisms for nucleotide flipping and base excision. Cell 1998; 95:249–258.

27. Doherty AJ, Serpell LC, Ponting CP. The helix-hairpin-helix DNA-binding motif: a structural basis for non-sequence-specific recognition of DNA. Nucleic Acids Res 1996; 24:2488–2497.

28. Hosfield DJ, Mol CD, Shen B, Tainer JA. Structure of the DNA repair and replication endonuclease and exonuclease FEN-1: coupling DNA and PCNA binding to FEN-1 activity. Cell 1998; 95:135–146.

29. Sugahara M, Mikawa T, Kumasaka T, Yamamoto M, Kato R, Fukuyama K, Inoue Y, Kuramitsu S. Crystal strucure of a repair enzyme of oxidatively damaged DNA, MutM (Fpg), from an extreme thermophile, Thermus thermophilus HB8. EMBO J 2000; 19: 3857–3869.

30. Serre L, Pereira de Jesus K, Boiteux S, Zelwer C, Castaing B. Crystal structure of the Lactococcus lactis formamidopyrimidine-DNA glycosylase bound to an abasic site analouge-containing DNA. EMBO J 2002; 21:2854–2865.

31. Fromme JC, Verdine GL. DNA lesion recognition by the bacterial repair enzyme MutM. J Biol Chem 2003.

32. Hazra TK, Izumi T, Boldogh I, Imhoff B, Kow YW, Jaruga P, Dizdaroglu M, Mitra S. Identification and characterization of a human DNA glycosylase for repair of modified bases in oxidatively damaged DNA. Proc Natl Acad Sci USA 2002; 99:3523–3528.

33. Hazra TK, Izumi T, Venkataraman R, Kow YW, Dizdaroglu M, Mitra S. Characterization of a novel 8-oxoguanine-DNA glycosylase activity in *Escherichia coli* and identification of the enzyme as endonuclease VIII. J Biol Chem 2000; 275:27762–27767.

34. Takao M, Kanno SI, Kobayashi K, Zhang QM, Yonei S, Van Der Horst GT, Yasui A. A Back-up Glycosylase in Nth1 Knock-out Mice Is a Functional Nei (Endonuclease VIII) Homologue. J Biol Chem 2002; 277:42205–42213.

35. Radicella JP, Dherin C, Desmaze C, Fox MS, Boiteux S. Cloning and characterization of hOGG1, a human homolog of the OGG1 gene of Saccharomyces cerevisiae. Proc Natl Acad Sci USA 1997; 94:8010–8015.

36. Ikeda S, Biswas T, Roy R, Izumi T, Boldogh I, Kurosky A, Sarker AH, Seki S, Mitra S. Purification and characterization of human NTH1, a homolog of *Escherichia coli* endonuclease III. Direct identification of Lys-212 as the active nucleophilic residue. J Biol Chem 1998; 273:21585–21593.

37. Hazra TK, Kow YW, Hatahet Z, Imhoff B, Boldogh I, Mokkapati SK, Mitra S, Izumi T. Identification and characterization of a novel human DNA glycosylase for repair of cytosine-derived lesions. J Biol Chem 2002; 277:30417–30420.

38. Bjoras M, Luna L, Johnsen B, Hoff E, Haug T, Rognes T, Seeberg E. Opposite base-dependent reactions of a human base excision repair enzyme on DNA containing 7, 8-dihydro-8-oxoguanine and abasic sites. EMBO J 1997; 16:6314–6322.

39. Bjoras M, Seeberg E, Luna L, Pearl LH, Barrett TE. Reciprocal "flipping" underlies substrate recognition and catalytic activation by the human 8-oxo-guanine DNA glycosylase. J Mol Biol 2002; 317:171–177.

40. Zharkov DO, Gilboa R, Yagil I, Kycia JH, Gerchman SE, Shoham G, Grollman AP. Role for lysine 142 in the excision of adenine from A:G mispairs by MutY DNA glycosylase of *Escherichia coli*. Biochemistry 2000; 39:14768–14778.

41. Michaels ML, Miller JH. The GO system protects organisms from the mutagenic effect of the spontaneous lesion 8-hydroxyguanine (7,8-dihydro-8-oxoguanine). J Bacteriol 1992; 174:6321–6325.

42. Abeygunawardana C, Weber DJ, Gittis AG, Frick DN, Lin J, Miller AF, Bessman MJ, Mildvan AS. Solution structure of the MutT enzyme, a nucleoside triphosphate pyro-phosphohydrolase. Biochemistry 1995; 34:14997–15005.

43. Noll DM, Gogos A, Granek JA, Clarke ND. The C-terminal domain of the adenine-DNA glycosylase MutY confers specificity for 8-oxoguanine adenine mispairs and may have evolved from MutT, an 8-oxo-dGTPase. Biochemistry 1999; 38:6374–6379.

44. Volk DE, House PG, Thiviyanathan V, Luxon BA, Zhang S, Lloyd RS, Gorenstein DG. Structural similarities between MutT and the C-terminal domain of MutY. Biochemistry 2000; 39:7331–7336.

45. Bernards AS, Miller JK, Bao KK, Wong I. Flipping duplex DNA inside out: a double base-flipping reaction mechanism by *Escherichia coli* MutY adenine glycosylase. J Biol Chem 2002; 277:20960–20964.

46. Wist E, Unhjem O, Krokan H. Accumulation of small fragments of DNA in isolated HeLa cell nuclei due to transient incorporation of dUMP. Biochim Biophys Acta 1978; 520:253–270.

47. Tye BK, Nyman PO, Lehman IR, Hochhauser S, Weiss B. Transient accumulation of Okazaki fragments as a result of uracil incorporation into nascent DNA. Proc Natl Acad Sci USA 1977; 74:154–157.

48. Nilsen H, Otterlei M, Haug T, Solum K, Nagelhus TA, Skorpen F, Krokan HE. Nuclear and mitochondrial uracil-DNA glycosylases are generated by alternative spli-cing and transcription from different positions in the UNG gene. Nucleic Acids Res 1997; 25:750–755.

49. Savva R, McAuley-Hecht K, Brown T, Pearl L. The structural basis of specific base-excision repair by uracil-DNA glycosylase. Nature 1995; 373:487–493.

50. Slupphaug G, Mol CD, Kavli B, Arvai AS, Krokan HE, Tainer JA. A nucleotide-flipping mechanism from the structure of human uracil-DNA glycosylase bound to DNA. Nature 1996; 384:87–92.

51. Xiao G, Tordova M, Jagadeesh J, Drohat AC, Stivers JT, Gilliland GL. Crystal struc-ture of *Escherichia coli* uracil DNA glycosylase and its complexes with uracil and glycerol: structure and glycosylase mechanism revisited. Proteins 1999; 35:13–24.

52. Leiros I, Moe E, Lanes O, Smalas AO, Willassen NP. The structure of uracil-DNA glycosylase from Atlantic cod (Gadus morhua) reveals cold-adaptation features. Acta Crystallogr D Biol Crystallogr 2003; 59:1357–1365.

53. Parikh SS, Mol CD, Slupphaug G, Bharati S, Krokan HE, Tainer JA. Base excision repair initiation revealed by crystal structures and binding kinetics of human uracil-DNA glycosylase with DNA. EMBO J 1998; 17:5214–5226.

54. Wong I, Lundquist AJ, Bernards AS, Mosbaugh DW. Presteady-state analysis of a single catalytic turnover by *Escherichia coli* uracil-DNA glycosylase reveals a "pinch-pull-push" mechanism. J Biol Chem 2002; 277:19424–19432.

55. Pearl LH. Structure and function in the uracil-DNA glycosylase superfamily. Mutat Res 2000; 460:165–181.

56. Brown TC, Jiricny J. A specific mismatch repair event protects mammalian cells from loss of 5-methylcytosine. Cell 1987; 50:945–950.

57. Hardeland U, Bentele M, Lettieri T, Steinacher R, Jiricny J, Schar P. Thymine DNA glycosylase. Prog Nucleic Acid Res Mol Biol 2001; 68:235–253.

58. Gallinari P, Jiricny J. A new class of uracil-DNA glycosylases related to human thymine-DNA glycosylase. Nature 1996; 383:735–738.

59. Hang B, Medina M, Fraenkel-Conrat H, Singer B. A 55-kDa protein isolated from human cells shows DNA glycosylase activity toward 3,N4-ethenocytosine and the G/T mismatch. Proc Natl Acad Sci USA 1998; 95:13561–13566.

60. Barrett TE, Savva R, Panayotou G, Barlow T, Brown T, Jiricny J, Pearl LH. Crystal structure of a G:T/U mismatch-specific DNA glycosylase: mismatch recognition by complementary-strand interactions. Cell 1998; 92:117–129.

61. Barrett TE, Scharer OD, Savva R, Brown T, Jiricny J, Verdine GL, Pearl LH. Crystal structure of a thwarted mismatch glycosylase DNA repair complex. EMBO J 1999; 18:6599–6609.

62. Haushalter KA, Todd Stukenberg MW, Kirschner MW, Verdine GL. Identification of a new uracil-DNA glycosylase family by expression cloning using synthetic inhibitors. Curr Biol 1999; 9:174–185.

63. Kavli B, Sundheim O, Akbari M, Otterlei M, Nilsen H, Skorpen F, Aas PA, Hagen L, Krokan HE, Slupphaug G. hUNG2 Is the Major Repair Enzyme for Removal of Uracil from U:A Matches, U:G Mismatches, and U in Single-stranded DNA, with hSMUG1 as a Broad Specificity Backup. J Biol Chem 2002; 277:39926–39936.

64. Boorstein RJ, Cummings A Jr., Marenstein DR, Chan MK, Ma Y, Neubert TA, Brown SM, Teebor GW. Definitive identification of mammalian 5-hydroxymethyluracil DNA N-glycosylase activity as SMUG1. J Biol Chem 2001; 276:41991–41997.

65. Wibley JE, Waters TR, Haushalter K, Verdine GL, Pearl LH. Structure and specificity of the vertebrate anti-mutator uracil-DNA glycosylase SMUG1. Mol Cell 2003; 11: 1647–1659.

66. Hoseki J, Okamoto A, Masui R, Shibata T, Inoue Y, Yokoyama S, Kuramitsu S. Crystal structure of a family 4 uracil-DNA glycosylase from Thermus thermophilus HB8. J Mol Biol 2003; 333:515–526.

67. Petronzelli F, Riccio A, Markham GD, Seeholzer SH, Genuardi M, Karbowski M, Yeung AT, Matsumoto Y, Bellacosa A. Investigation of the substrate spectrum of the human mismatch-specific DNA N-glycosylase MED1 (MBD4): fundamental role of the catalytic domain. J Cell Physiol 2000; 185:473–480.

68. Parikh SS, Walcher G, Jones GD, Slupphaug G, Krokan HE, Blackburn GM, Tainer JA. Uracil-DNA glycosylase-DNA substrate and product structures: conformational strain promotes catalytic efficiency by coupled stereoelectronic effects. Proc Natl Acad Sci USA 2000; 97:5083–5088.

69. Ohki I, Shimotake N, Fujita N, Nakao M, Shirakawa M. Solution structure of the methyl-CpG-binding domain of the methylation-dependent transcriptional repressor MBD1. EMBO J 1999; 18:6653–6661.

70. Ohki I, Shimotake N, Fujita N, Jee J, Ikegami T, Nakao M, Shirakawa M. Solution structure of the methyl-CpG binding domain of human MBD1 in complex with methylated DNA. Cell 2001; 105:487–497.

71. Wakefield RI, Smith BO, Nan X, Free A, Soteriou A, Uhrin D, Bird AP, Barlow PN. The solution structure of the domain from MeCP2 that binds to methylated DNA. J Mol Biol 1999; 291:1055–1065.

72. Wu P, Qiu C, Sohail A, Zhang X, Bhagwat AS, Cheng X. Mismatch repair in methylated DNA. Structure and activity of the mismatch-specific thymine glycosylase domain of methyl-CpG-binding protein MBD4. J Biol Chem 2003; 278:5285–5291.

73. Huffman JL, Li H, White RH, Tainer JA. Structural basis for recognition and catalysis by the bifunctional dCTP deaminase and dUTPase from Methanococcus jannaschii. J Mol Biol 2003; 331:885–896.

74. Mol CD, Harris JM, McIntosh EM, Tainer JA. Human dUTP pyrophosphatase: uracil recognition by a beta hairpin and active sites formed by three separate subunits. Structure 1996; 4:1077–1092.

75. Bjornberg O, Neuhard J, Nyman PO. A bifunctional dCTP deaminase-dUTP nucleotidohydrolase from the hyperthermophilic archeon Methanococcus jannaschii. J Biol Chem 2003.

76. Dauter Z, Persson R, Rosengren AM, Nyman PO, Wilson KS, Cedergren-Zeppezauer ES. Crystal structure of dUTPase from equine infectious anaemia virus; active site metal binding in a substrate analogue complex. J Mol Biol 1999; 285:655–673.

77. Cedergren-Zeppezauer ES, Larsson G, Nyman PO, Dauter Z, Wilson KS. Crystal structure of a dUTPase. Nature 1992; 355:740–743.

78. Prasad GS, Stura EA, McRee DE, Laco GS, Hasselkus-Light C, Elder JH, Stout CD. Crystal structure of dUTP pyrophosphatase from feline immunodeficiency virus. Protein Sci 1996; 5:2429–2437.

79. Daniels DS, Mol CD, Arvai AS, Kanugula S, Pegg AE, Tainer JA. Active and alkylated human AGT structures: a novel zinc site, inhibitor and extrahelical base binding. EMBO J 2000; 19:1719–1730.

80. Hashimoto H, Inoue T, Nishioka M, Fujiwara S, Takagi M, Imanaka T, Kai Y. Hyperthermostable protein structure maintained by intra and inter-helix ion-pairs in archaeal O6-methylguanine-DNA methyltransferase. J Mol Biol 1999; 292:707–716.

81. Lin Y, Dotsch V, Wintner T, Peariso K, Myers LC, Penner-Hahn JE, Verdine GL, Wagner G. Structural basis for the functional switch of the E. coli Ada protein. Biochemistry 2001; 40:4261–4271.

82. Moore MH, Gulbis JM, Dodson EJ, Demple B, Moody PC. Crystal structure of a suicidal DNA repair protein: the Ada O6-methylguanine-DNA methyltransferase from E. coli. EMBO J 1994; 13:1495–1501.

83. Trewick SC, Henshaw TF, Hausinger RP, Lindahl T, Sedgwick B. Oxidative demethylation by *Escherichia coli* AlkB directly reverts DNA base damage. Nature 2002; 419:174–178.

84. Falnes PO, Johansen RF, Seeberg E. AlkB-mediated oxidative demethylation reverses DNA damage in *Escherichia coli*. Nature 2002; 419:178–182.

85. Aravind L, Koonin EV. The DNA-repair protein AlkB, EGL-9, and leprecan define new families of 2-oxoglutarate- and iron-dependent dioxygenases. Genome Biol 2001; 2: RESEAR CH0007.

86. Aas PA, Otterlei M, Falnes PO, Vagbo CB, Skorpen F, Akbari M, Sundheim O, Bjoras M, Slupphaug G, Seeberg E, Krokan HE. Human and bacterial oxidative demethylases repair alkylation damage in both RNA and DNA. Nature 2003; 421: 859–863.

87. Lau AY, Wyatt MD, Glassner BJ, Samson LD, Ellenberger T. Molecular basis for discriminating between normal and damaged bases by the human alkyladenine glycosylase, AAG. Proc Natl Acad Sci USA 2000; 97:13573–13578.

88. Brennan RG, Matthews BW. The helix-turn-helix DNA binding motif. J Biol Chem 1989; 264:1903–1906.

89. Lindahl T. Repair of intrinsic DNA lesions. Mutat Res 1990; 238:305–311.

90. Barzilay G, Hickson ID. Structure and function of apurinic/apyrimidinic endonucleases. Bioessays 1995; 17:713–719.

91. Demple B, Harrison L. Repair of oxidative damage to DNA: enzymology and biology. Annu Rev Biochem 1994; 63:915–948.

92. Ramotar D. The apurinic-apyrimidinic endonuclease IV family of DNA repair enzymes. Biochem Cell Biol 1997; 75:327–336.

93. Ljungquist S, Lindahl T, Howard-Flanders P. Methyl methane sulfonate-sensitive mutant of *Escherichia coli* deficient in an endonuclease specific for apurinic sites in deoxyribonucleic acid. J Bacteriol 1976; 126:646–653.

94. Yajko DM, Weiss B. Mutations simultaneously affecting endonuclease II and exonuclease III in *Escherichia coli*. Proc Natl Acad Sci USA 1975; 72:688–692.

95. Ljungquist S. A new endonuclease from *Escherichia coli* acting at apurinic sites in DNA. J Biol Chem 1977; 252:2808–2814.

96. Xanthoudakis S, Smeyne RJ, Wallace JD, Curran T. The redox/DNA repair protein, Ref-1, is essential for early embryonic development in mice. Proc Natl Acad Sci USA 1996; 93:8919–8923.

97. Bennett RA, Wilson DM, 3rd, Wong D, Demple B. Interaction of human apurinic endonuclease and DNA polymerase beta in the base excision repair pathway. Proc Natl Acad Sci USA 1997; 94:7166–7169.

98. Vidal AE, Boiteux S, Hickson ID, Radicella JP. XRCC1 coordinates the initial and late stages of DNA abasic site repair through protein-protein interactions. EMBO J 2001; 20:6530–6539.

99. Dianova, II, Bohr VA, Dianov GL. Interaction of human AP endonuclease 1 with flap endonuclease 1 and proliferating cell nuclear antigen involved in long-patch base excision repair. Biochemistry 2001; 40:12639–12644.

100. Ranalli TA, Tom S, Bambara RA. AP endonuclease 1 coordinates flap endonuclease 1 and DNA ligase I activity in long patch base excision repair. J Biol Chem 2002; 27:27.

101. Xanthoudakis S, Miao G, Wang F, Pan YC, Curran T. Redox activation of Fos-Jun DNA binding activity is mediated by a DNA repair enzyme. EMBO J 1992; 11: 3323–3335.

102. Tell G, Zecca A, Pellizzari L, Spessotto P, Colombatti A, Kelley MR, Damante G, Pucillo C. An 'environment to nucleus' signaling system operates in B lymphocytes: redox status modulates BSAP/Pax-5 activation through Ref-1 nuclear translocation. Nucleic Acids Res 2000; 28:1099–1105.

103. Tell G, Pellizzari L, Cimarosti D, Pucillo C, Damante G. Ref-1 controls pax-8 DNA-binding activity. Biochem Biophys Res Commun 1998; 252:178–183.

104. Jayaraman L, Murthy KG, Zhu C, Curran T, Xanthoudakis S, Prives C. Identification of redox/repair protein Ref-1 as a potent activator of p53. Genes Dev 1997; 11:558–570.

105. Mol CD, Kuo CF, Thayer MM, Cunningham RP, Tainer JA. Structure and function of the multifunctional DNA-repair enzyme exonuclease III. Nature 1995; 374:381–386.

106. Mol CD, Izumi T, Mitra S, Tainer JA. DNA-bound structures and mutants reveal abasic DNA binding by APE1 and DNA repair coordination. Nature 2000; 403:451–456.

107. Levin JD, Shapiro R, Demple B. Metalloenzymes in DNA repair. *Escherichia coli* endonuclease IV and Saccharomyces cerevisiae Apn1. J Biol Chem 1991; 266:22893–22898.

108. Hosfield DJ, Guan Y, Haas BJ, Cunningham RP, Tainer JA. Structure of the DNA repair enzyme endonuclease IV and its DNA complex: double-nucleotide flipping at abasic sites and three-metal-ion catalysis. Cell 1999; 98:397–408.

109. Haas BJ, Sandigursky M, Tainer JA, Franklin WA, Cunningham RP. Purification and characterization of Thermotoga maritima endonuclease IV, a thermostable apurinic/apyrimidinic endonuclease and 3'-repair diesterase. J Bacteriol 1999; 181:2834–2839.

110. Ide H, Tedzuka K, Shimzu H, Kimura Y, Purmal AA, Wallace SS, Kow YW. Alpha-deoxyadenosine, a major anoxic radiolysis product of adenine in DNA, is a substrate for *Escherichia coli* endonuclease IV. Biochemistry 1994; 33:7842–7847.

111. Ischenko AA, Saparbaev MK. Alternative nucleotide incision repair pathway for oxidative DNA damage. Nature 2002; 415:183–187.

112. Ocampo MT, Chaung W, Marenstein DR, Chan MK, Altamirano A, Basu AK, Boorstein RJ, Cunningham RP, Teebor GW. Targeted deletion of mNth1 reveals a novel DNA repair enzyme activity. Mol Cell Biol 2002; 22:6111–6121.

113. Blaisdell JO, Wallace SS. Abortive base-excision repair of radiation-induced clustered DNA lesions in *Escherichia coli*. Proc Natl Acad Sci USA 2001; 98:7426–7430.

15
Repair of Oxidized Bases

Yoke Wah Kow
Department of Radiation Oncology, Division of Cancer Biology, Emory University School of Medicine, Atlanta, Georgia, U.S.A.

1. BIOLOGICAL CONSEQUENCES OF OXIDATIVE DAMAGE

Oxidative damage to DNA constitutes a major portion of the endogenous DNA damage in a cell. Most of the oxidative lesions are the result of reactive oxygen species (ROS) interacting with DNA bases and the deoxyribose moiety (1,2). Reactive oxygen species, in particular superoxide radicals, are generated in mitochondria as a result of incomplete reduction of oxygen during oxidative phosphorylation (3). Superoxide is then converted to other forms of ROS including hydrogen peroxide and hydroxyl radicals (Fig. 1) (1,4,5). Reactive oxygen species are also generated when cells are exposed to exogenous agents such as redox chemicals (6), UV [in particular high doses of UVA (7)] radiation, and ionizing radiation (1,2). Singlet oxygen, another highly ROS, is generated via photoreactions in the presence of photoactive pigments or dyes (8,9). The majority of oxidative DNA damage is the result of hydroxyl radicals' attack on DNA; under physiological condition, hydroxyl radical reacts with DNA at almost diffusion-controlled rates ($10^9 \, M^{-1} \, sec^{-1}$–$10^{10} \, M^{-1} \, sec^{-1}$) (1,2).

Reactions of hydroxyl radicals with DNA bases generate a wide spectrum of DNA base modifications. The major stable base oxidation products are the hydroxylation products of the 5, 6 double bond of pyrimidines and the C-8 hydroxylation products of purines (1,2). The initial hydroxyl radical adducts of purine and pyrimidines can further react and lead to the formation of ring saturation, contraction, or fragmentation of DNA bases. Figs. 2 and 3 list the structures of a number of commonly identifiable DNA base lesions. Oxidation of DNA bases also leads to increased instability of the *N*-glycosidic bond and to the formation of apurinic/apyrimidinic (AP) sites in DNA. However, AP sites are generated predominantly as a result of hydroxyl radical interaction with the deoxyribose moiety (1,2), abstraction of hydrogen, or the addition of hydroxyl radical to the deoxyribose ring, leading to the formation of various kinds of modification of the deoxyribose moiety (1,2). Many of the damaged deoxyribose species are characterized by substantial decrease in the stability the *N*-glycosidic bond, thus generating various forms of oxidized AP site and DNA-strand breaks containing modified sugar residues or phosphoryl termini (1,2).

A. Oxidative Phosphorylation:

$$O_2 \xrightarrow{\quad e^- \quad} O_2^{\cdot -}$$

B. Harber-Weiss reaction

$$2O_2^{\cdot -} + 2H^+ \longrightarrow H_2O_2 + O_2$$

$$O_2^{\cdot -} + Fe(III) \longrightarrow O_2 + Fe(III)$$

$$H_2O_2 + Fe(III) \longrightarrow HO^{\cdot} + OH^- + Fe(III)$$

$$O_2^{\cdot -} + H_2O_2 \longrightarrow HO^{\cdot} + OH^- + O_2$$

C. Fenton Reaction:

$$H_2O_2 + Fe(II) \longrightarrow HO^{\cdot} + OH^- + Fe(III)$$

D. Radiolysis of water

$$H_2O \longrightarrow H_2O^+ + H_2O^* + e_{aq}^-$$

$$H_2O^* \longrightarrow HO^{\cdot} + H^{\cdot}$$

$$H_2O^+ + H_2O \longrightarrow H_3O^+ + HO^{\cdot}$$

Figure 1 Formation of ROS.

Oxidized DNA bases and AP sites can produce significant deleterious effect to cells. They can be blocks to DNA replication and thus lead to increased cellular lethality. They can also be bypassed by DNA polymerases and, when the lesion bypass is mutagenic, will result in increased mutability. The biological consequences of many oxidized base lesions have been studied extensively in *Eschericia coli*, yeast, and human cells (10–24). Thymine glycol is generally a strong block to DNA replication, and unrepaired thymine glycol leads to cellular lethality (10–12). It has been estimated that it takes about 12 thymine glycols per double-stranded phage Φx-174 DNA to constitute a lethal event when the phage containing thymine glycols is transfected into a wild-type *E. coli* host (24,25). However, in single-stranded DNA, a single thymine glycol is sufficient to constitute a block to replication, as demonstrated in in vivo transformation and in vitro replication assays (26–28). These data thus suggest the importance of cellular repair of oxidative DNA damage for cellular survival. It is important to point out that thymine glycol can be bypassed at low frequency, and the bypass usually results in the insertion of A opposite thymine glycol (26). Thymine glycol is thus poorly mutagenic.

Other products of thymine oxidation, such as hydantoin, urea residues, and so on, exhibit similar lethality as thymine glycol when examined in wild-type *E. coli* cells (24–26,29). Thymine oxidation products are generally poorly mutagenic, and lesion bypass occurring at sites of thymine oxidation is usually not mutagenic; this is simply because the repair polymerase tends to incorporate A opposite oxidized thymine lesions (24,25). However, under certain circumstances, apparently determined by the sequence context where the lesion is embedded, the lesion bypass of thymine

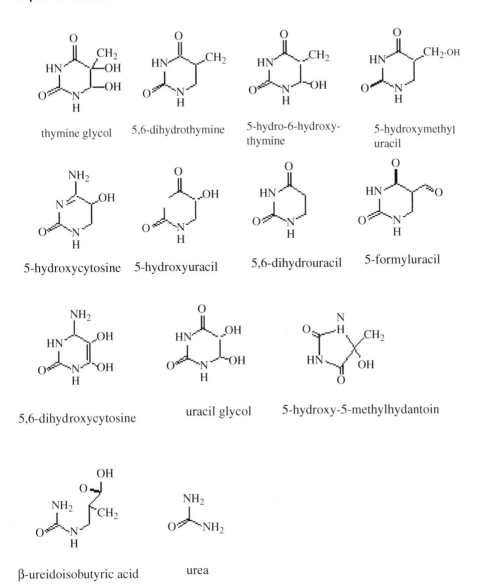

Figure 2 Structures of oxidized pyrimidines.

oxidation products can be mutagenic (11). Oxidation products of cytosine such as 5-hydroxyuracil, uracil glycol, and 5-hydroxycytosine are also blocks to DNA replication; when bypassed, 5-hydroxyuracil and uracil glycol are highly mutagenic because 5-hydroxyuracil pairs with an A, thus resulting in C to T mutations (30).

The major purine oxidation products generated by ROS include 8-oxoG, 8-oxoA, formamidopyrimide-dA (FaPy-A), and formamidopyrimidine-dG (FaPy-G). 8-oxoG has been shown to lead to a stall in DNA polymerase progression or DNA replication (31). When the DNA polymerase reaction or DNA replication resumes, a high proportion of the bypassed 8-oxoG lesions pair with As, leading to high frequencies of G to T mutations (32–34). In contrast, 8-oxoA is only poorly mutagenic (31–36). Despite the fact that 8-oxoA and 8-oxoG are structurally similar, 8-oxoG is prone to further oxidation and 8-oxoA is chemically more stable.

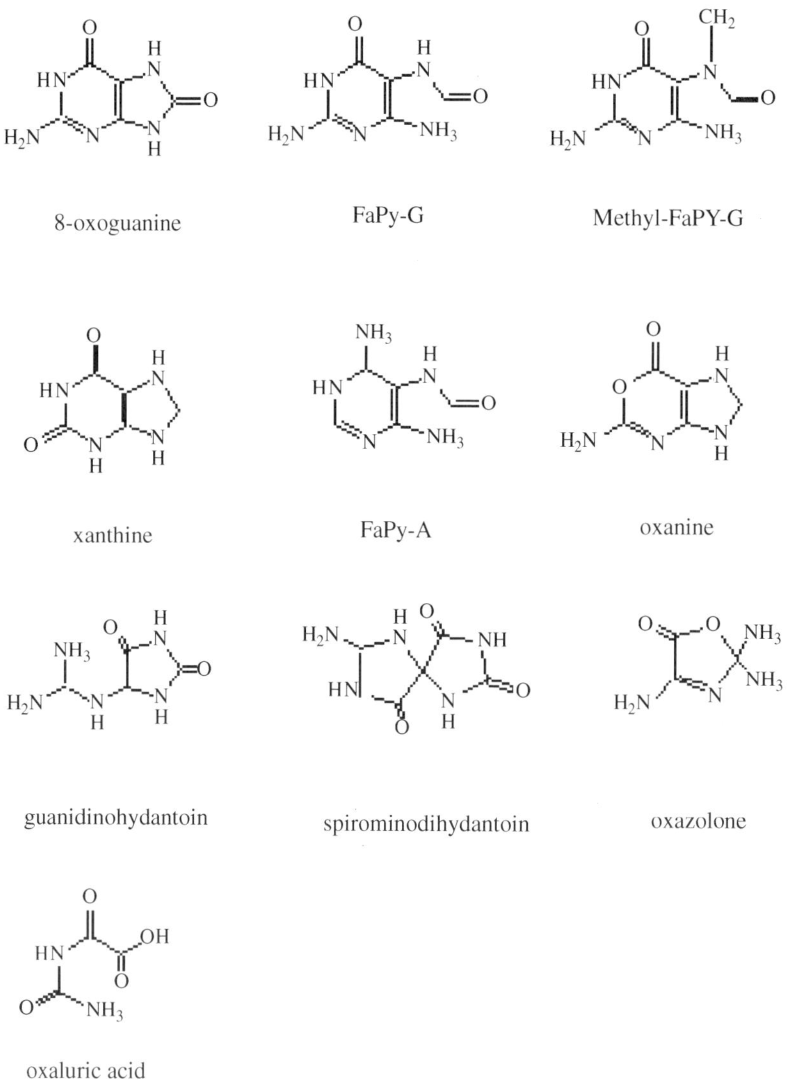

8-oxoguanine   FaPy-G   Methyl-FaPY-G

xanthine   FaPy-A   oxanine

guanidinohydantoin spirominodihydantoin oxazolone

oxaluric acid

Figure 3 Structures of oxidized purines.

The one-electron oxidation products of 8-oxoG are guanidinohydantoin and spiro-iminodihydantoin (37,38). Similar to 8-oxoG, these oxidation products are highly mutagenic because of their abilities to pair with A or G (37,39). However, further oxidation of 8-oxoG, under conditions where singlet oxygen is present, generates predominantly oxaluric acid (40). Oxaluric acid was shown to be a block to DNA replication; when it is bypassed, it can lead to GC to TA and GC to CG mutations (41).

2. MAJOR REPAIR ENZYMES THAT RECOGNIZE OXIDATIVE BASE DAMAGE

Despite the fact that ROS generate a wide spectrum of oxidized DNA bases, there are in fact only four major cellular repair enzymes that recognize these chemical modifications: endonucleases III and VIII, formamidopyrimidine *N*-glycosylase (Fpg), and oxoguanine *N*-glycosylase (Ogg). Endonucleases III and VIII are the two major enzymes that recognize predominantly oxidized pyrimidines, whereas Fpg and Ogg recognize predominantly oxidized purines. However, the in vitro substrate recognition by these enzymes appears to be quite relaxed. For example, endonuclease III and its homologs have been shown to also recognize oxidized purines such as oxoG (42), and Fapy was shown to recognize some oxidized pyrimidines such as 5-hydroxyuracil and dihydrouracil (43), albeit at rather low efficiencies when compared with their more efficiently recognized substrates. These glycosylases belong to two major families of repair enzymes: the HhH-GPD superfamily that include the endonuclease III and Ogg, and the helix-2 turn-helix (H-2T-H) family that include the *E. coli* Fpg, endonuclease VIII, and human Neil1 and Neil2. In contrast to simple glycosylases that recognize alkylation base damage or uracil, these glycosylases are bifunctional, containing both an *N*-glycosylase and an AP lyase activity. Despite being structurally quite different, these glycosylases possess a DNA-binding domain that binds and interacts with a duplex DNA and a damage recognition domain that binds specifically to the damage. The binding of the glycosylase to the base damage usually results in conformational changes to both DNA and protein, leading to the flipping of the damaged base into the lesion binding pocket of the repair enzyme (44–46). Despite the fact that HhH-GPD and H-2T-H families of glycosylases are structurally dissimilar and exhibit major differences in substrate specificities, they employ a common strategy in their catalysis: the reactions carried out by these enzymes that lead to the release of damaged base and the cleavage of the phosphodiester bond 3′ to the lesion are remarkably similar. In general, these glycosylases possess lysine (for HhH-GPD family) (44–46) or proline residues (for H-2T-H family) (44–46) in the active site that can form a covalent Schiff base (iminium) intermediate with the C1′ of the deoxyribose. The nucleophilicity of the lysine residue is enhanced substantially by a nearby glutamic acid (or basic residue) that is in juxtaposition to deprotonate the positively charged lysine residue. During the formation of the covalent iminium intermediate, the damaged base is released (the glycosylase reaction). The formation of the covalent iminium intermediate leads to substantial increase in the acidity of the C-2 hydrogen, which can then be readily abstracted by other nearby basic residues and leads to the subsequent cleavage of the 3′ phosphodiester bond (β-elimination reaction) (47). This will generate a 4-hydroxy-2-pentenal moiety at the 3′ terminus of the nick (the AP lyase activity) (47). Endonuclease III and Ogg are examples of repair enzymes that can carry out the β-elimination reaction after the *N*-glycosylase reaction (44,45). Endonuclease VIII, Fpg, human Neil1 and Neil2 (44), however, can catalyze a second β-elimination reaction, generating DNA nicks that contain 3′ and 5′ phosphate terminiaccompanying the release of a modified sugar moiety as 4-oxopent-e-enal (44,45,47).

2.1. Endonuclease III

Escherichia coli endonuclease III was first partially purified on the basis of its ability to recognize X-irradiated DNA and was called x-ray endonuclease (48). It was later shown to be identical to endonuclease III, a repair enzyme that was identified as an endonuclease that recognized a broad spectrum of oxidized pyrimidines (49). The enzyme was shown to be a DNA glycosylase that has an associated AP lyase activity that can nick DNA containing an AP site by a β-elimination reaction. Endonuclease

III is present in most cells examined, and orthologs of *E. coli* endonuclease III have been cloned, overexpressed, and characterized including *Saccharomyces cerevisiae*, *Schizasaccharomyces pombe*, mouse, human, and many other organisms.

Escherichia coli endonuclease III recognizes predominantly oxidized pyrimidines. Oxidized pyrimidines that have been shown to be substrates for endonuclease III included thymine glycols, 5,6-dihydrothymine, 5,6-dihydrouracil, 5-hydroxyuracil, 5-hydroxycytosine, 5-hydroxyl 5-methyl hydantoin, formylurea, and urea (49). 5-Formyl and 5-hydroxymethyluracil are also recognized by endonuclease III (50). A surprising result was observed by the Yonei's laboratory (42) showing that *E. coli* endonuclease III is able to excise oxoG when it is opposite G. A oxodG:dG pair could occur in DNA only during misincorporation of dG opposite the unrepaired oxoG or misincorporation of oxoG opposite dG from the damaged triphosphate, 8-oxodGTP. The in vitro recognition was supported by the observed increase G–C to C–G mutation frequency in triple *E. coli* mutant *nth mutM nei*. The increase is greater than either *mutM nei*, *mutM nth*, or *nth nei* double mutant (42). Human Nth was also shown to have a weak activity on oxoG when it is paired with G (42). Similar opposite base effect on substrate recognition was observed earlier for human endonuclease III; the enzyme exhibits a much higher activity on hydroxycytosine or AP site when the lesion is opposite G. The opposite base discrimination exhibited by endonuclease III appears to be affected by the presence of Mg^{++} (52). In addition to exhibiting opposite base discrimination, the removal of thymine glycol by endonuclease III is also stereospecific (53). Endonuclease III efficiently removes the *cis*-5*R*,6*S*-thymine glycol isomers but removes *cis*-5*S*,6*R*-thymine glycol isomers at an extremely slow rate (53). The selective removal of a particular stereoisomer can potentially lead to an accumulation of the lesion even in wild-type cells and impart significant biological consequences.

Ntg2, the yeast homolog of *E. coli* endonuclease III, also recognizes 8-oxoG (54); in contrast, Ntg1 appears to only recognize pyrimidine oxidation products (55). Oxidation products of 8-oxoG, guanidinohydantoin and spiroiminodihydantoin, are also substrates for the *E. coli* enzyme endonuclease III (38). Interestingly, these oxidation products of 8-oxoG are also recognized by endonuclease VIII and Fpg and Ogg (38,39,56). Oxaluric acid, another oxidation product of 8-oxoG is also recognized by both *E. coli* endonuclease III and Fpg (40,41). It is interesting to note that compared with 5-hydroxycytosine, oxaluric acid appears to be a much better substrate for *E. coli* endonuclease III (41). Pyrimidine ring opened product of 1,N^6-ethenoadenine was also reported to be recognized by endonuclease III (57,58). Ethenoadenine is formed as a result of lipid peroxidation (59).

High expression of endonuclease III was shown to lead to partial relief of the alkylation sensitivity of *E. coli* alkA mutant cells. These data suggest that endonuclease III might also recognize a certain class of alkylation damage that is substrate of alkA protein. Despite the fact that the nature of the alkylation products is not known, it is likely that they will be simple alkylation products. Evidently, the recognition of these alkylation lesions by endonuclease III is rather poor, as only a high level of cellular endonuclease III is required to impart a partial complementation of alkA deficiency (60).

Escherichia coli endonuclease III exhibits an active *N*-glycosylase and a very robust AP lyase activity. In general, little or no intermediary AP site is observed in an endonuclease III reaction. The *N*-glycosylase and AP lyase activity thus appears to be concerted in *E. coli* endonuclease III. In contrast, the AP lyase activity of eukaryotic endonuclease III, such as human Nth1 and yeast Ntg, are much slower than their *N*-glycosylase activities (45). The addition of the Y-box binding protein-1 was shown to lead to an increase in both the *N*-glycosylase and AP endonuclease activity (61). The increase in the AP lyase activity by Y-box binding protein-1 was thought to be due to the shifting of the equilibrium between the covalent iminium intermediate and the noncovalent AP aldehyde. Human AP endonuclease (APE1) was shown to increase significantly the *N*-glycosylase activity of human Nth1 (62). The addition of APE1 leads to an increased dissociation of hNTH1 from the intermediary AP site, resulting in the abrogation of AP lyase activity and thus an increase in turnover of the enzyme (62). It is interesting to note that hNTH1 binds more tightly to an AP site when the lesion is opposite to G than when it is opposite to A (62). The opposite base specificity in binding strength is in agreement with the opposite base effect on substrate recognition. The activity of the human Nth1 was also reported to be enhanced by the presence of XPG protein, a component of the human nucleotide excision complex (63,64). Interestingly, the N-terminal protein sequence of hNTH appears to be important for dimerization of the human Nth (65). Apparently, dimerization of human Nth1 leads to a faster release of Nth1 from the nick product and thus a higher observed catalytic activity of human Nth1. The dimerization of hNTH1 might be an important means for the modulation of the Nth activity in human cells.

Despite the fact that the endonuclease III exhibits robust activity on DNA containing an AP site, endonuclease III from *E. coli*, yeast, and human are inefficient in the removal of 2-deoxyribonolactone (66), a C-2 oxidized AP site that is generated by ionizing radiation (1,2) and neocarzinostatin (67). Instead of leading to a β-elimination reaction and the cleavage of phosphodiester backbone, endonuclease III forms a cross-link with 2-deoxyribonolactone when it binds to DNA containing 2-deoxyribonolactone. However, the covalent dead-end endonuclease III–DNA complex appears to be formed only at relatively high concentrations of endonuclease III. Furthermore, 2-deoxyribonolactone is recognized efficiently by both exonuclease III and endonuclease IV; the likelihood that such a dead-end complex will be generated in vivo is poor.

Most endonuclease III contains a [4Fe–4S] cluster. The [4Fe–4S] cluster of endonuclease III has been shown to be stable and resistant to both oxidation and reduction (68). However, it was recently demonstrated that this iron–sulfur center can be readily modified by nitric oxide (NO), forming a dinitrosyl iron complex. Nitric oxide modification of endonuclease III leads to the loss in both N-glycosylase and AP endonuclease activities (69). Interestingly, NO modification appears to be reversible both in vitro and in vivo, with full restoration of activity.

2.2. Formamidopyrimidine *N*-glycosylase

Escherichia coli Fpg was originally identified as an enzyme that removes alkylated forms of formamidopyrimidine (Me-FaPy) products (70). It was later shown to remove 8-oxoguanine from X-irradiated DNA (44,71). Formamidopyrimidine *N*-glycosylase was cloned, overexpressed, and purified in *E. coli*, and the enzyme was shown to recognize predominantly oxoG and the pyrimidine products of purines (44). However, recently, Fpg enzyme was also shown to recognize oxidized pyrimidine products such as uracil glycol, 5-hydroxycytosine, 5-hydroxyuracil, hydantoins, and urea (43,44). However, Fpg is significantly more active on DNA containing 8-oxoG and FaPy than DNA containing oxidized pyrimidines (43). Both the in vitro biochemical and in vivo genetic data provide ample evidence that the in vivo substrate for Fpg is 8-oxoG and FaPy derived from either adenine or guanine. It is, therefore, doubtful that Fpg protein contributes significantly to the repair of oxidized pyrimidines in vivo. It is interesting to note that a similar product, 8-oxoA, is not recognized by Fpg protein. Similar to endonuclease III, the activity of Fpg is also significantly affected by the nature of the base opposite oxoG, and Fpg is much more efficient in the removal of oxoG when opposite C (44,70,71). Formamidopyrimidine N-glycosylase removes oxoG at a reasonable rate when it is opposite C or T; however, the rate of oxoG removal by Fpg is extremely low when oxoG is opposite A. This is indeed quite interesting because oxoG:A pair is generated when oxoG is bypass. MutY, a mismatch DNA glycosylase, was shown to remove the misincorporated A opposite oxoG (72). The inadequacy of Fpg to remove oxoG from oxoG:A pair thus prevents the untimely removal of the base damage that can lead to the fixation of misincorporated base into the genome. Such a mutation-avoiding system has been extensively studied in E. coli (73). In addition to the N-glycosylase activity, Fpg possesses a robust AP lyase activity that catalyzes a β,δ-elimination reaction. The AP lyase activity of Fpg also exhibits opposite base preference, with AP:C pair being the best substrate (71). Interestingly, Fpg was shown to have the ability to remove the deoxyribose-phosphate residue located on the 5′ terminus [5′-deoxyribosephosphate (5′-dRP)] of DNA (74,75). Such a terminus is generated when an AP site is acted upon by an AP endonuclease such as endonuclease IV or exonuclease III (see Figs. 4 and 5) (See Chapter 20). Whether the ability of Fpg to remove dRP has any biological role is still not fully understood. Other base oxidation products reported to be recognized by Fpg include the pyrimidine ring opened product of 1,N6-ethenoadenine (57,58).

Three-dimensional structures of Fpg have been obtained for a number of bacterial Fpg including that of E. coli (44). The C-terminal α-helix domain containing the four-cysteine zinc finger motif has been identified as the primary DNA-binding motif. Another structural element that is important for DNA binding is also located at the C-terminal domain, the H-2T-H motif (44). In contrast to the H-h-H-GPD endonuclease III family of glycosylases that employ an internal lysine residue as the nucleophile, Fpg utilizes the N-terminal proline residue as the active site nucleophile (44). It is interesting to point out that Fpg sequence homolog is not present in eukaryotes. However, functional homolog of Fpg, oxoguanine glycosylase (Ogg1), is present in all eukaryotes cells (22,45,46). Recently, human repair proteins, hNeil1 and hNeil2, have been cloned and characterized. Despite the fact that hNeil1 and hNeil2 are structurally similar to the bacterial Fpg, functionally these proteins are more similar to E. coli endonuclease VIII, a DNA glycosylase that recognized predominantly oxidized pyrimidines (44,76–78). Ogg1 recognizes 8-oxoG as the predominantly substrate. As discussed earlier, the Ogg1 family shares no sequence similarities to Fpg. This family of repair proteins shares significant homology with the Nth family and is a member of HhH-GPD superfamily of repair proteins. Yeast Ogg1 was cloned by complementation assay using the E. coli fpg mutY mutator strain and also by reverse genetics via the covalent trapping of the protein with the substrates. Interestingly, yeast Ntg2 (yeast endonuclease III) and the endonuclease III homolog of Methanobacterium thermoautotrophicum also recognizes 8-oxoG (79). In human, 8-oxoG is predominantly removed by Ogg1 (22,46). The human Neil1 protein also exhibits a limited ability to recognize 8-oxoG (76).

The Ogg family of proteins also shows substrate specificities that are similar to the bacterial Fpg proteins and recognize predominantly oxoG opposite C, G, or T and exhibit extremely poor activity when oxoG is opposite A. Ogg1 also recognize

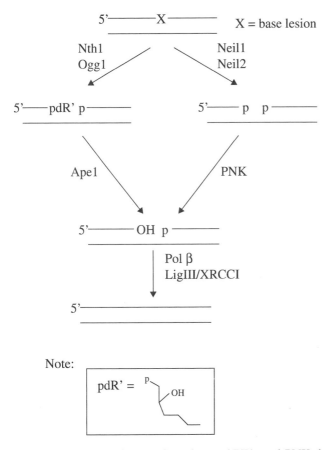

Figure 4 Base excision repair pathway: APE1- and PNK-dependent subpathways.

Fapy-G and Fapy-A (22,44–46). However, it is interesting to note that Ogg1 can recognize 8-oxoA at a low rate when the lesion is opposite C (80,81). In contrast to the Fpg, the AP lyase activity of Ogg1 catalyzes a β-elimination reaction (45). In general, the AP lyase activity of Ogg1 is much lower when compared with its *N*-glycosylase activity. These enzymes bind to uncleaved AP sites and dissociate slowly from the reaction products. The *N*-glycosylase activity of human Ogg1 is stimulated in the presence of Ape1 (82,83). The role of Ape1 in stimulating the activity of Ogg1 is thought to displace Ogg1 from the AP site or its nicked product, thus leading to an increase in the turnover rate of Ogg1 (83). Extensive description of the ability of AP endonuclease and other proteins to stimulate the glycosylase activity can be found in Chapter 3.

Despite differences in substrate specificity, the reactions catalyzed by Fpg and Ogg1 are remarkably similar to that of endonuclease III. The nucleophilic attack by the N-terminal proline (Fpg) or internal lysine residue (Ogg1) on C1 represents the initial step of the *N*-glycosylase reaction. The subsequent cleavage of the *N*-glycosydic bond and the concomitant formation of the intermediate covalent Schiff base intermediate are common among these glycosylases. Following the formation of the Schiff base intermediate, in a number of concerted steps, the covalent DNA–protein complex can yield either a 3′ 4-hydroxypentenal (Ogg1) or lead to the gen-

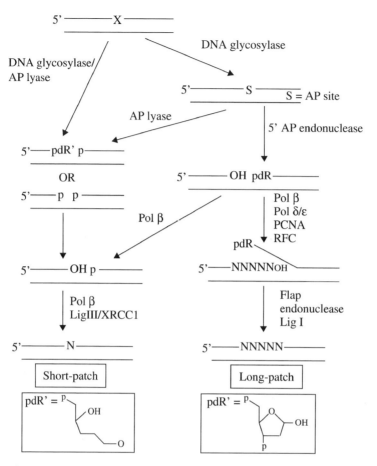

Figure 5 Base excision repair pathway: short-patch and long-patch subpathways.

eration of a 3′ phosphate (Fpg) and the release of the modified sugar moiety as 4-oxopent-e-enal. Extensive biochemical experiments and structural analysis (45,46) have yielded significant insight for understanding each of the steps of the reaction and the amino-acid residues involved in each of the reaction steps. Excellent reviews are available in the literature and highly recommended for readers who are interested in an in-depth understanding of the mechanism of action of these DNA glycosylases (44–46).

2.3. Endonuclease VIII

Endonuclease VIII was identified as a backup enzyme for the removal of oxidized pyrimidine in *E. coli* (84). Functionally, it is almost identical to endonuclease III and exhibits similar substrate specificity (84). In addition to recognizing a wide spectrum of oxidized pyrimidines, endonuclease VIII also recognizes 8-oxoG (85). In contrast to *E. coli* endonuclease III, which shows significant stereoselectivity in the recognition of thymine glycols, endonuclease VIII recognizes both the $5S,6R$ and $5R,6S$ stereoisomers with equal efficiency (53). Endonuclease VIII shows significant sequence homology to Fpg protein and is structurally similar to Fpg (44). Similar to Fpg, endonuclease VIII has an associated AP lyase activity that catalyzes an elimination reaction. Eukaryotic homologs of endonuclease VIII have been cloned and identified in human (hNeil1 and hNeil2) and mouse. Interestingly, *S. cerevisiae* appears to be lacking the structural homolog of both Fpg and endonuclease VIII.

3. REPAIR PATHWAYS FOR OXIDATIVE DNA DAMAGE

3.1. Base Excision Repair

It has been estimated that approximately 20,000 oxidative DNA damages are formed per human cell per day (86). The level of oxidative lesions increases substantially when cells are under oxidative stress or exposed to exogenously agents such as ionizing radiation and chemical oxidants. Many of these oxidative base damages influence biological endpoints such as cytotoxicity and mutagenicity significantly; it is therefore important for cells to be equipped with systems that can efficiently recognize and repair these damages. In general, oxidative damages are repaired predominantly via the base excision repair (BER) pathway. In BER, oxidative lesions are recognized by DNA glycosylases. Many of the DNA glycosylases that recognize oxidative damage also possess AP lyase activities that cleave the phosphodiester bond 3' to the AP site. Endonuclease III, endonuclease VIII, Fpg, and Ogg are the major enzymes that recognize oxidized bases. Other DNA glycosylases, such as uracil DNA glycosylase, do not have an associated AP lyase activity. The biochemical properties of these glycosylases have been discussed in detail in an earlier section of this chapter and elsewhere in this book. It is clear that most, if not all, glycosylases that recognize oxidized base have an associated AP lyase activity, and thus the removal of the oxidized base by these DNA glycosylases will generate a one base gap containing either 5' phosphate/3' 4-hydroxypentenal termini or 5' phosphate/ 3' phosphate termini. The former DNA gap is generated by DNA glycosylases that are associated with an AP lyase that catalyzes a β-elimination reaction (endonuclease III and Ogg). The latter is generated by DNA glycosylases that are associated with an AP lyase that catalyzes a β,δ-elimination reaction (endonuclease VIII, Fpg, Neil1, and Neil2). In bacteria, the one-base-gapped DNA is then further processed by a 5' AP endonuclease such as endonuclease IV or exonuclease III (48), which will remove these 3' moieties (both the phosphate and the 4-hydroxypentenal group), generating a 3' hydroxyl end. The repair process is then completed by the participation of a repair DNA polymerase and DNA ligase.

However, the processing of a gapped DNA generated in human cells is a bit more complicated. Owing to the apparent lack of 3' phosphatase activity in human APE (87) (see Chapter 20), the 3' termini generated by hNeil1 and hNeil2 can not be processed by human APE. The 3' phosphate is thought to be processed by polynucleotide kinase (PNK) which is present in abundance in mammalian cells (88,89). Therefore, as demonstrated in Mitra's laboratory, gapped DNA generated by human repair glycosylases is processed via two subpathways, an APE1- and a PNK-dependent pathways to generate a one-base-gapped DNA that contains a 3' hydroxyl terminus that is suitable for repair synthesis and DNA ligation (Fig. 4) (Sankar Mitra, personal communications; Lee et al. manuscript submitted).

The nature of the residues generated in a nick after the initial processing of the base lesions by *N*-glycosylases not only direct the BER pathway to proceed via the APE1-dependent or PNK-dependent pathway, but will also influence which repair polymerase will participate in subsequent repair synthesis and thus the size of the repair patch. When a nicked or AP site is generated after the initiation of the repair of base lesions, the nick or AP site is then further repaired via two alternative pathways, involving either the replacement of one nucleotide (short patch) or of 2–16 nucleotides (long patch) at the lesion site (Fig. 5). The short patch human BER pathway is very similar to the simple bacteria BER pathway and requires basically four major proteins: the human AP endonuclease APE1, DNA polymerase β (Pol β), and the DNA ligase III/XRCC1 heterodimer or DNA ligase I (see Ref. 90 for review and Chapter 3). When an AP site is acted upon by APE1, the enzyme hydrolyzes the phosphodiester bond 5′ to the AP site, leading to the generation of 3′-hydroxyl and 5′-dRP termini. In addition to the DNA polymerase activity, Pol β also has an intrinsic 5′ deoxyribosephosphatase (dRPase) activity (91). Therefore, upon binding of Pol β to the nicked site after the APE1 activity, Pol β will catalyze the polymerization reaction and at the same time will also remove the 5′-dRP residue. This will essentially result in a one-nucleotide replacement at the lesion site, and the resulting nick can then be efficiently ligated by DNA ligase III/XRCC1 heterodimer or DNA ligase I (90).

In contrast, long-patch BER involves polymerases other than Pol β. dRPase is intrinsic only to Pol β. When a polymerase other than Pol β is involved in the repair of a nick, the 5′-dRP will have to be removed by a process that involves additional enzymes before ligation can occur. It has been demonstrated in human cells that, when Pol δ/ε is the repair polymerase, strand displacement synthesis occurs, generating a 5′ flap that contains the dRP. This 5′ flap is removed by the Flap endonuclease I (FEN1), thus removing the dRP residue along with several additional nucleotides (usually between 2 and 10 nucleotides) (92). Therefore, the involvement of FEN1 is unique to the long-patch BER pathway. Other accessory proteins such as PCNA and RFC were also shown to be essential for the long-patch BER pathway (see Chapter 3). It is interesting to note that, in the presence of FEN1, Pol β can also initiate strand displacement synthesis, and under this condition, long-patch BER will result (93). Other proteins such as PARP also influence Pol β-dependent BER to adopt a long-patch repair mode. The stimulation of long-patch BER by PARP was suggested to operate under conditions where cellular ATP levels are severely compromised (94).

3.2. Role of Nucleotide Excision Repair in Oxidative DNA Damage Removal

As indicated earlier, the majority of the base lesions generated by ROS is recognized by DNA glycosylases and is thus repaired via the BER pathway. However, some base lesions such as 8-oxoguanine, thymine glycols, and AP sites were reported to be also recognized by the nucleotide excision repair pathway that involves a multiple protein complex (95–97). Most of the base lesions listed in Figs. 2 and 3 do not confer much distortion to the local DNA structure surrounding the base lesion. However, ROS also generate lesions that can result in significant local distortion to the DNA structure. Under anaerobic or hypoxic condition, hydroxyl radical interaction with purines can lead to the generation of 5′,8-cyclodeoxyadenosine and 5′,8-cyclodeoxyguanosine (98,99). These base lesions are the result of abstraction of the hydrogen from the C-5′ position of 2-deoxyribose by hydroxyl radicals, and in the absence of oxygen, the C-5′ sugar radical can add to C-8 of adenine or guanine to generate a cyclopurine deoxyribonucleoside. Studies from various laboratories indicated that these lesions are not recognized by any of the known DNA glycosylases; however, these lesions were efficiently recognized by the nucleotide excision repair proteins in *E. coli* and in human cells (100,101).

In addition to reacting with DNA, ROS also interact with membrane lipids generating a spectrum of highly reactive lipid peroxidation products such as malondialdehyde, hydroxynonenal, and other short-chain lipid enals (59). These highly reactive enals can react with DNA bases generating a spectrum of DNA adduct (59). Some of these, like ethenodC and ethenodA, are recognized by mismatch uracil glycosylase (MUG) (102,103) and 3-methyladenine DNA glycosylase (Alk A) (104,105) respectively, and these lesions are repaired via the BER pathway. Other adducts such as priopional-dG and 4-hydroxynonenal adducts are recognized by the nucleotide excision repair nuclease complex (106), further demonstrating the importance of nucleotide excision repair in the repair of certain types of oxidative DNA damage.

4. CONCLUSIONS

The extensive work from many laboratories has demonstrated that many repair mechanisms are involved in maintaining the stability of the genetic material: BER, nucleotide excision repair, direct reversal of DNA damage, and double-strand break repair. DNA damage tolerance pathway such as lesion bypass and recombination are also important in helping cells to cope with unpaired DNA lesion to allow survival, even at a risk that might lead to increased mutability. Studies in yeast demonstrate that these repair pathways do not operate independent of each other. Many of these base lesions are channeled through other repair pathways when the major pathway handling these lesions is compromised, suggesting a complex interplay and signaling among these repair pathways to achieve an efficient repair of the DNA damage and recover from the initial insult imparted by DNA damaging agents (106). Defects in some of these repair pathways have been shown to associate with one or more specific human diseases. Additionally, the repair of damaged DNA is also intimately associated with many distinct cellular processes such as DNA replication, DNA recombination, cell cycle checkpoint arrest, and other basic cellular mechanisms further underscoring the fundamental importance of the repair process for maintaining the genomic stability.

ACKNOWLEDGMENTS

The author would like to thank Cheryl Bowie for critically reading of the manuscript. This work is supported by grants received from NIH and NIEHS (CA 90860 and P01 ES11163).

REFERENCES

1. Von Sonntag C. The Chemical Basis of Radiation Biology. London: Taylor and Francis, 1987.
2. Huttermann J, Kuhnlein W, Teoule R. Effects of Ionizing Radiation on DNA: Physical, Chemical, and Biological Aspects. Berlin, Heidelberg, New York: Springer-Verlag, 1978.
3. Esposito LA, Melov S, Panov A, Cottrell B, Wallace DC. Proc Natl Acad Sci USA 1999; 96:4820–4825.

4. Haber F, Weiss J. Meth Phys Soc 1934; 147:942–946.
5. Bveris A. Medicine 1988; 58:350–356.
6. Aust SD, Chignell CF, Bray TM, Kalyanaraman B, Mason RP. Toxicol Appl Pharmacol 1993; 120:168–178.
7. Douki T, Reynaud-Angelin A, Cadet J, Sage E. Biochemistry 2003; 42:9221–9226.
8. Mehrdad Z, Noll A, Grabner EW, Schmidt R. Photochem Photobiol Sci 2002; 1: 263–269.
9. Eckardt T, Hagen V, Schade B, Schmidt R, Schweitzer C, Bendig J. J Org Chem 2002; 67:703–710.
10. Hayes RC, Petrullo LA, Huang HM, Wallace SS, LeClerc JE. J Mol Biol 1988; 201:239–246.
11. Basu AK, Loechler EL, Leadon SA, Essigmann JM. Proc Natl Acad Sci USA 1989; 86:7677–7681.
12. Ide H, Petrullo LA, Hatahet Z, Wallace SS. J Biol Chem 1991; 266:1469–1477.
13. Bregeon D, Doddridge ZA, You HJ, Weiss B, Doetsch PW. Mol Cell 2003; 12:959–970.
14. Hailer-Morrison MK, Kotler JM, Martin BD, Sugden KD. Biochemistry 2003; 42:9761–9770.
15. Sekiguchi M, Tsuzuki T. Oncogene 2002; 21:8895–8904.
16. Henderson PT, Delaney JC, Gu F, Tannenbaum SR, Essigmann JM. Biochemistry 2002; 41:914–921.
17. Wood ML, Esteve A, Morningstar ML, Kuziemko GM, Essigmann JM. Nucleic Acids Res 1992; 20:6023–6032.
18. Wood ML, Dizdaroglu M, Gajewski E, Essigmann JM. Biochemistry 1990; 29: 7024–7032.
19. Boiteux S, Gellon L, Guibourt N. Free Rad Biol Med 2002; 32:1244–1253.
20. Schulz I, Mahler HC, Boiteux S, Epe B. Mutat Res 2000; 461:145–156.
21. Pastoriza Gallego M, Sarasin A. Biochimie 2003; 85:1073–1082.
22. Fortini P, Pascucci B, Parlanti E, D'Errico M, Simonelli V, Dogliotti E. Mutat Res 2003; 531:127–139.
23. Please provide missing details.
24. Laspia MF, Wallace SS. J Mol Biol 1989; 207:53–60.
25. Laspia MF, Wallace SS. J Bacteriol 1988; 170:3359–3366.
26. Ide H, Kow YW, Wallace SS. Nucleic Acids Res 1985; 13:8035–8052.
27. Rouet P, Essigmann JM. Nucleic Acids Res 1985; 45:6113–6118.
28. Hayes RC, LeClerc JE. Nucleic Acids Res 1986; 14:1045–1061.
29. Evans J, Maccabee M, Hatahet Z, Courcelle J, Bockrath R, Ide H, Wallace SS. Mutat Res 1993; 299:147–156.
30. Kreutzer DA, Essigmann JM. Proc Natl Acad Sci USA 1998; 95:3578–3582.
31. Miller J, Grollman AP. Biochemistry 1997; 36:15336–15342.
32. Zhang Y, Yuan F, Wu X, Rechkoblit O, Taylor JS, Geacintov NE, Wang Z. Nucleic Acids Res 2000; 28:4717–4724.
33. Avkin S, Livneh Z. Mutat Res 2002; 510:81–90.
34. Haracska L, Prakash L, Prakash L. Proc Natl Acad Sci USA 2002; 99:16000–16005.
35. Kamiyama H, Miura H, Murata-Kamiyama N, Ishikawa H, et al. Nucleic Acids Res 1995; 23:2893–2898.
36. Guschbauer W, Duplaa AM, Guy A, Teoule R, Fazakerley GV. Nucleic Acids Res 1991; 19:1753–1761.
37. Henderson PT, Delaney JC, Muller JG, Neeley WL, Tannenbaum SR, Burrows CJ, Essigmann JM. Biochemistry 2003; 42:9257–9262.
38. Hazra TK, Muller JG, Manuel RC, Burrows CJ, Lloyd RS, Mitra S. Nucleic Acids Res 2001; 29:1967–1974.
39. Kornyushyna O, Burrows CJ. Biochemistry 2003; 42:13008–13018.
40. Duarte V, Gasparutto D, Yamaguchi LF, Ravanat JL, Martinez GR, Medeiros MHG, Di Mascio P, Cadet J. J Chem Soc 2000; 122:12622–12628.

41. Duarte V, Gasparutto D, Jaquinod M, Ravanat J, Cadet J. Chem Res Toxicol 2001; 14:46–53.
42. Matsumoto Y, Zhang QM, Takao M, Yasui A, Yonei S. Nucleic Acids Res 2001; 29:1975–1981.
43. Zaika IE, Perlow RA, Matz E, Broyde S, Gilboa R, Grollman AP, Zharkov DO. J Biol Chem 2003; 279:4849–4861.
44. Zharkov DO, Shoham G, Grollman AP. DNA Repair 2003; 2:839–862.
45. Stivers JT, Jiang YL. Chem Rev 2003; 103:2729–2759.
46. Denver DR, Swenson SL, Lynch M. Mol Bol Evol 2003; 20:1603–1611.
47. Rabow L, Venkataraman R, Kow YW. Prog Nucleic Acid Res Mol Biol 2001; 68: 223–234.
48. Wallace SS. Environ Mutagen 1983; 5:769–788.
49. Wallace SS. In: Scandalios J, ed. Oxidative Stress and the Molecular Biology of Antioxidant Defenses. Cold Spring Harbor, New York: Cold Spring Harbor Laboratory Press, 1997:49–90.
50. Masaki H, Yonei S, Sugiyama H, Kino K, Yamamoto K, Zhang QM. Nucleic Acids Res 2003; 31:1191–1196.
51. Miyabe I, Zhang QM, Kino K, Sugiyama H, Takao M, Yasui A, Yonei S. Nucleic Acids Res 2002; 30:3443–3448.
52. Eide L, Luna L, Gustad EC, Henderson PT, Essigmann JM, Demple B, Seeberg E. Biochemistry 2001; 40:6653–6659.
53. Miller H, Fernandes AS, Zaika E, McTigue MM, Torres MC, Wente M, Iden CR, Grollman AP. Nucleic Acids Res 2004; 32:338–345.
54. Kim JE, You HJ, Choi JY, Doetsch PW, Kim JS, Chung MH. Biochem Biophys Res Commun 2001; 285:1186–1191.
55. You HJ, Swanson RL, Harrington C, Corbett AH, Jinks-Robertson S, Senturker S, Wallace SS, Boiteux S, Dizdaroglu M, Doetsch PW. Biochemistry 1999; 38: 11298–11306.
56. Leipold MD, Workman H, Muller JG, Burrows CJ, David SS. Biochemistry 2003; 42:11373–11381.
57. Speina E, Ciesla JM, Wojcik J, Bajek M, Kusmierek JT, Tudek B. J Biol Chem 2001; 276:21821–21827.
58. Bajek M, Ciesla JM, Tudek B. DNA Repair (Amst) 2002; 1:251–257.
59. Marnett LJ. Carcinogenesis 2000; 21:361–370.
60. Eide L, Fosberg E, Hoff E, Seeberg E. FEBS Lett 2001; 491:59–62.
61. Marenstein DR, Ocampo MT, Chan MK, Altamirano A, Basu AK, Boorstein RJ, Cunningham RP, Teebor GW. J Biol Chem 2001; 276:21242–21249.
62. Marenstein DR, Chan MK, Altamirano A, Basu AK, Boorstein RJ, Cunningham RP, Teebor GW. J Biol Chem 2003; 278:9005–9012.
63. Bessho T. Nucleic Acids Res 1999; 27:979–983.
64. Klungland A, Hoss M, Gunz D, Constantinou A, Clarkson SG, Doetsch PW, Bolton PH, Wood RD, Lindahl T. Mol Cell 1999; 3:33–42.
65. Liu X, Choudhury S, Roy R. J Biol Chem 2003; 278:50061–50069.
66. Hashimoto M, Greenberg MM, Kow YW, Hwang JT, Cunningham RP. J Am Chem Soc 2001; 123:3161–3162.
67. Kappen LS, Chen CQ, Goldberg IH. Biochemistry 1988; 27:4331–4337.
68. Kuo CF, McRee DE, Cunningham RP, Tainer JA. J Mol Biol 1992; 227:347–351.
69. Rogers PA, Eide L, Klungland A, Ding H. DNA Repair (Amst) 2003; 2:809–817.
70. Chetsanga CJ, Lindahl T. Nucleic Acids Res 1979; 6:3673–3684.
71. Castaing B, Fourrey JL, Hervouet N, Thomas M, Boiteux S, Zelwer C. Nucleic Acids Res 1999; 27:608–615.
72. Fromme JC, Banerjee A, Huang SJ, Verdine GL. Nature 2004; 427:652–656.
73. Michaels ML, Miller JH. J Bacteriol 1992; 174:6321–6325.

74. Graves RJ, Felzenszwalb I, Laval J, O'Connor TR. J Biol Chem 1992; 267: 14429–14435.
75. Piersen CE, McCullough AK, Lloyd RS. Mutat Res 2000; 459:43–53.
76. Hazra TK, Izumi T, Boldogh I, Imhoff B, Kow YW, Jaruga P, Dizdaroglu M, Mitra S. Proc Natl Acad Sci USA 2002; 99:3523–3528.
77. Hazra TK, Kow YW, Hatahet Z, Imhoff B, Boldogh I, Mokkapati SK, Mitra S, Izumi T. J Biol Chem 2002; 277:30417–30420.
78. Hazra TK, Izumi T, Kow YW, Mitra S. Carcinogenesis 2003; 24:155–157.
79. Back JH, Chung JH, Park YI, Kim KS, Han YS. DNA Repair (Amst) 2003; 2:455–470.
80. Jensen A, Calvayrac G, Karahalil B, Bohr VA, Stevnsner T. J Biol Chem 2003; 278:19541–19548.
81. Girard PM, D'Ham C, Cadet J, Boiteux S. Carcinogenesis 1998; 19:1299–1305.
82. Hill JW, Hazra TK, Izumi T, Mitra S. Nucleic Acids Res 2001; 29:430–438.
83. Vidal AE, Hickson ID, Boiteux S, Radicella JP. Nucleic Acids Res 2001; 29:1285–1292.
84. Jiang D, Hatahet Z, Blaisdell JO, Melamede RJ, Wallace SS. J Bacteriol 1997; 179:3773–3782.
85. Hazra TK, Izumi T, Venkataraman R, Kow YW, Dizdaroglu M, Mitra S. J Biol Chem 2000; 275:27762–27767.
86. Ames BN, Shigenaga MK. Ann NY Acad Sci 1992; 663:85–96.
87. Izumi T, Mitra S. Carcinogenesis 1998; 19:525–527.
88. Karimi-Bushen F, Lee J, Tomkinson AE, Weinfeld M. Nucleic Acids Res 1998; 26:4395–4400.
89. Karimi-Busheri F, Daly G, Robins P, Canas B, Pappin DJ, Sgouros J, Miller GG, Fakhrai H, Davis EM, Le Beau MM, Weinfeld M. J Biol Chem 1999; 274:24187–24194.
90. Dogliotti E, Fortini P, Pascucci B, Parlanti E. Prog Nucleic Acid Res Mol Biol 2001; 68:3–27.
91. Podlutsky AJ, Dianova II, Wilson SH, Bohr VA, Dianov GL. Biochemistry 2001; 40(3):809–813.
92. Hiraoka LR, Harrington JJ, Gerhard DS, Lieber MR, Hsieh CL. Genomics 1995; 25:220–225.
93. Prasad R, Dianov GL, Bohr VA, Wilson SH. J Biol Chem 2000; 275:4460–4466.
94. Petermann E, Ziegler M, Oei SL. DNA Repair (Amst) 2003; 2:1101–1114.
95. Kow YW, Wallace SS, Van Houten B. Mutat Res 1990; 235:147–156.
96. Czeczot H, Tudek B, Lambert B, Laval J, Boiteux S. J Bacteriol 1991; 173:3419–3424.
97. Van Houten B. Microbiol Rev 1990; 54:18–51.
98. Fuciarelli AF, Miller GG, Raleigh JA. Radiat Res 1985; 104:272–283.
99. Jaruga P, Birincioglu M, Rodriguez H, Dizdaroglu M. Biochemistry 2002; 41: 3703–3711.
100. Brooks PJ, Wise DS, Berry DA, Kosmoski JV, Smerdon MJ, Somers RL, Mackie H, Spoonde AY, Ackerman EJ, Coleman K, Tarone RE, Robbins JH. J Biol Chem 2000; 275:22355–22362.
101. Kuraoka I, Bender C, Romieu A, Cadet J, Wood RD, Lindahl T. Proc Natl Acad Sci USA 2000; 97:3832–3837.
102. Saparbaev M, Laval J. Proc Natl Acad Sci USA 1998; 95:8508–8513.
103. Hang B, Downing G, Guliaev AB, Singer B. Biochemistry 2002; 41:2158–2165.
104. Saparbaev M, Kleibl K, Laval J. Nucleic Acids Res 1995; 23:3750–3755.
105. Dosanjhv MK, Roy R, Mitra S, Singer B. Biochemistry 1994; 33:1624–1628.
106. Johnson KA, Fink SP, Marnett LJ. J Biol Chem 1997; 272:11434–11438.
107. Doetsch PW, Morey NJ, Swanson RL, Jinks-Robertson S. Prog Nucleic Acid Res Mol Biol 2001; 68:29–39.

16

Recognition of Alkylating Agent Damage in DNA

Timothy R. O'Connor

Biology Department, Beckman Research Institute, City of Hope National Medical Center, Duarte, California, U.S.A.

1. MODIFICATION OF DNA BY SMALL ALKYLATING AGENTS

Alkylating agents from both exogenous and endogenous sources constantly modify DNA (1–12). Alkylating agents are found in plants, foods, and are produced industrially. Sources of endogenous alkylation include nitrosating agents and lipoperoxidation (1,3,12,13). The sites of modification on DNA bases are the same for all alkylating agents (Fig. 1a) and include all the exocyclic nitrogens and oxygens. Ring nitrogens without hydrogens (N7-G, N3-A, N1-A, N3-C, N7-A) are also targets for alkylation. In addition to alterations of DNA bases, the phosphodiester linkages can be converted to phosphotriesters (Fig. 1b). Although the modification sites are the same, the percentage of modification at each site depends on the specific alkylating agent. The damage resulting from bulky alkylating agents is generally removed by NER, but many less bulky modifications of DNA (groups with formula weights ca. 40 g/mol) are removed by at least five different types of repair. Prior to discussion of the mechanisms of recognition, a brief introduction regarding the generation of these damage sites is appropriate.

1.1. S_N1 vs. S_N2

Mechanisms of alkylation are separated into two pathways, either S_N1 or S_N2 based on the kinetics of the alkylation reaction (Fig. 2). Unimolecular nucleophilic substitution involves the formation of an intermediate carbonium ion prior to its attack on DNA (first-order kinetics). Therefore, the rate-limiting step for the reaction is the formation of a carbonium ion. By contrast, a reaction involving an S_N2 mechanism is dependent on both the alkylating agent and its target (second-order kinetics). Examples of some S_N1 and S_N2 alkylating agents are presented in Table 1. Differences in kinetics result in changes in the relative ratio of modifications at different sites on DNA (Table 2). Agents that react via an S_N1 mechanism tend to produce increased amounts of modified oxygens compared to agents that react via an S_N2 mechanism.

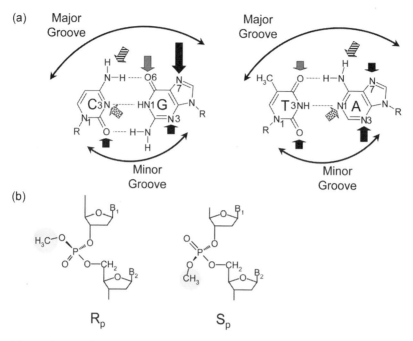

Figure 1 (a) Sites of alkylation damage on DNA bases. Gray arrows indicate the sites modified mainly by S_N1 reagents, black arrows by S_N2 agents, and spotted arrows by S_N2 agents in ss-DNA. The black striped arrows indicate exocyclic amino groups that are important in the formation of cyclized DNA base adducts. The locations of the major and minor grooves of DNA are indicated. R is the attachment of the base to the deoxyribose and the phosphodiester backbone. (b) Modified phosphodiester positions. Two isomers are possible when a phosphodiester is modified, either Rp or Sp. S_N1 reagents generally form more phosphotriester products than S_N2 reagents.

1.2. Direct and Metabolically Activated Alkylating Agents

Some alkylating agents react directly with DNA and do not require any activation to modify DNA. Many agents, including a number of carcinogens, require activation by the cytochrome P450 system to generate reactive species capable of modifying DNA. Table 1 lists some examples of direct acting alkylating agents; Table 3 lists some examples of metabolically activated agents.

1.3. Bifunctional Alkylating Agents

The alkylating agents shown in Table 1 and most of the repair discussed in this chapter will deal with monofunctional agents. Nonetheless, a number of alkylating agents exist that can react in a bifunctional manner to form cyclized DNA bases. Exocyclic amino groups and ring nitrogens can react with those bifunctional agents to permit cyclization. Those adducts present a problem for DNA polymerases to use as a template since the base pairing positions are occupied (Fig. 3). Much data have been accumulated on repair of potentially mutagenic etheno adducts formed by molecules such as chloracetaldehyde, a decomposition product of vinyl chloride. Such bifunctional alkylating agents can be generated from vinyl chloride. Although the etheno adducts have been most studied, cyclized adducts are also observed for larger alkylating agents (9). The structures formed with five member rings are generally

(a) S_N1

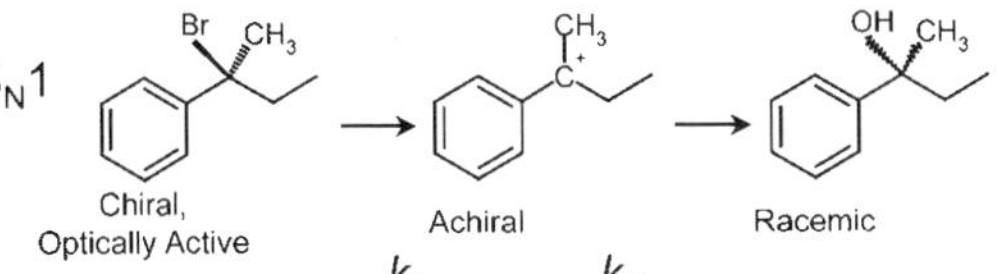

Chiral,
Optically Active   Achiral   Racemic

$$A + H_2O \xrightarrow{k_1} B \xrightarrow{k_2} C$$

$$\frac{d[C]}{dt} = k_1[B] \quad (\text{First Order, } k_1 \text{ units sec}^{-1})$$

(b) S_N2

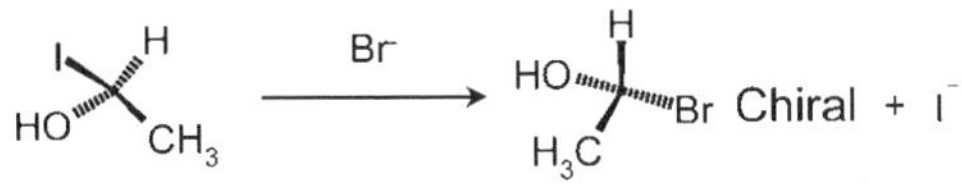

Chiral + I⁻

$$A + B \xrightarrow{k} C + D$$

$$\frac{d[C]}{dt} = k[A][B] \quad (\text{Second Order, k units M}^{-1}\text{sec}^{-1})$$

Figure 2 (a) Example of an S_N1 reaction. S_N1 reactions proceed with the formation of a carbonium ion intermediate that is rate-limiting. Thus the kinetics of the reaction is dependent only on the formation of the intermediate. The products are a racemic mixture at chiral centers since the planar intermediate can be attacked on either side. (b) Example of an S_N2 reaction. S_N2 reactions proceed by direct attack by the nucleophile. Both reactants are required to describe the kinetics, making the reaction second order. Since a transition state is formed with the chiral center, chirality is maintained.

removed by DNA glycosylases. Formation of cyclic ring products is not limited to five member rings. An example of a six member ring product is presented in Fig 3c as a point of comparison. The repair system or systems implicated in removal of six member ring structures is still not clear.

1.4. Phosphotriesters

Alkylating agents can also form phosphotriesters (Fig. 1b). In RNA, but not in DNA, phosphotriester formation can lead to single-strand breaks. Therefore, these modifications are believed to be relatively innocuous. There is a reduced rate of synthesis in polynucleotides with phosphotriesters (14–17), but this does not result in significantly increased mutation rates.

2. DNA REPAIR SYSTEMS FOR REMOVAL OF ALKYLATING AGENT DAMAGE

To date, five DNA repair systems in cells have been linked to the recognition of small alkylating agent DNA damage (Fig. 4):

1. O6-alkylguanine-methyltransferases.
2. α-ketoglutarate-Fe(II)-dependent oxygenases.
3. BER via DNA glycosylases.
4. NER.
5. Phosphotriester transfer (Ada N terminal) (only in prokaryotes).

Table 1 Examples of S_N1 and S_N2 Alkylating Agents: None of These Agents Require Activation to Alkylate DNA

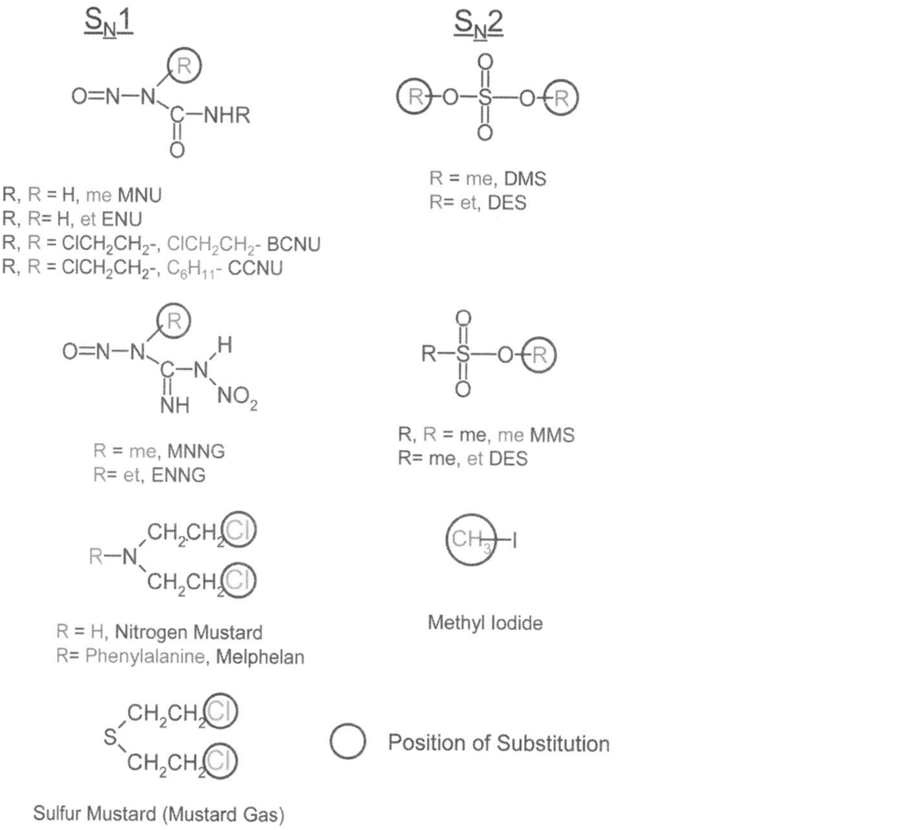

$\underline{S_N1}$

O=N—N(R)C(=O)—NHR

R, R = H, me MNU
R, R= H, et ENU
R, R = ClCH$_2$CH$_2$-, ClCH$_2$CH$_2$- BCNU
R, R = ClCH$_2$CH$_2$-, C$_6$H$_{11}$- CCNU

O=N—N(R)—C(=NH)—N(H)(NO$_2$)

R = me, MNNG
R= et, ENNG

R—N(CH$_2$CH$_2$Cl)(CH$_2$CH$_2$Cl)

R = H, Nitrogen Mustard
R= Phenylalanine, Melphelan

S(CH$_2$CH$_2$Cl)(CH$_2$CH$_2$Cl)

Sulfur Mustard (Mustard Gas)

$\underline{S_N2}$

(R)—O—S(=O)$_2$—O—(R)

R = me, DMS
R= et, DES

R—S(=O)$_2$—O—(R)

R, R = me, me MMS
R= me, et DES

(CH$_3$)—I

Methyl Iodide

◯ Position of Substitution

All repair mechanisms for elimination of DNA alkylation damage will be described in this chapter except that of NER, which is described elsewhere in this volume. Only the overlap between BER and NER will be discussed in this chapter. A single type of phosphodiester repair has been described. Except for phosphotriester repair, these systems are found in every cell and are important in removing alkylation damage. Tables 4 and 5 list *E. coli* and human genes involved in protecting cells from alkylating agent damage along with some of their properties. It is noteworthy that all the genes in this chapter code for rather small proteins less than 50 kDa in molecular mass.

3. O6-ALKYLGUANINE DNA METHYLTRANSFERASES—AGTs

O6-meG adducts in DNA are extremely mutagenic (18,19), but also block DNA polymerase extension that is generally associated with cytotoxicity (20,21). There was a debate in the literature regarding the relative cytotoxicity and mutagenicity of the O6-meG adduct which has been resolved by the realization that this lesion has both characteristics (22). Therefore removal of O6-meG from DNA both reduces toxicity and mutagenicity resulting from treatment with S_N1 type alkylating agents

Table 2 Alkylation Damage Observed in Vitro on DNA at Different Positions when Exposed to S_N1 or S_N2 Type Alkylating Agents

Alkylation Site	DMS (S_N2)	MNU (S_N1)	ENU (S_N1)
Adenine			
N1	1.9	1.3	0.2
N3	18	9	4.0
N7	1.9	1.7	0.3
Guanine			
N3	1.1	0.8	0.6
O6	0.2	6.3	7.8
N7	74	67	11.5
Thymine	—		
O2	—	0.1	7.4
N3	—	0.3	0.8
O4		0.4	2.5
Cytosine			
O2	(nd)	0.1	3.5
N3	(<2)	0.6	0.2
Diester	—		
		17	57

Source: Adapted from Ref. 138.

such as MNNG. The two adducts associated with repair by AGTs are O6-meG and O4-meT. Transition mutations of the type G:C→A:T are observed by a failure to repair O6-meG in DNA prior to replication, whereas a failure to repair O4-meT results in T:A→C:G transition mutations

The recognition of the O6-meG and O4-meT is accomplished by a highly conserved amino acid structure found in the transferases, -PCHR-, that includes a cysteine to which the methyl group is irreversibly transferred (Fig. 5) (23). Many derivatives of AGTs have been identified (Fig. 6). In repair, AGTs are unique in that the removal of each damaged base uses a single protein in a stoichiometric or suicide

Table 3 Examples of Small Alkylating Agents Requiring Enzymatic Activation: Arrows Correspond to One or More Steps Needed to Activate the Agents[a]

Dialkylnitrosamines → Alkyldiazonium ion

Vinyl Chloride → Chloracetaldehyde

NDMAOAc → Methyldiazonium ion

[a]This list is not exhaustive.

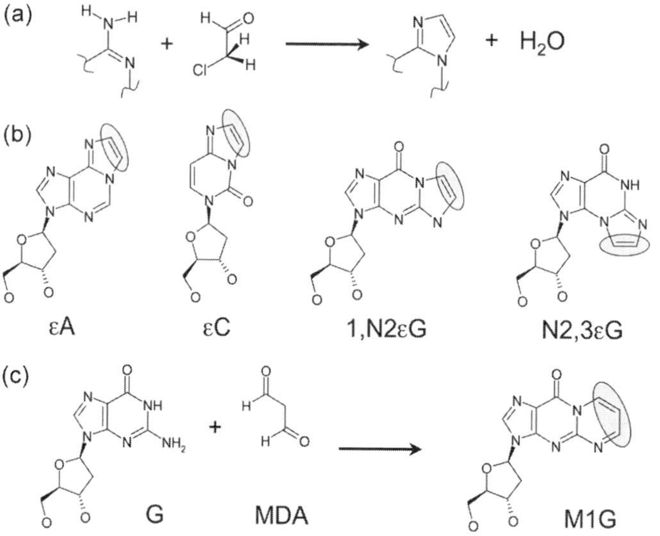

Figure 3 Cyclic ring adducts. The position of the cyclic adducts is highlighted by gray shading. (a) Modification of DNA producing five member cyclic etheno DNA adducts. (b) Etheno adducts of DNA bases. (c) Propeno M1G adduct formed by reaction of G with malondialdehyde (MDA).

reaction (24). Thus, one AGT molecule is responsible for the removal of one O6-meG adduct. This suggests that the enzyme has an important role in the elimination of DNA damage, since cells expend an enormous amount of energy to synthesize a protein that functions only once to remove each damaged base. Since repair is effected by a direct reversal of DNA damage, unlike repair based on DNA synthesis, this repair is error free.

All AGTs eliminate alkyl damage by a second order reaction, i.e., alkyl removal is dependent on the concentration of the AGT and the substrate. Elimination of the O6-meG is preferred compared to elimination of O4-meT, but the efficiency of the O6-meG:O4-meT removal varies with respect to species (25–27). In general, O4-meT removal is most efficient in bacteria and least efficient in mammalian cells (Table 4).

3.1. Procaryotic AGTs

3.1.1. Ogt

E. coli constitutively produces Ogt that has a single domain involved in transfer of methyl groups from O6-meG and O4-meT in DNA to Ogt (28,29). Another gene, *ada*, that codes for an AGT removing O6-meG, is also found in *E. coli*. The *ogt* gene was identified based on the fact that there was residual O6-meG activity in *E. coli* strains deficient in *ada*, this other AGT. In addition to the residual activity, strains deficient in both ada and ogt have reduced survival to MNNG and increased spontaneous mutation frequencies (Fig. 7) (30). It should be emphasized that the increase in mutation frequency is spontaneous, i.e., no external alkylating agent treatment is needed to produce a much larger mutation frequency than that observed in wild type cells. Ogt removes O6-meG at about the same rate as Ada, but Ogt is 173 and 84 times more efficient at abstraction of the alkyl moieties of O6-etG and O4-meT, than Ada respectively (Table 6) (31). Similar to Ada, a preference for repair at GG or CG

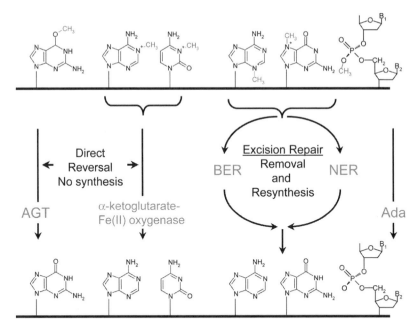

Figure 4 DNA repair systems implicated in repair of alkylation damage. The top line presents examples of adducts removed by each one of the systems. AGT is direct reversal by alkyltransferases to reform the original base both O6-meG and O4-meT are repaired by this system. The α-ketoglutarate-Fe(II) oxygenase repairs 1-meA and 3-meC damage also in a reversible manner. Both 3-meA and 7-meG are removed prior to resynthesis in the excision repair system (both BER and NER). The last system repairs phosphotriesters and is only known to occur via a reversal of DNA damage. Only bacterial proteins that have the N-terminal region similar to that of Ada (bacterial AGT) repair phosphotriesters. No mammalian equivalent is known. Modifications are shaded in gray.

sequences with the damage at the 3'G was suggested by three-fold fewer mutations than at AG or TG sequences (32). The differences in mutations observed were eliminated when higher doses of EMS were used, indicating that sequence context effects for Ogt repair are concentration dependent.

3.1.2. Ada

The *ada* gene is inducible following low-level exposure of *E. coli* to DNA alkylating agents and was cloned based on its capacity to complement *E. coli ada*-cells (33). Ada is 39 kDa and is composed of a 20 kDa N-terminal domain and a 19 kDa C-terminal domain that are involved in the removal of alkylation damage by a transferase mechanism (34) (Fig. 8). AGTs that have two domain structures repairing phosphotriesters and O-methylated bases like Ada (O6-meG and O4-meT) are found only in bacterial systems (35). The two domain structure is generally associated with inducible repair upon exposure to alkylating agents (36).

The 19 kDa C-terminal domain of Ada is responsible for the transfer of methyl groups from O6-meG and O4-meT. Ada-catalyzed methyl group removal from O6-meG is more efficient than O4-meT removal, but O4-meT removal is not as efficient as for Ogt (Table 6). The crystal structure of the C domain of Ada has shown that the binding in the active site of the enzyme is limited to relatively small molecules

Table 4 *E. coli* Genes Involved in Repair of Alkylating Agent Damage

Protein	Protein size (kDa)	Chromosomal location	Inducible expression	Substrates	Mutant phenotype
Ogt	19.2	1,398,260 → 1,397,745	No	O6-meG, O4-meT	Sensitive to alkylating agents
Ada	39.3	2,308,425 → 2,307,361	Yes	O6-meG, O4-meT	Sensitive to alkylating agents, hypermutagenic
AlkB	24.1	2,307,361 → 2,306,711	Yes	1-meA, 3-meC, ssDNA, RNA	Sensitive to alkylating agents
Tag	21.1	3,710,721 → 3,711,284	No	3-meA, 3-meG	Slightly sensitive to alkylating agents
AlkA	31.4	2,145,562 → 2,144,714	Yes	3-meA, 3-meG, 7-meG, Hx, X, εA,	Slightly sensitive to alkylating agents
Mug	18.7	3,213,114 → 3,212,608	No	εC, U/G	?

Table 5 Human Proteins Eliminating Alkylating Agent Damage from DNA

Protein	Protein size (kDa)	Human chromosome	Gene size (kbp)	Substrates	Mouse model	ES cell mutant phenotype	Animal mutant phenotype
MGMT	21.6	10q26	230	O6-meG >> O4-meT	Mgmt	Extremely sensitive to alkylating agents, increased mutations	Alkylation sensitive, Increased mutations
ABH1	43.9	14q24	36	?	Abh1	?	?
ABH2	29.3	12q 24.12	5.3	1-meA, 3-meC, dsDNA	Abh2	?	?
ABH3	33.4	11p11	39	1-meA, 3-meC, ssDNA, RNA	Abh3	?	?
MPG	32.2	16p13	6.5	3-meA, 3-meG, 7-meG, Hx, X, εA, 3-etA, 7-etG	Aag/Apng	ES cells sensitive to alkylating agents	Viable, fertile
TDG	46.0	12q24.1	23	εC, U/G, T/G	Tdg		Embryonic lethal

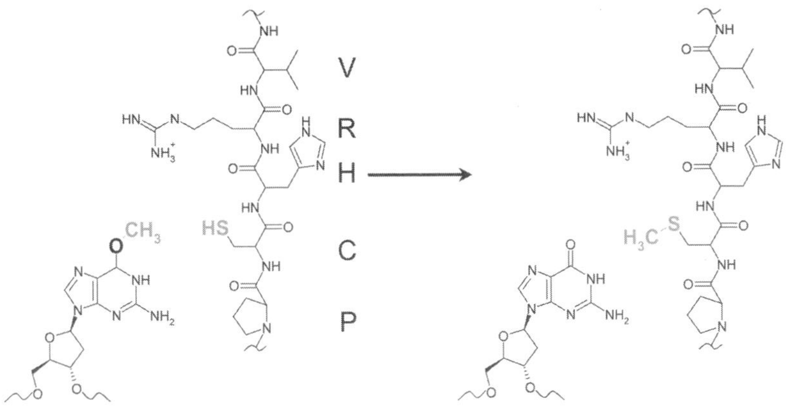

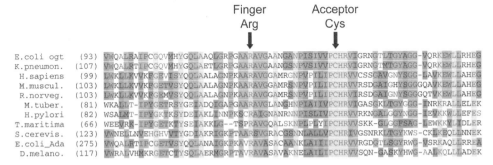

Figure 5 Mechanism for repair of O6-meG by the -PCHR-conserved sequence found in all AGTs. The methyl or alkyl group on the base is irreversibly transferred to the active site Cys that has a conserved consensus sequence of PCHRV/I. The methyl group transferred and the acceptor Cys are shaded in gray.

and excludes some larger molecules such as O6-BzG that serve as substrates or inhibitors of the mammalian enzymes (37). A structural study of the C-terminal domain of Ada indicated that the protein binds to its substrate using a base flipping mechanism (38), as observed for DNA glycosylases. The base flipping mechanism is assisted by an "arginine finger" to move the damaged base out of the helix (Fig. 9) (39). (Base flipping is discussed in detail elsewhere in this volume.)

The sequence preference of Ada has been examined indirectly using a forward mutagenesis assay to show that there is a bias for the formation of G:C→A:T transition mutations in GG sequences with the mutation at the 3' position (40). Therefore, it would appear that repair is poorer in the 3'G in GG sequences. A structural bias also has been directly demonstrated. Ada removes the methyl group from O6-meG in B-DNA in poly d(G-5meC) but does not remove the methyl group from the same polymer when changed to left-handed Z-DNA (41). The presence of such a

```
                Finger              Acceptor
                 Arg                  Cys
                  ↓                    ↓
E.coli ogt   (93)  VWQALRAIPCGQVMHYGQLAAQLGRPGAARAVGAANGANPISIVVPCHRVIGRNGTLTGYAGG-VQRKEWLLRHEG
K.pneumon.  (107)  VWQALRTIPCGQVMHYGQLAETLGRPGAARAVGAANGSNPVSIVVPCHRVIGRNGTMTGYAGG-VQRKEWLLRHEG
H.sapiens    (99)  LWKLLKVVKFCEVISYQQLAALAGNPKAARAVGGAMRSNPVPILIPCHRVVCSSCAVGNYSGG-LAVKEWLLAHEG
M.muscul.   (103)  LWKLLKVVKFGETVSYQQLAALAGNPKAARAVGGAMRGNPVPILIPCHRVVCSSCAVGNYSGG-LAVKEWLLAHEG
R.norveg.   (103)  LWKLLKVVKFGEMVSYQQLAALAGNPKAARAVGGAMRSNPVPILIPCHRVIRSDGAIGNYSGGGQTVKEWLLAHEG
M.tuber.     (81)  WKALLT-IPYGETRSYGEIADQIGAPGAARAVGLANGHNPIAIIVPCHRVIGASGKLTGYGGG-INRKRALLELEK
H.pylori     (82)  WSALMT-IPYGKTKSYDEIAKLINNPKSCRAIGNANRNNPISLIVPCHRVVRKNGALGGYNGG-IEVKKWLLEFES
T.maritima   (66)  WEEVRK-IPYGETKTYSEIAKKLG--TSPRAVGQALSKNPLPLYIPCHRVVSKK-GLGGFSAG-LEWKKYLIDLER
S.cerevis.  (123)  VWNELLNVEHGHVVTYGDIAKRIGKPTAARSVGRACGSNNLALLVPCHRIVGSNRKLTGYKWS-CKLKEQLLNNEK
E.coli_Ada  (275)  VWQALRTIPCGETVSYQQLANAIGKPKAVRAVASACAANKLAIIIPCHRVVRGDGTLSGYRWG-VSRKAQLLRREA
D.melano.   (117)  VWRALVHMKRGETCTYSQLAERMGRPTAVRAVASAVAKNELAILIPCHRVVSQN-GASKYHWG-AALKQLLLADEK
```

E. coli ogt: 2108312A:, Klebsiella pneumon.: NP_299875:, H.sapiens: NP_002403, Mus musculus: NP_032624, Rattus norvegicus: NP_036993, M.tubercul.: NP_215832, H.pylori: NP_223336, T.maritima: NP_228695, S.cerevis.: NP_010081, D.melano.: NP_477366

Figure 6 Alignment of conserved regions of various AGT protein sequences. The universally conserved sequences are shown in yellow, less conserved in blue, and the least conserved in green. The Cys that accepts the alkyl group and the Arg finger that helps flip the base out of the DNA helix to facilitate transfer are indicated. (*See color insert.*)

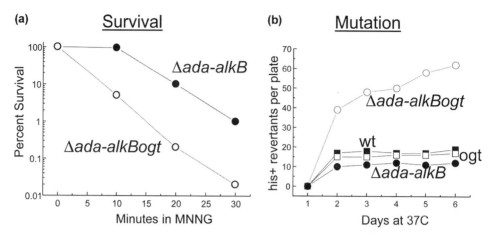

Figure 7 Biological properties of *E. coli ada* and *ogt* mutations (a) An *ada-alkB ogt* mutant is more sensitive to MNNG than the Δ*ada-alkB* mutant. (b) An *ada alkB ogt* mutant has a higher spontaneous mutation frequency than wild type (wt) or an *ada-alkB* mutant or an ogt mutant. The *ada* and *alkB* loci are contiguous with expression of the *alkB* gene dependent on expression of the ada gene (forming an operon). *Source*: Adapted from Ref. 30.

structural motif in DNA could therefore contribute to an increased mutation load on an organism from a failure to repair O6-meG damage.

3.1.3. Phosphotriester Repair by E. coli Ada

The 20 kDa N terminal fragment of *E. coli* Ada is associated with repair of the Sp isomer (Fig. 1) phosphotriester as discussed above (Fig. 8). However, until present, repair of these adducts has not been considered as critical except as a signal for the adaptive response. This is because phosphotriesters in DNA neither introduce strand breaks nor have miscoding properties that result in lethality or mutations. Bacterial cells have used this hitherto innocuous adduct to signal the initiation of the adaptive response to alkylating agents. Some other bacteria, e.g., B. subtilis, have similar mechanisms for the elimination of phosphotriesters. However, there is no parallel to the adaptive response to alkylating agents in mammalian cells or yeast and no means to repair phosphotriesters has been elucidated. No significant repair or biological consequence has as yet been associated with the formation of phosphotriesters in mammalian cells.

As mentioned, Ada serves as both a transcriptional activator and a DNA repair enzyme (42). The 20 kDa N-terminal domain removes methyl groups from the Sp phosphotriester binding to Cys at position 38 (43–45). This transfer of a

Table 6 Second-Order Rate Constants for the Excision of O6-meG and O4-meT by Selected AGTs

Protein	$k_{O6\text{-meG}}(M^{-1}s^{-1})$	$k_{O4\text{-meT}}(M^{-1}s^{-1})$	$k_{O6\text{-meG}}/k_{O4\text{-meT}}$	References
Ogt	2.9×10^7	2.1×10^5	138	31
Ada	2.5×10^7	2.5×10^3	1.0×10^4	31, 243–245
hMGMT	3.1×10^7	3.0×10^3	1.0×10^4	26, 246, 247
mMgmt	3.7×10^6	–	–	246
rMgmt	1.6×10^7	8.0×10^4	2.0×10^2	26

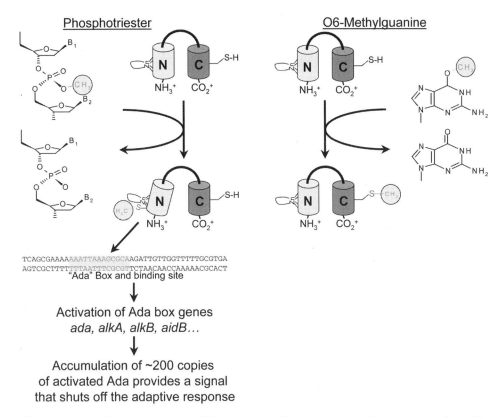

TCAGCGAAAAAAATTAAAGCGCAAGATTGTTGGTTTTTGCGTGA
AGTCGCTTTTTTTTAATTTCGCGTTCTAACAACCAAAAACGCACT
"Ada" Box and binding site

Activation of Ada box genes
ada, alkA, alkB, aidB...

Accumulation of ~200 copies
of activated Ada provides a signal
that shuts off the adaptive response

Figure 8 Two domain structure of *E. coli* Ada and activation by alkylation at position 38. Ada has an N-terminal domain with a Zn finger that transfers an alkyl group from only the Sp isomer to the protein that activates Ada as a transcription factor. Ada then can activate the transcription of other genes including itself and alkA. The Ada box for binding of the Ada as a transcription activator is highlighted in gray. Once more than 200 copies of the activated protein exist in cells, the adaptive response is shut off. The C-terminal domain is responsible for the transfer of alkyl groups from O6 of G or O4 of T. *Source*: Adapted from Ref. 250.

methyl group from the Sp phosphotriester isomer serves as the signal activating the adaptive response to alkylating agents (Fig. 8) (43). The C38 can also be directly alkylated by treatment with S_N2 type reactants, e.g., methyl iodide (46,47). When position 38 is methylated, Ada induces the expression of other genes including *aidB* and *alkA* during the adaptive response (43,48,49). The adaptive response to alkylating agents will be discussed elsewhere in this volume. The N domain of Ada has a Zn finger structure that could have a role in binding to DNA (50). Accumulation of more than 200 molecules of the activated N domain Ada inhibits the transcriptionally activated form of Ada and a consequent reduction in the adaptive response (51).

3.2. Eucaryotic AGTs

All eucaryotic AGTs isolated catalyze only the transfer of alkyl groups from O-alkylated bases. No enzyme similar to the N-terminal domain of the bacterial Ada has been identified. Because of its impact on human health, the most extensively studied AGT is hMGMT. This protein was first isolated based on complementation of either bacterial or human cells deficient in AGT activity and by use of probes

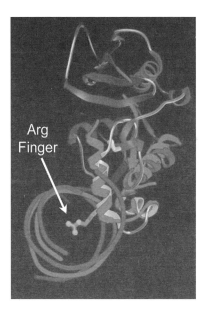

Figure 9 Structure of hMGMT, the human AGT. The Arg finger that assists base flipping in AGTs is shown in blue in the helix. *Source*: Adapted from Ref. 58. *(See color insert.)*

constructed from the active site of the bovine enzyme (52–55). hMGMT does not efficiently remove O4-meT compared to the Ogt and Ada found in *E. coli* (Table 6).

To probe the active site architecture of hMGMT in the absence of a crystallographic or NMR structure, a number of site directed mutants were constructed based both on homology and on random mutagenesis. The methyl group acceptor site of the enzyme was identified as Cys145 by construction of a C145A mutant that eliminates alkyl group transfer, but permits substrate binding (Fig. 5) (56). Based on homology to the bacterial Ada, hMGMT was proposed to move the target base out of the helix using a base flipping mechanism to find its substrate. An R128A mutant of hMGMT supported that hypothesis (see the chapter on base flipping) (57) and confirmed by structures of MGMT (38,58). In fact, the Arg at position 128 in hMGMT is conserved in analogous AGTs demonstrating the universal use of base flipping in AGTs (Fig. 5 and 9). Directed evolution was used to prepare mutants around the active site of MGMT to obtain enzymes that recognize O6-meG or O4-meT better than those found in nature (59–62). However, there is some controversy with respect to the efficient removal of the O6-meG and O4-meT by the generated mutants in different *E. coli* strains (27). The crystallization of hMGMT has permitted the identification of a Zn in the protein that increases the activity in vitro (63). The Zn does not increase binding to the DNA, but does evoke a small conformational change in hMGMT that could assist in stabilizing the structure and could help influence its regulation.

While many DNA repair proteins have a specific requirement for double-stranded DNA, hMGMT can bind to ssDNA (64). The hMGMT interacts with a 16-mer having a unique O6-meG with a protein:oligo ratio of 4:1, indicating that there is a cooperative interaction of the hMGMT with DNA, i.e., the binding of the first protein facilitates the binding of subsequent hMGMTs. Binding of hMGMT is accompanied by a structural change that has also been detected by crystallographic means using an O6-BzG to fit into the active site.

Some data on the sequence specificity of hMGMT have appeared in the literature. Three double-stranded dodecamers were alkylated with [^{3}H]-MNU to introduce both O6-meG and N7-meG (65). The oligos were subjected to reaction with cell extracts from the human colon cancer line HT-29 to assay for hMGMT activity. In these experiments, the HT-29 extracts excised the methyl group of O6-meG faster when preceded by a C than when preceded by a G, suggesting that there is sequence specificity to the rate of methyl group removal by hMGMT. More recent data have indicated that there is a position effect for excision of O6-BzG from DNA. In the context 5'-d[TGGGG(O6-BzG)G]-3', O6-BzG is more poorly repaired by hMGMT than is 5'-d[T(O6-BzG)GGGG]-3' and 5'-d[TGG(b(6)G)GGG]-3' (66). This difference in removal extended to 16-mer oligonucleotides. Changes in DNA structure also have an effect on excision by rMgmt. Similar to the experiments using the Ada protein, when extracts from rat hepatoma H4 cells are used to remove O6-meG from [^{3}H]-CH$_3$poly d(G-5meC), the extracts remove the methylated base damage in B-DNA, but are unable to excise the O6-meG from Z-DNA (67). As for the bacterial enzyme, this could have the consequence of increased mutagenesis in regions of Z-DNA found in mammalian cells.

One question that has been raised repeatedly is the reason that cells would expend resources to use a protein such as hMGMT a single time. One possible reason is that the modified protein could be used as a signal. In fact, MGMT also prevents transcriptional activation of the ERα (Fig. 10) (68). MGMT binds to ERα following the removal of an alkyl group. This interaction abrogates the binding

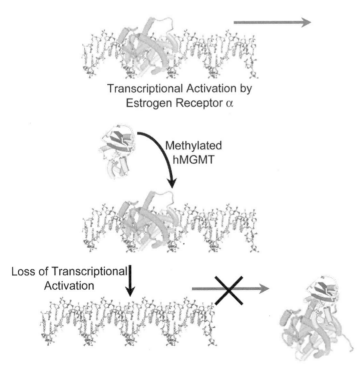

Figure 10 Illustration of transcriptional inactivation of estrogen receptor α (ERα) by hMGMT. The proteins and the DNA are not drawn to scale or for accuracy.

of ERα to the steroid receptor co-activator-1 that would result in the decreased production of transcripts from genes involved in cell proliferation. This would serve as a signal transduction from the O6-meG that would transfer the methyl group to MGMT, changing the conformation of the protein, and enabling MGMT to bind to ERα. That complex between MGMT and ERα would block transcription and subsequently replication. Thus, the repair reaction is only the initial step in transducing signals that would arrest cellular proliferation following exposure to alkylating agents and the use of a single protein for each repair reaction is amortized.

A specific inhibitor of hMGMT was developed using O6-BzG (Fig. 11) (69,70). The identification of O6-BzG as an inhibitor addressed a potential concern in chemotherapy: how tumor cells could be rendered more sensitive to chemotherapeutic agents. Administration of O6-BzG depletes cells of hMGMT, allowing more alkylation of the O6 position of G. This, in turn, increases the sensitivity of cells to a given alkylating agent. The cytotoxicity is observed via apoptosis. Other inhibitors of hMGMT have also been developed, but, as yet, the most widely used is O6-BzG (71). There is a significant difference in O6-BzG inhibition of hMGMT activity when compared to the *E. coli* orthologs. The bacterial AGTs (Ada and Ogt) are not significantly inhibited by O6-BzG, whereas the human enzyme is inhibited by O6-BzG even in cells (72). Even among mammalian AGTs, there are differences in inhibition by O6-BzG. For example, the mouse AGT enzyme, mMgmt, is more resistant to O6-BzG than the rat, rMgmt. But hMGMT is still more sensitive to depletion of its activity by O6-BzG than either of the rodent AGTs. The in vitro study of O6-BzG as an inhibitor of hMGMT has led to clinical trials (73,74).

Some hMGMT mutants could confer O6-BzG resistance to human cells. These mutants could have either beneficial or detrimental effects therapeutically depending

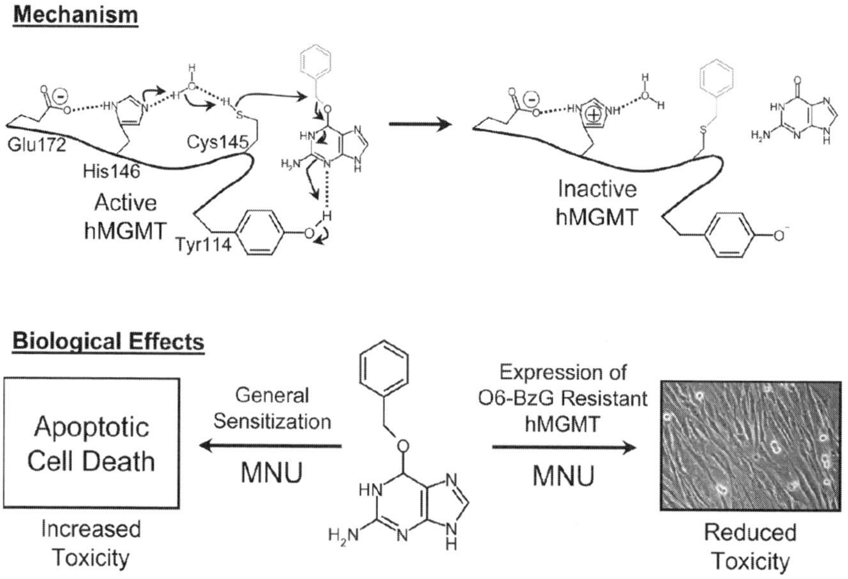

Mechanism

Biological Effects

Figure 11 O6-BzG inhibition of the activity of hMGMT. *Top*: Mechanism of O6-BzG modification of hMGMT. *Source*: Adapted from Ref. 58. *Bottom*: Biological effects of O6-BzG. O6-BzG depletes cellular levels of hMGMT and renders the cells more sensitive to alkylating agents than cells not exposed to O6-BzG.

on whether the mutation is natural or introduced during therapy. If some individuals are resistant to O6-BzG, therapy would not be successful. Alternatively, transfer and expression of hMGMT mutants resistant to O6-BzG in cell lineages extremely sensitive to alkylating agents (e.g., bone marrow) could reduce cell killing that would allow the use of higher doses of alkylating agents during chemotherapy that would eliminate tumors in other cell lineages. Some polymorphisms generating mutant hMGMTs exist in human populations (27,75–78), but their widespread existence in human populations and the risk they pose is unknown. The presence of such variants in a patient would signal that the use of O6-BzG in a treatment regimen would not result in an outcome different from the alkylating agent alone. To identify hMGMT mutants that could protect cells from alkylating agent cytotoxicity, directed evolution was used to isolate a mutant hMGMT, V139F/P140R/L142M, that is resistant to O6-BzG (61). Another in vitro selection process produced at least four mutants that were more resistant to BCNU than wild-type hMGMT when tested in human K562 cells (59). The mutations conferring resistance to O6-BzG were just outside the conserved PCHRV sequence. Despite the theoretical promise for the use of O6-BzG clinically, the precise conditions and protocols for its use remain to be established. A major problem associated with O6-BzG is that it will inactivate hMGMT in tumor and normal cells, rendering them both sensitive to DNA alkylating agents. This lack of specificity may present a problem for use of O6-BzG (79). Nonetheless the use of O6-BzG as an inhibitor of hMGMT has opened up a field of study that could result in enhancement of chemotherapeutic drugs. It is important to recognize that chemotherapeutic drugs are still the major treatment option in most forms of cancer.

Once hMGMT has accepted the methyl group, no further reactions are catalyzed, so the protein must be eliminated. Therefore, the fate of hMGMT following inactivation has also been explored. The degradation of hMGMT is a ubiquitination-dependent process whether the protein is inactivated by O6-BzG, BCNU, or NO (80). Several NO generating agents inactivate hMGMT (81,82). The position of nitrosation on hMGMT in this case was at C145, also the methyl group acceptor residue, in the active site. Thus the degradation of either methylated or nitroasated hMGMT should be similar. Following the inactivation, the amount of ubiquitination increased 2 hr posttreatment in the presence of MG132 an inhibitor of the 26S proteasome. Degradation of the C145A mutant was not observed, which confirmed the C145 as the modification site.

In some transformed human cell lines, it was demonstrated that there was no hMGMT activity and this was termed a mer- or mex-phenotype (83–86). Cells with this phenotype are extremely sensitive to S_N1 alkylating agents such as MNU compared to lines of similar lineage that are mer+. Moreover, expression of the 3′end of the *E. coli ada* gene coding for the activity reversing the O6-meG modification renders the mer− cells resistant to S_N1 alkylating agents (87–90). Treatment of cells with deazacytosine, an inhibitor of DNA methylases, restores a mer+ phenotype (91,92). This indicates that the mer− phenotype is caused by epigenetic silencing of the hMGMT promoter by enzymatic methylation at the 5 position of cytosine in CpG sites (93–96). Subsequently, hot spots for methylation were noted in the hMGMT promoter region and methylation was demonstrated to occur with a change in chromatin structure (91,92,95). The silencing has been shown to be a frequent occurrence in human cancers, particularly hepatocarcinoma (97–100). All of these data attest to the importance of hMGMT in the repair of damage inflicted on cells by alkylating agents.

The reports above demonstrate the central role of hMGMT in both cell survival and mutation avoidance. Despite the toxicity of mammalian cells in the absence of hMGMT, MGMT-mediated toxcity requires another repair system. It was noted that mer− cells (deficient in MGMT) that were also defective in long patch MMR were more resistant to alkylating agents than corresponding cell lines that expressed MMR genes or lines complemented with the defective MMR genes (101–103). MMR is absolutely necessary for cells to manifest sensitivity to alkylating agents. Complementation of cells with the defective mismatch repair genes re-establishes sensitivity to alkylating agents (104,105). Therefore, this presents questions concerning the function of other modified bases in cellular toxicity and the contribution of alkylation damage to the formation of MMR mutations that could render the cells more resistant to lethal adducts. A mouse model system with Mgmt and Mlh1 (a MMR gene) deficiencies constructed to investigate methylation tolerance in mammalian cells and a whole animal model is discussed below (105–107). Although tolerance to alkylating agents is increased in these methylation tolerant cells, mutation frequency is also increased. Thus, methylation tolerance can have a significant impact with respect to treatment of individuals using chemotherapy, because many tumors have defective MMR systems. The flawed MMR system could result in drug resistance to alkylating agents along with an increase in mutations that could induce secondary tumors. One model for cell killing via apoptosis in these cells is based on the model shown in Fig. 12. If an O6-meG adduct is present in replicating DNA, the MMR system can be involved in abortive processing of the lesion. If the adduct is not removed by hMGMT because it is on the template strand, it remains in the template DNA allowing the MMR system to recognize the adduct in a cyclic manner eventually leading to cell death via abortive repair (108).

The construction of murine mutant cell lines and animal models either deficient or overexpressing AGTs has clarified the role of AGTs in protecting cells or an organism from DNA damage. Mouse ES cells that are Mgmt−/− are highly susceptible to apoptotic cell death when exposed to MNNG (109). Mgmt−/− mice are approximately 10-fold more sensitive to MNU treatment than were wild type or heterozygotes (110). Mice that are deficient in Mgmt are extremely sensitive to nitrosourea and alkylating agents used in chemotherapy (111,112). Major damage to hematopoetic tissues is observed in both the wild-type and mutant mice, but the Mgmt−/− mice show damage at much lower doses than the wild-type mice. As discussed above, mammalian cells deficient in AGTs are extremely sensitive to killing by alkylating agents, but lethality can be significantly reduced by the loss of MMR. Mice with homozygous deletions of Mgmt and Mlh1 (Mlh1 is a component of MMR) exhibit phenotypes that permit the separation of lethality and mutagenesis, and mimic the phenotype of cells with methylation tolerance (106,107). Immortalized mouse cell lines which are either homo- or heterozygous for Mgmt or Mlh1 have been constructed and shown to be models that mimic the human system (Fig. 12) (106). Despite this similarity, the difference between mouse and human cell lines demonstrates that methylation tolerance is much greater in human cell lines compared to mouse cell lines (113). In the animal, similar effects are observed as those in cells, but effects with implications in chemotherapy are observed. Mice that are Mgmt−/−Mlh1−/− survive MNU treatment as well as wild-type mice, but develop an increased number of tumors, suggesting that although initial toxicity was avoided, the number of mutations introduced increased. In humans, this indicates that if individuals are deficient in MMR, the mutation frequency would increase following chemotherapy, without toxicity; that increase in mutation frequency could lead to secondary tumors. In contrast to the Mlh1−/− mice, the Mgmt−/−Mlh1+/+ mice have decreased thymus size and a reduced number of cells

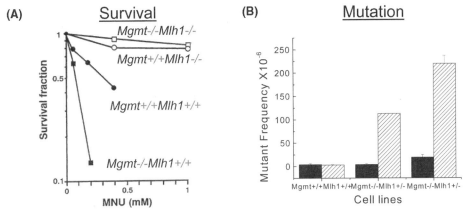

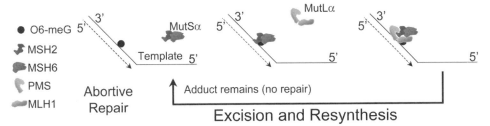

Figure 12 Methylation tolerance in immortalized mouse homo and heterozygotes for Mgmt and Mlh1 (106). (a) Survival of Mgmt and Mlh1 mutants with respect to MNU treatment. Note that even Mgmt−/− homozygotes survive better than Mgmt+/+ homozygotes in the absence of MMR (Mlh1−/−). (b) Increased mutation rates in hetero and homozygotes of Mlh1. (c) Model for methylation sensitivity in cells deficient in MMR. Cells that normally have MMR, but deficient in AGT mediated repair would abortively process an O6-meG adduct. Repeated cycles of repair would lead to cell death. However, without the AGT and MMR the toxicity of such lesions would be avoided, but not their mutagenicity. *Source*: Adapted from Ref. 106.

in bone marrow following MNU exposure (107). Bone marrow is one of the most sensitive sites to alkylating agents. The use of an animal model for methylation tolerance should be useful in elucidating the mechanism for resistance to alkylating agents observed in human cells (22,105,114,115). Other work has demonstrated that overexpression of AGTs protects animals from both the mutagenic and cytotoxic effects of alkylating agent damage and that expression of Mgmt correlates with animal survival (116–119).

4. AlkB—2-OXOGLUTARATE-DEPENDENT FE(II)-DEPENDENT OXYGENASES

Two of the systems for elimination of alkylating agent damage effect repair without concomitant unscheduled DNA synthesis and are characterized as error-free. AGT repair was discussed in the previous section. The other error-free repair system involved in alkylation damage repair is that of AlkB or α-ketoglutarate-Fe(II) oxygenases. The function of AlkB remained a puzzle for many years. The sensitivity of *E. coli alkB* mutants to S_N2 alkylating agent damage was known for almost

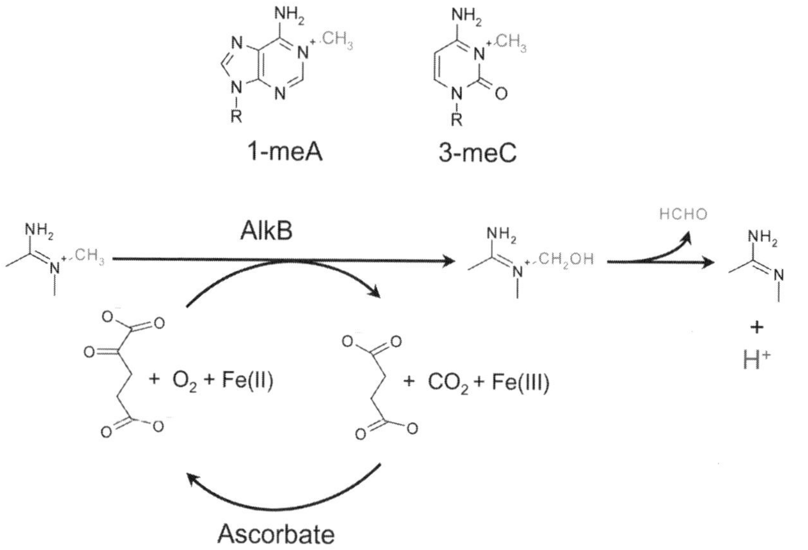

Figure 13 Mechanism of action of α-ketoglutarate/Fe(II) oxygenases. Two modified bases removed by AlkB, 1-meA and 3-meC, are shown at the top of the figure. The gray indicates the modifications removed. In the initial steps, the methyl group is converted to a hydroxymethyl group. The hydroxymethyl group is removed by AlkB as formaldehyde. Ascorbate helps to regenerate the Fe(II) center of AlkB. *Source*: Adapted from Refs. 124,125,131.

20 years without a mechanism to explain its protection (120). Transfection of *alkB* on a mammalian expression vector into mammalian cells conferred resistance to alkylating agents, suggesting that AlkB acts independently in repairing damage (121). Subsequently, more recent work demonstrated that ss phages modified with S_N2 alkylating agents, such as DMS, had significantly reduced survival compared to double-stranded phages (122). The function of AlkB was initially suggested by bioinformatics as a 2-oxoglutarate-Fe(II)-dependent oxygenase (123), but the actual substrates for AlkB were not revealed. Recently, two groups have used that bioinformatics result coupled with the observation of sensitivity of ssDNA to S_N2 reactions in *alkB* deficient strains (122) to show that AlkB is responsible for the removal of 1-meA and 3-meC from DNA (124,125). Both 1-meA and 3-meC are modifications that occupy base pairing positions (Fig. 1) and would halt DNA synthesis if not eliminated. This accounts for the toxicity observed for *E. coli alkB* mutants when presented with ssDNA modified by S_N2 agents. In genomic DNA, such modifications at the 1 position of A and at the 3 position of C would be more prevalent in ss-DNA regions compared to ds-DNA regions. Since single-stranded regions could exist at or near DNA replication forks, those regions would be particularly sensitive to the absence of AlkB type repair. In the reaction mechanism, AlkBs release formaldehyde that permits the formation of the initial base without repair synthesis (Fig. 13).

4.1 Procaryotic 2-Oxoglutarate-Dependent Fe(II)-Dependent Oxygenases

AlkB removes 1-meA and 3-meC from both DNA and RNA. In addition to 1-meA and 3-meC, the *E. coli* enzyme can eliminate ethyl adducts at the same positions producing acetaldehyde instead of formaldehyde (126). AlkB was cloned based on complementation of *alkB* mutants using a genomic library (120,127,128) (Fig. 14). Prior to sequencing of the *E. coli* genome, the *alkB* gene was located 3' to the ada gene responsible for the adaptive response to alkylating agents (127). Other significant data have shown that AlkB functions to remove damage efficiently from modified ss phages and is not active on DNA containing 3-meA (122,129). A bioinformatics search identified a family of structures including AlkB that placed AlkB and other similar potential repair proteins from different organisms in the α-ketoglutarate-Fe(II) oxygenase family (123) (Fig. 15). A putative secondary structure could also be modeled from the similarity of the α-ketoglutarate-Fe(II) oxygenases (Fig. 16) (123,130). The AlkB protein was purified and substrates prepared using ss polymers or ss phage treated with [^{3}H]-MNU (124,125,128). This generated a significant number of sites in the substrate that were modified at N1-A or N3-C.

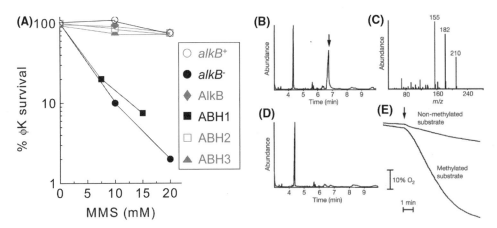

Figure 14 Complementation of *E. coli alkB* mutant phenotype by overexpression of *E. coli* alkB, human ABH1, ABH2, or ABH3 and reversal of damage, release of formaldehyde, and consumption of oxygen during AlkB mediated repair. (a) A host cell reactivation assay of alkylating agent damaged φK ss phage DNA was monitored in several *E. coli* strains. Two different types of plasmids were used for overexpression, but little difference is observed using the different constructs and different *alkB* deficient strains. For illustrative purposes, these are included in a single figure. The strains used were the *alkB22* derivative of *E. coli* BL21(DE3) as *alkB⁻* (for the ABH1 complementation) and AB1157 as the *alkB⁺* or the *alkB22* strain for AlkB-deficient strains for the rest of the series. The *alkB⁻* and *alkB⁺* designations have the pGEX6P-1 plasmid. The AlkB designation is the alkB-strain hosting the pBAR32 plasmid that has the alkB gene expressed. The ABH1 designation is the alkB-strain hosting the ABH1 cDNA in the pET15 plasmid. The ABH2 designation is the alkB-strain hosting the ABH2 cDNA in a GST fusion plasmid (pGST.ABH2). The ABH3 designation is the alkB-strain hosting the ABH3 cDNA in a GST fusion plasmid (pGST.ABH3). (b) Methylated oligo(dA)·oligo(dT) was incubated with AlkB. The products were derivatized with PFPH, an agent used to volatilize products for GC/MS analysis, and analyzed by gas chromatography mass spectrometry. Gas chromatography traces are shown of derivatized products generated when the DNA was methylated (b) or not methylated (c). (d) Mass spectrum of the product indicated by the arrow in a. (e) Consumption of O$_2$ during incubation of AlkB with either methylated or non-methylated oligo(dA)·oligo(dT) was determined by using an oxygen electrode. AlkB (9mM) was added at the time point indicated by the arrow. *Source*: Adapted from Ref 125.

```
H.sapiens ABH1 (133) KHMSKEETQDLWEQSKEFLRYKEATKRPRSLLEKLRWTVGYHYN-WDSKKYSADHYTPFPSDLGFLSEQVAAACGFEDFRAEGLINY
E.coli          (39) DVASQSPFRQMVTPGGYTMSVAMTNCGHLG------WTTHRQGIYSPIDPQINKWPAMPQSFHNLCQRAATAAGYPDFQPDACINR
S.cerevisiae   (180) VKSHKNLFPTLTEQIIQHNINQDFTESTYDEDYVFSSIWANFMEGLINHYLEKVIVYSEMKVCQQLYKPMMKIISLYN-EYNELMVKS
S.pombe         (60) ESVQRNIINNVPKELSIYGSGKQSHIY--------------------IPFPAHINCLNDYIPSDFKQRLWKGQDA---EAIIMQV
A.thaliana     (116) FQQSWTFFDYLDKHIPWTRPTIRVFGRSCLQPRDTCYVASSLTAIVYSGYRPTSYSWDDFPPLKEILDATYKVLPGS--RFNSLLLNR
H.sapiens ABH3  (99) VKEADWILEQLCQDVPWKQRTGIREDT--YQQPRLTAWVGELPYTYSRITMEPNE-HWHPVLRTLKNRIEENT-GH--TFNSLLCNL
M.musculus1     (99) LKEADWILEQLCKDVPWKQRMGIREDVT--YPQPRLTAWYGELPYTYSRITMEPNE-HWLPVLWTLKSRIEENT-SH--TFNSLLCNF
H.sapiens ABH2  (75) KAEADEIFQEIEKEVEYFTGALARVQVFGKWHSVPRKQATYGDAGLTYFSGLETLSKPWIPVLERIRDHVSGVT-GQ--TFNFVLINR
M.musculus2     (53) KAEADIFRELEQEVEYFTGALAKVQVFGKWHSVPRKQATYGDAGLTYFSGLLTPKPWVPVLERVRDRCEVT-GQ--TFNFVLVNR
R.norvegicus    (53) KAEADQIFRELEQEVEYFTGALAKVQVFGKWHSVPRKQATYGDAGLTYFSGLLTPKPWIPVLERVRDQVCRVT-GQ--TFNFVLVNR

H.sapiens ABH1 (222) YRLDSTLGIHVDRSELDHSKPLL-----SFSFGQSAIFLLGGLQRDEAPTA--------------------M
E.coli         (122) YAPGAKLSLHQDKDEPDLRAPIV----SVSLGLPAIFQFGGLKRNDPLKR-----------------L
S.cerevisiae   (268) EKNGFLPSLQDSENVQGKEKESKDDAVSQERLERAQKLMQAREDIPKTISKELTLLSEMYSTLSADEQDYELDEFVCCAEEYIELEY
S.pombe        (123) YNPG-DGIIPHKLEMFGDG--------VAIFSFLSNTTMIFTHPELKLKS----------------KI
A.thaliana     (203) YKGASDYVAWHADDEKIYGPTPE----IASVSFGCERDFVIKKKDEESSQGKTGDSG------PAKKRLKRSSREDQQSL
H.sapiens ABH3 (181) YRNEKDSVDWHSDDEPSLGRCPI----IASLSFGATRTEMRKKPPPEENG---------------DYTYVERVKI
M.musculus1    (181) YRDEKDSVDWHSDDEPSLGSCPV----IASLSFGATRTEMRKKPPPEENG---------------DYTYVERVKI
H.sapiens ABH2 (161) YKDGCDHIGEHRDERELAPGSP-----IASVSFGACRDFVFRHKDSRGKS---------------PSRRVAVVRL
M.musculus2    (139) YKDGCDHIGEHRDERELAPGSP-----IASVSFGACRDFIFRHKDSRGKR---------------PRRTVEVVRL
R.norvegicus   (139) YKDGCDHIGEHRDERELAPGSP-----IASVSFGACRDILFRHKDSRGKR---------------PRRAVEVVRL

H.sapiens ABH1 (269) FMHSGDIMIMSGFSRLLNHAVPRVLPNPEGEGLPHCLEAPLPAVLPRDSMVEPCSMEDWQVCASYLKTARVNMTVRQVLATDQNF
E.coli         (169) LLEHGDVVVWGESRLFYHGIQPLKAGFHPLT------I-----------DCRYNLTFRQAGKKE--
S.cerevisiae   (358) LPALVDVLFANCGINNFWKIMIVLEPFFYYIEDVGGDDDEDNVDN-S----------EGDEESLLSRNVEGDDNVV
S.pombe        (167) RLEKGSLLDMSTARYDWFHEIPFRAGDWVMNDGEEKWVSR----------SQRLSVTMRRIIENHVFG
A.thaliana     (274) TLKHGSLLVRGYTQRDWIHSVPRAKAEGT--------------------RINLTFRLVI
H.sapiens ABH3 (238) PLDHGTLLINEGAIQADWQHRVPKEYHSREP-------------RVNLTFRTVYPDPRGA
M.musculus1    (238) PLDHGTLLINEGAIQADWQHRVPKEYHSRQP-------------RVNLTFRTVYPDPRGA
H.sapiens ABH2 (217) PLAHGSLLMNHPINTHWYHSLPVRKKVLAP-------------RVNLTFRKILTKK--
M.musculus2    (195) QLAHGSLLMNNPPINTHWYHSLPIRKRVLAP------------RVNLTFRKILPTKK--
R.norvegicus   (195) QLAHGSLLMNNPINTHWYHSLPIRKRVLAP-------------RINLTFRKILPTKK--

ABH1: AAH25787, E.coli: NP_416716, S. cerevisiae: NP_012848, S.pombe: NP_565530, A.thaliana: NP_594941,
ABH3: NP_631917, M.musculus1: XP_130317, ABH2: XP_058581, M.musculus2: XP_132383, R.norvegicus: XP_222273
```

Figure 15 Sequence homology of AlkB derivatives. Sequences were aligned using Clustal W software with Vector NTI. The universally conserved sequences are shown in yellow, less conserved in blue, and the least conserved in green. The sequence reference number in GenBank is indicated at the bottom of the figure. Note that the conservation with *E. coli* AlkB among the higher eucaryotes is minimal. (*See color insert.*)

Enzymatic assays using those substrates were performed and the products separated using HPLC to demonstrate the loss of 1-meA or 3-meC. Release of HCHO (formaldehyde) was shown using a chemical reaction (124,125). The removal of 1-meA or 3-meC is favored using ssDNA. There are at least two co-factors that were noted initially, i.e., α-ketoglutarate and Fe(II). Ascorbate also can play a role to help convert the Fe(III) to Fe(II) to regenerate AlkB (131). The turnover number for AlkB has been estimated at ~0.2 min^{-1} that is similar to that of many DNA repair enzymes (124). The putative active site predicted by sequence alignment includes H131, D133, and H187 (132) (Fig. 17). Mutation of one of the sites to a C and the modification of the substrate to provide a disulfide crosslink showed that all of the mutant proteins can form attached products with the DNA (Fig. 17). Therefore, H131, D133, and H187 are in the active site of the AlkB. Further detail on the structure of AlkB awaits the elucidation of a crystal structure of the protein and with its bound substrate and co-factors.

4.2. Eucaryotic 2-Oxoglutarate-Dependent Fe(II)-Dependent Oxygenases

Derivatives of AlkB have been identified in a number of eucaryotic organisms (Fig. 15). Originally, a potential AlkB deriviative, ABH (now ABH1), was identified from human EST clones as having sequence homology with AlkB (133). However, the

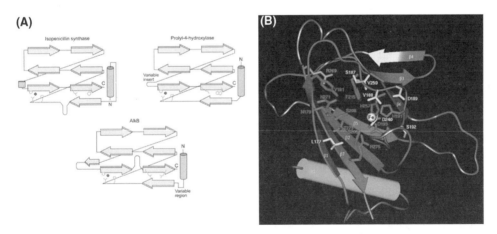

Figure 16 Structural model for AlkB. (A) Topological diagrams for three members of the 2OG-Fe(II) dioxygenase superfamily. The diagrams are based on the experimentally determined structures for *E. nidulans* isopenicillin N synthase (PDB: 1ips) and structural models of prolyl-4-hydroxylase and AlkB. The amino acid residues of the active site and the Fe(II) ion are shown. (B) Three-dimensional model for the core region of ABH3. By using the known crystal structures of the three αKG-dependent DNA dioxygenases as templates and the alignment, a three-dimensional model was constructed for the core of ABH3 by using the program 3D-JIGSAW. The overall quality of side-chain packing and stereochemistry of the final model were checked by using program QUANTA 2000 (Accelrys, Orsay, France). The conserved α-helix is colored cyan. The β-strands, forming the jelly roll structure, are colored orange. The approximate location of the Fe(II) ion is represented as a white sphere. Residues involved in metal binding in AlkB identified by disulfide cross-linking. *Source*: Adapted from Refs. 123,130,132,251. (*See color insert.*)

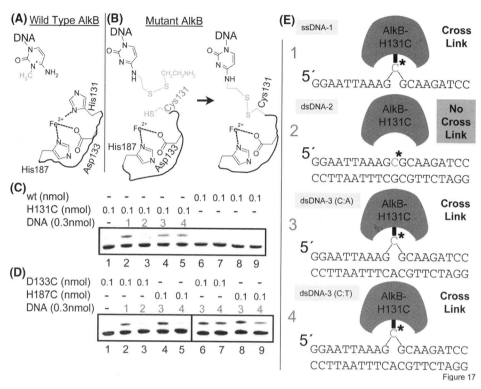

Figure 17 Residues involved in metal binding in AlkB identified by disulfide cross-linking. (A) Proposed substrate binding in the putative AlkB active site. (B) Replacement of one of the three putative metal-binding ligand His131 to Cys131 and introduction of a thiol-tether in DNA (positions in e with the C* indicate the position of the base with the thiol-tether) provide the partners for the formation of a disulfide cross-link. (C) Oligonucleotides used in cross-linking. (D) SDS gel analysis of the cross-linking reactions between H131C AlkB and the DNAs shown in (E). Lane 1 is a size standard for AlkB. Disulfide cross-linking between the protein and DNA results in the appearance of a new band having retarded mobility. Lanes 2–5: Cross-linking results between H131C AlkB and DNAs shown in (E) after 24 hr of incubation at 4°C. Lanes 5–9: Controls with wild-type AlkB. (D) SDS gel analysis of the cross-linking reactions between D133C AlkB and DNAs in e and H187C AlkB and DNAs in e. The reactions were analyzed after 24 hr of incubation at 4°C. (E) Models for the cross-linking reactions along with the substrates bound to AlkA. The C* with a heavy line to the AlkB representation indicates that a cross-link has formed. The only instance where the AlkB does not cross link is when the DNA has a G/C base pair and therefore remains double-stranded. *Source*: Adapted from Ref. 132.

ABH1 protein did not have any AlkB activity and no increased survival was observed upon complementation of an *alkB* deficient *E. coli* strain with ABH1 cDNA (130,134). Thus, ABH1 has no ascribed function as an AlkB ortholog. Further searching of the human genome sequence revealed ABH2 and ABH3, both human AlkB homologs. ABH2 is located adjacent to the human UNG gene (134). When purified, both ABH2 and ABH3 have activity in the enzymatic assay and complement *alkB*-deficient *E. coli* strains (Fig. 14). The three AlkB related genes are found on different chromosomes (Table 5). Both AlkB and ABH3 prefer ssDNA as a substrate, whereas ABH2 has a preference for double-stranded DNA.

ABH3 and AlkB also repair alkylated RNAs, whereas ABH2 does not. Although AlkB can use even 1-medATP as a substrate, the ABH2 and ABH3 do not function on substrates shorter than three nucleotides (126).

In transient transfection experiments, ABH2 and ABH3 were localized in the nucleus, and in S phase, ABH2 co-localizes with PCNA, suggesting that ABH2 is present near replication forks (Fig. 18) (134). ABH3, on the other hand, is sometimes localized to the nucleoli and in the nucleoplasm. The substrate specificity of ABH3 and its location predict a role for the repair of DNA and RNA in the process of transcription. Although this is a promising start in the study of mammalian AlkB homologs, much work remains with respect to the biochemistry and biology of these proteins.

The reversal of damage catalyzed by AlkB and its human homologues presents an interesting target for chemotherapy. Similar to that of hMGMT, ABH2 and ABH3 act independently and perform repair in an error-free manner. The fact that the prokaryotic AlkB enzyme increases survival when expressed in human cells also supports its use as a potential target for chemotherapy (121). Some inhibitors of AlkB have been identified that could eventually be used for ABH2 and ABH3 (131). In the future, work on the biological role of ABH1–3 will be required not only

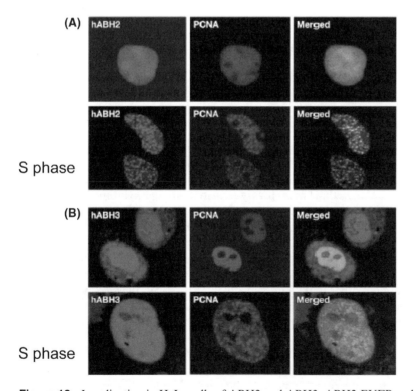

Figure 18 Localization in HeLa cells of ABH2 and ABH3. ABH2-EYFP and ABH3-EYFP fusion proteins were constructed and transfected into HeLa cells that were either outside or in S phase cells with respect to a PCNA-ECFP fusion protein (134). (A) Localization of hABH2 (hABH2–EYFP) and PCNA (ECFP–PCNA) in co-transfected cycling living cells outside S phase (no label) and in S phase. (B) Localization of hABH3 (hABH3–EYFP) and PCNA (ECFP–PCNA) in cells outside S phase (no label) and in S phase. *Source*: Adapted from Ref. 134. (*See color insert.*)

to understand their function in human cells, but to exploit those results for therapeutic purposes.

5. DNA GLYCOSYLASES—BASE EXCISION REPAIR

DNA glycosylases initiate BER on DNA damaged by alkylating agents. The BER pathway has been described at several points in this volume, so only the recognition and the specific DNA glycosylases involved in repair of alkylation damage will be discussed. The DNA glycosylases involved in this process can be broadly divided into those excising 3-meA (methylpurine-DNA glycosylases) and those excising εC (thymine-DNA glycosylases).

5.1. 3-Methyladenine-DNA Glycosylases

These DNA glycosylases have at least three identified classes. All the proteins in this group remove 3-meA, a toxic adduct, since it occupies a position involved in base pairing that blocks extension by DNA polymerases (135). These DNA glycosylases were among the first DNA glycosylases examined biochemically (136,137). Some of the 3-meA-DNA glycosylases, especially those found in mammalian systems, also remove other damaged bases, including Hx and εA. This group of enzymes removes a wide range of modified bases, a number of which are indicated in Fig. 19. Interestingly, 3-meA-DNA glycosylase orthologues are found even in thermophilic organisms, despite the fact that 3-meA has a half-life of under 30 min at 80°C at neutral pH (138).

5.1.1. *Procaryotic and S. cerevisiae 3-Methyladenine DNA Glycosylases*

Tag 3-Methyladenine-DNA Glycosylases. The *tag* gene codes for a DNA glycosylase that excises both 3-meA and 3-meG and there are several examples of proteins similar to it (Fig. 20). The *tag* gene was cloned by complementation of *E. coli tagalkA* mutants that were highly sensitive to methylating agents (139) (Fig. 21). The gene has been overexpressed and the corresponding protein purified to homogeneity (140–143). Some inhibitors of the enzyme include 3-ethyl, propyl, butyl, and benzyl adenine modifications (144). The structure of Tag was solved using NMR and it has also been included in the helix–hairpin–helix HhH family including Nth and AlkA proteins (145) (Fig. 22). This was unexpected, based on its lack of sequence homology to other HhH family members. Moreover, the enzyme is a metalloprotein with a unique Zn structural feature that has been termed a Zn snap (146), an arrangement that permits the enzyme to be organized into a HhH structure to recognize the damaged DNA base by chelating a Zn ion using Cys and His residues at both the N and C termini (147).

In vivo, *tag* is constitutively expressed in *E. coli*. Overexpression of *tag* generally reduces the toxicity of S_N2 methylating agents in *tagalkA* deficient *E. coli* (Fig. 21) (148). Moreover, compared to other 3-meA-DNA glycosylases, Tag excises the lowest number of unmodified bases suggested by the fact that DNA transfected into cells following treatment with Tag produces the lowest mutation frequencies (149,150). In Chinese hamster fibroblast cells treated with either MMS or EMS, Tag expression reduced mutagenicity (151) and 3-meA residues did not contribute significantly to observed Hprt mutations (152).

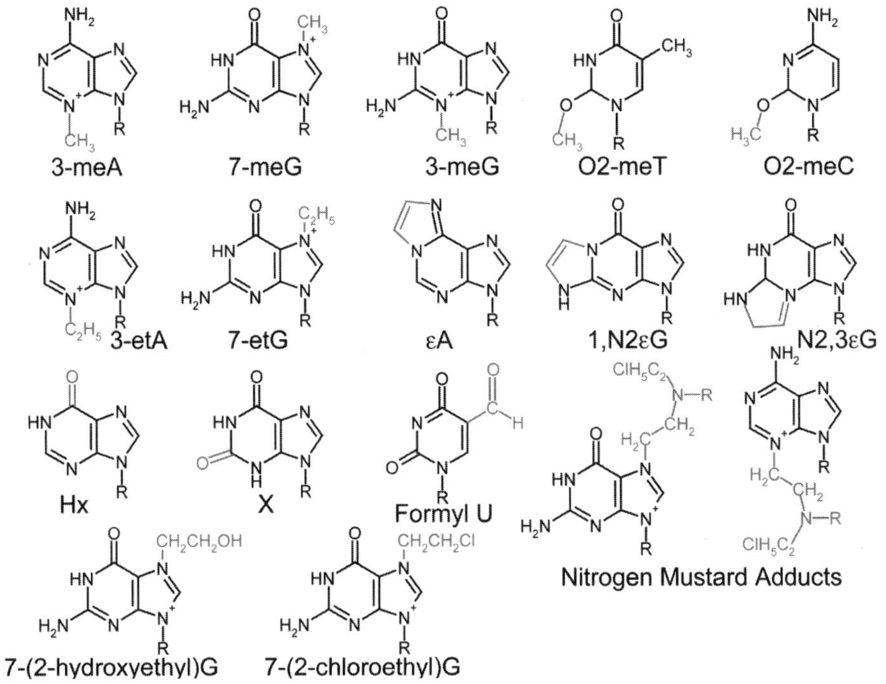

Figure 19 Different substrates for 3-meA-DNA glycosylases. These substrates are recognized by at least one of the 3-meA-DNA glycosylases. The shaded parts indicate the adduct. From the top left: 3-meA is recognized by all the enzymes of this class. 7-meG is recognized by AlkA, MPG. 3-meG is recognized by Tag, AlkA, MPG. O2meT and O2-meC are recognized by AlkA and most probably by MPG. 3-etA and 7-etG are recognized by AlkA and MPG. εA is recognized by AlkA (although not efficiently) and MPG. 1,N2εG is recognized by MPG. N2,3εG is removed by AlkA. Hx is recognized by AlkA (although not efficiently) and MPG. X is recognized by AlkA and MPG. Formyl U is recognized by AlkA. Nitrogen mustard adducts are recognized by AlkA. 7-(2-hydroxyethyl)G and 7-(2-chloroethyl)G are recognized by AlkA. *Source*: Adapted from Ref. 192.

AlkA 3-Methyladenine-DNA Glycosylases. The *alkA* gene is under the control of Ada and *alkA* is induced during the adaptive response to alkylating agents. Otherwise the protein is maintained at low levels in *E.coli*. The *alkA* gene was cloned by rescue of the sequence in *tagada* or *alkA* mutants, in the same manner as that of *tag* (143,153,154). A number of proteins that are mainly prokaryotic share sequence homology with AlkA (Fig. 23) (155,156). The structure of AlkA has been determined alone and complexed with a substrate (157–159). It is classed with Nth (Endo III) in the HhH or helix–hairpin–helix family. In fact, the structure is surprisingly close in fold and active site location to that of Nth, but AlkA has no Fe-S cluster. AlkA has a large cleft in which the active site is located. This cleft allows the approach of large substrates and as with other DNA glycosylases, AlkA flips the base out of the DNA helix prior to cleavage. The base that is flipped out (the modified base) is stabilized by stacking interactions with Trp272 and a number of aromatic residues in the active site region. When the substrate binds to ds-DNA, a distinct 66° bend is introduced that widens the DNA minor groove. The Glu238 residue abstracts a proton from water providing a base to attack the glycosylic bond releasing the modified base (Fig. 24).

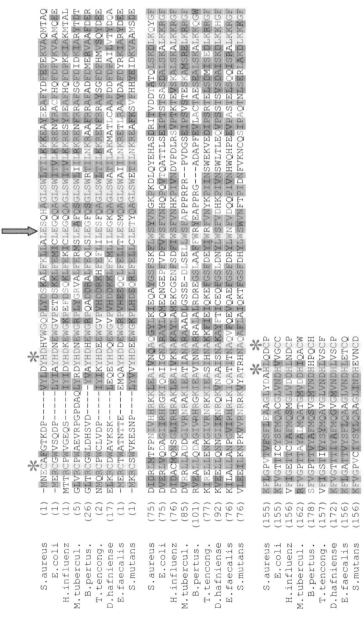

Figure 20 Alignment of Tag sequences from different organisms. The universally conserved sequences are shown in blue, and the least conserved in yellow, less conserved sequences are shown in yellow, less conserved in green. Sequences were aligned using ClustalW with Vector NTI software. The sequence reference number in GenBank is indicated at the bottom of the figure. The red asterisks indicate the residues involved in the Zn snap and the catalytic Glu38 in *E. coli* is indicated. (*See color insert.*)

S.aureus: CAB60736, E.coli: NP_418005, H.influenz: NP_438814, M.tubercul.: NP_215726, B.pertus.: NP_881431,
T.tencong.: NP_621791, D.hafniense: ZP_00100284, E.faecalis: NP_816662, S.mutans: NP_722376

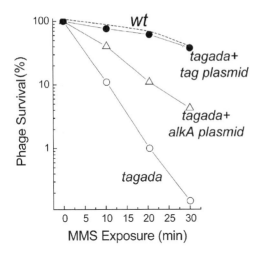

Figure 21 Complementation of tagada deficient *E. coli* by expression of the tag and alkA genes. The tagada strain is deficient in production of AlkA since the alkA gene must be induced by the transcriptional activation of Ada. The dashed line at the top represents survival in a wild type strain. *Source*: Adapted from Ref. 143.

As suggested by its structure, AlkA has a wide substrate range that includes 3-meA, 7-meG, but also Hx, εA, and nitrogen mustard modified bases (Table 7). AlkA even removes ring-opened pyrimidine products suggesting, a similarity in the active sites among AlkA, Nth, and Fpg (Fig. 24) (160). However, despite the broad substrate range, the preferred substrate for AlkA, based on steady-state kinetics, is 3-meA.

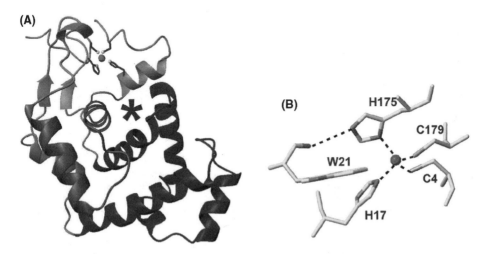

Figure 22 Structure of Tag and the Zn snap from an NMR solution structure. (A) Three-dimensional structure of Tag including restraints to a Zn^{2+} ion. The signature HhH structural motif is highlighted in green, and the remaining elements of the conserved helical domain of the HhH glycosylase fold are colored blue. The structure elements that are unique to Tag are shown in orange-red, and the 3-methyladenine binding pocket is indicated with an asterisk. (B) Tetrahedral Zn^{2+} binding site of Tag. The residue numbers are indicated. *Source*: Adapted from Ref. 146. (*See color insert*.)

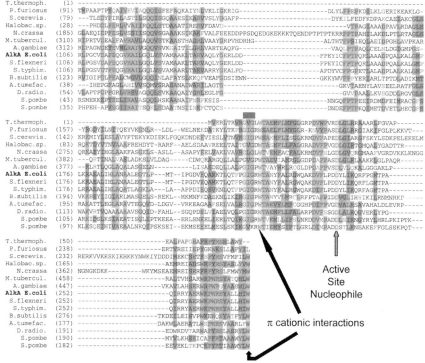

Thermus thermophilus: BAA88678, Pyrococcus furiosus: NP_578240., S.cerevis.: NP_011069, Halobac.sp.: NP_280175, N.crassa: XP_331330,
M.tubercul.: NP_215833, A.gambiae: XP_306313, AlkA E.coli: NP_416572, S.flexneri: NP_707962, S.typhim.: NP_461069, B.subtilis: NP_388061,
A.tumefac.: NP_357024, D.radio.: NP_296303, S.pombe: NP_593991, S.pombe: NP_595869

Figure 23 Alignment of AlkA sequences from different organisms. The universally conserved sequences are shown in yellow, less conserved in blue, and the least conserved in green. Sequences were aligned using ClustalW with Vector NTI software. The sequence reference number in GenBank is indicated at the bottom of the figure. The GV/IG sequence indicative of the HhH sequence, the active site nucleophile position, and some of the residues stabilizing the π interactions with the substrate are indicated. (*See color insert.*)

The size of DNA adducts recognized by AlkA has been investigated in vitro and in vivo for adducts formed by a monopyrrole-containing distamycin A analog. The enzyme has a favored excision site at the 3′ purine in TTTTGPu. However, the di- and tripyrrole derivatives of distamycin A were poorly recognized (161). There is also evidence that adducts between 4–5 carbons are repaired, presumably by AlkA (162).

AlkA discriminates in base pairing with Hx, but only by about a factor of 2 with the Hx/T excision (formed by deamination of A) being the most efficient (163). The excision of Hx is also dependent on sequence context. Changing the sequence context alters excision of Hx:G pairs to a point where there is a five-fold difference, but only in one of the sequence contexts examined (164). In contrast to the differences for the Hx excision, AlkA does not excise εA with any base pair preference (165).

Scanning the DNA for damage is the role of DNA glycosylases, but these enzymes must distinguish damage among a great background of normal DNA bases. Often less than 1 base in 10^4–10^7 normal bases is damaged. The difference in interaction between damaged substrates and DNA for DNA glycosylases is often

a. AlkA

b. Nth (Endonuclease III)

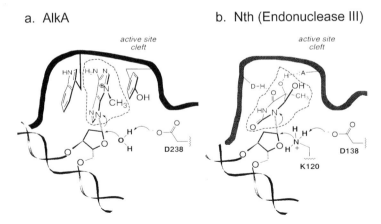

Figure 24 Extrahelical recognition and excision of aberrant DNA bases by AlkA and Nth (Endonuclease III). The bases excised are surrounded by a dashed line. The common location of a catalytically essential, conserved aspartic acid residue in the active site clefts of AlkA (a, Asp-238) and Nth (b, Asp-138) suggests the two proteins may use a common mode of nucleophilic activation. In the case of AlkA, a monofunctional glycosylase, Asp-238 is envisioned to deprotonate water, activating it for attack on the glycosidic bond of the substrate. In the case of Endo III, the Asp-138 deprotonates the ε-NH3 1 group of Lys-120, which attacks the glycosidic bond to form a covalent enzyme-substrate intermediate. The active site cleft of AlkA is rich in electron-donating aromatic residues (here illustrated in (a) as a Trp and a Tyr residue), which are ideally suited to interact with electron-deficient bases through π-DA interactions. On the other hand, the corresponding cleft in Endo III contains mostly hydrophilic residues, which may interact with substrate bases such as thymine glycol (shown in (b)) through either direct or water-mediated hydrogen-bonding. *Source*: Adapted from Ref. 158.

Table 7 Substrates of Different Methylpurine-DNA Glycosylases—Tag, AlkA, and MPG

Protein	Substrate	k_{cat} (sec^{-1})	K_m (nM)	k_{cat}/K_m (M^{-1} s^{-1})	K_{Dapp} (nM)	References
Tag	3-meA	—	0.14	—	—	137
AlkA	3-meA	4×10^{-3}	8	5×10^5	nd	163, 248
hMPG	3-meA	1.8×10^{-1}	8	2.3×10^7	nd	184
AlkA	7-meG		11		nd	137
hMPG	7-meG	5.8×10^{-3}	25	2.3×10^5	nd	184
AlkA	εA	2×10^{-5}	800	3×10^0	—	165
hMPG	εA		24		—	165
hMPG	εG	2.9×10^{-5}	25	1.9×10^3		194
AlkA	Hx	1×10^{-5}	420	2.4×10^1		163
hMPG	Hx	1.6×10^{-3}	11	1.4×10^5	8.6	186
AlkA	X	2×10^{-4}	53	3.8×10^3	—	185, 249
hMPG	X	9.1×10^{-4}	13	7.0×10^4		185
AlkA	THF	—	—	—	45	158
AlkA	Pyr	—	—	—	0.016	158
hMPG	AP	—	—	—	1.6	186
hMPG	rAP	—	—	—	8.6	186

approximately only 10^3 or less (166). This could lead to problems in excision of many normal DNA bases (150,167). Apparently the ability to accommodate many substrates in its cleft has also helped the AlkA protein to receive the distinction of being the DNA glycosylase excising the highest percentage of unmodified bases from DNA (150). Such removal of normal bases can result in repair synthesis that produces more mutations and this is observed when AlkA or other 3-meA-DNA glycosylases are overproduced in *E. coli.* (149,150). But the DNA glycosylase producing the highest mutation rate is AlkA. It is also interesting that although overexpression of *alkA* in cells should protect them from alkylation damage, at high levels the cells are sensitized to killing (148). One way to limit removal of normal DNA bases and potential cell killing is to use AlkA only when necessary and that is the solution that *E. coli* has evolved (168). AlkA is produced at low levels and induced only during the adaptive response limiting any mutation that could occur due to excision of normal bases. Deletion of the two 3-meA-DNA glycosylases in *E. coli* results in sensitivity to MMS approximately 10^5 times that of wild-type *E. coli* (169). This presence of 3-meA in *E. coli* elicits the SOS response.

Little biochemical data have been accumulated on Mag (170), the *S. cerevisiae* 3-meA-DNA glycosylase, so the discussion concerning yeast will focus on results obtained from yeast genetics involving Mag. *S. cerevisiae* provides an example of the equilibrium in cells that protect against DNA mutation, especially with respect to Mag, the 3-meA-DNA glycosylase similar to AlkA (171–173). Mag and Apn1, the AP endonuclease in the second step of BER in *S. cerevisiae*, work in concert. If Mag is overexpressed, the spontaneous mutation rate increases, but if Mag expression is reduced by generating MagApn1 mutants, the spontaneous mutation rate is reduced (Fig. 25). Thus, an increase in unrepaired AP sites can lead to an increase in mutation rate in cells that are only exposed to endogenous DNA damage or the spontaneous mutation rate. The generation of such mutations in yeast requires the REV1/REV3/REV7 pathway for lesion bypass. These experiments establish that endogenous DNA damage must be eliminated in lower eucaryotic as well as prokaryotic cells to prevent spontaneous mutations.

5.1.2. Eucaryotic MPG 3-Methyladenine DNA Glycosylases

A series of 3-meA-DNA glycosylases have been isolated generally by complementation of strains of *E. coli* deficient in *tagalkA* (Fig. 26). Although functionally similar to AlkA, the members of this series have no sequence homology to AlkA. These proteins are found in many organisms including *A. thaliana* (plant), mouse, rat, and humans (Fig. 27) (174–178) and are grouped as MPG or methylpurine-DNA glycosylases.

The hMPG protein (also known as AAG and ANPG) adopts a structure that is different from that of AlkA. The HhH that is found in almost all DNA glycosylases does not exist in hMPG (179). The Tyr (Fig. 28) in the β hairpin structure protrudes into the minor groove of DNA and stabilizes the helix after the damaged base is extruded into the protein. There is a bend of 22° that is introduced upon binding of the protein to the DNA (179). A crystal structure of a bound oligonucleotide with an εA substrate was obtained showing the positions of the amino acids around the base without cleavage of the glycosylic bond (Fig. 28). This is exceptional, since the enzyme was functional and the crystals were obtained at room temperature. One potential explanation for the possibility to obtain this MPG–εA structure is that chemistry of excision for εA is much slower than for Hx in single turnover kinetic

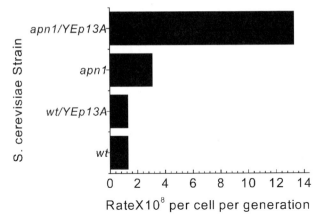

Figure 25 Spontaneous mutation rates in cells expressing high levels of the MAG in *S. cerevisiae* MagApn1 are increased when MAG is overexpressed in *Apn* mutants. Increased production of MAG results in increased amounts of abasic sites that are not processed in the *Apn* mutants. Rate was measured for DBY747 (wild type, wt), WX9105 (dapn1::HIS3), DBY747/Yep13A, and WX9105/Yep13A. Values were the average of three experiments in which all four strains were tested simultaneously. The rate in WX9105/Yep13A is significantly different from that of the three other strains P = 0.01, and the rate is significantly different from that of wild type cells ($P = 0.01$). *Source*: Adapted from Ref. 172.

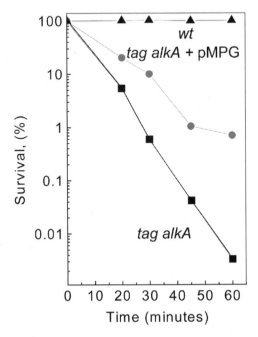

Figure 26 Expression of MPG cDNA in *E. coli tagalkA* mutants increases survival. Time is minutes in 0.05% MMS in solution prior to plating. The strains are AB1157 (wt), MV1932 (*tag alkA*), and the *tagalkA* strain complemented with a plasmid harboring the hMPG cDNA MV1932 + pP5-3 (*tagalkA* + pMPG). *Source*: Adapted from Ref. 175.

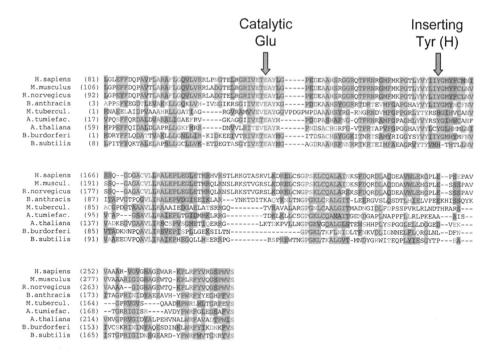

Figure 27 Alignment of MPG sequences from different organisms. The universally conserved sequences are shown in yellow, less conserved in blue, and the least conserved in green. Sequences were aligned using ClustalW with Vector NTI software. The sequence reference number in GenBank is indicated at the bottom of the figure. The residues for the human MPG for the catalytic Glu120 and the Tyr157 that inserts in the DNA helix to stabilize the structure are indicated in the figure. *Source*: From Ref. 252. (*See color insert.*)

experiments. Those experiments reflect the chemistry of the active site (180). As the damaged base is flipped out, it is placed into a binding pocket that makes van der Waals contact with the base to stabilize the interaction prior to cleavage of the glycosylic bond. There are at least nine residues that have interactions with the εA to stabilize the flipped out base in the structure (Fig. 28). A number of those bases surrounding the modified base stabilize it outside the helix using stacking interactions. Examination of thermal stability with normal and εA bases demonstrated that the initial recognition of substrate by MPG is dependent on the capacity of the base to be flipped out of the helix (181). Site directed mutants at the active site of hMPG are consistent with the Tyr157 inserting into the helix to stabilize the hole left by the flipped out damaged base (182). Once flipped out of the helix and in the binding pocket, the glycosylic bond of the modified base is severed by E120 abstraction of a proton from water in the active site and attack of the OH^- on the glycosylic bond.

MPG and associated proteins have enzymatic constants similar to those of AlkA for excision of methylated bases including 3-meA and 7-meG (175–178,183, 184) (Table 7). Excision of 3-meA is favored compared to that of 7-meG by about 100-fold (Fig. 29). MPG releases a number of modified bases, most overlapping with those of AlkA, but there are differences in efficiency and in some cases, excision is not observed. The turnover numbers are extremely slow compared to those observed

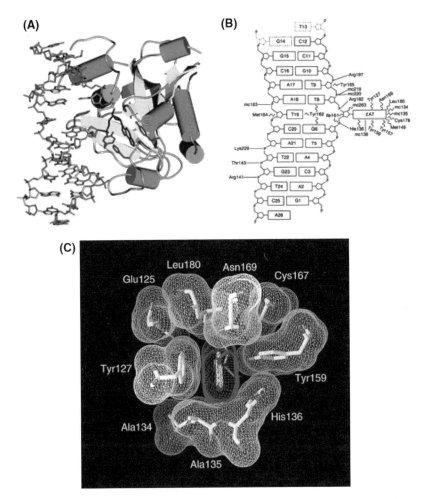

Figure 28 Crystal structure and binding pocket of hMPG. (A) Crystal structure of the E125Q AAGY′A–DNA complex. The εA base (black) is flipped into the protein active site to stack between Tyr-122 on one side and His-131 and Tyr-154 on the other (shown in purple). Tyr-157 intercalates between the bases that flank the flipped-out εA, filling the abasic gap in the DNA. (B) Schematic diagram of contacts between AAG and the εA–DNA. The flipped-out εA base (labeled εA7) participates in a hydrogen-bonding interaction with the main chain amide of His-131 (solid line labeled "mc136") and many van der Waals interactions (wavy lines) with residues of the active site. Hydrogen bonds and salt bridges (solid lines) with the DNA backbone anchor the protein to DNA. The nucleotides T13 and G14 (dashed outlines) are not visible in the electron density. The differences in numbering are due to the different exon 1 splicing in MPG. The full length protein is generally accepted to be the 293 amino acids and a splicing variant 298 amino acids (175,252). (C) The εA (purple surface) fits into a pocket with the DNA helix oriented vertically behind the plane of the diagram. Met149 and Cys178 make additional contacts that are not shown. *Source*: Adapted from Ref. 253. (*See color insert.*)

for many enzymes. For example, MPG excises both Hx and X from DNA with an efficiency close to that of 7-meG, but excision of Hx and X from DNA by AlkA is less efficient (163,164,184–186). Even though MPG does not remove adducts such as those formed by nitrogen mustards, MPG does eliminate some relatively bulky adducts, εA and εG (165,187–194). The only difference in the substrates recognized

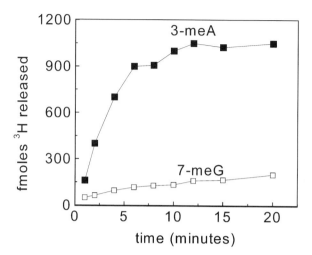

Figure 29 In vitro hMPG catalyzed excision of 3-meA and 7-meG by MPG. A substrate containing ³H labeled 3-meA and 7-meG in calf thymus DNA was incubated in the presence of homogeneous MPG and the products were separated using HPLC. The closed squares represent 3-meA excision and the open squares 7-meG excision. *Source*: Adapted from Ref. 184.

by the series of MPG proteins is that mMPG removes 8-oxoG, whereas hMPG does not remove 8-oxoG (195).

Binding of a DNA glycosylase to an adduct is not sufficient to indicate that the adduct is excised by the glycosylase. MPG protein binds to intrastrand cross-links of cis-Pt in an oligonucleotide with dissociation constants of ~100 nM for various constructs (196). That is not significantly different from the dissociation constants observed for MPG interacting with other adducts (162). Despite the interaction, no excision is observed. Even in the presence of an oligonucleotide with a unique εA, the bound MPG is not released from the cis-Pt cross-link (196). This type of interaction could trap proteins on DNA and result in eventual cell death if not removed.

It is worth noting that the majority of substrates have been studied in vitro and the relevance of MPG removal in vivo has not been established. Only repair of 3-meA and 7-meG in vivo have been studied in mammalian cells (197–201). Some substrates of DNA glycosylases are also removed by other DNA repair systems. For example, an enzyme that exists in *E. coli* and mouse, EndoV, is an endonuclease that will remove Hx and X from DNA. It is unclear how MPG and EndoV function in vivo to eliminate Hx and X (202–205).

hMPG discriminates different base pairs depending on the adduct and the pair. Hx is readily excised from Hx/T pairs, but is poorly excised from Hx/C pairs (164,206). By manipulation of the base binding pocket of hMPG using site-directed mutagenesis, it is possible to alter the specificity of the enzyme for different base pairs (182). These differences could contribute to hot and cold spots for mutations based on the efficiency of repair.

Sequence context dependent excision by MPG has been observed for specific substrates and in randomly damaged DNA fragments. First-order rate constants for excision as a function of sequence suggest that base removal by mMPG is sequence context dependent. MPG-catalyzed excision of Hx is more sequence context dependent than excision of εA (207). The presence of flanking G or C bases (e.g., GG or CC) on either side of εA results in a distinct sequence context effect that favors excision of εA from a DNA sequence with increased stability. In sequences with A or T bases (e.g., AA or TT) on either side of the εA, cleavage is much slower. Examination of cleavage of Hx from four sequences revealed approximately a five-fold difference in sequence context for excision of Hx when paired with T (164). However, it is unclear whether the sequence context differences observed are linked to steady-state kinetic parameters. In cells, the initial velocity or first-order rate constants may more accurately reflect the in vivo situation, since steady-state conditions are probably not maintained. The sequence context excision by MPG was studied in vitro for approximately 150 base positions in the human PGK-1 promoter and first exon (201). The first-order rate constants differed by a factor of 180 (Fig. 30). This suggests that under conditions similar to those for excision in cells, there is a broad range for removal of bases. The association of 7-meG in vitro excision and in vivo repair rates suggests that that the removal of the base is the rate-limiting step for repair in vivo. Although excision of 7-meG is rate-limiting in vivo, 3-meA excision is not rate-limiting in vivo as determined using mouse embryonic stem cells (208).

Repair of 3-meA and 7-meG has also been examined in mammalian cells. This repair was initially linked only to BER. In contrast to repair of some types of adducts via BER and NER, there is no difference in the removal of 3-meA and 7-meG when examined at the global, gene, and transcribed gene levels in mammalian cells (197–199,209–212). However, at nucleotide resolution in normal human cells, there is a striking heterogeneity of repair of 3-meA and 7-meG adducts. Some 7-meG adducts are repaired in less than 3 hr, whereas others remain even at 64 hr.

MPG has been used to sensitize human tumor cells to DNA damaging agents (213). By both nuclear and mitochondrial targeting, the survival of MPG overexpressing

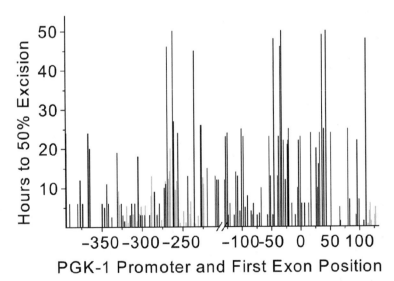

Figure 30 Variation of in vitro 7-meG excision from the human PGK-1 promoter region and first exon as a function of base position. Note the wide range of times for the different positions. Plasmid DNA containing the promoter and the first exon of the human PGK-1 gene which was 3′ end labeled with ^{32}P were incubated with MPG for periods of time up to 48 hr. DNA was reacted with Nth to introduce breaks at abasic sites and first-order rate constants for excision were determined. Black lines represent G positions on the non-transcribed strand and gray lines represent positions on the transcribed strand. *Source*: Adapted from Ref. 201.

cells to alkylating agents is reduced. Mitochondrial targeting of MPG resulted in apoptosis at lower doses of alkylating agents than for nuclear-targeted MPG producing cells to undergo apoptosis. This suggests that mitochondrial-targeted MPG could be used to augment the effects of alkylating agents used in chemotherapy.

As for the AGT series, the use of targeted deletions in mouse embryonic stem cells has permitted the construction of models for repair in mammalian cells and in a whole mouse model. For MPG, at least two groups have reported targeted deletions in mice for the Aag (Apng) genes (188,208,214). The Aag−/− cells are more sensitive to alkylating agent damage than the wild-type or the Aag+/− mutant. The use of Aag−/− cells and a unique methylating agent Me-lex (a compound specifically modifying the 3 position of A) demonstrated that damage, in this case 3-meA, serves as an induction signal for P53 (215). In other work, the Aag−/− lines were used to demonstrate that the only DNA glycosylase activity that excises Hx is produced by the Aag gene (188,214).

The fact that DNA glycosylases remove adducts relatively slowly in vitro, but rapidly in vivo suggests that there are most likely protein–protein interactions that facilitate repair. One complex has been found in the recognition of lesions in BER—that of MPG with HRAD23B (216) (Fig. 31). HRAD23 also complexes with XPC in the initial step of damage recognition for NER. The role of this complex in vivo remains to be defined, but could provide a link between BER and NER recognition. It is possible that a number of other complexes are formed with MPG during repair or in other processes. These complexes could have the effect of tranducing signals related to damage into transcriptional messages that can inhibit or enhance transcription depending on the protein and damage levels in cells. One example is the interaction of MPG with MBD1 (217). MBD1 is a methyl-CpG binding protein that binds to MPG. The corresponding complex silences gene transcription. Exposure to alkylating agents releases the MBD1–MPG complex and the specific binding

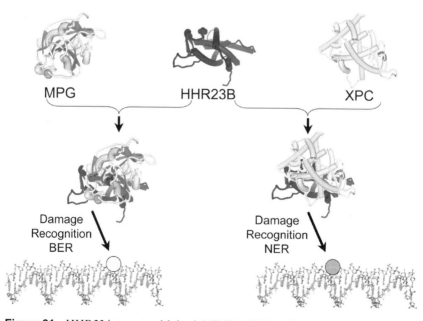

Figure 31 HHR23 interacts with both MPG in BER and XPC in NER. The proteins and the DNA are not drawn to scale or for accuracy and are only for illustrative purposes.

of MBD1–MPG to methylated promoters is restored following repair. Thus, MBD1 can function to store MPG and that the MPG can serve as a signal for damage to DNA in chromatin.

5.1.3. Other 3-meA-DNA Glycosylases

There are at least three other 3-meA-DNA glycosylases that have been examined that do not fit into the existing classification of AlkA type proteins. One was identified using homology searches as an AlkA in archaea (218). The archaea protein AfalkA from the thermophile Archaeoglobus fulgidus is unusual because it is specific for 3-meA and 7-meG. It does not remove Hx, foU, or εA, which are all substrates for E.coli AlkA. In DNA, 3-meA has limited stability even at room temperature and would not necessarily be removed faster than loss due to thermal instability. Another thermophile 3-meA-DNA glycosylase was discovered in Thermotoga maritime, MpgII, with similar characteristics to that of the AfalkA (219). The most unusual characteristic of this protein is that it was identified in a bioinformatics scan as Nth-like. The last 3-meA-DNA glycosylase in this different category was also found in a scan for Nth-like proteins in *Heliobacter pyloris* a causative agent of stomach ulcers. There was no alignment with AlkA protein, but similar to the case of the T. *maritima* and A. *fulgidus* enzymes, there is a specificity for 3-meA and 7-meG (220).

5.2. Thymine-DNA Glycosylases

Thymine-DNA glycosylases were first described based on their capacity to excise T from G-T mismatches. This class of DNA glycosylases also removes U from DNA. However, subsequent work demonstrated that these glycosylases are more efficient at the removal of εC from DNA. In fact, these DNA glycosylases are at least 1–2 orders of magnitude more efficient at removal of εC adducts than for U or T. A number of these enzymes have been identified in different organisms (Fig. 32). Since these proteins will be discussed elsewhere in this volume with respect to uracil-DNA glycosylases, this section will focus on the excision of εC and εG from DNA.

5.2.1. Procaryotic Thymine-DNA Glycosylases Excising εC

E.coli Mug was discovered after the cloning of TDG by a sequence search of the bacterium. Mug excises U from dsDNA, but also excises εC and εG adducts from dsDNA (194,221–226). The crystal structure of Mug (Fig. 33) reveals an architecture similar to that of uracil-DNA glycosylase, even though there is only minimal sequence homology (225) (Fig. 32). Excision of εC by Mug is actually greater than that of U by Mug (194). Examination of excision of εC in E.coli *mug* mutants demonstrates that this mutagenic base is eliminated only by Mug. Two other alkylation products generating cyclic adducts are also removed by Mug in vitro, εG and hydroxymethylεC. Although Mug has been reported to excise U, it seems that the principal role of this glycosylase is to eliminate εC adducts (226,227).

5.2.2. Eucaryotic Thymine-DNA Glycosylases Excising εC

Thymine-DNA glycosylase or TDG was first isolated based on its ability to release thymine from G-T mispairs (228,229). A crystal structure of a thermal stable TDG has been reported (230). Subsequently, it was found that TDG releases εC is much more rapidly than that of T from G/T mispairs, indicating that the actual in vivo

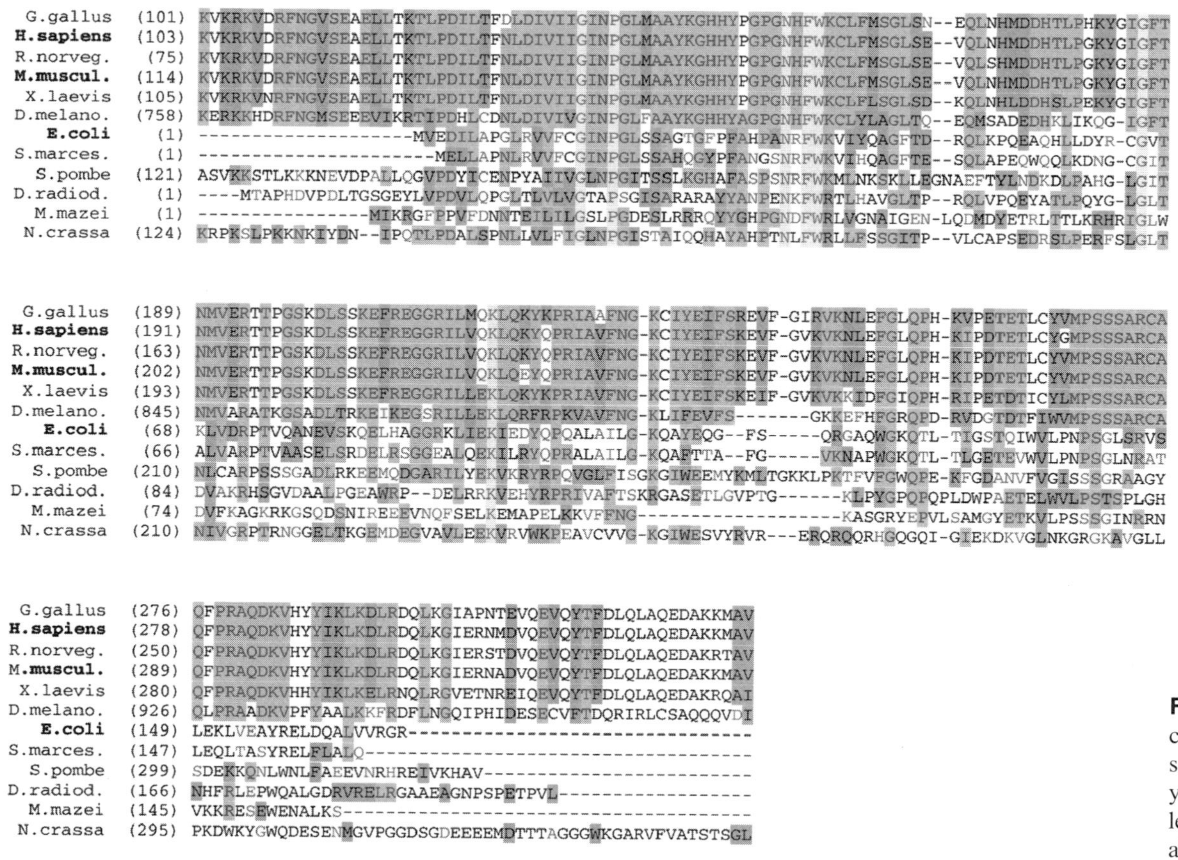

Figure 32 Alignment of TDG sequences from different organisms. The universally conserved sequences are shown in yellow, less conserved in blue, and the least conserved in green. Sequences were aligned using ClustalW with Vector NTI software. The sequence reference number in GenBank is indicated at the bottom of the figure. (*See color insert.*)

G.gallus: AAF14308, H.sapiens: NP_003202, R.norveg.: NP_446181, M.muscul.: NP_766140, X.laevis: AAH41496, D.melano.:NP_651925, E.coli MUG:NP_417540, S.marces.: P43343, S.pombe: NP_588515, D.radiodur: NP_294438, M.mazei: NP_633675, N.crassa: XP_329288

Mug HIV1-UDG

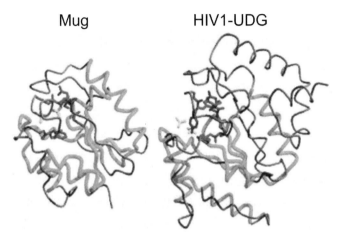

Figure 33 Crystal structure of *E. coli* Mug compared to that of the HIV1 UDG. The secondary structure elements identified as equivalent are highlighted in cyan, residues contributing to the binding pockets in each molecule are shown in red, and the bound sulphate ions observed in native crystals of both enzymes are shown in yellow. *Source*: Adapted from Ref. 225. (*See color insert.*)

substrate for TDG is εC (227,231). In addition to the excision of T and εC, TDG also removes from U and hydroxymethyl U from DNA (232). Interestingly, although εG is excised by Mug, εG is not excised by TDG (194).

Kinetic studies indicate that TDG binds tightly to the abasic sites generated during its reaction and that the time for release is on the order of hours (233,234) (Fig. 34). Thus, steady-state kinetic parameters do not hold. The release of substrate can be facilitated, however, by the use of APE, an endonuclease specific for abasic sites.

TDG has been found to have a number of partners, including SUMO, RXR, ERα, and XPC-HHR23B (235–239) (Fig. 35). Nuclear receptors such as RXR and ERα both interact with TDG to stimulate transcription (235,239). In contrast, MGMT complexes with ERα to repress transcription (68). Moreover, TDG interacts with the transcriptional co-activator CBP/p300 to increase its transcriptional activity. CBP also acetylates TDG favoring its release from the complex and the recruitment of APE. Sumoylation is the process of conjugation of SUMO, a small ubiquitin like monomer, onto proteins. The sumoylation of TDG potentiates the turnover of TDG in the presence of APE. The role of these interactions in the excision of εC or T from G/T mispairs in vivo has yet to be demonstrated.

6. NUCLEOTIDE EXCISION REPAIR

Much small alkylated base damage is removed by BER, but it has also been shown that the NER system can help to eliminate this type of DNA damage. The extent to which NER functions has yet to be established, but at present repair is believed to occur in much the same manner as for that of bulky adducts. The role of this important repair system and its mechanism has been discussed in detail in other chapters. In addition to NER, incomplete BER repair intermediates can also be processed by homologous recombination (240).

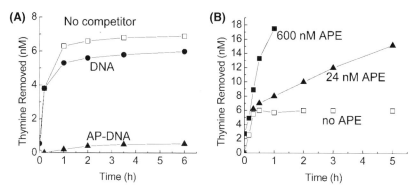

Figure 34 TDG excision from G/T mispairs. (a) Preincubation of TDG with DNA containing an abasic site strongly inhibits its reaction with DNA containing a G/T mismatch. 7 nM thymine DNA glycosylase was preincubated with either 10 nM CpG/AP DNA (▲) or 10 nM CpG/C DNA (●) in reaction buffer for 30 min. At the end of the preincubation, 10 nM ^{32} P-labeled CpG/T DNA was then added to each and the amount of thymine removal monitored by chromatography. For comparison, the ^{32}P-labeled CpG/T DNA was added directly to 7 nM thymine DNA glycosylase in the absence of competitor and without preincubation (open squares). (b) TDG excision from a G/T mismatch is enhanced in the presence of human APE. Concentration dependence of the effect of APE on the glycosylase reaction with a G/T mismatch. G/T DNA (20 nM) was incubated with thymine DNA glycosylase (6 nM) in 2 mM magnesium in the absence of or in the presence of APE at the concentrations as indicated. *Source*: Adapted from Refs. 233,234.

6.1. Procaryotic NER

The role of NER in repair of large adducts by the UvrABC pathway is well established, but it is now evident that NER can also play a role in repair of 3-meA adducts that are formed specifically by the minor groove alkylating agent Me-lex (241).

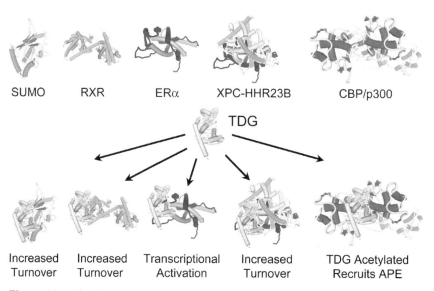

Figure 35 Reactions of TDG with different proteins. The effects of the interactions are listed at the bottom of the figure. The proteins and the DNA are not drawn to scale.

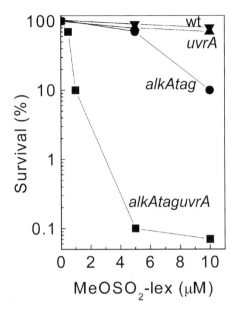

Figure 36 Toxicity of Me-lex with respect to BER and NER in different *E. coli* strains. *Source*: Adapted from Ref. 241.

At low concentrations, tagalkA or uvrA mutants are relatively insensitive to Me-lex, while the triple mutant tagalkAuvrA is almost 1000 times more sensitive (Fig. 36). One other study has suggested that there is a crossover with respect to adduct size that has a role in the use of the NER and BER systems (162). Adducts larger than

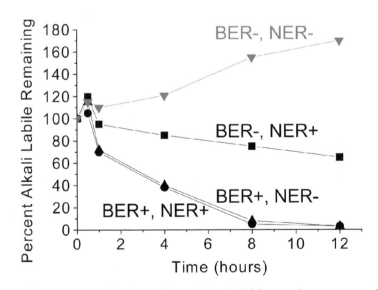

Figure 37 Repair time of 3-meA in the Dhfr gene in spontaneously transformed mouse embryonic fibroblasts as a function of BER and NER. The Aag+/− or Xpa+/− heterozygotes show no deficiency in either BER or NER, respectively. Aag codes for the MPG in mouse and Xpa is involved in damage recognition in NER. The Aag−/−Xpa−/− (BER-, NER-) line is highlighted in gray. *Source*: Adapted from Refs. 200,242.

ethylated bases have a significant NER component to their repair, whereas a majority of repair occurs via BER for methyl and ethyl adducts.

6.2. Eucaryotic NER

Excision of methylated bases is generally considered to occur via BER. For the human system, there is correlation between 7-meG excision in vitro and in vivo repair (201). Therefore, it was surprising when in knock out mouse cells deficient in the MPG (Aag−/−) 7-meA excision was as fast as in Aag+/+ (242), despite the absence of the BER pathway. To address this, four MEF cell lines with combined mutations-Aag+/-Xpa+/−, Aag+/−Xpa−/−, Aag−/−Xpa+/−, Aag−/−Xpa−/− were constructed using Aag and Xpa (200). Xpa is involved in the recognition step of NER and the use of the heterozygous cells is justified by results that show that repair is identical with a single allele of either Aag or Xpa. Repair of 7-meG is almost the same in either Xpa−/− or Aag−/− cells. However, in cells with homozygous deletions in both genes simultaneously Aag−/−Xpa−/−, repair of 7-meG is not observed (Fig. 37) (200). This is significant because it suggests that there is overlap of at least the two repair pathways for adducts that were previously believed to be repaired only by BER. This highlights the necessity to examine the role of multiple DNA repair pathways in adduct elimination.

ACKNOWLEDGMENTS

The author would like to thank Gerald Holmquist and Steven Bates for critical reading of the manuscript and the National Institutes of Health for funding during the preparation of the manuscript.

ABBREVIATIONS

AAG—alkyladenine-DNA glycosylase, ANPG, MPG
ABH—human AlkB homologue ABH1
ABH2—human AlkB homologue 2
ABH3—human AlkB homologue 3
Ada—*E. coli* adaptive response and inducible O6-alkylguanine-DNA methyltransferase
A. gambiae—*Anopheles gambiae*
AGT—The general class of O6-alkylguanine-DNA methyltransferases
AidB—gene inducible during the adaptive response
ANPG—alkyl-*N*-purine-DNA glycosylase, AAG, MPG
AlkA—3-methyladenine-DNA glycosylase II
AlkB—α-ketooxoglutarate-Fe(II) dependent oxygenase
A. thaliana—*Arabadopsis thaliana*
A. tumefac.—*Agrobacterium tumefaciens*
B. anthracis—*Bacillus anthracis*
B. burdorferi—*Bacillus burdorferi*
BCNU—bis-chloroethylnitrosourea
BER—base excision repair
B. pertus.—*Bordetella pertussis*
B. subtilis—*Bacillus subtilis*

D. hafniense—Desulfitobacterium hafniense
D. melano—Drosophila melanogaster
DMS—dimethylsulfate
D. radiodur.—Deinococcus radiodurans
ds—double-stranded
E. coli—Escherichia coli
E. faecalis—Enterococcus faecalis
EMS—ethylmethane sulfonate
ENU—ethylnitrosourea
ERα—estrogen receptor alpha
et—ethyl
εA—$1,N_6$-ethenoadenine
εC—$3,N_4$-ethenocytosine
1,N2εG—$1,N_2$-ethenoguanine
N2,3εG—$N_{2,3}$-ethenoguanine
foU—formyl uracil
Fpg—formamidopyrimidine-DNA glycosylase MutM
G. gallus—Gallus gallus
Halobac.sp.—Halobacterium sp.
H. influenz.—Haemophilus influenzae
H. pylori—Helicobacter pylori
H. sapiens—Homo sapiens
Klebsiella pneumon.—Klebsiella pneumoniae
Me-lex—[1-methyl-4-[3-(methanesulfonyl)propanamido)pyrrole 2-carboxami-
    do]pyrrole-2-carboxiamido]propane
me—methyl
hMGMT—human MGMT
hMPG—human MPG
MGMT—O6-methylguaninine-DNA methyltransferase
M. mazei—Methanosarcina mazei Goe1
MMR—long patch mismatch repair
MMS—methyl methane sulfonate
mMgmt—mouse MGMT
M. muscul.—Mus musculus
mMPG—mouse MPG
MPG—methylpurine-DNA glycosylase, AAG, ANPG
MNU—methylnitrosourea
M. tubercul.—Mycobacterium tuberculosis
Mug—mismatch repair uracil-DNA glycosylase
nd—not determined
NDMAOAc—N-nitroso(acetoxymethyl) methylamine
NER—nucleotide excision repair
N. crassa—Neurospora crassa
NO—nitric oxide
Nth—product of nth gene, endonuclease III
P. furiosus—Pyrococcus furiosus
rMgmt—rat MGMT
rMpg—rat MPG, Apdg
R. norveg.—Rattus norvegicus
Rp—R phosphotriester stereoisomer

S. aureus—Staphlococcus aureus
S. cerevis. or *S.cerevisiae—Saccharomyces cerevisiae*
S. flexneri—Shigella flexneri
S. typhim.—Salmonella typhimurium
S. maracas.—Serratia maracescens
S. mutans—Streptococcus mutans
Sp—S phophotriester stereoisomer
S. pombe-Schizosaccharomyces pombe
ss—single-stranded
S_N1—substitution nucleophilic unimolecular
S_N2—substitution nucleophilic bimolecular
Tag—3-methyladenine-DNA glycosylase I
TDG—thymine-DNA glycosylase
T. maritima—Thermotoga maritime
T. tencong.– Thermoanaerobacter tengcongensis
T. thermoph.—Thermus thermophilus
wt—wild type
X—xanthine
X. laevis—Xenopus laevis

REFERENCES

1. Marnett LJ. Mutat Res 1999; 424:83–95.
2. Bartsch H, Nair J, Velic I. Eur J Cancer Prev 1997; 6:529–534.
3. Marnett LJ. IARC Sci Publ 1999:17–27.
4. Zhao C, Hemminki K. Carcinogenesis 2002; 23:307–310.
5. Povey AC. Toxicol Pathol 2000; 28:405–414.
6. Hemminki K, Koskinen M, Rajaniemi H, Zhao C. Regul Toxicol Pharmacol 2000; 32:264–275.
7. Jackson AL, Loeb LA. Mutat Res 2001; 477:7–21.
8. Epe B. Biol Chem 2002; 383:467–475.
9. Marnett LJ. Toxicology 2002; 181–182:219–222.
10. Marnett LJ, Riggins JN, West JD. J Clin Invest 2003; 111:583–593.
11. Lindahl T, Barnes DE. Cold Spring Harb Symp Quant Biol 2000; 65:127–133.
12. Taverna P, Sedgwick B. J Bacteriol 1996; 178:5105–5111.
13. Sedgwick B. Carcinogenesis 1997; 18:1561–1567.
14. Sobol RW, Henderson EE, Kon N, Shao J, Hitzges P, Mordechai E, Reichenbach NL, Charubala R, Schirmeister H, Pfleiderer W, et al. J Biol Chem 1995; 270:5963–5978.
15. Yashiki T, Yamana K, Nunota K, Negishi K. Nucl Acids Symp Ser 1992; 197–198.
16. Yashiki T, Yamana K, Nishijima Y, Negishi K. Nucl Acids Symp Ser 1991; 25–26.
17. Miller PS, Chandrasegaran S, Dow DL, Pulford SM, Kan LS. Biochemistry 1982; 21:5468–5474.
18. Green CL, Loechler EL, Fowler KW, Essigmann JM. Proc Natl Acad Sci USA 1984; 81:13–17.
19. Loechler EL, Green CL, Essigmann JM. Proc Natl Acad Sci USA 1984; 81:6271–6275.
20. Voigt JM, Topal MD. Carcinogenesis 1995; 16:1775–1782.
21. Reha-Krantz LJ, Nonay RL, Day RS, Wilson SH. J Biol Chem 1996; 271:20088–20095.
22. Bignami M, O'Driscoll M, Aquilina G, Karran P. Mutat Res 2000; 462:71–82.
23. Ihara K, Kawate H, Chueh LL, Hayakawa H, Sekiguchi M. Mol Gen Genet 1994; 243:379–389.
24. Lindahl T, Demple B, Robins P. EMBO J 1982; 1:1359–1363.

25. Samson L, Han S, Marquis JC, Rasmussen LJ. Carcinogenesis 1997; 18:919–924.
26. Zak P, Kleibl K, Laval F. J Biol Chem 1994; 269:730–733.
27. Pegg AE. Mutat Res 2000; 462:83–100.
28. Potter PM, Wilkinson MC, Fitton J, Carr FJ, Brennand J, Cooper DP, Margison GP. Nucl Acids Res 1987; 15:9177–9193.
29. Rebeck GW, Coons S, Carroll P, Samson L. Proc Natl Acad Sci USA 1988; 85: 3039–3043.
30. Rebeck GW, Samson L. J Bacteriol 1991; 173:2068–2076.
31. Wilkinson MC, Potter PM, Cawkwell L, Georgiadis P, Patel D, Swann PF, Margison GP. Nucl Acids Res 1989; 17:8475–8484.
32. Vidal A, Abril N, Pueyo C. Carcinogenesis 1995; 16:817–821.
33. Sedgwick B. Mol Gen Genet 1983; 191:466–472.
34. Demple B, Sedgwick B, Robins P, Totty N, Waterfield MD, Lindahl T. Proc Natl Acad Sci USA 1985; 82:2688–2692.
35. Teo IA. Mutat Res 1987; 183:123–127.
36. Sedgwick B, Lindahl T. Oncogene 2002; 21:8886–8894.
37. Moore MH, Gulbis JM, Dodson EJ, Demple B, Moody PC. EMBO J 1994; 13: 1495–1501.
38. Verdemato PE, Brannigan JA, Damblon C, Zuccotto F, Moody PC, Lian LY. Nucl Acids Res 2000; 28:3710–3718.
39. Daniels DS, Tainer JA. Mutat Res 2000; 460:151–163.
40. Roldan-Arjona T, Luque-Romero FL, Ariza RR, Jurado J, Pueyo C. Mol Carcinog 1994; 9:200–209.
41. Boiteux S, Costa de Oliveira R, Laval J. J Biol Chem 1985; 260:8711–8715.
42. Teo I, Sedgwick B, Demple B, Li B, Lindahl T. EMBO J 1984; 3:2151–2157.
43. Teo I, Sedgwick B, Kilpatrick MW, McCarthy TV, Lindahl T. Cell 1986; 45:315–324.
44. Lindahl T, Sedgwick B, Sekiguchi M, Nakabeppu Y. Annu Rev Biochem 1988; 57: 133–157.
45. Norman DP, Chung SJ, Verdine GL. Biochemistry 2003; 42:1564–1572.
46. Takahashi K, Kawazoe Y. Biochem Biophys Res Commun 1987; 144:447–453.
47. Takahashi K, Kawazoe Y, Sakumi K, Nakabeppu Y, Sekiguchi M. J Biol Chem 1988; 263:13490–13492.
48. Landini P, Volkert MR. J Bacteriol 2000; 182:6543–6549.
49. Sakumi K, Sekiguchi M. J Mol Biol 1989; 205:373–385.
50. Myers LC, Terranova MP, Nash HM, Markus MA, Verdine GL. Biochemistry 1992; 31:4541–4547.
51. Saget BM, Walker GC. Proc Natl Acad Sci USA 1994; 91:9730–9734.
52. Hayakawa H, Koike G, Sekiguchi M. J Mol Biol 1990; 213:739–747.
53. Tano K, Shiota S, Collier J, Foote RS, Mitra S. Proc Natl Acad Sci USA 1990; 87: 686–690.
54. Rydberg B, Spurr N, Karran P. J Biol Chem 1990; 265:9563–9569.
55. Rydberg B, Hall J, Karran P. Nucl Acids Res 1990; 18:17–21.
56. Crone TM, Pegg AE. Cancer Res 1993; 53:4750–4753.
57. Crone TM, Goodtzova K, Pegg AE. Mutat Res 1996; 363:15–25.
58. Daniels DS, Mol CD, Arvai AS, Kanugula S, Pegg AE, Tainer JA. EMBO J 2000; 19:1719–1730.
59. Davis BM, Encell LP, Zielske SP, Christians FC, Liu L, Friebert SE, Loeb LA, Gerson SL. Proc Natl Acad Sci USA 2001; 98:4950–4954.
60. Christians FC, Loeb LA. Proc Natl Acad Sci USA 1996; 93:6124–6128.
61. Christians FC, Dawson BJ, Coates MM, Loeb LA. Cancer Res 1997; 57:2007–2012.
62. Encell LP, Loeb LA. Biochemistry 1999; 38:12097–12103.
63. Rasimas JJ, Kanugula S, Dalessio PM, Ropson IJ, Fried MG, Pegg AE. Biochemistry 2003; 42:980–990.
64. Fried MG, Kanugula S, Bromberg JL, Pegg AE. Biochemistry 1996; 35:15295–15301.

65. Dolan ME, Oplinger M, Pegg AE. Carcinogenesis 1988; 9:2139–2143.
66. Luu KX, Kanugula S, Pegg AE, Pauly GT, Moschel RC. Biochemistry 2002; 41: 8689–8697.
67. Boiteux S, Laval F. Carcinogenesis 1985; 6:805–807.
68. Teo AK, Oh HK, Ali RB, Li BF. Mol Cell Biol 2001; 21:7105–7114.
69. Dolan ME, Mitchell RB, Mummert C, Moschel RC, Pegg AE. Cancer Res 1991; 51:3367–3372.
70. Dolan ME, Moschel RC, Pegg AE. Proc Natl Acad Sci USA 1990; 87:5368–5372.
71. Middleton MR, Kelly J, Thatcher N, Donnelly DJ, McElhinney RS, McMurry TB, McCormick JE, Margison GP. Int J Cancer 2000; 85:248–252.
72. Dolan ME, Pegg AE, Dumenco LL, Moschel RC, Gerson SL. Carcinogenesis 1991; 12:2305–2309.
73. Dolan ME, Posner M, Karrison T, Radosta J, Steinberg G, Bertucci D, Vujasin L, Ratain MJ. Clin Cancer Res 2002; 8:2519–2523.
74. Gerson SL. J Clin Oncol 2002; 20:2388–2399.
75. Xu-Welliver M, Leitao J, Kanugula S, Meehan WJ, Pegg AE. Biochem Pharmacol 1999; 58:1279–1285.
76. Wu MH, Lohrbach KE, Olopade OI, Kokkinakis DM, Friedman HS, Dolan ME. Clin Cancer Res 1999; 5:209–213.
77. Loktionova NA, Pegg AE. Biochem Pharmacol 2002; 63:1431–1442.
78. Rusin M, Samojedny A, Harris CC, Chorazy M. Hum Mutat 1999; 14:269–270.
79. Middleton MR, Margison GP. Lancet Oncol 2003; 4:37–44.
80. Srivenugopal KS, Yuan XH, Friedman HS, Ali-Osman F. Biochemistry 1996; 35: 1328–1334.
81. Laval F, Wink DA. Carcinogenesis 1994; 15:443–447.
82. Liu L, Xu-Welliver M, Kanugula S, Pegg AE. Cancer Res 2002; 62:3037–3043.
83. Sklar R, Strauss B. Nature 1981; 289:417–420.
84. Day RS, 3rd, Ziolkowski CH. Nature 1979; 279:797–799.
85. Day RS III, Ziolkowski CH, Scudiero DA, Meyer SA, Lubiniecki AS, Girardi AJ, Galloway SM, Bynum GD. Nature 1980; 288:724–727.
86. Yarosh DB, Scudiero D, Ziolkowski CH, Rhim JS, Day RS III. Carcinogenesis 1984; 5:627–633.
87. Brennand J, Margison GP. Carcinogenesis 1986; 7:2081–2084.
88. Brennand J, Margison GP. Proc Natl Acad Sci USA 1986; 83:6292–6296.
89. Hall J, Kataoka H, Stephenson C, Karran P. Carcinogenesis 1988; 9:1587–1593.
90. Samson L, Derfler B, Waldstein EA. Proc Natl Acad Sci USA 1986; 83:5607–5610.
91. Qian XC, Brent TP. Cancer Res 1997; 57:3672–3677.
92. Bhakat KK, Mitra S. Carcinogenesis 2003; 24:1337–1345.
93. Wang Y, Kato T, Ayaki H, Ishizaki K, Tano K, Mitra S, Ikenaga M. Mutat Res 1992; 273:221–230.
94. Cairns-Smith S, Karran P. Cancer Res 1992; 52:5257–5263.
95. Watts GS, Pieper RO, Costello JF, Peng YM, Dalton WS, Futscher BW. Mol Cell Biol 1997; 17:5612–5619.
96. von Wronski MA, Harris LC, Tano K, Mitra S, Bigner DD, Brent TP. Oncol Res 1992; 4:167–174.
97. Esteller M, Hamilton SR, Burger PC, Baylin SB, Herman JG. Cancer Res 1999; 59:793–797.
98. Esteller M, Risques RA, Toyota M, Capella G, Moreno V, Peinado MA, Baylin SB, Herman JG. Cancer Res 2001; 61:4689–4692.
99. Matsukura S, Soejima H, Nakagawachi T, Yakushiji H, Ogawa A, Fukuhara M, Miyazaki K, Nakabeppu Y, Sekiguchi M, Mukai T. Br J Cancer 2003; 88:521–529.
100. Zhang YJ, Chen Y, Ahsan H, Lunn RM, Lee PH, Chen CJ, Santella RM. Int J Cancer 2003; 103:440–444.
101. Kat A, Thilly WG, Fang WH, Longley MJ, Li GM, Modrich P. Proc Natl Acad Sci USA 1993; 90:6424–6428.

102. Griffin S, Branch P, Xu YZ, Karran P. Biochemistry 1994; 33:4787–4793.
103. Branch P, Aquilina G, Bignami M, Karran P. Nature 1993; 362:652–654.
104. Umar A, Koi M, Risinger JI, Glaab WE, Tindall KR, Kolodner RD, Boland CR, Barrett JC, Kunkel TA. Cancer Res 1997; 57:3949–3955.
105. Koi M, Umar A, Chauhan DP, Cherian SP, Carethers JM, Kunkel TA, Boland CR. Cancer Res 1994; 54:4308–4312.
106. Takagi Y, Takahashi M, Sanada M, Ito R, Yamaizumi M, Sekiguchi M. DNA Repair (Amst) 2003; 2:1135–1146.
107. Kawate H, Sakumi K, Tsuzuki T, Nakatsuru Y, Ishikawa T, Takahashi S, Takano H, Noda T, Sekiguchi M. Proc Natl Acad Sci USA 1998; 95:5116–5120.
108. Karran P. Carcinogenesis 2001; 22:1931–1937.
109. Tominaga Y, Tsuzuki T, Shiraishi A, Kawate H, Sekiguchi M. Carcinogenesis 1997; 18:889–896.
110. Sakumi K, Shiraishi A, Shimizu S, Tsuzuki T, Ishikawa T, Sekiguchi M. Cancer Res 1997; 57:2415–2418.
111. Glassner BJ, Weeda G, Allan JM, Broekhof JL, Carls NH, Donker I, Engelward BP, Hampson RJ, Hersmus R, Hickman MJ, Roth RB, Warren HB, Wu MM, Hoeijmakers JH, Samson LD. Mutagenesis 1999; 14:339–347.
112. Shiraishi A, Sakumi K, Sekiguchi M. Carcinogenesis 2000; 21:1879–1883.
113. Humbert O, Fiumicino S, Aquilina G, Branch P, Oda S, Zijno A, Karran P, Bignami M. Carcinogenesis 1999; 20:205–214.
114. Branch P, Hampson R, Karran P. Cancer Res 1995; 55:2304–2309.
115. Karran P, Hampson R. Cancer Surv 1996; 28:69–85.
116. Allay E, Veigl M, Gerson SL. Oncogene 1999; 18:3783–3787.
117. Gerson SL, Zaidi NH, Dumenco LL, Allay E, Fan CY, Liu L, O'Connor PJ. Mutat Res 1994; 307:541–555.
118. Reese JS, Allay E, Gerson SL. Oncogene 2001; 20:5258–5263.
119. Reese JS, Qin X, Ballas CB, Sekiguchi M, Gerson SL. J Hematother Stem Cell Res 2001; 10:115–123.
120. Kataoka H, Yamamoto Y, Sekiguchi M. J Bacteriol 1983; 153:1301–1307.
121. Chen BJ, Carroll P, Samson L. J Bacteriol 1994; 176:6255–6261.
122. Dinglay S, Trewick SC, Lindahl T, Sedgwick B. Genes Dev 2000; 14:2097–2105.
123. Aravind L, Koonin EV. Genome Biol 2, RESEARCH0007, 2001.
124. Falnes PO, Johansen RF, Seeberg E. Nature 2002; 419:178–182.
125. Trewick SC, Henshaw TF, Hausinger RP, Lindahl T, Sedgwick B. Nature 2002; 419:174–178.
126. Koivisto P, Duncan T, Lindahl T, Sedgwick B. J Biol Chem 2003; 278:44348–44354.
127. Kataoka H, Sekiguchi M. Mol Gen Genet 1985; 198:263–269.
128. Kondo H, Nakabeppu Y, Kataoka H, Kuhara S, Kawabata S, Sekiguchi M. J Biol Chem 1986; 261:15772–15777.
129. Dinglay S, Gold B, Sedgwick B. Mutat Res 1998; 407:109–116.
130. Duncan T, Trewick SC, Koivisto P, Bates PA, Lindahl T, Sedgwick B. Proc Natl Acad Sci USA 2002; 99:16660–16665.
131. Welford RW, Schlemminger I, McNeill LA, Hewitson KS, Schofield CJ. J Biol Chem 2003; 278:10157–10161.
132. Mishina Y, He C. J Am Chem Soc 2003; 125:8730–8731.
133. Wei YF, Carter KC, Wang RP, Shell BK. Nucl Acids Res 1996; 24:931–937.
134. Aas PA, Otterlei M, Falnes PO, Vagbo CB, Skorpen F, Akbari M, Sundheim O, Bjoras M, Slupphaug G, Seeberg E, Krokan HE. Nature 2003; 421:859–863.
135. Larson K, Sahm J, Shenkar R, Strauss B. Mutat Res 1985; 150:77–84.
136. Riazuddin S, Lindahl T. Biochemistry 1978; 17:2110–2118.
137. Thomas L, Yang CH, Goldthwait DA. Biochemistry 1982; 21:1162–1169.
138. Singer B, Grunberger D. Molecular Biology of Mutagens and Carcinogens, New York: Plenum, 1983.

139. Karran P, Lindahl T, Ofsteng I, Evensen GB, Seeberg E. J Mol Biol 1980; 140:101–127.
140. Sakumi K, Nakabeppu Y, Yamamoto Y, Kawabata S, Iwanaga S, Sekiguchi M. J Biol Chem 1986; 261:15761–15766.
141. Bjelland S, Seeberg E. Nucl Acids Res 1987; 15:2787–2801.
142. Steinum AL, Seeberg E. Nucl Acids Res 1986; 14:3763–3772.
143. Clarke ND, Kvaal M, Seeberg E. Mol Gen Genet 1984; 197:368–372.
144. Tudek B, Van Zeeland AA, Kusmierek JT, Laval J. Mutat Res 1998; 407:169–176.
145. Drohat AC, Kwon K, Krosky DJ, Stivers JT. Nat Struct Biol 2002; 9:659–664.
146. Kwon K, Cao C, Stivers JT. J Biol Chem 2003; 278:19442–19446.
147. Cao C, Kwon K, Jiang YL, Drohat AC, Stivers JT. J Biol Chem 2003; 278:48012–48020.
148. Kaasen I, Evensen G, Seeberg E. J Bacteriol 1986; 168:642–647.
149. Posnick LM, Samson LD. J Bacteriol 1999; 181:6763–6771.
150. Berdal KG, Johansen RF, Seeberg E. EMBO J 1998; 17:363–367.
151. Klungland A, Bjoras M, Hoff E, Seeberg E. Nucl Acids Res 1994; 22:1670–1674.
152. Klungland A, Laake K, Hoff E, Seeberg E. Carcinogenesis 1995; 16:1281–1285.
153. Nakabeppu Y, Kondo H, Sekiguchi M. J Biol Chem 1984; 259:13723–13729.
154. Nakabeppu Y, Miyata T, Kondo H, Iwanaga S, Sekiguchi M. J Biol Chem 1984; 259:13730–13736.
155. Chen J, Derfler B, Maskati A, Samson L. Proc Natl Acad Sci USA 1989; 86:7961–7965.
156. Memisoglu A, Samson L. Gene 1996; 177:229–235.
157. Hollis T, Ichikawa Y, Ellenberger T. EMBO J 2000; 19:758–766.
158. Labahn J, Scharer OD, Long A, Ezaz-Nikpay K, Verdine GL, Ellenberger TE. Cell 1996; 86:321–329.
159. Yamagata Y, Kato M, Odawara K, Tokuno Y, Nakashima Y, Matsushima N, Yasumura K, Tomita K, Ihara K, Fujii Y, Nakabeppu Y, Sekiguchi M, Fujii S. Cell 1996; 86:311–319.
160. Privezentzev CV, Saparbaev M, Sambandam A, Greenberg MM, Laval J. Biochemistry 2000; 39:14263–14268.
161. Brooks N, McHugh PJ, Lee M, Hartley JA. Anticancer Drug Des 1999; 14:11–18.
162. Bouziane M, Miao F, Ye N, Holmquist G, Chyzak G, O'Connor TR. Acta Biochim Pol 1998; 45:191–202.
163. Saparbaev M, Laval J. Proc Natl Acad Sci USA 1994; 91:5873–5877.
164. Saparbaev M, Mani JC, Laval J. Nucl Acids Res 2000; 28:1332–1339.
165. Saparbaev M, Kleibl K, Laval J. Nucl Acids Res 1995; 23:3750–3755.
166. Castaing B, Geiger A, Seliger H, Nehls P, Laval J, Zelwer C, Boiteux S. Nucl Acids Res 1993; 21:2899–2905.
167. Holmquist GP. Mutat Res 1998; 400:59–68.
168. Evensen G, Seeberg E. Nature 1982; 296:773–775.
169. Boiteux S, Huisman O, Laval J. EMBO J 1984; 3:2569–2573.
170. Bjoras M, Klungland A, Johansen RF, Seeberg E. Biochemistry 1995; 34:4577–4582.
171. Glassner BJ, Rasmussen LJ, Najarian MT, Posnick LM, Samson LD. Proc Natl Acad Sci USA 1998; 95:9997–10002.
172. Xiao W, Samson L. Proc Natl Acad Sci USA 1993; 90:2117–2121.
173. Xiao W, Chow BL, Hanna M, Doetsch PW. Mutat Res 2001; 487:137–147.
174. Santerre A, Britt AB. Proc Natl Acad Sci USA 1994; 91:2240–2244.
175. Samson L, Derfler B, Boosalis M, Call K. Proc Natl Acad Sci USA 1991; 88:9127–9131.
176. O'Connor TR, Laval J. Biochem Biophys Res Commun 1991; 176:1170–1177.
177. O'Connor TR, Laval F. EMBO J 1990; 9:3337–3342.
178. Chakravarti D, Ibeanu GC, Tano K, Mitra S. J Biol Chem 1991; 266:15710–15715.
179. Lau AY, Scharer OD, Samson L, Verdine GL, Ellenberger T. Cell 1998; 95:249–258.
180. Abner CW, Lau AY, Ellenberger T, Bloom LB. J Biol Chem 2001; 276:13379–13387.
181. Biswas T, Clos LJ II, SantaLucia J Jr, Mitra S, Roy R. J Mol Biol 2002; 320:503–513.
182. Vallur AC, Feller JA, Abner CW, Tran RK, Bloom LB. J Biol Chem 2002; 277:31673–31678.

183. Engelward BP, Boosalis MS, Chen BJ, Deng Z, Siciliano MJ, Samson LD. Carcinogenesis 1993; 14:175–181.
184. TR O'Connor. Nucl Acids Res 1993; 21:5561–5569.
185. Wuenschell GE, O'Connor TR, Termini J. Biochemistry 2003; 42:3608–3616.
186. Miao F, Bouziane M, O'Connor TR. Nucl Acids Res 1998; 26:4034–4041.
187. Hang B, Chenna A, Rao S, Singer B. Carcinogenesis 1996; 17:155–157.
188. Hang B, Singer B, Margison GP, Elder RH. Proc Natl Acad Sci USA 1997; 94: 12869–12874.
189. Dosanjh MK, Roy R, Mitra S, Singer B. Biochemistry 1994; 33:1624–1628.
190. Singer B, Antoccia A, Basu AK, Dosanjh MK, Fraenkel-Conrat H, Gallagher PE, Kusmierek JT, Qiu ZH, Rydberg B. Proc Natl Acad Sci USA 1992; 89:9386–9390.
191. Dosanjh MK, Chenna A, Kim E, Fraenkel-Conrat H, Samson L, Singer B. Proc Natl Acad Sci USA 1994; 91:1024–1028.
192. Singer B, Hang B. Chem Res Toxicol 1997; 10:713–732.
193. Saparbaev M, Laval J. IARC Sci Publ 1999:249–261.
194. Saparbaev M, Langouet S, Privezentzev CV, Guengerich FP, Cai H, Elder RH, Laval J. J Biol Chem 2002; 277:26987–26993.
195. Bessho T, Roy R, Yamamoto K, Kasai H, Nishimura S, Tano K, Mitra S. Proc Natl Acad Sci USA 1993; 90:8901–8904.
196. Kartalou M, Samson LD, Essigmann JM. Biochemistry 2000; 39:8032–8038.
197. Scicchitano DA, Hanawalt PC. Proc Natl Acad Sci USA 1989; 86:3050–3054.
198. Scicchitano DA, Hanawalt PC. Mutat Res 1990; 233:31–37.
199. Scicchitano DA, Hanawalt PC. Environ Health Perspect 1992; 98:45–51.
200. Plosky B, Samson L, Engelward BP, Gold B, Schlaen B, Millas T, Magnotti M, Schor J, Scicchitano DA. DNA Repair (Amst) 2002; 1:683–696.
201. Ye N, Holmquist GP, O'Connor TR. J Mol Biol 1998; 284:269–285.
202. Yao M, Kow YW. J Biol Chem 1996; 271:30672–30676.
203. He B, Qing H, Kow YW. Mutat Res 2000; 459:109–114.
204. Kow YW. Free Radic Biol Med 2002; 33:886–893.
205. Moe A, Ringvoll J, Nordstrand LM, Eide L, Bjoras M, Seeberg E, Rognes T, Klungland A. Nucl Acids Res 2003; 31:3893–3900.
206. Wyatt MD, Samson LD. Carcinogenesis 2000; 21:901–908.
207. Connor EE, Wyatt MD. Chem Biol 2002; 9:1033–1041.
208. Engelward BP, Dreslin A, Christensen J, Huszar D, Kurahara C, Samson L. EMBO J 1996; 15:945–952.
209. Wang W, Sitaram A, Scicchitano DA. Biochemistry 1995; 34:1798–1804.
210. Bartlett JD, Scicchitano DA, Robison SH. Mutat Res 1991; 255:247–256.
211. Hartshorn JN, Scicchitano DA, Robison SH. Basic Life Sci 1990; 53:233–249.
212. May A, Nairn RS, Okumoto DS, Wassermann K, Stevnsner T, Jones JC, Bohr VA. J Biol Chem 1993; 268:1650–1657.
213. Fishel ML, Seo YR, Smith ML, Kelley MR. Cancer Res 2003; 63:608–615.
214. Engelward BP, Weeda G, Wyatt MD, Broekhof JL, de Wit J, Donker I, Allan JM, Gold B, Hoeijmakers JH, Samson LD. Proc Natl Acad Sci USA 1997; 94:13087–13092.
215. Allan JM, Engelward BP, Dreslin AJ, Wyatt MD, Tomasz M, Samson LD. Cancer Res 1998; 58:3965–3973.
216. Miao F, Bouziane M, Dammann R, Masutani C, Hanaoka F, Pfeifer G, O'Connor TR. J Biol Chem 2000; 275:28433–28438.
217. Watanabe S, Ichimura T, Fujita N, Tsuruzoe S, Ohki I, Shirakawa M, Kawasuji M, Nakao M. Proc Natl Acad Sci USA 2003; 100:12859–12864.
218. Birkeland NK, Anensen H, Knaevelsrud I, Kristoffersen W, Bjoras M, Robb FT, Klungland A, Bjelland S. Biochemistry 2002; 41:12697–12705.
219. Begley TJ, Haas BJ, Noel J, Shekhtman A, Williams WA, Cunningham RP. Curr Biol 1999; 9:653–656.

220. O'Rourke EJ, Chevalier C, Boiteux S, Labigne A, Ielpi L, Radicella JP. J Biol Chem 2000; 275:20077–20083.
221. Liu P, Burdzy A, Sowers LC. DNA Repair (Amst) 2003; 2:199–210.
222. Valinluck V, Liu P, Burdzy A, Ryu J, Sowers LC. Chem Res Toxicol 2002; 15: 1595–1601.
223. Liu P, Burdzy A, Sowers LC. Chem Res Toxicol 2002; 15:1001–1009.
224. Barrett TE, Scharer OD, Savva R, Brown T, Jiricny J, Verdine GL, Pearl LH. Embo J 1999; 18:6599–6609.
225. Barrett TE, Savva R, Panayotou G, Barlow T, Brown T, Jiricny J, Pearl LH. Cell 1998; 92:117–129.
226. Lutsenko E, Bhagwat AS. J Biol Chem 1999; 274:31034–31038.
227. Saparbaev M, Laval J. Proc Natl Acad Sci USA 1998; 95:8508–8513.
228. Gallinari P, Jiricny J. Nature 1996; 383:735–738.
229. Neddermann P, Gallinari P, Lettieri T, Schmid D, Truong O, Hsuan JJ, Wiebauer K, Jiricny J. J Biol Chem 1996; 271:12767–12774.
230. Mol CD, Arvai AS, Begley TJ, Cunningham RP, Tainer JA. J Mol Biol 2002; 315: 373–384.
231. Hang B, Medina M, Fraenkel-Conrat H, Singer B. Proc Natl Acad Sci USA 1998; 95:13561–13566.
232. Hang B, Downing G, Guliaev AB, Singer B. Biochemistry 2002; 41:2158–2165.
233. Waters TR, Gallinari P, Jiricny J, Swann PF. J Biol Chem 1999; 274:67–74.
234. Waters TR, Swann PF. J Biol Chem 1998; 273:20007–20014.
235. Um S, Harbers M, Benecke A, Pierrat B, Losson R, Chambon P. J Biol Chem 1998; 273:20728–20736.
236. Tini M, Benecke A, Um SJ, Torchia J, Evans RM, Chambon P. Mol Cell 2002; 9: 265–277.
237. Shimizu Y, Iwai S, Hanaoka F, Sugasawa K. EMBO J 2003; 22:164–173.
238. Hardeland U, Steinacher R, Jiricny J, Schar P. EMBO J 2002; 21:1456–1464.
239. Chen D, Lucey MJ, Phoenix F, Lopez-Garcia J, Hart SM, Losson R, Buluwela L, Coombes RC, Chambon P, Schar P, Ali S. J Biol Chem 2003; 278:38586–38592.
240. Sobol RW, Kartalou M, Almeida KH, Joyce DF, Engelward BP, Horton JK, Prasad R, Samson LD, Wilson SH. J Biol Chem 2003; 278:39951–39959.
241. Shah D, Kelly J, Zhang Y, Dande P, Martinez J, Ortiz G, Fronza G, Tran H, Soto AM, Marky L, Gold B. Biochemistry 2001; 40:1796–1803.
242. Smith SA, Engelward BP. Nucl Acids Res 2000; 28:3294–3300.
243. Spratt TE, Wu JD, Levy DE, Kanugula S, Pegg AE. Biochemistry 1999; 38:6801–6806.
244. Graves RJ, Li BF, Swann PF. Carcinogenesis 1989; 10:661–666.
245. Bhattacharyya D, Tano K, Bunick GJ, Uberbacher EC, Behnke WD, Mitra S. Nucl Acids Res 1988; 16:6397–6410.
246. Roy R, Shiota S, Kennel SJ, Raha R, von Wronski M, Brent TP, Mitra S. Carcinogenesis 1995; 16:405–411.
247. Chan CL, Wu Z, Ciardelli T, Eastman A, Bresnick E. Arch Biochem Biophys 1993; 300:193–200.
248. Bjelland S, Birkeland NK, Benneche T, Volden G, Seeberg E. J Biol Chem 1994; 269:30489–30495.
249. Terato H, Masaoka A, Asagoshi K, Honsho A, Ohyama Y, Suzuki T, Yamada M, Makino K, Yamamoto K, Ide H. Nucl Acids Res 2002; 30:4975–4984.
250. Friedberg EC, Walker GC, Siede W. DNA Repair and Mutagenesis. Washington, DC: ASM Press, 1995.
251. Bates PA, Sternberg MJ. Proteins 1999; (suppl 3):47–54.
252. Vickers MA, Vyas P, Harris PC, Simmons DL, Higgs DR. Proc Natl Acad Sci USA 1993; 90:3437–3441.
253. Lau AY, Wyatt MD, Glassner BJ, Samson LD, Ellenberger T. Proc Natl Acad Sci USA 2000; 97:13573–13578.

17

Deaminated Bases in DNA

Bernard Weiss
Department of Pathology, Emory University School of Medicine, Atlanta,
Georgia, U.S.A.

1. INTRODUCTION

Because the exocyclic amino groups of the DNA bases mediate base pairing, their loss is mutagenic. The amino groups are subject to attack by normal intracellular compounds as well as by environmental agents, and the types of enzymes that are involved in their repair are universal. This chapter focuses on the enzymatic repair of deaminated purines in DNA, specifically hypoxanthine (deaminated adenine) and xanthine (deaminated guanine). The pyrimidines cytosine and 5-methylcytosine are deaminated by many of the same agents as the purines, and so the causes and consequences of their deamination are also discussed. Their repair, however, is covered in separate chapters (Chapters 1, 14, and 22). 5-Hydroxyuracil (derived from cytosine) and 5-hydroxymethyluracil (derived from 5-methylcytosine or thymidine) are oxidation products and are separately covered (Chapter 15). The repair of deaminated bases in DNA, including deaminated oxidation products, has been previously reviewed (1).

2. LESIONS AND THEIR CONSEQUENCES

Deamination alters the base-pairing properties of nucleotides in DNA (Fig. 1). For example, when cytosine, which normally pairs with guanine, is deaminated to form uracil, it pairs with adenine on subsequent replication. The adenine then pairs with thymine, so that the overall result of the deamination is a C:G → T:A transition mutation. Similarly, adenine is deaminated to hypoxanthine (the base in deoxyinosine), which pairs preferentially with cytosine, thereby producing an A:T → G:C transition. The situation is less clear with deaminated guanine (xanthine). Xanthine is a poor subject for base pairing because it exists as the ionized enol at neutral pH. Poly-X does not pair detectably with Poly-C, and it pairs very poorly with homopolymers of the other nucleotides (2). Studies with purified DNA polymerases indicate that xanthine in the template is a hindrance but not a block to DNA replication and that it leads more frequently to the misincorporation of thymine than of guanine or adenine (3–5). Therefore, the deamination of guanine may primarily produce

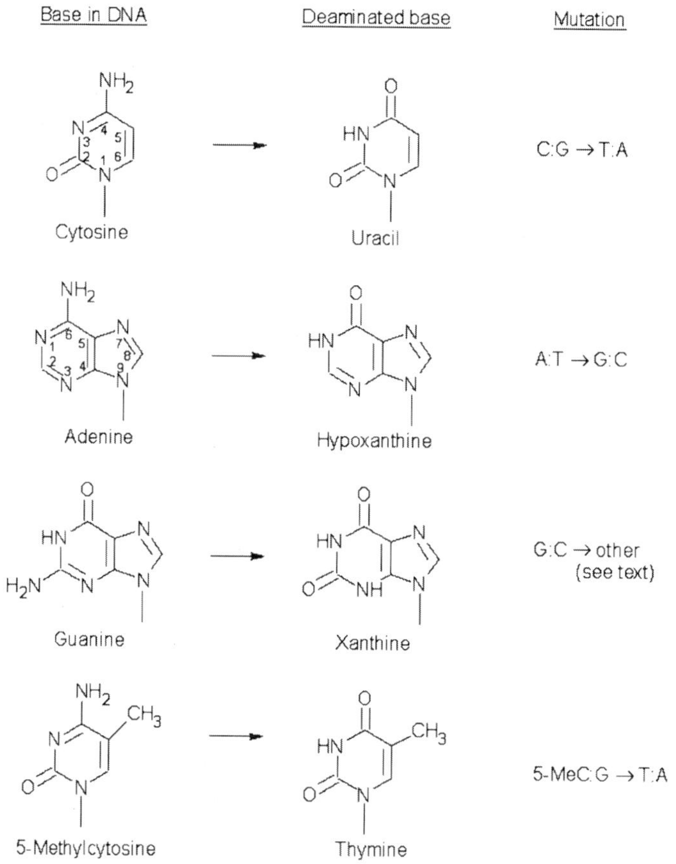

Figure 1 The consequences of base deamination in DNA.

G:C → A:T transitions, a suggestion that has been supported by in vivo evidence as well: experiments with an *Escherichia coli* mutant lacking endonuclease V (a xanthine-specific DNase) suggested that thymine is incorporated opposite xanthine (6). However, it is not known if this result is organism specific; i.e., does it reflect the chemical nature of the nucleobases or the specificity of the DNA polymerases and accessory proteins involved in repair, replication, or translesion bypass?

The deamination of 5-methylcytosine, a common naturally occurring modified base, is a special case because it results in thymine, a normal base. All of the other deaminations result in bases that do not occur naturally in DNA and are therefore easily recognized by specific repair enzymes. Sites of DNA cytosine methylation are thus hotspots for mutation in *E. coli* (7) as well as in mammalian cells (8), and removal of the resulting thymine must depend on the enzymatic recognition of the resulting T:G base pair rather than on the altered base alone (Chapter 15).

3.   DEAMINATING AGENTS

3.1.   Water

The exocyclic amines in the nucleobases are susceptible to spontaneous (hydrolytic) deamination. It is estimated that in the double-stranded DNA of a rat liver cell, there are about 1700 deamination events per day per 10^{10} bases; the rate in single-stranded regions is about 25-fold greater (9). The most common spontaneous deamination event in DNA is that of cytosine to uracil (10). Adenine deamination occurs at about 2% the rate of cytosine deamination (11), and the rate of guanine deamination is even lower (12). Although hydrolytic deamination of DNA purines is less frequent than that of DNA cytosine, it is nevertheless a significant potential cause of mutation.

3.2. Nitrosating Agents

Nitrosating compounds are widespread deaminating agents (13). They are produced in nature as byproducts of the metabolism of nitrogen oxides or of ammonia. The major ones are dinitrogen trioxide (N_2O_3 or nitrous anhydride) and nitrosamines (RNHN=O). N_2O_3 may be formed reversibly from nitrite ion via the dehydration of molecular nitrous acid (HNO_2), or it may be formed by the auto-oxidation of nitric oxide ($NO^\bullet$). N_2O_3 may then react with primary amines to form nitrosamines (which are also nitrosating agents). It may also react with secondary amines and amides to form alkylating agents, which are themselves potent mutagens. The reaction of N_2O_3 or a primary nitrosamine with a nucleobase amine produces first an unstable N-nitroso derivative and then an unstable diazonium compound that hydrolyzes to form the deaminated base, releasing N_2:

$$-NH_2 \rightarrow -NHN = O \rightarrow -N^+N \rightarrow -OH$$

In bacteria such as *E. coli*, nitrate and nitrite are the preferred electron acceptors when oxygen is limiting for growth (14). Nitrite and $NO^\bullet$ are the first reduction products of nitrate. Whereas nitrite is an obligatory free intermediate, it is a poor direct source for N_2O_3. Unlike nitrite, $NO^\bullet$ can readily form N_2O_3, but it is only a latent intermediate. Nitrite is a poor direct source of N_2O_3 because at neutral pH, it forms very little undissociated HNO_2 (the immediate precursor for nitrous anhydride) because HNO_2 has a low pK. On the other hand, NO_2^- can be reduced to $NO^\bullet$, which is readily auto-oxidized to N_2O_3 (15). However, in *E. coli*, $NO^\bullet$ is mostly bound to the major nitrite reductase, and it is therefore only a latent intermediate in the reduction of nitrate to ammonia. Furthermore, this nitrite reductase is induced only under O_2-limiting conditions, but O_2 would be needed to generate N_2O_3 from $NO^\bullet$. Some additional $NO^\bullet$ may also be produced by the chemical reduction of NO_2^- by Fe^{2+} (16) or by formate (17). Recent experiments suggest that main direct cause of mutagenic nitrosative deamination in *E. coli* is $NO^\bullet$ that forms during hypoxic nitrate/nitrite respiration and that is subsequently auto-oxidized. An endonuclease V mutant (which is unable to repair deaminated adenine in DNA) has a high rate of nitrate-induced A:T → G:C mutations, but this mutator effect was eliminated by a deletion of the gene for the major nitrite reductase (B. Weiss, unpublished results). Therefore, $NO^\bullet$ rather than NO_2^- is the more proximate source of most intracellular N_2O_3. However, extracellular $NO^\bullet$ may be produced by other pathways, two of which have been substantiated in *E. coli*: (i) an aberrant reaction in which nitr*ate* reductase reduces nitrite (17,18), and (ii) the reduction of nitrite by a periplasmic cytochromal nitrite reductase (19). Only hypoxic conditions are needed to induce the nitrate and nitrite reductases; they need not be anaerobic. Thus, nitrosative deamination of DNA occurs even when nitrate-supplemented bacterial cultures are grown to saturation in open flasks (20).

Although mammals have no nitrate/nitrite respiration, we are continuously exposed to nitrosating agents. NO•, which is produced mainly by the enzymatic oxidation of arginine, is a widespread chemical messenger in many tissues. It is also generated by enteric bacteria, and it is produced by activated phagocytes as an antibacterial agent. We are exposed to mutagenic nitrogen oxides as pollutants in water and in air, in acid rain, smog, and tobacco smoke. Nitrate in fertilizers is converted to nitrite by bacteria in streams. Nitrite is used to preserve the color of meat products, and nitrosamines are formed during the cooking of meat at high temperature. Despite the prevalence of these major endogenous and environmental mutagens, their study has been relatively neglected when one considers the voluminous literature on reactive oxygen species.

Guanine is the most susceptible of the nucleobases to nitrosative deamination. When DNA in solution is exposed to nitrous acid, guanine, cytosine, and adenine are deaminated in the approximate ratio of 4:2:1 (21). Similarly, NO• at neutral pH deaminates guanine in double-stranded oligonucleotides at twice the rate of cytosine (adenine was not measured) (22), and NO• attacked mainly guanine residues in the DNA of intact mitochondria (23). As with almost all agents that attack the bases in DNA, single-stranded targets are far more susceptible than double-stranded ones (22). Based on its behavior in acid, xanthine in DNA was assumed to be highly susceptible to spontaneous depurination. However, at neutral pH, it is only a little less stable than guanine (5).

Nitrosative damage also results in other lesions that are not strictly deaminated bases, although their formation involves deamination reactions. These are oxanine, and interstrand cross-links. Oxanine is formed by a deamination of guanine accompanied by a rearrangement to result in a guanine than contains an oxygen atom in place of the N1 imino group in its six-membered ring (24). Appreciable quantities are formed only at very low pH, and therefore little is known of its physiological significance or mode of repair. Interstrand cross-links are formed between diagonally opposite purines. A diazonium derivative of one purine reacts with free amino group of the other, releasing N_2 and forming an imino cross-link between the two rings. A 2–2 G/G bridge is the most common event, whereas the 2–6 G/A bridge is a minor lesion (25). Oxanine–protein cross-links have also been described (26). Presumably, the cross-linking lesions are handled by repair systems that recognize DNA cross-links and bulky adducts and not by those that recognize deaminated bases.

3.3. Other Agents

Bisulfite converts cytosine to uracil by transiently forming an addition product from which the amino group is readily hydrolyzed (9). The reaction occurs only at unpaired cytosine residues, i.e., in single-stranded rather than in double-stranded DNA. The deamination of 5-methylcytosine occurs at less than 1% the rate of cytosine deamination. Bisulfite is produced in cells as an intermediary metabolite in the breakdown of sulfur-containing amino acids. However, mutagenesis has been demonstrated in vivo only with extremely high external concentrations (1M).

Some agents that are primarily oxidants, such as peroxynitrite and ferric nitrilotriacetate, have been reported to produce deaminated purines in vivo (27,28).

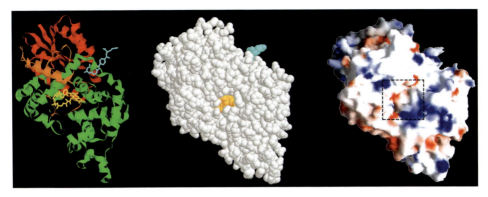

Figure 5–3 Molecular structure of *E. coli* CPD photolyase as determined by x-ray crystallography. (See p. 101.)

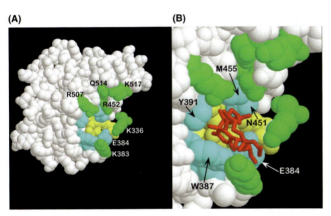

(A)

(B)

Figure 5–5 Space filling model of the *C*-terminal domain of *S. cerevisiae* photolyase. (See p. 103.)

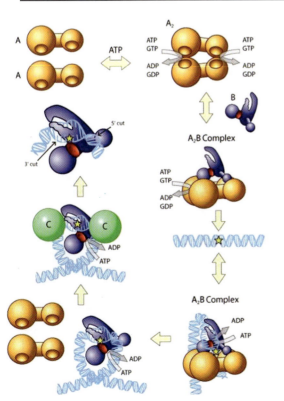

Figure 6–1 Damage detection, verification, and incision by the prokaryotic NER system. (See p. 112.)

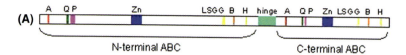

(A) N-terminal ABC — C-terminal ABC

(B)

```
                      A
NtEcA    24 DKLIVVTGLSGSGKSSLAFDTLYAEGQRRYVESLSA----YARQFLSLMEKPDVDHIEGL
NtBcA    24 GKLVVLTGLSGSGKSSLAFDTIYAEGQRRYVESLSA----YARQFLGQMEKPDVDAIEGL
CtEcA   633 GLFTCITGVSGSGKSTLINDTLFPIAQRQLNG---------ATIAEPAPYRDIQGLEHF
CtBcA   630 GTFVAVTGVSGSGKSTLVNEVLYKALAQKLH----------RAKAKPGEHRDIRGLEHL
HisP     32 GDVISIIGSSGSGKSTFLRCINFLEKPSEGAIIVNGQNINLVRDKDGQLKVADKNQLRLL
MJ0796   31 GEFVSIMGPSGSGKSTMLNIIGCLDKPTEGEVYIDN-------IKTN--DLDDDELTKIR
                Q    PRS/ENI
NtEcA    81 SP-AISIEQKSTSHNPRSTVGTITEIHDYLRLLFAR--Zn---RLKFLVNVGLNYLTLSR
NtBcA    81 SP-AISIDQKTTSRNPRSTVGTVTEIYDYLRLLFAR--Zn---RLGFLQNVGLDYLTLSR
CtEcA   684 DK-VIDIDQSPIGRTPRSNPATYTGVFTPVRELFAG--Zn---KLQTLMDVGLTYIRLGQ
CtBcA   680 DK-VIDIDQSPIGRTPRSNPATYTGVFDDIRDVFAS--Zn---KLETLYDVGLGYMKLGQ
HisP     92 RTRLTMVFQHFNLWSHMTVLENVMEAPIQVLGLSKHDA--RERALKYLAKVGIDERAQGK
MJ0796   82 RDKIGFVFQQFNLIPLLTALENVELPLIFKYRGAMSGEERRKRALECLKMAELEERFANH
              LSGG                              B
NtEcA   483 SAETLSGGEAQRIRLASQIGAGLVG-VMYVLDEPSIGLHQRDNERLLGTLIHLRDL-GNT
NtBcA   481 SAGTLSGGEAQRLATQIGSRLTG-VLYVLDEPSIGLHQRDNDRLIATLKSMRDL-GNT
CtEcA   826 SATTLSGGEAQRVKLARELSKRGTGQTLYILDEPTTGLHFADIQQLLDVLHKLRDQ-GNT
CtBcA   822 PATTLSGGEAQRVKLAAELHRRSNGRTLYILDEPTTGLHVDDIARLLDVLHRLVDN-GDT
HisP    150 YPVHLSGGQQQRVSIARALAMEPDV---LF-DEPTSALDPELVGEVLRIMQQLAEE-GKT
MJ0796  142 KPNQLSGGQQQRVAIARALANNPPI---IL-ADEPTGALDSKTGEKIMQLLKKLNEEDGKT
              H
NtEcA   167 VIVVEHDEDAIRAADHVIDIGPGAGVHGG--------------------
NtBcA   167 LIVVEHDEDTMLAADYLIDIGPGAGIHGG--------------------
CtEcA   163 IVVIEHNLDVIKTADWIVDLGPEGGSGGGEILVSGTPETVAECEASHTA
CtBcA   161 VLVIEHNLDVIKTADYIIDLGPEGGDRGG--------------------
HisP    206 MVVVTHEMGFARHVSSHVIFLHQGKIEEEG--------------------
MJ0796  199 VVVVTHD INVARFGERI IYLKDG--------------------
```

(C)

```
Tth_N    82 AISIDQKTTSHNPRSTVGTVTEIHDYLRLLFAR 114
Drd_N    82 AISIDQKTTSHNPRSTVGTVTEIHDYLRLLFAR 114
Bsu_N    84 AISIDQKTTSRNPRSTVGTVTEIYDYLRLLYAR 116
Eco_N    82 AISIDQKTTSRNPRSTVGTVTEIYDYLRLLFAR 114
Bpe_N    82 AISIEQKSAGHNPRSTVGTITEIHDYLRLLYAR 114
Tth_C   687 VIEIDQSPIGRTPRSNPATYTGVFDEIRDLFAK 719
Drd_C   723 VIEIDQSPIGRTPRSNPATYTGVFTEIRDLFTR 755
Bsu_C   683 VIDIDQAPIGRTPRSNPATYTGVFDDIRDVFAQ 715
Eco_C   681 VIDIDQSPIGRTPRSNPATYTGVFDDIRDVFAS 713
Bpe_C   691 TISVDQSPIGRTPRSNPATYTGLFTPIRELFAG 723
```

Figure 6–2 UvrA. (See p. 116.)

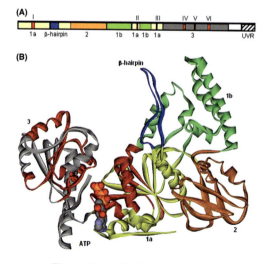

Figure 6–3 UvrB. (See p. 118.)

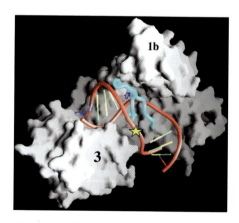

Figure 6–4 Hypothetical space filling model of UvrB bound to DNA. (See p. 119.)

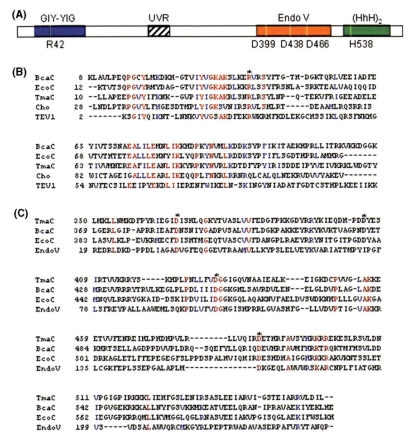

(A)

GIY-YIG   UVR   Endo V   (HhH)₂

R42   D399 D438 D466   H538

(B)

```
BcaC    8 KLAVLPEQPGCYLMKDKH-GTVIYVGKAKSLKERVRSYFTG-TH-DGKTQRLVEEIADFE
EcoC   12 --KTVTSQPGVYRMYDAG-GTVIYVGKAKDLKKRLSSYFRSNLA-SRKTEALVAQIQQID
TmaC   10 --LLAPEEPGVYIFKNK--GVPIYIGKAKRLSNRLRSYLNP--Q-TEKVFRIGEEADELE
Cho    28 -LNDLPTRPGUYLFHGESDTMPLYIGKSVNIRSRVLSHLRT-----DEAAMLRQSRRIS
TEV1    2 -------KSGIYQIKNT-LNNKVYVGSAKDFEKRWKRHFKDLEKGCHSSIKLQRSFNKHG

BcaC   65 YIVTSSNAEALILEMNLIKKHDPKYNVMLKDDKSYPFIKITAEKHPRLLITRKVKKDGGK
EcoC   68 VTVTHTETEALLLEHNYIKLYQPRYNVLLRDDKSYPFIFLSGDTHPRLAMHRG-------
TmaC   63 TIVVMNEREAFILEANLIKKYRPKYNVRLKDTDFYPYIRISDDEIPYVEIVKRKLWDGTY
Cho    82 WICTAGEIGALLLEARLIKEQQPLFNKRLRRNRQLCALQLNEKRVDVVYAKEV-------
TEV1   54 NVFECSILEEIPYEKDLIIERENFWIKELN-SKINGYNIADATFGDTCSTHPLKEEIIKK
```

(C)

```
TmaC  350 LMKLLNMKDFPYRIEGIDISHLQGKYTVASLVVFEDGFPKKGDYRRYKIEQDH-PDDYES
BcaC  369 LGERLGIP-APRRIEAFDNSNIYGADPVSALVVFLDGKPAKKEYRKYKVKTVAGPNDYET
EcoC  383 LASVLKLP-EVKRMECFDISHTMGEQTVASCVVFDANGPLRAEYRRYNITGITPGDDYAA
EndoV  19 REDRLDKD-PPDLIAGADVGFEQGGEVTRAAMVLLKYPSLELVEYKVARIATTMPYIPGF

TmaC  409 IRTVVKRRYS-----KHPLPNLLFVDGGIGQVNAAIEALK----EIGKDCPVVG-LAKKE
BcaC  428 MREVVRRRYTRVLKEGLPLPDLIIIDGGKGHLSAVRDVLEN---ELGLDVPLAG-LAKDE
EcoC  442 MNQVLRRRYGKAID-DSKIPDVILIDGGKGQLAQAKNVFAELDVSWDKNHPLLLGVAKGA
EndoV  78 LSFREYPALLAAWEMLSQKPDLVFVDGHGISHPRRLGVASHFG--LLVDVPTIG-VAKKR

TmaC  459 ETVVFENREIHLPHDHPVLR--------LLVQIRDETHRFAVSYHRKRREKESLRSVLDN
BcaC  484 KHRTSELLAGDPPPDVVPLDRQ--SQEFYLLQRIQDEVHRFAVMFHRKTRQKTMFHSVLDD
EcoC  501 DRKAGLETLFFEPEGEGFSLPPDSPALHVIQHIRDESHDHAIGGHRKKRAKVKNTSSLET
EndoV 135 LCGKFEPLSSEPGALAPLM--------------DKGEQLAWVWRSKARCNPLFIATGHR

TmaC  511 VPGIGPIRKKKLIEHFGSLENIRSASLEEIARVI-GSTEIARRVLDIL--
BcaC  542 IPGVGEKRKKALLNYFGSVKKMKEATVEELQRAN-IPRAVAEKIYEKLHE
EcoC  562 IEGVGPKRRQMLLKYMGGLQGLRNASVEEIAKVPGISQGLAEKIFWSLKH
EndoV 199 VS-----VDSALAWVQRCMKGYRLPEPTRWADAVASERPAFVRYTANQP-
```

Figure 6–5   UvrC. (See p. 120.)

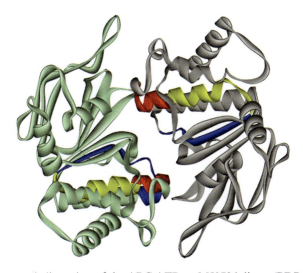

Figure 6–6   Helix–turn–helix region of the ABC ATPase MJ0796 dimer (PDB 1L2T). (See p. 125.)

3

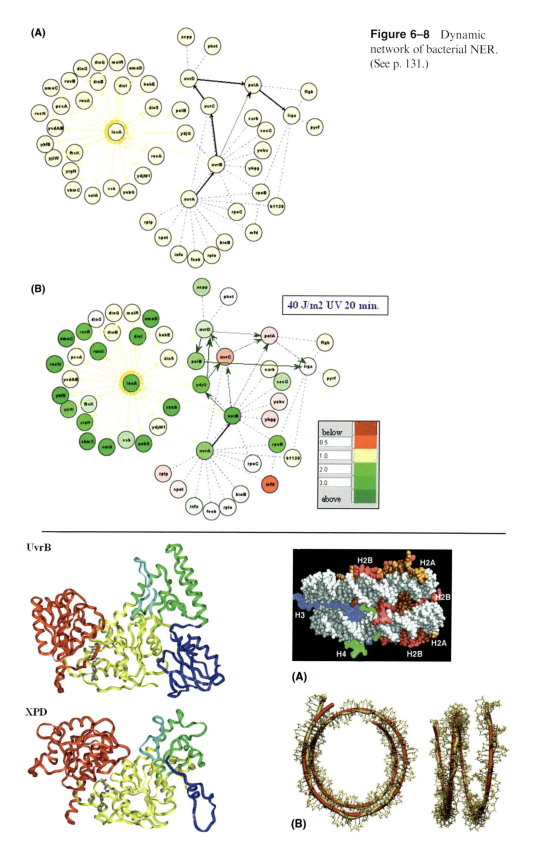

Figure 6–8 Dynamic network of bacterial NER. (See p. 131.)

40 J/m2 UV 20 min.

below
0.5
1.0
2.0
3.0
above

UvrB

XPD

Figure 6–9 Model of XPD using UvrB as a template. (See p. 132.)

H2B   H2A
H2B
H3
H2A
H4   H2B

(A)

(B)

Figure 10–1 Structural features of nucleosomes. (See p. 202.)

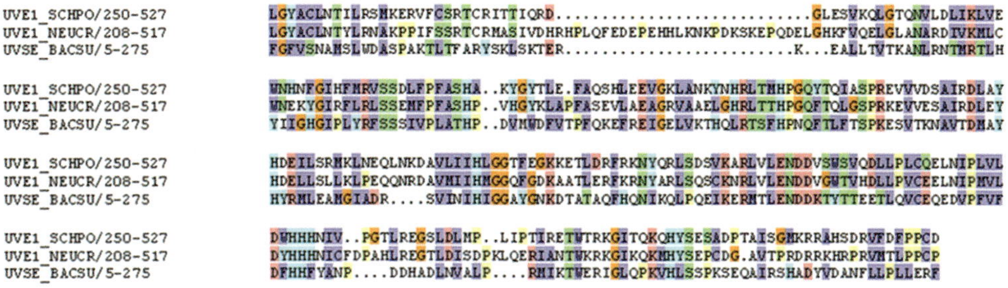

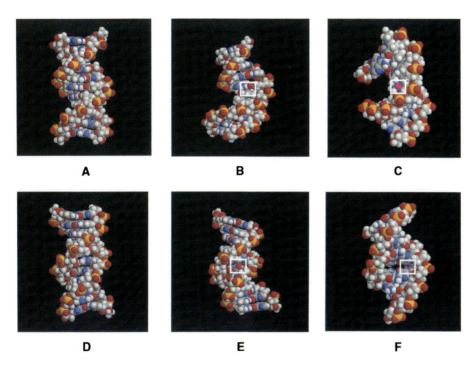

<pre>
UVE1_SCHPO/250-527   LGYACLNTILRSHKERVFCSRTCRITTIQRD........................GLESVKQLGTQNVLDLIKLVE
UVE1_NEUCR/208-517   LGYACLNTYLRNAKPPIFSSRTCRMASIVDHRHPLQFEDEPEHHLKNKPDKSKEPQDELGHKFVQELGLANARDIVKMLC
UVSE_BACSU/5-275   FGFVSNAMSLWDASPAKTLTFARYSKLSKTER....................K...EALLTVTKANLRNTHRTLH

UVE1_SCHPO/250-527   WNHNFGIHFMRVSSDLFPFASHA..KYGYTLE.FAQSHLEEVGKLANKYNHRLTHHPGQYTQIASPREVVVDSAIRDLAY
UVE1_NEUCR/208-517   WNEKYGIRFLRLSSEMFPFASHP..VHGYKLAPFASEVLAEAGRVAAELGHRLTTHPGQFTQLGSPRKEVVESAIRDLEY
UVSE_BACSU/5-275   YIIGHGIPLYRFSSSIVPLATHP..DVMWDFVTPFQKEFREIGELVKTHQLRTSFHPNQFTLFTSPKESVTKNAVTDHAY

UVE1_SCHPO/250-527   HDEILSRNKLNEQLNKDAVLIIHLGGTFEGKKETLDRFRKNYQRLSDSVKARLVLENDDVSWSVQDLLPLCQELNIPLVL
UVE1_NEUCR/208-517   HDELLSLLKLPEQQNRDAVHIIHMGGQFGDKAATLERFKRNYARLSQSCKNRLVLENDDVGWTVHDLLPVCEELNIPHVL
UVSE_BACSU/5-275   HYRHLEAHGIADR....SVINIHIGGAYGNKDTATAQFHQNIKQLPQEIKERHTLENDDKTYTTEETLQVCEQEDWPFVF

UVE1_SCHPO/250-527   DWHHHNIV..PGTLREGSLDLMP..LIPTIRETWTRKGITQKQHYSESADPTAISGHKRRAHSDRVFDFPPCD
UVE1_NEUCR/208-517   DYHHHNICFDPAHLREGTLDISDPKLQERIANTWKRKGIKQKMHYSEPCDG.AVTPRDRRKHRPRVHTLPPCP
UVSE_BACSU/5-275   DFHHFYANP....DDHADLNVALP....RMIKTWERIGLQPKVHLSSPKSEQAIRSHADYVDANFLLPLLERF
</pre>

Figure 11–2 *Schizosaccharomyces pombe* UVDE conserved region sequence alignments with *N. crassa* and *B. subtilis* functional homologs. (See p. 225.)

Figure 11–5 Duplex DNA structural distortion comparisons of oligomers. (See p. 228.)

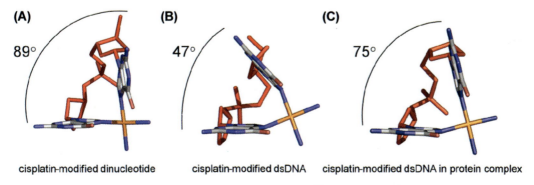

cisplatin-modified dinucleotide   cisplatin-modified dsDNA   cisplatin-modified dsDNA in protein complex

Figure 12–2 Structures of single-stranded *cis*-{[Pt(NH$_3$)$_2$]}$^{2+}$ platinated dinucleotide. (See p. 241.)

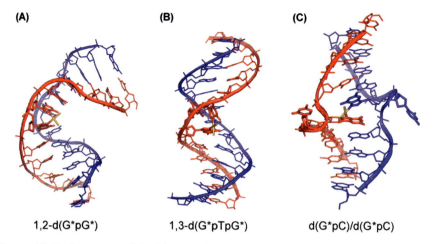

(A)

1,2-d(G*pG*)

(B)

1,3-d(G*pTpG*)

(C)

d(G*pC)/d(G*pC)

Figure 12–3 Structures of double-stranded DNA modified with cisplatin. (See p. 242.)

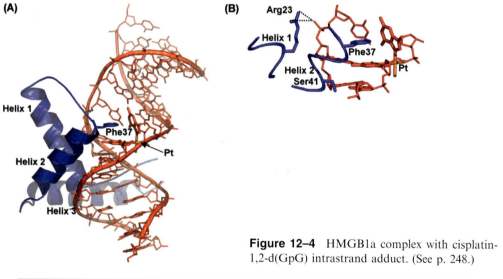

(A)

Helix 1

Helix 2

Phe37

Pt

Helix 3

(B)

Arg23

Helix 1

Phe37

Pt

Helix 2

Ser41

Figure 12–4 HMGB1a complex with cisplatin-1,2-d(GpG) intrastrand adduct. (See p. 248.)

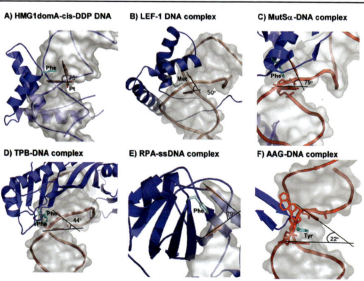

A) HMG1domA-cis-DDP DNA

B) LEF-1 DNA complex

C) MutSα-DNA complex

D) TPB-DNA complex

E) RPA-ssDNA complex

F) AAG-DNA complex

Figure 12–5 Comparison of complexes of minor groove DNA-binding proteins. (See p. 249.)

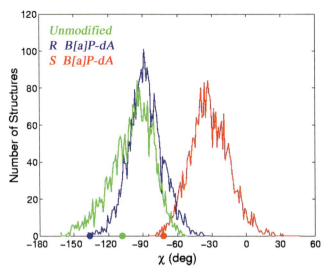

Figure 13–10 Glycosidic torsion angle χ of the B[*a*]P-modified adenine (See p. 282.)

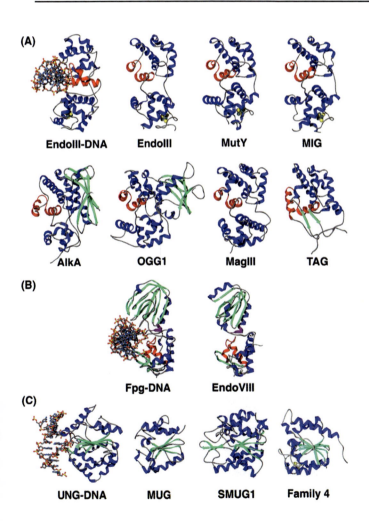

(A)

EndoIII-DNA   EndoIII   MutY   MIG

AlkA   OGG1   MagIII   TAG

(B)

Fpg-DNA   EndoVIII

(C)

UNG-DNA   MUG   SMUG1   Family 4

Figure 14–1 Structures of HhH (A), H$_2$TH (B), UDG (C) family members. (See p. 300.)

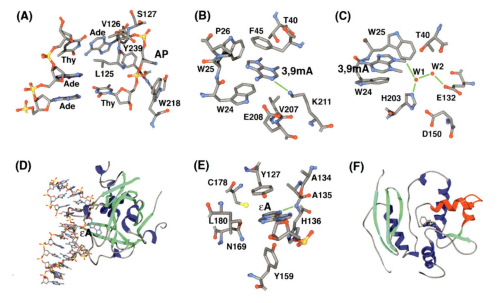

Figure 14–4 Alkylation damage repair proteins. (See p. 311.)

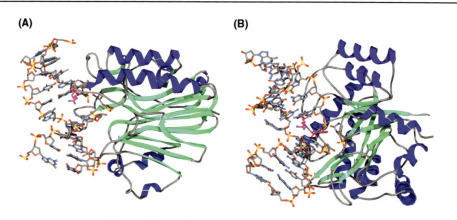

Figure 14–5 AP endonucleases. (See p. 314.)

Finger
Arg

Acceptor
Cys

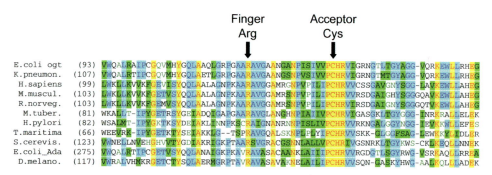

E. coli ogt: 2108312A:, Klebsiella pneumon.: NP_299875:, H.sapiens: NP_002403, Mus musculus: NP_032624, Rattus norvegicus: NP_036993, M.tubercul.: NP_215832, H.pylori: NP_223336, T.maritima: NP_228695, S.cerevis.: NP_010081, D.melano.: NP_477366

Figure 16–6 Alignment of conserved regions of various AGT protein sequences. (See p. 347.)

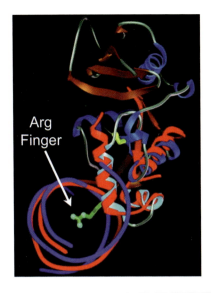

Figure 16–9 Structure of hMGMT. (See p. 350.)

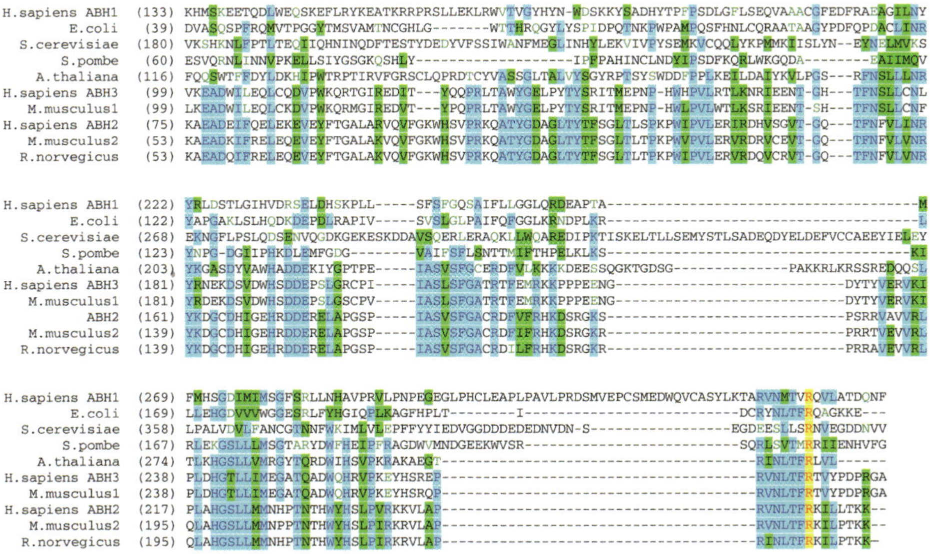

```
H.sapiens ABH1  (133) KHMSKEETQDLWEQSKEFLRYKEATKRRPRSLLEKLRWVTVGYHYN-WDSKKYSADHYTPFFSDLGFLSEQVAAACGFEDFRAEAGILNY
       E.coli    (39) DVASQSPFRQMVTPGGYTMSVAMTNCGHLG-------WTTHRQGYLYSPIDPQTNKPWPAMEQSFHNLCQRAATAAGYPDFQPDACLINR
  S.cerevisiae  (180) VKSHKNLFPTLTEQIIQHNINQDFTESTYDEDYVFSSIWANFMEGLINHYLEKVIVPYSEMKVCQQLYKPMMKIISLYN--EYNELMVKS
       S.pombe   (60) ESVQRNLINNVPKELLSIYGSGKQSHLY-----------------IPFPAHINCLNDYIPSDFKQRLWKGQDA-------EAIIMQV
   A.thaliana   (116) FQQSWTFFDYLDKHIPWTRPTIRVFGRSCLQPRDTCYVASSGLTALVYSGYRPTSYSWDDFFPLKEILDALYKVLPGS---RFNSLLLNR
H.sapiens ABH3   (99) VKEADWILEQLCQDVPWKQRTGIREDIT---YQQPRLTAWYGELPYTYSRITMEPNP-HWHPVLRTLKNRIEENT-GH---TFNSLLCNL
   M.musculus1   (99) LKEADWILEQLCKDVPWKQRMGIREDVT---YPQPRLTAWYGELPYTYSRITMEPNP-HWLPVLWTLKSRIEENT-SH---TFNSLLCNF
H.sapiens ABH2   (75) KAEADEIFQELEKEVEYFTGALARVQVFGKWHSVPRKQATYGDAGLTYTFSGLTLSPKPWIPVLERIRDHVSGVT-GQ---TFNFVLINR
   M.musculus2   (53) KAEADKIFRELEQEVEYFTGALAKVQVFGKWHSVPRKQATYGDAGLTYTFSGLTLTPKPWVPVLERVRDRVCEVT-GQ---TFNFVLVNR
  R.norvegicus   (53) KAEADQIFRELEQEVEYFTGALAKVQVFGKWHSVPRKQATYGDAGLTYTFSGLTLTPKPWIPVLERVRDQVCRVT-GQ---TFNFVLVNR

H.sapiens ABH1  (222) YRLDSTLGIHVDRSELDHSKPLL-----SFSFGQSAIFLLGGLQRDEAPTA--------------------------------------M
       E.coli   (122) YAPGAKLSLHQDKDEPDLRAPIV-----SVSLGLPAIFQFGGLKRNDPLKR--------------------------------------L
  S.cerevisiae  (268) EKNGFLPSLQDSENVQGDKGEKESKDDAVSQERLERAQKLLWQAREDIPKTISKELTLLSEMYSTLSADEQDYELDEFVCCAEEYIELEY
       S.pombe  (123) YNPG-DGIIPHKDLEMFGDG--------VAIFSFLSNTTMIFTHPELKLKS-------------------------------------KI
   A.thaliana   (203) YKGASDYVAWHADDEKIYGPTPE-----IASVSFGCERDFVLKKKKDEESSQGKTGDSG--------------PAKKRLKRSSREDQQSL
H.sapiens ABH3  (181) YRNEKDSVDWHSDDEPSLGRCPI-----IASLSFGATRTTEMRKKPPPEENG--------------------------DYTYVERVKI
   M.musculus1  (181) YRDEKDSVDWHSDDEPSLGSCPV-----IASLSFGATRTTEMRKKPPPEENG--------------------------DYTYVERVKI
          ABH2  (161) YKDGCDHIGEHRDDERELAPGSP-----IASVSFGACRDFVFRHKDSRGKS--------------------------------PSRRVAVVRL
   M.musculus2  (139) YKDGCDHIGEHRDDERELAPGSP-----IASVSFGACRDFIFRHKDSRGKR--------------------------------PRRTVEVVRL
  R.norvegicus  (139) YKDGCDHIGEHRDDERELAPGSP-----IASVSFGACRDILFRHKDSRGKR--------------------------------PRRAVEVVRL

H.sapiens ABH1  (269) FMHSGDIMTMSGFSRLLNHAVPRVLPNPEGEGLPHCLEAPLPAVLPRDSMVEPCSMEDWQVCASYLKTARVNMIVRQVLATDQNF
       E.coli   (169) LLEHGDVVVWGGESRLFVHGIQPLKAGFHPLT-------I-------------------------DCRYNLTFRQAGKKE---
  S.cerevisiae  (358) LPALVDVLFANCGTNNFWKIMLVLEPFFYYIEDVGGDDDEDEDNVDN-S----------------EGDEESLLSRNVEGDDNVV
       S.pombe  (167) RLEKGSLLLMSGTARYDWFHEIPFRAGDWVMNDGEEKWVSR----------------------SQRLSVTMRRIIENHVFG
   A.thaliana   (274) TLKHGSLLVMRGYTQRDWIHSVPKRAKAEGT----------------------------------RINLTFRLVL------
H.sapiens ABH3  (238) PLDHGTLLIMEGATQADWQHRVPKEYHSREP----------------------------------RVNLTFRTVYPDPRGA
   M.musculus1  (238) PLDHGTLLIMEGATQADWQHRVPKEYHSRQP----------------------------------RVNLTFRTVYPDPRGA
H.sapiens ABH2  (217) PLAHGSLLMMNHPTNTHWYHSLPVRKKVLAP----------------------------------RVNLTFRKILLTKK--
   M.musculus2  (195) QLAHGSLLMMNNPPTNTHWYHSLPIRKRVLAP---------------------------------RVNLTFRKILPTKK--
  R.norvegicus  (195) QLAHGSLLMMNHPTNTHWYHSLPIRKRVLAP----------------------------------RINLTFRKILPTKK-
```

ABH1: AAH25787, E.coli: NP_416716, S. cerevisiae: NP_012848, S.pombe: NP_594941, A.thaliana: NP_565530, ABH3: NP_631917, M.musculus1: XP_130317, ABH2: XP_058581, M.musculus2: XP_132383, R.norvegicus: XP_222273

Figure 16–15 Sequence homology of AlkB derivatives. (See p. 358.)

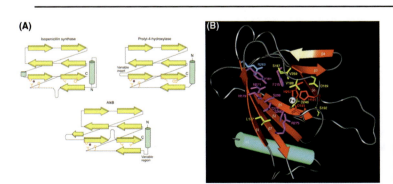

Figure 16–16 Structural model for AlkB. (See p. 359.)

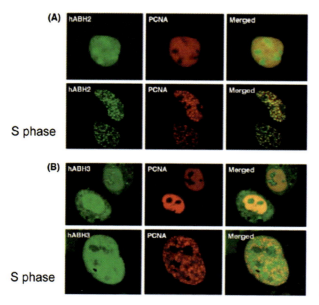

Figure 16–18 Localization in HeLa cells of fusion proteins. (See p. 361.)

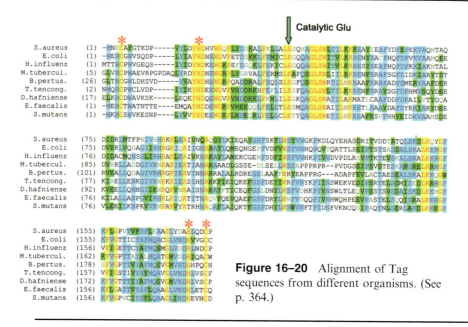

Figure 16–20 Alignment of Tag sequences from different organisms. (See p. 364.)

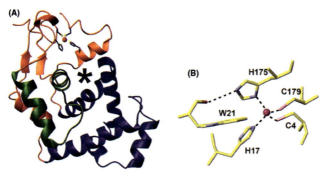

Figure 16–22 Structure of Tag and the Zn snap from an NMR solution structure. (See p. 365.)

Figure 16–23 Alignment of AlkA sequences from different organisms. (See p. 366.)

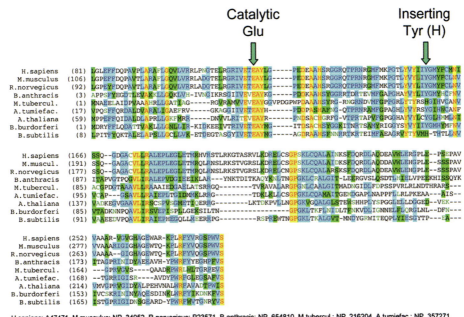

Figure 16–27 Alignment of MPG sequences from different organisms. (See p. 370.)

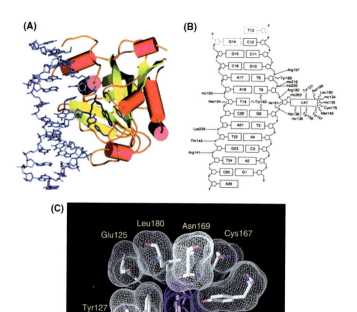

Figure 16–28 Crystal structure and binding pocket of hMPG. (See p. 371.)

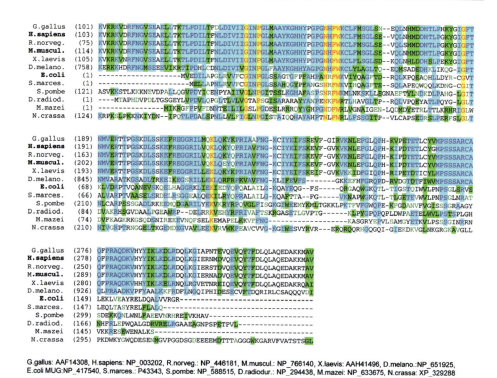

G.gallus: AAF14308, H.sapiens: NP_003202, R.norveg.: NP_446181, M.muscul.: NP_766140, X.laevis: AAH41496, D.melano.:NP_651925, E.coli MUG:NP_417540, S.marces.: P43343, S.pombe: NP_588515, D.radiodur.: NP_294438, M.mazei: NP_633675, N.crassa: XP_329288

Figure 16–32 Alignment of TDG sequences from different organisms. (See p. 376.)

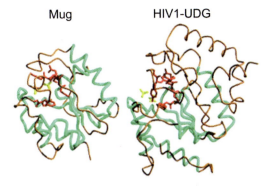

Mug   HIV1-UDG

Figure 16–33 Crystal structure of *E. coli* Mug compared to that of the HIV1 UDG. (See p. 377.)

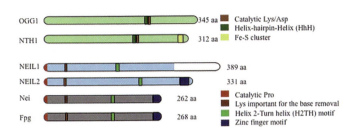

Figure 18–1 Human DNA glycosylases/AP lyases for oxidized bases. (See p. 405.)

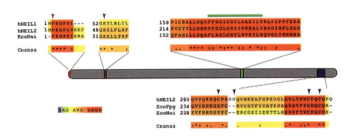

Figure 18–2 Sequence alignment of NEIL1/NEIL2 and *E. coli* Nei/Fpg. (See p. 407.)

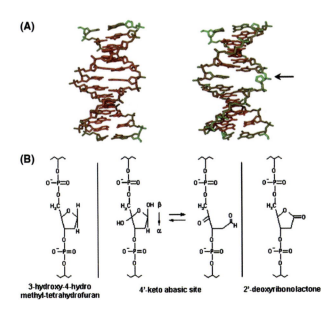

(A)

(B)

3-hydroxy-4-hydro methyl-tetrahydrofuran   4'-keto abasic site   2'-deoxyribonolactone

Figure 19–2 AP-DNA dynamics and the major AP site chemical forms studied by NMR. (See p. 424.)

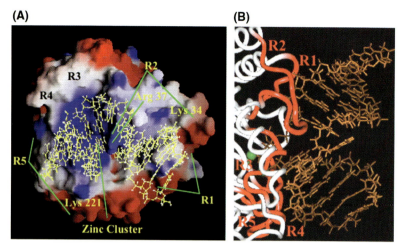

Figure 19–3 A molecular image of the endonuclease IV–DNA complex. (See p. 427.)

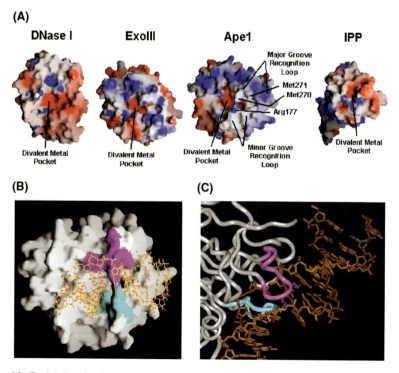

Figure 19–5 Molecular image of the Ape1 protein alone and in complex with AP-DNA. (See p. 429.)

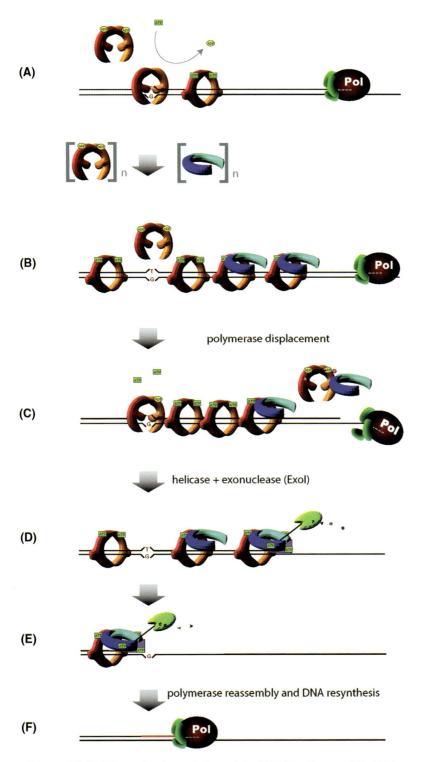

Figure 21–3 The molecular switch model of MMR. (See pp. 472–473.)

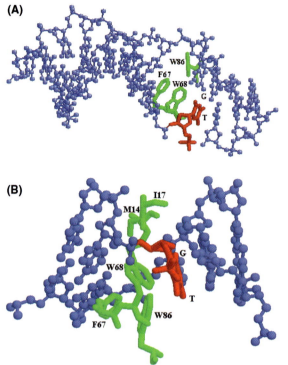

(A)

W86
F67
W68
G
T

(B)

I17
M14
W68
G
T
W86
F67

Figure 22–3 Protein–DNA recognition by Vsr. (See p. 487.)

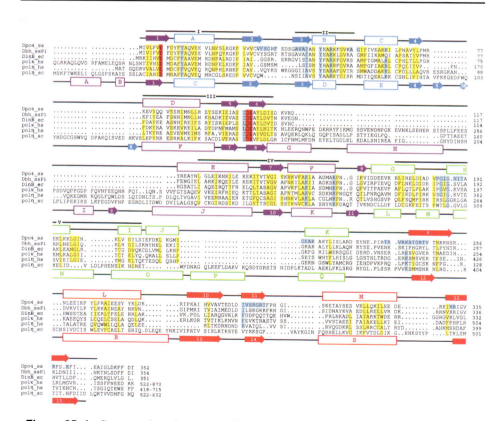

Figure 25–1 Structure-based sequence alignment of Y-family polymerases. (See p. 530.)

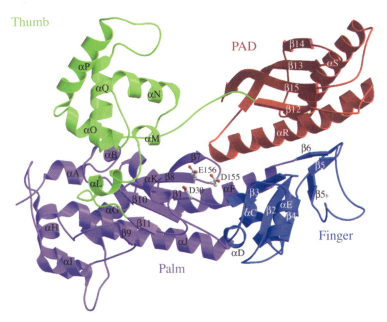

Figure 25–2 Overall structure of polη from *S. cerevisiae*. (See p. 531.)

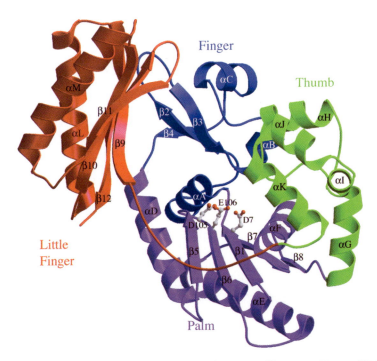

Figure 25–3 Overall structure of Dpo4 from *S. solfataricus*. (See p. 533.)

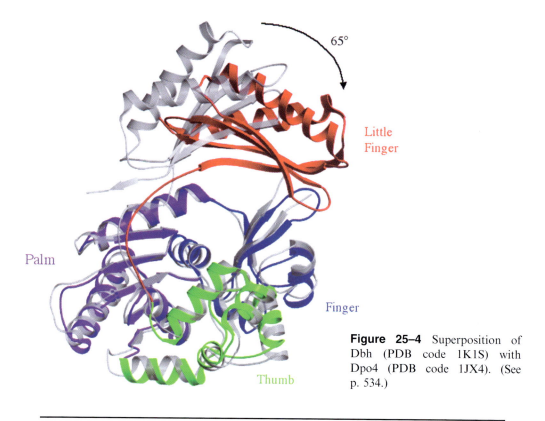

Figure 25–4 Superposition of Dbh (PDB code 1K1S) with Dpo4 (PDB code 1JX4). (See p. 534.)

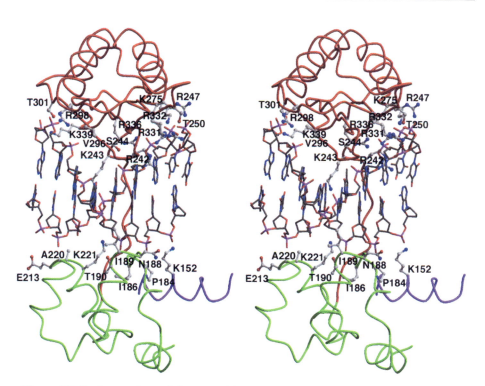

Figure 25–5 Stereo view, of the overall interaction of Dpo4 with DNA. (See p. 536.)

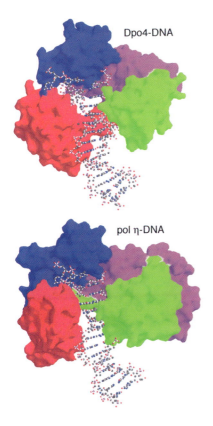

Dpo4-DNA

polη-DNA

Figure 25–6 Space filling model of Dpo4 with DNA (top) and polη (bottom). (See p. 537.)

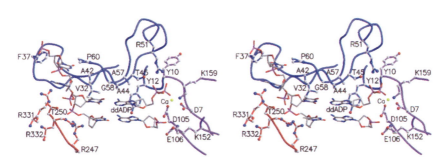

Figure 25–7 Active site of Dpo4. (See p. 538.)

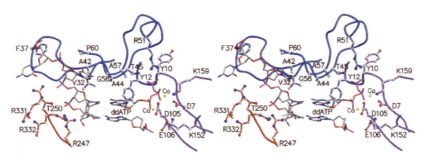

Figure 25–8 Active site of Dpo4 with a thymidine dimer. (See p. 539.)

19

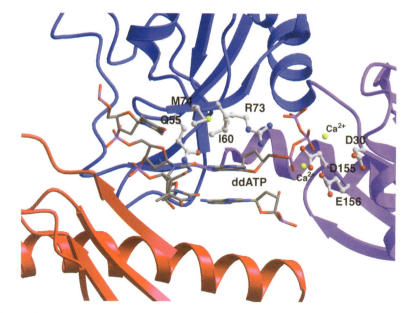

Figure 25–9 Active site of polη in the presence of a thymidine dimer. (See p. 542.)

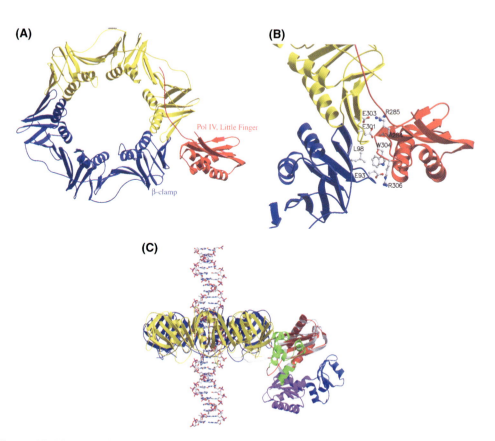

Figure 25–10 Little finger domain of *E. coli* Pol IV bound to the β-clamp processivity factor. (See p. 544.)

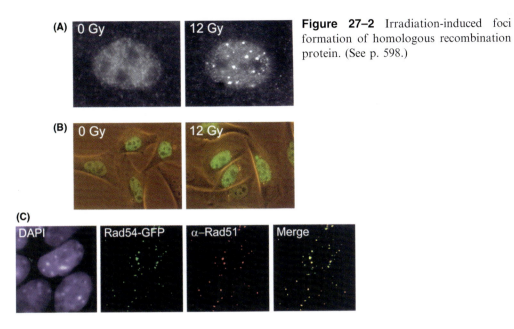

Figure 27–2 Irradiation-induced foci formation of homologous recombination protein. (See p. 598.)

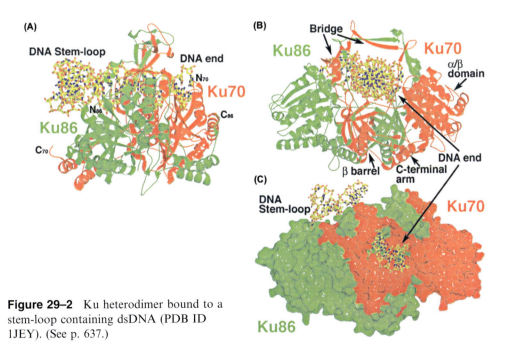

Figure 29–2 Ku heterodimer bound to a stem-loop containing dsDNA (PDB ID 1JEY). (See p. 637.)

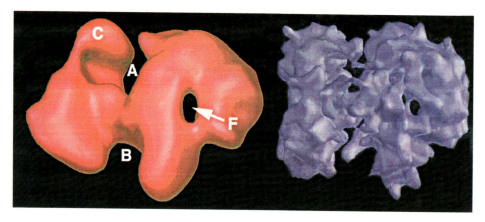

Figure 29–3 Structure of DNA-PK$_{cs}$. (See p. 651.)

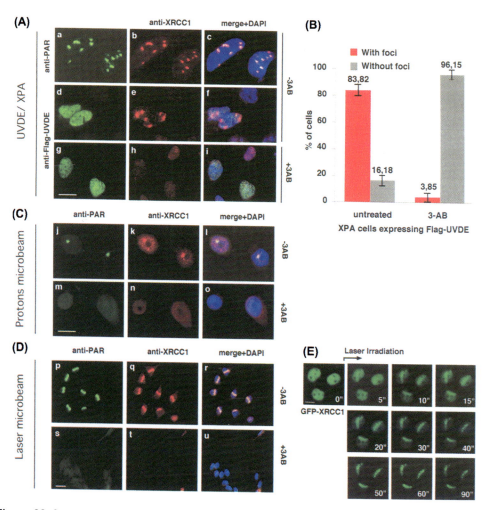

Figure 33–6 XRCCl colocalizes with PAR synthesis at nuclear foci induced by local UV-C irradiation in Flag-UVDE expressing XPA cells. (See p. 746.)

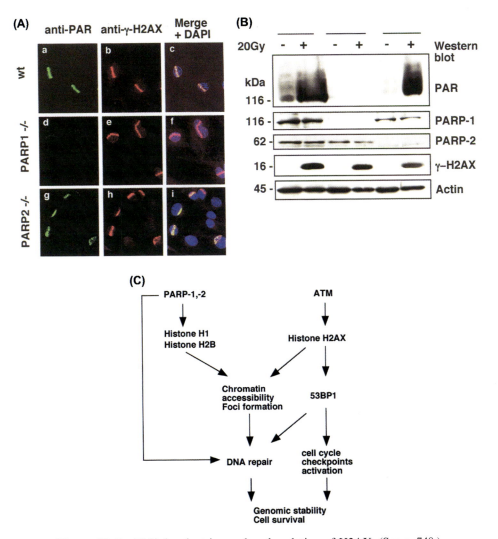

Figure 33–7 DNA breaks trigger phosphorylation of H2AX. (See p. 748.)

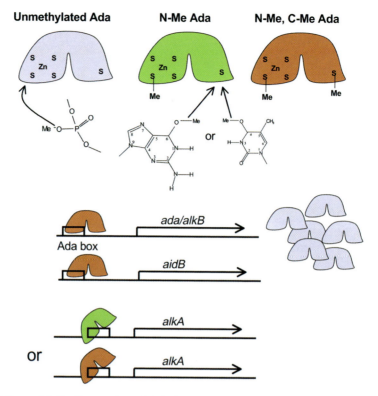

Figure 34–2 Conversion of Ada into a transcription factor. (See p. 761.)

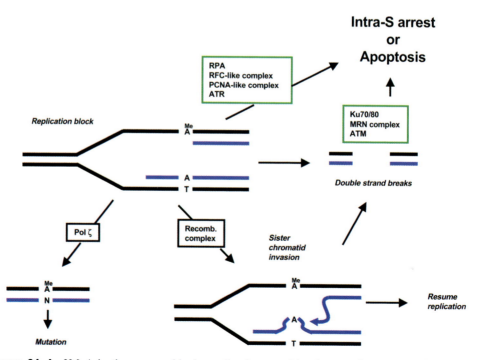

Figure 34–4 3MeA in the genome blocks replication, resulting in mutations, recombination, single-strand gaps, and DSBs. (See p. 771.)

4. ENDONUCLEASE V, AN ENZYME SPECIFIC FOR DEAMINATED PURINES

Endonuclease V (Endo V) of *E. coli* is the prototype for a ubiquitous enzyme that recognizes deaminated adenine and guanine (i.e., hypoxanthine and xanthine) in DNA. It is not to be confused with endonuclease V of bacteriophage T4, to which it is unrelated. It cleaves the second phosphodiester bond 3' to the deaminated base, leaving a 3'-hydroxyl and a 5'-phosphoryl end group. It is thus thought to initiate an excision repair pathway in which a $3' \rightarrow 5'$ exonuclease must remove the lesion. The enzyme also attacks many other DNA substrates that do not contain deaminated bases, but which have duplex regions that are adjacent to unpaired bases.

When first discovered (29,30), Endo V was described as an enzyme that attacks DNAs that are damaged by a variety of agents, untreated single-stranded DNA, and DNA that contained large amounts of uracil in place of thymine. It seemed likely that the enzyme recognizes lesions that result in unpaired regions next to duplex ones. Single-stranded DNA, for example, contains such regions due to intrastrand base pairing, and uracil substitution results in reduced base stacking. The enzyme was later rediscovered as a deoxyinosine 3' endonuclease (31) that was shown to be encoded by the same gene as Endo V, the *nfi* gene of *E. coli* (32,33). Endo V was then found to have similar activity on xanthine in DNA (6,34). Studies with synthetic oligonucleotide substrates (33,35) confirmed its specificity for regions of altered secondary structure. Endo V attacked DNAs containing: base mismatches, urea glycosides, and AP sites. When Mn^{2+} was substituted for Mg^{2+} in the reactions, it cut duplex DNA adjacent to deletion/substitution loops, hairpins, and flaps, and its activity was increased at uracil residues and base mismatches. It is difficult to see how one substrate binding site in this relatively small (24.9 kDa) protein could recognize all of the substrates. For example, hypoxanthine and xanthine are not recognized in the same way as are mismatched bases. Hypoxanthine is recognized even when it is stably base paired with cytosine or when it is in single-stranded DNA, and although xanthine is never stably base paired, it is always the xanthine-containing strand that is cleaved when it is opposite any other base.

Four apparently unrelated features in DNA are detected by the enzyme: (i) hypoxanthine in single- or double-stranded DNA, (ii) xanthine in double-stranded DNA, (iii) uracil opposite any base (at high enzyme levels), and (iv) unpaired or mismatched bases. Yao and Kow (36) proposed a model to at least explain the ability of the enzyme to recognize the different deaminated bases through common features. The enzyme must bind to the 6-keto group of deoxyinosine and deoxyxanthosine. The 4-keto group of deoxyuridine serves the same function, and it is in almost the same relative position in DNA, but there is no imidazole ring between it and the glycosylic bond. Because of this lateral displacement of its keto group with respect to that of the purines, it is a relatively poor substrate. In accordance with this theory, deoxynebularine (purine deoxynucleoside), which contains no 6-keto group, binds poorly to the enzyme. To explain why some bases that have the appropriate keto groups are not recognized by Endo V, it was postulated that enzyme binding is blocked by other ring substituents, such as the 5-methyl group of thymine and the 2-amino group of guanine. In discussing Endo V from *Thematoga maritima*, Huang et al. (37) extended this model to cover the enzyme's action on AP sites and base mismatches as well. In addition to the aforementioned unfavorable and favorable base contacts (to which they added the N7 of purine), they suggest that distortions caused by base-flipping and non-Watson–Crick pairing may also affect substrate recognition.

Alanine substitutions at nine sites were used to identify regions involved in substrate recognition and cleavage by the *Thermatoga* enzyme (38). The altered amino acids were those conserved in orthologs of Endo V and of UvrC, a protein with which Endo V shares a large region of homology. Each of the mutations affected the activity or specificity of the enzyme. Separate sites were found that diversely affected the enzyme's recognition of hypoxanthine, uracil, and AP sites in DNA, the ability of the enzyme to bind to hypoxanthine in the absence of metal ions, and the ability of the enzyme to remain bound to a deoxyinosine-containing substrate after cleavage. We must await correlation with data on crystal structure for the results to be more meaningful.

We are still left with the problem of explaining how the enzyme also recognizes mismatched bases and larger unpaired regions. We might hastily conclude that there is more than one substrate recognition site in the protein or that the other activities are contaminants, except for an additional observation: the cleavage site in the DNA is always two phosphodiester bonds 3′ to the recognized landmark, whether it be a hypoxanthine, xanthine, uracil, mismatched base, the base of a stem-loop structure, the end of a deletion/substitution loop, or the base of a flap. This fixed spatial relationship between a substrate feature and the cleavage site suggests that there is a similar relationship between the corresponding binding site and catalytic site of the enzyme. In that case, there should be only one substrate recognition site despite the disparate nature of the substrates. Thus, we are left with a paradox. In addition, there are two other features of the enzyme that must eventually be explained. (i) Some base mismatches are attacked better than others; for example, C/C mismatches are barely recognized. (ii) When there is a mismatch near one end of an oligonucleotide duplex, cleavage occurs near the mismatched base that is closer to the 5′ end of its strand (35). This second property suggests that the enzyme may gain access to the DNA duplex by threading in from the 5′ end of a strand, but the enzyme also cuts (although with low efficiency) circular single-stranded molecules (29,30), which of course have no free ends. In cells, free 5′ ends exist in Okazaki fragments, which gave rise to the hypothesis, as yet unproven, that Endo V might specifically repair the lagging strand during DNA synthesis (35). This property should also have the useful effect of preventing the enzyme from cutting the leading strand near the gaps at replication forks, which would produce potentially lethal double-strand breaks.

Despite its many substrates, the only demonstrated function for Endo V of *E. coli* in vivo is in the repair of hypoxanthine and xanthine in DNA. Thus, *nfi* (Endo V) mutants display an increased frequency of A:T → G:C and G:C → AT transition mutations when exposed to nitrous acid (6) or when grown under hypoxic conditions in the presence of nitrate or nitrite (20); and whereas single-stranded and uracil-containing DNAs are substrates for the purified enzyme, an *nfi* mutation does not affect the plating efficiency of M13, a single-stranded DNA phage, nor does it further enhance the growth of a uracil-containing λ bacteriophage on an *ung* (uracil-DNA glycosylase) mutant (39). Perhaps the possible formation of complexes between Endo V and other proteins narrows its specificity in vivo.

The binding of *E. coli* Endo V to its substrates was studied by gel mobility shift and DNase I protection assays (40). Stable binding required a duplex region 5′ to the deoxyinosine. Two molecules of Endo V were bound. The first protected 4 to 5 nucleotides 5′ to the deoxyinosine, and the second protected at least 13 nucleotides 3′ to it. The affinity of the enzyme for the nicked product was the same as that for the intact substrate, indicating that the enzyme must have the unique property of remaining bound to its substrate after cleavage. By analogy with nonenzymatic proteins that are involved in DNA repair, it was postulated that Endo V might remain on the substrate to recruit other enzymes such as an endonuclease cleaving 5' to the deoxyinosine. In vitro, there is no repeated turnover of the enzyme on hypoxanthine-containing substrates. Therefore, in vivo, it would have to be displaced, probably by enzymes involved in subsequent repair steps. There is a DNase I-hypersensitive site in the protein/DNA complex, 2 to 3 nucleotides 3' to the nick, which might provide an entry point for such an enzyme, perhaps a helicase. A homologous enzyme from *Thermatoga maritima* also remained bound to deoxyinosine-containing duplexes after nicking but turned over repeatedly on other substrates (37). A second molecule of enzyme was bound more weakly than that of *E. coli* unless the substrate contained a dI/dI mismatch, suggesting that each protein molecule covers only one strand.

Homologs of Endo V are found in species as diverse as bacteria and man and form a superfamily (41). A homolog is notably absent in *Saccharomyces cerevisiae*, although one does exist in *Schizosaccharomyces pombe*. The conservation of sequences is extraordinary: the human and *E. coli* proteins have 32% identity. There is also significant identity of Endo V homologs to a portion of UvrC of *E. coli* containing a conserved domain that catalyzes the 5' incision event by UvrC.

So far, the enzyme has been purified from the following sources of cloned genes: *E. coli* (36), the eubacterial thermophile *Thermatoga maritima* (37), the archaeal hyperthermophile *Archaeoglobus fulgidus* (42), and mouse (43). Although the enzymes from different organisms vary in their range of substrate specificities, they all cleave DNA at the second phosphodiester 3' to a hypoxanthine in DNA. The enzyme from *Thermatoga maritima* has activities similar to that of *E. coli*; it recognizes deoxyinosine, deoxyuridine, AP sites, and base mismatches. Other known substrates for the *E. coli* enzyme were not tested. Contrary to results with *E. coli* Endo V, both strands were cleaved at a mismatch, without terminus dependence. In addition, it was shown to cleave a duplex directly opposite a nick produced near deoxyinosine, resulting in a double-strand break, but a contaminating activity has not been ruled out. Mouse Endo V, a 37-kDa protein, has a smaller substrate range (43). It recognizes hypoxanthine (better in double- than in single-stranded DNA) and uracil (but only in single-stranded DNA). Although it did not cleave at C/C mismatches (a poor substrate for the *E. coli* enzyme as well), AP sites or 5' flap structures at the levels used, the enzyme preparation tested was of extraordinarily low specific activity toward its preferred substrate. Other mismatches and DNA xanthine were not tested. The enzyme from *Archaeoglobus fulgidus* possesses only the deoxyinosine endonuclease activity; it lacks the ability to cleave at deoxyxanthosine, an A/A mismatch, or a 5' flap. These findings suggested that a primordial enzyme may have had only that one activity. Indeed, studies of reaction rates and binding constants indicate that deoxyinosine-containing DNA is the preferred substrate for all of the purified Endo V homologs.

It is difficult to assess the relative rates of activity of Endo V-like enzymes on their various substrates in vitro. In most experiments, rate measurements were not performed, and the enzyme concentrations were equal to or greater than those of the substrate. Large amounts of enzyme were used because the enzyme does not detach itself from its best substrate, a deoxyinosine-containing duplex. Therefore, one molecule of enzyme can cleave no more than one bond. This is true at least for the two homologs for which binding has been studied, namely, those from *E. coli* and *Thermatoga*. On all of its other substrates, however, Endo V can turn over many times during the usual reaction periods (see Ref. 37, for example, for time course comparisons). Therefore, in vitro measurements of deoxyinosine endonuclease activity are grossly underestimated when compared to other activities of the enzyme. We might obtain a more accurate measurement of relative rates by examining single-turnover kinetics for the various substrates, as has been done for the thymine DNA glycosylases (44,45). However, in vivo, the rate-limiting step in repair may not be hydrolysis but rather the displacement of the enzyme from the cleavage site in hypoxanthine-containing DNA.

It is possible, therefore, that the minor side activities of the enzyme, which we observe in vitro, are relatively even smaller in vivo. Perhaps that is why, for example, Endo V does not appear to be used for the repair of uracil-containing DNA in *E. coli* and that despite its ability to act on diverse substrates, its only documented biological role has been in the repair of deaminated purines in DNA (39). The chief value of the other activities may lie not in their biological roles but in what they may tell us someday about how the enzyme recognizes its substrates.

5. HYPOXANTHINE/ALKYLPURINE DNA GLYCOSYLASES

5.1. Two Classes of Enzyme

Hypoxanthine DNA glycosylases catalyze the hydrolysis of the glycosylic bond between the altered base and deoxyribose in the DNA, thereby releasing free hypoxanthine and leaving behind an AP site. The preferred substrate for these enzymes is not hypoxanthine but rather alkylated purine in DNA. This section will dwell on the properties of these enzymes as they relate to the repair of deaminated purines. For details of their general properties, especially with respect to their action on alkylated substrates, see the chapters by O'Connor and Tainer in this volume.

The known hypoxanthine DNA glycosylases fall into two classes; one belongs to microorganisms and the other to higher eukaryotes. The first class is exemplified by the AlkA protein (3-methyladenine glycosylase II) of *E. coli*, which has a relatively weak hypoxanthine DNA glycosylase activity. It is not to be confused with 3-methyladenine DNA glycosylase I (the Tag protein) of *E. coli*, which has no apparent activity on hypoxanthine in DNA. Homologs have been found in other bacteria, in fission yeast, in budding yeast, and in the thermophilic archaeon *Archaeoglobus fulgidus* (46–48). In yeast, it is referred to as the MAG (methyladenine glycosylase) protein. A common 9-amino acid consensus has been identified (47). The second class of hypoxanthine DNA glycosylase is exemplified by the mammalian alkyladenine glycosylase, which has also been referred to as the AAG protein and as methylpurine glycosylase (MPG). The plant *Arabidopsis thaliana* also has a paralog belonging to this class (47). Although the bacterial and mammalian enzymes have similar substrate ranges and require base flipping to get the damaged base into the active site, they are otherwise unrelated. They have no significant homology, they use different amino acids to replace the flipped-out base, they have different DNA-binding motifs, they have different consensus motifs, and they probably employ different catalytic mechanisms. Whereas the bacterial AlkA enzyme has a relatively low specificity for hypoxanthine in DNA, the mammalian AAG enzymes recognize hypoxanthine almost as well as they recognize some alkylpurines. With respect to its hypoxanthine DNA glycosylase activity, the specific activity of the purified human enzyme is about 120 times that of *E. coli* AlkA (46). Another important difference, and one that may be of physiological significance, is that AlkA is relatively indifferent to the base opposite hypoxanthine (46), whereas AAG is highly specific for the hypoxanthine in dI/dT (49,50).

Both enzymes hydrolyze N^3-methyladenine, N^7-methyladenine, and N^7-methyl-guanine from duplex DNA. The mammalian enzyme also rapidly attacks $1,N^6$-ethenoadenine and hypoxanthine. Unfortunately, there are no comprehensive kinetic data comparing all of the substrates, for which they would have to be within the same randomly selected sequence and tested opposite each of the four normal bases. Therefore, reported differences in substrate specificity are largely qualitative.

5.2. AlkA Protein (3-Methyladenine Glycosylase II) of *E. coli*

5.2.1. History

Hypoxanthine-DNA glycosylase activity was described by Karran and Lindahl (51) in *E. coli* in 1978. In 1994, Saparbaev and Laval (46) discovered a hypoxanthine-DNA glycosylase associated with AlkA (3-methyladenine glycosylase II) of *E. coli*. That the two activities belonged to the same protein was established by co-purification, co-mutability, co-induction by an alkylating agent, and overproduction by a plasmid containing the *alkA* gene. Several findings suggested that this hypoxanthine-DNA glycosylase is the same as that described earlier. It was purified by a similar scheme, it has the same molecular weight (about 30 kDa), and an *alkA* mutant lacked measurable hypoxanthine glycosylase activity. The activity was not affected by a mutation in *tag*, the gene for 3-methyladenine glycosylase I. An apparently different hypoxanthine-DNA glycosylase activity was reported by Harosh and Sperling (52), but it is now believed to be due to the sequential action of a DNase and enzymes that break down the dIMP released (49). Therefore, at present, the only known hypoxanthine-DNA glycosylase in *E. coli* is the AlkA protein.

5.2.2. Properties

The enzyme cleaves at an almost equal rate deoxyinosine that is paired with any other deoxynucleoside (46). It has no measurable activity on single-stranded substrates, and it is inactive on dX/dT (46,51). A lack of activity on N^6-methyladenine and O^6-methylguanine signified that the enzyme did not merely recognize any purines that like hypoxanthine were altered in the 6th position (51). The enzyme also attacks *N*-alkylpurines and *O*-methyl pyrimidines in addition to hypoxanthine. In fact, the hypoxanthine glycosylase activity of the AlkA protein appears to be a weak side activity of the enzyme. Under standard reaction conditions, its activity on 3-methyladenine residues in DNA is about 800 times greater than on hypoxanthine and there is a 50,000-fold difference in k_{cat}/K_m for the two substrates (46). Its activity on 7-methylguanine residues is over 100-fold that on hypoxanthine (53). Therefore, the hypoxanthine glycosylase activity, which is so weak that it could not be measured accurately in crude extracts (51), may have little relative significance in DNA repair, especially when compared to the activity of Endo V. In fact, an *alkA* mutation, either alone or in combination with an *nfi* mutation, had little if any effect on nitrous acid-induced A:T $\rightarrow$ G:C mutations (6,54). However, the nitrous acid treatments had to be performed on stationary phase cells because they are not killed by the pH used. A more reasonable test of *alkA* function would be in cells that are growing anaerobically with nitrate and that are then exposed to air. It is during this shift from nitrate to oxygen metabolism that (i) N_2O_3 is formed by the auto-oxidation of accumulated NO•, (ii) DNA adenine is nitrosatively deaminated, and (iii) endogenous alkylating agents may be formed by the nitrosation of inter-mediary metabolites. Alkylating agents can induce AlkA at least 7- to 10-fold (55,56) and perhaps as much as 75-fold (57). Therefore, under these growth condi-tions, which the cell is likely to encounter in nature, it is conceivable that AlkA might become as important as Endo V in the repair of hypoxanthine in DNA. This hypoth-esis remains to be tested.

A study by Terato et al. (58) suggested that the enzyme might play an active role in the excision of xanthine and possibly oxanosine from DNA. The activity (V_{max}/K_m) of purified AlkA on xanthine in DNA was one-fifth that on N^7-methyl-guanine. An increase in xanthine glycosylase activity was detected in a crude extract of wild-type cells that were treated with N-methyl-N'-nitro-N-nitrosoguanidine, but not in that of an *alkA* mutant. Wuenschell et al. (5) found that xanthine could be excised from X:C pairs by AlkA and mouse AAG (as well as by endonuclease III and formaminopyrimidine glycosylase of *E. coli*): however, incomplete digestion by each was obtained after 2 h of incubation with a 50-fold molar ratio of enzyme to substrate.

End-product inhibition by hypoxanthine was discovered and shown to be due to the binding of hypoxanthine within the active site (59). This inhibition was pos-tulated to serve a purpose in vivo, that of modulating the activity of the enzyme to keep pace with the repair of the resulting AP sites so as to prevent their lethal accumulation.

5.3. Human Alkyladenine (Alkylpurine) Glycosylase (hAAG)

Mammalian AAG proteins have been purified from human, mouse, rat, and calf thy-mus (43,46,49). In contrast to the *E. coli* enzyme, the human enzyme is almost as active on hypoxanthine in DNA as it is on some of its alkylated substrates (60). It is therefore more likely to be important in the repair of deaminated adenine than AlkA is. The enzyme activity is cell-cycle dependent (61). It preferentially hydrolyzes deoxyinosine in a dI/dT mismatch (49,50), which is the product of adenine deamina-tion, and not in normally paired dI:dC, which would result from the subsequent replication of dI/dT. Therefore, AAG could only participate in the repair of deami-nated adenine residues before replication. After replication, Endo V could still oper-ate on the lesion, but then it would be too late to prevent a heritable mutation from being established.

5.3.1. Structural Basis for Substrate Specificity

AlkA and AAG have both been co-crystallized with nonhydrolyzable substrate ana-logs. Although their catalytic mechanisms are now mostly understood, the basis for their substrate specificities is still largely conjectural (62, and this volume, chapter by Tainer). Briefly, the enzyme binds to a small region of the DNA. The altered base flips into the active site pocket of the enzyme and is stabilized in its extrahelical posi-tion through replacement by an amino acid that fills the resulting space in the minor groove. This base flipping would be impeded by stable base pairing, which explains why dI/dT in DNA is a better substrate for AAG than is dI:dC. Hydrolysis depends on the proximity of the glycosylic bond to a catalytic site containing a bound water molecule. To explain substrate specificity, it was postulated (62) that the fit and alignment of the base within the active site are affected by (i) positive charges produced by alkylation, (ii) H-bonding to the O^6 position of the purine, and (iii) steric hindrance. Thus, hypoxanthine works because it has an O^6 group. Although guanine also has an O^6 group, it is sterically hindered by an N^2 amino group that is not present in hypoxanthine. The explanation for N^7-methylguanine, however, is not straightforward. Like guanine, N^7-methylguanine also has an N^2-amino group that should prevent a close fit at the active site, but it is nevertheless a substrate, presumably because its alkylation compensates for that disadvantage. Thus, there are no absolute rules.

6. ENDONUCLEASE VIII OF *E. COLI*

Endo VIII is mainly a glycosylase/endolyase for oxidized pyrimidines in DNA. Cleavage was detected at xanthine residues in DNA when it was incubated in great molar excess over the DNA substrate (58). Its activity (V_{max}/K_m) on xanthine in DNA was about 50 times lower than that on thymine glycol, its preferred substrate. Despite this apparently poor activity in vitro, the biological significance of this activity was documented with the finding that the *alkA nei* (Endo VIII) double mutant, was about twice as sensitive to killing by nitrous acid than the wild type or the single mutants.

7. CONCLUSIONS

Endo V, AlkA, and Endo VIII each has a broad specificity, and each has been found to attack additional substrates if it is used at substrate-level concentrations. Are such side reactions of any importance? Repair enzymes must distinguish abnormal features of DNA from normal ones. When the enzyme is designed to recognize one feature (uracil, for example) the task of discrimination is relatively easy. However, the enzymes described in this chapter each recognize sets of disparate substrates the structural similarities of which are not immediately obvious. Because of their broad specificity, they are bound occasionally to cleave unintended substrates. This situation has evolved for the sake of genetic economy; one enzyme can handle several types of lesions, and this ability more than compensates for an occasional mistake. However, in the laboratory, it means that if we add enough of these enzymes to almost any DNA substrate, we might see a reaction in vitro that is of little consequence in vivo or that might be due to a contaminating enzyme of higher specificity. This point was driven home by Berdal et al. (63), who discovered that AlkA and similar eukaryotic glycosylases can even remove normal bases from DNA. Therefore, we must turn to in vivo studies. A delineation of these repair pathways must ultimately come through the study of mutants.

REFERENCES

1. Kow YW. Free Radical Biol Med 2002; 33:886–893.
2. Michelson AM, Monny C. Biochim Biophys Acta 1966; 129:460–474.
3. Eritja R, Horowitz DM, Walker PA, Ziehler-Martin JP, Boosalis MS, Goodman MF, Itakura K, Kaplan BE. Nucl Acids Res 1986; 14:8135–8153.

4. Kamiya H, Sakaguchi T, Murata N, Fujimuro M, Miura H, Ishikawa H, Shimizu M, Inoue H, Nishimura S, Matsukage A, et al. Chem Pharmaceut Bullet 1992; 40: 2792–2795.
5. Wuenschell GE, O'Connor TR, Termini J. Biochemistry 2003; 42:3608–3616.
6. Schouten KA, Weiss B. Mutat Res 1999; 435:245–254.
7. Duncan BK, Miller JH. Nature 1980; 287:560–561.
8. Tomso DJ, Bell DA. J Mol Biol 2003; 327:303–308.
9. Singer B, Grunberger D. Molecular Biology of Mutagens and Carcinogens. New York: Plenum Press, 1983.
10. Lindahl T, Nyberg B. Biochemistry 1974; 13:3405–3410.
11. Karran P, Lindahl T. Biochemistry 1980; 19:6005–6011.
12. Lindahl T. Prog Nucl Acid Res Mol Biol 1979; 22:135–192.
13. Williams DLH. Nitrosation. Cambridge: Cambridge Univ Press, 1988.
14. Gennis RB, Stewart V. In: Neidhardt FC, Curtiss R III, Ingraham JL, Lin ECC, Low KB, Magasanik B, Reznikoff WS, Riley M, Schaechter M, Umbarger HE, eds. *Escherichia Coli* and Salmonella: Cellular and Molecular Biology. 2nd ed. Washinghton, DC: ASM Press, 1996:217–261.
15. Ji XB, Hollocher TC. Appl Environ Microbiol 1988; 54:1791–1794.
16. Brons HJ, Hagen WR, Zehnder AJ. Arch Microbiol 1991; 155:341–347.
17. Metheringham R, Cole JA. Microbiology 1997; 143:2647–2656.
18. Ralt D, Wishnok JS, Fitts R, Tannenbaum SR. J Bacteriol 1988; 170:359–364.
19. Calmels S, Ohshima H, Henry Y, Bartsch H. Carcinogenesis 1996; 17:533–536.
20. Weiss B. Mut Res 2001; 461:301–309.
21. Shapiro R, Pohl SH. Biochemistry 1968; 7:448–455.
22. Caulfield JL, Wishnok JS, Tannenbaum SR. J Biol Chem 1998; 273:12689–12695.
23. Grishko VI, Druzhyna N, LeDoux SP, Wilson GL. Nucl Acids Res 1999; 27:4510–4516.
24. Suzuki T, Ide H, Yamada M, Endo N, Kanaori K, Tajima K, Morii T, Makino K. Nucl Acids Res 2000; 28:544–551.
25. Harwood EA, Hopkins PB, Sigurdsson ST. J Org Chem 2000; 65:2959–2964.
26. Nakano T, Terato H, Asagoshi K, Masaoka A, Mukuta M, Ohyama Y, Suzuki T, Makino K, Ide H. J Biol Chem 2003; 278:25264–25272.
27. Spencer JP, Wong J, Jenner A, Aruoma OI, Cross CE, Halliwell B. Chem Res Toxicol 1996; 9:1152–1158.
28. Toyokuni S, Mori T, Dizdaroglu M. Int J Cancer 1994; 57:123–128.
29. Gates FT, Linn S. J Biol Chem 1977; 252:2802–2807.
30. Demple B, Linn S. J Biol Chem 1982; 257:2848–2855.
31. Yao M, Hatahet Z, Melamede RJ, Kow YW. J Biol Chem 1994; 269:16260–16268.
32. Guo G, Ding Y, Weiss B. J Bacteriol 1997; 179:310–316.
33. Yao M, Kow YW. J Biol Chem 1996; 271:30672–30676.
34. He B, Qing H, Kow YW. Mut Res 2000; 459:109–114.
35. Yao M, Kow YW. J Biol Chem 1994; 269:31390–31396.
36. Yao M, Kow YW. J Biol Chem 1997; 272:30774–30779.
37. Huang J, Lu J, Barany F, Cao W. Biochemistry 2001; 40:8738–8748.
38. Huang J, Lu J, Barany F, Cao W. Biochemistry 2002; 41:8342–8350.
39. Guo G, Weiss B. J Bacteriol 1998; 180:46–51.
40. Yao M, Kow YW. J Biol Chem 1995; 270:28609–28616.
41. Aravind L, Walker DR, Koonin EV. Nucl Acids Res 1999; 27:1223–1242.
42. Liu J, He B, Qing H, Kow YW. Mut Res 2000; 461:169–177.
43. Moe A, Ringvoll J, Nordstrand LM, Eide L, Bjoras M, Seeberg E, Rognes T, Klungland A. Nucl Acids Res 2003; 31:3893–3900.
44. O'Neill RJ, Vorob'eva OV, Shahbakhti H, Zmuda E, Bhagwat AS, Baldwin GS. J Biol Chem 2003; 278:20526–20532.
45. Abu M, Waters TR. J Biol Chem 2003; 278:8739–8744.
46. Saparbaev M, Laval J. Proc Natl Acad Sci USA 1994; 91:5873–5877.

47. Memisoglu A, Samson L. Gene 1996; 177:229–235.
48. Birkeland NK, Anensen H, Knaevelsrud I, Kristoffersen W, Bjoras M, Robb FT, Klungland A, Bjelland S. Biochemistry 2002; 41:12697–12705.
49. Dianov G, Lindahl T. Nucl Acids Res 1991; 19:3829–3833.
50. Abner CW, Lau AY, Ellenberger T, Bloom LB. J Biol Chem 2001; 276:13379–13387.
51. Karran P, Lindahl T. J Biol Chem 1978; 253:5877–5879.
52. Harosh I, Sperling J. J Biol Chem 1988; 263:3328–3334.
53. Gasparutto D, Dherin C, Boiteux S, Cadet J. DNA Repair 2002; 1:437–447.
54. Sidorkina O, Saparbaev M, Laval J. Mutagenesis 1997; 12:23–28.
55. Karran P, Hjelmgren T, Lindahl T. Nature 1982; 296:770–773.
56. Evensen G, Seeberg E. Nature 1982; 296:773–775.
57. Nakabeppu Y, Miyata T, Kondo H, Iwanaga S, Sekiguchi M. J Biol Chem 1984; 259:13730–13736.
58. Terato H, Masaoka A, Asagoshi K, Honsho A, Ohyama Y, Suzuki T, Yamada M, Makino K, Yamamoto K, Ide H. Nucl Acids Res 2002; 30:4975–4984.
59. Teale M, Symersky J, DeLucas L. Bioconjugate Chem 2002; 13:403–407.
60. Asaeda A, Ide H, Asagoshi K, Matsuyama S, Tano K, Murakami A, Takamori Y, Kubo K. Biochemistry 2000; 39:1959–1965.
61. Bouziane M, Miao F, Bates SE, Somsouk L, Sang BC, Denissenko M O'Connor TR. Mut Res 2000; 461:15–29.
62. Hollis T, Lau A, Ellenberger T. Prog Nucl Acids Res Mol Biol 2001; 68:305–314.
63. Berdal KG, Johansen RF, Seeberg E. EMBO J 1998; 17:363–367.

18

New Paradigms for DNA Base Excision Repair in Mammals

Sankar Mitra, Lee R. Wiederhold, Hong Dou, Tadahide Izumi, and Tapas K. Hazra
Sealy Center for Molecular Science and Department of Human Biological Chemistry and Genetics, University of Texas Medical Branch, Galveston, Texas, U.S.A.

1. INTRODUCTION

In addition to strand breaks, a wide variety of oxidized bases are continuously generated in the genomes of all aerobic organisms as a result of reaction with reactive oxygen species (ROS) invariably generated as by-products of respiration (1,2). In mammals, ROS are also produced during inflammation and infection. For example, NADPH oxidase in activated neutrophils recruited to the site of infection produces $O_2^{\Gamma\bullet}$, which undergoes superoxide dismutases catalyzed dismutation to H_2O_2 (3). In the presence of Fe^{2+} or Cu^{2+}, H_2O_2 is reduced via the Fenton reaction to the $OH\bullet$ radical, the most reactive ROS of all. Hydrogen perodixe and $O_2^{\Gamma\bullet}$ could react between themselves to generate the $OH\bullet$ radical via the Haber reaction (4).

The oxidatively damaged bases induced by ROS are often mutagenic and/or toxic, and these base lesions are repaired primarily via the base excision repair (BER) pathway. The BER is initiated with excision of the damaged base by ubiquitous DNA glycosylases. Most of these glycosylases, which appear to number fewer than a dozen in both *E. coli* and mammalian cells, have broad and sometimes overlapping substrate range (5). Monofunctional DNA glycosylases activate a water molecule to function as a nucleophile in attacking the *N*-glycosyl bond. The damaged base is thus hydrolyzed, resulting in the formation of an abasic (AP) site (6). On the other hand, the oxidized base-specific DNA glycosylases utilize the α-imino group of an N-terminal Pro or an ε-amino group of an internal Lys as the active site nucleophile and act as AP lyases. Thus after base excision, these enzymes cleave the phosphodiester bond at the resulting abasic (AP) site. These DNA glycosylase/AP lyases are divided into the Nth (endonuclease III) and Fpg/Nei families (Fapy DNA glycosylase/endonuclease VIII). The active site nucleophile residues in both families form Schiff base intermediates with the C'_1 aldehyde of the free deoxyribose after excision of the damaged base. The Schiff base may be hydrolyzed to release the enzyme for its turnover, in which case an AP site is generated, as with monofunctional DNA glycosylases. Alternatively, the Schiff base permits a DNA lyase reaction with β or successive β and δ elimination (6,7). The β elimination reaction generates a DNA strand break at the AP site containing 5′ phosphate and 3′ α,β unsaturated deoxyribose. Subsequent δ elimination causes the second strand break at the 3′ site, and produces 3′ phosphate and 5′ phosphate termini, along with a gap due to the missing deoxynucleoside (8–10). Nth, which carries out β elimination, has no sequence homology with Fpg or Nei. Also, Fpg has extensive homology with Nei, and both carry out βδ elimination (5,11). However, Fpg prefers as substrates 8-oxoguanine and ring-opened purines, namely formamidopyrimidines (Fapy), while Nei was discovered as a glycosylase specific for oxidized pyrimidines (12). In both cases, the completion of repair requires the removal of the 3′ blocking group so that the 3′ OH could act as the primer terminus for repair synthesis by a DNA polymerase which uses the uninterrupted and undamaged complementary DNA strand as the template. In the absence of exonucleolytic removal of additional nucleotides, the 1-nucleotide gap will then be filled in order to produce the substrate for a DNA ligase. The ligase completes the BER process by sealing the nick (13).

2.   OXIDIZED BASE-SPECIFIC GLYCOSYLASES IN *E. COLI* AND MAMMALS

The conservation of DNA repair processes in general, and DNA BER in particular, ranges from prokaryotes to eukaryotes. Although the primary sequence and structures of the cognate enzymes in BER have significantly diverged between *E. coli* and mammals, the functional similarities of the early enzymes in BER, (DNA glycosylases and AP-endonuclease, APE), are quite remarkable. It was therefore surprising that while the *E. coli* genome encodes three DNA glycosylases specific for oxidized bases (Nth, Fpg, and Nei), only two enzymes NTH1 (the ortholog of Nth) and 8-oxoguanine-DNA glycosylase (OGG1) were identified and characterized in various mammalian cells (14–16). NTH1 and OGG1 have distinct preference for specific oxidized base lesions. NTH1 prefers oxidized pyrimidines, whereas OGG1 prefers oxidized purines. More interestingly, both of the enzymes belong to the Nth family, based on the conservation of structural motifs and reaction mechanism. Both utilize an internal Lys residue as the active site nucleophile and carry out β elimination (14,17). An Asp residue, 18–20 amino acid residues downstream from the active site Lys present in both, is essential for activity and believed to function as a proton donor to the N–C glycosyl bond. This protonation facilitates release of the base and formation of the Schiff base with the Lys residue (17). All Nth family members have a helix–hairpin–helix motif, although conservation at the primary sequence level is not extensive (Fig. 1) (18).

In contrast to the presence of Nth orthologs in eukaryotes including yeast (19), the Nei/Fpg type enzymes were observed only in prokaryotes. *E. coli* Fpg and Nei have significant sequence homology between themselves (Fig. 1), including sequence identity in key motifs. *E. coli* Nei and Fpg have distinct substrate preferences, although both of them are active on some common substrates, e.g., 5-hydroxyuracil (5-OHU) and dihydrouracil (20). More interestingly, Nei and Nth share many common substrates, including oxidation products of pyrimidines (21). The substrate preferences of these DNA glycosylases pose an interesting challenge because not only do these enzymes utilize multiple base lesions of distinct chemical structures, but homologs such as Fpg and Nei recognize disparate bases, while heterologous enzymes (Nei and Nth) excise several common bases. The molecular bases of substrate recognition by these DNA glycosylases require elucidation of X-ray crystallographic structures

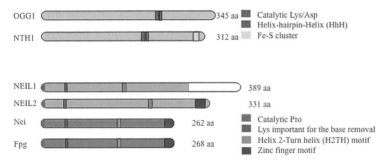

Figure 1 Human DNA glycosylases/AP lyases for oxidized bases. OGG1 and NTH1 carry out β elimination, whereas NEILs cleave AP sites in DNA strands via βδ elimination. Conserved domains and essential residues are indicated as colored boxes. *E. coli* Nei and Fpg are shown for comparison. The C-terminal domain of NEIL1 (shown in white) is dispensable for activity. (*See color insert.*)

of these enzymes bound to diverse substrates and their products. Some success has recently been achieved in solving the structures of bacterial Fpg and Nei bound to an AP site analog (22,23).

2.1. Discovery of Nei Orthologs in Mammalian Cells

In view of the presence of mammalian orthologs of many bacterial DNA glycosylases, it was surprising that no ortholog of Nei/Fpg was identified in the eukaryotes until recently. The database compiled as a result of the Human Genome Project provided an exciting clue about the existence of Nei/Fpg-like genes in the human genome. We and others had been searching for Nei/Fpg-like proteins based on the conserved N-terminal motif PEGP present in all prokaryotic members of the Nei/Fpg family, along with some key internal sequences and conserved residues. We originally identified a cDNA whose open reading frame encodes a 62 kD polypeptide. Although the molecular mass of the candidate DNA glycosylase is much higher than that of a common DNA glycosylase (the 25–40 kD size range), we cloned the cDNA and expressed the recombinant protein in *E. coli*. However, we could not detect any DNA glycosylase nor AP lyase activity (T. Izumi and T. K. Hazra, unpublished). Thus the cellular function of this protein, which we subsequently named NEIL (Nei-like) 3, remains obscure. More recently, we identified two other candidate orthologs of Nei/Fpg in both human and mouse genome databases (24–26). We expressed the wild-type human proteins in *E. coli* and tested for their DNA glycosylase activity. It should be noted in this context that the Fpg/Nei family of DNA glycosylases is distinct from other DNA glycosylases by having N-terminal Pro as the active site. This precludes the approach of expressing N-terminal fusion polypeptides for ease of purification of active recombinant proteins. We then decided to express the wild-type proteins in "Codon Plus©" *E. coli*, in which mammalian proteins with rare codons are translated more efficiently (24,25).

Once we established that these two recombinant proteins have AP lyase activity, we named them NEIL1 and NEIL2. We substituted NEIL for the initial name of NEH (Nei homolog) on the recommendation of the human genome organization (HUGO). Although NEIL1 and NEIL2 have some sequence similarity to both *E. coli* Fpg and Nei, we believed that Nei-like is a better name for these enzymes

(24–26). Subsequent to our publications on the characterization of these enzymes, other groups have also accepted these names (27–29).

2.2. Comparative Enzymatic Properties of NEIL1 and NEIL2

Our initial studies showed that both NEIL1 and NEIL2 function as DNA glycosylase/AP lyases and carry out βδ elimination on AP sites. NEIL1 prefers Fapys as substrates, while NEIL2 excises almost exclusively oxidation products of C, namely, 5-OHU and 5-hydrocytosine (25). Other investigators have subsequently shown that thymine glycol (TG), generated from thymine by the OH• radical, is also a substrate of NEIL1 (27). It is interesting that TG is a substrate of both Nth and Nei (28,29). However, there is a significant difference in the activity of these enzymes for the two diastereoisomers of TG (28,30). It is also surprising that 8-oxoG, a major base lesion induced by ROS and often used as a marker of oxidative stress, is not a good substrate for either NEIL enzyme. On the other hand, many pyrimidine lesions, including dihydrouracil and 5-OHU, are good substrates for both enzymes. As expected, NEILs share many substrates with NTH1. NEIL1 is the only mammalian enzyme identified so far which excises Fapy A. In this respect it is similar to Fpg and not OGG1, even though unlike Fpg or OGG1 it has rather weak activity in excising 8-oxoG (24).

Although our results on NEILs were subsequently confirmed or extended by others, DNA glycosylase activity of NEIL2 was not reported to be significant by other investigators (30). One possible explanation for this discrepancy is that NEIL2 is quite unstable, as we had initially reported. In fact, we use a high concentration of an osmolyte, trimethylamine N-oxide (TMAO) or glycerol, for storing the active enzyme (25). We also routinely use glycerol in the assay mixtures.

2.3. Structural Differences of NEIL1 and NEIL2

Sequences comparison clearly shows very little similarity between NEIL1 and NEIL2, other than the presence of the conserved N-terminal PEGP motif and Lys (K51 for NEIL1 and K49 for NEIL2; 27). In fact, identification of NEIL1 was serendipitous, because this protein does not have the signature Zn finger motif present in Fpg/Nei. In this respect, NEIL2 is closer to Nei/Fpg than to NEIL1. The presence of a Zn finger motif in NEIL2 was suspected from the presence of three Cys and one His residues in the conserved C-terminal region of this protein (Fig. 2). Although this potential CHCC type Zn finger motif is distinct from the C4 type Zn finger present in Nei/Fpg, we have now confirmed that NEIL2 does have a bound Zn atom coordinated by the candidate residues. The Zn finger is essential for maintaining structural integrity of the polypeptide, because mutations in the coordinating Cys or His not only abolish the DNA-binding activity of the protein, but also grossly alter its secondary structure (as indicated from CD spectra; Das, A. and Hazra, T. K., unpublished).

2.4. Distinct Requirements for NEIL-Initiated BER

The discovery of NEIL1 and NEIL2 has raised the issue of the subsequent steps in repair initiated by these enzymes. These are the only enzymes identified so far in mammalian cells which generate 3' phosphate termini after AP lyase reaction.

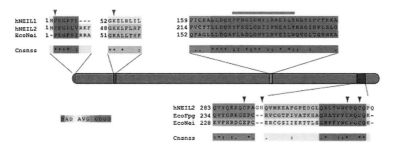

Figure 2 Sequence alignment of NEIL1/NEIL2 and *E. coli* Nei/Fpg. The domains critical for the glycosylase activity are aligned. The central bar denotes NEIL2 polypeptide with domains containing the essential Pro1 after removal of the N-terminal Met (red). Other conserved residues and motifs are Lys (brown), H2TH (green), and Zn-finger (dark blue). The sequence alignments of hNEIL1/hNEIL2/Nei are shown for the conserved domains, except for the Zn-finger motifs with hNEIL2/Fpg/Nei. The alignment was carried out using "T-coffee" (http://www.ch.embnet.org/software/TCoffee.html) with the homology scores indicated in color (blue to red). Positions from N-termini are numerically shown in the primary sequences. Critical side chains and boxes are also indicated with triangles (Pro and Lys; CHCC type Zn-finger) or a bar (H2TH) along with the alignments. The consensus rows (Cons) depicts identical match (∗), chemically similar residues (":" and "."). (*See color insert.*)

The 3′ end cleaning of AP lyase products is a key step in the repair of oxidized bases in DNA (31,32). In *E. coli*, APEs are responsible for removing all types of 3′ blocking ends. *E. coli* expresses two major APEs, Xth and Nfo (33). Xth, which accounts for most of the bacterial APE activity, was in fact discovered as a DNA 3′ phosphatase/exonuclease (34). It is also used as a common reagent in recombinant DNA techniques for its 3′ exonuclease activity, which removes not only the 3′ α,β unsaturated deoxyribose, the product of β elimination reaction, but also other types of 3′ blocking residues, including oxidation products. Nfo lacks the 3′ exonuclease activity, but is still efficient in hydrolyzing both free 3′ phosphates or 3′ phosphates attached to the deoxyribose fragments from the DNA (35).

Thus APEs are characterized by two distinct types of intrinsic activities. They act as both endonucleases and as 3′ phosphodiesterases. In all cases they generate a 3′ OH terminus which is essential for the DNA polymerase to utilize as a primer for DNA synthesis.

Unlike in *E. coli* and both fission and budding yeasts, all of which express two distinct APEs of Xth and Nfo types (33,36–38), the mammalian cells have so far been shown to express only one APE, named APE1. A homolog of APE1 was cloned from the mammalian cells several years ago (39). Although this protein shows strong homology with the yeast APN2, which does have authentic AP-endonuclease activity, no enzymatic activity could be detected in purified mammalian APE2 (Wiederhold et al., unpublished results).

APE1 belongs to the Xth group, with which it shares significant sequence identity (33). Surprisingly, in spite of broad sequence conservation, the human APE1 and *E. coli* Xth profoundly differ in substrate preference and activity. While both enzymes have robust endonuclease activity for AP sites, Xth has an even higher specific activity as a 3′ exonuclease/phosphatase (36). The human APE1, on the other hand, has robust 3′ phosphodiesterase activity in removing 3′ phospho α,β unsaturated deoxy ribose and 3′ phosphoglycolate. However, unlike Xth, APE1 was shown to have very weak DNA 3′ phosphatase activity, although it was never carefully characterized. Early studies with the human APE1, partially purified from HeLa cells, showed that the specific activity for 3′ phosphate removal was significantly less than that for AP site cleavage (40).

In any event, the issue regarding the involvement of APE1 in removing DNA 3′ phosphate during BER was moot before 2002, because OGG1 and NTH1, the only oxidized base-specific DNA glycosylase/AP lyases known in mammalian cells, generate AP sites and β elimination products which are efficiently removed by APE1.

2.5. Polynucleotide Kinase is the Predominant DNA 3′ Phosphatase in Mammalian Cells

The discovery of polynucleotide kinase (PNK), which transfers phosphate from ATP to 5′ OH terminus of DNA (and RNA) was a critical landmark in biomedical research, and heralded the era of recombinant DNA technology (41). The first PNK was found to be encoded by colophage T4, and it was immediately evident that *E. coli* itself does not possess this activity. Also, PNK is particularly useful as a reagent for manipulating recombinant DNA and 5′ terminal labeling of oligo nucleotide with ^{32}P. What was not previously publicized is the intrinsic 3′ phosphatase activity of this enzyme (42). W. Verly's lab discovered the presence of PNK activity in the nuclear extract of rat liver, the activity was then extensively purified and its enzymatic activities characterized. The mammalian PNK like the T4 encoded enzyme was found to have dual activities (10). It is also a potent DNA 3′ phosphatase in addition to its ability to transfer phosphate to the 5′ termini of DNA. The kinase and phosphatase activities are localized in two distinct domains in both T4 and mammalian PNKs (43,44). Habraken and Verley's results (10) indicated that PNK accounts for some 90% of the liver chromatin-associated DNA 3′ phosphatase activity in liver cells.

We tested the possibility that PNK could function in NEIL-initiated repair of an oxidized base in an in vitro reconstituted systems. These results confirm the earlier results of Habraken and Verly, and support our contention that PNK could provide the DNA 3′ phosphatase activity in vivo, not only because of its high catalytic specificity, but also its comparable cellular abundance to APE1 (data not shown). The catalytic specificity of APE1 for the DNA 3′ phosphodeoxyribose derivative is similar to that of PNK for DNA 3′ phosphate (L. Wiederhold, unpublished experiment). Thus, there is a clear dichotomy in the functions of APE1 and PNK in mammalian cells. Like in *E. coli*, APE1 acts in mammals as an endonuclease for AP sites by cleaving the DNA strand to produce 3′ OH and also as a phosphodiesterase to hydrolyze 3′ α,β unsaturated deoxyribose. However, PNK is required for removal of DNA 3′ phosphate.

2.6. In Vivo Functions of NEIL1 and NEIL2

Despite the accumulation of a large body of information regarding the BER enzymes in vitro, the in vivo activities of DNA glycosylases in the complex mosaic of cellular functions remain largely unexplored. The presence of a multitude of DNA glycosylases with broad and overlapping substrate ranges suggests that these enzymes provide back-up functions for one another in vivo. This appears to be particularly important for oxidized bases, which are continuously generated due to endogenous ROS. Homozygous null mice lacking OGG1 or NTH1 have recently been generated, and these animals have no phenotype even during maturity (45–48). Because the mutagenic and toxic ROS-induced base lesions are continuously generated, the lack of toxic or mutagenic response in the absence of NTH1 or OGG1 strongly suggests the presence of other repair enzymes for oxidative damage. Thus the NEILs are excellent candidates for fulfilling this role.

The dispensability of OGG1 in the repair of 8-oxoG was also shown in a different type of study in cultured cells. The repair of 8-oxoG was monitored in mouse fibroblasts transiently transfected with a plasmid in which the base lesion was inserted at a single site in either the transcribed or nontranscribed strand. 8-oxoG was found to be repaired at a much higher rate when present in the transcribed strand of the plasmid than when incorporated in the nontranscribed strand. Furthermore, 8-oxoG repair in the transcribed strand was as efficient in OGG1-null cells as in wild-type control cells (49,50). These results are consistent with the possibility that an additional repair system, presumably a second DNA glycosylase, is responsible for repair of 8-oxoG specifically from the transcribed strand. Furthermore, these studies strongly support the possibility that oxidized bases are also subject to transcription-coupled repair (TCR), which was first characterized as a subpathway of NER of bulky adducts (51). XPG, a $3'$ endonuclease is involved in cleaving the damage-containing DNA segment during NER (52). However, XPG, a very large polypeptide essential for cell survival has multiple functions in vivo. Cooper and her colleagues showed that TCR of 8-oxoG requires XPG in the plasmid system (49). More critically, the endonuclease activity of XPG is not required for this function. Thus XPG may provide a structural function in TCR of 8-oxoG.

2.7. Preference of NEIL1 and NEIL2 for Substrate Lesions in DNA Bubble Structures

Most DNA glycosylases are unable to excise base lesions from single-stranded DNA. This is perhaps expected, because base excision and strand cleavage at the resulting AP site by oxidized base-specific DNA glycosylase/AP lyases will produce a product which could be repaired in the absence of a complementary strand. It was therefore surprising that Takao et al. (29) first reported that NEIL1 could excise oxidized base lesion from single-stranded DNA. We then speculated that the true substrate of NEIL1 is a DNA bubble structure and not single-stranded DNA. Indeed, we observed that both NEIL1 and NEIL2 are highly efficient in excising the substrate base lesions when placed in the middle of unpaired sequences in duplex oligos (53); the activities are comparable those for ssDNA. In contrast, in confirmation of previous studies both OGG1 and NTH1, the other two oxidized base-specific DNA glycosylases, were unable to excise their substrate lesions either from bubble DNA or single-stranded DNA (53). As already mentioned, 5-OHU is a common substrate for NEIL1, NEIL2, and NTH1. 5-OHU-containing oligos representing both duplex and bubble DNA substrates could be generated in addition to the single-stranded oligo itself. Figure 3 shows that both NEIL1 and NEIL2 were active with all three substrates, while NTH1 could utilize only the duplex DNA. Interestingly, NEIL2 was much more active with the bubble DNA relative to duplex or single-stranded DNA, while NEIL1 had comparable activity with duplex and bubble DNA (53).

We then tested for 8-oxoG excision by NEIL1 from oligos of different structures. Our earlier studies showed that NEIL1 has weak 8-oxoG excision activity. However, this activity was found to be significantly higher when the base lesion was placed within a base unrepaired region. NEIL1 was then compared with

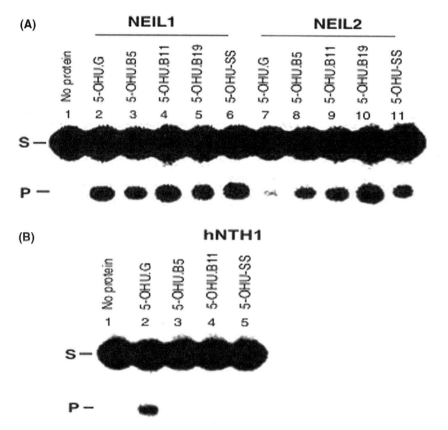

Figure 3 Activity assay of NEIL1, NEIL2, and NTH1 with substrates (500 nM) in different structures. Identical 5-OHU-containing oligo strands were used as is (5-OHU-ss), annealed with a complementary strand containing G opposite 5-OHU (5-OHU·G), or with a noncomplementary strand to generate B5, B11, or B19 bubbles flanked by duplex sequences, as described previously. A, relative activity of NEIL1 and NEIL2. B, activity of NTH1 (100 nM) was measured as in A. *S*, substrate; *P*, product. *Source*: From Ref. 53.

OGG1, the major 8-oxoG repairing enzyme. Again OGG1 was able to excise 8-oxoG present only in the duplex oligo (Fig. 4). We also examined the effect of the length of unpaired region on the base excision activity of NEIL1 and NEIL2. Figure 3 shows that the activity increased with increasing length of the bubble. However, these studies were not extensive and the activity may be affected by other factors, e.g., the intrastrand secondary structure of the single-stranded region which would be determined by its sequence. Furthermore, the impact of the lesion site on the enzyme activity relative to the unrepaired sequence is unknown, and needs to be systematically investigated.

2.8. Affinity of NEIL1 and NEIL2 for Bubble Structure

The results described so far suggested that NEIL1 and NEIL2 have intrinsic affinity for single-stranded or bubble structures. We measured the binding of these enzymes to both undamaged oligos and 5-OHU-containing oligos with single-stranded duplex and bubble structures. All of the oligos have identical sequences except for the 5-OHU which was substituted fir C in the normal DNA. This avoided any potential

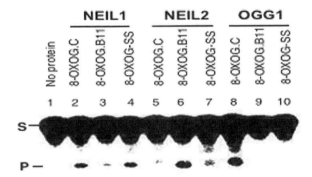

Figure 4 Relative activities of NEIL1, NEIL2, and OGG1 in excision of 8-oxoG from duplex and bubble-containing oligos. An 8-oxoG-containing oligo (500 nM) of the same sequence as the 5-OHU oligo was used as such (ss) or was annealed with appropriate complementary strands to generate an 8-oxoG·C-containing duplex or 8-oxoG·B11 bubble oligo. The activities of NEIL1 (40 nM), NEIL2 (60 nM), and OGG1 (20 nM) were measured as described previously. *Source*: From Ref. 53.

impact of sequence context on the binding. We used wild-type enzymes with normal DNA and inactive mutants of NEIL1 and NEIL2 with the damaged oligo in the electrophoretic mobility shift analysis (EMSA) to measure the affinities (53). Because DNA glycosylases do not require any cofactor for their enzymatic reactions, active NEIL1 and NEIL2 will continue to cleave the substrate oligos during incubation for EMSA, which make the quantitative measurement unreliable. Pro1 (after cleavage of the initiator Met) is the active site nucleophile for both NEIL1 and NEIL2. We first planned to use an Ala1 mutant of NEIL for EMSA, based on the assumption that such a mutation will inactivate the protein without causing significant structural perturbation. However, we observed that the Ala1 mutant of NEIL1 shows a major structural perturbation, based on its chromatographic behavior (unpublished observation) and hence may not be a good surrogate of wild-type NEIL1. We therefore decided to use Lys 53 and Lys 51 mutants of NEIL1 and NEIL2, respectively. These Lys residues are conserved in all Fpg/Nei family members (27), although the role of this Lys residue in the enzyme reaction is not fully understood. An *E. coli* Fpg mutant containing Gln at this conserved Lys site was found to have lost base excision activity, but retained AP lyase activity (54). We generated a Lys → Leu mutation in NEIL1 and Lys → Arg in NEIL2 by site-directed mutagenesis and purified the recombinant proteins to homogeneity. These mutants were found to have lost both base excision and AP lyase activities (unpublished observation). In any case, we used the mutant enzymes for the binding studies with 5-OHU containing oligos. We made the surprising observation that neither wild-type nor mutant enzymes showed significant affinity for single-stranded oligos with or without a damaged base, so the binding constants could not be accurately determined. However, the affinity for duplex and bubble DNA could be measured. Also, EMSA with both proteins showed formation of two distinct complexes, at higher enzyme concentrations. Assuming that the larger complex contained two enzyme molecules/oligo molecule, we fitted the binding isotherm to an equation derived for similar studies with a restriction enzyme (55,56). Table 1 shows the calculated affinity constants of wild type and mutant enzymes for undamaged and damaged DNA, respectively. It is evident that both NEIL1 and NEIL2 have intrinsic affinity for bubble DNA, even

Table 1 Affinity of WT and Mutant NEILs for DNA

K_d app(nM)	Duplex	Bubble (B11)
NEIL1[a]	714 ± 76	32.2 ± 2.7
NEIL2[a]	833 ± 34	119 ± 6
NEIL1(K53L)[b]	1428 ± 142	87 ± 16
NEIL2(K49R)[b]	5000 ± 100	286 ± 24

[a]WT enzymes and nondamaged oligo duplex and bubbles were used.
[b]Mutant enzymes and 5-OHU-containing oligos were used.
Source: From Ref. 53.

in the absence of a damaged base lesion. In the case of NEIL2, the binding affinity for the bubble structure is about an order of magnitude higher than for the duplex oligo of the same sequence. Although a similar trend in affinity of the mutant NEIL1 and NEIL2 was observed for the 5-OHU containing oligo with highest affinity for the bubble substrates, the binding affinities were much lower. The opposite was expected because the enzymes should have higher affinity for the substrate lesion than normal bases. It thus appears likely that subtle structural perturbation due to point mutation in these proteins have decreased the overall binding affinity.

In any event, our results raise the interesting possibility that while NEIL1 and NEIL2 prefer binding to single-stranded DNA sequence, they require a duplex region for initial binding. Thus the single-stranded oligos do not stably bind to these enzymes. The NEILs again show very distinctive behavior compared to OGG1 and NTH1, and possibly other DNA glycosylases. The mechanism of substrate recognition by DNA glycosylases as a prerequisite to catalysis is not completely understood. Both kinetic and structural studies suggest a general model of initial interaction of the enzyme with duplex DNA, followed by scanning the DNA via a "push–pinch–pull" mode until the enzyme recognizes and binds a substrate lesion with much higher affinity, which is followed by catalysis (57). In this model, the enzyme always translocates along the duplex DNA. However, in the case of NEILs this model needs to be modified in that after initial nonspecific binding to a duplex region, the enzymes translocate along the DNA backbone but stably bind to the substrate base lesion located in a single-stranded region. X-ray crystallographic structures of several DNA glycosylases complexed with duplex DNA substrate mimics have recently been elucidated, providing significant insight into about the interaction between the DNA base residues and phosphate and specific interaction with amino acid residues (23,58). It will be interesting to solve similar structure of NEILs bound to bubble DNA substrate, in which interactions should be distinct from those of OGG1.

2.9. Potential Role of NEILs in Transcription and Replication-Associated Repair

The unexpected ability of NEILs to excise base lesions from an unrepaired region in duplex DNA suggests that the preferred substrates of these enzymes are indeed bubble structures in vivo. Although in our in vitro studies the stable bubble structure in substrate oligos was generated using noncomplementary base sequences, similar bubble structures are transiently generated in vivo by localized unwinding of the duplex DNA during transcription and DNA replication. In the case of eukaryotic transcription, the transcription factor TFIIH, a component of the RNA polymerase

II (Pol II) homoenzyme, has intrinsic DNA helicase activity, and presumably unwinds the duplex DNA template ahead of the nascent RNA chain. The separation from the complementary DNA strands facilitates copying of the transcribed strand by the RNA polymerase. As RNA synthesis continues, the Pol II complex with the bound nascent RNA chain moves along the DNA, and the transcribed strand reforms the duplex DNA behind the enzyme. The length of the RNA•DNA hybrid, which is an obligatory intermediate during RNA synthesis, has been variously estimated from a few to about 10 base pairs (59). The structure of the hybrid is an R-loop which is different from the bubble structure formed ahead of the nascent RNA chain. The structure of yeast RNA Pol II has recently been solved by X-ray crystallography; it appears that RNA Pol II covers bases on the template, while bases ahead of the 3′ OH terminus of the RNA chain remain in an unpaired state or as a transient bubble (60).

In the case of DNA replication, both strands serve as templates for replicative DNA polymerase complexes. However, one strand (the same as transcribed) is copied continuously and the other discontinuously because of the single polarity of 5′→3′ chain elongation by both RNA and DNA polymerases. At the same time, it is now clear that the two DNA strands are replicated almost simultaneously because the DNA polymerase acts as a dimer and the template DNA assumes a "trombone" shape in order to accommodate synthesis of both nascent chains in the 5′ → 3′ direction, while one template strand is of 3′→5′ and the other 5′→3′ orientation. Furthermore, the discontinuous strand synthesis requires repeated de novo priming for synthesis of Okazaki fragments. In any event, the strands needed to be separated ahead of replication in the same way during transcription although both strands then serve as templates.

Many DNA helicases have been characterized in both prokaryotes and eukaryotes. In mammalian cells, however, the identity of helicase(s) involved in replication-associated DNA unwinding is not clear. It is possible that several such helicases including the Mcm 4/6/7 complex may be involved (61). Regardless of the precise mechanism and identity of helicases, it is clear that DNA ahead of the replication form assumes a bubble structure containing unpaired bases.

2.10. Role of NEILs in Transcription-Coupled Repair

Transcription-coupled repair (TCR) was discovered by Hanawalt and his colleagues in the mid-1980s as a subpathway of NER (51). They observed that UV-induced pyrimidine photoproducts were preferentially repaired in transcriptionally active genes relative to the lesions formed in nontranscribed sequences, repair of which was named global genome repair, or GGR (62). While the key components of the repair process are common to both GGR and TCR, some additional factors are unique. Although it has not been reproduced in an in vitro assay using a reconstituted system, it is generally accepted that TCR is triggered when the transcription complex stalls at a bulky adduct in the template strand that completely blocks transcription. The stalled polymerase complex at a lesion site generates a signal for regression of the enzyme complex and for repair of the lesion by assembling the TCR complex. In this process, TFIIH and TFIID are components of both the transcription machinery and the TCR complex. The details about the identity of the proteins in the complex and the signaling for sequential steps in assembly are still not completely understood.

Hanawalt's group also showed that TCR occurs selectively in the transcribed strand and the repair of the nontranscribed strand occurs comparable to GGR. The teleological basis for TCR is obvious. The bulk of the mammalian genome does not have any apparent function and most of the cells in the adult are terminally differentiated and nondividing. Thus the persistence of bulky adducts in the absence of repair have no consequence in these noncoding sequences. On the other hand, the damage in the transcribed sequences has to be repaired for synthesis of wild-type proteins in order to maintain turnover of cellular proteins.

The discovery of TCR also explains the "mouse paradox," which was discovered decades ago when repair of UV photoproducts was first examined. Mouse cells were found to function normally in spite of inefficient removal of pyrimidine photoproducts from their genomes compared to efficient removal of these DNA lesions from the human cell DNA. It was later discovered, however, that the TCR in human and mouse cells after UV irradiation is comparable (63). In contrast, the mouse cells have relatively inefficient GGR relative to the human cells. Because the transcriptionally inactive sequences constitute the bulk of the mammalian genome, the overall repair of UV damage is much less in the mouse cells than in human cells.

2.11. TCR of Oxidatively Damaged Bases in the Mammalian Genome

All experimental evidence supports the idea that oxidatively damaged bases are repaired primarily via the BER pathway. Furthermore, most of the lesions unlike the bulky adduct do not block transcription in vitro. Thus it was first thought that BER may not include TCR of oxidized bases. In fact, an early study showed that methylated bases, e.g., 7-methylguanine, are not preferentially repaired in the active gene sequences (64). On the other hand, based on the teleological argument we should expect preferential repair of mutagenic base adducts such as 8-oxoguanine or 5-hydroxyuracil in the transcribed sequences. While these lesions would not block transcription, mutant mRNA and thus mutant proteins will be generated if these lesions remain unrepaired. Such "transcriptional mutagenesis" was indeed demonstrated elegantly in a model system (65).

Indeed, there are several independent lines of evidence supporting TCR of oxidized bases in mammalian cells. First, NTH1 and OGG1 previously, characterized in mammalian cells, are likely to be responsible in repair of the bulk of oxidatively damaged bases in vivo. This is consistent with the observed accumulation of 8-oxoG in the genome of OGG1-null mouse cells and tissues (45,46). However, OGG1 is not essential for repair of 8-oxoG in the transcribed strand of a transfected plasmid. Thus repair of 8-oxoG and possibly other oxidatively damaged bases may not utilize OGG1 or NTH1. One possibility is that the TCR subpathway of NER is also utilized in repairing oxidized bases. Although there is some evidence that 8-oxoG could be repaired via the NER pathway (66), no follow-up study to support the role of NER in oxidized base repair has been published. On the other hand, accumulation of 8-oxoG in OGG1-deficient cells provides strong support for the role of BER in repairing 8-oxoG. At the same time, the same proteins involved in TCR of bulky adducts, including XPG, Cockayne syndrome B (CSB) and BRCA1 and 2, have also been implicated in the repair of 8-oxoG in mammalian cells (49,67,68).

We should stress here that for repair of oxidized bases via the TCR subpathway, the analogy with repair of bulky adducts is not valid because 8-oxoG (and other similar oxidized base lesions) does not block transcription (69). We have therefore developed a model for transcription-coupled repair of oxidized bases including 8-oxoG, which still involves the BER pathway. Based on the preference of NEILs for a DNA bubble substrate, we propose the following scenario for transcription-coupled BER. The transcription complex carrying out RNA synthesis encounters an oxidized base lesion in the transient transcription bubble ahead of the growing RNA chain. Based on X-ray crystallographic studies of the yeast RNA Pol II complex, the size of the bubble appears to be about nine nucleotides (60). NEIL excises the damaged bases and cleaves the DNA strand due to its AP lyase activity. The strand breaks blocks transcription, followed by regression of the transcription complex and collapse of the bubble structure. The reformed duplex DNA with the single base gap is then repaired, followed by resumption of RNA synthesis (Fig. 5). Thus, unlike bulky adducts which directly block transcription and trigger TCR, the oxidized bases trigger TCR indirectly after the action of NEILs. The fact both NEILs, (but not OGG1 and NTH1) utilize bubble substrates is consistent with this model.

2.12. Role of NEIL1 in Replication-Associated Repair

Finally, we would like to hypothesize about the specialized role of NEIL1 in replication-associated repair. Such a possibility was raised by our earlier observation that NEIL1 is induced during the S-phase (24), and that it, but not NEIL2, stably interacts with proliferating cell nuclear antigen (PCNA; H. Dou unpublished). PCNA, often used as a marker of S-phase cells, acts as a sliding clamp on the replicating DNA and stimulates the activities of multiple proteins involved in DNA replication, including DNA polymerase δ and flap endonuclease 1 (FEN1, which removes the RNA primers of Okazaki fragment).

We propose that NEIL1, by its ability to repair oxidized base lesions in a single-stranded region, acts as a "cow-catcher" in the DNA replication machinery in surveillance of DNA lesions ahead of the polymerase complex. A lesion in the single-stranded sequence generated by a DNA helicase will be recognized and excised by NEIL1. The resulting strand break will interrupt daughter strand synthesis complementary to the damaged strand. However, because the leading and lagging strands are coordinately synthesized by a dimeric DNA polymerase complex (70–72), interruption in synthesis of one parental strand will stall the polymerase complex, which may then regress. This allows reannealing of separated strands and repair of the strand break induced by NEIL1, analogous to the final stages of TCR initiated by NEIL2 (Fig. 5).

Although there is no evidence for preferential repair of oxidized bases in the prereplicated genome, it is tempting to speculate that continuous and insidious production of oxidative base damage in the genome warrants availability of a multitude of protective repair processes in order to main genomic fidelity. In extending this scenario, we propose that replication-associated repair includes both preferential repair of the template strand ahead of the replication fork via the BER pathway and repair of misincorporated bases in the nascent DNA strand via the mismatch repair (MMR) pathway. The role of MMR in nascent strand repair has already been established (73).

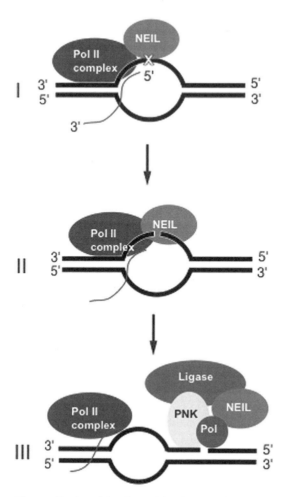

Figure 5 A minimal model of NEIL2's proposed role in transcription-coupled repair. I. NEIL2 acts as a cow-catcher ahead of the transcription complex moving along the template strand recognizes an oxidized base lesion in the transient bubble. II. NEIL2 excises the lesion and produces a single-strand break. III. The bubble collapses after regression of the transcription complex; NEIL2 recruits BER proteins to repair the nick and allows transcription to resume.

3. CONCLUSIONS

It should be evident from this short review that repair of oxidized bases via the BER pathway is far more complex than was originally thought. The BER enzymes were among the first repair enzymes to be thoroughly characterized and the overall pathway delineated (74). However, the presence of a multitude of subpathways in these processes, including distinct and additional requirement for other proteins including BRCA1, BRCA2, CSB, and XPG, and the subtleties of repair regulation and signaling are now being gradually recognized. It is safe to predict that we have not identified all of the stable interactions among BER proteins, specifically those between DNA glycosylases and proteins involved in other DNA transactions including replication and transcription, or even other DNA repair pathways, namely NER and

MMR. We have also begun to identify covalent modification of these DNA glyco-sylases and other BER proteins (75). Development of a comprehensive picture of BER complexes in a dynamic fashion will remain a major challenge for many years to come.

ACKNOWLEDGMENTS

The research described in this review was partially supported by the U.S. Public Health grants R01 CA53791, CA81063, ES08457, CA098664, and also by P01 CA92584 grant and P30 ES05576. The authors would like to thank Dr. R. Roy and current members of the Mitra lab for reagents and other help, and Dr. David Konkel for critically reading the manuscript.

ABBREVIATIONS

AP—abasic (apurinic, apyrididinic)
BER—base excision repair
DHU—dihydrouracil
8-oxoG—8-oxo 7,8 dihydroguanine
EMSA—electrophoretic mobility shift analysis
5-OHU—5-hydroxyuracil
Fapy—formamido pyrimidine
Fpg—Fapy-DNA glycoylase
GGR—global genome repair
Nei—endonuclease VIII
NEIL—Nei-like
NER—nucleotide excision repair
Nfo—endonuclease IV
Nth—endonuclease III
NTH—Nth homolog
OGG—8-oxoguanine-DNA glycosylase
PNK—polynucleotide kinase
Polβ—DNA polymerase β
ROS—reactive oxygen species
TCR—transcription coupled repair
TG—thymine glycol
Xth—exonuclease III

REFERENCES

1. Gotz ME, Kunig G, Riederer P, Youdim MB. Pharmacol Ther 1994; 63:37–122.
2. Grisham MB, Granger DN, Lefer DJ. Free Radic Biol Med 1998; 25:404–433.
3. Tan AS, Berridge MV. J Immunol Methods 2000; 238:59–68.
4. Karam LR, Simic MG. Environ Health Perspect 1989; 82:37–41.
5. Krokan HE, Standal R, Slupphaug G. Biochem J 1997; 325:1–16.
6. Dodson ML, Michaels ML, Lloyd RS. J Biol Chem 1994; 269:32709–32712.
7. Tchou J, Grollman AP. J Biol Chem 1995; 270:11671–11677.

8. Bailly V, Verly WG, O'Connor T, Laval J. Biochem J 1989; 262:581–589.
9. Habraken Y, Verly WG. FEBS Lett 1983; 160:46–50.
10. Habraken Y, Verly WG. Eur J Biochem 1988; 171:59–66.
11. Bhagwat M, Gerlt JA. Biochemistry 1996; 35:659–665.
12. Blaisdell JO, Hatahet Z, Wallace SS. J Bacteriol 1999; 181:6396–6402.
13. Singhal RK, Prasad R, Wilson SH. J Biol Chem 1995; 270:949–957.
14. Ikeda S, Biswas T, Roy R, Izumi T, Boldogh I, Kurosky A, Sarker AH, Seki S, Mitra S. J Biol Chem 1998; 273:21585–21593.
15. Lu R, Nash HM, Verdine GL. Curr Biol 1997; 7:397–407.
16. Krokan HE, Nilsen H, Skorpen F, Otterlei M, Slupphaug G. FEBS Lett 2000; 476:73–77.
17. Nash HM, Lu R, Lane WS, Verdine GL. Chem Biol 1997; 4:693–702.
18. Thayer MM, Ahern H, Xing D, Cunningham RP, Tainer JA. EMBO J 1995; 14:4108–4120.
19. Girard PM, Boiteux S. Biochimie 1997; 79:559–566.
20. Hatahet Z, Kow YW, Purmal AA, Cunningham RP, Wallace SS. J Biol Chem 1994; 269:18814–18820.
21. Dizdaroglu M, Laval J, Boiteux S. Biochemistry 1993; 32:12105–12111.
22. Gilboa R, Zharkov DO, Golan G, Fernandes AS, Gerchman SE, Matz E, Kycia JH, Grollman AP, Shoham G. J Biol Chem 2002; 277:19811–19816.
23. Zharkov DO, Golan G, Gilboa R, Fernandes AS, Gerchman SE, Kycia JH, Rieger RA, Grollman AP, Shoham G. EMBO J 2002; 21:789–800.
24. Hazra TK, Izumi T, Boldogh I, Imhoff B, Kow YW, Jaruga P, Dizdaroglu M, Mitra S. Proc Natl Acad Sci USA 2002; 99:3523–3528.
25. Hazra TK, Kow YW, Hatahet Z, Imhoff B, Boldogh I, Mokkkapati SK, Mitra S, Izumi T. J Biol Chem ; 277:30417–30420.
26. Hazra TK, Izumi T, Kow YW, Mitra S. Carcinogenesis 2003; 24:155–157.
27. Bandaru V, Sunkara S, Wallace SS, Bond JP. DNA Repair (Amst) 2002; 1:517–529.
28. Rosenquist TA, Zaika E, Fernandes AS, Zharkov DO, Miller H, Grollman AP. DNA Repair (Amst) ; 2:581–591.
29. Takao M, Kanno S, Kobayashi K, Zhang QM, Yonei S, van der Horst GT, Yasui A. J Biol Chem 2002; 277:42205–42213.
30. Katafuchi A, Nakano T, Masaoka A, Terato H, Iwai S, Hanaoka F, Ide H. J Biol Chem. In press.
31. Mitra S, Izumi T, Boldogh I, Bhakat KK, Hill JW, Hazra TK. Free Radic Biol Med 2002; 33:15–28.
32. Izumi LR, Wiederhold Roy G, Roy R, Jaiswal A, Bhakat KK, Mitra S, Hazra TK. Toxicology 2003; 193:43–65.
33. Demple B, Harrison L. Annu Rev Biochem 1994; 63:915–948.
34. Richardson CC, Lehman IR, Kornberg A. J Biol Chem 1964; 239:251–258.
35. Levin JD, Johnson AW, Demple B. J Biol Chem 1988; 263:8066–8071.
36. Bennett RA. Mol Cell Biol 1999; 19:1800–1809.
37. Johnson, Torres-Ramos CA, Izumi T, Mitra S, Prakash S, Prakash L. Genes Dev 1998; 12:3137–3143.
38. B Ribar B, Izumi T, Mitra S. Nucleic Acids Res 2004; 32:115–126.
39. Hadi MZ, Wilson DM. Environ Mol Mutagen 2000; 36:312–324.
40. Chen DS, Herman T, Demple B. Nucleic Acids Res 1991; 19:5907–5914.
41. Weiss B, Richardson CC. Proc Natl Acad Sci USA 1967; 57:1021–1028.
42. Amitsur M, Levitz R, Kaufman G. EMBO J 1987; 6:2499–2503.
43. Mani RS, Karimi Busheri F, Cass CE, Weinfeld M. Biochemistry 2001; 40:12967–12973.
44. Wang LK, Shurman S. EMBO J 2001; 21:3873–3880.
45. Klungland A, Roswell I, Hollenbach S, Larsen E, Daly G, Epe B, Seeberg E, Lindahl T, Barnes DE. Proc Natl Acad Sci USA 1999; 96:13300–13305.
46. Minowa O, Arai T, Hirano M, Monden Y, Nakai S, Fukuda M, Itoh M, Takano H, Hippou Y, Aburatani H, Masumura K, Nohmi T, Nishimura S, Noda T. Proc Natl Acad Sci USA 2000; 97:4156–4161.

47. Ocampo MT, Chaung W, Marenstein DR, Chan MK, Altamirano A, Basu AK, Boorstein RJ, Cunningham RP, Teebor GW. Mol Cell Biol 2002; 22:6111–6121.
48. Takao M, Kanno S, Shiromoto T, Hasegawa R, Ide H, Ikeda S, Sarker AH, Seki S, Xing JZ, Le XC, Weinfeld M, Kobayashi K, Miyazaki J, Muijtjens M, Hoeijmakers JHJ, van der Horst G, Yasui A. EMBO J 2002; 21:3486–3493.
49. Le Page F, Kwoh EE, Avrutskaya A, Gentil A, Leadon SA, Sarasin A, Cooper PK. Cell 2000; 101:159–171.
50. Le Page F, Klungland A, Barnes DE, Sarasin A, Boiteux S. Proc Natl Acad Sci USA 2000; 97:8397–8402.
51. Mellon I, Spivak G, Hanawalt PC. Cell 1987; 51:241–249.
52. O'Donovan A, Davies AA, Moggs JG, West SC, Wood RD. Nature 1994; 371:432–435.
53. Dou H, Mitra S, Hazra TK. J Biol Chem 2003; 278:49679–49684.
54. Sparabaev M, Sidorkina OM, Jurado J, Privezentzev CV, Greenber MM, Laval J. Environ Mol Mutagen 2002; 39:10–17.
55. Taylor JD, Badcoe IG, Clarke AR, Halford SE. Biochemistry 1991; 30:8743–8753.
56. Vallone PM, Benight, AS. Biochemistry 2000; 39:7835–7846.
57. Parikh SS, Putnam CD, Tainer JA. Mutat Res 2000; 460:183–199.
58. Fromme JC, Verdine GL. EMBO J 2003; 22:3461–3471.
59. Shilatifard A, Conaway RC, Conaway JW. Annu Rev Biochem 2003; 72:693–715.
60. Gnatt AL, Cramer P, Fu J, Bushnell DA, Kornberg RD. Science 2001; 292:1876–1882.
61. You Z, Ishimi Y, Mizuno T, Sugasawa K, Hanaoka F, Masai H. EMBO J 2003; 22:6148–6160.
62. Bowman KK, Sicard DM, Ford JM, Hanawalt PC. Mol Carcinog 2000; 29:17–24.
63. Hanawalt PC. Environ Mol Mutagen 2001; 38:89–96.
64. Plosky B, Samson L, Engelward BP, Gold B, Schlaen B, Millas T, Magnotti M, Schor J, Scicchitano DA. DNA Repair (Amst) 2002; 1:683–696.
65. Bregeon D, Doddridge ZA, You HJ, Weiss B, Doetsch PW. Mol Cell 2003; 12:959–970.
66. Reardon JT, Bessho T, Kung HC, Bolton PH, Sancar A. Proc Natl Acad Sci USA: 94: 9463–9468.
67. Tuo J, Muftuoglu M, Chen C, Jaruga P, Selzer RR, Brosh RM, Rodriguez H, Dizdaroglu M, Bohr VA. J Biol Chem 2001; 276:45772–45779.
68. Le Page F, Randrianarison V, Marot D, Cabannes J, Perricaudet M, Feunteun J, Sarasin A. Cancer Res 2000; 60:5548–5552.
69. Tornaletti S, Maeda LS, Lloyd DR, Reins D, Hanawalt PC. J Biol Chem 2001; 276:45367–45371.
70. Gerik J, Li X, Pautz A, Burgers PMJ. J Biol Chem 1998; 273:19747–19755.
71. Burgers PMJ, Gerik KJJ. Biol Chem 1998; 273:19756–19762.
72. Zuo V Bermudez, Zhang G, Kelman Z, Hurwitz J. J Biol Chem 2000; 275:5153–5162.
73. Modrich P. J Biol Chem 1997; 272:24727–24730.
74. Friedberg EC, Walker, Siede W. DNA Repair and Mutagenesis. Washington, DC: ASM Press, 1995.
75. Bhakat KK, Izumi T, Yang SH, Hazra TK, Mitra S. EMBO J 2003; 22:6299–6309.

19

Recognition and Repair of Abasic Sites

David M. Wilson III
Laboratory of Molecular Gerontology, GRC, National Institute on Aging, Baltimore, Maryland, U.S.A.

David F. Lowry
Macromolecular Structure and Dynamics, Pacific Northwest National Laboratory, Richland, Washington, U.S.A.

It was initially assumed that our genetic material, i.e., DNA, being the master blueprint of the cell, would retain a high degree of fidelity. However, it was soon recognized that DNA is in fact a dynamic molecule, subject to constant molecular change (1). For example, early studies found that DNA is prone to considerable base loss, where the extent of depurination varies inversely with pH and where elevated temperature promotes depurination through an acid-catalyzed hydrolytic reaction (2,3). Because heat had previously been shown to promote genetic mutations at physiological pH (4), the biological significance of base loss was postulated early on. In this chapter, we cover the mechanisms of AP site formation, the biological consequences of unrepaired AP lesions in DNA, the structural and dynamical consequences of AP sites when present in duplex DNA, and the mechanisms by which cells recognize and process abasic damage in an effort to avoid genome instability, cellular lethality, and ultimately human disease.

1. AP SITE FORMATION AND BIOLOGICAL IMPACT

Owing to the intrinsic chemical instability of the N-glycosylic bond linking the base to the deoxyribose sugar of DNA, AP sites are estimated to form spontaneously at a rate of $\sim$10,000 times per human genome per day under normal physiological conditions, with purine loss making up a majority of the events (5–7). In addition, chemical modification of the base moiety such as nonenzymatic alkylation or oxidation (as well as conditions such as heat or low pH) can accelerate hydrolytic base loss as much as six orders of magnitude (8,9). Moreover, many base damages, including 8-oxoguanine, thymine glycol, and 3-methyladenine, are substrates for DNA glycosylases, which excise target base lesions from DNA as the first step in the base excision repair (BER) pathway (Chapter Tainer within). Thus, AP sites (Fig. 1) make up a large portion of the naturally occurring damage in any genome. However, due to

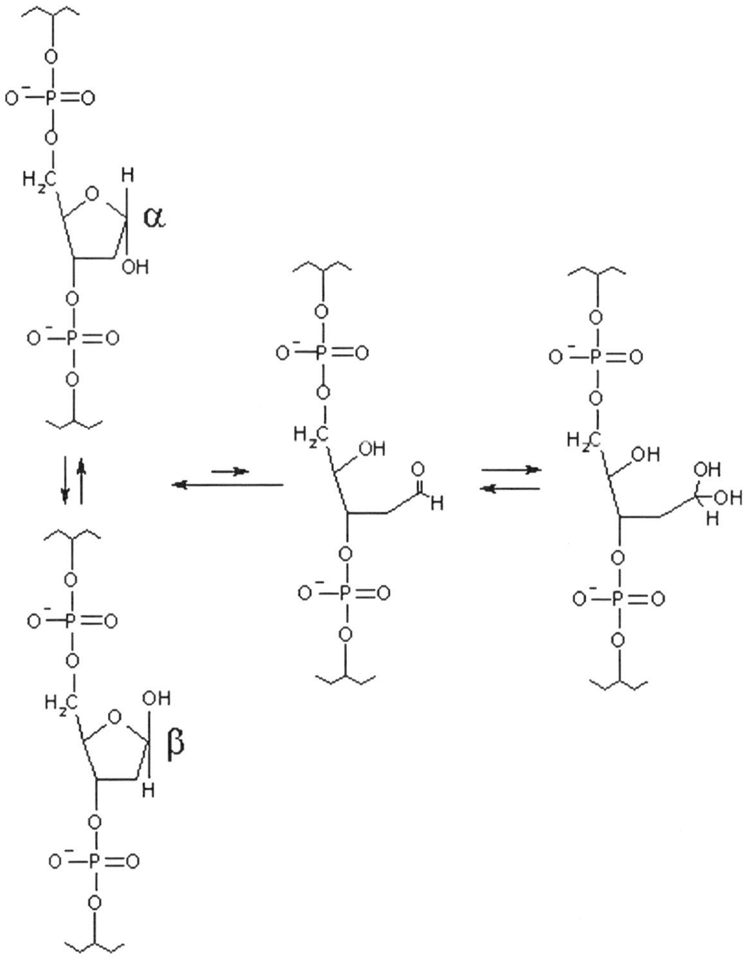

Figure 1 Chemical structures of the natural 2′-deoxyribose AP site. Hydrolytic AP sites exist primarily (at ~99%) in the ring-closed conformation in one of two racemeric hemiacetal forms, α (top left) or β (bottom left). The racemeric species are in an equilibrium mixture. Reduction of the ring-closed AP site can produce a ring-opened aldehyde form (middle). This ring-opened form is susceptible to hydration, generating a hydrated aldehyde AP site (right). *Source*: From Ref. 43.

limitations in current experimental methods, the precise steady-state level of AP lesions remains unknown, although estimates suggest <0.67 per 10^6 nucleotides in the human IMR90 fibroblast genome (10).

As noncoding lesions lack the instructional information of the base moiety, if left unrepaired in DNA (see in what follows for repair mechanisms), AP sites pose both cytotoxic and mutagenic threats to the cell (8,11). Most commonly, AP lesions block DNA or RNA polymerase progression, and in this capacity, hinder successful completion of chromosome replication (causing collapse of the replication fork and the formation of DNA double-strand breaks) or gene transcription, respectively. These biological outcomes can promote cellular defects or, in extreme cases, cell death.

Although AP sites typically block highly accurate (i.e., high fidelity) replicative polymerases, in certain situations they can serve as mutagenic templates.

For example, in *Escherichia coli*, arrest of DNA replication and the consequent formation of DNA single-stranded regions—events most commonly associated with exposure to DNA-damaging agents—activate the SOS (survival) response (12,13). Three DNA polymerases, PolII, PolIV, and PolV (a heterotrimer composed of one UmuC and two UmuD' molecules), are upregulated as part of this response, with the latter two being error-prone translesion enzymes. Thus, to facilitate propagation and avoid cell death, the damage-induced SOS response promotes mutagenic bypass synthesis and, in turn, genomic instability (with fidelity of DNA synthesis being decreased ~100-fold). The combined results of in vitro and in vivo analyses suggest that PolV is responsible for AP site translesion synthesis and cell survival (12,14,15), preferentially inserting adenine opposite the lesion [the so-called "A" rule (16)].

In higher eukaryotes, AP site (or DNA damage in general) "tolerance" appears to be more complex, as at least nine different error-prone bypass polymerases have been identified in humans to date (13,17). In addition, there does not appear to be a clearly defined SOS-type response in vertebrates. Perhaps consistent with this, the mutational spectrum of AP site-containing DNA transfected into a mammalian cell varies from study to study (18–22). Moreover, in vitro bypass studies using AP site-containing templates indicate that the mutagenic profile for AP damage is dependent on the polymerase utilized and the auxiliary proteins present (23–39). Thus, whether there exists a damage-specific mutagenic response (and an accompanying, specific mutagenic profile for AP sites) requires further analysis of the biochemical and biological workings of the mammalian translesion DNA polymerases. Above all, are these proteins linked to the DNA replication and/or repair machinery, and if so, how? For a more thorough review of the eukaryotic bypass DNA polymerases, see Chapter 24.

2. AP-DNA DYNAMICS AND STRUCTURE

DNA structure affects processes such as DNA repair, replication, and recombination, and events typically initiated by precise DNA binding and/or recognition. Thus, in the realm of understanding AP site repair, we first describe the thermodynamic and structural features of abasic DNA.

Ultraviolet spectroscopic and calorimetric studies reveal that incorporation of an abasic site into duplex DNA dramatically reduces thermal and thermodynamic duplex stability (40). Moreover, structural studies involving nuclear magnetic resonance (NMR) indicate that, under most experimental conditions and in nearly all sequence contexts, AP-DNA, regardless of the AP site chemistry, generally retains canonical right-handed B-form, with no severe bending, kinking, or backbone distortions (41–44). Nevertheless, NMR has revealed increased dynamical movements associated with AP-DNA, typically within two base pairs of the abasic site (Fig. 2A). Fluorescence studies, molecular mechanic calculations, and molecular dynamic simulations have detected similar increased dynamical movement (flexibility) around the abasic lesion, not observed in undamaged DNA (45–47). Some of these structural deviations include altered torsion angles of the phosphodiester backbone, decreased base stacking, and additional local conformational variations described in what follows.

To date, the primary AP site forms studied include the 2'-deoxyribose abasic site (Fig. 1), the tetrahydrofuran AP site analog, the 4'-keto abasic lesion, and the 2'-deoxyribonolactone AP damage (Fig. 2B). These AP site chemistries have been

(A)

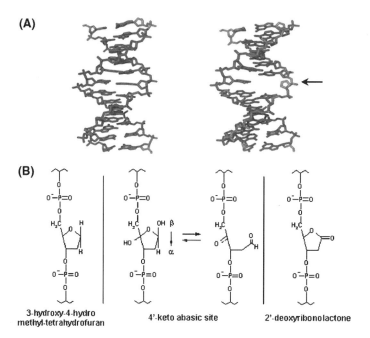

(B)

3-hydroxy-4-hydro
methyl-tetrahydrofuran   4'-keto abasic site   2'-deoxyribonolactone

Figure 2 AP-DNA dynamics and the major AP site chemical forms studied by NMR. (A) Average DNA structures from molecular dynamic simulations of identical dodecamer duplexes with either no abasic site (left) or an abasic site opposite an orphan C base (right). Shown are all nonhydrogen atoms (as capped-sticks), colored by the degree of internal movement where blue indicates the largest movement (5.0 Å) and red the smallest (0.7 Å). An arrow points to the abasic sugar. Notice that atoms near the abasic site show more movement (increased dynamics, as indicated by the more green coloration) relative to the same atoms in the control DNA, and that the "melted" nature of the region surrounding the C-orphan abasic site is visible. (These average structures were created as described in Ref. 47.) (B) Chemical composition of the major AP site forms studied by structural techniques. See text for details. (*See color insert.*)

found to affect the local structure of AP-DNA differently (43,44,48,49). For instance, the tetrahydrofuran residue lacks the 1'-hydroxyl group present in the natural 2'-deoxyribose AP site and thus cannot assume specific hydrogen-bonded conformations achieved by the hydrolytic AP lesion (43,50–52). In addition, the natural AP site exists in multiple forms (Fig. 1), and each of these chemical species can produce unique local structural distortions around the abasic lesion. Specifically, when the sugar is in the β hemiacetal configuration, the deoxyribose remains within the helix, whereas when the sugar is in the α configuration, the deoxyribose exists in an extra-helical conformation (43,44). α and β anomers of the 4'-keto abasic lesion also exist in an extrahelical form (49). Notably, the structural effects of the less-frequent ring-opened aldehyde and hydrated aldehyde AP site forms (Fig. 1) are currently unknown.

In addition to AP site chemistry, the sequence context surrounding the lesion can have a profound impact on duplex stability and the local structure of AP-DNA (50,51,53–58). Most notably, the solvent accessibility of both the abasic site and the base positioned opposite depends mainly on the nature of the opposed ("orphan") nucleotide and the interaction energies of the bases flanking the AP site. As a general rule of thumb, oligonucleotides containing AP sites opposed by a purine conserve right-handed DNA geometry, with the orphan base positioned within the helix, whereas oligonucleotide duplexes harboring a pyrimidine opposite exist in alternative forms, namely, with the abasic sugar and sometimes the pyrimidine opposite taking on extrahelical conformations (depending on the stacking energetics of the flanking bases).

In addition to these local deviations, more dramatic structural alterations have been observed in AP-DNA. For instance, Coppel et al. (54) found that, while abasic DNA retained a right-handed native form with the thymine base opposite stacked within the helix, the AP site induced a kink in the duplex of ~30°. This kink allowed for a bifurcated hydrogen bond to be formed between the unpaired deoxythymidine orphan and the base 5′ to the AP site. Similar kinking (or increased flexibility) and bifurcated hydrogen bonding have also been observed in molecular dynamic simulations (47). Moreover, both computational (47) and NMR studies (54) have revealed (a) unique abasic site movements (including partial flipping out), (b) A-DNA tendencies around the lesion, and (c) destabilization of hydrogen bonding throughout the AP-DNA duplex, but most prominently around the AP site. In a separate study, an AP site positioned within a dA tract, a sequence that normally creates a stable curvature in duplex DNA, was found to alter the stability of this structure up to four base pairs away from the lesion, albeit in an AP site configuration-dependent manner (59).

Thus, in total, current data suggest that AP-DNA typically takes on a normal canonical B-form structure, with increased dynamics around the AP site, where the severity of the conformational rearrangements depends primarily on the surrounding nucleotide sequence context and the AP site chemistry. How nucleotide sequence context and AP site chemistry affect AP site repair (see in what follows) and AP site-directed mutagenesis (see earlier) needs to be more explicitly determined, particularly in mammalian systems (60,61). In future structural analyses, incorporation of restraints from just recently developed reduced dipolar coupling experiments— restraints that are apparently necessary to diminish the force-field dependence of final structures—will help reveal more subtle, but consistent, deviations from B-DNA.

X-ray crystallography has generated a number of DNA–enzyme cocomplex structures. Although the DNA bound by a repair enzyme does not necessarily represent an unbound structure, it may represent a conformation that is attainable to the free, unbound damaged DNA. As will be discussed in more detail in Section 4, this structural "accessibility" (or tendency) may contribute to the specificity of the repair enzyme. Structures of damaged or product DNA in complex with an AP site repair protein (see in what follows) reveal a DNA that is kinked at the abasic site and compressed at the major groove, features likely unique to these complexes. Moreover, in certain protein–DNA structures, the abasic sugar and sometimes the opposite base are unstacked and flipped out of the helix. Although, in general, these severely altered AP-DNA structures are not observed in the structural studies of DNA alone, the propensity of AP-DNA towards increased dynamics/flexibility is likely a key element in targeted enzyme recognition.

3. AP ENDONUCLEASES

To counteract the cytotoxic or mutagenic potential of AP lesions, organisms rely primarily on a class of proteins termed AP endonucleases. These enzymes first recognize AP sites in DNA and then catalyze incision of the DNA backbone, initiating a cascade of events that leads to damage removal and ultimately genome restoration (see Section 5 for more details). AP endonucleases have thus far been divided into two major families designated after the two predominant AP endonucleases of *Escherichia coli*: endonuclease IV (nfo) and exonuclease III (xth) (62).

   To date, mammalian homologs to only *E. coli* exonuclease III have been found, with the two human proteins, Ape1 and Ape2, making up separate subfamilies of the exonuclease III family (Table 1). Whereas it is clear that Ape1 comprises the majority (if not all) of the AP endonuclease activity in mammalian cells, the contribution of Ape2 to DNA repair and AP site processing, in particular, remains in question (63,64). Next, we describe how AP endonucleases target AP sites in DNA and execute strand incision, two key steps in initiating AP damage repair. For a more extensive discussion of the physical properties and various biochemical activities of the prokaryotic and eukaryotic exonuclease III and endonuclease IV counterparts, see Refs. 11, 62, 65, 66.

4. AP SITE RECOGNITION AND PROCESSING

Crystal structures of *E. coli* endonuclease IV, a $\sim$30 kDa Zn^{2+}-dependent AP endonuclease, and human Ape1, a $\sim$35 kDa Mg^{2+}-dependent nuclease, in complex with AP-DNA have been solved (67,68). These x-ray structures provide a detailed molecular picture of how representatives from the two major AP endonuclease families (Table 1) likely recognize and incise AP-DNA.

4.1. Endonuclease IV Family

In the case of *E. coli* endonuclease IV (67), which is a single-domain protein built upon a conserved $\alpha8\beta8$ TIM (originally identified in *t*riose phosphate *iso*merase)

Table 1 Distribution of the Major AP Endonuclease Family Members

Taxonomy	Exonuclease III family		Endonuclease IV family
Eukaryota			
Homo sapiens	Ape1	Ape2	?
Mus musculus	Apex1	Apex2	?
Caenorhabditis elegans	Exo3	?	Apn1
Drosophila melanogaster	Rrp1	—	—
Arabidopsis thaliana	Arp, 3_T29H11_60	4_T19 K4_180	—
Schizosaccharomyces pombe	—	Apn2/Eth1	Apn1
Saccharomyces cerevisiae	—	Apn2/Eth1	Apn1
Archaea			
Methanobacterium thermoautotrophicum	B69126	—	G69001
Bacteria			
Escherichia coli	Exonuclease III (Xth)	—	Endonuclease IV (Nfo)

Note: "—" denotes that a sequence homolog was not detected in the complete genome sequence. "?" denotes that it is currently unknown whether a functional homolog exists (although unlikely).
Source: For additional information, see Refs. 11, 138, 139, and references within.

barrel fold, specificity for AP-DNA is conveyed by five unique DNA recognition loops (the so-called R loops). These loops, which emanate from the C-terminal end of the central β barrel, form the walls of a positively charged crescent-shaped groove that specifically complements the negatively charged phosphate backbone of double-stranded DNA (Fig. 3). As revealed by the endonuclease IV–DNA cocrystal structures (67), three of the protein loops contact the DNA phosphate backbone directly through specific side chain interactions while the other two loops (R2 and R3) both bind the DNA phosphate backbone and provide key base contacts at the AP site. Through these physical contacts, endonuclease IV is able to flip the AP site and the nucleotide opposite out of the duplex and bend the DNA ~90°, substrate conformational changes that are necessary for targeted recognition and ultimately catalysis. Notably, unlike the human Ape1 protein (see in what follows), which maintains a rigid, preformed recognition pocket, the DNA binding surface of endonuclease IV undergoes significant local conformational changes that permit the two recognition loops (R2 and R3) to penetrate the DNA minor groove.

Upon specific complex formation, endonuclease IV cuts the AP site-containing DNA strand immediately 5′ to the lesion. The crystal structures of endonuclease IV bound to either intact AP-DNA or the incised DNA product reveal a trinuclear divalent metal catalytic active site center consisting of three Zn^{2+} ions (67). As seen with other zinc cluster-containing enzymes, most notably P1 nuclease from *Penicillium citrinum* and phospholipase C from *Bacillus cereus* (69–73), these divalent ions appear to facilitate phosphodiester bond cleavage. In particular, Zn1 and Zn2 of endonuclease IV have been proposed to deprotonate a bridging water molecule, creating

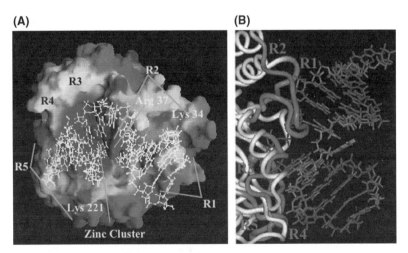

Figure 3 A molecular image of the endonuclease IV–DNA complex. (A) Endonuclease IV–AP DNA cocrystal model with the surface rendered protein and zinc cluster colored according to electrostatic potential. DNA is rendered as a ball/stick model. The 5 R-loops of the TIM-like β-barrel are indicated (Grasp v1.3). These loops are designated as follows: R1, β1–α1 (residues 8–13); R2, β2–α2 (33–45); R3, β3–α3 (70–78); R4, β5–α5 (147–152); and R5, β7–α7 (220–240). Conserved basic residues in loops R2 and R5 are noted, along with the location of the zinc cluster. (B) A closer view of the endonuclease IV–AP DNA interaction, highlighting the R-loops and zinc atoms (Insight II 98.0). Note the severe kink in the DNA substrate and the minor groove penetration by the R2 and R3 loops. *Source*: From Ref. 67. (*See color insert.*)

the active site nucleophile required for hydrolysis (Fig. 4). Moreover, each of the three Zn^{2+} ions neutralizes the negative charge of the abasic 5'-phosphate via direct ligation, rendering the phosphorus atom susceptible to nucleophilic substitution. Upon strand cleavage, Zn3 stabilizes the developing negative charge of the 3'-terminal hydroxyl-leaving group through a direct interaction. Thus, all three metal ions participate in this highly tuned reaction scheme. Interestingly, while TIM barrel fold-containing proteins exist throughout evolution (74,75), functional homologs to *E. coli* endonuclease IV have not been identified in mammals.

4.2. Exonuclease III Family

As seen with endonuclease IV, Ape1 is built upon a conserved protein fold that has acquired substrate specificity via specific structural adaptations, i.e., through the acquisition of unique loop elements (Fig. 5A). Ape1 is a member of the α/β superfamily of phosphohydrolases, which includes exonuclease III, inositol 5'-phosphatases, and sphingomyelinases (76–78). These proteins retain a common four-layered α/β-sandwich fold, yet exhibit highly divergent substrate specificities, ranging from phospholipids, to nucleic acids, to polypeptides. The cocrystal structure of Ape1 bound to

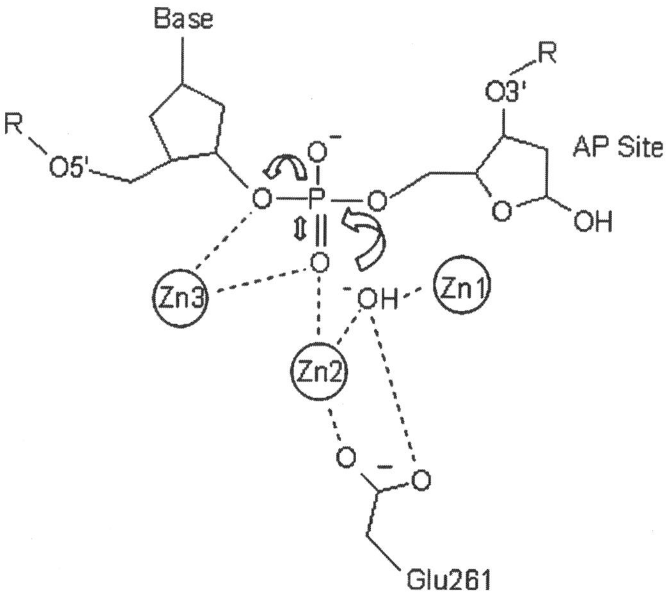

Figure 4 Structure-based reaction mechanism for endonuclease IV. The high-resolution x-ray structures of endonuclease IV in complex with intact and incised DNA suggest a three-metal-ion mechanism for phosphodiester hydrolysis, similar to that of other trinuclear Zn center enzymes (see text). Charge neutralization of the phosphate group, which renders the phosphorus atom susceptible to nucleophilic substitution, is achieved by interaction with all three Zn^{2+} ions. Glu261, which is a Zn2 ligand, may assist in orienting and activating the attacking nucleophile. As the transition state collapses to the reaction product, the stereochemical configuration of the phosphate is inverted and the developing negative charge at O3' is stabilized by interactions with Zn3. The 3' hydroxyl and 5' phosphate oxygen positions observed in the endonuclease IV–DNA complex suggest a mechanism in which the hydrolytic reaction proceeds through a pentaco-ordinate transition state where the unesterified phosphate oxygen that bridges Zn2 and Zn3 remains bound to its cognate metal ions. *Source*: From Ref. 67.

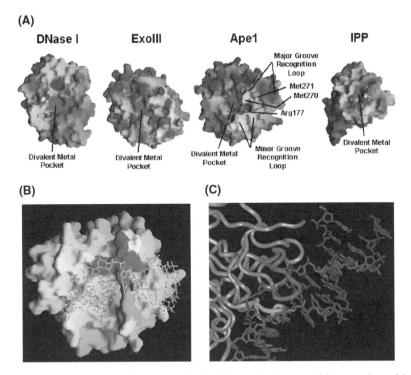

Figure 5 Molecular image of the Ape1 protein alone and in complex with AP-DNA. (A) Grasp surface rendering of four members of the α/β sandwich superfamily of phosphohydrolases (Grasp v1.3). Note that Ape1 (like the homologous exonuclease III protein), compared with DNase I and IPP, has unique loops that form an arch over the substrate-binding pocket (which is identical to the divalent metal binding pocket). In addition, note that exonuclease III and Ape1 have more positive electrostatic potential on their surface than DNase I, perhaps suggesting that exonuclease III and Ape1 are more suited to scan DNA via weak nonspecific ionic interactions with the phosphodiester backbone. (B) A global view of the Ape1–DNA interaction (Grasp v1.3). Major groove (which includes Met270 and Met271) and minor groove (which includes Arg177) loops are colored magenta and cyan, respectively. DNA is shown in yellow/orange. (C) A closer view of the unique Ape1 recognition loops (colored as in panel B) and conserved basic sidechains (noted in panel B) within these loops (Insight II 98.0). Note the kinking of the DNA. *Source*: From Ref. 79. (*See color insert.*)

AP-DNA (68) reveals that site-specific recognition is mediated largely by two unique loop elements, which harbor amino acid residues that penetrate the DNA helix through the minor (Met270 and Met271) and major (Arg177) grooves (Figs. 5B and C). In addition, Hadi et al. (63) showed that the active site "hydrophobic" core, which is found to sequester the abasic lesion in the Ape1–DNA complex (68), is another determinant in the nuclease specificity of the exonuclease III-like proteins. These enzyme–DNA interactions, as well as specific contacts along the DNA phosphate backbone, permit the endonuclease to bend the AP-DNA substrate $\sim35°$, "flip out" the abasic sugar, and kink the DNA helical axis $\sim5\text{Å}$ immediately around the AP site. In total, the crystal data (68), in combination with a series of biochemical studies (reviewed in Ref. 11), suggest that AP site targeting is facilitated via a protein-induced strain mechanism (an energetic process), where the substrate undergoes a significant, as well as distinct, conformational change upon Ape1 recognition. To locate AP sites amongst the largely undamaged DNA, work of Carey and Strauss (79) suggests that Ape1 scans in a quasiprocessive manner (i.e., about 300 nucleotides at a time), presumably probing DNA for regions that display increased flexibility.

Although the matter of substrate recognition by Ape1 appears generally resolved, how the enzyme catalyzes phosphodiester bond cleavage is more uncertain (reviewed in Refs. 11, 66, 80, 81). In the initial papers reporting the crystal structures of unbound exonuclease III and Ape1 [i.e., without DNA (82,83)], a single divalent metal ion was found chelated within the active site of each enzyme by the analogous amino acid residues (principally Asp70 and Glu96 of Ape1; only the human residues in Section 4.2 are listed). Thus, it was proposed that exonuclease III and Ape1 catalyze strand incision via a single metal ion-assisted reaction. In this hydrolytic reaction, the divalent ion was suggested to orient the scissile phosphate, stabilize the pentacovalent transition state intermediate, and/or polarize the target phosphate bond. An active site histidine residue (His309), which forms a catalytic "network" with an aspartate (Asp283) and a water molecule, acts as the general base in the reaction, abstracting a proton from water to generate the hydroxyl ion that carries out nucleophilic attack on the phosphorous atom. A separate active site aspartate (Asp210) was proposed to protonate the O3′ leaving group, based on active site proximity and the biochemical properties of certain site-specific mutants (83,84). However, subsequent crystal structures of Ape1 bound in pseudo-Michaelis and DNA product complexes suggested an alternative reaction scheme, in which His309 acts to orient and polarize the scissile phosphate bond and Asp210 functions as the general base (68). In this reaction format, the single active site divalent metal ion functions to stabilize both the transition state intermediate and the 3′ leaving group.

A potential concern with the initial Ape1 structures was that crystallization was performed under acidic conditions, where the enzyme is generally inactive (85,86). Thus, a crystal form of Ape1 was grown at neutral pH, and this new structure, in the absence of DNA, was found to have two lead ions bound within its active site (86). These findings prompted the proposal of a two metal reaction scheme, similar to that of the DNA PolI Klenow fragment (87). However, it was duly noted that Pb^{2+} is an "unproductive" metal in which Ape1 is unable to carry out strand incision in its presence. In fact, subsequent analysis revealed that lead is inhibitory to Ape1 activity in vitro, raising doubt about the proposed two metal reaction scheme (88).

To more precisely interrogate the two metal reaction format, the pK_a value of Ape1 His309, a likely chelating residue for the second metal ion (86), was determined using NMR spectroscopy. Surprisingly, the pK_a was >9.0 pH units (88). Thus, the reaction scheme for Ape1 does not likely involve a second metal bound directly to the His309–Asp283 pair (Fig. 6). However, the metal may bind to the pair indirectly through a water molecule—as is the case in the crystal model of alkaline phosphatase (89)—but a lack of significant proton or nitrogen chemical shift change in Ape1 upon addition of Mg^{2+} would appear to rule this out (88). Alternatively, His309 may bind directly to Mg^{2+} via a lone pair with partial charge built up on the imidazole ring (90). A partially charged ring, in this case, could lead fortuitously to spectral frequencies nearly identical to a charged protonated histidine. In order to distinguish between these possibilities, future studies, such as direct or indirect spectroscopy of His309 Nε2 nucleus and direct spectroscopy with $^{25}Mg^{2+}$, are underway.

It is important to note that while the mechanisms of recognition and catalysis are generally conserved among the members of each AP endonuclease family (Table 1), the members within each family have acquired subtle, but significant, differences in their substrate specificities. As an example, the exonuclease III protein, in

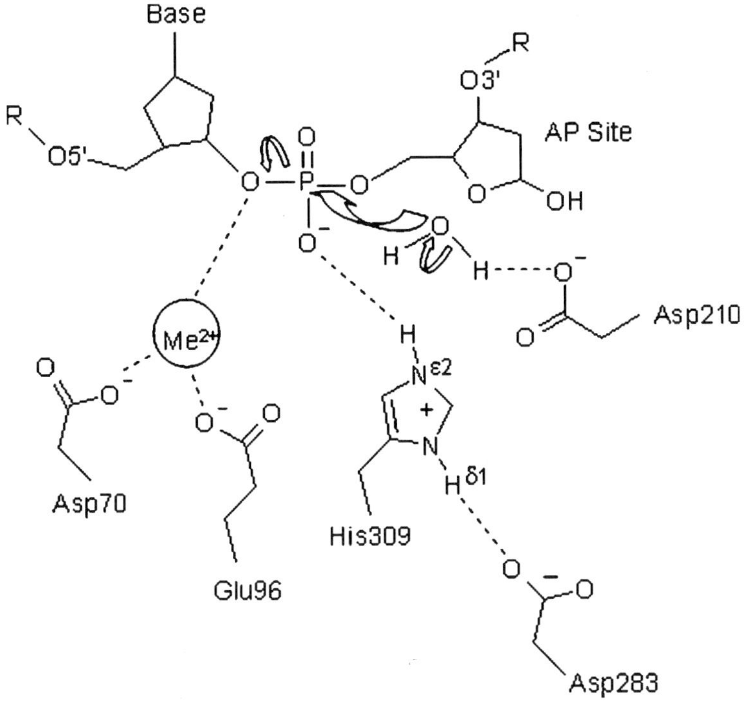

Figure 6 Current model for the Ape1 catalytic reaction. Recent Ape1 substrate and product complexes suggest a reaction mechanism in which the target phosphate is polarized about the scissile bond and attacked by an aspartate-activated hydroxyl nucleophile; the metal ion stabilizes the transition state and the leaving group of the phosphodiester cleavage reaction. In this structure-based reaction scheme, which agrees with known mutagenesis and NMR spectroscopy results (see text), the conserved Asp283/His309 pair makes a direct hydrogen bond from His309 Nε2 to the remaining O1P atom. These Ape1–phosphate contacts orient and polarize the scissile P-O3′ bond with the Asp210 side chain favorably aligned to activate an attacking nucleophile. Positioned by hydrogen bonds with the backbone amide of Asn212 and Asn68 Nδ2 (not shown), the pK_a of the buried Asp210 is probably elevated, and along with the Asp283/His309 phosphate interaction, likely results in the observed maximal Ape1 activity at pH 7.5, with a pH 6.7–9 activity profile. Both Asp308 (not shown) and Asp283 play a role in positioning and establishing the pK_a of His309. The divalent metal ion may facilitate the O3′ leaving group either by direct ligation or through a water in the first hydration shell of an Mg^{2+} ion. Further investigation is currently underway. *Source*: From Refs. 68, 140.

addition to being a major AP endonuclease, is also a prominent 3′–5′ exonuclease in *E. coli* (91,92). Biochemical analysis has revealed that these two functional activities of exonuclease III are present at roughly equal levels. In contrast, the human AP endonuclease counterpart, Ape1, while being an extremely potent AP site incision enzyme (roughly 10-fold better than Exonuclease III), displays a comparatively poor 3′–5′ exonuclease activity, which is present at ∼0.03% of its AP endonuclease activity (93). However, the 3′–5′ exonuclease activity of Ape1 has been found to be influenced by reaction conditions, the terminal base pairing (Ape1 is more effective on 3′ mismatches), the 3′- and 5′-terminal chemistry (e.g., a phosphate vs. a hydroxyl or a 5′-deoxyribose phosphate end), and the surrounding nucleotide sequence context (63,93–97). The protein elements that dictate 3′-nuclease efficiency remain largely unknown, although there is evidence that the "hydrophobic" pocket plays some role in this capacity (63).

5. AP SITE REPAIR IN GENERAL

To place the earlier-mentioned discussions into a broader context, it is important to emphasize that AP endonucleases (and many of the BER proteins) act co-operatively with other members of the BER pathway to complete repair of abasic DNA (Fig. 7). For instance, initiating DNA glycosylases have been argued to exhibit "co-ordination" with AP endonucleases—although the specific results have not always been consistent from report to report (98–105). That is, the high affinity of DNA glycosylases for their AP site product (more specifically, their slow dissociation) has been proposed to allow this enzyme–DNA complex not only to protect

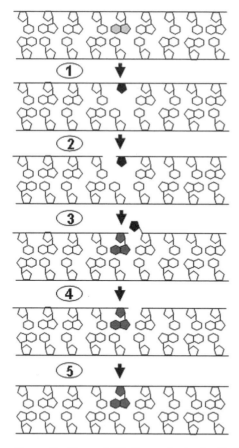

Figure 7 Steps of AP site repair. As detailed in the text, AP sites are formed by spontaneous, chemical-induced or enzyme-catalyzed hydrolytic base loss/excision (step 1). AP sites are then repaired as follows (with the major corrective enzyme of humans indicated in parentheses): step 2, incision 5′ to the abasic residue by an AP endonuclease (Ape1); step 3, gap-filling by a DNA polymerase (Polβ); step 4, excision of the 5′-deoxyribose phosphate residue (Polβ); and step 5, ligation of the remaining single-strand nick (DNA ligase 1 or an Xrcc1/Ligase 3 complex). Additional processing related to single-strand breaks is presented in Chapter 33.

the lesion from spontaneous strand cleavage but also to "recruit" AP endonucleases to the site of the damage. However, Ape1 incision activity is not stimulated by the addition of a DNA glycosylase. Conversely, the base removal activity of many DNA glycosylases has been reported to be stimulated by Ape1. This activation is likely the result of the AP endonuclease promoting dissociation of the DNA glycosylase from its AP site product (enhancing the dissociation constant), thus allowing the enzyme to act on another damage-containing substrate (i.e., turnover). Therefore, it is unclear whether there exists a "true" co-ordination between DNA glycosylases and AP endonucleases, where two proteins co-operate to facilitate or expedite the repair process.

In addition to the purported glycosylase–AP endonuclease co-ordination, in vitro biochemical studies indicate that human Ape1 can stimulate the dRp excision activity of DNA polymerase β (106), presumably through a "passing the baton" mechanism (68,107). In this process, the first enzyme's product is transitioned into a more favorable substrate for the second protein. This co-ordinated product-substrate "hand-off" appears to result in a more efficient BER process. However, in vivo demonstration of the legitimacy of this structure/function-based model is currently lacking, as is supporting biological evidence for many of the reported in vitro BER "interactions". For a more thorough review of the BER process, see Ref. 108 and references within.

Although the predominant repair system for AP sites is the AP endonuclease-driven BER pathway, these lesions can be processed effectively by alternative DNA repair responses in vivo, particularly in the absence of AP endonuclease activity (109–111). In particular, when the major AP endonuclease genes of yeast (*apn1* and *apn2/eth1*) are deleted, these mutant cells still repair AP sites with notable efficiency via one of three distinct pathways (Fig. 8).

As determined through complex genetic analysis in yeast, the first (in no particular order) of the back-up AP site repair pathways is the AP lyase-directed response. In this process, a multifunctional (or bifunctional) DNA glycosylase initiates the repair cascade by cleaving the DNA backbone 3' to the AP site lesion (via its AP lyase function), leaving behind a normal 5'-phosphate group and a nonconventional 3'-abasic fragment (110). In *Saccharomyces cerevisiae*, this 3' end, which is typically processed by the 3'-repair diesterase activity of an AP endonuclease, is then processed (in the absence of *apn1* and *apn2*) by either the Rad1/Rad10 or Mus81/Mms4 nuclease complex (112,113). Upon removal of the 3'-blocking terminus by a 3'-flap endonuclease complex, BER-like gap filling and DNA ligation would presumably proceed normally.

Recently, the Ntg1 and Ntg2 glycosylase/AP lyase proteins of *S. cerevisiae* were found to exhibit similar in vitro catalytic efficiencies (i.e., k_{cat}/K_M) for incising at AP sites as Apn1, the major AP endonuclease of yeast (114). These biochemical data, in combination with the yeast genetic studies mentioned earlier, have prompted the suggestion that 3' AP lyase-directed repair is as biologically critical as the 5' AP endonuclease-driven pathway (Fig. 8). However, whether this concept will hold true in mammals is presently unclear, as many of the repair enzymes and processing mechanisms differ substantially between yeast and humans, particularly in BER (115).

As a second alternative for AP site repair, nucleotide excision repair (NER) has been found to correct AP lesions. In particular, biochemical studies, using purified bacterial or human proteins, have revealed that reconstituted NER systems can bind and/or process AP site damages (116,117). Moreover, yeast studies have found that complex mutant strains, where an NER deficiency has been introduced into cells

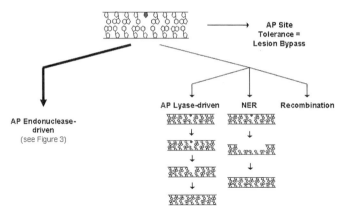

Figure 8 Pathways for repairing AP sites in DNA. In addition to the major AP site repair pathway, which involves incision 5′ to the AP site by an AP endonuclease (see Fig. 7), three alternative AP site corrective pathways exist. As discussed in the text, the first (in no particular order) involves 3′ incision to the AP site by a DNA glycosylase/AP lyase, 3′-terminus "clean-up", gap filling, and ligation. The second is NER, a pathway typically associated with the repair of helix-distorting DNA lesions, including bulky base adducts. Nucleotide excision repair involves recognition, incision of the damage-containing strand on both sides of the lesion, subsequent removal of this DNA fragment, gap filling, and DNA ligation. The last AP site repair pathway entails homologous recombination, a pathway that utilizes an intact (undamaged) homologous chromosome for accurate genetic exchange (not depicted). In some circumstances, to avoid cell death, organisms may engage error-prone polymerases for lesion bypass.

lacking the major AP endonuclease and/or AP lyase activities, exhibit a more pronounced sensitivity to oxidizing and alkylating agents, relative to comparable NER-proficient strains (110,118). Analysis of chromosomal DNA isolated from these complex mutants (via alkaline sucrose gradient sedimentation) suggested that the unrepaired DNA intermediates are indeed abasic sites, with the rate of removal of AP lesions in AP endonuclease/NER-deficient strains being markedly slower than the related single mutants (118). Thus, it is likely that NER offers a back-up system for the repair of AP sites in all organisms. It has been proposed that the neurological abnormalities associated with NER-defective patients (i.e., those suffering from the human disease xeroderma pigmentosum) may arise, at least in part, from the inability of cells to efficiently repair oxidative DNA damage, such as AP sites (119–122).

The third mechanism that copes with unrepaired AP lesions involves homologous recombination. Studies in bacteria and yeast have found that accumulation of AP sites and/or other BER DNA intermediates makes cells dependent on homologous recombinational repair for survival (110,111). Furthermore, upregulation of DNA glycosylases in either yeast or mammalian cell lines leads to an induction of homologous recombination and/or sensitizes DSB repair-deficient cells to the effects of BER intermediates (111,123–125). In *Schizosaccharomyces pombe*, the alkylation sensitivity of strains defective in recombinational repair can be counteracted by deleting the 3-methyladenine DNA glycosylase gene, suggesting that certain BER intermediates (such as abasic sites and subsequent strand break products) can be channeled into recombinational repair (126). Which precise DNA intermediate(s) is most recombinogenic is still under active investigation.

In addition to the three corrective processes described earlier (i.e., AP lyase-directed repair, NER, or recombination; Fig. 8), which typically operate to retain genetic integrity, AP sites can be "tolerated" by invoking an error-prone (mutagenic) response (see Section 1 for further discussion) (110,111). This process, which incorporates a translesion DNA polymerase, sacrifices fidelity in favor of survival of the cell by permitting mutagenic bypass synthesis (see Chapter 24).

Finally, our current understanding of the biology associated with the mammalian Ape1 protein is briefly discussed. Because a genetically defined mutant cell has yet to be created, the precise quantitative contributions of the mammalian enzyme remain unclear. Nonetheless, antisense and mouse knockout studies have revealed some important features of this protein. In particular, targeted antisense experiments in mammalian cell lines have found that reduced Ape1 levels associate with increased cellular sensitivity to the monofunctional alkylating agent MMS and several oxidizing agents (such as hydrogen peroxide, menadione, paraquat, and ionizing radiation) but not to ultraviolet irradiation, which creates bulky DNA adducts typically repaired by NER (127–129). Bacteria or yeast deficient in AP endonuclease function exhibit a parallel DNA-damaging agent hypersensitivity profile, with alkylating agent sensitivity being the most pronounced and the most consistently observed (130–132). These results are in line with the notion that AP endonucleases function primarily as part of a pathway that repairs oxidized and alkylated bases (Fig. 7), as well as various DNA intermediates (e.g., AP sites and single-strand breaks) formed either directly or during the BER process. It is noteworthy that the severity of the various DNA-damaging agent sensitivities differs measurably among the different organisms, likely due to their disparate AP endonuclease profiles (Table 1) and their differing compensatory responses.

Null mice are inviable, indicating an essential role for the Ape1 protein in embryonic development (133,134). Although heterozygous animals appear to grow normally, these animals exhibit increased susceptibility to oxidative stress (as revealed by elevated oxidative stress markers), as well as potentially reduced survival and increased carcinogenesis (135). These observations suggest that haploinsufficiency of Ape1 can play a role in disease susceptibility, most likely in an exposure-dependent manner. Notably, *S. cerevisiae* strains devoid in the major AP endonuclease gene *apn1* display a marked increase in spontaneous mutagenesis, further arguing for a role of AP endonucleases in maintaining genetic integrity and averting disease-related outcomes (132). However, to date, individuals harboring polymorphic alleles that encode dramatically reduced-function Ape1 molecules have not been identified in the human population (136,137). Thus, whether Ape1 is tied to human diseases, namely cancer, neurodegeneration, or premature aging phenotypes, remains unclear.

ACKNOWLEDGMENTS

This chapter was written in a personal capacity and does not represent the opinions of the NIH, DHHS, or the Federal Government. The authors thank Drs. Bevin Engelward (Massachusetts Institute of Technology), Carlos de los Santos (State University of New York, Stony Brook), and Phyllis Strauss (Northeastern University) for helpful comments on this chapter and Dr. Daniel Barsky (Lawrence Livermore National Laboratory) for assistance with the figures. This effort was supported by the Biological and Environmental Research Program, U.S. Department of Energy (Interagency Agreement No. DE-AI02–02ER63424 to DMWIII).

REFERENCES

1. Friedberg EC. Repair DNA. 1st ed. New York: W.H. Freeman, 1985.
2. Roger M, Hotchkiss R. Selective heat inactivation of pneumococcal transforming deoxyribonucleate. Proc Natl Acad Sci USA 1961; 47:653–669.
3. Greer S, Zamenhof S. Studies on depurination of DNA by heat. J Mol Biol 1962; 4: 123–141.
4. Zamenhof S, Greer S. Heat as an agent producing high frequency of mutations and unstable genes in Escherichia coli. Nature 1958; 182:611–613.
5. Lindahl T, Nyberg B. Rate of depurination of native deoxyribonucleic acid. Biochemistry 1972; 11:3610–3618.
6. Lindahl T, Karlstrom O. Heat-induced depyrimidination of deoxyribonucleic acid in neutral solution. Biochemistry 1973; 12:5151–5154.
7. Nakamura J, Walker VE, Upton PB, Chiang SY, Kow YW, Swenberg JA. Highly sensitive apurinic/apyrimidinic site assay can detect spontaneous and chemically induced depurination under physiological conditions. Cancer Res 1998; 58:222–225.
8. Loeb LA, Preston BD. Mutagenesis by apurinic/apyrimidinic sites. Annu Rev Genet 1986; 20:201–230.
9. Lindahl T. Instability and decay of the primary structure of DNA. Nature 1993; 362:709–715.
10. Atamna H, Cheung I, Ames BN. A method for detecting abasic sites in living cells: age-dependent changes in base excision repair. Proc Natl Acad Sci USA 2000; 97: 686–691.
11. Wilson III DM, Barsky D. The major human abasic endonuclease: formation, consequences and repair of abasic lesions in DNA. Mutat Res 2001; 485:283–307.
12. Gonzalez M, Woodgate R. The "tale" of UmuD and its role in SOS mutagenesis. Bioessays 2002; 24:141–148.
13. Goodman MF. Error-prone repair DNA polymerases in prokaryotes and eukaryotes. Annu Rev Biochem 2002; 71:17–50.
14. Sutton MD, Smith BT, Godoy VG, Walker GC. The SOS response: recent insights into umuDC-dependent mutagenesis and DNA damage tolerance. Annu Rev Genet 2000; 34:479–497.
15. Pham P, Rangarajan S, Woodgate R, Goodman MF. Roles of DNA polymerases V and II in SOS-induced error-prone and error-free repair in Escherichia coli. Proc Natl Acad Sci USA 2001; 98:8350–8354.
16. Strauss BS. The "A" rule revisited: polymerases as determinants of mutational specificity. DNA Repair (Amst) 2002; 1:125–135.
17. Friedberg EC, Wagner R, Radman M. Specialized DNA polymerases, cellular survival, and the genesis of mutations. Science 2002; 296:1627–1630.
18. Klinedinst DK, Drinkwater NR. Mutagenesis by apurinic sites in normal and ataxia telangiectasia human lymphoblastoid cells. Mol Carcinog 1992; 6:32–42.
19. Kamiya H, Suzuki M, Komatsu Y, Miura H, Kikuchi K, Sakaguchi T, Murata N, Masutani C, Hanaoka F, Ohtsuka E. An abasic site analogue activates a c-Ha-ras gene by a point mutation at modified and adjacent positions. Nucleic Acids Res 1992; 20:4409–4415.
20. Cabral Neto JB, Cabral RE, Margot A, Le Page F, Sarasin A, Gentil A. Coding properties of a unique apurinic/apyrimidinic site replicated in mammalian cells. J Mol Biol 1994; 240:416–420.
21. Takeshita M, Eisenberg W. Mechanism of mutation on DNA templates containing synthetic abasic sites: study with a double strand vector. Nucleic Acids Res 1994; 22: 1897–1902.
22. Avkin S, Adar S, Blander G, Livneh Z. Quantitative measurement of translesion replication in human cells: evidence for bypass of abasic sites by a replicative DNA polymerase. Proc Natl Acad Sci USA 2002; 99:3764–3769.

23. Kunkel TA, Schaaper RM, Loeb LA. Depurination-induced infidelity of deoxyribonucleic acid synthesis with purified deoxyribonucleic acid replication proteins in vitro. Biochemistry 1983; 22:2378–2384.

24. Takeshita M, Chang CN, Johnson F, Will S, Grollman AP. Oligodeoxynucleotides containing synthetic abasic sites. Model substrates for DNA polymerases and apurinic/apyrimidinic endonucleases. J Biol Chem 1987; 262:10171–10179.

25. Efrati E, Tocco G, Eritja R, Wilson SH, Goodman MF. Abasic translesion synthesis by DNA polymerase beta violates the "A-rule". Novel types of nucleotide incorporation by human DNA polymerase beta at an abasic lesion in different sequence contexts. J Biol Chem 1997; 272:2559–2569.

26. Mozzherin DJ, Shibutani S, Tan CK, Downey KM, Fisher PA. Proliferating cell nuclear antigen promotes DNA synthesis past template lesions by mammalian DNA polymerase delta. Proc Natl Acad Sci USA 1997; 94:6126–6131.

27. Daube SS, Tomer G, Livneh Z. Translesion replication by DNA polymerase delta depends on processivity accessory proteins and differs in specificity from DNA polymerase beta. Biochemistry 2000; 39:348–355.

28. Johnson RE, Prakash S, Prakash L. The human DINB1 gene encodes the DNA polymerase Poltheta. Proc Natl Acad Sci USA 2000; 97:3838–3843.

29. Masutani C, Kusumoto R, Iwai S, Hanaoka F. Mechanisms of accurate translesion synthesis by human DNA polymerase eta. EMBO J 2000; 19:3100–3109.

30. Ohashi E, Ogi T, Kusumoto R, Iwai S, Masutani C, Hanaoka F, Ohmori H. Error-prone bypass of certain DNA lesions by the human DNA polymerase kappa. Genes Dev 2000; 14:1589–1594.

31. Zhang Y, Yuan F, Wu X, Rechkoblit O, Taylor JS, Geacintov NE, Wang Z. Error-prone lesion bypass by human DNA polymerase eta. Nucleic Acids Res 2000; 28:4717–4724.

32. Zhang Y, Yuan F, Wu X, Wang M, Rechkoblit O, Taylor JS, Geacintov NE, Wang Z. Error-free and error-prone lesion bypass by human DNA polymerase kappa in vitro. Nucleic Acids Res 2000; 28:4138–4146.

33. Zhang Y, Yuan F, Wu X, Wang Z. Preferential incorporation of G opposite template T by the low-fidelity human DNA polymerase iota. Mol Cell Biol 2000; 20:7099–7108.

34. Zhang Y, Yuan F, Wu X, Taylor JS, Wang Z. Response of human DNA polymerase iota to DNA lesions. Nucleic Acids Res 2001; 29:928–935.

35. Haracska L, Kondratick CM, Unk I, Prakash S, Prakash L. Interaction with PCNA is essential for yeast DNA polymerase eta function. Mol Cell 2001; 8:407–415.

36. Haracska L, Unk I, Johnson RE, Phillips BB, Hurwitz J, Prakash L, Prakash S. Stimulation of DNA synthesis activity of human DNA polymerase kappa by PCNA. Mol Cell Biol 2002; 22:784–791.

37. Maga G, Villani G, Ramadan K, Shevelev I, Tanguy LG, Blanco L, Blanca G, Spadari S, Hubscher U. Human DNA polymerase lambda functionally and physically interacts with proliferating cell nuclear antigen in normal and translesion DNA synthesis. J Biol Chem 2002; 277:48434–48440.

38. Maga G, Shevelev I, Ramadan K, Spadari S, Hubscher U. DNA polymerase theta purified from human cells is a high-fidelity enzyme. J Mol Biol 2002; 319:359–369.

39. Zhang Y, Wu X, Guo D, Rechkoblit O, Taylor JS, Geacintov NE, Wang Z. Lesion bypass activities of human DNA polymerase mu. J Biol Chem 2002; 277:44582–44587.

40. Vesnaver G, Chang CN, Eisenberg M, Grollman AP, Breslauer KJ. Influence of abasic and anucleosidic sites on the stability, conformation, and melting behavior of a DNA duplex: correlations of thermodynamic and structural data. Proc Natl Acad Sci USA 1989; 86:3614–3618.

41. Cuniasse P, Sowers LC, Eritja R, Kaplan B, Goodman MF, Cognet JA, LeBret M, Guschlbauer W, Fazakerley GV. An abasic site in DNA. Solution conformation determined by proton NMR and molecular mechanics calculations. Nucleic Acids Res 1987; 15:8003–8022.

42. Kalnik MW, Chang CN, Grollman AP, Patel DJ. NMR studies of abasic sites in DNA duplexes: deoxyadenosine stacks into the helix opposite the cyclic analogue of 2-deoxyribose. Biochemistry 1988; 27:924–931.

43. Withka JM, Wilde JA, Bolton PH, Mazumder A, Gerlt JA. Characterization of conformational features of DNA heteroduplexes containing aldehydic abasic sites. Biochemistry 1991; 30:9931–9940.

44. Goljer I, Kumar S, Bolton PH. Refined solution structure of a DNA heteroduplex containing an aldehydic abasic site. J Biol Chem 1995; 270:22980–22987.

45. Stivers JT. 2-Aminopurine fluorescence studies of base stacking interactions at abasic sites in DNA: metal-ion and base sequence effects. Nucleic Acids Res 1998; 26: 3837–3844.

46. Ayadi L, Coulombeau C, Lavery R. The impact of abasic sites on DNA flexibility. J Biomol Struct Dyn 2000; 17:645–653.

47. Barsky D, Foloppe N, Ahmadia S, Wilson DM III, MacKerell AD Jr. New insights into the structure of abasic DNA from molecular dynamics simulations. Nucleic Acids Res 2000; 28:2613–2626.

48. Jourdan M, Garcia J, Defrancq E, Kotera M, Lhomme J. 2'-deoxyribonolactone lesion in DNA: refined solution structure determined by nuclear magnetic resonance and molecular modeling. Biochemistry 1999; 38:3985–3995.

49. Hoehn ST, Turner CJ, Stubbe J. Solution structure of an oligonucleotide containing an abasic site: evidence for an unusual deoxyribose conformation. Nucleic Acids Res 2001; 29:3413–3423.

50. Goljer I, Withka JM, Kao JY, Bolton PH. Effects of the presence of an aldehydic abasic site on the thermal stability and rates of helix opening and closing of duplex DNA. Biochemistry 1992; 31:11614–11619.

51. Singh MP, Hill GC, Peoc'h D, Rayner B, Imbach JL, Lown JW. High-field NMR and restrained molecular modeling studies on a DNA heteroduplex containing a modified apurinic abasic site in the form of covalently linked 9-aminoellipticine. Biochemistry 1994; 33:10271–10285.

52. Lin Z, Hung KN, Grollman AP, de los SC. Solution structure of duplex DNA containing an extrahelical abasic site analog determined by NMR spectroscopy and molecular dynamics. Nucleic Acids Res 1998; 26:2385–2391.

53. Cuniasse P, Fazakerley GV, Guschlbauer W, Kaplan BE, Sowers LC. The abasic site as a challenge to DNA polymerase. A nuclear magnetic resonance study of G, C and T opposite a model abasic site. J Mol Biol 1990; 213:303–314.

54. Coppel Y, Berthet N, Coulombeau C, Garcia J, Lhomme J. Solution conformation of an abasic DNA undecamer duplex d(CGCACXCACGC) x d(GCGTGTGTGCG): the unpaired thymine stacks inside the helix. Biochemistry 1997; 36:4817–4830.

55. Beger RD, Bolton PH. Structures of apurinic and apyrimidinic sites in duplex DNAs. J Biol Chem 1998; 273:15565–15573.

56. Gelfand CA, Plum GE, Grollman AP, Johnson F, Breslauer KJ. Thermodynamic consequences of an abasic lesion in duplex DNA are strongly dependent on base sequence. Biochemistry 1998; 37:7321–7327.

57. Sagi J, Guliaev AB, Singer B. 15-mer DNA duplexes containing an abasic site are thermodynamically more stable with adjacent purines than with pyrimidines. Biochemistry 2001; 40:3859–3868.

58. Rachofsky EL, Seibert E, Stivers JT, Osman R, Ross JB. Conformation and dynamics of abasic sites in DNA investigated by time-resolved fluorescence of 2-aminopurine. Biochemistry 2001; 40:957–967.

59. Wang KY, Parker SA, Goljer I, Bolton PH. Solution structure of a duplex DNA with an abasic site in a dA tract. Biochemistry 1997; 36:11629–11639.

60. Sagi J, Hang B, Singer B. Sequence-dependent repair of synthetic AP sites in 15-mer and 35-mer oligonucleotides: role of thermodynamic stability imposed by neighbor bases. Chem Res Toxicol 1999; 12:917–923.

61. Otsuka C, Sanadai S, Hata Y, Okuto H, Noskov VN, Loakes D, Negishi K. Difference between deoxyribose- and tetrahydrofuran-type abasic sites in the in vivo mutagenic responses in yeast. Nucleic Acids Res 2002; 30:5129–5135.

62. Demple B, Harrison L. Repair of oxidative damage to DNA: enzymology and biology. Annu Rev Biochem 1994; 63:915–948.

63. Hadi MZ, Ginalski K, Nguyen LH, Wilson DM III. Determinants in nuclease specificity of Ape1 and Ape2, human homologues of Escherichia coli exonuclease III. J Mol Biol 2002; 316:853–866.

64. Ide Y, Tsuchimoto D, Tominaga Y, Iwamoto Y, Nakabeppu YCharacterization of the genomic structure and expression of the mouse Apex2 geneGenomics 2003; 81: 47–57.

65. Wilson DM III, Engelward BP, Samson L. In: Nickoloff JA, Heokstra MF, eds. DNA Damage and Repair, Volume 1: DNA Repair in Prokaryotes and Lower Eukaryotes. Totowa, NJ: Humana Press Inc., 1998:29–64.

66. Mol CD, Hosfield DJ, Tainer JA. Abasic site recognition by two apurinic/apyrimidinic endonuclease families in DNA base excision repair: the $3'$ ends justify the means. Mutat Res 2000; 460:211–229.

67. Hosfield DJ, Guan Y, Haas BJ, Cunningham RP, Tainer JA. Structure of the DNA repair enzyme endonuclease IV and its DNA complex: double-nucleotide flipping at abasic sites and three-metal-ion catalysis. Cell 1999; 98:397–408.

68. Mol CD, Izumi T, Mitra S, Tainer JA. DNA-bound structures and mutants reveal abasic DNA binding by APE1 and DNA repair coordination. Nature 2000; 403: 451–456.

69. Hough E, Hansen LK, Birknes B, Jynge K, Hansen S, Hordvik A, Little C, Dodson E, Derewenda Z. High-resolution (1.5 A) crystal structure of phospholipase C from Bacillus cereus. Nature 1989; 338:357–360.

70. Volbeda A, Lahm A, Sakiyama F, Suck D. Crystal structure of Penicillium citrinum P1 nuclease at 2.8 A resolution. EMBO J 1991; 10:1607–1618.

71. Hansen S, Hansen LK, Hough E. Crystal structures of phosphate, iodide and iodate-inhibited phospholipase C from Bacillus cereus and structural investigations of the binding of reaction products and a substrate analogue. J Mol Biol 1992; 225: 543–549.

72. Hansen S, Hansen LK, Hough E. The crystal structure of tris-inhibited phospholipase C from Bacillus cereus at 1.9 A resolution. The nature of the metal ion in site 2. J Mol Biol 1993; 231:870–876.

73. Romier C, Dominguez R, Lahm A, Dahl O, Suck D. Recognition of single-stranded DNA by nuclease P1: high resolution crystal structures of complexes with substrate analogs. Proteins 1998; 32:414–424.

74. Aravind L, Walker DR, Koonin EV. Conserved domains in DNA repair proteins and evolution of repair systems. Nucleic Acids Res 1999; 27:1223–1242.

75. Nagano N, Orengo CA, Thornton JM. One fold with many functions: the evolutionary relationships between TIM barrel families based on their sequences, structures and functions. J Mol Biol 2002; 321:741–765.

76. Dlakic M. Functionally unrelated signalling proteins contain a fold similar to Mg2+-dependent endonucleases. Trends Biochem Sci 2000; 25:272–273.

77. Whisstock JC, Romero S, Gurung R, Nandurkar H, Ooms LM, Bottomley SP, Mitchell CA. The inositol polyphosphate 5-phosphatases and the apurinic/apyrimidinic base excision repair endonucleases share a common mechanism for catalysis. J Biol Chem 2000; 275:37055–37061.

78. Tsujishita Y, Guo S, Stolz LE, York JD, Hurley JH. Specificity determinants in phosphoinositide dephosphorylation: crystal structure of an archetypal inositol polyphosphate 5-phosphatase. Cell 2001; 105:379–389.

79. Carey DC, Strauss PR. Human apurinic/apyrimidinic endonuclease is processive. Biochemistry 1999; 38:16553–16560.

80. Strauss PR, O'Regan NE. In: Nickoloff JA, Hoekstra MF, eds. DNA Damage and Repair, Volume III: Advances from Phage to Humans. Totowa, NJ: Humana Press Inc, 2001:43–85.

81. Hickson ID, Gorman MA, Freemont PS. In: Nickoloff JA, Hoekstra MF, eds. DNA Damage and Repair, Volume III: Advances from Phage to Humans. Totowa, NJ: Humana Press Inc., 2001:87–105.

82. Mol CD, Kuo CF, Thayer MM, Cunningham RP, Tainer JA. Structure and function of the multifunctional DNA-repair enzyme exonuclease III. Nature 1995; 374:381–386.

83. Gorman MA, Morera S, Rothwell DG, de La FE, Mol CD, Tainer JA, Hickson ID, Freemont PS. The crystal structure of the human DNA repair endonuclease HAP1 suggests the recognition of extra-helical deoxyribose at DNA abasic sites. EMBO J 1997; 16:6548–6558.

84. Erzberger JP, Wilson DM III. The role of Mg2+ and specific amino acid residues in the catalytic reaction of the major human abasic endonuclease: new insights from EDTA-resistant incision of acyclic abasic site analogs and site-directed mutagenesis. J Mol Biol 1999; 290:447–457.

85. Kane CM, Linn S. Purification and characterization of an apurinic/apyrimidinic endonuclease from HeLa cells. J Biol Chem 1981; 256:3405–3414.

86. Beernink PT, Segelke BW, Hadi MZ, Erzberger JP, Wilson DM III, Rupp B. Two divalent metal ions in the active site of a new crystal form of human apurinic/apyrimidinic endonuclease, Ape1: implications for the catalytic mechanism. J Mol Biol 2001; 307:1023–1034.

87. Beese LS, Steitz TA. Structural basis for the 3'-5' exonuclease activity of Escherichia coli DNA polymerase I: a two metal ion mechanism. EMBO J 1991; 10:25–33.

88. Lowry DF, Hoyt DW, Khazi FA, Bagu J, Lindsey AG, Wilson DM III. Investigation of the role of the histidine-aspartate pair in the human exonuclease III-like abasic endonuclease, Ape1. J Mol Biol 2003; 329:311–322.

89. Stec B, Holtz KM, Kantrowitz ER. A revised mechanism for the alkaline phosphatase reaction involving three metal ions. J Mol Biol 2000; 299:1303–1311.

90. Alia JM, Soede-Huijbregts C, Baldus M, Raap J, Lugtenburg J, Gast P, van Gorkom HJ, Hoff AJ, de Groot HJM. Ultrahigh field MAS NMR dipolar correlation spectroscopy of the histidine residues in light-harvesting complex II from photosynthetic bacteria reveals partial internal charge transfer in the B850/His complex. J Amer Chem Soc 2001; 123:4803–4809.

91. Weiss B. Endonuclease II of Escherichia coli is exonuclease III. J Biol Chem 1976; 251:1896–1901.

92. Demple B, Johnson A, Fung D. Exonuclease III and endonuclease IV remove 3' blocks from DNA synthesis primers in H2O2-damaged Escherichia coli. Proc Natl Acad Sci USA 1986; 83:7731–7735.

93. Wilson DM III, Takeshita M, Grollman AP, Demple B. Incision activity of human apurinic endonuclease (Ape) at abasic site analogs in DNA. J Biol Chem 1995; 270: 16002–16007.

94. Seki S, Hatsushika M, Watanabe S, Akiyama K, Nagao K, Tsutsui K. cDNA cloning, sequencing, expression and possible domain structure of human APEX nuclease homologous to Escherichia coli exonuclease III. Biochim Biophys Acta 1992; 1131:287–299.

95. Chou KM, Cheng YC. An exonucleolytic activity of human apurinic/apyrimidinic endonuclease on 3' mispaired DNA. Nature 2002; 415:655–659.

96. Wong D, DeMott MS, Demple B. Modulation of the 3'->5'-Exonuclease Activity of Human Apurinic Endonuclease (Ape1) by Its 5'-incised Abasic DNA Product. J Biol Chem 2003; 278:36242–36249.

97. Wilson DM III. Properties of and substrate determinants for the exonuclease activity of human apurinic endonuclease Ape1. J Mol Biol 2003; 330:1027–1037.

98. Parikh SS, Mol CD, Slupphaug G, Bharati S, Krokan HE, Tainer JA. Base excision repair initiation revealed by crystal structures and binding kinetics of human uracil-DNA glycosylase with DNA. EMBO J 1998; 17:5214–5226.

99. Waters TR, Gallinari P, Jiricny J, Swann PF. Human thymine DNA glycosylase binds to apurinic sites in DNA but is displaced by human apurinic endonuclease 1. J Biol Chem 1999; 274:67–74.

100. Privezentzev CV, Saparbaev M, Laval J. The HAP1 protein stimulates the turnover of human mismatch-specific thymine-DNA-glycosylase to process 3,N(4)-ethenocytosine residues. Mutat Res 480–481; 2001:277–284.

101. Hill JW, Hazra TK, Izumi T, Mitra S. Stimulation of human 8-oxoguanine-DNA glycosylase by AP-endonuclease: potential coordination of the initial steps in base excision repair. Nucleic Acids Res 2001; 29:430–438.

102. Vidal AE, Hickson ID, Boiteux S, Radicella JP. Mechanism of stimulation of the DNA glycosylase activity of hOGG1 by the major human AP endonuclease: bypass of the AP lyase activity step. Nucleic Acids Res 2001; 29:1285–1292.

103. Saitoh T, Shinmura K, Yamaguchi S, Tani M, Seki S, Murakami H, Nojima Y, Yokota J. Enhancement of OGG1 protein AP lyase activity by increase of APEX protein. Mutat Res 2001; 486:31–40.

104. Yang H, Clendenin WM, Wong D, Demple B, Slupska MM, Chiang JH, Miller JH. Enhanced activity of adenine-DNA glycosylase (Myh) by apurinic/apyrimidinic endonuclease (Ape1) in mammalian base excision repair of an A/GO mismatch. Nucleic Acids Res 2001; 29:743–752.

105. Marenstein DR, Chan MK, Altamirano A, Basu AK, Boorstein RJ, Cunningham RP, Teebor GW. Substrate specificity of human endonuclease III (hNTH1). Effect of human APE1 on hNTH1 activity. J Biol Chem 2003; 278:9005–9012.

106. Bennett RA, Wilson DM III, Wong D, Demple B. Interaction of human apurinic endonuclease and DNA polymerase beta in the base excision repair pathway. Proc Natl Acad Sci USA 1997; 94:7166–7169.

107. Wilson SH, Kunkel TA. Passing the baton in base excision repair. Nat Struct Biol 2000; 7:176–178.

108. Wilson DM III, Sofinowski TM, McNeill DR. Repair mechanisms for oxidative DNA damage. Front Biosci 2003; 8:d963–d981.

109. Xiao W, Chow BL. Synergism between yeast nucleotide and base excision repair pathways in the protection against DNA methylation damage. Curr Genet 1998; 33:92–99.

110. Swanson RL, Morey NJ, Doetsch PW, Jinks-Robertson S. Overlapping specificities of base excision repair, nucleotide excision repair, recombination, and translesion synthesis pathways for DNA base damage in Saccharomyces cerevisiae. Mol Cell Biol 1999; 19: 2929–2935.

111. Otterlei M, Kavli B, Standal R, Skjelbred C, Bharati S, Krokan HE. Repair of chromosomal abasic sites in vivo involves at least three different repair pathways. EMBO J 2000; 19:5542–5551.

112. Guillet M, Boiteux S. Endogenous DNA abasic sites cause cell death in the absence of Apn1, Apn2 and Rad1/Rad10 in Saccharomyces cerevisiae. EMBO J 2002; 21: 2833–2841.

113. Karumbati AS, Deshpande RA, Jilani A, Vance JR, Ramotar D, Wilson TE. The role of yeast DNA 3'-phosphatase Tpp1 and rad1/Rad10 endonuclease in processing spontaneous and induced base lesions. J Biol Chem 2003; 278:31434–31443.

114. Meadows KL, Song B, Doetsch PW. Characterization of AP lyase activities of Saccharomyces cerevisiae Ntg1p and Ntg2p: implications for biological function. Nucleic Acids Res 2003; 31:5560–5567.

115. Kelley MR, Kow YW, Wilson DM III. Disparity between DNA base excision repair in yeast and mammals: translational implications. Cancer Res 2003; 63:549–554.

116. Snowden A, Kow YW, Van Houten B. Damage repertoire of the Escherichia coli UvrABC nuclease complex includes abasic sites, base-damage analogues, and lesions containing adjacent $5'$ or $3'$ nicks. Biochemistry 1990; 29:7251–7259.

117. Huang JC, Hsu DS, Kazantsev A, Sancar A. Substrate spectrum of human excinuclease: repair of abasic sites, methylated bases, mismatches, and bulky adducts. Proc Natl Acad Sci USA 1994; 91:12213–12217.

118. Torres-Ramos CA, Johnson RE, Prakash L, Prakash S. Evidence for the involvement of nucleotide excision repair in the removal of abasic sites in yeast. Mol Cell Biol 2000; 20: 3522–3528.

119. Reardon JT, Bessho T, Kung HC, Bolton PH, Sancar A. In vitro repair of oxidative DNA damage by human nucleotide excision repair system: possible explanation for neurodegeneration in xeroderma pigmentosum patients. Proc Natl Acad Sci USA 1997; 94:9463–9468.

120. Lipinski LJ, Hoehr N, Mazur SJ, Dianov GL, Senturker S, Dizdaroglu M, Bohr VA. Repair of oxidative DNA base lesions induced by fluorescent light is defective in xeroderma pigmentosum group A cells. Nucleic Acids Res 1999; 27:3153–3158.

121. Kuraoka I, Bender C, Romieu A, Cadet J, Wood RD, Lindahl T. Removal of oxygen free-radical-induced 5',8-purine cyclodeoxynucleosides from DNA by the nucleotide excision-repair pathway in human cells. Proc Natl Acad Sci USA 2000; 97:3832–3837.

122. Brooks PJ, Wise DS, Berry DA, Kosmoski JV, Smerdon MJ, Somers RL, Mackie H, Spoonde AY, Ackerman EJ, Coleman K, Tarone RE, Robbins JH. The oxidative DNA lesion 8,5'-(S)-cyclo-2'-deoxyadenosine is repaired by the nucleotide excision repair pathway and blocks gene expression in mammalian cells. J Biol Chem 2000; 275:22355–22362.

123. Coquerelle T, Dosch J, Kaina B. Overexpression of N-methylpurine-DNA glycosylase in Chinese hamster ovary cells renders them more sensitive to the production of chromosomal aberrations by methylating agents—a case of imbalanced DNA repair. Mutat Res 1995; 336:9–17.

124. Posnick LM, Samson LD. Imbalanced base excision repair increases spontaneous mutation and alkylation sensitivity in Escherichia coli. J Bacteriol 1999; 181:6763–6771.

125. Hendricks CA, Razlog M, Matsuguchi T, Goyal A, Brock AL, Engelward BP. The S. cerevisiae Mag1 3-methyladenine DNA glycosylase modulates susceptibility to homologous recombination. DNA Repair (Amst) 2002; 1:645–659.

126. Memisoglu A, Samson L. Contribution of base excision repair, nucleotide excision repair, and DNA recombination to alkylation resistance of the fission yeast Schizosaccharomyces pombe. J Bacteriol 2000; 182:2104–2112.

127. Chen DS, Olkowski ZL. Biological responses of human apurinic endonuclease to radiation-induced DNA damage. Ann N Y Acad Sci 1994; 726:306–308.

128. Ono Y, Furuta T, Ohmoto T, Akiyama K, Seki S. Stable expression in rat glioma cells of sense and antisense nucleic acids to a human multifunctional DNA repair enzyme, APEX nuclease. Mutat Res 1994; 315:55–63.

129. Walker LJ, Craig RB, Harris AL, Hickson ID. A role for the human DNA repair enzyme HAP1 in cellular protection against DNA damaging agents and hypoxic stress. Nucleic Acids Res 1994; 22:4884–4889.

130. Demple B, Halbrook J, Linn S. Escherichia coli xth mutants are hypersensitive to hydrogen peroxide. J Bacteriol 1983; 153:1079–1082.

131. Cunningham RP, Saporito SM, Spitzer SG, Weiss B. Endonuclease IV (nfo) mutant of Escherichia coli. J Bacteriol 1986; 168:1120–1127.

132. Ramotar D, Popoff SC, Gralla EB, Demple B. Cellular role of yeast Apn1 apurinic endonuclease/3'-diesterase: repair of oxidative and alkylation DNA damage and control of spontaneous mutation. Mol Cell Biol 1991; 11:4537–4544.

133. Xanthoudakis S, Smeyne RJ, Wallace JD, Curran T. The redox/DNA repair protein, Ref-1, is essential for early embryonic development in mice. Proc Natl Acad Sci USA 1996; 93:8919–8923.

134. Ludwig DL, MacInnes MA, Takiguchi Y, Purtymun PE, Henrie M, Flannery M, Meneses J, Pedersen RA, Chen DJ. A murine AP-endonuclease gene-targeted deficiency with post-implantation embryonic progression and ionizing radiation sensitivity. Mutat Res 1998; 409:17–29.

135. Meira LB, Devaraj S, Kisby GE, Burns DK, Daniel RL, Hammer RE, Grundy S, Jialal I, Friedberg EC. Heterozygosity for the mouse Apex gene results in phenotypes associated with oxidative stress. Cancer Res 2001; 61:5552–5557.

136. Hadi MZ, Coleman MA, Fidelis K, Mohrenweiser HW, Wilson DM III. Functional characterization of Ape1 variants identified in the human population. Nucleic Acids Res 2000; 28:3871–3879.

137. Mohrenweiser HW, Wilson DM III, Jones IM. Challenges and complexities in estimating both the functional impact and the disease risk associated with the extensive genetic variation in human DNA repair genes. Mutat Res 2003; 526:93–125.

138. Ramotar D. The apurinic-apyrimidinic endonuclease IV family of DNA repair enzymes. Biochem Cell Biol 1997; 75:327–336.

139. The Arabidopsis Genome Initiative. Analysis of the genome sequence of the flowering plant Arabidopsis thaliana. Nature 2000; 408:796–815.

140. Lucas JA, Masuda Y, Bennett RA, Strauss NS, Strauss PR. Single-turnover analysis of mutant human apurinic/apyrimidinic endonuclease. Biochemistry 1999; 38:4958–4964.

20

Oxidative Mitochondrial DNA Damage Resistance and Repair

Gerald S. Shadel
Department of Pathology, Yale University School of Medicine, New Haven, Connecticut, U.S.A.

1. INTRODUCTION

In most eukaryotic cells, the DNA genome is housed not only in the nucleus but also in cytoplasmic organelles called mitochondria. The mitochondrion represents a major crossroads of metabolism, being the site for hundreds of essential reactions including those involved in the tricarboxylic acid cycle, fatty acid oxidation, and pathways for the biosynthesis of heme, amino acids, steroid hormones, and many other biochemicals. Of course, mitochondria also perform oxidative phosphorylation (OXPHOS), the process through which the energy derived from the final oxidation of major foodstuffs is efficiently converted to cellular ATP. It is here where the essential function of mitochondrial DNA (mtDNA) is primarily realized, because mtDNA encodes essential protein subunits of the OXPHOS system. Therefore, mutations in mtDNA can result in reduced OXPHOS capacity, declined ATP production, and subsequent loss of proper cell function. While the decline in OXPHOS capacity alone is certainly a very severe negative consequence, the deleterious effects of mitochondrial dysfunction are compounded further by the fact that mitochondria are also a major source of reactive oxygen species (ROS) that can damage cellular components, the production of which is often enhanced under conditions where OXPHOS is disabled. Finally, mitochondria are intricately intertwined with the cellular apoptotic pathways that control programmed cell death. Thus, mtDNA mutations and mitochondrial dysfunction can lead to the purposeful or, in some cases, untimely loss of cells from a population or tissue, again potentially exacerbating the cellular effects of mitochondrial dysfunction.

In this chapter, I first describe the general features of human mtDNA and its involvement in human disease and aging, and then review what is known about pathways and mechanisms that exist to resist and repair of oxidative mtDNA damage in mammalian cells. I close by summarizing how recent studies in the budding yeast, *Saccharomyces cerevisiae*, have increased our overall understanding of mtDNA damage resistance pathways.

2. GENERAL FEATURES OF HUMAN mtDNA

2.1. Gene Content and Expression

Human mtDNA is a circular 16,569 base pair molecule which encodes essential protein subunits of the mitochondrial OXPHOS complexes, as well as two rRNAs (16S and 12S) and 22 tRNAs required to translate them. Specifically, mtDNA encodes 13 mRNAs: seven (*ND1-3*, *ND4*, *ND4L*, *ND5*, and *ND6*), encoding subunits of NADH dehydrogenase, or complex I; four (*COX1-3* and *CYTB*) encoding subunits of cytochome c oxidase, or complex III; and two (*ATP6* and *ATP8*) encoding subunits of the ATP synthase, or complex IV. The mRNAs are translated on mitochondrial ribosomes that are comprised of ~80 ribosomal proteins (encoded by nuclear genes and imported into the mitochondrion). An interesting consequence of the fact that all human mitochondrial mRNAs encode integral membrane proteins is that mitochondrial translation occurs primary, if not exclusively, at the inner mitochondrial membrane on membrane-associated ribosomes (1).

In addition to the 13 mtDNA-encoded proteins, ~1000 mitochondrial proteins are encoded by nuclear genes, translated on cytoplasmic ribosomes, and imported into the organelle. This includes the remainder of the OXPHOS subunits (~70 proteins), metabolic enzymes, and proteins dedicated to the expression and maintenance of mtDNA, including mtDNA polymerase (polγ), mtRNA polymerase, transcription factors, and DNA repair enzymes.

The mtDNA molecule contains a major non-coding control site called the displacement-loop (D-loop) region, which contains several regulatory loci, including two primary transcription promoters (one for each strand) and the heavy-strand origin (O_H) of mtDNA replication (Fig. 1). Expression of mitochondrial genes begins with transcription from each of the promoters and requires a dedicated mtRNA polymerase (*HPOLRMT*) (2), the DNA-binding transcription factor, h-mtTFA (*TFAM*) (3), and two RNA methyltransferase-related mitochondrial transcription factors, h-mtTFB1 (*TFB1M*) (4) and h-mtTFB2 (*TFB2M*) (5). Interestingly h-mtTFB1 has been shown to have rRNA methyltransferase activity and appears therefore to be a dual-function protein (6). The two large mitochondrial transcripts are polycistronic with the mRNA and rRNA genes usually flanked by tRNAs. Therefore, tRNA excision is a major RNA processing event required to liberate mature RNA species in mitochondria (7). In addition, transcription and RNA processing events are required for mtDNA replication (see Sec. 2.2).

2.2. mtDNA Replication, Copy Number, and Organization

A detailed description of the classical model of mtDNA replication in mammalian cells has been reviewed (8), therefore, only a brief summary is included here. Replication occurs by an asynchonous, strand-displacement mechanism involving two origins of replication termed O_H and O_L that are physically separated on the molecule (Fig. 1). Initiation of the leading strand at O_H is primed by transcription from the light-strand promoter (LSP) and involves the formation of a stable RNA/DNA hybrid at the origin (9) that is processed to form the requisite 3′ hydroxyl group used for extension by DNA polγ. This forms a unidirectional replication fork that proceeds to approximately two-thirds of the distance around the circular mtDNA molecule to O_L, displacing the parental heavy (H) strand. Once O_L is displaced as a single strand, priming of the light (L) mtDNA strand occurs by a mitochondrial DNA primase. Completion of the partially synthesized L and H mtDNA

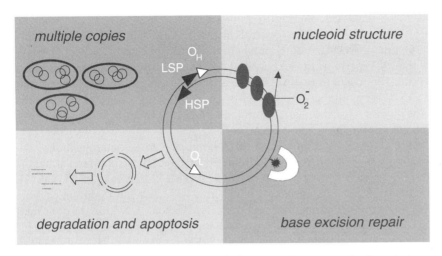

Figure 1 Summary of oxidative mtDNA damage resistance mechanisms in human cells. A schematic representation of the circular human mitochondrial genome is shown in the center of the figure. The two divergent transcription promoters (L-strand promoter, LSP and H-strand promoter, HSP) and the physically separated origins of replication (origin of H-strand, O_H and origin of L-strand, O_L) are indicated by black and open arrowheads, respectively. Each shaded quadrant represents a known or postulated mode of oxidative mtDNA damage resistance. Clockwise, starting from the upper left, quadrant 1 indicates that the multi-copy nature of the mtDNA genome (100–1000 copies per cell) can provide resistance by allowing genetic complementation between damaged or mutated molecules to provide a "buffer" of protection. Multiple mitochondria (indicated as ovals) per cell and multiple mtDNA molecules (circles) per mitochondrion contribute to the overall mtDNA copy number. Quadrant 2 indicates that mtDNA is not naked, but rather packaged by proteins into nucleoids that likely provide shielding from oxidative damage in much the same way as nucleosomes protect nuclear DNA. Shown is a ROS (superoxide) being denied access (bent arrow) to the mtDNA by a DNA-binding protein (gray oval; e.g., h-mtTFA). Quadrant 3 represents the presence of BER pathways present in mitochondria to remove oxidatively damaged bases from mtDNA. The star depicts an oxidatively damaged base in mtDNA that is being recognized by a BER DNA glycosylase (open u-shape; e.g., hNTH1). Finally, quadrant 4 represents removal of damaged molecules from the population by intracellular degradation of damaged mitochondria and therefore mtDNA (depicted schematically as broken circles that progress to degraded linear fragments) or elimination of damaged cells (and hence damaged mtDNA) from a tissue or population by apoptosis.

strands, ligation, resolution of two daughter molecules, and introduction of negative supercoils complete the replication process (10). Recently evidence consistent with a more standard, coupled leading and lagging strand mode of mtDNA replication has been proposed (11), suggesting that other sites on the displaced H strand, in addition to O_L, may be capable of priming L-strand synthesis. While this has been interpreted by some to indicate an entirely different mode of mtDNA replication than that described above, this remains controversial and, at this point, appears to represent an interesting refinement of the original model, rather than evidence for an entirely new mechanism.

Generally, there are many mitochondria per cell and multiple mtDNA molecules per mitochondrion, with both parameters varying dramatically depending on the cell or tissue type (12,13). Therefore, the mitochondrial genome is a multi-copy genome, present at 100–10,000 copies/cell. The number of mtDNA molecules per cell appears to be regulated, leading to a characteristic copy number for each cell type. While the mechanisms that control mtDNA copy number remain largely unknown (14), one provocative study points to the ability of cells to limit or "count" the mass of mtDNA, perhaps through availability or sensing of mitochondrial deoxyribonucleoside triphosphate pools (15). Regardless of the precise mechanism that governs mtDNA copy number, the impact of the multi-copy nature of this genetic system on the phenotypic expression of mtDNA mutations in mitochondrial diseases is dramatic and will be discussed in Sec. 3.

The major units of organization of mtDNA appear to be heterogeneous protein/DNA complexes called nucleoids, visualized as punctate structures within the mitochondrial matrix when mtDNA is stained with fluorescent dyes (16,17). Nucleoids have been most extensively studied in *S. cerevisiae* and contain at least 20 different proteins, including the abundant HMG-box, DNA-binding protein Abf2p (18), complexed with 3–4 mtDNA molecules (19). Recently, several investigators have reported on the protein composition (20) and dynamics (21,22) of human mtDNA nucleoids, which like their yeast counterparts are composed of multiple proteins, including mitochondrial single-strand DNA-binding protein and h-mtTFA (the human ortholog of Abf2p), as well as multiple (~5–10) mtDNA molecules. Documentation of this type of nucleoid structure should dispel the commonly and mistakenly propagated myth that mtDNA is largely "naked" in vivo.

2.3. Involvement of mtDNA Mutations in Human Disease

As discussed above, mitochondria are essential for normal cell function and viability because of the multitude of biochemical reactions that occur within the organelles as well as their role in mediating cellular signaling processes such as apoptosis and calcium homeostasis. However, the placement of mitochondria at this critical intersection of metabolism comes at a considerable cost, because mitochondrial dysfunction disrupts multiple cellular processes that, in turn, affect multiple tissues and organ systems, resulting in disease. At present, there are greater than one hundred known human diseases caused by mutations in nuclear genes that encode mitochondrial gene products (23). Most of these involve dysfunction of metabolic enzymes that must carry out their reactions in the mitochondrial compartment. However, an interesting subset includes those required for expression and replication of mtDNA. For example, mutations in polγ, the mtDNA helicase twinkle, and several genes that regulate mitochondrial nucleotide pools cause mtDNA depletion and deletions that lead to either myopathy and/or progressive external ophthalmoplegia (24).

In addition to the many nuclear gene mutations that cause mitochondrial disorders, a breakthrough in our understanding of mitochondrial disease came with the first demonstration in 1988 that mutations in mtDNA likewise can be pathogenic (25,26). It is now recognized that many multi-system degenerative diseases (mitochondrial myopathies) affecting heart, muscle, the nervous system, and other major organs are caused by mtDNA mutations that result from reduced or aberrant mitochondrial gene expression and OXPHOS (27,28). Because mtDNA is exclusively maternally inherited in humans, these diseases are diagnosed, in large part, by this characteristic inheritance pattern. However, because of the multi-copy nature of the mitochondrial genome, pathogenic mutations almost always coexist with wild-type (or other types of mutant) molecules, a condition called heteroplasmy (as opposed to homoplasmy, where all mtDNA molecules are identical). Heteroplasmy results in remarkably complex genetics and clinical manifestations of mtDNA-based diseases, including incomplete penetrance due to random genetic drift within tissues and the existence of tissue-specific thresholds of OXPHOS dysfunction. In addition, because of the prominent nuclear involvement in mitochondrial function, the nuclear genetic background of individuals further complicates the inheritance and clinical presentation of mtDNA-based diseases.

2.4. Mitochondria as a Source and Target of Cellular Oxidative Damage in Normal and Aging Cells

As noted above, the advantages of mitochondrial aerobic metabolism do not come without cost. During OXPHOS, reactive oxygen species (ROS) such as superoxide and hydrogen peroxide are produced as side reactions (29). The ROS are normally kept at low levels through the action of protective enzymes, but certain conditions, including inhibition of the respiratory chain, can lead to enhanced ROS production and cellular oxidative stress (30). Current models for mtDNA-derived oxidative damage suggest that impairment of mitochondrial respiration can create a self-perpetuating cycle of oxidative stress that leads to mtDNA damage, nuclear DNA damage, cellular dysfunction, apoptosis, and disease (31). The cornerstone of these models is oxidative damage-induced inactivation or mutagenesis of mtDNA due to its proximity to, or direct physical association with (32), the inner membrane OXPHOS complexes. Accumulation of mtDNA mutations is predicted to cause a reduction of mitochondrial gene expression, promoting a further reduction in respiration capacity, enhanced ROS production and more mtDNA damage.

Though still somewhat controversial, a decline in mitochondrial respiration with age has long been noted (33) and mtDNA mutations have been shown to accumulate in normal aging tissues (34). These observations are the foundation of numerous related theories (35) that invoke a cycle of oxidative stress, similar to that described above, ultimately leading to age-related declines in cellular functions. This scenario is commonly referred to as either the "Mitochondrial Theory of Aging" (36) or the "Free Radical Theory of Aging" (37). Most recently, specific mutations in the mtDNA D-loop region have been found to accumulate preferentially in older individuals, indicating for the first time the potential involvement of specific mtDNA mutations in human aging (38).

3. OXIDATIVE mtDNA DAMAGE RESISTANCE AND REPAIR

The mutation rate of mammalian mtDNA is estimated to be as much as 10-fold greater than that of nuclear DNA (39), promoting its use as both a "molecular clock" for evolutionary studies (40) and a forensic tool (41). Speculation by many investigators has attributed this high mutation rate to many factors unique to the environment to which mtDNA is exposed, including proximity to the ROS-generating OXPHOS complexes, lack of a compact nucleosome structure to protect the DNA, and a paucity of DNA repair mechanisms. In addition, mitochondria are susceptible to environmental agents, many of which target the organelle preferentially and damage mtDNA and/or inhibit respiration. These include ultraviolet light, the known carcinogens N-methyl-N-nitrosourea, N-nitrosodimethylamine and cigarette smoke, and certain DNA-damaging agents, like polyaromatic hydrocarbons and cisplatin (reviewed in Ref. 42). Furthermore, replication of mtDNA is inhibited by certain antibacterial and antiviral drugs that result in depletion of mtDNA in cells and mitochondrial disease-like side effects (43). By inhibiting mtDNA expression and/or respiration, these and probably many yet-to-be-identified environmental factors likely promote and exacerbate mitochondrial-mediated oxidative stress in cells.

Initially, most of the studies on mtDNA damage focused primarily on demonstrating that specific types of damage induced by endogenous, chemical or environmental agents can occur in mtDNA and whether or not the specific DNA lesions introduced can be repaired in a given time-frame relative to the same damage in nuclear DNA (reviewed in Ref. 44). While providing evidence that some DNA repair pathways exist in mitochondria, these studies did not address directly the number or type of pathways available to resist mtDNA damage. Here, I will evaluate mechanisms that are known or predicted to provide protection to mtDNA from oxidative damage, according to more recent and direct studies (summarized in Fig. 1).

3.1. Resistance to Oxidative mtDNA Damage: Packaging, Copy Number, and Degradation

The packaging of nuclear DNA into nucleosomes and higher order chromatin by histones and non-histone DNA-binding proteins is generally believed to protect the DNA from oxidative and other types of damage (45). While appealing conceptually, relatively few studies indicate that this is indeed the case. Certainly, a more likely scenario is that this form of protection depends both on the dynamics of the nucleosome structure at specific chromosomal sites as well as the particular DNA-damaging agent involved. Nonetheless, it remains likely that, in a general sense, nuclear DNA is insulated to a significant degree by its packaging into higher order protein/DNA complexes. This is an important consideration when one speculates whether mtDNA is likewise protected via its packaging into nucleoids. Since first postulated by Ames and co-workers in 1988 (46), the concept that mtDNA is subject to increased damage and subsequent mutagenesis because it is not protected by histones has been widely accepted and promoted in the literature. While it is clear that a histone octamer-based nucleosome structure does not exist in mitochondria, mtDNA is indeed bound to a significant degree by proteins in the context of nucleoids. Interestingly, a common component of the mitochondrial nucleoid in both yeast (18) and mammals (20) is the HMG-box DNA-binding factor mtTFA, which is related to HMG1, a nuclear non-histone chromosomal protein (3). Altogether, these data indicate that mtDNA, like nuclear DNA, is organized in a manner that could protect it from damage. Whether protection of mtDNA provided by proteins present in nucleoids is similar to that provided to nuclear DNA by the nucleosome remains an important unanswered question.

The high copy number of mtDNA, in principle, provides another level of protection from mtDNA damage. This is because with multiple molecules per organelle, and certainly with hundreds or thousands per cell, the inherent capacity to maintain a significant number of fully functional molecules per cell even in the face significant oxidative damage remains high. In addition, the potential for *trans*-complementation of mutations in mtDNA molecules further increases the ability to contend with a high mtDNA mutation load. This phenomenon almost certainly underlies much of the threshold effect observed in mtDNA-based disease (27,28). However, despite the logical foundation of this "copy number protection" hypothesis, whether copy number indeed protects cells from mtDNA damage has yet to be addressed directly. Several human diseases (24) and mouse gene knock-out animals (47) result in mtDNA depletion. Analysis of oxidative mtDNA damage in cell lines derived from these could provide a model system for analyzing this issue.

A third mechanism to resist oxidative mtDNA damage by means of selective degradation of damaged mitochondria has been proposed (48). Mitochondria, and therefore mtDNA molecules (49), are constantly undergoing a cycle of biogenesis and turnover, the latter of which involves the process of lysosome-mediated autophagy (50,51). While very little is known regarding the influence of autophagy in the elimination of oxidatively damaged mtDNA, a formal possibility is that mitochondria with extensively damaged mtDNA are somehow identified, eliminated from the cellular population, and replaced using the "surviving" undamaged organelles. The ability of cells to synthesize mtDNA throughout the cell cycle (52) and in non-dividing cells (48) is compatible with this idea.

Finally, the ultimate mtDNA damage resistance pathway may indeed be cell death by apoptosis. It is known that cytochrome c and other apoptosis-inducing factors are released from mitochondria (53). In addition, many pro- and *anti*-apoptotic proteins reside in mitochondria or are translocated there under conditions that impact cell survival. Thus, mitochondria are clearly at the crossroads of important signals that can induce or inhibit programmed cell death. While it remains unclear whether excessive oxidative damage to mtDNA or mitochondrial proteins promote cell death, several studies indeed suggest that this may be the case (54–57). In this way, cells in a tissue or population that are severely compromised for mitochondrial function can be eliminated rather than allowed to continue to negatively influence tissue function. If this is the case, purposefully imbalancing mtDNA repair systems to allow over-accumulation of oxidative damage is one promising way to make cancer cells more susceptible to cell death by chemotherapeutic agents (58,59), making mtDNA a potential target for cancer therapy. However, this may be a slippery slope given the recent findings of homoplasmic mtDNA mutations in human tumors (60,61), suggesting that mtDNA mutagenesis, perhaps as a result of oxidative damage, may also play a direct role in tumor initiation or progression in certain tissues.

3.2. Base Excision Repair of Oxidative mtDNA Damage

One of the earliest examinations of whether pathways exist to repair mammalian mtDNA revealed that UV-induced thymine dimers were not efficiently, if at all, removed from mtDNA in cultured mouse and human cells (62). This was the first indication that mitochondria have an abbreviated repertoire for repair of mtDNA and, in particular, appeared to lack nucleotide excision repair (NER) capacity to remove bulky DNA lesions, an observation that has been verified in subsequent studies (42). More recently, however, it has been definitively shown that mammalian mitochondria have, at the very least, base excision repair (BER) pathways for the removal of certain less bulky lesions, like those characteristic of spontaneous or induced oxidative DNA damage. This has been shown in a variety of ways including repair studies in cultured mammalian cells, biochemical analysis of mitochondrial extracts for specific BER activities, and demonstration of localization of known BER proteins to the organelle (31,42,44,50). The significance of other DNA repair pathways in human mitochondria is less clear. Recently, putative mismatch repair (MMR) activity has been detected in rat liver mitochondria (63), perhaps similar to that observed in yeast a number of years ago (64), yet definitive proof of this concept has not yet surfaced. In addition, NER-like (64) and photoreactivation (65,66) pathways have been reported in yeast, but these remain of questionable relevance in human cells. For these reasons, I will focus on reviewing the mammalian BER pathway as it relates to the negotiation of oxidative mtDNA damage.

The process of BER is initiated by cadre of relatively specific DNA glycosylase enzymes that, as a group, recognize a wide variety of DNA base damages. After binding to a lesion, the *N*-glycosylase activity of these enzymes cleaves the glycosidic bond to remove the damaged base, creating an apurinic or apyrimidinic (abasic) site. The resulting abasic site is then cleaved by an apurinic/apyrimidinic (AP) endonuclease. The process is completed by processing of the 5′-end by a deoxyribose phophodiesterase (dRPase) which creates a 5′-hydroxyl end at the nick, allowing the one nucleotide gap to be filled in by a DNA polymerase and then sealed by a DNA ligase. It has been reported that the all the requisite activities for this conventional pathway of BER exist in mitochondria of higher organisms. For example, the basic components of BER werc first analyzed from *Xenopus* mitochondria by Pinz and Bogenhagen (67). Others have shown subsequently that similar activities and/ or proteins known to be involved in BER in the nucleus are also localized to mammalian mitochondria (68). It is noted that several of these proteins have multiple activities. For example, some DNA glycosylases have an associated AP lyase activity that is able to cleave the abasic site left by their DNA glycosylase activity, thus creating an alternative sub-pathway for completion of the BER process. In addition, both mtDNA ligase III (68) and polγ (69) have weak associated dRPase activities that may also play specialized roles in mitochondrial BER. All of the mammalian mitochondrial BER proteins, like most mitochondrial proteins, are encoded in the nucleus and imported into the organelle. Most these are isoforms of the same enzymes used in nuclear BER that are derived by alternative splicing or other mechanisms that allow differential localization of the enzymes. Interestingly, a recent report suggests that there are age-related deficiencies in the import of certain DNA repair proteins into mitochondria that could contribute to declines in oxidative mtDNA repair with age (70).

With regard to oxidative mtDNA damage specifically, isoforms of the 8-oxoguanine DNA glycosylase (hOGG1) and the ortholog of *E. coli* endonuclease III glycosylase (hNTH) are targeted to mitochondria in human cells (71,72), where they initiate repair of oxidatively damaged purines (e.g., 8-oxoguanine) and pyrimidines (e.g., thymine glycol), respectively. Recent studies in cultured cells from wild-type and knock-out mice have confirmed the role of these enzymes in mitochondrial BER (73,74) and also suggest the existence of additional glycosylase activities in mitochondria that can repair oxidized pyrimidines such as 5-hydroxycytosine and 5-hydroxyuracil. In summary, mitochondria harbor DNA glycosylases and other BER proteins that provide significant protection against oxidative mtDNA damage, which is likely the most abundant type of DNA damage encountered by the organelle under normal circumstances.

4. NEW LESSONS ABOUT OXIDATIVE mtDNA DAMAGE FROM THE BUDDING YEAST, *S. CEREVISIAE*, GENETIC MODEL SYSTEM

The yeast, *S. cerevisiae*, has long served as model system for studying mitochondrial genome regulation as well as other fundamental aspects of mitochondrial biology (75). The primary advantage of this organism for these purposes is that it is a facultative anaerobe that can survive in the absence of mitochondrial respiration if a fermentable carbon source is available. Yeast strains that are respiration deficient exhibit a mitochondrial "petite" phenotype, that is, easily scored as impaired growth on non-fermentable carbon sources (e.g., glycerol and ethanol) or by a red/white colony color difference in certain strain backgrounds. The ability to isolate petite mutations has led to the identification of many nuclear genes involved in mitochondrial respiration, gene expression, and mtDNA maintenance. While significant differences exist between yeast and vertebrates, which certainly must be kept in mind when employing this organism as a model, fundamental aspects of mitochondrial gene expression and mtDNA replication are conserved from yeast to humans (75).

Some of the earliest studies of mtDNA repair and stability were performed in *S. cerevisiae* (reviewed in Ref. 76). These studies elucidated many of the basic principles we now embrace with regard to the mutagenesis and repair of mtDNA in human cells (8), including the inherent susceptibility of mtDNA to spontaneous and induced damage and mutagenesis, the influence of many nuclear gene products on mtDNA stability and repair, and the generation of large deletions as a common mtDNA mutagenic event in cells. These studies also pointed to some salient differences between yeast and human cells, perhaps most notably, the presence of extensive genetic recombination in yeast mitochondria and some ability to repair ultraviolet-induced mtDNA damage. However, more recent studies in yeast have led to the identification and characterization of many conserved factors and pathways involved in mtDNA oxidative damage resistance and repair that will be summarized in the following sections.

4.1. BER and Overlapping Oxidative mtDNA Damage Resistance Pathways

First, as already discussed for human cells, studies in yeast have confirmed the existence of and elucidated further the BER system in mitochondria for the removal of oxidative mtDNA damage. For example, the yeast homolog of the human hOGG1 DNA glycosylase, Ogg1p, is localized to mitochondria and deletion of its gene causes an increase in petite mutant formation consistent with a role in repairing spontaneous mtDNA damage (77). In addition, there are two homologs of the human NTH1 gene in yeast, *NTG1* and *NTG2*. Ntg2p is localized in the nucleus, while Ntg1p is localized both to the nucleus and mitochondria (78,79). Though the predicted role for Ntg1p in repairing oxidative mtDNA damage has been elucidated (80), it is noteworthy that deletion of this gene does not result in a significant increase in mitochondrial petite mutants. With regard to mitochondria, these observations indicate the presence of additional activities or pathways that can fulfill, bypass, or compensate for the lack of Ntg1p in the organelle. Consistent with this concept is the observation that the increase in petite formation rate in an *ogg1* null strain is rescued by deletion of *NTG1*, pointing to pathways other than BER compensating for the reduced capacity to handle oxidative damage in these strains (77). Finally, another recent study (80) shed additional light into this issue by demonstrating that deletion of either *PIF1*, encoding a DNA helicase located in the nucleus and mitochondria, or *ABF2*, encoding an abundant mtDNA-binding protein (the yeast ortholog of h-mtTFA) results in a synthetic increase in petite mutant induction in combination with an *ntg1* null mutation. Because each of these proteins has a documented role in mtDNA recombination (81,82), and since recombination has been implicated in mtDNA repair in yeast (83), this study also provided evidence that it is not the recombination function of these proteins per se that facilitates resistance to mtDNA damage, but rather novel functions such as mtDNA packaging and shielding (Abf2p) or regulation of mtDNA replication to facilitate repair (Pif1p). These observations may be more directly relevant to mitochondrial resistance pathways in human cells because recombination is not at all prevalent in mammalian mitochondria (84) and homologs of Pif1p and Abf2p exist in humans.

Other components of the mitochondrial BER pathway have been identified in *S. cerevisiae* including uracil DNA glycosylase (*UNG1*; 85), AP endonuclease (*APN1*; 86), and DNA ligase (*CDC9*; 87). In all three cases, the respective proteins are localized both to the nucleus and mitochondria. However, the localization mechanism of Apn1p is apparently unique in that it requires another protein factor, Pir1p, to facilitate its targeting to mitochondria (86), perhaps pointing to a novel mode of regulation for certain mtDNA repair proteins.

4.2. Other Factors Involved in mtDNA Maintenance and Repair

Genetic analysis of *S. cerevisiae* has also led to the identification of new players involved in mtDNA maintenance. Those with potential impact on oxidative mtDNA damage resistance and relevance to humans are considered here. The first of these is *MSH1*, which encodes a homolog of bacterial MutS, that is involved in DNA mismatch repair and is localized exclusively to mitochondria (64). Deletion of this gene results in mtDNA point mutagenesis and rearrangements, consistent with a role in DNA repair, although other functions in mtDNA maintenance have been postulated as well (88). The identification of a putative mismatch repair system in yeast mitochondria has prompted a search for similar activities in human cells. Biochemical evidence for such a pathway has been reported (63), but direct visualization of mitochondrial localization of any known mismatch repair proteins in mammalian cells has not yet surfaced. The second is the product of the damage-inducible 7 (*DIN7*) gene, a novel nuclease of the Xeroderma pigmentosum group G (XPG) family of proteins that is localized exclusively to mitochondria in *S. cerevisiae* (89). Overexpression of Din7p results in enhanced mtDNA point mutagenesis, and deletion of the gene rescues the petite phenotype of a *dun1* strain that is deficient in the induction of the ribonucleotide reductase genes (90), suggesting a novel role for Din7 in mtDNA repair or recombination. It will be of particular interest to see if members of this protein family localize to and function in human mitochondria. The final two examples emphasize the role of the mitochondrial environment in mtDNA stability and cellular oxidative stress. Two genes, *ATM1* (91) and *YHF1* (92,93), are involved in the metabolism of iron–sulphur clusters and matrix iron homeostasis (94). Deletion of either gene results in increased mtDNA mutagenesis, presumably due to enhanced oxidative stress from increased levels of free matrix iron, a known catalyst of ROS generation. Interest in these proteins is heightened by the fact that both have known homologs in humans (95). In fact, the human human homolog of *YFH1* is *FRA-TAXIN*, the gene that causes the human neurodegenerative disease, Friedreich's ataxia.

4.3. Influence of Mitochondrial Function on Nuclear Genome Stability

As already discussed for human cells, the production of ROS during mitochondrial respiration has been clearly demonstrated in *S. cerevisiae* and shown to have deleterious consequences under a variety of conditions (96–98). While this is likely the major influence of mitochondrial function on cellular oxidative stress, several recent studies have uncovered a much more complex relationship between mitochondrial function, cellular oxidative stress, and nuclear genome integrity. For example, strains devoid of mtDNA (rho°) do not have a functional respiratory chain and therefore generally produce fewer ROS. Under many circumstances, rho° strains exhibit reduced oxidative stress, again pointing to the deleterious effects of mitochondrial ROS production (99,100). However, rho° strains have been reported more recently to result in a nuclear mutator phenotype that is dependent on error-prone nuclear DNA polymerases, despite their reduced production of ROS (101). In a similar study, an increase in nuclear mutations that restored a frameshift mutation reporter allele was observed in rho° yeast cells (102). These observations suggest that previously unrecognized lines of communication exist between the nucleus and mitochondria that do not involve mitochondrial-derived ROS. Finally, a fascinating interaction between the yeast nuclear and mitochondrial genomes was recently uncovered in two reports (103,104) that show convincingly that double-strand breaks in yeast nuclear DNA are repaired by short patches of mtDNA. Thus, not only are there ROS-dependent and ROS-independent signaling pathways and consequences that connect the nuclear and mitochondrial compartments functionally, but also complex physical exchanges between these two normally compartmentalized genomes.

5. CONCLUSIONS AND NEW HORIZONS

Detailed studies of mitochondrial DNA damage resistance and repair over the past decade have greatly enhanced our knowledge of these pathways and also clarified some early misconceptions regarding mtDNA. First, the early seminal experiments that led subsequent investigators to conclude that DNA repair was altogether absent in mitochondria have been extended to show that, while a classical NER pathway does not operate in human mitochondria, efficient BER pathways for the repair of oxidative and other types of DNA damage to the mitochondrial genome are present and accounted for. Second, the concept of a "naked" mitochondrial genome that is inherently susceptible to DNA damage is also clearly incorrect. Packaging of mtDNA into nucleoids by conserved DNA-binding proteins no doubt provides some protection to oxidative damage. And finally, the idea that the nuclear and mitochondrial genomes are compartmentalized and largely independent has been challenged by recent findings in model systems such as *S. cerevisiae*. However, much remains to be learned and some outstanding questions in the field with regard to oxidative stress and mtDNA remain. What is the full complement of mtDNA damage resistance pathways present in human mitochondria and can these be manipulated in some way to protect cells from the deleterious effects of ROS, or perhaps even to make cancer cells more susceptible to therapeutics? What is the molecular and cellular basis for the many ways in which mtDNA damage and mutagenesis cause and contribute to human pathology and aging? And finally, what are the signaling pathways that orchestrate the complex interplay between the nuclear and mitochondrial genomes? Only through continued research on these and related questions will the full impact of mtDNA mutations and mitochondrial dysfunction on human health be realized.

REFERENCES

1. Liu M, Spremulli L. Interaction of mammalian mitochondrial ribosomes with the inner membrane. J Biol Chem 2000; 275:29400–29406.

2. Tiranti V, Savoia A, Forti F, D'Apolito MF, Centra M, Rocchi M, Zeviani M. Identification of the gene encoding the human mitochondrial RNA polymerase (h-mtRPOL) by cyberscreening of the Expressed Sequence Tags database. Hum Mol Genet 1997; 6:615–625.

3. Parisi MA, Clayton DA. Similarity of human mitochondrial transcription factor 1 to high mobility group proteins. Science 1991; 252:965–969.

4. McCulloch V, Seidel-Rogol BL, Shadel GS. A human mitochondrial transcription factor is related to RNA adenine methyltransferases and binds S-adenosylmethionine. Mol Cell Biol 2002; 22:1116–1125.

5. Falkenberg M, Gaspari M, Rantanen A, Trifunovic A, Larsson NG, Gustafsson CM. Mitochondrial transcription factors B1 and B2 activate transcription of human mtDNA. Nat Genet 2002; 31:289–294.

6. Seidel-Rogol BL, McCulloch V, Shadel GS. Human mitochondrial transcription factor B1 methylates ribosomal RNA at a conserved stem-loop. Nat Genet 2003; 33:23–24.

7. Ojala D, Montoya J, Attardi G. tRNA punctuation model of RNA processing in human mitochondria. Nature 1981; 290:470–474.

8. Shadel GS, Clayton DA. Mitochondrial DNA maintenance in vertebrates. Annu Rev Biochem 1997; 66:409–435.

9. Lee DY, Clayton DA. Initiation of mitochondrial DNA replication by transcription and R-loop processing. J Biol Chem 1998; 273:30614–30621.

10. Clayton DA. Replication of animal mitochondrial DNA. Cell 1982; 28:693–705.

11. Holt IJ, Lorimer HE, Jacobs HT. Coupled leading- and lagging-strand synthesis of mammalian mitochondrial DNA. Cell 2000; 100:515–524.

12. Michaels GS, Hauswirth WW, Laipis PJ. Mitochondrial DNA copy number in bovine oocytes and somatic cells. Dev Biol 1982; 94:246–251.

13. Veltri KL, Espiritu M, Singh G. Distinct genomic copy number in mitochondria of different mammalian organs. J Cell Physiol 1990; 143:160–164.

14. Moraes CT. What regulates mitochondrial DNA copy number in animal cells? Trends Genet 2001; 17:199–205.

15. Tang Y, Schon EA, Wilichowski E, Vazquez-Memije ME, Davidson E, King MP. Rearrangements of human mitochondrial DNA (mtDNA): new insights into the regulation of mtDNA copy number and gene expression. Mol Biol Cell 2000; 11:1471–1485.

16. Miyakawa I, Aoi H, Sando N, Kuroiwa T. Fluorescence microscopic studies of mitochondrial nucleoids during meiosis and sporulation in the yeast, *Saccharomyces cerevisiae*. J Cell Sci 1984; 66:21–38.

17. Satoh M, Kuroiwa T. Organization of multiple nucleoids and DNA molecules in mitochondria of a human cell. Exp Cell Res 1991; 196:137–140.

18. Kaufman BA, Newman SM, Hallberg RL, Slaughter CA, Perlman PS, Butow RA. In organello formaldehyde crosslinking of proteins to mtDNA: identification of bifunctional proteins. Proc Natl Acad Sci USA 2000; 97:7772–7777.

19. Miyakawa I, Sando N, Kawano S, Nakamura S, Kuroiwa T. Isolation of morphologically intact mitochondrial nucleoids from the yeast, *Saccharomyces cerevisiae*. J Cell Sci 1987; 88:431–439.

20. Bogenhagen DF, Wang Y, Shen EL, Kobayashi R. Protein components of mitochondrial DNA nucleoids in higher eukaryotes. Mol Cell Proteomics 2003; 2:1205–1216.

21. Alam TI, Kanki T, Muta T, Ukaji K, Abe Y, Nakayama H, Takio K, Hamasaki N, Kang D. Human mitochondrial DNA is packaged with TFAM. Nucleic Acids Res 2003; 31:1640–1645.

22. Garrido N, Griparic L, Jokitalo E, Wartiovaara J, van der Bliek AM, Spelbrink JN. Composition and dynamics of human mitochondrial nucleoids. Mol Biol Cell 2003; 14:1583–1596.

23. Zeviani M. The expanding spectrum of nuclear gene mutations in mitochondrial disorders. Semin Cell Dev Biol 2001; 12:407–416.

24. Elpeleg O, Mandel H, Saada A. Depletion of the other genome–mitochondrial DNA depletion syndromes in humans. J Mol Med 2002; 80:389–396.

25. Holt IJ, Harding AE, Morgan-Hughes JA. Deletions of muscle mitochondrial DNA in patients with mitochondrial myopathies. Nature 1988; 331:717–719.

26. Wallace DC, Singh G, Lott MT, Hodge JA, Schurr TG, Lezza AM, Elsas LJ II, Nikoskelainen EK. Mitochondrial DNA mutation associated with Leber's hereditary optic neuropathy. Science 1988; 242:1427–1430.

27. Wallace DC. Mitochondrial diseases in man and mouse. Science 1999; 283:1482–1488.

28. DiMauro S, Schon EA. Mitochondrial DNA mutations in human disease. Am J Med Genet 2001; 106:18–26.

29. Boveris A, Chance B. The mitochondrial generation of hydrogen peroxide. General properties and effect of hyperbaric oxygen. Biochem J 1973; 134:707–716.

30. Esposito LA, Melov S, Panov A, Cottrell BA, Wallace DC. Mitochondrial disease in mouse results in increased oxidative stress. Proc Natl Acad Sci USA 1999; 96: 4820–4825.

31. Mandavilli BS, Santos JH, Van Houten B. Mitochondrial DNA repair and aging. Mutat Res 2002; 509:127–151.

32. Albring M, Griffith J, Attardi G. Association of a protein structure of probable membrane derivation with HeLa cell mitochondrial DNA near its origin of replication. Proc Natl Acad Sci USA 1977; 74:1348–1352.

33. Miquel J, Economos AC, Fleming J, Johnson JE Jr. Mitochondrial role in cell aging. Exp Gerontol 1980; 15:575–591.

34. Chomyn A, Attardi G. MtDNA mutations in aging and apoptosis. Biochem Biophys Res Commun 2003; 304:519–529.

35. Miquel J. An update on the oxygen stress-mitochondrial mutation theory of aging: genetic and evolutionary implications. Exp Gerontol 1998; 33:113–126.

36. Linnane AW, Marzuki S, Ozawa T, Tanaka M. Mitochondrial DNA mutations as an important contributor to ageing and degenerative diseases. Lancet 1989; 1:642–645.

37. Harman D. The aging process. Proc Natl Acad Sci USA 1981; 78:7124–7128.

38. Wang Y, Michikawa Y, Mallidis C, Bai Y, Woodhouse L, Yarasheski KE, Miller CA, Askanas V, Engel WK, Bhasin S, Attardi G. Muscle-specific mutations accumulate with aging in critical human mtDNA control sites for replication. Proc Natl Acad Sci USA 2001; 98:4022–4027.

39. Brown WM, George M Jr, Wilson AC. Rapid evolution of animal mitochondrial DNA. Proc Natl Acad Sci USA 1979; 76:1967–1971.

40. Wallace DC. Mitochondrial DNA sequence variation in human evolution and disease. Proc Natl Acad Sci USA 1994; 91:8739–8746.

41. Budowle B, Allard MW, Wilson MR, Chakraborty R. Forensics and mitochondrial DNA: applications, debates, and foundations. Annu Rev Genomics Hum Genet 2003; 4:119–141.

42. Croteau DL, Stierum RH, Bohr VA. Mitochondrial DNA repair pathways. Mutat Res 1999; 434:137–148.

43. Lewis W, Dalakas MC. Mitochondrial toxicity of antiviral drugs. Nat Med 1995; 1: 417–422.

44. LeDoux SP, Driggers WJ, Hollensworth BS, Wilson GL. Repair of alkylation and oxidative damage in mitochondrial DNA. Mutat Res 1999; 434:149–159.

45. Enright HU, Miller WJ, Hebbel RP. Nucleosomal histone protein protects DNA from iron-mediated damage. Nucleic Acids Res 1992; 20:3341–3346.

46. Richter C, Park JW, Ames BN. Normal oxidative damage to mitochondrial and nuclear DNA is extensive. Proc Natl Acad Sci USA 1988; 85:6465–6467.

47. Larsson NG, Wang J, Wilhelmsson H, Oldfors A, Rustin P, Lewandoski M, Barsh GS, Clayton DA. Mitochondrial transcription factor A is necessary for mtDNA maintenance and embryogenesis in mice. Nat Genet 1998; 18:231–236.

48. Kopsidas G, Kovalenko SA, Heffernan DR, Yarovaya N, Kramarova L, Stojanovski D, Borg J, Islam MM, Caragounis A, Linnane AW. Tissue mitochondrial DNA changes. A stochastic system. Ann NY Acad Sci 2000; 908:226–243.

49. Gross NJ, Getz GS, Rabinowitz M. Apparent turnover of mitochondrial deoxyribonucleic acid and mitochondrial phospholipids in the tissues of the rat. J Biol Chem 1969; 244:1552–1562.

50. Kang D, Hamasaki N. Maintenance of mitochondrial DNA integrity: repair and degradation. Curr Genet 2002; 41:311–322.

51. Marzella L, Ahlberg J, Glaumann H. Isolation of autophagic vacuoles from rat liver: morphological and biochemical characterization. J Cell Biol 1982; 93:144–154.

52. Bogenhagen D, Clayton DA. Mouse L cell mitochondrial DNA molecules are selected randomly for replication throughout the cell cycle. Cell 1977; 11:719–727.

53. Newmeyer DD, Ferguson-Miller S. Mitochondria: releasing power for life and unleashing the machineries of death. Cell 2003; 112:481–490.

54. Santos JH, Hunakova L, Chen Y, Bortner C, Van Houten B. Cell sorting experiments link persistent mitochondrial DNA damage with loss of mitochondrial membrane potential and apoptotic cell death. J Biol Chem 2003; 278:1728–1734.

55. Mirabella M, Di Giovanni S, Silvestri G, Tonali P, Servidei S. Apoptosis in mitochondrial encephalomyopathies with mitochondrial DNA mutations: a potential pathogenic mechanism. Brain 2000; 123:93–104.

56. Esteve JM, Mompo J, Garcia de la Asuncion J, Sastre J, Asensi M, Boix J, Vina JR, Vina J, Pallardo FV. Oxidative damage to mitochondrial DNA and glutathione oxidation in apoptosis: studies in vivo and in vitro. FASEB J 1999; 13:1055–1064.

57. Yoneda M, Katsumata K, Hayakawa M, Tanaka M, Ozawa T. Oxygen stress induces an apoptotic cell death associated with fragmentation of mitochondrial genome. Biochem Biophys Res Commun 1995; 209:723–729.

58. Fishel ML, Seo YR, Smith ML, Kelley MR. Imbalancing the DNA base excision repair pathway in the mitochondria; targeting and overexpressing N-methylpurine DNA glycosylase in mitochondria leads to enhanced cell killing. Cancer Res 2003; 63:608–615.

59. Shokolenko IN, Alexeyev MF, Robertson FM, LeDoux SP, Wilson GL. The expression of Exonuclease III from E. coli in mitochondria of breast cancer cells diminishes mitochondrial DNA repair capacity and cell survival after oxidative stress. DNA Repair (Amst) 2003; 2:471–482.

60. Fliss MS, Usadel H, Caballero OL, Wu L, Buta MR, Eleff SM, Jen J, Sidransky D. Facile detection of mitochondrial DNA mutations in tumors and bodily fluids. Science 2000; 287:2017–2019.

61. Polyak K, Li Y, Zhu H, Lengauer C, Willson JK, Markowitz SD, Trush MA, Kinzler KW, Vogelstein B. Somatic mutations of the mitochondrial genome in human colorectal tumours. Nat Genet 1998; 20:291–293.

62. Clayton DA, Doda JN, Friedberg EC. The absence of a pyrimidine dimer repair mechanism in mammalian mitochondria. Proc Natl Acad Sci USA 1974; 71:2777–2781.

63. Mason PA, Matheson EC, Hall AG, Lightowlers RN. Mismatch repair activity in mammalian mitochondria. Nucleic Acids Res 2003; 31:1052–1058.

64. Reenan RA, Kolodner RD. Characterization of insertion mutations in the Saccharomyces cerevisiae MSH1 and MSH2 genes: evidence for separate mitochondrial and nuclear functions. Genetics 1992; 132:975–985.

65. Pasupathy K, Pradhan DS. Evidence for excision repair in promitochondrial DNA of anaerobic cells of Saccharomyces cerevisiae. Mutat Res 1992; 273:281–288.

66. Yasui A, Yajima H, Kobayashi T, Eker AP, Oikawa A. Mitochondrial DNA repair by photolyase. Mutat Res 1992; 273:231–236.

67. Pinz KG, Bogenhagen DF. Efficient repair of abasic sites in DNA by mitochondrial enzymes. Mol Cell Biol 1998; 18:1257–1265.

68. Bogenhagen DF, Pinz KG, Perez-Jannotti RM. Enzymology of mitochondrial base excision repair. Prog Nucleic Acid Res Mol Biol 2001; 68:257–271.

69. Longley MJ, Prasad R, Srivastava DK, Wilson SH, Copeland WC. Identification of 5'-deoxyribose phosphate lyase activity in human DNA polymerase gamma and its role in mitochondrial base excision repair in vitro. Proc Natl Acad Sci USA 1998; 95:12244–12248.

70. Szczesny B, Hazra TK, Papaconstantinou J, Mitra S, Boldogh I. Age-dependent deficiency in import of mitochondrial DNA glycosylases required for repair of oxidatively damaged bases. Proc Natl Acad Sci USA 2003; 100:10670–10675.

71. Nishioka K, Ohtsubo T, Oda H, Fujiwara T, Kang D, Sugimachi K, Nakabeppu Y. Expression and differential intracellular localization of two major forms of human 8-oxoguanine DNA glycosylase encoded by alternatively spliced OGG1 mRNAs. Mol Biol Cell 1999; 10:1637–1652.

72. Takao M, Aburatani H, Kobayashi K, Yasui A. Mitochondrial targeting of human DNA glycosylases for repair of oxidative DNA damage. Nucleic Acids Res 1998; 26:2917–2922.

73. de Souza-Pinto NC, Eide L, Hogue BA, Thybo T, Stevnsner T, Seeberg E, Klungland A, Bohr VA. Repair of 8-oxodeoxyguanosine lesions in mitochondrial dna depends on the oxoguanine dna glycosylase (OGG1) gene and 8-oxoguanine accumulates in the mitochondrial dna of OGG1-defective mice. Cancer Res 2001; 61:5378–5381.

74. Karahalil B, de Souza-Pinto NC, Parsons JL, Elder RH, Bohr VA. Compromised incision of oxidized pyrimidines in liver mitochondria of mice deficient in NTH1 and OGG1 glycosylases. J Biol Chem 2003; 278:33701–33707.

75. Shadel GS. Yeast as a model for human mtDNA replication. Am J Hum Genet 1999; 65:1230–1237.

76. Moustacchi E, Heude M. Mutagenesis and repair in yeast mitochondrial DNA. Basic Life Sci 1982; 20:273–301.

77. Singh KK, Sigala B, Sikder HA, Schwimmer C. Inactivation of Saccharomyces cerevisiae OGG1 DNA repair gene leads to an increased frequency of mitochondrial mutants. Nucleic Acids Res 2001; 29:1381–1388.

78. Alseth I, Eide L, Pirovano M, Rognes T, Seeberg E, Bjoras M. The Saccharomyces cerevisiae homologues of endonuclease III from Escherichia coli, Ntg1 and Ntg2, are both required for efficient repair of spontaneous and induced oxidative DNA damage in yeast. Mol Cell Biol 1999; 19:3779–3787.

79. You HJ, Swanson RL, Harrington C, Corbett AH, Jinks-Robertson S, Senturker S, Wallace SS, Boiteux S, Dizdaroglu M, Doetsch PW. Saccharomyces cerevisiae Ntg1p and Ntg2p: broad specificity N-glycosylases for the repair of oxidative DNA damage in the nucleus and mitochondria. Biochemistry 1999; 38:11298–11306.

80. O'Rourke TW, Doudican NA, Mackereth MD, Doetsch PW, Shadel GS. Mitochondrial dysfunction due to oxidative mitochondrial DNA damage is reduced through cooperative actions of diverse proteins. Mol Cell Biol 2002; 22:4086–4093.

81. Foury F, Lahaye A. Cloning and sequencing of the PIF gene involved in repair and recombination of yeast mitochondrial DNA. EMBO J 1987; 6:1441–1449.

82. Zelenaya-Troitskaya O, Newman SM, Okamoto K, Perlman PS, Butow RA. Functions of the high mobility group protein, Abf2p, in mitochondrial DNA segregation, recombination and copy number in Saccharomyces cerevisiae. Genetics 1998; 148:1763–1776.

83. Ling F, Morioka H, Ohtsuka E, Shibata T. A role for MHR1, a gene required for mitochondrial genetic recombination, in the repair of damage spontaneously introduced in yeast mtDNA. Nucleic Acids Res 2000; 28:4956–4963.

84. Eyre-Walker A, Awadalla P. Does human mtDNA recombine? J Mol Evol 2001; 53:430–435.

85. Chatterjee A, Singh KK. Uracil-DNA glycosylase-deficient yeast exhibit a mitochondrial mutator phenotype. Nucleic Acids Res 2001; 29:4935–4940.

86. Vongsamphanh R, Fortier PK, Ramotar D. Pir1p mediates translocation of the yeast Apn1p endonuclease into the mitochondria to maintain genomic stability. Mol Cell Biol 2001; 21:1647–1655.

87. Donahue SL, Corner BE, Bordone L, Campbell C. Mitochondrial DNA ligase function in Saccharomyces cerevisiae. Nucleic Acids Res 2001; 29:1582–1589.

88. Mason PA, Lightowlers RN. Related Articles, Links Abstract Why do mammalian mitochondria possess a mismatch repair activity? FEBS Lett 2003; 554:6–9.

89. Fikus MU, Mieczkowski PA, Koprowski P, Rytka J, Sledziewska-Gojska E, Ciesla Z. The product of the DNA damage-inducible gene of Saccharomyces cerevisiae, DIN7, specifically functions in mitochondria. Genetics 2000; 154:73–81.

90. Koprowski P, Fikus MU, Dzierzbicki P, Mieczkowski P, Lazowska J, Ciesla Z. Enhanced expression of the DNA damage-inducible gene DIN7 results in increased mutagenesis of mitochondrial DNA in Saccharomyces cerevisiae. Mol Genet Genomics 2003; 269:632–639.

91. Senbongi H, Ling F, Shibata T. A mutation in a mitochondrial ABC transporter results in mitochondrial dysfunction through oxidative damage of mitochondrial DNA. Mol Gen Genet 1999; 262:426–436.

92. Babcock M, de Silva D, Oaks R, Davis-Kaplan S, Jiralerspong S, Montermini L, Pandolfo M, Kaplan J. Regulation of mitochondrial iron accumulation by Yfh1p, a putative homolog of frataxin. Science 1997; 276:1709–1712.

93. Wilson RB, Roof DM. Respiratory deficiency due to loss of mitochondrial DNA in yeast lacking the frataxin homologue. Nat Genet 1997; 16:352–357.

94. Rotig A, de Lonlay P, Chretien D, Foury F, Koenig M, Sidi D, Munnich A, Rustin P. Aconitase and mitochondrial iron-sulphur protein deficiency in Friedreich ataxia. Nat Genet 1997; 17:215–217.

95. Mitsuhashi N, Miki T, Senbongi H, Yokoi N, Yano H, Miyazaki M, Nakajima N, Iwanaga T, Yokoyama Y, Shibata T, Seino S. MTABC3, a novel mitochondrial ATP-binding cassette protein involved in iron homeostasis. J Biol Chem 2000; 275:17536–17540.

96. Barros MH, Netto LE, Kowaltowski AJ. H(2)O(2) generation in Saccharomyces cerevisiae respiratory pet mutants: effect of cytochrome c. Free Radic Biol Med 2003; 35:179–188.

97. Laun P, Pichova A, Madeo F, Fuchs J, Ellinger A, Kohlwein S, Dawes I, Frohlich KU, Breitenbach M. Aged mother cells of Saccharomyces cerevisiae show markers of oxidative stress and apoptosis. Mol Microbiol 2001; 39:1166–1173.

98. Longo VD, Gralla EB, Valentine JS. Superoxide dismutase activity is essential for stationary phase survival in Saccharomyces cerevisiae. Mitochondrial production of toxic oxygen species in vivo. J Biol Chem 1996; 271:12275–12280.

99. Davermann D, Martinez M, McKoy J, Patel N, Averbeck D, Moore CW. Impaired mitochondrial function protects against free radical-mediated cell death. Radic Biol Med 2002; 33:1209–1220.

100. Machida K, Tanaka T, Fujita K, Taniguchi M. Farnesol-induced generation of reactive oxygen species via indirect inhibition of the mitochondrial electron transport chain in the yeast Saccharomyces cerevisiae. J Bacteriol 1998; 180:4460–4465.

101. Rasmussen AK, Chatterjee A, Rasmussen LJ, Singh KK. Mitochondria-mediated nuclear mutator phenotype in Saccharomyces cerevisiae. Nucleic Acids Res 2003; 31:3909–3917.

102. Heidenreich E, Wintersberger U. Replication-dependent and selection-induced mutations in respiration-competent and respiration-deficient strains of Saccharomyces cerevisiae. Mol Gen Genet 1998; 260:395–400.

103. Ricchetti M, Fairhead C, Dujon B. Mitochondrial DNA repairs double-strand breaks in yeast chromosomes. Nature 1999; 402:96–100.

104. Yu X, Gabriel A. Patching broken chromosomes with extranuclear cellular DNA. Mol Cell 1999; 4:873–881.

Part IV

Mismatch Repair

This section of the book deals with DNA mismatch repair (MMR), a process vital to all organisms for maintaining genomic stability. MMR contributes to the high degree of fidelity of DNA replication and participates in other important events where correction of base–base mispairs or other types of damaged base–base mispairing may occur. Chapter 21 provides a focused, biochemically oriented review of this process with an emphasis on bacterial MMR components together with appropriate comparisons to their eukaryotic homologs. The currently debated models for MMR are presented here for the reader to compare and evaluate.

Chapter 22–the second (and final) chapter of this short section–is a brief review of another, and less well-understood, mismatch-repair system in *Escherichia coli* termed the very short patch (VSP) repair system mediated by the Vsr protein. A biochemical focus as well as its relationship to the general MMR system in bacteria is presented.

21

Mechanism of DNA Mismatch Repair from Bacteria to Human

Samir Acharya and Richard Fishel
Department of Molecular Virology, Immunology and Medical Genetics,
Ohio State University, Columbus, Ohio, U.S.A.

1. INTRODUCTION

Cells have evolved efficient, integrated pathways for detecting and avoiding genetic catastrophe. Events that have the potential to develop into such a fate include genomic lesions resulting from exposure to DNA damaging agents, physical and chemical stress, aberrant recombination events or replication-induced DNA biosynthetic errors. Left unresolved, such damage might result in mutability, genomic instability and/or eventual cell death. Specialized DNA repair pathways—nucleotide excision repair, base excision repair, the recombination pathways (nonhomologous end joining and double strand break repair), and mismatch repair (MMR)—have evolved to correct these errors. Typically, specific damage-recognition proteins initiate the repair process followed by the coordinated assembly of multiprotein complexes that repair the damage. Loss of MMR results in a mutator phenotype and is closely associated with hereditary nonpolyposis colorectal cancer (HNPCC) (1,2). Understanding the basis for this link has propelled the research on MMR and a considerable genetic, structural, and biochemical knowledge has accumulated. However, the precise mechanism of MMR is fervently debated. Here, we review the biochemical history of the MMR field and the various mechanisms proposed, and discuss the mechanism that appears most consistent with the biochemical observations.

1.1. Introduction to the Mechanism—The Bacterial Paradigm

The existence of MMR was postulated to explain the phenomenon of gene conversion in lower eukaryotes and bromouracil-mediated mutagenesis in bacteria (3,4). The connection between mutability and MMR was subsequently established in *Pneumococcus* and *Escherichia coli* and involved the function of "mutator" genes (5–11). The MMR-specific genes were originally identified in Gram-negative bacteria and include the mutator genes, MutS, MutL, MutH, and MutU (UvrD). MutS and MutL genes are conserved throughout evolution with multiple homologues in eukaryotes (Table 1).

Table 1 MutS and MutL Homologues

E. coli	Yeast	Human		
		Locus	Functional heterodimers	Function (substrate specificity)
MutS	MSH1	None		
	MSH2	hMSH2 2p21-16.3		
	MSH3	hMSH3 5q11-q12	hMSH2–hMSH6	MMR (base–base; oxidative lesions; small insertion/deletions loops)
	MSH4	hMSH4 1p31	hMSH2–hMSH3	MMR (base–base; insertion/deletions)
	MSH5	hMSH5 6p22.3-21.3	hMSH4–hMSH5	Meiosis (Holliday junctions)
	MSH6	hMSH6 2p21-16.3		
MutL	MLH1	hMLH1 3p21.3	hMLH1–hPMS2	MMR (base–base; oxidative lesions; small insertion/deletions loops)
	PMS1	hPMS2 7p22	hMLH1–hPMS1	MMR (base–base; insertion/deletions loops)
	MLH2	hPMS1 2p31.33	hMLH1–hMLH3	Meiosis (Holliday junctions)
	MLH3	hMLH3 14q24.3		

However, outside of Gram-negative bacteria, MutH or its homologues have not yet been identified.

The MMR pathway corrects mismatches in DNA that arise primarily from replicative errors and recombination between heterologous regions of DNA (12–14). The DNA adenine methylation (Dam)-instructed MMR pathway is the most studied pathway of MMR in *E. coli* and involves the functions of MutS, MutL, and MutH. MutS has been identified as a mismatch recognition protein that binds a variety of mismatches (15,16). MutL acts as a molecular chaperone connecting MutS to the downstream effectors in the pathway—the hemi-methylated GATC-specific endonuclease MutH and the helicase UvrD (17–20). Interaction of the MutS–MutL complex with MutH triggers its endonuclease activity resulting in the generation of a single strand break on the transiently unmethylated nascent strand (18,21,22). This activity sets up a strand-specific directional repair system wherein UvrD, in conjunction with MutS and MutL, catalyzes DNA unwinding from the strand break towards the mismatch (20,22,23). Exonucleolytic degradation of the unwound strand by one of four exonucleases (ExoX, RecJ, ExoVIII, and ExoI) creates a gap that specifically spans the entire region between the strand break and the mismatch and can extend up to 1 kb in length (21,24–29). The replicative polymerase machinery (Pol III holoenzyme), the single-strand binding protein (SSB) and DNA ligase complete the repair process (Fig. 1).

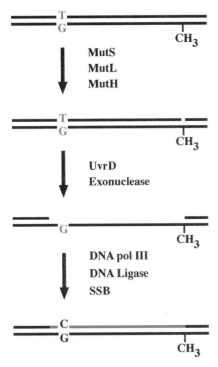

Figure 1 Bacterial MMR. The Dam-instructed MMR in *E. coli* involves the functions of MutS, MutL, MutH, UvrD and one of four exonucleases (see text). MutS and MutL initiate the repair process and activate MutH to incise a hemi-methylated GATC site. The helicase UvrD unwinds the strand at the incision towards the mismatch in a MutS–MutL-dependent manner. Exonuclease action on the unwound strand results in a gap that spans the DNA between the mismatch and the strand break. DNA polIII, SSB, and DNA ligase repair the gap.

1.2. The Eukaryotic MutS and MutL Homologues

The prokaryotic MutS and MutL proteins function as homodimers, and repair a variety of mismatches ranging from single nucleotide mismatches to insertion deletion loops. The eukaryotic MutS homologues (MSH) and MutL homologues (MLH) function as heterodimers and are specific to the kinds of mismatches being repaired (30–38). Single nucleotide mismatches are primarily repaired via the MSH2–MSH6 heterodimer and the MLH1–PMS1 heterodimer (hMLH1–hPMS2 in human; Table 1). Insertion–deletion loop repair involves primarily MSH2–MSH3 and MLH1–MLH2 heterodimers (hMLH1–hPMS1 in human; Table 1). Consequently, mutations in human MSH2 or MLH1 correlate with lack of repair in vivo and in vitro and account for the majority of HNPCC cases (1,39–44). These mutations affect the functions of these proteins including heterodimer formation, mismatch recognition, and interaction with downstream effectors. The cells that lack these proteins also acquire resistance to DNA damaging agents and may have an altered apoptotic response (45–51). The association of these mutations with a predisposition of the cells to tumorigenesis underscores the importance of mismatch repair in the regulation of cell survival and its fate (37,51).

2. BIOCHEMISTRY OF MISMATCH REPAIR PROTEINS

Central to the mechanism of mismatch repair is the processing of adenine nucleotides by MutS and MutL proteins. Adenine nucleotides modulate the first step of MMR, recognition and binding of MutS to a mismatch, as well as subsequent events until the repair is completed. MutS and MutL proteins and their eukaryotic homologues contain consensus ATP binding/hydrolysis motifs—the Walker Box in MutS and the Bergerat fold in MutL—that are conserved amongst several nucleotide-binding proteins (52–54). The Walker box domain lies in the C-terminus region of the MutS family of proteins and the Bergerat fold is in the N-terminus of the MutL family. Much of the debate on MMR mechanism centers on the use of ATP binding and hydrolysis by these proteins in carrying out their individual functions.

2.1. Mismatch-Dependent Stimulation of ATP Hydrolysis by MutS Proteins

Bacterial MutS and its eukaryotic homologues MSH2–MSH6 and MSH2–MSH3 display a weak steady state ATP hydrolase (ATPase) activity that is uniquely stimulated by mismatch-containing DNA (55–61). Interestingly, even perfectly paired duplex DNA is capable of stimulating the ATPase activity of these proteins. However, ATPase activity in the presence of mismatch-containing DNA is 4–5-fold more than in the absence of DNA and 2–3-fold more than that in the presence of perfectly paired duplex DNA. This stimulation of the ATPase activity is critically dependent on the concentration of salt and peaks in the physiological range of 100–160 mM. The differential stimulation of ATPase by mismatch-containing DNA and a perfectly paired duplex DNA is not apparent at low salt concentrations (at or below 50 mM) or at high salt concentrations (at or above 200 mM) (58,60,61). Furthermore, the range of salt concentration for maximal differentiation in ATPase activity between a mismatch-containing DNA and a perfectly paired duplex DNA closely correlates with the salt profile observed for MMR in vitro (56,58,62). Perhaps the most intriguing aspect of the ATPase cycle of all MutS proteins studied to date is the fact that the rate-limiting step is the exchange of ADP for ATP and not the cleavage of the terminal phosphate (55,61). The rate of exchange is enhanced significantly in the presence of a mismatch in the DNA. Different mismatches provoke this exchange with varying efficiencies that correlate closely with their stimulation of ATP hydrolysis by MutS proteins and their repair in vivo (57,58,62–67). Thus, mismatches act as ADP→ATP exchange factors similar to the guanine nucleotide exchange factors that trigger a signaling cascade in the G-protein signaling pathways (68).

2.2. Mode of Mismatch Binding and Recognition by MutS Proteins

MutS and its eukaryotic homologues, MSH2–MSH6 or MSH2–MSH3, perform two important functions in the initiation of mismatch repair: recognition of a mismatch in the context of a large excess of perfectly paired duplex DNA, and mismatch-dependent recruitment of, or "relay of signal to," downstream effectors. They have the unique ability to recognize and repair a variety of mismatches in the DNA albeit with different affinities. The mismatch binding domain resides in the N-terminus of the protein whereas binding competence is modulated by the adenine nucleotide bound at the C-terminal domain of the protein. MutS proteins undergo subtle conformational changes depending on the specific nucleotide bound (69,70). These conformational states define the function performed by the protein. The ADP-bound conformation of MutS proteins is competent for stable mismatch binding and recognition. The ATP-bound conformation facilitates the disassociation of MutS proteins from the mismatch and subsequent translocation along the adjoining DNA (61,69,71). A consequence of the rapid ADP→ATP exchange at the mismatch followed by ATP-induced dissociation from the mismatch is the iterative loading of several MutS proteins on the DNA (61,69). The crystal structures of the bacterial MutS proteins in the presence of mismatch-containing DNA confirm the ADP-bound MutS as the mismatch binding conformation (72). Electron microscopy and biochemical studies suggest that prokaryotic and eukaryotic MutS proteins can translocate at least 1–2 kb away from the mismatch in the presence of ATP (21,61,69,73). Moreover, the ATP-bound conformation is also required for subsequent functions e.g., interaction with the MutL proteins. The switch from one conformation to another thereby outlines a commitment by the MutS proteins to performing one function (mismatch recognition/binding: ADP-bound conformation) or another (translocation on the DNA and interaction with downstream effectors: ATP-bound conformation). The unique ability of the mismatch to provoke ADP→ATP exchange imparts an elegant means of regulating this separation of functions.

The crystal structures of bacterial MutS protein bound to several mismatch-containing DNA molecules, as well as the individual structures and biophysical studies of various mismatch-containing DNA molecules, suggest that subtle conformational changes in both DNA and protein are involved in the basic recognition of a mismatch and additionally specify the determinants of mismatch binding (59,72,74,75). An obvious insight from these studies has been the asymmetry displayed by bacterial MutS dimer in mismatch binding. Only one monomer, which retains the bound ADP, contacts the mismatch. The other monomer is free of nucleotide. Both monomers maintain limited nonspecific contacts with the DNA backbone in a clamp-like structure. Such asymmetric recognition of a mismatch is also conserved among eukaryotic MSH proteins; in the MSH2–MSH6 heterodimer, MSH6 contacts the mismatch (59,76). Asymmetry is further manifest in adenine nucleotide binding amongst MutS proteins with one monomer displaying a higher affinity for binding the nucleotide and consequently there appears to be a sequential hydrolysis of the bound ATP (77–79). Taken together, these studies suggest a sequential mechanism of ADP release and ADP→ATP exchange provoked by the mismatch.

Among the amino acids that are critical for stable mismatch binding by MutS are the conserved phenylalanine and glutamate in the Phe-X-Glu motif in the N-terminus domain (80–82). A general theme of binding emerging from the bacterial and eukaryotic studies entails the mismatched-base specific stacking with the conserved phenylalanine and associated H-bonding with the conserved glutamate residue (59,80). Such an interaction leads to local unwinding or destabilization of neighboring base pairs (83). Interestingly, with all the mismatches studied, MutS binding results in a 60° bend in the DNA around the mismatch (75). The mismatches themselves possess varying degrees of local flexibility that is different from perfectly paired duplex DNA (75,83). This has led to the suggestion that local flexibility of the DNA might prove to be a critical requisite for stable mismatch binding and discrimination by MutS. It also underscores the importance of the sequence context around the mismatch for recognition. Together, these studies indicate that the kinetics of local DNA distortion and nucleotide exchange are likely to both play a role in determining the efficiency of MMR.

2.3. Role of ATP Hydrolysis by MutS Proteins

The role of ATP hydrolysis in the initiation of mismatch repair is controversial. The role ranges from its proposed use in propelling the movement of MutS or MutS–MutL complexes along the DNA (akin to the action of multimeric helicases) to scanning for mismatches by MutS proteins on the DNA to recycling of MutS proteins from the DNA (84–86). Several observations point to its function in turnover of MutS proteins from DNA (61,69). First, MutS or MutS–MutL complexes are able to traverse at least 1 kb of DNA following exchange at the mismatch by the nonhydrolyzable analogue, ATPγMS. Second, the stimulation of ATPase of MutS proteins by mismatch-containing DNA is dependent on the availability of free DNA ends in the substrate. Double-stranded DNA ends also contribute to the exchange of ADP for ATP and account for the modest stimulation of ATP hydrolysis by MutS proteins in the presence of perfectly paired duplex DNA substrates. When the DNA substrate is circular or has blocked ends, nucleotide exchange is abolished in the presence of perfectly paired duplex DNA—but not in the presence of DNA containing a mismatch. Free DNA ends are a very rare event in vivo and are processed by different pathways e.g., nonhomologous end joining or double-strand break repair. Thus, perfectly paired duplexes do not appear to possess inherent structural features that would promote successive cycles of exchange of ADP for ATP, a step essential for continuous steady state ATP hydrolysis. Third, the ATP-bound MutS clamps require MutL to turnover from end-blocked DNA (see below) and ATP hydrolysis by MutS is intricately associated with that turnover. The presence of a nonhydrolyzable analogue of ATP, ATPγS, prevents turnover and results in stable MutS clamps on the DNA. These observations have redefined the role of ATP hydrolysis by MutS proteins in MMR. Hydrolysis enables recycling of the mismatch binding proficient state of MutS—the ADP-bound conformation. Furthermore, the translocation of the MutS proteins on the DNA occurs by diffusion and is independent of ATP hydrolysis. ATP hydrolysis-independent diffusion of protein complexes on DNA and the use of nucleotide hydrolysis for recycling of the active protein conformation are not unique to the MMR system. Analogies exist in the G-protein signaling mechanisms, clamp unloaders in replication and the SbcCD mediated processing of double-stranded DNA ends (68,87,88).

2.4. Function of MutL—A Second Switch in Mismatch Repair

Like MutS proteins, the MutL proteins from bacteria to human share several similarities in function. They also undergo nucleotide-dependent conformational changes and are the first effectors to interact with the ATP-bound MutS (89–96). Nucleotide binding by the MutL proteins is not essential for their interaction with MutS proteins and they do not interact with the ADP-bound conformation of MutS on the mismatch-containing DNA (61,96–98). In addition to their role as molecular chaperones that connect the MutS proteins to the downstream protein machinery in MMR, studies on bacterial MutL have revealed an additional function that these proteins perform: they assist in the turnover of stable ATP-bound MutS clamps from the DNA by provoking ATP hydrolysis (61). Furthermore, MutL acts as a protein-induced molecular switch in bacterial MMR. Interaction with MutH, the immediate downstream effector in Gram-negative bacteria, provokes ATP binding by MutL and in turn activates the hemi-methylated GATC-endonuclease activity of MutH. The coordinated formation of two adenine nucleotide-dependent switches—mismatch provoked MutS switch and protein induced MutL switch—thus appears to be a recurring feature of MMR.

2.5. The Excision Reaction in Mismatch Repair

Steps subsequent to mismatch recognition involve the combined action of helicases and exonucleases resulting in a gap. The creation of a precise unidirectional excision tract that initiates from a strand-break downstream of a mismatch and proceeds towards the mismatch is a hallmark of MMR that is conserved from bacteria to human (64–66). Both MutS and MutL proteins are essential for the excision reaction and they enhance the processivity of the helicase/exonuclease action from a strand break towards the mismatch (22,99). The eukaryotic $5'\rightarrow 3'$ exonuclease, ExoI is involved in MMR and interacts with both MSH2 and MLH1 (100–104). Human ExoI displays a predominant $5'\rightarrow 3'$ exonuclease activity as well as a $3'\rightarrow 5'$ exonuclease activity that initiates from a strand-break (99,105). The activation and processivity of the exonuclease activity from a strand break is enhanced in the presence of the human MutS heterodimer hMSH2–hMSH6. Both hMSH2–hMSH6 and hMLH1–hPMS2 are required for the creation of a $3'$-excision tract. hMLH1–hPMS2 appears dispensable for the $5'$-excision tract, although it is necessary for enhancing specificity of excision on the mismatch-containing duplex. The nature of the excision products formed appears to be discrete, indicative of pause/restart sites during this process. Similar to the prokaryotic system, removal of the mismatch signal abolishes further excision beyond 100–150 bp upstream of the mismatch. These remarkable biochemical similarities between the prokaryotic and the eukaryotic MutS and MutL proteins and the overall excision reaction point to a common mechanism of mismatch repair mediated by these proteins (64–66,106,107).

3. MECHANISM OF MISMATCH REPAIR

3.1. The Static *Trans*-Activation Model

Two major mechanistic schemes have been proposed that incorporate the aforementioned biochemical findings (Fig. 2). These schemes differ in the mode of signal transduction between a mismatch site and the downstream signal—a strand break on the incorrect strand—that initiates the excision process. Signal transduction in trans, as proposed from the crystal structure of bacterial MutS, suggests assembly of the mismatch repair complex at the mismatch site followed by a search through space (trans) by the MutS–MutL complex for the downstream strand break (85). The static *trans*-activation model proposes that MutS uses ATP to scan the DNA for mismatches and interaction with a mismatch results in a long-lived MutS–ATP intermediate at mismatch. This prolonged lifetime of such an intermediate enables assembly, at the mismatch site, of the MMR complex that can search for downstream signals. Such a search is independent of the length of the intervening DNA as DNA looping achieves accessibility to the strand break. The apparent randomness of the search makes it difficult to envisage the specificity and unidirectionality of the excision process, by this mode of signal transduction. A static assembly of a single mismatch-repair complex at or near the mismatch site is contrary to the observed

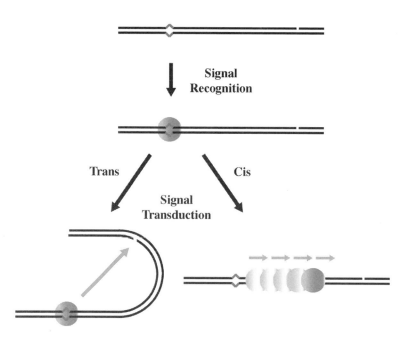

Figure 2 Signal transduction modes between a mismatch and the downstream strand break. Following recognition of a mismatch, the MMR machinery can access the downstream signal (strand break) either directly (trans) via DNA looping or by tracking along the DNA (cis). Access to the strand break initiates the excision phase of MMR.

ATP-dependent dynamism of the MSH or MSH–MLH complexes on mismatch-containing DNA in vitro. While the available evidence does not favor a static MMR complex at or near a mismatch, it does not eliminate the possibility of *trans*-activation of the excision reaction presented by this model and more studies are required to test the hypothesis.

3.2. The Hydrolysis-Dependent Translocation Model

In contrast to the trans mode, a *cis*-mode of signal transduction invokes tracking of the repair components along the DNA to the downstream strand break. According to this scheme, a mismatch site should act as a nucleation site for the assembly of the MMR components followed by translocation. The observed biochemical properties of MSH and MLH proteins agree well with such a mechanism and two models, the hydrolysis-dependent translocation model and the molecular switch model, have been proposed. The hydrolysis-dependent translocation model proposes the assembly of a MSH–MLH complex that translocates along the DNA fueled by hydrolysis of ATP (13,84). The MSH–MLH complex thus acts as a signal transducer for the mismatch signal and its interaction with the excision machinery initiates a directional excision reaction. Though compelling, the role of ATP hydrolysis in propelling the movement of the MSH or MSH–MLH complexes along the DNA is contrary to the reported biochemical observations (see above) and remains controversial. MSH proteins are unable to exchange ADP for ATP in the absence of mismatch—a condition that would not be encountered once they have dissociated from the mismatch to the adjoining duplex DNA. Moreover, the visualization of MutS–MutL foci associated with a moving DNA polymerase in bacteria suggests the existence of not one, but multiple MSH or MSH–MLH complexes in vivo (108).

3.3. The Molecular Switch Model

An alternative mechanism for mismatch repair, the molecular switch model incorporates the observed dynamism and redundancy of MMR (Fig. 3) (61). The model proposes that the mismatch is only transiently associated with the ADP-bound MSH proteins and provokes rapid exchange of ADP for ATP. The ATP-bound MSH clamps move away from the mismatch allowing another round of binding and exchange. This results in iterative loading of multiple MSH clamps that diffuse freely along the DNA adjoining the mismatch and interact with MLH proteins (61,69,109). Multiple clamps outline the DNA around a mismatch and serve as a marker that can explain the observed precision and directionality of the subsequent excision event. The MSH–MLH clamps provide a platform for the further loading of downstream effectors in the MMR process. Probable candidates that may productively interact with the MSH–MLH clamps are helicases, exonucleases, and replication components that would complete the assembly of a mismatch "repairosome" at a pre-existing or nascent strand break on the newly replicated strand. Directed helicase/exonuclease activity is stimulated by the interaction with the MSH–MLH clamps. It is proposed that this complex has a distinct half-life on the DNA and ATP hydrolysis by the MSH proteins results in disassociation of the "repairosome." However, the existence of a continuous stream of redundant MSH–MLH clamps along the DNA (emanating from the mismatch and proceeding in a direction opposing the exonuclease direction) ensures reassembly and another round of processive unwinding/degradation of the mismatch-containing strand. A consequence of MSH clamp turnover and reassembly would be pauses in the exonucleolytic phase of the reaction and result in the discrete nature of products, as observed in vitro (99). The degradation of the DNA strand beyond the mismatch removes the signal for loading of more MSH clamps and thus prevents reloading of the repairosome once it has disassociated. The outcome of such a coordinated process would be a precise and unidirectional excision event that encompasses the DNA between a strand break and a mismatch. The additional function of MLH proteins in facilitating turnover of MutS clamps ensures timely removal of any remaining clamps on the DNA.

3.4. Downstream Effectors in MMR

The last stage that completes the process of mismatch repair is the resynthesis of the resultant gap in the DNA. Several studies have revealed the components of the replication machinery that interact with MMR components. β-clamp in bacteria and its eukaryotic homologue, PCNA, were amongst the first to be discovered (110–112). PCNA has been suggested to play a significant role in MMR both during DNA synthesis as well at an earlier step. It forms strong associations with MSH3 and MSH6 (110,113,114). Such associations are presumed to enhance the specificity of the MSH2–MSH6 and MSH2–MSH3 complexes for binding mispairs (115,116). While a human MMR-specific helicase is yet to be discovered, potential candidates that have been considered for such a function include the Bloom helicase, BLM or the Werner helicase, WRN (117,118). BLM is implicated in resolution of replication forks and interacts strongly with MLH1 (117,119). However, cell extracts deficient in BLM or lymphoblastoid cells deficient in WRN do not show MMR defects,

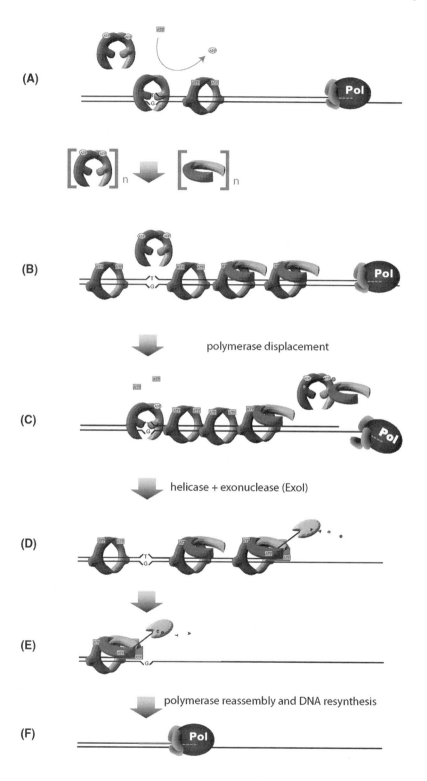

Figure 3 (*Caption on facing page*)

suggesting the involvement of other helicases, a redundant helicase function or no helicase requirement in humans (99, 117–119). It is likely that such interactions with replication machinery components possess the potential to displace or alter the function of the existing replication machinery.

3.5. The MMR-Replication Conundrum

One of the major questions in the field is how the MMR process coordinates with the ongoing DNA replication machinery. The close association of the MMR components with the replication apparatus is evident from the co-localization of the MutS–MutL foci with the polymerase foci upon induction of DNA damage in bacteria (108). Prokaryotic studies with T4 and T7 polymerase replication systems and *E. coli* DNA Pol III have given clues as to how the MMR machinery "catches up" with the forward moving replication system (120). The observed necessity for dimerization of the two polymerases corresponding to the lagging and the leading strand is critical in this regard (121). Constant ATP hydrolysis-dependent recycling of the lagging strand polymerase enables the polymerase to reload on the new primer synthesized by primase at the base of the replication fork (122). The need for dimerization creates a situation wherein the leading strand polymerase must pause on the DNA until the reassembly of the lagging strand polymerase on the new primer is accomplished. Dimerization of the two polymerases then resumes DNA synthesis on both strands. The recycling of the polymerase on the lagging strand and pausing on the leading strand sets up conditions wherein the MMR machinery could access either a free DNA end on the lagging strand or polymerase components on the leading strand prior to polymerase reassembly. Consequent displacement or realignment of the leading strand polymerase as a result of interaction with the MMR components would expose the free end on the leading DNA strand to helicase/exonuclease proteins resulting in degradation of the nick-containing strand, followed by repair of the mismatch. The presence of multiple MSH–MLH clamps along the

---

Figure 3 *(Facing page)* The molecular switch model of MMR. (A) MSH proteins function as mismatch sensors that recognize mismatches in the ADP-bound state. Mismatches provoke ADP→ATP exchange and the protein switches to the ATP-bound sliding-clamp form. Iterative binding and exchange results in multiple clamps that diffuse along the DNA. (B) The MLH proteins associate with the ATP-bound sliding clamps. Multiple MSH–MLH complexes outline the repair tract and interact with the downstream effectors. (C) Interaction of the MSH–MLH complex with the polymerase machinery is proposed to displace or realign the polymerase and expose the newly replicated single-strand end. MLH proteins also enhance recycling of MSH clamps from the DNA via ATP hydrolysis. (D) Interaction of MLH proteins with helicases/exonucleases results in the formation of a ternary MSH–MLH–helicase/exonucleases complex at the exposed strand-break. Such an interaction is proposed to activate ATP binding by the MLH proteins and enhance unwinding of the DNA from the exposed strand-break towards the mismatch. Concomitant activation of the exonuclease (ExoI) results in degradation of the unwound strand. The MSH–MLH–helicase/ExoI complex has a distinct half-life and dissociates at defined intervals. The requirement of a helicase is not clear in humans and ExoI in the presence of MSH-MLH clamps is sufficient for directed degradation of the nascent strand from the strand-break towards the mismatch (see text). The presence of multiple MSH–MLH clamps ensures reassembly of the complex and degradation resumes. (E) The degradation of the unwound strand continues beyond the mismatch. Subsequent dissociation results in an exposed 3'-end and (F) The polymerase reassembly at the exposed 3'-end complete the repair process. (*See color insert.*)

length of DNA behind a replication tract provides a signal for constant activation/reassembly of helicase/exonuclease proteins (see above). As the strand is degraded beyond the mismatch, this signal is lost and subsequent disassociation of the MSH–MLH clamps is an indication for the polymerase machinery to realign (or reassemble afresh at the site) and synthesize the new strand. For replication-induced mismatches created on the lagging strand, the window of opportunity for MMR to access a strand break constantly presents itself when the polymerase is being recycled and the strand break is thereby potentially exposed. However, on the leading strand, access to a strand break depends entirely on the probability of the MMR system to "catch" the paused polymerase in the time interval between the loading of the lagging strand polymerase and its subsequent dimerization with the leading strand polymerase. This raises the possibility that repair on the leading strand should be less frequent than that on the lagging strand. Indeed, studies in bacterial systems and yeast have demonstrated that mutation frequencies vary between the strands with the leading strand having a higher mutation frequency than the lagging strand (123,124). Higher processivity of the polymerase on the leading strand has been suggested as the probable cause for the observed higher mutation frequency (125). In addition to polymerase processivity, differential accessibility of the MMR machinery to a strand break may further enhance the observed mutability on the leading strand (126).

4. IMPLICATIONS

Several studies have suggested that mismatch repair proteins, MSH2 and MLH1 are involved in DNA-damage-dependent signaling (48,127–133). The existence of a threshold level of multiple clamps on the DNA offers a means for MSH proteins to act as damage sensors and thereby act as signaling proteins for the activation of MMR as well as apoptotic pathways (49,134). Such a mechanism of signal transduction may explain the resistance of MSH2 and MLH1 deficient cells to DNA damaging agents like MNNG. Drug resistance in these cells is proposed to be due to a lack of damage sensor and/or transducer (MSH or MSH–MLH clamps) signaling to the apoptotic pathway. Thus, lack of MMR in these cells would not only result in an increased mutation frequency but also give a selective advantage to such cells by allowing them to escape DNA-damage induced apoptosis. Likewise, a persistence of active MSH or MSH–MLH clamps on the DNA (due to defective turnover as a result of mutations in MSH proteins or absence of/mutant MLH proteins) would provide a constant "amplified" signal for other pathways to salvage the DNA or lead to apoptosis. Furthermore, the molecular switch model provides a means of regulation of MMR. The equivalence in the amount of the MSH and MLH proteins in the cell is evident and implied from the model. Any change in the relative amounts or modification of either of the proteins would skew the signaling sensor (MSH or MSH–MLH clamps) and alter the threshold level signaling to apoptosis, in addition to having an impact of mismatch repair.

The functions of another MutS heterodimer (MSH4–MSH5) and MutL heterodimer (MLH1–MLH3) are as yet unknown. These proteins act specifically during meiosis and are important for the successful progression of meiosis (135–144). The proposed mechanism of MMR and the implicit roles of the MSH and MLH proteins in signal transfer provides a framework to understand their function during meiosis, where the primary form of DNA "aberration," recognized by MSH4–MSH5 is a Holliday junction (145,146). Further studies on the basic mechanism of MMR will aid in our understanding of the links between the MSH and/or MLH proteins and the various damage signaling pathways in a cell.

ACKNOWLEDGMENTS

The authors are indebted to Drs. Kristine Yoder, Vicki Whitehall, Christoph Schmutte, Michael McIlhatton, and Christopher Heinen for critical reading of the manuscript and invaluable editorial changes.

REFERENCES

1. Muller A, Fishel R. Mismatch repair and the hereditary non-polyposis colorectal cancer syndrome (HNPCC). Cancer Invest 2002; 20:102–109.
2. Peltomaki P. Role of DNA mismatch repair defects in the pathogenesis of human cancer. J Clin Oncol 2003; 21:1174–1179.
3. Holliday RA. A mechanism for gene conversion in fungi. Genet Res 1964; 5:282–304.
4. Witkin EM. Pure clones of lactose negative mutants obtained in *Escherichia coli* after treatment with 5-bromouracil. J Mol Biol 1964; 8:610–613.
5. Treffers HP, Spinelli V, Belser NO. A factor (or mutator gene) influencing mutation rates in *Escherichia coli*. Proc Natl Acad Sci USA 1954; 40:1064–1071.
6. Siegel EC, Bryson V. Mutator gene of *Escherichia coli* B. J Bacteriol 1967; 94:38–47.
7. Tiraby J-G, Fox MS. Marker discrimination in transformation and mutation of pneumococcus. Proc Natl Acad Sci USA 1973; 70:3541–3545.
8. Nevers P, Spatz HC. *Escherichia coli* mutants uvr D and uvr E deficient in gene conversion of lambda-heteroduplexes. Mol Gen Genet 1975; 139:233–243.
9. Wildenberg J, Meselson M. Mismatch repair in heteroduplex DNA. Proc Natl Acad Sci USA 1975; 72:2202–2206.
10. Rydberg B. Bromouracil mutagenesis in *Escherichia coli* evidence for involvement of mismatch repair. Mol Gen Genet 1977; 152:19–28.
11. Rydberg B. Bromouracil mutagenesis and mismatch repair in mutator strains of *Escherichia coli*. Mutat Res 1978; 52:11–24.
12. Radman M, Wagner R. Mismatch repair in *Escherichia coli*. Annu Rev Genet 1986; 20:523–538.
13. Modrich P. DNA Mismatch Correction. Annu Rev Biochem 1987; 56:435–466.
14. Modrich P. Methyl-directed DNA mismatch correction. J Biol Chem 1989; 264:6597–6600.
15. Su S-S, Modrich P. *Escherichia coli* mutS-encoded protein binds to mismatched DNA base pairs. Proc Natl Acad Sci USA 1986; 83:5057–5061.
16. Su SS, Lahue RS, Au KG, Modrich P. Mispair specificity of methyl-directed DNA mismatch correction in vitro.[erratum appears in J Biol Chem 1988 Aug 5;263(22):11015]. J Biol Chem 1988; 263:6829–6835.
17. Grilley M, Welsh KM, Su SS, Modrich P. Isolation and characterization of the *Escherichia coli* mutL gene product. J Biol Chem 1989; 264:1000–1004.
18. Au KG, Welsh K, Modrich P. Initiation of methyl-directed mismatch repair. J Biol Chem 1992; 267:12142–12148.
19. Hall MC, Jordan JR, Matson SW. Evidence for a physical interaction between the *Escherichia coli* methyl-directed mismatch repair proteins MutL and UvrD. EMBO J 1998; 17:1535–1541.
20. Hall MC, Matson SW. The *Escherichia coli* MutL protein physically interacts with MutH and stimulates the MutH-associated endonuclease activity. J Biol Chem 1999; 274:1306–1312.

21. Dao V, Modrich P. Mismatch-, MutS-, MutL-, and Helicase II-dependent Unwinding from the Single-strand Break of an Incised Heteroduplex. J Biol Chem 1998; 273: 9202–9207.

22. Yamaguchi M, Dao V, Modrich P. MutS and MutL activate DNA helicase II in a mismatch-dependent manner. J Biol Chem 1998; 273:9197–9201.

23. Mechanic LE, Frankel BA, Matson SW. *Escherichia coli* MutL loads DNA helicase II onto DNA. J Biol Chem 2000; 275:38337–38346.

24. Wagner R Jr, Meselson M. Repair tracts in mismatched DNA heteroduplexes. Proc Natl Acad Sci USA 1976; 73:4135–4139.

25. Laengle-Rouault F, Maenhaut-Michel G, Radman M. GATC sequence and mismatch repair in *Escherichia coli*. EMBO J 1986; 5:2009–2013.

26. Bruni R, Martin D, Jiricny J. d(GATC) sequences influence *Escherichia coli* mismatch repair in a distance-dependent manner from positions both upstream and downstream of the mismatch. Nucleic Acids Res 1988; 16:4875–4890.

27. Cooper DL, Lahue RS, Modrich P. Methyl-directed mismatch repair is bidirectional. J Biol Chem 1993; 268:11823–11829.

28. Viswanathan M, Burdett V, Baitinger C, Modrich P, Lovett ST. Redundant exonuclease involvement in *Escherichia coli* methyl-directed mismatch repair. J Biol Chem 2001; 276:31053–31058.

29. Burdett V, Baitinger C, Viswanathan M, Lovett ST, Modrich P. In vivo requirement for RecJ, ExoVII, ExoI, and ExoX in methyl-directed mismatch repair. Proc Natl Acad Sci USA 2001; 98:6765–6770.

30. Prolla TA, Pang Q, Alani E, Kolodner RD, Liskay RM. MLH1, PMS1, and MSH2 interactions during the initiation of DNA mismatch repair in yeast. Science 1994; 265:1091–1093.

31. Drummond JT, Li GM, Longley MJ, Modrich P. Isolation of an hMSH2-p160 heterodimer that restores DNA mismatch repair to tumor cells[comment]. Science 1995; 268:1909–1912.

32. Acharya S, Wilson T, Gradia S, Kane MF, Guerrette S, Marsischky GT, Kolodner R, Fishel R. hMSH2 forms specific mispair-binding complexes with hMSH3 and hMSH6. Proc Natl Acad Sci USA 1996; 93:13629–13634.

33. Habraken Y, Sung P, Prakash L, Prakash S. Binding of insertion/deletion DNA mismatches by the heterodimer of yeast mismatch repair proteins MSH2 and MSH3. Curr Biol 1996; 6:1185–1187.

34. Iaccarino I, Palombo F, Drummond J, Totty NF, Hsuan JJ, Modrich P, Jiricny J. MSH6, a *Saccharomyces cerevisiae* protein that binds to mismatches as a heterodimer with MSH2. Curr Biol 1996; 6:484–486.

35. Marsischky GT, Filosi N, Kane MF, Kolodner R. Redundancy of *Saccharomyces cerevisiae* MSH3 and MSH6 in MSH2-dependent mismatch repair. Genes Dev 1996; 10:407–420.

36. Palombo F, Iaccarino I, Nakajima E, Ikejima M, Shimada T, Jiricny J. hMutSbeta, a heterodimer of hMSH2 and hMSH3, binds to insertion/deletion loops in DNA. Curr Biol 1996; 6:1181–1184.

37. Fishel R, Wilson T. MutS homologs in mammalian cells. Curr Opin Genet Dev 1997; 7:105–113.

38. Kolodner RD, Marsischky GT. Eukaryotic DNA mismatch repair. Curr Opin Genet Dev 1999; 9:89–96.

39. Fishel R, Lescoe MK, Rao MR, Copeland NG, Jenkins NA, Garber J, Kane M, Kolodner R. The human mutator gene homolog MSH2 and its association with hereditary nonpolyposis colon cancer[erratum appears in Cell 1994; 77(1):167]. Cell 1993; 75:1027–1038.

40. Leach FS, Nicolaides NC, Papadopoulos N, Liu B, Jen J, Parsons R, Peltomaki P, Sistonen P, Aaltonen LA, Nystrom-Lahti M, et al. Mutations of a mutS homolog in hereditary nonpolyposis colorectal cancer. Cell 1993; 75:1215–1225.

41. Nicolaides NC, Papadopoulos N, Liu B, Wei YF, Carter KC, Ruben SM, Rosen CA, Haseltine WA, Fleischmann RD, Fraser CM, et al. Mutations of two PMS homologues in hereditary nonpolyposis colon cancer. Nature 1994; 371:75–80.

42. Bronner CE, Baker SM, Morrison PT, Warren G, Smith LG, Lescoe MK, Kane M, Earabino C, Lipford J, Lindblom A, et al. Mutation in the DNA mismatch repair gene homologue hMLH1 is associated with hereditary non-polyposis colon cancer. Nature 1994; 368:258–261.

43. Papadopoulos N, Nicolaides NC, Liu B, Parsons R, Lengauer C, Palombo F, D'Arrigo A, Markowitz S, Willson JK, Kinzler KW, et al. Mutations of GTBP in genetically unstable cells[comment]. Science 1995; 268:1915–1917.

44. Thibodeau SN, French AJ, Roche PC, Cunningham JM, Tester DJ, Lindor NM, Moslein G, Baker SM, Liskay RM, Burgart LJ, Honchel R, Halling KC. Altered expression of hMSH2 and hMLH1 in tumors with microsatellite instability and genetic alterations in mismatch repair genes. Cancer Res 1996; 56:4836–4840.

45. Hawn MT, Umar A, Carethers JM, Marra G, Kunkel TA, Boland CR, Koi M. Evidence for a connection between the mismatch repair system and the G2 cell cycle checkpoint. Cancer Res 1995; 55:3721–3725.

46. Branch P, Hampson R, Karran P. DNA mismatch binding defects, DNA damage tolerance, and mutator phenotypes in human colorectal carcinoma cell lines. Cancer Res 1995; 55:2304–2309.

47. Gong JG, Costanzo A, Yang HQ, Melino G, Kaelin WG Jr, Levrero M, Wang JY. The tyrosine kinase c-Abl regulates p73 in apoptotic response to cisplatin-induced DNA damage.[comment]. Nature 1999; 399:806–809.

48. Hickman MJ, Samson LD. Role of DNA mismatch repair and p53 in signaling induction of apoptosis by alkylating agents. Proc Natl Acad Sci USA 1999; 96:10764–10769.

49. Zhang H, Richards B, Wilson T, Lloyd M, Cranston A, Thorburn A, Fishel R, Meuth M. Apoptosis induced by overexpression of hMSH2 or hMLH1. Cancer Res 1999; 59: 3021–3027.

50. Karran P. Mechanisms of tolerance to DNA damaging therapeutic drugs. Carcinogenesis 2001; 22:1931–1937.

51. Fishel R. The selection for mismatch repair defects in hereditary nonpolyposis colorectal cancer: revising the mutator hypothesis. Cancer Res 2001; 61:7369–7374.

52. Walker JE, Saraste M, Runswick MJ, Gay NJ. Distantly related sequences in the alpha- and beta-subunits of ATP synthase, myosin, kinases and other ATP-requiring enzymes and a common nucleotide binding fold. EMBO J 1982; 1:945–951.

53. Bergerat A, de Massy B, Gadelle D, Varoutas PC, Nicolas A, Forterre P. An atypical topoisomerase II from Archaea with implications for meiotic recombination[comment]. Nature 1997; 386:414–417.

54. Mushegian AR, Bassett DE Jr, Boguski MS, Bork P, Koonin EV. Positionally cloned human disease genes: Patterns of evolutionary conservation and functional motifs. Proc Natl Acad Sci USA 1997; 94:5831–5836.

55. Gradia S, Acharya S, Fishel R. The human mismatch recognition complex hMSH2-hMSH6 functions as a novel molecular switch. Cell 1997; 91:995–1005.

56. Blackwell LJ, Bjornson KP, Modrich P. DNA-dependent activation of the hMutSalpha ATPase. J Biol Chem 1998; 273:32049–32054.

57. Wilson T, Guerrette S, Fishel R. Dissociation of mismatch recognition and ATPase activity by hMSH2-hMSH3. J Biol Chem 1999; 274:21659–21644.

58. Gradia S, Acharya S, Fishel R. The role of mismatched nucleotides in activating the hMSH2-hMSH6 molecular switch. J Biol Chem 2000; 275:3922–3930.

59. Dufner P, Marra G, Raschle M, Jiricny J. Mismatch recognition and DNA-dependent stimulation of the ATPase activity of hMutSalpha is abolished by a single mutation in the hMSH6 subunit. J Biol Chem 2000; 275:36550–36555.

60. Hess MT, Gupta RD, Kolodner RD. Dominant *Saccharomyces cerevisiae* msh6 mutations cause increased mispair binding and decreased dissociation from mispairs by Msh2-Msh6 in the presence of ATP. J Biol Chem 2002; 277:25545–25553.

61. Acharya S, Foster PL, Brooks P, Fishel R. The coordinated functions of the *E. coli* MutS and MutL proteins in mismatch repair. Mol Cell 2003; 12:233–246.

62. Su S, Lahue R, Au K, Modrich P. Mispair specificity of methyl-directed DNA mismatch correction in vitro [published erratum appears in J Biol Chem 1988 Aug 5;263(22):11015]. J Biol Chem 1988; 263:6829–6835.

63. Bishop DK, Andersen J, Kolodner RD. Specificity of mismatch repair following transformation of *Saccharomyces cerevisiae* with heteroduplex plasmid DNA. Proc Natl Acad Sci USA 1989; 86:3713–3717.

64. Holmes J Jr, Clark S, Modrich P. Strand-specific mismatch correction in nuclear extracts of human and Drosophila melanogaster cell lines. Proc Natl Acad Sci USA 1990; 87:5837–5841.

65. Fang WH, Modrich P. Human strand-specific mismatch repair occurs by a bidirectional mechanism similar to that of the bacterial reaction. J Biol Chem 1993; 268:11838–11844.

66. Thomas DC, Roberts JD, Kunkel TA. Heteroduplex repair in extracts of human HeLa cells. J Biol Chem 1991; 266:3744–37451.

67. Genschel J, Littman SJ, Drummond JT, Modrich P. Isolation of MutSbeta from human cells and comparison of the mismatch repair specificities of MutSbeta and MutSalpha [erratum appears in J Biol Chem 1998; 273(41):27034]. J Biol Chem 1998; 273: 19895–19901.

68. Sprang SR. G protein mechanisms: insights from structural analysis. Annu Rev Biochem 1997; 66:639–678.

69. Gradia S, Subramanian D, Wilson T, Acharya S, Makhov A, Griffith J, Fishel R. hMSH2-hMSH6 forms a hydrolysis-independent sliding clamp on mismatched DNA. Mol Cell 1999; 3:255–261.

70. Iaccarino I, Marra G, Dufner P, Jiricny J. Mutation in the magnesium binding site of hMSH6 disables the hMutSalpha sliding clamp from translocating along DNA. J Biol Chem 2000; 275:2080–2086.

71. Blackwell LJ, Bjornson KP, Allen DJ, Modrich P. Distinct MutS DNA-binding modes that are differentially modulated by ATP binding and hydrolysis. J Biol Chem 2001; 276:34339–34347.

72. Lamers MH, Perrakis A, Enzlin JH, Winterwerp HH, de Wind N, Sixma TK. The crystal structure of DNA mismatch repair protein MutS binding to a G x T mismatch. Nature 2000; 407:711–717.

73. Allen DJ, Makhov A, Grilley M, Taylor J, Thresher R, Modrich P, Griffith JD. MutS mediates heteroduplex loop formation by a translocation mechanism. EMBO J 1997; 16:4467–4476.

74. Obmolova G, Ban C, Hsieh P, Yang W. Crystal structures of mismatch repair protein MutS and its complex with a substrate DNA. Nature 2000; 407:703–710.

75. Natrajan G, Lamers MH, Enzlin JH, Winterwerp HHK, Perrakis A, Sixma TK. Structures of *Escherichia coli* DNA mismatch repair enzyme MutS in complex with different mismatches: a common recognition mode for diverse substrates. Nucleic Acids Res 2003; 31:4814–4821.

76. Drotschmann K, Yang W, Brownewell FE, Kool ET, Kunkel TA. Asymmetric recognition of DNA local distortion. Structure-based functional studies of eukaryotic Msh2-Msh6. J Biol Chem 2001; 276:46225–46229.

77. Drotschmann K, Yang W, Kunkel TA. Evidence for sequential action of two ATPase active sites in yeast Msh2-Msh6. DNA Repair 2002; 1:743–753.

78. Lamers MH, Winterwerp HH, Sixma TK. The alternating ATPase domains of MutS control DNA mismatch repair. EMBO J 2003; 22:746–756.

79. Bjornson KP, Modrich P. Differential and simultaneous adenosine di- and triphosphate binding by MutS. J Biol Chem 2003; 278:18557–18562.

80. Malkov VA, Biswas I, Camerini-Otero RD, Hsieh P. Photocross-linking of the NH2-terminal region of Taq MutS protein to the major groove of a heteroduplex DNA. J Biol Chem 1997; 272:23811–23817.

81. Yamamoto A, Schofield MJ, Biswas I, Hsieh P. Requirement for Phe36 for DNA binding and mismatch repair by *Escherichia coli* MutS protein. Nucleic Acids Res 2000; 28:3564–3569.

82. Schofield MJ, Brownewell FE, Nayak S, Du C, Kool ET, Hsieh P. The Phe-X-Glu DNA binding motif of MutS. The role of hydrogen bonding in mismatch recognition. J Biol Chem 2001; 276:45505–45508.

83. Isaacs RJ, Rayens WS, Spielmann HP. Structural differences in the NOE-derived structure of G-T mismatched DNA relative to normal DNA are correlated with differences in (13)C relaxation-based internal dynamics. J Mol Biol 2002; 319:191–207.

84. Blackwell LJ, Martik D, Bjornson KP, Bjornson ES, Modrich P. Nucleotide-promoted Release of hMutSalpha from Heteroduplex DNA Is Consistent with an ATP-dependent Translocation Mechanism. J Biol Chem 1998; 273:32055–32062.

85. Junop MS, Obmolova G, Rausch K, Hsieh P, Yang W. Composite active site of an ABC ATPase: MutS uses ATP to verify mismatch recognition and authorize DNA repair. Mol Cell 2001; 7:1–12.

86. Fishel R. Mismatch repair, molecular switches, and signal transduction. Genes Dev 1998; 12:2096–2101.

87. Bertram JG, Bloom LB, Hingorani MM, Beechem JM, O'Donnell M, Goodman MF. Molecular Mechanism and Energetics of Clamp Assembly in *Escherichia coli*. The role of ATP Hydrolysis when gamma complex loads beta on DNA. J Biol Chem 2000; 275: 28413–28420.

88. Connelly JC, de Leau ES, Leach DRF. Nucleolytic processing of a protein-bound DNA end by the *E. coli* SbcCD (MR) complex. DNA Repair 2003; 2:795–807.

89. Ban C, Yang W. Crystal structure and ATPase activity of MutL: implications for DNA repair and mutagenesis. Cell 1998; 95:541–552.

90. Ban C, Junop M, Yang W. Transformation of MutL by ATP binding and hydrolysis: a switch in DNA mismatch repair. Cell 1999; 97:85–97.

91. Galio L, Bouquet C, Brooks P. ATP hydrolysis-dependent formation of a dynamic ternary nucleoprotein complex with MutS and MutL. Nucleic Acids Res 1999; 27:2325–2331.

92. Spampinato C, Modrich P. The MutL ATPase is required for mismatch repair. J Biol Chem 2000; 275:9863–9869.

93. Tran PT, Liskay RM. Functional studies on the candidate ATPase domains of *Saccharomyces cerevisiae* MutLalpha. Mol Cell Biol 2000; 20:6390–6398.

94. Schofield MJ, Nayak S, Scott TH, Du C, Hsieh P. Interaction of *Escherichia coli* MutS and MutL at a DNA mismatch. J Biol Chem 2001; 276:28291–28299.

95. Tomer G, Buermeyer AB, Nguyen MM, Liskay RM. Contribution of human mlh1 and pms2 ATPase activities to DNA mismatch repair. J Biol Chem 2002; 277:21801–21809.

96. Raschle M, Dufner P, Marra G, Jiricny J. Mutations within the hMLH1 and hPMS2 subunits of the human MutLalpha mismatch repair factor affect its ATPase activity, but not its ability to interact with hMutSalpha. J Biol Chem 2002; 277:21810–21820.

97. Habraken Y, Sung P, Prakash L, Prakash S. ATP-dependent Assembly of a Ternary Complex Consisting of a DNA Mismatch and the Yeast MSH2-MSH6 and MLH1-PMS1 Protein Complexes. J Biol Chem 1998; 273:9837–9841.

98. Selmane T, Schofield MJ, Nayak S, Du C, Hsieh P. Formation of a DNA Mismatch Repair Complex Mediated by ATP. J Mol Biol 2003; 334:949–965.

99. Genschel J, Modrich P. Mechanism of 5′-Directed Excision in Human Mismatch Repair. Mol Cell 2003; 12:1077–1086.

100. Tishkoff DX, Boerger AL, Bertrand P, Filosi N, Gaida GM, Kane MF, Kolodner RD. Identification and characterization of *Saccharomyces cerevisiae* EXO1, a gene encoding an exonuclease that interacts with†MSH2. PNAS 1997; 94:7487–7492.

101. Tishkoff DX, Amin NS, Viars CS, Arden KC, Kolodner RD. Identification of a human gene encoding a homologue of *Saccharomyces cerevisiae* EXO1, an exonuclease implicated in mismatch repair and recombination. Cancer Res 1998; 58:5027–5031.

102. Schmutte C, Marinescu RC, Sadoff MM, Guerrette S, Overhauser J, Fishel R. Human exonuclease I interacts with the mismatch repair protein hMSH2. Cancer Res 1998; 58:4537–4542.

103. Schmutte C, Sadoff MM, Shim KS, Acharya S, Fishel R. The interaction of DNA mismatch repair proteins with human exonuclease I. J Biol Chem 2001; 276:33011–33018.

104. Tran PT, Simon JA, Liskay RM. Interactions of Exo1p with components of MutLalpha in Saccharomyces cerevisiae. Proc Natl Acad Sci USA 2001; 98:9760–9765.

105. Genschel J, Bazemore LR, Modrich P. Human exonuclease I is required for 5′ and 3′ mismatch repair. J Biol Chem 2002; 277:13302–13311.

106. Modrich P. Mismatch repair, genetic stability and tumour avoidance. Philos Transact R Soc London Ser B: Biol Sci 1995; 347:89–95.

107. Modrich P. Strand-specific mismatch repair in mammalian cells. J Biol Chem 1997; 272:24727–24730.

108. Smith BT, Grossman AD, Walker GC. Visualization of mismatch repair in bacterial cells. Mol Cell 2001; 8:1197–1206.

109. Heinen CD, Wilson T, Mazurek A, Berardini M, Butz C, Fishel R. HNPCC mutations in hMSH2 result in reduced hMSH2-hMSH6 molecular switch functions. Cancer Cell 2002; 1:469–478.

110. Umar A, Buermeyer AB, Simon JA, Thomas DC, Clark AB, Liskay RM, Kunkel TA. Requirement for PCNA in DNA mismatch repair at a step preceding DNA resynthesis. Cell 1996; 87:65–73.

111. Gu L, Hong Y, McCulloch S, Watanabe H, Li GM. ATP-dependent interaction of human mismatch repair proteins and dual role of PCNA in mismatch repair. Nucleic Acids Res 1998; 26:1173–1178.

112. Lopez de Saro FJ, O'Donnell M. Interaction of the beta sliding clamp with MutS, ligase, and DNA polymerase I. Proc Natl Acad Sci USA 2001; 98:8376–8380.

113. Clark AB, Valle F, Drotschmann K, Gary RK, Kunkel TA. Functional interaction of proliferating cell nuclear antigen with MSH2-MSH6 and MSH2-MSH3 complexes. J Biol Chem 2000; 275:36498–36501.

114. Kleczkowska HE, Marra G, Lettieri T, Jiricny J. hMSH3 and hMSH6 interact with PCNA and colocalize with it to replication foci. Genes Dev 2001; 15:724–736.

115. Flores-Rozas H, Clark D, Kolodner RD. Proliferating cell nuclear antigen and Msh2p-Msh6p interact to form an active mispair recognition complex. Nat Genet 2000; 26: 375–378.

116. Lau PJ, Kolodner RD. Transfer of the MSH2-MSH6 complex from proliferating cell nuclear antigen to mispaired bases in DNA. J Biol Chem 2003; 278:14–17.

117. Pedrazzi G, Perrera C, Blaser H, Kuster P, Marra G, Davies SL, Ryu GH, Freire R, Hickson ID, Jiricny J, Stagljar I. Direct association of Bloom's syndrome gene product with the human mismatch repair protein MLH1. Nucleic Acids Res 2001; 29:4378–4386.

118. Bennett SE, Umar A, Oshima J, Monnat RJ Jr, Kunkel TA. Mismatch repair in extracts of Werner syndrome cell lines. Cancer Res 1997; 57:2956–2960.

119. Langland G, Kordich J, Creaney J, Goss KH, Lillard-Wetherell K, Bebenek K, Kunkel TA, Groden J. The Bloom's syndrome protein (BLM) interacts with MLH1 but is not required for DNA mismatch repair. J Biol Chem 2001; 276:30031–30035.

120. Benkovic SJ, Valentine AM, Salinas F. Replisome-medicated DNA replication. Annu Rev Biochem 2001; 70:181–208.

121. Ishmael FT, Trakselis MA, Benkovic SJ. Protein-Protein Interactions in the Bacteriophage T4 Replisome. The leading strand holoenzyme is physically linked to the lagging strand holoenzyme and the primosome. J Biol Chem 2003; 278:3145–3152.

122. Li X, Marians KJ. Two Distinct Triggers for Cycling of the Lagging Strand Polymerase at the Replication Fork. J Biol Chem 2000; 275:34757–34765.

123. Fijalkowska IJ, Jonczyk P, Tkaczyk MM, Bialoskorska M, Schaaper RM. Unequal fidelity of leading strand and lagging strand DNA replication on the *Escherichia coli* chromosome. PNAS 1998; 95:10020–10025.

124. Pavlov YI, Newlon CS, Kunkel TA. Yeast origins establish a strand bias for replicational mutagenesis. Mol Cell 2002; 10:207–213.

125. Mozzherin DJ, McConnell M, Jasko MV, Krayevsky AA, Tan C-K, Downey KM, Fisher PA. Proliferating Cell Nuclear Antigen Promotes Misincorporation Catalyzed by Calf Thymus DNA Polymerase delta. J Biol Chem 1996; 271:31711–31717.

126. Pavlov YI, Mian IM, Kunkel TA. Evidence for Preferential Mismatch Repair of Lagging Strand DNA Replication Errors in Yeast. Curr Biol 2003; 13:744–748.

127. Nehme A, Baskaran R, Aebi S, Fink D, Nebel S, Cenni B, Wang JY, Howell SB, Christen RD. Differential induction of c-Jun NH2-terminal kinase and c-Abl kinase in DNA mismatch repair-proficient and -deficient cells exposed to cisplatin. Cancer Res 1997; 57:3253–3257.

128. Davis TW, Wilson-Van Patten C, Meyers M, Kunugi KA, Cuthill S, Reznikoff C, Garces C, Boland CR, Kinsella TJ, Fishel R, Boothman DA. Defective expression of the DNA mismatch repair protein, MLH1, alters G2-M cell cycle checkpoint arrest following ionizing radiation. Cancer Res 1998; 58:767–778.

129. Yan T, Schupp JE, Hwang HS, Wagner MW, Berry SE, Strickfaden S, Veigl ML, Sedwick WD, Boothman DA, Kinsella TJ. Loss of DNA mismatch repair imparts defective cdc2 signaling and G(2) arrest responses without altering survival after ionizing radiation. Cancer Res 2001; 61:8290–8297.

130. Lan Z, Sever-Chroneos Z, Strobeck MW, Park CH, Baskaran R, Edelmann W, Leone G, Knudsen ES. DNA damage invokes mismatch repair-dependent cyclin D1 attenuation and retinoblastoma signaling pathways to inhibit CDK2. J Biol Chem 2002; 277:8372–8381.

131. Brown KD, Rathi A, Kamath R, Beardsley DI, Zhan Q, Mannino JL, Baskaran R. The mismatch repair system is required for S-phase checkpoint activation. Nat Genet 2003; 33:80–84.

132. Bellacosa A. Functional interactions and signaling properties of mammalian DNA mismatch repair proteins. Cell Death Differ 2001; 8:1076–1092.

133. Bernstein C, Bernstein H, Payne CM, Garewal H. DNA repair/pro-apoptotic dual-role proteins in five major DNA repair pathways: fail-safe protection against carcinogenesis. Mutat Res 2002; 511:145–178.

134. Fishel R. Signaling mismatch repair in cancer. Nat Med 1999; 5:1239–1241.

135. Ross-Macdonald P, Roeder GS. Mutation of a meiosis-specific MutS homolog decreases crossing over but not mismatch correction. Cell 1994; 79:1069–1080.

136. Hollingsworth NM, Ponte L, Halsey C. MSH5, a novel MutS homolog, facilitates meiotic reciprocal recombination between homologs in *Saccharomyces cerevisiae* but not mismatch repair. Genes Dev 1995; 9:1728–1739.

137. Pochart P, Woltering D, Hollingsworth NM. Conserved properties between functionally distinct MutS homologs in yeast. J Biol Chem 1997; 272:30345–30349.

138. de Vries SS, Baart EB, Dekker M, Siezen A, de Rooij DG, de Boer P, te Riele H. Mouse MutS-like protein Msh5 is required for proper chromosome synapsis in male and female meiosis. Genes Dev 1999; 13:523–531.

139. Wang TF, Kleckner N, Hunter N. Functional specificity of MutL homologs in yeast: evidence for three Mlh1-based heterocomplexes with distinct roles during meiosis in recombination and mismatch correction[comment]. Proc Nat Acad Sci USA 1999; 96:13914–13919.

140. Zickler D, Kleckner N. Meiotic chromosomes: integrating structure and function. Annu Rev Genet 1999: 33.

141. Kneitz B, Cohen PE, Avdievich E, Zhu L, Kane MF, Hou H Jr, Kolodner RD, Kucherlapati R, Pollard JW, Edelmann W. MutS homolog 4 localization to meiotic chromosomes is required for chromosome pairing during meiosis in male and female mice. Genes Dev 2000; 14:1085–1097.

142. Novak JE, Ross-Macdonald PB, Roeder GS. The budding yeast Msh4 protein functions in chromosome synapsis and the regulation of crossover distribution. Genetics 2001; 158:1013–1025.

143. Santucci-Darmanin S, Neyton S, Lespinasse F, Saunieres A, Gaudray P, Paquis-Flucklinger V. The DNA mismatch-repair MLH3 protein interacts with MSH4 in meiotic cells, supporting a role for this MutL homolog in mammalian meiotic recombination. Hum Mol Genet 2002; 11:1697–1706.

144. Lipkin SM, Moens PB, Wang V, Lenzi M, Shanmugarajah D, Gilgeous A, Thomas J, Cheng J, Touchman JW, Green ED, Schwartzberg P, Collins FS, Cohen PE. Meiotic arrest and aneuploidy in MLH3-deficient mice. Nat Genet 2002; 31:385–390.

145. Lenzi ML, Smith J, Snowden T, Kim M, Fishel R, Poulos BK, Cohen PE. Extreme heterogeneity in the molecular events leading to the establishment of chiasmata during meiosis I in human oocytes. Am J Hum Genet, 2005; 76:112–127. Epub 2004: 2022.

146. Snowden T, Acharya S, Butz C, Berardini M, Fishel R. hMSH4-hMSH5 recognizes Holliday Junctions and forms a meiosis-specific sliding clamp that embraces homologous chromosomes. Mol Cell 2004; 15:437–451.

22

Interaction of the *Escherichia coli* Vsr with DNA and Mismatch Repair Proteins

Ashok S. Bhagwat
Department of Chemistry, Wayne State University, Detroit, Michigan, U.S.A.

Bernard Connolly
School of Cell and Molecular Biosciences, University of Newcastle, Newcastle Upon Tyne, U.K.

Escherichia coli protein Vsr is a sequence-specific, mismatch-specific endonuclease that is involved in the maintenance of cytosine methylation in DNA and mutation avoidance. This chapter surveys what is known about the structure of this protein, how it finds the substrate T•G mismatch in DNA and its catalytic mechanism. A summary is also presented concerning how two accessory proteins, MutS and MutL, may help Vsr perform its nuclease function and help prevent mutations.

1. INTRODUCTION

Methylation of cytosine at position 5 (C5 methylation) is frequently used by organisms ranging from bacteria to mammals as a physical or epigenetic tag that allows them to distinguish DNAs from different sources. A wide variety of biological phenomena including restriction-modification, modified-cytosine restriction, gene silencing, epigenetic inheritance, and stimulation of an immune response use C5 methylation of DNA. However, this use of methylation comes with a serious drawback: an increase in C to T mutations due to the deamination of 5-methylcytosine (5meC) to thymine. Consequently, all organisms that contain C5 methylation are expected to also contain processes that reduce such mutations. The best-known example of such a repair process is from *E. coli* and is called very short-patch (VSP) repair. This process depends on the action of a sequence-specific mismatch-specific endonuclease, Vsr, and its mode of interaction with its substrate DNA is reviewed here.

The key intermediate in the pathway for 5meC to thymine mutations is a T•G mismatch created by the deamination of 5meC. Vsr prevents the fixing of this mismatch as a T•A pair by nicking the DNA immediately upstream of the mispair. Although the subsequent steps in the pathway have not been studied biochemically, genetic evidence suggests that DNA polymerase I performs nick translation to replace the offending T with a C and DNA ligase seals the nick creating a short (<10 nt) repair patch (Fig. 1). VSP repair occurs predominantly within 5'-CTWGG/CCWGG (W is A or T) creating 5'-CCWGG/CCWGG sequences which are substrates for the *E. coli* DNA cytosine methyltransferase, Dcm. Thus while the principal biological function of VSP repair is to prevent 5meC to T mutations, its overall effect on the enterobacterial genomes is to maintain Dcm sites. VSP repair has been reviewed previously (1) and details regarding its discovery, genetic requirements, and biological function can be found therein. This chapter will focus largely on the interaction of Vsr with DNA.

The ability of Vsr to prevent mutations has been confirmed in exponentially growing and stationary phase *E. coli*. In exponentially growing wild-type bacteria, VSP repair is not very active and methylated cytosines within Dcm sites are hotspots for C to T mutations (2). However, these sites become even hotter spots for mutations in a *vsr* background (2,3). In contrast, stationary phase *E. coli* do not accumulate 5meC to T mutations unless Vsr is inactive (4).

There are multiple reasons why VSP repair in exponentially growing cells is relatively inefficient. They include lower levels of Vsr in cells, DNA replication outpacing repair and its competition with the general mismatch repair (MMR) in cells. The latter process repairs replication errors that escape proof-reading and overlaps with VSP repair in its genetic requirements (see below). Regardless of its limitations, Vsr is an important antimutator protein in enteric bacteria and is unusual among DNA repair enzymes in terms of its ability to integrate the tasks of DNA damage-recognition and strand cleavage into a single protein. For example, in MMR, these tasks are performed by the coordinated action of three separate proteins, MutS, MutL, and MutH.

Although the relationship between VSP repair and the avoidance of mutations promoted by cytosine methylation is clear in *E. coli*, this has not been established in any other organism. Sequence homologs of Vsr can be found in a wide variety of bacteria using BLAST, PSI-BLAST, and other sequence alignment engines, but the only homolog for which biochemical characterization of the protein has been done is from *Geobacillus stearothermophilus* (5). In this case, the protein does nick

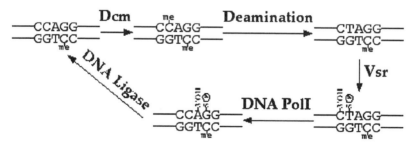

Figure 1 DNA methylation, 5meC deamination and VSP repair. The steps involved in the maintenance of cytosine methylation in *E. coli* DNA are shown. CCWGG sequences in DNA are methylated by Dcm and either 5meC at the Dcm site may suffer deamination to give rise to a T•G mispair. Vsr recognizes this abnormality and hydrolyzes the phosphodiester linkage immediately upstream of the offending T. DNA polymerase is expected to perform a nick translation reaction replacing the T with a C. DNA ligase completes repair by sealing the nick. The patch of repair is small (<10 nt) and is arbitrarily shown to be 2 nt.

the DNA immediately upstream of a T•G mismatch, but the relationship of this activity to mutation avoidance is less clear.

2. STRUCTURE OF Vsr

A number of structures, determined by X-ray crystallography, are available for Vsr. An initial structure of the free enzyme, at 1.8 Å resolution, indicated a single domain comprising a central β-sheet buttressed by three α-helices and a structural Zn^{2+} atom co-ordinated by three cysteines and a histidine (6). The main chain positions of Vsr demonstrated homology to type II restriction endonucleases, a class of enzymes whose folding results in a defined location of three key amino acids present in a catalytic motif, $DX_{(6-20)}(D/E) XK$ (7). With restriction endonucleases, the conserved acidic amino acids are responsible for co-ordinating essential Mg^{2+} ions which provide acid/base catalysis and transition state stabilisation; the lysine probably activates the water involved in the hydrolytic reaction. Although Vsr contains an aspartate (D51) at the position corresponding to the first conserved residue, the second and third amino acids are replaced by phenylalanine (F62) and histidine (H64). Site-directed mutagenesis confirmed the critical nature of D51, but H64 appeared to be less important, mutants showing reduced but appreciable activity (8).

The DNA hydrolysis mechanism of Vsr, together with the identification of catalytic amino acids, only became apparent with the solution of two Vsr–DNA complexes (8,9). Both structures represent Vsr–product complexes; the first arising from cutting of the oligodeoxynucleotide during crystallisation (8), the second due to the use of a self-assembling nucleic acid that formed a nicked product complex (9). As shown in Fig. 2, Vsr requires two Mg^{2+} atoms for DNA hydrolysis and both are liganded by the critical acidic residue D51. The peptide carbonyl oxygen of T63 (which is at a location near to the second conserved acidic residue in the restriction endonuclease $DX_{(6-20)}(D/E) XK$ motif) is also a metal ligand. A key histidine H69 (which is in the vicinity H64 and so occupies a site near to but distinct from the third conserved amino acid in the $DX_{(6-20)}(D/E) XK$ motif) is bound to the product phosphate, but assumed to act as a base to activate the attacking water molecule in the substrate complex. Other important amino acids either bind to Mg^{2+} co-ordinated water molecules (E25 and H64) or, in the case of D97, hydrogen bond to H69, thereby increasing its pK_a value. The most likely Vsr mechanism involves in line attack of OH^-, generated by deprotonation of a water molecule using H69, with the two Mg^{2+} ions and their associated water molecules stabilizing the transition state and providing acidic catalysis to labilize the scissile bond (10). Similar mechanisms are used by a number of nucleases, although the precise disposition and roles of key amino acids and the two metal ions varies between different enzymes (7).

Both Vsr–DNA structures provided clues concerning the mode of recognition of the T•G mismatch. Perhaps the most remarkable feature is DNA intercalation of three aromatic amino acids, F67, W68, and W86, between the T•G mismatch and the adjacent, central A•T base-pair (Fig. 3A). The penetrating aromatic amino acids take part in stacking interactions with the nucleic acid, F67 with the central A•T base-pair, W68 with the mispaired thymine and W86 with two sugars on the strand opposite this thymine. Insertion, via the major groove, of this hydrophobic wedge causes considerable DNA distortion with separation of the mismatch and the neighbouring base-pair, opening of the major groove by 57° and bending of the DNA duplex by 44°. Two amino acids, M14 and I17, insert into the minor groove opposite

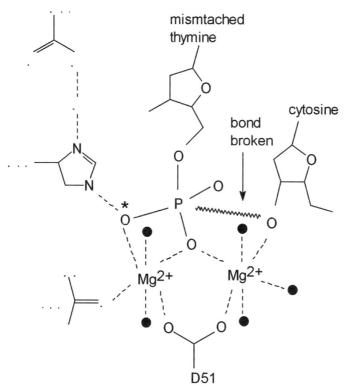

Figure 2 Metal ions and amino acids involved in DNA hydrolysis by Vsr. The two essential Mg^{2+} ions, together with their ligands are illustrated. Solid circles represent Mg^{2+} bound water molecules. The oxygen indicted with an asterisk probably derives from the water molecule that attacks the phosphodiester bond. The bond broken during the Vsr reaction is also illustrated. *Source*: Adapted from Ref. 8.

the three aromatic amino acids (Fig. 3B) and contribute to the overall distortion of the nucleic acid (8). The T•G mismatch forms a wobble base-pair and so presents a unique array of hydrogen bond donor/acceptors in the major groove; hydrogen bonds are observed between guanine and both K89 and M14 (through the major and minor grooves, respectively) and cytosine and N93 (through the major groove). However, mismatch base recognition does not solely result from these hydrogen-bonding interactions; additionally shape complementarity is observed between the protein and the nucleic acid, particularly the unusual wobble structure of the T•G mismatch and the perturbed adjacent base-pairs. Much of this shape recognition arises from the intercalated hydrophobic wedge and the insertion of M14 and I17 into the minor groove.

The role of the CTWGG recognition sequence was also clarified by the Vsr–DNA structures. One difference in the two structures concerns the nature of the oligodeoxynucleotides used; the first contained a G•C base-pair at the fourth position (8), the second G•5meC (9). As mentioned above, the Dcm methyltransferase acts on 5′-CCWGG/CCWGG duplexes, adding methyl groups to the second cytosine in both strands. Deamination of one of these 5-methyl-cytosines gives a T•G mismatch with a G•5meC base-pair at the symmetry-related site. With the oligodeoxynucleotide containing a T•G mismatch in an un-methylated context, a number of polar

(A)

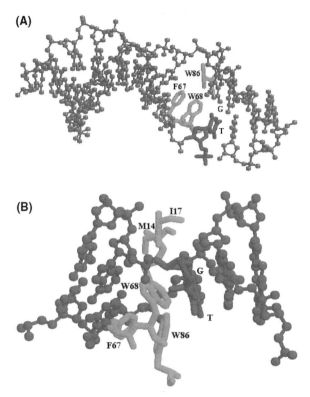

(B)

Figure 3 Protein–DNA recognition by Vsr. (A) Intercalation of Vsr amino acids F67, W68, and W86 into DNA between the T•G mismatch (mismatched T shown in red) and the central T•A base-pair and (B) Insertion of the hydrophobic Vsr amino acids M14 and I17 into the DNA minor groove in the vicinity of the T•G mismatch (mismatched T shown in red). The intercalating amino acids are also illustrated. Diagrams have been derived using PDB accession 1CWO. *Source*: Adapted from Ref. 8. (*See color insert.*)

interactions were seen between the protein and all the bases comprising the Vsr recognition sequence, accounting for the sequence preference. However, the degree of direct interaction between Vsr and its target site was limited and the toleration of a central A•T (as in this structure) or a T•A base-pair, together with the critical nature of the fourth G•C base-pair (which is sparsely contacted) were not obvious. Using the physiologically more relevant DNA possessing a T•G mismatch in a hemi-methylated context, yielded further insights. The self-assembling oligodeoxynucleotide results in alternating protein-bound and free Vsr target sites, allowing an investigation of the DNA in both states. It was observed that the central A•T base-pair did not exist as a usual Watson–Crick base-pair but adopted the Hoogsteen form. Remarkably, this unusual conformation was seen with both Vsr-bound and free DNA, suggesting the Hoogsteen base-pair is an intrinsic property of the nucleic acid sequence, rather than a consequence of binding to the protein. The stabilisation of the central Hoogsteen A•T base-pair was explained by the unusual nature of both flanking base-pairs. Should the A•T base-pair retain the Watson–Crick conformation, the wobble T•G mismatch to one side results in poor base stacking; the G•5meC base-pair to the other side leads to a steric clash between the methyl groups of 5meC and thymine within the A•T pair. Flipping the A•T into

Hoogsteen conformation both restores base stacking and relieves steric strain. Although structures containing the alternate central T•A base-pair have yet to be determined, it was proposed that the Watson–Crick form would be retained as the thymine would be adequately stacked with the mismatched guanosine and be far enough from the 5meC to eliminate steric problems. Thus the requirement for a central A•T or T•A is explained by alternating between Hoogsteen and Watson–Crick forms, the fourth G•C is critical as it allows cytosine methylation.

In summary, the central three bases of the Vsr sequence are highly atypical comprising a wobble T•G mismatch, a Hoogsteen base-pair (at least in some circumstances) and G•5meC. While recognition does involve typical polar interactions to the base-pair edges (direct readout), shape recognition of the unusual triumvirate is just as important.

2.1. Biochemistry of Vsr

Although 5′-CTWGG/C5meCWGG is the best substrate for Vsr, two lines of evidence show that VSP repair acts on a broader set of mismatches. Lieb and colleagues found that VSP repair interfered with recombination not only at the Dcm site, but also at sites that included 4 out of 5 and 3 out of 5 base pairs within the CCWGG site (11,12). For example, if recombination between different strains of bacteriophage λ creates a 5′-NTAGG/5′-CCTGN′ intermediate, it is subject to VSP repair. Similarly, intermediates with 5′-CTAGN/5′-N′CTGG and 5′-NTAGN/5′-N′CTGN′ are also subject to some amount of VSP repair.

This dissonance between the sequence specificities of Dcm and Vsr has had an impact on the nucleotide composition of the *E. coli* genome. Comparison of observed frequencies of tetranucleotides of the form 5′-TWGG, 5′-CTWG (and their complements), with frequencies predicted from the observed frequencies of component di- and tri- nucleotides shows that the tetranucleotide sequences are underrepresented in the *E. coli* genome (13,14). Concurrently, CWGG and CCWG sequences, and their complements, are significantly overrepresented in the genome.

Biochemical investigations on the selectivity of Vsr have complemented these genetic and computational studies. Interestingly, there is a direct correlation between the efficiency with which Vsr nicks various T•G mismatches and the extent of underrepresentation of the corresponding T-containing sequences in the genome (15). This analysis shows that sequence preferences of Vsr have shaped significantly the composition of the *E. coli* genome.

Comparison of various fluorescently labeled oligonucleotide DNAs as substrates showed that, as expected, CTWGG (paired with CCWGG) sequences were the best substrates (15). However, changes to one and even two bases were relatively well tolerated. Sequences NT(A/T) GG and CT(A/T) GN had relative rates (expressed as second order rate constants for hydrolysis) approximately 20%–60% of the canonical substrate and NT(A/T) GN as well as CT(N) GG also turned over.

An alternative approach used single turnover rate constants (k_{st}) and equilibrium dissociation constants (K_D), allowing Vsr selectivity to be expressed as a specificity constant k_{st}/K_D (16,17). Although, Vsr was capable of binding and hydrolysing a T•G mismatch, and less efficiently U•G, embedded in the (unmethylated) Dcm sequence but, in agreement with structural data, better substrates were found when the symmetry-related cytosine in the fourth G•C base pair was replaced by 5meC. For all substrates, specificity arose at both the binding (K_D) and catalytic

(k_{st}) stages and, similarly to many restriction endonucleases (7), binding was dependent on a divalent metal ion. In general, single base-pair changes to the Dcm sequence were tolerated and, while the numerical values for site preference sometimes differed from the study referred to above, it is clear that base substitutions at the first, third, and fifth positions give Vsr substrates. The fourth G•C (or better G•5meC) base pair appeared to be the most critical.

Finally a study has probed the recognition of the T•G mismatch itself (18). While T•G and U•G produced the most efficient substrates, a number of alterations including modified bases and abasic sites at both mismatch partners, were accepted. In conclusion, all in vitro biochemical studies indicate that Vsr prefers 5'-CTWGG/ 5'-C(5meC) WGG duplexes, but that the enzyme does not show extreme specificity, certainly very much less than for the type II restriction endonucleases (7). Numerous changes to the target site (both at the mismatch and the flanking bases) give reasonable substrates. In general, these investigations agree with structural data which indicates a paucity of direct protein–DNA contacts and general shape recognition, expected to be more tolerant of DNA modification.

2.2. Interaction Between MMR and VSP Repair

The relationship between VSP repair and the general mismatch repair in *E. coli* is complex and has been reviewed recently (19). Although it is poorly understood, this interaction illuminates both VSP repair and MMR, and will be briefly discussed here. The MMR pathway targets base–base mismatches (including T•G) and small insertion/deletion loops and the key protein in MMR that binds the mismatches and loops is MutS. Consequently, MutS can potentially compete with Vsr for mismatch binding, but is known to aid Vsr in an unknown way (20–22). MutH, a sequence-specific, adenine methylation-specific endonuclease is required in MMR to create a nick near the mismatch. This protein shares some similarities with Vsr in its structure (23) and is not required for VSP repair. However, MutL, a protein that communicates between MutS and MutH, enhances VSP repair. Additional observations that help construct a model for the role of MutL and MutS in VSP repair include: (a) overproduction of either MutL or MutS in cells inhibits VSP repair (24), (b) overproduction of Vsr inhibits MMR (25), and (c) Vsr and MutL directly interact with each other (26). The way in which *E. coli* appears to avoid a constant competition between MMR and VSP repair for the substrate of the latter pathway is as follows: it maintains high levels of MutH, MutL, and MutS proteins and low levels of Vsr during the exponential phase of cell growth. This allows repair of replication errors by MMR without significant interference from Vsr. At the end of the *E. coli* life cycle, the stationary phase, cells maintain low levels of MMR proteins, but high levels of Vsr. This assures that MMR does not interfere with VSP repair. This makes biological sense because while the former process is not needed during the stationary phase, the latter process is needed to correct steadily accumulating thymine from deaminated 5meC residues.

It is clear from the crystal structures of Vsr and MutS with DNA that the two proteins are unlikely to bind simultaneously to the T•G mismatch. If so, MutS must assist Vsr in mismatch binding in some indirect way. One possibility is that MutS creates a specialized DNA structure in the vicinity of the mismatch to promote Vsr binding. Alternately, MutS may bring MutL to Vsr and the resulting interaction of MutL with Vsr may stimulate the endonucleolytic activity of the latter protein. There is some genetic and biochemical evidence to support an interaction between

MutL and Vsr (26,27); however, the data do not directly support or eliminate either model. It remains possible that MutS and MutL play a very different role in VSP repair than described here.

3. SUMMARY AND CONCLUDING REMARKS

Vsr is part of a specialized DNA repair process that reduces mutations caused by the deamination of methylated cytosines. Although, it recognizes T•G mismatches in DNA, it does not resemble MutS in primary sequence, its 3D structure and many aspects of its interactions with DNA. Instead, it is structurally similar to MutH and some restriction endonucleases, and also shares many of the catalytic features of restriction enzymes. VSP repair has evolved a close and well-nuanced relationship with the general mismatch repair proteins in order to assure that the two processes do not significantly interfere with each other.

REFERENCES

1. Lieb M, Bhagwat AS. Very short patch repair: reducing the cost of cytosine methylation. Mol Microbiol 1996; 20:467–473.
2. Lieb M. Spontaneous mutation at a 5-methylcytosine hotspot is prevented by very short patch (VSP) mismatch repair. Genetics 1991; 128:23–27.
3. Lutsenko E, Bhagwat AS. Principal causes of hot spots for cytosine to thymine mutations at sites of cytosine methylation in growing cells. A model, its experimental support and implications. Mutat Res 1999; 437:11–20.
4. Lieb M, Rehmat S. 5-methylcytosine is not a mutation hot spot in nondividing *Escherichia coli*. Proc Natl Acad Sci USA 1997; 94:940–945.
5. Laging M, Lindner E, Fritz HJ, Kramer W. Repair of hydrolytic DNA deamination damage in thermophilic bacteria: cloning and characterization of a vsr endonuclease homolog from Bacillus stearothermophilus. Nucleic Acids Res 2003; 31:1913–1920.
6. Tsutakawa SE, Muto T, Kawate T, Jingami H, Kunishima N, Ariyoshi M, Kohda D, Nakagawa M, Morikawa K. Crystallographic and functional studies of very short patch repair endonuclease. Mol Cell 1999; 3:621–628.
7. Pingoud A, Jeltsch A. Structure and function of type II restriction endonucleases. Nucleic Acids Res 2001; 29:3705–3727.
8. Tsutakawa SE, Jingami H, Morikawa K. Recognition of a TG mismatch: the crystal structure of very short patch repair endonucleases in complex with a DNA duplex. Cell 1999; 99:615–623.
9. Bunting KA, Roe SM, Headley A, Brown T, Savva R, Pearl LH. Crystal structure of the *Escherichia coli* dcm very-short-patch DNA repair endonuclease bound to its reaction product-site in a DNA superhelix. Nucleic Acids Res 2003; 31:1633–1639.
10. Tsutakawa SE, Morikawa K. The structural basis of damaged DNA recognition and endonucleolytic cleavage for very short patch repair endonuclease. Nucleic Acids Res 2001; 29:3775–3783.
11. Lieb M. Specific mismatch correction in bacteriophage lambda crosses by very short patch repair. Mol Gen Genet 1983; 191:118–125.
12. Lieb M, Allen E, Read D. Very short patch repair in phage lambda: repair sites and length of repair tracts. Genetics 1986; 114:1041–1060.
13. Bhagwat AS, McClelland M. DNA mismatch correction by very short patch repair may have altered the abundance of oligonucleotides in the *E. coli* genome. Nucleic Acids Res 1992; 20:1663–1668.

14. Merkl R, Kröger M, Rice P, Fritz HJ. Statistical evaluation and biological interpretation of non-random abundance in the *E. coli* K-12 genome of tetra- and pentanucleotide sequences related to VSP DNA mismatch repair. Nucleic Acids Res 1992; 20:1657–1662.

15. Gläsner W, Merkl R, Schellenberger V, Fritz HJ. Substrate preferences of vsr DNA mismatch endonuclease and their consequences for the evolution of the *Escherichia coli* K-12 Genome. J Mol Biol 1995; 245:1–7.

16. Gonzalez-Nicieza R, Turner DP, Connolly BA. DNA binding and cleavage selectivity of the *Escherichia coli* DNA G:T-mismatch endonuclease (vsr protein). J Mol Biol 2001; 310:501–508.

17. Turner DP, Connolly BA. Interaction of the *E. coli* DNA G:T-mismatch endonuclease (vsr protein) with oligonucleotides containing its target sequence. J Mol Biol 2000; 304:765–778.

18. Fox KR, Allinson SL, Sahagun-Krause H, Brown T. Recognition of GT mismatches by vsr mismatch endonuclease. Nucleic Acids Res 2000; 28:2535–2540.

19. Bhagwat AS, Lieb M. Cooperation and competition in mismatch repair: very short-patch repair and methyl-directed mismatch repair in *Escherichia coli*. Mol Microbiol 2002; 44:1421–1428.

20. Jones M, Wagner R, Radman M. Mismatch repair of deaminated 5-methyl-cytosine. J Mol Biol 1987; 194:155–159.

21. Lieb M. Bacterial genes mutL, mutS, and dcm participate in repair of mismatches at 5-methylcytosine sites. J Bacteriol 1987; 169:5241–5246.

22. Zell R, Fritz HJ. DNA mismatch-repair in *Escherichia coli* counteracting the hydrolytic deamination of 5-methyl-cytosine residues. Embo J 1987; 6:1809–1815.

23. Ban C, Yang W. Structural basis for MutH activation in *E.coli* mismatch repair and relationship of MutH to restriction endonucleases. Embo J 1998; 17:1526–1534.

24. Lieb M, Rehmat S, Bhagwat AS. Interaction of MutS and Vsr: some dominant-negative mutS mutations that disable methyladenine-directed mismatch repair are active in very-short- patch repair. J Bacteriol 2001; 183:6487–6490.

25. Doiron KM, Viau S, Koutroumanis M, Cupples CG. Overexpression of vsr in Escherichia coli is mutagenic. J Bacteriol 1996; 178:4294–4296.

26. Mansour CA, Doiron KM, Cupples CG. Characterization of functional interactions among the *Escherichia coli* mismatch repair proteins using a bacterial two-hybrid assay. Mutat Res 2001; 485:331–338.

27. Drotschmann K, Aronshtam A, Fritz HJ, Marinus MG. The *Escherichia coli* MutL protein stimulates binding of vsr and MutS to heteroduplex DNA. Nucleic Acids Res 1998; 26:948–953.

Part V

Replication and Bypass of DNA Lesions

DNA damage can be repaired or tolerated. The tolerance mechanism is of special importance to permit completion of replication in the presence of unrepaired damage. Not long ago, it became clear that numerous DNA polymerases exist in pro- and eukaryotes that are designed to replicate with reduced fidelity in order to accommodate base damage in the template. Such "bypass polymerases" can replace a high-fidelity replicative polymerase that cannot synthesize past an offending altered base. Several questions arise that are within the scope of the topic of damage recognition. How is the situation of unrepaired base damage that has resulted in polymerase stalling being recognized? How does the cell decide which polymerase to use and when to switch back to high-fidelity replication? What are the structural features and molecular mechanisms that enable a bypass polymerase to accept a distorted template?

Chapter 23 and 24 review the basic findings in pro- and eukaryotes, respectively.

Next, Chapter 25 provides an indepth view of the structural properties of eukaryotic bypass polymerases.

Finally, the fascinating topic of pathway choice necessitated by replication-stalling base damage is discussed in Chapter 26 on the basis of insights from budding yeast.

23

Mechanism of Translesion DNA Synthesis in *Escherichia coli*

Zvi Livneh, Ayelet Maor-Shoshani, Moshe Goldsmith,
Gali Arad, Ayal Hendel, and Lior Izhar
Department of Biological Chemistry, Weizmann Institute of Science, Rehovot, Israel

1. INTRODUCTION

The need to transmit genetic information from parent to daughter cells in a precise manner is fulfilled by high-fidelity DNA replication. The main components in this machinery are high-fidelity replicative DNA polymerases. The high precision of these replication machines is achieved by a tight fit into the active site of the Watson–Crick A:T and G:C base pairs, along with specific hydrogen bonds and other stabilizing interactions, and a $3' \rightarrow 5'$ exonuclease proofreading activity (1–3). This high-precision and finely tuned mechanism of action of replicative DNA polymerases leads to inhibition of DNA synthesis upon encounter with lesions in DNA, whose chemistry and architecture strongly deviate from that of the four canonical nucleotides. These modified template nucleotides cannot be accommodated in the active site of replicative DNA polymerases, or else they cause structural or conformational changes that greatly reduce the catalytic efficiency and/or fidelity of the polymerase (4–6). Why are there DNA lesions left in DNA despite the presence and operation of multiple DNA repair mechanisms that act before replication commences? The two simplest reasons are (1) DNA repair mechanisms, like any other biological mechanism, are not 100% efficient and (2) some lesions may form in DNA during replication.

Inhibition of replication at DNA lesions leads to the formation of single-stranded gaps in the daughter strands (reviewed in Ref. 6). These gaps must be filled in for replication to be completed, such that cells can proceed and divide. In addition, any nick in the ssDNA region will cause a double-strand break (DSB), thereby converting a local lesion (a modified base) to a chromosomal lesion (DSB), which is much more severe in its biological consequences (7–9). Therefore, organisms from *Escherichia coli* to humans utilize DNA damage tolerance mechanisms, which function to fill in the gaps, despite the presence of lesions (7). This will enable both completion of replication and a second attempt of error-free DNA repair in order to eliminate the damaged base from DNA. There are at least two mechanisms for the repair of damaged gaps in DNA: translesion DNA synthesis (TLS), also termed translesion replication, and gap-filling homologous recombination repair (GFRR).

In GFRR, the gaps are filled in by transferring complementary DNA segments from the intact sister chromatids and by pairing them with the damaged ssDNA. Such a mechanism was proposed in the late 1960s (10,11), and direct evidence for its operation in *E. coli* was recently presented (12). The two main features of this process are the requirement for an intact homologous DNA and its error-free nature. If the ssDNA is broken resulting a DSB, homologous recombination can function to repair the broken DNA (13). In TLS replication, gaps are filled in by specialized DNA polymerases, which are capable to replicate across DNA lesions. Because this process involves insertion of a nucleotide opposite a damaged base, where normal base pairing is usually compromised, it is inherently mutagenic. In fact, TLS is responsible for most mutations caused by DNA damaging agents such as sunlight and a variety of chemical carcinogens (14–16).

Two other DNA damage tolerance mechanisms were proposed for rescuing replication stalled at DNA lesions. In the replication fork regression model, after encountering a lesion the fork regresses, thereby leading the lesion back to a double-stranded configuration (17). This allows a second attempt of error-free repair, after which replication can proceed. The second model is the strand-switching model (copy choice replication), in which the newly synthesized DNA strand that is blocked at a lesion switches to the intact sister chromatid, is extended by copying the complementary strand, and then returns back and patches the gap (18). This mechanism too is fundamentally nonmutagenic. This chapter will not deal with nonmutagenic DNA damage tolerance mechanisms. It will be devoted to TLS in *E. coli*.

2. TRANSLESION DNA SYNTHESIS AND THE SOS RESPONSE

Translesion DNA synthesis in *E. coli* is part of the Internation distress signal (SOS) response to DNA damage. This response, which is described in detail in another chapter of this book, involves 30–40 genes that function to increase cell survival under environmental conditions that cause DNA damage. SOS genes are subjected to negative regulation at the transcriptional level by a common repressor (LexA), which binds to a consensus sequence in the promoter region (SOS box, LexA-binding site). Induction of these genes occurs following exposure to DNA damaging agents, such as UV light, and is effected via autocleavage of the LexA repressor promoted by activated RecA. The latter, known for its role as the main recombinase in *E. coli*, acts in SOS induction as a positive transcriptional regulator, in a role totally distinct from its role in strand pairing and exchange during homologous recombination (7).

Among the SOS genes, an operon consisting two genes, *umuD* and *umuC*, was found to be specifically required for mutagenesis caused by DNA damaging agents such as UV light (19,20). It was later discovered that UmuD is cleaved to a shorter form, termed UmuD′, which is the active form in mutagenesis (reviewed in Ref. 21). This gene dependence highlights the surprising fact that UV mutagenesis is not merely a byproduct of replication through miscoding DNA lesions, but rather a genetically regulated process, which requires specific proteins. From the isolation of their mutants in 1977, it took more than 20 years until the activity of UmuC and UmuD was discovered. Originally, it was believed that UmuD′ and UmuC are accessory proteins, which modulate the activity of DNA polymerase III, thereby enabling it to bypass otherwise blocking lesions. This hypothesis turned out to be incorrect. Eventually, the Umu proteins were purified and the TLS reaction was reconstituted independently in two laboratories (22,23), leading to the exciting and unexpected discovery that UmuC is a novel type of a DNA polymerase, which is specialized for replication across DNA lesions (24,25). It is now clear that TLS by specialized lesion bypass polymerases is a universal pathway conserved from *E. coli* to humans. Many of the polymerases involved in TLS, including pol V, belong to the new Y family of DNA polymerases (26). However, it should be kept in mind that DNA polymerases that belong to other families are also involved in TLS (27).

3. OVERVIEW ON POL V

Genetic analysis performed over the years has indicated that UV mutagenesis, most of which occurs by TLS, requires the *umuD* and *umuC* genes, and in addition *recA* and *polC (dnaE)*, encoding the α (catalytic) subunit of pol III (for reviews, see Refs. 6, 7, 21). Three of these proteins, UmuC, UmuD′, and RecA, were required in the purified reconstituted TLS system. In addition, the single-strand binding (SSB) protein was required in the in vitro reaction. Pol III, however, was dispensable in the in vitro reaction, and that led to the discovery that UmuC is a DNA polymerase by itself, and it was termed pol V (24,25).

Pol V by itself is a very weak DNA polymerase, which is incapable of bypassing DNA lesions. This is true for the UmuC protein with or without UmuD′ (25), as well as the UmuD′$_2$C complex (24). However, in the presence of both RecA and SSB, pol V becomes a lesion-bypass DNA polymerase (24,25), which is further stimulated by the processivity subunits of pol III, such as the β subunit DNA sliding clamp and the γ complex clamp loader (28,29). Pol V has low fidelity during synthesis on undamaged DNA (28,30), and it lacks a proofreading 3′→5′ exonuclease activity, as expected from a mutagenic polymerase, which needs to form 'mispairs' of damaged and normal bases during TLS. The processivity of pol V is low, ranging from approximately 3 in the presence of RecA and SSB, up to 14–18 in the presence of RecA, SSB, and the processivity proteins (28,29). This is consistent with a need to synthesize only short stretches of DNA during the bypass reaction, before being replaced by the replicative high-fidelity high-processivity pol III holoenzyme.

4. FIDELITY OF POL V

The fidelity of pol V on undamaged DNA templates is 10^{-3}–10^{-4} errors/nucleotide replicated (28,30). This is 100–1000-fold lower than the fidelity of the replicative pol III holoenzyme (30–32). DNA sequence analysis of pol V errors showed that pol V produces all types of errors, including frameshifts and base substitutions. However, it has a propensity to produce transversion mutations, namely, changing a pyrimidine into a purine or vice versa. Pol V forms transversions at a frequency up to 300-fold higher than pol III holoenzyme (30). It should be noted that the base–base mismatches produced by pol V during synthesis on undamaged DNA are potential substrates for the mismatch repair system (33,34). This system corrects primarily mismatches that are precursors for transition mutations, because these are the types of mistakes that are made by the replicative polymerase, pol III holoenzyme (31,32). The repair of mismatches that are precursors of transversions is 20-fold less effective that transition mismatches (34–36). This implies that pol V generates mismatches that are largely immune to mismatch repair and yield mutations. This provides the mechanistic basis for the phenomenon of untargeted SOS mutagenesis, also termed

SOS mutator activity (37). In essence, this phenomenon involves the formation of mutations in undamaged regions of DNA in cells in which the SOS response was induced. The process is *umuDC-* and *recA*-dependent and yields primarily transversion mutations (34,38,39). On the basis of the fidelity of pol V, untargeted mutagenesis can be explained by the activity of pol V under SOS conditions in undamaged regions of the *E. coli* chromosome. It should be noted that there is a second branch of untargeted mutagenesis that depends on pol IV rather than pol V (39,40).

5. LESION BYPASS BY POL V

The most remarkable property of pol V is its ability to replicate through a variety of DNA lesions, some of which are extremely challenging. In general, TLS involves at least two steps: (i) insertion of a nucleotide opposite the lesion and (ii) extension past the lesion. Pol V carries out both steps, leading to complete bypass of a variety of lesions, including an abasic site, a T–T cyclobutyl pyrimidine dimmer (CPD), and a 6–4 T–T adduct (24,25,28). When replicating an abasic site, pol V inserts primarily purine nucleotides opposite the abasic site, with a preference for A over G (25,28). When replicating a T–T CPD, pol V inserts primarily two correct A nucleotides; however, when replicating a 6–4 T–T adduct, pol V prefers to insert G opposite the 3′ T of the 6–4 adduct. Insertion opposite the 5′ T is primarily the correct A nucleotide (28). Overall, these results are consistent with the in vivo mutagenic specificity of these DNA lesions when assayed in DNA constructs carrying site-specific lesions (22,41–43). What enables pol V to bypass lesions that severely block "classical" DNA polymerase? The crystal structure of pol V is unknown yet. However, on the basis of several structures of other Y family DNA polymerases (44–49), the ability to bypass lesions can be explained, at least in part by a structural design of the active site that includes a spacious template-binding pocket, which can accommodate two template nucleotides (e.g., for a T–T CPD), a much weaker geometrical fit to its substrates, and an overall flexibility that might allow accommodation of bulky lesions. This permissiveness comes at the expense of fidelity, which is low even during synthesis on undamaged DNA as discussed earlier. It should be noted though that all the structures known to date were obtained with TLS polymerases that perform bypass unassisted by other proteins, unlike pol V, suggesting that there might be significant differences in structure.

A remarkable example for the bypass capacity of pol V was its ability to replicate across a stretch of three or 12 methylene residues, inserted into a segment of ssDNA in a gapped plasmid (50). The $-(CH_2)_3-$ and $-(CH_2)_{12}-$ inserts span lengths corresponding to the lengths of approximately one and 2.5 nucleotides, respectively. These artificial "lesions" share no common features with DNA. They lack the base, sugar, and phosphate moieties of DNA, they are not charged, and they are hydrophobic. Remarkably, pol V was able to bypass these hydrocarbon "lesions" both in vitro and in vivo with efficiencies similar to that of an abasic site. Bypass across the $-(CH_2)_{12}-$ insert occurred primarily by looping out the hydrocarbon chain, followed by polymerase hopping to the nucleotide located 5′ to the insert, and continued synthesis. This is functionally an editing reaction, which leads to restoration of the original DNA sequence (50). More remarkable, however, was the finding that pol V is able to insert one or even two nucleotides opposite the M12 insert. This is not due to a misalignment mechanism, in which downstream nucleotides are used as a template for the synthesis of one to two nucleotides, and then realignment places those nucleotides opposite the lesion (50). Can pol V use the $-(CH_2)_{12}-$ insert as a "template"? The chemical structure of this insert makes such a possibility unlikely. Instead, it was suggested that when facing such a highly abnormal segment in the template strand, pol V switches locally to a template-independent mode of synthesis. It should be stressed that pol V has no terminal transferase activity and does require a template. In other words, its local switching to a template-independent mode requires contact with the DNA flanking the non-DNA segment (50). A somewhat similar behavior of local loss of template direction was recently observed with human pol μ (51) and may be a general property common to many lesion-bypass DNA polymerases.

6. ACCESSORY PROTEINS ARE REQUIRED FOR LESION BYPASS BY POL V

As described earlier, pol V on its own is a very weak DNA polymerase, which is unable to bypass DNA lesions. Bypass activity requires also RecA and SSB and is stimulated by the β subunit processivity clamp and the γ complex clamp loader. This complexity is in striking contrast to purified eukaryotic lesion bypass DNA polymerases, which can carry out highly effective in vitro TLS on their own, without the need for any accessory proteins. Why then does pol V need RecA and SSB for lesion bypass?

To understand the mechanistic requirement for RecA and SSB in pol V-catalyzed TLS, one needs to consider the cellular events that lead to the formation of damaged gaps in DNA. As the replication fork progresses, SSB binds the unwound single-strand regions in a co-operative fashion. The SSB–ssDNA complex is composed of SSB tetramers (each binding a stretch of 35 nucleotides), with ssDNA wrapped around SSB. The stretching out of ssDNA around SSB serves several functions, including prevention of reannealing and melting out of secondary structures in DNA, thereby preparing it for replication by pol III holoenzyme (52). On the basis of this scenario, when the replisome is blocked at a lesion, the unwound region including the lesion is wrapped around SSB. This is the situation that is the starting point of the events that will eventually lead to TLS by pol V (Fig. 1, stage 1). Soon after replication arrest at a lesion, SSB is displaced from DNA by the RecA protein (Fig. 1, stage 2). RecA is constitutively present in the cell at an estimated number of 5000 molecules, and it binds the DNA in a mode totally different from that of SSB. Each RecA monomer binds a stretch of three nucleotides, and binding is highly co-operative. However, unlike SSB, RecA forms a helical sleeve, which engulfs the DNA (a nucleoprotein filament; Fig. 1, stage 2) (53). This serves two immediate purposes: (i) it provides physical protection to the DNA, previously exposed on the surface of SSB molecules, and (ii) the RecA nucleoprotein filament activates the SOS response, by promoting the autocleavage of free LexA. This cleavage reduces the concentration of free LexA, shifting the equilibrium from the DNA-bound state of LexA to the free state, thereby causing derepressing of SOS genes. The *umuDC* operon is induced relatively late in the SOS response (54,55), suggesting that error-free gap-filling mechanisms like GFRR are preferred, before resorting to the mutagenic TLS.

Once the *umuDC* operon is induced, the UmuD is processed to its active derivative UmuD′, in a reaction promoted by the RecA nucleoprotein filament (similar to the autocleavage of LexA), and together with UmuC it forms pol V, which is ready for the gap-filling reaction. This means that pol V has to act on a DNA

1. Arrest of DNA replication at a blocking lesion

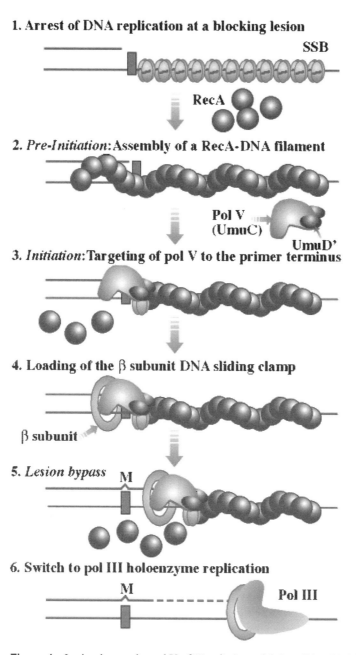

2. *Pre-Initiation*: Assembly of a RecA-DNA filament

3. *Initiation*: Targeting of pol V to the primer terminus

4. Loading of the β subunit DNA sliding clamp

5. *Lesion bypass*

6. Switch to pol III holoenzyme replication

Figure 1 Lesion bypass by pol V of *E. coli*. A model describing TDS by pol V in the presence of RecA, SSB, the β subunit processivity clamp, and the γ complex clamp loader. See text for details. *Source*: From Ref. 15.

region that is already covered by RecA (56). In fact, the RecA serves to target pol V to the primer-template lesion junction (28,56,57) (Fig. 1, stage 3). This sequence of events has the advantage that it limits the activity of pol V to sites of DNA damage, where a RecA nucleoprotein has assembled, and not on undamaged regions on DNA. This serves to minimize undesired mutagenic synthesis by pol

V on undamaged DNA regions, providing additional regulation on the activity of pol V. Interestingly, purified pol V strongly binds naked ssDNA regardless of the presence of DNA lesions (58). Thus, pol V will not be able to perform DNA synthesis on naked DNA because it will be trapped by the strong and nonproductive interactions with ssDNA, preventing it from finding its way to the primer terminus. The targeting to the primer-template–lesion junction is facilitated by interactions between UmuD' and RecA (57) and between UmuC and the primer template. There is no evidence for a direct interaction between UmuC and RecA. In summary, it is possible that pol V evolved to act on RecA-coated ssDNA, as part of a regulatory process that limits its activity to primers terminated at blocking template lesions. This model provides an explanation for the requirement of RecA, but why then is SSB required?

Single-strand binding increases the affinity of pol V to the primer template (59). This is somewhat puzzling in light of the fact that RecA binds DNA stronger than SSB (53,60). A possible hint is provided by the finding that ATPγS, a poorly hydrolysable ATP analog that strongly increases binding of RecA to DNA, severely inhibits the initiation of synthesis by pol V but affects elongation to a much lower extent (Ref. 56 and Goldsmith, M. and Livneh, Z., unpublished results). This suggests that while RecA targets pol V to the primer terminus, once pol V is there, RecA monomers must dissociate to allow stable binding of pol V and initiation of DNA synthesis (Fig. 1, stage 3). Once bound to the primer terminus, pol V can proceed with DNA synthesis and bypass (Fig. 1, stages 4 and 5) even in the presence of ATPγS. This stage of facilitating local RecA dissociation might be aided by SSB (59). Recall that RecA binding is dynamic, with monomers dissociating and reassociating at a rapid rate. In addition, although RecA–ssDNA binding is more stable than SSB–ssDNA binding, SSB binding is faster (53,60). Therefore, it is possible that SSB displaces RecA from the template strand locally at the vicinity of the primer terminus, and this enables the exposure of the DNA for productive pol V binding.

7. OTHER DNA POLYMERASES INVOLVED IN TLS IN *E. COLI*

Of the five known DNA polymerases in *E. coli*, three are SOS inducible: pol II, pol IV, and pol V. Using an episome-based assay system, it was reported that all these three DNA polymerases are involved in TLS in *E. coli* (61). However, while there is ample evidence for the role of pol V in TLS (reviewed in Ref. 7), the evidence for the involvement of pol II and pol IV is scarce. In fact, it appears that TLS in *E. coli* is carried out primarily by one generalized TLS polymerase, pol V. The other two SOS polymerases, pol II and pol IV, seem to function primarily in processes of recovery from DNA damage other than TLS: pol II in resumption of replication after DNA damage [replication restart (62)] and pol IV in stationary phase mutagenesis (63). Particularly enigmatic is pol IV, which shows different behavior in vitro and in vivo. When assayed in vitro, purified pol IV, in the presence of the β subunit and γ complex, exhibits efficient bypass across several lesions, including an abasic site (64) and the $-(CH_2)_3-$ and $-(CH_2)_{12}-$ "lesions" (50), with efficiencies comparable to that of pol V. However, when assayed in vivo, pol IV does not appear to participate in the bypass across these lesions in any significant way (50,64). It is possible that the main function of pol IV is as a mutase (e.g., in stationary mutations), and that the in vitro ability to bypass lesions is a byproduct of its low fidelity, with little in vivo significance. Alternatively, pol IV may act in vivo with pol V, to bypass a subset of lesions that pol V has a hard time to deal with. In those cases, pol V may carry out the insertion step, whereas pol IV may carry out the extension step.

The replicative pol III holoenzyme is often blocked at DNA lesions. However, some lesions are weak blockers (e.g., 8-oxoguanine) and may therefore be bypassed by the replicative polymerase at least to some extent (65). In fact, it was shown that the β subunit can endow pol III with high bypass efficiency across an abasic site in vitro under some conditions (66), and that this might have in vivo significance under conditions when the proofreading activity of pol III is compromised (67).

A surprising finding was the presence of multiple pol V homologs on native con-jugative plasmids (68). These plasmids are of broad host specificity and frequently carry multiple antibiotic-resistance genes. The most extensively studied one is encoded by the *MucA'* and *MucB* genes (69) and termed pol RI (70). Pol RI bypasses lesion in a manner similar to pol V, requiring both RecA and SSB for bypass (70). *Escherichia coli* cells harboring plasmids with *MucB* and *MucA* exhibit higher UV resistance and higher UV mutability, and it was suggested that they have a survival effect higher than pol V. Other plasmidic homologs include RumAB and ImpABC (68,71,72). It was suggested that plasmidic pol V-homologs act to increase adaptation and survival of the plasmids in a variety of hosts, and that this might have a role in the spreading of antibiotics resistance among bacterial pathogens (70).

8. IN VIVO ROLE OF TLS

Escherichia coli cells lacking pol V show slightly reduced UV survival and greatly reduced UV mutagenesis. This led to two different views on the in vivo role of pol V-promoted TLS in *E. coli* (37,73,74). According to the "repair at a price" view, TLS is a DNA repair mechanism that functions to increase cell survival when other mechanisms have failed or are unable to act. This may occur when two closely opposed lesions that have escaped repair block replication on both daughter stands (75). This gives rise to overlapping daughter strand gaps, which can be filled in only by TLS. The price of this gap filling is an increase in mutation frequency, as TLS across modified nucleotides is inherently mutagenic. This view is supported by the lower UV survival of cells lacking pol V (7). The finding that the $-(CH_2)_{12}-$ "lesion" cannot be repaired in *E. coli* by excision repair, but can be bypassed by pol V is also consistent with such a view (50). The other view is that TLS is a process of induced mutagenesis that functions to increase fitness by allowing faster adaptation to hostile environments (e.g., with DNA damaging agents) (76,77). A third hypothesis, which combines the two former views, is that TLS evolved first as a generic, primitive "repair" mechanism to overcome replication barriers by read-through. As more sophisticated DNA repair mechanisms evolved (e.g., base excision repair, nucleotide excision repair, mismatch repair, recombinational repair), the significance of the repair aspect of TLS decreased, and its mutagenesis aspect increased, keeping it from being lost during evolution (15). Of course, these two functions are not mutually exclusive. Pol V-catalyzed TLS may act as a mutator to facilitate adaptation, but it can also act as a backup generic primitive replication-rescue operation, to deal with unrepaired lesions in DNA.

REFERENCES

1. Echols H, Goodman MF. Fidelity mechanisms in DNA replication. Annu Rev Biochem 1991; 60:477–511.
2. Kunkel TA, Bebenek K. DNA replication fidelity. Annu Rev Biochem 2000; 69:497–526.
3. Kool ET. Active site tightness and substrate fit in DNA replication. Annu Rev Biochem 2002; 71:191–219.
4. Moore PD, Bose KK, Rabkin SD, Strauss BS. Sites of termination of *in vitro* DNA synthesis on ultraviolet- and N-acetylaminofluorene-treated $\phi \times 174$ templates by prokaryotic and eukaryotic DNA polymerases. Proc Natl Acad Sci USA 1981; 78:110–114.
5. Shwartz H, Livneh Z. Dynamics of termination during *in vitro* replication of ultraviolet-irradiated DNA with DNA polymerase III holoenzyme of *Escherichia coli*. J Biol Chem 1987; 262:10518–10523.
6. Livneh Z, Cohen-Fix O, Skaliter R, Elizur T, Crit CRC. Replication of damaged DNA and the molecular mechanism of ultraviolet light mutagenesis. Rev Biochem Mol Biol 1993; 28:465–513.
7. Friedberg EC, Walker GC, Siede W. DNA Repair and Mutagenesis. Washington, DC: ASM Press, 1995.
8. Michel B, Ehrlich SD, Uzest M. DNA double-strand breaks caused by replication arrest. EMBO J 1997; 16:430–438.
9. Khanna KK, Jackson SP. DNA double-strand breaks: signaling, repair and the cancer connection. Nat Genet 2001; 27:247–254.
10. Rupp WD, Howard-Flanders P. Discontinuities in the DNA synthesized in an excision-defective strain of *Escherichia coli* following ultraviolet irradiation. J Mol Biol 1968; 31:291–304.
11. Rupp WD, Wilde CE III, Reno DL, Howard-Flanders P. Exchanges between DNA strands in ultraviolet-irradiated *Escherichia coli*. J Mol Biol 1971; 61:25–44.
12. Berdichevsky A, Izhar L, Livneh Z. Error-free recombinational repair predominates over mutagenic translesion replication in *E. coli*. Mol Cell 2002; 10:917–924.
13. Michel B. Replication fork arrest and DNA recombination. Trends Biochem Sci 2000; 25:173–178.
14. Goodman MF. Coping with replication 'train wrecks' in *Escherichia coli* using Pol V, Pol II and RecA proteins. Trends Biochem Sci 2000; 25:189–195.
15. Livneh Z. DNA damage control by novel DNA polymerases: translesion replication and mutagenesis. J Biol Chem 2001; 276:25639–25642.
16. Prakash S, Prakash L. Translesion DNA synthesis in eukaryotes: A one- or two-polymerase affair. Genes Dev 2002; 16:1872–1883.
17. Courcelle J, Donaldson JR, Chow KH, Courcelle CT. DNA damage-induced replication fork regression and processing in Escherichia coli. Science 2003; 299:1064–1067.
18. Higgins NP, Kato K, Strauss B. A model for replication repair in mammalian cells. J Mol Biol 1976; 101:417–425.
19. Kato T, Shinoura Y. Isolation and characterization of mutants of *Escherichia coli* deficient in induction of mutagenesis by ultraviolet light. Mol Gen Genet 1977; 156:121–131.
20. Steinborn G. Uvm mutants of Escherichia coli K12 deficient in UV mutagenesis. I. Isolation of uvm mutants and their phenotypical characterization in DNA repair and mutagenesis. Mol Gen Genet 1978; 165:87–93.
21. Walker GC. SOS-regulated proteins in translesion DNA synthesis and mutagenesis. Trends Biochem Sci 1995; 20:416–420.
22. Reuven NB, Tomer G, Livneh Z. The mutagenesis proteins UmuD' and UmuC prevent lethal frameshifts while increasing base substitution mutations. Mol Cell 1998; 2:191–199.
23. Tang M, Bruck I, Eritja R, Turner J, Frank EG, Woodgate R, O'Donnell M, Goodman MF. Biochemical basis of SOS-induced mutagenesis in *Escherichia coli*: Reconstitution of in vitro lesion bypass dependent on the UmuD'$_2$C mutagenic complex and RecA protein. Proc Natl Acad Sci USA 1998; 95:9755–9760.

24. Tang M, Shen X, Frank EG, O'Donnell M, Woodgate R, Goodman MF. UmuD'$_2$C is an error-prone DNA polymerase, *Escherichia coli* pol V. Proc Natl Acad Sci USA 1999; 96:8919–8924.

25. Reuven NB, Arad G, Maor-Shoshani A, Livneh Z. The mutagenesis protein UmuC is a DNA polymerase activated by UmuD', RecA and SSB, and specialized for translesion replication. J Biol Chem 1999; 274:31763–31766.

26. Ohmori H, Friedberg EC, Fuchs RPP, Goodman MF, Hanaoka F, Hinkle D, Kunkel TA, Lawrence CW, Livneh Z, Nohmi T, Prakash L, Prakash S, Todo T, Walker GC, Wang Z, Woodgate R. The Y-family of DNA polymerases. Mol Cell 2001; 8:7–8.

27. Nelson JR, Lawrence CW, Hinkle DC. Thymine-thymine dimer bypass by yeast DNA polymerase ζ. Science 1996; 272:1646–1649.

28. Tang M, Pham P, Shen X, Taylor JS, O'Donnell M, Woodgate R, Goodman MF. Roles of *E. coli* DNA polymerases IV and V in lesion-targeted and untargeted SOS mutagenesis. Nature 2000; 404:1014–1018.

29. Maor-Shoshani A, Livneh Z. Analysis of the stimulation of DNA polymerase V of *Escherichia coli* by processivity proteins. Biochemistry 2002; 41:14438–14446.

30. Maor-Shoshani A, Reuven NB, Tomer G, Livneh Z. Highly mutagenic replication of undamaged DNA by DNA polymerase V (UmuC) provides a mechanistic basis for SOS untargeted mutagenesis. Proc Natl Acad Sci USA 2000; 97:565–570.

31. Pham PT, Olson MW, McHenry CS, Schaaper RM. The base substitution and frameshift fidelity of *Escherichia coli* DNA polymerase III holoenzyme *in vitro*. J Biol Chem 1998; 273:23575–23584.

32. Fujii S, Akiyama M, Aoki K, Sugaya Y, Higuchi K, Hiraoka M, Miki Y, Saitoh N, Yoshiyama K, Ihara K, Seki M, Ohtsubo E, Maki H. DNA replication errors produced by the replicative apparatus of *Escherichia coli*. J Mol Biol 1999; 289:835–850.

33. Caillet-Fauquet P, Maenhaut-Michel G, Radman M. SOS mutator effect in *E. coli* mutants deficient in mismatch correction. EMBO J 1984; 3:707–712.

34. Fijalkowska IJ, Dunn RL, Schaaper RM. Genetic requirements and mutational specificity of the *Escherichia coli* mutator activity. J Bacteriol 1997; 179:7435–7445.

35. Schaaper RM, Dunn RL. Spectra of spontaneous mutations in *Escherichia coli* strains defective in mismatch correction: the nature of in vivo DNA replication errors. Proc Natl Acad Sci USA 1987; 84:6220–6224.

36. Schaaper RM. Base selection, proofreading, and mismatch repair during DNA replication in *Escherichia coli*. J Biol Chem 1993; 268:23762–23765.

37. Witkin EM, Wermundsen IE. Targeted and untargeted mutagenesis by various inducers of SOS functions in *E. coli*. Cold Spring Harb Symp Quant Biol 1979; 43:881–886.

38. Miller JH, Low KB. Specificity of mutagenesis resulting from the induction of the SOS system in the absence of mutagenic treatment. Cell 1984; 37:675–682.

39. Caillet FP, Maenhaut MG. Nature of the SOS mutator activity: genetic characterization of untargeted mutagenesis in *Escherichia coli*. Mol Gen Genet 1988; 213:491–498.

40. Brotcorne-Lannoye A, Maenhaut-Michel G. Role of RecA protein in untargeted UV mutagenesis of bacteriophage l: Evidence for the requirement of the *dinB* gene. Proc Natl Acad Sci USA 1986; 83:3904–3908.

41. Banerjee SK, Christensen RB, Lawrence CW, LeClerc JE. Frequency and spectrum of mutations produced by a single *cis-syn* thymine-thymine cyclobutane dimer in a single-stranded vector. Proc Natl Acad Sci USA 1988; 85:8141–8145.

42. Lawrence CW, Borden A, Banerjee SK, LeClerc JE. Mutation frequency and spectrum resulting from a single abasic site in a single-stranded vector. Nucleic Acids Res 1990; 18: 2153–2157.

43. LeClerc JE, Borden A, Lawrence CW. The thymine-thymine pyrimidine-pyrimidone(6-4) ultraviolet light photoproduct is highly mutagenic and specifically induces 3' thymine-to-cytosine transitions in *Escherichia coli*. Proc Natl Acad Sci USA 1991; 88: 9685–9689.

44. Trincao J, Johnson RE, Escalante CR, Prakash S, Prakash L, Aggarwal AK. Structure of the catalytic core of *S. cerevisiae* DNA polymerase η: implications for translesion DNA synthesis. Mol Cell 2001; 8:417–426.

45. Silvian LF, Toth EA, Pham P, Goodman MF, Ellenberger T. Crystal structure of a DinB family error-prone DNA polymerase from Sulfolobus solfataricus. Nat Struct Biol 2001; 8:984–989.

46. Zhou BL, Pata JD, Steitz TA. Crystal structure of a DinB lesion bypass DNA polymerase catalytic fragment reveals a classic polymerase catalytic domain. Mol Cell 2001; 8:427–437.

47. Ling H, Boudsocq F, Woodgate R, Yang W. Crystal structure of a Y-family DNA polymerase in action. A mechanism for error-prone and lesion-bypass replication. Cell 2001; 107:91–102.

48. Ling H, Boudsocq F, Plosky BS, Woodgate R, Yang W. Replication of a *cis-syn* thymine dimer at atomic resolution. Nature 2003; 424:1083–1087.

49. Ling H, Boudsocq F, Woodgate R, Yang W. Snapshots of replication through an abasic lesion; structural basis for base substitutions and frameshifts. Mol Cell 2004; 13:751–762.

50. Maor-Shoshani A, Ben-Ari V, Livneh Z. Lesion bypass DNA polymerases replicate across non-DNA segments. Proc Natl Acad Sci USA 2003; 100:14760–14765.

51. Covo S, Blanco L, Livneh Z. Lesion bypass by human DNA polymerase μ reveals a template-dependent sequence-independent nucleotidyl transferase activity. J Biol Chem 2004; 279:859–865.

52. Lohman TM, Ferrari ME. *Escherichia coli* single-stranded DNA binding protein: Multiple DNA-binding modes and cooperativities. Annu Rev Biochem 1994; 63:527–570.

53. Roca AI, Cox MM, Crit CRC. The RecA protein: Structure and function. Rev Biochem Mol Biol 1990; 25:415–456.

54. Sommer S, Bailone A, Devoret R. The appearance of the UmuD′C protein complex in *Escherichia coli* switches repair from homologous recombination to SOS mutagenesis. Mol Microbiol 1993; 10:963–971.

55. Ronen M, Rosenberg R, Shraiman B, Alon U. Assigning numbers to the arrows: parameterizing a gene regulation network by using accurate expression kinetics. Proc Natl Acad Sci USA 2002; 99:10555–10560.

56. Reuven NB, Arad G, Stasiak AZ, Stasiak A, Livneh Z. Lesion bypass by the *Escherichia coli* DNA polymerase V requires assembly of a RecA nucleoprotein filament. J Biol Chem 2001; 276:5511–5517.

57. Frank EG, Hauser J, Levine AS, Woodgate R. Targeting of the UmuD, UmuD′, and MucA′ mutagenesis proteins to DNA by RecA protein. Proc Natl Acad Sci USA 1993; 90:8169–8173.

58. Bruck I, Woodgate R, McEntee K, Goodman MF. Purification of a soluble UmuD′C complex from *Escherichia coli*. Cooperative binding of UmuD′C to single-stranded DNA. J Biol Chem 1996; 271:10767–10774.

59. Pham P, Bertram JG, O'Donnell M, Woodgate R, Goodman MF. A model for SOS-lesion-targeted mutations in *Escherichia coli*. Nature 2001; 409:366–370.

60. Roca AI, Cox MM. RecA protein: Structure, function, and role in recombinational DNA repair. Prog Nucleic Acids Res Mol Biol 1997; 56:129–223.

61. Napolitano R, Janel-Bintz R, Wagner J, Fuchs RPP. All three SOS-inducible DNA polymerases (Pol II, Pol IV, and Pol V) are involved in induced mutagenesis. EMBO J 2000; 19:6259–6265.

62. Rangarajan S, Woodgate R, Goodman MF. A phenotype for enigmatic DNA polymerase II: a pivotal role for pol II in replication restart in UV-irradiated *Escherichia coli*. Proc Natl Acad Sci USA 1999; 96:9224–9229.

63. McKenzie GJ, Lee P, Lombardo MJ, Hastings PJ, Rosenberg SM. SOS mutator DNA polymerase IV functions in adaptive mutation and not adaptive amplification. Mol Cell 2001; 7:571–579.

64. Maor-Shoshani A, Hayashi K, Ohmori H, Livneh Z. Analysis of translesion replication across an abasic site by DNA polymerase IV of *Escherichia coli*. DNA Repair 2003; 2:1227–1238.

65. Freisinger E, Grollman AP, Miller H, Kisker C. Lesion (in)tolerance reveals insights into DNA replication fidelity. EMBO J 2004; 23:1494–1505.

66. Tomer G, Reuven NB, Livneh Z. The b subunit sliding DNA clamp is responsible for unassisted mutagenic translesion replication by DNA polymerase III holoenzyme. Proc Natl Acad Sci USA 1998; 95:14106–14111.

67. Vandewiele D, Borden A, O'Grady PI, Woodgate R, Lawrence CW. Efficient translesion replication in the absence of *Escherichia coli* Umu proteins and 3′-5′ exonuclease proof-reading function. Proc Natl Acad Sci USA 1998; 95:15519–15124.

68. Woodgate R, Sedgwick SG. Mutagenesis induced by bacterial UmuDC proteins and their plasmid homologues. Mol Microbiol 1992; 6:2213–2218.

69. Perry KL, Walker GC. Identification of plasmid (pKM101)-coded proteins involved in mutagenesis and UV resistance. Nature 1982; 300:278–281.

70. Goldsmith M, Sarov-Blat L, Livneh Z. Plasmid-encoded MucB protein is a DNA polymerase (pol RI) specialized for lesion bypass in the presence of MucA′, RecA, and SSB. Proc Natl Acad Sci USA 2000; 97:11227–11231.

71. Woodgate R, Singh M, Kulaeva OI, Frank EG, Levine AS, Koch WH. Isolation and characterization of novel plasmid-encoded UmuDC mutants. J Bacteriol 1994; 176: 5011–5021.

72. Szekeres EJ, Woodgate R, Lawrence CW. Substitution of mucAB or rumAB for UmuDC alters the relative frequencies of the two classes of mutations induced by a site-specific T-T cyclobutane dimer and the efficiency of translesion DNA synthesis. J Bacteriol 1996; 178:2559–2563.

73. Radman M. SOS repair hypothesis: phenomenology of an inducible DNA repair which is accompanied by mutagenesis. In: Hanawalt P, Setlow RB, eds. Molecular Mechanisms for Repair of DNA. New York: Plenum Press, 1975:355–367.

74. Echols H. SOS functions, cancer, and inducible evolution. Cell 1981; 25:1–2.

75. Sedgwick SG. Misrepair of overlapping daughter strand gaps as a possible mechanism for UV induced mutagenesis in *uvr* strains of *Escherichia coli*: a general model for induced mutagenesis by misrepair (SOS repair) of closely spaced DNA lesions. Mutat Res 1976; 41:185–200.

76. Radman M. Enzymes of evolutionary change. Nature 1999; 401:866–869.

77. Yeiser B, Pepper ED, Goodman MF, Finkel SE. SOS-induced DNA polymerases enhance long-term survival and evolutionary fitness. Proc Natl Acad Sci USA 2002; 99:8737–8741.

24

Mechanism of Bypass Polymerases in Eukaryotes

Zhigang Wang
Graduate Center for Toxicology, University of Kentucky, Lexington, Kentucky, U.S.A.

1. INTRODUCTION

Cells contain extraordinarily efficient replication apparatus to duplicate their genomes during cell division. Faithfully copying the long stretches of cellular DNA demands extraordinarily high efficiency and fidelity for DNA synthesis. The replicative DNA polymerases (Pol) δ and ε are indeed highly efficient and accurate in copying DNA. This high efficiency and accuracy owes in part to their ability to tightly grab onto the template DNA and stringently select the correct nucleotide at the geometrically well-confined active site. The $3' \rightarrow 5'$ proofreading exonuclease activity of both Polδ and Polε adds another level of accuracy by removing an incorrectly incorporated nucleotide. Such a tight fit is well suited for high efficient and accurate replication from DNA templates of normal structure and chemical compositions. DNA damage such as chemical modifications of the bases would disrupt this intimate match between the replicative polymerases and the DNA template. In the presence of DNA damage, the replicative polymerases become ineffective, as they have evolved so effectively and specifically to deal with the normal DNA template. Specialized DNA polymerases are required to copy the damaged sites of the DNA template. The challenge for these specialized polymerases is not high efficiency and accuracy, but merely being able to copy the damaged sites. Consequently, these polymerases had evolved to possess very different biochemical properties as compared to the replicative polymerases.

Damage to DNA is unavoidable because of its reactive chemical components. Cells rely on DNA repair and cell-cycle checkpoint control for defense against DNA damage. In multicell organisms, apoptosis is additionally employed to remove excessively damaged cells. Nevertheless, these cellular defense systems do not function with perfection, leading to some persistent DNA lesions during replication. Several factors further promote persistence of DNA lesions during replication, including: (a) high levels of damage, (b) poorly repaired lesions, (c) inefficiently repaired genomic regions, and (d) damage sustained in the S phase of the cell cycle. Cells have evolved a sophisticated system called damage tolerance to respond to the unrepaired DNA lesions during replication. It allows cells to replicate their genomes in the presence of DNA damage that would normally block the replicative polymerases. This system tolerates, rather than removes, DNA damage. After replication, the tolerated lesions are then subject to removal by DNA repair systems. In eukaryotes, damage tolerance consists of at least two mechanisms: error-free postreplication repair (template switching) and translesion synthesis (Fig. 1). The term postreplication repair originated from yeast experiments employing alkaline sucrose gradient to examine DNA synthesis following UV radiation of the cell (1,2). Immediately after UV radiation, smaller DNA fragments are generated as detected by the alkaline sucrose gradient. With extended incubation time, these smaller DNA fragments are converted to large fragments that are normally detected without DNA damage. This cellular response requires at least Rad6, Rad18, Rad5, Mms2–Ubc13 complex, PCNA, and Polδ, and is an error-free mechanism of damage tolerance (reviewed in Ref. 3). Thus, template switching is a preferred term to describe this mechanism of damage tolerance in order to avoid the somewhat misleading "repair" description.

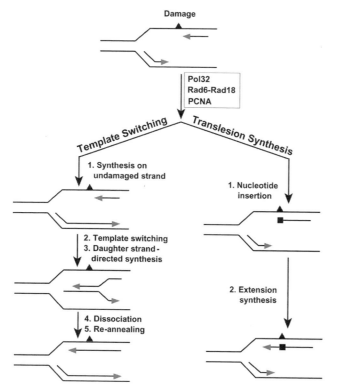

Figure 1 Models of two damage tolerance mechanisms. At the lesion site, template switching (the left pathway) uses the newly synthesized daughter strand as the template for DNA synthesis, thus, bypassing the lesion in an error-free manner. Template switching is likely a molecular mechanism responsible for the phenomenon of error-free postreplication repair that involves Rad5, Mms2, and Ubc13. In contrast, translesion synthesis (the right pathway) directly copies the damaged site on the template. Consequently, mutations, shown as a square, are often generated opposite the lesion. At early steps of damage tolerance, the Pol32 subunit of Polδ, the Rad6–Rad18 complex, and PCNA may be involved in recognition of the stalled replication complex and subsequently recruiting other proteins for template switching or translesion synthesis.

Template switching in mammalian cells was originally proposed by Higgins et al. in 1976 (4). The molecular details of template switching remain mostly unknown in eukaryotes, largely due to lack of biochemical studies. Nevertheless, a conceptual model is shown in Fig. 1. In this model, when DNA synthesis is blocked by a template lesion, synthesis on the undamaged template strand may continue to a limited extent. Then, by using the newly synthesized daughter strand as the template (template switching), the lesion-blocked DNA synthesis can proceed further downstream beyond the site corresponding to the lesion. Following dissociation of the two newly synthesized daughter strands, each is reannealed to its original parental strand, thus bypassing the damaged site on the parental DNA strand (Fig. 1). Because template switching avoids copying the damaged site of the DNA template, the result is error-free tolerance of the damage. In contrast, translesion synthesis directly copies the damaged site of the template. Consequently, mutations are often produced at sites of the damage (Fig. 1).

2. CONCEPTS OF TRANSLESION SYNTHESIS

Translesion synthesis is the cellular process that directly copies damaged sites of the template during DNA synthesis. It consists of nucleotide insertion opposite the lesion and extension synthesis from opposite the lesion (Fig. 1). According to this definition, a DNA polymerase that performs the insertion step, the extension step, or both qualifies as a translesion polymerase. The term lesion bypass has also been interchangeably used with translesion synthesis in the literature.

Depending on the accuracy of nucleotide insertion opposite the DNA damage, translesion synthesis is further divided into two categories: error free and error prone. Error-free translesion synthesis predominantly inserts the correct nucleotide opposite the lesion. Thus, it is a mutation-avoiding mechanism that suppresses DNA damage-induced mutagenesis. Error-prone translesion synthesis frequently inserts an incorrect nucleotide opposite the lesion. Thus, it is a mutation-generating mechanism that promotes DNA damage-induced mutagenesis. In cells, error-prone translesion synthesis constitutes the major mechanism of base damage-induced mutagenesis in cells and is thus of great interest to the field of mutagenesis.

Error free vs. error prone is a relative description for the accuracy of translesion synthesis. As translesion polymerases copy undamaged DNA with very low fidelity, how can translesion synthesis be error free? The key lies in the fact that copying undamaged genome during replication involves a vast amounts of template nucleotides, $\sim 3 \times 10^9$ bp in the human genome, whereas only a very tiny fraction of the genome is encountered by translesion synthesis. Considering a polymerase with a 10^{-2} synthesis fidelity per base, if this polymerase were to copy the whole genome, the errors would have been enormous ($10^{-2} \times 3 \times 10^9 = 3 \times 10^7$ per human genome!), obviously an error-prone result. Assuming the same fidelity of this polymerase in copying a specific DNA lesion, if 50 of such lesions were left unrepaired during replication, translesion synthesis by this same polymerase would not produce a single mutation at the lesion sites, obviously an error-free result. Therefore, error free or error prone is meaningful only when the comparison is made within the replication domain or the translesion synthesis domain. Sometimes, it may not be obvious to distinguish between error free and error prone on the basis of in vitro biochemical analysis of a polymerase in response to a specific lesion. The ultimate distinction between these two modes of translesion synthesis in cells can be made through genetic analysis. If the polymerase activity suppresses the lesion-induced mutagenesis, then it is error free. If the polymerase activity promotes the lesion-induced mutagenesis, then it is error prone.

It is not clear how far extension synthesis is needed beyond the lesion before normal replication can resume. Nevertheless, as indicated by in vitro biochemical experiments (5), extension synthesis of at least a few nucleotides is required for the switch from translesion synthesis to normal replicative synthesis to take place. In response to a mispaired A opposite a template *trans-anti*-benzo[*a*]pyrene-7,8-dihydrodiol-9,10-epoxide (BPDE)-N^2-dG adduct, yeast Polδ-catalyzed DNA synthesis, without the interference of its $3' \rightarrow 5'$ proofreading exonuclease activity, is possible only when the primer ends at least six nucleotides beyond the lesion site (5). Therefore, when translesion synthesis efficiency is considered, both the nucleotide insertion and the extension synthesis steps will have to be considered. Depending on the specific lesion and the polymerase, either the insertion or the extension can become a rate-limiting step in translesion synthesis. Furthermore, translesion synthesis may be completed by one polymerase or it may require two polymerases in which one catalyzes the insertion step and the other performs extension synthesis.

3. TRANSLESION POLYMERASES

A DNA polymerase that performs nucleotide insertion opposite the lesion and/or extension synthesis from opposite the lesion is referred to as translesion polymerase or bypass polymerase. Multiple translesion polymerases have been discovered. During translesion synthesis, the active site of the polymerase must be able to accommodate the damaged template base. As there are various types of base lesions that differ drastically in chemistry and structure, it may well be an evolutionary consequence that a group of translesion polymerases is devoted to copy the damaged sites of DNA, with each polymerase possessing different lesion specificities. The involvement of multiple polymerases makes the process of translesion synthesis highly complex.

3.1. Polζ

Studies on translesion synthesis in eukaryotes began over 30 years ago in the yeast *Saccharomyces cerevisiae* by traditional genetics (6). Several yeast mutant strains were isolated in which the wild-type versions of their affected genes are required for UV-induced mutagenesis (6–12). Among them are the *REV1* (required for *reversion* mutation), *REV3*, and *REV7* genes. When the *REV3* gene was cloned from *S. cerevisiae* in 1989, it became clear that this gene codes for a DNA polymerase (13). Unlike the replicative polymerases α, δ, and ε, *REV3* gene is not essential for growth (13), although they all belong to the B family of DNA polymerases (14). Therefore, it was widely believed that Rev3 is a specialized polymerase specifically devoted for translesion synthesis. Seven years later, it was experimentally demonstrated that Rev3 is indeed the catalytic subunit of Polζ and that Polζ is capable of translesion synthesis in vitro (15). Thus, Polζ was established as the first translesion polymerase in eukaryotes. The small Rev7 protein (29 kDa) forms a tight complex with Rev3 and is considered as a noncatalytic subunit of Polζ (15).

Purified yeast Polζ (the Rev3–Rev7 complex) is able to perform limited nucleotide insertions opposite a template TT (6–4) photoproduct, AAF-dG adduct, and (+) or (−) -*trans-anti*-BPDE-N^2-dG adduct (16,17). Furthermore, Polζ also catalyzes extension synthesis from opposite many types of lesions with varying efficiencies, including an AP site, *cis–syn* TT dimer, (6–4) photoproduct, AAF-dG adduct, (+) or (–)-*trans-anti*-BPDE-N^2-dG adduct, and an acrolein-derived dG adduct (16–21). Therefore, it has been proposed that Polζ functions both as an insertion polymerase and as an extension polymerase (16). It appears that the extension activity of Polζ is versatile. Thus, it is believed that Polζ is a major extension polymerase during translesion synthesis in eukaryotes (16,20). It should be noted, however, that in vitro extension synthesis by purified yeast Polζ is rather inefficient in most cases (16,22). Hence, it is possible that there may exist other factors that stimulate the translesion synthesis activity of Polζ in cells. Additionally, the Y family DNA polymerases also participate in translesion synthesis, catalyzing the nucleotide insertion step, the extension step, or both.

On the basis of the yeast Rev3 protein sequence, the human full-length *REV3* cDNA was cloned (23,24). The polymerase domains of these two proteins are well conserved (23,24). Human REV7 protein is much less conserved and was cloned by the yeast two-hybrid screening for interaction with human REV3 (25). REV3 (353 kDa) is one of the largest proteins in human cells, making it highly difficult to produce the recombinant protein for in vitro biochemical studies. In addition to its C-terminal polymerase domain, human REV3 protein contains a large N-terminal region, accounting for ~3/4 of the protein (23,24). The function of this large N-terminal region remains completely unknown. It is possible that this N-terminal region may be involved in protein–protein interactions during the recruitment of Polζ to sites of DNA damage during translesion synthesis. Suppressing *REV3* gene expression by antisense RNA in cultured human or mouse cells results in significant reduction of UV-induced mutagenesis (23,26), thus suggesting that mammalian Polζ also plays an important role in error-prone translesion synthesis of UV lesions, as is the case in yeast. Surprisingly, Polζ-knockout mice are embryonic lethal (27–30). The failed embryos exhibited increased double-strand breaks and massive apoptosis. It was suggested that DNA double-strand breaks accumulate at sites of unreplicated DNA lesions in the absence of Polζ, leading to excessive apoptosis that results in lethality (30). Furthermore, a Rev3$^{-/-}$ cell line could not be established from the early embryos of the Polζ-knockout mice (30), suggesting that Polζ is essential for long-term cell survival. This is in contrast to the yeast and chicken cells, in which deletion of the *REV3* gene is compatible with cell survival under normal growth conditions (13,31). Whether Polζ is similarly essential for cell growth in humans, however, should not be simply extrapolated from the mouse results.

Yeast genetic studies have clearly indicated that Polζ does not work alone in error-prone translesion synthesis. Additional proteins function together with Polζ in a biochemical pathway to achieve error-prone translesion synthesis. This is referred to as the Polζ mutagenesis pathway. The other cloned genes in the Polζ mutagenesis pathway are *RAD6*, *RAD18*, and *REV1* (32–35). Recent studies indicate that the Pol32 subunit of Polδ and PCNA are also involved in the Polζ mutagenesis pathway (36,37). While Rev1 is a member of the Y family of DNA polymerases (38), the Rad6–Rad18 complex possesses a ubiquitin-conjugating activity and is thought to function at an early step of the Polζ mutagenesis pathway (39,40). Human homologs of *RAD6*, *RAD18*, and *REV1* have been cloned (41–45). Moreover, the human *RAD18*, *REV1*, and *REV3* genes are ubiquitously expressed in various tissues examined (24,42,44), supporting the notion that the Polζ pathway may represent a major mutagenesis pathway in humans.

3.2. The Y Family of DNA Polymerases

In 1996, Nelson et al. (18) discovered that the yeast Rev1 protein is a DNA template-dependent deoxycytidyl (dCMP) transferase. This milestone discovery laid a foundation for the discovery of the Y family of DNA polymerases 3 years later. Shortly after the Rev1 biochemical studies, the yeast *RAD30* gene was cloned as a homolog of the *E. coli dinB* and *umuC* (46,47). As Rev1 protein shares sequence homology with Rad30 protein, it was suspected that Rad30 might possess a similar dCMP transferase. Experiments attempting to test this prediction led to the surprising discovery that Rad30 protein is in fact a novel DNA polymerase, designated Polη, which is capable of error-free translesion synthesis across from a template *cis–syn* TT dimer, a major DNA damage induced by UV radiation (48). Meanwhile, using a traditional biochemical approach, Masutani et al. (49) found that the human xeroderma pigmentosum variant (XPV) protein is a DNA polymerase capable of error-free translesion synthesis across from a template *cis–syn* TT dimer. The fact that XPV protein is encoded by a human *RAD30* gene was then quickly established (50,51). Search for a human homolog of the yeast *RAD30* gene, which had began prior to the discovery of Polη, resulted in two additional homologs: *DINB1* coding for Polκ and *RAD30B* coding for Polι (52–54). Thus, the Y family of DNA polymerases was discovered. The Y family members in humans consist of Polη, Polκ, Polι, and REV1 (38). Remarkably, all of these human genes were published in 1999 (44,50–54).

Subsequent studies revealed an intrinsic biochemical activity common to the Y family of DNA polymerases: translesion synthesis. Hence, it is widely believed that a major cellular function of the Y family of DNA polymerases is translesion synthesis. The Y family polymerases share several biochemical properties: (a) lacking $3' \rightarrow 5'$ proofreading exonuclease activity; (b) synthesizing DNA in a more or less distributive manner; (c) capable of both error-free and error-prone translesion synthesis, depending on the lesion; and (d) synthesizing DNA from undamaged templates with extraordinarily low fidelity. These features are well suited for the task of a translesion polymerase.

The low fidelity nature of the Y family polymerases probably reflects an inevitable consequence of their biological function in translesion synthesis. It was speculated that a Y family polymerase contains a loose and flexible active site such that a damaged template base can be accommodated for translesion synthesis (55–57). This concept is indeed supported by crystal structures of yeast Polη and Dpo4, a Polκ homolog in *Sulfolobus solfataricus* (58–60). Consequently, when copying undamaged DNA, the active site of a Y family polymerase would loose the stringent geometry constraints that characterize highly accurate Watson–Crick base pairing, resulting in extraordinarily low fidelity DNA synthesis. Such infidelity reaches to an extreme extent for Polι when copying undamaged template T, opposite which G is preferred by 3–10-fold over the correct A (20,61,62). In the case of REV1, the only template base that is copied by the Watson–Crick base-pairing rule is G (18,63,64). Therefore, REV1 acts as a DNA polymerase on repeating template G sequences (63,65). Amazingly, REV1 recognizes undamaged template A, C, and T and several damaged bases and preferentially inserts a C opposite each without an exception (18,63–65). In this case, the rules of geometry constraints and hydrogen bonding that govern nucleotide selection during normal DNA synthesis simply do not apply. Hence, REV1 acts more as a dCMP transferase than a polymerase.

Owing to their low fidelity nature, the Y family polymerases must be excluded from normal DNA replication in order to maintain genomic stability. Conceivably, these specialized polymerases are accessible only to the damaged sites on the template through a recruiting mechanism. Additionally, these polymerases may be regulated transcriptionally and post-transcriptionally. The Polη expression in yeast is indeed inducible by UV radiation (46,47). Polη also contains a nuclear localization sequence in its C-terminal region. Deleting this region does not affect its polymerase activity but renders Polη biologically inactive in response to UV radiation (66,67). Thus, protein truncation at the C-terminus may represent an important control mechanism.

3.3. Polμ

Polμ is a newly discovered member of the X family polymerases (68,69). Human Polμ is highly prone to frameshift DNA synthesis (70). At single-nucleotide repeat sequences, DNA synthesis by human Polμ in vitro is mainly mediated by a deletion mechanism due to primer-template realignment prior to synthesis (70). Furthermore, when the primer 3' end contains one or a few mismatches, human Polμ can promote microhomology search and microhomology pairing through primer-template realignment (70). It has been proposed that Polμ may play an important role in nonhomologous end joining for double-strand break repair (70,71).

In vitro, purified human Polμ is capable of translesion synthesis across from a template 8-oxoguanine, AP site, $1,N^6$-ethenoadenine, AAF-dG, ($\pm$)-*trans-anti*-benzo[a]pyrene-N^2-dG, and cisplatin (72,73). However, unlike Polζ and the Y family polymerases, translesion synthesis by Polμ is achieved mainly through a deletion mechanism (72,73). Thus, Polμ may simply loop out the lesion, avoiding directly copying the damaged template base during translesion synthesis. This is possible because of the extraordinary ability of Polμ in promoting primer-template realignment. Surprisingly, however, human Polμ is capable of error-free translesion synthesis in response to a template *cis–syn* TT dimer by incorporating AA opposite the lesion (72). In contrast, a template (6–4) photoproduct completely blocks human Polμ (72). It was proposed that, due to covalent linkage between the two thymine bases, the TT dimer and the TT (6–4) photoproduct may not be flexible enough to allow loop-out by human Polμ. Unlike the TT dimer, the TT (6–4) photoproduct may be too distorting to DNA structure to allow Polμ for nucleotide insertion opposite the lesion (72). The in vivo significance of the translesion synthesis activity of Polμ remains to be determined.

4. MECHANISTIC MODELS OF TRANSLESION SYNTHESIS

Although exciting progress has been made in the last few years in the field of translesion synthesis in eukaryotes, several key issues need to be clearly defined, which include the followings. How is the stalled replication apparatus at a lesion site handed over to the damage tolerance system? How does the cell choose translesion synthesis vs. template switching? How are translesion polymerases recruited to the lesion site? How is the translesion apparatus handed back to the replication apparatus following translesion synthesis? Does translesion synthesis differ between the leading strand and the lagging strand during replication? Nevertheless, what is clear is that multiple translesion polymerases are involved, and it appears that multiple mechanisms exist when copying the lesion site by the polymerases.

In the simplest case, one polymerase inserts a nucleotide opposite the lesion. Then, the same polymerase extends the synthesis from opposite the lesion. This constitutes the one-polymerase two-step mechanism (Fig. 2A). Examples of this mode of translesion synthesis include the bypass of a TT dimer by Polη (48,49) and the bypass of a (–)-*trans-anti*-benzo[*a*]pyrene-N^2-dG by Polκ (56). In a more complex scheme, following nucleotide insertion opposite the lesion by one polymerase, subsequent extension synthesis is catalyzed by another polymerase. This constitutes the two-polymerase two-step mechanism (Fig. 2B). The two-polymerase two-step model of translesion synthesis was based on the studies of Nelson et al. (18) and Yuan et al. (19), where Rev1–Polζ and Polη–Polζ co-operation, respectively, were demonstrated for translesion synthesis of AP sites in vitro. Additional examples of two-polymerase co-operation include Polι–Polζ (20), REV1–Polκ (63), and Polη–Polκ (5). Although Polζ was the first polymerase believed to play an important role in extension synthesis by the two-polymerase two-step mechanism, Polκ is also capable of extension synthesis from opposite certain types of lesions in vitro (5,63,74,75). Thus, it is likely that multiple extension polymerases may be involved in translesion synthesis in cells.

The choice of one-polymerase two-step vs. two-polymerase two-step mechanism in cells most likely depends on the specific type of lesions. Apparently, efficient bypass of a lesion by a single polymerase, as TT dimer bypass by Polη, is exceptional. Thus, translesion synthesis of most types of lesions likely involves the two-polymerase two-step mechanism. It is possible that some lesions are bypassed by both mechanisms of translesion synthesis, where a fraction of the bypass involves a single polymerase while the remaining bypass requires the combination of two different polymerases. Hence, in vivo translesion synthesis would often involve the participation of multiple polymerases. Such a multiple-polymerase mode of translesion synthesis has been observed in yeast cells (17,76,77). Physical interactions

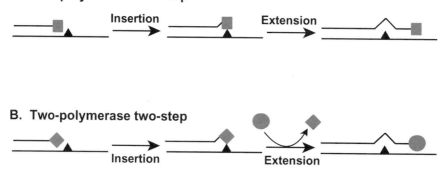

Figure 2 Mechanistic models of translesion synthesis. (A) The one-polymerase two-step model. The nucleotide insertion step and the extension step are catalyzed by the same translesion polymerase (gray rectangle). (B) The two-polymerase two-step model. A polymerase (gray diamond) inserts one nucleotide opposite the lesion at the insertion step. Subsequently, a different translesion polymerase (gray circle) replaces the insertion polymerase and catalyzes the extension synthesis from opposite the lesion. Polζ, Polη, and Polκ may function as the extension polymerase, depending on the specific lesion. The DNA lesion is shown as a triangle on the template.

between the C-terminal region of mouse REV1 and mouse Polη, Polκ, Polι, and REV7, respectively, have been observed (78). The functional significance of these interactions, however, is not clear.

5. TRANSLESION SYNTHESIS OF VARIOUS DNA DAMAGE IN EUKARYOTES

Since the discovery of the Y family DNA polymerases, translesion synthesis has been extensively studied using in vitro biochemistry. Biochemical approach is an extremely powerful tool and has yielded detailed molecular information and models of translesion synthesis. However, biochemistry cannot precisely duplicate the in vivo condition. Therefore, biochemical results need to be validated by in vivo genetics. By combining biochemical and genetic approaches, insightful information about translesion synthesis in cells is emerging. Our understanding on translesion synthesis of several selected lesions is summarized in what follows.

5.1. UV Photoproducts

The major DNA lesions of UV radiation are cyclobutane pyrimidine dimers (CPDs) and (6–4) photoproducts. Among the UV lesions, *cis–syn* TT dimers and TT (6–4) photoproducts are quite stable and have therefore been extensively studied. A template *cis–syn* TT dimer is efficiently bypassed by Polη by inserting AA opposite the lesion (48,49), which is the hallmark of Polη. The importance of Polη-catalyzed error-free translesion synthesis of TT dimers in human health is underscored by the hereditary disease XPV that results from inactivation of Polη. Without Polη, XPV cells become hypersensitive to and hypermutable by UV radiation (79–81). Consequently, XPV patients exhibit photosensitivity and a predisposition to skin cancer (82). Although the molecular defect of XPV is very different from that of other XP patients (XPA, XPB, XPC, XPD, XPE, XPF, and XPG), who are deficient in nucleotide excision repair, the clinical manifestation of the diseases is quite similar (82). This is not surprising because defect in either Polη or nucleotide excision repair results in a common problem: genomic overload of TT dimers and perhaps other CPDs for error-prone translesion synthesis by other bypass polymerases during replication. The result is predictable: elevated cytotoxicity and mutagenesis induced by the UV component of the sunlight, which constitute the cellular bases of XP diseases.

The 3′ T of the *cis–syn* TT dimer, the first T encountered by the polymerase, completely blocks human Polκ and yeast Polζ in vitro (16,20,56). In contrast, Polζ is able to efficiently insert the correct A opposite the 5′ T of the dimer and subsequently extends the synthesis beyond the lesion (16). Polκ is also able to perform extension synthesis from a G opposite the 3′ T of the TT dimer (74). Opposite a template TT dimer, human Polι has a very limited activity, preferentially inserting a T and less frequently a G opposite the 3′ T of the lesion (83,84). This activity, albeit inefficient, may make a significant contribution to TT dimer-induced mutagenesis. Extension synthesis by Polι from opposite the 3′ T of the dimer, however, is undetectable (83) or very inefficient (84). The extension synthesis following Polι-catalyzed insertion may thus involve Polζ and Polκ. Consistent with a role of Polκ in extension synthesis of TT dimers, Polκ-knockout mouse cells are slightly sensitive to UV radiation (85,86). The unresponsiveness of Polζ and Polκ and the inefficient response of

Polι to the 3′ T of TT dimers likely contribute to the UV-sensitive phenotype of Polη-deficient cells.

In contrast to TT dimers, TT (6–4) photoproducts cannot be bypassed by Polη alone in vitro (49,87). However, Polη is able to insert a G opposite the 3′ T of the TT (6–4) photoproduct before aborting DNA synthesis (87). The resulting intermediate of translesion synthesis is a substrate for extension synthesis by Polζ (16). Co-ordination between these two polymerases could therefore achieve bypass of TT (6–4) photoproducts by the two-polymerase two-step mechanism of translesion synthesis. This indeed occurs in yeast cells and is the major mechanism of G misinsertion opposite the 3′ T of TT (6–4) photoproducts, leading to T→C transition mutations (76,77). Polζ also possesses limited nucleotide insertion activity opposite the TT (6–4) photoproduct, frequently misinserting a T opposite the 3′ T but predominantly inserting the correct A opposite the 5′ T of the lesion (16). This explains the observation that, in yeast cells lacking Polη, the mutation spectrum at the presumed TT (6–4) photoproducts was altered from predominant T→C to predominant T→A mutations as a result of T insertion opposite the 3′ T of the lesion (76). In mammalian cells, TT (6–4) photoproducts also mainly induce T→C transition mutations as a result of G misinsertion opposite the 3′ T of the lesion (88). Given the yeast results, a major role of Polη in T→C mutations induced by TT (6–4) photoproducts is anticipated in mammals.

With respect to UV mutagenesis, Polη plays two opposing functions: suppressing mutagenesis in response to TT dimers and promoting mutagenesis in response to TT (6–4) photoproducts. Then, why are XPV cells lacking Polη hypermutable by UV radiation? Two key factors are probably responsible for this. First, the yield of TT (6–4) photoproducts is significantly lower than TT dimers, especially for sunlight in which UV-A and UV-B dominate and the short wave UV-C has been largely filtered out by the ozone layer of the atmosphere. Secondly and perhaps more importantly, TT (6–4) photoproducts are rapidly removed from DNA by nucleotide excision repair (89). In contrast, TT dimers are poorly repaired, especially in the nontranscribed strand of an active gene and in the transcriptionally silent regions of the genome (90,91). Hence, TT dimers, rather than TT (6–4) photoproducts, are much more prevalent UV lesions left unrepaired during replication following cell exposure to the sunlight. Consequently, the error-free translesion synthesis of Polη predominates its biological function in response to UV radiation. In lower eukaryotes, CPDs can be alternatively repaired by a photolyase. Unfortunately, this enzyme was lost during evolution to mammals. This fact further underscores our dependence on Pol η in protection against the cytotoxic and mutagenic effects of UV radiation. Without this simple translesion polymerase, life becomes very hazardous under the sun.

Genetic analyses indicate that Rev1 is required for UV-induced mutagenesis (45,92,93). However, Rev1 is unable to respond to a template TT dimer and a template (6–4) photoproduct in vitro (63,92). Thus, it has been proposed that Rev1 may play a noncatalytic function in error-prone translesion synthesis of UV lesions (92,93). At the present time, there is no clue as to what noncatalytic function Rev1 might play.

5.2. AP Sites

AP sites are a major type of spontaneous DNA lesions, although they can also be induced by many environmental agents. AP sites are noncoding. Translesion synthesis of an AP site by any polymerase simply cannot be error free. Therefore, AP sites are highly mutagenic. In *E. coli*, translesion synthesis of an AP site results in preferential incorporation of an A opposite the lesion, leading to the "A rule" hypothesis (94). Such a strong bias opposite an AP site does not appear to be the case in mammals. Using a plasmid mutagenesis system, insertions of A, C, and T opposite an AP site were detected at similar frequencies in simian COS cells (95–98). In a different study in COS cells, preferential A incorporation was observed (99). In another study using human lymphoblastoid cells, preferential G incorporation was reported (100).

In yeast cells, however, C is predominantly inserted opposite an AP site, which is dependent on the Rev1 and Polζ function in yeast cells (101). The Rev1 dCMP transferase is efficient in inserting a C opposite an AP site in vitro, but it cannot catalyze extension synthesis from opposite the lesion (18). The combination of Rev1 and Polζ, however, results in bypass of the AP site in vitro (18). Hence, this two-polymerase two-step action of Rev1–Polζ constitutes the major mechanism of translesion synthesis of AP sites in yeast cells. Because the human REV1 also efficiently inserts a C opposite an AP site (63), a similar REV1–Polζ mechanism for translesion synthesis of AP sites is most likely operational in humans as well. However, in contrast to yeast, human cells additionally contain the Y family polymerases Polι and Polκ (14,38). Human Polι efficiently inserts a G and less frequently a T opposite an AP site in vitro (83). Extension synthesis was not observed (83). Human Polκ preferentially inserts an A opposite an AP site in vitro (56,102). Efficient extension synthesis can be achieved by Polκ through a –1 deletion mechanism if the next template base is a T (56,102). It is possible that REV1, Polι, and Polκ all participate in nucleotide insertions during translesion synthesis of AP sites in human cells, resulting in insertions of C, G, A, and T opposite the lesions. This hypothesis needs to be experimentally tested by genetic experiments.

Yeast Polη catalyzes G and less frequently A insertion opposite an AP site in vitro (19). Purified human Polη, however, prefers A than G for insertion opposite an AP site (87,103). Because G insertion was not observed in yeast cells under normal conditions (101), Polη may not directly catalyze nucleotide insertions during translesion synthesis of AP sites in vivo. Other functions of Polη in translesion synthesis of AP site, however, cannot be excluded at the present time.

5.3. 8-Oxoguanine

8-Oxoguanine is a major product of oxidative damage in DNA. It has long been recognized as a miscoding lesion (104). 8-Oxoguanine can pair either with a C or an A (104). Unlike most other DNA lesions, 8-oxoguanine is not a strong blocker to DNA polymerases. Not surprisingly, the Y family polymerases and Polζ all can perform translesion synthesis in response to a template 8-oxoguanine. The probability of C vs. A insertion opposite 8-oxoguanine depends on the specific DNA polymerase. C is strongly preferred by yeast Polη and human Polι (19,83,105) but is only slightly preferred by human Polη (87). A is preferred by human Polκ and yeast Polζ (56,106). REV1 inserts a C opposite 8-oxoguanine but is unable to catalyze subsequent extension synthesis (63). Human Polι can only extend from opposite the lesion by one nucleotide (83). In contrast, extension synthesis can be efficiently catalyzed by Polη, Polκ, and Polζ (56,87,106). Despite these detailed in vitro biochemical studies, in cells, however, it is not known whether all of these polymerases are involved in translesion synthesis of 8-oxoguanine or to what extent each of the polymerases contribute to the bypass of this lesion. Because of the ubiquitous occurrence of 8-oxoguanine and its mutagenic nature, studies are eagerly awaited to uncover the polymerase(s) responsible for error-prone translesion synthesis of this lesion in cells.

5.4. AAF-dG Adducts

N-2-acetylaminofluorene (AAF) is an aromatic amine and is a known carcinogen. Upon bioactivation in cells, N-acetoxy-2-acetylaminofluorene is formed as an ultimate carcinogen of AAF. N-acetoxy-2-acetylaminofluorene readily attacks DNA, forming covalent AAF adduct at the C8 position of guanine as the major DNA lesion (107). The AAF DNA adducts are mutagenic. Based on forward mutation assays, the major mutations induced by AAF DNA adducts are frameshift mutations in yeast (108) and human cells (109). In yeast, the Polζ mutagenesis pathway plays a major role in error-prone translesion synthesis of AAF-dG adducts (110,111). Consistent with these in vivo results, yeast Polζ is able to perform limited translesion synthesis across from an AAF-dG adduct in vitro, misinserting a G opposite the lesion (16). Furthermore, extension synthesis from opposite the AAF-dG adduct was also detected (16). However, the strong inhibitory effect of AAF-dG adduct on purified yeast Polζ is evident (16). Although REV1 is able to insert a C opposite several types of DNA lesions, its dCMP transferase is essentially unresponsive to a template AAF-dG adduct (63,112). Instead, a noncatalytic role of yeast Rev1 in stimulating Polζ-catalyzed translesion synthesis of AAG-dG adducts was observed (112). Human Polκ is capable of error-prone translesion synthesis in vitro, inserting T or C at similar frequencies and A at a lower frequency opposite the AAF-dG adduct (56,113). Whether Polκ plays an important role in error-prone translesion synthesis in cells, however, remains to be determined.

Efficient error-free nucleotide insertion opposite an AAF-dG adduct can be catalyzed by Polη in vitro (19,103). The human Polη is more efficient in subsequent extension synthesis when compared with the yeast Polη (19,103). If the error-free translesion synthesis activity of Polη is utilized in cells in response to AAF-dG adducts, this polymerase would function to suppress AAF-induced mutagenesis. Using a plasmid-based mutagenesis assay in yeast cells, Bresson and Fuchs (77) reported that Polη participates in error-free bypass of a site-specific AAF-guanine within two sequence contexts (77). Surprisingly, Polη was found to be additionally required for frameshift mutagenesis at these two sequence contexts (77). Hence, it is still unknown about the contribution of Polη to AAF-induced mutagenesis. Does Polη act in cells to suppress or promote AAF-induced mutations, or neither? Definitive genetic experiments are needed to answer this question. Opposite a template AAF-dG adduct, human Polι is able to insert predominantly a C in vitro (83). Subsequent extension synthesis, however, was not observed (83). It is not known whether this intrinsic biochemical activity of Polι is employed for translesion synthesis of AAF-dG adducts in vivo.

5.5. BPDE-dG Adducts

Benzo[a]pyrene belongs to the class of polycyclic aromatic hydrocarbon compounds. It is produced by the incomplete combustion of organic materials and is therefore commonly found in the environment. In cells, benzo[a]pyrene is metabolically activated to highly reactive bay region dihydrodiol epoxide derivatives (114,115). Among the benzo[a]pyrene metabolites, ($\pm$)-$anti$-BPDE reacts with DNA mainly at the N^2 position of guanine, forming stereoisomeric bulky adducts

(+)-*trans-anti*-BPDE-N^2-dG, (+)-*cis-anti*-BPDE-N^2-dG, (−)-*trans-anti*-BPDE-N^2-dG, and (−)-*cis-anti*-BPDE-N^2-dG (115,116). In cells, the major DNA adduct derived from benzo[*a*]pyrene is (+)-*trans-anti*-BPDE-N^2-dG (115).

Zhang et al. (56) discovered that human Polκ efficiently performs error-free translesion synthesis across from the (−)-*trans-anti*-BPDE-N^2-dG adduct in vitro. Later, it was reported that human Polκ also catalyzes error-free translesion synthesis across from the (+)-*trans-anti*-BPDE-N^2-dG adduct, although less efficiently compared with the (−)-stereoisomeric adduct (22,75). Some sequence contexts dramatically affect the efficiency of Polκ-catalyzed translesion synthesis across from the (+)-*trans-anti*-BPDE-N^2-dG adduct (117). This illustrates the importance of sequence contexts in translesion synthesis and damage-induced mutagenesis. Very likely, sequence contexts make a significant contribution to mutation hot spots in cells. The biochemical results predict that Polκ may function to suppress BPDE-induced mutagenesis in cells. Indeed, an important role of Polκ in suppressing BPDE-induced mutagenesis has been established recently by Ogi et al. (85) using Polκ-knockout mouse cells. Purified human REV1 mainly inserts the correct C opposite (±)-*trans-anti*-BPDE-N^2-dG adducts (63). Extension synthesis cannot be performed by REV1 but can be effectively catalyzed by human Polκ in vitro (63,75). Thus, it was proposed that Polκ might additionally function as an extension polymerase in cells during bypass of BPDE-dG adducts by multiple translesion polymerases (75).

In contrast, human Polη performs error-prone translesion synthesis opposite (+)- and (−)-*trans-anti*-BPDE-N^2-dG DNA adducts in vitro, and it is more active in response to the former isomeric lesion (5,22,87,118). In yeast cells, Polη indeed plays a role in error-pone translesion synthesis of BPDE DNA lesions (17). The majority of BPDE-induced mutagenesis, however, is generated through a Polζ mechanism in yeast cells (17). The in vivo function of Polζ in BPDE mutagenesis is supported by in vitro biochemical results showing that this polymerase is able to perform limited translesion synthesis across from the (+)- and (−)-*trans-anti*-BPDE-N^2-dG DNA adducts with predominant G misinsertion opposite the lesions (17). Li et al. (119) reported that (±)-*anti*-BPDE-induced mutagenesis is largely abolished in cultured human cells expressing *hREV3* antisense RNA. Thus, a major role of Polζ in (±)-*anti*-BPDE-induced mutagenesis is apparently conserved from yeast to humans.

5.6. Cisplatin

Cisplatin [*cis*-diamminedichloroplatinum(II)] is a clinically used chemotherapeutic agent in the treatment of a variety of human cancers (120). Cisplatin can form intrastrand crosslinks at GG (65%), AG (25%), and GNG (6%) sequences in DNA and minor interstrand crosslinks at GC/CG (2%) (121). The anticancer activity of cisplatin is believed to result from cytotoxicity induced by cisplatin DNA adducts (120,121). Not surprisingly, cisplatin itself is a mutagen and carcinogen. G→T and A→T transversions appear to be the main mutations induced by cisplatin (122–125). DNA structural distortion by cisplatin damage remarkably resembles that observed in DNA duplexes bound by the HMG-domain proteins (126), which explains the efficient binding of cisplatin-damaged DNA by HMG proteins (121).

Little is known about the mechanism of cisplatin-induced mutagenesis in eukaryotic cells. Presumably, error-prone translesion synthesis is involved. In vitro, human Polη can effectively perform translesion synthesis across from a template cisplatin-GG adduct, inserting mainly the correct C opposite the damaged 3′ G (103,127). Opposite the 5′ damaged G, the correct C was predominantly inserted in one study (127), whereas A was preferentially inserted in another study (103). Because different sequence contexts were used in these studies, the discrepancy most likely reflects a strong sequence context effect on nucleotide selection by Polη in response to the damaged 5′ G of the cisplatin adduct. Nevertheless, bypass of a template cisplatin-GG adduct by Polη is likely to be predominantly error-free. This is because extension by Polη from the correctly paired primer 3′ end opposite the lesion is much more efficient than from a mismatched end (103,128). At some sequence contexts, deletions were also observed during in vitro translesion synthesis of cisplatin-GG adducts by human Polη (129).

6. IMPORTANCE OF TRANSLESION SYNTHESIS IN EUKARYOTIC BIOLOGY

At the cellular level, translesion synthesis can enhance cell survival in the presence of DNA damage. The contribution of translesion synthesis to cell survival is rather modest because of the template switching mechanism that functions in parallel in response to unrepaired DNA damage during replication (Fig. 1). In contrast, error-prone translesion synthesis is the major mechanism of base damage-induced mutagenesis. Many mutations may be harmful to an individual. For example, mutations can cause hereditary diseases and cancer. However, in the context of the whole species, mutations are essential for evolution and adaptation. Under genotoxic stress conditions, the genome becomes more flexible in the sense that it is more prone to permanent change through error-prone translesion synthesis. This would help a species to survive and subsequently proliferate in a changing environment. Hence, translesion synthesis is critical for the long-term survival of a species. On the other hand, mutagenesis is the very basis of most, if not all, human cancers. Clinical intervention to inhibit error-prone translesion synthesis is predicted to result in cancer prevention (93,130,131). This is an essentially untouched area in cancer research. If this novel anticancer strategy turns into reality, the benefits to human health will be enormous.

In higher eukaryotes, recent studies have implicated translesion synthesis in two more biological processes: embryo development and somatic hypermutation. Without Polζ, the mouse embryo cannot complete its development (27–30). The simplest interpretation is that mouse embryo development depends on Polζ-catalyzed translesion synthesis of spontaneous DNA lesions such that rapid cell divisions can take place during embryogenesis. Somatic hypermutation is a key step in the development of immunoglobulin genes to generate diverse antibodies. It introduces mainly point mutations into the V region of Ig genes at a rate of 10^{-3}–10^{-4}/bp per generation, which is ~10^6-fold higher than the spontaneous mutation rate in the rest of the genome (132). Somatic hypermutation requires the AID protein, which is able to deaminate C in single-stranded DNA in vitro (133,134). The controlled cytosine deamination in the targeted DNA regions results in uracil formation, which can be removed by a uracil-DNA glycosylase, leaving AP sites in DNA. Thus, it is believed that translesion synthesis and localized DNA synthesis involving an error-prone Y family DNA polymerase are important mechanisms of somatic hypermutation. Experimental results have been reported supporting the involvement of these translesion polymerases in somatic hypermutation: Polζ (135,136), REV1 (137), Polι (138), and Polη (139–141). It is likely that these polymerases function with a large degree of redundancy such that missing one of them will not eliminate the entire somatic hypermutation. Such functional redundancy may provide an explanation for the hypermutation-proficient phenotype of the 129 mouse strain that naturally lacks Polι (142). Generations after the natural nonsense mutation in the Polι gene in the 129 strain, the animal may have well adapted to rely on the other polymerases for somatic hypermutation, leading to a fully functional hypermutation system in the absence of Polι. It appears that Polκ is not involved in somatic hypermutation (86).

Translesion synthesis is clearly a fundamental aspect of life. We have witnessed a period of great leaps in the field of translesion synthesis in recent years. Nevertheless, much more needs to be learnt. It is breathtaking to think about the possibility that one day we may be able to drastically change human lives through intervention of translesion synthesis.

REFERENCES

1. Prakash L. Characterization of postreplication repair in *Saccharomyces cerevisiae* and effects of *rad6, rad18, rev3* and *rad52* mutations. Mol Gen Genet 1981; 184:471–478.
2. di Caprio L, Cox BS. DNA synthesis in UV-irradiated yeast. Mutat Res 1981; 82:69–85.
3. Broomfield S, Hryciw T, Xiao W. DNA postreplication repair and mutagenesis in *Saccharomyces cerevisiae*. Mutat Res 2001; 486:167–184.
4. Higgins NP, Kato K, Strauss B. A model for replication repair in mammalian cells. J Mol Biol 1976; 101:417–425.
5. Zhang Y, Wu X, Guo D, Rechkoblit O, Geacintov NE, Wang Z. Two-step error-prone bypass of the (+)- and (−)-*trans-anti*-BPDE-N_2-dG adducts by human DNA polymerases η and κ. Mutat Res 2002; 510:23–35.
6. Lemontt JF. Mutants of yeast defective in mutation induction by ultraviolet light. Genetics 1971; 68:21–33.
7. Cox BS, Parry JM. The isolation, genetics and survival characteristics of ultraviolet light-sensitive mutants in yeast. Mutat Res 1968; 6:37–55.
8. Lawrence CW, Stewart JW, Sherman F, Christensen R. Specificity and frequency of ultraviolet-induced reversion of an iso-1-cytochrome c ochre mutant in radiation-sensitive strains of yeast. J Mol Biol 1974; 85:137–162.
9. Lawrence CW, Christensen R. UV mutagenesis in radiation-sensitive strains of yeast. Genetics 1976; 82:207–232.
10. Cassier-Chauvat C, Fabre F. A similar defect in UV-induced mutagenesis conferred by the rad6 and rad18 mutations of *Saccharomyces cerevisiae*. Mutat Res 1991; 254:247–253.
11. Lawrence CW, Krauss BR, Christensen RB. New mutations affecting induced mutagenesis in yeast. Mutat Res 1985; 150:211–216.
12. Lawrence CW, Das G, Christensen RB. REV7, a new gene concerned with UV mutagenesis in yeast. Mol Gen Genet 1985; 200:80–85.
13. Morrison A, Christensen RB, Alley J, Beck AK, Bernstine EG, Lemontt JF, Lawrence CW. REV3, a *Saccharomyces cerevisiae* gene whose function is required for induced mutagenesis, is predicted to encode a nonessential DNA polymerase. J Bacteriol 1989; 171:5659–5667.
14. Burgers PM, Koonin EV, Bruford E, Blanco L, Burtis KC, Christman MF, Copeland WC, Friedberg EC, Hanaoka F, Hinkle DC, Lawrence CW, Nakanishi M, Ohmori H, Prakash L, Prakash S, Reynaud CA, Sugino A, Todo T, Wang Z, Weill JC, Woodgate R.

Eukaryotic DNA polymerases: proposal for a revised nomenclature. J Biol Chem 2001; 276:43487–43490.

15. Nelson JR, Lawrence CW, Hinkle DC. Thymine-thymine dimer bypass by yeast DNA polymerase ζ. Science 1996; 272:1646–1649.

16. Guo D, Wu X, Rajpal DK, Taylor J-S, Wang Z. Translesion synthesis by yeast DNA polymerase ζ from templates containing lesions of ultraviolet radiation and acetylamino-fluorene. Nucleic Acids Res 2001; 29:2875–2883.

17. Xie Z, Braithwaite E, Guo D, Bo Z, Geacintov NE, Wang Z. Mutagenesis of benzo[*a*]-pyrene diol epoxide in yeast: Requirement for DNA polymerase ζ and involvement of DNA polymerase η. Biochemistry 2003; 42:11253–11262.

18. Nelson JR, Lawrence CW, Hinkle DC. Deoxycytidyl transferase activity of yeast REV1 protein. Nature 1996; 382:729–731.

19. Yuan F, Zhang Y, Rajpal DK, Wu X, Guo D, Wang M, Taylor J-S, Wang Z. Specificity of DNA lesion bypass by the yeast DNA polymerase η. J Biol Chem 2000; 275: 8233–8239.

20. Johnson RE, Washington MT, Haracska L, Prakash S, Prakash L. Eukaryotic polymerase ι and ζ act sequentially to bypass DNA lesions. Nature 2000; 406:1015–1019.

21. Yang IY, Miller H, Wang Z, Frank EG, Ohmori H, Hanaoka F, Moriya M. Mammalian translesion DNA synthesis across an acrolein-derived deoxyguanosine adduct. Participation of DNA polymerase η in error- prone synthesis in human cells. J Biol Chem 2003; 278:13989–13994.

22. Rechkoblit O, Zhang Y, Guo D, Wang Z, Amin S, Krzeminsky J, Louneva N, Geacintov NE. *trans*-Lesion synthesis past bulky benzo[*a*]pyrene diol epoxide N^2-dG and N^6-dA lesions catalyzed by DNA bypass polymerases. J Biol Chem 2002; 277:30488–30494.

23. Gibbs PE, McGregor WG, Maher VM, Nisson P, Lawrence CW. A human homolog of the *Saccharomyces cerevisiae REV3* gene, which encodes the catalytic subunit of DNA polymerase ζ. Proc Natl Acad Sci USA 1998; 95:6876–6880.

24. Lin W, Wu X, Wang Z. A full-length cDNA of hREV3 is predicted to encode DNA polymerase ζ for damage-induced mutagenesis is humans. Mutat Res 1999; 433:89–98.

25. Murakumo Y, Roth T, Ishii H, Rasio D, Numata S, Croce CM, Fishel R. A human REV7 homolog that interacts with the polymerase ζ catalytic subunit hREV3 and the spindle assembly checkpoint protein hMAD2. J Biol Chem 2000; 275:4391–4397.

26. Diaz M, Watson NB, Turkington G, Verkoczy LK, Klinman NR, McGregor WG. Decreased frequency and highly aberrant spectrum of ultraviolet-induced mutations in the *hprt* gene of mouse fibroblasts expressing antisense RNA to DNA polymerase ζ. Mol Cancer Res 2003; 1:836–847.

27. Wittschieben J, Shivji MK, Lalani E, Jacobs MA, Marini F, Gearhart PJ, Rosewell I, Stamp G, Wood RD. Disruption of the developmentally regulated *rev3l* gene causes embryonic lethality. Curr Biol 2000; 10:1217–1220.

28. Bemark M, Khamlichi AA, Davies SL, Neuberger MS. Disruption of mouse polymerase ζ (*Rev3*) leads to embryonic lethality and impairs blastocyst development *in vitro*. Curr Biol 2000; 10:1213–1216.

29. Esposito G, Godindagger I, Klein U, Yaspo M, Cumano A, Rajewsky K. Disruption of the *REv3l*-encoded catalytic subunit of polymerase ζ in mice results in early embryonic lethality. Curr Biol 2000; 10:1221–1224.

30. Van Sloun PP, Varlet I, Sonneveld E, Boei JJ, Romeijn RJ, Eeken JC, De Wind N. Involvement of mouse Rev3 in tolerance of endogenous and exogenous DNA damage. Mol Cell Biol 2002; 22:2159–2169.

31. Sonoda E, Okada T, Zhao GY, Tateishi S, Araki K, Yamaizumi M, Yagi T, Verkaik NS, van Gent DC, Takata M, Takeda S. Multiple roles of Rev3, the catalytic subunit of polζ in maintaining genome stability in vertebrates. EMBO J 2003; 22:3188–3197.

32. Reynolds P, Weber S, Prakash L. RAD6 gene of *Saccharomyces cerevisiae* encodes a protein containing a tract of 13 consecutive asparatates. Proc Natl Acad Sci USA 1985; 82:168–172.

33. Chanet R, Magana-Schwencke N, Fabre F. Potential DNA-binding domains in the RAD18 gene product of *Saccharomyces cerevisiae*. Gene 1988; 74:543–547.

34. Jones JS, Weber S, Prakash L. The *Saccharomyces cerevisiae* RAD18 gene encodes a protein that contains potential zinc finger domains for nucleic acid binding and a putative nucleotide binding sequence. Nucleic Acids Res 1988; 16:7119–7131.

35. Larimer FW, Perry JR, Hardigree AA. The REV1 gene of *Saccharomyces cerevisiae*: isolation, sequence, and functional analysis. J Bacteriol 1989; 171:230–237.

36. Huang ME, de Calignon A, Nicolas A, Galibert F. *POL32*, a subunit of the *Saccharomyces cerevisiae* DNA polymerase δ, defines a link between DNA replication and the mutagenic bypass repair pathway. Curr Genet 2000; 38:178–187.

37. Stelter P, Ulrich HD. Control of spontaneous and damage-induced mutagenesis by SUMO and ubiquitin conjugation. Nature 2003; 425:188–191.

38. Ohmori H, Friedberg EC, Fuchs RPP, Goodman MF, Hanaoka F, Hinkle D, Kunkel TA, Lawrence CW, Livneh Z, Nohmi T, Prakash L, Prakash S, Todo T, Walker GC, Wang Z, Woodgate R. The Y-family of DNA polymerases. Mol Cell 2001; 8:7–8.

39. Bailly V, Lamb J, Sung P, Prakash S, Prakash L. Specific complex formation between yeast RAD6 and RAD18 proteins: a potential mechanism for targeting RAD6 ubiquitin-conjugating activity to DNA damage sites. Genes Dev 1994; 8:811–820.

40. Bailly V, Lauder S, Prakash S, Prakash L. Yeast DNA repair proteins Rad6 and Rad18 form a heterodimer that has ubiquitin conjugating, DNA binding, and ATP hydrolytic activities. J Biol Chem 1997; 272:23360–23365.

41. Koken MHM, Reynolds P, Jaspers-Dekker I, Prakash L, Prakash S, Bootsma D, Hoeijmakers JHJ. Structural and functional conservation of two human homologs of the yeast DNA repair gene *RAD6*. Proc Natl Acad Sci USA 1991; 88:8865–8869.

42. Xin H, Lin W, Sumanasekera W, Zhang Y, Wu X, Wang Z. The human *RAD18* gene product interacts with HHR6A and HHR6B. Nucleic Acids Res 2000; 28:2847–2854.

43. Tateishi S, Sakuraba Y, Masuyama S, Inoue H, Yamaizumi M. Dysfunction of human Rad18 results in defective postreplication repair and hypersensitivity to multiple mutagens. Proc Natl Acad Sci USA 2000; 97:7927–7932.

44. Lin W, Xin H, Zhang Y, Wu X, Yuan F, Wang Z. The human *REV1* gene codes for a DNA template-dependent dCMP transferase. Nucleic Acids Res 1999; 27:4468–4475.

45. Gibbs PE, Wang XD, Li Z, McManus TP, McGregor WG, Lawrence CW, Maher VM. The function of the human homolog of *saccharomyces cerevisiae REV1* is required for mutagenesis induced by UV light. Proc Natl Acad Sci USA 2000; 97:4186–4191.

46. McDonald JP, Levine AS, Woodgate R. The *Saccharomyces cerevisiae RAD30* gene, a homologue of *Escherichia coli din B* and *umuC*, is DNA damage inducible and functions in a novel error-free postreplication repair mechanism. Genetics 1997; 147:1557–1568.

47. Roush AA, Suarez M, Friedberg EC, Radman M, Siede W. Deletion of the *Saccharomyces cerevisiae* gene *RAD30* encoding an *Escherichia coli DinB* homolog confers UV radiation sensitivity and altered mutability. Mol Gen Genet 1998; 257:686–692.

48. Johnson RE, Prakash S, Prakash L. Efficient bypass of a thymine-thymine dimer by yeast DNA polymerase, Polη. Science 1999; 283:1001–1004.

49. Masutani C, Araki M, Yamada A, Kusumoto R, Nogimori T, Maekawa T, Iwai S, Hanaoka F. Xeroderma pigmentosum variant (XP-V) correcting protein from HeLa cells has a thymine dimer bypass DNA polymerase activity. EMBO J 1999; 18: 3491–3501.

50. Masutani C, Kusumoto R, Yamada A, Dohmae N, Yokoi M, Yuasa M, Araki M, Iwai S, Takio K, Hanaoka F. The XPV (xeroderma pigmentosum variant) gene encodes human DNA polymerase η. Nature 1999; 399:700–704.

51. Johnson RE, Kondratick CM, Prakash S, Prakash L. hRAD30 mutations in the variant form of xeroderma pigmentosum. Science 1999; 285:263–265.

52. Gerlach VL, Aravind L, Gotway G, Schultz RA, Koonin EV, Friedberg EC. Human and mouse homologs of *Escherichia coli* DinB (DNA polymerase IV), members of the UmuC/DinB superfamily. Proc Natl Acad Sci USA 1999; 96:11922–11927.

53. Ogi T, Kato T Jr, Kato T, Ohmori H. Genes Cells 1999; 4:607–618.

54. McDonald JP, Rapic-Otrin V, Epstein JA, Broughton BC, Wang X, Lehmann AR, Wolgemuth DJ, Woodgate R. Novel human and mouse homologs of *Saccharomyces cerevisiae* DNA polymerase η. Genomics 1999; 60:20–30.

55. Washington MT, Johnson RE, Prakash S, Prakash L. Fidelity and processivity of *Saccharomyces cerevisiae* DNA polymerase η. J Biol Chem 1999; 274:36835–36838.

56. Zhang Y, Yuan F, Wu X, Wang M, Rechkoblit O, Taylor J-S, Geacintov NE, Wang Z. Error-free and error-prone lesion bypass by human DNA polymerase κ *in vitro*. Nucleic Acids Res 2000; 28:4138–4146.

57. Matsuda T, Bebenek K, Masutani C, Hanaoka F, Kunkel TA. Low fidelity DNA synthesis by human DNA polymerase η. Nature 2000; 404:1011–1013.

58. Trincao J, Johnson RE, Escalante CR, Prakash S, Prakash L, Aggarwal AK. Structure of the catalytic core of *S. cerevisiae* DNA polymerase η: implications for translesion DNA synthesis. Mol Cell 2001; 8:417–426.

59. Ling H, Boudsocq F, Woodgate R, Yang W. Crystal structure of a y-family DNA polymerase in action. a mechanism for error-prone and lesion-bypass replication. Cell 2001; 107:91–102.

60. Ling H, Boudsocq F, Plosky BS, Woodgate R, Yang W. Replication of a *cis-syn* thymine dimer at atomic resolution. Nature 2003; 424:1083–1087.

61. Zhang Y, Yuan F, Wu X, Wang Z. Preferential incorporation of G opposite template T by the low fidelity human DNA polymerase ι. Mol Cell Biol 2000; 20:7099–7108.

62. Tissier A, McDonald JP, Frank EG, Woodgate R. polι, a remarkably error-prone human DNA polymerase. Genes Dev 2000; 14:1642–1650.

63. Zhang Y, Wu X, Rechkoblit O, Geacintov NE, Taylor JS, Wang Z. Response of human REV1 to different DNA damage: preferential dCMP insertion opposite the lesion. Nucleic Acids Res 2002; 30:1630–1638.

64. Masuda Y, Takahashi M, Fukuda S, Sumii M, Kamiya K. Mechanisms of dCMP transferase reactions catalyzed by mouse Rev1 protein. J Biol Chem 2001; 277:3040–3046.

65. Haracska L, Prakash S, Prakash L. Yeast Rev1 protein is a G template-specific DNA polymerase. J Biol Chem 2002; 277:15546–15551.

66. Kannouche P, Broughton BC, Volker M, Hanaoka F, Mullenders LH, Lehmann AR. Domain structure, localization, and function of DNA polymerase η, defective in xeroderma pigmentosum variant cells. Genes Dev 2001; 15:158–172.

67. Kondratick CM, Washington MT, Prakash S, Prakash L. Acidic resudues critical for the activity and biological function of yeast DNA polymerase η. Mol Cell Biol 2001; 21: 2018–2025.

68. Dominguez O, Ruiz JF, Lain de Lera T, Garcia-Diaz M, Gonzalez MA, Kirchhoff T, Martinez AC, Bernad A, Blanco L. DNA polymerase mu (Pol μ), homologous to TdT, could act as a DNA mutator in eukaryotic cells. EMBO J 2000; 19:1731–1742.

69. Aoufouchi S, Flatter E, Dahan A, Faili A, Bertocci B, Storck S, Delbos F, Cocea L, Gupta N, Weill JC, Reynaud CA. Two novel human and mouse DNA polymerases of the polX family. Nucleic Acids Res 2000; 28:3684–3693.

70. Zhang Y, Wu X, Yuan F, Xie Z, Wang Z. Highly frequent frameshift DNA synthesis by human DNA polymerase μ. Mol Cell Biol 2001; 21:7995–8006.

71. Mahajan KN, Nick McElhinny SA, Mitchell BS, Ramsden DA. Association of DNA polymerase μ (pol μ) with Ku and ligase IV: role for pol μ in end-joining double-strand break repair. Mol Cell Biol 2002; 22:5194–5202.

72. Zhang Y, Wu X, Guo D, Rechkoblit O, Taylor JS, Geacintov NE, Wang Z. Lesion bypass activities of human DNA polymerase μ. J Biol Chem 2002; 277:44582–44587.

73. Havener JM, McElhinny SA, Bassett E, Gauger M, Ramsden DA, Chaney SG. Translesion synthesis past platinum DNA adducts by human DNA polymerase η. Biochemistry 2003; 42:1777–1788.

74. Washington MT, Johnson RE, Prakash L, Prakash S. Human *DINB1*-encoded DNA polymerase κ is a promiscuous extender of mispaired primer termini. Proc Natl Acad Sci USA 2002; 99:1910–1914.

75. Zhang Y, Wu X, Guo D, Rechkoblit O, Wang Z. Activities of human DNA polyemrase κ in response to the major benzo[*a*]pyrene DNA adduct: error-free bypass and extension synthesis from opposite the lesion. DNA Repair 2002; 1:559–569.

76. Zhang H, Siede W. UV-induced T->C transition at a TT photoproduct site is dependent on *Saccharomyces cerevisiae* polymerase η *in vivo*. Nucleic Acids Res 2002; 30:1262–1267.

77. Bresson A, Fuchs RP. Lesion bypass in yeast cells: Pol η participates in a multi-DNA polymerase process. EMBO J 2002; 21:3881–3887.

78. Guo C, Fischhaber PL, Luk-Paszyc MJ, Masuda Y, Zhou J, Kamiya K, Kisker C, Friedberg EC. Mouse Rev1 protein interacts with multiple DNA polymerases involved in translesion DNA synthesis. EMBO J 2003; 22:6621–6630.

79. Maher VM, Ouellette LM, Curren RD, McCormick JJ. Frequency of ultraviolet light-induced mutations is higher in exeroderma pimentosum variant cells than in normal human cells. Nature 1976; 261:593–595.

80. Wang YC, Maher VM, McCormick JJ. Xeroderma pigmentosum variant cells are less likely than normal cells to incorporate dAMP opposite photoproducts during replication of UV-irradiated plasmids. Proc Natl Acad Sci USA 1991; 88:7810–7814.

81. McGregor WG, Wei D, Maher VM, McCormick JJ. Abnormal, error-prone bypass of photoproducts by xeroderma pigmentosum variant cell extracts results in extreme strand bias for the kinds of mutations induced by UV light. Mol Cell Biol 1999; 19:147–154.

82. Cleaver JE, Kraemer KH. Scriver CR, Beaudet AL, Sly WS, Valle D eds. The Metabolic Basis of Inherited Disease. 6th ed. New York: McGraw-Hill Book Co., 1989: 2949–2971.

83. Zhang Y, Yuan F, Wu X, Taylor JS, Wang Z. Response of human DNA polymerase ι to DNA lesions. Nucleic Acids Res 2001; 29:928–935.

84. Tissier A, Frank EG, McDonald JP, Iwai S, Hanaoka F, Woodgate R. Misinsertion and bypass of thymine-thymine dimers by human DNA polymerase ι. EMBO J 2000; 19:5259–5266.

85. Ogi T, Shinkai Y, Tanaka K, Ohmori H. Polκ protects mammalian cells against the lethal and mutagenic effects of benzo[a]pyrene. Proc Natl Acad Sci USA 2002; 99: 15548–15553.

86. Schenten D, Gerlach VL, Guo C, Velasco-Miguel S, Hladik CL, White CL, Friedberg EC, Rajewsky K, Esposito G. DNA polymerase κ deficiency does not affect somatic hypermutation in mice. Eur J Immunol 2002; 32:3152–3160.

87. Zhang Y, Yuan F, Wu X, Rechkoblit O, Taylor J-S, Geacintov NE, Wang Z. Error-prone lesion bypass by human DNA polymerase η. Nucleic Acids Res 2000; 28: 4717–4724.

88. Kamiya H, Iwai S, Kasai H. The (6-4) photoproduct of thymine-thymine induces targeted substitution mutations in mammalian cells. Nucleic Acids Res 1998; 26: 2611–2617.

89. Mitchell DL, Brash DE, Nairn RS. Rapid repair kinetics of pyrimidine(6-4)pyrimidone photoproducts in human cells are due to excision rather than conformational change. Nucleic Acids Res 1990; 18:963–971.

90. Bohr VA, Smith CA, Okomuto DS, Hanawalt PC. DNA repair in an active gene: removal of pyrimidine dimers from the DHFR gene of CHO cells is much more efficient than in the genome overall. Cell 1985; 40:359–369.

91. Mellon I, Spivak G, Hanawalt PC. Selective removal of transcription-blocking DNA damage from the transcribed strand of the mammalian DHFR gene. Cell 1987; 51: 241–249.

92. Nelson JR, Gibbs PE, Nowicka AM, Hinkle DC, Lawrence CW. Evidence for a second function for *Saccharomyces cerevisiae* Rev1p. Mol Microbiol 2000; 37:549–554.

93. Rajpal DK, Wu X, Wang Z. Alteration of ultraviolet-induced mutagenesis in yeast though molecular modulation of the *REV3* and *REV7* gene expression. Mutat Res 2000; 461:133–143.

94. Strauss BS. The 'A rule' of mutagen specificity: a consequence of DNA polymerase bypass of non-instructional lesions? Bioessays 1991; 13:79–84.

95. Gentil A, Renault G, Madzak C, Margot A, Cabral-Neto JB, Vasseur JJ, Rayner B, Imbach JL, Sarasin A. Mutagenic properties of a unique abasic site in mammalian cells. Biochem Biophys Res Commun 1990; 173:704–710.

96. Gentil A, Cabral-Neto JB, Mariage-Samson R, Margot A, Imbach JL, Rayner B, Sarasin A. Mutagenicity of a unique apurinic/apyrimidinic site in mammalian cells. J Mol Biol 1992; 227:981–984.

97. Neto JB, Gentil A, Cabral RE, Sarasin A. Mutation spectrum of heat-induced abasic sites on a single-stranded shuttle vector replicated in mammalian cells. J Biol Chem 1992; 267:19718–19723.

98. Cabral Neto JB, Cabral RE, Margot A, Le Page F, Sarasin A, Gentil A. Coding properties of a unique apurinic/apyrimidinic site replicated in mammalian cells. J Mol Biol 1994; 240:416–420.

99. Takeshita M, Eisenberg W. Mechanism of mutation on DNA templates containing synthetic abasic sites: study with a double strand vector. Nucleic Acids Res 1994; 22: 1897–1902.

100. Klinedinst DK, Drinkwater NR. Mutagenesis by apurinic sites in normal and ataxia telangiectasia human lymphoblastoid cells. Mol Carcinog 1992; 6:32–42.

101. Gibbs PE, Lawrence CW. Novel mutagenic properties of abasic sites in *Saccharomyces cerevisiae*. J Mol Biol 1995; 251:229–236.

102. Ohashi E, Ogi T, Kusumoto R, Iwai S, Masutani C, Hanaoka F, Ohmori H. Error-prone bypass of certain DNA lesions by the human DNA polymerase κ. Genes Dev 2000; 14:1589–1594.

103. Masutani C, Kusumoto R, Iwai S, Hanaoka F. Mechanisms of accurate translesion synthesis by human DNA polymerase η. EMBO J 2000; 19:3100–3109.

104. Shibutani S, Takeshita M, Grollman AP. Insertion of specific bases during DNA synthesis past the oxidation- damaged base 8-oxodG. Nature 1991; 349:431–434.

105. Haracska L, Yu SL, Johnson RE, Prakash L, Prakash S. Efficient and accurate replication in the presence of 7,8-dihydro-8-oxoguanine by DNA polymerase η. Nat Genet 2000; 25:458–461.

106. Haracska L, Prakash S, Prakash L. Yeast DNA polymerase ζ is an efficient extender of primer ends opposite from 7,8-dihydro-8-Oxoguanine and O^6-methylguanine. Mol Cell Biol 2003; 23:1453–1459.

107. Friedberg EC, Walker GC, Siede W. DNA Repair and Mutagenesis. Washington, DC: American Society of Microbiology Press, 1995.

108. Roy A, Fuchs RP. Mutational spectrum induced in *Saccharomyces cerevisiae* by the carcinogen *N*-2-acetylaminofluorene. Mol Gen Genet 1994; 245:69–77.

109. Ogawa HI, Kimura H, Koya M, Higuchi H, Kato T. *N*-acetoxy-*N*-acetyl-2-aminofluorene-induced mutation spectrum in a human *hprt* cDNA shuttle vector integrated into mammalian cells. Carcinogenesis 1993; 14:2245–2250.

110. Baynton K, Bresson-Roy A, Fuchs RP. Distinct roles for Rev1p and Rev7p during translesion synthesis in *Saccharomyces cerevisiae*. Mol Microbiol 1999; 34:124–133.

111. Baynton K, Bresson-Roy A, Fuchs RP. Analysis of damage tolerance pathways in *Saccharomyces cerevisiae*: a requirement for Rev3 DNA polymerase in translesion synthesis. Mol Cell biol 1998; 18:960–966.

112. Guo D, Xie Z, Shen H, Bo Z, Wang Z. Translesion synthesis of acetylaminofluorene-dG adducts by DNA polymerase ζ is stimulated by yeast Rev1 protein. Nucleic Acids Res 2004; 32:1122–1130.

113. Gerlach VL, Feaver WJ, Fischhaber PL, Friedberg EC. Purification and characterization of polκ, a DNA polymerase encoded by the human *DINB1* gene. J Biol CheDINB1m 2001; 276:92–98.

114. Phillips DH, Grover PL. Polycyclic hydrocarbon activation: bay regions and beyond. Drug Metab Rev 1994; 26:443–467.

115. Peltonen K, Dipple A. Polycyclic aromatic hydrocarbons: chemistry of DNA adduct formation. J Occup Environ Med 1995; 37:52–58.

116. Cheng SC, Hilton BD, Roman JM, Dipple A. DNA adducts from carcinogenic and noncarcinogenic enantiomers of benzo[*a*]pyrene dihydrodiol epoxide. Chem Res Toxicol 1989; 2:334–340.

117. Huang X, Kolbanovsky A, Wu X, Zhang Y, Wang Z, Zhuang P, Amin S, Geacintov NE. Effects of base sequence context on translesion synthesis past a bulky (+)-*trans-anti*-B[a]P-N^2-dG lesion catalyzed by the Y-family polymerase pol κ. Biochemistry 2003; 42:2456–2466.

118. Chiapperino D, Kroth H, Kramarczuk IH, Sayer JM, Masutani C, Hanaoka F, Jerina DM, Cheh AM. Preferential misincorporation of purine nucleotides by human DNA polymerase η opposite benzo[*a*]pyrene 7,8-diol 9,10-epoxide deoxyguanosine adducts. J Biol Chem 2002; 277:11765–11771.

119. Li Z, Zhang H, McManus TP, McCormick JJ, Lawrence CW, Maher VM. hREV3 is essential for error-prone translesion synthesis past UV or benzo[*a*]pyrene diol epoxide-induced DNA lesions in human fibroblasts. Mutat Res 2002; 510:71–80.

120. Colvin OM. In: Holland JF, Bast RC Jr, Morton DL, Frei E III, Kufe DW, Weichselbaum RR eds. Cancer Medicine. 4th ed. Baltimore: Williams & Wilkins, 1997; 1:949–975.

121. Trimmer EE, Essigmann JM. Cisplatin. Essays Biochem 1999; 34:191–211.

122. Burnouf D, Gauthier C, Chottard JC, Fuchs RP. Single d(ApG)/cis-diamminedichloroplatinum(II) adduct-induced mutagenesis in *Escherichia coli*. Proc Natl Acad Sci USA 1990; 87:6087–6091.

123. Pillaire MJ, Villani G, Hoffmann JS, Mazard AM, Defais M. Characterization and localization of *cis*-diamminedichloro-platinum(II) adducts on a purified oligonucleotide containing the codons 12 and 13 of *H-ras* proto-oncogene. Nucleic Acids Res 1992; 20:6473–6479.

124. Pillaire MJ, Hoffmann JS, Defais M, Villani G. Replication of DNA containing cisplatin lesions and its mutagenic consequences. Biochimie 1995; 77:803–807.

125. Bradley LJ, Yarema KJ, Lippard SJ, Essigmann JM. Mutagenicity and genotoxicity of the major DNA adduct of the antitumor drug cis-diamminedichloroplatinum(II). Biochemistry 1993; 32:982–988.

126. Gelasco A, Lippard SJ. NMR solution structure of a DNA dodecamer duplex containing a *cis*-diammineplatinum(II) d(GpG) intrastrand cross-link, the major adduct of the anticancer drug cisplatin. Biochemistry 1998; 37:9230–9239.

127. Vaisman A, Masutani C, Hanaoka F, Chaney SG. Efficient translesion replication past oxaliplatin and cisplatin GpG adducts by human DNA polymerase η. Biochemistry 2000; 39:4575–4580.

128. Bassett E, Vaisman A, Havener JM, Masutani C, Hanaoka F, Chaney SG. Efficiency of extension of mismatched primer termini across from cisplatin and oxaliplatin adducts by human DNA polymerases β and η *in vitro*. Biochemistry 2003; 42:14197–14206.

129. Bassett E, Vaisman A, Tropea KA, McCall CM, Masutani C, Hanaoka F, Chaney SG. Frameshifts and deletions during *in vitro* translesion synthesis past Pt-DNA adducts by DNA polymerases β and η. DNA Repair 2002; 1:1003–1016.

130. Lawrence CW, Hinkle DC. DNA ζ and the control of DNA damage induced mutagenesis in eukaryotes. Cancer Surv 1996; 28:21–31.

131. Wang Z. DNA damage-induced mutagenesis: a novel target for cancer prevention. Mol Interv 2001; 1:269–281.

132. Wagner SD, Neuberger MS. Somatic hypermutation of immunoglobulin genes. Annu Rev Immunol 1996; 14:441–457.

133. Bransteitter R, Pham P, Scharff MD, Goodman MF. Activation-induced cytidine deaminase deaminates deoxycytidine on single-stranded DNA but requires the action of RNase. Proc Natl Acad Sci USA 2003; 100:4102–4107.

134. Pham P, Bransteitter R, Petruska J, Goodman MF. Processive AID-catalysed cytosine deamination on single-stranded DNA simulates somatic hypermutation. Nature 2003; 424:103–107.

135. Zan H, Komori A, Li Z, Cerutti A, Schaffer A, Flajnik MF, Diaz M, Casali P. The translesion DNA polymerase ζ plays a major role in Ig and bcl-6 somatic hypermutation. Immunity 2001; 14:643–653.

136. Diaz M, Verkoczy LK, Flajnik MF, Klinman NR. Decreased frequency of somatic hypermutation and impaired affinity maturation but intact germinal center formation in mice expressing antisense RNA to DNA polymerase ζ. J Immunol 2001; 167:327–335.

137. Simpson LJ, Sale JE. Rev1 is essential for DNA damage tolerance and non-templated immunoglobulin gene mutation in a vertebrate cell line. EMBO J 2003; 22:1654–1664.

138. Faili A, Aoufouchi S, Flatter E, Gueranger Q, Reynaud CA, Weill JC. Induction of somatic hypermutation in immunoglublulin genes is dependent on DNA polymerase ι. Nature 2002; 419:944–947.

139. Zeng X, Winter DB, Kasmer C, Kraemer KH, Lehmann AR, Gearhart PJ. DNA polymerase η is an A-T mutator in somatic hypermutation of immunoglublulin variable genes. Nat Immunol 2001; 2:537–541.

140. Yavuz S, Yavuz AS, Kraemer KH, Lipsky PE. J Immunol 2002; 169:3825–3830.

141. Pavlov YI, Rogozin IB, Galkin AP, Aksenova AY, Hanaoka F, Rada C, Kunkel TA. Correlation of somatic hypermutation specificity and A-T base pair substitution errors by DNA polymerase η during copying of a mouse immunoglobulin κ light chain transgene. Proc Natl Acad Sci USA 2002; 99:9954–9959.

142. McDonald JP, Frank EG, Plosky BS, Rogozin IB, Masutani C, Hanaoka F, Woodgate R, Gearhart PJ. 129-derived strains of mice are deficient in DNA polymerase ι and have normal immunoglublulin hypermutation. J Exp Med 2003; 198:635–643.

25

Structural Features of Bypass Polymerases

Caroline Kisker

*Department of Pharmacological Sciences, Center for Structural Biology, State
University of New York at Stony Brook, Stony Brook, New York, U.S.A.*

1. INTRODUCTION

A block in DNA synthesis can have serious consequences for the cell. The recently
discovered Y-family of DNA polymerases allows the cell to bypass specific lesions,
which lead to a block for replicative polymerases and thereby permits cell survival
(1). Y-family DNA polymerases replicate DNA in a distributive manner, which
allows them to bypass the lesion, but ensures that a high-fidelity DNA polymerase
is able to continue replication as soon as the block has been overcome. Despite
the fact that these new enzymes are classified as DNA polymerases, they share little
sequence homology to the five other DNA polymerase families, but have significant
sequence homology within their family (Fig. 1) and contain five conserved sequence
motifs, I–V (2). Additional residues mostly located towards the C-terminal end are
highly variable and may be required for protein–protein interactions or other additional functions. Currently, the Y-family is divided into four subfamilies, which can
be described as the 1. UmuC or pol V family that has only been identified in prokaryotes; 2. the Din B or pol IV family with members in bacteria, eukaryotes, and
archaea presenting the most widely distributed family and the 3. Rev1 and 4.
Rad30 families that are found exclusively in eukaryotes (2). All members of the
Y-family lack a 3′–5′ exonuclease activity and are characterized as low-fidelity and
low-processivity polymerases. Shortly after the identification of the Y-family of
DNA polymerases, the first structures of members within this family were solved
and currently two of the four subfamilies have been structurally characterized,
(3–8) allowing the first detailed insight into their ability of lesion bypass and
providing the structural basis of their low fidelity and processivity.

1.1. Overall Structure of *Saccharomyces cerevisiae* polη

Both human and yeast polη misincorporate nucleotides *in vitro* with a high rate
(9,10), with an intrinsic error rate of $\sim 1/22$ for the human enzyme (11), and support
TLS across diverse types of base damage that typically cause replicative arrest (9,10).
The crystal structure of the first 513 amino acids of the 632 residue-containing DNA
polymerase polη from *S. cerevisiae* provided first insights into the Rad30 subfamily

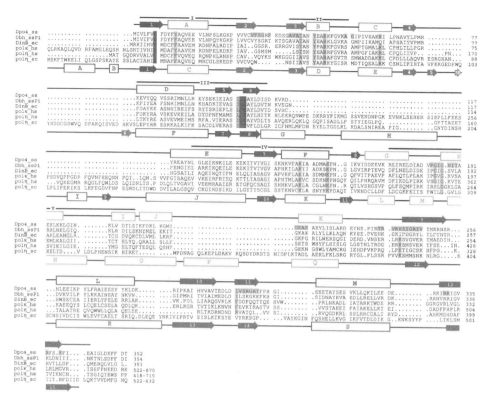

Figure 1 Structure-based sequence alignment of Y-family polymerases. Dpo4 from *S. solfataricus P2* (Dpo4_ss) (AAK42588), Dbh from *S. acidocaldarius* (Dbh_ssP1) (T46875), DinB/pol IV from *E. coli* (DinB_ec) (Q47155), human polκ/DINB1 (polκ_hs) (XP_048874), human polη/RAD30A (polη_hs) (AF158185_1), and polη/RAD30 from *S. cerevisiae* (polη_sc) (18158626). The alignment is based on the structures of polη_cs, Dpo4 and Dbh. Secondary structure elements are colored according to the different domains with magenta for the palm, blue for the fingers, green for the thumb and red for the PAD or little finger domains. Open rectangles indicate α-helices and arrows β-strands, secondary structure elements were defined for all structures using PROMOTIF. The assignment above the sequence corresponds to the Dpo4 structure and below the sequences to the polη structure. In the polη structure, an additional β-strand was identified in PROMOTIF that was not assigned before. To keep the same assignment as in Ref. 4, this additional strand was named 5B. The five conserved motifs in Y-family DNA polymerases are indicated by roman numerals above the alignment. The three invariant carboxylates required for catalysis are highlighted in red. Other conserved residues are highlighted in yellow. Residues interacting with the DNA are highlighted in blue. *Source*: Adapted from Refs. 3,4,6,7,53. (*See color insert.*)

of Y-family DNA polymerases. This fragment has identical properties with respect to DNA replication and bypass activity compared to the full-length enzyme (12). The C-terminal residues that are not present in the structure are not highly conserved within this subfamily. As seen in the high-fidelity DNA polymerases, the architecture of polη can be described as resembling a right hand with a palm, finger, and thumb domain. In addition to these domains, it also contains an additional domain which has been described as the palm-associated domain or PAD (Fig. 2).

The palm domain can be divided into a large and a small subdomain. The large subdomain contains a mixed six-stranded β-sheet (β1, β7–β11) that is flanked by two

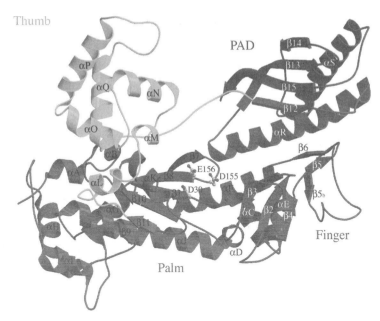

Figure 2 Overall structure of polη from *S. cerevisiae*. The domains are color coded as in Fig. 1 (palm in magenta, finger in blue, thumb in green and PAD in red), β-strands are numbered, and α-helices labeled with letters. The three active site carboxylates, Asp 30, Asp 155, and Glu 156, are shown in a ball-and-stick representation. (*See color insert.*)

long α-helices (αF and αJ) on the bottom and a shorter α-helix (αK) on the top. The top face of the β-sheet provides the base for protein–DNA interactions. The positions of the three conserved active site carboxylates, Asp 30, Asp 155, and Glu 156, is reminiscent of the position seen in the catalytically important carboxylates of the high-fidelity DNA polymerases (Fig. 2). Asp 30 is located in the central β-strand (β1) of the six-stranded β-sheet. The other two carboxylates, Asp 155 and Glu 156, are in close proximity and are part of a neighboring β-hairpin. Structural homology extends past these carboxylates and almost the entire large subdomain can be superimposed onto the palm domains of A-family DNA polymerases such as T7 DNA polymerase (PDB code 1T7P) with residues from four of the six β-strands (β1, β7, β8, β10) and all three α-helices (αF, αJ, and αK) of the large sub-domain resulting in an rms deviation of 1.8 Å for 61 Cα's. The three carboxylates of polη, Asp 30, Asp 155, and Glu 156, align with Asp 475, Asp 654, and Glu 655 of T7 DNA polymerase (13). Superposition with B-family DNA polymerases such as RB69 DNA polymerase (PDB code 1IH7) (14,15) yields an rms deviation of 1.9 Å for 67 Cα's and the catalytically important residues Asp 411 and Asp623 of RB69 align with Asp 30 and Asp 155 of polη. The overall structural similarity of the palm domains of polη and the A- and B-family polymerases points to a similarity of the different motifs that are characteristic within these families. Motifs A and C (16) in the high-fidelity polymerases directly correspond to motifs I and III in polη, forming β-strand β1 which contains Asp 30 and the hairpin that contains the two other carboxylates required for catalysis. Motif IV is located towards the C-terminal end of the palm domain within helices αJ, αK, and β-strand β10 and has two differ-ent functions: First it is structural and important for packing helices αJ and αK against the β-sheet of the palm domain. Second, other basic residues such as Lys 272 and Lys 279 were proposed (4) to contact the primer strand as seen for Arg 452 and His 704 in T7 DNA polymerase (13).

The small subdomain of the palm domain is entirely α-helical and contains 5 α-helices (αA, αB, αG, αH, and αI). Sequence comparisons reveal that this subdomain may have a different fold in human polη since the first a-helix (αA) is not present at all in the human ortholog and the fourth helix (αH) would be shorter based on current sequence alignments (Fig. 1). Since it is located close to helices αJ and αF at the bottom of the β-sheet, it is unlikely to be involved in protein–DNA interactions, but may be important for interactions with other proteins.

The thumb domain of polη shares very limited structural homology with the thumb domains of A- or B-family DNA polymerases. It is stubby and much smaller compared to the corresponding domains in the high-fidelity polymerases and is entirely α-helical (αL-αQ). It is comprised of residues 285–395 and is located C-terminal to the palm domain. The structure of this domain is related to the thumb of human DNA polymerase β. Superposition of 26 Cα atoms within this domain with the thumb domain of pol β (PDB code 9ICW) (17) reveals an rms deviation of 2.2 Å. The conserved region includes residues of motif V, which are located in helices αL to αO and points toward the DNA binding cleft, but the structural homology extends beyond this conserved motif.

The fingers contain two small β-sheets (β2–β4 and β5–β6) and three short α-helices (αC–αE), and can also be described as a small and stubby domain compared to other DNA polymerases. The second β-sheet formed by β-strands β5, β5B, and β6 may have a different fold in human polη since only β-strand β5 is present in the human enzyme based on sequence alignments, strands β5B and β6 are absent in human polη. The finger domains within the other DNA polymerase families are mostly α-helical, but more importantly contain the so-called O-helix, which plays a critical role with respect to the fidelity of the enzyme since it tightly interacts with the incoming nucleotide and ensures formation of the correct Watson–Crick base pair (18). The finger domain in polη does not contain an analogous helix, which may be one of the reasons for the reduced fidelity of this enzyme. Motif II is located within the finger domain and contains the conserved YXAR sequence (Fig. 1) that is located in helix αD of the finger domain. This motif is similar to motif B found in the A-family DNA polymerases, where the conserved Tyr and Arg could function analogously to Arg 518 and His 506 of T7 DNA polymerase interacting with the incoming nucleotide (19).

In addition to the palm, finger, and thumb domains, polη contains the PAD. This domain is joined to the thumb by a flexible linker that positions the PAD opposite the thumb domain (Fig. 2) with the DNA binding site in between both domains. To accommodate this position, the 16 amino acids of the linker region form a 30 Å long bridge over the inner surface of the palm domain. The PAD domain contains residues 395–508 and can be described as a four-stranded antiparallel β-sheet (β12–β15), which is flanked by two antiparallel α-helices located opposite to the DNA binding surface. The PAD and palm domains are related both in terms of their structure and their mode of DNA interaction. In both cases, the helices form part of the outer surface of the domain, whereas one side of the β-sheet points towards the DNA binding groove.

An important consequence of the smaller finger and thumb domains of polη is the small surface area provided for DNA binding through these two domains and the palm domain. This reduction in surface area is compensated by the additional area provided by the PAD domain. Taken together, similar values for the DNA binding surface as seen in A- or B-family DNA polymerases are realized with all four domains (4).

1.2 Structure of the *Sulfolobus* DinB Family Members from *Sulfolobus solfataricus*

The *Escherichia coli* DinB protein (DNA polIV) is required for spontaneous muta-genesis in *E. coli* and generates –1 frame shift mutations when extending mispaired primer/templates *in vitro*. The enzyme has also been implicated in translesion synth-esis (20). The crystal structure of an N-terminal fragment of a structural ortholog of the *E. coli* DinB protein from the archaebacterium *S. acidocaldarius P1*, designated Dbh (for DinB homolog), (Fig. 1) has been determined (3). This fragment was iden-tified after chymotrypsin cleavage and contains residues 2–216 of the 354 amino acid full-length protein. The fragment from *S. acidocaldarius* retains the catalytic domain of the protein and includes the five conserved motifs that are characteristic for Y-family DNA polymerases. At the same time, the structure of the full-length protein from *S. acidocaldarius* has been determined (7). Finally, the structure of a DinB homolog from the P2 strain of *S. solfataricus P2*, named Dpo4 (for DNA polymerase IV) (Figs. 1 and 3), has been solved in the presence of DNA (6).

The three structures of the two archaeal polymerases are very similar and can be superimposed with rms deviations of 1.1 Å for 197 Cα's of the finger and palm domain between Dpo4 and the short fragment of Dbh and 1.5 Å rms deviation

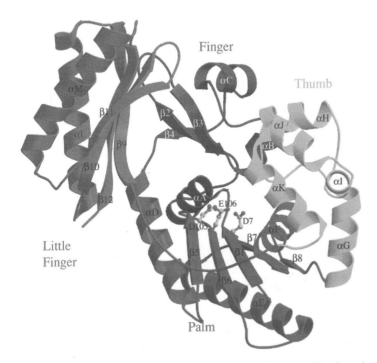

Figure 3 Overall structure of Dpo4 from *S. solfataricus*. The domains are color coded as in Fig. 1 (palm in magenta, finger in blue, thumb in green and little finger in red), β-strands are numbered and α-helices are assigned letters. The three active site carboxylates, Asp 7, Asp 105, and Glu 106, are shown as a ball-and-stick representation. (*See color insert.*)

for 228 Cα's of the palm, finger, and thumb domains between Dpo4 and Dbh (Fig. 4). The structure of Dpo4 will be discussed here as a representative member of the DinB family. The basic architecture of Dpo4 is very similar to the polη structure and retains the classic polymerase fold (Fig. 3) with the palm, finger, and thumb domains. In addition to these three domains, Dpo4 also contains the additional PAD domain, which was named "little finger" for Dpo4 (6) or "wrist" for Dbh (7).

The palm domain contains a central mixed five-stranded β-sheet surrounded by two long α-helices on one side of the β-sheet and an additional α-helix following the 7th β-strand. The catalytic triad formed by residues Asp 7, Asp 105, and Glu 106 is positioned as described above for polη, with Asp 7 on β-strand β1 and Asp 105 and Glu 106 in the adjacent β-hairpin. The additional α-helical small subdomain observed in polη is not present in Dpo4.

The thumb domain and the fingers domain of Dpo4 are also small and stubby. The thumb domain is entirely α-helical and contains 5 α-helices (αG–αK). The fingers contain one β-sheet formed by β-strands (β2–β4), that is, flanked by three short α-helices (αB and αC). The fingers in polη are bigger in comparison to Dpo4, since they contain an insertion allowing them to form a second small β-sheet after β-strand β4.

The little finger contains a central four-stranded antiparallel β-sheet and two long anti-parallel α-helices on one side of the β-sheet. Interestingly, the little finger domain of

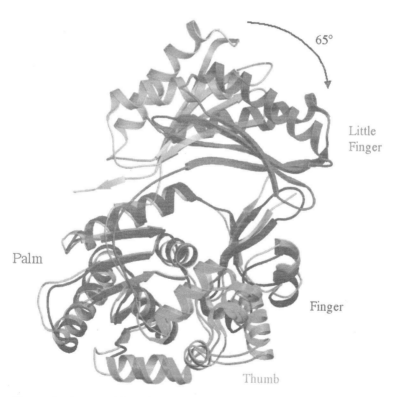

Figure 4 Superposition of Dbh (PDB code 1K1S) with Dpo4 (PDB code 1JX4). Dpo4 is color coded as in Fig. 3, Dbh is shown in gray. The arrow indicates the movement of the little finger domain toward the finger domain. (*See color insert.*)

Dpo4 and the corresponding wrist domain of Dbh are rotated by approximately 65° relative to each other (Fig. 4) and therefore adopt different positions with respect to the finger domains. Dpo4 approaches the finger domain much closer so that the Cα's of residues Lys 252 and Arg 36 are only 4.6 Å apart, whereas the closest approach between Dbh and the finger domain is between the Cα's of residues Leu 293 and Gly 35 featuring a distance of 12.3 Å. For the full-length Dbh and Dpo4, it was only possible to obtain crystals in the presence of DNA (6,7), but in the case of the Dbh structure it was not possible to locate the primer–template strands, whereas the DNA is resolved in the Dpo4 structure. The difference in position of the little finger/wrist domains is not surprising, since it is connected to the thumb by a long flexible linker containing about 13 amino acids. The flexibility of this domain may also point to additional functions other than binding to DNA.

2. DNA SYNTHESIS BY THE DINB FAMILY MEMBERS FROM THE *SULFOLOBUS* GENUS

In DNA polymerase families A, B, X, and reverse transcriptase, the catalytic complex formed between the protein, DNA, and incoming nucleotide results in large conformational changes (18). The different structures of the B-family DNA polymerase RB69 (14,15,21) reveal that the thumb interacts with the minor groove of the duplex DNA upon binding and forms multiple contacts with the DNA. The finger domain undergoes the largest conformational changes during DNA polymerization, and allows the polymerase to alternate between its "open" and "closed" conformations. In the absence of DNA, the enzyme is in the "open" conformation, characterized by a rotation of the fingers domain by 60° away from the palm domain (15). Upon nucleotide binding, the finger domain "closes" the active site, ensuring correct Watson–Crick base pair formation (14).

Superpositions of the N-terminal fragment of Dbh and the full-length Dbh structures with Dpo4 in a ternary complex with primer/template DNA and an incoming ddADP suggest that the induced fit mechanism observed in all other DNA polymerase families is not required for this archaeal Y-family polymerase. The structure of the short fragment of Dbh (3) was obtained in the absence of DNA and therefore represents the apo structure. Superposition of the apo structure with the ternary complex of Dpo4 clearly shows that the finger and the thumb domain undergo almost no conformational change upon DNA binding. One reason for the lack of movement of the finger domain could be the position of Tyr 10, which is the first residue of helix αA and directly connects the palm and the finger domain. The side chain of Tyr 10 stacks between residues of the finger domain Gln 14 and Tyr 48 and thereby locks the position of the finger domain in place. The little finger domain, in contrast, undergoes a large conformational change as seen by comparison of Dpo4 and the full-length Dbh structure (Fig. 4). Upon DNA binding, the little finger domain approaches the finger and closes the gap that is present in the Dbh structure. This conformational change of the little finger domain nevertheless cannot be compared to the closing of the fingers in the other DNA polymerase families, since the active site remains relatively open and permits the presence of two adjacent template bases into the active site (5,6). For polη, it has been suggested that the role of the little finger domain could be analogous to that seen for thioredoxin and T7 polymerase. The two proteins form a tight complex with DNA (22,23) and prevent dissociation of the primer/template DNA during catalysis (4). The primer–template

DNA maintains its standard B-form around the active site and does not adopt the A-form as seen in the A-family of DNA polymerases (19,24–26). Adopting A-form DNA causes a widening of the minor groove, making it more shallow and therefore more accessible. In B-family DNA polymerases like RB69 (14), however, the primer–template also maintains the B-form and in polymerase β this has been observed as well (14,27,28).

The thumb and the little finger bind to eight bases of the double-stranded DNA and form hydrogen bonds and van der Waals interactions with the backbone, primarily with the phosphodiester moieties (Fig. 5). The thumb binds into the minor groove and interacts with both the primer and the template strand through the *N*-termini of α-helices αH and αK. The little finger fits nicely with one side of the β-sheet into the major groove and β-strands β9 and β11, which form the rim on either side of the β-sheet, interact with both DNA strands. In addition, β-strand β12, next to β-strand β9, contacts the DNA through several arginines (Fig. 5) and Lys 152 from the palm domain approaches the DNA backbone from the same side as the thumb domain. The flexible linker between the thumb and the little finger domain can be readily cleaved by trypsin in the absence of DNA, releasing the little finger from the catalytic core and resulting in lower and less processive polymerase activity *in vitro*. In the presence of DNA, the protein is protected from proteolytic cleavage, suggesting that the little finger facilitates stable binding of the polymerase

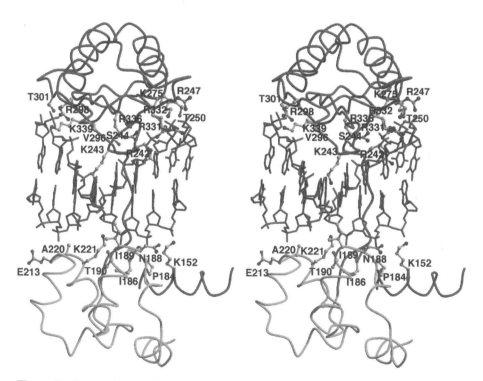

Figure 5 Stereo view, of the overall interaction of Dpo4 with DNA. For a better view only the thumb, little finger, and helix αF of the palm domains are shown as Cα traces. The thumb is located below the DNA and shown in green, whereas the little finger domain is shown above the DNA and is drawn in red. Residues that form van der Waals interactions or hydrogen bonds to the DNA are shown in a ball-and-stick representation. The DNA is shown in all-bonds representation. (*See color insert.*)

to substrate DNA (6). The linker contributes several positively charged residues that mostly interact with the backbone of the template strand. It remains an open question whether this reorientation affects processivity and fidelity or is primarily involved in clamping, as suggested in the polη structure (4). The remaining five base pairs of the DNA are solvent exposed. An interesting feature is the formation of a large crevice between the little finger domain and the catalytic core (Fig. 6). The minor groove of the primer/template DNA in the active site faces the gap between

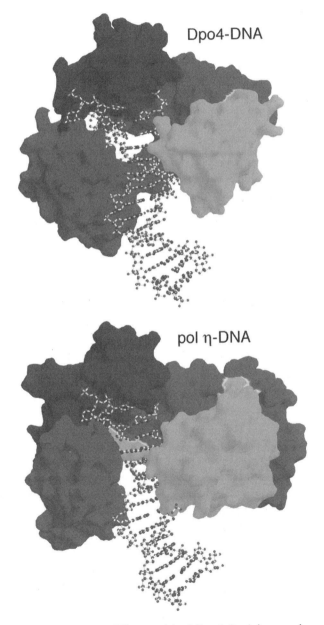

Figure 6 Space filling model of Dpo4 (top) in complex with DNA and color coded as in Fig. 2 and polη (bottom) after superposition of the individual domains onto the ternary Dpo4 complex. (*See color insert.*)

the finger and the little finger domains and thereby becomes solvent accessible. In other DNA polymerases, the minor groove is much more restricted through close contacts with the protein and less solvent accessible (14,19).

2.1. The Active Site of Dpo4

The three carboxylates, Asp 7, Asp 105, and Glu 106, are absolutely conserved (Fig. 1) among the Y-family polymerases. In the structure with non-damaged DNA, only one metal ion was observed, which has been characterized as a Ca^{2+} ion (6). The Ca^{2+} ion is coordinated to Asp 7 and Asp 105, the main chain oxygen of Phe 8, the non-bridging oxygens of the α and β phosphates of the incoming nucleotide and a water molecule (Fig. 7). In this structure, the presence of the second metal ion may have been prohibited because the α-phosphate of the ddADP is located too far away to permit catalysis. However, in the structures that were obtained in the presence of a cyclobutane pyrimidine dimer, ddATP was bound instead of ddADP and both metal ions are present (Fig. 8). The two metals are about 3.9 Å apart and form multiple interactions with the protein and the incoming nucleotide. The first Ca^{2+} ion coordinates with the carboxylates of Asp 105 and the non-bridging oxygens of the α and the β phosphate and to four water molecules. The second Ca^{2+} ion coordinates with Asp 7, Asp 105, the main chain oxygen of Phe 8, to a non-bridging oxygen of the β-phosphate and two water molecules. This arrangement is highly similar to other DNA polymerases and is in agreement with the proposed two metal ion mechanism for the phosphoryl transfer reaction (29,30).

Mainly small amino acids build the walls of the active site (Fig. 7). Residues from the finger domain, Val 32, Ala 42, and Gly 58, interact with the template base from the 5′ end and from the minor groove. Residues 44–58, which are part of motif II and contain the conserved YXAR sequence, form a lid on the incoming nucleotide and interact with the base and the phosphodiester moiety of the incoming nucleotide. Ala 44 and Ala 57 are in proximity to the adenine of the ddADP, while Thr 45 and Arg 51 form hydrogen bonds to the phosphates. The palm domain forms part of the active site as well and contributes residues such as the catalytic carboxylates Asp 7 and Asp 105, but also Lys 159 and the main chain nitrogen of Tyr 10, which form

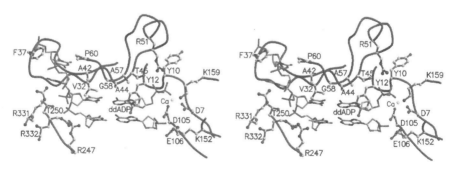

Figure 7 A view into the active site of Dpo4. Residues from the finger (blue), palm (magenta), and little finger domain (red) are shown. Amino acids in close proximity to the DNA are shown in a ball-and-stick representation. The Ca^{2+} ion is shown as a yellow sphere. The primer/template DNA is shown in all-bonds representation with the 3′ end of the primer strand, the incoming nucleotide (ddADP), and the template strand on the opposite side. (*See color insert.*)

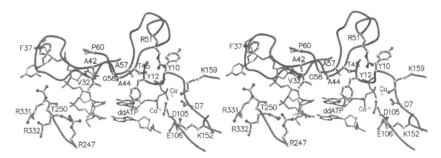

Figure 8 The active site of Dpo4 with a thymidine dimer. (Same view as Fig. 5.) The template strand contains a thymidine dimer. The 5′ thymine of the CPD is base-paired with ddATP. The two Ca^{2+} ions present in this structure are shown as yellow spheres. (*See color insert.*)

interactions with the non-bridging oxygens of the β- and γ-phosphate, respectively. Tyr 12 stacks against the ribose of the incoming nucleotide and would not permit the presence of the 2′ hydroxyl. The active site of Dpo4 is thus primarily formed by small and uncharged residues and the replicating base pair is not restricted towards the major or minor groove. These features are in contrast to the high-fidelity DNA polymerase, which mostly position large amino acids close to the replicating base pair and form tight interactions on the minor groove side, being only permissive toward the major groove (14,19,26,31). The most intriguing feature of this Y-family DNA polymerase is the possibility to accommodate two bases in the active site. In an attempt to obtain a G-G mismatch, Ling et al. showed (6) that two bases are accommodated in the active site to avoid the formation of the mismatch; the incoming ddGTP rather formed a base, pair with the next template base a C. This example of template slippage may also indicate how an abasic site is replicated in DinB-like polymerases. It has been shown that all DinB-like polymerases are capable of replicating past an abasic site (32–34) but often result in –1 frameshift mutations, thus, the newly obtained daughter strand would be one nucleotide shorter. In the case of the abasic site, the situation could be similar as seen in the structure harboring two bases in the active site, the abasic site would be "skipped" and the next template base would pair with the incoming nucleotide.

2.2 Lesion Bypass

Dpo4 is so far the only DinB-like DNA polymerase identified that is able to bypass a thymine dimer (35). Ling et al. (5) captured the structures of Dpo4 in complex with a *cis-syn* cyclobutane pyrimidine dimer (CPD) bound to the active site of Dpo4 with the 3′ and the 5′ thymine of the CPD base paired to an incoming ddATP. Superpositions of the unmodified DNA-containing structures on the CPD-containing structure with the 3′ thymine base paired to ddATP revealed rms deviations of 0.72 Å (PDB code 1JXL) and 0.84 Å (PDB code 1JX4) for 342 Cα atoms. The overall structures of the protein in the presence of undamaged DNA and with the CPD bound are therefore very similar and the biggest shifts of only 1–2 Å were observed in the thumb and little finger domains. These shifts lead also to a shift of the DNA, but the pivot of this movement is located around the catalytic residues Asp 7 and Asp 105, causing the resulting protein–DNA interactions to be similar irrespective of the templating base. The ability to accommodate two bases in the active site of the polymerase is essential for a covalently linked lesion like CPD. In the structure with undamaged DNA, it has already been shown that the active site of Dpo4 is wide enough to accommodate two bases and is thereby able to avoid a mismatch. In the structure with the 3′ thymine of the CPD located opposite the incoming nucleotide, both thymines are accommodated in the active site and the conformation of the 3′ thymine allows the formation of Watson–Crick hydrogen bonds with the incoming ddATP, even though both bases are not entirely coplanar. The distances between O4 and N6 and N3 and N1 of the thymine and the adenine base are 2.6 and 2.7 Å, respectively. A similar conformation was observed in the structure of a DNA decamer containing a CPD, where the 3′ thymine forms Watson–Crick base pairs with the opposing adenine (36). The 5′ thymine, due to its covalent bond to the 3′ thymine, is only positioned 2.7 Å away from the main chain oxygen of Gly 58. Interestingly, this glycine is not conserved among the Y-family DNA polymerases, but its Cα carbon is positioned such that a bigger side chain would point towards the major groove of the DNA and would not limit the space provided in the active site.

To accommodate the 5′ thymine of the CPD in the active site, the finger domain moves towards the palm domain and the little finger approaches the major groove by about 1 Å in comparison to the complex formed when the 3′ thymine of the CPD is positioned opposite the incoming nucleotide. Due to steric constraints within the CPD, it is not possible for the 5′ thymine to form a Watson–Crick base pair with the incoming ddATP. While the 3′ thymine forms a canonical Watson–Crick base pair with the adenine at the 3′ end of the primer, the 5′ thymine would be required to rotate by 36° and rise by 3.4 Å relative to the neighboring base pair to maintain standard B-form geometry. To accommodate the 5′ thymine in the active site the ribose of the 5′ thymine approaches Val 32 and thereby shortens the distance to this region of the finger domain by approximately 2.3 Å (O4 of the ribose from the 5′ thymine opposite the incoming nucleotide is at a distance of 4.3 Å to Cγ1 of Val 32, whereas the O4 of the ribose in the 3′ thymine opposite ddATP is positioned 6.6 Å away from the same atom.). The 5′ thymine does not enter the binding pocket as far as the 3′ thymine, which leads to an increase in the distance between the base and Gly 58 by about 1.2 Å and the base protrudes into the space, which would normally be occupied by the incoming nucleotide. To accommodate both bases and still maintain the conformation required for catalysis, the incoming ddATP has to adopt the *syn* conformation rather than *anti* as required for the formation of a Watson–Crick base pair. This arrangement is also advantageous since it does not require a backbone distortion as seen in the crystal structure of a CPD-containing decamer, which maintains Watson–Crick hydrogen bonds for the 5′ thymine but the hydrogen bond between N6 and O4 of the adenine and the 5′ thymine becomes too long and therefore only two instead of three hydrogen bonds remain (36). In the *syn* conformation, the incoming nucleotide forms a Hoogsteen base pair with the 5′ thymine. N6 of the adenine and O4 of the thymine form a hydrogen bond at a distance of 3.1 Å and N7 of the adenine forms a hydrogen bond with N3 of the thymine at a distance of 2.6 Å.

Analysis of mutations caused by replication across a CPD within other Y-family DNA polymerases such as UmuD′₂C and polη have shown that primarily T to C and T to A mutations take place at the 3′ thymine of the CPD but not at the 5′ thymine (37,38). Assuming that both polymerases form similar interactions with the 3′ and 5′ thymines of CPD, the steric restraints causing the formation of a Hoogsteen base pair between the 5′ thymine and the incoming nucleotide may explain why only the 3' thymine is prone to mutations. If a Hoogsteen base pair would be formed with an incoming dGTP, only one instead of two hydrogen bonds could be formed, which would disfavor this interaction. At the 3' thymine, however, space is less restricted and wobble base pairs seem feasible.

Very recently, two structures of Dpo4 in complex with a primer/template DNA containing a benzo[a]pyrene diol epoxide-adenine adduct were reported (8). The two structures contain the adduct in the primer-extension complex, i.e., the adduct is base paired to a dT at the 3' end of the primer strand and the incoming dATP forms a base pair with the templating base next to the adduct, a dT. Interestingly, the adduct adopts two very distinctive conformations. In the first complex, the benzo[a]pyrene intercalates at the 5' side of the adenine between the lesion containing and the replicating base pair. Even though this DNA conformation may be energetically favored, it would form a high barrier to extend synthesis beyond this lesion. The overall structure of the polymerase in this complex remains similar to the ones described previously. However, the 3' OH of the primer strand and the α-phosphate of the incoming nucleotide are more than 10 Å apart due to the intercalation of the benzo[a]pyrene, thereby making the nucleotidyl transfer reaction impossible. The incoming dATP adopts the *syn* conformation, which allows the formation of a Hoogsteen base pair with the dT of the template strand and aromatic stacking with the benzo[a]pyrene. A reoccurring observation is the conformation of the triphosphate in this structure, which positions the γ phosphate at the position of the α phosphate, the β phosphate at the position of the γ phosphate and the α phosphate at the position of the β phosphate, compared to the catalytically active conformation of the triphosphate.

In the second complex, the benzo[a]pyrene swings out into the major groove and one side of the polycyclic aromatic hydrocarbons is almost perpendicular to the plane of the base pairs. This causes a move of the adenine base that is covalently attached to the adduct towards the major groove by almost 2 Å. Even though this conformation of the DNA is energetically less favored and is therefore less likely to be observed, it does not block DNA synthesis in Dpo4 and allows the modified base and the replicating base pairs to adopt a similar conformation as seen in undamaged DNA. The incoming nucleotide and the 3' hydroxyl group of the primer strand and the α phosphate are less than 5 Å apart, indicating that the nucleotidyl transfer reaction can take place.

Due to the small finger and thumb domains of the Y-family polymerases, the interactions towards the minor and major groove of the DNA are much more limited in comparison to the high-fidelity DNA polymerases. A bulky lesion like benzo[a]-pyrene should therefore be accommodated as long as it adopts the conformation seen in the second complex, where it is swung out into the major groove. A consequence of this movement, as discussed above, is the concomitant shift of the adenine base that is covalently bound to the benzo[a]pyrene. This shift could explain the preferential incorporation of dAMP opposite this adduct as seen for Dpo4 (8), since the exocyclic amino group of the incoming dAMP would reach the displaced adenine better than the O4 of thymine.

3. DNA BINDING AND LESION BYPASS IN POLη

So far, no structural information is available for polη in complex with DNA. Therefore, the catalytic domain of polη (PDB code 1JIH) was modeled onto the ternary complex of Dpo4 containing the thymidine dimer (PDB code 1PM0). After superposition of the conserved palm domains, a rotation of the thumb, finger, and little finger domain (PAD domain in polη) was required to obtain similar positions of the corresponding domains as seen in the Dpo4 structure, thereby leading to a closer approach to the DNA. This is in contrast to the Dpo4 structure where only the little finger seems to undergo a conformational change upon DNA binding, and no induced fit as seen in high-fidelity DNA polymerases takes place. The thumb domain, the PAD domain, and the finger domain were rotated by 31°, 45°, and 19°, respectively, to approach the DNA in a similar way as seen in the Dpo4 structure. In the resulting model, almost the entire primer/template DNA fits well and the thymidine dimer can be accommodated in the active site (Figs. 6 and 9). Only a loop region (residues 361–366) in the thumb domain approaches the DNA too closely but a minor movement of this loop would avoid any steric clashes.

Polη replicates very efficiently past a thymidine dimer (39,40). Prior to the knowledge about the structures of the ternary Dpo4 complexes, the authors of the polη structure modeled a ternary complex by superposition of the polη structure with the T7 ternary complex (4). They also suggested that the few putative contacts between polη and DNA may accommodate two nucleotides in the active site, thereby explaining the ability of polη to bypass thymine dimers (4). Studies by the Prakash lab have suggested that polη is not highly selective for the incorporation of the correct nucleotide and that the active site would differ significantly compared to other high-fidelity DNA polymerases. Despite this inaccuracy, an induced fit mechanism is proposed to selectively incorporate the correct nucleotide (41), which is in agreement with the model that suggests conformational changes of the finger

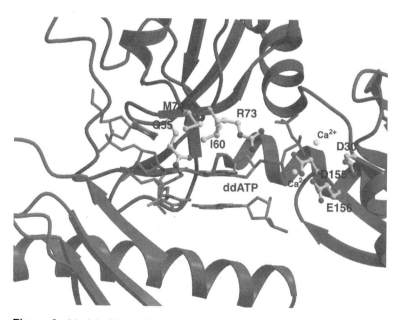

Figure 9 Model of the active site of polη in the presence of a thymidine dimer. The domains are color coded as in Fig. 2. Residues Gln 55, Ile 60, Arg 73, and Met 74 have been proposed to enhance the efficiency for bypass of the CPD. The active site residues, Asp 30, Asp 155, and Glu156, are shown as well, while the Ca^{2+} ions are shown as yellow spheres. *Source*: From Ref. 5. (*See color insert.*)

and thumb domains as well as the PAD domain, with the former also being observed in high-fidelity DNA polymerases.

A comparison between the putative ternary complex of polη and the structure of Dpo4 in complex with DNA reveals significant differences in the active site (Figs. 7–9). Several side chains in polη are larger and therefore shield the active site more efficiently from the solvent. Gly 58 in Dpo4 is replaced by Met 74 in yeast polη and by a serine in human polη. This side chain, as well as the neighboring Arg 73, could shield the incoming nucleotide from above, whereas residues of the finger and palm domains shield it from the opposite side. Arg 73, which replaces Ala 57, could also form charged interactions with the phosphates of the incoming nucleotide and thereby support the correct orientation of the incoming nucleotide. Gln 55 in yeast polη replaces Val 32 of Dpo4 and Ile 60 replaces Ala 44. Ile 60 is located in close proximity to Arg 73 and tightens the active site. Gln 55 is located in close proximity to the thymidine dimer and approaches the base more closely than the shorter valine in Dpo4. Trp 56 in polη replaces Phe 33 of Dpo4 and forms a wall against the backbone of the template strand. Based on this model, the active site is therefore much less solvent accessible compared to Dpo4 and more importantly, the incoming nucleotide cannot access the active site without an opening of the polymerase. The suggested mechanism of an induced fit to incorporate the correct nucleotide seems therefore very likely.

4. RECRUITMENT OF Y-FAMILY DNA POLYMERASES

Another important issue relates to the effects of interactions between the Y-family DNA polymerases and known replication accessory proteins, as well as possible novel proteins. The recent structure of the little finger domain of *Escherichia coli* Pol IV and the β-clamp processivity factor has shown how a bacterial Y-family DNA polymerase can interact with the β-clamp and how the interactions can orient the polymerase so that it can switch from an inactive orientation to an active orientation, thereby allowing a close approach to the DNA (42). This scheme would permit strict regulation of the Y-family DNA polymerase so that replication would proceed via pol IV if a blocking lesion is encountered, but in all other cases, a high-fidelity DNA polymerase would interact with the DNA. All five DNA polymerases from *E. coli* have been shown to interact with the β-clamp (43–47) and three human Y-family DNA polymerases are known to interact with PCNA (48,49). In the case of Pol IV, two regions within the little finger domain are important for the interaction with the β-clamp: First, the extended C-terminal tail of the little finger domain fits into a hydrophobic channel on the surface of the β-clamp that is formed by the second and the third domain of the β-clamp (Fig. 10A). The very C-terminal five residues of pol IV are critical for its lesion bypass and mutagenesis activity. Second, surface loops of the little finger domain interact with surface residues of the eight-stranded β-sheet formed between the N-terminal domain of one subunit and the C-terminal domain of another subunit (Fig. 10B). Two-thirds of the interface is provided through this second interaction.

In the Dbh and the Dpo4 structures, the last 10 and 11 residues, were disordered respectively. This is not surprising since these residues may only adopt a defined structure in the presence of the accessory subunit to fit snugly into the hydrophobic channel formed on the surface of the β-clamp. Superposition of the little finger domain of pol IV and Dpo4 reveals an rms deviation of 1.85 Å for 90

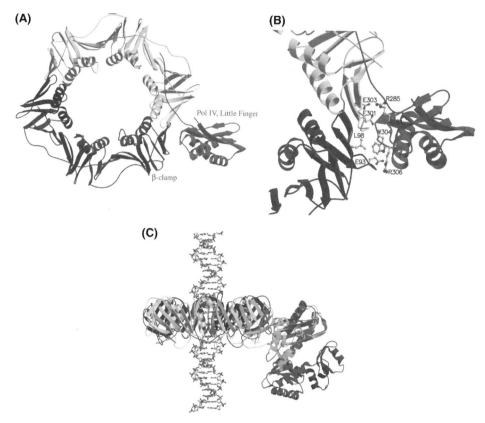

Figure 10 (A) The little finger domain of *E. coli* Pol IV bound to the β-clamp processivity factor. The β-clamp dimer is shown in yellow and blue. The little finger domain of Pol IV is shown in red. The extended C-terminus of pol IV, interacting with the β-clamp, is shown as a Cα trace (B) The interface between the little finger domain of pol IV and the β-clamp, which keeps the Y-family DNA polymerase in the "locked down" complex. Leu 98 of the β-clamp forms hydrophobic interactions with Val 303, Trp 304, and Pro 305 (not shown) of pol IV. Towards the edges of this interface, Arg 285, Arg 306, and the main chain oxygen of Ala 284 from pol IV form hydrogen bonds or salt bridges with Glu 93, Glu 303, and Glu 301 of the β-clamp. (C) Model of the Y-family DNA polymerase bound to the β-clamp in the "locked-down" position. The little finger domain of Dpo4 was superimposed onto the pol IV domain. The little finger of pol IV is shown in grey and Dpo4 is color coded as in Fig. 3. The DNA is shown in all-bonds representation. In this position, the polymerase would be located too far away to access the primer/template DNA and therefore represents the inactive state in which the replicative polymerase could interact with the DNA. (*See color insert*.)

Cα-atoms, indicating that the overall fold of this domain is very similar despite the lack of sequence identity. Based on the superposition of the little finger domains of pol IV and Dpo4, it is possible to predict how the entire polymerase interacts with the β-clamp (Fig. 10C). Assuming that the DNA is passing perpendicularly through the center of the β-clamp, the polymerase would not be able to approach the DNA as long as both interfaces of the polymerase interact with the β-clamp. This structure therefore presents the inactive complex, referred to as the "locked-down" complex by the authors (42). To obtain a "tethered" complex, which is able to approach the DNA, it would be necessary to disrupt the protein–protein interactions between the surface loops of the little finger domain and the β-clamp. The polymerase could then "swing" towards the DNA and perform translesion synthesis. The swinging mechanism would require main chain rotations of residues that are located between the C-terminus of pol IV and the core of the little finger domain. Switching between a tightly-bound inactive complex and a more loosely-bound active complex is an attractive model for a Y-family DNA polymerase and fulfills several requirements. The translesion DNA polymerase is in close proximity to the DNA and can replace the high-fidelity DNA polymerase if a lesion is encountered and provides a block in DNA synthesis. On the other hand, the Y-family DNA polymerase should not compete with a high-fidelity DNA polymerase for access to the DNA. It is, therefore, reasonable to assume that the inactive complex involves a thermodynamically stable complex, representing the DNA un-bound and inactive form, thereby shifting the equilibrium towards the interaction of the high-fidelity DNA polymerase with the DNA. Since the sequence of the little finger domain is not highly conserved among the Y-family DNA polymerases, it is unclear if a similar mode of interaction is also used by other polymerases – an issue that awaits further analysis.

5. LESION SPECIFICITY OF THE Y-FAMILY DNA POLYMERASES

Despite the knowledge that has been gained through the recent discovery of the Y-family DNA polymerases and the subsequent structural characterization of some members within this family, several questions remain unresolved. What are the factors that confer specificity to bypass certain lesions while others cause a block? Why is one lesion accurately bypassed while others cause misincorporation? Polη, for example, incorporates two A's correctly opposite cis–syn thymidine dimers (50), but preferentially misincorporates G opposite the 3′ T in TT (6–4) photoproducts (51). Pol V shows a similar discrimination (52), suggesting that the difference in conformation of the *cis-syn* thymidine dimer and of the 6–4 photoproducts and how it can be accommodated in the active site of the polymerase determines the accuracy of translesion bypass. Based on the Dpo4 structure in complex with DNA, it can be assumed that in addition to the catalytic carboxylates originating from the palm domain, residues in the finger domain, mainly the stretch covering residues 30–60, are responsible for the formation of the active site. This region of the finger domain contains the conserved YXAR sequence, but the residues that have been described to be in close proximity to the replicating base pair are located N- and C-terminal to this motif. While some of these residues like Arg 51, which is forming hydrogen bonds to the phosphates of the incoming nucleotide, is conserved, other residues are not conserved at all and based on the current sequence alignments even the number of residues responsible for this part of the finger domain seems to vary among the Y-family DNA polymerases, suggesting that the diversity allows different substrate specificity.

    Even though structural information is now available for some of the Y-family DNA polymerases, only two of the four subfamilies have been characterized so far. The Rev 1 family and the UmuC family await further analysis and may shed more light on the various functions of this new intriguing family of DNA polymerases.

ACKNOWLEDGMENTS

I would like to thank James J. Truglio for help with Figures 1 and 6 and Hermann Schindelin and Holly Miller for critical reading of the manuscript.

REFERENCES

1. Friedberg EC, Fischhaber PL, Kisker C. Error-prone DNA polymerases: novel structures and the benefits of infidelity. Cell 2001; 107:9–12.
2. Ohmori H, Friedberg EC, Fuchs RP, Goodman MF, Hanaoka F, Hinkle D, Kunkel TA, Lawrence CW, Livneh Z, Nohmi T, Prakash L, Prakash S, Todo T, Walker GC, Wang Z, Woodgate R. The Y-family of DNA polymerases. Mol Cell 2001; 8:7–8.
3. Zhou BL, Pata JD, Steitz TA. Crystal structure of a DinB lesion bypass DNA polymerase catalytic fragment reveals a classic polymerase catalytic domain. Mol Cell 2001; 8:427–437.
4. Trincao J, Johnson RE, Escalante CR, Prakash S, Prakash L, Aggarwal AK. Structure of the catalytic core of *S. cerevisiae* DNA polymerase eta: implications for translesion DNA synthesis. Mol Cell 2001; 8:417–426.
5. Ling H, Boudsocq F, Plosky BS, Woodgate R, Yang W. Replication of a cis-syn thymine dimer at atomic resolution. Nature 2003; 424:1083–1087.
6. Ling H, Boudsocq F, Woodgate R, Yang W. Crystal structure of a Y-family DNA polymerase in action: a mechanism for error-prone and lesion-bypass replication. Cell 2001; 107:91–102.
7. Silvian LF, Toth EA, Pham P, Goodman MF, Ellenberger T. Crystal structure of a DinB family error-prone DNA polymerase from Sulfolobus solfataricus. Nat Struct Biol 2001; 8:984–989.
8. Ling H, Sayer JM, Plosky BS, Yagi H, Boudsocq F, Woodgate R, Jerina DM, Yang W. Crystal structure of a benzo[a]pyrene diol epoxide adduct in a ternary complex with a DNA polymerase. Proc Natl Acad Sci USA 2004; 101:2265–2269.
9. Masutani C, Kusumoto R, Yamada A, Yuasa M, Araki M, Nogimori T, Yokoi M, Eki T, Iwai S, Hanaoka F. Xeroderma pigmentosum variant: from a human genetic disorder to a novel DNA polymerase. Cold Spring Harb Symp Quant Biol 2000; 65:71–80.
10. Prakash S, Johnson RE, Washington MT, Haracska L, Kondratick CM, Prakash L. Role of yeast and human DNA polymerase eta in error-free replication of damaged DNA. Cold Spring Harb Symp Quant Biol 2000; 65:51–59.
11. Matsuda T, Bebenek K, Masutani C, Hanaoka F, Kunkel TA. Low fidelity DNA synthesis by human DNA polymerase-eta. Nature 2000; 404:1011–1013.
12. Kondratick CM, Washington MT, Prakash S, Prakash L. Acidic residues critical for the activity and biological function of yeast DNA polymerase eta. Mol Cell Biol 2001; 21: 2018–2025.
13. Doublie S, Tabor S, Long AM, Richardson CC, Ellenberger T. Crystal structure of a bacteriophage T7 DNA replication complex at 2.2 A resolution. Nature 1998; 391: 251–258.
14. Franklin MC, Wang J, Steitz TA. Structure of the replicating complex of a pol alpha family DNA polymerase. Cell 2001; 105:657–667.
15. Wang J, Sattar AK, Wang CC, Karam JD, Konigsberg WH, Steitz TA. Crystal structure of a pol alpha family replication DNA polymerase from bacteriophage RB69. Cell 1997; 89:1087–1099.
16. Delarue M, Poch O, Tordo N, Moras D, Argos P. An attempt to unify the structure of polymerases. Protein Eng 1990; 3:461–467.

17. Pelletier H, Sawaya MR, Wolfle W, Wilson SH, Kraut J. Crystal structures of human DNA polymerase beta complexed with DNA: implications for catalytic mechanism, processivity, and fidelity. Biochemistry 1996; 35:12742–12761.

18. Doublie S, Sawaya MR, Ellenberger T. An open and closed case for all polymerases. Structure Fold Des 1999; 7:R31–R35.

19. Doublié S, Tabor S, Long AM, Richardson CC, Ellenberger T. Crystal structure of a bacteriophage T7 DNA replication complex at 2.2 A resolution. Nature 1998; 391: 251–258.

20. Napolitano R, Janel-Bintz R, Wagner J, Fuchs RP. All three SOS-inducible DNA polymerases (Pol II, Pol IV and Pol V) are involved in induced mutagenesis. EMBO J 2000; 19:6259–6265.

21. Shamoo Y, Steitz TA. Building a replisome from interacting pieces: sliding clamp complexed to a peptide from DNA polymerase and a polymerase editing complex. Cell 1999; 99:155–166.

22. Huber HE, Tabor S, Richardson CC. Escherichia coli thioredoxin stabilizes complexes of bacteriophage T7 DNA polymerase and primed templates. J Biol Chem 1987; 262:16224–16232.

23. Modrich P, Richardson CC. Bacteriophage T7 Deoxyribonucleic acid replication in vitro. A protein of Escherichia coli required for bacteriophage T7 DNA polymerase activity. J Biol Chem 1975; 250:5508–5514.

24. Eom SH, Wang J, Steitz TA. Structure of Taq ploymerase with DNA at the polymerase active site. Nature 1996; 382:278–281.

25. Kiefer JR, Mao C, Braman JC, Beese LS. Visualizing DNA replication in a catalytically active Bacillus DNA polymerase crystal. Nature 1998; 391:304–307.

26. Li Y, Korolev S, Waksman G. Crystal structures of open and closed forms of binary and ternary complexes of the large fragment of Thermus aquaticus DNA polymerase I: structural basis for nucleotide incorporation. EMBO J 1998; 17:7514–7525.

27. Pelletier H, Sawaya MR, Kumar A, Wilson SH, Kraut J. Structures of ternary complexes of rat DNA polymerase beta, a DNA template-primer, and ddCTP. Science 1994; 264:1891–1903.

28. Sawaya MR, Prasad R, Wilson SH, Kraut J, Pelletier H. Crystal structures of human DNA polymerase beta complexed with gapped and nicked DNA: evidence for an induced fit mechanism. Biochemistry 1997; 36:11205–11215.

29. Steitz TA. DNA- and RNA-dependent DNA polymerases. Curr Opin Struct Biol 1993; 3:31–38.

30. Beese LS, Steitz TA. Structural basis for the 3′-5′exonuclease activity of Escherichia coli DNA polymerase I: a two metal ion mechanism. EMBO J 1991; 10:25–33.

31. Huang H, Chopra R, Verdine GL, Harrison SC. Structure of a covalently trapped catalytic complex of HIV-1 reverse transcriptase: implications for drug resistance. Science 1998; 282:1669–1675.

32. Ohashi E, Ogi T, Kusumoto R, Iwai S, Masutani C, Hanaoka F, Ohmori H. Error-prone bypass of certain DNA lesions by the human DNA polymerase kappa. Genes Dev 2000; 14:1589–1594.

33. Wagner J, Gruz P, Kim SR, Yamada M, Matsui K, Fuchs RP, Nohmi T. The dinB gene encodes a novel E. coli DNA polymerase, DNA pol IV, involved in mutagenesis. Mol Cell 1999; 4:281–286.

34. Zhang Y, Yuan F, Wu X, Wang M, Rechkoblit O, Taylor JS, Geacintov NE, Wang Z. Error-free and error-prone lesion bypass by human DNA polymerase kappa in vitro. Nucleic Acids Res 2000; 28:4138–4146.

35. Boudsocq F, Iwai S, Hanaoka F, Woodgate R. Sulfolobus solfataricus P2 DNA polymerase IV (Dpo4): an archaeal DinB- like DNA polymerase with lesion-bypass properties akin to eukaryotic poleta. Nucleic Acids Res 2001; 29:4607–4616.

36. Park H, Zhang K, Ren Y, Nadji S, Sinha N, Taylor JS, Kang C. Crystal structure of a DNA decamer containing a cis-syn thymine dimer. Proc Natl Acad Sci USA 2002; 99:15965–15970.

37. Szekeres EZ Jr, Woodgate R, Lawrence CW. Substitution of mucAB or rumAB for umuDC alters the relative frequencies of the two classes of mutations induced by a site-specific T-T cyclobutane dimer and the efficiency of translesion DNA synthesis J Bacteriol 1996; 178:2559–2563.

38. Zhang H, Siede W. UV-induced T→C transition at a TT photoproduct site is dependent on Saccharomyces cerevisiae polymerase eta in vivo. Nucleic Acids Res 2002; 30:1262–1267.

39. Johnson RE, Washington MT, Prakash S, Prakash L. Fidelity of human DNA polymerase eta. J Biol Chem 2000; 275:7447–7450.

40. Masutani C, Kusumoto R, Iwai S, Hanaoka F. Mechanisms of accurate translesion synthesis by human DNA polymerase eta. EMBO J 2000; 19:3100–3109.

41. Washington MT, Prakash L, Prakash S. Yeast DNA polymerase eta utilizes an induced-fit mechanism of nucleotide incorporation. Cell 2001; 107:917–927.

42. Bunting KA, Roe SM, Pearl LH. Structural basis for recruitment of translesion DNA polymerase Pol IV/DinB to the beta-clamp. EMBO J 2003; 22:5883–5892.

43. Hughes AJ Jr, Bryan SK, Chen H, Moses RE, McHenry CS. Escherichia coli DNA polymerase II is stimulated by DNA polymerase III holoenzyme auxiliary subunits. J Biol Chem 1991; 266:4568–4573.

44. Naktinis V, Turner J, O'Donnell M. A molecular switch in a replication machine defined by an internal competition for protein rings. Cell 1996; 84:137–145.

45. Wagner J, Fujii S, Gruz P, Nohmi T, Fuchs RP. The beta clamp targets DNA polymerase IV to DNA and strongly increases its processivity. EMBO Rep 2000; 1:484–488.

46. Lopez de Saro FJ, O'Donnell M. Interaction of the beta sliding clamp with MutS, ligase, and DNA polymerase I. Proc Natl Acad Sci USA 2001; 98:8376–8380.

47. Sutton MD, Farrow MF, Burton BM, Walker GC. Genetic interactions between the Escherichia coli umuDC gene products and the beta processivity clamp of the replicative DNA polymerase. J Bacteriol 2001; 183:2897–2909.

48. Haracska L, Johnson RE, Unk I, Phillips B, Hurwitz J, Prakash L, Prakash S. Physical and functional interactions of human DNA polymerase eta with PCNA. Mol Cell Biol 2001; 21:7199–7206.

49. Haracska L, Unk I, Johnson RE, Phillips BB, Hurwitz J, Prakash L, Prakash S. Stimulation of DNA synthesis activity of human DNA polymerase kappa by PCNA. Mol Cell Biol 2002; 22:784–791.

50. Johnson RE, Prakash S, Prakash L. Efficient bypass of a thymine-thymine dimer by yeast DNA polymerase, Poleta. Science 1999; 283:1001–1004.

51. Johnson RE, Haracska L, Prakash S, Prakash L. Role of DNA polymerase zeta in the bypass of a (6-4) TT photoproduct. Mol Cell Biol 2001; 21:3558–3563.

52. Tang M, Pham P, Shen X, Taylor JS, O'Donnell M, Woodgate R, Goodman MF. Roles of E. coli DNA polymerases IV and V in lesion-targeted and untargeted SOS mutagenesis. Nature 2000; 404:1014–1018.

53. Hutchinson EG, Thornton JM. PROMOTIF–a program to identify and analyze structural motifs in proteins. Protein Sci 1996; 5:212–220.

26

Regulation of Damage Tolerance by the *RAD6* Pathway

Helle D. Ulrich

Cancer Research UK, Clare Hall Laboratories, South Mimms, U.K.

1. INTRODUCTION

Tolerance to replication-blocking lesions in the DNA is of vital importance for a cell's resistance to genotoxic agents. Within double-stranded (ds) DNA, excision repair systems efficiently remove various types of damage to bases and nucleotides, and the information encoded by the complementary strand is used to correctly restore the original information (1). In single-stranded (ss) DNA regions, however, which arise during the duplication of the genome, these repair systems cannot operate due to the absence of an instructive template. As a consequence, unrepaired lesions act as "road blocks" for the replication machinery during S phase, because the active sites of replicative DNA polymerases, streamlined for accurate and processive DNA synthesis, do not accommodate distorted template structures (2,3). It is generally believed that this situation poses a minor problem on the lagging strand, where DNA replication can in principle resume by the initiation of a new Okazaki fragment, leaving a small gap opposite the damaged site. On the other hand, a blocked polymerase on the leading strand causes a permanent stalling of the replication fork, which would elicit a checkpoint response leading to cell cycle arrest and ultimately cell death if the lesion could not be circumvented or passed.

Damage bypass mechanisms that allow the completion of replication in the presence of DNA lesions are therefore critical for the survival and proliferation of a cell. Accordingly, a number of different systems appear to be operating in any given organism, each designed to allow replication forks to pass over sites of damage. Importantly, these bypass systems differ markedly in the accuracy with which they fill the position opposite a lesion. In addition to a damage avoidance pathway that acts in a virtually error-free manner (4,5), most organisms employ several specialized polymerases capable of using damaged templates for replication with reduced fidelity (2,6–10). Due to their activity, DNA-damaging agents generally induce mutations, even at concentrations too low to have a pronounced effect on cell survival. The carcinogenic potential of genotoxic agents can thus be attributed not primarily to the inflicted damage itself, but rather to its mutagenic processing by damage-tolerant polymerases. A tight control over the choice between error-free and error-prone pathways is therefore of vital importance for the prevention of potentially harmful mutations and the maintenance of overall genome stability, and it is not surprising to find an intricate machinery for the regulation of damage bypass activity in eukaryotic cells.

The regulatory system responsible for the choice of damage bypass mechanism in eukaryotes is called the *RAD6* pathway after its most prominent member in yeast (11). While most of its components were cloned decades ago, it has emerged only recently that this group of enzymes controls damage bypass by means of posttranslational protein modification, in particular by the small, highly conserved ubiquitin. Ubiquitin is a protein of 76 amino acids common to eukaryotes and most prominent for its function in the targeting of short-lived proteins for regulated degradation by the 26S proteasome, a large intracellular protease (for reviews see Refs. 12–14). Potential substrates are marked for destruction by the attachment of a multimeric chain of ubiquitin molecules in an intricate conjugation reaction that usually requires a cascade of at least three different enzymes (Fig. 1). In contrast to multi-ubiquitination, the attachment of a single ubiquitin moiety conveys entirely different, proteasome-independent signals, ranging from the inititation of endocytosis and intracellular vesicle transport to the regulation of chromatin structure and transcription (15–17). In the context of DNA damage tolerance, distinct functions, both apparently unrelated to proteolysis, have been revealed for multi- and monoubiquitination by components of the *RAD6* pathway.

This chapter briefly summarizes the mechanisms of damage bypass and their consequences for damage-induced and spontaneous mutagenesis before describing in detail the components of the *RAD6* pathway, their genetic and physical interactions, their relations to the ubiquitin system, and our present knowledge about their mechanism of action. As both the genetic and biochemical details of the *RAD6* pathway have been best characterized in the yeast *Saccharomyces cerevisiae*, the review is focused largely on this organism, but mentions parallels and differences to other organisms, including mammals, where they are relevant.

2. MECHANISMS OF DAMAGE BYPASS

2.1. Translesion Synthesis

The simplest way to continue DNA replication in the presence of a lesion on the template strand is to ignore it and try to polymerize across the site of damage. Naturally, this strategy, which is called translesion synthesis (TLS), poses several problems for the cell: replicative DNA polymerases are highly efficient and accurate enzymes whose catalytic centers have evolved to fit an unperturbed template and primer terminus along with the matching deoxynucleoside triphosphate complementary to the next position on the template. Any unphysiological change, ranging from bulky adducts and backbone distortions to small base modifications altering their chemical properties or coding capacity, will therefore present an obstacle to the processive activity of the replicative enzyme, causing polymerization to stall. Similarly, abasic sites, frequently arising by spontaneous hydrolysis, do not serve as a template for polymerase δ (Polδ), the principal replicative DNA polymerase in eukaryotes. Thus, most organisms harbor specialized polymerases with more relaxed catalytic centers, which are less demanding with respect to the template DNA and therefore capable of inserting nucleotides opposite a variety of abnormal structures (3,6–10). If the replicative polymerase is transiently exchanged for one of the damage-tolerant enzymes,

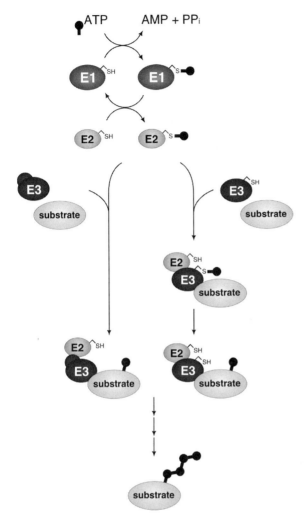

Figure 1 Mechanism of ubiquitin conjugation. Attachment of ubiquitin to its substrate proteins is mediated by a cascade of enzymes: an active-site cysteine residue within ubiquitin-activating enzyme (E1) undergoes a covalent thioester linkage with the C-terminal carboxy group of ubiquitin in an energy-dependent reaction involving a ubiquitin-AMP intermediate (not shown). Ubiquitin, now in its activated form, is then transferred in an energy-neutral manner to the active-site cysteine of a ubiquitin-conjugating enzyme (E2). The E2, aided by a ubiquitin protein ligase (E3), mediates the transfer of ubiquitin to the substrate protein. This may occur by one of two different mechanisms, depending on the class of E3 involved: RING E3s harbor a specialized form of a zinc finger, the RING domain, which in many cases contacts the E2 and is essential for ubiquitin transfer. This class of E3 stimulates conjugation by mediating the contact between the E2 and the substrate in a noncovalent manner (left branch). HECT E3s are characterized by a domain that harbors another conserved cysteine. This residue takes part in the thioester cascade and takes over the activated ubiquitin from the E2 before it is transferred to the substrate in an additional step (right branch). Coupling results in an isopeptide bond between ubiquitin's C-terminus and the ε-amino group of an internal lysine residue on the substrate protein. Repeated conjugation of ubiquitin, using internal lysine residues of the preceeding ubiquitin moieties, then leads to the formation of multimeric chains. While most organisms encode a single E1 for activation of the total cellular ubiquitin pool, multiple E2s with varying subcellular locations and an even larger number of E3s of distinct properties and substrate specificities provide for a combinatorial system of high selectivity and flexibility.

the lesion can thus be overcome and processive replication can resume (Fig. 2a). Depending on the type of lesion, however, the genetic information of the damaged template may of course be obscured or even completely lost, as in the case of an abasic site. Therefore, the newly synthesized sequence opposite the lesion may differ

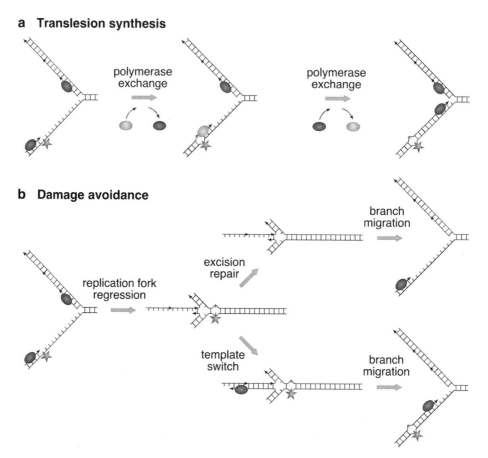

Figure 2 Pathways of damage bypass. (a) Translesion synthesis. A lesion in the leading strand (star symbol) causes the stalling of the replicative polymerase (dark grey oval). The enzyme is exchanged for a damage-tolerant polymerase (light grey oval), which can accommodate distorted templates in its active site and inserts nucleotides opposite the lesion. After polymerization of a few nucleotides, the damage-tolerant enzyme is replaced again by the replicative polymerase. Direction of DNA polymerization is indicated by arrowheads. (b) Damage avoidance. A stalled replication fork can be overcome in an error-free manner by an unwinding of the lagging strand and the reannealing of the original duplex (branch migration). When the replication fork has regressed far enough, the two newly synthesized strands can anneal with each other, resulting in a Holliday structure with the lesion now in its original double-stranded context. The damage may now be removed by an excision repair system, and branch migration in the opposite direction can restore the replication fork structure (top). Alternatively, the stalled primer terminus of the leading strand may be extended, using the newly synthesized strand of the sister chromatid as a template (bottom). When this structure is resolved by branch migration, the lesion now resides in a double-stranded region and can again be repaired in an error-free manner by an excision repair system. In both cases, the intermediate branched structure can also be resolved by cleavage of the junction, followed by homologous recombination.

from the original, resulting in a mutation that will be fixed in the genome once the lesion—now in a double-stranded region—is removed by an excision repair system. Furthermore, the employment of a translesion polymerase creates an additional problem intrinsic to its ability to accommodate distorted templates: since the supreme accuracy of a replicative polymerase is a consequence of an exact substrate fit to its catalytic center, damage-tolerant polymerases are in turn less discriminatory even on undamaged templates, incorporating mismatched nucleotides with a frequency significantly higher than that observed for replicative polymerases. In summary, the elevated mutation rates caused by the action of damage-tolerant polymerases have a two-fold origin: on one hand, the loss of information associated with the lesion itself and, on the other hand, the high rates of nucleotide misincorporation even on undamaged template DNA.

The recent discovery of a growing number of translesion polymerases in higher organisms has caused a surge of biochemical investigations into the properties and specificities of this class of enzymes (see Chapters 23 and 25). There appears to be significant variation in the activities of individual polymerases toward different types of lesions, and while some of them are specialized in the insertion of individual nucleotides opposite a lesion, but cannot extend the resulting mismatched primer terminus, others exhibit complementary abilities, suggesting that different translesion polymerases may frequently cooperate in the bypass of a single lesion. A feature common to all the damage-tolerant polymerases appears to be their low degree of processivity. In fact, considering the potential harm that inaccurate replication over long stretches of DNA could cause, processivity would be one of the least desirable properties of a bypass enzyme. Thus, the danger of inducing mutations during TLS might be minimized by keeping a tight control over the activities of the bypass polymerases and allowing them to incorporate only a few nucleotides before the replicative polymerase takes over again.

2.2. Damage Avoidance

An alternative strategy that altogether avoids the use of the damaged region as a template for DNA synthesis takes advantage of the genetic information encoded by the original complementary strand to restore the sequence opposite the lesion in an error-free manner. In bacteria, many of the components involved in this pathway have been identified, and the process is beginning to be unravelled in molecular detail (4,5,18). In eukaryotes, meanwhile, mechanistic aspects are not at all understood, even though it is now generally accepted that a damage avoidance system does exist. Several nonexclusive mechanisms are being discussed for the bypass of a lesion in the leading strand, each involving a temporary reversal of the replication fork movement (Fig. 2b). Fork regression is believed to be achieved by the unwinding of both newly synthesized daughter strands by means of helicases and the reannealing of the original duplex. The newly synthesized strands may then also anneal with each other, in effect resulting in a Holliday junction. Structures like this have been observed in response to DNA damage in bacteria (19,20) as well as eukaryotes (21), although in the latter case firm evidence for their physiological significance is still missing. Nevertheless, by the creation of a four-way junction, which has been dubbed "chicken foot" (20), the lesion would be brought back into a double-stranded sequence and could be removed by an excision repair system, thus correctly restoring the genetic information. The Holliday junction might then simply be resolved by branch migration, and replication could proceed again. Alternatively, the stalled 3′ end of the leading strand, now annealed to the newly synthesized sequence of the lagging strand, could act as a primer for renewed DNA synthesis on the new template. Again, branch migration can resolve the Holliday structure, resulting in a Y-shaped replication fork with the 3′ end of the leading strand now extending beyond the lesion. A resolution of the four-way junction by nucleolytic cleavage [using a Holliday junction resolvase (22)] and a subsequent reconstruction of the replication fork by a recombinational mechanism has also been discussed (23). A prerequisite for the strand switching model is the continuation of replication on the lagging strand upon arrest of the leading strand in order to provide the template for the leading strand primer. Evidence for this "overshoot" synthesis has been found in *Escherichia coli* (24) as well as mammalian systems (25). Obviously, some lesions, such as interstrand crosslinks or protein adducts involving both strands, would prevent any progression of one primer terminus beyond the other, as they would inhibit not only the polymerases but also the replicative helicases responsible for separating the template strands. In these cases, the only way to remove the damage would be a replication fork regression and removal of the damage before a restart of synthesis. In yeast and *E. coli*, this type of damage has been shown to be associated with elevated levels of recombination, although it is not at all clear how a recombinational event by itself could contribute to the removal of a fork block, other than facilitating its repair by clearance of the replisome.

3. THE *RAD6* PATHWAY

The significance of damage tolerance for the survival of a cell was recognized several decades ago in genetic experiments with collections of randomly isolated yeast mutants sensitive towards ultraviolet (UV) irradiation (26). Double mutant analysis was used to determine whether the effects of two individual mutants on survival were additive, in which case they were considered to belong to independently operating repair systems, or mutually dependent, suggesting an epistatic relationship and thus an involvement in the same pathway. Based on this phenotypic analysis, three major epistasis groups of DNA repair genes were identified and each named after a prominent member (1): the *RAD3* group, whose mutants are characterized by high sensitivity towards UV irradiation and various alkylating agents, was later found to encode components of the nucleotide excisison repair system (27); mutants of the *RAD52* group, highly sensitive towards ionizing radiation, turned out to be defective in the repair of double-strand breaks (DSBs) via homologous recombination (28); finally, mutants of the *RAD6* group showed varying sensitivities towards a variety of genotoxic agents that were additive to mutants of the *RAD3* group and proved to be phenotypically quite heterogeneous, as some of them appeared to affect not so much the efficiency, but rather the accuracy of DNA repair (11). It soon became clear that the *RAD6* group of genes is responsible for controlling both of the branches of damage tolerance: one class of mutants was found to contribute only little to overall resistance to damage, but to be unable to accumulate mutations upon treatment with DNA-damaging agents, consistent with a defect in error-prone TLS (29,30); the other class exhitibed defects in a recovery system called postreplication repair, which is highly important for survival, but has no effect on damage-induced mutagenesis and is therefore considered to be error-free (31–33). This activity is detected by the cell's ability to convert low molecular weight (MW) DNA synthesized on damaged templates (e.g., in excision repair defective mutants) to the high

MW form that is normally produced in the absence of damage, and it most likely reflects the action of the error-free damage avoidance system (34,35). Interestingly, some of the members of the *RAD6* group, including the *RAD6* gene itself, are required for both of the activities described above, suggesting that the *RAD6* pathway may act as a control system that regulates the balance between error-prone TLS and error-free damage avoidance (31–33,36–38).

Due to the pleiotropic phenotypes of its mutants, the *RAD6* pathway has received a number of different names: most often, it has been called error-prone repair due to its effects on damage-induced mutagenesis, or postreplication repair due to its influence on the restoration of the size of DNA synthesized on damaged templates. Both names are awkward because they do not reflect the regulatory nature of the pathway. First, not all its activities, not even concerning TLS, are mutagenic. Second, both TLS and the error-free system are believed to operate mainly during rather than after DNA synthesis. Finally, it is inaccurate to speak of a genuine repair system, since neither TLS nor damage avoidance actually removes the lesion from the DNA. Thus, damage bypass or damage tolerance is more appropriate terms and should be used (11).

Genetic relationships between the members of the *RAD6* pathway have been studied extensively, but a first insight into the mechanistic aspects of damage bypass has come more recently from the biochemical characterization of the core components and their physical interactions (39–41). According to their enzymatic functions, the majority of them falls into one of two classes: ubiquitin conjugation factors and damage-tolerant polymerases. While the action of the polymerases appears restricted to the TLS pathway, many of the ubiquitin conjugation factors contribute to both the mutagenic and the error-free bypass systems.

3.1. Members of the *RAD6* Pathway

3.1.1. Ubiquitin Conjugation Factors

RAD6: Mutants in the *RAD6* gene itself display the most severe and at the same time most pleiotropic phenotype among all the factors involved in damage tolerance. They are highly sensitive towards any type of DNA-damaging agent (11) and defective in both damage-induced mutagenesis (30,36) as well as postreplication repair (31,33) based on the synthesis of large MW DNA on damaged templates, indicating that *RAD6* is required for both mutagenic TLS and error-free damage avoidance. Mutants of *rad6* have elevated rates of spontaneous mutagenesis and mitotic recombination (32), and they are growth-retarded even in the absence of genotoxic agents. In addition, the *RAD6* gene affects several other aspects of metabolism apparently unrelated to DNA damage tolerance: mutants are defective in sporulation (32) and transcriptional silencing (42), but hyperactive with respect to Ty1 retrotransposition (43), suggesting a defect in chromatin metabolism. Finally, a temperature-sensitive cell cycle arrest has been reported in some genetic backgrounds (44,45).

Two homologs of *RAD6* have been identified in the human genome, the X-linked HHR6A and the autosomal HHR6B, which encode two highly similar proteins with an overall homology of 70% to the yeast Rad6p, but lacking its acidic C-terminal tail (46). Expression in yeast *rad6* mutants rescues their hypersensitivity against genotoxic agents, but not the sporulation and silencing phenotypes, which correlate with the absence of the acidic tail. Both mammalian *RAD6* homologs appear to participate in postreplicative damage avoidance; they are transcriptionally upregulated and recruited to chromatin in response to DNA damage (47). Nevertheless, they also appear to play a role in meiosis, as deletion of HHR6B in mice causes male sterility (48,49).

A link between DNA damage tolerance and the ubiquitin system was first established when Rad6p was discovered in a biochemical approach designed to identify components of the yeast ubiquitin conjugation machinery by means of affinity chromatography (50). Based on its ability to undergo a covalent thioester linkage to ubiquitin in the presence of ATP and ubiquitin-activating enzyme (see Fig. 1 for details), Rad6p was identified as a member of the family of ubiquitin-conjugating enzymes (E2). Mutations in the active-site cysteine, which is involved in the formation of the essential thioester intermediate and is conserved among all E2s, resulted in a phenotype indistinguishable from a deletion of the gene, indicating that the catalytic activity of Rad6p as a ubiquitin conjugation factor is a prerequisite for activity in DNA damage bypass (51). Until recently, however, the target proteins relevant to damage tolerance have remained obscure.

In contrast, a range of activities unrelated to the DNA damage response has been characterized in detail, explaining some of the pleiotropic phenotypes of *rad6* mutants. Soon after its isolation as an E2, Rad6p was found to mediate one of the best characterized pathways of protein degradation, the so-called "N-end rule" (52), which relates the in vivo half-life of a protein to the identity of its amino-terminal residue (53). In the context of this degradation pathway, Rad6p cooperates with the ubiquitin protein ligase (E3) Ubr1p, which is responsible for the recognition of substrates via their destabilizing N-terminal residues (54). In cooperation with Ubr1p, Rad6p produces multiubiquitin chains in which one ubiquitin moiety is linked to the following via its lysine (K) 48 (55). This type of linkage is known to be the principal chain geometry recognized by the 26S proteasome as a degradation signal (55,56). Thus, together with the E3 Ubr1p, Rad6p functions as a classical E2 in mediating the ubiquitination of short-lived proteins for proteasomal degradation. Its activity is not limited to N-end rule substrates, but extends to other short-lived proteins such as Gpa1p, the α subunit of a heterotrimeric G protein (57), and Cup9p, a transcriptional repressor responsible for the regulation of peptide import (58). Both proteins are recognized by the E3 Ubr1p, however not via their N-termini, but via a less well-characterized internal site.

In higher eukaryotes, monoubiquitination is one of the prominent histone modifications affecting chromatin structure and transcriptional activity (59,60). In fact, histones were the very first proteins found to be subject to modification by ubiquitin in mammalian cells (61). The observation that purified yeast Rad6p is able to ubiquitinate histones H2A and H2B in vitro (50,62) prompted an examination of histone modifications in yeast, and histone H2B was indeed found to be ubiquitinated in a Rad6p-dependent manner at a single lysine within its C-terminal tail (63). The absence of this modification correlated with the growth and sporulation defects observed in *rad6* mutants. Moreover, it was later shown that monoubiquitination of H2B is required for methylation of histone H3, which in turn is a prerequisite for telomeric gene silencing (64,65). In contrast to the damage bypass-related activities, histone modification by Rad6p requires the protein's highly acidic C-terminus. This observation suggested that—as in the in vitro reaction—no E3 enzyme would be required to mediate the interaction between the E2 and the basic histone substrate. However, a candidate E3, named Bre1p, was recently identified as a Rad6p-interacting protein that is required for H2B ubiquitination in vivo and recruits the E2 to the promoter regions of several inducible genes, where it contributes to transcriptional activation (66–68).

Ubr1p and Bre1p as well as Ubr2p, a distant Ubr1p homolog of unknown function that also interacts with Rad6p, share a feature that has proved to be a hallmark of one of the two known classes of E3 enzymes: the RING finger, a zinc-coordinating domain, which in E3s generally appears to function as an interaction module for the E2 (for review see Refs. 69–71). Although Ubr1p and Ubr2p in fact bind Rad6p via an adjacent domain, the Ubr1p RING finger is nevertheless essential for substrate ubiquitination via the N-end rule (72). Thus, it appears that Rad6p is able to cooperate with several E3s, each of which directs the E2 to different substrates according to its function in cellular metabolism. In the context of DNA damage bypass, Rad6p cooperates with yet another RING finger protein, the DNA-binding repair factor Rad18p.

RAD18: The phenotype of *rad18* mutants is somewhat less severe than that of *rad6* mutants and appears to be restricted to those aspects of *RAD6* function related to DNA damage tolerance. Thus, like *rad6*, *rad18* strains are highly sensitive to DNA-damaging agents, defective in damage-induced mutagenesis as well as error-free recovery, and hyperactive with respect to recombination and spontaneous mutation rates, but they do not share the sporulation or silencing defects of *rad6* cells (31,33,37,38,73). Homologs of *RAD18* have been identified in higher eukaryotes, and deletion of the gene in mouse embryonic fibroblasts and chicken B-lymphocyte lines leads to DNA damage hypersensitivity and increased levels of recombination and sister chromatid exchange (74–77).

RAD18 from baker's yeast was found to encode a protein of 487 amino acids, harboring an N-terminal RING domain, a second putative zinc-binding motif located centrally, and a region close to the C-terminus that is rich in basic residues (78,79). The protein was shown to bind DNA with a preference for ssDNA, a property that might be attributable to the presence of a SAP domain spanning amino acids 278–312 (80). Rad18p exists in a stable complex with Rad6p, and purification studies have shown that the protein might even require this interaction for stable folding (80,81). As in the case of Ubr1p, the RING domain does not mediate the contact to Rad6p in yeast; instead, the interaction depends on the basic motif close to the Rad18p C-terminus (82). In the mammalian system, however, the RING finger does in fact contact the E2 (75), and even in yeast this domain is required for biological activity, as mutants in which one of the conserved cysteines involved in zinc coordination is replaced by serine do not complement the *rad18* deletion with respect to DNA damage sensitivity (H. D. Ulrich, unpublished observations). In mammals, a RING finger mutant of Rad18p acts in a dominant negative way when overexpressed in cell culture (74). Thus, Rad18p exhibits all the features suggestive of an E3 enzyme that targets its cognate E2 Rad6p to DNA and is responsible for mediating damage bypass via TLS as well as error-free damage avoidance.

UBC13, MMS2: Identification of Rad6p as an E2 enzyme initially suggested an involvement of proteolysis in damage bypass, such as the removal of proteins from a stalled replication fork (83). However, this concept was complicated by the observation that multiubiquitin chains of a nonstandard topology, apparently unrelated to proteasomal signalling, participate in the *RAD6* pathway: yeast strains in which endogenous ubiquitin was replaced by a variant in which K63 was mutated to arginine displayed a UV sensitivity that fell into the *RAD6* epistasis group, suggesting that multiubiquitin chain formation via this lysine was important for Rad6p-dependent damage bypass (84). Accordingly, a prominent set of ubiquitin conjugates was absent in the *ubi(K63R)* strain when compared to wild-type (*wt*) ubiquitin. It turned out, however, that Rad6p was not the E2 responsible for their synthesis (84). The relevant E2 (or rather its mammalian homolog) was identified later by fractionation of cell extracts and purification of the activity responsible for the synthesis of K63-linked diubiquitin (85). Interestingly, both the mammalian and the yeast enzymes were found to be heterodimers, consisting of Ubc13p, a classical ubiquitin-conjugating enzyme with high homology to other E2 enzymes, and Mms2p, an E2-related protein (dubbed UEV, for ubiquitin-conjugating enzyme variant) whose sequence resembles that of genuine E2s, but lacks the conserved active-site cysteine characteristic of catalytically active E2s. In contrast to other E2s, which appear to function as monomers or homodimers, Ubc13p is active only when in complex with Mms2p (85). Moreover, unlike most other E2s the purified complex polymerizes free, unanchored multiubiquitin chains in vitro in the absence of a substrate. The crystal structures of the yeast and the human heterodimers reveal that the unusual ubiquitin chain geometry likely results from an Mms2p-mediated alignment of the acceptor ubiquitin in a position that places K63 in proximity to the catalytic cysteine of Ubc13p (86,87).

The ability of Ubc13p and Mms2p to synthesize multiubiquitin chains correlates with their activity in DNA damage tolerance, as mutations that inhibit chain synthesis in vitro also result in a hypersensitivity of the corresponding yeast strains to UV irradiation and chemical damage (39,85). Moreover, the defect of *ubc13* and *mms2* deletion mutants is similar and epistatic to the phenotype of the *ubi(K63R)* mutant (85), and in fact the *MMS2* gene had previously been isolated as a member of the *RAD6* pathway based on the complementation of a mutant sensitive to the alkylating agent methyl methane sulfonate (MMS) (88). Thus, Ubc13p and Mms2p function in the *RAD6* pathway by means of K63-linked multiubiquitin chain synthesis. Despite the presence of possible nuclear localization signals, in yeast both proteins reside largely in the cytoplasm and are redistributed to the nucleus in response to DNA damage (89). *MMS2* is required—like *RAD6* and *RAD18*—for the restoration of high MW DNA after replication of damaged templates (90). In contrast to *rad6* and *rad18* mutants, however, *ubc13* and *mms2* cells exhibit only moderate sensitivities to UV- and chemically induced DNA damage and virtually no sensitivity towards γ irradiation. Like *rad6* and *rad18*, they have elevated spontaneous mutation rates, but they do not show any defect in damage-induced mutagenesis. In fact, their damage sensitivity is synergistic to that of mutations defective in TLS. Taken together, these data place *UBC13* and *MMS2* into the error-free damage avoidance pathway and exclude a participation in TLS (85,88–91).

Interestingly, MMS2 belongs to a whole family of E2-related proteins (UEVs) found in a variety of different contexts, such as transcriptional activation and cell differentiation (92–96). While human hMMS2 has 50% homology to yeast Mms2p and complements the corresponding deletion mutant, the closely related CROC-1 proteins bear N-terminal extensions with no homology to the ubiquitin-conjugating enzymes and complement yeast *mms2* only when the extension is removed (95), indicating that in vivo they most likely fulfill other functions unrelated to DNA damage bypass.

RAD5: The role of *RAD5* within the *RAD6* epistasis group has remained somewhat more ambiguous. Mutants of *rad5* were originally characterized as defective in the UV-induced reversion of ochre alleles and hence named *rev2* (29), but were later shown to be proficient in damage-induced mutagenesis in most other reporter systems, based on amber, missense, and frameshift reversions or forward mutations

(73,97). Moreover, *rad5* and TLS-defective mutants show a synergistic increase in UV sensitivity (98), and the *RAD5* gene was found to be required for the postreplicative restoration of high MW DNA (90). *RAD5* is therefore now viewed as a component of the error-free damage avoidance system. Similar to *rad18*, the *rad5* deletion causes a hyperrecombination and hypermutator phenotype and a moderate sensitivity towards ionizing radiation (73,99). In most situations, the function of *RAD5* appears strictly dependent on the presence of *RAD18*; however, in some circumstances, for example in stationary phase cells, the effects of *rad18* and *rad5* mutants on damage sensitivity were found to be additive (73,100). In addition, genetic experiments have indicated an involvement in the repair of double-strand breaks, and a recent analysis of the *rad5* mutator phenotype has again suggested some contribution to TLS (99,101). Thus, although its main function appears to be in the error-free system, its position seems less restricted than that of *UBC13* or *MMS2*, and additional activities beyond the error-free branch are expected to emerge.

The ambiguous nature of *RAD5* function is reflected by the protein's sequence. *RAD5* encodes a protein of 134 kDa whose most prominent features are the seven conserved sequence motifs in the C-terminal half of the protein that place Rad5p into the SNF2/SWI2 family of DNA and RNA helicases as well as chromatin remodeling factors (102). Accordingly, the purified protein was found to exhibit DNA-binding and ssDNA-dependent ATPase activity, although helicase activity could not be demonstrated (103). Interestingly, however, mutations in the conserved Walker type A motif, which should affect ATP binding and hydrolysis, cause only a partial loss of function with respect to damage sensitivity (H. D. Ulrich, unpublished observations). A more severe phenotype is caused by mutation of the RING finger, which is located within the helicase-like domain (39). The N-terminal half of the protein has little homology to other known sequences, but harbors a leucine zipper-like heptad repeat preceded by a region rich in basic residues that may be important for protein–protein contacts (102). Its placement in the error-free damage avoidance pathway, where it cooperates with Ubc13p and Mms2p (see below), suggests that Rad5p contributes to ubiquitin conjugation as an E3 enzyme.

3.1.2. Damage-Tolerant Polymerases

REV3, REV7: Among all the factors that contribute to damage bypass by TLS, the damage-tolerant DNA polymerase ζ (Polζ), encoded in yeast by *REV3* and *REV7* (104), is arguably the one with the most significant impact on the overall fidelity of DNA replication in the presence of genotoxic agents (105). Deletion of *REV3*, which encodes the catalytic subunit of the polymerase, results in only a modest increase of DNA damage sensitivity, but a virtually complete loss of damage-induced mutagenesis, implying that Polζ is responsible for more than 95% of all base substitutions and frameshift mutations induced by DNA damage (106). In contrast to most damage-tolerant polymerases, Polζ does not belong to the Y family of polymerases (see Chapter 25), but is instead related to the replicative enzymes of the B family, although it lacks a 3′–5′ exonuclease (proofreading) activity. Its accuracy on undamaged templates is comparable to that of replicative polymerases in the absence of proofreading, but it is much less processive (104). Considering its involvement in TLS, the enzyme has a remarkably limited ability to insert nucleotides opposite a lesion. Instead, it very efficiently extends terminally mismatched primers resulting from natural mispairings or the presence of a lesion in the template strand

opposite the terminal nucleotide. Thus, it is now believed that Polζ probably cooperates with additional bypass polymerases, which would be responsible for the actual
incorporation of mismatched nucleotides (see below). Damage-induced mutations
would then result from the extension of the mispaired termini by Polζ.

In addition to its prominent role in damage-induced mutagenesis, Polζ also
significantly affects spontaneous mutation rates. Judging by the antimutator phenotype of *rev3* strains, the polymerase is responsible for 50–75% of all spontaneous
mutations during normal replication (107,108). Moreover, the enhanced mutation
frequencies associated with homologous recombination or high levels of transcription
and those resulting from defects in excision repair can be attributed to the activity of
Polζ (108–110). It appears unlikely that these effects can be explained entirely by the
Polζ-dependent bypass of unrepaired exogenous or spontaneous damage. Therefore,
it has been suggested that Polζ should rather be viewed as a replicative enzyme that is
used to overcome not only unrepaired lesions, but also other refractory structures
that obstruct replication fork progression, such as unedited terminal mismatches
or elements of secondary structure in the template DNA (105).

Epistasis analysis based upon the UV sensitivities of the respective strains has
classified *REV3* and *REV7* as members of the *RAD6* pathway, and *rad6* and *rad18*
mutants indeed share the defect in damage-induced mutagenesis with *rev3*, implying
that Rad6p is required for the function of Polζ during TLS. This notion, however,
stands in contrast to the observation that the spontaneous hypermutator phenotype
of *rad6* and *rad18* mutants is also largely *REV3* dependent (111). This apparently
paradoxical situation could be resolved if Polζ were activated for spontaneous mutagenesis in a manner independent of *RAD6* and *RAD18*. A possible explanation for
this dual mode of activation will be given below.

REV1: Like mutants in the *REV3* and *REV7* genes, *rev1* mutants were first
isolated in a screen for yeast mutants defective in UV-induced reversion of base
substitutions (29). The *REV1* gene encodes a protein with homology to the *E. coli*
UmuC and has now been classified as a member of the Y family of translesion polymerases. Biochemical assays with the purified protein have revealed a rather unusual
polymerase activity, as the enzyme prefers to insert a single dCMP opposite an abasic site (112). This deoxycytidyl (dC) transferase activity is less pronounced opposite
purines in the template strand, and nucleotides other than dC are not used at all. In
vitro, Rev1p stimulates bypass of an abasic site by Polζ, suggesting that the two
enzymes may similarly cooperate in vivo (112). However, *REV1* is also required
for the bypass of lesions other than abasic sites that do not involve incorporation
of dC, such as the TT(6–4) photoadduct. Moreover, a *rev1-1* allele was isolated that
is defective in the bypass of abasic sites despite a functional dC transferase activity.
Thus, Rev1p appears to have an additional function in TLS that is independent of its
dC transferase activity (113). It has recently been shown that the mouse homolog of
Rev1p can competitively interact with several other translesion polymerases, suggesting that it may act as a coordinator of bypass polymerases at the replication fork
(114). Although direct physical interactions have not been demonstrated in the *S.
cerevisiae* system, a similar scenario would nicely explain the additional function
observed for the yeast enzyme.

RAD30: Polymerase η (Polη), encoded in yeast by the *RAD30* gene, is related
to the *E. coli dinB* and *umuC* and thus represents another member of the recently
characterized Y family of bypass polymerases (98,115,116). Its in vitro properties
have been characterized in detail (117,118). In contrast to Polζ, its error-rate on
undamaged templates is extremely high, while the accuracy of lesion bypass varies

with the type of damage. For example, a cyclobutane thymine dimer is copied by Polη quite accurately, while bypass of the less frequent TT(6–4) photoproduct is mutagenic.

In humans, Polη is encoded by the *XPV* (xeroderma pigmentosum variant) gene (119). Mutations in *XPV* cause a strong predisposition for skin cancer, and the corresponding cells are defective in the bypass of UV lesions. In yeast, *rad30* mutants exhibit no obvious defect in damage-induced mutagenesis, but a modest, *REV3*-dependent enhancement of spontaneous mutation rates. With respect to UV sensitivity, *RAD6* and *RAD18* are epistatic to *RAD30*, but additive effects are observed in a *rad30 rad5* double mutant. Sensitivity towards chemical damage by the alkylating agent MMS or the radiomimetic drug 4-nitroquinoline oxide (4-NQO) is very moderate. Interestingly, combination of the *rad30* and the *rad5* mutants results in a synergistic increase in damage-induced mutagenesis. Taken together, these data indicate that the translesion polymerase is responsible for the bypass of mostly UV-induced damage in a relatively (but not absolutely) error-free manner and thus in effect contributes to the overall protection from damage-induced mutations (98,116). Considering its low fidelity on undamaged templates, Polη's modest effect on mutability is thus attributable to the enzyme's low processivity (120). In this context, it is important to note that the enzyme's activity is stimulated by proliferating cell nuclear antigen (PCNA), the processivity clamp for replicative polymerases, but PCNA does not enhance the processivity of Polη (121). Instead, it is believed that Polη dissociates from the primer terminus after insertion of one or two nucleotides opposite a damaged template, and a second polymerase is responsible for elongation.

3.2. Genetic and Physical Interactions Within the *RAD6* Pathway

The factors described above seem to constitute the core components of the *RAD6* pathway, and their genetic relationships, as summarized in Fig. 3, support the notion of two mechanistially different systems of damage bypass: the damage-tolerant polymerases carry out TLS in a more or less mutagenic fashion, depending on the intrinsic accuracy of the enzyme as well as the type of lesion; the dimeric E2 Ubc13p/Mms2p cooperates with the RING finger protein Rad5p in the error-free damage avoidance pathway; and both branches are controlled by the ubiquitin conjugation factors Rad6p and Rad18p. A number of additional genes have been implicated in *RAD6*-dependent damage tolerance. However, as their positions within the system outlined in Fig. 3 are less well defined, current models for their possible contributions to the *RAD6* pathway will be discussed separately in Section 6.

The genetic hierarchy within the *RAD6* pathway is paralleled by direct physical interactions between those components that are involved in ubiquitin conjugation (Fig. 4). A cooperation between Rad6p and Rad18p was suggested based on the stable interaction between the two proteins, by which the RING finger protein Rad18p recruits the E2 Rad6p to DNA (80). More recently, it was shown that Rad18p is capable of self-association while at the same time maintaining its interaction with Rad6p, thus resulting in the coordiation of two (or more) E2 molecules on the chromatin (89). Rad5p, on the other hand, associates with Ubc13p by means of its RING domain and in turn recruits the Ubc13p/Mms2p heterodimer to chromatin in response to DNA damage. A weak self-association of Rad5p has also been observed. Finally, a connection between the two E2–RING finger protein pairs is established by the mutual interaction of Rad18p and Rad5p. The domains involving

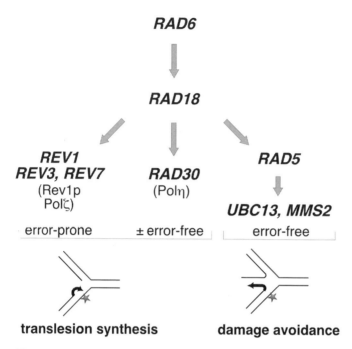

Figure 3 Genetic relationships within the *RAD6* pathway in yeast. Epistasis relations based on the damage sensitivities of the respective mutants are indicated by arrows. *RAD30*, *REV1*, *REV3*, and *REV7* encode damage-tolerant polymerases involved in translesion synthesis. Rev1p and Polζ, encoded by *REV3* and *REV7*, account for most of damage-induced mutagenesis. Bypass of UV lesions mediated by Polη, encoded by *RAD30*, is largely error free. *RAD5*, *UBC13*, and *MMS2* are components of the error-free damage avoidance systems. All three branches depend on the presence of *RAD6* and *RAD18*.

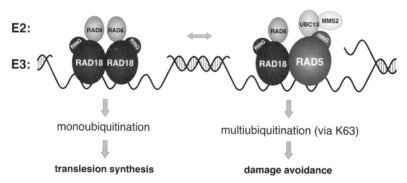

Figure 4 Physical interactions within the *RAD6* pathway. The DNA-binding RING finger proteins Rad18p and Rad5p recruit the two E2s, Rad6p, and the Ubc13p–Mms2p complex, to chromatin in response to DNA damage, resulting in different combinations of E2s and E3s. The Ubc13p–Mms2p complex (right) assembles K63-linked multiubiquitin chains, which are a prerequisite for error-free damage avoidance. Monoubiquitination by Rad6p and Rad18p (left) is required for translesion synthesis. *Source*: Adapted from Ref. 89.

this contact are distinct from those required for interaction with the respective E2. Thus, the RING finger proteins coordinate the assembly of the two different E2s, Rad6p and Ubc13p/Mms2p, in a single complex on the DNA. As the central domain within Rad18p that is responsible for contacting Rad5p overlaps with the region involved in self-association, and likewise the N-terminal domain within Rad5p mediates both the Rad5p–Rad5p and the Rad5p–Rad18p contacts, it is likely that homo- and heterodimerization are mutually exclusive, thus resulting in an equilibrium between different DNA-associated complexes as depicted in Fig. 4.

Although higher-order complexes have been isolated by coimmunoprecipitation (89), and the association of Rad18p with Ubc13p and Mms2p has been confirmed by a genome-wide pulldown experiment (122), the heteromeric complex involving all five components (Rad6p, Rad18p, Rad5p, Ubc13p, and Mms2p) has not been purified in its entirety. Most likely, the interactions between the individual components are too transient to withstand purification. Support for this notion comes from the analysis of individual contacts within the complex (39,123). For example, the Ubc13p–Mms2p interaction is characterized by fast association and dissociation kinetics (39), which prevents the formation of a stable complex despite a reasonably high affinity. Similarly, the Ubc13p–Rad5p contact appears weak or transient based on two-hybrid and coimmunoprecipitation experiments (39,89,123). However, tight complexes are apparently dispensable for biological function in damage bypass, as mutations in the Rad5p–Ubc13p or the Ubc13p–Mms2p interface cause only partial phenotypes that can be compensated by the overproduction of the mutant factors. Dynamic interactions between E2 and E3 enzymes have been found in other instances and may even be an integral feature of the ubiquitin conjugation reaction (124). Likewise, a rapid equilibrium between the complexes shown in Fig. 4 may be a prerequisite for the regulation of error-free vs. mutagenic damage bypass by the *RAD6* system (89).

When the physical interactions between the ubiquitin conjugation factors are correlated with the genetics of the *RAD6* pathway and the biochemical activities of the E2 enzymes involved, it becomes apparent that the heteromeric complex involving Rad6p and Ubc13p–Mms2p (Fig. 4, right) must mediate error-free damage avoidance via the synthesis of K63-linked multiubiquitin chains, while by default the Rad6p–Rad18p complex (Fig. 4, left), through monoubiquitination or K48-linked chains, would be responsible for TLS. The spontaneous hypermutator phenotypes of *rad5*, *ubc13*, or *mms2* mutants strongly support the notion that a dynamic equilibrium between the different complexes may regulate the balance between TLS and error-free damage avoidance, because the absence of these components would strongly favor mutagenic TLS mediated by the now over-abundant Rad6p–Rad18p complex. On a more speculative notion, additional functions of Rad5p independent of Ubc13p, Mms2p or even Rad18p (see above) might involve its self-associated form.

3.3. Proliferating Cell Nuclear Antigen Is a Target of the *RAD6* Pathway

Significant insight into how the chromatin-associated complexes described above regulate damage tolerance came from the identification of PCNA as a target for Rad6p-dependent ubiquitination (41). Proliferating cell nuclear antigen, encoded in yeast by the *POL30* gene, functions analogously to the bacterial β-clamp, the subunit of the replicative DNA polymerase that confers processivity to the polymerization reaction (125). Although eukaryotic PCNA—in contrast to the dimeric β-clamp—is a homotrimer, its overall six-fold geometry and the ring-shaped structure encircling the DNA closely resemble the bacterial protein (126). Eukaryotic PCNA stimulates the activities of the two replicative polymerases, δ and ε, but in addition it serves as a binding platform for a multitude of other factors involved in excision and mismatch repair, chromatin assembly, and cell cycle regulation (127). Interestingly, most of these contact PCNA via a conserved sequence motif that is often situated at the extreme N- or C-terminus of the protein. In light of the trimeric nature of PCNA, this set-up suggests that PCNA acts as a central communication site and signal integrator for the coordination of multiple events and factors at the replication fork, ranging from replication to repair and postreplicational chromatin assembly.

In yeast as well as mammals, PCNA is modified at a single conserved lysine residue, K164, with a K63-linked multiubiquitin chain in response to DNA damage (41). Modification requires the components of the error-free damage avoidance pathway, Rad6p, Rad18p, Rad5p, Ubc13p, and Mms2p. While ubiquitination is completely abolished in *rad6* and *rad18* mutants, deletion of *RAD5*, *UBC13*, or *MMS2*, or mutation of K63 of ubiquitin to arginine results in the accumulation of monoubiquitinated PCNA after DNA damage. This suggests that the conjugation reaction is a two-step process in which Rad6p and Rad18p attach the first ubiquitin moiety, which is then expanded to a multimeric chain by Ubc13p, Mms2p, and Rad5p, according to the unusual linkage specificity of the dimeric E2 (Fig. 5). Consistent

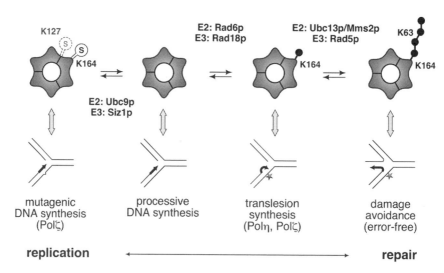

Figure 5 Consequences of PCNA modifications for DNA replication and mutagenesis. PCNA (ring symbol) is modified alternatively by ubiquitin (black symbols) or SUMO (white symbols). Acceptor lysines on PCNA (K127, K164) and the linkage of multiubiquitin chains (K63) as well as the conjugation factors (E2, E3) are indicated. The modification states of PCNA are correlated with different activities of the replication fork: unmodified PCNA promotes processive DNA synthesis by the replicative polymerases, multiubiquitination is a prerequisite for the error-free damage avoidance pathway, and monoubiquitination of PCNA in response to DNA damage activates the damage-tolerant polymerases Polη and Polζ for translesion synthesis. Polζ can be activated for mutagenic DNA synthesis during replication in the absence of damage by modification of PCNA with SUMO, possibly as a means to overcome refractory DNA template structures or unedited terminal mismatches. *Source*: Adapted from Ref. 40.

with an E3 function, both RING finger proteins, Rad18p and Rad5p, physically interact with the substrate protein, PCNA.

Genetic analysis has shown that multiubiquitination of PCNA is indeed the activity that mediates error-free damage avoidance in yeast (41): mutation of the acceptor lysine of PCNA to arginine, *pol30(K164R)*, results in a hypersensitivity to DNA-damaging agents that falls into the error-free branch of the *RAD6* epistasis group, while additive or synergistic effects are observed with mutants in the nucleotide excision or double-strand break repair systems. Consistent with a requirement for ubiquitination at K164, overepxression of the lysine mutant *pol30(K164R)* sensitizes the cells to MMS treatment. In contrast, overexpression of *wt* PCNA partially suppresses the hypersensitivity of *rad5*, *ubc13*, and *mms2* mutants, indicating that even monoubiquitinated PCNA can protect the cell from the harmful effects of DNA damage. While this observation does not answer the question whether monoubiquitinated PCNA is a biologically relevant species or merely a nonphysiological intermediate in multiubiquitination-deficient mutants, a distinct function of the monoubiquitinated form has been derived from the effect of PCNA modifications on TLS (40). Double mutant analysis has demonstrated that the activities of both damage-tolerant polymerases, Polζ and Polη, are dependent on the presence of K164 of PCNA, and the fact that the *UBC13*-dependent branch of the *RAD6* pathway acts independently of TLS implies that it is really the mono- and not the multiubiquitinated form of PCNA that is relevant in this process. Further evidence for this notion is the complete absence of damage-induced mutagenesis in the *pol30(K164R)* mutant, a response that depends almost exclusively on the activity of Polζ (see above) and does not require *UBC13*-dependent multiubiquitination (40).

In summary, identification of PCNA as a target for *RAD6*-dependent ubiquitination has uncovered two distinct and novel aspects of ubiquitin function (Fig. 5): multiubiquitination of PCNA elicits the error-free damage-avoidance response that likely involves a regression of the replication fork, and monoubiquitination of the same protein at the same lysine residue in turn activates the bypass polymerases η and ζ for TLS and damage-induced mutagenesis. Intriguingly, both PCNA modifications appear to convey a nonconventional, i.e., proteasome-independent signal. Monoubiquitination is generally not sufficient to target a protein to the 26S proteasome (55), and K63-linked multiubiquitin chains adopt a geometry that strongly differs from that of conventional, K48-linked chains that normally act as recognition signals for the 26S proteasome (128). As a note of caution, however, it is important to remark that there is no definitive proof yet that multiubiquitination of PCNA via K63-linkage really does not affect the stability of the modified protein. On one hand, the pool of ubiquitinated PCNA is much too small to detect a noticable reduction in the overall abundance of the protein in vivo, and on the other hand, when chemically coupled to an artificial substrate protein, a K63-linked multiubiquitin chain can indeed promote its in vitro recognition and degradation by the proteasome, albeit less effectively than a chain of K48-linkage (129). Moreover, mutants in catalytic subunits of the proteasome as well as in the proteasome maturation factor Ump1p were found to exhibit elevated spontaneous mutation rates and slight UV sensitivities genetically related to the *RAD6* pathway, raising the possibility that proteasomal activity may yet play a role in damage bypass (130).

As outlined above, the requirement of PCNA monoubiquitination for the activation of Polζ explains the defect in damage-induced mutagenesis observed in *rad6* and *rad18* mutants as well as the spontaneous Polζ-dependent hypermutator phenotype of *rad5*, *ubc13*, and *mms2*. It does not, however, provide a satisfying explanation for the fact that *rad6* and *rad18* mutants are also spontaneous Polζ-dependent hyper-mutators. In fact, if monoubiquitination of PCNA were a strict requirement for activation of Polζ, *rad6* and *rad18* mutants should exhibit lowered rather than elevated mutation rates, whether induced or spontaneous. As outlined below, however, the activation of the bypass polymerase turns out to be more complex than anticipated and shows different requirements under damage vs. nondamage conditions.

4. PROLIFERATING CELL NUCLEAR ANTIGEN MODIFICATION BY THE UBIQUITIN-LIKE PROTEIN SUMO

The action of the *RAD6* pathway in damage tolerance is complicated by the finding that ubiquitin is not the only modifier to be attached to PCNA. In fact, PCNA was first isolated as a substrate of the small ubiquitin-related modifier SUMO, and only when this modification was studied in detail the ubiquitinated forms were discovered (41).

SUMO, in yeast encoded by the *SMT3* gene, is a 12 kDa protein with moderate sequence and structural similarity to ubiquitin itself (reviewed in Ref. 31). In higher eukaryotes, three closely related forms have been identified, and according to the various approaches that have led to its discovery, SUMO has received several additional names (UBL, sentrin, PIC-1, GMP1). Together with at least two other distant relatives, ubiquitin and SUMO form a family of eukaryotic protein modification factors whose function is mediated by covalent attachment to distinct sets of target proteins (132). A common phylogenetic origin is suggested not only by the homology of the modifiers themselves, but also the similarities of their respective conjugation machineries. SUMO is activated by an E1 enzyme, a heterodimer of the two subunits Uba2p and Aos1p, which closely resemble the N- and C-terminal portions of the ubiquitin-specific E1, Uba1p, respectively (133). In contrast to the ubiquitin system, a single conjugating or E2 enzyme, Ubc9p, mediates the transfer of SUMO to the target proteins (134). In vivo, Ubc9p is aided in this process by a SUMO-specific ligase or E3. In yeast, two such enzymes, Siz1p and Siz2p, are known, while in higher eukaryotes their number is somewhat larger (135). Interestingly, most SUMO-specific ligases belong to the PIAS family, which share a domain related to the RING finger. The consequences of SUMO modification are much less well defined than those of ubiquitination. In some instances, SUMO conjugation has been shown to affect the localization of the target protein, in others it appears to modulate its activity, and some studies even suggest that it might act as an antagonist of ubiquitination in cases where the two modifiers compete for the same lysine on a common target protein (131). More and more proteins are being identified as substrates for both ubiquitination and SUMO modification. In contrast to ubiquitin, SUMO appears to be conjugated mainly in its monomeric form, although multimeric SUMO chains have been observed in vitro as well as in vivo (135,136).

SUMO modification of PCNA was discovered by means of bulk purification of SUMO conjugates in yeast and identification of the targets by mass spectrometry, although it is still unclear whether this modification occurs in mammalian cells as well (41). In yeast, low levels of modification are found in replicating cells during S phase, but not in G1, G2 and mitosis. In addition, extensive modification is observed when cells are treated with lethal amounts of the alkylating agent MMS. Apart from the SUMO-specific E2 Ubc9p, the ligase Siz1p was found to be required for SUMO conjugation to PCNA in vivo (Fig. 5). Proliferating cell nuclear antigen is modified primarily at K164, the same lysine that is also subject to ubiquitination (see above). In addition, modification is observed to a minor extent at K127, which—unlike K164—is part of a consensus motif that was found to serve as a SUMO attachment site in several other proteins, but is not conserved in the PCNA sequences of other species.

Although not all aspects of PCNA SUMO modification are fully understood to date, one of its functions during S phase became apparent upon analysis of spontaneous mutation rates in strains deficient in ubiquitin or SUMO conjugation to PCNA (40). Here—in contrast to damage-induced mutagenesis—the requirement for monoubiquitination of PCNA was not absolute, but could be substituted by SUMO modification. Overall, spontaneous mutation rates were found to depend on the balance between the different modification states of PCNA: whenever multi-ubiquitination was possible, mutation rates were close to *wt*. If, on the other hand, the monoubiquitinated or SUMO-modified forms predominated in the absence of multiubiquitin chains, elevated mutation rates were observed, and in cases where neither modification of PCNA was possible, mutation rates were lower than *wt*. Thus, both monoubiquitin and SUMO are capable of stimulating Polζ activity, with the difference that SUMO modification is not inducible by DNA damage and thus does not contribute to induced mutagenesis. This observation now explains the spontaneous hypermutator phenotype of *rad6* and *rad18* mutants by an activation of Polζ through PCNA SUMO modification in the absence of ubiquitination. It has been argued previously that the activity of Polζ during S phase in the absence of exogenous DNA damage serves not only for the bypass of spontaneous lesions, but also to overcome replication fork blocks caused by other refractory sequences such as secondary structures or unedited terminal mismatches (105). Activation of Polζ by SUMO modification of PCNA as part of the normal S phase would fulfill exactly this purpose.

Interestingly, however, SUMO modification of PCNA appears to have a negative effect on the cell's resistance towards DNA damage (41). It turns out that *pol30(K164R)* mutants, which are no longer able to ubiquitinate PCNA, are less sensitive to UV and alkylation damage than *pol30(K127/164R)* double mutants, which in addition have lost the option of SUMO modification. Moreover, deletion of the SUMO-specific ligase gene *SIZ1* alleviates the hypersensitivity of *rad18* and *rad5* mutants (40). Thus, the detrimental effect of SUMO modification of PCNA cannot be explained simply by a SUMO-dependent inhibition of ubiquitination, but instead its consequences are most visible in mutants defective in the *RAD6* pathway. There is currently no satisfying model to explain this phenomenon.

5. MECHANISTIC CONSIDERATIONS

The model shown in Fig. 5 depicts PCNA as a molecular switchboard that controls the mechanism of replication and damage bypass by means of distinct modification states. While it is attractive to invoke the PCNA modifications as a means to trigger alternative responses to DNA damage, several new questions arise from this scenario. On one hand, upstream signals must exist that determine which of the modifications is appropriate at what time, and these signals should control the activities of the respective conjugation enzymes. On the other hand, attachment of either monoubiquitin, multiubiquitin chains or SUMO to PCNA must elicit distinct cellular responses based on the recognition of these modifications by downstream effectors. In the following section, our current state of knowledge and highlights of important unanswered questions about the prerequisites and consequences of PCNA modifications are summarized.

5.1. Upstream Signals

SUMO modification of PCNA appears to be a regular event associated with normal S phase. Nevertheless, the signal that triggers the reaction is far from clear. Conjugation may simply be governed by the stage of the cell cycle. Considering that only a small portion of the PCNA pool is modified at any given moment, it appears more likely, however, that specific events, such as the stalling of a replication fork, would induce the modification at the relevant location. This scenario is supported by the hypermodification of PCNA upon treatment with lethal amounts of MMS (see above), but this in turn raises the question about the factors that convey the relevant signal. Likely candidates are the proteins involved in damage recognition and the checkpoint response (see Chapters 36, 37).

The same questions hold true for the ubiquitination of PCNA, although the requirements for this modification clearly differ from those that pertain to the SUMO system. As ubiquitinated PCNA is undetectable in cycling cells, it appears likely that DNA damage exceeding that of a stalled replication fork is a prerequisite for activation of the *RAD6* pathway. The observation that a PCNA mutant with a defect in trimerization is less efficiently modified than *wt* PCNA suggests that ubiquitination indeed takes place while PCNA is associated with DNA and not in the soluble phase. Again, damage recognition or checkpoint factors may participate in the recruitment of the conjugation machinery to the relevant sites. Alternatively, since both Rad18p and Rad5p are known as DNA-binding proteins with a preference for ssDNA, they could be directly involved in the recognition of the lesion.

Once ubiquitination has been triggered, the cell of course has the choice between mono- and multiubiquitination of PCNA. It is currently unknown whether these two modifications are freely interconvertible, i.e., whether monoubiquitinated PCNA can be converted into the multiubiquitinated form by means of Ubc13p, Mms2p, and Rad5p alone, or whether the two states are produced by distinct and dedicated complexes according to Fig. 4. In the former case, TLS and error-free damage avoidance could be attempted successively in a time-dependent manner. This would allow the cell to bypass the damage in a straightforward way by using damage-tolerant polymerases, while it would only have to resort to the more elaborate mechanism of replication fork regression if the lesion cannot be overcome by any of the bypass polymerases. In the second case, however, the choice between the two different bypass mechanisms would have to be made immediately according to the type of damage encountered. Availability of Rad5p, which appears to be present in the cell at rate-limiting concentrations, or the import of Ubc13p and Mms2p into the nucleus may influence the balance between mono- and multiubiquitination.

Finally, it is interesting to note that there is an interplay between the components of the ubiquitin and the SUMO conjugation machinery, as the SUMO-specific E2 Ubc9p was found to physically interact with the ubiquitin-specific E3s Rad18p and Rad5p (41). This interaction does not compete with the association of Rad18p and Rad5p with their cognate E2s (P. Stelter and H. D. Ulrich, unpublished results), suggesting that it may be of regulatory significance rather than a requirement for ubiquitin or SUMO conjugation. Although the consequences of these interactions are currently unknown, they may affect the choice between ubiquitin vs. SUMO modification of PCNA.

5.2. Activation of TLS

Several models can be envisioned for the activation of damage-tolerant polymerases by ubiquitinated or SUMO-modified PCNA (for a more detailed review, see Ref. 137). On one hand, recruitment of the bypass enzyme to the replication fork might be facilitated if the affinity of the polymerase to PCNA is enhanced by the modification. On the other hand, attachment of ubiquitin or SUMO might cause a dissociation of the replicative polymerase from the clamp, thereby allowing access of the damage-tolerant enzyme to the primer terminus in a more passive way. A third option would be the dissociation of PCNA, catalyzed by the clamp loader RF-C, upon modification (138). Inspection of the PCNA structure with respect to the site of modification does not give any clue as to the most plausible mechanism of polymerase exchange. As shown in Fig. 6, it turns out that K164 is situated at the periphery of the molecule, but pointing towards the "back" side of the molecule, which faces away from the polymerase-binding site. RF-C as well as most other PCNA-interacting factors also associate with regions of PCNA that do not overlap with the ubiquitin acceptor lysine (127). Thus, polymerase exchange is not expected to function via a simple displacment of associated molecules by competition with the modifier. However, an active recruitment of a bypass polymerase or the activation of the clamp loader by the modifier would also have to involve regions of PCNA that are spacially not immediately adjacent to K164. Finally, the exchange mechanism might differ depending on the polymerase involved. For example, Polη is known to require PCNA for activity in lesion bypass (121), which would favor a recruitment or displacement model over a complete removal of PCNA from the fork, while Polζ can apparently function in the absence of a clamp. On the other hand, Polζ can be

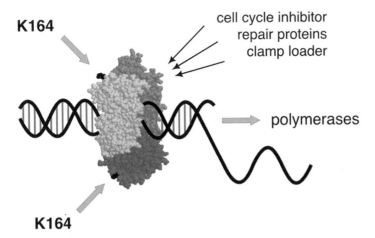

Figure 6 Sites of modifications and interactions on PCNA. In a schematic representation, the PCNA trimer is shown to encircle dsDNA at the site of a primer terminus. The subunits are depicted in different shades of grey. K164, the conserved acceptor lysine for ubiquitin and SUMO (marked in black), is situated at the periphery of the ring-shaped molecule, facing away from the surface that contacts polymerases as well as most other known PCNA-interacting proteins. *Source*: From Ref. 126.

activated by either ubiquitination or SUMO modification of PCNA (40). In light of the fact that the two modifiers differ significantly in their surface properties, it appears unlikely that they should equally contribute to the recruitment of the polymerase to the fork, thus favoring a displacement of the replicative polymerase over an active recruitment of the bypass enzyme by the modification. It may well turn out in the end that more than one of the above mechanisms contribute to polymerase exchange, but biochemical analysis is clearly needed to differentiate between the individual scenarios.

5.3. Activation of Damage Avoidance

In contrast to TLS by damage-tolerant polymerases, which has lately become an object of intense research, virtually nothing is known about the mechanistic details of the error-free damage avoidance pathway in eukaryotes. Most of the working models are derived from the situation observed in *E. coli* and thus give no clue as to the function of ubiquitination in the regression of a stalled replication fork or a template switch in DNA synthesis. Obviously, the restructuring of the replication fork is significantly more complicated than TLS and must involve additional components, such as helicases and a system that would facilitate strand invasion and homology search (4,5,18). A regulatory signal mediated by K63-linked multiubiquitin chains would most likely act upstream of the enzymes carrying out the actual bypass and might be mediated by proteins that recognize the unusual chain geometry.

While no such factors have been identified yet, linkage-specific chain recognition might be a more widely used concept, considering that K63-linked multiubiquitin chains are involved in a number of other aspects of cellular metabolism. In yeast, they were found to promote endocytosis when attached to plasma membrane transporters (139) and to contribute to ribosomal function as a modification of one of its subunits (140). In addition, there is evidence that K63 of ubiquitin may affect mitochondrial inheritance (141). In humans, K63-linked chains synthesized by the homologs of Ubc13p and Mms2p play an important role in tumor necrosis factor (TNF) -induced signalling and the inflammatory response (142–146). In this instance, K63-linked multiubiquitin chains are attached in a most likely autocatalytic fashion to the RING finger proteins TRAF6 and TRAF2. The modified proteins then activate multiple kinase pathways. An involvement of the 26S proteasome in this process could be excluded. It remains to be seen whether a common, proteasome-independent mechanism of K63-chain function may emerge from these observations.

6. INTERACTIONS OF THE *RAD6* PATHWAY WITH OTHER FACTORS

In addition to the core components described in Section 3.1, several other genes involved in DNA metabolism have been associated phenotypically with the *RAD6* pathway. However, the details of their contribution to damage bypass are much less clearly defined than those of the ubiquitin conjugation factors or bypass polymerases. While the majority appears to act in the complex damage avoidance pathway and may act as those accessory factors discussed in the previous section, some of them also appear to affect TLS. Here they are grouped according to their biochemical properties for a brief discussion of their relevance to the *RAD6* pathway.

6.1. Polymerases

POL3: When temperature-sensitive mutants of the catalytic subunits of the two essential DNA polymerases Polδ, encoded by *POL3*, and Polε, encoded by *POL2*, were examined in postreplication recovery assays for their ability to restore high MW DNA after replication of UV-damaged templates, Polδ, but not Polε, was found to be required for this process, suggesting that the major replicative enzyme, Polδ, is also responsible for the error-free damage avoidance pathway described above (147). Interestingly, another allele of *POL3*, *pol3-13*, was found to be somewhat deficient in damage-induced mutagenesis, implying a contribution of Polδ to TLS (148). In contrast, the function of Polε remains somewhat obscure. The enzyme is clearly present at replication forks and takes part in bulk DNA replication; its essential role, however, does not lie within the catalytic center, but instead in a domain that is believed to mediate signalling in damage recognition and the checkpoint response (149). The fact that a single temperature-sensitive mutant in the enzyme's catalytic subunit did not show a defect in the postreplicative recovery assay, however, does of course not rule out a contribution of Polε to PCNA-mediated damage bypass.

   POL32: *POL32* encodes a nonessential subunit of Polδ that directly contacts PCNA and is required for efficient DNA polymerization in vitro (150,151). It has been suggested that Pol32p stabilizes the Polδ–PCNA complex. Mutants of *POL32* are hypersensitive towards a range of genotoxic agents, implying that this subunit is of particular importance for the replication of damaged DNA. Interestingly, its phenotype appears to be due mainly to a defect in TLS, as *pol32* is epistatic with *rev3*, shows a similar defect in damage-induced mutagenesis and—like *rev3*—abolishes the spontaneous hypermutator phenotypes of *rad6* and *rad18* mutants (152). It is currently unknown whether Pol32p acts as part of the Polδ holoenzyme or as an independent entity in this context, and whether contacts to PCNA are required for this process. The recent finding that Pol32p directly interacts with the helicase Srs2p (152) indicates that the protein might acts as a communicator between the replication machinery and the components of the *RAD6* pathway.

6.2. Helicases and Strand-Annealing Factors

SRS2: The *SRS2* gene occupies a unique position within the *RAD6*-dependent bypass system, as deletion of the gene suppresses a substantial portion of the damage sensitivities of many *RAD6* pathway mutants. First identified as suppressors of *rad6* and *rad18*, *srs2* mutants were independently found in a screen for hyper-gene conversion mutants (*hpr5*) and as radiation-sensitive mutants (*radH*) in yeast (106,153,154). *SRS2* encodes a helicase with 3′–5′ polarity (155). Deletion of the gene causes only slight hypersensitivity to genotoxic agents, but strong UV-stimulated hyperrecombination. Genetic interactions with mutants of the *RAD52* group and a synthetic phenotype with another helicase gene involved in recombinational repair, *SGS1*, suggest that *SRS2* acts in an antirecombinogenic fashion to protect the cell from the lethal effects of aberrant recombination intermediates (156–158). Consistent with this idea, the Srs2p protein was recently shown to be capable of disrupting prerecombinogenic Rad51p filaments on ssDNA (159,160). The shortened life span of *srs2* mutants may also be attributable to hyperrecombination (161).

   Suppression of the phenotypes of *RAD6* pathway mutants is restricted to the error-free damage avoidance pathway and depends on the presence of a functional homologous recombination system (100,162,163). This dependence on the *RAD52* group has prompted the suggestion that the helicase acts upstream of the *RAD6* pathway and—consistent with its antirecombinogenic action—channels lesions into *RAD6*-dependent bypass. In its absence, stalled replication forks would have to be rescued by recombination.

Conflicting results have been reported about the role of *SRS2* in the mutagenic pathway. According to one of the earlier studies, *srs2* mutants are severely depressed in damage-induced forward mutagenesis (154). However, no significant defect was found by a reversion assay (164). Yet, *srs2* mutants suppress the spontaneous hyper-mutator phenotype of *mms2* cells to a large extent, indicating that the balance between error-free and error-prone replication is affected by the helicase (164).

Its effect on the ubiquitination of PCNA has not been examined in molecular terms. An action of *SRS2* upstream of the *RAD6* pathway might suggest that ubiqui-tination should be absent in the *srs2* mutant. Activation of Polζ, however, might still be mediated by SUMO modification of PCNA, which would of course influence mutation rates. In this context, it is interesting that deletion of *SRS2* has a similar effect on *RAD6* pathway mutants as deletion of the SUMO-specific ligase *SIZ1*: both mutations significantly suppress their hypersensitivities and partially or fully reduce their spontaneous hypermutator phenotypes. In a high-throughput two-hybrid screen, Srs2p was even found to physically interact with the yeast SUMO pro-tein (165). Thus, it is conceivable that the helicase Srs2p and the SUMO modification of PCNA cooperate in the regulation of *RAD6*-dependent bypass, possibly by means of repressing homologous recombination.

Finally, Srs2p may provide a link between the regulation of damage bypass and the checkpoint response, as *srs2* mutants were found to exhibit a defect in replication arrest in response to DNA damage during S phase (166). Moreover, Srs2p is required for full activation of Rad53p phosphorylation, and the protein is itself phosphory-lated in a damage- and checkpoint-dependent manner (167).

MPH1: *MPH1* encodes a protein with homology to the DEAH family of heli-cases and DNA-dependent ATPases. It was first characterized because of the strong hypermutator phenotype of the *mph1* deletion (168). This phenotype, in combination with a moderate sensitivity to genotoxic agents, suggests that the protein functions in an error-free bypass system that protects the cells from the effects of damage-induced mutations. Yet, *MPH1* appears to act independently of the members of the error-free *RAD6* pathway, as the phenotypes of the respective deletions were found to be additive. Instead, *MPH1* function depends on the components of homologous recombination (101). Interestingly, *rad5* mutants to some degree suppress the stong mutator phenotype of *mph1*, indicating that the two systems do communicate. Thus, its relationship to the *RAD6* pathway remains unclear; however, it is conceivable that the helicase could cooperate with the *RAD6*-dependent factors, either in the replication fork regression itself or in the resolution of the regressed fork by recom-bination proteins, as both events are required for the postulated mechanism of error-free damage avoidance.

MGS1: Mgs1p, another DNA-dependent ATPase, is related to *E. coli* RuvB, a Holliday junction branch migration protein, and accordingly possesses single-strand annealing activity (169). A connection to the *RAD6* pathway was established when *mgs1* mutants were found to be synthetically lethal with *rad6*, near lethal with *rad18*, and growth-retarded in combination with *rad5* (170). The growth defect of *mgs1 rad5* double mutant was shown to be due to problems during replication, as evident by a strong sensitivity to the replication fork inhibitor hydroxyurea. In addition, the double mutant displayed elevated levels of recombination. However, deletion of *MGS1* had no influence on the damage sensitivity of the *rad5* mutant. The lethal effects on *rad6* and *rad18* could be overcome by deletion of *SRS2*, again in a manner dependent on components of homologous recombination. Interestingly, overepxression of MGS1 caused lethality in combination with mutations in RF-C and PCNA (171). Thus, *MGS1* and the *RAD6* pathway apparently both contribute to genomic stability in a redundant manner during normal replication in the absence of exogenous damage, possibly by modulation of replication fork movement or branched intermediates that result from replication fork stalling.

6.3. Chromatin Components

CAF1: Most studies on the DNA damage response neglect the fact that bypass and repair must occur in the context of chromatin. Yet, chromatin components appear to play an important role in the maintenance of genome stability, as defects in chromatin assembly can lead to increased damage sensitivity. In yeast, deletion of *CAC1*, encoding a subunit of the chromatin assembly factor CAF-1, which couples DNA replication to histone deposition, results in a measurable UV sensitivity related to the *RAD6* pathway, as *rad6* and *rad18* mutants were found to be epistatic to *cac1* (172). Additive effects were observed with *rad5* as well as *rev3* mutants, and the *rad5 cac1* double mutant had increased levels of induced mutagenesis, suggesting that CAF-1 contributes to error-free rather than mutagenic bypass. However, the relationship between CAF-1 and PCNA ubiquitination has yet to be elucidated in molecular terms.

H2B: In addition to the chromatin assembly factor, the nucleosomes themselves apparently have an influence on damage tolerance. This was suggested by the identification of a UV-sensitive mutant of histone H2B, *htb1-3*, whose phenotype was genetically assigned to the *RAD6* epistasis group (173). Epistasis was found with *rad6*, *rad18*, and *rad5* mutants, but not with *rev1*, *rad30*, or *ubc13*, and UV-induced mutagenesis was normal, suggesting that *HTB1* functions in an error-free *RAD5*-mediated pathway, yet independent of multiubiquitin chain synthesis. Again, mechanistic details of the contribution of the nucleosome to damage tolerance and PCNA modifications remain unclear.

7. SUMMARY AND OUTLOOK

Until recently, the action of the *RAD6* pathway on damaged DNA has remained elusive. It was recognized to act as a regulatory system promoting tolerance to replication-blocking lesions rather than actual repair of the damage. However, the relevance of the ubiquitin system for this process was not at all understood. Moreover, it was unclear how two mechanistically totally distinct activities, translesion synthesis vs. error-free damage avoidance, could be controlled and balanced against each other by means of ubiquitination.

Insight into the mechanism of *RAD6*-dependent damage bypass has come recently from the identification of PCNA as a physiological target of ubiquitination and the assignment of distinct biological functions to its modified forms (40,41). The core components of the *RAD6* pathway are now recognized as ubiquitin conjugation factors that cooperate in the modification of PCNA with either monoubiquitin or multiubiquitin chains, depending on the combination of enzymes that assemble on the chromatin in response to DNA damage. While multiubiquitination of PCNA promotes the error-free damage avoidance pathway that is believed to involve a regression of the replication fork, the monoubiquitinated form of PCNA activates the damage-tolerant polymerases Polη and Polζ for TLS, thus contributing to damage-induced mutagenesis. Interestingly, modification of PCNA by the ubiquitin-like molecule SUMO can also stimulate Polζ in the absence of exogenous damage during DNA replication. Thus, PCNA acts as a multistate molecular switchboard that determines the composition and direction of the replicative machinery by means of distinct modifications. Other factors, such as helicases and chromatin components, which make up the enzymatic machinery for replication fork movement, are of course expected to contribute to damage bypass. However, PCNA appears to function as the central signal integrator in the control of damage tolerance, mutation rates and overall genome stability.

Yet, our picture of *RAD6*-dependent damage bypass is far from complete. On one hand, the upstream signalling pathways leading from the recognition of a lesion or the stalling of a replication fork to the appropriate modification of PCNA have yet to be elucidated. On the other hand, the consequences of the individual modifications for the activity of DNA polymerases, the direction of the replication fork or the recruitment of additional repair factors are poorly understood in molecular terms. Finally, although this regulatory system appears to be highly conserved among eukaryotes, no comparable system has been found in prokaryotes, despite the fact that the enzymatic machinery itself that is used for damage bypass closely resembles that of higher organisms. It remains to be seen whether prokaryotes have independently developed alternative strategies for regulating the activities of error-free and mutagenic bypass, or whether the *RAD6* pathway is a unique acquisition of eukaryotes, which appeared later in evolution as an addition to the bypass systems themselves, providing a more flexible means of modulation and control over the choice of damage bypass pathways.

REFERENCES

1. Friedberg EC, Walker GC, Siede W. DNA Repair and Mutagenesis. Washington, DC: American Society for Microbiology Press, 1995.
2. Goodman MF. Trends Biochem Sci 2000; 25:189–195.
3. Baynton K, Fuchs RP. Trends Biochem Sci 2000; 25:74–79.
4. McGlynn P, Lloyd RG. Nat Rev Mol Cell Biol 2002; 3:859–870.
5. Cox MM. Mutat Res 2002; 510:107–120.
6. Woodgate R. Genes Dev 1999; 13:2191–2195.
7. Pages V, Fuchs RP. Oncogene 2002; 21:8957–8966.
8. Lehmann AR. Mutat Res 2002; 509:23–34.
9. Sutton MD, Walker GC. Proc Natl Acad Sci USA 2001; 98:8342–8349.
10. Friedberg EC, Gerlach VL. Cell 1999; 98:413–416.
11. Lawrence C. Bioessays 1994; 16:253–258.
12. Hochstrasser M. Annu Rev Genet 1996; 30:405–439.
13. Hershko A, Ciechanover A. Annu Rev Biochem 1998; 67:425–479.
14. Peters JM, Harris JR, Finley D. Ubiquitin and the Biology of the Cell. New York, NY: Plenum Press, 1998.
15. Conaway RC, Brower CS, Conaway JW. Science 2002; 296:1254–1258.
16. Hicke L, Dunn R. Annu Rev Cell Dev Biol 2003; 19:141–172.
17. Ulrich HD. Eukaryot Cell 2002; 1:1–10.

18. Michel B, Flores MJ, Viguera E, Grompone G, Seigneur M, Bidnenko V. Proc Natl Acad Sci USA 2001; 98:8181–8188.
19. McGlynn P, Lloyd RG, Marians KJ. Proc Natl Acad Sci USA 2001; 98:8235–8240.
20. Postow L, Ullsperger C, Keller RW, Bustamante C, Vologodskii AV, Cozzarelli NR. J Biol Chem 2001; 276:2790–2796.
21. Sogo JM, Lopes M, Foiani M. Science 2002; 297:599–602.
22. Sharples GJ. Mol Microbiol 2001; 39:823–834.
23. Saintigny Y, Makienko K, Swanson C, Emond MJ, Monnat RJ Jr. Mol Cell Biol 2002; 22:6971–6978.
24. Pages V, Fuchs RP. Science 2003; 300:1300–1303.
25. Svoboda DL, Vos JM. Proc Natl Acad Sci USA 1995; 92:11975–11979.
26. Cox BS, Parry JM. Mutat Res 1968; 6:37–55.
27. Prakash S, Prakash L. Mutat Res 2000; 451:13–24.
28. Symington LS. Microbiol Mol Biol Rev 2002; 66:630–670, table of contents.
29. Lemontt JF. Genetics 1971; 68:21–33.
30. Lawrence CW, Christensen R. Genetics 1976; 82:207–232.
31. di Caprio L, Cox BS. Mutat Res 1981; 82:69–85.
32. Montelone BA, Prakash S, Prakash L. Mol Gen Genet 1981; 184:410–415.
33. Prakash L. Mol Gen Genet 1981; 184:471–478.
34. Ley RD. Photochem Photobiol 1973; 18:87–95.
35. Lehmann AR. J Mol Biol 1972; 66:319–337.
36. Prakash L. Genetics 1974; 78:1101–1118.
37. Armstrong JD, Chadee DN, Kunz BA. Mutat Res 1994; 315:281–293.
38. Cassier-Chauvat C, Fabre F. Mutat Res 1991; 254:247–253.
39. Ulrich HD. J Biol Chem 2003; 278:7051–7058.
40. Stelter P, Ulrich HD. Nature 2003; 425:188–191.
41. Hoege C, Pfander B, Moldovan GL, Pyrowolakis G, Jentsch S. Nature 2002; 419: 135–141.
42. Huang H, Kahana A, Gottschling DE, Prakash L, Liebman SW. Mol Cell Biol 1997; 17:6693–6699.
43. Picologlou S, Brown N, Liebman SW. Mol Cell Biol 1990; 10:1017–1022.
44. Ellison KS, Gwozd T, Prendergast JA, Paterson MC, Ellison MJ. J Biol Chem 1991; 266:24116–24120.
45. Raboy B, Marom A, Dor Y, Kulka RG. Mol Microbiol 1999; 32:729–739.
46. Koken MH, Reynolds P, Jaspers-Dekker I, Prakash L, Prakash S, Bootsma D, Hoeijmakers JH. Proc Natl Acad Sci USA 1991; 88:8865–8869.
47. Lyakhovich A, Shekhar MP. Oncogene 2004.
48. Koken MH, Hoogerbrugge JW, Jasper-Dekker I, de Wit J, Willemsen R, Roest HP, Grootegoed JA, Hoeijmakers JH. Dev Biol 1996; 173:119–132.
49. Roest HP, van Klaveren J, de Wit J, van Gurp CG, Koken MH, Vermey M, van Roijen JH, Hoogerbrugge JW, Vreeburg JT, Baarends WM, Bootsma D, Grootegoed JA, Hoeijmakers JH. Cell 1996; 86:799–810.
50. Jentsch S, McGrath JP, Varshavsky A. Nature 1987; 329:131–134.
51. Sung P, Prakash S, Prakash L. Proc Natl Acad Sci USA 1990; 87:2695–2699.
52. Dohmen RJ, Madura K, Bartel B, Varshavsky A. Proc Natl Acad Sci USA 1991; 88:7351–7355.
53. Varshavsky A, Turner G, Du F, Xie Y. Biol Chem 2000; 381:779–789.
54. Bartel B, Wunning I, Varshavsky A. Embo J 1990:93179–93189.
55. Chau V, Tobias JW, Bachmair A, Marriott D, Ecker DJ, Gonda DK, Varshavsky A. Science 1989; 243:1576–1583.
56. Finley D, Sadis S, Monia BP, Boucher P, Ecker DJ, Crooke ST, Chau V. Mol Cell Biol 1994; 14:5501–5509.
57. Madura K, Varshavsky A. Science 1994; 265:1454–1458.
58. Byrd C, Turner GC, Varshavsky A. Embo J 1998; 17:269–277.

59. Zhang Y. Genes Dev 2003; 17:2733–2740.
60. Jason LJ, Moore SC, Lewis JD, Lindsey G, Ausio J. Bioessays 2002; 24:166–174.
61. Goldknopf IL, Busch H. Proc Natl Acad Sci USA 1977; 74:864–868.
62. Sung P, Prakash S, Prakash L. Genes Dev 1988; 2:1476–1485.
63. Robzyk K, Recht J, Osley MA. Science 2000; 287:501–504.
64. Dover J, Schneider J, Tawiah-Boateng MA, Wood A, Dean K, Johnston M, Shilatifard A. J Biol Chem 2002; 277:28368–28371.
65. Sun ZW, Allis CD. Nature 2002; 418:104–108.
66. Kao CF, Hillyer C, Tsukuda T, Henry K, Berger S, Osley MA. Genes Dev 2004; 18:184–195.
67. Hwang WW, Venkatasubrahmanyam S, Ianculescu AG, Tong A, Boone C, Madhani HD. Mol Cell 2003; 11:261–266.
68. Wood A, Krogan NJ, Dover J, Schneider J, Heidt J, Boateng MA, Dean K, Golshani A, Zhang Y, Greenblatt JF, Johnston M, Shilatifard A. Mol Cell 2003; 11:267–274.
69. Joazeiro CA, Weissman AM. Cell 2000; 102:549–552.
70. Jackson PK, Eldridge AG, Freed E, Furstenthal L, Hsu JY, Kaiser BK, Reimann JD. Trends Cell Biol 2000; 10:429–439.
71. Ulrich HD. Curr Top Microbiol Immunol 2002; 268:137–174.
72. Xie Y, Varshavsky A. Embo J 1999; 18:6832–6844.
73. Liefshitz B, Steinlauf R, Friedl A, Eckardt-Schupp F, Kupiec M. Mutat Res 1998; 407:135–145.
74. Tateishi S, Niwa H, Miyazaki J, Fujimoto S, Inoue H, Yamaizumi M. Mol Cell Biol 2003; 23:474–481.
75. Tateishi S, Sakuraba Y, Masuyama S, Inoue H, Yamaizumi M. Proc Natl Acad Sci USA 2000; 97:7927–7932.
76. Yamashita YM, Okada T, Matsusaka T, Sonoda E, Zhao GY, Araki K, Tateishi S, Yamaizumi M, Takeda S. Embo J 2002; 21:5558–5566.
77. van der Laan R, Roest HP, Hoogerbrugge JW, Smit EM, Slater R, Baarends WM, Hoeijmakers JH, Grootegoed JA. Genomics 2000; 69:86–94.
78. Chanet R, Magana-Schwencke N, Fabre F. Gene 1988; 74:543–547.
79. Jones JS, Weber S, Prakash L. Nucleic Acids Res 1988; 16:7119–7131.
80. Bailly V, Lauder S, Prakash S, Prakash L. J Biol Chem 1997; 272:23360–23365.
81. Bailly V, Lamb J, Sung P, Prakash S, Prakash L. Genes Dev 1994; 8:811–820.
82. Bailly V, Prakash S, Prakash L. Mol Cell Biol 1997; 17:4536–4543.
83. Prakash L. Ann N Y Acad Sci 1994; 726:267–273.
84. Spence J, Sadis S, Haas AL, Finley D. Mol Cell Biol 1995; 15:1265–1273.
85. Hofmann RM, Pickart CM. Cell 1999; 96:645–653.
86. VanDemark AP, Hofmann RM, Tsui C, Pickart CM, Wolberger C. Cell 2001; 105:711–720.
87. Moraes TF, Edwards RA, McKenna S, Pastushok L, Xiao W, Glover JN, Ellison MJ. Nat Struct Biol 2001; 8:669–673.
88. Broomfield S, Chow BL, Xiao W. Proc Natl Acad Sci USA 1998; 95:5678–5683.
89. Ulrich HD, Jentsch S. Embo J 2000; 19:3388–3397.
90. Torres-Ramos CA, Prakash S, Prakash L. Mol Cell Biol 2002; 22:2419–2426.
91. Brusky J, Zhu Y, Xiao W. Curr Genet 2000; 37:168–174.
92. Fritsche J, Rehli M, Krause SW, Andreesen R, Kreutz M. Biochem Biophys Res Commun 1997; 235:407–412.
93. Sancho E, Vila MR, Sanchez-Pulido L, Lozano JJ, Paciucci R, Nadal M, Fox M, Harvey C, Bercovich B, Loukili N, Ciechanover A, Lin SL, Sanz F, Estivill X, Valencia A, Thomson TM. Mol Cell Biol 1998; 18:576–589.
94. Thomson TM, Khalid H, Lozano JJ, Sancho E, Arino J. FEBS Lett 1998; 423:49–52.
95. Xiao W, Lin SL, Broomfield S, Chow BL, Wei YF. Nucleic Acids Res 1998; 26:3908–3914.
96. Rothofsky ML, Lin SL. Gene 1997; 195:141–149.

97. Lawrence CW, Christensen RB. Genetics 1978; 90:213–226.
98. McDonald JP, Levine AS, Woodgate R. Genetics 1997; 147:1557–1568.
99. Ahne F, Jha B, Eckardt-Schupp F. Nucleic Acids Res 1997; 25:743–749.
100. Friedl AA, Liefshitz B, Steinlauf R, Kupiec M. Mutat Res 2001; 486:137–146.
101. Schurer KA, Rudolph C, Ulrich HD, Kramer W. Regulation of Damage Tolerance. Genetics 2004; 166:1673–1686.
102. Johnson RE, Henderson ST, Petes TD, Prakash S, Bankmann M, Prakash L. Mol Cell Biol 1992; 12:3807–3818.
103. Johnson RE, Prakash S, Prakash L. J Biol Chem 1994; 269:28,259–28,262.
104. Nelson JR, Lawrence CW, Hinkle DC. Science 1996; 272:1646–1649.
105. Lawrence CW, Maher VM. Philos Trans R Soc Lond B Biol Sci 2001; 356:41–46.
106. Lawrence CW, Christensen RB. J Bacteriol 1979; 139:866–876.
107. Quah SK, von Borstel RC, Hastings PJ. Genetics 1980; 96:819–839.
108. Roche H, Gietz RD, Kunz BA. Genetics 1994; 137:637–646.
109. Datta A, Jinks-Robertson S. Science 1995; 268:1616–1619.
110. Holbeck SL, Strathern JN. Genetics 1997; 147:1017–1024.
111. Roche H, Gietz RD, Kunz BA. Genetics 1995; 140:443–456.
112. Nelson JR, Lawrence CW, Hinkle DC. Nature 1996; 382:729–731.
113. Nelson JR, Gibbs PE, Nowicka AM, Hinkle DC, Lawrence CW. Mol Microbiol 2000; 37:549–554.
114. Guo C, Fischhaber PL, Luk-Paszyc MJ, Masuda Y, Zhou J, Kamiya K, Kisker C, Friedberg EC. Embo J 2003; 22:6621–6630.
115. Johnson RE, Prakash S, Prakash L. Science 1999; 283:1001–1004.
116. Roush AA, Suarez M, Friedberg EC, Radman M, Siede W. Mol Gen Genet 1998; 257:686–692.
117. Goodman MF. Annu Rev Biochem 2002; 71:17–50.
118. Hubscher U, Maga G, Spadari S. Annu Rev Biochem 2002; 71:133–163.
119. Masutani C, Kusumoto R, Yamada A, Dohmae N, Yokoi M, Yuasa M, Araki M, Iwai S, Takio K, Hanaoka F. Nature 1999; 399:700–704.
120. Washington MT, Johnson RE, Prakash S, Prakash L. J Biol Chem 1999; 274: 36835–36838.
121. Haracska L, Kondratick CM, Unk I, Prakash S, Prakash L. Mol Cell 2001; 8:407–415.
122. Ho Y, Gruhler A, Heilbut A, Bader GD, Moore L, Adams SL, Millar A, Taylor P, Bennett K, Boutilier K, Yang L, Wolting C, Donaldson I, Schandorff S, Shewnarane J, Vo M, Taggart J, Goudreault M, Muskat B, Alfarano C, Dewar D, Lin Z, Michalickova K, Willems AR, Sassi H, Nielsen PA, Rasmussen KJ, Andersen JR, Johansen LE, Hansen LH, Jespersen H, Podtelejnikov A, Nielsen E, Crawford J, Poulsen V, Sorensen BD, Matthiesen J, Hendrickson RC, Gleeson F, Pawson T, Moran MF, Durocher D, Mann M, Hogue CW, Figeys D, Tyers M. Nature 2002; 415:180–183.
123. McKenna S, Hu J, Moraes T, Xiao W, Ellison MJ, Spyracopoulos L. Biochemistry 2003; 42:7922–7930.
124. Deffenbaugh AE, Scaglione KM, Zhang L, Moore JM, Buranda T, Sklar LA, Skowyra D. Cell 2003; 114:611–622.
125. Jónsson ZO, Hübscher U. Bioessays 1997; 19:967–975.
126. Krishna TS, Kong XP, Gary S, Burgers PM, Kuriyan J. Cell 1994; 79:1233–1243.
127. Warbrick E. Bioessays 2000; 22:997–1006.
128. Varadan R, Assfalg M, Haririnia A, Raasi S, Pickart C, Fushman D. J Biol Chem 2004; 279:7055–7063.
129. Hofmann RM, Pickart CM. J Biol Chem 2001; 276:27,936–27,943.
130. Podlaska A, McIntyre J, Skoneczna A, Sledziewska-Gojska E. Mol Microbiol 2003; 49:1321–1332.
131. Melchior F. Annu Rev Cell Dev Biol 2000; 16:591–626.
132. Schwartz DC, Hochstrasser M. Trends Biochem Sci 2003; 28:321–328.
133. Johnson ES, Schwienhorst I, Dohmen RJ, Blobel G. Embo J 1997; 16:5509–5519.

134. Johnson ES, Blobel G. J Biol Chem 1997; 272:26799–26802.
135. Johnson ES, Gupta AA. Cell 2001; 106:735–744.
136. Bylebyl GR, Belichenko I, Johnson ES. J Biol Chem 2003; 278:44113–44120.
137. Ulrich HD. Cell Cycle 2004; 3:15–18.
138. Mossi R, Hubscher U. Eur J Biochem 1998; 254:209–216.
139. Galan JM, Haguenauer-Tsapis R. Embo J 1997; 16:5847–5854.
140. Spence J, Gali RR, Dittmar G, Sherman F, Karin M, Finley D. Cell 2000; 102:67–76.
141. Fisk HA, Yaffe MP. J Cell Biol 1999; 145:1199–1208.
142. Wang C, Deng L, Hong M, Akkaraju GR, Inoue J, Chen ZJ. Nature 2001; 412:346–351.
143. Kovalenko A, Chable-Bessia C, Cantarella G, Israel A, Wallach D, Courtois G. Nature 2003; 424:801–805.
144. Trompouki E, Hatzivassiliou E, Tsichritzis T, Farmer H, Ashworth A, Mosialos G. Nature 2003; 424:793–796.
145. Brummelkamp TR, Nijman SM, Dirac AM, Bernards R. Nature 2003; 424:797–801.
146. Deng L, Wang C, Spencer E, Yang L, Braun A, You J, Slaughter C, Pickart C, Chen ZJ. Cell 2000; 103:351–361.
147. Torres-Ramos CA, Prakash S, Prakash L. J Biol Chem 1997; 272:25,445–25,448.
148. Giot L, Chanet R, Simon M, Facca C, Faye G. Genetics 1997; 146:1239–1251.
149. Chilkova O, Jonsson BH, Johansson E. J Biol Chem 2003; 278:14,082–14,086.
150. Eissenberg JC, Ayyagari R, Gomes XV, Burgers PM. Mol Cell Biol 1997; 17:6367–6378.
151. Burgers PM, Gerik KJ. J Biol Chem 1998; 273:19756–19762.
152. Huang ME, de Calignon A, Nicolas A, Galibert F. Curr Genet 2000; 38:178–187.
153. Rong L, Palladino F, Aguilera A, Klein HL. Genetics 1991; 127:75–85.
154. Aboussekhra A, Chanet R, Zgaga Z, Cassier-Chauvat C, Heude M, Fabre F. Nucleic Acids Res 1989; 17:7211–7219.
155. Rong L, Klein HL. J Biol Chem 1993; 268:1252–1259.
156. Gangloff S, Soustelle C, Fabre F. Nat Genet 2000; 25:192–194.
157. Klein HL. Genetics 2001; 157:557–565.
158. Lee SK, Johnson RE, Yu SL, Prakash L, Prakash S. Science 1999; 286:2339–2342.
159. Krejci L, Van Komen S, Li Y, Villemain J, Reddy MS, Klein H, Ellenberger T, Sung P. Nature 2003; 423:305–309.
160. Veaute X, Jeusset J, Soustelle C, Kowalczykowski SC, Le Cam E, Fabre F. Nature 2003; 423:309–312.
161. McVey M, Kaeberlein M, Tissenbaum HA, Guarente L. Genetics 2001; 157:1531–1542.
162. Schiestl RH, Prakash S, Prakash L. Genetics 1990; 124:817–831.
163. Ulrich HD. Nucleic Acids Res 2001; 29:3487–3494.
164. Broomfield S, Xiao W. Nucleic Acids Res 2002; 30:732–739.
165. Ito T, Chiba T, Ozawa R, Yoshida M, Hattori M, Sakaki Y. Proc Natl Acad Sci USA 2001; 98:4569–4574.
166. Vaze MB, Pellicioli A, Lee SE, Ira G, Liberi G, Arbel-Eden A, Foiani M, Haber JE. Mol Cell 2002; 10:373–385.
167. Liberi G, Chiolo I, Pellicioli A, Lopes M, Plevani P, Muzi-Falconi M, Foiani M. Embo J 2000; 19:5027–5038.
168. Scheller J, Schurer A, Rudolph C, Hettwer S, Kramer W. Genetics 2000; 155:1069–1081.
169. Hishida T, Iwasaki H, Ohno T, Morishita T, Shinagawa H. Proc Natl Acad Sci USA 2001; 98:8283–8289.
170. Hishida T, Ohno T, Iwasaki H, Shinagawa H. Embo J 2002; 21:2019–2029.
171. Branzei D, Seki M, Onoda F, Enomoto T. Mol Genet Genomics 2002; 268:371–386.
172. Game JC, Kaufman PD. Genetics 1999; 151:485–497.
173. Martini EM, Keeney S, Osley MA. Genetics 2002; 160:1375–1387.

Part VI

DNA Strand Breaks

This section deals with the recognition and handling of strand breaks, primarily in eukaryotic organisms. DNA strand breaks are a frequent consequence of exogenous agents, such as ionizing radiation and radiomimetic chemicals, and of endogenous DNA transactions, such as replication or recombination. Especially double-strand breaks represent a type of lesion of high toxicity, with severe genetic consequences if unrepaired. The cell goes to great lengths to locate such lesions and to recruit various repair proteins.

One repair pathway exploits the existence of redundant sequence information on a homologous chromatid or chromosome—homologous recombination. Chapter 27 provides an overview of homologous recombination and discusses aspects of the dynamics of complex formation and damage recognition in the cellular context.

The alternative important pathway can be described formally as direct endjoining of nonhomologous ends. Immune system gene rearrangements depend heavily on this repair system which also allows the introduction of genetic variability. In this context, Chapter 28 reviews our current knowledge of the vertebrate system.

Next, Chapter 29 focuses on a major player in this system that recognizes and binds double-strand breaks in a sequence-independent manner and orchestrates downstream repair events—the Ku proteins and the DNA-dependent protein kinase.

The final step in this and many other repair pathways is DNA ligation; the various ligases and their mechanisms are discussed in Chapter 30.

Harder to classify is the next topic. The Rad50/Mre11/Nbs1 is an evolutionary conserved protein complex that plays multiple roles—in homologous recombination, end joining and checkpoint control. Chapter 31 provides an in-depth discussion and demonstrates how its unique structural features reflect its multiple functions.

In spite of all of these versatile proteins, the fast and highly sensitive process of initial double-strand break recognition remains puzzling. It has long been suspected that signals may emanate from chromatin alterations that may occur at a distance from the original damage. Chapter 32 summarizes our knowledge of mechanisms and significance of phosphorylation of H2AX, a fast-appearing modification that is detectable in the vicinity of a double-strand break and that seems to spread quickly over amazing distances.

Last but not least, we will have a look at single-strand breaks where poly(ADP) ribose polymerase provides an example of a highly abundant sensor that has been suspected to take part in a confusing array of activities. Chapter 33 discusses how structural features correlate with damage binding and activities. Additionally, it reviews our current level of understanding of its significance in repair.

Exchange of DNA strands between homologous DNA molecules via recombination ensures accurate genome duplication and preservation of genome integrity. Biochemical studies have provided insights into the molecular mechanisms by which homologous recombination proteins perform these essential tasks. More recent cell biological experiments are addressing the behavior of homologous recombination proteins in cells. The challenge ahead is to uncover the relationship between the individual biochemical activities of homologous recombination proteins and their coordinated action in the context of the living cell.

27

Biochemical and Cellular Aspects of Homologous Recombination

Lieneke van Veelen
Department of Radiation Oncology, Erasmus Medical Center-Daniel, Rotterdam, The Netherlands

Joanna Wesoly
Department of Cell Biology and Genetics, Erasmus Medical Center, Rotterdam, The Netherlands

Roland Kanaar
Departments of Radiation Oncology and Cell Biology and Genetics, Erasmus Medical Center, Rotterdam, The Netherlands

1. INTRODUCTION

Homologous recombination, the exchange of DNA sequences between two homologous DNA molecules, is essential for the preservation of genome integrity. It contributes to the repair of a wide range of DNA lesions, including DNA double-strand breaks (DSBs) and DNA interstrand crosslinks (1,2). In addition, homologous recombination plays a pivotal role in underpinning genome duplication, through its role in rebuilding DNA replication forks that have collapsed due to lesions in the template DNA (3). Homologous recombination is mediated by an extensive group of proteins that need to work together in a coordinated fashion. This cooperation is necessary to choreograph the complicated DNA gymnastics which is required to accurately restore DNA damage on one molecule using information of a second homologous DNA molecule.

An extensive number of biochemical studies on the enzymes that mediate homologous recombination have provided a number of working models of how the reaction can take place in the test tube (1,4). One important conclusion from these studies has been that the core of the process, homology recognition and DNA strand exchange is remarkably conserved throughout evolution. More recent cell biology studies have begun to address the behavior of homologous recombination proteins inside cells (5). The interesting challenge ahead is to link our understanding of the biochemical mechanisms of homologous recombination with its operation in the context of the living cell.

2. DNA DOUBLE-STRAND BREAK REPAIR THROUGH HOMOLOGOUS RECOMBINATION

The fundamentals of homologous recombination are highly conserved from phages to humans (6). For the sake of brevity, we consider here an example of one model for repair of a DSB by homologous recombination. More in depth discussions of different models for homologous recombination can be found in a number of extensive reviews (1,2,4). During DSB repair via homologous recombination, missing DNA is restored using the intact homologous sequence provided by the sister chromatid. In the early stage of the reaction, referred to as presynapsis, the DNA ends are processed into 3′ single-stranded overhang, by yet unidentified nucleases and/or helicases (Fig. 1). The single-stranded DNA tails are coated with a strand exchange protein to form a nucleoprotein filament (see below) that can recognize a homologous DNA sequence. During synapsis, the middle step of the recombination process, the nucleoprotein filament invades the homologous template DNA to form a joint heteroduplex molecule linking the broken end(s) and the undamaged template DNA. In the postsynaptic, or late stage of recombination, DNA polymerases restore the missing information and DNA ends are ligated. In this last step of the reaction, resolution of recombined molecules into separate DNA duplexes can be promoted by structure-specific endonucleases (7).

3. BIOCHEMICAL PROPERTIES OF HOMOLOGOUS RECOMBINATION PROTEINS

Below we discuss a number of proteins involved in homologous recombination, with emphasis on the proteins involved in synapsis, the central core reaction of homologous recombination, in which the joint molecule between the broken DNA and the intact repair template is established.

3.1. The Rad51 Protein

Rad51, conserved in all kingdoms of life, is a key protein in homologous recombination because it promotes homology recognition and DNA strand exchange. Biochemical studies have shown that Rad51 binds both single-stranded and double-stranded DNA (8). Its preferred substrate is single-stranded tailed duplex DNA, which resembles a DSB repair intermediate (9) (Fig. 1). Rad51 polymerizes on single-stranded DNA to form a nucleoprotein filament that is capable of recognizing homologous double-stranded DNA and promotes DNA strand exchange between the double-stranded template DNA and the Rad51-coated single-stranded DNA (10,11). Rad51-mediated joint molecule formation is stimulated by a number of accessory proteins; the single-stranded DNA binding protein RPA, Rad52, and Rad54 (10,11).

Cells from the yeast *Saccharomyces cerevisiae* that lack Rad51 are viable but display strongly reduced mitotic and meiotic recombination and are sensitive to ionizing radiation (12). In vertebrates, Rad51 is essential for cell proliferation. Depletion of Rad51 from chicken DT40 cells leads to accumulation of chromosomal abnormalities and cell death (13). Targeted disruption of Rad51 in mouse cells results in early embryonic lethality (14,15). Together these observations suggest that Rad51 plays an important role in proliferation processes.

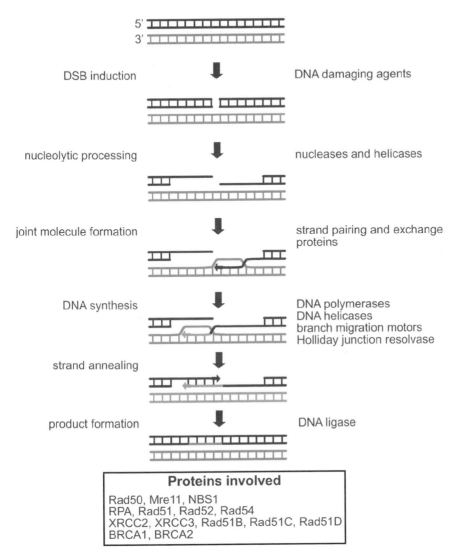

5'

3'

DSB induction   DNA damaging agents

nucleolytic processing   nucleases and helicases

joint molecule formation   strand pairing and exchange
                                       proteins

DNA synthesis   DNA polymerases
                                       DNA helicases
                                       branch migration motors
                                       Holliday junction resolvase

strand annealing

product formation   DNA ligase

Proteins involved

Rad50, Mre11, NBS1
RPA, Rad51, Rad52, Rad54
XRCC2, XRCC3, Rad51B, Rad51C, Rad51D
BRCA1, BRCA2

Figure 1 A model for DSB repair through homologous recombination. The black and gray double-stranded DNA, depicted as ladders, are homologous in sequence. A DSB can be generated by DNA damaging agents or replication of DNA containing a single-stranded break. The DSB is processed by the combined action of helicases and/or nucleases resulting in the generation of single-stranded DNA tails with a 3' overhang. The Mre11/Rad50/NBS1 complex has been implicated in this step, although its precise role is still unclear. The single-stranded DNA tail is bound by the Rad51 strand exchange protein to form a nucleoprotein filament. This filament can recognize homologous double-stranded DNA. DNA strand exchange generates a joint molecule between the homologous damaged DNA and undamaged DNA. In addition to Rad51, these steps require the coordinated action of the single-stranded DNA binding protein RPA (replication protein A), Rad52 and Rad54. The role of the five Rad51 paralogs, XRCC2, XRCC3, Rad51B, Rad51C, and Rad51D, as well as the function of the breast cancer susceptibility proteins Brca1 and Brca2 has not yet been defined in great detail. DNA synthesis, requiring a DNA polymerase, its accessory factors and a ligase, restores the missing information. Resolution of crossed DNA strands (Holliday junctions) by a resolvase yields two intact duplex DNAs. Only one pair of possible recombination products is depicted.

Rad51 interacts in vitro with a number of proteins involved in DSB repair, including RPA, Rad52, Rad54, and BRCA2. RPA is thought to remove the secondary structures on single-stranded DNA. BRCA2 was implicated in DSB repair recently and is thought to play a controlling role upstream of *Rad51* function in DNA strand exchange (16–21).

Both in yeast and in vertebrates paralogs of Rad51 have been identified. There are two mitotic Rad51 paralog proteins in *S. cerevisiae*; Rad55 and Rad57. They form a heterodimer that interacts with Rad51 and stimulates Rad51-mediated strand exchange. The recombination defect and ionizing radiation sensitivity of *Rad55* and *Rad57* mutants can be overcome by overexpression of Rad51 or Rad52 (22,23). In total, five paralogs have been discovered in mitotically dividing vertebrate cells; XRCC2, XRCC3, Rad51B, Rad51C, and Rad51D (24,25). The paralogs are present in two distinct complexes. One contains XRCC3 and Rad51C, while the other consists of XRCC2, Rad51B, Rad51C, and Rad51D (26,27). Disruption of the paralogs in chicken cells leads to chromosomal instability, moderately increased ionizing radiation sensitivity, significantly increased sensitivity to the cross-linking agent mitomycin C and it affects homologous recombination efficiency. Similarly to *Rad51*-deficient mice, targeted disruption of *Rad51B*, *Rad51D*, and *Xrcc2* results in embryonic lethality (24,25).

3.2. The Rad52 Protein

Rad52 is a central homologous recombination protein in the *S. cerevisiae*. *Rad52* mutants display the most severe recombination phenotype of all *RAD52* epistasis group mutants in yeast. The *Rad52* mutants are extremely sensitive to DNA damaging agents and almost completely deficient in all pathways of homology-mediated repair, including pathways that are independent of Rad51 (1,4). In vertebrates, *Rad52* mutants have only a two-fold decreased level of homologous recombination compared to wild type cells as measured by homologous gene targeting efficiency (28,29). It is possible that some of the functions of Rad52 in mammals can be taken over by Rad51 paralogs (6,30,31). Furthermore, although Rad52 homologs have been identified in the yeast *S. cerevisiae* and *Schizosaccharomyces pombe*, no such proteins have been identified in vertebrates (32,33).

In vitro Rad52 binds to single-stranded DNA, protects the ends from nucleolytic degradation and forms rings interacting with DNA (5). Rad52 also interacts with Rad51 and RPA and stimulates Rad51-mediated strand exchange by overcoming the inhibitory role of RPA (34,35).

3.3. The Rad54 Protein

Rad54 belongs to the SWI2/SNF2 family of proteins involved in many biological processes such as transcriptional activation and repression, destabilization of nucleosomes, DNA repair, and chromosome segregation (36). In general, these proteins function by modulating protein–DNA interactions. Rad54 is an important accessory factor for Rad51 (37). A number of biochemical characteristics of Rad54 have been well defined for different species ranging from yeast to humans. Rad54 is a double-stranded DNA-dependent ATPase with ability to change DNA topology and chromatin structure (38–41). Rad54 has been implicated to participate throughout the whole duration of the homologous recombination reaction by first stabilizing the Rad51 nucleoprotein filament, subsequently by stimulating Rad51-mediated joint molecule formation and chromatin remodeling. Finally, in the last stage of the reaction it could displace Rad51 from the product DNA (42).

Rad54-deficient mouse embryonic stem cells show increased sensitivity to ionizing radiation, mitomycin C and methanesulfonate and have a defect in homology-dependent DSB repair (42–44). Mice lacking Rad54 are viable (44). They are sensitive to the cross-linking agent mitomycin C (45). In S. cerevisiae, a homologue of RAD54–RDH54/TID1 has been identified (46,47). The proteins have similar biochemical properties (48,49). Yeast Rad54 and Tid1 promote Rad51-mediated joint molecule formation and have ability to modify DNA topology (48). Both Rad54 and Tid1 interact with Rad51 (46,48). In yeast, there is a functional overlap between both proteins, but Rad54 is more important in mitosis, where the sister chromatid is used as a template, while Tid1 is important in meiosis directing recombination towards the homologous chromosome (47,50,51). Recently, a human gene, termed Rad54B, sharing a significant homology to Rad54 has been isolated (52).

3.4. The Rad50/Mre11/NBS1 Protein Complex

The presence of Rad50/Mre11/NBS1 complex is required for proper functioning of DSB repair, although its role is still elusive (4) (see also Chapter 31). The complex consists of two proteins, Rad50 and Mre11, conserved from yeast to human, while the third subunit, NBS1 in mammals and Xrs2 in yeast, is less conserved at the amino acid level (53). In yeast, the complex is involved in nonhomologous DNA end joining, sister chromatid repair by homologous recombination, telomere maintenance and formation and processing of DSBs in meiosis (53). Biochemical analysis of Mre11 revealed its strand dissociation, strand annealing and 3'-5' exo/endo dsDNA nuclease activity properties (53). The Rad50/Mre11 complex has been shown to bind DNA ends and tether linear DNA molecules (54,55).

Conditional inactivation of Mre11 in chicken cells causes accumulation of chromosomal breaks, increased radiosensitivity, and reduced targeted integration frequencies (56). NBS1-deficient chicken cells display similar defects as the Mre11 knockout cells, with additional reduction of gene conversion levels and lower rates of sister chromatid exchanges (57). Mre11, Rad50, and NBS1 null mutations in mice lead to cellular and/or embryonic lethality indicating the importance of this complex for the function of the cell (58–60). Mutations in the Mre11 gene have been found in patients with ataxia telangiectasia-like disorder, while mutations in NBS1 cause the Nijmegen breakage syndrome (61,62). Cells derived from these patients display chromosomal instability and radioresistant DNA synthesis, which is an indication of a defective intra-S checkpoint. Indeed, mammalian Mre11 and NBS1 are phosphorylated by ATM, a cell cycle checkpoint protein that is crucial in the cellular response to DSBs in response to ionizing radiation (53). Ultraviolet light (UV), hydroxyurea or methylmethane sulfonate treatment also leads to phosphorylation of both proteins, most likely by the ATR kinase (63,64). The presence of the Rad50/Mre11/NBS1 complex is required for proper activation of checkpoints throughout all cell cycle phases (53). The complex could serve as a signal modifier by nucleolytic modification of the lesions in order to make them detectable by the checkpoint machinery.

3.5. The Brca1 and Brca2 Proteins

Mutations in the BRCA1 and BRCA2 breast cancer susceptibility genes predispose to breast, ovarian, prostate, and pancreatic cancer (65). Mouse Brca1- and

Brca2-deficient and human mutant cell lines display chromosomal instability and sensitivity to DNA damaging agents (66). Both proteins are required for homology-directed repair and gene targeting events. In comparison to wild type cells, gene targeting is respectively 20-fold and 2-fold decreased in Brca1 and Brca2 mutant cells (18,67). Similarly to *Rad51*, targeted disruption of *Brca1* and *Brca2* in mice leads to embryonic lethality, associated with a proliferation defect. This defect is partially suppressed by a p53 mutation (68–71). While Brca2 interacts directly with Rad51, the interaction between Brca1 and Rad51 appears to be indirect (21). Crystallographic data, characterizing the conserved BRC repeats and C-terminal single-strand DNA binding folds of Brca2 suggest that Brca2 can recruit Rad51 to a DSB and regulate the spatial distribution of Rad51 (20,72).

3.6. Other Proteins Involved in Homologous Recombination

Homologous recombination is a collection of complex processes. So far, not all proteins involved in these processes have been identified. The number of proteins known to participate in homologous recombination has significantly increased over the last years. The variety of substrates on which homologous recombination can act could explain the diversity of proteins required for its successful completion. Recently, a group of DNA helicases has been implicated in homologous recombination. Human homologs of the RecQ helicase in *E. coli*; Bloom, Werner, and Rothmund–Thomson proteins (BLM, WRN, and Recql 4, respectively), are thought to resolve abnormal replication structures after the replication forks stall or collapse (73). These proteins could also promote joint molecule formation and take part in the resolution of joint molecules. The three proteins are ATP-dependent 3′-5′ helicases which are able to unwind forked DNA structures and synthetic Holliday junctions in vitro (74,75). Mutations in BLM, WRN or the yeast RecQ homologue Sgs1, lead to chromosomal instability and an increased risk of tumor formation in patients (73). Patient derived cell lines accumulate abnormal replication intermediates (74). BLM mutant cells are characterized by hyperrecombination, visualized through increased numbers of sister chromatid exchanges. WRN mutant cells have increased levels of translocations and deletions (74). The BLM protein interacts with Rad51, Rad51D, and RPA; WRN protein with DNA-PK and RPA (76–81).

4. CELLULAR PROPERTIES OF HOMOLOGOUS RECOMBINATION PROTEINS

4.1. Local Accumulations of Proteins Involved in Homologous Recombination

In addition to unraveling the function of homologous recombination proteins in the test tube, it is also of great importance to understand the action mechanisms of these proteins in the context of the cell, where they have to function in chromatin and compete with other DNA metabolic processes. The response to DNA damage of a number of proteins involved in homologous recombination has been visualized inside cells using immunofluorescence (5). Many of the homologous recombination proteins studied to date, including Rad51, Rad52, Rad54, Brca1, Brca2, and Rad50/Mre11/NBS1 accumulate into subnuclear structures at sites of DNA damage (82–86). These subnuclear structures are referred to as foci. An overview of a number of proteins detected in foci is given in Table 1.

(*Text continued on page 597.*)

Table 1 Overview of Proteins Involved in Spontaneous or Induced Foci Formation

Protein	Spontaneous foci formation	Cell cycle dependency of spontaneous or induced foci	Damage-induced foci formation	Co-localization			
				Before treatment	After treatment		
					Complete	Partial	None
Core homologous recombination protein							
RPA	Yes	S-phase G2-phase	IR UV CPT	WRN BLM PML-NB Brca1 Rad51	WRN Rad51 Brca1 ATR	Brca1	BrdU[3]
Rad51	Yes	S-phase G2-phase	IR MMS MMC UV-C CPT HU Cisplatin Etoposide Tetracyclin	RPA Rad54B BLM Brca1 Fanc-D2 p53	RPA Rad54B γ-H2AX Rad54 PML-NB BrdU[4]	Rad52 PCNA WRN BLM phospho-p53 Brca2	Mre11 Rad50 Nbs1 MDC1
Rad52			IR MMS			Rad50 Rad51	
Rad54			IR		Rad51 Rad54B		
Rad54B	Yes		IR	Rad51	Rad51 Rad54	Brca1	
Rad51 paralogs	No		No				

(Continued)

Table 1 Overview of Proteins Involved in Spontaneous or Induced Foci Formation (*Continued*)

Protein	Spontaneous foci formation	Cell cycle dependency of spontaneous or induced foci	Damage-induced foci formation	Co-localization Before treatment	Complete	Partial	None
Brca proteins							
Brca1	Yes	S-phase	IR MMC HU UV MMS	Bard1 Cds1 Fanc-D2 Rad51 RPA	Bard1 γ-H2AX RPA Bach1 Fanc-D2 MDC1	Rad54B Nbs1 BLM PCNA Mre11 Rad50 Rad51	
Brca2			IR MMC HU MMS UV		Rad51 Fanc-C		
BARD1			HU	Brca1	Brca1	PCNA	
BACH1		S-phase G2-phase	HU		Brca1		
Mre11 complex							
Mre11	Yes	Staining pattern depends on cell cycle	IR HU CPT MMC	TRF1 BrdU PCNA PML-NB	Rad50 Nbs1 53BP1 PCNA Fanc-D2 γ-H2AX MDC1	Brca1 PML-NB p53 p21	Rad51

Rad50	Yes	No cell cycle dependency	IR HU CPT MMC MMS UV	TRF1/2 PML-NB	Mre11 Nbs1 γ-H2AX	Brca1 Rad52 PCNA PML-NB p53	Rad51
Nbs1	Yes		IR HU CPT MMC AS NCS	γ-H2AX TRF2 PML-NB	Mre11 Rad50 53BP1 Fanc-D2 γ-H2AX	Brca1 ATM	Rad51
MDC1	Yes[1] No[2]		IR UV Phleomycin		Mre11 Nbs1 γ-H2AX 53BP1 Brca1 Chk2		Rad51
DNA helicases WRN	Yes	S-phase	IR HU UVC CPT Etoposide Bleomycin 4NQO	BLM	RPA ATR	Rad51 BdrU	
BLM	Yes	S-phase	IR HU Etoposide	PML-NB Rad51 RPA phospho-p53 PCNA WRN	PML-NB Rad51 BrdU p53	Brca1 Mre11 Rad50 Rad51 RPA PCNA	

(Continued)

Table 1 Overview of Proteins Involved in Spontaneous or Induced Foci Formation (*Continued*)

Protein	Spontaneous foci formation	Cell cycle dependency of spontaneous or induced foci	Damage-induced foci formation	Co-localization			
				Before treatment	After treatment		
					Complete	Partial	None
Cell cycle proteins							
p21		G0-phase G1-phase	LET		Mre11		PCNA
p53	Rare	S-phase	IR APH HU	Rad51	Mre11 PML-NB BLM	Rad51 BrdU PCNA	
53BP1	Yes		IR MMC CPT 4NQO MMS Etoposide Neomycin VM26 UV HU IR[1] no[2]		γ-H2AX Mre11 Nbs1 MDC1		
CHK2			IR[1] NCS[1] no[2]		MDC1		
ATM	No		IR HU UVC CPT APH			γ-H2AX Nbs1	
ATR				PML-NB	RPA WRN Brca1		

Other						
H2AX	Yes	S-phase[g] All phases[h]	IR NCS UV HU CPT	Rad51 Rad50 Brca1 Mre11	Rad51 Rad50 Mre11 Nbs1 Brca1 53BP1 PML-NB p53 MDC1 Mre11	ATM PCNA
PCNA		S-phase	IR HU UV	Mre11 γ-H2AX	Mre11	Brca1 BLM Rad50 BARD1 BrdU γ-H2AX BLM p53
Fanc-D2	Yes	S-phase	IR MMC UV	Brca1 Mre11 Nbs1 Rad51	Brca1 Mre11 Nbs1	
PML-NB	Nuclear bodies		Nuclear bodies	Mre11	Mre11 γ-H2AX p53 Rad50	
Cds1	Yes		No foci upon IR	Brca1	Brca1	

(*Continued*)

Table 1 Overview of Proteins Involved in Spontaneous or Induced Foci Formation (*Continued*)

Protein	Absence	Foci formation influenced by a mutation					No influence on foci formation	Comments	References
		Decrease		Delay	Increase				
		No. of foci-positive cells	No. of foci per nucleus	No. of foci-positive cells	No. of foci-positive cells	No. of foci per nucleus			
Core homologous recombination proteins									
RPA	ATR				Brca1[a]	Brca1[a]	WRN ATM DNA-PK Brca2	Upon IR, not UV[a]	78,92–99
Rad51	Xrcc2 Xrcc3 Rad51B Rad51C Rad51D Brca1[1] Brca2[10] PML	LigaseIV WRN[b] Fanconi (all 8 groups)[5] Fanc-D2[6] Fanc-A/C/G[7] p21 c-Abl MSH2[b] Arg		Fanc-A/C/G[7] AT	ATM DNA-PK Rad54 BLM[9] MSH2[a] WRN[a] Xrcc4	ATM Xrcc4	p53 Brca1[2] Mre11 Nbs1 Rad52 H2AX Fanconi (all 8 groups except Fanc-D2) 53BP1 p53 BLM[6]	[1]100,101 [2]102,103 [3]83 [4]94 [5]104 [6]105 [7]106 [8]107 [9]94,108 [10]S-phase foci are present [a]spontaneous foci. [b]damage induced foci.	29,56,82,83,85, 88,92–94,96, 99–141
Rad52	c-Abl								88,117,142,143
Rad54									
Rad54B					ATM				88,112,113,116
Rad51 paralogs									113

Brca proteins

Protein					References
Brca1	H2AX, Brca1, 53BP1, MDC1	53BP1, MDC1	Fanc-C[1]	Nbs1[1], DNA-PK[2], ATM, BLM, Brca2, Fanc (all 8 groups)[2], Brca1	87,91,97,99, 101,102,104– 106,113,114, 119,125,132, 136,137,139, 141,144–157
Brca2					99,140,141
BARD1	Brca1				152,158
BACH1					149
Mre11 complex					
Mre11 — Nbs1, Mre11, Ku70, Ku80, BLM[c], ATRkd[10], Fanc-A/C/G[8,e], MSH2, H2AX, MDC1	ATM[6], Brca1[1], Fanc-D2[3]		LigaseIV	DNA-PK[1], p53[2], Brca1[2], ATM[5], Fanc (all 8 groups)[4], BLM[d], WRN[7], Fanc-D1[7], Fanc-D2[7], XP-F, 53BP1, ATR[9]	61,62,85,87, 102,104– 106,110,120, 135,138,148, 153,154,156, 159–173
Rad50 — Nbs1, H2AX	Brca1[1]			Brca2[1], Brca1[2]	61,102,103, 110,114,117, 120,134,142, 148,153,165, 166,168,170

Footnote legends:

Brca1: [1]106; [2]104

Mre11: [1]102; [2]87,148; [3]159; [4]104,105; [5]120,153,154,156; [6]110; [7]173; [8]106; [9]156; [10]153; [c]Upon HU, not IR; [d]Upon IR, not HU; [e]Upon MMC, not IR

Rad50: [1]102; [2]148

(Continued)

Table 1 Overview of Proteins Involved in Spontaneous or Induced Foci Formation (*Continued*)

Protein	Foci formation influenced by a mutation							Comments	References
	Decrease		Delay	Increase			No influence on foci formation		
	Absence	No. of foci-positive cells	No. of foci per nucleus	No. of foci-positive cells	No. of foci-positive cells	No. of foci per nucleus			
Nbs1	Mre11 Nbs1 Rag1/2 H2AX BLM[c] Fanc-C/G/A[e]	Brca1[1] MDC1	MDC1				Brca1[2] BLM Fanc-D1 Fanc-A/C/G[3,f] 53BP1 ATM	[1] 102 [2] 87,148 [3] 106 [c] Upon HU, not IR [d] Upon IR, not HU [e] Upon MMC, not IR [f] Upon IR, not MMC	61,62,87,91, 102,105,106, 120,125,136, 139,148,153, 154,157,161, 163,165,166, 170–177
MDC1	H2AX						ATM ATR Mre11 Nbs1 DNA-PK 53BP1 Chk2 Brca1	[1] 157 [2] 138	138,155,157, 178
DNA helicases									
WRN	WRN			ATRkd ATM ATR		Telomerase ATM ATR Telomerase	ATM DNA-PK p53		78,96,179,180 94,107,108,133, 148,153,179
BLM									

Protein					Notes	References
Cell cycle proteins						
p21						84,169
p53						107,133,168
53BP1	BLM, ATM, MDC1	MDC1	DNA-PK[g]	Nbs1, XPA, p53, ATM	[g] before treatment	125,138,151, 154,157,163, 172,181,182
H2AX	53BP1		ATM	Nbs1, ATM, p53, DNA-PK, MDC1		91,139,178
CHK2					[1] 139,178 · [2] 91 · Opposite results on Chk2 foci formation have been reported, depending on the antibody used	
ATM	MDC1	MDC1	Ligase IV	Nbs1[g]	[1] 155,176 · [2] 147,180	147,155,176, 180
ATR	ATRkd					98,147,180
Other						
H2AX ATR Top1[i]	DNA-PK, MDC1	MDC1	Nbs1[g], DNA-PK[g]	ATM, HUS1, 53BP1, Rad50, Mre11, Nbs1, MDC1	[g] untreated cells · [h] Upon IR · [i] Upon HU, CPT and UV	91,114,136– 139,151,155– 157,163,168, 170,174,176, 178,181,183

(Continued)

Table 1 Overview of Proteins Involved in Spontaneous or Induced Foci Formation (*Continued*)

		Foci formation influenced by a mutation							
		Decrease		Delay		Increase			
Protein	Absence	No. of foci-positive cells	No. of foci per nucleus	No. of foci-positive cells	No. of foci-positive cells	No. of foci per nucleus	No influence on foci formation	Comments	References
PCNA									107,148,151, 156,158,167, 169,184
Fannc-D2	Fanc-A/C/G Brca1						Mre11 Nbs1		132,150,173
PML-NB									168
Cds1									145

The table shows an overview of recombination protein known to form spontaneous or damage-induced subnulcear structures called foci in mitotically dividing cells. For this overview the literature from 1995 until August 2003 was surveyed. Co-localization with other proteins of interest and the influence of mutant or absent proteins on foci formation of a specific other protein is shown. In cases where contradictory results have been reported, that reference is given in the Comments column.

[a]Spontaneous foci.
[b]Damage-induced foci.
[c]Upon HU, not IR.
[d]Upon IR, not HU.
[e]Upon MMC, not IR.
[f]Upon IR, not MMC.
[g]Untreated cells.
[h]Upon IR.
[i]Upon HU, CPT, and UV.

Besides this accumulation at sites of induced DNA damage, proteins involved in homologous recombination can also be observed in foci in cells that have not been treated with exogenous DNA damaging agents. These so-called "spontaneous" foci occur specifically in S-phase cells. They represent the cytological manifestation of the link between DNA replication and homologous recombination. DSBs occur during DNA replication, for example when imperfections in the DNA template are encountered. This can lead to replication fork arrest and breakdown. The resulting DSB intermediates are acted upon by homologous recombination factors that rebuild a functional replication fork (3). Usually, these DNA replication associated foci have a similar appearance as the DNA damage-induced foci of the same protein. However, cells show less spontaneous foci per nucleus probably because there are less spontaneous DSBs than DNA damage-induced DSBs.

4.2. Detection of Foci by Antibodies in Chemically Fixed Cells

Immunostaining is a commonly used method of detecting nuclear foci of proteins of interest. After treatment of cells by damaging agents, cells are fixed, permeabilized, and foci can be detected by a fluorescently labeled antibody specific for the protein of interest. An example of ionizing radiation-induced Rad51 foci is displayed in Fig. 2A. Immunostaining experiments can be done for many cell types. The method is relatively simple and fast, though the results depend on various factors, such as the fixation and permeabilization techniques and the cell lines and antibodies used. This variability complicates the interpretation of published data on DNA damage-induced foci formation (87).

4.3. Detection of Foci by Expression of the Protein of Interest Tagged to a Fluorescent Group

Another approach to study foci formation is by stably transfecting the cDNA of the protein of interest tagged to a fluorescent group, such as one of a number of the spectral variants of the green fluorescent protein (GFP). After ascertaining that this tagged protein is functioning similarly as the endogenous protein, the transfected cells can be studied by fluorescence microscopy (Fig. 2B). In this manner, the behavior of the protein can be observed before and after damage induction. Foci of the tagged proteins are smaller in size than foci observed by immunostaining, though their number is usually not different (88). An advantage of studying foci formation by expression of the protein of interest is that it can be done in both living and fixed cells. The unfixed cells can even be used to study the dynamic behavior of the protein in time (see below). The main disadvantage of this technique is the fact that not all cell lines are suitable for stable transfection.

4.4. Colocalization of Proteins in Foci

DNA damage-induced formation of nuclear foci implicates the involvement of certain proteins during the process of DNA repair. Moreover, a possible cooperation of specific proteins in the repair of DSBs can be studied by investigating foci formation of two or more proteins at the same time in the same cell. Colocalization of foci suggests an association between the proteins of interest. They may be part of the same DNA repair complex or participate in the same cascade of proteins that are essential

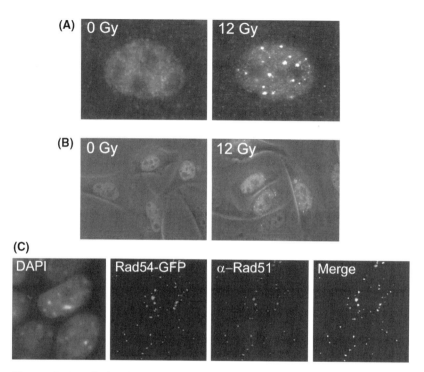

Figure 2 Irradiation-induced foci formation of homologous recombination protein. (A) Rad51 ionizing radiation-induced foci by antibody detection in fixed cells. Chinese hamster ovary (CHO) cells were irradiated with 12 Gy and chemically fixed after 2 hr. Immunostaining was performed using antibodies against Rad51. After ionizing radiation treatment Rad51 foci appear in the nucleus. (B) Rad52 ionizing radiation-induced foci detected by expression of Rad52-GFP in living cells. CHO cells expressing Rad52-GFP were irradiated with 12 Gy and investigated after 2 hr. Using a fluorescence microscope, ionizing radiation-induced Rad52 foci can be observed in living cells. (C) Colocalization of Rad51 and Rad54 ionizing radiation-induced foci. Wild type CHO cell lines expressing Rad54 fused to GFP were irradiated with 12 Gy and fixed at 2 hr after irradiation. Cells were counterstained with a Rad51 antibody. The first panel from the left shows the nuclei of the cells, visualized through DAPI staining. The second and third panels show ionizing radiation-induced Rad54 and Rad51 foci detected through the GFP signal and the fluorescently labeled antibody, respectively. The last panel shows the merged images, resulting in a yellow focus in case of Rad51 and Rad54 colocalization. For Rad51 and Rad54 the colocalization is virtually complete. (*See color insert.*)

for repair of DSBs. Colocalization of foci can be complete (Fig. 2C) or partial, which may provide information about the cooperation of the proteins. It is also possible that the two proteins of interest are mutually exclusive with respect to their presence in foci. An explanation for this phenomenon might be the recruitment of specific repair proteins at different stages of the cell cycle. An example is provided in Fig. 3. While ionizing radiation-induced Rad51 foci are observed in replicating cells, Mre11 foci are detected in cells outside of S-phase. Though colocalization experiments do not provide any information about the actual interaction of the proteins of interest, the results can lead to the suggestion whether specific proteins may or may not cooperate in the repair of DSBs.

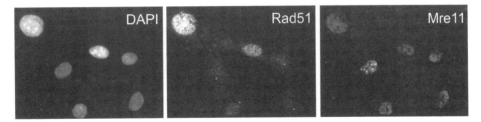

Figure 3 Rad51 and Mre11 ionizing radiation-induced foci formation depends on cell cycle stage. Primary human fibroblasts were irradiated with 12 Gy and incubated for 8 hr before fixation. Double immunostaining was performed using antibodies against Rad51 and Mre11. The nuclei were visualized by DAPI staining. A representative picture of each staining pattern is shown. Cells positive for Rad51 foci do usually lack Mre11 foci and vice versa. Note that cells which are positive for Rad51 IRIFs display a brighter DAPI staining, indicating that they are replicative cells.

4.5. Foci Formations in DNA Repair Deficient Mutant Cell Lines

In addition to colocalization experiments, DNA repair deficient mutant cell lines can be used to establish a possible cooperation between DSB repair proteins. Cell lines with a defect in one of the DNA repair proteins may show less or more spontaneous or damage-induced foci per nucleus, depending on the repair pathway that is diminished. An increase in number of foci might be due to the inability of the cell to repair the spontaneous occurring DSBs. A decrease in number of foci may occur in case the protein of interest, or one of its cooperating proteins, is not functioning properly. This may lead to impaired complex formation at the site of the DSB, thus preventing an accumulation of the protein of interest (Table 1). Interestingly, even though Rad51 foci do not form in, for example, the Rad51 paralog mutant cell lines in response to induced DNA damage, these mutant cell lines are capable of forming the DNA replication associated spontaneous Rad51 foci. Possibly, the accessory proteins to Rad51 are not absolutely required for Rad51 foci under all circumstances. However, given that most of the mutants are hypomorphic, the proteins still retain parts of their functions. Consistent with this idea is the finding that complete knockouts for these proteins result in embryonic lethality in mice (89).

4.6. Nuclear Dynamics of Homologous Recombination Proteins in Living Cells

Cell lines expressing the protein of interest tagged to a GFP spectral variant may be utilized to study the dynamic behavior of the protein in living cells using a confocal microscope. The principle of studying the dynamic behavior is based on the rate of recovery of the fluorescent signal of the protein in an area that has been bleached by a short laser pulse (Fig. 4A) (90). Fluorescence redistribution after photobleaching (FRAP) can be used to determine the diffusion rate of proteins by bleaching a small strip spanning the entire nucleus. Recovery of the fluorescence in the strip is monitored at specific time intervals. The kinetics with which the fluorescence intensity in the strip reaches the same intensity as the unbleached area relates to the diffusion rate of a protein (Fig. 4B). Fluorescence redistribution after photobleaching can also be used to study the residence times of specific proteins in the DNA damage-induced foci. In this case, a single focus is bleached and the time interval between bleaching and

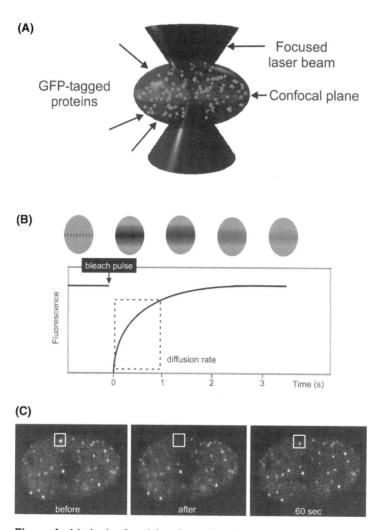

Figure 4 Methods of studying the nuclear dynamics of proteins. (A) The principle of fluorescence redistribution after photobleaching (FRAP). The diffusion of a protein and the fraction of mobile proteins can be determined by bleaching a small area in living cells expressing the protein of interest tagged to a fluorescent group using a focused laser beam. The recovery of fluorescence in the bleached are can be measured. This gives an indication of the mobility of proteins in the nucleus. Fluorescence redistribution after photobleaching analysis can be done for freely moving proteins (B) or proteins that are bound to DNA (C). (B) The principle of strip bleaching. This technique can be used to determine the diffusion rate of recombination proteins in living cells. A bleach pulse of 200 ms is given such that only a small strip spanning the entire nucleus will be bleached. After this bleach pulse the recovery of fluorescence in the strip is monitored at intervals of 100 ms. After a while the fluorescence intensity in the strip will have reached the same level as the unbleached area. This represents influx of the GFP-tagged proteins from the surroundings into the bleached area. In this way the diffusion rate of a GFP-tagged protein can be measured and plotted as is shown in the graph. (C) Fluorescence redistribution after photobleaching for Rad52-GFP ionizing radiation-induced foci. A similar analysis as described in (B) can be done for DNA damage-induced foci. A single focus in the nucleus is bleached and the time interval between bleaching and recovery of the focus is measured. The picture shows an example for a Rad52-GFP focus before bleaching, immediately after bleaching and after 60 sec. Recovery of the fluorescence at the damaged site can be observed within 60 sec. The time to recovery of the focus relates to the residence time of the protein at the damaged site. Various repair proteins have different residence times in DNA damage-induced foci.

recovery of fluorescence focus is measured. The time to recovery can be used to estimate the residence time of that specific protein at the damaged site (Fig. 4C) (88,91).

4.7. Implications of Analyses of Homologous Recombination Proteins in Living Cells

Investigation of the nuclear dynamics of DNA repair proteins has provided new insight in the recognition mechanisms of DNA damage and the interaction between mechanistically distinct DNA repair pathways. For example, Rad52 and Rad54 diffuse through the nucleus (88). Diffusion ensures that the proteins are everywhere in the nucleus all of the time which is a useful property of proteins that need to repair DSBs that can occur anywhere in the genome. Furthermore, diffusion rate measurements of Rad52 and Rad54 showed that these proteins have different diffusion rates before the induction of DNA damage (88). Because the two proteins diffuse through the nucleus independently of each other, they cannot be part of the same preassembled holo-complex in the absence of DNA damage. Possibly, the affinity of the Rad52 group proteins for the DSB site compared to intact DNA might be slightly increased. This difference in affinity ensures that the Rad52 group proteins will be immobilized for a longer time at the DSB site than at other sites in the genome, resulting in a local accumulation or focus at the site of DNA damage. The large local concentration of the different proteins ensures that the reaction that each of them mediate can be driven to completion. Performing repair of DNA lesions by diffusable proteins that are temporarily immobilized due to the encounter of sites of increased affinity has an important advantage over the use of preassembled holo-complexes. In situ assembly allows greater flexibility in the components of a DNA repair complex. Because different components can reversibly interact with the DNA damage-induced structure, the correct components required for repair of a specific lesion can be selected. The reversible interaction of proteins with DNA damage-induced foci alleviates the necessity of having to dissemble a DNA repair holo-complex that does not contain all of the specialized components required to repair the lesion it is associated with. Furthermore, in situ assembly allows exchange of components between different multistep DNA repair pathways. Multistep DNA repair pathways can mix and match components, instead of linear DNA repair pathways that each uses a defined set of enzymes. This cross talk is biologically significant because it will lead to an increase in the diversity of DNA lesions that can be repaired.

ABBREVIATIONS

APH—aphidicolin
AS—arsenic
CPT—Camptothecin
HU—hydroxyurea
IR—ionizing radiation
kd—kinase dead
LET—low energy transfer
MMC—mitomycin C
MMS—methylmethanesulfonate
NCS—neocarzinostatin
PML-NB—promyelocytic leukemia nuclear bodies

UV—ultraviolet radiation
VM26—teniposide
4NQO—4-nitroquinoline 1-oxide.

ACKNOWLEDGMENT

The authors wish to thank Jeroen Essers for kindly providing a number of images used in the figures.

REFERENCES

1. Symington LS. Microbiol Mol Biol Rev 2002; 66:630–670.
2. Dronkert ML, Kanaar R. Mutat Res 2001; 486:217–247.
3. Cox MM, Goodman MF, Kreuzer KN, Sherratt DJ, Sandler SJ, Marians KJ. Nature 2000; 404:37–41.
4. Paques F, Haber JE. Microbiol Mol Biol Rev 1999; 63:349–404.
5. West SC. Nat Rev Mol Cell Biol 2003; 4:435–445.
6. Modesti M, Kanaar R. Genome Biol 2001; 2:1014.1–1014.5.
7. Heyer WD, Ehmsen KT, Solinger JA. Trends Biochem Sci 2003; 28:548–557.
8. Baumann P, West SC. Trends Biochem Sci 1998; 23:247–251.
9. Mazin AV, Bornarth CJ, Solinger JA, Heyer WD, Kowalczykowski SC. Mol Cell 2000; 6:583–592.
10. Sung P, Trujillo KM, Van Komen S. Mutat Res 2000; 451:257–275.
11. Bianco PR, Tracy RB, Kowalczykowski SC. Front Biosci 1998; 3:570–603.
12. Game JC. Semin Cancer Biol 1993; 4:73–83.
13. Sonoda E, Sasaki MS, Buerstedde JM, Bezzubova O, Shinohara A, Ogawa H, Takata M, Yamaguchi-Iwai Y, Takeda S. Embo J 1998; 17:598–608.
14. Lim DS, Hasty P. Mol Cell Biol 1996; 16:7133–7143.
15. Tsuzuki T, Fujii Y, Sakumi K, Tominaga Y, Nakao K, Sekiguchi M, Matsushiro A, Yoshimura Y, Morita T. Proc Natl Acad Sci USA: 1996; 93:6236–6240.
16. Shinohara A, Ogawa T. Mutat Res 1999; 435:13–21.
17. Davies AA, Masson JY, McIlwraith MJ, Stasiak AZ, Stasiak A, Venkitaraman AR, West SC. Mol Cell 2001; 7:273–282.
18. Moynahan ME, Pierce AJ, Jasin M. Mol Cell 2001; 7:273–282.
19. Orelli BJ, Bishop DK. Breast Cancer Res 2001; 3:294–298.
20. Pellegrini L, Yu DS, Lo T, Anand S, Lee M, Blundell TL, Venkitaraman AR. Nature 2002; 420:287–293.
21. Venkitaraman AR. J Cell Sci 1995; 92:3591–3598.
22. Hays SL, Firmenich AA, Berg P. Proc Natl Acad Sci USA 1995; 92:6925–6929.
23. Johnson RD, Symington LS. Mol Cell Biol 1995; 15:4843–4850.
24. Thacker J. Trends Genet 1999; 15:166–168.
25. Schild D, Lio YC, Collins DW, Somondo T, Chen DJ. J Biol Chem 2000; 275: 16443–16449.
26. Masson JY, Tarsounas MC, Stasiak AZ, Stasiak A, Shah R, McIlwraith MJ, Benson FE, West SC. Genes Dev 2001; 15:3296–3307.
27. Sigurdsson S, S Van Komen, Bussen W, Schild D, Albala JS, Sung P. Genes Dev 2001; 15:3308–3318.
28. Rijkers T, J Van Den Ouweland, Morolli B, Rolink AG, Baarends WM, PP Van Sloun, Lohman PH, Pastink A. Mol Cell Biol 1998; 18:6423–6429.
29. Yamaguchi-Iwai Y, Sonoda E, Buerstedde JM, Bezzubova O, Morrison C, Takata M, Shinohara A, Takeda S. Mol Cell Biol 1998; 18:6430–6435.

30. Fujimori A, Tachiiri S, Sonoda E, Thompson LH, Dhar PK, Hiraoka M, Takeda S, Zhang Y, Reth M, Takata M. Embo J 2001; 20:5513–5520.

31. Sung P. Genes Dev 1997; 11:1111–1121.

32. van den Bosch M, Vreeken K, Zonneveld JB, Brandsma JA, Lombaerts M, Murray JM, Lohman PH, Pastink A. Mutat Res 2001; 461:311–323.

33. Bai Y, Symington LS. Genes Dev 1996; 10:2025–2037.

34. Sugiyama T, Kowalczykowski SC. J Biol Chem 2002; 277:31663–31672.

35. Hiom K. Curr Biol 1999; 9:446–448.

36. Pazin MJ, Kadonaga JT. Cell 1997; 88:737–740.

37. Tan RTL, Kanaar R, Wyman C. DNA Repair (Amst) 2003; 2:787–794.

38. Alexeev A, Mazin A, Kowalczykowski SC. Nat Struct Biol 2003; 10:182–186.

39. Alexiadis V, Kadonaga JT. Genes Dev 2002; 16:2767–2771.

40. Jaskelioff M, Van Komen S, Krebs JE, Sung P, Peterson CL. J Biol Chem 2003; 278:9212–9218.

41. Ristic D, Wyman C, Paulusma C, Kanaar R. Proc Natl Acad Sci USA 2001; 98: 8454–8460.

42. Solinger JA, Kiianitsa K, Heyer WD. Mol Cell 2002; 10:1175–1188.

43. Dronkert ML, Beverloo HB, Johnson RD, Hoeijmakers JH, Jasin M, Kanaar R. Mol Cell Biol 2000; 20:3147–3156.

44. Essers J, Hendriks RW, Swagemakers SM, Troelstra C, de Wit J, Bootsma D, Hoeijmakers JH, Kanaar R. Cell 1997; 89:195–204.

45. Essers J, van Steeg H, de Wit J, Swagemakers SM, Vermeij M, Hoeijmakers JH, Kanaar R. Embo J 2000; 19:1703–1710.

46. Dresser ME, Ewing DJ, Conrad MN, Dominguez AM, Barstead R, Jiang H, Kodadek T. Genetics 1997; 147:533–544.

47. Klein HL. Genetics 1997; 147:1533–1543.

48. Petukhova G, Sung P, Klein H. Genes Dev 2000; 14:2206–2215.

49. Tanaka K, Kagawa W, Kinebuchi T, Kurumizaka H, Miyagawa K. Nucleic Acids Res 2002; 30:1346–1353.

50. Arbel A, Zenvirth D, Simchen G. Embo J 1999; 18:2648–2658.

51. Shinohara M, Shita-Yamaguchi E, Buerstedde JM, Shinagawa H, Ogawa H, Shinohara A. Genetics 1997; 147:1545–1556.

52. Hiramoto T, Nakanishi T, Sumiyoshi T, Fukuda T, Matsuura S, Tauchi H, Komatsu K, Shibasaki Y, Inui H, Watatani M, Yasutomi M, Sumii K, Kajiyama G, Kamada N, Miyagawa K, Kamiya K. Oncogene 1999; 18:3422–3426.

53. Amours DD, Jackson SP. Nat Rev Mol Cell Biol 2002; 3:317–327.

54. Chen L, Trujillo K, Ramos W, Sung P, Tomkinson AE. Mol Cell 2001; 8:1105–1115.

55. de Jager M, van Noort J, van Gent DC, Dekker C, Kanaar R, Wyman C. Mol Cell 2001; 8:1129–1135.

56. Yamaguchi-Iwai Y, Sonoda E, Sasaki MS, Morrison C, Haraguchi T, Hiraoka Y, Yamashita YM, Yagi T, Takata M, Price C, Kakazu N, Takeda S. Embo J 1999; 18:6619–6629.

57. Tauchi H, Matsuura S, Kobayashi J, Sakamoto S, Komatsu K. Oncogene 2002; 21:8967–8980.

58. Zhu J, Petersen S, Tessarollo L, Nussenzweig A. Curr Biol 2001; 11:105–109.

59. Xiao Y, Weaver DT. Nucleic Acids Res 1997; 25:2985–2991.

60. Luo G, Yao MS, Bender CF, Mills M, Bladl AR, Bradley A, Petrini JH. Proc Natl Acad Sci USA 1999; 96:7376–7381.

61. Carney JP, Maser RS, Olivares H, Davis EM, M Le Beau, JR Yates III, Hays L, Morgan WF, Petrini JH. Cell 1998; 93:477–486.

62. Stewart GS, Maser RS, Stankovic T, Bressan DA, Kaplan MI, Jaspers NG, Raams A, Byrd PJ, Petrini JH, Taylor AM. Cell 1999; 99:577–587.

63. Lim DS, Kim ST, Xu B, Maser RS, Lin J, Petrini JH, Kastan MB. Nature 2000; 404:613–617.

64. Gatei M, Young D, Cerosaletti KM, A Desai-Mehta, Spring K, Kozlov S, Lavin MF, Gatti RA, Concannon P, Khanna K. Nat Genet 2000; 25:115–119.

65. Venkitaraman AR. Cell 2002; 108:171–182.

66. Scully R, Puget N, Vlasakova K. Oncogene 2000; 19:6176–618.

67. Moynahan ME, Chiu JW, Koller BH, Jasin M. Mol Cell 1999; 4:511–518.

68. Gowen LC, Johnson BL, Latour AM, Sulik KK, Koller BH. Nat Genet 1996; 12: 191–194.

69. Hakem R, de la Pompa JL, Sirard C, Mo R, Woo M, Hakem A, Wakeham A, Potter J, Reitmair A, Billia F, Firpo E, Hui CC, Roberts J, Rossant J, Mak TW. Cell 1996; 85:1009–1023.

70. Ludwig T, Chapman DL, Papaioannou VE, Efstratiadis A. Genes Dev 1997; 11: 1226–1241.

71. Suzuki A, de la Pompa JL, Hakem R, Elia A, Yoshida R, Mo R, Nishina H, Chuang T, Wakeham A, Itie A, Koo W, Billia P, Ho A, Fukumoto M, Hui CC, Mak TW. Genes Dev 1997; 11:1242–1252.

72. Yang H, Jeffrey PD, Miller J, Kinnucan E, Sun Y, Thoma NH, Zheng N, Chen PL, Lee WH, Pavletich NP. Science 2002; 297:1837–1848.

73. Bachrati CZ, Hickson ID. Biochem J 2002; 374:577–606.

74. Chakraverty RK, Hickson ID. Bioessays 1999; 21:286–294.

75. Karow JK, Wu L, Hickson ID. Curr Opin Genet Dev 2000; 10:32–38.

76. Braybrooke JP, Li JL, Wu L, Caple F, Benson FE, Hickson ID. J Biol Chem 2003; 15:48357–48366.

77. Brosh RM Jr, Li JL, Kenny MK, Karow JK, Cooper MP, Kureekattil RP, Hickson ID, Bohr VA. J Biol Chem 2000; 275:23500–23508.

78. Constantinou A, Tarsounas M, Karow JK, Brosh RM, Bohr VA, Hickson ID, West SC. EMBO Rep 2000; 1:80–84.

79. Cooper MP, Machwe A, Orren DK, Brosh RM, Ramsden D, Bohr VA. Genes Dev 2000; 14:907–912.

80. Wu L, Hickson ID. Science 2001; 292:229–230.

81. Xia SJ, Shammas MA, Shmookler Reis RJ. Mol Cell Biol 1997; 17:7151–7158.

82. Raderschall E, Golub EI, Haaf T. Proc Natl Acad Sci USA 1999; 96:1921–1926.

83. Haaf T, Golub EI, Reddy G, Radding CM, Ward DC. Proc Natl Acad Sci USA 1995; 92:2298–2302.

84. Jakob B, Scholz M, Taucher-Scholz G. Radiat Res 2000; 154:398–405.

85. Nelms BE, Maser RS, MacKay JF, Lagally MG, Petrini JH. Science 1998; 280:590–592.

86. Tashiro S, Walter J, Shinohara A, Kamada N, Cremer T. J Cell Biol 2000; 150:283–291.

87. Wu X, Petrini JH, Heine WF, Weaver DT, Livingston DM, Chen J. Science 2000; 289:11.

88. Essers J, Houtsmuller AB, van Veelen L, Paulusma C, Nigg AL, Pastink A, Vermeulen W, Hoeijmakers JH, Kanaar R. Embo J 2002; 21:2030–2037.

89. Thompson LH, Schild D. Mutat Res 2001; 477:131–153.

90. Houtsmuller AB, Vermeulen W. Histochem Cell Biol 2001; 115:13–21.

91. Lukas C, Falck J, Bartkova J, Bartek J, Lukas J. Nat Cell Biol 2003; 5:255–260.

92. Gasior SL, Wong AK, Kora Y, Shinohara A, Bishop DK. Genes Dev 1998; 12:2208–2221.

93. Golub EI, Gupta RC, Haaf T, Wold MS, Radding CM. Nucleic Acids Res 1998; 26:5388–5393.

94. Bischof O, Kim SH, Irving J, Beresten S, Ellis NA, Campisi J. J Cell Biol 2001; 153: 367–380.

95. MacPhail SH, Olive PL. Radiat Res 2001; 155:672–679.

96. Sakamoto S, Nishikawa K, Heo SJ, Goto M, Furuichi Y, Shimamoto A. Genes Cells 2001; 6:421–430.

97. Choudhary SK, Li R. J Cell Biochem 2002; 84:666–674.

98. Barr SM, Leung CG, Chang EE, Cimprich KA. Curr Biol 2003; 13:1047–1051.

99. Tarsounas M, Davies D, West SC. Oncogene 2003; 22:1115–1123.

100. Bhattacharyya A, Ear US, Koller BH, Weichselbaum RR, Bishop DK. J Biol Chem 2000; 275:23899–23903.

101. Huber LJ, Yang TW, Sarkisian CJ, Master SR, Deng CX, Chodosh LA. Mol Cell Biol 2001; 21:4005–4015.

102. Zhong Q, Chen CF, Li S, Chen Y, Wang CC, Xiao J, Chen PL, Sharp ZD, Lee WH. Science 1999; 285:747–750.

103. Yuan SS, Lee SY, Chen G, Song M, Tomlinson GE, Lee EY. Cancer Res 1999; 59:3547–3551.

104. Digweed M, Rothe S, Demuth I, Scholz R, Schindler D, Stumm M, Grompe M, Jordan A, Sperling K. Carcinogenesis 2002; 23:1121–1126.

105. Godthelp BC, Artwert F, Joenje H, Zdzienicka MZ. Oncogene 2002; 21:5002–5005.

106. Pichierri P, Averbeck D, Rosselli F. Hum Mol Genet 2002; 11:2531–2546.

107. Sengupta S, Linke SP, Pedeux R, Yang Q, Farnsworth J, Garfield SH, Valerie K, Shay JW, Ellis NA, Wasylyk B, Harris CC. Embo J 2003; 22:1210–1222.

108. Wu L, Davies SL, Levitt NC, Hickson ID. J Biol Chem 2001; 276:19375–19381.

109. Tashiro S, Kotomura N, Shinohara A, Tanaka K, Ueda K, Kamada N. Oncogene 1996; 12:2165–21670.

110. Maser RS, Monsen KJ, Nelms BE, Petrini JH. Mol Cell Biol 1997; 17:6087–6096.

111. Bishop DK, Ear U, Bhattacharyya A, Calderone C, Beckett M, Weichselbaum RR, Shinohara A. J Biol Chem 1998; 273:21482–21488.

112. Morrison C, Sonoda E, Takao N, Shinohara A, Yamamoto K, Takeda S. Embo J 2000; 19:463–471.

113. Tanaka K, Hiramoto T, Fukuda T, Miyagawa K. J Biol Chem 2000; 275:26316–26321.

114. Paull TT, Rogakou EP, Yamazaki V, Kirchgessner CU, Gellert M, Bonner WM. Curr Biol 2000; 10:886–895.

115. Morrison C, Shinohara A, Sonoda E, Yamaguchi-Iwai Y, Takata M, Weichselbaum RR, Takeda S. Mol Cell Biol 1998; 19:6891–6897.

116. Tan TL, Essers J, Citterio E, Swagemakers SM, de Wit J, Benson FE, Hoeijmakers JH, Kanaar R. Curr Biol 1999; 9:325–328.

117. Liu Y, Maizels N. EMBO Rep 2000; 1:85–90.

118. Takata M, Sasaki MS, Sonoda E, Fukushima T, Morrison C, Albala JS, Swagemakers SM, Kanaar R, Thompson LH, Takeda S. Mol Cell Biol 2000; 20:6476–6482.

119. Scully R, Chen J, Plug A, Xiao Y, Weaver D, Feunteun J, Ashley T, Livingston DM. Cell 1997; 88:265–275.

120. Mirzoeva OK, Petrini JH. Mol Cell Biol 2001; 21:281–288.

121. O'Regan P, Wilson C, Townsend S, Thacker J. J Biol Chem 2001; 276:22148–22153.

122. Pichierri P, Franchitto A, Mosesso P, Palitti F. Mol Biol Cell 2001; 12:2412–2421.

123. Pichierri P, Franchitto A, Piergentili R, Colussi C, Palitti F. Carcinogenesis 2001; 22: 1781–1787.

124. Takata M, Sasaki MS, Tachiiri S, Fukushima T, Sonoda E, Schild D, Thompson LH, Takeda S. Mol Cell Biol 2001; 21:2858–2866.

125. Celeste A, Petersen S, Romanienko PJ, Fernandez-Capetillo O, Chen HT, Sedelnikova OA, Reina-San-Martin B, Coppola V, Meffre E, Difilippantonio MJ, Redon C, Pilch DR, Olaru A, Eckhaus M, Camerini-Otero RD, Tessarollo L, Livak F, Manova K, Bonner WM, Nussenzweig MC, Nussenzweig A. Science 2002; 296:922–927.

126. Delacote F, Han M, Stamato TD, Jasin M, Lopez BS. Nucleic Acids Res 2002; 30: 3454–3463.

127. Godthelp BC, Wiegant WW, van Duijn-Goedhart A, Scharer OD, van Buul PP, Kanaar R, Zdzienicka MZ. Nucleic Acids Res 2002; 30:2172–2182.

128. Kraakman-van der Zwet M, Overkamp WJ, van Lange RE, Essers J, van Duijn-Goedhart A, Wiggers I, Swaminathan S, Van Buul PP, Errami A, Tan RT, Jaspers NG, Sharan SK, Kanaar R, Zdzienicka MZ. Mol Cell Biol 2002; 22:669–679.

129. Li Y, Shimizu H, Xiang SL, Maru Y, Takao N, Yamamoto K. Biochem Biophys Res Commun 2002; 299:697–702.

130. Liu N. J Biomed Biotechnol 2002; 2:106–113.
131. Raderschall E, Bazarov A, Cao J, Lurz R, Smith A, Mann W, Ropers HH, Sedivy JM, Golub EI, Fritz E, Haaf T. J Cell Sci 2002; 115:153–164.
132. Taniguchi T, Garcia-Higuera I, Andreassen PR, Gregory RC, Grompe M, D'Andrea AD. Blood 2002; 100:24114–24120.
133. Yang Q, Zhang R, Wang XW, Spillare EA, Linke SP, Subramanian D, Griffith JD, Li JL, Hickson ID, Shen JC, Loeb LA, Mazur SJ, Appella E, Brosh RM Jr, Karmakar P, Bohr VA, Harris CC. J Biol Chem 2002; 277:31980–31987.
134. Yuan SS, Chang HL, Lee EY. Mutat Res 2003; 525:85–92.
135. Franchitto A, Pichierri P, Piergentili R, Crescenzi M, Bignami M, Palitti F. Oncogene 2003; 22:2110–2120.
136. Petersen S, Casellas R, Reina-San-Martin B, Chen HT, Difilippantonio MJ, Wilson PC, Hanitsch L, Celeste A, Muramatsu M, Pilch DR, Redon C, Ried T, Bonner WM, Honjo T, Nussenzweig MC, Nussenzweig A. Nature 2001; 414:660–665.
137. Bassing CH, Chua KF, Sekiguchi J, Suh H, Whitlow SR, Fleming JC, Monroe BC, Ciccone DN, Yan C, Vlasakova K, Livingston DM, Ferguson DO, Scully R, Alt FW. Proc Natl Acad Sci USA 2002; 99:8173–8178.
138. Goldberg M, Stucki M, Falck J, D'Amours D, Rahman D, Pappin D, Bartek J, Jackson SP. Nature 2003; 421:952–956.
139. Wang B, Matsuoka S, Carpenter PB, Elledge SJ. Science 2002; 298:1435–1438.
140. Shin DS, Pellegrini L, Daniels DS, Yelent B, Craig L, Bates D, Yu DS, Shivji MK, Hitomi C, Arvai AS, Volkmann N, Tsuruta H, Blundell TL, Venkitaraman AR, Tainer JA. EMBO J 2003; 22:4566–4576.
141. Chen J, Silver DP, Walpita D, Cantor SB, Gazdar AF, Tomlinson G, Couch FJ, Weber BL, Ashley T, Livingston DM, Scully R. Mol Cell 1998; 2:317–328.
142. Liu Y, Li M, Lee EY, Maizels N. Curr Biol 1999; 9:975–978.
143. Kitao H, Yuan ZM. J Biol Chem 2002; 277:48944–48948.
144. Jin Y, Xu XL, Yang MC, Wei F, Ayi TC, Bowcock AM, Baer R. Proc Natl Acad Sci USA 1997; 94:12075–12080.
145. Lee JS, Collins KM, Brown AL, Lee CH, Chung JH. Nature 2000; 404:201–204.
146. Cortez D, Wang Y, Qin J, Elledge SJ. Science 1999; 286:1162–1166.
147. Tibbetts RS, Cortez D, Brumbaugh KM, Scully R, Livingston D, Elledge SJ, Abraham RT. Genes Dev 2000; 14:2989–3002.
148. Wang Y, Cortez D, Yazdi P, Neff N, Elledge SJ, Qin J. Genes Dev 2000; 14:927–939.
149. Cantor SB, Bell DW, Ganesan S, Kass EM, Drapkin R, Grossman S, Wahrer DC, Sgroi DC, Lane WS, Haber DA, Livingston DM. Cell 2000; 105:149–160.
150. Garcia-Higuera I, Taniguchi T, Ganesan S, Meyn MS, Timmers C, Hejna J, Grompe M, D'Andrea AD. Mol Cell 2001; 7:249–262.
151. Ward IM, Chen J. J Biol Chem 2001; 276:47759–47762.
152. Fabbro M, Rodriguez JA, Baer R, Henderson BR. J Biol Chem 2002; 277:21315–21324.
153. Franchitto A, Pichierri P. J Cell Biol 2002; 157:19–30.
154. Celeste A, Fernandez-Capetillo O, Kruhlak MJ, Pilch DR, Staudt DW, Lee A, Bonner RF, Bonner WM, Nussenzweig A. Nat Cell Biol 2003; 5:675–679.
155. Lou Z, Chini CC, K Minter-Dykhouse, Chen J. J Biol Chem 2003; 278:13599–13602.
156. Mirzoeva OK, Petrini JH. Mol Cancer Res 2003; 1:207–218.
157. Stewart GS, Wang B, Bignell CR, Taylor AM, Elledge SJ. Nature 2003; 421:961–966.
158. Scully R, Chen J, Ochs RL, Keegan K, Hoekstra M, Feunteun J, Livingston DM. Cell 1997; 90:425–435.
159. Digweed M, Demuth I, Rothe S, Scholz R, Jordan A, Grotzinger C, Schindler D, Grompe M, Sperling K. Oncogene 2002; 21:4873–4878.
160. Goedecke W, Eijpe M, Offenberg HH, van Aalderen M, Heyting C. Nat Genet 1999; 23:194–198.
161. Paull TT, Gellert M. Mol Cell 1998; 1:969–979.
162. Dong Z, Zhong Q, Chen PL. J Biol Chem 1999; 274:19513–19516.

163. Schultz LB, Chehab NH, Malikzay A, Halazonetis TD. J Cell Biol 2000; 151: 1381–1390.
164. Zhao S, Renthal W, Lee EY. Nucleic Acids Res 2002; 30:4815–4822.
165. Zhu XD, Kuster B, Mann M, Petrini JH, de Lange T. Nat Genet 2001; 25:347–352.
166. Desai-Mehta A, Cerosaletti KM, Concannon P. Mol Cell Biol 2001; 21:2184–2191.
167. Maser RS, Mirzoeva OK, Wells J, Olivares H, Williams BR, Zinkel RA, Farnham PJ, Petrini JH. Mol Cell Biol 2001; 21:6006–6016.
168. Carbone R, Pearson M, Minucci S, Pelicci PG. Oncogene 2002; 21:1633–1640.
169. Jakob B, Scholz M, G Taucher-Scholz. Int J Radiat Biol 2002; 78:75–88.
170. Furuta T, Takemura H, Liao ZY, Aune GJ, Redon C, Sedelnikova OA, Pilch DR, Rogakou EP, Celeste A, Chen HT, Nussenzweig A, Aladjem MI, Bonner WM, Pommier Y. J Biol Chem 2003; 278:20303–20312.
171. Lee JH, Xu B, Lee CH, Ahn JY, Song MS, Lee H, Canman CE, Lee JS, Kastan MB, Lim DS. Mol Cancer Res 2003; 1:674–681.
172. Anderson L, Henderson C, Adachi Y. Mol Cell Biol 2001; 21:1719–1729.
173. Nakanishi K, Taniguchi T, Ranganathan V, New HV, Moreau LA, Stotsky M, Mathew CG, Kastan MB, Weaver DT, D'Andrea AD. Nat Cell Biol 2002; 4:913–920.
174. Chen HT, Bhandoola A, Difilippantonio MJ, Zhu J, Brown MJ, Tai X, Rogakou EP, Brotz TM, Bonner WM, Ried T, Nussenzweig A. Science 2000; 290:1962–1965.
175. Zhao S, Weng YC, Yuan SS, Lin YT, Hsu HC, Lin SC, Gerbino E, Song MH, Zdzienicka MZ, Gatti RA, Shay JW, Ziv Y, Shiloh Y, Lee EY. Nature 2000; 405: 473–477.
176. Andegeko Y, Moyal L, Mittelman L, Tsarfaty I, Shiloh Y, Rotman G. J Biol Chem 2001; 276:38224–38230.
177. Yuan SS, Su JH, Hou MF, Yang FW, Zhao S, Lee EY. DNA Repair (Amst) 2002; 1:137–142.
178. Lou Z, Minter-Dykhouse K, Wu X, Chen J. Nature 2003; 421:957–961.
179. von Kobbe C, Karmakar P, Dawut L, Opresko P, Zeng X, Brosh RM Jr, Hickson ID, Bohr VA. J Biol Chem 2002; 277:22035–22044.
180. Pichierri P, Rosselli F, Franchitto A. Oncogene 2003; 22:1491–1500.
181. Rappold I, Iwabuchi K, Date T, Chen J. J Cell Biol 2001; 153:613–620.
182. Ward IM, Minn K, Jorda KG, Chen J. J Biol Chem 2003; 278:19579–19582.
183. Rothkamm K, Lobrich M. Proc Natl Acad Sci USA 2003; 100:5057–5062.
184. Balajee AS, Geard CR. Nucleic Acids Res 2001; 29:1341–1351.

28

The Mechanism of Vertebrate Nonhomologous DNA End Joining and Its Role in Immune System Gene Rearrangements

Michael R. Lieber, Yunmei Ma and Kefei Yu
USC Norris Comprehensive Cancer Center, University of Southern California Keck School of Medicine, Los Angeles, California, U.S.A.

Ulrich Pannicke and Klaus Schwarz
Department of Transfusion Medicine, Institute for Clinical Transfusion Medicine and Immunogenetics, University of Ulm, Ulm, Germany

1. INTRODUCTION

The vertebrate immune system utilizes two different double-strand DNA breakage strategies to permit the gene rearrangements necessary to generate the antigen receptor repertoire of lymphocytes. After those double-strand breaks have been created, the DNA joinings required to complete the process are carried out by the nonhomologous DNA end joining pathway, or NHEJ. The NHEJ pathway is present not only in lymphocytes, but in all eukaryotic cells ranging from yeast to humans. The NHEJ pathway is needed to repair these physiologic breaks, as well as challenging pathologic breaks that arise from ionizing radiation and oxidative damage to DNA.

2. ESSENTIAL ASPECTS OF VERTEBRATE NONHOMOLOGOUS DNA END JOINING (NHEJ)

Pathologic double-strand DNA (dsDNA) breaks arise when ionizing radiation in the environment passes near the double-helix (1). In addition, oxidative metabolism produces reactive oxygen species that can produce nicks or double-strand breaks in DNA (2,3). Replication across a nick can produce a dsDNA break. Because these pathologic events arise frequently (4), there is a need for pathways that can repair dsDNA breaks.

One pathway for the repair of dsDNA breaks is homologous recombination (5) (Fig. 1). This pathway requires that the cell be diploid for the region that requires

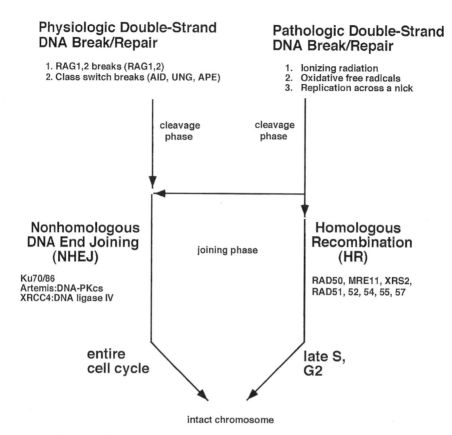

Figure 1 Physiologic and pathologic double-strand DNA breaks and their pathways for rejoining. Physiologic dsDNA breaks only occur in lymphocytes as part of V(D)J recombination and class switch recombination. RAG1 and 2 proteins along with HMG1 form a complex that cleaves sites in V(D)J recombination. The cleavage events in class switch recombination involve activation-induced deaminase (AID), uracil DNA glycosylase (UDG), and abasic endonuclease (APE1) (see Fig. 6). Pathologic DNA breaks can result from ionizing radiation, oxidative free radicals, and replication across a nick. Homologous recombination in vertebrates is limited to late S and G2 of the cell cycle. NHEJ can occur at any time during the cell cycle. The repair of physiologic DNA breaks relies on NHEJ. In V(D)J recombination, this is because the breaks occur during G1 of the cell cycle. For class switch recombination, the cell cycle dependence of the breaks has not been fully defined, but CSR does not occur in Ku null mice and is not normal in DNA-PKcs null mice.

repair. In vertebrate organisms, this pathway is restricted to late S and G_2 of the cell cycle (6). The second pathway, NHEJ, is active throughout the cell cycle, including late S and G_2 (6). Hence, NHEJ is thought to repair the majority of dsDNA breaks.

The chemical configuration at both pathologic and physiologic dsDNA breaks is highly diverse. The NHEJ pathway is ideally suited to handle the diverse chemical nature of dsDNA breaks. This is because the proteins and enzymes involved in NHEJ recognize DNA ends based on their structure rather than on their sequence. This theme is illustrated throughout the mechanistic steps of NHEJ.

The first step in NHEJ is thought to be the binding of Ku to the two DNA ends (see also Chapter 29) (Fig. 2). Ku is a heterodimer, consisting of Ku70 and 86 and has

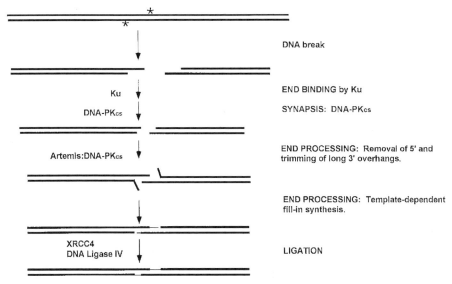

Figure 2 The steps and proteins in the vertebrate nonhomologous DNA end joining pathway. When a double-stranded DNA break occurs, the density of Ku in the nucleus and its high affinity are such that Ku is likely to bind before anything else occurs. The ends must be held in proximity to permit subsequent repair steps to proceed and to align the two ends. This step can be referred to as synapsis. Though Ku alone may not be able to achieve synapsis, there is some evidence that DNA-PKcs is capable of noncovalently holding on to each of two DNA ends. For complex DNA ends, as would be generated by ionizing radiation, there is often need for nucleolytic processing. If so, Artemis:DNA-PKcs are likely to carry out most, and perhaps all, such processing. For only a subset of DNA ends, end alignment can occur at chance sites of terminal microhomology of 1–4 nucleotides. This is an optional aspect, as NHEJ occurs regardless of microhomology. End processing not only includes nucleolytic removal of sections of the DNA termini, but also the filling in of gaps by polymerases. Ligation is the final step, and it requires a ligatable nick on each strand. Ligation in NHEJ is done by the XRCC4:DNA ligase IV complex. Though this diagram proposes the events at a complex pair of DNA ends (not blunt or incompatible), simpler events may occur at blunt or compatible ends or at ends that do not require a nuclease or a polymerase (see Fig. 7).

a toroidal structure. The 20 angstrom hole through the center of Ku permits it to thread onto any DNA end (7). Functional studies suggest that long single-stranded arms projecting off both strands at the DNA end (a pseudo-Y) are not an impediment to Ku (8). Based on this, one could conjecture that Ku may thread onto only one of the strands of such a pseudo-Y DNA end, and this may be understandable in light of the action of the nuclease that acts in the next step.

The second step is the recruitment of the nuclease. In vertebrate organisms, Artemis (9) and DNA-PKcs form a complex (10). Artemis appears to have a weak nuclease activity of its own, but this is limited to $5'$ to $3'$ exonuclease activity. DNA-PKcs has no nuclease activity but has serine/threonine protein kinase activity when bound to a DNA end (11). DNA-PKcs is the only protein kinase that requires a DNA end as a cofactor. Within eukaryotic cells, some fraction of the Artemis and DNA-PKcs populations of molecules are bound to one another in a very stable complex (10). Ku is able to recruit DNA-PKcs to DNA ends (12–14), and presumably this applies to the Artemis:DNA-PKcs complex. Once bound to a DNA end, DNA-PKcs phosphorylates itself and Artemis (10). One of these two phosphorylation targets (or perhaps both) permits the Artemis:DNA-PKcs complex to function as an endonuclease, allowing it to cleave both 5′ and 3′ overhangs of any length (10) (Fig. 3). Once trimming has occurred, the nuclease, Artemis:DNA-PKcs, may dissociate from the Ku:DNA complex (see below) to permit the next step.

In the third step, the XRCC4:DNA ligase IV complex is recruited to the DNA end by Ku (15,16). This complex ligates the ends (17–20). Additional details, points of uncertainty, and avenues for future work on NHEJ are discussed below. But first, we briefly summarize how NHEJ functions in physiologic double-strand breaks, specifically V(D)J recombination and class switch recombination, which normally only occur in lymphocytes.

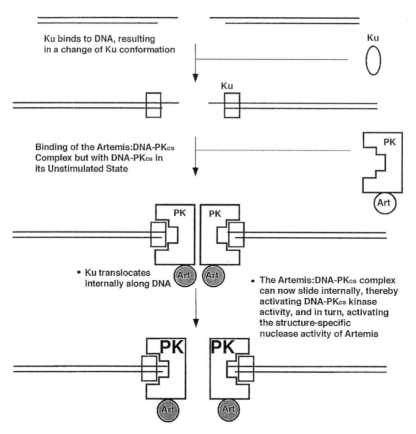

Figure 3 Physical model for the function of the Artemis:DNA-PKcs complex. Ku loads on to DNA at a double-stranded DNA end. This may result in a change in Ku conformation because now the Ku:DNA complex binds to the DNA-dependent protein kinase catalytic subunit (DNA-PKcs) with an affinity that is 100-fold higher than the affinity of DNA for DNA-PKcs directly. In vivo, Artemis is in complex with DNA-PKcs. Hence, it is likely that the recruitment of DNA-PKcs to the Ku:DNA end also recruits Artemis with it. Once DNA-PKcs contacts within the Ku:DNA end complex, it becomes active as a protein kinase. Artemis is a key phosphorylation target of this kinase activity. This permits the Artemis: DNA-PKcs complex to be active as an endonuclease.

3. OVERVIEW OF V(D)J RECOMBINATION AND ITS UTILIZATION OF NHEJ IN THE REJOINING PROCESS

V(D)J recombination is a specialized double-strand break and rejoining pathway that generates the exon that encodes the variable domain for immunogloblulins and T-cell receptors. Joining events in each pre-B and pre-T cell are the basis of the antigen-specific immune system of vertebrates (21). The RAG proteins, which generate the DNA breaks, are only expressed in early lymphoid progenitors. The ends are joined in a new configuration by NHEJ in the joining phase.

The RAG-dependent (DNA cleavage) phase begins when the RAG1, RAG2, and HMG1 proteins form a complex that binds at the recombination signal sequences (RSS, depicted as filled triangles) adjacent to each V, D, or J element (depicted as open rectangles in Fig. 4A). The RAG complex nicks the DNA at each signal (22,23). One nicked 12-signal (next to a V segment, for example, at the kappa light chain gene locus) and one nicked 23-signal (next to a J segment at the J segment of the kappa light chain locus) are brought into synapsis by the RAG complex. The 3′OH at each nick is then used as the nucleophile to attack the antiparallel strand at each site to generate a DNA hairpin at each V, D, or J element. The Ku protein is presumed to bind to some or all of the four DNA ends, perhaps thereby displacing the RAG complex from the two signal ends and allowing their ligation (Fig. 4B) (4,24). The Artemis:DNA-PKcs complex, probably recruited to the codings ends by Ku, opens the hairpinned V, D, or J ends (10). The Artemis:DNA-PKcs complex then acts endonucleolytically (and perhaps Artemis exonucleolytically) to trim the V, D, or J ends to variable extents, thereby contributing to junctional diversity (10). A template-dependent polymerase is likely to fill in gaps (Fig. 4C). If terminal deoxynucleotidyl transferase is present, as it is in B- and T-cell lymphoid progenitors, then template-independent DNA synthesis markedly contributes to the junctional diversity at the coding end. Finally, the XRCC4:DNA ligase IV complex carries out the ligation of the two DNA ends, as in all NHEJ (17–20) (see also Chapter 30).

The signal ends are joined together by the XRCC4:DNA ligase IV complex also (17). The lack of signal joint formation in Ku null cells may be because Ku is needed to recruit the ligase complex (15,16); additionally, it has been conjectured that Ku is needed to displace the RAGs (25). The signal end joining does not require the Artemis:DNA-PKcs complex, though some abnormalities of signal joints may be found in cells defective for this complex (26,27). This may be because blunt ends do not require nucleolytic processing and, hence, Ku and XRCC4:DNA ligase IV may be sufficient (see below).

The RAG complex almost certainly evolved from a transposase because the chemistry of how the RAGs cleave is indistinguishable from that of the well-characterized retrotransposons (28,29). These types of reactions begin with a nick at the end or at the edge of a recognition sequence. Then the 3′OH of the nick is used as a nucleophile in a transesterification reaction. For transposons, the nucleophile is used to attack anywhere in the genome. For the RAG complex, the nucleophile is used to attack the phosphodiester backbone directly opposite the nick, resulting in the hairpin. Some of the prokaryotic transposons also utilize a hairpin intermediate. Hence, the chemistry of the RAG cleavage is likely to have evolved from a transposase (22,23).

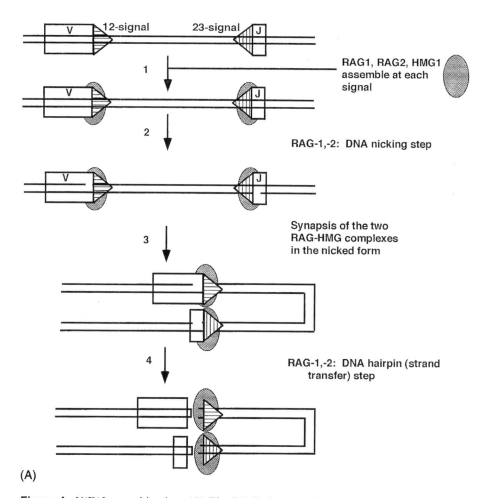

(A)

Figure 4 V(D)J recombination. (A) The RAG cleavage phase. RAG1, RAG2, and HMG1 form a complex that binds at the recombination signal sequences, often abbreviated RSS. The stoichiometry of this complex is still an active area of investigation but may be [(RAG1)2(RAG2)2HMG1]. The signal sequences consist of a palindromic heptamer and an AT-rich nonamer separated by either 12 or 23 bp (V coding end-CACAGTG-(12/23 bp) ACAAAAACC). Initially the RAG complex creates a nick adjacent to the coding end side of the heptamer. Then the two nicked species must be brought into synapsis before the nicked coding ends can be converted to double-strand breaks. The RAGs achieve this by using the 3'OH as a nucleophile to attack the opposite strand to create hairpinned coding ends at the end of the V, D, or J ends. (B) Displacement of the RAG complex and the hairpin opening step. Ku binds to one or both of the coding ends. Ku recruits the Artemis:DNA-PK$_{cs}$ complex. The Artemis:DNA-PK$_{cs}$ complex opens the hairpins. Ku binding to the signals displaces the RAG complex, thereby permitting the ligation of the two signal ends (see below). (C, see p. 616) Coding end processing and ligation of the ends. The two signal ends may be ligated by XRCC4:DNA ligase IV as soon as Ku displaces the RAGs. The two coding ends may be treated like any two DNA ends being processed by the full NHEJ pathway with the only exception being the participation of a template-independent polymerase called terminal transferase or TdT. Any nucleotide trimming is likely to be done by the Artemis:DNA-PKcs complex and gap fill-in (template-dependent) synthesis may be done by any of a number of polymerases. Polymerase μ has been shown to associate with the DNA:Ku:XRCC4:DNA ligase IV complex, and pol μ null mice have subtle alterations in the processing of some of their coding ends. The ligase for the coding ends is also the XRCC4:DNA ligase IV complex. The stoichiometry of the XRCC4:DNA ligase IV complex is not yet determined but has been conjectured to be four XRCC4 molecules and two DNA ligase IV molecules. *Source*: From Refs. 59,61,96.

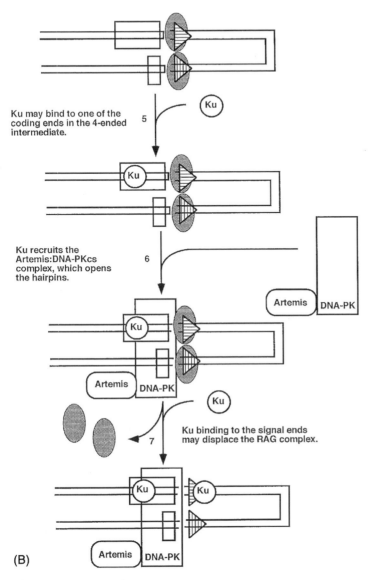

Ku may bind to one of the coding ends in the 4-ended intermediate.

5

Ku recruits the Artemis:DNA-PKcs complex, which opens the hairpins.

6

Ku binding to the signal ends may displace the RAG complex.

7

(B)

Figure 4 (*Continued*)

4. OVERVIEW OF IMMUNOGLOBULIN CLASS SWITCH RECOMBINATION AND ITS UTILIZATION OF NHEJ IN THE REJOINING PROCESS

In contrast to most DNA recombination events, class switch recombination (CSR) has no biological or enzymatic precedents for comparison. It does not involve homologous recombination. The target zone in CSR is distinctive for its extensive length (kilobases), defying the term site-specific. Therefore, the term regionally specific recombination was coined to refer to switch recombination (30). The switch sequences are 1 to 10 kb long, and the recombination points can be anywhere within them (31) (Fig. 5). Approximately 60% of the donor and acceptor switch breakpoint sequences show 1–4 nucleotides of terminal microhomology between the two ends at

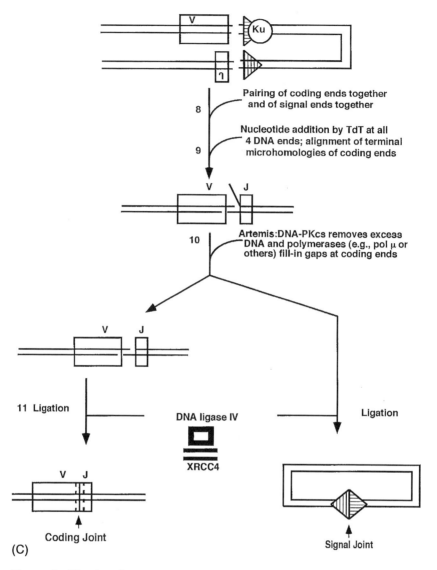

Figure 4 (*Continued*)

the recombination junction such that these few nucleotides can be assigned to either participating DNA end (32). The 1–4 nts of terminal microhomology in CSR contrasts with the more than 25 bp (and, more typically, thousands of bps) of homology involved in homologous recombination; this rules out any role for homologous recombination in CSR. The lack of terminal microhomology use in 40% of switch junctions is incompatible with any obligate role of variants of a single-strand annealing mechanism also. The NHEJ pathway is thought to be involved in the rejoining of the broken DNA ends within the switch region based on studies of mice lacking NHEJ components (33–36). Such studies are technically complex to interpret, and the requirement for DNA-PKcs has been contested (37). The need for nuclease components, such as DNA-PKcs and Artemis, in processes where the single-stranded DNA overhangs are undamaged could be optional, as we discuss below. Whether

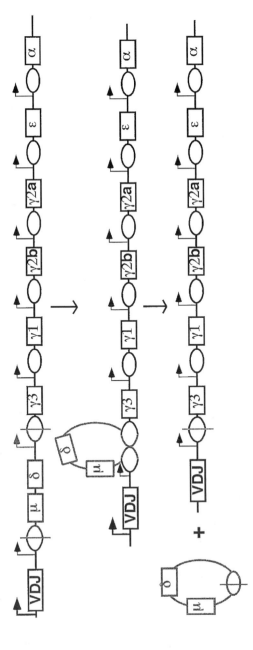

Figure 5 Diagram of class switch recombination. The rectangles represent exons (each set of constant regions has been simplified to a single rectangle), and the ovals represent class switch recombination (CSR) sequences. A single class switch recombination event utilizes the Sμ region and one of the downstream "acceptor" switch regions, in this case Sγ3. The deleted region is shown. The right-angle arrows are promoters. The largest right-angle arrow upstream of the VDJ exon is the actual promoter of the gene, and the smaller arrows represent sterile transcript promoters. The ends of the deleted region appear to be ligated, based on the detectability of such circles. *Source:* From Refs. 97,98.

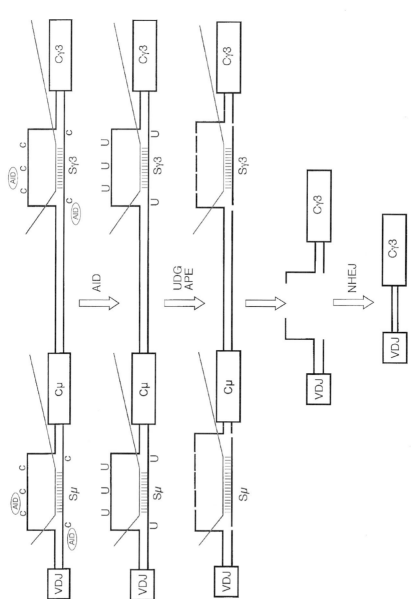

Figure 6 R-loop:deaminase model of class switch recombination. A putative CSR event to IgG3 is shown. Switch region RNA transcripts were paired with the C-rich DNA template strand to form an R-loop structure. The entire displaced G-rich DNA strand and part of the C-rich DNA at the edges of the R-loop are single-stranded and, therefore, serve as targets for AID. AID deaminates C residues located in the single-stranded region to convert them to uracil (U). UDG removes uracils in the DNA and leaves behind abasic (apyrimidinic) sites, which are cleaved by APE. The sum of nicks on both strands results in double-strand DNA breaks in the switch region, which are repaired by the NHEJ pathway to complete CSR. Coding regions (VDJ and constant region exons) are indicated by rectangles. Small dashes in the switch region indicate base pairing between the switch RNA transcript and the C-rich DNA template.

additional proteins, such as mismatch repair proteins and other nucleases, participate in the end joining has not been determined; null mice for some mismatch repair proteins and for exonuclease 1 have decreased CSR by up to two- to three-fold (38,39).

The switch regions are highly repetitive and G-rich on the non-template strand. The G residues are often clustered in groups of 2–5. The repeat lengths vary from 20 to 80 nt. A single recombination event appears to require both a donor and an acceptor switch region (32,40). The upstream or donor switch region is Sμ (Fig. 5). The downstream or acceptor switch region can be any of Sγ3, γ1, γ2b, γ2a, ε, or α in mouse and any of Sγ3, γ1 α1, γ2, γ4, ε, or α2 in human in that physical order along chromosome 12 or 14, respectively (Igδ and pseudogenes are excluded because no DNA recombination events involve these).

Upon transcription, an R-loop forms at the donor and acceptor switch regions in vivo, as it does in vitro (41) (Fig. 6). This R-loop forms at the switch regions because a G-rich RNA base pairs to a C-rich DNA more stably than the original duplex. The displaced DNA strand of the R-loop is single-stranded for lengths of over one kilobase in the chromosomes of stimulated B cells (41).

A cytidine deaminase called activation-induced deaminase (AID) acts on any single-stranded DNA portions of the R-loop (42). AID converts cytosine to uracil at regions of single-strandedness (43). This would be expected to generate several sites of U in the displaced G-rich strand and possibly at the edges of the R-loop on the C-rich DNA strand. Uracil glycosylase removes the uracil base from the DNA strands, creating an abasic site (44). Then apurinic/apyrimidinic endonuclease (APE) is likely to cut the phosphodiester backbone immediately 5′ of the abasic sites (44). This is suspected to result in the double-strand DNA breaks that are needed for the recombination, though this is still not proven.

The details of AID's single-stranded DNA substrate preferences provide additional insight into its role in CSR. It requires a single-stranded DNA length of more than 20 nts for significant activity (45), though duplex substrates containing 3nt bubbles are also suitable substrates (43). The R-loops formed at the chromosomal switch regions are of ample length for AID binding. However, the distance of the displaced G-rich DNA strand from the RNA:DNA helix may influence access by AID to the displaced single-strand (45,45a). AID has a sequence preference for acting at C's within WRC motifs (45,46). WRC motifs are quite frequent in class switch sequences, and this is probably important for the action of AID in CSR (45). Importantly, there are WRC sequences located across from one another on the two strands (45). This may turn out to be critical for the generation of double-strand DNA breaks during class switch recombination. Otherwise, action of AID might only generate nicks on the top and bottom strands, and if the nicks are not near one another, a double-strand break would not result. Hence, the evolution of the class switch sequences and of the AID sequence preference may have been closely coupled so as to generate end configurations that would be processed by NHEJ.

5. POINTS OF BIOCHEMICAL DETAIL IN THE NHEJ PATHWAY

There are two major facets to the NHEJ pathway that remain uncertain. First, how are the two DNA ends held in proximity—a feature sometimes called synapsis? Second, what polymerases carry out any template-dependent fill-in synthesis?

An intial study provided data that Ku could bring two DNA ends together (47). However, this observation remains unverified, despite efforts to duplicate it (48). More recently, evidence was provided that DNA-PKcs could synapse two ends (48). The synapsis by DNA-PKcs was substantial at very low ionic strengths (<30 mM monovalent salt) and at low temperatures (4°C), but raising the ionic strength or the temperature was sufficient to eliminate the apparent synapsis. Hence, additional work will be needed to determine if DNA-PKcs is capable of synapsis under conditions that are closer to physiologic.

Ku is likely to change conformation after binding to DNA. This is based on the fact that free Ku:DNA-PKcs complexes are not detectable in the absence of DNA ends (14,49–52). Moreover, DNA-PKcs is able to bind to Ku:DNA complexes when the DNA is only 18 bp (14), which is the footprint size of Ku (53). This means that most, if not all, of the initial contacts between the Ku:DNA complex and the incoming DNA-PKcs are on the Ku protein, rather than on the combination of DNA and Ku. The lack of Ku:DNA-PK$_{cs}$complexes in free solution and in the cell and the presence of DNA:Ku:DNA-PK$_{cs}$ complexes on 18 bp DNA would indicate that Ku has changed conformation once it has bound to DNA. Electron microscopy of DNA-PKcs in free solution vs. when bound to a DNA end shows that it also changes conformation (54). Therefore, in the assembly of the nuclease complex, Ku binds to the DNA end, changes conformation, recruits the Artemis:DNA-PK$_{cs}$complex, whereupon the DNA-PKcs changes conformation.

There is evidence that the DNA:Ku:DNA-PK$_{cs}$ complex also changes as DNA-PKcs phosphorylates itself (50,55). This autophosphorylation appears to down-modulate the kinase activity of DNA-PKcs. This may then allow reconfiguration of the complex to permit the NHEJ steps to go to completion (56).

Regarding template-dependent polymerases, there is considerable data in *S. cerevisiae* that POL4, a POL X family member, significantly contributes to, but is not exclusively responsible for, fill-in synthesis during NHEJ and that POL4 physically and functionally interacts with the DNA ligase IV complex in yeast (Lif1 and DNL4) (57,58). The POL X family members in vertebrates are polymerase beta, mu, lambda, and terminal transferase (58).

One study has shown that when Ku and the XRCC4:DNA ligase IV complex is bound at a DNA end (15,16), then polymerase μ is recruited (59). The polymerase μ knockout mouse and cells from it are not sensitive to ionizing radiation (60). However, the pol μ null mouse does have slightly shortened coding ends at the kappa light chain junctions (61). Immunodepletion of polymerase lambda from human crude extracts has been used in support of the possibility that this polymerase is also involved in NHEJ (62). Use of murine genetic null mice is one key approach that may shed light on which and how many template-dependent polymerases are involved. In addition, correlation of in vitro reconstitution (using purified components) with cellular NHEJ events is another approach that will help clarify this aspect.

As mentioned, simpler DNA end configurations may not require a nuclease and/or a polymerase in NHEJ (Fig. 7). Such simpler cases may simply require Ku and XRCC4:DNA ligase IV. The fact that a polymerase or a nuclease is not required for some particular pair of ends cannot be construed as denoting a separate end joining pathway (see below).

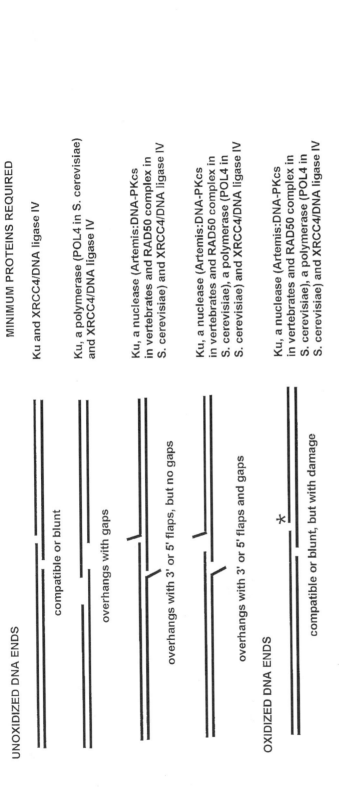

Figure 7 Simpler DNA end configurations may not require a nuclease or a polymerase. On the left side of the figure, various DNA end configurations are shown. Unoxidized DNA means that the nucleotides are normal. Oxidized DNA (at the asterisk) means that there is base or sugar oxidative damage, requiring some nuclease action for removal. The top DNA diagram is drawn as compatible, but the same minimal protein requirements would likely apply to blunt ends. The right side of the diagram lists the minimal proteins or activities that are conjectured to be required. As described in the text, POL4 is only responsible for part of the fill-in synthesis fin S. cerevisiae NHEJ, and it is unclear what other polymerases participate. The RAD50 complex refers to RAD50/Mre11/Xrs2. This figure reflects the versatility of an NHEJ pathway that begins with Ku and ends with XRCC4/DNA ligase IV, but for which a nuclease and a polymerase are necessary only when the ends are incompatible.

6. SPECIAL ASPECTS OF NHEJ AS IT RELATES TO V(D)J RECOMBINATION

One issue in V(D)J recombination concerns the transition from a RAG post-cleavage complex into two NHEJ events: signal joint formation and coding joint formation. There are limited firm data on this transition at the current time. It is known that the RAG complex binds to the two signal ends and the two coding ends in a post-cleavage complex (63,64). An argument can be made that Ku might displace the RAGs from the signal ends in order for the signal joint to form (25); however, there has been no direct experimental test of this hypothesis.

Another point of special interest concerns the role of terminal deoxynucleotidyl transferase (TdT) in the NHEJ phase of V(D)J recombination. TdT adds deoxyribonucleotides to the 3'OH of any single- or double-stranded DNA (65). It can add any of the four nucleotides, but appears to have the highest binding affinity for G, and therefore, incorporates G preferentially (66). Moreover, TdT appears to preferentially incorporate strings of pyrimidine nucleotides or of puririe nucleotides, consistent with stabilization of the incoming dNTP by stacking on the 3' base in the chain (67). In V(D)J recombination, the addition by TdT at the two coding ends adds enormously to the diversity of the immune system (30). Though the sole physiologic function of TdT is the junctional diversification at the coding joint, it is a point of biochemical interest that TdT adds nucleotides to the 3'OH of the signal ends as well (68). Some have suggested that Ku recruits TdT, based on the lack of junctional additions in the rare joints formed in Ku86 null cells (25). However, in Ku70 null cells, the rare joints can have TdT additions, indicating that TdT does not require Ku for its recruitment (69,70).

7. ARE THERE MULTIPLE NHEJ PATHWAYS?

The initial biochemical description of the XRCC4:DNA ligase IV complex demonstrated that DNA ligase IV alone could ligate blunt and compatible overhangs (17). Hence, the ligase complex alone could function as an "NHEJ pathway" when the ends are blunt or compatible. Given that pathologic causes (such as ionizing radiation) cause multiple local sites of oxidative damage to the bases and sugar moieties (1,2,71), it was assumed that XRCC4:DNA ligase complex would more typically function along with nucleolytic activities and polymerases.

The fact that the ligase complex can readily ligate blunt or compatible ends may explain some of the claims in the literature that there are multiple NHEJ pathways. The description of a Ku-independent pathway by one group using crude extracts could quite conceivably be a purely DNA ligase IV-dependent joining (72).

Similarly, description of a DNA-PKcs-independent, RAD50/Mre11/Nbs1-dependent NHEJ pathway (see Chapter 31) in the absence of genetic evidence to support such an alternative pathway in mammalian cells is of uncertain (73). Of particular note, cells deficient for Nbs1 or defective for RAD50 can support perfectly normal V(D)J recombination (74–76).

The role of some components has been evaluated by transfection of linearized plasmids into cell lines where the plasmids have 6 bp of perfect terminal microhomology (77). End joining in such cases may involve nucleolytic resection, annealing of the single-stranded overhangs at the sites of such microhomology, filling of the gaps by a polymerase, followed by ligation of the displaced nicks by DNA ligase I, based on the fact that some of these events are independent of DNA ligase IV; one could think of this as a variant of the single-strand annealing mechanism (SSA), though the specific protein requirements for this scheme may differ from the usual SSA pathway. Six base pairs of perfect microhomology would only very rarely be encountered in vivo. Hence, this is not an adequate basis for invoking multiple types of NHEJ. Consistent with this, more recently, these authors have shown that when breaks are generated within the cell, ligase IV is indeed required for the joining. It is only when linear DNA is transfected into the same cells that they observe a ligase IV-independent joining (78). The authors conclude that there are differences between the joining of ends generated in vivo and those introduced exogenously with exposed DNA termini on naked DNA. In addition to the bases for such differences noted in that study, there is the very likely possibility of the variant single-strand annealing mechanism noted above. The considerations here may apply to quite a few linear transfection studies and biochemical studies where variations of NHEJ have been invoked.

One study has described joining events in *S. cerevisiae* between two incompatible sites created by HO endonuclease (79). Thirty-five out of 46 joinings of such ends in yeast result in losses from one end of 53 or more bp. This is quite different from NHEJ and reflects a different pathway. It is dependent on RAD1 and the RAD50/Mre11/Xrs2 complex (79). This salvage form of joining may not be so commonly used. Most DNA ends in yeast and mammals do not suffer such large extents of nucleolytic loss. In mammalian cells, XP-F null patients are not particularly senstive to ionizing radiation, suggesting that such an RAD1-dependent pathway is not a major contributor to end joining (80). These events could be salvage attempts using a variation of single-strand annealing, as suggested (79). Hence, it may be premature to designate names for such pathways or to deem them to be variants of NHEJ (79). Nevertheless, these salvage pathways may explain the rejoining mechanism for the rare chromosomal translocations that arise in mammals lacking core NHEJ components.

Genetics is the strongest manner by which to assess the contribution of any particular component to a pathway. Genetic evidence in both yeast and mammals is strong that NHEJ requires DNA ligase IV and its partner protein, XRCC4 (or Lif1 in yeast) (17–20,81). Likewise, it is clear that NHEJ in yeast requires Ku (82–84). Ku-deficient mice, DNA-PKcs-deficient mice, Artemis-deficient humans and mice, XRCC4-deficient mice, and DNA ligase IV-deficient mice are sensitive to ionizing radiation, as are the cells from these organisms (85). If there are backup pathways for the NHEJ pathway, they do not appear to be very effective in vivo; otherwise, such pathways would compensate for the lack of NHEJ, and there would be no radiation sensitivity. In this regard, it should be noted that radiation-induced double-strand breaks are likely to require a nuclease, in this case Artemis:DNA-PKcs, to remove damaged overhangs. Blunt ends, as, for example, in signal joint formation of V(D)J recombination, do not require a nuclease for joining, and the genetics illustrates that neither Artemis nor DNA-PKcs are required for signal joint formation (see above). This illustrates that the core components of NHEJ are Ku and the ligase complex, XRCC4:DNA ligase IV, and that the nuclease participates in those ends that cannot be joined without resection. This view is consistent with the diverse array of end joining outcomes typically seen in mammalian NHEJ for the same ends.

8. NHEJ AND HUMAN DISEASE

In addition to their radiation-sensitivity, mice lacking NHEJ components also lack lymphocytes because they cannot complete V(D)J recombination. Deficiency of Artemis in humans results in the lack of lymphocytes and radiation-sensitivity for the same reason.

Mutations in DNA ligase IV in humans have been reported but these are presumably partial reductions in the level of active ligase IV activity rather than true nulls; otherwise, they would have resulted in fewer lymphocytes than they do and, if like the mice, would be expected to have been lethal in early life (86).

It will be interesting to learn if heterozygosity for some NHEJ proteins results in intermediate phenotypes. Chromosomal instability has been documented in the cells of mice heterozygous for Ku86 and DNA ligase IV (87), and inactivation of one Ku86 allele in a human cell line has a cell culture phenotype (88). Consistent with the chromosomal instability of NHEJ heterozygotes (87), heterozygosity for DNA ligase IV in mice results in increased tumorigenesis in mice null for ink4a/ arf (89). These initial studies indicate that heterozygosity for NHEJ components may be an important predictor of human disease.

9. FUTURE AVENUES OF STUDY OF THE NHEJ PATHWAY

In addition to clarification on the synapsis and polymerase points described above, future clarification on the role of chromatin will also be of interest. Are double-strand breaks that are in DNA which is wrapped around a nucleosome repaired in precisely the same way as internucleosomal double-strand breaks? Does a break in the DNA wrapped around a nucleosome require removal of the nucleosome in order for repair to proceed? In addition to the fine-structural chromatin aspects of NHEJ, there are more regional chromatin issues. One study has shown that sites of V(D)J recombination involve immunofluorescent foci of H2AX (90); however, the H2AX knockout mice carry out V(D)J recombination relatively normally (91,92). Moreover, additional work has shown that H2AX is dispensable in many ways for the repair of double-strand breaks (93) (see also Chapter 32).

Another area for future work involves the relationship between dsDNA breaks and the cell cycle. How many dsDNA breaks must there be in a mammalian cell before the cell cycle is halted? If repair is sufficiently rapid, then a single break may not be sufficient to halt the mammalian cell cycle, despite evidence for such an effect in *S. cerevisiae* (94).

Finally, are there additional components for NHEJ? Until biochemical reconstitution is achieved, this is difficult to rule out. One study reported that a patient was radiation-sensitive, but his cells appeared to be normal for all of the known components of NHEJ (95). This study did not fully rule out point mutations in DNA-PKcs that might influence its function or its interaction with Ku or Artemis. Hence, at the present time, there is no clear evidence for additional NHEJ components directly involved in the mammalian pathway.

REFERENCES

1. Ward JF. Cold Spring Harb Symp Quant Biol 2000; 65:377–382.
2. Imlay JA, Linn S. Science 1988; 240:1302–1309.

3. Karanjawala ZE, Murphy N, Hinton DR, Hsieh C-L, Lieber MR. Current Biol 2002; 12:397–402.
4. Lieber MR, Ma Y, Pannicke U, Schwarz K. Nature Rev Mol Cell Biol 2003; 4:712–720.
5. West SC. Nat Rev Mol Cell Biol 2003; 4:435–445.
6. Takata M, Sasaki MS, Sonoda E, Morrison C, Hashimoto M, Utsumi H, Yamaguchi-lwai Y, Shinohara A, Takeda S. EMBO J 1998; 17:5497–5508.
7. Walker JR, Corpina RA, Goldberg J. Nature 2001; 412:607–614.
8. Falzon M, Fewell J, Kuff EL. J Biol Chem 1993; 268:10546–10552.
9. Moshous D, Callebaut I, Chasseval Rd, Corneo B, Cavazzana-Calvo M, Diest FL, Tezcan I, Sanal O, Bertrand Y, Philippe N, Fischer A, PdVillartay J. Cell 2001; 105:177–186.
10. Ma Y, Pannicke U, Schwarz K, Lieber MR. Cell 2002; 108:781–794.
11. Anderson CW, Carter TH. The DNA-activated protein kinase-DNA-PK. In: Jessberger R, Lieber MR, eds. Molecular Analysis of DNA Rearrangements in the Immune system. Heidelberg: Springer-Verlag, 1996:91–112.
12. Gottlieb T, Jackson SP. Cell 1993; 72:131–142.
13. Yaneva M, Kowalewski T, Lieber MR. EMBO J 1997; 16:5098–5112.
14. West RB, Yaneva M, Lieber MR. Mol Cell Biol 1998; 18:5908–5920.
15. Chen L, Trujillo K, Sung P, Tomkinson AE. J Biol Chem 2000; 275:26196–26205.
16. NickMcElhinny SA, Snowden CM, McCarville J, Ramsden DA. Mol Cell Biol 2000; 20:2996–3003.
17. Grawunder U, Wilm M, Wu X, Kulesza P, Wilson TE, Mann M, Lieber MR. Nature 1997; 388:492–495.
18. Wilson TE, Grawunder U, Lieber MR. Nature 1997; 388:495–498.
19. Schar P, Herrmann G, Daly G, Lindahl T. Genes Dev 1997; 11:1912–1924.
20. Teo SH, Jackson SP. EMBO J 1997; 16:4788–4795.
21. Lewis SM. Adv Imm 1994; 56:27–150.
22. Gellert M. Annu Rev Biochem 2002; 71:101–132.
23. Fugmann SD, Al Lee, Shockett PE, Villey IJ, Schatz DG. Ann Rev Immunol 2000; 18:495–527.
24. Chu G. J Biol Chem 1997; 272:24097–24100.
25. Zhu C, Bogue MA, D-Lim S, Hasty P, Roth DB. Cell 1996; 86:379–389.
26. Nicolas N, Moshous D, Cavazzana-Calvo M, Papadoupoulo D, Chasseval Rd, Deist FL, Fischer A, Villartay J-Pd. J Exp Med 1998; 188:627–634.
27. Lieber MR, Hesse JE, Lewis S, Bosma GC, Rosenberg N, Mizuuchi K, Bosma MJ, Gellert M. Cell 1988; 55:7–16.
28. McBlane JF, Gent DCv, Ramsden DA, Romeo C, Cuomo CA, Gellert M, Oettinger MA. Cell 1995; 83:387–395.
29. vanGent DC, Mizuuchi K, Gellert M. Science 1996; 271:1592–1594.
30. Lieber MR. FASEB J 1991; 5:2934–2944.
31. Gritzmacher CA. Crit Rev Immunol 1989; 9:173–200.
32. Dunnick WA, Hertz GZ, Scappino L, Gritzmacher C. Nucl Acid Res 1993; 21:365–372.
33. Rolink A, Melchers F, Andersson J. Immunity 1996; 5:319–330.
34. Casellas R, Nussenzweig A, Wuerffel R, Pelanda R, Reichlin A, Suh H, Qin XF, Besmer E, Kenter A, Rajewsky K, Nussenzweig MC. EMBO J 1998; 17:2404–2411.
35. Manis JP, Dudley D, Kaylor L, Alt FW. Immunity 2002; 16:607–617.
36. Manis JP, Gu Y, Lansford R, Sonoda E, Ferrini R, Davidson L, Rajewsky K, Alt FW. J Exp Med 1998; 187:2081–2089.
37. Bosma GC, Kim J, Ulrich T, Fath DM, Cotticelli MG, Ruetsch NR, Radic MZ, Bosma MJ. J Exp Med 2002; 196:1483–1495.
38. Schrader CE, Vardo J, Stavnezer J. J Exp Med 2003; 197:1377–1383.
39. Bardwell PD, Woo CJ, Wei K, Li Z, Martin A, Sack SZ, Parris T, Edelmann W, Scharff MD. Nat Immunol 2004; 5:224–229.
40. Stavnezer J. Adv Immunol 1996; 61:79–146.

41. Yu K, Chedin F, C-Hsieh L, Wilson TE, Lieber MR. Nat Immunol 2003; 4:442–451.
42. Shinkura R, Tian M, Khuong C, Chua K, Pinaud E, Alt FW. Nature Immunol 2003; 4:435–441.
43. Bransteitter R, Pham P, Scharff MD, Goodman MF. Proc Natl Acad Sci 2003; 100:4102–4107.
44. DiNoia J, Neuberger MS. Nature 2002; 419:43–48.
45. Yu K, Huang FT, Lieber MR. J Biol Chem 2004; 279:6496–6500.
45a. Yu K, Roy D, Bayramyam M, Haworth IS, Lieber MR. (2005) Fine-structure analysis of activation-induced deaminase accessibility to class switch region R-loops. Mol Cell Biol 25:1730–1736.
46. Pham P, Bransteitter R, Petruska J, Goodman MF. Nature 2003; 424:103–107.
47. Ramsden DA, Gellert M. EMBO J 1998; 17:609–614.
48. DeFazio LG, Stansel RM, Griffith JD, Chu G. EMBO J 2002; 21:3192–3200.
49. Suwa A, Hirakata M, Takeda Y, Jesch S, Mimori T, Hardin JA. Proc Natl Acad Sci USA 1994; 91:6904–6908.
50. Chan DW, Lees-Miller SP. J Biol Chem 1996; 271:8936–8941.
51. Chan DW, Mody CH, Ting NS, Lees-Miller SP. Biochem Cell Biol 1996; 74:67–73.
52. Ma Ya. Lieber MR. J Biol Chem 2002; 277:10756–10759.
53. deVries E, vanDriel W, Bergsma WG, Amberg AC, vanderVliet PC. J Mol Biol 1989; 208:65–78.
54. Boskovic J, Rivera-Calzada A, Maman J, Chacon P, Willison K, Pearl L, Llorca O. EMBO J 2003; 22:5875–5882.
55. Merkle D, Douglas P, Moorhead GB, Leonenko Z, Yu Y, Cramb D, Bazett-Jones DP, Lees-Miller SP. Biochemistry 2002; 41:1206–1714.
56. Ding O, Reddy Y, Wang W, Woods T, Douglas P, Ramsden DA, Lees-Miller SP, Meek K. Mol Cell Biol 2003; 23:5836–5848.
57. Wilson T, Lieber MR. J Biol Chem 1999; 274:23599–23609.
58. Tseng HM, Tomkinson AE. J Biol Chem 2002; 277:45630–45637.
59. Mahajan KN, NickMcElhinny SA, Mitchell BS, Ramsden DA. Mol Cell Biol 2002; 22:5194–5202.
60. Bertocci B, Smet AD, Flatter E, Dahan A, Bories J, Landreau C, Weill JC, Reynaud CA. J Immunol 2002; 168:3702–3706.
61. Bertocci B, Smet AD, Berek C, Weill J-C, Reynaud C-A. Immunity 2003; 19:203–211.
62. Lee JW, Blanco L, Zhou T, M Garcia-Diaz, Bebenek K, Kunkel TA, Wang Z, Povirk LF. J Bio Chem 2003; 279:805–811.
63. Agrawal A, Schatz DG. Cell 1997; 89:43–53.
64. Hiom K, Gellert M. Cell 1997; 88:65–72.
65. Bollum FJ. J Biol Chem 1960; 235:2399–2403.
66. Kornberg A, Baker T. DNA Replication. 2nd ed. New York: W.H. Freeman, 1992.
67. Gauss GH, Lieber MR. Mol Cell Biol 1996; 16:258–269.
68. Lieber MR, Hesse JE, Mizuuchi K, Gellert M. Proc Natl Acad Sci 1988; 85:8588–8592.
69. Gu Y, Seidl K, Rathbun GA, Zhu C, Manis JP, vanderStoep N, Davidson L, Cheng HL, Sekiguchi J, Frank K, Stanhope-Baker P, Schlissel MS, Roth DB, Alt FW. Immunity 1997; 7:653–665.
70. Ouyang H, Nussenzweig A, Kurimasa A, Soares V, Li X, Cordon-Cardo C, Li W, Cheong N, Nussenzweig M, lliakis G, Chen DJ, Li GC. J Exp Med 1997; 186:921–929.
71. Imlay JA. Adv Microb Physiol 2002; 46:111–153.
72. Wang H, Perrault AR, Takeda Y, Qin W, Wang H, Lliakis G. Nucl Acids Res 2003; 31:5377–5388.
73. Udayakumar D, Bladen CL, Hudson FZ, Dynan WS. J Biol Chem 2003; 278:1631–1635.
74. Harfst E, Cooper S, Neubauer S, Distel L, Grawunder U. Mol Immunol 2000; 37:915–929.
75. Yeo TC, Xia D, Hassouneh S, Yang XO, Sabath DE, Sperling K, Gatti RA, Concannon P, Willerford DM. Mol Immunol 2000; 37:1131–1139.

76. Bender CF, Sikes ML, Sullivan R, Huye LE, Beau MML, Roth DB, Mirzoeva OK, Oltz EM, Petrini JH. Genes Dev 2002; 16:2237–2251.

77. Verkaik NS, Lange RE-v, Heemst Dv, Bruggenwirth H, Hoeijmakers JH, Zdzienicka MZ, Gent DCv. Eur J Immunol 2002; 32:701–709.

78. vanHeemst D, Brugmans L, Verkaik N, vanGent DC. DNA Repair 2004; 3:43–50.

79. Ma JL, Kim EM, Haber JE, Lee SE. Mol Cell Biol 2003; 23:8820–8828.

80. Friedberg EC, Walker GC, Siede W. DNA Repair and Mutagenesis. Washington, DC: ASM Press, 1995.

81. Frank KM, Sharpless NE, Gao Y, Sekiguchi YM, Ferguson DO, Zhu C, Manis JP, Homer J, DePinho R, Alt FW. Mol Cell 2000; 5:993–1002.

82. Milne GT, Jin S, Shannon KB, Weaver DT. Mol Cell Biol 1996; 16:4189–4198.

83. Boulton SJ, Jackson SP. EMBO J 1996; 15:5093–5103.

84. Moore JK, Haber JE. Mol Cell Biol 1996; 16:2164–2173.

85. Mills KD, Ferguson DO, Alt FW. Immunol Rev 2003; 194:77–95.

86. O'Driscoll M, Cerosaletti KM, Girard PM, Dai T, Stumm M, Kysela B, Hirsch B, Gennery A, Palmer SE, Seidel J, Gatti RA, Varon R, Oettinger MA, Neitzel H, Jeggo PA, Concannon P. Mol Cell 2001; 8:1175–1185.

87. Karanjawala ZE, Grawunder U, C-Hsieh L, Lieber MR. Curr Biol 1999; 9:1501–1504.

88. Li G, Nelsen C, Hendrickson EA. Proc Natl Acad Sci 2002; 99:832–837.

89. Sharpless NE, Ferguson DO, O'Hagen R, Castrillon D, Lee C, Farazi P, Alson S, Fleming J, Morton C, Frank K, Chin L, Alt FW, DePinho RA. Mol Cell 2001; 8: 1187–1196.

90. Chen HT, Bhandoola A, Difilippantonio MJ, Zhu J, Brown M, Tai X, Rogakou EP, Brotz TM, Bonner WM, Ried T, Nussenzweig A. Science 2000; 290:1962–1965.

91. Bassing CH, Suh H, Ferguson DO, Chua KF, Manis J, Eckersdorff M, Branson R, Lee C, Alt FW. Cell 2003; 114:359–370.

92. Celeste A, Difilippantonio S, Difilippantonio MJ, Fernandez-Capetillo O, Pilch DR, Sedelnikova OA, Eckhaus M, Ried T, Bonner WM, Nussenzweig A. Cell 2003; 114: 371–383.

93. Celeste A, Fernandez-Capetillo O, Kruhlak MJ, Pilch DR, Staudt DW, Lee A, Bonner RF, Bonner WM, Nussenzweig A. Nat Cell Biol 2003; 5:675–679.

94. Bennett CB, Lewis AL, Baldwin KK, Resnick MA. Proc Natl Acad Sci 1993; 90: 5613–5617.

95. Dai Y, Kysela B, Hanakahi LA, Manolis K, Riballo E, Stumm M, Harville TO, West SC, Oettinger MA, Jeggo PA. Proc Natl Acad Sci 2003; 100:2462–2467.

96. Modesti M, Junop MS, Ghirlando R, Mvd Rakt, Gellert M, Yang W, Kanaar R. J Mol Biol 2003; 334:215–228.

97. Matsuoka M, Yoshida K, Maeda T, Usuda S, Sakano H. Cell 1990; 62:135–142.

98. Iwasato T, Shimizu A, Honjo H, Yamagishi H. Cell 1990; 62:143–149.

29

Structural Aspects of Ku and the DNA-Dependent Protein Kinase Complex

Eric A. Hendrickson
University of Minnesota Medical School, Minneapolis,
Minnesota, U.S.A.

Joy L. Huffman and John A. Tainer
Department of Molecular Biology—MB4, Skaggs Institute for Chemical Biology and
The Scripps Research Institute, La Jolla, California, U.S.A.

1. INTRODUCTION

DNA-dependent protein kinase (DNA-PK) is a trimeric complex comprised of the DNA-PK catalytic subunit (DNA-PK$_{cs}$) and the autoantigens, Ku86 and Ku70. DNA-dependent protein kinase has attracted enormous interest because it is critically important for four distinct, but probably mechanistically overlapping, processes in vertebrate cells: (i) the generation of a functional immune system through lymphoid variable(diversity)joining [V(D)J] recombination, (ii) immunoglobulin isotype switch recombination, (iii) the repair of spontaneous and exogenously generated DNA double-strand breaks (DSBs) via nonhomologous end joining (NHEJ), and (iv) telomere length maintenance and function.

1.1. Variable (Diversity) Joining [V(D)J] Recombination

Variable (diversity) joining [V(D)J] recombination is a site-specific recombination process that is absolutely required for the generation of the vertebrate immune system (reviewed in Refs. 1,2). The process is initiated by the recombination activating genes-1 and -2 (RAG-1 and RAG-2, respectively) proteins, which form a lymphoid-restricted, heterodimeric, site-specific recombinase. These proteins initiate V(D)J recombination by introducing DNA DSBs into chromosomal DNA at specific recombination signal sequences that flank the elements to be recombined (Fig. 1A, *i*). The resolution of the resulting four broken DNA ends (Fig. 1A, *ii*) generally results in restoration of an intact chromosome along with the deletion of the intervening DNA via the formation of a circular DNA fragment in which the signal sequences are precisely joined in a head-to-head fashion (Fig. 1A, *iii*). In contrast to the lack of modification at the signal junction, the chromosomal junction is often marked by the random loss

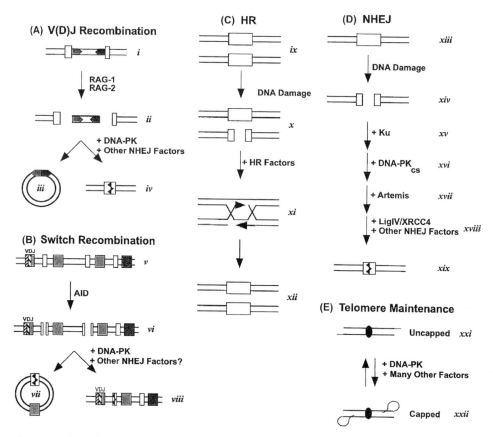

Figure 1 Drawing showing DNA-PK's involvement in a variety of cellular processes. (A) V(D)J recombination. The double line represents dsDNA. The open rectangles are V, D, or J elements. The black-filled arrowheads represent the *cis*-acting recombination signal sequences. The jagged arrowheads represent a repaired junction in which variable numbers of nucleotides have been deleted and/or inserted. (B) Switch recombination. The three, small hatched rectangles correspond to V, D, and J elements that have been recombined. The open rectangles correspond to switch elements. The rectangles represent various constant regions. The black-filled, jagged arrowheads are as in panel (A). (C) Homologous recombination (HR). The open rectangles represent homologous regions on sister chromatids or homologous chromosomes. The crossed lines represent a Holliday junction in which a strand from a broken end has invaded the intact duplex DNA. The solid arrowheads represent DNA synthesis. (D) Nonhomologous end joining (NHEJ). The open rectangle corresponds to a region where a DSB will occur. The jagged arrowheads are as in panel (A). (E) Telomere maintenance. The double line represents dsDNA. The thinner lines protruding from the ends represent the G-strand overhangs. The black oval corresponds to a centromere. The looped structures represent t-loops. In all panels, the vertical arrows are drawn implying a temporal order to each process, although in many cases the precise sequence of events is not known. In all panels, the italicized Roman numerals refer to steps in each pathway that are described in the text.

and/or addition of nucleotides (Fig. 1A, *iv*). Indeed, this imprecision is a general hallmark of most of the joins mediated by DNA-PK and NHEJ (Figs. 1A, *iv*; B, *vii* and *viii*; and D, *xix*). DNA-dependent protein kinase and other NHEJ factors are, with some unusual exceptions (3), essential for accurate V(D)J recombination and mutation of any of the three subunits of the DNA-PK complex invariably leads to profound immune deficiencies due to the inability to complete V(D)J recombination. For a comprehensive discussion of the mechanism of V(D)J recombination and the role that DNA-PK plays in it, the reader is referred to chapter 28.

1.2. Switch Recombination

Mutations of DNA-PK also deleteriously affect switch recombination (4–7) [reviewed by O'Driscoll and Jeggo (8)]. This is a B-cell-restricted, somatic DNA recombination process that temporally follows V(D)J recombination [reviewed by Manis et al. (9)]. Switch recombination allows a B-cell, after it has encountered an antigen, to change the heavy chain class of the antibody it synthesizes. This change results in the expression of an antibody with an unaltered antigen-binding specificity, but with different effector functions. Switch recombination is superficially similar to V(D)J recombination: i.e., a lymphoid-specific factor uses *cis*-linked signal sequences to induce predominately deletional recombination (Fig. 1B, *v*). The two processes, however, employ clearly distinct mechanisms. Thus, whereas V(D)J recombination utilizes a classical recombinase that binds and cleaves with nucleotide precision at short recombination signal sequences, switch recombination utilizes a B-cell-specific deaminase, activation-induced deaminase (AID) (10), which indirectly introduces ostensibly nonrecombinogenic nucleotide base modifications, randomly over large (1–10 kb) regions [Fig. 1B, *vi*; reviewed by Stavnezer and Bradley (11)]. To facilitate switch recombination, the AID-dependent modifications are first converted into nicks, which, in turn, are probably converted, by an as-of-yet undefined process and additional set of factors, into frank DNA DSBs (Fig. 1B, *vi*) (12). It is these DSBs that are likely to be the substrates for DNA-PK and other NHEJ factors that convert the breaks into a circular product (Fig. 1B, *vii*), which is lost from the cells and a repaired chromosome expressing the novel heavy chain gene (Fig. 1B, *viii*). For a more comprehensive review of switch recombination and the role that AID plays in it, see the recent review by Neuberger et al. (13).

1.3. Homologous Recombination

Cells that sustain DNA DSBs either via endogenous processes or through the exposure to exogenous DNA damaging agents can repair this damage through at least two distinct pathways: NHEJ or homologous recombination (HR) (Fig. 1C). Although HR appears to be completely DNA-PK independent, it is important to remember that HR likely competes with the DNA-PK-dependent NHEJ pathway for substrates during DNA DSB repair. In bacteria and in lower eukaryotes, the process of HR dominates. In HR (reviewed in Refs. 14,15), the broken DNA ends are resected to yield 3'-single-stranded DNA overhangs. These overhangs are then used in homology searches—facilitated by strand exchange proteins—to locate an appropriately undamaged segment of DNA, which is usually a sister chromatid or a homologous chromosome (Fig. 1C, *ix* and *x*). Following synapsis, and strand invasion, a displacement (D) loop is generated (Fig. 1C, *xi*). The invading strand is then extended by DNA synthesis to regain the genetic information that was lost. Double-strand break repair is ultimately completed as a gene conversion event by resolution of the cross-stranded intermediates with or without crossovers (Fig. 1C, *xii*). For a fuller description of the HR pathway and the gene products required for it, the reader is referred to the chapter 27.

1.4. Nonhomologous End Joining

The existence and usage of HR extends to higher eukaryotes. Thus, most of the mitotic and meiotic recombination that takes place in eukaryotic cells and the repair of DSBs in late S and G_2 phases of the cell cycle appears to be carried out by HR (reviewed in Refs. 16,17). In striking contrast to bacteria and lower eukaryotes, however, the bulk of DSB repair in higher eukaryotes proceeds more frequently by a process that does not require extended regions of homology. Specifically, mammalian cells have evolved a highly efficient ability to join nonhomologous DNA molecules together (18,19) (reviewed in Refs. 20,21). In their seminal work on gene targeting, Thomas and Capecchi (22) showed that although somatic mammalian cells can integrate linear duplex DNA into corresponding homologous chromosomal sequences using HR, the frequency with which recombination into nonhomologous sequences occurs is at least 1000-fold greater. This NHEJ pathway appears to be predominately active during the G_1/early S phase of the cell cycle (23,24), which is when V(D)J recombination is also known to occur (25,26). Given that NHEJ DSB repair is generally error-prone, the increased percentage of noncoding DNA in higher eukaryotes may have facilitated the evolution of this pathway. In summary, mammals are different from bacteria and lower eukaryotes in that DSB repair proceeds primarily through an NHEJ recombinational pathway.

While the many details of NHEJ remain to be worked out, a reasonable model is that following the introduction of a DNA DSB (Fig. 1D, *xiii* and *xiv*), the Ku heterodimer binds to the broken DNA ends to prevent unnecessary DNA degradation (27–29) and/or to juxtapose the ends (30–33) (Fig. 1D, *xv*). The binding of Ku to the free DNA ends recruits and activates DNA-PK$_{cs}$ (34,35) (Fig. 1D, *xvi*), which, in turn recruits and activates a nuclease, Artemis (36,37) [reviewed by Moshous et al. (38); Fig. 1D, *xvii*], to help trim the damaged DNA ends (Fig. 1D, *xvii*). DNA-PK$_{cs}$ may also help to synapse the free ends (39,40). The rejoining of the DNA DSB (Fig. 1D, *xix*) requires the recruitment (41) of DNA ligase IV (LigIV) (42,43), x-ray cross-complementing group (XRCC4) (44,45), and probably at least one additional factor (46) as well (Fig. 1D, *xviii*) (41,47–49).

It is also important to note that NHEJ appears to consist of at least two subpathways: the main, DNA-PK-dependent end-joining pathway described above and one mediated by microhomologies. Thus, mammalian cell lines defective for Ku86, DNA-PK$_{cs}$, LigIV, and XRCC4 are still proficient for some NHEJ, but the majority of repair products generated appear to have utilized microhomologies (32,50–53). The genes and mechanism required for the microhomology-directed subpathway are currently poorly defined, but there are some interesting clues. First, meiotic recombination defective 11 (Mre11), a 3'–5' exonuclease also required for homologous recombinational repair, has enzymatic properties consistent with it playing a role in microhomology-mediated NHEJ repair (54,55). Similarly, chromosomal translocations and DNA repair events in yeast cells deficient in MRE11 often lack microhomology at their junctions (56,57) although this lack of microhomology-mediated repair has yet to be demonstrated in human Mre11 patients (58). Alternatively, some genetically uncharacterized mammalian cell lines sensitive to DNA cross-linking agents appear to be specifically defective in the microhomology-directed subpathway (52). Intriguingly, cell lines derived from Fanconi anemia patients, which have long been known to be sensitive to DNA cross-linking agents, have been reported to be defective in end-joining activity (59,60). Thus, there appear to be DNA-PK-dependent and DNA-PK-independent (but perhaps Mre11/Fanconi anemia-dependent?) pathways (23,49,55,61–66) that utilize primarily direct end joining and microhomology-mediated end joining, respectively, to repair DSBs. This chapter focuses exclusively on DNA-PK-dependent activities.

1.5. Telomeres

Telomeres are the terminal structures of linear chromosomes. Telomeres perform at least two functions: (i) they allow for the replication of the ends of chromosomes and (ii) they stabilize chromosomes by keeping them from recombining with one another [reviewed by de Lange (67)]. Telomeric DNA consists of a repetitive motif of the general form $T_xA_yG_z$, which in mammals is $T_2A_1G_3$. At the ends of the chromosomes, the G-rich strand is often extended over the C-rich strand for a variable number of nucleotides (68) [reviewed by Blackburn (69); Fig. 1E, *xxi*]. This single-stranded extension is referred to as the G-strand overhang and at least some of the time, it folds back internally into the chromosome to form a "t-loop" (70) [reviewed by de Lange (71); Fig. 1E, *xxii*]. The (dys)regulation of telomeric DNA structure has been associated with immortalization (72), senescence/aging [reviewed by Stewart and Weinberg (73)] and tumorigenesis (reviewed in Refs. 74,75). While the precise role that telomeres play in these processes is still under intense investigation, it is clear that telomere dysregulation results in cataclysmic pathological consequences. Many of the genes involved in telomere biogenesis and stability have been identified and their characterization has led to the identification of additional genes leaving the field with a rich, yet complicated, picture. In yeast, for example, mutation of any of over 25 different genes can deleteriously affect telomere length and/or structure (76). The mammalian counterparts of some of these genes have been identified and these include the ribonucleoprotein complex consisting of telomerase reverse transcriptase (TERT) (77) and telomerase RNA component (TERC) (78) that is responsible for the synthesis of the $T_2A_1G_3$ repeat; telomere recognition factors 1 and 2 (TRF1 and TRF2, respectively) (79,80) that bind to the double-strand portion of the $T_2A_1G_3$ repeat and protection of telomeres 1 (Pot1) (81), that binds to the G-strand overhang. In addition—and superficially very paradoxically—a variety of DNA repair proteins, including hMre11, hRAD50, and particularly Ku are also associated, directly or indirectly, with telomeres [reviewed by Weaver (82)]. A current working model is that telomeres can exist in two conformations: "uncapped" (Fig. 1E, *xxi*) and "capped" (Fig. 1E, *xxii*). Telomeres are likely susceptible to modifying enzymes in the uncapped confirmation and resistant to them in the capped confirmation. DNA-PK appears to be associated predominately with the capped conformation (Fig. 1E, *xxii*), although, as detailed below, the precise role(s) that DNA-PK plays in telomere maintenance is not known (83). Lastly, it should be noted that in a small percentage of human cancer cells telomeres can be maintained by a telomerase-independent alternative lengthening of telomeres (ALT) pathway that appears to rely heavily on HR [reviewed by Reddel et al. (84)]. Given the rarity of the ALT pathway, however, it will not be addressed further in this review.

While no discussion of DNA-PK would be completely satisfactory without a significant understanding of the biology behind the abovementioned processes, the topics of lymphoid and homologous recombination as well as NHEJ will be covered in depth in other chapters in this book and not this one. Consequently, this review will focus primarily on the structure and function of the proteins themselves and of their emerging role in proper telomere maintenance.

2. THE KU AUTOANTIGEN

2.1. A Brief History

Ku is an ancient protein and orthologs can be found in some, but not all, bacterial species (85–87) as well as in certain bacteriophage genomes (88). In bacteria, Ku orthologs are often found in an operon with DNA ligase, suggesting that the DNA repair activity of Ku is also anciently conserved. The bacteriophage Mu ortholog of Ku, called Gam [short for *gamma*, as in the third (*alpha, beta, gamma*) gene identified that regulated bacteriophage λ recombination (89)] is essential for Mu replication and it binds on to the linear ends of the phage DNA to protect it from cellular nucleases during infection (90,91), suggesting perhaps the primordial origin of the telomeric maintenance functions of modern eukaryotic Ku. Bacterial Ku is encoded by a single gene, consists of a small protein of ~30 kDa and almost certainly functions as a homodimer. The basic biochemical activity of bacterial and phage Ku is indistinguishable from its eukaryotic counterpart and consists of the unique ability to bind onto the free ends of any linear, double-stranded DNA irrespective of the DNA's sequence (87,88). Later in evolution, a gene duplication event likely occurred [all eukaryotic organisms described to date have two (Ku70 and Ku86) Ku homologs] with both the primordial Ku70 and Ku86 genes subsequently gaining domains on their N- and C-terminus, probably in response to the additional demands of more complex telomere maintenance and vertebrate V(D)J recombination. An extensive phylogenic analysis of all the known Ku genes suggests that the ancestral eukaryotic Ku was Ku86 (92).

Eukaryotic Ku86 and Ku70 were so named because of their apparent molecular weights on protein gels. The larger polypeptide is interchangeably, confusingly, and incorrectly, referred to as either Ku86 or Ku80. The origin of these different monikers is obscure but probably resulted either from the over- or underestimation, respectively, of the protein's actual size (82,713 Da) (93). Thus, while "Ku83" would technically be more correct, for the purposes of this review the larger subunit will be referred to as Ku86. Ku was originally identified in humans in a clinical setting as an autoantigen recognized by the sera of patients afflicted with the autoimmune diseases systemic lupus erythematosus, scleroderma, polyomyositis, Sjorgen's syndrome, or a combination of these disorders (94). The antigen was termed "Ku" based upon the first two letters of the patient's last name whose sera was used in the initial clinical characterizations. The association of Ku with autoimmune disorders is still not understood. While it is conceivable that unique epitopes that promote the production of autoantibodies are induced when Ku binds onto the ends of DNA [this and other possibilities reviewed by Takeda and Dynan (95)], the subsequent two decades of research on Ku has led most researchers to conclude that Ku's association with these disorders is not directly linked to the protein's biochemical function. More likely it results from Ku being a moderately abundant protein (estimated at roughly half-a-million molecules per cell) (96,97), which makes it statistically a better candidate to become an autoantigen than most other (less abundant) cellular proteins.

Nonetheless, in the hope of understanding the pathogenic mechanism of these autoimmune diseases, Ku was extensively biochemically characterized long before its cellular role(s) was discovered. Ku has the unique ability to bind onto the free ends of any linear, double-stranded DNA irrespective of the DNA's sequence. This activity was first described using a filter-binding assay, where it was demonstrated that competition for filter binding by Ku was directly proportional to the number of dsDNA ends available (98). Subsequent investigators utilized what has become the traditional assay for Ku's DNA end binding (DEB) activity—an electrophoretic mobility shift assay (EMSA)—to confirm and extend these studies (99–102). Although Ku can bind to double-stranded oligonucleotides as short as 14–18 bp, optimum binding requires approximately 30 bp of dsDNA (33,102–104), a length that is certainly almost always available in vivo. Once bound, Ku has the unusual property of being able to translocate internally in an ATP-independent manner, thus freeing up the end for additional Ku binding. There is conflicting data as to whether the additional Ku binding is noncooperative (104) or cooperative (105). Regardless, the maximal number of Ku molecules that can be bound on a given DNA fragment is directly proportional to the length of that fragment. A simple—but very elegant—experiment in which multiple Ku molecules where loaded onto a linear plasmid DNA and then the ends of the plasmid were resealed by treating with DNA ligase demonstrated that Ku slides along the DNA with no preferential binding to specific sequences and that it is extremely resistant to dissociation (100). Moreover, the interpretation of this experiment correctly and presciently—a decade before the solution structure of the protein was actually solved—predicted that Ku must assume a ring-like shape through which the dsDNA is threaded. The absence of a dsDNA DEB activity, later shown to be Ku (27,106), in extracts derived from ionizing radiation (IR)-sensitive rodent cell lines was the first direct connection between Ku and DNA repair (107). Contemporaneously with these studies, Ku was identified as the DNA-binding subunit of a DNA-dependent serine–threonine protein kinase complex (i.e., DNA-PK) (34). When it was subsequently shown (108,109) that the kinase catalytic subunit (i.e., DNA-PK$_{cs}$) of DNA-PK was encoded by the severe combined immune deficient [scid (110)]/protein kinase DNA-activated catalytic subunit [PRKDC (111)] locus, a gene long known to control V(D)J recombination (112–115), a putative functional link between Ku and V(D)J recombination was established as well. Confirmation of Ku's role in both DNA repair and V(D)J recombination was demonstrated by functional complementation studies of the rodent IR-sensitive cell lines which indicated that they were defective in Ku86 expression (116–118) and by the identification of inactivating mutations in the corresponding endogenous Ku86 gene (119,120). Moreover, all these phenotypes have subsequently been recapitulated in mice containing targeted disruptions of either Ku86 (121,122) or Ku70 (123,124). Impressively, the Ku-deficient phenotypes have been elaborated in these murine lines to demonstrate additional—and somewhat unexpected—growth retardation defects (125), premature senescence (126), and a marked increase in chromosomal aberrations (127–129) with elevated telomeric fusions (130,131).

In summary, in 25 years, Ku has gone from an origin as an obscure factor associated with rheumatic disorders to its relatively exalted status as one of the premiere vertebrate DNA repair/recombination factors.

2.2. Structure

The determination of the crystal structure of a truncated Ku heterodimer (alone and in complex with dsDNA) (33) validated the general structure determined by neutron scattering (132) and significantly extended several decades of biochemical observations carried out with purified or recombinant proteins. The crystal structure demonstrated that each subunit of the Ku heterodimer consists of four domains with three of the domains conserved between the proteins: (i) an N-terminal α-helical domain, (ii) a central β-barrel domain with an extended bridge, and (iii) a C-terminal helical region (Fig. 2). In addition, each polypeptide has an unrelated extreme C-terminal region. Together, these polypeptides fashion a "donut" or ring-shaped protein that can encircle dsDNA (Fig. 2B). Ku has dyad symmetry with similar topological folds and resembles a high school or college class ring with the gem facing inward, where a person's finger would mimic the DNA. The central β-barrel domains constitute the primary regions for heterodimerization and for DNA binding. This interpretation is consistent with only the central β-barrel domain being conserved in the bacterial and bacteriophage Ku molecules, which are nonetheless capable of (homo)dimerizing and DNA end binding (86–88). The importance of this domain is also generally consistent with results obtained earlier from several independent laboratories using the yeast two-hybrid system and/or in vitro biochemistry. These studies suggested that the middle portion of each protein was important for both heterodimerization and DNA binding (133–137). Extending from the β barrels are bridge elements forged from two crossed β strands at the top of the ring band connected by three interleaved β hairpins, which are contributed asymmetrically from each protomer (Fig. 2B).

Heterodimerization occurs through an extensive interface and involves interdigitation of the two polypeptide chains (Fig. 2). The tight interdigitation of the heterodimer results in a buried surface of $\sim$9000 Å^2, indicating an extremely stable complex (138). This polypeptide interface is fundamentally different from other DNA binding ringlike structures (e.g., sliding clamps) that have been described, which generally consist of individual subunits, each of which forms only part of the ring that encircles the DNA (139,140). In the case of Ku, each monomer polypeptide fully encircles the DNA (Fig. 2B). Consequently, the disassembly of Ku from the DNA is likely to require more than just simple disassembly into monomers as is observed for some sliding clamplike proteins (141). Instead, this observation leads to the speculation that proteolysis of the thin bridge element might be the simplest mechanism for irreversible unloading (33). A hypothetical Ku protease might be similar to separase, a protease that cleaves the cohesin ring that surrounds sister chromatids prior to cell division [reviewed by Nasmyth (142)]. Importantly, and in stark contrast to bacterial Ku, the dimer interface between vertebrate Ku70 and Ku86, albeit extensive, involves very low sequence identity between the two subunits. The lack of identity probably prevents the formation of stable Ku70:Ku70 or Ku86:Ku86 homodimers. This hypothesis is consistent with the extreme instability of the nonmutated Ku70 subunit in a variety of Ku86 mutant cell lines and the corresponding instability of the nonmutated Ku86 subunit in Ku70-deficient cell lines (118,119,121,143). The function of the N-terminal α-helical domains and the C-terminal helical regions is less clear, and the best available evidence suggests that they are predominately involved in ancillary protein:protein interactions to facilitate DSB repair and telomere maintenance (e.g., see the section of separation-of-function mutations for telomere maintenance below). An additional possibility is that they help to orient the dimer. Thus, although the molecule has overall dyad symmetry, Ku binds DNA ends asymmetrically, with the Ku70 subunit most proximal to the free DNA end and Ku86 most distal (Figs. 2A and C). The Ku dimer asymmetry probably arises from a structural shift in the handle region, away from the dyad axis of the heterodimer and toward Ku70, and possibly from a steric or electrostatic block provided by the Ku70 N-terminal 33 amino acids, which are highly acidic

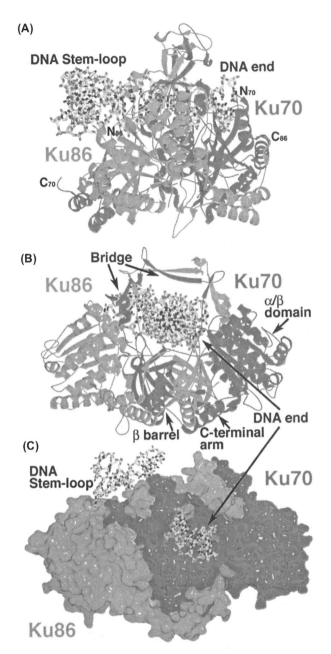

Figure 2 Structure of the Ku heterodimer bound to a stem-loop containing dsDNA (PDB ID 1JEY) (33). (A) Ribbon diagram showing the side view. Ku86 is in green and Ku70 in red, with their respective N- and C-terminus marked. The DNA is shown as balls and sticks. Yellow, red, blue, and orange correspond to carbon, oxygen, nitrogen, and phosphorus atoms, respectively. (B) Ribbon diagram of the Ku heterodimer structure viewed down the DNA end. Major structural domains are marked. (C) A space-filling model showing a side view and the tight interdigitation of the Ku subunits. (*See color insert.*)

and disordered in the crystal structure. This asymmetry of binding is significant as it may impart a preferred orientation for bridging of two DNA ends via Ku's interaction with other repair factors (see Section 2.4) and/or it may provide an energy barrier to Ku to keep it from sliding along the DNA, away from the end. This observation validates elegant biochemical experiments, which came to precisely the same conclusion using photoaffinity labeling of Ku with dsDNA substrates having only a single free end (144). One function the N- and C-terminal regions do not appear to provide is a dimer:dimer interface. Thus, some researchers have postulated that two Ku heterodimers may perform the bridging function that is an obligate necessity in DNA DSB repair (30,31,145,146). If Ku does indeed perform this function, there is nothing about the crystal structure to suggest how this might occur and/or what domains would facilitate such an interaction. Indeed, more recent reports have implicated DNA-PK$_{cs}$ (39,40) or LigIV:XRCC4 (147,148) rather than Ku, as a factor that can synapse free ends.

Comparison of the free and DNA-bound structures reveals a preformed, positively charged DNA binding site: In the crystal structure, DNA end binding by Ku encompassed approximately two full turns (i.e., ~20 bp) of the DNA (33), a fact in agreement with a bevy of biochemical experiments (98,99,102–104). The elements connecting the β barrels and bridge motifs lie just inside the major groove, and—perhaps not surprisingly—there are no specific DNA base contacts with the interior ring portion of Ku and few contacts with the sugar:phosphate backbone, implying that charge and shape are the major contributors to recognition of DNA ends. This lack of specific interactions almost certainly accounts for Ku's unique ability to bind onto virtually all DNA ends, irrespective of their sequence. Interestingly, several residues of Ku project into the internal portion of the ring and superficially appear to occlude the DNA. That occlusion does not occur suggests that the DNA takes a helical path through Ku, in essence much like a bolt threading itself onto a nut. Consequently, these interior Ku residues appear to be nestled into the major and minor groove contours of the DNA and are stabilized by a preponderance of positively charged residues. Lastly, the structure of Ku alone is virtually identical to the structure of Ku complexed with dsDNA, implying that Ku does not undergo significant conformational changes upon binding to DNA ends although binding seems to reduce the flexibility of some Ku regions.

Four additional structural features of Ku are worth noting. First, the loop or bridge that defines the dorsal portion of the central β-barrel domain is extremely narrow, whereas the ventral side encompasses the majority of this domain as well as the bulk of the N- and C-terminal helical domains (Fig. 2B). Consequently, significant portions of the DNA remain exposed to solvent on the top side, while the bulk of Ku cradles the DNA from underneath. Walker et al. (33) hypothesized that this would leave much of the DNA accessible to other DNA modifying enzymes [reviewed by Jones et al. (149)]. While this hypothesis is mechanistically appealing, experimental evidence supporting this view is lacking. Indeed, it is known from biochemical data that Ku is actually displaced internally ~10 bp upon the binding of DNA-PK$_{cs}$ (103) or ~30 bp by the binding of LigIV:XRCC4 (148) and is thus physically removed from the residues undergoing repair. Future studies will hopefully illuminate the precise geometry of the DNA-PK subunits as well as the additional accessory factors assembled on a broken DNA end (40).

Second, the extreme C-terminus of Ku70 is divergent from the C-terminus of Ku86. The structure of a bacterially expressed Ku70 C-terminal fragment had previously been determined by nuclear magnetic resonance (NMR) (150). Both the

NMR and crystal structures agreed that this region is predominately α-helical. More importantly, the Ku70 C-terminus encompasses a SAP (after motifs found in SAF-A/B, acinus and PIAS proteins) domain, which encodes a helix–turn–helix DNA binding motif (151). The Ku70 SAP motif appears to possess legitimate (albeit weak) DNA binding because it has been shown by independent laboratories using Southwestern blotting protocols that the Ku70 subunit by itself can bind to DNA (98,152–155). Importantly, the DNA recognition helix of the SAP domain is exposed to the solvent suggesting that in vivo it should be capable of interacting with DNA. Whether this domain accounts for some of the controversy surrounding site-specific binding by Ku (see below) warrants further investigation.

Third, the crystal structure of Ku was incomplete in one important regard: it utilized a Ku86 polypeptide that was missing, presumably for technical reasons to facilitate crystalization, the last ~20 kDa of its C-terminus. Walker et al. (33), however, did note that the last residue of the C-terminus of the truncated Ku86 in their crystal structure was situated proximal to the DNA end (Fig. 2A). Mechanistically, this is consistent with the demonstrations that the Ku86 C-terminus encodes a DNA-PK$_{cs}$ recruitment motif (156–160) and that overexpression of just the Ku86 C-terminus in mammalian cells results in a dominant negative IR sensitivity (161,162). Thus, if the C-terminus of full-length Ku86 is proximal to the free DNA end, it would be well situated to provide a landing pad to recruit DNA-PK$_{cs}$. Subsequently, the structure of a bacterially expressed C-terminal fragment of Ku86 was determined by NMR (163,164). This globular region consists of six α-helices connected by loops and does not bear significant structural homology to any proteins currently in the structural database. There is nothing particularly remarkable about this region, except for a couple of pockets of hydrophobic residues, which have been postulated to mediate protein:protein interactions (164). Unfortunately, in one important aspect the NMR structure was also incomplete because it was determined with a Ku86 polypeptide lacking 2.5 kDa (23 amino acids) from the extreme C-terminus. Since removal of even the C-terminal 12 amino acids of Ku86 will disrupt the interaction with DNA-PK$_{cs}$ (157), it is likely that the Harris et al. (164) NMR structure, like the Walker et al. crystal structure, lacks the critical DNA-PK$_{cs}$ interaction domain. When the entire C-terminal region was used for NMR, the polypeptide assumed a completely random coil conformation (164). Given the defined structure obtained with the truncated polypeptide, the authors cogently argued that in vivo, the binding of DNA-PK$_{cs}$ to Ku may cause an induced-fit transition in the C-terminus of Ku86 from a random coil to an α-helical structure. Future structural studies will certainly be focused on resolving this important domain/interaction. Lastly, it is also relevant to note here that the DNA-PK-defective form of Ku86 that Walker et al. crystallized was virtually identical to the C-terminally truncated Ku86 that has been repeatedly observed by immunoblotting in a bevy of mammalian cell lines (156,159,165–171). Careful manipulation of the cell extract conditions, however, demonstrated that this truncated form of Ku86 was generated by a leupeptin-sensitive, trypsin-like serine protease that is activated upon cell extraction (167,170). Thus, it is still unclear whether these C-terminally truncated forms of Ku86 are extraction artifacts or whether proteolysis of the C-terminal region of Ku86 is biologically relevant. Intriguingly, proteolysis of Ku86 appears to only take place on DNA-bound Ku (165). Since, the crystal structure of Ku was nearly identical in the presence and absence of DNA (33) and since Ku interacts only weakly with DNA-PK$_{cs}$ in the absence of dsDNA (172), this observation lends support to the contention that the C-terminus of Ku86 undergoes a conformational change (from protease insensitive to sensitive) when it interacts with DNA-PK$_{cs}$ (164). Purification and characterization of the Ku86 C-terminal protease would clearly be enlightening.

Fourth, almost every paper written about Ku contains an introductory sentence stating that Ku binds in a sequence nonspecific fashion onto the ends of dsDNA regardless of whether the ends are blunt-ended, contain 5'- or 3'-overhangs or are hairpinned. The structure of Ku is completely consistent with such claims. Unfortunately, many papers often go on to suggest that Ku can bind to nicks, gaps and/or other transitions between single-to-double-stranded DNA. With one exception, the structure of Ku suggests that such scenarios are highly unlikely. Telomeric DNA does contain a single (the G-strand overhang) -to-double (the telomere itself) strand transition (Fig. 1E) and it is probable that Ku binds there (173–176). In this instance, however, the proximal chromosome end would readily allow Ku access to this site. Indeed, when telomeric sequences are ectopically integrated at an internal location on a chromosome, Ku has no affinity for them (177). The evidence that Ku can bind to internal single-to-double-stranded transitions and/or nicks and gaps comes from studies in which annealed or ligated oligonucleotides or nicked or gapped plasmids were gel-purified and utilized for in vitro binding or DNA-PK kinase assays (101,102,111). In retrospect, it seems likely that some of these DNA substrate preparations may have been contaminated with DNA molecules containing frank double-stranded broken ends. Thus, Ku is not likely to bind to nicks, gaps or single-to-double-stranded transitions in vivo and DNA-PK is not activated by such substrates (178,179). Indeed, there is no compelling evidence to suggest that Ku plays a role in either base [reviewed by Fortini et al. (180)] or nucleotide excision repair [reviewed by Costa et al. (181)], where such DNA substrates are routinely generated.

2.3. Structural Implications for Ku's Role in Transcription and DNA Replication

The structure of Ku supports a role for Ku in DNA repair, recombination and telomere biology as an end-binding protein. In addition, however, there are many reports suggesting that Ku may also play a role in transcription and/or DNA replication. Thus, Ku has been shown to repress RNA polymerase I transcription (182) and to affect RNA polymerase II transcription reinitiation (183). Moreover, the literature is rich with reports demonstrating that Ku can facilitate, in the context of DNA-PK, the in vitro phosphorylation of a veritable bevy of transcription (e.g., SP1) (34) or DNA replication (e.g., RPA) (184) factors [exhaustively reviewed by Lees-Miller (185)]. The implication of all these studies has always been that said phosphorylation would positively or negatively alter the factor's ability to promote transcription or replication. Given, however, the structural requirement for free dsDNA ends for Ku binding (a situation that almost certainly does not exist in vivo during normal transcription and DNA replication), and the strong promiscuous in vitro ability of purified DNA-PK in the presence of dsDNA to phosphorylate virtually any protein, the biological significance of most of these studies awaits complementary experimental evidence [reviewed by Smith and Jackson (186)]. In fact, the bias that Ku/DNA-PK is irrelevant for general transcription or DNA replication is supported by genome-wide gene expression profiling studies on DNA-PK defective cell lines, which show minor differences in comparison to DNA-PK proficient lines

(187,188) and the lack of significant cell division defects in Ku/DNA-PK defective cell lines [reviewed by Zdzienicka (189)], respectively.

Similar dismissive arguments can be made for a role of Ku as a sequence-specific transcription/replication factor. Thus, Ku has repeatedly been reisolated as a sequence-specific transcription factor, e.g., nuclear factor IV (NF-IV) (99,190,191), proximal sequence element binding protein 1 (PSE1) (192,193); an Epstein Barr virus promoter binding protein (194); a herpes simplex virus promoter binding protein (195); an E_2F motif binding protein (196); osteocalcin fibroblast growth factor response element (OCFRE) binding factor (197); a collagen III promoter binding factor (198); transferrin receptor enhancer binding factor 1 (aka Ku86) and 2 (aka Ku70) (TREF1 and TREF2) (199); an NF-κB promoter binding factor (200,201); an octamer motif binding factor (202); glucocorticoid receptor enhancing factors 1 (aka Ku86) and 2 (aka Ku70) (GREF1 and GREF2) (203); CTC binding factor (CTCBF) (204); enhancer 1 binding factor (E_1BF) (205–207); a T-cell receptor enhancer binding protein (208); negative calcium response element binding protein (nCAREB) (209); glucose responsive proteins 78 kDa (aka Ku70) and 94 kDa (aka Ku86) (GRP78 and GRP94) promoter binding protein (210,211); a heat shock element binding factor (212,213); glycophorin B (GPB) promoter repressor protein (214) and negative response element 1 (NRE1) binding protein (215,216). Similarly, Ku has been identified by three independent laboratories as an origin of DNA replication binding protein: as origin binding factor 2 (OBF2) (217), a factor bound to the B48 origin (218), and as origin binding activity (OBA) (219,220). However, since the dizzying array of sequences to which these factors bind to have no common consensus motif and since most of these experiments were performed using crude whole cell extracts and EMSAs where Ku is known to be a major nonspecific contaminating activity (221), the biological significance of these studies is unclear. The most compelling cases can be made for NRE1 (215) and OBA (219) where the respective laboratories have taken painstaking care to show that sequence-specific Ku binding can occur even within the context of a circular plasmid. There are at least two possible explanations for this apparent paradox of Ku DNA binding in the absence of a free DNA end. One possibility is that the SAP domain in the Ku70 C-terminus is responsible for the binding. As noted above, it has been repeatedly demonstrated by Southwestern blotting protocols that this domain possesses DNA binding activity (98,152–155) and the crystal structure suggests that the SAP domain is exposed to the solvent (33) and is thus available for interaction with DNA. This hypothesis is not consistent, however, with the contention that it is the Ku86 subunit that constitutes the DNA binding component of OBA (219,220). Site-directed mutagenesis of the SAP region coupled with functional complementation studies could be used to address the relevance of this hypothesis directly. A second possibility, originally suggested by Walker et al. (33), is that the NRE1 and OBA DNA sequences may be prone to forming hairpins. This model is consistent with the observation that Ku, which normally has extremely weak ssDNA binding activity (101,102), can bind tightly to the NRE1 sequence when it is single-stranded (222). Since Ku binds tightly to hairpin DNA (100,102)—indeed, the crystal structure of Ku complexed to DNA was done essentially with hairpin DNA (33)—this hypothesis explains how Ku could, in a manner completely consistent with its normal structural requirements, bind in a pseudosequence-specific fashion to closed, circular DNA. If this hypothesis is correct, it will be incumbent upon future investigators to confirm the existence of hairpins at these sequences in vivo.

Lastly, it should be emphasized that there are two instances where the evidence for Ku's involvement in DNA replication is mounting. In both of these cases, however, the effects are likely to be indirect. Thus, the presence or absence of Ku has been shown to affect the reinitiation of replication following DNA damage (223,224). This phenomenon is likely due to the involvement of Ku in the repair of broken replication forks and/or by "capping" the end of a broken chromosome and preventing the disassembly of the replication machinery (224). On a tangential note, this later study in particular is potentially a paradigm shift because it emphasizes for the first time that Ku may cap the ends of DNA to keep factors on—that might otherwise slip off—the broken chromosome, rather than capping the ends of DNA to protect them from nucleases and/or pathological recombination reactions as is normally envisioned (225). Secondly, Ku-deficient yeast cells are altered in their ability to utilize telomeric origins (226) and in their ability to replicate telomeric DNA (227). As a telomere-binding protein (see below), however, Ku is known to effect the chromatin structure at the end of a telomere (56,228) and it is likely that at least the defects in telomeric origin utilization is an indirect effect of this function.

In summary, there is currently little compelling evidence to suggest that Ku plays an essential and/or significant role in transcription or DNA replication. If Ku does play a role in either process, it is likely to be an indirect outcome of Ku's normal end binding activity and/or its ability to interact with chromatin remodeling proteins (see below).

2.4. Structural Implications for the Functions of the N- and C-Terminus of Ku70 and Ku86

The crystal structure of human Ku and the existence of truncated prokaryotic homologs suggest that the central β-barrel domain is likely to be sufficient for dimerization and DNA binding. If this is true, it begs a larger question of what is the function of the N- and C-terminal helical domains and the unique C-terminal regions (henceforth collectively referred to as the auxiliary regions) that each of the eukaryotic subunits possesses? One possibility is that the auxiliary regions perform a regulatory function for the DNA binding domain. It has been known for almost two decades that Ku can be phosphorylated in vivo (184,229,230). Moreover, there is good biochemical evidence that autophosphorylation of Ku by DNA-PK promotes either the disassembly of the complex from DNA ends (147,172,231,232) or Ku's inward translocation along the DNA (233). Unfortunately, when the DNA-PK-mediated phosphorylation sites were mapped (Ku70 at serine 6; i.e., within the N-terminal α-helical domain and Ku86 at serines 577 and 580 and threonine 715; i.e., within the C-terminal helical and unique regions) they did not conform to canonical DNA-PK phosphorylation sites (231). Moreover, there have been no subsequent follow-up site-directed mutagenesis studies to examine the validity of these sites and thus their biological relevance is unknown. The possibility that the auxiliary regions of the Ku subunits can be modified by phosphorylation or other regulatory modifications—such as a recent report describing the acetylation of the C-terminus of Ku70 (234)—is an area of Ku research that deserves significantly more effort than it has received to date.

A second possible function for the auxiliary regions is that they serve as protein:protein interaction domains. In stark contrast to the paucity of literature surrounding regulatory modifications of Ku, the Ku interaction literature is rich with

Table 1 DNA-PK Interactions with Proteins That Are of Unknown or Questionable Biological Significance

	Reference
Ku70 with:	
Apolipoprotein J	Yang et al. (1999; 2000) (235,236)
HOXC4	Schild-Poulter et al. (2001) (237)
p21	Kumaravel et al. (1998) (238)
Vav	Romero et al. (1996) (239)
Ku86 with:	
Somatostatin	Le Romancer et al. (1994) (240)
Tyk2	Adam et al. (2000) (241)
Vav	Adam et al. (2000) (241)
Ku with:	
CD40	Morio et al. (1999) (242)
RBP-Jκ	Lim et al. (2004) (201)
REF1	Chung et al. (1996) (209)
TBP	Genersch et al. (1995) (204)
DNA-PK$_{cs}$ with:	
Lyn	Kumar et al. (1998) (243)
PKCδ	Bharti et al. (1998) (244)
DNA-PK with:	
EGFR	Bandyopadhyay et al. (1998) (245)
HSF1	Huang et al. (1997) (246)
NF45/NF90	Aoki et al. (1998) (247); Ting et al. (1998) (248)
Progesterone Receptor	Sartorius et al. (2000) (249)

reports. Indeed by almost any definition, Ku qualifies as a "sticky" protein. In yeast, where whole genome approaches have been applied, Ku70 is known to interact with at least nine and Ku86 with an astounding 64 other proteins (http://www.yeast genome.org). In mammals, over 40 discrete interactors have been described (e.g., see Tables 1 and 2). For some proteins, there is simply no obvious biological connection, and many of these reports have not been followed up (Table 1). For other proteins, however, there is significant direct or circumstantial evidence to suggest that the interaction is biologically meaningful (Table 2). While an exhaustive discussion of each interaction is prohibitive, certain interactions should be highlighted:

CBP, PCAF, GCN5, HP1α & Sir4: CREB binding protein (CBP), p300/CBP-associated factor (PCAF) and general control nonderepressible 5 (GCN5) all possess histone acetyltransferase activity [reviewed by Roth et al. (283)]. Heterochromatin protein 1α (HP1α) is a nonhistone protein that is tightly associated with heterochromatin (284) and telomeres and its expression prevents telomere fusions (285). Silent information regulator 4 (Sir4) is a component of the SIR complex that inhibits gene expression at mating-type loci and at telomeres by packaging DNA into heterochromatin (286). The common activity of these proteins is that they are either associated with heterochromatin (HP1α and Sir4) or are used to alter it (CBP, PCAF, and GCN5). Since all five of these proteins interact with Ku (Table 2), it is tempting to speculate that Ku may perform some of its functions by facilitating the formation and disassembly of heterochromatin (228). A simple scenario would be that

Table 2 DNA-PK Interactions with Proteins That Are of Likely Biological Significance

	Reference
Ku70 with:	
Bax	Sawada et al. (2003) (250,251)
CBP	Cohen et al. (2004) (234)
GCN5	Barlev et al. (1998) (252)
HP1α	Song et al. (2001) (253)
Mlp2	Galy et al. (2000) (254)
Mre11	Goedecke et al. (1999) (255)
PCAF	Cohen et al. (2004) (234)
Sir4	Tsukamoto et al. (1997) (256)
TRF2	Song et al. (2000) (257)
WRN	Karmakar et al. (2002) (258)
Ku86 with:	
DNA-PKcs	Gell and Jackson (1999) (157); Singleton et al. (1999) (158)
Sir4	Roy et al. (2004) (259)
WRN	Li and Comai (2000) (260); Li and Comai (2001) (261); Karmaker et al. (2002) (258)
Ku with:	
cAbl	Jin et al. (1997) (262); Kharbanda et al. (1997) (262); Kumaravel et al. (1998) (238)
LigIV/XRCC4	Leber et al. (1998) (264); Chen et al. (2000) (265); Nick McElhinny et al. (2000) (42); Hsu et al. (2002) (266); Kysela et al. (2003) (148)
PARP	Galande and Kohwi-Shigematsu (1999) (267); Sartorius et al. (2000) (249)
PCNA	Balajee and Geard (268)
TdT	Mahajan et al. (1999) (269)
Telomerase	Chai et al. (2002) (270)
TLC1	Peterson et al. (2001) (271); Stellwagen et al. (2003) (272)
TRF1	Hsu et al. (2002) (273)
VLPs	Downs and Jackson (1999) (274)
WRN	Cooper et al. (2000) (275); Orren et al. (2001) (276); Karmaker et al. (2002) (160); Li and Comai (2002) (277)
DNA-PK$_{cs}$ with:	
Artemis	Ma et al. (2002) (37)
Ku	Jin et al. (1997) (262)
LigIV/XRCC4	Chen et al. (2000) (265)
PP5	Wechsler et al. (278)
RPA	Shao et al. (1999) (279)
WRN	Yannone et al. (2001) (280)
XRCC4	Hsu et al. (2002) (266)
DNA-PK with:	
LigIV/XRCC4	Calsou et al. (2003) (147)
P53	Achanta et al. (2001) (281)
TdT	Mickelsen et al. (1999) (282)

immediately after a DSB Ku binds on to the free ends and by recruiting HP1α and/ or Sir4 establishes a heterochromatic state at the site of the break to protect it from unwanted modifying activities. Then, as the DNA repair response proceeds, Ku may recruit CBP, PCAF, or GCN5 to help open up the heterochromatin to let the appropriate DNA repair enzymes gain access to the broken end. A similar scenario could apply at telomeres, where Ku with the assistance of HP1α and/or Sir4 may keep the telomeres in a heterochromatic state—a condition that indeed normally exists in vivo [reviewed by Chan and Blackburn (76)]—until access to the telomere needs to be gained; during, for example, DNA replication (226). It should be emphasized that Ku does not need a telomere or even a DSB to establish a heterochromatic structure, since Ku was capable of inhibiting gene expression when it was artificially tethered to internal chromosomal sites (228). This observation suggests that the DNA binding domain (i.e., the central β-barrel region) of Ku is likely dispensable for this heterochromatin generating function. In the case of HP1α, the domain of Ku70 required for interaction was mapped to a large region [amino acid (AA) 200–385] encompassing both the N-terminal domain and the central β-barrel region (253). For GCN5, the interaction domain was shown to reside between AA349 and AA499, a region that encompasses the central β-barrel region and the C-terminal helical region. Since the interaction domains for HP1α and GCN5 do not overlap significantly, it is conceivable that Ku70 could interact with both proteins simultaneously. The situation for Sir4 is more complex as the original interaction was described as occurring with Ku70 (256). This interaction was recently confirmed, but Roy et al. (259) concluded that there was also a direct (and stronger) interaction with Ku86. Moreover, point mutations within Ku86 were isolated that disrupted the interaction with Sir4. These mutations (ca. AA420 in humans) map to the most N-terminal portion of the C-terminal arm, just outside the central β-barrel region. Interestingly, molecular modeling (these studies were done with yeast Ku86) of the mutations onto the structure of human Ku suggests that these residues are normally partially buried. The authors concluded that this region, therefore, may not constitute a direct Sir4 binding site for Ku86, but rather suggested that the mutations in this region caused a conformation change in either Ku86, or the heterodimer, that disrupted the interaction with Sir4 (259). Given the importance of the Ku:Sir4 interaction, it will be important in the future to determine the precise geometry of the interactions between Ku70, Ku86, and Sir4.

Mlp2: In yeast, interphase chromosomes cluster near the nuclear periphery (287). Mutation of either Ku70, Ku86, or myosin-like protein 2 (Mlp2) in yeast disrupts this clustering, resulting in chromosome mislocalization (254,288). Myosin-like protein 2 is a protein that associates with the nuclear pore complex (289) and makes physical contacts with Ku70 (Table 2) (254). Thus, it seems likely that telomeric Ku70, using Mlp2 as a bridging protein, tethers chromosomes to nuclear pores. It is unknown which domain of Ku70 interacts with Mlp2 and whether this interaction is applicable to higher eukaryotes—both areas of worthwhile future investigation.

TRF1, TRF2, Telomerase, and TLC1: As briefly referred to above and as described in more detail below, Ku and DNA-PK clearly play important roles in telomere regulation. In this capacity, it is not surprising that they interact with telomere-specific factors (Table 2). Thus, Ku70 has been shown to interact with telomere recognition factor 2 (TRF2) (257) and the Ku heterodimer has been shown to interact with telomere recognition factor 1 (TRF1) (273). TRF1 and TRR2 are factors that recognize and bind to the double-stranded $T_2A_1G_3$ repeats found at the end of chromosomes. Importantly, they are integral components of the "cap" that seems to render telomeres invisible to the DSB machinery [reviewed by de Lange (67)]. In the case of TRF2, the domain of Ku70 required for interaction was demonstrated to be AA200–385, encompassing both the N-terminal domain and the central β-barrel region (257). Which domain, or even which subunit, of Ku that interacts with TRF1 was not determined (273). It is currently unclear what the function of TRF1/2:Ku interactions might be. One possibility would be maintenance of the t-loop structure that appears critical to the capping function [reviewed by de Lange (67)]. As with the Ku:Sir4 interaction, the exact geometry of the interactions between Ku and TRF1/2 is worth refining given the almost certainty of the biological relevance of the interactions.

Telomerase (hTERT) is the reverse transcriptase required for the synthesis of new telomeres. In humans, a direct interaction between Ku and telomerase using coimmunoprecipitations has been reported (Table 2) (270). Which subunit of Ku is required for this interaction was not determined. This observation is, however, consistent with the demonstration that human cells containing reduced levels of Ku86 due to a gene targeting inactivation of a single allele have shorter telomeres, which can be partially complemented by the reintroduction of a Ku86 cDNA (176). Whether the short telomere phenotype observed in Ku86 heterozygous cells is due to a reduction in the recruiting of telomerase to the telomeric end—a view supported by recent work on Ku86 in yeast (175)—or whether it is due to an actual decrease in telomerase activity (e.g., see below) is unknown. Lastly, while the utility of Ku recruiting telomerase to a telomere end to facilitate telomere elongation should be obvious, it is important to note that such a function would—except as a last resort—be exceptionally detrimental for Ku that is localized to a DSB where such an event would facilitate the loss of all genetic information distal to the break (272) [reviewed by Williams and Lustig (290)]. How Ku limits its interaction with telomerase to predominately telomeric ends is unknown, but it is likely mediated by other telomere-associated factors.

TLC1 in yeast encodes the RNA component of the telomerase complex and Ku is known to physically interact with TLC1 (Table 2) (271,272). The ability of Ku to interact with a variety of RNAs in vitro had been previously demonstrated, although none of these RNAs was capable of activating the kinase function of DNA-PK (291). In the case of yeast, which lacks a true DNA-PK$_{cs}$ homolog—and hence a DNA-PK complex—this may not be as relevant. In higher organisms, if such a Ku:telomeric RNA interaction is conserved, one would need to posit that telomeric RNA may uniquely be able to activate DNA-PK and/or that this interaction only requires Ku. Recently, the interaction between Ku and TLC1 was clarified when a synthetic lethality screen was used to identify a 5AA substitution in the N-terminus of Ku86 (AA53) that abolished RNA binding activity (272). Importantly, this mutation did not affect heterodimerization or DNA binding by Ku (consistent with the mutation being outside the central β-barrel region) and it did not affect NHEJ activities. Thus, this represents a "separation-of-function" mutation in which Ku86's telomere functions can be separated from its DNA repair functions. Previously, a 25AA C-terminal truncation of yeast Ku70 had also been shown to affect telomere maintenance more than end joining (292). Moreover, although Ku is known to bind to the stem-loop portion of TLC1 (271), a quasihairpin-like structure, these data imply that the RNA binding domain (potentially the N-terminus of Ku86 or the C-terminus of Ku70) is distinctly different from the DNA binding domain. This hypothesis is supported by an additional study, which, utilizing a similar experimental strategy, isolated a series of point mutations that inactivated Ku86's telomeric activity without apparently altering its DNA repair activities (175).

Although binding of the mutant Ku proteins to TLC1 RNA was not directly measured in this study, it is likely that some or all of the mutations affected this interaction. Importantly, five of the mutations map to the N-terminal α-helical domain including one at AA57 (175), very near the insertion site described by Stellwagen et al. (272), again implicating this region as being important for Ku86:TLC1 interactions. An additional two mutations map to the most N-terminal portion of the C-terminal α-helical arm, just beyond the central β-barrel region. How these mutations affect Ku86:TLC1 interactions, or whether they are affecting some other telomeric function of Ku, is not clear. One final mutation was a very nonconservative serine to proline mutation within the base of the central β-barrel region. The amino acid sequence in this region is not well conserved between yeast and humans although in both species it resides within a β-sheet region that separates two α-helical regions. Why this mutation affected Ku86's telomeric and not its NHEJ functions is unclear, although a direct demonstration of DNA binding activity for this mutant was not assessed. Together, these studies [reviewed by Bertuch and Lundblad (293)] suggest that an attempt to determine the crystal structure of Ku complexed to telomeric RNA would be mechanistically very informative.

WRN: Werner's syndrome (WRN) protein is an exonuclease and a helicase that is mutated in patients with a premature aging syndrome [reviewed by Opresko (294)]. WRN protein interacts with Ku/DNA-PK (Table 2), but how it does so is currently a matter of more than slight confusion. Some laboratories have shown that WRN protein makes contacts with Ku70 (258), and/or Ku86 (258,260,277), with the Ku heterodimer but not DNA-PK$_{cs}$ (258,275,277), or with DNA-PK$_{cs}$ but not Ku (280), or potentially in a complex with all three proteins (160,280). There is less, albeit some, confusion as to what the functional significance of these interactions entails. There is relatively good agreement that the interaction of WRN with Ku stimulates the exonuclease activity of WRN (260,275,280). Moreover, it has been reported that the interaction of WRN with Ku alters WRN's normal 3′–5′ exonuclease activity so that it can now work on alternative substrates such as ssDNA and blunt and 3′-protruding ends (261). Ku also appears to alter WRN's nuclease activity in a way that permits WRN to degrade DNA containing 8-oxoguanine modifications (276). And although there is one report to the contrary (277), two independent laboratories have concluded that the subsequent phosphorylation of WRN by DNA-PK is inhibitory (160,280). A working model that explains some of these results is that Ku recruits WRN to the ends of DSBs (and possibly also telomeres) (294) where, by altering/modulating WRN's activity, it facilitates the removal of damaged or modified bases at the site of the break. After completion of DNA repair, DNA-PK might phosphorylate WRN to allow for the ligation step to proceed. This model is, however, not consistent with the lack of an overt IR sensitivity in WRN patients, a phenotype universal to mutation of other NHEJ genes. In summary, the mechanistic details and the biological significance of WRN:Ku interactions are currently unresolved, but given that this is a very active area of research, it is anticipated that new information will be shortly forthcoming.

LigIV:XRCC4: Since Ku binds onto broken dsDNA ends and the LigIV: XRCC4 heterodimer is the enzymatic complex required to rejoin those same ends, it is not surprising that the proteins have been demonstrated to interact (Table 2). Initially, the functional significance of this interaction was unclear as there were reports of Ku both inhibiting (265) and activating (146) the ligation reaction. More recent reports have clarified the issue and suggested that lower (and probably more physiologically relevant) concentrations of Ku activate the ligation reaction whereas higher concentrations, probably by mass action, are inhibitory (147,148). A model consistent with most of the available data is that Ku initially binds onto a free DNA end. The loading of DNA-PK$_{cs}$ onto Ku displaces the Ku internally about 10 base pairs (295) while simultaneously activating the kinase activity of DNA-PK. An attractive scenario is that subsequent autophosphorylation of the complex results in the release of DNA-PK$_{cs}$ (265), while Ku remains attached to the DNA (147). This would then allow LigIV:XRCC4 to bind onto the free end (147), possibly facilitated by a DNA:adenylated LigIV complex (148). Thus, it is quite likely that Ku and LigIV:XRCC4 lie adjacent to one another and that most of the interactions between them are facilitated by the end of the broken DNA (147,148). This model is consistent with the observation that LigIV:XRCC4 binding and ligation are inhibited when Ku is restricted to a DNA end and not allowed to translocate internally (148), but it is not consistent with the demonstration that the kinase activity of DNA-PK$_{cs}$ is dispensable for LigIV:XRCC4 recruitment (147). Juxtapositioning of the broken ends would probably be facilitated by a tetrameric double heterodimer of LigIV:XRCC4—a complex which is believed to exist in vivo (42). Again, the obvious biological importance of this reaction warrants further research into precisely how these proteins interact and how they may modulate their respective activities.

Bax: One of the most unexpected recent findings concerning Ku was the compelling demonstration that in normally growing mammalian cells Ku70 utilizes a small 5 amino acid region in its unique C-terminal domain to bind to and sequester Bax in the cytoplasm (250,251). Bax is a proapoptotic protein that, in cells undergoing apoptosis, translocates from the cytoplasm to mitochondria, where it induces the release of cytochrome c, which, in turn, activates a proteolytic cascade that results in cell death [reviewed by Daniel et al. (296)]. Overexpression of this 5 amino acid motif in cells is sufficient to bind Bax and suppress apoptosis while not interfering with Ku heterodimerization, as would be predicted from the structure. The binding site maps to the SAP domain, just in front of the recognition helix and appears to be exposed to the solvent and hence competent for interaction with Bax. More than any other work, this report potentially opens new doors into studies on alternative (non-DEB related) Ku functions. It has, for example, lead to the discovery that Ku70's interaction with Bax can be regulated by acetylation of the Ku70 C-terminus (234). Additional work in this area should prove quite interesting.

In summary, Ku makes direct interactions with many proteins. With the exception of Bax, all of these interactions are likely to result in Ku recruiting these proteins either to the sites of DSBs or to the ends of telomeres. While the precise regions of Ku required for these interactions is in most cases currently unknown, it is likely that many of them will reside in the auxiliary regions. The identification, structural characterization, and biological understanding of these interactions are a goal of paramount importance for the field.

3. DNA-PK$_{CS}$

3.1. A Brief History

DNA-PK$_{cs}$ [encoded by the rather obliquely named protein kinase DNA-activated catalytic subunit (Prkdc) gene; (111)] entered our scientific consciousness unbeknownst to most researchers under the guise of the moniker "scid" (severe combined immune deficient). In 1973, a strain of Arabian foals was described that presented with hypogammaglobulinemia and thymic hypoplasia; i.e., they were scid (297). Unfortunately, the horse is a rather expensive and experimentally unwieldy model system and consequently little progress was achieved on the molecular basis underlying this interesting phenotype. A decade later, a group led by Melvin Bosma identified a mouse that contained extremely low levels of serum immunoglobulin (110). By breeding brothers against sisters, a colony of mice were established that were subsequently shown to be functionally B- and T-cell deficient; i.e., they were also scid (110). Even armed with a tractable model system, progress on scid languished for several years until it was demonstrated that the underlying molecular cause of the B- and T-cell deficiency was a defect in V(D)J recombination (112). This observation enticed an army of molecular immunologists into the fray and resulted in the rapid identification that the scid mutation profoundly and directly affected V(D)J coding junction formation (113–115). Work over the next 5 years focused on clarifying this process and culminated with the observation that the block to V(D)J recombination in the scid mouse resulted in coding ends that were stable hairpins (298). Unexpectedly, but very importantly, it was also demonstrated that the scid mutation was pleiotropic and resulted not only in the well-characterized, lymphoid-restricted immune deficits, but in a ubiquitous IR sensitivity (299) that was later shown to be due to a deficiency in generalized DNA DSB repair (300,301). Thus, after a decade's worth of research, the scid mutation was known to deleteriously impact site-specific recombination and DNA DSB repair.

Contemporaneously with the work described above, unrelated researchers were adding mRNA into rabbit reticulocyte lysates and observing that this resulted in the phosphorylation of proteins within the lysate (302). These investigators swiftly discovered that their mRNA preparation was contaminated with dsDNA and that it was this latter nucleic acid, which was inducing the phosphorylation (302). Subsequent investigations demonstrated that this "dsDNA-activated" kinase activity was not present in lower eukaryotes but was present in extracts derived from most vertebrates, including humans, and was responsible for the phosphorylation of the Sp1 transcription factor (303–306). The biochemical characterization of the DNA-dependent Sp1 kinase revealed that it consisted of the DNA-end-binding, heterodimeric Ku protein and a large catalytic subunit (i.e., DNA-PK$_{cs}$) (34). When Ku was linked the following year to DNA repair through its absence in several IR-sensitive hamster cells lines (27,107,116,117), the obvious implication was that DNA-PK$_{cs}$ could also be involved. The *tour-de-force*, three-year undertaking to clone the cDNA for DNA-PK$_{cs}$ fulfilled this prediction when the cDNA was mapped to the murine scid locus (109), confirming that DNA-PK$_{cs}$ was encoded by the genetically defined scid (subsequently renamed Prkdc) locus (108). Several years later, it was demonstrated that the original strain of scid Arabian foals was also defective in DNA-PK$_{cs}$, thus bringing full circle two-plus decades of scid-related research (307). Interestingly, unlike Ku, which is conserved throughout evolution, homologs of DNA-PK$_{cs}$ appear to exist only in vertebrates and apparently have evolved with the development of the vertebrate immune system [reviewed by Smith and Jackson (186)]. A putative mosquito homolog of DNA-PK$_{cs}$ has recently been described (92), however, the fact that the mosquito Ku86 gene is missing the C-terminal DNA-PK$_{cs}$ interaction domain suggests that this gene may simply be a closely related member of the phosphatidylinositol-3 kinase (PI3K) family [reviewed by Smith and Jackson (186)], to which DNA-PK$_{cs}$ belongs and not a true DNA-PK$_{cs}$ ortholog.

The murine scid mutation was known to be "leaky," i.e., while younger animals were profoundly immune deficient, older animals were able to generate an, albeit severely limited, B- and T-cell repertoire (308). Early speculation suggested that this might be due to a hypomorphic mutation, resulting in the production of a crippled, but still functional protein. This hypothesis was not supported by the demonstration that the murine (309,310) and equine (307) mutations resulted in unstable, C-terminally truncated polypeptides. Moreover, the leaky phenotype was recapitulated in murine DNA-PK$_{cs}$ knockout strains (311–314). Thus, while DNA-PK$_{cs}$ is extremely important for lymphoid V(D)J recombination, some rearrangements can take place in its absence. These studies are consistent with a DNA-damage-inducible (3), DNA-PK-independent form of NHEJ (Section 1.4) that with quite low efficiency can substitute for DNA-PK (52). In addition, characterization of DNA-PK$_{cs}$-null animals has revealed that, as observed for each of the Ku subunits, this protein is apparently important for the regulation of telomere length (315–317) and results in gross chromosomal rearrangements when absent (130,318,319). Lastly, while the basic defects observed in Ku70 and Ku86 knockout animals are recapitulated in DNA-PK$_{cs}$-null animals, there are some subtle differences implying that Ku and DNA-PK$_{cs}$ may have functions independent of the DNA-PK complex.

In summary, the history of DNA-PK$_{cs}$ provides a classic example of how researchers working on basic problems in quite disparate fields can suddenly find themselves strange bedfellows. Moreover, three decades of work has elevated DNA-PK$_{cs}$ from a relatively obscure disorder in an esoteric strain of horses to its acknowledged status as a critical component of vertebrate lymphoid V(D)J recombination and of the cellular response to DSBs.

3.2. Structure

Human DNA-PK$_{cs}$ is a polypeptide of 465,482 Da (108) and because of its daunting size or its functionally implicated flexibility it has, to date, defied x-ray diffraction. Undeterred, three independent laboratories have attempted to glean as much information as possible concerning the structure of DNA-PK$_{cs}$. Technically, these attempts have consisted of utilizing cryoelectron microscopy, which first yielded a structure of 21 Å (angstroms) (320), which was subsequently refined to 17 Å (321) and then 15 Å (322), and two additional electron microscopic attempts which yielded structures resolved to 22 (323) and 30 Å (324), respectively. The latter study, while at the lowest resolution, was significant because it contained the first attempt to compare DNA bound and unbound structures. Although at ~20 Å only the gross, overall structure of a protein can be determined, these studies have been remarkably consistent in their findings and their interpretations. In particular, DNA-PK$_{cs}$ appears to be monomeric in solution and consists of a nonglobular head or crown separated by a deep channel from a globular base (Fig. 3). The head domain is particularly interesting since it appears to be hollow or cage-like (320,323). While the volume of the interior of the head domain of DNA-PK$_{cs}$ is theoretically large enough to accommodate the Ku heterodimer (320), the existence of multiple tunnels and folds almost certainly precludes this possibility as does the lack of an opening (there are three) large enough through which Ku could gain access to the interior (Fig. 3; e.g., opening "F"). Instead, Leuther et al. (323) have argued that the openings may accommodate a dsDNA end and that actual repair (i.e., ligation) may take place within the cavity. This model has been elaborated or modified to envision that the synapsis of two DNA-PK$_{cs}$ molecules (39) may facilitate the binding of dsDNAs to the channels (Fig. 3; clefts "A" and "B") separating the head and the base. The dsDNA would then be threaded as ssDNA into the cavity where single-strands

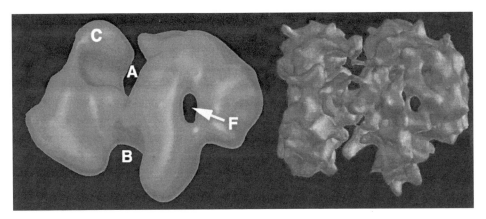

Figure 3 Structure of DNA-PK$_{cs}$. Cryoelectron microscopy 3D image reconstructions of DNA-PK$_{cs}$ from two different laboratories (shown in the same orientation) have revealed two large domains, termed the head (C; leftmost protein density) and the base (rightmost protein density), the latter of which encloses a large cavity. There are three openings to the cavity, only one of which (F) can be seen from this orientation while the other two openings (D and E) are on the back side of the molecule. The head and the base are separated by a deep cleft that is formed by two open channels (A and B). *Source*: On the left, the 22 Å resolution structure from Ref. 323. On the right, the 15 Å structure determined by single particle reconstruction from Ref. 322 reveals numerous additional surface features. Images were courtesy of Drs. Gilbert Chu and Phoebe Stewart. (*See color insert.*)

from two different ends would ultimately base pair (325,326). This model is consistent with the demonstration that dsDNA with single-stranded ends is the most potent activator for the DNA-PK kinase activity (326,327). On the other hand, this model is difficult to reconcile with the requirement for DNA-PK complex disassembly (see below), which apparently precedes the repair/ligation step (328). If DNA-PK$_{cs}$ needs to be released from the DNA end(s) in order to facilitate repair, it is unlikely that the ssDNA ends could be internally base paired within DNA-PK$_{cs}$.

A simplified model is suggested by the work of Boskovic et al. (324), who demonstrated that the deep channel that separates the head from the base domain is wide enough to accommodate dsDNA (Fig. 3, Cleft A). Moreover, the structure of dsDNA bound to DNA-PK$_{cs}$ suggested that the head and the base undergo significant conformational changes upon DNA binding and that they appear to clamp over the DNA. Stabilization of this closure was postulated to require about 15–20 bp of dsDNA—a length in very good agreement with the minimum length of DNA needed to activate the kinase (323,325,326)—such that it could make contact with both the head and the base domains. Whether or not any ssDNA ends could still be threaded into the internal cavity when DNA-PK$_{cs}$ was in this configuration was not addressed.

In summary, a high-resolution, x-ray diffraction-based structure of DNA-PK$_{cs}$ is not currently available and consequently the atomic details of the location and architecture of the various domains of the protein are unknown. The valuable structural data that are available at low resolution from electron microscopic image reconstructions are consistent with the ability of DNA-PK$_{cs}$ to bind/clamp onto the ends of DNA, where it undergoes significant conformational changes. Elucidation of how these conformational changes affect the catalytic core, DNA binding and interaction with other repair factors are critical for understanding the precise mechanism of NHEJ. This deficiency is perhaps the biggest challenge facing the DNA-PK field and it will certainly be the focus of much future research.

3.3. Domains

The paucity of detailed structural information concerning DNA-PK$_{cs}$ is exacerbated by the fact that much of the protein bears no homology to known structural and/or functional domains in the database. Several domains within the protein are, however, noteworthy. First, around AA1500–1550 is a hexaheptad repeat of leucines (Fig. 4). Heptad repeats of leucines are characteristic of a subset of coiled-coil proteins known as leucine zippers (329). Leucine zippers constitute protein:protein interaction domains and generally facilitate either homo- or heterodimerization with either self or other leucine zipper-containing proteins, respectively. Although there was some doubt due to thermodynamical/structural considerations as to whether the DNA-PK$_{cs}$ leucine repeat region formed a true leucine zipper (330,331), a leucine zipper-containing protein, C1D, was identified based upon its ability to interact with DNA-PK$_{cs}$ through this region (332). C1D, named for a phage clone and not as a biological acronym, is a tiny nuclear protein of 16 kDa that is tightly associated with chromatin (333). C1D contains no recognizable motif or domain other than its leucine zipper and C1D's biological function is unknown. The protein is, however, highly conserved in higher eukaryotes (334), implying that its function, whatever it may be, is likely to be conserved as well. Since C1D has DNA binding affinity, it was originally suggested that it may serve as a Ku-like surrogate in directing DNA-PK$_{cs}$ to DNA, perhaps at internal DNA sites (332). The fact that C1D homologs, however, exist in lower eukaryotes whose genomes do not encode DNA-PK$_{cs}$ homologs, suggests that this cannot be the only function of C1D. Recent studies using mutant C1D yeast strains have demonstrated that it unexpectedly may play roles in both NHEJ and HR (334). Even more confusing, however, was a report showing that the primary role of the yeast 25 kDa homolog of C1D, rRNA processing 47 protein (Rrp47p), is as a component of the rRNA exonucleolytic exosome (335). If the later report is verified, it is likely that the effects of C1D/Rdp47p mutations in yeast on DNA repair may be indirect. The functional significance of the DNA-PK$_{cs}$:C1D interaction clearly needs to be investigated in more detail and the generation of murine C1D knockout strains, if viable, should be informative. A second leucine zipper-containing protein, Ku86 autoantigen related protein-1

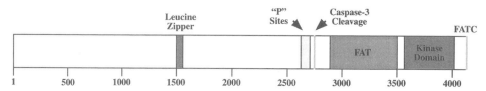

Figure 4 Schematic drawing of DNA-PKcs showing potentially biologically relevant domains. The protein is represented as an open rectangle and the approximate AA positions are shown below. As indicated, shaded rectangles represent the leucine zipper region and a region containing at least six autophosphorylation ("P") sites. Note that another autophosphorylation site at AA3205 is not within this region and is not shown. Additional rectangles correspond to the DEVDN motif cleaved by caspase-3 during apoptosis and to the FAT, kinase catalytic core and FATC domains, respectively. See the text for a fuller description of all these domains.

(KARP-1) was also postulated to interact with the DNA-PK$_{cs}$ leucine zipper domain (336). Results from directed two-hybrid interaction assays, however, using both of these protein domains have consistently been negative (EAH, unpublished observations) and thus the biological relevance of this interaction remains unknown. Moreover, an alternative leucine zipper partner KARP-1 binding protein 1 (KAB1) that does interact with the leucine zipper of KARP-1 has been described (337). The function of KAB1 is unknown although its overexpression seems to protect against oxidative stress, indicative of a role in DNA damage response (337). It would be informative to determine whether KAB1 interacts with DNA-PK$_{cs}$ as well. In summary, the lack of structural information about whether the DNA-PK$_{cs}$ leucine repeat region actually forms an alpha helical structure and/or whether this domain is even exposed on the surface of the protein precludes a rigorous assessment of its importance. Moreover, neither of the two proteins, C1D and KARP-1, that have been implicated as DNA-PK$_{cs}$ binding partners have been compellingly linked to NHEJ.

Second, around AA2600 is a series of six or more serine and threonine residues that are the targets of autophosphorylation (328,338–340) [reviewed by Lees-Miller and Meek (20); Fig. 4]. The preferred AA sequence for DNA-PK phosphorylation in vitro is a serine (S) or threonine (T) residue followed by glutamine (Q), i.e., an "S/T-Q" motif (341). Five of the seven identified sites correspond to this motif and four of these are conserved in all six of the sequenced DNA-PK$_{cs}$ homologs (339,340). Not surprisingly, the same four sites are phosphorylated in vivo in phosphatase inhibitor-treated cells (339). Autophosphorylation of DNA-PK$_{cs}$ is important since it has been known for over a decade (306) that self-modification by DNA-PK leads to disassembly—and hence, inactivation—of the complex (147,172,231,232). The functional significance of the putative autophosphorylation sites was demonstrated using alanine substitution mutagenesis for the respective serine or threonine residues and then testing the expression of the altered cDNA expression constructs for their ability to functionally rescue the ionizing radiation sensitivity of DNA-PK$_{cs}$-defective cell lines. In all cases, the expression of the alanine-substituted DNA-PK$_{cs}$ proteins resulted in the lack of complementation (328,338,340). Indeed, several mutations altered at multiple sites give rise to increased radiation sensitivity over the parental DNA-PK$_{cs}$-defective cell line, implying a dominant negative effect in which the mutated DNA-PK complex could bind to, but could not be disassembled from a DNA DSB end, thereby sequestering the break from other DNA DSB pathways. Where tested, the alanine-substitution mutants were also incapable of functionally complementing the V(D)J recombination defect of the DNA-PK$_{cs}$-null cells (328); a result consistent with the known role of DNA-PK$_{cs}$ in this process. These experiments unequivocally demonstrated the importance of autophosphorylation. Still, the story is likely incomplete as other kinases, in particular cellular Abelson (cAbl), are known to phosphorylate DNA-PK$_{cs}$ (263), although the location of these modifications and their biological relevance has not yet been established. Lastly, DNA-PK$_{cs}$ is an exceptionally large protein with an exceptionally long half-life ($\sim$5 days) (342) and thus energetically and mechanistically it makes much more sense to recycle old protein rather than resynthesize new protein. In this context, it is perhaps not surprising that the inhibitory autophosphorylation of DNA-PK$_{cs}$ can be reversed by the action of phosphatases, with protein phosphatase 1 (PP1) (232), a PP2A-like (230) and PP5 (278) phosphatases being implicated in the reversal.

Third, located at AA2709–2713 is the amino acid sequence DEVDN, which is a canonical caspase-3 cleavage site [reviewed by Shi (343); Fig. 4]. In cells undergoing apoptosis, DNA-PK$_{cs}$ (but not either Ku subunit) is proteolytically cleaved at this site by caspase-3 and inactivated (344–349). It is thus reasonable to predict that these residues (AA2709–2713) will be on the surface of the DNA-PK$_{cs}$ protein, where they would be exposed to the solvent and readily accessible to caspase-3. In addition, it seems reasonable to postulate that the inactivation of DNA-PK$_{cs}$ by caspase-3 facilitates apoptosis. Thus, one of the critical steps in apoptosis is the caspase-mediated activation of a DNase (350), which randomly cleaves chromosomes into smaller DNA fragments. The concomitant inactivation of DNA-PK$_{cs}$—and consequently DNA DSB repair—would clearly enhance the efficacy of the DNase and thus expedite the apoptotic process. Interestingly, poly(ADP-ribose)polymerase (PARP) is also a target of caspase-3 and it is known that cleavage of PARP separates its DNA binding domain from the catalytic domain, resulting in a dominant negative fragment capable of binding DNA ends, but not ribosylating (351). Determining whether caspase-3-mediated cleavage of DNA-PK$_{cs}$ results in a similar outcome for DNA-PK would be of mechanistic interest and potentially of biological importance.

Fourth, stretching from AA2908–3539 is a FAT (FRAPP, ATM, and TRAPP) domain (331) (Fig. 4). This large domain is conserved, albeit poorly (~16% sequence identity), in all phosphatidylinositol kinase-related family members. The function of this region is unknown, although recent work on the DNA-PK$_{cs}$ homolog, ataxia telangiectasia mutated (ATM), which is also a FAT domain-containing protein, has provided some tantalizing suggestions. In ATM, autophosphorylation within this domain appears to regulate the transition between a dimer (inactive) and a monomer (active) form of ATM (352). Thus, FAT may define a homodimerization domain. Interestingly, one of the seven autophosphorylation sites identified for DNA-PK$_{cs}$ (339) maps within the FAT domain and it would be intriguing to see if this site/domain regulates DNA-PK$_{cs}$ in a similar fashion. Alternatively or in addition, the FAT domain may facilitate interaction with Ku, as a region between AA3002 and AA3850 (which encompasses FAT and more C-terminal residues) of DNA-PK$_{cs}$ has been shown to be required for this interaction (262).

Fifth, DNA-PK$_{cs}$ and all other phosphatidylinositol kinase-related family members contain a well-conserved (34% sequence identity), small (35AA long) domain, FAT extreme C-terminal region (FATC), at their C-terminus (Fig. 4). Interestingly, this region is specifically deleted by a premature nonsense mutation in the original scid mouse, although the mutation also affects the protein's stability, such that the overall protein expression levels are greatly reduced (309). As with the FAT domain, the biological function of the FATC region remains unclear and to date no site-directed mutagenesis studies or FATC domain-interacting proteins have been described.

Sixth and last, the catalytic kinase domain resides between residues AA3645 and AA4049 (Fig. 4). A long-standing major bias in the field has been that the kinase activity of DNA-PK would be essential for its function. Indeed, all DNA-PK$_{cs}$-defective cell lines described to date are deficient in DNA-PK kinase activity and this always correlates with deficiencies in NHEJ, V(D)J recombination, switch recombination, and/or telomere maintenance [reviewed by Smith and Jackson (186)]. Complementary experiments in which cells acquired NHEJ-defective phenotypes when DNA-PK activity was abolished using specific inhibitors substantiated this bias (83,353–357). These studies have been elegantly augmented with functional complementation experiments from several laboratories in which either a wild-type DNA-PK$_{cs}$ cDNA or one containing a missense mutation inactivating the kinase catalytic core were introduced into DNA-PK$_{cs}$-defective cell lines. In every case, the wild-type cDNA was able to complement, whereas the mutated versions could not (354,358,359). Altogether, these studies unequivocally demonstrated the essential nature of the kinase activity for DNA-PK function. More recent studies, however, have shown that while kinase activity is essential, it is not sufficient. Thus, the DNA-PK$_{cs}$ constructs that contained the multiple serine or threonine to alanine mutations that were defective in autophosphorylation, still retained an unaltered kinase domain. Although there was no direct in vivo demonstration that the kinase domain of these proteins was still functional, the fact that cell lines expressing these constructs were able to open up V(D)J recombination-generated coding hairpins (which is thought to require the phosphorylation of Artemis by DNA-PK$_{cs}$) (37) implies that this was so. In spite of this functional kinase activity, however, completion of V(D)J recombination was blocked, because the DNA-PK$_{cs}$ protein was incapable of autophosphorylation (328). In their entirety, these experiments argue very strongly that the kinase activity of DNA-PK is necessary but not sufficient for V(D)J recombination and NHEJ. Moreover, the implication is that DNA-PK$_{cs}$ itself is perhaps the most critical downstream target of DNA-PK. Needless to say, a structural determination of the catalytic core of DNA-PK$_{cs}$ would be very informative. Insight into how DNA-PK$_{cs}$ is able to phosphorylate all of its putative substrates (e.g., Artemis, Ku, Lig4:XRCC-4, WRN) and itself (in this case presumably in *trans*?) would be greatly benefited by structural information.

4. DNA-PK, TELOMERES AND GENOMIC STABILITY

4.1. Introduction

Modern biology is replete with examples of genes and processes that are highly conserved throughout evolution. Ku is clearly one such gene and the pervasive existence of telomeres for the problem of chromosome end maintenance is a good example of a conserved process. Yet despite the near universal existence of Ku and telomeres, there appears to be an almost perverse idiosyncratic madness to the role that Ku plays in telomere biology and the variety of mechanisms that different species utilize to maintain their telomeres. Thus, although one of the most intensely researched areas concerning DNA-PK over the past 5 years has been its role in vertebrate telomere biology, and even though much elegant, compelling data have been generated, no clear picture of the role that Ku/DNA-PK plays in this process has emerged. The one thing that does appear to be true is that while the results obtained for a given species are—by and large—reproducible from laboratory to laboratory, great care must be taken in extrapolating the results from one species onto another. Since a complete discussion of DNA-PK and telomeres would require a separate review, this section will focus solely on three topics: DNA-PK's role in (i) telomere length maintenance, (ii) G-strand overhang maintenance, and (iii) chromosomal stability.

4.2. Telomere Length Maintenance

The first report implicating Ku and telomere length maintenance came from the laboratory of Thomas Petes, which showed that *Saccharomyces cerevisiae* strains defective in Ku70 had a shortened, but stable, telomere phenotype (360). This observation has been exhaustively reproduced in many independent laboratories and extended to mutant Ku86 *S. cerevisiae* strains as well (56,175,228,272,

292,361–367). *Schizosaccharomyces pombe* (177,368,369) and trypanosome (370) Ku-null cells also show telomere shortening. In contrast to these organisms, however, the majority of Ku70-null chicken DT40 cell lines exhibited telomeres of parental length although telomeric expansions were observed in some independent subclones (371). Different yet from the sporadic telomeric expansions seen in the chicken—and in sharp contrast to what was observed with either yeast species and trypanosomes—*Arabidopsis thaliana* Ku-null plants showed consistent, massive telomeric expansions (372–374). Adding enormous confusion to these already disparate results is the problem that in mice—sometimes using identical strains of animals—multiple contradictory studies have been published. Specifically, there are claims of slight telomere shortening in DNA-PK$_{cs}$ animals (375), significant telomeric shortening in Ku heterozygotes and severe telomere shortening in Ku-null animals (129), as well as claims of telomeric expansions—some slight (131,376) some large (315,316)—and/or no discernable effects (317–319) in Ku- or DNA-PK$_{cs}$-null cell lines and animals. Lastly, in human somatic cells, the inactivation of even a single allele of Ku86 resulted in dramatic telomere shortening (176). Moreover, human Ku86-null cells were not viable (377) and this appears to be due to massive telomere loss (176). In summary, the loss or deficiency of DNA-PK or Ku activity can result in either shorter, the same or longer telomeres and the cellular consequences range from no obvious effects to cell death.

In most single-cell eukaryotes, the data are compelling that a deficiency in Ku leads to telomere shortening. Importantly, in yeast, a back-up pathway exists that—in the absence of telomerase—uses HR and unequal sister chromatid exchange to maintain and often hyperextend telomeric ends (378,379) [reviewed by Lustig (380)]. The existence of a similar pathway, which has evolved to become the primary mechanism of telomere length maintenance, could explain the absence of significant phenotypic effects of Ku mutations in the chicken DT40 cell line. This hypothesis is consistent with DT40's known heavy reliance on HR to carry out most of its recombinational processes (381) [reviewed by Winding and Berchtold (382)]. Similarly, it is tempting to speculate that a related HR-mediated mechanism is responsible for the remarkable telomeric expansions observed in *A. thalania* Ku mutants. This hypothesis is, however, not consistent with two observations. First, the telomeric expansions in plants are observed in the first generation of Ku-null plants (372,373). In contrast, a lag time is required in yeast to activate the HR-mediated pathway because the telomeres must first become shortened. Second and more importantly, the HR-mediated pathways of telomere expansion, including those in humans [reviewed by Henson et al. (383)], are generally telomerase-independent, whereas in *A. thalania* the telomere expansions required telomerase (374). Thus, plants appear to employ a unique, Ku-dependent mechanism of telomere length maintenance that currently defies explanation. The situation in mice is perhaps not as confusing as it superficially appears to be. In the case of the DNA-PK$_{cs}$ knockout mouse lines, there have been reports of slight shortening (375), slight elongation (317), and no change (318,319), which are all consistent with DNA-PK$_{cs}$ (and therefore also DNA-PK) not playing a significant role in telomere length maintenance in the mouse. The two reports that appear to be in contradiction with these studies by demonstrating significant telomere expansions in DNA-PK$_{cs}$-defective cells are actually not directly comparable as they were carried out using a different mouse line: the original scid mouse (315,316). Thus, different genetic backgrounds are likely to account for the phenotypic differences observed in these animals (319) and they reinforce the contention that many genetic loci are involved in telomere length regulation. There has been a single report on telomere length in Ku70-deficient mice and this demonstrated that there was significant telomere shortening (129). The only major confusion concerns Ku86, where two independent groups using the same knockout mouse strain have come to diametrically opposed conclusions demonstrating either significant shortening (129) or slight expansions (131,376). While telomere shortening (129) is consistent with the results from lower eukaryotes, murine Ku70 and human Ku86-defective cell lines, the slight telomere expansions (131,376) are more similar to what has been reported for the chicken DT40 Ku86 mutant cell lines and for the DNA-PK$_{cs}$ knockout murine cell lines. This important disagreement, which can only be explained by technical differences in data interpretation, needs to be resolved before an accurate assessment of the role of Ku in murine telomere length maintenance can be addressed.

At the heart of the matter, in spite of the caveats discussed above, is the question of why does telomere length generally change in Ku/DNA-PK mutant cell lines/animals? In the case of telomere shortening, the shortening occurs too rapidly to be due to the normal passive loss of sequences caused by incomplete replication (363). Instead, a plausible explanation is that Ku binds onto the end of telomeres and protects them from ongoing nucleolytic attack (Fig. 5C). This hypothesis is completely consistent with Ku's known ability to bind to telomeric DNA (129, 173–175,259,384,385) and to protect DNA ends from nucleolytic degradation (27,156). Consistent with the structural requirements for a free DNA end, it should be reemphasized that Ku's binding to telomeres is certainly facilitated by the open end of the chromosome as Ku has no affinity for telomeric sequences that are located internally within a chromosome (177). Telomeric binding and protection are at best, however, only a partial explanation because certain Ku mutant strains nonetheless have shortened telomeres even when the mutated Ku protein is still able to bind onto telomeric DNA (175,272). In addition, if Ku "only" performed a role in blocking telomeric ends, then the reintroduction of Ku into Ku mutant strains should result simply in a stabilization of the already shortened telomeres. Instead, the reintroduction of a Ku86 cDNA to trypanosome (370) and yeast (361) Ku86-null strains or a Ku70 cDNA to a yeast Ku70-null strain (227) resulted in a complete restoration of the telomere length to parental levels. Similar effects have been observed in human cell lines, although the complementation is only partial (176). Together, these studies suggest that Ku must also play an active role in telomere elongation. One possibility is that Ku is required for telomerase biogenesis/activity via its interactions with the telomeric RNA component. This hypothesis—which was discussed in detail in Section 2.4—is consistent with the demonstration in yeast that Ku86 will directly bind to telomeric RNA (271,272) and can be indirectly inferred from additional yeast mutants that have telomere defects, but are still proficient for DNA binding (175). Alternatively, Ku may recruit telomerase to the growing end of a telomere (Fig. 5D). This model is supported by the observation that human Ku86 heterozygous cell lines have reduced levels of telomerase activity and that Ku86 interacts with human telomerase, but not telomeric RNA (270). This function appears to be specific for Ku and not DNA-PK since murine DNA-PK$_{cs}$-deficient cell lines have wild-type levels of telomerase activity (318). This would also be consistent with the apparent differences in telomere lengths between Ku and DNA-PK$_{cs}$ knockout mouse strains discussed above.

In summary, a model consistent with most of the available data is that Ku, mediated by its binding to the single-to-double stranded transition at a telomeric end, performs two functions: (i) it primarily blocks the end, protecting it from

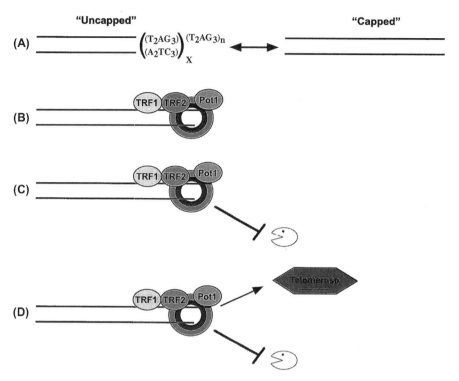

Figure 5 Illustrations showing Ku's speculative role in telomere length maintenance. (A) A cartoon showing only the telomeric DNA associated with "uncapped" and "capped" telomeric states. DNA is represented as a thick black line. The double-stranded, repetitive portion of the telomere and the G-strand overhang are shown. The left-hand panel shows the G-strand extended (uncapped) and the right hand panel shows a t-loop (capped). The double-headed arrow implies that these forms are interconvertible. (B) Some of the proteins known to associate with telomeric DNA. TRF1 and TRF2 are shown as ovals, respectively and they bind to the repetitive, double-stranded portion of the telomere. Pot1 binds to the G-strand overhang. The Ku heterodimer is shown as two rings. (C) One role for Ku is to inhibit nucleolytic degradation of the telomeric end. Nucleases are cartooned as a PacManTM. Inhibition is cartooned as a barred line. All other symbols are as in panel (B). (D) Ku may also help in the activation or recruit of telomerase. Telomerase is shown as a hexagon and the telomeric RNA is shown as extended horizontal line. Activation is cartooned as an arrow. The thickness of the lines is meant to imply their relative importance. All other symbols are as in panel (B).

nucleolytic attack and (ii) at least some of the time it may also recruit or activate telomerase to promote telomere elongation (Fig. 5D). It should also be noted that in the case of telomere shortening, it appears—somewhat paradoxically—that the resulting short telomere phenotype is stable. Thus, regardless of what Ku's actual function may be, it would have been reasonable to postulate that in Ku-defective cells the telomeres should become "ever shorter" until genomic catastrophe and/ or death ensued. Instead, in Ku-defective cells the telomeres shorten, but then are stable for innumerable generations (176,361,370). These data suggest strongly that there is a Ku-independent backup mechanism for protecting telomeres from progressing to the critically short length that promotes genomic catastrophe (386). Lastly, in the case of plants, Ku may act as a novel negative regulator of telomerase.

4.3. G-Strand Overhang Maintenance

All eukaryotic telomeres, with a few exceptions (387), end in a 3′-G-rich, single-stranded extension called the G-strand overhang (68,388–390) (Fig. 1). This structure, somewhat counterintuitively, is produced, not by the action of telomerase per se (391,392), but likely by the preferential resection of the C-strand (388,393,394). The nuclease responsible for the resection is not known. The trimeric complex of meiotic recombination defective 11/radiation sensitive 50/Nijmegen chromosome breakage syndrome 1 (Mre11/Rad50/Nbs1) is a nuclease that is known to be involved in telomere length maintenance (56,362) and was often implicated as the C-strand nuclease. More recent data, however, suggest that it is either exonuclease 1 (Exo1) (367,395,396) or a third (unidentified) nuclease (392). The G-overhang is bound by the sequence-specific, single-stranded binding protein, Pot1 (81), and at least some of the time it folds back internally into the chromosome to form a "t-loop" (70) [reviewed by de Lange (71)] probably mediated by the TRF1 and TRF2 proteins (Fig. 2B). Importantly, the integrity of the G-overhang, or lack thereof, has been implicated as being causal in senescence (397) and genomic instability (393). Lastly, the relationship between the length of the G-overhang and overall telomere length is currently unknown.

As observed with telomere length maintenance, the influence of Ku mutations on G-strand overhangs shows significant species variability. The laboratory of Raymund Wellinger was the first to demonstrate that mutations in Ku genes affected the G-strand overhang (384). Specifically, null mutations in *S. cerevisiae* of either Ku70 or Ku86 resulted in G-overhangs that were 2–5-fold longer (384). This phenotype has been repeatedly confirmed and elaborated in numerous independent laboratories (175,363,366). Similarly, in *S. pombe* (398) and *A. thaliana* (374) mutation of Ku70 results in longer G-strands, as does the heterozygous inactivation of Ku86 in human somatic cells (176). In contrast, however, mutation of Ku70 in chicken DT40 cells (371) and Ku86 (131) or DNA-PK$_{cs}$ (319,375) in the mouse does not result in any demonstrable changes in the G-strand overhang. The reader can be thankful that there are no reports (yet) of Ku mutations making G-overhangs shorter.

Mechanistically, the observation of elongated G-strand overhangs in Ku mutant cells can be accommodated by postulating that Ku binding to the end of a telomere specifically sterically blocks and protects the telomere from a C-strand nuclease rather than blocking the end from general nucleases as elaborated above (Section 4.2; Fig. 5C). The portion of Ku most likely to provide this activity would be the N-terminal α-helical domain of Ku70, which is most proximal to the free end (33,144), or the C-terminus of Ku86, which is also likely to reside near the free end (33). Presumably when Ku is reduced or absent, the C-strand gets resected more than normal, resulting in longer G-strand overhangs. The binding of Pot1 to the G-strand overhang would presumably protect this strand from nucleolytic attack, although this has not been directly demonstrated. Ultimately, if the G-overhang becomes too long it may break and/or be subject to nucleolytic attack itself, resulting in overall shorter telomeres (consistent with many of the telomere length phenotypes described above) with hyperextended G-overhangs. Alternatively, especially if the amount of Pot1 in a cell is limiting, the elongated G-strand may simply become uncovered. Since single-stranded DNA is thought to be a potent signal for the activation of DNA repair checkpoints (399), this would lead to cell cycle arrest or—if and when arrest failed—genomic instability. Consistent with this prediction, radiation-sensitive 53 (Rad53), an S-phase checkpoint kinase, is chronically activated in Ku-null yeast strains with elongated G-overhangs (366) and p53 [a transcription factor that plays a critical role in the mammalian DNA repair response; reviewed by Sharpless and DePinho (400)] levels are significantly elevated in asynchronously growing human somatic cells that are heterozygous for Ku86 (176,377). The hypothesis that Ku exerts its effects by regulating the amount of single-stranded telomeric DNA is also consistent with the observation that the length of the G-overhang can control the overall length of the telomere (401). Moreover, it is consistent with the observation that mutations of Exo1 will partially suppress the aberrant G-strand overhang phenotypes of certain Ku86 mutant yeast strains (367,395,396). It is not, however, consistent with a report demonstrating that the G-overhang is lost in cells undergoing replicative senescence (397) nor with the demonstration that the overexpression of telomerase or telomeric RNA will suppress the temperature-sensitive lethality of yeast Ku strains without suppressing the aberrant G-strand overhangs (366).

In summary, mutations of Ku, but not apparently DNA-PK$_{cs}$, generally result in aberrantly long G-overhangs. The generation and maintenance of the G-overhang, however, involves many enzymes and a better understanding of this process requires additional experimentation. The structural requirements for Ku suggest that it performs its role in this process by binding to the end of the telomere to protect it from C-strand-specific nucleases (Fig. 5C). When this does not occur, the resulting elongated G-strands signal the presence of damaged DNA and this leads to cell cycle arrest or genomic instability (see below).

4.4. Chromosomal Stability

It has been over 60 years since Barbara McClintock (402) demonstrated that chromosomes without telomeres undergo a very high rate of aberrant recombination, usually referred to as a breakage:fusion:bridge cycle [reviewed by Maser and DePinho (75)]. In the intervening decades, this observation has been elaborated to show that while the loss of telomeric sequences is key to this process, it is not essential (403). Instead, the concept of "capping," which nebulously consists of a multiprotein complex and potentially higher-order structures [reviewed by de Lange (67)], appears—at least in mammals—to be the essential feature of telomeres for limiting GCRs (gross chromosomal rearrangements) (Fig. 5A). Telomeric shortening, with the concomitant loss of protein: DNA interactions, is simply one of several ways in which the telomeric cap can be disrupted.

Yeast Ku mutants do not show any evidence of chromosomal instability when grown at normal temperatures, but at elevated temperatures they die rapidly, with all the hallmarks of telomeric rearrangements and genomic instability (56,227,363,364,384). *S. pombe* Ku mutant strains are also prone to GCRs although they ultimately stabilize their genomes by circularizing their chromosomes (368). In Ku70 mutant chicken DT40 cells (371), trypanosomes (370), and plants (374), there is no evidence for increased chromosomal fusions, again implicating compensating HR processes for the lack of phenotypic effects. In the mouse, there is (amazingly) unanimous agreement that mutations in Ku (130,131,273,404) and DNA-PK$_{cs}$ (130,318,319,375,405,406) result in greatly elevated levels of chromosomal telomere:telomere fusions and that this accounts, at least in part, for the premature senescence (121,126) associated with these animals and cell lines. Lastly, in human somatic cells heterozygous for Ku86, nearly one in four metaphases contains a GCR (176).

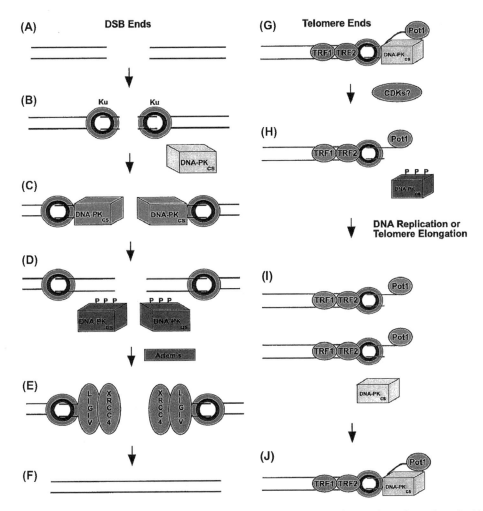

Figure 6 A model explaining the similar role of DNA-PK at DSBs and at telomeric ends. (A) Double-stranded DNA is shown as a pair of horizontal lines. (B) Ku (double rings) binds onto the terminal ends of the DSB. (C) DNA-PKcs in a neutral conformation (rectangle) binds to the dsDNA (displacing Ku internally along the DNA) and is activated (indicated by darker shading). (D) DNA-PKcs undergoes autophosphorylation (PPP), which inactivates DNA-PKcs (dark grey rectangle) and causes it to dissociate from the DNA. (E) The exposed dsDNA undergoes modifications, perhaps mediated by the nuclease Artemis (rectangle). Ultimately the ends become substrates for the LigIV:XRCC4 complex (gray ovals), which potentially keeps the two DNA ends in proximity to one another via dimerization. (F) The integrity of the DNA has been restored by NHEJ DSB repair. (G) Speculative structure of what a capped telomeric end might look like. TRF1 and TRF2 (gray ovals, respectively) are bound to the double-strand, repetitive portion of the telomere (gray horizontal lines). Pot1 (oval) is bound to the G-strand overhang, which is folded back into a t-loop. Ku is shown as the double rings. Perhaps because of the unusual conformation of the t-loop, the DNA-PKcs subunit is bound, but not activated (light grey rectangle). (H) To facilitate chromosome replication, a kinase [speculatively(?), CDKs (cyclin-dependent kinase; oval)] initiate the autophosphorylation (PPP) process in DNA-PKcs, which inactivates DNA-PKcs (dark grey rectangle) and causes it to dissociate from the DNA. (I) Following DNA replication, DNA-PKcs in a neutral conformation (light grey rectangle) is recruited back to the telomeric end. (J) The original structures shown in panel (G) is re-established.

The exceptions noted above notwithstanding, the sum of the available data suggests that the absence of Ku and DNA-PK$_{cs}$ leads to telomere "uncapping" and chromosomal instability (Fig. 5A). A possible explanation, at least for mammals, is that the absence of Ku or DNA-PK impinges deleteriously on TRF function. TRF1 and TRF2 are known to physically interact with Ku (257,273) (Fig. 5B) and the absence of at least TRF2 can result in increased chromosomal fusions without the obvious loss of telomeric sequences. Thus, even in the mouse, where the evidence for a role for Ku and especially DNA-PK in telomere length and/or G-strand overhang maintenance is weak or missing, respectively—and thus difficult to functionally assess—a lack of proper interaction with TRF2 could mimic the genetic loss of TRF2. In addition, we propose a—highly speculative—alternative model. Thus, the ostensibly paradoxical nature of NHEJ (i.e., that in the presence of DNA-PK DSBs get rejoined and in its absence they do not) and its involvement in telomere maintenance (i.e., that in the presence of DNA-PK telomeres do not get joined together and in its absence they do) has been recently discussed repeatedly and highlighted [reviewed by Lees-Miller and Meek (20)]. Most of these data can be readily explained if the assumption is made that DNA-PK is differentially activated by DNA DSBs and telomere ends. Thus, when a chromosome sustains a DNA DSB (Fig. 6A), Ku binds on to the ends (Fig. 6B). DNA-PK$_{cs}$ is subsequently—but transiently—recruited (Fig. 6C), displacing Ku internally along the chromosome (295) while simultaneously activating the kinase activity of DNA-PK. The subsequent autophosphorylation of the complex results in the release of DNA-PK$_{cs}$ (265), while Ku remains attached to the DNA (147) (Fig. 6D). This would allow nucleases like Artemis (37) access to the ends (Fig. 6D) and ultimately allow LigIV:XRCC4 to bind onto the free ends (147,148) (Fig. 6E) to facilitate repair of the break (Fig. 6F). In the case of "capped" telomere ends, we envision that DNA-PK$_{cs}$ is bound, but not activated (Fig. 6G). The seminal observation for this hypothesis is that when DNA-PK$_{cs}$ cannot or does not phosphorylate itself, it remains bound to DNA ends and, in a dominant-negative manner, acts to impede DNA end joining (328,338,340), which is precisely what it appears to be doing at a telomeric end. There is precedent for this possibility as DNA-PK is known to bind to RNA (291) and hairpin DNA (179) and yet these nucleic acids/structures fail to stimulate DNA-PK's kinase activity. Occasionally, of course, telomeres need to become uncapped, for example, during DNA replication. In this instance, we postulate that cyclin dependent kinases (CDKs) [reviewed by Nishitani and Lygerou (407)] may initiate the activation of DNA-PK$_{cs}$, which induces autophosphorylation and disassembly of the capped telomere (Fig. 6H). Following DNA replication and/or telomere elongation, a new DNA-PK$_{cs}$ subunit or one reactivated by dephosphorylation (230,232,278) is recruited to the telomere (Fig. 6I), where a capped structure is reestablished (Fig. 6J). This model explains how mutations of Ku, but not DNA-PK$_{cs}$, lead to telomeric shortening and hyperextended G-overhangs, while mutations of either subunit lead to genomic instability. This model clearly predicts that the structure of DNA-PK$_{cs}$ bound to a DNA DSB will be conformationally different from the structure of DNA-PK$_{cs}$ bound to a telomere.

5. SUMMARY

From their humble beginnings as either the apparently random target of autoantibodies (Ku70/Ku86) or as an obscure disease in an esoteric strain of horses

(DNA-PK$_{cs}$), these three proteins have shown themselves over the intervening three decades to be critical for a vast array of DNA metabolic processes. Besides the ongoing investigations into all of these processes, the recent observations that alterations in DNA-PK may also affect transposition (408,409) and retrotransposition (88,274,410–413) [reviewed by Friedl (414)] and be a potential biomarker for cancer predisposition (415,416) strongly suggests that the Golden Age of DNA-PK is by no means over.

ACKNOWLEDGMENTS

We thank Drs. Anja-Katrin Bielinsky and Dennis Livingston (University of Minnesota Medical School) for their comments and helpful discussions. The work on DNA break repair in the Tainer laboratory is supported by NIH Grants CA97209 and CA92584 and a grant from the Human Frontiers in Science Program. J. L. H. is supported by a NIH training grant (HL07781). The work on Ku and DNA-PK in the Hendrickson laboratory is supported by NIH grants HL 79559 and GM69576.

REFERENCES

1. Gellert M. V(D)J recombination: RAG proteins, repair factors, and regulation. Annu Rev Biochem 2002; 71:101–132.
2. Jung D, Alt FW. Unraveling V(D)J recombination. Insights into gene regulation. Cell 2004; 116:299–311.
3. Danska JS, Pflumio F, Williams CJ, Huner O, Dick JE, Guidos CJ. Rescue of T cell-specific V(D)J recombination in SCID mice by DNA-damaging agents. Science 1994; 266:450–455.
4. Rolink A, Melchers F, Andersson J. The SCID but not the RAG-2 gene product is required for S$_\mu$-S$_\varepsilon$e heavy chain class switching. Immunity 1996; 5:319–330.
5. Casellas R, Nussenzweig A, Wuerffel R, Pelanda R, Reichlin A, Suh H, Qin XF, Besmer E, Kenter A, Rajewsky K, Nussenzweig MC. Ku80 is required for immunoglobulin isotype switching. EMBO J 1998; 17:2404–2411.
6. Manis JP, Gu Y, Lansford R, Sonoda E, Ferrini R, Davidson L, Rajewsky K, Alt FW. Ku70 is required for late B cell development and immunoglobulin heavy chain class switching. J Exp Med 1998; 187:2081–2089.
7. Manis JP, Dudley D, Kaylor L, Alt FW. IgH class switch recombination to IgG$_1$ in DNA-PK$_{cs}$-deficient B cells. Immunity 2002; 16:607–617.
8. O'Driscoll M, Jeggo P. Immunological disorders and DNA repair. Mutat Res 2002; 509:109–126.
9. Manis JP, Tian M, Alt FW. Mechanism and control of class-switch recombination. Trends Immunol 2002; 23:31–39.
10. Muramatsu M, Sankaranand VS, Anant S, Sugai M, Kinoshita K, Davidson NO, Honjo T. Specific expression of activation-induced cytidine deaminase (AID), a novel member of the RNA-editing deaminase family in germinal center B cells. J Biol Chem 1999; 274:18470–18476.
11. Stavnezer J, Bradley SP. Does activation-induced deaminase initiate antibody diversification by DNA deamination? TIGS 2002; 18:541–543.
12. Wuerffel RA, Du J, Thompson RJ, Kenter AL. Ig Sγ3 DNA-specifc double strand breaks are induced in mitogen-activated B cells and are implicated in switch recombination. J Immunol 1997; 159:4139–4144.

13. Neuberger MS, Harris RS, Di Noia J, Petersen-Mahrt SK. Immunity through DNA deamination. Trends Biochem Sci 2003; 28:305–312.

14. Sonoda E, Takata M, Yamashita YM, Morrison C, Takeda S. Homologous DNA recombination in vertebrate cells. Proc Natl Acad Sci USA 2001; 98:8388–8394.

15. Thompson LH, Schild D. Homologous recombinational repair of DNA ensures mammalian chromosome stability. Mutat Res 2001; 477:131–153.

16. Hoeijmakers JH. Genome maintenance mechanisms for preventing cancer. Nature 2001; 411:366–374.

17. Jackson SP. Sensing and repairing DNA double-strand breaks. Carcinogenesis 2002; 23:687–696.

18. Wilson JH, Berget PB, Pipas JM. Somatic cells efficiently join unrelated DNA segments end-to-end. Mol Cell Biol 1982; 2:1258–1269.

19. Roth DB, Wilson JH. Relative rates of homologous and nonhomologous recombination in transfected DNA. Proc Natl Acad Sci USA 1985; 82:3355–3359.

20. Lees-Miller SP, Meek K. Repair of DNA double strand breaks by non-homologous end joining. Biochimie 2003; 85:1161–1173.

21. Lieber MR, Ma Y, Pannicke U, Schwarz K. Mechanism and regulation of human non-homologous DNA end-joining. Nat Rev Mol Cell Biol 2003; 4:712–720.

22. Thomas KR, Capecchi MR. Site-directed mutagenesis by gene targeting in mouse embryo-derived stem cells. Cell 1987; 51:503–512.

23. Lee SE, Mitchell RA, Cheng A, Hendrickson EA. Evidence for DNA-PK-dependent and -independent DNA double-strand break repair pathways in mammalian cells as a function of the cell cycle. Mol Cell Biol 1997; 17:1425–1433.

24. Takata M, Sasaki MS, Sonoda E, Morrison C, Hashimoto M, Utsumi H, Yamaguchi-Iwai Y, Shinohara A, Takeda S. Homologous recombination and non-homologous end-joining pathways of DNA double-strand break repair have overlapping roles in the maintenance of chromosomal integrity in vertebrate cells. EMBO J 1998; 17: 5497–5508.

25. Schlissel M, Constantinescu A, Morrow T, Baxter M, Peng A. Double-strand signal sequence breaks in V(D)J recombination are blunt, 5′-phosphorylated, RAG-dependent, and cell cycle regulated. Genes Dev 1993; 7:2520–2532.

26. Lin WC, Desiderio S. Cell cycle regulation of V(D)J recombination-activating protein RAG-2. Proc Natl Acad Sci USA 1994; 91:2733–2737.

27. Getts RC, Stamato TD. Absence of a Ku-like DNA end binding activity in the *xrs* double-strand DNA repair-deficient mutant. J Biol Chem 1994; 269:15981–15984.

28. Liang F, Romanienko PJ, Weaver DT, Jeggo PA, Jasin M. Chromosomal double-strand break repair in Ku80-deficient cells. Proc Natl Acad Sci USA 1996; 93:8929–8933.

29. Inamdar KV, Yu Y, Povirk LF. Resistance of 3′-phosphoglycolate DNA ends to digestion by mammalian DNase III. Radiat Res 2002; 157:306–311.

30. Bliss TM, Lane DP. Ku selectively transfers between DNA molecules with homologous ends. J Biol Chem 1997; 272:5765–5773.

31. Pang D, Yoo S, Dynan WS, Jung M, Dritschilo A. Ku proteins join DNA fragments as shown by atomic force microscopy. Cancer Res 1997; 57:1412–1415.

32. Kabotyanski EB, Gomelsky L, Han JO, Stamato TD, Roth DB. Double-strand break repair in Ku86- and XRCC4-deficient cells. Nucl Acids Res 1998; 26:5333–5342.

33. Walker JR, Corpina RA, Goldberg J. Structure of the Ku heterodimer bound to DNA and its implications for double-strand break repair. Nature 2001; 412:607–614.

34. Gottlieb TM, Jackson SP. The DNA-dependent protein kinase: requirement for DNA ends and association with Ku antigen. Cell 1993; 72:131–142.

35. Suwa A, Hirakata M, Takeda Y, Jesch SA, Mimori T, Hardin JA. DNA-dependent protein kinase (Ku protein-p350 complex) assembles on double-stranded DNA. Proc Natl Acad Sci USA 1994; 91:6904–6908.

36. Moshous D, Callebaut I, de Chasseval R, Corneo B, Cavazzana-Calvo M, Le Deist F, Tezcan I, Sanal O, Bertrand Y, Philippe N, Fischer A, de Villartay JP. Artemis, a novel

DNA double-strand break repair/V(D)J recombination protein, is mutated in human severe combined immune deficiency. Cell 2001; 105:177–186.

37. Ma Y, Pannicke U, Schwarz K, Lieber MR. Hairpin opening and overhang processing by an Artemis/DNA-dependent protein kinase complex in nonhomologous end joining and V(D)J recombination. Cell 2002; 108:781–794.

38. Moshous D, Callebaut I, de Chasseval R, Poinsignon C, Villey I, Fischer A, de Villartay JP. The V(D)J recombination/DNA repair factor artemis belongs to the metallo-beta-lactamase family and constitutes a critical developmental checkpoint of the lymphoid system. Ann NY Acad Sci 2003; 987:150–157.

39. DeFazio LG, Stansel RM, Griffith JD, Chu G. Synapsis of DNA ends by DNA-dependent protein kinase. EMBO J 2002; 21:3192–3200.

40. Weterings E, Verkaik NS, Bruggenwirth HT, Hoeijmakers JH, van Gent DC. The role of DNA dependent protein kinase in synapsis of DNA ends. Nucl Acids Res 2003; 31:7238–7246.

41. Critchlow SE, Bowater RP, Jackson SP. Mammalian DNA double-strand break repair protein XRCC4 interacts with DNA ligase IV. Curr Biol 1997; 7:588–598.

42. Nick McElhinny SA, Snowden CM, McCarville J, Ramsden DA. Ku recruits the XRCC4-ligase IV complex to DNA ends. Mol Cell Biol 2000; 20:2996–3003.

43. Teo SH, Jackson SP. Lif1p targets the DNA ligase Lig4p to sites of DNA double-strand breaks. Curr Biol 2000; 10:165–168.

44. Li Z, Otevrel T, Gao Y, Cheng HL, Seed B, Stamato TD, Taccioli GE, Alt FW. The XRCC-4 gene encodes a novel protein involved in DNA double-strand break repair and V(D)J recombination. Cell 1995; 83:1079–1089.

45. Gao Y, Sun Y, Frank KM, Dikkes P, Fujiwara Y, Seidl KJ, Sekiguchi JM, Rathbun GA, Swat W, Wang J, Bronson RT, Malynn BA, Bryans M, Zhu C, Chaudhuri J, Davidson L, Ferrini R, Stamato T, Orkin SH, Greenberg ME, Alt FW. A critical role for DNA end-joining proteins in both lymphogenesis and neurogenesis. Cell 1998; 95:891–902.

46. Dai Y, Kysela B, Hanakahi LA, Manolis K, Riballo E, Stumm M, Harville TO, West SC, Oettinger MA, Jeggo PA. Nonhomologous end joining and V(D)J recombination require an additional factor. Proc Natl Acad Sci USA 2003; 100:2462–2467.

47. Grawunder U, Wilm M, Wu X, Kulesza P, Wilson TE, Mann M, Lieber MR. Activity of DNA ligase IV stimulated by complex formation with XRCC4 protein in mammalian cells. Nature 1997; 388:492–495.

48. Grawunder U, Zimmer D, Fugmann S, Schwarz K, Lieber MR. DNA ligase IV is essential for V(D)J recombination and DNA double-strand break repair in human precursor lymphocytes. Mol Cell 1998; 2:477–484.

49. Wang H, Zeng ZC, Bui TA, Sonoda E, Takata M, Takeda S, Iliakis G. Efficient rejoining of radiation-induced DNA double-strand breaks in vertebrate cells deficient in genes of the RAD52 epistasis group. Oncogene 2001; 20:2212–2224.

50. Feldmann E, Schmiemann V, Goedecke W, Reichenberger S, Pfeiffer P. DNA double-strand break repair in cell-free extracts from Ku80-deficient cells: implications for Ku serving as an alignment factor in non-homologous DNA end joining. Nucl Acids Res 2000; 28:2585–2596.

51. Chen S, Inamdar KV, Pfeiffer P, Feldmann E, Hannah MF, Yu Y, Lee JW, Zhou T, Lees-Miller SP, Povirk LF. Accurate *in vitro* end joining of a DNA double strand break with partially cohesive 3′-overhangs and 3′-phosphoglycolate termini: effect of Ku on repair fidelity. J Biol Chem 2001; 276:24323–24330.

52. Verkaik NS, Esveldt-van Lange RE, van Heemst D, Bruggenwirth HT, Hoeijmakers JH, Zdzienicka MZ, van Gent DC. Different types of V(D)J recombination and end-joining defects in DNA double-strand break repair mutant mammalian cells. Eur J Immunol 2002; 32:701–709.

53. Wang H, Perrault AR, Takeda Y, Qin W, Iliakis G. Biochemical evidence for Ku-independent backup pathways of NHEJ. Nucl Acids Res 2003; 31:5377–5388.

54. Paull TT, Gellert M. A mechanistic basis for Mre11-directed DNA joining at micro-homologies. Proc Natl Acad Sci USA 2000; 97:6409–6414.

55. Ma JL, Kim EM, Haber JE, Lee SE. Yeast Mre11 and Rad1 proteins define a Ku-independent mechanism to repair double-strand breaks lacking overlapping end sequences. Mol Cell Biol 2003; 23:8820–8828.

56. Boulton SJ, Jackson SP. Components of the Ku-dependent non-homologous end-joining pathway are involved in telomeric length maintenance and telomeric silencing. EMBO J 1998; 17:1819–1828.

57. Chen C, Kolodner RD. Gross chromosomal rearrangements in *Saccharomyces cerevisiae* replication and recombination defective mutants. Nat Genet 1999; 23:81–85.

58. Stewart GS, Maser RS, Stankovic T, Bressan DA, Kaplan MI, Jaspers NG, Raams A, Byrd PJ, Petrini JH, Taylor AM. The DNA double-strand break repair gene hMRE11 is mutated in individuals with an *ataxia-telangiectasia*-like disorder. Cell 1999; 99:577–587.

59. Escarceller M, Buchwald M, Singleton BK, Jeggo PA, Jackson SP, Moustacchi E, Papadopoulo D. Fanconi anemia C gene product plays a role in the fidelity of blunt DNA end-joining. J Mol Biol 1998; 279:375–385.

60. Lundberg R, Mavinakere M, Campbell C. Deficient DNA end joining activity in extracts from *Fanconi anemia* fibroblasts. J Biol Chem 2001; 276:9543–9549.

61. Tzung TY, Runger TM. Reduced joining of DNA double strand breaks with an abnormal mutation spectrum in rodent mutants of DNA-PK$_{cs}$ and Ku80. Int J Radiat Biol 1998; 73:469–474.

62. Cheong N, Perrault AR, Wang H, Wachsberger P, Mammen P, Jackson I, Iliakis G. DNA-PK-independent rejoining of DNA double-strand breaks in human cell extracts *in vitro*. Int J Radiat Biol 1999; 75:67–81.

63. Labhart P. Ku-dependent nonhomologous DNA end joining in Xenopus egg extracts. Mol Cell Biol 1999; 19:2585–2593.

64. Wang H, Zeng ZC, Perrault AR, Cheng X, Qin W, Iliakis G. Genetic evidence for the involvement of DNA ligase IV in the DNA-PK-dependent pathway of non-homologous end joining in mammalian cells. Nucl Acids Res 2001; 29:1653–1660.

65. Udayakumar D, Bladen CL, Hudson FZ, Dynan WS. Distinct pathways of nonhomologous end joining that are differentially regulated by DNA-dependent protein kinase-mediated phosphorylation. J Biol Chem 2003; 278:41631–41635.

66. Yu X, Gabriel A. Ku-dependent and Ku-independent end-joining pathways lead to chromosomal rearrangements during double-strand break repair in *Saccharomyces cerevisiae*. Genetics 2003; 163:843–856.

67. de Lange T. Protection of mammalian telomeres. Oncogene 2002; 21:532–540.

68. McElligott R, Wellinger RJ. The terminal DNA structure of mammalian chromosomes. EMBO J 1997; 16:3705–3714.

69. Blackburn EH. Switching and signaling at the telomere. Cell 2001; 106:661–673.

70. Griffith JD, Comeau L, Rosenfield S, Stansel RM, Bianchi A, Moss H, de Lange T. Mammalian telomeres end in a large duplex loop. Cell 1999; 97:503–514.

71. de Lange T. Cell biology. Telomere capping–one strand fits all. Science 2001; 292: 1075–1076.

72. Hahn WC, Counter CM, Lundberg AS, Beijersbergen RL, Brooks MW, Weinberg RA. Creation of human tumour cells with defined genetic elements. Nature 1999; 400: 464–468.

73. Stewart SA, Weinberg RA. Senescence: does it all happen at the end? Oncogene 2002; 21:627–630.

74. Greider CW. Telomerase activation. One step on the road to cancer? TIGS 1999; 15:109–112.

75. Maser RS, DePinho RA. Connecting chromosomes, crisis, and cancer. Science 2002; 297:565–569.

76. Chan SW, Blackburn EH. New ways not to make ends meet: telomerase, DNA damage proteins and heterochromatin. Oncogene 2002; 21:553–563.

77. Greider CW, Blackburn EH. Identification of a specific telomere terminal transferase activity in *Tetrahymena* extracts. Cell 1985; 43:405–413.

78. Blasco MA, Funk W, Villeponteau B, Greider CW. Functional characterization and developmental regulation of mouse telomerase RNA. Science 1995; 269:1267–1270.

79. Chong L, van Steensel B, Broccoli D, Erdjument-Bromage H, Hanish J, Tempst P, de Lange T. A human telomeric protein. Science 1995; 270:1663–1667.

80. Broccoli D, Smogorzewska A, Chong L, de Lange T. Human telomeres contain two distinct Myb-related proteins, TRF1 and TRF2. Nat Genet 1997; 17:231–235.

81. Baumann P, Cech TR. Pot1, the putative telomere end-binding protein in fission yeast and humans. Science 2001; 292:1171–1175.

82. Weaver DT. Telomeres: moonlighting by DNA repair proteins. Curr Biol 1998; 8: R492–R494.

83. Bailey SM, Brenneman MA, Halbrook J, Nickoloff JA, Ullrich RL, Goodwin EH. The kinase activity of DNA-PK is requried to protect mammalian telomeres. DNA Repair 2004; 3:225–233.

84. Reddel RR, Bryan TM, Colgin LM, Perrem KT, Yeager TR. Alternative lengthening of telomeres in human cells. Radiat Res 2001; 155:194–200.

85. Aravind L, Koonin EV. Prokaryotic homologs of the eukaryotic DNA-end-binding protein Ku, novel domains in the Ku protein and prediction of a prokaryotic double-strand break repair system. Genome Res 2001; 11:1365–1374.

86. Doherty AJ, Jackson SP, Weller GR. Identification of bacterial homologues of the Ku DNA repair proteins. FEBS Lett 2001; 500:186–188.

87. Weller GR, Kysela B, Roy R, Tonkin LM, Scanlan E, Della M, Devine SK, Day JP, Wilkinson A, d'Adda di Fagagna F, Devine KM, Bowater RP, Jeggo PA, Jackson SP, Doherty AJ. Identification of a DNA nonhomologous end-joining complex in bacteria. Science 2002; 297:1686–1689.

88. d'Adda di Fagagna F, Weller GR, Doherty AJ, Jackson SP. The Gam protein of bacteriophage Mu is an orthologue of eukaryotic Ku. EMBO Rep 2003; 4:47–52.

89. Zissler J, Signer E, Schaefer F. The bacteriophage lambda. In: Hershey AD, ed. The Bacteriophage *Lambda*. Cold Spring Harbor, NY: Cold Spring Harbor Press, 1971:455–468.

90. Williams JG, Radding CM. Partial purification and properties of an exonuclease inhibitor induced by bacteriophage Mu-1. J Virol 1981; 39:548–558.

91. Akroyd J, Symonds N. Localization of the *gam* gene of bacteriophage Mu and characterisation of the gene product. Gene 1986; 49:273–282.

92. Dore AS, Drake AC, Brewerton SC, Blundell TL. Identification of DNA-PK in the arthropods. Evidence for the ancient ancestry of vertebrate non-homologous end-joining. DNA Repair 2004; 3:33–41.

93. Mimori T, Ohosone Y, Hama N, Suwa A, Akizuki M, Homma M, Griffith AJ, Hardin JA. Isolation and characterization of cDNA encoding the 80-kDa subunit protein of the human autoantigen Ku (p70/p80) recognized by autoantibodies from patients with scleroderma-polymyositis overlap syndrome. Proc Natl Acad Sci USA 1990; 87: 1777–1781.

94. Mimori T, Akizuki M, Yamagata H, Inada S, Yoshida S, Homma M. Characterization of a high molecular weight acidic nuclear protein recognized by autoantibodies in sera from patients with polymyositis-scleroderma overlap. J Clin Invest 1981; 68:611–620.

95. Takeda Y, Dynan WS. Autoantibodies against DNA double-strand break repair proteins. Front Biosci 2001; 6:D1412–D1422.

96. Mimori T, Hardin JA, Steitz JA. Characterization of the DNA-binding protein antigen Ku recognized by autoantibodies from patients with rheumatic disorders. J Biol Chem 1986; 261:2274–2278.

97. Tuteja N, Tuteja R, Ochem A, Taneja P, Huang NW, Simoncsits A, Susic S, Rahman K, Marusic L, Chen J, Zhang J, Wang S, Pongor S, Falaschi A. Human DNA helicase II: a novel DNA unwinding enzyme identified as the Ku autoantigen. EMBO J 1994; 13: 4991–5001.

98. Mimori T, Hardin JA. Mechanism of interaction between Ku protein and DNA. J Biol Chem 1986; 261:10375–10379.

99. de Vries E, van Driel W, Bergsma WG, Arnberg AC, van der Vliet PC. HeLa nuclear protein recognizing DNA termini and translocating on DNA forming a regular DNA-multimeric protein complex. J Mol Biol 1989; 208:65–78.

100. Paillard S, Strauss F. Analysis of the mechanism of interaction of simian Ku protein with DNA. Nucl Acids Res 1991; 19:5619–5624.

101. Blier PR, Griffith AJ, Craft J, Hardin JA. Binding of Ku protein to DNA. Measurement of affinity for ends and demonstration of binding to nicks. J Biol Chem 1993; 268:7594–7601.

102. Falzon M, Fewell JW, Kuff EL. EBP-80, a transcription factor closely resembling the human autoantigen Ku, recognizes single- to double-strand transitions in DNA. J Biol Chem 1993; 268:10546–10552.

103. Yoo S, Dynan WS. Geometry of a complex formed by double strand break repair proteins at a single DNA end: recruitment of DNA-PK$_{cs}$ induces inward translocation of Ku protein. Nucl Acids Res 1999; 27:4679–4686.

104. Arosio D, Cui S, Ortega C, Chovanec M, Di Marco S, Baldini G, Falaschi A, Vindigni A. Studies on the mode of Ku interaction with DNA. J Biol Chem 2002; 277:9741–9748.

105. Taghva A, Ma Y, Lieber MR. Analysis of the kinetic and equilibrium binding of Ku protein to DNA. J Theor Biol 2002; 214:85–97.

106. Rathmell WK, Chu G. Involvement of the Ku autoantigen in the cellular response to DNA double-strand breaks. Proc Natl Acad Sci USA 1994; 91:7623–7627.

107. Rathmell WK, Chu G. A DNA end-binding factor involved in double-strand break repair and V(D)J recombination. Mol Cell Biol 1994; 14:4741–4748.

108. Hartley KO, Gell D, Smith GC, Zhang H, Divecha N, Connelly MA, Admon A, Lees-Miller SP, Anderson CW, Jackson SP. DNA-dependent protein kinase catalytic subunit: a relative of phosphatidylinositol 3-kinase and the ataxia telangiectasia gene product. Cell 1995; 82:849–856.

109. Sipley JD, Menninger JC, Hartley KO, Ward DC, Jackson SP, Anderson CW. Gene for the catalytic subunit of the human DNA-activated protein kinase maps to the site of the *XRCC7* gene on chromosome 8. Proc Natl Acad Sci USA 1995; 92:7515–7519.

110. Bosma GC, Custer RP, Bosma MJ. A severe combined immunodeficiency mutation in the mouse. Nature 1983; 301:527–530.

111. Morozov VE, Falzon M, Anderson CW, Kuff EL. DNA-dependent protein kinase is activated by nicks and larger single-stranded gaps. J Biol Chem 1994; 269:16684–16688.

112. Schuler W, Weiler IJ, Schuler A, Phillips RA, Rosenberg N, Mak TW, Kearney JF, Perry RP, Bosma MJ. Rearrangement of antigen receptor genes is defective in mice with severe combined immune deficiency. Cell 1986; 46:963–972.

113. Hendrickson EA, Schatz DG, Weaver DT. The *scid* gene encodes a *trans*-acting factor that mediates the rejoining event of Ig gene rearrangement. Genes Dev 1988; 2:817–829.

114. Lieber MR, Hesse JE, Lewis S, Bosma GC, Rosenberg N, Mizuuchi K, Bosma MJ, Gellert M. The defect in murine severe combined immune deficiency: joining of signal sequences but not coding segments in V(D)J recombination. Cell 1988; 55:7–16.

115. Malynn BA, Blackwell TK, Fulop GM, Rathbun GA, Furley AJ, Ferrier P, Heinke LB, Phillips RA, Yancopoulos GD, Alt FW. The *scid* defect affects the final step of the immunoglobulin VDJ recombinase mechanism. Cell 1988; 54:453–460.

116. Smider V, Rathmell WK, Lieber MR, Chu G. Restoration of X-ray resistance and V(D)J recombination in mutant cells by Ku cDNA. Science 1994; 266:288–291.

117. Taccioli GE, Gottlieb TM, Blunt T, Priestley A, Demengeot J, Mizuta R, Lehmann AR, Alt FW, Jackson SP, Jeggo PA. Ku80: product of the XRCC5 gene and its role in DNA repair and V(D)J recombination. Science 1994; 265:1442–1445.

118. Boubnov NV, Hall KT, Wills Z, Lee SE, He DM, Benjamin DM, Pulaski CR, Band H, Reeves W, Hendrickson EA, Weaver DT. Complementation of the ionizing radiation

sensitivity, DNA end binding, and V(D)J recombination defects of double-strand break repair mutants by the p86 Ku autoantigen. Proc Natl Acad Sci USA 1995; 92:890–894.

119. Errami A, Smider V, Rathmell WK, He DM, Hendrickson EA, Zdzienicka MZ, Chu G. Ku86 defines the genetic defect and restores X-ray resistance and V(D)J recombination to complementation group 5 hamster cell mutants. Mol Cell Biol 1996; 16:1519–1526.

120. Singleton BK, Priestley A, Steingrimsdottir H, Gell D, Blunt T, Jackson SP, Lehmann AR, Jeggo PA. Molecular and biochemical characterization of *xrs* mutants defective in Ku80. Mol Cell Biol 1997; 17:1264–1273.

121. Nussenzweig A, Chen C, da Costa Soares V, Sanchez M, Sokol K, Nussenzweig MC, Li GC. Requirement for Ku80 in growth and immunoglobulin V(D)J recombination. Nature 1996; 382:551–555.

122. Zhu C, Bogue MA, Lim DS, Hasty P, Roth DB. Ku86-deficient mice exhibit severe combined immunodeficiency and defective processing of V(D)J recombination intermediates. Cell 1996; 86:379–389.

123. Ouyang H, Nussenzweig A, Kurimasa A, Soares VC, Li X, Cordon-Cardo C, Li W, Cheong N, Nussenzweig M, Iliakis G, Chen DJ, Li GC. Ku70 is required for DNA repair but not for T cell antigen receptor gene recombination *in vivo*. J Exp Med 1997; 186:921–929.

124. Li GC, Ouyang H, Li X, Nagasawa H, Little JB, Chen DJ, Ling CC, Fuks Z, Cordon-Cardo C. Ku70: a candidate tumor suppressor gene for murine T cell lymphoma. Mol Cell 1998; 2:1–8.

125. Nussenzweig A, Sokol K, Burgman P, Li L, Li GC. Hypersensitivity of *Ku80*-deficient cell lines and mice to DNA damage: the effects of ionizing radiation on growth, survival, and development. Proc Natl Acad Sci USA 1997; 94:13588–13593.

126. Vogel H, Lim DS, Karsenty G, Finegold M, Hasty P. Deletion of Ku86 causes early onset of senescence in mice. Proc Natl Acad Sci USA 1999; 96:10770–10775.

127. Karanjawala ZE, Grawunder U, Hsieh CL, Lieber MR. The nonhomologous DNA end joining pathway is important for chromosome stability in primary fibroblasts. Curr Biol 1999; 9:1501–1504.

128. Difilippantonio MJ, Zhu J, Chen HT, Meffre E, Nussenzweig MC, Max EE, Ried T, Nussenzweig A. DNA repair protein Ku80 suppresses chromosomal aberrations and malignant transformation. Nature 2000; 404:510–514.

129. d'Adda di Fagagna F, Hande MP, Tong WM, Roth D, Lansdorp PM, Wang ZQ, Jackson SP. Effects of DNA nonhomologous end-joining factors on telomere length and chromosomal stability in mammalian cells. Curr Biol 2001; 11:1192–1196.

130. Bailey SM, Meyne J, Chen DJ, Kurimasa A, Li GC, Lehnert BE, Goodwin EH. DNA double-strand break repair proteins are required to cap the ends of mammalian chromosomes. Proc Natl Acad Sci USA 1999; 96:14899–14904.

131. Samper E, Goytisolo FA, Slijepcevic P, van Buul PP, Blasco MA. Mammalian Ku86 protein prevents telomeric fusions independently of the length of TTAGGG repeats and the G-strand overhang. EMBO Rep 2000; 1:244–252.

132. Zhao J, Wang J, Chen DJ, Peterson SR, Trewhella J. The solution structure of the DNA double-stranded break repair protein Ku and its complex with DNA: a neutron contrast variation study. Biochemistry 1999; 38:2152–2159.

133. Wu X, Lieber MR. Protein-protein and protein-DNA interaction regions within the DNA end-binding protein Ku70-Ku86. Mol Cell Biol 1996; 16:5186–5193.

134. Osipovich O, Durum SK, Muegge K. Defining the minimal domain of Ku80 for interaction with Ku70. J Biol Chem 1997; 272:27259–27265.

135. Cary RB, Chen F, Shen Z, Chen DJ. A central region of Ku80 mediates interaction with Ku70 *in vivo*. Nucl Acids Res 1998; 26:974–979.

136. Wang J, Dong X, Reeves WH. A model for Ku heterodimer assembly and interaction with DNA. Implications for the function of Ku antigen. J Biol Chem 1998; 273:31068–31074.

137. Wang J, Dong X, Myung K, Hendrickson EA, Reeves WH. Identification of two domains of the p70 Ku protein mediating dimerization with p80 and DNA binding. J Biol Chem 1998; 273:842–848.

138. Lo Conte L, Chothia C, Janin J. The atomic structure of protein-protein recognition sites. J Mol Biol 1999; 285:2177–2198.

139. Krishna TS, Kong XP, Gary S, Burgers PM, Kuriyan J. Crystal structure of the eukaryotic DNA polymerase processivity factor PCNA. Cell 1994; 79:1233–1243.

140. Moarefi I, Jeruzalmi D, Turner J, O'Donnell M, Kuriyan J. Crystal structure of the DNA polymerase processivity factor of T4 bacteriophage. J Mol Biol 2000; 296:1215–1223.

141. Stewart J, Hingorani MM, Kelman Z, O'Donnell M. Mechanism of β clamp opening by the δ subunit of *Escherichia coli* DNA polymerase III holoenzyme. J Biol Chem 2001; 276:19182–19189.

142. Nasmyth K. Segregating sister genomes: the molecular biology of chromosome separation. Science 2002; 297:559–565.

143. Gu Y, Jin S, Gao Y, Weaver DT, Alt FW. Ku70-deficient embryonic stem cells have increased ionizing radiosensitivity, defective DNA end-binding activity, and inability to support V(D)J recombination. Proc Natl Acad Sci USA 1997; 94:8076–8081.

144. Yoo S, Kimzey A, Dynan WS. Photocross-linking of an oriented DNA repair complex. Ku bound at a single DNA end. J Biol Chem 1999; 274:20034–20039.

145. Cary RB, Peterson SR, Wang J, Bear DG, Bradbury EM, Chen DJ. DNA looping by Ku and the DNA-dependent protein kinase. Proc Natl Acad Sci USA 1997; 94:4267–4272.

146. Ramsden DA, Gellert M. Ku protein stimulates DNA end joining by mammalian DNA ligases: a direct role for Ku in repair of DNA double-strand breaks. EMBO J 1998; 17:609–614.

147. Calsou P, Delteil C, Frit P, Drouet J, Salles B. Coordinated assembly of Ku and p460 subunits of the DNA-dependent protein kinase on DNA ends is necessary for XRCC4-ligase IV recruitment. J Mol Biol 2003; 326:93–103.

148. Kysela B, Doherty AJ, Chovanec M, Stiff T, Ameer-Beg SM, Vojnovic B, Girard PM, Jeggo PA. Ku stimulation of DNA ligase IV-dependent ligation requires inward movement along the DNA molecule. J Biol Chem 2003; 278:22466–22474.

149. Jones JM, Gellert M, Yang W. A Ku bridge over broken DNA. Structure 2001; 9:881–884.

150. Zhang Z, Zhu L, Lin D, Chen F, Chen DJ, Chen Y. The three-dimensional structure of the C-terminal DNA-binding domain of human Ku70. J Biol Chem 2001; 276:38231–38236.

151. Aravind L, Koonin EV. SAP - a putative DNA-binding motif involved in chromosomal organization. Trends Biochem Sci 2000; 25:112–114.

152. Allaway GP, Vivino AA, Kohn LD, Notkins AL, Prabhakar BS. Characterization of the 70KDA component of the human Ku autoantigen expressed in insect cell nuclei using a recombinant baculovirus vector. Biochem Biophys Res Commun 1990; 168:747–755.

153. Abu-Elheiga L, Yaneva M. Antigenic determinants of the 70-kDa subunit of the Ku autoantigen. Clin Immunol Immunopathol 1992; 64:145–152.

154. Zhang WW, Yaneva M. On the mechanisms of Ku protein binding to DNA. Biochem Biophys Res Commun 1992; 186:574–579.

155. Jacoby DB, Wensink PC. Yolk protein factor 1 is a *Drosophila* homolog of Ku, the DNA-binding subunit of a DNA-dependent protein kinase from humans. J Biol Chem 1994; 269:11484–11491.

156. Han Z, Johnston C, Reeves WH, Carter T, Wyche JH, Hendrickson EA. Characterization of a Ku86 variant protein that results in altered DNA binding and diminished DNA-dependent protein kinase activity. J Biol Chem 1996; 271:14098–14104.

157. Gell D, Jackson SP. Mapping of protein-protein interactions within the DNA-dependent protein kinase complex. Nucl Acids Res 1999; 27:3494–3502.

158. Singleton BK, Torres-Arzayus MI, Rottinghaus ST, Taccioli GE, Jeggo PA. The C terminus of Ku80 activates the DNA-dependent protein kinase catalytic subunit. Mol Cell Biol 1999; 19:3267–3277.

159. Tai YT, Teoh G, Lin B, Davies FE, Chauhan D, Treon SP, Raje N, Hideshima T, Shima Y, Podar K, Anderson KC. Ku86 variant expression and function in multiple myeloma cells is associated with increased sensitivity to DNA damage. J Immunol 2000; 165:6347–6355.

160. Karmakar P, Piotrowski J, Brosh RM Jr, Sommers JA, Miller SP, Cheng WH, Snowden CM, Ramsden DA, Bohr VA. Werner protein is a target of DNA-dependent protein kinase *in vivo* and *in vitro*, and its catalytic activities are regulated by phosphorylation. J Biol Chem 2002; 277:18291–18302.

161. Marangoni E, Foray N, O'Driscoll M, Douc-Rasy S, Bernier J, Bourhis J, Jeggo P. A Ku80 fragment with dominant negative activity imparts a radiosensitive phenotype to CHO-K1 cells. Nucl Acids Res 2000; 28:4778–4782.

162. Kim CH, Park SJ, Lee SH. A targeted inhibition of DNA-dependent protein kinase sensitizes breast cancer cells following ionizing radiation. J Pharmacol Exp Ther 2002; 303:753–759.

163. Harris R, Maman JD, Hinks JA, Sankar A, Pearl LH, Driscoll PC. Backbone ^{1}H, ^{13}C, and ^{15}N resonance assignments for the C-terminal region of Ku86 (Ku86CTR). J Biomol NMR 2002; 22:373–374.

164. Harris R, Esposito D, Sankar A, Maman JD, Hinks JA, Pearl LH, Driscoll PC. The 3D solution structure of the C-terminal region of Ku86 (Ku86CTR). J Mol Biol 2004; 335:573–582.

165. Paillard S, Strauss F. Site-specific proteolytic cleavage of Ku protein bound to DNA. Proteins 1993; 15:330–337.

166. Muller C, Dusseau C, Calsou P, Salles B. Human normal peripheral blood B-lymphocytes are deficient in DNA-dependent protein kinase activity due to the expression of a variant form of the Ku86 protein. Oncogene 1998; 16:1553–1560.

167. Jeng YW, Chao HC, Chiu CF, Chou WG. Senescent human fibroblasts have elevated Ku86 proteolytic cleavage activity. Mutat Res 1999; 435:225–232.

168. Coffey G, Campbell C. An alternate form of Ku80 is required for DNA end-binding activity in mammalian mitochondria. Nucl Acids Res 2000; 28:3793–3800.

169. Sallmyr A, Henriksson G, Fukushima S, Bredberg A. Ku protein in human T and B lymphocytes: full length functional form and signs of degradation. Biochim Biophys Acta 2001; 1538:305–312.

170. Sallmyr A, Du L, Bredberg A. An inducible Ku86-degrading serine protease in human cells. Biochim Biophys Acta 2002; 1593:57–68.

171. Tai YT, Podar K, Kraeft SK, Wang F, Young G, Lin B, Gupta D, Chen LB, Anderson KC. Translocation of Ku86/Ku70 to the multiple myeloma cell membrane: functional implications. Exp Hematol 2002; 30:212–220.

172. Chan DW, Lees-Miller SP. The DNA-dependent protein kinase is inactivated by autophosphorylation of the catalytic subunit. J Biol Chem 1996; 271:8936–8941.

173. Hsu HL, Gilley D, Blackburn EH, Chen DJ. Ku is associated with the telomere in mammals. Proc Natl Acad Sci USA 1999; 96:12454–12458.

174. Martin SG, Laroche T, Suka N, Grunstein M, Gasser SM. Relocalization of telomeric Ku and SIR proteins in response to DNA strand breaks in yeast. Cell 1999; 97:621–633.

175. Bertuch AA, Lundblad V. The Ku heterodimer performs separable activities at double-strand breaks and chromosome termini. Mol Cell Biol 2003; 23:8202–8215.

176. Myung K, Ghosh G, Fattah FJ, Li G, Kim H, Dutia A, Pak E, Smith S, Hendrickson EA. Regulation of telomere length and suppression of genomic instability in human somatic cells by Ku86. Mol Cell Biol 2004; 24:5050–5059.

177. Miyoshi T, Sadaie M, Kanoh J, Ishikawa F. Telomeric DNA ends are essential for the localization of Ku at telomeres in fission yeast. J Biol Chem 2003; 278:1924–1931.

178. Weinfeld M, Chaudhry MA, D'Amours D, Pelletier JD, Poirier GG, Povirk LF, Lees-Miller SP. Interaction of DNA-dependent protein kinase and poly(ADP-ribose) polymerase with radiation-induced DNA strand breaks. Radiat Res 1997; 148:22–28.

179. Smider V, Rathmell WK, Brown G, Lewis S, Chu G. Failure of hairpin-ended and nicked DNA to activate DNA-dependent protein kinase: implications for V(D)J recombination. Mol Cell Biol 1998; 18:6853–6858.

180. Fortini P, Pascucci B, Parlanti E, D'Errico M, Simonelli V, Dogliotti E. The base excision repair: mechanisms and its relevance for cancer susceptibility. Biochimie 2003; 85: 1053–1071.

181. Costa RM, Chigancas V, da Silva Galhardo R, Carvalho H, Menck CF. The eukaryotic nucleotide excision repair pathway. Biochimie 2003; 85:1083–1099.

182. Labhart P. DNA-dependent protein kinase specifically represses promoter-directed transcription initiation by RNA polymerase I. Proc Natl Acad Sci USA 1995; 92: 2934–2938.

183. Woodard RL, Anderson MG, Dynan WS. Nuclear extracts lacking DNA-dependent protein kinase are deficient in multiple round transcription. J Biol Chem 1999; 274:478–485.

184. Boubnov NV, Weaver DT. *Scid* cells are deficient in Ku and replication protein A phosphorylation by the DNA-dependent protein kinase. Mol Cell Biol 1995; 15:5700–5706.

185. Lees-Miller SP. The DNA-dependent protein kinase, DNA-PK: 10 years and no ends in sight. Biochem Cell Biol 1996; 74:503–512.

186. Smith GC, Jackson SP. The DNA-dependent protein kinase. Genes Dev 1999; 13: 916–934.

187. Galloway AM, Allalunis-Turner J. cDNA expression array analysis of DNA repair genes in human glioma cells that lack or express DNA-PK. Radiat Res 2000; 154: 609–615.

188. Bryntesson F, Regan JC, Jeggo PA, Taccioli GE, Hubank M. Analysis of gene transcription in cells lacking DNA-PK activity. Radiat Res 2001; 156:167–176.

189. Zdzienicka MZ. Mammalian X ray sensitive mutants: a tool for the elucidation of the cellular response to ionizing radiation. Cancer Surv 1996; 28:281–293.

190. Stuiver MH, Coenjaerts FE, van der Vliet PC. The autoantigen Ku is indistinguishable from NF IV, a protein forming multimeric protein-DNA complexes. J Exp Med 1990; 172:1049–1054.

191. Stuiver MH, Celis JE, van der Vliet PC. Identification of nuclear factor IV/Ku autoantigen in a human 2D-gel protein database. Modification of the large subunit depends on cellular proliferation. FEBS Lett 1991; 282:189–192.

192. Gunderson SI, Knuth MW, Burgess RR. The human U1 snRNA promoter correctly initiates transcription *in vitro* and is activated by PSE1. Genes Dev 1990; 4:2048–2060.

193. Knuth MW, Gunderson SI, Thompson NE, Strasheim LA, Burgess RR. Purification and characterization of proximal sequence element-binding protein 1, a transcription activating protein related to Ku and TREF that binds the proximal sequence element of the human U1 promoter. J Biol Chem 1990; 265:17911–17920.

194. Shieh B, Schultz J, Guinness M, Lacy J. Regulation of the human IgE receptor (Fc εRII/CD23) by Epstein-Barr virus (EBV): Ku autoantigen binds specifically to an EBV-responsive enhancer of CD23. Int Immunol 1997; 9:1885–1895.

195. Petroski MD, Wagner EK. Purification and characterization of a cellular protein that binds to the downstream activation sequence of the strict late U_L38 promoter of herpes simplex virus type 1. J Virol 1998; 72:8181–8190.

196. Pedley J, Pettit A, Parsons PG. Inhibition of Ku autoantigen binding activity to the E_2F motif after ultraviolet B irradiation of melanocytic cells. Melanoma Res 1998; 8:471–481.

197. Willis DM, Loewy AP, Charlton-Kachigian N, Shao JS, Ornitz DM, Towler DA. Regulation of *osteocalcin* gene expression by a novel Ku antigen transcription factor complex. J Biol Chem 2002; 277:37280–37291.

198. Giampuzzi M, Botti G, Di Duca M, Arata L, Ghiggeri G, Gusmano R, Ravazzolo R, Di Donato A. Lysyl oxidase activates the transcription activity of human collagene III promoter. Possible involvement of Ku antigen. J Biol Chem 2000; 275: 36341–36349.

199. Roberts MR, Miskimins WK, Ruddle FH. Nuclear proteins TREF1 and TREF2 bind to the transcriptional control element of the transferrin receptor gene and appear to be associated as a heterodimer. Cell Regul 1989; 1:151–164.

200. Lim JW, Kim H, Kim KH. Expression of Ku70 and Ku80 mediated by NF-κB and cyclooxygenase-2 is related to proliferation of human gastric cancer cells. J Biol Chem 2002; 277:46093–46100.

201. Lim JW, Kim H, Kim KH. The Ku antigen-recombination signal-binding protein J_k complex binds to the nuclear factor-κB p50 promoter and acts as a positive regulator of p50 expression in human gastric cancer cells. J Biol Chem 2004; 279:231–237.

202. May G, Sutton C, Gould H. Purification and characterization of Ku-2, an octamer-binding protein related to the autoantigen Ku. J Biol Chem 1991; 266:3052–3059.

203. Warriar N, Page N, Govindan MV. Expression of human glucocorticoid receptor gene and interaction of nuclear proteins with the transcriptional control element. J Biol Chem 1996; 271:18662–18671.

204. Genersch E, Eckerskorn C, Lottspeich F, Herzog C, Kuhn K, Poschl E. Purification of the sequence-specific transcription factor CTCBF, involved in the control of human collagen IV genes: subunits with homology to Ku antigen. EMBO J 1995; 14:791–800.

205. Hoff CM, Jacob ST. Characterization of the factor E_1BF from a rat hepatoma that modulates ribosomal RNA gene transcription and its relationship to the human Ku autoantigen. Biochem Biophys Res Commun 1993; 190:747–753.

206. Niu H, Jacob ST. Enhancer 1 binding factor (E_1BF), a Ku-related protein, is a growth-regulated RNA polymerase I transcription factor: association of a repressor activity with purified E_1BF from serum-deprived cells. Proc Natl Acad Sci USA 1994; 91:9101–9105.

207. Niu H, Zhang J, Jacob ST. E_1BF/Ku interacts physically and functionally with the core promoter binding factor CPBF and promotes the basal transcription of rat and human ribosomal RNA genes. Gene Expr 1995; 4:111–124.

208. Messier H, Fuller T, Mangal S, Brickner H, Igarashi S, Gaikwad J, Fotedar R, Fotedar A. p70 lupus autoantigen binds the enhancer of the T-cell receptor β-chain gene. Proc Natl Acad Sci USA 1993; 90:2685–2689.

209. Chung U, Igarashi T, Nishishita T, Iwanari H, Iwamatsu A, Suwa A, Mimori T, Hata K, Ebisu S, Ogata E, Fujita T, Okazaki T. The interaction between Ku antigen and REF1 protein mediates negative gene regulation by extracellular calcium. J Biol Chem 1996; 271:8593–8598.

210. Alexandre S, Nakaki T, Vanhamme L, Lee AS. A binding site for the cyclic adenosine 3′,5′-monophosphate-response element-binding protein as a regulatory element in the grp78 promoter. Mol Endocrinol 1991; 5:1862–1872.

211. Liu ES, Lee AS. Common sets of nuclear factors binding to the conserved promoter sequence motif of two coordinately regulated ER protein genes, GRP78 and GRP94. Nucl Acids Res 1991; 19:5425–5431.

212. Kim D, Ouyang H, Yang SH, Nussenzweig A, Burgman P, Li GC. A constitutive heat shock element-binding factor is immunologically identical to the Ku autoantigen. J Biol Chem 1995; 270:15277–15284.

213. Li GC, Yang SH, Kim D, Nussenzweig A, Ouyang H, Wei J, Burgman P, Li L. Suppression of heat-induced hsp70 expression by the 70-kDa subunit of the human Ku autoantigen. Proc Natl Acad Sci USA 1995; 92:4512–4516.

214. Camara-Clayette V, Thomas D, Rahuel C, Barbey R, Cartron JP, Bertrand O. The repressor which binds the -75 GATA motif of the GPB promoter contains Ku70 as the DNA binding subunit. Nucl Acids Res 1999; 27:1656–1663.

215. Giffin W, Torrance H, Rodda DJ, Prefontaine GG, Pope L, Hache RJ. Sequence-specific DNA binding by Ku autoantigen and its effects on transcription. Nature 1996; 380:265–268.

216. Jeanson L, Mouscadet JF. Ku represses the HIV-1 transcription: identification of a putative Ku binding site homologous to the mouse mammary tumor virus NRE1 sequence in the HIV-1 long terminal repeat. J Biol Chem 2002; 277:4918–4924.

217. Shakibai N, Kumar V, Eisenberg S. The Ku-like protein from *Saccharomyces cerevisiae* is required *in vitro* for the assembly of a stable multiprotein complex at a eukaryotic origin of replication. Proc Natl Acad Sci USA 1996; 93:11569–11574.

218. Sordas C, Toth EC, Marusic L, Ochem A, Patthy A, Pongor S, Giacca M, Falaschi A. Interactions of USF and Ku antigen with a human DNA region containing a replication origin. Nucl Acids Res 1993; 21:3257–3263.

219. Ruiz MT, Matheos D, Price GB, Zannis-Hadjopoulos M. OBA/Ku86: DNA binding specificity and involvement in mammalian DNA replication. Mol Biol Cell 1999; 10:567–580.

220. Novac O, Matheos D, Araujo FD, Price GB, Zannis-Hadjopoulos M. *In vivo* association of Ku with mammalian origins of DNA replication. Mol Biol Cell 2001; 12: 3386–3401.

221. Klug J. Ku autoantigen is a potential major cause of nonspecific bands in electrophoretic mobility shift assays. Biotechniques 1997; 22:212–216.

222. Torrance H, Giffin W, Rodda DJ, Pope L, Hache RJ. Sequence-specific binding of Ku autoantigen to single-stranded DNA. J Biol Chem 1998; 273:20810–20819.

223. Lundin C, Erixon K, Arnaudeau C, Schultz N, Jenssen D, Meuth M, Helleday T. Different roles for nonhomologous end joining and homologous recombination following replication arrest in mammalian cells. Mol Cell Biol 2002; 22:5869–5878.

224. Park SJ, Ciccone SL, Freie B, Kurimasa A, Chen DJ, Li GC, Clapp DW, Lee SH. A positive role for the Ku complex in DNA replication following strand break damage in mammals. J Biol Chem 2004; 279:6046–6055.

225. Frit P, Li RY, Arzel D, Salles B, Calsou P. Ku entry into DNA inhibits inward DNA transactions *in vitro*. J Biol Chem 2000; 275:35684–35691.

226. Cosgrove AJ, Nieduszynski CA, Donaldson AD. Ku complex controls the replication time of DNA in telomere regions. Genes Dev 2002; 16:2485–2490.

227. Gravel S, Wellinger RJ. Maintenance of double-stranded telomeric repeats as the critical determinant for cell viability in yeast cells lacking Ku. Mol Cell Biol 2002; 22: 2182–2193.

228. Mishra K, Shore D. Yeast Ku protein plays a direct role in telomeric silencing and counteracts inhibition by rif proteins. Curr Biol 1999; 9:1123–1126.

229. Yaneva M, Busch H. A 10S particle released from deoxyribonuclease-sensitive regions of HeLa cell nuclei contains the 86-kilodalton-70-kilodalton protein complex. Biochemistry 1986; 25:5057–5063.

230. Douglas P, Moorhead GB, Ye R, Lees-Miller SP. Protein phosphatases regulate DNA-dependent protein kinase activity. J Biol Chem 2001; 276:18992–18998.

231. Chan DW, Ye R, Veillette CJ, Lees-Miller SP. DNA-dependent protein kinase phosphorylation sites in Ku 70/80 heterodimer. Biochemistry 1999; 38:1819–1828.

232. Merkle D, Douglas P, Moorhead GB, Leonenko Z, Yu Y, Cramb D, Bazett-Jones DP, Lees-Miller SP. The DNA-dependent protein kinase interacts with DNA to form a protein-DNA complex that is disrupted by phosphorylation. Biochemistry 2002; 41: 12706–12714.

233. Calsou P, Frit P, Humbert O, Muller C, Chen DJ, Salles B. The DNA-dependent protein kinase catalytic activity regulates DNA end processing by means of Ku entry into DNA. J Biol Chem 1999; 274:7848–7856.

234. Cohen HY, Lavu S, Bitterman KJ, Hekking B, Imahiyerobo TA, Miller C, Frye R, Ploegh H, Kessler BM, Sinclair DA. Acetylation of the C terminus of Ku70 by CBP and PCAF controls Bax-mediated apoptosis. Mol Cell 2004; 13:627–638.

235. Yang CR, Yeh S, Leskov K, Odegaard E, Hsu HL, Chang C, Kinsella TJ, Chen DJ, Boothman DA. Isolation of Ku70-binding proteins (KUBs). Nucl Acids Res 1999; 27:2165–2174.

236. Yang CR, Leskov K, Hosley-Eberlein K, Criswell T, Pink JJ, Kinsella TJ, Boothman DA. Nuclear clusterin/XIP8, an x-ray-induced Ku70-binding protein that signals cell death. Proc Natl Acad Sci USA 2000; 97:5907–5912.

237. Schild-Poulter C, Pope L, Giffin W, Kochan JC, Ngsee JK, Traykova-Andonova M, Hache RJ. The binding of Ku antigen to homeodomain proteins promotes their phosphorylation by DNA-dependent protein kinase. J Biol Chem 2001; 276:16848–16856.

238. Kumaravel TS, Bharathy K, Kudoh S, Tanaka K, Kamada N. Expression, localization and functional interactions of Ku70 subunit of DNA-PK in peripheral lymphocytes and Nalm-19 cells after irradiation. Int J Radiat Biol 1998; 74:481–489.

239. Romero F, Dargemont C, Pozo F, Reeves WH, Camonis J, Gisselbrecht S, Fischer S. p95vav associates with the nuclear protein Ku-70. Mol Cell Biol 1996; 16:37–44.

240. Le Romancer M, Reyl-Desmars F, Cherifi Y, Pigeon C, Bottari S, Meyer O, Lewin MJ. The 86-kDa subunit of autoantigen Ku is a somatostatin receptor regulating protein phosphatase-2A activity. J Biol Chem 1994; 269:17464–17468.

241. Adam L, Bandyopadhyay D, Kumar R. Interferon-α signaling promotes nucleus-to-cytoplasmic redistribution of p95Vav, and formation of a multisubunit complex involving Vav, Ku80, and Tyk2. Biochem Biophys Res Commun 2000; 267:692–696.

242. Morio T, Hanissian SH, Bacharier LB, Teraoka H, Nonoyama S, Seki M, Kondo J, Nakano H, Lee SK, Geha RS, Yata J. Ku in the cytoplasm associates with CD40 in human B cells and translocates into the nucleus following incubation with IL-4 and anti-CD40 mAb. Immunity 1999; 11:339–348.

243. Kumar S, Pandey P, Bharti A, Jin S, Weichselbaum R, Weaver D, Kufe D, Kharbanda S. Regulation of DNA-dependent protein kinase by the Lyn tyrosine kinase. J Biol Chem 1998; 273:25654–25658.

244. Bharti A, Kraeft SK, Gounder M, Pandey P, Jin S, Yuan ZM, Lees-Miller SP, Weichselbaum R, Weaver D, Chen LB, Kufe D, Kharbanda S. Inactivation of DNA-dependent protein kinase by protein kinase Cδ: implications for apoptosis. Mol Cell Biol 1998; 18:6719–6728.

245. Bandyopadhyay D, Mandal M, Adam L, Mendelsohn J, Kumar R. Physical interaction between epidermal growth factor receptor and DNA-dependent protein kinase in mammalian cells. J Biol Chem 1998; 273:1568–1573.

246. Huang J, Nueda A, Yoo S, Dynan WS. Heat shock transcription factor 1 binds selectively *in vitro* to Ku protein and the catalytic subunit of the DNA-dependent protein kinase. J Biol Chem 1997; 272:26009–26016.

247. Aoki Y, Zhao G, Qiu D, Shi L, Kao PN. CsA-sensitive purine-box transcriptional regulator in bronchial epithelial cells contains NF45, NF90, and Ku. Am J Physiol Cell Physiol 1998; 275:L1164–L1172.

248. Ting NS, Kao PN, Chan DW, Lintott LG, Lees-Miller SP. DNA-dependent protein kinase interacts with antigen receptor response element binding proteins NF90 and NF45. J Biol Chem 1998; 273:2136–2145.

249. Sartorius CA, Takimoto GS, Richer JK, Tung L, Horwitz KB. Association of the Ku autoantigen/DNA-dependent protein kinase holoenzyme and poly(ADP-ribose) polymerase with the DNA binding domain of progesterone receptors. J Mol Endocrinol 2000; 24:165–182.

250. Sawada M, Hayes P, Matsuyama S. Cytoprotective membrane-permeable peptides designed from the Bax-binding domain of Ku70. Nat Cell Biol 2003; 5:352–357.

251. Sawada M, Sun W, Hayes P, Leskov K, Boothman DA, Matsuyama S. Ku70 suppresses the apoptotic translocation of Bax to mitochondria. Nat Cell Biol 2003; 5:320–329.

252. Barlev NA, Poltoratsky V, Owen-Hughes T, Ying C, Liu L, Workman JL, Berger SL. Repression of GCN5 histone acetyltransferase activity via bromodomain-mediated

binding and phosphorylation by the Ku-DNA-dependent protein kinase complex. Mol Cell Biol 1998; 18:1349–1358.

253. Song K, Jung Y, Jung D, Lee I. Human Ku70 interacts with heterochromatin protein 1α. J Biol Chem 2001; 276:8321–8327.

254. Galy V, Olivo-Marin JC, Scherthan H, Doye V, Rascalou N, Nehrbass U. Nuclear pore complexes in the organization of silent telomeric chromatin. Nature 2000; 403:108–112.

255. Goedecke W, Eijpe M, Offenberg HH, van Aalderen M, Heyting C. Mre11 and Ku70 interact in somatic cells, but are differentially expressed in early meiosis. Nat Genet 1999; 23:194–198.

256. Tsukamoto Y, Kato J, Ikeda H. Silencing factors participate in DNA repair and recombination in *Saccharomyces* cerevisiae. Nature 1997; 388:900–903.

257. Song K, Jung D, Jung Y, Lee SG, Lee I. Interaction of human Ku70 with TRF2. FEBS Lett 2000; 481:81–85.

258. Karmakar P, Snowden CM, Ramsden DA, Bohr VA. Ku heterodimer binds to both ends of the Werner protein and functional interaction occurs at the Werner N-terminus. Nucl Acids Res 2002; 30:3583–3591.

259. Roy R, Meier B, McAinsh AD, Feldmann HM, Jackson SP. Separation-of-function mutants of yeast Ku80 reveal a Yku80p-Sir4p interaction involved in telomeric silencing. J Biol Chem 2004; 279:86–94.

260. Li B, Comai L. Functional interaction between Ku and the werner syndrome protein in DNA end processing. J Biol Chem 2000; 275:28349–28352.

261. Li B, Comai L. Requirements for the nucleolytic processing of DNA ends by the Werner syndrome protein-Ku70/80 complex. J Biol Chem 2001; 276:9896–9902.

262. Jin S, Kharbanda S, Mayer B, Kufe D, Weaver DT. Binding of Ku and c-Abl at the kinase homology region of DNA-dependent protein kinase catalytic subunit. J Biol Chem 1997; 272:24763–24766.

263. Kharbanda S, Pandey P, Jin S, Inoue S, Bharti A, Yuan ZM, Weichselbaum R, Weaver D, Kufe D. Functional interaction between DNA-PK and c-Abl in response to DNA damage. Nature 1997; 386:732–735.

264. Leber R, Wise TW, Mizuta R, Meek K. The XRCC4 gene product is a target for and interacts with the DNA-dependent protein kinase. J Biol Chem 1998; 273:1794–1801.

265. Chen L, Trujillo K, Sung P, Tomkinson AE. Interactions of the DNA ligase IV-XRCC4 complex with DNA ends and the DNA-dependent protein kinase. J Biol Chem 2000; 275:26196–26205.

266. Hsu HL, Yannone SM, Chen DJ. Defining interactions between DNA-PK and ligase IV/XRCC4. DNA Repair 2002; 1:225–235.

267. Galande S, Kohwi-Shigematsu T. Poly(ADP-ribose) polymerase and Ku autoantigen form a complex and synergistically bind to matrix attachment sequences. J Biol Chem 1999; 274:20521–20528.

268. Balajee AS, Geard CR. Chromatin-bound PCNA complex formation triggered by DNA damage occurs independent of the ATM gene product in human cells. Nucl Acids Res 2001; 29:1341–1351.

269. Mahajan KN, Gangi-Peterson L, Sorscher DH, Wang J, Gathy KN, Mahajan NP, Reeves WH, Mitchell BS. Association of terminal deoxynucleotidyl transferase with Ku. Proc Natl Acad Sci USA 1999; 96:13926–13931.

270. Chai W, Ford LP, Lenertz L, Wright WE, Shay JW. Human Ku70/80 associates physically with telomerase through interaction with hTERT. J Biol Chem 2002; 277:47242–47247.

271. Peterson SE, Stellwagen AE, Diede SJ, Singer MS, Haimberger ZW, Johnson CO, Tzoneva M, Gottschling DE. The function of a stem-loop in telomerase RNA is linked to the DNA repair protein Ku. Nat Genet 2001; 27:64–67.

272. Stellwagen AE, Haimberger ZW, Veatch JR, Gottschling DE. Ku interacts with telomerase RNA to promote telomere addition at native and broken chromosome ends. Genes Dev 2003; 17:2384–2395.

273. Hsu HL, Gilley D, Galande SA, Hande MP, Allen B, Kim SH, Li GC, Campisi J, Kohwi-Shigematsu T, Chen DJ. Ku acts in a unique way at the mammalian telomere to prevent end joining. Genes Dev 2000; 14:2807–2812.

274. Downs JA, Jackson SP. Involvement of DNA end-binding protein Ku in Ty element retrotransposition. Mol Cell Biol 1999; 19:6260–6268.

275. Cooper MP, Machwe A, Orren DK, Brosh RM, Ramsden D, Bohr VA. Ku complex interacts with and stimulates the Werner protein. Genes Dev 2000; 14:907–912.

276. Orren DK, Machwe A, Karmakar P, Piotrowski J, Cooper MP, Bohr VA. A functional interaction of Ku with Werner exonuclease facilitates digestion of damaged DNA. Nucl Acids Res 2001; 29:1926–1934.

277. Li B, Comai L. Displacement of DNA-PK$_{cs}$ from DNA ends by the Werner syndrome protein. Nucl Acids Res 2002; 30:3653–3661.

278. Wechsler T, Chen BP, Harper R, Morotomi-Yano K, Huang BC, Meek K, Cleaver JE, Chen DJ, Wabl M. DNA-PK$_{cs}$ function regulated specifically by protein phosphatase 5. Proc Natl Acad Sci USA 2004; 101:1247–1252.

279. Shao RG, Cao CX, Zhang H, Kohn KW, Wold MS, Pommier Y. Replication-mediated DNA damage by camptothecin induces phosphorylation of RPA by DNA-dependent protein kinase and dissociates RPA:DNA-PK complexes. EMBO J 1999; 18:1397–1406.

280. Yannone SM, Roy S, Chan DW, Murphy MB, Huang S, Campisi J, Chen DJ. Werner syndrome protein is regulated and phosphorylated by DNA-dependent protein kinase. J Biol Chem 2001; 276:38242–38248.

281. Achanta G, Pelicano H, Feng L, Plunkett W, Huang P. Interaction of p53 and DNA-PK in response to nucleoside analogues: potential role as a sensor complex for DNA damage. Cancer Res 2001; 61:8723–8729.

282. Mickelsen S, Snyder C, Trujillo K, Bogue M, Roth DB, Meek K. Modulation of terminal deoxynucleotidyltransferase activity by the DNA-dependent protein kinase. J Immunol 1999; 163:834–843.

283. Roth SY, Denu JM, Allis CD. Histone acetyltransferases. Annu Rev Biochem 2001; 70:81–120.

284. Eissenberg JC, James TC, Foster-Hartnett DM, Hartnett T, Ngan V, Elgin SC. Mutation in a heterochromatin-specific chromosomal protein is associated with suppression of position-effect variegation in *Drosophila melanogaster*. Proc Natl Acad Sci USA 1990; 87:9923–9927.

285. Fanti L, Giovinazzo G, Berloco M, Pimpinelli S. The heterochromatin protein 1 prevents telomere fusions in *Drosophila*. Mol Cell 1998; 2:527–538.

286. Aparicio OM, Billington BL, Gottschling DE. Modifiers of position effect are shared between telomeric and silent mating-type loci in *S. cerevisiae*. Cell 1991; 66:1279–1287.

287. Gotta M, Laroche T, Formenton A, Maillet L, Scherthan H, Gasser SM. The clustering of telomeres and colocalization with Rap1, Sir3, and Sir4 proteins in wild-type *Saccharomyces cerevisiae*. J Cell Biol 1996; 134:1349–1363.

288. Laroche T, Martin SG, Gotta M, Gorham HC, Pryde FE, Louis EJ, Gasser SM. Mutation of yeast Ku genes disrupts the subnuclear organization of telomeres. Curr Biol 1998; 8:653–656.

289. Strambio-de-Castillia C, Blobel G, Rout MP. Proteins connecting the nuclear pore complex with the nuclear interior. J Cell Biol 1999; 144:839–855.

290. Williams B, Lustig AJ. The paradoxical relationship between NHEJ and telomeric fusion. Mol Cell 2003; 11:1125–1126.

291. Yoo S, Dynan WS. Characterization of the RNA binding properties of Ku protein. Biochemistry 1998; 37:1336–1343.

292. Driller L, Wellinger RJ, Larrivee M, Kremmer E, Jaklin S, Feldmann HM. A short C-terminal domain of Yku70p is essential for telomere maintenance. J Biol Chem 2000; 275:24921–24927.

293. Bertuch AA, Lundblad V. Which end: dissecting Ku's function at telomeres and double-strand breaks. Genes Dev 2003; 17:2347–2350.

294. Opresko PL, Cheng WH, von Kobbe C, Harrigan JA, Bohr VA. Werner syndrome and the function of the Werner protein; what they can teach us about the molecular aging process. Carcinogenesis 2003; 24:791–802.

295. Dynan WS, Yoo S. Interaction of Ku protein and DNA-dependent protein kinase catalytic subunit with nucleic acids. Nucl Acids Res 1998; 26:1551–1559.

296. Daniel PT, Schulze-Osthoff K, Belka C, Guner D. Guardians of cell death: the Bcl-2 family proteins. Essays Biochem 2003; 39:73–88.

297. McGuire TC, Poppie MJ. Hypogammaglobulinemia and thymic hypoplasia in horses: a primary combined immunodeficiency disorder. Infect Immun 1973; 8:272–277.

298. Roth DB, Menetski JP, Nakajima PB, Bosma MJ, Gellert M. V(D)J recombination: broken DNA molecules with covalently sealed (hairpin) coding ends in *scid* mouse thymocytes. Cell 1992; 70:983–991.

299. Fulop GM, Phillips RA. The *scid* mutation in mice causes a general defect in DNA repair. Nature 1990; 347:479–482.

300. Biedermann KA, Sun JR, Giaccia AJ, Tosto LM, Brown JM. *scid* mutation in mice confers hypersensitivity to ionizing radiation and a deficiency in DNA double-strand break repair. Proc Natl Acad Sci USA 1991; 88:1394–1397.

301. Hendrickson EA, Qin XQ, Bump EA, Schatz DG, Oettinger M, Weaver DT. A link between double-strand break-related repair and V(D)J recombination: the *scid* mutation. Proc Natl Acad Sci USA 1991; 88:4061–4065.

302. Walker AI, Hunt T, Jackson RJ, Anderson CW. Double-stranded DNA induces the phosphorylation of several proteins including the 90 000 mol. wt. heat-shock protein in animal cell extracts. EMBO J 1985; 4:139–145.

303. Carter TH, Kopman CR, James CB. DNA-stimulated protein phosphorylation in HeLa whole cell and nuclear extracts. Biochem Biophys Res Commun 1988; 157:535–540.

304. Carter T, Vancurova I, Sun I, Lou W, DeLeon S. A DNA-activated protein kinase from HeLa cell nuclei. Mol Cell Biol 1990; 10:6460–6471.

305. Jackson SP, MacDonald JJ, Lees-Miller S, Tjian R. GC box binding induces phosphorylation of Sp1 by a DNA-dependent protein kinase. Cell 1990; 63:155–165.

306. Lees-Miller SP, Chen YR, Anderson CW. Human cells contain a DNA-activated protein kinase that phosphorylates simian virus 40 T antigen, mouse p53, and the human Ku autoantigen. Mol Cell Biol 1990; 10:6472–6481.

307. Shin EK, Perryman LE, Meek K. A kinase-negative mutation of DNA-PK$_{CS}$ in equine SCID results in defective coding and signal joint formation. J Immunol 1997; 158:3565–3569.

308. Bosma GC, Fried M, Custer RP, Carroll A, Gibson DM, Bosma MJ. Evidence of functional lymphocytes in some (leaky) *scid* mice. J Exp Med 1988; 167:1016–1033.

309. Danska JS, Holland DP, Mariathasan S, Williams KM, Guidos CJ. Biochemical and genetic defects in the DNA-dependent protein kinase in murine *scid* lymphocytes. Mol Cell Biol 1996; 16:5507–5517.

310. Araki R, Fujimori A, Hamatani K, Mita K, Saito T, Mori M, Fukumura R, Morimyo M, Muto M, Itoh M, Tatsumi K, Abe M. Nonsense mutation at Tyr-4046 in the DNA-dependent protein kinase catalytic subunit of severe combined immune deficiency mice. Proc Natl Acad Sci USA 1997; 94:2438–2443.

311. Jhappan C, Morse HC, 3rd, Fleischmann RD, Gottesman MM, Merlino G. DNA-PKcs: a T-cell tumour suppressor encoded at the mouse *scid* locus. Nat Genet 1997; 17:483–486.

312. Bogue MA, Jhappan C, Roth DB. Analysis of variable (diversity) joining recombination in DNA-dependent protein kinase (DNA-PK)-deficient mice reveals DNA-PK-independent pathways for both signal and coding joint formation. Proc Natl Acad Sci USA 1998; 95:15559–15564.

313. Gao Y, Chaudhuri J, Zhu C, Davidson L, Weaver DT, Alt FW. A targeted DNA-PKcs-null mutation reveals DNA-PK-independent functions for KU in V(D)J recombination. Immunity 1998; 9:367–376.

314. Taccioli GE, Amatucci AG, Beamish HJ, Gell D, Xiang XH, Torres Arzayus MI, Priestley A, Jackson SP, Marshak Rothstein A, Jeggo PA, Herrera VL. Targeted disruption of the catalytic subunit of the DNA-PK gene in mice confers severe combined immunodeficiency and radiosensitivity. Immunity 1998; 9:355–366.

315. Slijepcevic P, Hande MP, Bouffler SD, Lansdorp P, Bryant PE. Telomere length, chromatin structure and chromosome fusigenic potential. Chromosoma 1997; 106:413–421.

316. Hande P, Slijepcevic P, Silver A, Bouffler S, van Buul P, Bryant P, Lansdorp P. Elongated telomeres in *scid* mice. Genomics 1999; 56:221–223.

317. Espejel S, Franco S, Sgura A, Gae D, Bailey SM, Taccioli GE, Blasco MA. Functional interaction between DNA-PK$_{cs}$ and telomerase in telomere length maintenance. EMBO J 2002; 21:6275–6287.

318. Gilley D, Tanaka H, Hande MP, Kurimasa A, Li GC, Oshimura M, Chen DJ. DNA-PK$_{cs}$ is critical for telomere capping. Proc Natl Acad Sci USA 2001; 98:15084–15088.

319. Goytisolo FA, Samper E, Edmonson S, Taccioli GE, Blasco MA. The absence of the DNA-dependent protein kinase catalytic subunit in mice results in anaphase bridges and in increased telomeric fusions with normal telomere length and G-strand overhang. Mol Cell Biol 2001; 21:3642–3651.

320. Chiu CY, Cary RB, Chen DJ, Peterson SR, Stewart PL. Cryo-EM imaging of the catalytic subunit of the DNA-dependent protein kinase. J Mol Biol 1998; 284: 1075–1081.

321. Stewart PL, Cary RB, Peterson SR, Chiu CY. Digitally collected cryo-electron micrographs for single particle reconstruction. Microsc Res Tech 2000; 49:224–232.

322. Stewart PL, Chiu CY, Haley DA, Kong LB, Schlessman JL. Review: resolution issues in single-particle reconstruction. J Struct Biol 1999; 128:58–64.

323. Leuther KK, Hammarsten O, Kornberg RD, Chu G. Structure of DNA-dependent protein kinase: implications for its regulation by DNA. EMBO J 1999; 18:1114–1123.

324. Boskovic J, Rivera-Calzada A, Maman JD, Chacon P, Willison KR, Pearl LH, Llorca O. Visualization of DNA-induced conformational changes in the DNA repair kinase DNA-PK$_{cs}$. EMBO J 2003; 22:5875–5882.

325. Hammarsten O, DeFazio LG, Chu G. Activation of DNA-dependent protein kinase by single-stranded DNA ends. J Biol Chem 2000; 275:1541–1550.

326. Martensson S, Hammarsten O. DNA-dependent protein kinase catalytic subunit. Structural requirements for kinase activation by DNA ends. J Biol Chem 2002; 277: 3020–3029.

327. Martensson S, Nygren J, Osheroff N, Hammarsten O. Activation of the DNA-dependent protein kinase by drug-induced and radiation-induced DNA strand breaks. Radiat Res 2003; 160:291–301.

328. Ding Q, Reddy YV, Wang W, Woods T, Douglas P, Ramsden DA, Lees-Miller SP, Meek K. Autophosphorylation of the catalytic subunit of the DNA-dependent protein kinase is required for efficient end processing during DNA double-strand break repair. Mol Cell Biol 2003; 23:5836–5848.

329. Landschulz WH, Johnson PF, McKnight SL. The leucine zipper; a hypothetical structure common to a new class of DNA binding proteins. Science 1988; 240:1759–1764.

330. Bornberg-Bauer E, Rivals E, Vingron M. Computational approaches to identify leucine zippers. Nucl Acids Res 1998; 26:2740–2746.

331. Bosotti R, Isacchi A, Sonnhammer EL. FAT: a novel domain in PIK-related kinases. Trends Biochem Sci 2000; 25:225–227.

332. Yavuzer U, Smith GC, Bliss T, Werner D, Jackson SP. DNA end-independent activation of DNA-PK mediated via association with the DNA-binding protein C1D. Genes Dev 1998; 12:2188–2199.

333. Nehls P, Keck T, Greferath R, Spiess E, Glaser T, Rothbarth K, Stammer H, Werner D. cDNA cloning, recombinant expression and characterization of polypeptides with exceptional DNA affinity. Nucl Acids Res 1998; 26:1160–1166.

334. Erdemir T, Bilican B, Cagatay T, Goding CR, Yavuzer U. *Saccharomyces cerevisiae* C1D is implicated in both non-homologous DNA end joining and homologous recombination. Mol Microbiol 2002; 46:947–957.

335. Mitchell P, Petfalski E, Houalla R, Podtelejnikov A, Mann M, Tollervey D. Rrp47p is an exosome-associated protein required for the 3′ processing of stable RNAs. Mol Cell Biol 2003; 23:6982–6992.

336. Myung K, He DM, Lee SE, Hendrickson EA. KARP-1: a novel leucine zipper protein expressed from the Ku86 autoantigen locus is implicated in the control of DNA-dependent protein kinase activity. EMBO J 1997; 16:3172–3184.

337. Do E, Taira E, Irie Y, Gan Y, Tanaka H, Kuo CH, Miki N. Molecular cloning and characterization of rKAB1, which interacts with KARP-1, localizes in the nucleus and protects cells against oxidative death. Mol Cell Biochem 2003; 248:77–83.

338. Chan DW, Chen BP, Prithivirajsingh S, Kurimasa A, Story MD, Qin J, Chen DJ. Autophosphorylation of the DNA-dependent protein kinase catalytic subunit is required for rejoining of DNA double-strand breaks. Genes Dev 2002; 16:2333–2338.

339. Douglas P, Sapkota GP, Morrice N, Yu Y, Goodarzi AA, Merkle D, Meek K, Alessi DR, Lees-Miller SP. Identification of *in vitro* and *in vivo* phosphorylation sites in the catalytic subunit of the DNA-dependent protein kinase. Biochem J 2002; 368:243–251.

340. Soubeyrand S, Pope L, Pakuts B, Hache RJ. Threonines 2638/2647 in DNA-PK are essential for cellular resistance to ionizing radiation. Cancer Res 2003; 63:1198–2201.

341. Lees-Miller SP, Sakaguchi K, Ullrich SJ, Appella E, Anderson CW. Human DNA-activated protein kinase phosphorylates serines 15 and 37 in the amino-terminal trans-activation domain of human p53. Mol Cell Biol 1992; 12:5041–5049.

342. Ajmani AK, Satoh M, Reap E, Cohen PL, Reeves WH. Absence of autoantigen Ku in mature human neutrophils and human promyelocytic leukemia line (HL-60) cells and lymphocytes undergoing apoptosis. J Exp Med 1995; 181:2049–2058.

343. Shi Y. Mechanisms of caspase activation and inhibition during apoptosis. Mol Cell 2002; 9:459–470.

344. Casciola-Rosen LA, Anhalt GJ, Rosen A. DNA-dependent protein kinase is one of a subset of autoantigens specifically cleaved early during apoptosis. J Exp Med 1995; 182:1625–1634.

345. Han Z, Malik N, Carter T, Reeves WH, Wyche JH, Hendrickson EA. DNA-dependent protein kinase is a target for a CPP32-like apoptotic protease. J Biol Chem 1996; 271:25035–25040.

346. Le Romancer M, Cosulich SC, Jackson SP, Clarke PR. Cleavage and inactivation of DNA-dependent protein kinase catalytic subunit during apoptosis in Xenopus egg extracts. J Cell Sci 1996; 109:3121–3127.

347. Song Q, Lees-Miller SP, Kumar S, Zhang Z, Chan DW, Smith GC, Jackson SP, Alnemri ES, Litwack G, Khanna KK, Lavin MF. DNA-dependent protein kinase catalytic subunit: a target for an ICE-like protease in apoptosis. EMBO J 1996; 15:3238–3246.

348. Teraoka H, Yumoto Y, Watanabe F, Tsukada K, Suwa A, Enari M, Nagata S. CPP32/Yama/apopain cleaves the catalytic component of DNA-dependent protein kinase in the holoenzyme. FEBS Lett 1996; 393:1–6.

349. McConnell KR, Dynan WS, Hardin JA. The DNA-dependent protein kinase catalytic subunit (p460) is cleaved during Fas-mediated apoptosis in Jurkat cells. J Immunol 1997; 158:2083–2089.

350. Enari M, Sakahira H, Yokoyama H, Okawa K, Iwamatsu A, Nagata S. A caspase-activated DNase that degrades DNA during apoptosis, and its inhibitor ICAD. Nature 1998; 391:43–50.

351. Kaufmann SH, Desnoyers S, Ottaviano Y, Davidson NE, Poirier GG. Specific proteolytic cleavage of poly(ADP-ribose) polymerase: an early marker of chemotherapy-induced apoptosis. Cancer Res 1993; 53:3976–3985.

352. Bakkenist CJ, Kastan MB. DNA damage activates ATM through intermolecular autophosphorylation and dimer dissociation. Nature 2003; 421:499–506.

353. DiBiase SJ, Zeng ZC, Chen R, Hyslop T, Curran WJ, Jr., Iliakis G. DNA-dependent protein kinase stimulates an independently active, nonhomologous, end-joining apparatus. Cancer Res 2000; 60:1245–1253.

354. Kienker LJ, Shin EK, Meek K. Both V(D)J recombination and radioresistance require DNA-PK kinase activity, though minimal levels suffice for V(D)J recombination. Nucl Acids Res 2000; 28:2752–2761.

355. Kim SH, Um JH, Dong-Won B, Kwon BH, Kim DW, Chung BS, Kang CD. Potentiation of chemosensitivity in multidrug-resistant human leukemia CEM cells by inhibition of DNA-dependent protein kinase using wortmannin. Leuk Res 2000; 24:917–925.

356. Kashishian A, Douangpanya H, Clark D, Schlachter ST, Eary CT, Schiro JG, Huang H, Burgess LE, Kesicki EA, Halbrook J. DNA-dependent protein kinase inhibitors as drug candidates for the treatment of cancer. Mol Cancer Ther 2003; 2:1257–1264.

357. Ismail IH, Martensson S, Moshinsky D, Rice A, Tang C, Howlett A, McMahon G, Hammarsten O. SU11752 inhibits the DNA-dependent protein kinase and DNA double-strand break repair resulting in ionizing radiation sensitization. Oncogene 2004; 23:873–882.

358. Kurimasa A, Kumano S, Boubnov NV, Story MD, Tung CS, Peterson SR, Chen DJ. Requirement for the kinase activity of human DNA-dependent protein kinase catalytic subunit in DNA strand break rejoining. Mol Cell Biol 1999; 19:3877–3884.

359. Beamish HJ, Jessberger R, Riballo E, Priestley A, Blunt T, Kysela B, Jeggo PA. The C-terminal conserved domain of DNA-PK$_{cs}$, missing in the SCID mouse, is required for kinase activity. Nucl Acids Res 2000; 28:1506–1513.

360. Porter SE, Greenwell PW, Ritchie KB, Petes TD. The DNA-binding protein Hdf1p (a putative Ku homologue) is required for maintaining normal telomere length in *Saccharomyces cerevisiae*. Nucl Acids Res 1996; 24:582–585.

361. Boulton SJ, Jackson SP. Identification of a *Saccharomyces cerevisiae* Ku80 homologue: roles in DNA double strand break rejoining and in telomeric maintenance. Nucl Acids Res 1996; 24:4639–4648.

362. Nugent CI, Bosco G, Ross LO, Evans SK, Salinger AP, Moore JK, Haber JE, Lundblad V. Telomere maintenance is dependent on activities required for end repair of double-strand breaks. Curr Biol 1998; 8:657–660.

363. Polotnianka RM, Li J, Lustig AJ. The yeast Ku heterodimer is essential for protection of the telomere against nucleolytic and recombinational activities. Curr Biol 1998; 8:831–834.

364. Fellerhoff B, Eckardt-Schupp F, Friedl AA. Subtelomeric repeat amplification is associated with growth at elevated temperature in yku70 mutants of *Saccharomyces cerevisiae*. Genetics 2000; 154:1039–1051.

365. Grandin N, Damon C, Charbonneau M. Cdc13 cooperates with the yeast Ku proteins and Stn1 to regulate telomerase recruitment. Mol Cell Biol 2000; 20:8397–8408.

366. Teo SH, Jackson SP. Telomerase subunit overexpression suppresses telomere-specific checkpoint activation in the yeast *yku80* mutant. EMBO Rep 2001; 2:197–202.

367. Bertuch AA, Lundblad V. EXO1 contributes to telomere maintenance in both telomerase-proficient and telomerase-deficient *Saccharomyces cerevisiae*. Genetics 2004; 166:1651–1659.

368. Baumann P, Cech TR. Protection of telomeres by the Ku protein in fission yeast. Mol Biol Cell 2000; 11:3265–3275.

369. Manolis KG, Nimmo ER, Hartsuiker E, Carr AM, Jeggo PA, Allshire RC. Novel functional requirements for non-homologous DNA end joining in *Schizosaccharomyces pombe*. EMBO J 2001; 20:210–221.

370. Conway C, McCulloch R, Ginger ML, Robinson NP, Browitt A, Barry JD. Ku is important for telomere maintenance, but not for differential expression of telomeric VSG genes, in African trypanosomes. J Biol Chem 2002; 277:21269–21277.

371. Wei C, Skopp R, Takata M, Takeda S, Price CM. Effects of double-strand break repair proteins on vertebrate telomere structure. Nucl Acids Res 2002; 30:2862–2870.

372. Bundock P, van Attikum H, Hooykaas P. Increased telomere length and hypersensitivity to DNA damaging agents in an *Arabidopsis* KU70 mutant. Nucl Acids Res 2002; 30:3395–3400.

373. Riha K, Watson JM, Parkey J, Shippen DE. Telomere length deregulation and enhanced sensitivity to genotoxic stress in *Arabidopsis* mutants deficient in Ku70. EMBO J 2002; 21:2819–2826.

374. Riha K, Shippen DE. Ku is required for telomeric C-rich strand maintenance but not for end-to-end chromosome fusions in *Arabidopsis*. Proc Natl Acad Sci USA 2003; 100:611–615.

375. Jaco I, Munoz P, Goytisolo F, Wesoly J, Bailey S, Taccioli G, Blasco MA. Role of mammalian Rad54 in telomere length maintenance. Mol Cell Biol 2003; 23: 5572–5580.

376. Espejel S, Franco S, Rodriguez-Perales S, Bouffler SD, Cigudosa JC, Blasco MA. Mammalian Ku86 mediates chromosomal fusions and apoptosis caused by critically short telomeres. EMBO J 2002; 21:2207–2219.

377. Li G, Nelsen C, Hendrickson EA. Ku86 is essential in human somatic cells. Proc Natl Acad Sci USA 2002; 99:832–837.

378. Lundblad V, Blackburn EH. An alternative pathway for yeast telomere maintenance rescues *est1*-senescence. Cell 1993; 73:347–360.

379. Teng SC, Zakian VA. Telomere-telomere recombination is an efficient bypass pathway for telomere maintenance in *Saccharomyces cerevisiae*. Mol Cell Biol 1999; 19: 8083–8093.

380. Lustig AJ. Clues to catastrophic telomere loss in mammals from yeast telomere rapid deletion. Nat Rev Genet 2003; 4:916–923.

381. Buerstedde JM, Takeda S. Increased ratio of targeted to random integration after transfection of chicken B cell lines. Cell 1991; 67:179–188.

382. Winding P, Berchtold MW. The chicken B cell line DT40: a novel tool for gene disruption experiments. J Immunol Meth 2001; 249:1–16.

383. Henson JD, Neumann AA, Yeager TR, Reddel RR. Alternative lengthening of telomeres in mammalian cells. Oncogene 2002; 21:598–610.

384. Gravel S, Larrivee M, Labrecque P, Wellinger RJ. Yeast Ku as a regulator of chromosomal DNA end structure. Science 1998; 280:741–744.

385. Bianchi A, de Lange T. Ku binds telomeric DNA *in vitro*. J Biol Chem 1999; 274: 21223–21227.

386. Hemann MT, Strong MA, Hao LY, Greider CW. The shortest telomere, not average telomere length, is critical for cell viability and chromosome stability. Cell 2001; 107: 67–77.

387. Mason JM, Biessmann H. The unusual telomeres of *Drosophila*. Trends Genet 1995; 11:58–62.

388. Makarov VL, Hirose Y, Langmore JP. Long G tails at both ends of human chromosomes suggest a C strand degradation mechanism for telomere shortening. Cell 1997; 88:657–666.

389. Wright WE, Tesmer VM, Huffman KE, Levene SD, Shay JW. Normal human chromosomes have long G-rich telomeric overhangs at one end. Genes Dev 1997; 11:2801–2809.

390. Jacob NK, Skopp R, Price CM. G-overhang dynamics at *Tetrahymena telomeres*. EMBO J 2001; 20:4299–4308.

391. Hemann MT, Greider CW. G-strand overhangs on telomeres in telomerase-deficient mouse cells. Nucl Acids Res 1999; 27:3964–3969.

392. Tomita K, Matsuura A, Caspari T, Carr AM, Akamatsu Y, Iwasaki H, Mizuno K, Ohta K, Uritani M, Ushimaru T, Yoshinaga K, Ueno M. Competition between the

Rad50 complex and the Ku heterodimer reveals a role for Exo1 in processing double-strand breaks but not telomeres. Mol Cell Biol 2003; 23:5186–5197.

393. Hackett JA, Greider CW. End resection initiates genomic instability in the absence of telomerase. Mol Cell Biol 2003; 23:8450–8461.

394. Jacob NK, Kirk KE, Price CM. Generation of telomeric G strand overhangs involves both G and C strand cleavage. Mol Cell 2003; 11:1021–1032.

395. Maringele L, Lydall D. Genes Dev 2002; 16:1919–1933.

396. Maringele L, Lydall D. Structural Aspects of Kuy. Genetics 2004; 166:1641–1649.

397. Stewart SA, Ben-Porath I, Carey VJ, O'Connor BF, Hahn WC, Weinberg RA. Erosion of the telomeric single-strand overhang at replicative senescence. Nat Genet 2003; 33:492–496.

398. Kibe T, Tomita K, Matsuura A, Izawa D, Kodaira T, Ushimaru T, Uritani M, Ueno M. Fission yeast Rhp51 is required for the maintenance of telomere structure in the absence of the Ku heterodimer. Nucl Acids Res 2003; 31:5054–5063.

399. Lee SE, Moore JK, Holmes A, Umezu K, Kolodner RD, Haber JE. *Saccharomyces* Ku70, mre11/rad50 and RPA proteins regulate adaptation to G_2/M arrest after DNA damage. Cell 1998; 94:399–409.

400. Sharpless NE, DePinho RA. Telomeres, stem cells, senescence, and cancer. J Clin Invest 2004; 113:160–168.

401. Huffman KE, Levene SD, Tesmer VM, Shay JW, Wright WE. Telomere shortening is proportional to the size of the G-rich telomeric 3′-overhang. J Biol Chem 2000; 275:19719–19722.

402. McClintock B. The behavior in successive nuclear divisions of a chromosome broken at meiosis. Proc Natl Acad Sci USA 1939; 25:405–416.

403. van Steensel B, Smogorzewska A, de Lange T. TRF2 protects human telomeres from end-to-end fusions. Cell 1998; 92:401–413.

404. Difilippantonio MJ, Petersen S, Chen HT, Johnson R, Jasin M, Kanaar R, Ried T, Nussenzweig A. Evidence for replicative repair of DNA double-strand breaks leading to oncogenic translocation and gene amplification. J Exp Med 2002; 196:469–480.

405. Bailey SM, Cornforth MN, Kurimasa A, Chen DJ, Goodwin EH. Strand-specific post-replicative processing of mammalian telomeres. Science 2001; 293:2462–2465.

406. Rebuzzini P, Lisa A, Giulotto E, Mondello C. Chromosomal end-to-end fusions in immortalized mouse embryonic fibroblasts deficient in the DNA-dependent protein kinase catalytic subunit. Cancer Lett 2004; 203:79–86.

407. Nishitani H, Lygerou Z. Control of DNA replication licensing in a cell cycle. Genes Cells 2002; 7:523–534.

408. van Attikum H, Bundock P, Hooykaas PJ. Non-homologous end-joining proteins are required for *Agrobacterium* T-DNA integration. EMBO J 2001; 20:6550–6558.

409. Yant SR, Kay MA. Nonhomologous-end-joining factors regulate DNA repair fidelity during *Sleeping Beauty* element transposition in mammalian cells. Mol Cell Biol 2003; 23:8505–8518.

410. Daniel R, Katz RA, Skalka AM. A role for DNA-PK in retroviral DNA integration. Science 1999; 284:644–647.

411. Baekelandt V, Claeys A, Cherepanov P, De Clercq E, De Strooper B, Nuttin B, Debyser Z. DNA-dependent protein kinase is not required for efficient lentivirus integration. J Virol 2000; 74:11278–11285.

412. Li L, Olvera JM, Yoder KE, Mitchell RS, Butler SL, Lieber M, Martin SL, Bushman FD. Role of the non-homologous DNA end joining pathway in the early steps of retroviral infection. EMBO J 2001; 20:3272–3281.

413. Kilzer JM, Stracker T, Beitzel B, Meek K, Weitzman M, Bushman FD. Roles of host cell factors in circularization of retroviral DNA. Virology 2003; 314:460–467.

414. Friedl AA. Ku and the stability of the genome. J Biomed Biotechnol 2002; 2:61–65.

415. Kettunen E, Nissen AM, Ollikainen T, Taavitsainen M, Tapper J, Mattson K, Linnainmaa K, Knuutila S, El-Rifai W. Gene expression profiling of malignant mesothelioma cell lines: cDNA array study. Int J Cancer 2001; 91:492–496.

416. Kitahara O, Katagiri T, Tsunoda T, Harima Y, Nakamura Y. Classification of sensitivity or resistance of cervical cancers to ionizing radiation according to expression profiles of 62 genes selected by cDNA microarray analysis. Neoplasia 2002; 4:295–303.

30

Cellular Functions of Mammalian DNA Ligases

John B. Leppard
Department of Molecular Medicine, Institute of Biotechnology, The University of Texas Health Science Center, San Antonio, Texas, U.S.A.

Julie Della-Maria Goetz, Teresa A. Motycka, Zhiwan Dong, Wei Song, Hui-Min Tseng, Sangeetha Vijayakumar, and Alan E. Tomkinson
Radiation Oncology Research Laboratory, Department of Radiation Oncology and Greenebaum Cancer Center, University of Maryland School of Medicine, Baltimore, Maryland, U.S.A.

1. INTRODUCTION

The integrity of the phosphodiester backbone of DNA is disrupted as a consequence of normal DNA metabolism. During DNA replication, the lagging strand is synthesized as a series of short Okazaki fragments that must be linked together to generate an intact strand. In addition, there are DNA rearrangements such as V(D) J recombination in mammals and mating type switching in yeast that are initiated by a site-specific DNA double strand break (DSB) that must be rejoined to restore genomic integrity. DNA is also subject to attack by endogenous and exogenous DNA damaging agents. Under these circumstances, DNA strand breaks may be introduced either directly by the action of the agent such as ionizing radiation or as a consequence removal of the DNA lesion by the DNA excision repair pathways. Irrespective of how they are caused, it is critical for genomic stability that DNA strand breaks are efficiently joined. In the cell, this task is performed by a group of enzymes known as DNA ligases. As expected from the involvement of DNA strand breaks in many different DNA transactions, cells mutated in a DNA ligase-encoding gene often exhibit a pleiotropic phenotype (1–4).

For many years, it was thought that simple prokaryotes such as *E. coli* only possessed a single species of DNA ligase. Recent DNA sequence analysis of prokaryotic organisms has revealed the presence of more than one DNA ligase gene per genome and discovered a novel family of DNA ligases that also have conserved nuclease and primase domains (5,6). By contrast, the presence of multiple DNA ligase genes in mammals was predicted based on the different biochemical properties of DNA ligase activities purified from mammalian cell extracts (7). Three human genes encoding DNA ligases, *LIG1*, *LIG3*, and *LIG4*, have been identified (8–10).

685

Notably, inherited mutations in both the *LIG1* and *LIG4* genes have been linked with human syndromes (11–13).

Although the polypeptides encoded by DNA ligase genes vary greatly in length, each of these gene products contains a conserved region that corresponds to the catalytic domain (14). The smaller enzymes such as the *Chlorella* DNA ligase appear to represent a minimal catalytic domain that has an intrinsic ability to recognize DNA nicks (15,16). In the larger enzymes, it is assumed that the unique sequences flanking the catalytic domain are involved in specific protein–protein interactions that target these enzymes to their site of action in vivo. In this chapter, we will first briefly describe the reaction mechanism and structure of the DNA ligase catalytic domain. The final section will focus on mammalian DNA ligases, their protein partners and how these interactions direct the specific participation of the different species of DNA ligase in DNA replication, DNA repair, and recombination pathways.

2. REACTION MECHANISM

DNA ligases can be divided into two groups depending on whether they use ATP or NAD as the cofactor in the ligation reaction. The laboratory of Dr. I. Robert Lehman played a major role in elucidating the three-step reaction catalyzed by the NAD-dependent DNA ligase of *Escherichia coli* DNA ligase of *E. coli* and the ATP-dependent DNA ligase encoded by bacteriophage T4 (4). In the first step of the reaction, the DNA ligase interacts with the nucelotide cofactor to form a covalent enzyme–adenylate complex. By determining the chemical sensitivity of the DNA ligase–adenylate complex, it was deduced that the AMP moiety was attached to a lysine residue by a phosphoramidite bond (17). The identification of the active site lysine residue by the sequencing of an adenylylated tryptic peptide from bovine DNA ligase I led to the definition of an active site motif, KXDGXR, that is diagnostic for DNA ligases (18).

In the first two steps of the ligation reaction, DNA ligases utilize essentially the same reaction mechanism as a larger family of enzymes known as nucleotidyl transferases that includes RNA ligases and mRNA capping enzymes. Based on alignments of the aniino acid sequences of these enzymes, five conserved motifs, in addition to the active site motif, were defined as key elements in the nucleotidyl transferase reaction mechanism (19). Although the nucleotidyl tranferases all transfer the covalently linked NMP group to a polynucleotide acceptor, the nature of this acceptor differs. In the case of DNA ligases, the AMP group is usually transferred to the 5′ terminus of a nick in duplex DNA that has 3′ hydroxyl and 5′ phosphate termini to form a covalent DNA–adenylate intermediate. In the final step of the ligation reaction, the nonadenylated form of DNA ligase interacts with the DNA adenylate intermediate in a reaction involving nucleophilic attack by the 3′ hydroxyl group to catalyze phosphodiester bond formation with the concomitant release of AMP.

3. DNA LIGASE STRUCTURE

Based on amino acid alignments, it appeared that the ATP-dependent DNA ligases contain a conserved catalytic domain. In accord with this notion, partial proteolysis of mammalian DNA ligase I generated a relatively protease-resistant fragment that retained catalytic activity (20). The structure of T7 DNA ligase determined by X-ray crystallography revealed that the catalytic domain contained two subdomains, one of which formed the enzyme–AMP complex (21,22). Notably, the larger adenylation subdomain and smaller oligomer binding (OB)-fold subdomain, form a cleft with the active site lysine at the bottom of the cleft and the other conserved motifs characteristic of nucleotidyl transferases lining the surfaces of the cleft (19,22). Further insights into the molecular architecture of the DNA ligase catalytic domain have come from additional structures determined by X-ray crystallography of the ATP-dependent *Chlorella* DNA ligase (23,24) and the NAD-dependent Bst and Tfi DNA ligases (25,26). Despite the low level of the amino acid sequence homology, the shape and structure of the catalytic domains of the ATP- and NAD-dependent are remarkably similar (14).

Although the crystal structures of the DNA ligases mentioned above and of the *Chlorella* mRNA capping enzyme (27) provide only a snapshot of the enzyme conformation, they do provide some evidence for the dynamic changes that may occur during the ligation reaction. For example, in the crystals formed by the *Chlorella* mRNA capping enzyme, the subdomains were in two different arrangements corresponding to open and closed conformations (27). Based on these structures, it was suggested that the open conformation allows nucleotide access and binding. After nucleotide binding, the enzyme changes to the closed conformation, in which it cleaves the nucleotide cofactor to form the covalent enzyme–AMP complex. The recent structures of the *Chlorella* DNA ligase–AMP complex revealed a change in the structure of the putative DNA binding site in the OB subdomain (23,24). Thus, it appears that, when the enzyme–AMP complex adopts the open conformation, pyrophosphate is released and the DNA binding site is now accessible. At the present time, there is no available structure of a DNA ligase complexed with its DNA substrate. However, analysis of DNA–enzyme complexes during the repair of DNA base lesions has shown that the DNA is bent to an increasing extent in the active sites of the sequentially acting base excision repair enzymes. Based on these studies, the nicked DNA duplex in the ligase active site may be bent to an even greater degree than it is in the DNA polymerase active site.

4. MAMMALIAN DNA LIGASES

A combination of biochemical and molecular biology approaches led to the cloning of three human genes encoding DNA ligases, *LIG1*, *LIG3*, and *LIG4* (8–10). Although considerable progress has been made in determining the cellular functions of the enzymes encoded by these genes, this analysis has been hampered by the lack of a simple eukaryotic model that possesses this repertoire of enzymes and is amenable to genetic manipulation. For example, the yeast, *Saccharomyces cerevisiae*, which has proven to be useful model in many instances, lacks a homolog of the mammalian *LIG3* gene. This raises the possibility that the mammalian *LIG3* gene products function in DNA metabolic pathways that are unique to higher eukaryotes. Furthermore, the *LIG3* gene products may have taken over functions that are carried out by the *LIG1* and *LIG4* gene products in lower eukaryotes.

4.1. *LIG1* Gene and Products

Mammalian DNA ligase I is 919 amino acids in length and shares more than 50% amino acid identity with the replicative yeast DNA ligase, Cdc9, but this homology is primarily restricted to the catalytic domains of these enzymes (8). Expression of the C-terminal catalytic domain of mammalian DNA ligase I complements the temperature sensitive phenotype of a *cdc9* mutant yeast strain (8). The N-terminus of DNA ligase I contains a nuclear localization signal (NLS) as well as regions responsible for protein–protein interactions discussed below. Mammalian DNA ligase I is localized to the nucleus and exhibits the same punctate distribution pattern in S phase cells as other DNA replication proteins such as DNA polymerase α (28,29). These nuclear foci visualized by immunofluorescence are presumed to be sites of DNA synthesis, implicating DNA ligase I in this process.

DNA ligase I has been identified as a component of a 21S DNA replication complex purified from cultured human cells (30) and as a component of a DNA base excision repair (BER) complex purified from bovine testes (31). Using affinity chromatography, it has been shown that DNA ligase I interacts directly with proliferating cell nuclear antigen (PCNA) (32), a homotrimeric ring protein that acts as a processivity factor for the replicative DNA polymerase, Polδ. There are a growing number of PCNA interacting proteins, many of which bind to the interdomain connector loop of PCNA using a conserved motif (33). The N-terminal PCNA binding motif of DNA ligase I is for targeting to replication foci and for efficient Okazaki fragment joining and completion of long patch base excision repair in vitro (34,35). Elegant studies by the Montecucco laboratory have identified a series of DNA ligase I phosphorylation sites that are modified during cell cycle progression and appear to regulate the association with PCNA (36,37).

There are contradictory reports as to whether the interaction with PCNA stimulates joining by DNA ligase I (32,38,39). Further studies are required to resolve this issue and to determine whether other factors can bind to a PCNA homotrimer at the same time as DNA ligase I. In contrast to the DNA ligase I–PCNA interaction, considerably less is known about the biological significance of the interaction between DNA ligase I and DNA Polβ that occurs within the BER complex purified from bovine testes. This interaction, which occurs between the noncatalytic N-terminal region of DNA ligase I and the N-terminal dRPase domain of Polβ, stimulates DNA joining by DNA ligase I presumably because Polβ recruits DNA ligase I to the site of gap-filling DNA synthesis (40,41).

Insights into the biological roles of DNA ligase I have come from the 46BR cell line, the only DNA ligase I-deficient human cell line currently available. These cells were derived from a patient who exhibited growth retardation, UV sensitivity, and severe immunodeficiencies, and died at age 19 of lymphoma (42,43). Sequence analysis of the *LIG1* alleles revealed that this patient was a compound heterozygote (13). The mutation in one allele (Glu-566 to Lys) inactivates the active site of the enzyme whereas the other mutant allele (Arg-771 to Trp) results in an enzyme with a 20-fold decrease in activity (13). Further analysis of the Arg-771 to Trp mutant version of DNA ligase I demonstrated that both enzyme-and DNA–AMP intermediates accumulate in vitro and in vivo (44). Transfection of 46BR cells with wild type DNA ligase I cDNA corrects both the defect in Okazaki fragment processing and the hypersensitivity to killing by DNA alkylating agents (35,43–46). Notably, this complementation is dependent upon the interaction between DNA ligase I and PCNA, demonstrating the biological significance of this interaction (35). More recently, the mutation causing the Arg-771 to Trp change in DNA ligase I has been reproduced in a mouse model (47). The mutant mice exhibited elevated genomic instability in the spleen and a predisposition to cancer. However, no defect in DNA repair was observed suggesting that the phenotype may largely be caused by the accumulation of DNA replication intermediates. This is consistent with data from a *LIG1* null mouse that survived until embryonic day 16.5 (48,49). These embryos exhibited a severe defect in fetal erythropoiesis, a process characterized by high levels of DNA replication.

4.2. *LIG3* Gene and Products

Human cDNAs encoding DNA ligase III were isolated by two independent approaches. First, by searching an EST database with a conserved peptide sequence from the C-terminal end of the DNA ligase I catalytic domain, two cDNAs encoding DNA ligases were identified (10). Full-length cDNAs were then isolated from a HeLa cDNA library, which were designated *LIG3* and *LIG4* (see below). In parallel, fragments of human DNA ligase III cDNA were amplified by the polymerase chain reaction (PCR) using degenerate oligonucleotide primers based on peptide sequences from purified bovine DNA ligase III (9). The PCR products were then used as probes to isolate cDNA sequences from a human testis cDNA library leading to construction of the full-length cDNA. Interestingly, these two independent approaches isolated full-length DNA ligase III cDNAs with different 3′ ends. The C-terminal 77 amino acids of the protein encoded by the HeLa cDNA are replaced with 17 unrelated amino acids in the polypeptide encoded by the testis cDNA. The 103 kDa enzyme and the 96 kDa enzyme are 922 and 862 amino acids in length, respectively (9,10). The presence of consensus splice donor and acceptor sequences at the point of divergence of these two cDNAs suggested that they are alternatively spliced mRNAs from the *LIG3* gene. This notion was verified when the exons encoding the different C-termini were mapped in the *LIG3* gene (50).

The two exons, designated α and β encode the C-termini of the 103 kDa DNA ligase IIIα and the 96 kDa DNA ligase IIIβ, respectively. The splicing of the *LIG3* gene is regulated in a tissue- and cell-type specific manner resulting in the ubiquitous expression of DNA ligase IIIβ (9,50), and the germ cell-specific expression of DNA ligase 111β (9,50). This may be achieved by silencing elements within the gene as both exonic and intronic silencers have been identified for the β exon (51).

In addition to the alternatively spliced transcripts of the *LIG3* gene, a mitochondrial form of DNA ligase III is generated by alternative translation initiation. The nuclear form is encoded from an internal in-frame ATG within the DNA ligase IIIα open reading frame, whereas translation of the mitochondrial form starts at the first in-frame ATG codon (52). The additional N-terminal region contains a 35 amino acid mitochondrial leader sequence that targets this form of DNA ligase III to mitochondria.

Although the amino acid sequences of the *LIG3* gene products share homology with the catalytic domain of DNA ligase I, they are more closely related to the DNA ligases encoded by vaccinia and other cytoplasmic pox viruses (9). Vaccinia DNA ligase is not required for viral DNA replication but deletion of the DNA ligase gene makes the virus more sensitive to UV light when grown in cultured cells and less virulent in vivo (53). Although the relationship with the pox virus DNA ligases has not provided insights into the cellular functions of the *LIG3* gene products, the cloning of the *LIG3* gene has helped to clarify several reports in the literature describing species of DNA ligase that appeared to be distinct from DNA ligase I. A 70 kDa polypeptide was designated as DNA ligase II because it had catalytic properties that distinguished it from DNA ligase I (54). Later, a 100 kDa DNA ligase was detected in extracts of rat liver and bovine thymus and described as a high molecular weight form of DNA ligase II (55) and DNA ligase III (56), respectively. Peptide mapping and partial amino acid sequences of DNA ligase II and the 100 kDa DNA ligase (DNA ligase III) demonstrated that these polypeptides were closely related and suggested that they are probably encoded by the same gene (57–59). This conclusion is supported by the absence of additional DNA ligase encoding genes in the completed sequence of the human genome. Despite the failure to generate an active 70 kDa fragment from the 100 kDa DNA ligase by proteolysis in vitro, it seems likely that DNA ligase II is generated from DNA ligase III by proteolysis in cell extracts.

A unique feature of all isoforms of DNA ligase III is the presence of a putative zinc finger at the amino terminus (10). The motif shares approximately 40% identity with the two tandemly arrayed zinc fingers of the type $CX_2CX_{28-30}HX_2C$ found at the amino terminus of poly(ADP-ribose) polymerase-1 (PARP-1) (see also Chapter 33). Like the PARP-1 zinc fingers, the DNA ligase III zinc finger allows the enzyme to bind to DNA single-strand and double-strand breaks (60,61). However, the zinc finger is not required for DNA ligase activity in vitro or in vivo (61).

A recent study described a direct physical interaction between PARP-1 and DNA ligase III that is mediated by the region adjacent to the DNA ligase III zinc finger (60). This suggested that there may be a functional interaction between the zinc finger DNA binding domains of PARP-1 and DNA ligase III. In response to DNA damaging agents that cause DNA single-strand breaks, PARP-1 binds to the breaks and its polymerase activity is activated. Under these circumstances, PARP-1 utilizes NAD to synthesize poly(ADP-ribose)(PAR) polymers on proteins including itself (62). DNA ligase III binds directly to PAR polymers and preferentially associates with automodified PARP-1 both in vitro and in vivo (60). Although the DNA ligase III zinc finger is not required for binding to PAR, it significantly enhances the ability of DNA ligase III to join DNA single-strand breaks in the presence of PAR (60). Thus, it appears that DNA ligase III is recruited into the vicinity of DNA single strand breaks by binding to poly(ADP-ribosylated) PARP-1 and then uses its zinc finger to locate the DNA single-strand break despite the network of negatively charged PAR polymers.

The DNA repair protein encoded by the X-ray repair cross complementing-1 (*XRCC1*) gene, was the first partner of DNA ligase III identified (63). This interaction is mediated through the C-terminal BRCT domain of XRCC1 and the unique C-terminal BRCT domain of DNA ligase IIIα (50). The *XRCC1* gene was initially detected in a screen of mutagenized AA8 Chinese hamster ovary (CHO) cell lines for mutants exhibiting hypersensitivity to the alkyating agent, ethyl methanesulfonate (EMS) (64,65). The levels of DNA ligase IIIα protein are reduced in the *xrcc1* mutant cells, indicating that complex formation with XRCC1 stabilizes DNA ligase IIIα and that *xrcc1* mutant cells are functionally DNA ligase III-deficient (66).

The most obvious defect in *xrcc1* -mutant cells is in the rejoining of single-strand breaks caused either directly by ionizing radiation or indirectly by repair of DNA alkylation damage by BER (64). The XRCC1 polypeptide has no known catalytic activity, but it binds to nicked and gapped DNA (67) and several other DNA repair proteins, including PARP-1 (68), PARP-2 (69), Polβ (70,71), polynucleotide kinase (PNK) (72), and APE1 (73), suggesting that it is a scaffolding factor involved in the assembly of multiprotein repair complexes. Based on their functional interaction, it has been assumed that DNA ligase IIIα and XRCC1 function together in BER and the repair of DNA single-strand breaks. However, recent studies have provided evidence that XRCC1 functions independently of DNA ligase IIIα in some repair pathways (74,75). In vitro reconstitution studies with DNA ligase IIIα have been limited because the DNA ligase IIIα/XRCCl complex, which is likely to be the active factor in BER and DNA single-strand break repair, has not yet been overexpressed and purified. For example, although XRCC1 inhibits strand displacement DNA synthesis by Pol β, and uracil–DNA glycosylase, apurinic/apyrimidinic endonuclease, Pol β and DNA ligase IIIα can efficiently repair a uracil-containing substrate without XRCC1 (71).

The role of DNA ligase IIIα in mitochondrial DNA metabolism is as yet poorly defined. Reduction of DNA ligase IIIα levels by an antisense approach impairs mitochondrial function indicating that mtDNA ligase III plays a key role in mitochondrial genome maintenance (76). Intriguingly, XRCC1 has not been detected in mitochondria suggesting that mtDNA ligase III functions independently of XRCC1 in mitochondrial DNA metabolism and may have other protein partners in this organelle (77). At the present time it is not known whether mtDNA ligase III functions in mitochondrial DNA replication and/or DNA repair. Repair of AP sites has been reconstituted with enzymes purified from *X. laevis* mitochondria including the *Xenopus* homolog of mammalian mtDNA ligase III (78). Given the relatively large quantities of ROS generated by aerobic metabolism in mitochondria, it is likely that the repair of oxidative lesions by BER and DNA single-strand break repair pathways, possibly involving mtDNA ligase III, will be critical to protect the mitochondrial genome.

Unlike DNA ligase IIIα, DNA ligase IIIβ is relatively stable when overexpressed in insect cells and in *E. coli* (61). Currently, no protein partners of the unique C-terminus of DNA ligase IIIβ have been identified. As previously mentioned, the DNA ligase IIIβ transcript is specific to germ cells. The DNA ligase IIIβ transcript can be detected early in the pachytene phase of male germ cell development, reaches a peak in late pachytene spermatocytes before declining in round spermatids (61). This expression profile suggests that the DNA ligase IIIβ isoform may be involved in meiotic recombination or post-meiotic DNA repair. The development of immunological reagents that can distinguish between the α and β isoforms of DNA ligase III would allow these proteins to be simultaneously visualized in male germ cells.

As mentioned above, *xrcc1* mutant CHO cells are effectively DNA ligase II-deficient. More recently, the generation of an *xrcc1*-null mouse model has revealed that *XRCC1* plays an essential role in embryonic development (79). Embryos with a homozygous deletion of *XRCC1* arrested at embryonic day 6.5 exhibiting increased cell death and an increase in unrepaired DNA strand breaks. Cells derived from mutant embryos are hypersensitive to DNA damaging agents, similar to other *xrcc1*-mutant cell lines. Interestingly, a double knockout of *XRCC1* and *Trp53* did not restore viability indicating that the mutant embryos are dying by a p53-independent mechanism (79). Given the DNA ligase III-independent functions of XRCC1 and the multiple *LIG3* gene products, the generation of *LIG3*-mutant cell lines and animals will greatly facilitate the elucidation of the biological roles of the *LIG3* gene products.

4.3. *LIG4* Gene and Products

As mentioned above, the *LIG4* gene was cloned by searching an EST database with a conserved peptide sequence from the catalytic domain of DNA ligase I (10). The single exon encoding DNA ligase encodes a polypeptide that is 911 amino acids in length with the catalytic domain located in the N-terminal half of DNA ligase IV and two BRCT domains in the C-terminal region. Similar to DNA ligase IIIα, the stability and activity of human DNA ligase IV is dependent upon complex formation with another DNA repair protein. DNA ligase IV forms a stable complex with XRCC4 (80–82) but, in this case, it is the region linking the two BRCT domains in DNA ligase IV that interacts with XRCC4 (83). Although XRCC4 forms dimers and tetramers, tetramerization of XRCC4 and interaction with DNA ligase IV are mutually exclusive (84–86). Based on these studies, it appears that the DNA ligase IV/XRCC4 complex contains one DNA ligase IV molecule and an XRCC4 dimer. There is, however, evidence for higher order complexes that contain more than one molecule of DNA ligase IV (87).

A common feature of most models for the repair of DSBs by NHEJ is a protein factor that bridges the DNA ends thereby bringing them together for processing and ligation. DNA ligase IV/XRCC4 complex has been overexpressed in and purified from insect cells (87–89). As expected, this complex binds to DNA ends and also has a weak DNA end-bridging activity (88). In addition, there are functional interactions with both the Ku70/Ku80 and DNA-PKcs subunits of the DNA-dependent protein kinase (DNA–PK) (88,89). Interestingly, DNA–PKcs changes the type of ligation product generated by DNA ligase IV/XRCC4 with linear cohesive-ended DNA fragments from circles to linear multimers (88). A recent study showing that DNA–PKcs has end-bridging activity provides a molecular explanation for its effect on the ligation reaction (90).

Like DNA ligase I, DNA ligase IV also has an *S. cerevisiae* homologue, Dnl4, which is 944 amino acids in length (91–93). However, there is no yeast homolog of DNA-PKcs. In this organism, the Rad50/Mrell/Xrs2 complex brings DNA ends together and also interacts with the Dnl4/Lifl complex, the functional homolog of mammalian DNA ligase IV/XRCC4 (94). At physiological salt concentrations, yeast but not mammalian Ku70/Ku80 heterodimer greatly enhances intermolecular DNA joining by Rad50/Mrell/Xrs2 and Dnl4/Lifl, suggesting that there are functional interactions between the yeast NHEJ factors at DNA ends (94).

As mentioned above, the ends of most in vivo DSBs will require processing prior to joining. In mammalian cells, the recently discovered Artemis nuclease, which forms a complex with DNA–PKcs, is likely to contribute to the nucleolytic processing of DNA ends (95). Similarly, in yeast, there is genetic evidence linking the DNA structure-specific endonuclease, FEN-1 (Rad27) with certain end processing events in NHEJ (96,97). It is likely that a fraction of the end joining events will also involve gap-filling DNA synthesis. In yeast, genetic studies linked Pol4, a member of the Pol X family of DNA polymerases, with gap-filling during NHEJ (96). The participation of Pol4 in the NHEJ pathway appears to be mediated, at least in part, by an interaction with the Dnl4 subunit of the Dnl4/Lifl complex that significantly enhances gap-filling DNA synthesis by Pol4 (98). In mammalian cells, there is biochemical evidence linking the Pol X family members, Pol Mu and terminal transferase with DNA synthesis in NHEJ and V(D)J recombination, respectively (99).

The first human cell line deficient in DNA ligase IV, 180BR, was derived from a patient with acute lymphoblastic leukemia who had a severe and eventually fatal response to radiation therapy (12,100). These cells are hypersensitive to ionizing radiation and DNA double-strand break repair is markedly reduced. Sequence analysis of *LIG4* in 180BR revealed a point mutation which changes the arginine in the active site motif KXDGXR into a histidine. Analogous to the case with the mutant form of DNA ligase I in 46BR cells, enzyme-AMP intermediate formation by DNA ligase IV is severely affected in 180BR cells (12). As alluded to above, the DNA ligase IV/XRCC4 complex is also required for the completion of V(D)J recombination (81,82), a site-specific recombination mechanism that generates a diverse repertoire of antibodies and T-cell receptors by rearranging the immunoglobulin genes. Although 180BR cells exhibit relatively normal levels of V(D)J recombination, the fidelity at signal joints is defective (12,100). Importantly, expression of wild type DNA ligase IV in these cells alleviates all the phenotypes of 180BR cells confirming the role of the enzyme in DSB repair and V(D)J recombination (12).

Recently, other mutations in the human *LIG4* gene have been identified in patients diagnosed with Nijmegen breakage syndrome (NBS) (11). NBS and a similar disorder, ataxia–telangiectasia (AT), are characterized by cells which have a defect in the response to DSBs. The mutation responsible for NBS is in the *NBS1* gene, the functional homologue of yeast *XRS2* (101,102). NBSl, like Xrs2 in yeast, is found in a complex with hRad50 and hMrell. As described above, the yeast counterpart to this complex stimulates DNA end-joining by Dnl4/Lifl (94). In contrast to yeast, the evidence linking the hRad50/Mre11/Nbsl complex with mammalian NHEJ is less convincing. Nonetheless, the similarity in clinical symptoms conferred by mutations in either the *NBS1* or *LIG4* genes (11) and the observation that partially purified fractions containing hRad50/hMre/Nbsl stimulate end joining by DNA ligase IV/XRCC4 (103) argue that, under certain circumstances, the hRad50/ hMrell/Nbsl complex does contribute to the repair of DSBs by NHEJ in mammals.

Homozygous deletion of the mouse *LIG4* gene causes late embryonic lethality (104,105) associated with extensive programmed cell death in the central nervous system (104). In addition, lymphopoiesis is blocked in the mutant embryos and V(D)J recombination does not occur in fibroblasts derived from the mutant embryos (106). Deletion of the genes encoding either Ataxia–telangiectasia mutated (ATM) or p53 in a *LIG4$^{-/-}$* background masks the phenotype caused by *LIG4* deletion (106,107) suggesting that the persistence of DSBs leads to apoptotic cell death in DNA ligase IV-deficient mice. However, the double knock-out mice do exhibit pre-disposition to cancer and often die from proB lymphomas (108). Moreover, the loss of even one allele of *LIG4* results in an increase in nonlymphoid tumorigenesis that presumably reflects less efficient repair of DSBs by NHEJ (109).

5. CELLULAR FUNCTIONS OF DNA LIGASE

Below we will summarise our current understanding of the involvement of the different species of eukaryotic DNA ligase in various cellular DNA transactions.

5.1. DNA Replication

There is compelling evidence linking DNA ligase I with DNA replication. This includes colocalization with replication foci (28), copurification with a 21S replication complex (30), a functional interaction with PCNA (34,35) and the defect in Okazaki fragment joining in the DNA ligase I deficient cell line 46BR (35). As expected for a key DNA replication enzyme, *CDC9*, the *LIG1* homolog in yeast, is an essential gene (2). Similarly, it was found that deletion of both *LIG1* alleles in a mouse embryonic stem cell caused lethality unless a cDNA encoding DNA ligase I was ectopically expressed (110).

Thus, it was surprising that homozygous deletion of the last 5 exons of the mouse *LIG1* gene encoding the C-terminal 174 amino acids was compatible with embryo formation and normal development until day 10.5 (48,49). These observations suggest that, whilst DNA ligase I is the main replicative DNA ligase, there are certain circumstances in which one of the other DNA ligases can substitute for DNA ligase I in DNA replication. In support of this notion, it was reported that another mammalian DNA ligase activity, in addition to DNA ligase I, was able to efficiently join Okazaki fragments when DNA replication was reconstituted with purified factors (111). Thus, it is possible that, in the absence of DNA ligase I, unlinked newly synthesized Okazaki fragments may be removed by strand displacement DNA synthesis. The net result of this would be longer fragments on the lagging strand, requiring fewer ligation events. In this scenario, it is possible that PARP-1 recruits the DNA ligase IIIα/XRCCl complex to join the single strand interruptions in the lagging strand.

5.2. DNA Excision Repair

5.2.1. DNA Mismatch Repair

Currently, there is no direct evidence linking a particular DNA ligase with DNA mismatch repair. Given the involvement of DNA replication proteins such as PCNA and Pol δ in this pathway (112,113), it seems that DNA ligase I is likely to be the enzyme that usually completes mismatch repair events.

5.2.2. DNA Nucleotide Excision Repair

Based on a combination of genetic and biochemical approaches, there is compelling evidence that the yeast homolog of DNA ligase I, Cdc9, completes nucleotide excision repair (NER) events (114). This observation plus the involvement of PCNA, replicative DNA polymerases and replication factor C in the repair synthesis step (115) strongly suggests that DNA ligase I functions in mammalian NER. However, the DNA ligase I-deficient 46BR human cells are only weakly sensitive to UV irradiation (43) and mouse fibroblasts with either the same mutation or a null mutation do not exhibit significantly increased sensitivity to UV light (47–49). Thus, it is possible that one of the other DNA ligases, such as DNA ligase IIIα/XRCCl can substitute for DNA ligase I in mammalian NER. In addition, it is also possible that different DNA ligases participate in the genome and transcription-coupled subpathways of NER.

5.2.3. DNA Base Excision Repair

Based on a combination of genetic and biochemical approaches, there is compelling evidence that the yeast homolog of DNA ligase I, Cdc9, completes base excision repair (BER) events (114). However, BER is more complex in mammalian cells. There are two subpathways that can be distinguished based on the length of the DNA repair synthesis tract. In addition there appears to be a transcription-coupled subpathway for certain base lesions.

Studies with the DNA ligase I- and DNA ligase III-deficient cell lines have provided evidence that DNA ligase I functions in the long patch BER pathway whereas DNA ligase IIIα/XRCCl completes the short patch BER subpathway (35,116,117). Because of the involvement of the replication factors such PCNA and FEN-1 in long patch BER, it seems likely that this subpathway, which was defined in vitro, corresponds to replication-associated BER whereas short patch BER may be a housekeeping repair pathway that functions in non-dividing and proliferating cells. The identity of the DNA ligase that participates in transcription-coupled repair of base lesions is not known.

5.3 DNA Strand Break Repair

5.3.1. DNA Single Strand Break Repair

In mammalian cells, it appears that the major pathway for repairing DNA single-strand breaks involves PARP-1, PARP-2, polynucleotide kinase, XRCC1, and DNA ligase IIIα (60,72,118,119). Presumably this pathway repairs breaks caused directly by agents such as ionizing radiation. In addition, it is conceivable that it acts as a back-up pathway to repair DNA single-strand breaks generated when incision events exceed the capacity of subsequent steps in DNA excision repair pathways. Similarly, the single-strand break repair pathway may join strand breaks in the lagging strand caused by the occasional failure to join Okazaki fragments during DNA replication, thereby preventing sister chromatid exchanges.

As stated previously, it is likely that PARP-l acts as the primary recognition factor for these lesions (62). The activation of PARP-1 catalytic activity by binding to DNA single strand breaks may not only serve to recruit other repair proteins such as DNA ligase IIIα/XRCCl but may also signal the presence of DNA damage to the network of genomic surveillance pathways.

5.3.2. DNA Double Strand Break Repair

The repair of DSBs can occur by processes which require long tracts of DNA sequence homology or by processes in which the DNA ends are simply brought together and joined in the absence of substantial DNA sequence homology (120). Although both these processes, recombinational repair and nonhomologous end-joining (NHEJ), occur in mammalian cells, their relative contribution to cell survival changes during progression through the cell cycle. NHEJ appears to be a house-keeping pathway that operates in nondividing and proliferating cells but is particularly important for the repair of DSBs in non dividing cells and in the Gl and early S phases of the cell cycle. By contrast, recombinational repair becomes more important for DSB repair as the cell progresses into late S and the G2 phase of the cell cycle. This appears to be a consequence of the availability of the sister chromatid to act as the homologous template in the recombinational repair pathway.

At the present time, the identity of the DNA ligase(s) that completes recombinational repair events is unknown. By contrast, there is compelling evidence that DNA ligase IV is a key player in the major NHEJ pathway (11,12,81). There is, however, evidence for the existence of additional end joining pathways in mammalian cells that may involve other DNA ligases (121).

5.4. Immunoglobulin Gene Rearrangement

In higher eukaryotes, the immune system plays a critical protective role that involves the recognition of a wide array of diverse foreign antigens. To achieve this, a wide repertoire of immunoglobulins and T cell receptors must be generated. The rearrangement of immunoglobulin genes, somatic hypermutataion, and class switching all contribute to the diversity of the immune response (see also Chapter 28). At the present time there in no compelling evidence for the involvement of a particular DNA ligase in either somatic hypermutation or class switching. In contrast, both DNA ligase I and DNA ligase IV have been linked with the completion of immunoglobulin gene rearrangement by V(D)J recombination (81,105,121,122). These events are initiated by the introduction of site-specific DSBs by the lymphoid-specific Ragl/Rag2 endonuclease and then completed by the core factors such as DNA–PK that repair DSBs by NHEJ. Thus, it was surprising that, in a cell free system, DNA ligase I was identified as the factor that was required to efficiently join DSBs generated by Ragl/Rag2 (122). Subsequent studies with mutant cell lines and animals have provided compelling evidence that DNA ligase IV/XRCC4 completes V(D)J recombination in vivo (81,82,105,108).

5.5. Meiosis

A unique feature of meiosis is the homologous recombination events initiated by programmed DSBs that are required for segregation of chromosomes at the first meiotic division and introduce genetic diversity into the gametes. At the present time, the identity of the DNA ligase(s) that completes meiotic recombinational repair events is unknown. Based on genetic studies in yeast (123), DNA ligase I is the most obvious candidate. However, DNA ligase IIIα and β mRNAs are both highly expressed when the recombinational events are occurring (50). Finally, it is possible that DNA ligase IIIβ participates in specialized DNA repair pathways in haploid gametes.

5.6. Mitochondrial DNA Metabolism

All the DNA replication, repair, and recombination pathways discussed above act on the nuclear genome. Considerably less is known about the DNA transactions that occur within the mitichondria. The role of a DNA ligase in the replication of the mitochondrial and nuclear genomes will be different because of differences both in the structure of these genomes and their mechanisms of DNA replication. Whilst there is convincing evidence for the repair of certain DNA lesions by BER (124), the existence of other excision repair and strand-break repair pathways has not been definitively established.

In mammalian cells, it appears that the *LIG3* gene encodes the mitochondrial DNA ligase (52,76). Since *S. cerevisiae* also has mitochondria but lacks a *LIG3* gene homolog, this implied that either the *CDC9* or *DNL4* gene would encode the mitochondrial DNA ligase in this organism. Elegant studies by the Stirling laboratory showed that the *CDC9* gene generates both mitochondrial and nuclear enzymes by an alternative translation initiation mechanism similar to the one that regulates expression of the mammalian *LIG3* gene and that the mitochondrial form of Cdc9 is essential for mitochondria (125).

ACKNOWLEDGMENTS

Studies in A.E.T.'s laboratory were supported by grants from the U.S. Department of Health and Human Services (R01GM47251, RO1GM57479, PO1 CA81020, and PO1 CA92584). T.A.M is supported by the Training Program in DNA Repair Mechanisms (T32 CA86800).

REFERENCES

1. Engler MJ, Richardson CC. In: Boyer PD, ed. The Enzymes XV. New York: Academic Press, 1982.
2. Johnston LH, Naysmith KA. *Saccharomyces cerevisiae* cell cycle mutant *cdc9* is defective in DNA ligase. Nature 1978; 274:891–893.
3. Johnston LH. The DNA repair capability of *cdc9*, the *Saccharomyces cerevisiae* mutant defective in DNA ligase. Mol Gen Genet 1979; 170:89–92.
4. Lehman IR. DNA ligase: structure, mechanism and function. Science 1974; 186:79–797.
5. Sriskanda V, Shuman S. A second NAD(+)-dependent DNA ligase (LigB) in *Escherichia coli*. Nucleic Acids Res 2001; 29:4930–4934.
6. Weller GR, Doherty AJ. A family of DNA repair ligases in bacteria? FEBS Lett 2001; 505:340–342.
7. Lindahl T, Barnes DE. Mammalian DNA ligases. Ann Rev Biochem 1992; 61:251–281.
8. Barnes DE, Johnston LH, Kodama K, Tomkinson AE, Lasko DD, Lindahl T. Human DNA ligase I cDNA: Cloning and functional expression in *Saccharomyces cerevisiae*. Proc Natl Acad Sci USA 1990; 87:6679–6683.
9. Chen J, Tomkinson AE, Ramos W, Mackey ZB, Danehower S, Walter CA, Schultz RA, Besterman JM, Husain I. Mammalian DNA ligase III: Molecular Cloning, Chromosomal Localization, and Expression in Spermatocytes undergoing Meiotic Recombination. Mol Cell Biol 1992; 15:5412–5422.
10. Wei Y-F, Robins P, Carter K, Caldecott K, Pappin DJC, Yu G-L, Wang R-P, Shell BK, Nash RA, Schar P, Barnes DE, Haseltine WA, Lindahl T. Molecular Cloning and Expression of Human cDNAs Encoding a Novel DNA Ligase IV and DNA Ligase III, an Enzyme Active in DNA Repair and Genetic Recombination. Mol Cell Biol 1995; 15:3206–3216.
11. O'Driscoll M, Cerosaletti KM, Girard PM, Dai Y, Stumm M, Kysela B, Hirsch B, Gennery A, Palmer SE, Seidel J, Gatti RA, Varon R, Oettinger MA, Neitzel H, Jeggo PA, Concannon P. DNA ligase IV mutations in patients exhibiting developmental delay and immunodeficiency. Mol Cell 2001; 8:1175–1185.
12. Riballo E, Critchlow SE, Teo S-H, Doherty AJ, Priestly A, Broughton B, Kysela B, Beamish H, Plowman N, Arlett CF, Lehmann AR, Jackson SP, Jeggo PA. Identification of a defect in DNA ligase IV in a radiosensitive leukemia patient. Curr Biol 1999; 9:699–702.
13. Barnes DE, Tomkinson AE, Lehmann AR, Webster ADB, Lindahl T. Mutation in the DNA ligase I gene of an individual with immunodeficiencies and cellular hypersensitivity to DNA damaging agents. Cell 1992; 69:495–503.
14. Doherty AJ, Suh SW. Structural and Mechanistic conservation in DNA ligases. Nucleic Acids Res 2000; 28:4051–4058.
15. Sriskanda V, Shuman S. *Chlorella* virus DNA ligase: nick recognition and mutational analysis. Nucleic Acids Res 1998; 26:525–531.
16. Ho CK, Van Etten JL, Shuman S. Characterization of an ATP-dependent DNA ligase encoded by *Chlorella* virus PBCV-1. J Virol 1997; 71:1931–1937.
17. Gumport RI, Lehman IR. Structure of the DNA ligase-adenylate Intermediate: Lysine-linked Adenosine monophosphoramidite. Proc Natl Acad Sci USA 1971; 68:2559–2563.
18. Tomkinson AE, Totty NF, Ginsburg M, Lindahl T. Location of the active site for enzyme-adenylate formation in DNA ligases. Proc Natl Acad Sci USA 1991; 88:400–404.
19. Shuman S, Schwer B. RNA capping enzyme and DNA ligase: a superfamily of covalent nucleotidyl transferases. Mol Microbiol 1995; 17:405–410.
20. Tomkinson AE, Lasko DD, Daly G, Lindahl T. Mammalian DNA ligases. Catalytic domain and size of DNA ligase I. J Biol Chem 1990; 265:12611–12617.
21. Doherty AJ, Wigley DB. Fuctional domains of an ATP-dependent DNA ligase. J Mol Biol 1999; 285:63–71.

22. Subramanya HS, Doherty AJ, Ashford SR, Wigley DB. Crystal structure of an ATP-dependent DNA ligase from bacteriophage T7. Cell 1996; 85:607–615.

23. Odell M, Malinina L, Sriskanda V, Teplova M, Shuman S. Analysis of the DNA joining repertoire of Chlorella virus DNA ligase and a new crystal structure of the ligase-adenylate intermediate. Nucleic Acids Res 2003; 31:5090–5100.

24. Odell M, Sriskanda V, Shuman S, Nikolov DB. Crystal structure of eukaryotic DNA ligase-adenylate illuminates the mechanism of nick sensing and strand joining. Mol Cell 2000; 6:1183–1193.

25. Singleton MR, Hakansson K, Timson DJ, Wigley DB. Structure of the adenylation domain of an NAD (+)-dependent DNA ligase. Struct Fold Des 1999; 7:35–42.

26. Lee JY, Chang C, Song HKJM, Yang YK, Kim HK, Kwon ST, Suh SW. Crystal Structure of NAD(+)-dependent DNA ligase: modular architecture and functional implications. EMBO J 2000; 19:1119–1129.

27. Hakansson K, Doherty AJ, Shuman S, Wigley DB. X-ray crystallography reveals a large conformational change during guanyl transfer by mRNA capping enzymes. Cell 1997; 89:545–553.

28. Lasko DD, Tomkinson AE, Lindahl T. Mammalian DNA ligases. Biosynthesis and intracellular localization of DNA ligase I. J Biol Chem 1990; 265:12618–12622.

29. Montecucco A, Savini E, Weighardt F, Rossi R, Ciarrocchi G, Villa A, Biamonti G. The N-terminal domain of human DNA ligase I contains the nuclear localization signal and directs the enzyme to sites of DNA replication. EMBO J 1995; 14:5379–5386.

30. Li C, Goodchild J, Baril EF. DNA ligase I is associated with the 21S complex of enzymes for DNA synthesis in HeLa cellls. Nucleic Acids Res 1994; 22:632–638.

31. Prasad R, Singhal RK, Srivastava DK, Molina JT, Tomkinson AE, Wilson SH. Specific interaction of DNA polymerase β and DNA ligase I in a multiprotein base excision repair complex from bovine testis. J Biol Chem 1996; 271:16000–16007.

32. Levin DS, Bai W, Tomkinson AE. An Interaction between DNA Ligase I and Proliferating Cell Nuclear Antigen: Implications for Okazaki Fragment Synthesis and joining. Proc Natl Acad Sci USA 1997; 94:12863–12868.

33. Warbrick E. PCNA binding through a conserved motif. Bioessays 1998; 20:195–199.

34. Montecucco A, Rossi R, Levin DS, Gary R, Park MS, Motycka TA, Ciarrocchi G, Villa A, Biamonti G, Tomkinson AE. DNA ligase I is recruited to sites of DNA replication by an interaction with proliferating cell nuclear antigen: identification of a common targeting mechanism for the assembly of replication factories. EMBO J 1998; 17:3786–3795.

35. Levin DS, McKenna A, Motycka T, Matsumoto Y, Tomkinson AE. Interaction between PCNA and DNA ligase I is critical for joining of Okazaki fragments and long-patch base-excision repair. Curr Biol 2000; 10:919–922.

36. Ferrari G, Rossi R, Arosio D, Vindigni A, Biamonti G, Montecucco A. cell cycle-dependent phosphorylation of human DNA ligase I at the cyclin-dependent kinase sites. J Biol Chem 2003; 278:37761–37777.

37. Rossi R, Villa A, Negri C, Scovassi I, Ciarrochi G, Biamonti G, Montecucco A. The replication factory targeting sequence/PCNA binding site is required in G(1) to control the phosphorylation status of DNA ligase I. EMBO J 1999; 18:5745–5754.

38. Jonsson Z, Hindges R, Hubscher U. Regulation of DNA replication and repair proteins through interaction with the front side of proliferating cell nuclear antigen. EMBO J 1998; 17:2412–2425.

39. Tom S, Henricksen LA, Park MS, Bambara RA. DNA ligase I and proliferating cell nuclear antigen form a functional complex. J Biol Chem 2001; 276:24817–24825.

40. Tomkinson AE, Chen L, Dong Z, Leppard JB, Levin DS, Mackey ZB, Motycka TA. Completion of Base Excision Repair by Mammalian DNA Ligases. Prog Nucleic Acid Res Mol Biol 2001; 68:151–164.

41. Dimitriadis EK, Prasad R, Vaske MK, Chen L, Tomkinson AE, Lewis MS, Wilson SH. Thermodynamics of DNA ligase I trimerization and association with DNA polymerase β. J Biol Chem 1998; 272:20540–20550.

42. Webster ADB, Barnes DE, Arlett CF, Lehmann AR, Lindahl T. Growth retardation and immunodeficiency in a patient with mutations in the DNA ligase I gene. Lancet 1992; 339:1508–1509.

43. Webster D, Arlett CF, Harcourt SA, Teo IA, Henderson L. A new syndrome of immuno-deficiency and increased cellular sensitivity to DNA damaging agents. In: Bridges BA. Harnden DG, eds. Ataxia telangiectasia—A Cellular and Molecular link between Cancer. Neuropathology and Immune Deficiency. John Wiley and sons Ltd 1982: 379–386.

44. Prigent C, Satoh MS, Daly G, Barnes DE, Lindahl T. Aberrant DNA Repair and DNA Replication due to an inherited defect in Human DNA ligase I. Mol Cell Biol 1994; 14:310–317.

45. Teo IA, Arlett CF, Harcourt SA, Priestly A, Broughton BC. Multiple hypersensitivity to mutagens in a cell strain (46BR) derived from a patient with immuno-deficiencies. Mut Res 1983; 107:371–386.

46. Somia NV, Jessop JK, Melton DW. Phenotypic correction of a human cell line (46BR) with aberrant DNA ligase I. Mut Res 1993; 294:51–58.

47. Harrison C, Ketchen AM, Redhead NJ, O'Sullivan MJ, Melton DW. Replication fail-ure, genome instability and increased cancer susceptibility in mice with a point mutation in the DNA ligase I gene. Cancer Res 2002; 62:4065–4074.

48. Bentley DJ, Harrison C, Ketchen AM, Redhead NJKS, Waterfall M, Ansell JD, Melton DW. DNA ligase I null cells show normal DNA repair activity but altered DNA replication and reduced genome stability. J Cell Sci 2002; 115:1551–1561.

49. Bentley DJ, Selfridge J, Millar JK, Samuel K, Hole N, Ansell JD, Melton DW. DNA ligase I is required for fetal liver erythropoiesis but is not essential for mammalian cell viability. Nat Gene 1996; 13:489–491.

50. Mackey ZB, Ramos W, Levin DS, Walter CA, McCarrey JR, Tomkinson AE. An alter-native splicing event, which occurs in mouse pachytene spermatocytes, generates a form of DNA ligase III with distinct biochemical properties that may function in meiotic recombination. Mol Cell Biol 1996; 17:989–998.

51. Chew SL, Baginsky L, Eperon IC. An exonic splicing silencer in the testes-specific DNA ligase III beta exon. Nucleic Acid Res 2000; 28:402–410.

52. Lakshmipathy U, Campbell C. The human DNA ligase III gene encodes nuclear and mitochondrial proteins. Mol Cell Biol 1999; 19:3869–3876.

53. Kerr SM, Johnston LH, Odell M, Duncan SA, Law KM, Smith GL. Vaccinia virus DNA ligase complements Saccharomyces cdc9, localizes in cytoplasmic factories and affects virulence and virus sensitivity to DNA damaging agents. EMBO J 1991; 10: 4343–4350.

54. Arrand JE, Willis AE, Goldsmith I, Lindahl T. Different substrate specifities of the two DNA ligases of Mammalian cells. J Biol Chem 1986; 261:9079–9082.

55. Elder RH, Rossignol J-M. DNA ligases from rat liver. Purification and partial charac-terization of two molecular forms. Biochem 1990; 29:6009–6017.

56. Tomkinson AE, Roberts E, Daly G, Totty NF, Lindahl T. Three distinct DNA ligases in mammalian cells. J Biol Chem 1991; 286:21728–21735.

57. Wang YC, Burkhart WA, Mackey ZB, Moyer MB, Ramos W, Husain I, Chen J, Besterman JM, Tomkinson AE. Mammalian DNA ligase II is highly homologous with vaccinia DNA ligase; identification of the DNA ligase II active site for enzyme-adeny-late formation. J Biol Chem 1994; 269:31923–31928.

58. Roberts E, Nash RA, Robins P, Lindahl T. Different active sites of mammalian DNA ligases I and II. J Biol Chem 1994; 269:3789–3792.

59. Husain I, Tomkinson AE, Burkhart WA, Moyer MB, Ramos W, Mackey ZB, Besterman JM, Chen J. Purification and characterization of DNA ligase III from bovine testes. J Biol Chem 1995; 270:9683–9690.

60. Leppard J, Dong Z, Mackey ZB, Tomkinson AE. Physical and functional interaction between DNA ligase III α and Poly(ADP-ribose) polymerase 1 in DNA single-strand break repair. Mol Cell Biol 2003; 23:5919–5927.

61. Mackey ZB, Niedergang C, Murcia JM, Leppard J, Au K, Chen J, de Murcia G, Tomkinson AE. DNA ligase III is recruited to DNA strand breaks by a zinc finger motif homologous to that of poly (ADP-ribose) polymerase. Identification of two functionally distinct DNA binding regions within DNA ligase III. J Biol Chem 1999; 274: 21679–21687.

62. de Murcia G, de Murcia JM. Poly(ADP) ribose polymerase: a molecular nick sensor. Trends Biochem Sci 1994; 19:172–176.

63. Caldecott KW, McKeown CK, Tucker JD, Ljunquist S, Thompson LH. An interaction between the Mammalian DNA Repair protein XRCC1 and DNA ligase III. Mol Cell Biol 1994; 14:68–76.

64. Thompson LH, Brookman KW, Dillehay LE, Carrano AV, Mazrimas JA, Mooney CL, Minkler JL. A CHO-cell strain having hypersensitivity to mutagens, a defect in strand break repair, and an extraordinary baseline frequency of sister chromatid exchange. Mut Res 1982; 95:247–254.

65. Thompson LH, Brookman KW, Jones NJ, Allen SA, Carrano AV. Molecular Cloning of the human XRCC1 gene, which corrects defective DNA strand break repair and sister chromatid exchange. Mol Cell Biol 1990; 10:6160–6171.

66. Caldecott KW, McKeown CK, Tucker JD, Stanker L, Thompson LH. Characterization of the Xrcc1-DNA ligase III complex in vitro and its absence from mutant hamster cells. Nucleic Acids Res 1996; 23:4836–4843.

67. Marintchev A, Mullen M, Maciejewski MW, Pan B, Gryk MR, Mullen GP. Solution structure of the single-strand break repair protein XRCC1 N-terminal domain. Nat Struct Biol 1999; 6:884–893.

68. Masson M, Niedergang C, Schreiebr V, Muller S, Menissier de Murcia J, de Murcia G. XRCC1 is specifically associated with poly(ADP-ribose) polymerase and negatively regulates its activity following DNA damage. Mol Cell Biol 1998; 18:3563–3571.

69. Schreiber V, Ame JC, Dolle P, Schultz I, Rinaldi B, Fraulob V, Menissier-de Murcia J, de Murcia G. Poly(ADP-ribose) polymerase -2 (PARP-2) is required for efficient base excision repair in association with PARP-1 and XRCC1. J Biol Chem 2002; 277: 23028–23036.

70. Caldecott KW, Aoufouchi S, Johnson P, Shall S. XRCC1 polypeptide interacts with DNA polymerase β and possibly poly (ADP-ribose) polymerase, and DNA ligase III is a novel molecular nick-sensor in vitro. Nucleic Acids Res 1996; 24: 4387–4394.

71. Kubota Y, Nah RA, Klungland A, Schar P, Barnes DE, Lindahl T. Reconstitution of DNA base excision-repair with purified human proteins: interaction between DNA polymerase β and the XRCC1 protein. EMBO J 1996; 15:6662–6670.

72. Whitehouse CJ, Taylor RM, Thistlethwaite A, Zhang H, Karimi-Busheri F, Lasko DD, Weineld M, Caldecott KW. XRCC1 stimulates human polynucleotide kinase at damaged DNA termini and accelerates DNA single-strand break repair. Cell 2001; 104:107–117.

73. Vidal AE, Boiteux S, Hickson ID, Radicella JP. XRCC1 coordinates the initial and late stages of DNA abasic site repair though protein-protein interactions. EMBO J 2001; 20:6530–6539.

74. Moore DJ, Taylor RM, Clements P, Caldecott KW. Mutation of a BRCT domain selectively disrupts DNA single-strand break repair in noncycling Chinese hamster ovary cells. Proc Natl Acad Sci USA 2000; 97:13649–13654.

75. Taylor RM, Moore DJ JW, Johnson P, Caldecott KW. A cell cycle-specific requirement for the XRCC1 BRCT II domain during mammalian DNA strand break repair. Mol Cell Biol 2000; 20:735–740.

76. Lakshmipathy U, Campbell C. Antisense-mediated decrease in DNA ligase III expression results in reduced mitochondrial DNA integrity. Nucleic Acids Res 2001; 29: 668–676.

77. Lakshmipathy U, Campbell C. Mitochondrial DNA ligase III function is independent of Xrccl. Nucleic Acids Res 2000; 28:3880–3886.

78. Pinz KG, Bogenhagen DF. Efficient repair of abasic sites in DNA by mitochondrial enzymes. Mol Cell Biol 1998; 18:1257–1265.

79. Tebbs RS, Flannery ML, Meneses JJ, Hartmann A, Tucker JD, Thompson LH, Cleaver JE, Pedersen RA. Requirement for the Xrccl DNA base excision repair gene during ealry mouse development. Dev Biol 1999; 208:513–529.

80. Grawunder U, Wilm M, Wu X, Kulesza P, Wilson TE, Mann ML, Lieber MR. Activity of DNA ligase IV stimulated by complex formation with XRCC4 protein in mammalian cells. Nature 1997; 388:492–495.

81. Grawunder U, Zimmer D, Fugmann S, Schwartz K, Lieber M. DNA ligase IV is essential for V(D)J recombination and DNA double-strand break repair in human precursor lymphocytes. Mol Cell 1998; 2:477–484.

82. Grawunder D, Zimmer D, Kulezwa P, Lieber MR. Requirement for interaction of XRCC4 with DNA ligase IV for wild type V(D)J recombination and DNA double-strand break repair. J Biol Chem 1998; 273:24708–24714.

83. Grawunder U, Zimmer D, Leiber MR. DNA ligase IV binds to XRCC4 via a motif located between rather than within its BRCT domains. Curr Biol 1998; 8:873–876.

84. Modesti M, Junop MS, Ghirlando R, van de Rakdt M, Gellert M, Yang W, Kanaar R. Tetramerization and DNA ligase IV interaction of the DNA doublestrand break repair protein XRCC4 are mutually exclusive. J Mol Biol 2003; 334:215–228.

85. Junop S, Modesti M, Guarne A, Ghirlando G, Gellert M, Yang W. Crystal Structure of the Xrcc4 DNA Repair protein and implications for end joining. EMBO J 2000; 19:5962–5970.

86. Sibanda BL, Critchlow SE, Begun J, Pei XY, Jackson SP, Blundell TL, Pellegrini L. Crystal structure of Xrcc4-DNA ligase IV complex. Nat Struct Biol 2001; 8:1015–1019.

87. Lee K-J, Huang J, Takeda Y, Dynan WS. DNA ligase IV and XRCC4 form a stable mixed tetramer that functions synergistically with other repair factors in a cell-free end-joining system. J Biol Chem 2000; 275:34787–34796.

88. Chen L, Trujillo K, Sung P, Tomkinson AE. Interactions of the DNA ligase IV-XRCC4 complex with DNA ends and the DNA-dependent protein kinase. J Biol Chem 2000; 275:26196–26205.

89. Nick McElhinny SA, Snowden CM, McCarville J, Ramsden D. Ku recruits the XRCC4-ligase IV-complex to DNA ends. Mol Cell Biol. 2000; 20:2996–3003.

90. DeFazio LG, Stansel RM, Griffith JD, Chu G. Synapsis of DNA ends by DNA-dependent protein kinase. EMBO J 2002; 21:3192–3200.

91. Schar P, Herrman G, Daly G, Lindahl T. A newly identified DNA ligase of *Saccharomyces cerevisiae* involved in RAD52-independent repair of DNA double-strand breaks. Genes Dev 1997; 11:1912–1924.

92. Teo SH, Jackson SP. Identification of *Saccharomyces cerevisiae* DNA ligase IV: involvement in DNA double strand break repair. EMBO J 1997; 16:4788–4795.

93. Wilson TE, Grawunder U, Lieber MR. Yeast DNA ligase IV mediates non-homologous DNA end joining. Nature 1997; 388:495–498.

94. Chen L, Trujillo K, Ramos W, Sung P, Tomkinson AE. Promotion of Dn14-catalysed DNA end-joining by the Rad50/Mre11/Xrs2 and Hdf1/Hdf2 complexes. Mol Cell 2001; 8:1105–1115.

95. Ma PU, Schartz K, Lieber MR. Hairpin opening and overhang processing by an Artemis/DNA-dependent protein kinase complex in nonhomologous end joining and V(D)J recombination. 2002; 208:781–794.

96. Wilson TE, Lieber MR. Efficient processing of DNA ends during yeast nonhomologous end joining. J Biol Chem 1999; 274:23599–23609.

97. Wu X, Wilson TE, Lieber MR. A role for FEN-1 in nonhomologous DNA end joining: the order of strand annealing and nucleolytic processing events. Proc Natl Acad Sci USA 1999; 96:1303–1308.

98. Tseng H-M, Tomkinson AE. A physical and functional interaction between Yeast Pol4 and Dnl4-Lif1 links DNA synthesis and ligation in non-homologous end joining. J Biol Chem 2002; 277:45630–45637.

99. Mahajan K, Nick McElhinney SA, Mitchell BS, Ramsden DA. Association of DNA polymerase μ (pol μ) with Ku and Ligase IV: Role for pol μ in end-joining double-strand break repair. Mol Cell Biol 2002; 22:5194–5202.

100. Badie C, Goodhardt M, Waugh A, Doyen N, Foray N, Calsou P, Singleton B, Gell D, Salles B, Jeggo P, Arlett C, Malaise EP. A double-strand break defective fibroblast cell line (180BR) derived from a radiosensitive patient represents a new mutant phenotype. Cancer Res 1997; 57:4600–4607.

101. Varon R, Vissinga C, Platzer M, Cerosaletti KM, Chrzanowska KH, Saar K, Beckmann G, Seemanova E, Cooper PR, Nowak NJ, Stumm M, Weemaes CM, Gatti RA, Wilson RK, Digweed M, Rosenthal A, Sperling K, Concannon P, Reis A. Nibrin, a novel DNA double-strand break repair protein, is involved in Nijmegen breakage syndrome. Cell 1998; 93:467–476.

102. Carney JP, Maser RS, Olivares H, Davis EM, Le Beau L, Yates JR, Hays L, Morgan WF, Petrini JHJ. The hMre11/hRad50 complex and Nijmegen breakage syndrome: linkage of double-strand break repair to the cellular DNA damage response. Cell 1998; 93:477–486.

103. Huang J, Dynan WS. Reconstitution of mammalian DNA double-strand break end-joining reveals a requirement for an Mre11/Rad50/NBS-containing fraction. Nucleic Acids Res 2002; 30:667–674.

104. Barnes DE, Stamp G, Rosewell I, Denzel A, Lindahl T. Targeted disruption of the gene encoding DNA ligase IV leads to lethality in embryonic mice. Curr Biol 1998; 8:1395–1398.

105. Frank KM, Sekiguchi JM, Seidl KJ, Swat W, Rathbun GA, Cheng HL, Davidson L, Kangaloo L, Alt FW. Late embryonic lethality and impaired V(D)J recombination in mice lacking DNA ligase IV. Nature 1998; 396:173–176.

106. Frank KM, Sharpless NE, Gao Y, Sekiguchi JM, Ferguson DO, Zhu C, Manis JP, Horner J, DePinho RA, Alt FW. DNA liagase IV deficiency in mice leads to defective neurogenesis and embryonic lethality via the p53 pathway. Mol Cell 2000; 5:993–1002.

107. Lee Y, Barnes DE, Lindahl T, McKinnon PJ. Defective neurogenesis resulting from DNA ligase IV deficiency requires Atm. Genes Dev 2000; 14:2576–2580.

108. Ferguson DO, Sekiguchi JM, Chang S, Frank KM, Gao Y, DePinho RA, Alt FW. The nonhomologous end-joining pathway of DNA repair is required for genomic stability and the supression of translocations. Proc Natl Acad Sci USA 2000; 97:6630–6633.

109. Sharpless NE, Ferguson DO, O'Hagan RC, Castrillon DH, Lee C, Farzi PA, Alson S, Fleming J, Morton CC, Frank K, Chin L, Alt FW, DePinho RA. Impaired nonhomologous end-joining provokes soft tissue sarcomas harboring chromosomal translocations, amplifications and deletions. Mol Cell 2001; 8:1187–1196.

110. Petrini JHJ, Xiao Y, Weaver DT. DNA ligase I mediates essential functions in mammalian cells. Mol Cell Biol 1995; 15:4303–4308.

111. Waga S, Bauer G, Stillman B. Reconstitution of complete SV40 replication with purified replication factors. J Biol Chem 1994; 269:10923–10934.

112. Longley MJ, Pierce AJ, Modrich P. DNA polymerase delta is required for human mismatch repair in vitro. J Biol Chem 1997; 272:10917–10921.

113. Umar A, Buermeyer AB, Simon JA, Thomas DC, Clark AB, Liskay RM, Kunkel TA. Requirement for PCNA in DNA mismatch repair at a step preceding DNA resynthesis. Cell 1996; 87:65–73.

114. Wu X, Braithwaite E, Wang Z. DNA ligation during excision repair in yeast cell-free extracts specifically catalyzed by the *CDC9* gene product. Biochem 1999; 38:2628–2635.

115. Aboussekhra A, Biggerstaff M, Shivji MKK, Vilpo JA, Moncollin V, Podust VN, Protic M, Hubscher U, Egly J-M, Wood RD. Mammalian DNA nucleotide excision repair reconstituted with purified protein components. Cell 1995; 80:859–868.

116. Frosina G, Fortini P, Rossi O, Carrozzino F, Raspaglio G, Cox LS, Dane DP, Abbondandolo A, Dogliotti E. Two pathways of base excision repair in mammalian cells. J Biol Chem 1996; 271:9573–9578.

117. Matsumoto Y, Kim K, Hurwitz J, Gary R, Levin DS, Tomkinson AE, Park M. Reconstitution of proliferating Cell Nuclear Antigen-dependent repair of apurinic/apyrimidinic sites with purified human proteins. J Biol Chem 1999; 274:33703–33708.

118. Menissier de Murcia J, Ricoul M, Tartier L, Niedergang C, Huber A, Dantzer F, Schreiber V, Ame JC, Dierich A, LeMur M, Sabatier L, Chambon P, de Murcia G. Functional interaction between PARP-1 and PARP-2 in chromosomal instability and embryonic development in mouse. EMBO J 2003; 22:2253–2263.

119. Okano S, Lan L, Caldecott KW, Mori T, Yasui A. Spatial and temporal responses to single-strand breaks in human cells. Mol Cell Biol 2003; 23:3974–3981.

120. Krejci L, Chen L, Van Komen S, Sung P, Tomkinson A. Mending the break: two repair machines in eukaryotes. Prog Nucleic Acid Res Mol Biol 2003; 74:159–201.

121. Baumann P, West SC. DNA end-joining catalyzed by human cell-free extracts. Proc Natl Acad Sci USA 1998; 95:14066–14070.

122. Ramsden DA, Paull TT, Gellert M. Cell-free V(D)J recombination. Nature 1997; 388:488–491.

123. Ramos W, Liu G, Giroux CN, Tomkinson AE. Biochemical and genetic characterization of the DNA ligase encoded by the *Saccharomyces cerevsiae* open reading frame YOR005c, a homolog of mammalian DNA ligase IV. Nucleic Acids Res 1998; 26:5676–5683.

124. Mandavilli BS, Santos JB, Van Houten B. Mitochondrial DNA repair and aging. Mut Res 2002; 509:127–151.

125. Wilier M, Rainey M, Pullen T, Stirling CJ. The yeast *CDC9* gene encodes both a nuclear and mitochondrial form of DNA ligase I. Curr Biol 1999; 9:1085–1094.

31

The Mre11/Rad50/Nbs1 Complex

Karl-Peter Hopfner
Gene Center, University of Munich, Munich, Germany

1. INTRODUCTION

DNA double-strand breaks (DSBs) are one of the most cytotoxic forms of DNA damage and disrupt the genomic integrity of a cell (1–5). DNA double-strand breaks can occur as products of ionizing radiation and genotoxic agents (6). However, the most frequent source of DSBs is DNA replication (7,8). Finally, DSBs are also generated by enzymatic activities, such as Spo11 in meiotic DNA recombination (9), Rag1/Rag2 in the generation of antibody and T-cell receptor diversity (10), and HO endonuclease in yeast mating-type switching (11). In these processes, DSBs are important cell physiological intermediates.

A single unrepaired DSB can arrest or kill a cell (12). Misrepaired DSBs can result in chromosomal aberrations and cancer (13). The highly cytotoxic nature of DSBs prompts repair by one of several pathways. The major DSB repair pathways are homologous recombination (HR) and nonhomologous end joining (NHEJ). In HR, the DSBs are resected in $5' \rightarrow 3'$ direction to form $3'$ single-strand DNA (ss DNA) tails. These tails pair with the homologous DNA segment of the sister chromatid (mitotic DNA repair) or with the equivalent segment of the homologous chromosome [meiotic recombination (Mre)] and repair proceeds by DNA synthesis without loss of genetic information. In NHEJ, on the contrary, the two broken ends are aligned and directly religated (14). As DSBs are frequently processed before ligation, NHEJ often results in loss of genetic information and is potentially mutagenic (15).

A key factor in the cellular response to DSB is the heterotrimeric Mre11/Rad50/Nbs1 complex (termed Mre11 complex in the following) (Fig. 1). In this chapter, Hopner focus on the emerging structural and functional role of this complex in DSB metabolism with an emphasis on the structural biochemistry of the Mre11 complex. Excellent recent reviews covering cell biological and genetic aspects of the Mre11 complex can be found in Refs. 16–19.

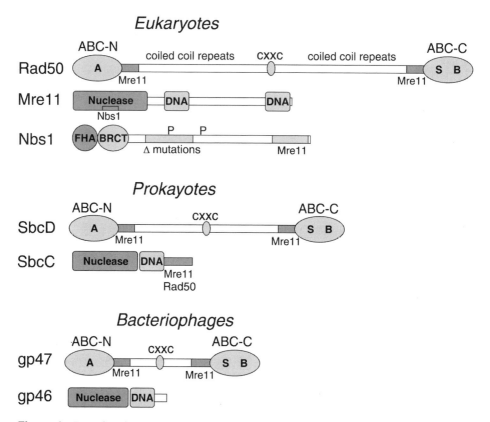

Figure 1 Functional domains of Mre11 complex proteins. The domain structure and architecture of Mre11 and Rad50 are highly conserved in different organisms. Nbs1 is only found in eukaryotes. Rad50 contains a bipartite ABC ATPase domain consisting of an N-terminal (ABC-N) and a C-terminal (ABC-C) segment. ABC-N harbors the Walker A motif (A), ABC-C harbors the Walker B (B) and signature motif (S). The ABC segments are the ends of a long heptad-repeat sequence (eukaryotes: ~900 amino acids; prokaryotes: ~600 amino acids; bacteriophages: ~300 amino acids) that folds into antiparallel coiled-coil domain. Adjacent to the ABC segments are Mre11-binding sites (indicated). The center of the heptad repeats contains a highly conserved Cys-X-X-Cys (CXXC) motif that forms a metal binding coiled-coil dimerization motif. Mre11 possesses a conserved phosphoesterase domain at the N-terminus that harbors the nuclease active site and a potential interaction loop with Nbs1 (in eukaryotes). The central and C-terminal regions contain additional DNA-binding sites, as well as a binding site, for Rad50 and Mre11–Mre11 dimerization (indicated). Nbs1 contains a forkhead associated (FHA) and breast cancer associated C-terminus (BRCT) domain at its N-terminus. These domains mediate interaction with phosphoproteins. The C-terminal region of Nbs1 contains a putative interaction site with Mre11. The central region of Nbs1/Xrs2 (shaded box) contains checkpoint phosphorylation sites (P) and mutations that are found in Nijmegen breakage syndrome (NBS) (including nonsense mutations Δ).

2. THE Mre11 COMPLEX

2.1. Identification of the Mre11 Complex

Genes involved in DSB repair were identified by *Saccharomyces cerevisiae* mutants that were sensitive to radiation (Rad, Xrs) and/or had defects in Mre (20–22). Three of the identified genes, Mre11, Rad50, and Xrs2, led to a similar phenotype: defects in meiosis along with a hyper-recombination phenotype and methyl methanesulfonate (a DSB-inducing agent) sensitivity in mitosis. These genes were subsequently found to form an epistasis group (along with the other HR genes) and to reside in a multiprotein complex (Mre11 complex) (23).

Homologs of Mre11 and Rad50 were readily identified and characterized in other eukaryotes, including humans, chicken, and *Schizosaccharomyces pombe* (24–26). In contrast, the identification of functional homologs of *S. cerevisiae* Xrs2 in other eukaryotes was difficult. Elegant work showed that human Mre11/ Rad50 copurifies with a third subunit of molecular mass 95 kDa (3). This protein was identified as the gene product of the *nbs1* gene (Nbs1), which is mutated in the recessive chromosome instability disorder Nijmegen breakage syndrome (NBS) (27,28). Surprisingly, it turned out that Nbs1 shares only limited homology to Xrs2. With the exception of conserved regions in the N-terminal part of the protein, it is not established if Xrs2/Nbs1 are true structural homologs. Recently, the Nbs1/ Xrs2 homolog of *S. pombe* was identified, again with limited sequence homology to Nbs1/Xrs2 (29,30). However, evidently all Xrs2/Nbs1 homologs are involved in HR and the S-phase checkpoint response to DSBs, suggesting they are a conserved feature of the eukaryotic Mre11 complex (3,30–33).

2.2. The Mre11 Complex Is a Central Player in the Cellular Response to DSBs

Genetic studies indicated that the Mre11 complex is required for genomic stability (4,34). In yeast, null mutations of the Mre11 are highly radiation sensitive and are deficient in Mre (35). In higher eukaryotes, null mutations in components of the Mre11 complex are even lethal (34). Hypomorphic mutations in Nbs1 and Mre11 (mutations that reduce but not abolish activity) cause the genome instability/cancer predisposition diseases such as NBS and Ataxia telangiectasia like disease (ATLD), respectively (3,5,27,28). These diseases show also similar cellular phenotypes with defects in the S-phase checkpoint and chromosome instability. Mutations in human Rad50 have been found in a patient with NBS and several separation of function mutations in yeast Rad50 (termed Rad50S mutations) showed hypomorphic phenotypes with defects in meiosis but relatively normal mitotic DSB repair (35). A Rad50S mutation has recently been introduced in the mouse (36). Rad50$^{S/S}$ mouse is viable but in general short lived, with partial embryonic lethality and cancer susceptibility. Taken together, the analysis of mutations in Mre11 complex proteins in yeast and mammals shows that even minor disturbances of the Mre11 complex activity have profound effects on the integrity of the genome, underlining the central role of this complex in the maintenance of the genetic information.

The similarity of diseases caused by hypomorphic mutations in Mre11 complex genes (ATLD and NBS) to ataxia telangiectasia (AT), which is caused by mutations in the large DNA damage checkpoint kinase ATM (Ataxia *t*elangiectasia *m*utated), indicated that the Mre11 complex is involved in the ATM-mediated DNA damage checkpoint response (37). Recent genetic and cell biological studies suggest that an important role of the Mre11 complex in eukaryotes is the activation of ATM-dependent cellular response to DSBs. In this process, the Mre11 complex probably forms a damage sensor for DSBs (see in what follows).

2.3. The Mre11 Complex Is an Ancient Damage Sensor

Mre11 and Rad50 are well conserved in evolution (Fig. 1) (38–40). Two *Escherichia coli* proteins, SbcC and SbcD, resemble Mre11 and Rad50 in both sequence and biochemical activity (41). SbcC and SbcD process DNA secondary structures and in bacterial replication. Additional homologs of Mre11 and Rad50 are found virtually in all eubacteria and archaea (39). The precise biological function of the archaeal homologs has not been revealed, but these proteins possess biochemical and structural features similar to the eukaryotic and *E. coli* proteins (42). Finally, homologs of Mre11 and Rad50 are even found in certain bacteriophages (T4 genes 46 and 47) that replicate via a break-induced (recombination like) replication mechanism (40,43,44). The particular distribution of Mre11/Rad50 homologs in certain bacteriophages, along with the biochemistry of bacterial and eukaryotic homologs, suggests that an evolutionary conserved role of the Mrc11 complex could be in the restart of replication forks by a break-induced replication mechanism.

The unusual evolutionary conservation suggests that Mre11/Rad50 complex is an ancient DSB detection/repair factor that probably coevolved with recombination processes. Likely, Mre11 and Rad50 form the evolutionary conserved, functional core of the Mre11 complex. Nbs1/Xrs2 has not been discovered outside eukaryotes and this component is much less sequence conserved in eukaryotes than Mre11 and Rad50. Thus, Nbs1/Xrs2 probably joined Mre11/Rad50 in eukaryotes with the evolution of cell-cycle checkpoints and adoption of new roles of the Mre11 complex in monitoring genome integrity.

2.4. The Mre11 Complex Is a Multisubunit Machine

The 150 kDa Rad50 consists of a bipartite ATP-binding cassette (ABC)-type ATPase domain with a 300–900 amino acid long (depending on organism) heptad-repeat insertion (Fig. 1). The ABC segments from the N- and C-termini of the Rad50 protein assemble into a single ABC ATPase domain (Fig. 2) (45). The 300–900 long heptad-repeat regions (bacteriophage–human) between the terminal ATPase segments fold into a 20–60 nm long coiled-coil domain that protrudes from the ABC domain as revealed by electron microscopy and scanning force spectroscopy of Rad50/Mre11 complexes (46–49). Although the length of the coiled-coil domain can vary among different taxonomic kingdoms, this striking architectural feature is conserved in all family members. For two regions of the coiled-coil domain, important protein–protein interaction functions have been assigned. The region directly adjacent to the ABC ATPase domain forms an Mre11-binding site on Rad50 (48). This close connection of Mre11 with the Rad50 ATPase domain suggests the Mre11 nuclease and Rad50 ATPase domain form a single molecular machine that processes and binds DNA ends (Fig. 2b). The precise center of the heptad-repeat sequences contains a conserved Cys-X-X-Cys (CXXC) motif. In the three-dimensional structure of the coiled-coil domain, this CXXC motif is located at the apex of the coiled coil, i.e., the coiled-coil end distal from the ABC ATPase domain (Fig. 2b and c). The CXXC sequence has been recognized early on as potential protein–protein interaction motif, as this spacing of two conserved cysteins is found in other macromolecular interaction motifs such as zinc fingers (41). In fact, recent structural and functional data suggest that the CXXC motif is a key element in architectural functions of the Mre11 complex: two coiled-coil domains of Rad50 can dimerize via a metal-mediated interaction of their CXXC motifs (50). These linked

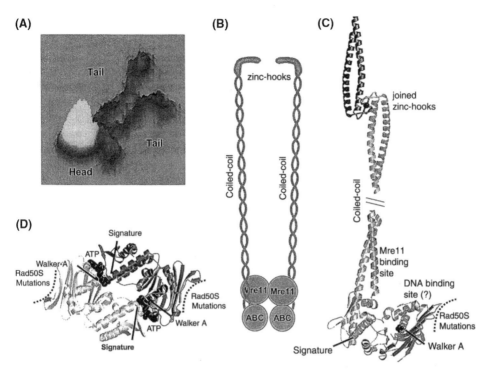

Figure 2 Structure of the core Mre11/Rad50 complex. This figure summarizes our current understanding of the functional topology of the Mre11 core complex. (a) Scanning force micrograph of Mre11/Rad50 reveals a globular head domain and two protruding coiled-coil domain, consisting with a (Mre11/Rad50)$_2$ complex (generously provided by Martijn De Jager and Claire Wyman). (b) Model of the core (Mre11/Rad50)$_2$ complex. (c) Ribbon model of high-resolution crystal structures of Rad50 functional domains. (Bottom) The Rad50 ATPase domain (with part of the coiled-coil) belongs to the ABC ATPases with characteristic Walker A and signature motifs. The coiled-coil region adjacent to the ABC domain contains an Mre11-binding site, indicating that Mre11 and Rad50 form a compact DNA processing machine. (Top) The apex of the Rad50 coiled-coil harbors the CXXC coiled-coil dimerization motif (zinchook). The zinchooks from two Rad50 coiled-coil domains (depicted as light and dark gray) join to form a composite zinc-binding site. The resulting joined Rad50 molecules form large structures that are well suited to link DNA ends or sister chromatids in recombination events. Rad50S mutations cluster at the surface (dashed line) near DNA and Mre11-binding sites. (d) The Rad50 ABC domains form a molecular machine that undergoes conformational changes in response to ATP binding. ATP induces a rotation in the ABC domains of Rad50 and engages two Rad50 ABC domains by creating an ABC–ATP$_2$–ABC sandwich. ATP is bond by the Walker A and signature motifs of opposing ABC domains. This molecular switching drives the nuclease activity of Mre11 and promotes DNA binding to the Mre11 complex.

coiled-coil domains create large Rad50 multimeric structures that are well suited to directly linking DNA ends or sister chromatids together (see in what follows).

  The nuclease Mre11 has a highly conserved N-terminal phosphodiesterase domain followed by a DNA-binding/active site capping domain (Fig. 1) (48). The phosphodiesterase domain harbors the nuclease active site of Mre11 and interaction sites for Xrs2/Nbs1 and possibly for Rad50 (5,51). The function of the C-terminal regions of Mre11 is less clear, but it contains interaction sites for Rad50,

Mre11–Mre11 dimer formation and DNA (apart from the phosphodiesterase domain). As Mre11 appears to interact with Rad50, Nbs1/Xrs2, and with itself, it can be viewed as architectural core of the Mre11 complex.

The structure and sequence motifs of Nbs1/Xrs2 are the least understood among the Mre11 complex proteins. Nbs1/Xrs2 probably mediates protein–protein interactions of the Mre11 complex with other repair/signaling factors at the break. Nbs1 has an N-terminal forkhead associated domain followed by a breast cancer associated C-terminus domain. Both domains mediate protein–protein interactions by binding to phosphopeptides (52–54). As phosphorylation is a key modification in the DNA damage checkpoint response, Nbs1 could be involved in targeting the Mre11 complex to damaged sites, to attract or interact with other repair or checkpoint factors to damaged sites (55), or to interact with chromatin components such as histone H2AX at DSBs (56).

3. CELLULAR BIOCHEMISTRY OF THE Mre11 COMPLEX

Research in many laboratories revealed a surprising variety of functions of the Mre11 complex in DNA end metabolism (57). The Mre11 complex has functions in virtually all aspects of DSB metabolism, including DSB detection (58), DSB processing (43,59–61), HR and meiosis (34,35,62,63), NHEJ (64–67), telomere maintenance (51,68–72), cell-cycle checkpoint response (3,31,73), and sister-chromatid cohesion (74). This kaleidoscope of functions suggested that the Mre11 complex has not only enzymatic activities in the metabolism of DNA ends, but also architectural roles, e.g., in the cohesion of DNA ends and sister chromatids. Such a dual enzymatic and architectural function is supported by the unusual domain structure and multiprotein architecture of Mre11 complex (see in what follows).

3.1 DNA Processing Activity

Although the Mre11 complex is a key factor in the cellular response paths to DSBs, the biochemical activity of Mre11 complex proteins and their role in DSB repair are somewhat puzzling. Biochemically, the Mre11 complex is an ATP-driven nuclease that degrades ssDNA endonucleolytically and double-strand DNA (dsDNA) and hairpins exonucleolytically (42,61,75–77). It was early suggested that this nuclease activity of Mre11 generates the 3′ tails in HR. However, with the exception of phage gp46/47, all Mre11/Rad50 proteins assayed degrade DNA in 3′→5′ direction, generating 5′ overhangs instead of the 3′ tails (48,61,77,78). In addition, in nuclease deficient Mre11 mutant *S. cerevisiae* strains, end resection in HR is slowed down but not completely abolished, and HR, NHEJ, and telomere maintenance function normally (51,59,62). Thus, the nuclease activity of Mre11 is probably not directly responsible for generation of the long 3′ tails in HR and even might be a minor function of the complex in mitotic and replication-linked functions. There is mounting evidence that the nuclease activity of Mre11 is involved in the processing of DNA secondary structures and misfolded DNA ends that arise during replication, at repeat structures and at DNA breaks (79–83).

The 3′ resection activity of the Mre11 complex could be important to remove damaged nucleotides from the DNA end that arise in the event of chemical and physical damage to DNA. Such a function would be important in order to allow extension of the new cleaned 3′ end by DNA polymerases in subsequent repair steps.

The role of ATP hydrolysis in the $3' \rightarrow 5'$ nuclease activity is not fully established. ATP hydrolysis is clearly linked to the main function of the complex and seems to be required for prokaryotic complex or stimulates the dsDNA-directed nuclease activities of the eukaryotic complex. Moreover, ATP binding to Rad50 can melt DNA structures and ends, modify DNA-binding activity and specificity of the Mre11 complex, and promote processing of hairpins (82–84). Yet the ATP-hydrolysis rate is too slow to drive a highly processive enzyme, and new data suggest that ATP binding to Rad50 could be only required primarily to load the complex on DNA, analogously to the related cohesion complex in chromosome interactions (85,86).

The nuclease activity of Mre11 complex is required for removal of covalently bound Spo11 (which generates meiotic DSBs) from the 5′ end of meiotic DSBs (59). A similar role in removing a covalently attached protein from a DNA end has been observed in adenovirus infected cells, where the Mre11 complex removes the adenovirus terminal protein from viral DNA ends (87). Leach and colleagues (80) recently showed that *E. coli* SbcC and SbcD could remove a protein from the 5′ end of DNA in vitro, by the generation of a DSB close to the end. Thus, the capability to remove proteins from DNA ends (possibly by the introduction of DSBs) seems to be an evolutionary preserved feature of the complex. Combined with the earlier-mentioned nuclease activities, it appears that the ATP-driven nuclease activity of the Mre11 complex can act virtually on all types of DNA ends to generate a clean processed end.

3.2. Damage Sensor and Checkpoint Functions

Nbs1/Xrs2 links the Mre11 complex to the checkpoint control in response to DSBs. This response is co-ordinated by the large kinase ATM. ATM becomes activated in the presence of DNA damage and other DNA-associated cellular stress phenomena and induces a phosphorylation cascade that mediates the cellular response mechanisms (88). ATM resides in the nucleus as inactive dimer and becomes active in the presence of DNA damage by autophosphorylation, monomerization and relocation to the sites of the damage (89). A key function of the eukaryotic Mre11 complex is to activate the S- and G2/M-checkpoint response to DSBs, by participating in the activation of ATM, possibly in conjunction with other checkpoint factors such as the mediator of damage checkpoint 1 (MDC1) (31,90–96). In this process, Nbs1 is phosporylated by ATM, a modification that is required for the intra S-checkpoint. Conversely, ATM activation requires the Mre11 complex and the Nbs1 phosphorylation, indicating an intimate functional connection of ATM and the Mre11 complex in the cellular response pathways to DSBs.

The precise mechanism of the role of the Mre11 complex in checkpoint activation is currently under intense research, but several lines of evidence suggest that the Mre11 complex could function as a primary sensor of DNA breaks in this process (18). Judged from the appearance of foci upon induction of DSBs, the Mre11 complex is an early component at the actual sites of the break (58). In addition, the Mre11 complex is intimately connected with replication, the primary site of DSB occurrence. Experiments with cell-free *Xenopus* egg extracts showed that in the absence of Mre11 complex, DNA replication resulted in many DSBs (7). In *E. coli*, SbcC and SbcD are required to process hairpins that arise during lagging-strand synthesis at inverted repeats, and propagation of inverted repeats in eukaryotes requires the Mre11 complex (60,79). In T4 and T5 phages, gp46/47 is required for recombination-induced replication (44,97,98). In eukaryotes, Nbs1 binds to E2F transcription factors near replication origins (99), and the Mre11 complex colocalizes with replication forks and is loaded onto nascent sister-chromatids in S-phase (100).

The role of Nbs1 protein–protein interaction and ATM-dependent Nbs1 phosphorylation and its consequences for the architecture and activity of the Mre11 complex are unclear. The phosphorylation might induce a structural switch that could alter the stoichiometry of the complex in a mechanism similar to that proposed for ATM activation. Nevertheless, besides mediating interaction with other checkpoint proteins and chromatin components at DSBs, Nbs1/Xrs2 could be directly involved in DNA damage recognition and processing by the Mre11 complex. Nbs1 modulates the ATP-driven nuclease activity of human Mre11/Rad50 and is required for the hairpin-processing activity of the complex (82). Recently, it was found that Xrs2 binds to DNA secondary structures and targets Mre11/Rad50 to DSBs (101). These data suggest that Nbs1/Xrs2 could be directly involved in DNA damage detection functions of the Mre11 complex. In particular, this direct DNA-binding activity could be important for activation of the DNA damage checkpoint.

3.3. Architectural Functions of the Mre11 Complex in Joining DSBs

Perhaps, the most important function of the Mre11 complex in genome integrity maintenance is a structural function in joining DNA ends and/or sister chromatids. This emerging structural function is encoded in the unusual sequence and structure of Mre11 complex proteins and the striking three-dimensional architecture of the Mre11 complex (Figs. 1 and 2).

Electron microscopy and scanning force spectroscopy of human and bacterial Mre11 complexes revealed important insights in the overall of architecture of the Mre11 complex and immediately suggested mechanisms for structural functions of the complex (46–49,102). In electron micrographs and scanning force images, the Mre11 complex appears as a lollipop-like structure with a global head and two long tails, consistent with a $Mre11_2/Rad50_2$ stoichiometry for the full prokaryotic and core eukaryotic (without Nbs1) complex (Fig. 2a). The globular head consists of two Rad50-ABC domains and the Mre11 dimer in a heterotetrameric architecture (Fig. 2b). The tails consist of the Rad50 coiled-coil domains. The head is the major interaction site for DNA and harbors the ATP-driven nuclease activity of the complex. This head probably also interacts with Nbs1/Xrs2, as judged from yeast mutagenesis data. The precise interaction of Mre11 with Nbs1 and the architecture of the full eukaryotic complex have not been revealed yet. Recent ultracentrifugation data, however, suggest that the stoichiometry of the eukaryotic Mre11 complex is probably $Rad50_2/Mre11_2/Nbs1_2$ (103). The close interaction of Mre11, Nbs1, and the Rad50-ABC domain suggests that these components probably form a single DNA recognition surface that detects and processes breaks in both repair and checkpoint activation. A model for the eukaryotic complex consistent with current data is shown in Fig. 3

The two coiled-coil domains of Rad50 protrude from the globular head in an almost parallel direction (Figs. 2 and 3) (45,46,49). At the apex of these coiled-coil domains are the intriguing hook structures, formed by the conserved CXXC motif (50). This hook can evidently join two Rad50 coiled-coils by forming an interlocked hook:zinc:hook bridge (Fig. 2c). In electron micrographs and scanning force images, two different species of joined Mre11 complexes are seen (Fig. 3a and b) (49,50).

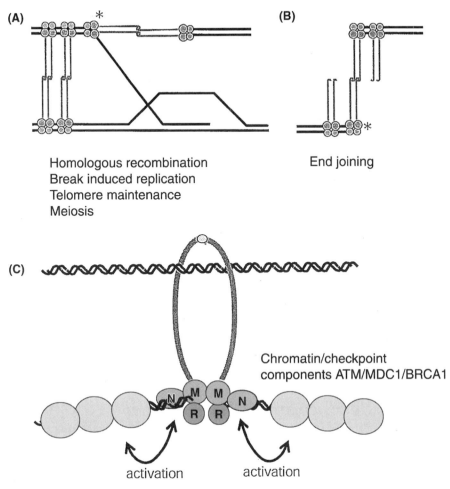

Figure 3 Architectural functions of the Mre11 complex in DSB repair (a,b). The Mre11 complex could directly link sister chromatids and DNA ends in recombination and end joining events by creating intermolecularly joined zinc-hooks. In this model, two DNA-binding head regions consisting of Mre11 (M) and the Rad50 ABC domain (A) are bound to opposite DNA ends of sister chromatids. This architectural role explains both nuclease-independent functions in joining DSBs and sister chromatids and nuclease-dependent functions (star) in processing DNA ends in DNA-end metabolism. (c) Alternative model in which two coiled-coil domains from the same $Mre11_2/Rad50_2$ complex are joined to form a large ring structure. This structure is analogous to the ring model proposed for the related cohesion complex. In principle, a single $Mre11_2/Rad50_2$ complex could link the sister chromatid to both DNA ends utilizing both potential Rad50 and Mre11 DNA-binding sites in the Mre11/Rad50 heterotetramer. Nbs1 binds to Mre11 in a 1:1 stoichiometry and, based on recent observations, could participate in the activation and interaction with ATM, MDC1 and/or BRCA1, and other chromatin components, as well as in DNA/DSB binding. *Source*: Adapted from Ref. 50.

First, the hook can join the two coiled-coils from a single $Mre11_2/Rad50_2$ complex (intramolecular joining), thereby forming a large loop (Fig. 3b). This loop is closed at one end by the global head and at the other end by the interlocked hooks. This structure is of remarkable similarity to the ring model proposed for cohesion (104) and provides a compelling mechanism for an architectural function of the Mre11 complex in recombinational sister chromatid interaction. Upon formation of these ring structures, the coiled-coil necessarily has to bend or kink. Recent scanning force microscopy (SFM) studies show that Rad50 coiled-coil domains have specific regions of high flexibility (105). These regions are colocalized with sequences of low coiled-coil propensities and could represent specifically designed hinge regions to allow for conformational flexibility of the coiled-coil domain.

An additional geometry of joined Mre11 complexes is also frequently found in electron micrographs and scanning force images (Fig. 3a). Upon addition of zinc, hetero-octameric $Mre11_2/Rad50_2$:$Mre11_2/Rad50_2$ complexes are formed (50). In this particular architecture, coiled coils from two different $Mre11_2/Rad50_2$ are joined by zinc-mediated hook association (intermolecular joining). At present, it is not clear which of the observed species (tetrameric or octameric) is physiological or whether both species have functions in DNA end metabolism. However, both species have the requisite structural features to efficiently connect DNA ends or sister chromatids.

Besides the capability of Rad50 coiled-coil ends to form joined complexes, the head domains can probably also multimerize. This multimerization is seen in SFM images and occurs at DNA ends (49). Presumable one complex directly binds to the end as damage sensor. Other complexes associate with this terminal complex along the DNA strand. The physiological basis of this multimerization is not fully established, but it might form a molecular velcro to efficiently join DSBs. The in vitro clustering of Mre11 complexes at DSBs could represent a molecular view of a small focus.

4.   STRUCTURAL BIOCHEMISTRY OF THE Mre11 COMPLEX

4.1.   Mre11 Nuclease

High-resolution structural data are available for archaebacterial Rad50 ABC domain structures in both ATP bound and nucleotide free conformations (Fig. 3d), and for the catalytic domain of Mre11 (Fig. 4a) (45,48). Mre11 is a nuclease with an unusually broad biochemical spectrum of activities. In addition, Mre11 can stimulate DNA ligase activity and sense microhomologies during simultaneous processing of two ends (106). Latter two activities are probably explained by biochemical findings, showing that Mre11 forms dimers and might process two DNA ends simultaneously (46,48,107,108).

How can we reconcile the broad nuclease specificity of Mre11 (exo/endo nuclease, hairpin, and terminal protein processing) in a single structural mechanism? The crystal structure of archaeal Mre11 gave some possible answers to this question. The nuclease core of Mre11 is a two-domain protein, consisting of a calcineurine protein phosphatase-like domain that contains the nuclease active site and a small capping domain that controls active site access (Fig. 4a). The capping domain presumably provides some DNA-binding specificity (48). The apparent similar active site geometry of Mre11 and calcineurine-like protein phosphatases suggests that the nuclease activity of Mre11 is mechanistically similar to the two metal-dependent phosphoesterase mechanism of protein phosphatases.

The active site of Mre11 binds two manganese ions at five conserved phosphodiesterase motifs (all Mre11 species tested have a strict requirement for manganese as metal cofactor). The active site is situated at a shallow groove at the interface between the phosphodiesterase domain and the putative DNA-binding domain.

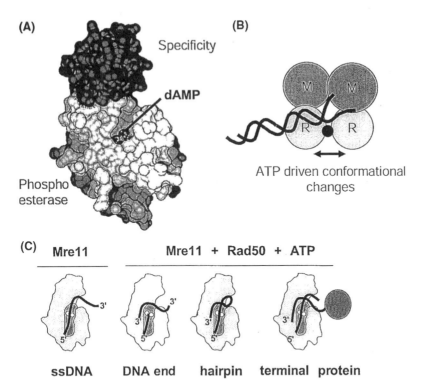

Figure 4 Mechanism of DNA damage detection and processing. (a) Molecular surface representation of the high-resolution crystal structure of archaeal Mre11 reveals a two-domain protein consisting of a phosphoesterase domain (light gray) and a specificity domain (dark gray). The active site is located at the domain interface and is formed by phosphoesterase motifs that co-ordinate two manganese ions (not shown). A dAMP molecule is bound to the catalytic manganese ions at the active-site depression via its phosphate suggesting that Mre11 mainly contacts the DNA backbone and requires DNA melting or deformation prior to metal-dependent cleavage. (b) Proposed mechanism for ATP-driven nuclease activity of the Mre11 complex. The Rad50 dimer (R) uses ATP (black sphere) driven conformational changes (arrow) to bind and possibly unwind DNA for cleavage by the Mre11 dimer (M). The coiled coils are omitted. (c) Unified model for multiple Mre11 nuclease activities. Single-strand DNA is readily cleaved by Mre11 because it is flexible enough to bind to the Mre11 active site metals. DNA ends, hairpins, or protein-obstructed ends need melting or unwinding by Rad50 and ATP to bind to the Mre11 active site metals for cleavage. *Source*: Adapted from Ref. 48.

In the crystal structure of an Mre11/deoxyadenosine monophosphate (dAMP) complex, the directionality of the bound dAMP is consistent with the $3' \rightarrow 5'$ nuclease direction of Mre11 that is observed biochemically. The active site structure of Mre11 revealed a putative mode of interaction with both dsDNA and ssDNA. DNA is predominantly recognized via its backbone, as expected for a nonspecific nuclease. Interestingly, docking experiments with B-DNA suggest that the geometry of the active site allows binding of flexible ssDNA but prevents direct binding of B-DNA to the active site metals. This specific architecture indicates that Mre11 needs Rad50 and ATP to unfold or melt DNA ends, hairpins, or obstructed ends in order to bind to the active site metals for cleavage. In fact, such an ATP-driven DNA melting activity has been biochemically observed, suggesting that one role of ATP binding to Rad50 could be to unfold DNA ends and hairpins (Figs. 4b and c) (82).

4.2. The Mre11/Rad50 DNA End Detection/Processing Machine

Crystallographic analysis of the ATP-bound and nucleotide-free Rad50 ATPase domain revealed major conformational changes that are well suited to drive DNA binding and assist nucleolytic processing of blocked/misfolded DNA ends (Fig. 4b) (45). Rad50 possesses an ABC ATPase domain, which consists of two lobes and is a highly conserved molecular engine in ABC transporter and DNA damage detection enzymes, such as the nucleotide excision repair protein UvrA, and the mismatch repair enzyme MutS/MSH (39). In general, these domains possess a unique structural switching mechanism that enables them to translate chemical binding energy of ATP into protein conformational changes (109). These conserved protein conformational changes are transmitted to interacting partners of the ABC domain and drive the enzymatic function of the ABC complex.

ATP binding drives the structural changes in the ABC domain of Rad50 by a two-step process. First, ATP binding induces a structural switch in the ABC domain that rotates the two lobes of the ABC domain $\sim 30°$ with respect to each other. This rotation is promoted by binding of ATP to Walker A and B motifs on one lobe, followed by binding to a conserved glutamine (Q-loop) on the other lobe to both the ATP γ-phosphate and the Mg^{++} ion. The intradomain rotation then allows two ABC domains to dimerization/engage. In this ATP-bound Rad50 ABC dimer, ATP binds in the ABC dimer interface, sandwiched between the Walker A and B motifs of one ABC domain and the signature/C motif of the other ABC domain (Fig. 3d). The signature motif, which is highly conserved in all ABC enzymes, specifically binds the γ-phosphate in trans and, therefore, drives ABC dimer engagement. Hydrolysis of ATP disrupts the interaction of both subunits and disengages the ABC-domain dimer. This mechanism is not unique to Rad50 but probably ubiquitously conserved in all ABC enzymes, including UvrA and MutS/MSH.

Because ATP binding to Rad50 modulates its interaction with DNA, the ATP-driven conformational change is probably an important stereochemical event in the damage recognition functions of the complex (35,45,80). Both the isolated ABC domains of Rad50 and the full Rad50 polypeptide can directly bind to DNA in an ATP-dependent manner, suggesting that ATP-driven structural changes modulate or form a DNA-binding site on the Rad50 ABC domain (Fig. 4b). Furthermore, in human Mre11/Rad50 complexes, ATP binding alters the preference of the Mre11 complex for various DNA ends (84). The precise location of the DNA-binding surface of Rad50 and the Mre11/Rad50 complex has not been established, but surface charge and molecular docking indicate that DNA binds at the surface next to the root of the Rad50 coiled-coils (48). This region is particularly interesting because it is located adjacent to the Mre11-binding site (coiled-coil), suggesting that ATP could modulate a combined Rad50/Mre11 DNA-binding site (Fig. 2c and d).

In the immediate vicinity of the putative DNA/Mre11 binding region of Rad50 is the surface cluster of the yeast Rad50S mutations (Fig. 2c and d). Thus, the effect of the Rad50S mutations could be the modulation of the DNA recognition properties of the Rad50/Mre11 DNA-processing machinery. Such a modulation could severely influence the capability of Rad50/Mre11 to process Spo11-linked DNA ends, the observed phenotype in rad50S yeast strains. Sister chromatid or

DNA-end cohesion activity of Mre11 complex, on the other hand, could be still functional as the coiled-coil loops or linkers are intact.

5. UNIFIED MODEL, CONCLUSIONS, AND OUTLOOK

In the past years, research from many laboratories began to provide a coherent picture of the central role of the Mre11 complex in the DSB response pathway. In this central role, the Mre11 complex has at least three critical functions. First, the Mre11 complex is a primary DNA damage recognition factor at DSBs, single DNA ends arising from stalled replication forks and otherwise misfolded or obstructed DNA. Using ATP, the Mre11 complex probably binds to these ends/structures and induces a signal that directly or indirectly assists the activation of the ATM kinase and checkpoint signaling.

Second, the ATP-driven nuclease activity is probably involved to process these misfolded/blocked DNA ends or to degrade secondary structures/hairpins that arise during replication. This function might be important to create a "common intermediate" that funnels heterogeneity arising at a break in a processed, clean DNA end. The processed end could be resected by another nuclease in HR, extended by repair/replication polymerases in HR or break-induced replication fork restart. In addition, ssDNA arising from processed or resected DNA ends could initiate the DNA damage checkpoint.

Finally, the perhaps most important function of the Mre11 is to link DNA ends with each other or DNA ends to sister chromatids, a pre-eminent step prior to NHEJ, HR, or rescue of blocked replication forks. The zinc-hook structure at the coiled-coil apex suggests a particular model how Rad50 probably achieves this linker function: Rad50 coiled-coil domains create large intramolecular loops or intermolecular bridges that could easily entrap or otherwise join sister chromatids to DNA ends.

Thus, all the diverse functions of the Mre11 complex could be elegantly explained by unified model such as depicted in Figs. 3a and b. The combined DNA-binding surface, comprising the Rad50 ATPase domain, Mre11, and possibly also Nbs1/Xrs2, could be involved in DNA damage detection, DNA processing, and checkpoint activation, possibly modulated by structural switching in the Rad50-ABC domain. Such a tethering of DNA ends to sister chromatids might be the ultimate role of the Mre11 complex, as this structural intermediate is found in most DSB-associated events, including telomeres and D-loops. It would further explain why Mre11/Rad50 homologs are only found in bacteriophages with a break-induced (D-loop) replication mechanism and why the Mre11 complex is intimately connected to replication in eukaryotes and prokaryotes.

Although we begin to have a coherent picture of the mechanistic role of the Mre11 complex in DNA end metabolism, several key questions remain to be addressed. The role of ATP in the Rad50/Mre11 complex is not fully understood and it is not clear whether the energy provided by ATP binding to Rad50 serves primarily to architecturally cohere DNA ends or sister chromatids or to provide the energy to drive the nuclease activity of Mre11. In addition, it remains to be seen how Mre11 and especially Rad50 interact with DNA in order to understand the function of the DNA-processing head. The greatest challenge and least understood function associated with the Mre11 complex is the mechanism of checkpoint activation. We need to understand the architecture of the Nbs1 interaction with Rad50/Mre11 and the functional and structural consequences of Nbs1 and Mre11 phosphorylations. Especially, this future challenge requires a combined cell-biological, biochemical, and structural effort over the next years.

ACKNOWLEDGMENTS

The author apologizes to all colleagues whose key contributions could not be cited due to space restrictions and focus. KPH is supported by a grant from the Deutsche Forschungsgemeinschaft and the EMBO Young Investigator Award.

REFERENCES

1. Game JC. Semin Cancer Biol 1993; 4:73–83.
2. Zdzienicka MZ. Cancer Surv 1996; 28:281–293.
3. Carney JP, Maser RS, Olivares H, Davis EM, Le Beau M, Yates JR III, Hays L, Morgan WF, Petrini JH. Cell 1998; 93:477–486.
4. Luo G, Yao MS, Bender CF, Mills M, Bladl AR, Bradley A, Petrini JH. Proc Natl Acad Sci USA 1999; 96:7376–7381.
5. Stewart GS, Maser RS, Stankovic T, Bressan DA, Kaplan MI, Jaspers NG, Raams A, Byrd PJ, Petrini JH, Taylor AM. Cell 1999; 99:577–587.
6. Ward JF. Prog Nucleic Acid Res Mol Biol 1988; 35:95–125.
7. Costanzo V, Robertson K, Bibikova M, Kim E, Grieco D, Gottesman M, Carroll D, Gautier J. Mol Cell 2001; 8:137–147.
8. Kuzminov A. Proc Natl Acad Sci USA 2001; 98:8461–8468.
9. Keeney S. Curr Top Dev Biol 2001; 52:1–53.
10. Gellert M. Annu Rev Biochem 2002; 71:101–132.
11. Haber JE. Annu Rev Genet 1998; 32:561–599.
12. Lee SE, Pellicioli A, Demeter J, Vaze MP, Gasch AP, Malkova A, Brown PO, Botstein D, Stearns T, Foiani M, Haber JE. Cold Spring Harb Symp Quant Biol 200; 65:303–314.
13. Chen C, Kolodner RD. Nat Genet 1999; 23:81–85.
14. Lewis LK, Resnick MA. Mutat Res 2000; 451:71–89.
15. Lieber MR, Ma Y, Pannicke U, Schwarz K. Nat Rev Mol Cell Biol 2003; 4:712–720.
16. D'Amours D, Jackson SP. Nat Rev Mol Cell Biol 2002; 3:317–327.
17. van den Bosch M, Bree RT, Lowndes NF. EMBO Rep 2003; 4:844–849.
18. Petrini JH, Stracker TH. Trends Cell Biol 2003; 13:458–462.
19. de Jager M, Kanaar R. Genes Dev 2002; 16:2173–2178.
20. Kupiec M, Simchen G. Mol Gen Genet 1984; 193:525–531.
21. Ivanov EL, Korolev VG, Fabre F. Genetics 1992; 132:651–664.
22. Ajimura M, Leem SH, Ogawa H. Genetics 1993; 133:51–66.
23. Johzuka K, Ogawa H. Genetics 1995; 139:1521–1532.
24. Petrini JH, Walsh ME, DiMare C, Chen XN, Korenberg JR, Weaver DT. Genomics 1995; 29:80–86.
25. Tavassoli M, Shayeghi M, Nasim A, Watts FZ. Nucleic Acids Res 1995; 23:383–388.
26. Dolganov GM, Maser RS, Novikov A, Tosto L, Chong S, Bressan DA, Petrini JH. Mol Cell Biol 1996; 16:4832–4841.
27. Varon R, Vissinga C, Platzer M, Cerosaletti KM, Chrzanowska KH, Saar K, Beckmann G, Seemanova E, Cooper PR, Nowak NJ, Stumm M, Weemaes CM, Gatti RA, Wilson RK, Digweed M, Rosenthal A, Sperling K, Concannon P, Reis A. Cell 1998; 93:467–476.
28. Matsuura S, Tauchi H, Nakamura A, Kondo N, Sakamoto S, Endo S, Smeets D, Solder B, Belohradsky BH, Der Kaloustian VM, Oshimura M, Isomura M, Nakamura Y, Komatsu K. Nat Genet 1998; 19:179–181.

29. Ueno M, Nakazaki T, Akamatsu Y, Watanabe K, Tomita K, Lindsay HD, Shinagawa H, Iwasaki H. Mol Cell Biol 2003; 23:6553–6563.

30. Chahwan C, Nakamura TM, Sivakumar S, Russell P, Rhind N. Mol Cell Biol 2003; 23:6564–6573.

31. D'Amours D, Jackson SP. Genes Dev 2001; 15:2238–2249.

32. Grenon M, Gilbert C, Lowndes NF. Nat Cell Biol 2001; 3:844–847.

33. Lim DS, Kim ST, Xu B, Maser RS, Lin J, Petrini JH, Kastan MB. Nature 2000; 404:613–617.

34. Yamaguchi-Iwai Y, Sonoda E, Sasaki MS, Morrison C, Haraguchi T, Hiraoka Y, Yamashita YM, Yagi T, Takata M, Price C, Kakazu N, Takeda S. EMBO J 1999; 18:6619–6629.

35. Alani E, Padmore R, Kleckner N. Cell 1990; 61:419–436.

36. Bender CF, Sikes ML, Sullivan R, Huye LE, Le Beau MM, Roth DB, Mirzoeva OK, Oltz EM, Petrini JH. Genes Dev 2002; 16:2237–2251.

37. Petrini JH. Curr Opin Cell Biol 2000; 12:293–296.

38. Gorbalenya AE, Koonin EV. J Mol Biol 1990; 213:583–591.

39. Aravind L, Walker DR, Koonin EV. Nucleic Acids Res 1999; 27:1223–1242.

40. Blinov VM, Koonin EV, Gorbalenya AE, Kaliman AV, Kryukov VM. FEBS Lett 1989; 252:47–52.

41. Sharples GJ, Leach DR. Mol Microbiol 1995; 17:1215–1217.

42. Hopfner KP, Karcher A, Shin D, Fairley C, Tainer JA, Carney JP. J Bacteriol 2000; 182:6036–6041.

43. Mickelson C, Wiberg JS. J Virol 1981; 40:65–77.

44. Kreuzer KN. Trends Biochem Sci 2000; 25:165–173.

45. Hopfner KP, Karcher A, Shin DS, Craig L, Arthur LM, Carney JP, Tainer JA. Cell 2000; 101:789–800.

46. Anderson DE, Trujillo KM, Sung P, Erickson HP. J Biol Chem 2001; 276:37027–37033.

47. Connelly JC, Kirkham LA, Leach DR. Proc Natl Acad Sci USA 1998; 95:7969–7974.

48. Hopfner KP, Karcher A, Craig L, Woo TT, Carney JP, Tainer JA. Cell 2001; 105: 473–485.

49. de Jager M, van Noort J, van Gent DC, Dekker C, Kanaar R, Wyman C. Mol Cell 2001; 8:1129–1135.

50. Hopfner KP, Craig L, Moncalian G, Zinkel RA, Usui T, Owen BA, Karcher A, Henderson B, Bodmer JL, McMurray CT, Carney JP, Petrini JH, Tainer JA. Nature 2002; 418:562–566.

51. Chamankhah M, Fontanie T, Xiao W. Genetics 2000; 155:569–576.

52. Durocher D, Henckel J, Fersht AR, Jackson SP. Mol Cell 1999; 4:387–394.

53. Yu X, Chini CC, He M, Mer G, Chen J. Science 2003; 302:639–642.

54. Manke IA, Lowery DM, Nguyen A, Yaffe MB. Science 2003; 302:636–639.

55. Bradbury JM, Jackson SP. Biochem Soc Trans 2003; 31:40–44.

56. Kobayashi J, Tauchi H, Sakamoto S, Nakamura A, Morishima K, Matsuura S, Kobayashi T, Tamai K, Tanimoto K, Komatsu K. Curr Biol 2002; 12:1846–1851.

57. Haber JE. Cell 1998; 95:583–586.

58. Mirzoeva OK, Petrini JH. Mol Cell Biol 2001; 21:281–288.

59. Moreau S, Ferguson JR, Symington LS. Mol Cell Biol 1999; 19:556–566.

60. Connelly JC, Leach DR. Genes Cells 1996; 1:285–291.

61. Trujillo KM, Yuan SS, Lee EY, Sung P. J Biol Chem 1998; 273:21447–21450.

62. Bressan DA, Baxter BK, Petrini JH. Mol Cell Biol 1999; 19:7681–7687.

63. Tauchi H, Kobayashi J, Morishima K, van Gent DC, Shiraishi T, Verkaik NS, vanHeems D, Ito E, Nakamura A, Sonoda E, Takata M, Takeda S, Matsuura S, Komatsu K. Nature 2002; 420:93–98.

64. Tsukamoto Y, Kato J, Ikeda H. Genetics 1996; 142:383–391.

65. Milne GT, Jin S, Shannon KB, Weaver DT. Mol Cell Biol 1996; 16:4189–4198.

66. Moore JK, Haber JE. Mol Cell Biol 1996; 16:2164–2173.

67. Udayakumar D, Bladen CL, Hudson FZ, Dynan WS. J Biol Chem 2003.
68. Kironmai KM, Muniyappa K. Genes Cells 1997; 2:443–455.
69. Boulton SJ, Jackson SP. EMBO J 1998; 17:1819–1828.
70. Teng SC, Chang J, McCowan B, Zakian VA. Mol Cell 2000; 6:947–952.
71. Zhu XD, Kuster B, Mann M, Petrini JH, de Lange T. Nat Genet 2000; 25: 347–352.
72. Lundblad V. Oncogene 2002; 21:522–531.
73. Zhao S, Weng YC, Yuan SS, Lin YT, Hsu HC, Lin SC, Gerbino E, Song MH, Zdzienicka MZ, Gatti RA, Shay JW, Ziv Y, Shiloh Y, Lee EY. Nature 2000; 405:473–477.
74. Hartsuiker E, Vaessen E, Carr AM, Kohli J. EMBO J 2001; 20:6660–6671.
75. Connelly JC, de Leau ES, Okely EA, Leach DR. J Biol Chem 1997; 272:19819–19826.
76. Ohta K, Nicolas A, Furuse M, Nabetani A, Ogawa H, T Shibata. Proc Natl Acad Sci USA 1998; 95:646–651.
77. Paull TT, Gellert M. Mol Cell 1998; 1:969–979.
78. Connelly JC, de Leau ES, Leach DR. Nucleic Acids Res 1999; 27:1039–1046.
79. Lobachev KS, Gordenin DA, Resnick MA. Cell 2002; 108:183–193.
80. Connelly JC, de Leau ES, Leach DR. DNA Repair (Amst) 2003; 2:795–807.
81. Farah JA, Hartsuiker E, Mizuno K, Ohta K, Smith GR. Genetics 2002; 161:461–468.
82. Paull TT, Gellert M. Genes Dev 1999; 13:1276–1288.
83. Trujillo KM, Sung P. J Biol Chem 2001; 276:35458–35464.
84. de Jager M, Wyman C, van Gent DC, Kanaar R. Nucleic Acids Res 2002; 30: 4425–4431.
85. Arumugam P, Gruber S, Tanaka K, Haering CH, Mechtler K, Nasmyth K. Curr Biol 2003; 13:1941–1953.
86. Weitzer S, Lehane C, Uhlmann F. Curr Biol 2003; 13:1930–1940.
87. Stracker TH, Carson CT, Weitzman MD. Nature 2002; 418:348–352.
88. Shiloh Y. Nat Rev Cancer 2003; 3:155–168.
89. Bakkenist CJ, Kastan MB. Nature 2003; 421:499–506.
90. Nakada D, Matsumoto K, Sugimoto K. Genes Dev 2003; 17:1957–1962.
91. Goldberg M, Stucki M, Falck J, D'Amours D, Rahman D, Pappin D, Bartek J, Jackson SP. Nature 2003; 421:952–956.
92. Stewart GS, Wang B, Bignell CR, Taylor AM, Elledge SJ. Nature 2003; 421:961–966.
93. Lou Z, Minter-Dykhouse K, Wu X, Chen J. Nature 2003; 421:957–961.
94. Usui T, Ogawa H, Petrini JH. Mol Cell 2001; 7:1255–1266.
95. Gatei M, Young D, Cerosaletti KM, Desai-Mehta A, Spring K, Kozlov S, Lavin MF, Gatti RA, Concannon P, Khanna K. Nat Genet 2000; 25:115–119.
96. Carson CT, Schwartz RA, Stracker TH, Lilley CE, Lee DV, Weitzman MD. Embo J 2003; 22:6610–6620.
97. Bleuit JS, Xu H, Ma Y, Wang T, Liu J, Morrical SW. Proc Natl Acad Sci USA 2001; 98:8298–8305.
98. George JW, Stohr BA, Tomso DJ, Kreuzer KN. Proc Natl Acad Sci USA 2001; 98:8290–8297.
99. Maser RS, Mirzoeva OK, Wells J, Olivares H, Williams BR, Zinkel RA, Farnham PJ, Petrini JH. Mol Cell Biol 2001; 21:6006–6016.
100. Mirzoeva OK, Petrini JH. Mol Cancer Res 2003; 1:207–218.
101. Trujillo KM, Roh DH, Chen L, Van Komen S, Tomkinson A, Sung P. J Biol Chem 2003; 278:48957–48964.
102. Chen L, Trujillo K, Ramos W, Sung P, Tomkinson AE. Mol Cell 2001; 8:1105–1115.
103. Lee JH, Ghirlando R, Bhaskara V, Hoffmeyer MR, Gu J, Paull TT. J Biol Chem 2003; 278:45171–45181.
104. Haering CH, Lowe J, Hochwagen A, Nasmyth K. Mol Cell 2002; 9:773–788.
105. van Noort J, van Der Heijden T, de Jager M, Wyman C, Kanaar R, Dekker C. Proc Natl Acad Sci USA 2003; 100:7581–7586.

106. Paull TT, Gellert M. Proc Natl Acad Sci USA 2000; 97:6409–6414.
107. Nairz K, Klein F. Genes Dev 1997; 11:2272–2290.
108. Chamankhah M, Wei YF, Xiao W. Gene 1998; 225:107–116.
109. Hopfner KP, Tainer JA. Curr Opin Struct Biol 2003; 13:249–255.

32

Histone γ-H2AX Involvement in DNA Double-Strand Break Repair Pathways

Nikolaos A. A. Balatsos
Institute of Molecular Biology and Genetics, B.S.R.C. Alexander Fleming, and Department of Biochemistry and Biotechnology, School of Health Sciences, University of Thessaly, Varkiza, Greece

Emmy P. Rogakou
Institute of Molecular Biology and Genetics, B.S.R.C. Alexander Fleming, Varkiza, Greece

1. INTRODUCTION

It was not long ago that chromatin was widely regarded as adding more "structure" to DNA, but not "function." Today, it is broadly perceived that chromatin modifications occur locally, in a regulated manner, to create different microenvironments along the chromatin fiber. Accordingly, these chromatin-governed microenvironments regulate—positively or negatively—the interaction between DNA and other protein players. Histone tails play a key role in these mechanisms. They provide the substrate, usually a stretch of a few aminoacids that extend out from the nucleosome towards the nucleoplasm, to accommodate post-translational modifications (1). Recently, it has been shown that this general theme applies also in the case of double-strand breaks introduced into cellular DNA.

H2AX is a mammalian variant that belongs to the H2A histone family (2) and its topography in nucleosomes is the same as for the other H2A family members (3). The H2AX protein sequence is almost identical to the H2A1 in mammals, except for the C-tail—in H2AX the C-terminus is longer and forms a characteristic SQ motif. The SQ motif is conserved in different species but resides on different histone variants, all members of the H2A family. In mammals, *Xenopus laevis* and *Tetrahymena thermophila*, the SQ motif resides on H2AX, in *Drosophila melanogaster* on H2AvD, and in *Saccharomyces cerevisiae* and *Saccharomyces pombe* on H2A1 and H2A2. It was only recently discovered that loss of chromosomal integrity from DNA double-strand breaks results in a specific serine phosphorylation at the conserved SQ motif, and this cellular response is conserved from yeast to mammals. Since H2A histone molecules have phosphorylation sites other than the SQ motif, this SQ phosphorylation is denoted "γ-phosphorylation" for simplicity and clarity. In cases where it is necessary to indistinguishably denote the H2A orthologoue that host the SQ motif, the H2A(X) term is used, and the SQ phosphorylation is indicated with a γ- before the relevant H2A homologue (2,4).

2. FORMATION AND DETECTION OF γ-PHOSPHORYLATION

2.1 Ionizing Radiation Induced γ-Phosphorylation

It was first demonstrated on AUT–AUC high resolution two-dimensional gels that H2AX is phosphorylated specifically on serine 139 (γ-H2AX), when mammalian cell cultures or mice are exposed to ionizing radiation (IR) (5). Up to date, it has been demonstrated that γ-H2A(X) in extracts from different species subjected to AUT–AUC gel electrophoresis resolve as distinct spots that migrate faster that the non-phosphorylated H2A(X) components on AUT–AUC gels (5,6). Following IR, γ-H2A(X) spots appear rapidly, within 1 min, half-maximal amounts are attained in less than 10 min, a plateau is reached within 15–60 min, and a slow decrease follows during 180 min (5). When immunocytochemistry methods are applied to irradiated cells, and cell specimens are observed under the fluorescent microscope, γ-H2AX forms large, bright, and discrete foci at a random distribution throughout the nucleus but not within the nucleoli area. Foci pattern formation follows fast kinetics; γ-H2AX foci appear as small and numerous within 1–3 min, become fewer in number but larger and better defined at 15 min, stay steady in size and number between 15–60 min, decrease in number at 180 min, and eventually almost disappear at 24–48 hr (7,8). A variety of different normal and cancer cell lines, as well as living organisms respond by the formation of γ-H2A(X) to lethal and nonlethal amounts of IR. Remarkably, this cellular reaction is conserved in evolution. The serine residue of the conserved SQ motif in C-terminus of the relevant H2A homologue becomes also phosphorylated in response to IR in all eucaryotes examined so far, e.g., in *Muntiacus muntiacus* (deer muntjak), *X. laevis*, *D. melanogaster*, and *S. cerevisiae* (5,7,9).

2.2 γ-Phosphorylation Is Induced by Double-Strand Breaks Generated by Ionizing Radiation, Particle Emission and Radiomimetic Drugs

Ionizing radiation produces multiple damaged sites (MDS) that contain base and sugar alterations as well as single-strand breaks, along with double-strand breaks. A series of experiments attribute the γ-H2AX formation to the double-strand breaks rather than to other types of DNA damage. In cells that are incubated with bleomycin, a chemical agent that introduces double- and single-strand breaks by nonradiolytic mechanism, γ-H2AX formation is apparent (5,10). Likewise, when [125]IdU is incorporated into DNA,[125]I, which is a Augen electron emitter with a very short-range radiation effect, generates double-strand breaks that lead to the formation of γ-H2AX (11). On the contrary, neither treatment with H_2O_2, that produces hydroxyl radicals and DNA single-strand breaks at a low spatial distribution, nor low-dose irradiation with ultraviolet A light (350 nm) induces γ-H2AX formation (5). When different drugs are administrated to the chinese hamster V79 cell line, γ-H2AX formation measured by flow cytometry can serve as a semi-quantitative indicator for clonogenic cell kill (12). According to a specific protocol, in cells sensitized with BrdU and Hoechst 33258 dye and then subjected to UV light, double-strand breaks are generated and γ-H2AX formation is apparent (5). Along the same line, a UVA laser beam that is narrow enough to run through a fraction of a cell nucleus, can demonstrate a precise γ-H2AX localization to the sites of DNA double-strand breaks. When cells sensitized with BrdU and Hoechst 33258 dye are subjected to a 390 nm UV laser-beam irradiation along a predetermined course, a track of double-strand breaks that runs through the nucleus is created, and γ-H2AX forms precisely along this track as shown by immunocytochemistry (7,8).

2.3. Double-Strand Breaks Generated as Intermediates in DNA Metabolic Processes Induce γ-Phosphorylation

In addition to DNA double-strand breaks generated by irradiation or chemicals, there are also DNA metabolic intermediates that are generated: (I) indirectly, in the process of DNA lesions (single stranded or others) being converted to double-strand breaks in succeeding phases of the cell cycle or subsequent DNA repair steps, (II) directly during highly specialized cellular processes such as V(D)J recombination, class switching, meiotic recombination, and apoptosis.

When cells are subjected to topoisomerase I or topoisomerase II inhibitors, covalent bonds between the topoisomerase molecules and DNA are stabilized, and the so called "cleavage complexes" are created. Cleavage complexes generated by topo II inhibitors affect both DNA strands and introduce double-strand breaks directly, whereas topo I inhibitors affect one DNA strand by introducing directly single-strand breaks. Consequently, topo II inhibitors induce γ-H2AX throughout the cell cycle (13), and topo I inhibitors induce γ-H2AX in S phase, as a result of collisions of the "cleavage complexes" with DNA replication forks, where single-strand breaks are converted into irreversible double-strand breaks (14). Acidic pH values within the cell are an additional mechanism to generate topo II mediated double-strand breaks, which also induce γ-phosphorylation in animal models (15).

Yeast strains that lack the C-terminus of the H2A, exhibit hypersensitivity when exposed to phleomycin—a radiomimetic drug—but not UV radiation (16,6). Strikingly, the same strains exhibit increased sensitivity to methyl methane-sulphonate (MMS), but not to ethyl methane-sulphonate (EMS), both being alkylating agents. Although not well understood, MMS may cause the appearance of double-strand breaks in the yeast DNA (16) indirectly, by a mechanism that is mediated by topoisomerases (6).

Telomeres extend the ends of chromosomes, with a characteristic nucleotide repeat forming a single stranded 3′ protruding overhang that is capped to prevent recognition by DNA processing enzymes. Uncapped telomeres that have been created through inhibition of the capping protein TRF2, resemble double-strand breaks and are extensively marked with 2AX (17). In human fibroblasts that exhibit telomerase-dependent senescence, telomeric γ-H2AX foci are increased in number and extent more than 270 kilobases inward of the chromosome end (18). Senescence related γ-H2AX foci colocalize with 53BP1, MDC1, NBS1, and the phosphorylated form of SMC1, associating DNA repair mechanisms with cellular senescence (18). Apart the telomeric γ-H2AX foci, there is another age-related, sub-population of γ-H2AX foci, identified in vitro aging cell lines, as well as in old-mouse organs. These foci are cryptogenic, and they do not follow repair kinetics as the irradiation-induced foci, suggesting that they may represent DNA lesions with unrepairable DSBs. Whether these unrepairable DSBs have a causal role in aging, remains to be shown (19).

During V(D)J recombination, RAG-mediated cleavage generates double-strand-breaks between immunoglobulins and T cell receptor loci. In developing thymocytes, γ-H2AX forms nuclear foci that colocalize with the T cell receptor α locus, as determined by immunofluorescence in situ hybridization. This result also demonstrates that immunocytochemistry of γ-H2AX is a very powerful tool as it detects the presence of only one double-strand break per mammalian genome (20). During the maturation of the immune system, one type of immunoglobulin heavy-chain constant region is replaced by another via class switch recombination (CSR) (see also Chapter 28). Activation-induced cytidine deaminase (AID) is an enzyme that either operates directly to produce nicks on both strands, or activates an endonuclease, resulting in double-strand breaks in switch regions (sμ) to facilitate CSR. γ-H2AX foci colocalize with IgH in wt but not in AID−/− mice, indicating that γ-H2AX forms at sites of CSR and is dependent on AID activity (21). B cells lacking H2AX show impaired CSR, but somatic hypermutation (SHM) is unaffected suggesting that the processing of DNA lesions leading to SHM is fundamentally different from CSR (22).

In meiotic recombination, γ-H2AX formation follows the Spo11 appearance along the leptotene-chromosomes. Spo11, a topoisomerase-related protein, introduces double-strand breaks in the meiotic chromosomes to initiate meiotic recombination. γ-H2AX formation is dependent on the Spo11 activity as demonstrated in Spo11−/− mice. Surprisingly, γ-H2AX signal that is located in the sex body is apparent also in Spo11−/− mice, indicating the existence of Spo11 independent double-strand breaks (23). H2AX−/− mice are infertile due to specific X–Y chromatin-related malfunctions.

During programmed cell death, also known as apoptosis, caspases orchestrate a cascade of cell reactions that result in the typical apoptotic morphology including DNA fragmentation and the formation of apoptotic bodies. Concomitant with the activation of the caspase activated DNAse (CAD or DFF) by caspases is the induction of γ-H2AX extensively throughout the apoptotic nucleus (24). The significance of γ-phosphorylation in a cellular process that is part of the execution phase of apoptosis is not well understood yet.

3. γ-PHOSPHORYLATION OF H2A(X) SPANS MEGABASE-LONG DOMAINS IN CHROMATIN

Up to date, the accumulated evidence is in accordance with the model proposed by the W.M. Bonner lab that H2AX molecules are distributed randomly through-out chromatin (with the possible exception of the nucleoli area), become γ-phosphorylated as they are incorporated into nucleosomes, and are not exchanged upon phosphorylation or dephosphorylation (5,7,25). In mammals, the percentage of H2AX with respect to total H2A is different between cell lines, spanning from 2.5% to 30% whereas in yeast the orthologue H2A comprises 95% of the whole complement. Stoichiometric analysis reveals that the same percentage of H2AX molecules becomes γ-phosphorylated per double-strand break in different cell lines or species. One of the intriguing implications of this stoichiometry is that megabase-long regions of chromatin are adjusted to the break sites. This model depicts a biological amplification mechanism where one double-strand break induces the γ-H2AX modification of thousand of nucleosomes, along megabase-long domains of chromatin. As γ-H2AX formation is not restricted to any phase of the cell cycle, visual confirmation of this model is provided by immunocytochemistry of irradiated cells throughout the cell cycle. Under the confocal microscope, γ-H2AX foci appear as large, roughly spherical conformations in interface and G2 premitotic condensed chromatin, and they adopt a band-like conformation in mitotic cells. Mitotic *M. muntiacus* chromosomes that undergo mitosis after irradiation, demonstrate that γ-phosphorylated, megabase long domains of chromatin are adjusted next to the break site in broken mitotic chromosomes (7). Variation of the γ-H2AX signal intensity per cell, measured by flow cytometry, has been reported as a function of cell-cycle phase (26). Immunoprecipitation experiments with γ-H2AX antibody revealed that γ-formation at the sites of double-strand break extents longer than several kilobases in yeast (R. Shroff and M. Lichten, personal communication, 2003).

4. KINASES INVOLVED IN γ-PHOSPHORYLATION OF H2A(X) HISTONE FAMILY

Phosphorylation of H2AX at serine 139 is governed by multiple kinases that are members of the phosphatidylinositol-3 family (PI3), namely ATM (ataxia telangiectasia mutated), ATR (ATM and Rad3 related), and DNA-PK (DNA-depended protein kinase) (5,8).

In *S. cerevisiae* Tel1 and Mec1, the yeast homologues to ATM and ATR, respectively, are also involved in γ-phosphorylation of H2A, which is the yeast homologue to H2AX. When yeast strains that bear deletions in tel1 or/and mec1 genes are subjected to MMS treatment that involves generation of putative double-strand breaks, γ-phosphorylation of H2A is impaired. A low-level signal is present in the mec1 or tel1 null single mutants, but not detectable signal is observed in tel1/mec1 double null. The above results indicate that both kinases are involved in γ-H2A phosphorylation, and they have overlapping roles (16,6). In addition, immunoprecipitated mec1 from yeast cell extracts can phosphorylate a C-terminal yeast H2A peptide in vitro, indicating that ser129 of the SQ motif of the yeast H2A is a direct target for mec1 (16).

In human cells, ATM seems to be the major kinase that controls γ-phosphorylation. ATM knockout cells exhibit only minimal γ-H2AX focus formation that can be further eliminated by low-concentration treatment of wortmannin, indicating perhaps a redundant role of DNA-PK and/or ATR (27). Low-dose IR activates ATM to γ-phosphorylate H2AX, whereas at higher doses other kinases substitute (28). ATM is also implicated in γ-H2AX formation during meiotic recombination as indicated in ATM−/− mice (29). In accordance with the in vivo experiments, immunoprecipitated ATM phosphorylates the SQ motif of H2AX on serine 139 in vitro (27). Fluorescent microscopy of ATM colocalizaton at the sites of double-strand breaks has been problematic because of the abundance of the former molecules throughout the nucleus. Retention of the ATM molecules at double-strand breaks and colocalization with γ-H2AX is shown only after in situ extraction of the unbound ATM molecules before immunocytochemistry (30). Although ATM and ATR are known to phosphorylate several common targets including H2AX, it has been shown in several cases that the two kinases pivot activities. Under hypoxic conditions, H2AX is γ-phosphorylated in an ATR-dependent manner (31). In S phase of the cell cycle, ATR is the kinase to take over γ-phosphorylation, in response to replication arrest and the consequent generation of double-strand breaks (32), or upon formation of topoisomerase I cleavage complexes after collision to DNA replication forks (14). The role of DNA-PK in γ-phosphorylation of H2AX is not well understood yet. Although crude extracts of DNA-PK phosphorylate the SQ motif of H2AX on serine 139 in vitro, several rodent cell lines that are deficient either in the DNA-PK catalytic subunit or the Ku antigen (see Chapter 29), exhibit no detectable defects in γ-phosphorylation of H2AX, (5,27). Nevertheless, it has been reported that DNA-PK plays a redundant role in several cases as shown in astrocytoma M059J cell line (5,8), and upon formation of topoisomerase I cleavage complexes at replication forks in S phase (14). By fluorescence microscopy, DNA-PK immuno-cytochemistry reveals a diffuse pattern throughout nucleus in both nonirradiated and irradiated cells. However, upon ionizing irradiation, DNA-PK catalytic subunit becomes auto-phosphorylated on threonine 2609 and colocalizes with γ-H2AX in distinct foci. Autophosphorylation of the DNA-dependent protein kinase catalytic subunit is required for rejoining of DNA double-strand breaks (33) but the involvement of γ-H2AX in this mechanism is not yet clear.

5. RECRUITMENT OF REPAIR FACTORS TO γ-PHOSPHORYLATED CHROMATIN

It has been demonstrated that there is a time-depended sequential assembly of repair factors and signal mediators on γ-H2AX positive repair foci. γ-H2AX foci appear within minutes after irradiation, and recruit other factors that participate either in direct DNA repair-homologous recombination (HR) and nonhomologous end-joining (NHEJ)- or in transducing the signal further to arrest the cell cycle or to initiate apoptosis.

Colocalization of γ-H2AX foci with the MRN complex (Mre11–Rad50–Nbs1) is evident in IR-induced foci. Following irradiation, Rad50 foci appear over a period of several hours and colocalize significantly with the already present γ-H2AX foci (8). It has been reported that Nbs1 physically interacts with γ-H2AX via its FHA/BRCT domain to relocalize the MRN complex in the vicinity of DNA damage (34). Moreover, IR in human fibroblasts induces the stable association of Mre11 and PML with p53, linking H2AX with both, DNA repair and checkpoint cell-cycle responses (35). In S phase, the generation of DNA double-strand breaks by topoisomerase I cleavage complex (14), or replication forks arrested by UV damage in virus transformed XP-V cells, induces colocalization of MRN complex with γ-H2AX (36,37). It is important to mention though, that in contrast to the IR-induced colocalization discussed above, Mre11 complex that is deposited onto S-phase chromatin at replication forks exhibits limited colocalization with γ-H2AX in nonirradiated cells (38).

In human cells, it is postulated that in S–G2 phase of the cell cycle, DNA damage is sensed by the hRad9–hRad1–Hus1 complex. hRad9 is shown to colocalize with topBP1(topoisomerase II beta-binding protein 1) and γ-H2AX at sites of double-strand breaks, in an S-phase depended fashion (39). TopBP1 is a protein that contains eight BRCT domains and interacts with DNA polymerase ε which participates in damage synthesis. Experimental evidence suggest that hRad9 recruits topBP1 to the damage sites by interaction of its C-terminal amino acids, and afterwards, topBP1 acts as a mediator and transduces the signal to other players involved in cell-cycle arrest (39).

BRCA1, another central player in DNA repair that is required for homologous recombination and DNA damage-induced S and G2/M phase arrest, also colocalizes with γ-H2AX foci with kinetics faster than the MRN complex (8). In

H2AX$-/-$ ES cells, as well as in MEFs and B cells, BRCA1 focus formation is impaired, contributing to the genomic instability that is evident in H2AX knockout phenotype (40–42). BRCA1 is dispensable for γ-phosphorylation of H2AX but is required for ATM- and ATR-dependent downstream phosphorylation of p53, c-Jun, Nbs1, and Chk2 (43). In addition to signal transduction, a role of BRCA1 in ubiquitination has been reported. The RING finger of BRCA1 confers ubiquitin ligase activity, and when complexed with BRCA1-associated ring domain protein (BARD1) can facilitate Ubc5c-mediated monoubiquitination of histone H2A/ H2AX in vitro (44). Whether γ-H2AX ubiquitination occurs in vivo and plays a role in DNA repair is not yet clear.

It has recently been demonstrated that the tumor suppressor p53 binding protein 1 (53BP1) regulates the BRCA1 phosphorylation. 53BP1 is a putative mammalian homologue to scRad9 that forms foci upon introduction of double-strand breaks into DNA. 53BP1 foci colocalize with γ-H2AX and exhibits kinetics parallel to γ-H2AX foci formation (45). In vitro assays revealed that a region upstream of the 53BP1 C-terminus binds to phosphorylated but not unphosphorylated H2AX (46). 53BP1 becomes hyperphosphorylated on its N terminus in an ATM-depended manner in response to IR, and mediates DNA damage-signaling pathways in mammalian cells (46,47,28). In H2AX$-/-$ cells, 53BP1 phosphorylation levels are reduced by 40% and 53BP1 foci formation is severely compromised but not eliminated.

Also known as KIAA0170 or NFBD1, the mediator of DNA damage checkpoint protein 1 (MDC1) is a novel protein that contains a forkhead-associated (FHA) domain, and two BRCA1 carboxy-terminal (BRCT) domains (48,49). In response to IR, MDC1 is rapidly recruited to γ-H2AX foci by its FHA domain (50–52), where it promotes the other DNA players to relocalize, and subsequently mediates further cellular responses to activate intra-S and G2/M phase cell cycle (48,49). MDC1 exhibits physical interaction with MRN complex, 53BP1, ATM, and γ-H2AX and its activity is placed upsream of 53BP1, BRCA1, and MRN complex, as well as of Chk2 and Chk1 phosphorylation, but downstream of γ-H2AX focus formation as indicated in H2AX$-/-$ MEFs (48–50,53). Down regulation of MDC1 abolishes the relocalization and hyperphosphorylation of BRCA1 following low-dose irradiation, suggesting the existence of redundant phosphorylation pathways (53). DNA damage induces hyperphosphorylation of MDC1 in a PI3-family dependent manner (48,49), and this hyperphosphorylation is also partially affected in H2AX$-/-$ MEFs (49). Remarkably, MDC1 affects also γ-phosphorylation levels of H2AX, indicating the existence of a positive feed-back loop mechanism (49,53).

6. MODELS AND SPECULATIONS ABOUT THE BIOLOGICAL ROLE OF γ-H2AX FOCI

A great development of our current understanding about the biological role of γ-H2AX modified chromatin comes from knock-out and knock-in models. The H2A(X)$-/-$ phenotype affects in a pleotropic way the organism that bares it, and although some aspects have been brought up earlier in this chapter, a more complete recapitulation follows. Though, for the purpose of this manuscript, the focus of the following discussion will be on DNA repair related features.

S. cerevisiae studies regarding the H2A function have been performed in genetically modified strains that either lack the C-tail of the H2A or carry mutated forms of H2A C-terminus where the serine 129 has been substituted with alanine or glutamic acid. Elimination of the critical phosphorylation site S129 results in sensitivity to inroduction of double-strand breaks by phleomycine, MMS, restriction enzymes, and TopoI (16,6). Complementation experiments revealed that phosphorylation at S129 is implicated in NHEJ rather than HR in yeast. In the mutated strain where serine 129 is substituted by glutamic acid to mimic constitutive phosphorylation, chromatin structure becomes relaxed as shown by chromatin-digestion (16).

Mice that lack the H2AX gene are viable and are characterized by sensitivity to IR, growth retardation and premature senescence, immune deficiency, impaired cell-cycle arrest, genomic instability, and male sterility (41,40). Combined deficiency in H2AX and p53, to diminish induction of apoptosis in mice, results in a more severe phenotype including the occurrence of lymphoid and solid tumors. In addition, the number of chromosomal breaks, fragments, and fusions decline from H2AX+/+ to H2AX+/− and further to H2AX−/− mice cells, when crossed to p53 deficient mice, indicating haploinsufficiency of the H2AX gene (54,55). Further, although V(D)J recombination products are largely intact in H2AX−/−, there is a two-fold reduction in the absolute number of lymphocytes in H2AX−/− mice (54,55). V(D)J recombination mechanism resembles that of nonhomologous end joining (NHEJ) path in DNA repair, as it requires the ubiquitously expressed DNA-PK, XRCC4, Ligase 4, and Artemis proteins and is restricted in G0/G1 phase of the cell cycle (see also Chapter 29). In H2AX−/−p53−/− mice, the resulting tumorigenesis is caused either by failure to correctly repair RAG-mediated DSBs by NHEJ, or via other spontaneous double-strand breaks that might arise during DNA replication (55). H2AX−/− embryonic stem cells (ES) are more sensitive to mitomycin C that induces interstrand crosslinks (40). Although the puzzle is still incomplete, it is perceived that repair of interstrand crosslinks is probably mediated by homologous recombination. Along the same line, gene-targeting efficiency is severely reduced, indicating that γ-H2AX has a critical role in homologous recombination in mammals (41).

While H2AX does not appear to be a vital component of double-strand repair pathway, it seems to assist both HR and NHEJ in mammals. On the molecular level, initial migration of MNR, 53BP1, BRCA1 to IRIF (IR-induced foci) is not totally abrogated in H2AX−/− cells, but further accumulation is diminished (42). A model that accommodates the experimental evidence build up so far has been proposed by A. Nussenzweig lab. γ-H2AX does not constitute the primary signal that is required for the redistribution of repair complexes to damaged chromatin, but functions as a platform to concentrate repair factors to the vicinity of DNA lesions and promote interactions between multicomponent complexes. The accumulation of repair and signaling factors in close proximity to a double-strand break would facilitate an amplification step of signal transduction and check-point pathways particularly in the case where low number of foci per nucleus are present in cells (42,56). This model is also consisted with the finding that H2AX−/− cells exhibit reduced ability to arrest the cell cycle only at low-doses of IR. When only a few double-strand breaks are generated in the nucleus, the DNA repair factors that are modified to transduce the signal are limited, and signal amplification at IRIF becomes essential (28). On the other hand, the haploinsufficiency of H2AX+/− mice, that is based on dosage dependence of H2AX gene, could be explained with the hypothesis that chromatin of H2AX+/− cells that comprises of sparsely γ-H2AX-containing nucleosomes, would not mediate efficient concentration of soluble DNA repair factors on IRIF and further amplification of the signal (28,56).

The importance of the phosphate group of γ-H2AX in recruitment of the other repair factors to IRIF has been demonstrated in knock-in experiments. In cell lines where the H2AX gene is genetically modified by substituting S136/139 with alanine or glutamic acid, IR-induced foci fail to form, and cells exhibit sensitivity to IR (42). It seems that the phosphate group on S136/139 is essential for interactions with other repair players and cannot be substituted.

Chromatin movements upon DSB lesions have been implicated in DNA repair, as to brink loose DNA-ends together. When cells are irradiated with α particles, and damaged chromosome-domain movements are followed utilizing anti-γ-H2AX as a probe, foci come together to form clusters in a subset of lesions. It has been proposed that the observed foci-clustering is facilitated by an adhesion process that takes place between the MRE 11 complex and the γ-H2AX (57).

Structural changes in chromatin may also play a critical role in repair processes. It has been reported that in yeast, the substitution of S129 to glutamic acid induces relaxation of chromatin structure that would facilitate the accessibility of repair factors to the lesion. On the contrary, γ-H2AX in mammals is correlated with chromatin condensation as in meiotic recombination (29) and apoptosis (24). Whether in IRIF-γ–H2AX induces chromatin relaxation to facilitate recruitment of repair factors, or chromatin condensation to provide a mechanistic way to condense the repair factors for subsequent phosphorylation and autophosphorylation steps that lead to signal amplification remains to be experimentally approached.

Along with the H2AX-γ-phosphorylation, there are several other chromatin modifications, specific to DSB, that have been recently identified. In *S. cerevisiae*, the linker histone Hho1p is inhibitory to DNA repair by HR, as well as to the recombination-dependent mechanism of telomere maintenance (58). Interestingly, histone H4 acetylation, that has a well-documented role in gene transcription, has been recently linked to DSB repair. In budding yeast, substitution of all four tail H4 lysines with glutamines cause a pronounced defect in a "replication-coupled" pathway (59) where in *S. cerevisiae*, lysine 16 of histone H4 becomes deacetylated during NHEJ, in a Sin3p-dependent manner (60).

Lysine 16 of histone H4 becomes deacetylated during NHEJ, in a Sin3p-dependent manner (60).

At the present time, chromatin is arising as a sophisticated player in DNA repair pathways. Although γH2AX plays a critical role in the biology of double strand break repair is not the only histone variant and/or histone modification to occur in response to DSB. Other chromatin modifications, have been identified up-to-day, to occur in parallel. However, the list is far from complete yet. Further chromatin studies are expected to brink into light specific chromatin modifications, that functioning in combination may modulate chromatin structure and/or chromatin cross talk. A new histone code related to DNA repair is expected to add critical pieces to our understanding in how different cell responses are fine-tuned in respect to different DNA lesions.

ACKNOWLEDGMENTS

We are grateful to Dr. W. M. Bonner for insightful discussion on chromatin and DNA double-strand break related topics. We also thank Dr. A. Nussenzweig for critical discussion on DNA-repair signaling topics.

REFERENCES

1. Strahl BD, Allis CD. The language of covalent histone modifications. Nature 2000; 403:41–45.
2. Redon C, Pilch D, Rogakou E, Sedelnikova O, Newrock K, Bonner W. Histone H2A variants H2AX and H2AZ. Curr Opin Genet Dev 2002; 12:162–169.
3. Luger K, Mader AW, Richmond RK, Sargent DF, Richmond TJ. Crystal structure of the nucleosome core particle at 2.8A resolution. Nature 1997; 389:251–260.
4. Hatch CL, Bonner WM, Moudrianakis EN. Minor histone 2A variants and ubiquinated forms in the native H2A:H2B dimer. Science 1983; 221:468–470.
5. Rogakou EP, Pilch DR, Orr AH, Ivanova VS, Bonner WM. DNA double-stranded breaks induce histone H2AX phosphorylation on serine 139. J Biol Chem 1998; 273: 5858–5868.
6. Redon C, Pilch DR, Rogakou EP, Orr AH, Lowndes NF, Bonner WM. Yeast histone 2A serine 129 is essential for the efficient repair of checkpoint-blind DNA damage. EMBO Rep 2003; 4:678–684.
7. Rogakou EP, Boon C, Redon C, Bonner WM. Megabase chromatin domains involved in DNA double-strand breaks in vivo. J Cell Biol 1999; 146:905–916.
8. Paull TT, Rogakou EP, Yamazaki V, Kirchgessner CU, Gellert M, Bonner WM. A critical role for histone H2AX in recruitment of repair factors to nuclear foci after DNA damage. Curr Biol 2000; 10:886–895.
9. Madigan JP, Chotkowski HL, Glaser RL. DNA double-strand break-induced phosphorylation of Drosophila histone variant H2Av helps prevent radiation-induced apoptosis. Nucleic Acids Res 2002; 30:3698–3705.
10. Tomilin NV, Solovjeva LV, Svetlova MP, Pleskach NM, Zalenskaya IA, Yau PM, Bradbury EM. Visualization of focal nuclear sites of DNA repair synthesis induced by bleomycin in human cells. Radiat Res 2001; 156:347–354.
11. Sedelnikova OA, Rogakou EP, Panyutin IG, Bonner WM. Quantitative detection of (125)IdU-induced DNA double-strand breaks with gamma-H2AX antibody. Radiat Res 2002; 158:486–492.
12. Banath JP, Olive PL. Expression of phosphorylated histone H2AX an a surrogate of cell killing by drugs that create DNA double-strand breaks. Cancer Res 2003; 63:4347–4350.
13. Huang X, Traganos F, Darzynkiewicz Z. DNA Damage Induced by DNA Topoisomerase I-and Topoisomerase II Inhibitors Detected by Histone H2AX Phosphorylation in Relation to the Cell Cycle Phase and Apoptosis. Cell Cycle 2003; 2:614–619.
14. Furuta T, Takemura H, Liao ZY, Aune GJ, Redon C, Sedelnikova OA, Pilch DR, Rogakou EP, Celeste A, Chen HT, Nussenzweig A, Aladjem MI, Bonner WM, Pommier Y. Phosphorylation of histone H2AX and activation of Mre11, Rad50, and Nbs1 in response to replication-dependent DNA double-strand breaks induced by mammalian DNA topoisomerase I cleavage complexes. J Biol Chem 2003; 278:20303–20312.
15. Xiao H, Li TK, Yang JM, Liu LF. Acidic pH induces topoisomerase II-mediated DNA damage. Proc Natl Acad Sci USA 2003; 100:5205–5210.
16. Downs JA, Lowndes NF, Jackson SP. A role for Saccharomyces cerevisiae histone H2A in DNA repair. Nature 2000; 408:1001–1004.
17. Takai H, Smogorzewska A, de Lange T. DNA damage foci at dysfunctional telomeres. Curr Biol 2003; 13:1549–1556.
18. d'Adda di Fagagna F, Reaper PM, Clay-Farrace L, Fiegler H, Carr P, Von Zglinicki T, Saretzki G, Carter NP, Jackson SP. A DNA damage checkpoint response in telomere-initiated senescence. Nature 2003; 426:194–198.
19. Sedelnikova OA, Horikawa I, Zimonjic DB, Popescu NC, Bonner WM, Barrett JC. Senescing human cells and ageing mice accumulate DNA lesions with unrepairable double-strand breaks. Nat Cell Biol 2004; 6:168–170.

20. Chen HT, Bhandoola A, Difilippantonio MJ, Zhu J, Brown MJ, Tai X, Rogakou EP, Brotz TM, Bonner WM, Ried T, Nussenzweig A. Response to RAG-mediated VDJ cleavage by NBS1 and gamma-H2AX. Science 2000; 290:1962–1965.

21. Petersen S, Casellas R, Reina-San-Martin B, Chen HT, Difilippantonio MJ, Wilson PC, Hanitsch L, Celeste A, Muramatsu M, Pilch DR, Redon C, Ried T, Bonner WM, Honjo T, Nussenzweig MC, Nussenzweig A. AID is required to initiate Nbs1/gamma-H2AX focus formation and mutations at sites of class switching. Nature 2001; 414:660–665.

22. Reina-San-Martin B, Difilippantonio S, Hanitsch L, Masilamani RF, Nussenzweig A, Nussenzweig MC. H2AX is required for recombination between immunoglobulin switch regions but not for intra-switch region recombination or somatic hypermutation. J Exp Med 2003; 197:1767–1778.

23. Mahadevaiah SK, Turner JM, Baudat F, Rogakou EP, de Boer P, Blanco-Rodriguez J, Jasin M, Keeney S, Bonner WM, Burgoyne PS. Recombinational DNA double-strand breaks in mice precede synapsis. Nat Genet 2001; 27:271–276.

24. Rogakou EP, Nieves-Neira W, Boon C, Pommier Y, Bonner WM. Initiation of DNA fragmentation during apoptosis induces phosphorylation of H2AX histone at serine 139. J Biol Chem 2000; 275:9390–9395.

25. Siino JS, Nazarov IB, Svetlova MP, Solovjeva LV, Adamson RH, Zalenskaya IA, Yau PM, Bradbury EM, Tomilin NV. Photobleaching of GFP-labeled H2AX in chromatin: H2AX has low diffusional mobility in the nucleus. Biochem Biophys Res Commun 2002; 297:1318–1323.

26. MacPhail SH, Banath JP, Yu Y, Chu E, Olive PL. Cell cycle-dependent expression of phosphorylated histone H2AX: reduced expression in unirradiated but not X-irradiated G1-phase cells. Radiat Res 2003; 159:759–767.

27. Burma S, Chen BP, Murphy M, Kurimasa A, Chen DJ. ATM phosphorylates histone H2AX in response to DNA double-strand breaks. J Biol Chem 2001; 276: 42462–42467.

28. Fernandez-Capetillo O, Chen HT, Celeste A, Ward I, Romanienko PJ, Morales JC, Naka K, Xia Z, Camerini-Otero RD, Motoyama N, Carpenter PB, Bonner WM, Chen J, Nussenzweig A. DNA damage-induced G2-M checkpoint activation by histone H2AX and 53BP1. Nat Cell Biol 2002; 4:993–997.

29. Fernandez-Capetillo O, Mahadevaiah SK, Celeste A, Romanienko PJ, Camerini-Otero RD, Bonner WM, Manova K, Burgoyne P, Nussenzweig A. H2AX is required for chromatin remodeling and inactivation of sex chromosomes in male mouse meiosis. Dev Cell 2003; 4:497–508.

30. Andegeko Y, Moyal L, Mittelman L, Tsarfaty I, Shiloh Y, Rotman G. Nuclear retention of ATM at sites of DNA double strand breaks. J Biol Chem 2001; 276:38224–38230.

31. Hammond EM, Dorie MJ, Giaccia AJ. ATR/ATM targets are phosphorylated by ATR in response to hypoxia and ATM in response to reoxygenation. J Biol Chem 2003; 278:12207–12213.

32. Ward IM, Chen J. Histone H2AX is phosphorylated in an ATR-dependent manner in response to replicational stress. J Biol Chem 2001; 276:47759–47762.

33. Chan DW, Chen BP, Prithivirajsingh S, Kurimasa A, Story MD, Qin J, Chen DJ. Autophosphorylation of the DNA-dependent protein kinase catalytic subunit is required for rejoining of DNA double-strand breaks. Genes Dev 2002; 16:2333–2338.

34. Kobayashi J, Tauchi H, Sakamoto S, Nakamura A, Morishima K, Matsuura S, Kobayashi T, Tamai K, Tanimoto K, Komatsu K. NBS1 localizes to gamma-H2AX foci through interaction with the FHA/BRCT domain. Curr Biol 2002; 12:1846–1851.

35. Carbone R, Pearson M, Minucci S, Pelicci PG. PML NBs associate with the hMre11 complex and p53 at sites of irradiation induced DNA damage. Oncogene 2002; 21:1633–1640.

36. Limoli CL, Laposa R, Cleaver JE. DNA replication arrest in XP variant cells after UV exposure is diverted into an Mre11-dependent recombination pathway by the kinase inhibitor wortmannin. Mutat Res 2002; 510:121–129.

37. Limoli CL, Giedzinski E, Bonner WM, Cleaver JE. UV-induced replication arrest in the xeroderma pigmentosum variant leads to DNA double-strand breaks, gamma -H2AX formation, and Mre11 relocalization. Proc Natl Acad Sci USA 2002; 99:233–238.

38. Mirzoeva OK, Petrini JH. DNA replication-dependent nuclear dynamics of the Mre11 complex. Mol Cancer Res 2003; 1:207–218.

39. Greer DA, Besley BD, Kennedy KB, Davey S. hRad9 rapidly blinds DNA containing double-strand breaks and is required for damage-dependent topoisomerase II beta binding protein 1 focus formation. Cancer Res 2003; 63:4829–4835.

40. Bassing CH, Chua KF, Sekiguchi J, Suh H, Whitlow SR, Fleming JC, Monroe BC, Ciccone DN, Yan C, Vlasakova K, Livingston DM, Ferguson DO, Scully R, Alt FW. Increased ionizing radiation sensitivity and genomic instability in the absence of histone H2AX. Proc Natl Acad Sci USA 2002; 99:8173–8178.

41. Celeste A, Petersen S, Romanienko PJ, Fernandez-Capetillo O, Chen HT, Sedelnikova OA, Reina-San-Martin B, Coppola V, Meffre E, Difilippantonio MJ, Redon C, Pilch DR, Olaru A, Eckhaus M, Camerini-Otero RD, Tessarollo L, Livak F, Manova K, Bonner WM, Nussenzweig MC, Nussenzweig A. Genomic instability in mice lacking histone H2AX. Science 2002; 296:922–927.

42. Celeste A, Fernandez-Capetillo O, Kruhlak MJ, Pilch DR, Staudt DW, Lee A, Bonner RF, Bonner WM, Nussenzweig A. Histone H2AX phosphorylation is dispensable for the initial recognition of DNA breaks. Nat Cell Biol 2003; 5:675–679.

43. Foray N, Marot D, Gabriel A, Randrianarison V, Carr AM, Perricaudet M, Ashworth A, Jeggo P. A subset of ATM-and ATR-dependent phosphorylation events requires the BRCA1 protein. Embo J 2003; 22:2860–2871.

44. Chen A, Kleiman FE, Manley JL, Ouchi T, Pan ZQ. Autoubiquitination of the BRAC1 *BARD1 RING ubiquitin ligase. J Biol Chem 2002; 277:22085–22092.

45. Schultz LB, Chehab NH, Malikzay A, Halazonetis TD. p53 binding protein 1 (53BP1) is an early participant in the cellular response to DNA double-strand breaks. J Cell Biol 2000; 151:1381–1390.

46. Ward IM, Minn K, Jorda KG, Chen J. Accumulation of checkpoint protein 53BP1 at DNA breaks involves its binding to phosphorylated histone H2AX. J Biol Chem 2003; 278:19579–19582.

47. Rappold I, Iwabuchi K, Date T, Chen J. Tumor suppressor p53 binding protein 1 (53BP1) is involved in DNA damage-signaling pathways. J Cell Biol 2001; 153:613–620.

48. Goldberg M, Stucki M, Falck J, D'Amours D, Rahman D, Pappin D, Bartek J, Jackson SP. MDC1 is required for the intra-S-phase DNA damage checkpoint. Nature 2003; 421:952–956.

49. Stewart GS, Wang B, Bignell CR, Taylor AM, Elledge SJ. MDC1 is a mediator of the mammalian DNA damage checkpoint. Nature 2003; 421:961–966.

50. Peng A, Chen PL. NFBD1, like 53BP1, is an early and redundant transducer mediating Chk2 phosphorylation in response to DNA damage. J Biol Chem 2003; 278:8873–8876.

51. Xu X, Stern DF. NFBD1/KIAA0170 is a chromatin-associated protein involved in DNA damage signaling pathways. J Biol Chem 2003; 278:8795–8803.

52. Shang YL, Bodero AJ, Chen PL. NFBD1, a novel nuclear protein with signature motifs of FHA and BRCT, and an internal 41-amino acid repeat sequence, is an early participant in DNA damage response. J Biol Chem 2003; 278:6323–6329.

53. Lou Z, Chini CCS, Minter-Dykhouse K, Chen J. Mediator of DNA Damage Checkpoint Protein 1 Regulates BRCA1 Localization and Phosphorylation in DNA Damage Checkpoint Control. J Biol Chem 2003; 278:13599–13602.

54. Bassing CH, Suh H, Ferguson DO, Chua KF, Manis J, Eckersdorff M, Gleason M, Bronson R, Lee C, Alt FW. Histone H2AX: a dosage-dependent suppressor of oncogenic translocations and tumors. Cell 2003; 114:359–370.

55. Celeste A, Difilippantonio S, Difilippantonio MJ, Fernandez-Capetillo O, Pilch DR, Sedelnikova OA, Eckhaus M, Ried T, Bonner WM, Nussenzweig A. H2AX haploinsufficiency modifies genomic stability and tumor susceptibility. Cell 2003; 114:371–383.

56. Fernandez-Capetillo O, Celeste A, Nussenzweig A. Focusing on Foci: H2AX and the Recruitment of DNA-Damage Response Factors. Cell Cycle 2003; 2:426–427.

57. Aten JA, Stap J, Krawczyk PM, van Oven CH, Hoebe RA, Essers J, Kanaar R. Dynamics of DNA double-strand breaks revealed by clustering of damaged chromosome domains. Science 2004; 303:92–95.

58. Downs JA, Kosmidou E, Morgan A, Jackson SP. Suppression of homologous recombination by the Saccharomyces cerevisiae linker histone. Mol Cell 2003; 11:1685–1692.

59. Bird AW, Yu DY, Pray-Grant MG, Qiu Q, Harmon KE, Megee PC, Grant PA, Smith MM, Christman MF. Acetylation of histone H4 by Esa1 is required for DNA double-strand break repair. Nature 2002; 419:411–415.

60. Jazayeri A, McAinsh AD, Jackson SP. Saccharomyces cerevisiae Sin3p facilitates DNA double-strand break repair. Proc Natl Acad Sci USA 2004; 101:1644–1649.

33

DNA Strand-Break Recognition, Signaling, and Resolution: The Role of Poly(ADP-Ribose) Polymerases-1 and -2

Emmanuelle Pion
Unité 7034 du CNRS, Laboratoire de Pharmacologie et Physico-chimie des Interactions Cellulaires et Moléculaires, Faculté de Pharmacie, Université Louis Pasteur, Illkirch, France

Catherine Spenlehauer, Laurence Tartier, Jean-Christophe Amé, Françoise Dantzer, Valérie Schreiber, Josiane Ménissier-de Murcia, and Gilbert de Murcia
Unité 9003 du CNRS, Ecole Supérieure de Biotechnologie de Strasbourg, Illkirch, France

Gérard Gradwohl
Unité INSERM U682, Strasbourg, France

1. BACKGROUND

Activation of DNA damage-dependent poly(ADP-ribose) polymerases (PARP-1, -2) is an immediate cellular reaction to DNA strand breakage as induced by alkylating agents, ionizing radiation, or oxidants. The resulting formation of protein-bound poly(ADP-ribose) (PAR) facilitates survival of proliferating cells under conditions of DNA damage probably via its contribution to DNA base excision repair and single-strand break repair. The role of PARP-1 during the repair of single-strand breaks in mammalian cells is now better understood, not only in vitro where PARP-1 activity facilitates the whole process, but also in living cells where PAR synthesis at sites of DNA lesions plays an essential role in the recruitment kinetics of one key player: X-ray cross-complementing factor 1. This review summarizes our present knowledge of a cellular response pathway to DNA damage specific for higher eukaryotes.

2. INTRODUCTION

The presence of DNA strand breaks in the cells of higher eukaryotes activates signal transduction pathways that rapidly trigger cell-cycle arrest and repair mechanisms, leading ultimately to cell survival or programed cell death. Central to pathways that

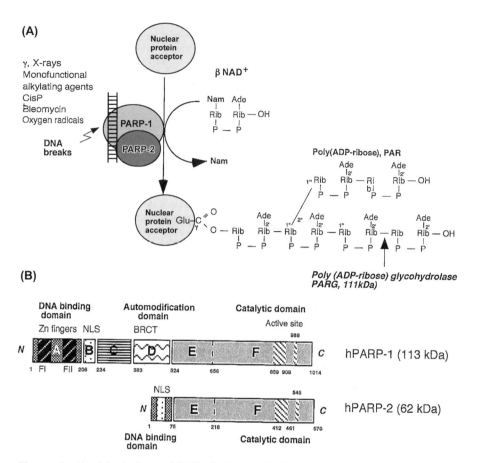

Figure 1 (A) Metabolism of PAR during DNA damage and repair induced by various genotoxins. (B) Schematic representation of the domain structure of the human PAR polymerases-1 and -2. *Source*: Adapted from Ref. 3.

maintain genomic integrity is the modification of histones and nuclear proteins by ADP-ribose polymers catalyzed by poly(ADP-ribose) polymerases (PARPs). The resulting formation of protein-bound poly(ADP-ribose) (PAR) facilitates survival of proliferating cells under conditions of DNA damage owing to chromatin structure opening and/or recruitment of DNA repair enzymes and factors (1–3). Poly (ADP-ribose) polymerases enzymes now constitute a large family of 18 proteins, encoded by different genes, and displaying a conserved catalytic domain (4). Among them, PARP-1 (113 kDa), the founding member, and PARP-2 (62 kDa) are the sole enzymes whose catalytic activity is immediately stimulated by DNA strandbreaks, suggesting that they are involved in the cellular response to DNA damage (5,6). At a site of DNA breakage, PARP-1 and PARP-2 catalyze the transfer of the ADP-ribose moiety from the respiratory coenzyme NAD^+ to a limited number of acceptor proteins involved in chromatin architecture and in DNA metabolism (Fig. 1A). Poly(ADP-ribosyl)ation of nuclear proteins establishes de facto a molecular link between DNA damage and chromatin modification and appears to be an obligatory step of a detection/signaling pathway leading ultimately to the resolution of strand-break interruptions. In certain pathophysiological conditions, this protecting function is dramatically activated leading to cell death, tissue damage, and organ failure (see Ref. 7 for review). This chapter mainly concentrates on the DNA strand-break recognition properties of PARP-1; occasionally, the DNA-binding properties of PARP-2 will be mentioned.

3. NICK SENSOR FUNCTION OF PARP-1

PARP-1 (113 kDa) is a highly conserved multifunctional enzyme whose enzymatic activity is stimulated more than 500-fold upon binding to DNA strand breaks. Its modular structure comprises four main distinct regions as in what follows.

i.   The N-terminal DNA-binding domain (DBD) that bears two zinc fingers acting as a molecular nick-sensor (Fig. 1B);

ii. A module of 38 amino acids, containing a bipartite motif of the form KRK-x_{11}- KKKSKK, which constitutes the nuclear homing sequence of PARP-1. This short region is not only recognized by the nuclear transport machinery, but also contains a proteolytic cleavage site of caspase-3 (Fig. 1B) localized in the sequence 211-DEVD-214.

iii. The central automodification domain, containing a BRCA1 C-terminus (BRCT) motif and auto-poly(ADP-ribosyl)ation sites, which are implicated in the regulation of PARP-1–DNA interactions.

iv. The carboxy-terminal region of PARP-1 (Fig. 1A, domain F) that bears all the different catalytic activities associated with the full-length enzyme: NAD^+ hydrolysis and initiation, elongation, branching, and termination of ADP-ribose polymers. This basal activity of PARP-1 is independent of the presence of DNA breaks. Domain F, by far the most evolutionarily conserved region, contains a block of 50 amino acids (aa 859–908) representing "the PARP signature" virtually unchanged from human to plants that turned out to be the PARP catalytic site. Interestingly, this region is folded in a β–α–loop–β–α motif structurally similar to the NAD^+-binding fold of several ADP-ribosylating bacterial toxins like diphtheria toxin, (see Ref. 3 for review).

Two characteristic properties of PARP-1 place this enzyme at early steps of the repair process, most probably downstream the action of DNA glycosylases and/or APE-1:

i.   The N-terminal region of PARP-1 that acts as a sensor of single-strand breaks (SSBs) (8,9).

ii. The bending of the nicked substrate that generates a distorted structure which in turn is recognized by the next enzyme in the base excision repair (BER) or single-strand break repair (SSBR) pathways. PARP-1 exploits the flexibility created by a sugar-phosphate backbone interruption in DNA to bend the nicked duplex by an angle of about 100°. The characteristic V-shaped conformation of the PARP-1-nicked-DNA complex, as visualized by dark-field electron microscopy (Fig. 2 A and B), presumably favors the formation of an active PARP-1 dimer ready to process its substrate NAD^+ for PAR synthesis (10).

The PARP-1 DBD interacts with one-and-a-half turns of the double helix, protecting seven nucleotides each side of the break, irrespective of the nucleotide

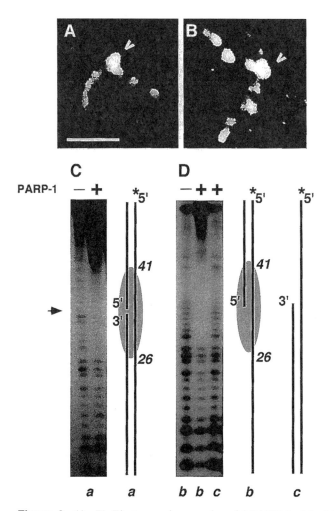

Figure 2 (A, B) Electron micrographs of hPARP-1-nicked DNA complexes. PARP-1 is located at the apex of a 139 bp DNA duplex containing an SSB at position 69. The bars represent 50 nm. (C) DNAse I footprinting of hPARP-1 bound to a 66 bp DNA duplex containing an SSB located at position 33. PARP-1 protects the nucleotides 26–41 (DNA probe a) on the continuous strand. (D) hPARP-1 binds to a 5′ recessed end (DNA probe b) and protects the double-to-single-strand junction. No protection is observed with a DNA probe containing a 3′ recessed end (probe c).

sequence in a DNase I footprinting assay where the protein is immobilized onto nitrocellulose and the 66 bp nicked-DNA probe is free to interact (9). The binding of PARP-1 or its DBD is dependent upon the co-ordination of two zinc atoms to the Cys and His that structurally organize the fingers (11,12). In fact, only the 5′ end of the nick is recognized (Fig. 2D) because PARP-1 has little affinity for a DNA probe with a 3′ recessed end that also behaves as a poor activator (E. Pion, unpublished results). Therefore, PARP-1 binds to a double-strand-to-single-strand junction, a structure functionally relevant of BER intermediates but present at telomeric ends. Using a photoaffinity labeling probe, a similar result has been obtained by Lavrik et al. (13) who identified not only PARP-1 but also FEN-1 and DNA pol-β cross-linked to a DNA probe bearing a sugar phosphate at the 5′ margin of a nick.

The PARP-1 DBD contains a repeated sequence (aa 2–97 and 106–207) encoding a zinc-finger motif of the form $Cx_2Cx_{28,30}Hx_2C$ strictly conserved from human to the plant *Arabidopsis thaliana* (Fig. 3A and B) suggesting that it might have arisen by sequence divergence of a duplicated primordial element, evolving to form two independently folded zinc-containing domains (FI, FII) (14). Interestingly, a sequence of 97 residues homologous to finger FI is present in the N-terminal domain of the human DNA ligase III protein (15,16) and is repeated three times in the nick-sensing DNA 3'-phosphoesterase from *A. thaliana* (17,18), an enzyme also involved in the resolution of DNA interruptions.

In order to assign a role to each of the hPARP-1 zinc fingers, we have over expressed and purified hPARP-1 mutants deleted in either FI or FII. These mutants were tested both for enzymatic activity and for their capacity to bind to the nicked DNA probe described earlier. As shown in Fig. 4A, the deletion of either FI or FII maintains a specific recognition of the DNA interruption located at position 33. However, some contacts are lost in each case: the nucleotides 26–28 or the nucleotides 44–46 become accessible to DNase I when FI and FII are, respectively, deleted (Fig. 4B).

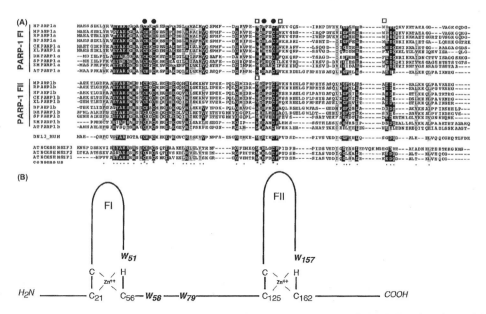

Figure 3 (A) The PARP-1-like zinc-finger family: Comparison of the deduced amino-acid sequences of the amino-terminal region of DNA ligase III (dnl_3hum, accession number P49916) and the three zinc fingers of the nick-sensing DNA 3'-phosphoesterase from *A. thaliana* (ATNCKSENZF-1, -2, -3, accession number AF453835) with the N-terminal zinc-finger region of PARP-1 from human (hPARP-1, accession number P09874), mouse (mPARP-1, accession number P1 1103), rat (rPARP-1, accession number P27008), bovine (bPARP-1, accession number P18493), chicken (ckPARP-1, accession number P26446), *Xenopus laevis* (xlPARP-1, accession number P31669), *Drosophila melanogaster* (dmPARP-1, accession number P35875), *Stichocorys peregrina* (spPARP-1, accession number D16482), *Zea mays* (zmPARP-1, accession number AF093627) and *A. thaliana* (atPARP-1, accession number AJ131705). Identical amino-acid residues are boxed in black. Conserved substitutions are indicated in grey. The closed circles point to the Cys and His involved in the co-ordination of zinc, the open squares point to the Trp residues. (B) PARP-1 DBD structure of the N-terminal binding domain of hPARP-1 (residues 1–234). The DBD is drawn to show the two zinc-co-ordinated fingers (FI, FII). The four trypto-phans: W51, W58, W79, W157 are indicated in bold italic capitals. *Source*: From Ref. 14.

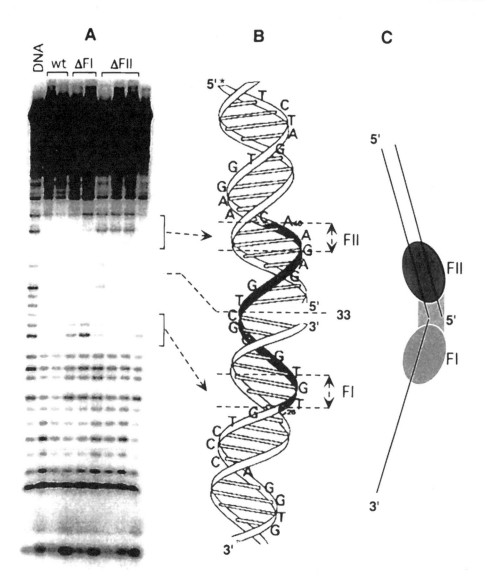

Figure 4 hPARP-1 zinc finger–DNA contact sites. (A). DNase I footprinting of wild-type and zinc-finger deletion mutants bound to a 66 bp DNA duplex containing an SSB located at position 33. (B) Nucleotide-numbers are reported relative to the 5'-labeled end of the DNA probe. DNA regions protected by FI and FII are indicated with brackets. (C) Schematic representation of a hPARP-1 DBD monomer interacting with the continuous strand at a junction between double-strand and single-strand DNA.

These results demonstrate that each zinc finger contacts the continuous strand in a symmetric manner with respect to the nick, FII being positioned at the 5' end. Interestingly, both deletion mutants display a strongly reduced enzymatic activity, indicating that each zinc finger plays a critical role in the stimulation of the enzymatic activity by DNA ends. The schematic model presented in Fig. 4C integrates the results obtained with the various DNA probes and PARP-1 mutants and tentatively represents a hPARP-1 monomer interacting with a double-strand-to-single-strand DNA junction.

The alignment of the PARP-1-like zinc-finger family (Fig. 3A) also points to the strict conservation of a Trp residue (at positions 51 and 157 in FI and FII, respectively,) which is also present in DNA ligase III as well as in each repeat finger of the DNA 3'-phosphoesterase. Because of their high sensitivity to even minor changes in the physicochemical environment, Trp residues constitute suitable and powerful intrinsic fluorescence probes for investigating the interaction of proteins with various ligands and, notably, nucleic acids. The interaction between hPARP-1 DBD and a double-stranded oligonucleotide bearing a 5'-recessed end was investigated by monitoring tryptophan fluorescence change upon addition of increasing oligonucleotide amounts. A systematic fluorescence decrease is observed until a plateau signal is reached at high DNA concentration (Fig. 5A, inset). Saturation is obtained with 0.5 equivalents of oligonucleotide indicating that two proteins are involved in the final complex. Accordingly, the DBD function was lost in the mutant Trp51Ala (E. Pion, unpublished results).

Fluorescence anisotropy is a suitable parameter for measuring the association of macromolecules, as it varies in response to a change in size and shape of the rotating molecules. Upon addition of increasing amounts of oligonucleotide, a remarkable increase of anisotropy was observed (Fig. 5B). The stoichiometry of binding 2:1 was confirmed (Figure 5B, inset). Because experimental evidences showed that hPARP-1 DBD is not a dimer in solution, the data were interpreted according to an "all-or-none" reaction for the binding of hPARP-1 DBD to DNA with the following reaction scheme (14):

$$P + P + N \overset{K}{\rightleftharpoons} PPN$$

Applying global analysis to fluorescence intensity and anisotropy data, we obtained a second-order binding constant $K = 1.5 \times 10^{14}\ M^{-2}$. This feature is typical for proteins that exist as a monomer and bind the DNA site as a dimer (19,20). Together, these results indicate that PARP-1 mainly recognizes the 5'-margin of a nick and dimerizes during binding to the DNA-damage site. In this process, the conserved Trp51 seems to play a critical role in DNA interaction and at least one tryptophan residue is involved in the stacking interaction with DNA bases.

4. DUAL ROLE OF DNA-DAMAGE INDUCED PAR SYNTHESIS: BREAK SIGNALING AND RECRUITMENT OF XRCC1

X-ray cross-complementing factor 1 (XRCCl) is a key factor involved in BER and SSBR, genomic stability, and embryonic viability (21,22). It interacts and modulates the function of enzymes involved in these pathways including APE-1, DNA pol-β, DNA ligase III, Polynucleotide kinase, and OGG1. The previously described functional interaction between XRCCl and PARP-1 (23) or PARP-2 (5) prompted us to examine the subcellular distribution of XRCCl comparatively to the spatial distribution of PAR in locally damaged cells. For this purpose, we set up three independent approaches to evaluate the biological function of PAR synthesis in response to locally produced DNA strand breaks.

First, we applied the finding by Okano et al. (24) that the transient expression of UV damage endonuclease (UVDE) in nucleotide excision repair deficient cells allows the introduction of SSBs, 5' to the UV-induced Cyclobutane pyrimidine dimers and 6–4 photoproducts in response to UV-C irradiation. We irradiated UVDE-expressing Xeroderma pigmentosum group A (XPA) cells through Isopore filters (25) to concentrate the breaks in defined nuclear volumes (3 μm in diameter), thus permitting a spatial and temporal analysis of XRCCl. Under these experimental conditions (Fig. 6A) discrete foci of PAR (panel a) colocalized with XRCCl foci (panel b) in cells that expressed the construct Flag-UVDE (panel d) but not in non-transfected cells (panel d–f). When XPA[Flag-UVDE] cells were treated with the PARP inhibitor 3-aminobenzamide (3-AB) (panel g–i) neither the relocalization of XRCCl into foci nor the synthesis of PAR (data not shown) could be detected. Figure 6B gives a quantitative estimation of UVDE-expressing cells displaying foci of XRCCl whether treated or not with 3-AB prior to UV-C irradiation. These results

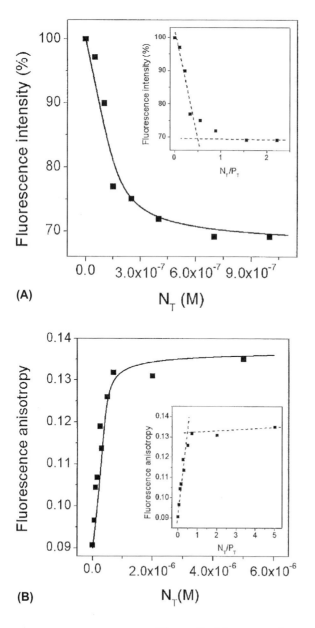

(A)

(B)

Figure 5 (*Caption on facing page*)

demonstrate that the relocalization of XRCCl from a homogeneous distribution in the nucleoplasm to nuclear foci critically depends upon local PAR synthesis at DNA breaks.

The microbeam facility, which has been developed at the Gray Cancer Institute (London), now allows microirradiation with charged particles of distinct cell compartments with a 2 μm beam spot (26). In a second approach, HeLa or V79 cells were subjected to microbeam irradiation with a number of protons equivalent to a dose of 10 Gy per nucleus (27). As shown in Fig. 6C, irradiation with 10 Gy was enough to obtain detectable PAR foci (panel j) in the nucleus of most irradiated cells and cell types. Again, foci of PAR, 2–3 μm in diameter, were observed in irradiated nuclei, which colocalized with the foci of XRCCl. Similarly to the earlier-mentioned results, the recruitment of XRCCl is conditioned by the PAR synthesis because no foci were found when cells were pretreated with a PARP inhibitor.

To follow the recruitment of XRCCl at DNA damage sites, we finally adopted the UV-A laser microirradiation technique developed by Rogakou et al. (28) that allows the introduction of DNA breaks in cells briefly incubated with the DNA intercalating dye Hoechst 33258. Given the size of the microbeam, the net advantage of this technique is to selectively target organelles and defined cellular compartments, taking the nonirradiated cellular zone as a negative control. As shown in Fig. 6D, microbeam-irradiated HeLa cells, immediately fixed and processed for immunofluorescence, display PAR synthesis along the laser path through the nuclei (panel p). Under these experimental conditions, the recruitment of XRCC1 was also observed at DNA breaks marked by PAR synthesis (panel q). When HeLa cells were pretreated with 3-AB prior to laser irradiation, neither PAR synthesis (panel r) nor XRCC1 recruitment (panel t) could be observed, in agreement with the earlier-mentioned results.

To further characterize the rapid PAR-dependent recruitment of XRCC1 at DNA breaks, we microirradiated HeLa cells expressing green fluorescent protein

---

Figure 5 (*Facing page*) (A) Titration curve for binding of hPARP-1 DBD to a 5′ recessed DNA end obtained with fluorescence intensity data. Fluorescence titrations were performed by adding increasing amounts of oligonucleotide to a fixed amount of protein in 50 mM Tris–HCl pH 8, 100 mM NaCl, and 1 mM DTT. The various molar ratios of oligonucleotide to protein were prepared as separate solutions. The excitation wavelength was set at 295 nm. The solid line corresponds to the fit of the fluorescence data using the binding constant determined from the simultaneous global analysis. The inset shows the dependence of fluorescence intensity of hPARP-1 DBD on the DNA concentration. The concentration of hPARP-1 DBD was 0.45 μM. The linear parts of the binding curve are fitted and extrapolated separately with a linear curve fitting routine (dashed lines). The intersection indicates that the saturating DNA:protein is 0.5. (B) Titration curve for binding of hPARP-1 DBD to a 5′ recessed DNA end obtained with fluorescence anisotropy data. Steady state anisotropy measurements were performed with a T-format SLM 8000 spectrofluorometer at 20°C. The emitted light was monitored through 350 nm interference filters (Schott). A device built in-house ensured the automatic rotation of the excitation polarizer. Increasing amounts of DNA were added to 1 μM hPARP-1 DBD under the same conditions as described earlier. The solid line corresponds to the fit of the anisotropy data using binding constant determined from the simultaneous global analysis. The inset shows the dependence of the fluorescence anisotropy of hPARP-1 DBD on the DNA concentration. The linear parts of the binding curve are fitted and extrapolated separately with a linear curve fitting routine (dashed lines). The intersection indicates that the saturating DNA:protein is 0.5. *Source*: From Ref. 14.

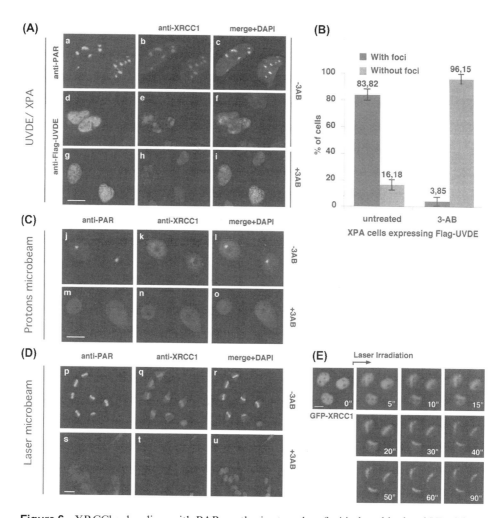

Figure 6 XRCCl colocalizes with PAR synthesis at nuclear foci induced by local UV-C irradiation in Flag-UVDE expressing XPA cells. (A) Xeroderma pigmentosum group A cells, transfected to transiently express Flag-UVDE, were irradiated with UV-C [$20\,J/m^2$] through 3 μm Isopore filters. In transfected cells, UV-C damaged areas are detected by immunofluorescent labeling of PAR (green, panel a) synthesized in response to DNA breakage. XRCCl foci (red, panel b) are detected at sites of DNA breaks marked and colocalized with PAR foci (panel c). XRCCl foci are detected only in transfected cells (green, panel d). Transfected cells (green, panel g) treated with the PARP inhibitor 3-AB at 10 mM before and during irradiation do not show XRCCl accumulation following UV-C irradiation. Bars indicate 10 μm. (B) Quantification of Flag-UVDE transfected cells displaying or not a foci distribution of XRCCl at sites of local DNA breaks after UV-C irradiation. Counting was performed on cells either untreated or treated with the PARP inhibitor 3-AB (10 mM) before and during irradiation. (C) Recruitment of XRCCl (red, panel k) to PAR foci (green, panel j) in V79 cells following a proton microbeam irradiation of 20 Gy. Neither PAR nor XRCCl recruitment in visible when V79 cells were pretreated with 3-AB before irradiation. (D) Laser-induced DNA damage triggers the local recruitment of XRCCl (red, panel q) dependent on PAR synthesis (green, panel p). UV-A light was delivered by a 337 nm laser targeting HeLa cells labeled with Hoechst dye 33258 and treated or not with the PARP inhibitor 3-AB (10 mM) before and during irradiation (panels s and t). Merged images of damaged nuclei stained with 4′6-diamidino-2-phenylindol (DAPI) are indicated (panels r and u). (E) Real-time recruitment of XRCCl is followed in HeLa cells transiently expressing GFP–XRCCl. Pictures captured every 5 sec indicate a gradual accumulation of GFP–XRCCl along the laser path within 10–15 sec. (*See color insert.*)

(GFP)–tagged XRCC1 or GFP alone as a control. Time-lapse experiments were performed and images were taken every 2–5 sec (Fig. 6E). Within 15 sec, a fast accumulation of GFP–XRCC1 was observed along the laser path. A similar result was obtained using a construct expressing the GFP–BRCT1 domain of XRCC1, known as the interacting interface with PARP-1 and PARP-2 (C. Spenlehauer, unpublished results). No accumulation was obtained using HeLa cells expressing GFP alone nor when cells were pre-treated with the PARP inhibitor NU1025 thus confirming the dynamics of XRCC1 relocalization to PAR foci in the nuclei of living cells.

Our results, in agreement with recent studies from Okano et al. (29) and El-Khamisy et al. (30), reveal that the immediate synthesis of PAR constitutes an initiating event in a damaged cell. This triggers the subsequent co-ordination of DNA strand-break detection and signaling to the SSBR pathway, the whole process being performed in <15 sec as shown in this study. PARP-1 appears to be particularly well suited to perform both functions: the efficient sensing of breaks is mediated by its zinc-finger domain, whereas their immediate signaling relies on the DNA strand-break-dependent synthesis of ADP-ribose polymers that, in turn, attracts XRCC1 to the damage sites. Although not essential in vitro, XRCC1 as a scaffold protein with no known enzymatic function plays a critical role in the co-ordinated handling of the damage DNA from one repair enzyme to the next in the BER pathway (31). The absence of PARP-1 or the inhibition of its enzymatic activity prevents the dynamic recruitment of XRCC1, thus explaining the important delay in strand-break rejoining that causes a severe DNA repair defect in PARP-1$^{-/-}$ damaged cells (32,33) or in damaged cells treated with 3-AB (34). It is worth mentioning that the repair defect observed in PARP-2$^{-/-}$ damaged cells (5) cannot be attributed to a lack of XRCC1 recruitment, as this step still occurred following microirradiation (not shown). In that case, the repair deficiency could be related to the impossibility to form PARP-1/PARP-2 heterodimers or to an unknown defect in a subsequent step in the repair pathway.

5. NO CROSS-TALK BETWEEN PAR SYNTHESIS AND γ-H2AX FORMATION IN RESPONSE TO DNA-STRAND BREAK INJURY

Poly(ADP-ribosyl)ation of histones and nuclear proteins and phosphorylation of histone H2AX on Ser 139 (35) are two epigenetic marks induced by DNA strand breaks that contribute to the histone code in damaged chromatin. Therefore, a possible connection between these two modifications was tested using the laser microbeam system. We found that both modifications, colocalized along the laser path, appear as two separate DNA damage signaling pathways because phosphorylation of H2AX occurred in response to laser microbeam irradiation, independently of the PARP-1 or PARP-2 status (Fig. 7A). Similarly, PARP-1- and PARP-2-deficient cell lines treated with a dose of 20 Gy displayed a strong induction of histone H2AX phosphorylation comparable with that observed in wild-type cells (Fig. 7B) thus confirming the absence of connection between poly (ADP-ribosyl)ation of nuclear proteins (especially histones) and phosphorylation of histone H2AX, two post-translational modifications of nuclear proteins induced by IR.

Changes in chromatin structure emanating from DNA breaks are probably the most initiating events in the damage response. They are catalyzed by several

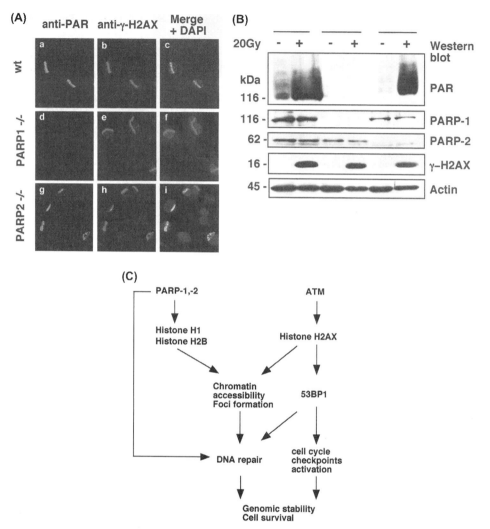

Figure 7 (A) Laser-induced DNA breaks trigger the phosphorylation of H2AX independently of PAR synthesis and PARP-1 or PARP-2 status. Immortalized wild-type, PARP-1$^{-/-}$ and PARP-2$^{-/-}$ mouse embryonic fibroblasts (MEFs) were locally irradiated with UV-A laser microbeam as described in Fig. 6. Cells were immediately fixed and process for immunostaining using anti-PAR (green, panels a, d, g) and anti-γ-H2AX (Panels red, b, e, h). Images are merged in panels c, f, i. (B) Induction of poly(ADP-ribosyl)ation by PARPs and phosphorylation of H2AX in wild-type PARP-1$^{-/-}$ and PARP-2$^{-/-}$ 3T3 cells, 30 min following 20 Gy irradiation. Each cell line was characterized in terms of PARP-1 and PARP-2 content. Antiactin was used for loading control. (C) Schematic representation of the DNA strand-break signaling and processing pathway. The pathway is triggered by IR resulting in SSB and DSB that immediately initiate two post-translational modifications of histones and nuclear proteins catalyzed by PARP-1/-2 and ATM. Poly(ADP-ribosyl)ation of histones H1 and H2B increases chromatin accessibility at the actual site of DNA damage thus promoting DNA repair. Phosphorylation of histone H2AX by ATM occurs at or near the DSB and is required for the phosphorylation of 53BP1 that participates to nuclear foci organization and the subsequent recruitment of several ATM downstream targets. *Source*: From Refs. 26,34,44,45. (*See color insert.*)

independent pathways: ATM seems to be involved predominantly in responding to double-strand break (DSB) (36,37), but PARP-1 responds to SSB and DSBs containing a 5′ recessed end. Interestingly, the simultaneous inactivation of either PARP-1 and ATM or PARP-2 and ATM genes in mice leads to early embryonic lethality suggesting the absolute necessity to maintain at least one of these two pathways, especially during early development (38). Therefore, both modifications appear to constitute a part of the histone code translating in that case the occurrence of a break into molecular signals emanating from the damaged chromatin to facilitate DNA repair and cell survival (Fig. 7C). It is interesting to note that the PAR signal persists during 1 hr maximum after DNA damage, whereas large γ-H2AX foci are still visible 8 hr after IR, probably signaling irreparable or slowly repaired DSBs lesions (39).

6. CONCLUSIONS AND FUTURE PROSPECTS

Several characteristics of PARP-1 behavior upon DNA damage classify this fascinating enzyme as a component of the early response to DNA strand breaks. As shown in this review, PARP-1 fulfills several key functions immediately following an interruption of the sugar-phosphate backbone: (i) efficient sensing of the break; (ii) translation and amplification of the damage signal in a post-translational modification of histones H1 and H2B leading to chromatin structure relaxation and to DNA accessibility; and (iii) immediate or concomitant recruitment of XRCC1 to the break. The domain structure of PARP-1 contains a recognition module (zinc fingers), a protein–protein interacting interface (BRCT) present in other enzymes and factors involved in the maintenance of genomic integrity, and a catalytic domain capable of fast signaling that makes it particularly well adapted to these tasks. Conversely, its reactivity at breaks renders PARP-1 a risky cellular factor when the genome is to be degraded during cell death. To limit futile DNA repair and to preserve the NAD and ATP pools, PARP-1 must be inactivated by a caspase-dependent cleavage (40,41). A quite different situation occurs during necrosis or in caspase-independent cell death (chromatinolysis) where PARP-1 is instrumentalized by the apoptosis-inducing factor that generates high amounts of PAR thus leading to an inflammatory response (42).

Three methods introducing local DNA breaks have been described in the present review; they all lead to a local and robust PAR synthesis at nuclear foci or along the laser path. Our results highlight a strict causal relationship between PAR synthesis and the recruitment of XRCC1, in agreement with the studies of Okano et al. (29) and El-Khamisy (30) that documented a link between PAR synthesis and XRCC1 accumulation at DNA damage foci. Here, we extend this conclusion in time-lapse experiments that demonstrate the kinetics of XRCC1 recruitment mediated by its BRCT1 module in response to laser microbeam irradiation. To our knowledge, this recruitment event at DNA breaks constitutes the first example of a relocalization of a SSBR factor, within seconds, in living cells. The physiological role of PAR previously identified in vitro as an attracting molecule for various DNA repair proteins is, therefore, validated.

PARP-1 could well be the first on breaks; however; it is not yet known how it finds the lesion inside a damaged nucleus. There are apparently enough PARP-1 molecules distributed in the nucleoplasm to cope with a repairable level of breaks. Moreover, no recruitment of PARP-1 to the damage sites could be detected either in living cells expressing a dsRed-PARP-1 fusion protein or in fixed cells immunostained with a specific anti-PARP antibody.

However, the function of PARP-1 in DNA damage and repair is not only restricted to the initial damage recognition step in the SSBR pathway, but also occurs during the repair of damaged bases by BER. Activation of polymer synthesis likely follows the incision step of the sugar-phosphate backbone by DNA glycosylase–AP lyases or APE-1. We have shown that the polymerization step of the long patch repair (LPR) pathway was mainly affected in PARP-1-deficient cells (43). Moreover, PARP-1 is associated to DNA pol-β (13,43) and efficiently binds to the repair intermediates containing a flap 5′-abasic site that are formed before subpathway choice, leading to either short patch repair (SPR) or LPR (13,44). In addition, PARP-1, together with FEN-1, stimulates strand displacement synthesis by DNA pol-β leading then to LPR. Under these conditions, the presence of the dRP group is supposed to trigger the LPR pathway following the PARP-1 recruitment.

The ligation step is also concerned by the presence of PARP-1, as it was demonstrated that a direct physical interaction between PARP-1 and DNA ligase III occurs via the region of aminoacids immediately adjacent to the N-terminal zinc finger of DNA ligase III (45). It was further demonstrated that DNA ligase III binds also directly to PAR and to poly(ADP-ribosyl)ated PARP-1, leading to a increase of DNA joining. Again, this provides a mechanism for the recruitment of the DNA ligase III–XRCC-1 complex at DNA breaks. Therefore, the presence of PARP-1 and/or PAR throughout the BER/SSBR process, in interaction with XRCC1 and DNA ligase III, none of them being found in lower eukaryotes, speaks for a unique role in maintaining genome stability in complex organisms.

ACKNOWLEDGMENTS

The authors are grateful to Didier Hentsch (IGBMC, Illkirch), Jean-Christophe Laval (Leica Microsystems, France), and Kevin Prise (Gray Laboratory, London), for their help with the laser microbeam and the proton microbeam irradiation. This work was supported by funds from the Centre National de la Recherche Scientifique, the Association pour la Recherche Contre le Cancer, Electricité de France, Ligue Nationale contre le Cancer, Commissariat à l'Energie Atomique, Ligue Contre le Cancer Région Alsace, Association Régionale pour l'Enseignement et la Recherche Scientifique et Technologique, and Université Louis Pasteur.

REFERENCES

1. de Murcia G, Ménissier-de Murcia J. Poly(ADP-ribose) polymerase: a molecular nick-sensor. Trends Biochem Sci 1994; 19:172–176.
2. D'e Amours D, Desnoyers S, D'Silva I, Poirier GG. Poly(ADP–ribosyl)ation reaction in the regulation of nuclear functions. Biochem J 1999; 342(Pt 2):249–268.
3. de Murcia G, Shall S. eds. From DNA damage and stress signalling to cell death: Poly (ADP-ribosylation) reactions. Oxford: Oxford University Press, 2000.
4. Ame JC, Spenlehauer C, de Murcia G. The PARP superfamily. Bioessays 2004; 26: 882–893.

5. Ame JC, Schreiber V, Fraulob V, Dolle P, de Murcia G, Niedergang CP. A bidirectional promoter connects the poly(ADP-ribose) polymerase 2 (PARP-2) gene to the gene for RNase P RNA. structure and expression of the mouse PARP-2 gene. J Biol Chem 2001; 276:11092–11099.

6. Schreiber V, Ame JC, Dolle P, Schultz I, Rinaldi B, Fraulob V, Menissier-de Murcia J, de Murcia G. Poly(ADP-ribose) polymerase-2 (PARP-2) is required for efficient base excision DNA repair in association with PARP-1 and XRCC1. J Biol Chem 2002; 277:23028–23036.

7. Burkle A. Physiology and pathophysiology of poly(ADP-ribosyl)ation. Bioessays 2001; 23:795–806.

8. Mazen A, Menissier-de Murcia J, Molinete M, Simonin F, Gradwohl G, Poirier G, de Murcia G. Poly(ADP-ribose)polymerase: a novel finger protein. Nucleic Acids Res 1989; 17:4689–4698.

9. Menissier-de Murcia J, Molinete M, Gradwohl G, Simonin F, de Murcia G. Zinc-binding domain of poly(ADP-ribose) polymerase participates in the recognition of single strand breaks on DNA. J Mol Biol 1989; 210:229–233.

10. Le Cam E, Fack F, Menissier-de Murcia J, Cognet JA, Barbin A, Sarantoglou V, Revet B, Delain E, de Murcia G. Conformational analysis of a 139 basepair DNA fragment containing a single-stranded break and its interaction with human poly(ADP-ribose) polymerase. J Mol Biol 1994; 235:1062–1071.

11. Gradwohl G, Menissier de Murcia JM, Molinete M, Simonin F, Koken M, Hoeijmakers JH, de Murcia G. The second zinc-finger domain of poly(ADP-ribose) polymerase polymerase determines specificity for single-stranded breaks in DNA. Proc Natl Acad Sci USA 1990; 87:2990–2994.

12. Molinete M, Vermeulen W, Burkle A, Menissier-de Murcia J, Kupper JH, Hoeijmakers JH, de Murcia G. Overproduction of the poly(ADP-ribose) polymerase DNA-binding domain blocks alkylation-induced DNA repair synthesis in mammalian cells. EMBO J 1993; 12:2109–2117.

13. Lavrik OI, Prasad R, Sobol RW, Horton JK, Ackerman EJ, Wilson SH. Photoaffinity labeling of mouse fibroblast enzymes by a base excision repair intermediate. Evidence for the role of poly(ADP-ribose) polymerase-1 in DNA repair. J Biol Chem 2001; 276:25541–25548.

14. Pion E, Bombarda E, Stiegler P, Ullmann GM, Mely Y, de Murcia G, Gerard, D. Poly (ADP-ribose) polymerase-1 dimerizes at a 5' recessed DNA end in vitro a fluorescence study. Biochemistry 2003; 42:12409–12417.

15. Chen J, Tomkinson AE, Ramos W, Mackey ZB, Danehower S, Walter CA, Schultz RA, Besterman JM, Husain I. Mammalian DNA ligase III: molecular cloning, chromosomal localization, and expression in spermatocytes undergoing meiotic recombination. Mol Cell Biol 1995; 15:5412–5422.

16. Mackey ZB, Niedergang C, Murcia JM, Leppard J, Au K, Chen J, de Murcia G, Tomkinson AE. DNA ligase III is recruited to DNA strand breaks by a zinc finger motif homologous to that of poly(ADP-ribose) polymerase. Identification of two functionally distinct DNA binding regions within DNA ligase III. J Biol Chem 1999; 274: 21679–21687.

17. Petrucco S, Volpi G, Bolchi A, Rivetti C, Ottonello S. A nick-sensing DNA 3'-repair enzyme from Arabidopsis. J Biol Chem 2002; 277:23675–23683.

18. Petrucco S. Sensing DNA damage by PARP-like fingers. Nucleic Acids Res 2003; 31:6689–6699.

19. Pomerantz JL, Wolfe SA, Pabo CO. Structure-based design of a dimeric zinc finger protein. Biochemistry 1998; 37:965–970.

20. Wang BS, Pabo CO. Dimerization of zinc fingers mediated by peptides evolved in vitro from random sequences. Proc Natl Acad Sci USA 1999; 96:9568–9573.

21. Caldecott KW. Mammalian DNA single-strand break repair: an X-ra(y)ted affair. Bioessays 2001; 23:447–455.

22. Caldecott KW. XRCC1 and DNA strand break repair. DNA Repair (Amst) 2003; 2: 955–969.

23. Masson M, Niedergang C, Schreiber V, Muller S, Menissier-de Murcia J, de Murcia G. XRCC1 is specifically associated with poly(ADP-ribose) polymerase and negatively regulates its activity following DNA damage. Mol Cell Biol 1998; 18:3563–3571.

24. Okano S, Kanno S, Nakajima S, Yasui A. Cellular responses and repair of single-strand breaks introduced by UV damage endonuclease in mammalian cells. J Biol Chem 2000; 275:32635–32641.

25. Mone MJ, Volker M, Nikaido O, Mullenders LH, van Zeeland AA, Verschure PJ, Manders EM, van Driel R. Local UV-induced DNA damage in cell nuclei results in local transcription inhibition. EMBO Rep 2001; 2:1013–1017.

26. Prise KM, Belyakov OV, Folkard M, Michael BD. Studies of bystander effects in human fibroblasts using a charged particle microbeam. Int J Radiat Biol 1998; 74:793–798.

27. Tartier L, Spenlehauer C, Newman HC, Folkard M, Prise KM, Michael BD, Menissier-de Murcia J, de Murcia G. Local DNA damage by proton microbeam irradiation induces poly(ADP-ribose) synthesis in mammalian cells. Mutagenesis 2003; 18:411–416.

28. Rogakou EP, Boon C, Redon C, Bonner WM. Megabase chromatin domains involved in DNA double-strand breaks in vivo. J Cell Biol 1999; 146:905–916.

29. Okano S, Lan L, Caldecott KW, Mori T, Yasui A. Spatial and temporal cellular responses to single-strand breaks in human cells. Mol Cell Biol 2003; 23:3974–3981.

30. El-Khamisy SF, Masutani M, Suzuki H, Caldecott KW. A requirement for PARP-1 for the assembly or stability of XRCC1 nuclear foci at sites of oxidative DNA damage. Nucleic Acids Res 2003; 31:5526–5533.

31. Tainer JA. Structural implications of BER enzymes: dragons dancing–the structural biology of DNA base excision repair. Prog Nucleic Acid Res Mol Biol 2001; 68:299–304.

32. Ménissier de Murcia J, Niedergang C, Trucco C, Ricoul M, Dutrillaux B, Mark M, Olivier FJ, Masson M, Dierich A, LeMeur M, Walztinger C, Chambon P, de Murcia G. Requirement of poly(ADP-ribose) polymerase in recovery from DNA damage in mice and in cells. Proc Natl Acad Sci USA 1997; 94:7303–7307.

33. Trucco C, Oliver FJ, de Murcia G, Ménissier-de Murcia J. DNA repair defect in poly (ADP-ribose) polymerase-deficient cell lines. Nucleic Acids Res 1998; 26:2644–2649.

34. Durkacz BW, Omidiji O, Gray DA, Shall S. (ADP-ribose)n participates in DNA excision repair. Nature 1980; 283:593–596.

35. Fernandez-Capetillo O, Chen HT, Celeste A, Ward I, Romanienko PJ, Morales JC, Naka K, Xia Z, Camerini-Otero RD, Motoyama N, Carpenter PB, Bonner WM, Chen J, Nussenzweig A. DNA damage-induced G2-M checkpoint activation by histone H2AX and 53BP1. Nat Cell Biol 2002; 4:993–997.

36. Shiloh Y, Kastan MB. ATM: genome stability, neuronal development, and cancer cross paths. Adv Cancer Res 2001; 83:209–254.

37. Kastan MB, Lim DS, Kim ST, Yang D. ATM–a key determinant of multiple cellular responses to irradiation. Acta Oncol 2001; 40:686–688.

38. Huber A, Bai P, de Murcia JM, de Murcia G. PARP-1, PARP-2 and ATM in the DNA damage response: functional synergy in mouse development. DNA Repair (Amst) 2004; 3:1103–1108.

39. Petrini JH, Stracker TH. The cellular response to DNA double-strand breaks: defining the sensors and mediators. Trends Cell Biol 2003; 13:458–462.

40. Tewari M, Quan LT, O'Rourke K, Desnoyers S, Zeng Z, Beidler DR, Poirier GG, Salvesen GS, Dixit VM. Yama/CPP32 beta, a mammalian homolog of CED-3, is a Crm-A-inhibitable protease that cleaves the death substrate poly(ADP-ribose) polymerase. Cell 1995; 81:801–809.

41. D'Amours D, Germain M, Orth K, Dixit VM, Poirier GG. Proteolysis of poly(ADP-ribose) polymerase by caspase 3: kinetics of cleavage of mono(ADP-ribosyl)ated and DNA-bound substrates. Radiat Res 1998; 150:3–10.

42. Yu SW, Wang H, Poitras MF, Coombs C, Bowers WJ, Federoff HJ, Poirier GG, Dawson TM, Dawson VL. Mediation of poly(ADP-ribose) polymerase-1-dependent cell death by apoptosis-inducing factor. Science 2002; 297:259–263.

43. Dantzer F, de La Rubia G, Ménissier-De Murcia J, Hostomsky Z, de Murcia G, Schreiber V. Base excision repair is impaired in mammalian cells lacking Poly(ADP-ribose) polymerase-1. Biochemistry 2000; 39:7559–7569.

44. Prasad R, Lavrik OI, Kim SJ, Kedar P, Yang XP, Vande Berg BJ, Wilson SH. DNA polymerase beta -mediated long patch base excision repair. Poly(ADP-ribose) polymerase-1 stimulates strand displacement DNA synthesis. J Biol Chem 2001; 276: 32411–32414.

45. Leppard JB, Dong Z, Mackey ZB, Tomkinson AE. Physical and functional interaction between DNA ligase IIIalpha and poly(ADP-Ribose) polymerase 1 in DNA single-strand break repair. Mol Cell Biol 2003; 23:5919–5927.

46. Poirier GG, de Murcia G, Jongstra-Bilen J, Niedergang C, Mandel P. Poly(ADP-ribosyl) ation of polynucleosomes causes relaxation of chromatin structure. Proc Natl Acad Sci USA 1982; 79:3423–3427.

47. de Murcia G, Huletsky A, Poirier GG. Modulation of chromatin structure by poly(ADP-ribosyl)ation. Biochem Cell Biol 1988; 66:626–635.

Part VII

Perception of DNA Damage for Initiating Regulatory Responses

DNA damage elicits a variety of regulatory responses in pro- and eukaryotic organisms. These include modifications of the transcriptional make-up as well as alterations of cell cycle parameters, all suspected to benefit the cell's repair capacity. At the beginning of the signal transduction processes involved stands damage recognition and assessment—directly or possibly indirectly, through sensing of some secondary consequence. The sensors used for regulatory responses and damage repair are not necessarily the same.

First, we will revisit alkylation damage. Chapter 34 discusses damage responses in pro- and eukaryotes with emphasis on regulatory signaling.

As reviewed in Chapter 35, we move on to the *Escherichia coli* SOS system, a coordinated transcriptional response with similarities to what is known as cell cycle checkpoints in eukaryotes.

Eukaryotic checkpoint responses are the topic of the last two chapters. Chapter 36 discusses the molecular events of replication-independent damage sensing that initiates eukaryotic checkpoint arrests and associated responses.

The complex damage signaling that occurs when DNA damage interferes with replication is discussed in the last chapter, Chapter 37.

34

Cellular and Molecular Responses to Alkylation Damage in DNA

James M. Bugni and Leona D. Samson
Department of Biological Engineering, Massachusetts Institute of Technology, Cambridge, Massachusetts, U.S.A.

1. INTRODUCTION

Organisms have evolved a variety of responses to DNA damage. This chapter describes regulatory responses that are initiated by DNA alkylation damage (see Chapter 16), with a focus on the initiation of events from specific methyl adducts.

1.1. Exposure to Alkylating Agents and Relevance to Disease

Chemicals that cause aberrant alkylation of cellular macromolecules are ubiquitous, and are produced as natural byproducts of cellular metabolism. One endogenous source of alkylating species is the nitrosation of amines which occurs enzymatically in enteric bacteria (1). Alkylation can also result indirectly from the reactive species produced upon oxidative damage-induced lipid peroxidation in the cell membrane, the major target of oxygen radicals (2). As one might expect, DNA lesions associated with this reactivity are present at higher levels in states of chronic inflammation. There are also numerous exogenous sources of alkylating agents as they are present in foods, cigarette smoke, and fuel combustion products. A highly abundant class of environmental alkylating toxicants, the methyl halides, are produced by the burning or decaying of biological materials, and are also present in commonly used industrial solvents. In addition, certain chemotherapeutic agents act by producing reactive alkylating species. It is, therefore, easy to see why an understanding of the mechanisms of toxicity and the cellular responses to these agents is important, because they are widespread in our environment and because of their widespread use in cancer treatment.

The evolutionary conservation of repair mechanisms directed towards alkyl damage in DNA suggests an important role of these pathways in preventing spontaneous as well as induced disease. Indeed, mice with genes disrupted in major pathways for DNA alkylation repair do show minor spontaneous abnormal phenotypes, and these DNA alkylation repair-deficient animals display greatly increased susceptibility to cancer and other diseases upon exposure to a variety of alkylating agents.

Similarly, alkylation repair deficiencies have been found in a variety of sporadic human cancers suggesting a major role of alkylation repair in preventing spontaneous or environmentally induced cancer (3).

1.2. Types of Alkylating Agents and DNA Adducts Created

Alkylating agents can be grouped mechanistically by their mode of action. Monofunctional methylating agents (i.e., those that possess a single reactive alkyl group) are classified as S_N1 agents, S_N2 agents, and methyl radicals. S_N1 agents act in a two-step reaction by first producing the nucleophilic methyl carbonium ion that then attacks DNA and other molecules. Typical S_N1 agents include methylnitrosourea (MNU) and N-methyl-N'-nitro-N-nitrosoguanidine (MNNG). These chemicals yield methyl adducts on ring nitrogens with the most prevalent adduct being N7-methylguanine (7MeG). S_N1 agents also methylate oxygens on the phosphodiester backbone, as well as exocyclic base oxygens yielding adducts that include methyl phosphotriesters (MePTs) and the highly mutagenic O^6-methylguanine (O^6MeG) and O^4-methylthymine (O^4MeT) (Fig. 1, Table 1) (4,5). The S_N2 alkylating agents generate the reactive group upon direct contact with DNA, and common S_N2 alkylating species include methylmethane sulfonate and methyl halides. Like the S_N1 agents, the major induced by S_N2 agents is 7MeG, but unlike the S_N1 alkylators, oxygens are minor targets for attack (5) (Fig. 1, Table 1). S_N2 agents can also form N1-methyladenine (1MeA) and N3-methylcytosine (3MeC) in single-stranded DNA, regions that are less susceptible to attack in dsDNA because they participate in base

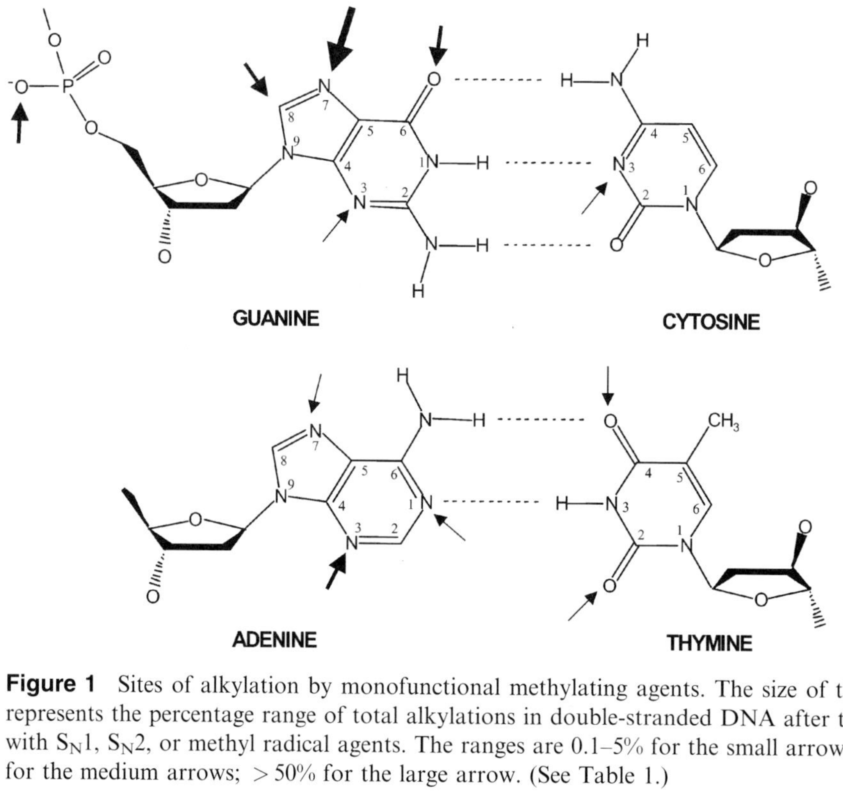

Figure 1 Sites of alkylation by monofunctional methylating agents. The size of the arrow represents the percentage range of total alkylations in double-stranded DNA after treatment with S_N1, S_N2, or methyl radical agents. The ranges are 0.1–5% for the small arrows; 5–50% for the medium arrows; >50% for the large arrow. (See Table 1.)

Table 1 Percentages of Methyl Adducts Formed in DNA by S_N1, S_N2, and Methyl Radical Agents

	Approximate %		
	S_N1[a]	S_N2[a]	Radicals[b]
1MeA	0.9	2	
3MeA	8	11	10
7MeA	2	2	
O^4MeT	0.7		
O^2MeT	0.1		
3MeC	0.5		
O^6MeG	6	0.3	
3MeG	0.6	0.6	
7MeG	66	81	20
8MeG			70
Me Phospho-triestersr	12	0.8	

[a]Percentages are from Beranek et al. (5). The S_N1 agent used was MNU, and the S_N2 agent was MMS.
[b]Percentages were calculated from Hix and Augusto (9). 8MeG, 7MeG, and 3MeA were the only adducts detected after treatment with the methyl radical producing agent tertiary butyl hydroperoxide. The relative amounts of these adducts formed are highly dependent on metals and chelators in the reaction.

pairing (6). Recently, S_N2 agents have been synthesized to target reactive alkyl groups at specific regions of DNA. One such compound, methyl lexitropsin (MeLex), produces ≫90% of its adducts at the N3 position of adenine in AT-rich regions of the genome (7,8). Compounds of this type have been developed for clinical applications but can be used experimentally to dissect the mechanism whereby a specific chemical adduct elicits various biological responses. Another class of methylating chemicals include radicals that are also present as environmental toxicants and used as experimental carcinogens. Methyl radicals yield similar adducts as S_N1 and S_N2 alkylators, but one adduct that is specific to the radicals is C8-methylguanine (9) (Fig. 1, Table 1). This chapter will focus on biological and regulatory responses that result from exposure to S_N1 and S_N2 methylating agents.

1.3. Responses to Alkylation Damage: A Prospectus

How is alkylation damage recognized and converted into cellular signals? One major response to alkylation in DNA is the upregulation of proteins that repair damage, and a description of the well-characterized adaptive response in *Escherichia coli* is presented in the context of how methyl adducts directly modify a sensor protein to initiate a transcriptional signal. Eukaryotes have evolved alternative systems to respond to DNA damage, such as the activation of cell cycle checkpoints which allow time for restorative processes to occur. Furthermore, multicellular organisms have evolved apoptosis as a response to DNA damage; apparently in certain circumstances it is better for a single cell to be removed from the viable population than to persist with mutations or chromosomal aberrations. In terms of these alternative responses we describe the cellular signals generated by the adducts O^6MeG and 3MeA. Direct recognition of these methyl adducts may be sufficient to initiate cellular signals, or alternatively, the response may result indirectly from repair intermediates or aberrant replication.

2. THE *E. COLI* ADAPTIVE RESPONSE: TRANSLATING METHYL DNA ADDUCTS INTO A TRANSCRIPTIONAL SIGNAL

2.1. Nature of the Adaptive Response

E. coli exposed to low doses of MNNG acquire resistance to the toxic and mutagenic effects of alkylating agents (10–13). Early work focused on distinguishing the adaptive response from the SOS response (14,15). Once that was established, it was shown that the adaptive response leads to increased activities for the removal of 3MeA and O^6MeG from DNA (16–18). *E. coli* mutants were identified that failed to acquire enhanced activity for the removal of 3MeA, or that failed to acquire enhanced activities for the removal of both 3MeA and O^6MeG activities, but initially, no mutants were identified that failed to acquire O^6MeG removal alone. The explanation for this phenomenon was revealed upon cloning of the *ada* gene, a gene that complemented the defect in both kinds of repair activities (19). Thus, whenever the Ada protein is present it does not just confer Ada activity, it confers AlkA, AklB, and AidB activity also. Ada was shown to function directly in DNA repair and in the transcriptional upregulation of four genes including the *ada* gene itself (20). Three of the regulated genes (*ada*, *alkB*, and *alkA*) encode three different DNA repair activities, and the fourth gene, *aidB*, encodes an activity that likely prevents damage from endogenous nitrosamines (21). Ada, a DNA repair methyltransferase (MTase), and AlkB, an oxidative demethylase, repair damage by very different direct reversal mechanisms, whereas AlkA is a DNA glycosylase that acts as part of the base excision repair (BER) process.

2.2. Ada Protein Structure and Function

The Ada protein turns out to be a bifunctional DNA repair MTase that serves as both an upstream sensor of damage and a downstream activator of transcription. The protein recognizes alkyl damage in the form of MePTs, O^6MeG, and O^4MeT, and repair of these adducts converts Ada to an active transcription factor.

2.2.1. Methyltransferase Functions of the Ada Protein

The 39 kd Ada protein consists of an N-terminal 20 kd domain (NAda20) and a C-terminal 19 kd domain (CAda19). These domains are tethered by a 10 amino acid hinge region that is susceptible to proteolytic cleavage (22). NAda20 specifically repairs the Sp diastereomers of MePTs (23,24), and CAda19 repairs O^6MeG and to a lesser extent O^4MeT (25). Both CAda19 and NAda20 transfer methyl groups to reactive cysteine residues, but their catalytic mechanisms are quite different. CAda19 transfers a methyl group from the O^6 of guanine or the O^4 of thymine to Cys321 (26). The active site cysteine acceptor is buried inside the protein (27), and access to the methyl group is gained by "flipping" the adducted nucleotide out of the helix and deep into the active site pocket (28). In contrast, NAda20 has four cysteines that coordinate a Zn++ ion in the active site to facilitate methyl transfer to Cys38 (not Cys69 as was previously thought) (29,30); mutations in any of the four coordinating cysteines have no adverse effect on protein folding or Zn++ binding, but each one is required for transferase activity (31). Thus, these non-acceptor residues likely contribute to the reactivity of Cys38 through the Zn++ ion. Reactions in both NAda20 and CAda19 are irreversible and expend one domain for each methyl group transferred. However, NAda20 has been modified by site-directed mutagenesis of the acceptor residue, a Cys38Gly mutant, to produce a protein capable of regenerating the unmethylated version, functioning as a true enzyme in vitro using methane thiol as an acceptor (32).

2.2.2. Mechanisms for Activating the Response

Upon methylation of Cys38, Ada is activated as a transcription factor. The Cys-38 methylated Ada protein has high-affinity binding for a specific DNA sequence known as the Ada box. Ada boxes are found as part of the promoters of the *ada/alkB* operon, as well as with the promoters of the *alkA* and *aidB* genes (33). The mechanism of activation at the *ada/alkB* and *aidB* promoters is distinct from the mechanism at the *alkA* promoter (Fig. 2) (13,34). At the *ada/alkB* and *aidB* promoters, methylation in the N-terminal domain enhances binding to the promoter, and the residues in the C terminus are required for transactivation. Methylation at Cys361 induces a conformational change that exposes residues that interact with σ^{70} and subsequently recruit RNA polymerase. These residues can also be exposed in partial truncation mutants and in a Cys361Ala mutant (35,36). In contrast, at

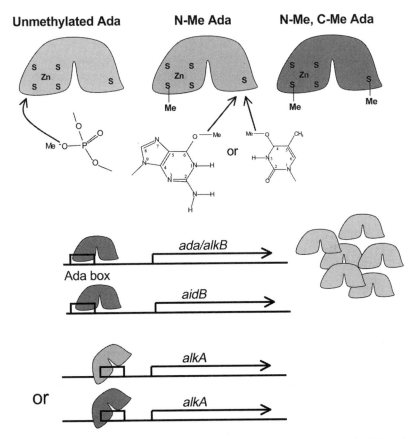

Figure 2 Conversion of Ada into a transcription factor. The N-terminal domain of Ada repairs MePTs, whereas the C-terminal domain repairs O^6MeG or O^4MeT. Methylations at both domains of Ada (red) are required to activate transcription at the *ada/alkB* and *aidB* promoters, leading to increased expression of unmethylated Ada (blue). Methylation at the N-terminal domain of Ada (green) is sufficient to activate *alkA* transcription. (*See color insert.*)

the *alkA* promoter, residues in the N-terminal domain interact directly with σ^{70} and with the alpha subunit of RNA pol (37). Therefore, at this promoter only methylation at Cys38 is required for binding the Ada box and recruiting the polymerase, both components of the transcriptional activation. At the *alkA* promoter, the C-terminal domain is dispensable, and transcription is not affected by Cys321 methylation, nor by proteolytic cleavage in the hinge region (38). The use of different Ada residues for recruiting the same σ^{70} factor likely results from the location of the Ada box relative to the transcriptional start site. Furthermore, these promoter-dependent mechanisms for activating transcription are inherently associated with different downstream mechanisms for downregulating transcription when the adaptive response is no longer needed. Ada has some minor affinity for the *ada/alkB* promoter in the absence of methylation at Cys38, a low-affinity interaction that becomes biologically significant when the protein is overexpressed. However, the C-terminal domain requires methylation in order to transactivate. Thus, when Ada protein continues to be produced after all of the damage has been repaired, the unmethylated Ada protein actually inhibits transcription at the *ada/alkB* promoter, and such inhibition acts to downregulate *ada/alkB* expression upon the completion of DNA repair. At the *alkA* promoter, downregulation may result passively from dilution of the methylated Ada protein, or possibly through an interaction with the stationary phase factor σ^{s} (13).

The transformation of Ada into a transcriptional activator can also be achieved by direct methylation of the protein by the $S_{N}2$-type alkylating methyl halides (39,40). In non-adapted *E. coli* there are approximately 2 Ada protein molecules per cell (41,42), and consequently, direct alkylation in a non-induced state is an unlikely event. However, it is thought that low levels of O-alkylations can upregulate Ada protein to a point at which direct alkylation becomes significant for further upregulation of the response. An adaptive response to methyl halides may be an important evolutionary response, given the environmental abundance of these chemicals and the bacteria that produce them.

In the response to alkyl damage, it is important to question what form of damage is being sensed. For induction of the adaptive response to alkylating agents, the DNA damage can be sensed in two different ways: for activation of *alkA*, Ada must sense adducts in the sugar phosphate backbone (MePTs), but for the induction of *ada/alkB* and *aidB*, Ada must also sense methylation at exocyclic base oxygens (O^{6}MeG and O^{4}MeT). Alternatively, as with activation by the methyl halides, the damaging agents themselves can activate the response independently of adducted DNA, by direct methylation of active site cysteine residues.

2.3. Genes Upregulated in the Adaptive Response

The DNA repair genes upregulated as part of the *E. coli* adaptive response, plus their orthologs and paralogs are discussed below in terms of substrate preferences for alkyl adducts.

2.3.1. DNA Repair Methyltransferases

The enhanced ability of adapted *E. coli* to remove O^{6}MeG from their genomes (16) ultimately led to the identification of Ada as the first DNA repair MTase (25). Non-adapted *E. coli* show constitutive levels of O^{6}MeG DNA MTase activity that is produced by the *ogt* gene (43,44). Ogt has considerable sequence similarity with CAda19, and has the ability to suppress both spontaneous and alkylation-induced mutations in bacteria (45). Subsequently, the *S. cerevisiae* homolog *MGT1* was cloned based on its ability to complement the alkylation sensitivity of Ada-deficient *E. coli* (46). The human homolog *MGMT* was identified by its ability to complement MTase-deficient cells and by screening cDNAs with an MTase-specific probe (47–49). Like Ada and Ogt in *E. coli*, Mgt1 suppresses spontaneous and alkylation-induced mutation in yeast, and MGMT appears to suppress C→T transitions (but not the overall mutation rate) in mammalian cells (50,51). MTases have been identified in several bacterial species and across biological kingdoms (52). These homologs have in common the absolutely conserved Pro-Cis-His-Arg sequence containing the cysteine acceptor, as well as significant conservation in a DNA recognition helix involved in "flipping" adducted nucleotides out of the helix (28).

S_N1 methylating agents will produce the O^6MeG adduct at 10- to 100-fold higher levels than O^4MeT (5,53). However, O^4MeT is also mutagenic because of its capacity to mispair with G during replication. Although the relative abundance of the O^4MeT adduct is low, its repair is likely to be biologically important (54). With regard to substrate preferences, Ada can repair both O^6MeG and O^4MeT, but it has significantly greater affinity for O^6MeG (55). The constitutively expressed Ogt will also repair both adducts, but in contrast to Ada, it actually has a slightly greater affinity for O^4MeT than O^6MeG. The yeast and mammalian homologs Mgt1 and MGMT, can act on O^4MeT but both have higher affinities for O^6MeG (55).

2.3.2. Methylpurine Glycosylases and BER

Glycosylases that repair alkylated bases are represented by different subfamilies of enzymes, which may have overlapping substrate specificities (56). In *E. coli* the alkyladenine DNA glycosylase, AlkA, is upregulated in the adaptive response and preferentially recognizes alkylated and deaminated bases (Table 2). AlkA actually removes 3MeG, a less common alkyl lesion, slightly better than 3MeA (57). Tag, a constitutively expressed glycosylase, on the other hand has only been shown to repair 3-methylpurines with a strong preference for 3MeA over 3MeG. Although AlkA and Tag have overlapping substrate specificities, they are in separate subfamilies within the helix-hairpin-helix (HhH) glycosylase superfamily (58). The methyl purine glycosylase in *S. cerevisiae* (Mag1) is related to AlkA (59,60), but the only known methyl purine glycosylase in mammals (AAG, also known as MPG or ANPG) is not an HhH glycosylase (61,62). Although these enzymes represent different classes of glycosylases, AAG can be considered a functional analog of AlkA and Mag1 based on their similar, broad substrate ranges (Table 2) (56).

Understanding the complete repair of *N*-methylpurines requires some explanation of the BER process. The BER is initiated by recognition of relatively small base lesions in DNA by a variety of DNA glycosylases. This enzyme family as a whole recognizes a variety of damaged bases resulting from deamination, alkylation, or oxidation, as well misincorporated base analogs (63). Glycosylases recognize structural distortions in the helix, and similar to the MTases, gain better access to the glycosylic bond by "flipping" the nucleotide with the target base out of the helix. Glycosylases hydrolyze the bond between the base and the sugar leaving an abasic (or apurinic/apyrimidinic, AP) site with the backbone structure otherwise intact. An AP endonuclease catalyzes nick formation 5′ to the AP site, at which point short- or long-patch repair ensues. In short-patch repair, Polβ extends a single base from the 3′ OH with its polymerase activity, and removes the 5′ sugar phosphate with

Table 2 Biochemical Activity of 3MeA Glycosylases Toward Other Modified or Normal Bases

	Tag	AlkA	Magl	AAG	Aag
3MeA	+	+	+	+	+
3MeG	+	+	+	+	+
7MeG	−	+	+	+	+
O^2MeT	−	+			
O^2MeC	−	+			
7CeG		+	+		
7HEG		+	+		
7EthoxyG		+			
εA	−	+	+	+	+
εG				+	
8oxoG				|	+
Hx	−	+	+	+	+
A		+			+
G	−	+	+	+	+
T		+			
C		+			

"+" denotes any measurable activity, and "−" denotes no measurable activity. Tag and AlkA are from *E. coli*, Magl from *S. cerevisiae*, AAG from human, and Aag from mouse.
Abbreviations: 3MeA, 3-methyladenine; 3MeG, 3-methylguanine; 7MeG, 7-methylguanine, O^2MeT, O^2-methylthymine; O^2MeC, O^2-methylcytosine; 7CeG, 7-chloroethylguanine; 7HEG, 7-hydroxyethylguanine; εA, ethenoadenine; εG, ethenoguanine; 8oxoG, 8-oxoguanine; Hx, hypoxanthine

its deoxyribosephosphate (dRP) lyase activity located at a separate active site. XRCC1 then facilitates an interaction between Polβ and DNA ligase III to seal the nick. In long-patch repair, Polδ or ε extend from the 3′OH displacing 2–6 nucleotides beyond the 5′ sugar phosphate, with the assistance of RFC and PCNA; the resulting flap is resolved by the FEN1 endonuclease and DNA ligase I. It is not clear how the decision is made to undergo short- or long-patch repair, but it may be dependent on the ability of the lyase to remove the 5′ dRP moiety (64).

2.4. Oxidative Demethylases

The induction of AlkB as part of the adaptive response and its contribution to alkylation resistance were known for several years before the activity of this enzyme was finally characterized (11,65). *E. coli alkB* mutants are extremely sensitive to the cytotoxic effects of alkylating agents, and this hypersensitivity is significantly greater for S_N2 agents than S_N1 agents (66). Clues to the activity of AlkB came from studies on the ability of wild-type and *alkB* mutant *E. coli* to reactivate alkylated phage DNA. Compared to wild-type *E. coli, alkB* mutants are defective in the reactivation MMS-treated single-stranded phages (M13, f1, G4), but the defect is not nearly as pronounced for alkylated duplex M13 or lambda (67). Subsequently, AlkB was shown to dealkylate 1MeA and 3MeC via an oxidative demethylation reaction (65,68). Although repair of these lesions occurs in both single and double-stranded DNA (68), these lesions are generated predominantly in single-stranded DNA by S_N2 alkylation. The mechanism by which AlkB dealkylates adducts was forecasted by similarities in the predicted AlkB protein structure to Fe^{++}- containing oxygenases

(69). AlkB homologs have now been identified in humans but to date no homologous genes have been found in yeast. The human gene products ABH2 and ABH3 (for AlkB homologs) have the same substrates as AlkB and also can dealkylate 1-ethyladenine (70). Interestingly, ABH3 and AlkB (but not ABH2) can dealkylate RNA in addition to DNA (71). Five additional putative AlkB homologs were recently identified in the human genome, and their functional characterization is hopefully forthcoming (72).

1MeA and 3MeC are predicted to present major blocks to DNA replication, and the importance of these lesions in cellular toxicity is clearly apparent in *E. coli*. Human cells deficient in AlkB homologs have not been described, but the importance of these lesions in cytotoxicity can be inferred from the fact that expression of the *E. coli* AlkB protein in human cells confers resistance to S_N2 alkylating agents (73). Whether the unrepaired AlkB substrate adducts in DNA or RNA can initiate cellular signals is not known, but the recent cloning and characterization of the ABH homologs can now be exploited to examine the effects of 1MeA and 3MeC in RNA and DNA on mutagenesis, toxicity, apoptosis, and cell signaling.

2.5. The Adaptive Response in Eukaryotes

The adaptive response is well conserved across several bacterial species (74), but in eukaryotes the story is more complicated. Some studies showed that when mammalian cells are pretreated with non-cytotoxic doses of alkylating agents there is no effect on subsequent alkylation-induced survival or mutagenesis (75,76). In contrast, other studies showed that chronic sublethal pretreatment with alkylating agents results in resistance to their mutagenic, cytotoxic, and clastogenic effects of subsequent high dose exposures (77–80). These differences are likely dependent on cell type. In a rat hepatoma line, resistance appears to result in part from the induction of MTase activity, but unlike the prokaryotic response there is no induction of DNA glycosylase activity (80,81). Also MTase activity can be induced by a variety of DNA damaging agents and is not specific to alkylating agents (82) ruling out the possibility that methylated MTases serve as the sensor and activator of the response, as in *E. coli*. Moreover, there has been no discovery of a mammalian MTase that repairs MePTs, and thus, any transcriptional upregulation in response to alkylating agents in eukaryotic cells, necessarily differs mechanistically from the adaptive response in *E. coli*.

Despite the fact that the mammalian adaptive response clearly differs from that in *E. coli*, the use of a MTase repair protein to generate a cellular signal upon alkylation may indeed occur. In a recent report Teo et al. (83) showed that direct alkylation of MGMT by methyl iodide or by O^6-benzylguanine stimulates an interaction with the estrogen receptor (ER). This physical interaction was associated with an inhibition of estrogen-dependent mitotic stimulation of ER-positive breast cancer cell lines. Alkylation of MGMT causes a structural rearrangement (84) which probably allows ER to interact with an LXXLL motif in MGMT. Mutations in this domain abrogate the ER/MGMT interaction, and ER is known to contact this same motif in its usual partner, the transcriptional enhancer SRC-1. The structural rearrangement that occurs upon alkylation of MGMT was shown to result in ubiquitin-mediated degradation of the protein (85), but it is possible that the interaction between MGMT and ER stabilizes MGMT. This interaction may serve as a model for cell signaling and transcriptional regulation in response to alkylation damage in mammalian cells.

3. CELLULAR RESPONSES TO O^6MeG

3.1. Biochemical Basis for Mutagenicity and Toxicity

The mutagenic capacity of O^6MeG lies in its ability to direct the misincorporation of thymine during replication (86–88). Although cytosine is still the best partner for O^6MeG with regard to Watson–Crick base pairing, pairing with either C or T both cause structural distortions in the helix (89,90). The frequency of inserting a T opposite this adducted base can be 7-fold higher than insertion of C, and the difference is dependent on sequence context (91). Furthermore, O^6MeG does not cause an overt block to in vitro replication using E. coli Pol I or AMV RT (92), but it can slow the rate of polymerization (93). Thus, it is possible that the evolution of responses (sometimes drastic) to this lesion have occurred to prevent GC→AT transitions. The cytotoxic cellular responses to O^6MeG turn out to be dependent upon the ability of DNA mismatch repair (MMR) complexes to recognize base pairs containing these adducts.

3.2. Background on MMR

The major function of MMR is to correct replication errors that escape proofreading at the replication fork. In bacteria, MMR is carried out by the gene products MutS, MutL, MutH, DNA helicase II, a series of exonucleases, ss DNA binding protein, and DNA Pol III (94). MutS recognizes a base mispair and recruits MutL and MutH. Repair correction occurs in the daughter strand of replication, and MutH recognizes this strand by linking the mispair with an adjacent hemi-methylated GATC sequence. A methyl adenine in the parental strand at the GATC site serves as a signal for MutH to create a nick in the unmethylated daughter strand. GATC methylation is catalyzed by the Dam methylase, and such methylation is absent in *dam⁻* E. coli. The daughter strand is degraded from the nicked GATC sequence toward and past the mispair, with the aid of the MutH helicase and the appropriate exonuclease. The resulting gap is filled in by DNA Pol III and sealed by a DNA ligase.

Human MMR is initiated by MutS homologs (MSH) and MutL homologs (MLH and PMS2). MSH2 heterodimerizes with MSH6 forming the MutSα complex, or with MSH3, forming the MutSβ complex. MutSα binds mispaired bases and +1/ −1 insertion/deletions, whereas the MutSβ heterodimer binds longer extrahelical loops. Upon recognition of the mispaired or misaligned sequences these complexes recruit a MLH1/PMS2 heterodimer, termed MutLα to the site of damage. From a nick that is either 5′ or 3′ of the mispair, exonuclease I (EXOI), with its 5′→3′ or 3′→5′ exonuclease activity, degrades towards and past the mispaired sequence; the gap is subsequently filled in by pol δ or ε and sealed by a DNA ligase. How the daughter strand is discriminated in mammalian cells is not known, but it probably results from an intimate association of MMR with the replication machinery such that the daughter strand is distinguishable (95).

3.3. MMR Recognition of Methylated Base Pairs

The interaction of MutSα with an O^6MeG-containing base pair was demonstrated by gel shifting oligonucleotides containing basepairs with this adduct. Both O^6MeG:C and O^6MeG:T show significantly greater affinity to MutSα than a G:C pair, with the O^6MeG:T oligo having slightly greater affinity than the O^6MeG:C-containing oligo (96,97). However, the oligos containing O^6MeG were not nearly as preferred as the oligo containing an unmodified G:T mispair (96–98). The binding of MutSα to a mispair activates its inherent ATPase activity, and this activity was shown to correlate perfectly with binding affinities for the O^6-methyl adducts (96). It is not surprising that MutSα would recognize both O^6MeG:C and O^6MeG:T pairs, given a model for mismatch recognition where MutSα tracks the double helix and is able to kink the DNA when there is a weakness in the duplex (99,100); O^6MeG-containing base pairs would create such a weakness (89,90). However, in an in vitro study of MMR activity in cell extracts demonstrated that while O^6MeG:T-containing oligos were efficiently repaired, the activity towards O^6MeG:C-containing oligos was undetectable (101). That MMR is not elicited at O^6MeG:C base pairs, at least in vitro is an important point for the models of cell signaling presented later in Sec. 4.5.

3.4. Biological Effects on Mutagenicity and Toxicity in *E. coli*

In general, the relative contributions of specific methyl adducts to cellular toxicity and mutagenesis have been examined using two complementary methodologies. The comparison of S_N1 to S_N2 alkylating agents is informative in that S_N1 agents produce significant *O* alkylations while S_N2 agents do not. However, as both agents produce a variety of overlapping *N*-alkylations, this approach has its limitations. A complementary approach is to compare toxicity and mutagenicity in isogenic strains with or without specific repair capacities. Using these approaches, early studies led to the designation of 3MeA as a cytotoxic lesion and O^6MeG as a mutagenic lesion. In agreement with this distinction, S_N1 agents are significantly more mutagenic than S_N2 agents, producing mostly GC→AT transitions (53). Furthermore AlkA protects against the cytotoxic effects of MNNG, whereas MTases protect against mutagenic effects. However, the generalization of O^6MeG as simply mutagenic in *E. coli* is an oversimplification, and its toxic effects are apparent. An *ogt* deletion enhances the toxicity of MNNG on an *ada⁻* but not an *ada⁺* background, and overexpression of *ogt* or *ada* confers resistance to MNNG (45). As with mammalian cells, the level of cytotoxicity caused by this lesion is determined largely by MMR function.

While, MNNG causes cytotoxicity in bacteria, the level of its effect is grossly modulated by MMR function (102). In *dam⁻* bacteria DNA mismatches are recognized, but their processing is defective due to the lack of daughter strand discrimination at hemimethylated GATC sites. These strains are extraordinarily sensitive to MNNG-induced toxicity but not mutagenesis. Furthermore, the cytotoxic effect is rescued by mutations in either *mutL* or *mutS*, again with no effect on mutagenesis (102). Altogether, these results suggested that the toxic effect of MNNG resulted from processing of O^6MeG by MMR. The detailed model proposed that the MMR pathway excises T opposite O^6MeG. Subsequently, DNA polymerase III reinserts T opposite the adducted base, and the process begins a new. It was originally proposed that in wild-type *E. coli,* the process continues until de novo methylation occurs on the daughter strand. This methylation would inhibit mismatch correction, but in *dam⁻* strains the signal would never be generated causing toxicity through continued repeated rounds of "futile repair". A later modification to the model was made when it was shown that MutH will nick both strands at a GATC in the absence of a strand-discriminating methyl A signal (103). This model better explains the extreme toxicity in *E. coli* because indiscriminate mismatch correction would directly cause lethal double strand breaks (DSBs). The general model of "futile repair" was later adopted to explain MMR-dependent toxicity in mammalian cells (104).

3.5. Mutagenicity and Toxicity in Human Cells

The elucidation of the cytotoxic effects of O^6MeG was aided by the identification of human cell lines that lacked the ability to repair this adduct. Human tumor and SV40-transformed cell lines that were unable to support the growth of MNNG-treated adenovirus were designated Mer⁻ (for methyl repair) (105), and concurrently characterized lymphoblastoid cell lines incapable of removing O^6MeG from their DNA were designated Mex⁻ (for methyl excision) (106). The Mer⁻ and Mex⁻ phenotypes are the same, and the underlying cause is a deficiency in MGMT that dramatically enhances the killing effects of S_N1 alkylating agents. In fact, complementation of the Mer⁻ phenotype led to the eventual cloning of *MGMT* (47), but it was expression of the bacterial Ada protein in human cells that convicted O^6MeG as a major cytotoxic lesion. Expression of Ada confers resistance to killing effects of alkylating agents and completely corrects the Mer⁻ phenotype (107–109).

Growing Mer⁻ or Mex⁻ cells in MNNG led to the selection of resistant clones, yet these revertants in most cases did not display a correction of the MTase deficiency, i.e., they had not acquired increased ability to remove methyl lesions from DNA, they nonetheless acquired resistance to the killing effects of alkylation (104,110). Because this resistance was not associated with increased DNA repair, the more appropriate designation of alkylation "tolerance" was used. Thus, cells sustain just as much *O*-alkyl damage, the damage is still mutagenic, but some other response (or lack of a response) to the damage determines the absence of lethality of this adduct. It was later shown that loss of DNA mismatch repair represents the major pathway through which MTase-deficient cells acquire alkylation tolerance (111–113).

3.5.1. MMR-Dependent Responses

The MMR complexes mediate several cellular responses to alkylating agents including cell cycle arrest and apoptosis (113). Exposure of mammalian cells to S_N1 alkylating agents can cause an arrest at the first G2/M transition post-treatment (104,114–117). The induction of this response is associated with an increase in p53 and cdc2p levels, even though the arrest does not appear to be p53 dependent (115). Of relevance here is the fact that cell cycle arrest and apoptosis are dependent on the presence of functional MMR.

MTase-deficient cells that acquire tolerance through loss of MMR incur less chromosome damage in response to S_N1 alkylating agents. Treatment of synchronized cells with MNNG or MMS causes strand breaks in the nascent strand that present at the first mitosis after treatment, even though the alkylation damage is distributed on both DNA strands (118). It was suggested that the greater frequency of breaks that occur with MNNG compared to MMS results from O^6MeG lesions. In HeLa cells, S_N1 alkylating agents cause chromosome damage that is apparent in the second cell cycle after treatment (119). This damage results from O^6MeG as inferred from the fact that MTase-deficient cells are more sensitive to S_N1 alkylation-induced SCEs (120), and that correction of the MTase deficiency with Ada expression significantly reduces the number of SCEs (109). Similar studies have shown a correlation of the MTase deficiency with sensitivity to alkylation-induced SCEs, where the acquisition of tolerance did not affect the frequency of these structures (110,121). However, in many cases where tolerance was shown to be caused by loss of mismatch repair, the sensitivity to alkylation-induced SCEs and chromosome aberrations was dramatically reduced (122–124). The appearance of SCEs and chromosome aberrations at the second S-phase after treatment with S_N1 alkylating agents differs from the timecourse for generating chromosome damage by other agents such as UV, gamma irradiation where this damage is apparent after the first post-treatment S-phase. Interestingly, loss of MMR dramatically reduces chromosome aberrations generated in the second S-phase, but has little effect on the frequency of aberrations produced in the first S-phase (123).

Human lymphoblastoid cells treated with MNNG display an increase in cells with a sub G1 DNA content (125). Later, it was shown that this response is truly apoptosis and is caused mostly by O^6MeG (126,127). Furthermore, like the G1/M arrest and the rise in SCEs, the apoptotic response to O^6MeG is mediated by MMR (128). The physical recognition of O^6MeG adducts by a MMR complex, and the models for how this interaction initiates cellular signaling cascades are presented below.

The presence of O^6MeG in the genome is not deterministic, i.e., some cells will undergo apoptosis, while other cells persist with mutations or chromosomal damage (120). In general, the MMR-dependent responses to O^6MeG can be summarized as follows. First, cells undergo an arrest at the first G2/M boundary. As cells overcome this boundary and enter the next S-phase there is a dramatic increase in sister chromatid exchanges. Concurrent with the second replication cycle, a fraction of cells undergo apoptosis. Models for the generation of apoptotic and checkpoint signals are presented below, with regard to this particular sequence of events.

3.5.2. Models for Activating Cellular Signals

One model posits that MMR proteins interact with the O^6MeG-containing base pairs, and upon binding directly generate apoptotic and checkpoint signals (Fig. 3). Mismatch recognition causes a conformational change in MutSα, and such changes may confer an ability to interact with downstream effector proteins (129). The ability of MMR to facilitate the activity of a known checkpoint effector, ATM, provides some indirect support of this model (130). However, the most compelling evidence for a direct signaling model was recently provided by studies employing a point mutant of Msh2 in which mismatch recognition is genetically separated from repair activity (131). The mouse *Msh2* gene was replaced with a variant coding protein that lacks ATPase activity. This mutant protein is capable of recognizing mismatches in DNA but cannot elicit repair of the mismatch. Nonetheless, mutant cells retained the ability to induce apoptosis in response to a variety of DNA damaging agents including MNNG. These studies did not examine effects on SCEs and chromosome aberrations, but nonetheless, the results strongly argue that at least part of the apoptotic signal is generated upon direct recognition of an O^6MeG-containing base pair.

Other models for MMR-dependent initiation of checkpoint and apoptotic signals parallel the futile repair model described above for *E. coli* (Fig. 3) (102,104). In this model, T is inserted opposite O^6MeG in the first S-phase after treatment. Then, the repeated excision/incorporation of T opposite O^6MeG leads to persistent single-strand gaps, some of which may persist as cells exit the S-phase. As replication forks encounter these gaps in the subsequent replication cycle, toxic and recombinogenic DSBs are generated (120,124). This model explains the increased SCEs and gross chromosomal rearrangements that are witnessed at the second S-phase after treatment, something that the direct signaling model simply does not account for. Of course, the two models are not mutually exclusive.

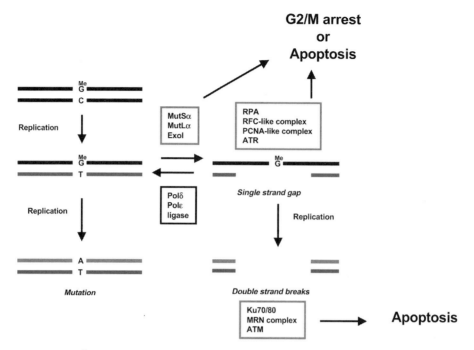

Figure 3 O^6MeG in the genome can lead to mutations and chromosomal aberrations that likely result from single-strand gaps or DSBs. All the protein complexes are boxed, and MMR complexes of MutSα, MutLα, and ExoI (boxed red) may initiate cellular signals upon direct binding of O^6MeG. Single-strand gaps and DSBs are secondary structures that result from O6MeG and are recognized protein complexes (boxed green) that signal through the ATM or ATR kinases. These secondary structures may be important for signaling regulatory endpoints.

The incorporation of T opposite O^6MeG is the primary event in initiating a MMR-dependent response. Although one study showed no detectable MMR activity on O^6MeG:C pairs (101), other studies showed an interaction between MutS or MutSα and O^6MeG:C-containing oligos (96,132). If MMR acted on O^6MeG:C pairs then one might expect the appearance of SCEs after the first replication cycle as forks encounter excision tracts. This prediction is clearly inconsistent with the observed results (123). However, if MMR complexes trailed the replication fork then recognition of O^6MeG-containing base pairs would not occur until after replication.(129). In such a circumstance, recognition of the O^6MeG:T pair (or less the less frequent O^6MeG:C pair) would still not occur until after replication. The trailing of polymerases by mismatch repair complexes has also been proposed to explain strand discrimination in mammalian cells (95).

Subsequent to the futile cycle of excision and synthesis, checkpoint and apoptotic signals could be generated by one or both of the following: (i) single-strand gaps that emerge from the first replication cycle; (ii) DSBs generated as replication forks encounter repair tracts (133). Single-strand gaps that result from MMR intermediates may be recognized by RPA, PCNA-like, and RFC-like complexes (134). These complexes initiate checkpoint responses in yeast and mammals by recruiting the ATR kinase. Alternatively, DSBs are potent activators of cell cycle checkpoints and apoptosis, and can initiate this signal through the ATM kinase.

Activation through these multiple pathways are not mutually exclusive and perhaps all are required for the full complement of biological responses. In summary, the primary lesion O^6MeG is "sensed" by the MMR complex, and the ensuing cellular responses may result from direct signaling upon recognition of the lesion, or downstream signaling upon recognition of single-strand gaps and DSBs.

4. CELLULAR RESPONSES TO 3MeA

3MeA can be a highly toxic lesion in bacteria, yeast, and mammalian cells (8). This specific adduct has been shown to cause S-phase arrest and apoptosis in some mammalian cells (Fig. 4), but appears to be innocuous in other cells. The complex responses to 3MeA can result directly from the lesion or indirectly from intermediates in the base excision repair process.

4.1. Biochemical Basis for Toxicity

After 7MeG, 3MeA is the most common adduct formed by S_N1 and S_N2 alkylating agents (Table 1) (5). When double-stranded DNA is treated with either S_N1 or S_N2 alkylating agents, 3MeA is the major replication blocking lesion for bacterial and viral polymerases (92). Modification at the N3 position of purines is known to dramatically retard DNA replication. The N3 of purines is a crucial hydrogen bond

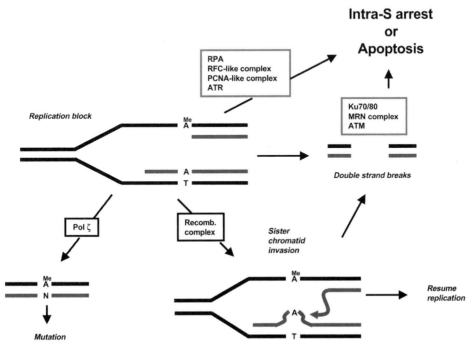

Figure 4 3MeA in the genome presents a strong block to replication, and this replication block can result in mutations, recombination, single-strand gaps, and DSBs. The protein complexes that directly or indirectly respond to 3MeA are boxed. Those complexes (boxed green) that signal regulatory endpoints through the ATM or ATR kinases likely recognize DNA structures that result a block in replication. (*See color insert.*)

acceptor for the highly conserved arginine in Pol I family polymerases (135–138). The absolute requirement for this crucial interaction for efficient polymerase activity explains how a relatively minor modification can have such drastic consequences for DNA replication. Moreover, in light of this, it is not surprising that all organisms have glycosylases for repairing this lesion.

4.2. Effect of 3MeA on Toxicity and Mutagenesis

4.2.1. Direct Responses to 3MeA

The only *E. coli* glycosylases known to repair 3MeA are Tag, which has a strong specificity for 3MeA, and AlkA, which repairs 3MeA plus a variety of other lesions. Deletion of either *tag* or *alkA* causes modest alkylation sensitivity, while deletion of both genes causes dramatic sensitivity (18,139). Tag can repair both 3MeA and 3MeG. However, because 3MeA is the preferred substrate, and because 3MeG is an extremely rare adduct following S_N2 alkylation, these studies strongly suggest that 3MeA is the major toxic lesion. Similarly, deletion mutants of *Mag1*, the only glycosylase known to repair 3MeA in *S. cerevisiae*, is highly sensitive to MMS. On the other hand, deletion of the only known 3MeA glycosylase in mice, Aag, produces an unpredictable phenotype; some *Aag* null cells become alkylation sensitive, some have no phenotype, and others become alkylation resistant (see below).

   Removal of 3MeA by DNA glycosylases is the major mechanism for repairing 3MeA. However, NER, homologous recombination, and translesion synthesis represent other pathways that act on this lesion. Genetic evidence in *S. cerevisiae* suggests that this lesion can be repaired by NER (140). On a *MAG1*-deficient background, deletion of the NER gene *RAD4*, enhances MMS-induced cytotoxicity and recombination, and both endpoints are reduced to the wild-type level by overexpressing Tag to remove specifically 3Me purine lesions. Also, toxicity and recombination rates are enhanced by deleting the translesion polymerase, Pol ζ, with a *rev3* mutation, suggesting that functional translesion polymerases serve to reduce the toxic effects of 3MeA and prevent recombination. In other words, while 3MeA can block the replicative polymerase to cause cell death, the Pol ζ translesion bypass polymerase can rescue the replication block and prevent cytotoxicity. However, Pol ζ would be expected to enhance the mutagenic effects of 3MeA, analogous to the tolerance phenotype in response to O^6MeG. Indeed, *E. coli tag* mutants are sensitive to alkylation-induced mutagenesis only when the SOS response is induced, a condition in *E. coli* wherein the translesion polymerases are upregulated (141). While BER represents the major pathway for repairing 3MeA, in BER-deficient cells, the NER, recombination, and translesion synthesis pathways can also respond to the primary lesion (Fig. 4).

4.2.2. Cellular Responses to DNA Repair Intermediates

Abasic sites, nicked DNA, and short single-strand gaps represent intermediates in the BER process that are known to have toxic consequences. Similar to 3MeA, abasic sites present strong blocks to in vitro replication, and can be mutagenic due to the activity of translesion bypass polymerases (92). The effect of abasic sites has been studied in vivo by altering the relative expression of DNA glycosylases and AP endonucleases. AP endonuclease-deficient bacteria and yeast have high spontaneous and DNA damage-induced mutation rates, demonstrating the detrimental effects of unprocessed abasic sites (142–144). Deletion of *MAG1* in AP-endonuclease-deficient yeast suppresses this mutagenicity, but the importance of its activity on 3MeA in this circumstance is not clear because Mag1 can also remove a variety of other alkylated bases, as well as deaminated and, at a low rate, normal bases (51,144,145). The corollary of this is that overexpression of Mag1 causes an increase in spontaneous mutagenesis even in a wild-type background, and this mutator phenotype is dramatically enhanced on an AP endonuclease-deficient background. The Mag1-enhanced spontaneous mutation rate is suppressed by overexpressing AP endonuclease, demonstrating that the generation of abasic sites by Mag1 is required for its mutator effects (146–148). In contrast to AAG and Mag1, overexpression of Tag does not significantly alter the mutation rate in AP endonuclease-deficient bacteria or yeast (146). Thus, because Tag is specific for 3Me purines, the abasic sites that result from 3MeA removal probably do not contribute significantly to spontaneous mutation; exactly which Mag1 substrates do contribute is not yet clear. Expression of the human 3MeA glycosylase, AAG, also confers a mutator phenotype in wild-type yeast, and part of the mutational spectrum is an enhancement of $+1/-1$ insertion/deletions (149). This spectrum may have a human parallel in that high protein levels of AAG are associated with microsatellite instability in chronically inflamed regions in the colon. Indeed, increased AAG expression in the human lymphoblastoid cell line, K562, induces a modest microsatellite instability phenotype.

The cytotoxicity of 3MeA has been examined in mammalian cells in the following ways: (i) by targeted disruption of the mouse *Aag* gene, which revealed that Aag is perhaps the only 3MeA glycosylase and that the absence of Aag can create alkylation sensitivity; and (ii) by the fact that MeLex, which generates almost exclusively 3MeA DNA lesions ($\gg$90% of the adducts in DNA) is cytotoxic in a variety of mammalian cell types (150,151). Embryonic stem cells deficient in Aag ($Aag-/-$) are sensitive to MMS and MeLex, and the difference in toxicity between $Aag-/-$ and $Aag+/+$ is greater for the MeLex compound. In embryonic stem cells, it is apparent that 3MeA is the major cytotoxic lesion, and cell death is accompanied or preceded by a production of SCEs and gross chromosomal aberrations. This result is consistent with the recombinogenic effects of this lesion in yeast (140). However, the Aag deficiency has little effect on the alkylation sensitivity in mouse embryo fibroblasts presumably because 3MeA lesions are efficiently bypassed in these cells. Finally, the Aag deficiency actually confers alkylation resistance to mouse bone marrow cells (BMCs) (152). In addition to MMS, *Aag-/-* BMCs were also resistant to MeLex suggesting that an intermediate in the repair of 3MeA is responsible for toxicity in the wild-type cells and that even mild levels of Aag can be detrimental to certain cells at times of alkylation stress (152).

4.3. Initiation of Checkpoints and Apoptosis

S_N2 alkylating agents cause S-phase arrest and apoptosis in ES cells and an intra-S-phase checkpoint in *S. cerevisiae*. Similar to the response to O^6MeG, we can consider the following mechanisms for initiating a signal: (i) direct recognition of the adducted base; (ii) recognition of repair intermediates; or (iii) direct inhibition of DNA replication. The preponderance of evidence suggests that 3MeA generates cellular signals not through direct recognition or processing to BER intermediates. Rather, these cellular signals are likely generated by 3MeA in its capacity to block replication (Fig. 4).

4.3.1. Responses to 3MeA in ES cells

In ES cells, S-phase arrest and apoptosis occur in response to MeLex, and the effects are more pronounced in *Aag*−/− than wild-type cells. Therefore, arrest and apoptosis most likely result from 3MeA lesions (7). As Aag is the only known mammalian glycosylase that repairs 3MeA, recognition of this lesion by a glycosylase and its subsequent processing to repair intermediates are not required to generate checkpoint and apoptotic signals, i.e., the 3MeA lesion itself causes the cellular responses in these cells. Cell cycle arrest and apoptosis that are induced by this lesion probably result directly from its capacity to block to DNA replication, but the specific mechanisms have not yet been examined in detail. Indeed, it is formally possible that blocking RNA polymerases by 3MeA may signal cell cycle arrest and apoptosis. MMS-induced S-phase arrest has been examined in *Xenopus* extracts and in detail in *S. cerevisiae*. While these responses have not been shown to directly result from 3MeA lesions, they are informative as to the process of arrest.

4.3.2. Response to MMS in Yeast

S. cerevisiae treated with MMS will arrest in the first S-phase after treatment, and this intra-S checkpoint is dependent on the checkpoint kinase proteins Mec1 and Rad53 (The yeast homologs of the mammalian proteins ATR and CHK2.) € see. also Chapter 37). When either of these proteins are inactive, cells treated with MMS will progress through S-phase at a normal rate as if no DNA damage were present (153,154). Furthermore, S-phase progression continues to appear normal even when the *Mec1* mutation is combined with a *Mag1* mutation, where 3MeA adducts persist (Samson, unpublished). These results demonstrate that while 3MeA blocks replication in vitro, it does not prolong S-phase in checkpoint mutants (153,154). Thus, the intra-S checkpoint is dependent on the generation of a cellular signal and does not result solely from the physical block to replication at 3MeA DNA lesions. Studies using *Xenopus* extracts support these conclusions, as MMS treatment of extracts results in a diffusible signal that inhibits the recruitment of PCNA to a replication fork (155,156). Mec1 and Rad53 activate a variety of "effector" proteins that inhibit replication at preformed replication forks, stabilize these forks, and inhibit firing from late origins of replication (134,157). Furthermore, the generation of a Mec1/Rad53-dependent signal requires the initiation of S-phase. Only after replication has begun are these checkpoint proteins activated (157).

The physical block to DNA replication by a 3MeA lesion does not directly alter the duration of S-phase in checkpoint mutants, but it is a likely candidate for activating the checkpoint signal. The processing of adducted bases to abasic sites is not required for S-phase arrest because *Mag1* deletion mutants fully activate the checkpoint (154). This result is consistent with other studies suggesting that the extent of single-stranded DNA generated in a BER intermediate is insufficient to activate a checkpoint through binding by RPA (158). However, this result does not mean that abasic sites are insignificant in wild-type yeast, as a variety of blocks to replication will activate the intra-S checkpoint (159). A discussion of the mechanisms by which replication blocks activate Mec1 is beyond the scope of this chapter, but it may have to do with recognition of large tracts of single-stranded DNA by RPA, RFC-like, and PCNA-like factors, or a signal generated by Pol ε (134) (see Chapters 36, 37).

It is possible that the same factors responsible for the intra S-phase checkpoint are responsible for part of the apoptotic signal in mammalian cells. The mammalian homolog of Mec1, ATR, can phosphorylate p53, and p53 is stabilized in response to

3MeA in ES cells (7). Alternatively, an apoptotic signal may be generated by the production of DSBs and a subsequent ATM-dependent response. DSBs could be generated in the first S-phase by nicking of single-strand gaps after breakdown of a replication fork or by the processing of recombination intermediates. DSBs could also be generated in the second S-phase after treatment, as replication forks encounter regions of single-stranded DNA.

5. GENOME-WIDE ANALYSIS OF RESPONSES TO ALKYLATING AGENTS

Several genome-wide techniques are currently available for querying global responses to alkylation damage. Using *S. cerevisiae* as a model organism, attempts have been made to discover novel genes and pathways that respond to alkylation damage using two powerful, genomic-scale techniques. One of these techniques involves microarray hybridization to measure gene expression changes, and the other technique involves high-throughput phenotypic analysis of single gene deletion mutants.

Gene expression changes have been measured in yeast in response to MMS (160–162). In one study it was found that ∼325 of the ∼6200 genes in the yeast genome are upregulated over 4-fold after a low dose treatment for 1 hr (160). Longer treatments with higher doses activates ∼2000 genes (162). It was previously known that Mag1 expression is elevated in response to MMS, and transcriptional profiling identified Mag1 as well as other transcripts representing DNA repair genes upregulated after MMS treatment. Additionally, transcripts representing protein degradation genes were highly represented among MMS-upregulated genes, and this response probably reflects the importance of removing damaged proteins that are modified by alkylating agents. Interestingly, several MMS-responsive genes (e.g., the BER gene *MAG1*, the NER gene *RAD23*, and a proteasome subunit gene *PRE2*) are regulated by the proteasome-associated protein Rpn4, which likely binds to a URS2 sequence motif common in the promoters of these genes (162). Thus, the mechanisms for upregulating repair genes and proteolytic processing genes in response to MMS are similar. How the Rpn4 transcription factor is activated in response to MMS remains to be determined.

Using the adaptive response as an analogy, one might expect all organisms to activate "damage control" genes in response to alkylating agents in order to protect against the toxic and mutagenic effects of these compounds. It would follow that deletion mutants of these genes would render cells hypersensitive to alkylation damage. In order to identify genes and protein networks that are important for alkylation resistance, MMS-sensitivity has been assayed on a genome-wide scale using haploid libraries of single gene yeast deletion mutants (163,164). One study found that one-fourth of the yeast deletion mutants tested were hypersensitive to MMS (163). It might be expected that genes upregulated in response to MMS are correlated with deletion mutants of those genes being sensitive to damage, yet this correlation was not observed. That is, a gene being transcriptionally activated in response to MMS has no predictive value as to whether its deletion mutant is sensitized to the killing effects of MMS. However, there were certain DNA repair genes and proteasome-associated genes that did overlap in term of being upregulated by MMS and hypersensitive to MMS when deleted.

The importance of proteasome-associated networks in the response to MMS is one phenomenon that underscores an important difference between these genome-wide screens and the other studies that have been described in this chapter. That is, alkylating agents are able to attack a variety of cellular macromolecules (4), and the regulatory networks that respond to alkylation damage need not be associated with damage in DNA. In many cases it is difficult to determine the types of macromolecular damage that cause, for example, a particular transcriptional response. Nonetheless, by using specific DNA repair mutants one may be able to apply genomic scale technologies to identify global cellular responses that result from specific lesions in DNA.

6. CONCLUSIONS

Cells display a variety of responses to the presence of methylation damage in their DNA. The adaptive response in *E. coli* is a perfect example of a cellular response generated by direct recognition of a DNA adduct. While the adaptive response is not conserved in eukaryotes, increasingly DNA repair proteins are being found to interact with proteins involved in transcriptional regulation. These interactions may be important for transforming DNA damage into a transcriptional response. In contrast to the adaptive response and its analogous mechanisms, the MMR-dependent response to O^6MeG probably results from a combination of direct recognition of the adducted base as well as processing of this lesion to repair intermediates. Finally, the responses to 3MeA can be explained by the ability of this lesion to block replication. Monofunctional methylation damage may seem a limited class of damage within the broad category of alkylation damage. However, the regulatory responses described in this chapter are useful examples to explain cellular responses to a variety of damaging agents.

REFERENCES

1. Taverna P, Sedgwick B. J Bacteriol 1996; 178:5105–5111.
2. Marnett LJ, Plastaras JP. Trends Genet 2001; 17:214–221.
3. Esteller M, Herman JG. Oncogene 2004; 23:1–8.
4. Lawley PD, Thatcher CJ. J Biochem 1970; 116:693–707.
5. Beranek DT, Weis CC, Swenson DH. Carcinogenesis 1980; 1:595–606.
6. Bodell WJ, Singer B. Biochemistry 1979; 18:2860–2863.
7. Engelward BP, Allan JM, Dreslin AJ, Kelly JD, Wu MM, Gold B, Samson LD. J Biol Chem 1998; 273:5412–5418.
8. Fronza G, Gold B. J Cell Biochem 2004; 91:250–257.
9. Hix S, Augusto O. Chem Biol Interact 1999; 118:141–149.
10. Samson L, Cairns J. Nature 1977; 267:281–283.
11. Sedgwick B, Lindahl T. Oncogene 2002; 21:8886–8894.
12. Lindahl T, Sedgwick B, Sekiguchi M, Nakabeppu Y. Annu Rev Biochem 1988; 57: 133–157.
13. Landini P, Volkert MR. J Bacteriol 2000; 182:6543–6549.
14. Jeggo P, Defais M, Samson L, Schendel P. Mol Gen Genet 1978; 162:299–305.
15. Jeggo P, Defais TM, Samson L, Schendel P. Mol Gen Genet 1977; 157:1–9.
16. Schendel PF, Robins PE. Proc Natl Acad Sci USA 1978; 75:6017–6020.
17. Karran P, Hjelmgren T, Lindahl T. Nature 1982; 296:770–773.
18. Evensen G, Seeberg E. Nature 1982; 296:773–775.
19. Sedgwick B. Mol Gen Genet 1983; 191:466–472.

20. Demple B, Sedgwick B, Robins P, Totty N, Waterfield MD, Lindahl T. Proc Natl Acad Sci USA 1985; 82:2688–2692.
21. Landini P, Hajec LI, Volkert MR. J Bacteriol 1994; 176:6583–6589.
22. Sedgwick B, Robins P, Totty N, Lindahl T. J Biol Chem 1988; 263:4430–4443.
23. Hamblin MR, Potter BV. FEBS Lett 1985; 189:315–317.
24. McCarthy TV, Lindahl T. Nucleic Acids Res 1985; 13:2683–2698.
25. Teo I, Sedgwick B, Demple B, Li B, Lindahl T. EMBO J 1984; 3:2151–2157.
26. Takano K, Nakabeppu Y, Sekiguchi M. J Mol Biol 1988; 201:261–271.
27. Moore MH, Gulbis JM, Dodson EJ, Demple B, Moody PC. EMBO J 1994; 13:1495–1501.
28. Daniels DS, Tainer JA. Mutat Res 2000; 460:151–163.
29. He C, Verdine GL. Chem Biol 2002; 9:1297–1303.
30. Lin Y, Dotsch V, Wintner T, Peariso K, Myers LC, Penner-Hahn JE, Verdine GL, Wagner G. Biochemistry 2001; 40:4261–4271.
31. Sun LJ, Yim CK, Verdine GL. Biochemistry 2001; 40:11596–11603.
32. He C, Wei H, Verdine GL. J Am Chem Soc 2003; 125:1450–1451.
33. Teo I, Sedgwick B, Kilpatrick MW, McCarthy TV, Lindahl T. Cell 1986; 45:315–324.
34. Nakabeppu Y, Sekiguchi M. Proc Natl Acad Sci USA 1986; 83:6297–6301.
35. Taketomi A, Nakabeppu Y, Ihara K, Hart DJ, Furuichi M, Sekiguchi M. Mol Gen Genet 1996; 250:523–532.
36. Lemotte PK, Walker GC. J Bacteriol 1985; 161:888–895.
37. Landini P, Busby SJ. J Bacteriol 1999; 181:524–529.
38. Akimaru H, Sakumi K, Yoshikai T, Anai M, Sekiguchi M. J Mol Biol 1990; 216:261–273.
39. Takahashi K, Kawazoe Y. Mutat Res 1987; 180:163–169.
40. Vaughan P, Lindahl T, Sedgwick B. Mutat Res 1993; 293:249–257.
41. Rebeck GW, Smith CM, Goad DL, Samson L. J Bacteriol 1989; 171:4563–4568.
42. Vaughan P, Sedgwick B, Hall J, Gannon J, Lindahl T. Carcinogenesis 1991; 12:263–268.
43. Potter PM, Wilkinson MC, Fitton J, Carr FJ, Brennand J, Cooper DP, Margison GP. Nucleic Acids Res 1987; 15:9177–9193.
44. Rebeck GW, Coons S, Carroll P, Samson L. Proc Natl Acad Sci USA 1988; 85:3039–3043.
45. Rebeck GW, Samson L. J Bacteriol 1991; 173:2068–2076.
46. Xiao W, Derfler B, Chen J, Samson L. EMBO J 1991; 10:2179–2186.
47. Hayakawa H, Koike G, Sekiguchi M. J Mol Biol 1990; 213:739–747.
48. Tano K, Shiota S, Collier J, Foote RS, Mitra S. Proc Natl Acad Sci USA 1990; 87:686–690.
49. Rydberg B, Spurr N, Karran P. J Biol Chem 1990; 265:9563–9569.
50. Aquilina G, Biondo R, Dogliotti E, Meuth M, Bignami M. Cancer Res 1992; 52:6471–6475.
51. Xiao W, Samson L. Proc Natl Acad Sci USA 1993; 90:2117–2121.
52. Samson L. Mol Microbiol 1992; 6:825–831.
53. Richardson KK, Richardson FC, Crosby RM, Swenberg JA, Skopek TR. Proc Natl Acad Sci USA 1987; 84:344–348.
54. Preston BD, Singer B, Loeb LA. Proc Natl Acad Sci USA 1986; 83:8501–8505.
55. Sassanfar M, Dosanjh MK, Essigmann JM, Samson L. J Biol Chem 1991; 266:2767–2771.
56. Krokan HE, Standal R, Slupphaug G. Biochem J 1997; 325(Pt 1):1–16.
57. Bjelland S, Bjoras M, Seeberg E. Nucleic Acids Res 1993; 21:2045–2049.
58. Kwon K, Cao C, Stivers JT. J Biol Chem 2003; 278:19442–19446.
59. Bjoras M, Klungland A, Johansen RF, Seeberg E. Biochemistry 1995; 34:4577–4582.
60. Chen J, Derfler B, Samson L. EMBO J 1990; 9:4569–4575.
61. Denver DR, Swenson SL, Lynch M. Mol Biol Evol 2003; 20:1603–1611.
62. Hollis T, Lau A, Ellenberger T. Mutat Res 2000; 460:201–210.

63. Scharer OD, Jiricny J. Bioessays 2001; 23:270–281.
64. Dianov GL, Sleeth KM, Dianova II, Allinson SL. Mutat Res 2003; 531:157–163.
65. Falnes PO, Johansen RF, Seeberg E. Nature 2002; 419:178–182.
66. Kataoka H, Yamamoto Y, Sekiguchi M. J Bacteriol 1983; 153:1301–1307.
67. Dinglay S, Trewick SC, Lindahl T, Sedgwick B. Genes Dev 2000; 14:2097–2105.
68. Trewick SC, Henshaw TF, Hausinger RP, Lindahl T, Sedgwick B. Nature 2002; 419:174–178.
69. Aravind L, Koonin EV. Genome Biol 2: RESEARCH0007, 2001.
70. Duncan T, Trewick SC, Koivisto P, Bates PA, Lindahl T, Sedgwick B. Proc Natl Acad Sci USA 2002; 99:16660–16665.
71. Aas PA, Otterlei M, Falnes PO, Vagbo CB, Skorpen F, Akbari M, Sundheim O, Bjoras M, Slupphaug G, Seeberg E, Krokan HE. Nature 2003; 421:859–863.
72. Kurowski MA, Bhagwat AS, Papaj G, Bujnicki JM. BMC Genomics 2003; 4:48.
73. Chen BJ, Carroll P, Samson L. J Bacteriol 1994; 176:6255–6261.
74. Sedgwick B, Vaughan P. Mutat Res 1991; 250:211–221.
75. Frosina G, Bonatti S, Abbondandolo A. Mutat Res 1984; 129:243–250.
76. Karran P, Arlett CF, Broughton BC. Biochimie 1982; 64:717–721.
77. Samson L, Schwartz JL. Nature 1980; 287:861–863.
78. Samson L, Schwartz JL. Basic Life Sci 1983; 23:291–309.
79. Kaina B. Mutat Res 1982; 93:195–211.
80. Laval F, Laval J. Proc Natl Acad Sci USA 1984; 81:1062–1066.
81. Laval F. Biochimie 1985; 67:361–364.
82. Cooper DP, O'Connor PJ, Margison GP. Cancer Res 1982; 42:4203–4209.
83. Teo AK, Oh HK, Ali RB, Li BF. Mol Cell Biol 2001; 21:7105–7114.
84. Daniels DS, Mol CD, Arvai AS, Kanugula S, Pegg AE, Tainer JA. EMBO J 2000; 19:1719–1730.
85. Xu-Welliver M, Pegg AE. Carcinogenesis 2002; 23:823–830.
86. Snow ET, Foote RS, Mitra S. J Biol Chem 1984; 259:8095–8100.
87. Eadie JS, Conrad M, Toorchen D, Topal MD. Nature 1984; 308:201–203.
88. Abbott PJ, Saffhill R. Biochim Biophys Acta 1979; 562:51–61.
89. Patel DJ, Shapiro L, Kozlowski SA, Gaffney BL, Jones RA. Biochemistry 1986; 25:1027–1036.
90. Patel DJ, Shapiro L, Kozlowski SA, Gaffney BL, Jones RA. Biochemistry 1986; 25:1036–1042.
91. Singer B, Chavez F, Goodman MF, Essigmann JM, Dosanjh MK. Proc Natl Acad Sci USA 1989; 86:8271–8274.
92. Larson K, Sahm J, Shenkar R, Strauss B. Mutat Res 1985; 150:77–84.
93. Dosanjh MK, Galeros G, Goodman MF, Singer B. Biochemistry 1991; 30:11595–11599.
94. Modrich P, Lahue R. Annu Rev Biochem 1996; 65:101–133.
95. Umar A, Buermeyer AB, Simon JA, Thomas DC, Clark AB, Liskay RM, Kunkel TA. Cell 1996; 87:65–73.
96. Berardini M, Mazurek A, Fishel R. J Biol Chem 2000; 275:27851–27857.
97. Duckett DR, Drummond JT, Murchie AI, Reardon JT, Sancar A, Lilley DM, Modrich P. Proc Natl Acad Sci USA 1996; 93:6443–6447.
98. Waters TR, Swann PF. Biochemistry 1997; 36:2501–2506.
99. Wang H, Yang Y, Schofield MJ, Du C, Fridman Y, Lee SD, Larson ED, Drummond JT, Alani E, Hsieh P, Erie DA. Proc Natl Acad Sci USA 2003; 100:14822–14827.
100. Natrajan G, Lamers MH, Enzlin JH, Winterwerp HH, Perrakis A, Sixma TK. Nucleic Acids Res 2003; 31:4814–4821.
101. Griffin S, Branch P, Xu YZ, Karran P. Biochemistry 1994; 33:4787–4793.
102. Karran P, Marinus MG. Nature 1982; 296:868–869.
103. Au KG, Welsh K, Modrich P. J Biol Chem 1992; 267:12142–12148.
104. Goldmacher VS, Cuzick Jr RA, Thilly WG. J Biol Chem 1986; 261:12462–12471.

105. Day III RS, Ziolkowski CH, Scudiero DA, Meyer SA, Lubiniecki AS, Girardi AJ, Galloway SM, Bynum GD. Nature 1980; 288:724–727.
106. Sklar R, Strauss B. Nature 1981; 289:417–420.
107. Kataoka H, Hall J, Karran P. EMBO J 1986; 5:3195–3200.
108. Brennand J, Margison GP. Proc Natl Acad Sci USA 1986; 83:6292–6296.
109. Samson L, Derfler B, Waldstein EA. Proc Natl Acad Sci USA 1986; 83:5607–5610.
110. Samson L, Linn S. Carcinogenesis 1987; 8:227–230.
111. Branch P, Aquilina G, Bignami M, Karran P. Nature 1993; 362:652–654.
112. Kat A, Thilly WG, Fang WH, Longley MJ, Li GM, Modrich P. Proc Natl Acad Sci USA 1993; 90:6424–6428.
113. Bignami M, O'Driscoll M, Aquilina G, Karran P. Mutat Res 2000; 462:71–82.
114. D'Atri S, Tentori L, Lacal PM, Graziani G, Pagani E, Benincasa E, Zambruno G, Bonmassar E, Jiricny J. Mol Pharmacol 1998; 54:334–341.
115. Cejka P, Stojic L, Mojas N, Russell AM, Heinimann K, Cannavo E, di Pietro M, Marra G, Jiricny J. EMBO J 2003; 22:2245–2254.
116. Hirose Y, Berger MS, Pieper RO. Cancer Res 2001; 61:1957–1963.
117. Carethers JM, Hawn MT, Chauhan DP, Luce MC, Marra G, Koi M, Boland CR. J Clin Invest 1996; 98:199–206.
118. Schwartz JL. Mutat Res 1989; 216:111–118.
119. Plant JE, Roberts JJ. Chem Biol Interact 1971; 3:337–342.
120. Rasouli-Nia A, Sibghat U, Mirzayans R, Paterson MC, Day III RS. Mutat Res 1994; 314:99–113.
121. Goth-Goldstein R, Hughes M. Mutat Res 1987; 184:139–146.
122. Aquilina G, Giammarioli AM, Zijno A, Di Muccio A, Dogliotti E, Bignami M. Cancer Res 1990; 50:4248–4253.
123. Galloway SM, Greenwood SK, Hill RB, Bradt CI, Bean CL. Mutat Res 1995; 346:231–245.
124. Kaina B, Ziouta A, Ochs K, Coquerelle T. Mutat Res 1997; 381:227–241.
125. Black KA, McFarland RD, Grisham JW, Smith GJ. Am J Pathol 1989; 134:53–61.
126. Meikrantz W, Bergom MA, Memisoglu A, Samson L. Carcinogenesis 1998; 19:369–372.
127. Tominaga Y, suzuki TT, Shiraishi A, Kawate H, Sekiguchi M. Carcinogenesis 1997; 18:889–896.
128. Hickman MJ, Samson LD. Proc Natl Acad Sci USA 1999; 96:10764–10769.
129. Fishel R. Nat Med 1999; 5:1239–1241.
130. Brown KD, Rathi A, Kamath R, Beardsley DI, Zhan Q, Mannino JL, Baskaran R. Nat Genet 2003; 33:80–84.
131. Lin DP, Wang Y, Scherer SJ, Clark AB, Yang K, Avdievich E, Jin B, Werling U, Parris T, Kurihara N, Umar A, Kucherlapati R, Lipkin M, Kunkel TA, Edelmann W. Cancer Res 2004; 64:517–522.
132. Rasmussen LJ, Samson L. Carcinogenesis 1996; 17:2085–2088.
133. Kaina B. Biochem Pharmacol 2003; 66:1547–1554.
134. Nyberg KA, Michelson RJ, Putnam CW, Weinert TA. Annu Rev Genet 2002; 36:617–656.
135. Eom SH, Wang J, Steitz TA. Nature 1996; 382:278–281.
136. Pelletier H, Sawaya MR, Kumar A, Wilson SH, Kraut J. Science 1994; 264:1891–1903.
137. Doublie S, Tabor S, Long AM, Richardson CC, Ellenberger T. Nature 1998; 391:251–258.
138. Spratt TE. Biochemistry 1997; 36:13292–13297.
139. Clarke ND, Kvaal M, Seeberg E. Mol Gen Genet 1984; 197:368–372.
140. Hendricks CA, Razlog M, Matsuguchi T, Goyal A, Brock AL, Engelward BP. DNA Repair (Amst) 2002; 1:645–659.
141. Chaudhuri I, Essigmann JM. Carcinogenesis 1991; 12:2283–2289.
142. Cunningham RP, Weiss B. Proc Natl Acad Sci USA 1985; 82:474–478.

143. Ramotar D, Popoff SC, Gralla EB, Demple B. Mol Cell Biol 1991; 11:4537–4544.
144. Kunz BA, Henson ES, Roche H, Ramotar D, Nunoshiba T, Demple B. Proc Natl Acad Sci USA 1994; 91:8165–8169.
145. Berdal KG, Bjoras M, Bjelland S, Seeberg E. EMBO J 1990; 9:4563–4568.
146. Posnick LM, Samson LD. J Bacteriol 1999; 181:6763–6771.
147. Xiao W, Chow BL, Hanna M, Doetsch PW. Mutat Res 2001; 487:137–147.
148. Glassner BJ, Rasmussen LJ, Najarian MT, Posnick LM, Samson LD. Proc Natl Acad Sci USA 1998; 95:9997–10002.
149. Hofseth LJ, Khan MA, Ambrose M, Nikolayeva O, Xu-Welliver M, Kartalou M, Hussain SP, Roth RB, Zhou X, Mechanic LE, Zurer I, Rotter V, Samson LD, Harris CC. J Clin Invest 2003; 112:1887–1894.
150. Engelward BP, Dreslin A, Christensen J, Huszar D, Kurahara C, Samson L. EMBO J 1996; 15:945–952.
151. Allan JM, Engelward BP, Dreslin AJ, Wyatt MD, Tomasz M, Samson LD. Cancer Res 1998; 58:3965–3973.
152. Roth RB, Samson LD. Cancer Res 2002; 62:656–660.
153. Paulovich AG, Hartwell LH. Cell 1995; 82:841–847.
154. Paulovich AG, Margulies RU, Garvik BM, Hartwell LH. Genetics 1997; 145:45–62.
155. Stokes MP, Van Hatten R, Lindsay HD, Michael WM. J Cell Biol 2002; 158:863–872.
156. Stokes MP, Michael WM. J Cell Biol 2003; 163:245–255.
157. Tercero JA, Longhese MP, Diffley JF. Mol Cell 2003; 11:1323–1336.
158. Leroy C, Mann C, Marsolier MC. EMBO J 2001; 20:2896–2906.
159. Donaldson AD, Blow JJ. Curr Biol 2001; 11:R979–R982.
160. Jelinsky SA, Samson LD. Proc Natl Acad Sci USA 1999; 96:1486–1491.
161. Gasch AP, Huang M, Metzner S, Botstein D, Elledge SJ, Brown PO. Mol Biol Cell 2001; 12:2987–3003.
162. Jelinsky SA, Estep P, Church GM, Samson LD. Mol Cell Biol 2000; 20:8157–8167.
163. Begley TJ, Jelinsky SA, Samson LD. Cold Spring Harb Symp Quant Biol 2000; 65: 383–393.
164. Chang M, Bellaoui M, Boone C, Brown GW. Proc Natl Acad Sci USA 2002; 99: 16934–16939.

35

Damage Signals Triggering the *Escherichia coli* SOS Response

Mark D. Sutton
Department of Biochemistry, School of Medicine and Biomedical Sciences, University at Buffalo, State University of New York, Buffalo, New York, U.S.A.

1. INTRODUCTION

Failure of a cell to accurately replicate and/or repair its DNA can lead to mutations, disease, and in extreme cases, even death. Thus, it is by no means trivial that the genetic information in all cells is subject to continuous assault by DNA damaging agents. Although many DNA damaging agents exist in the environment, some, such as reactive oxygen species, are produced as metabolic byproducts by cells themselves. As a result of this chronic threat, all organisms have evolved numerous important mechanisms by which to repair or tolerate DNA damage.

In *Escherichia coli*, most accurate as well as potentially mutagenic DNA repair functions are coordinately regulated as part of the global SOS response. This chapter discusses our current understanding of mechanisms leading to induction of the SOS response, as well the SOS-regulated mechanisms for repairing or tolerating the DNA damage, which is necessary in order for the SOS response to be turned off.

2. THE *E. COLI* SOS RESPONSE

E. coli has evolved a variety of mechanisms with which to deal with DNA damage. These mechanisms can be divided into four general classes, including: (i) direct reversal; (ii) excision repair; (iii) RecA-dependent recombination-mediated repair; and (iv) DNA damage tolerance via a specialized mode of DNA replication termed translesion DNA synthesis (TLS) (reviewed in Ref. 1). Some of these responses to DNA damage are constitutive, meaning that the proteins that function in these pathways are present continuously within the cell, and are functionally active for repair regardless of whether or not the cell has experienced DNA damage. However, most repair and damage tolerance functions in *E. coli* are not designed to act constitutively, but instead are coordinately regulated as part of the global SOS response, so named after the international distress signal (2).

Although a variety of different agents are capable of damaging DNA, only those that block progression of the replication fork induce the SOS response (reviewed in Refs. 1, 3–5) Thus, the main function of the *E. coli* SOS response is to delay cell division until such time as the cell has repaired and or tolerated the damage incurred to its DNA (reviewed in Ref. 1). To accomplish this task, the SOS response coordinately regulates the expression of a variety of different genes, many of whose products act in regulation of cell division, DNA replication, DNA repair, or DNA damage tolerance.

2.1. The SOS Regulon: Roles of the *recA*⁺ and *lexA*⁺ Genes

In the *E. coli* SOS response, the expression of more than 40 unlinked genes is induced following replication-blocking DNA damage in a *recA*- and *lexA*-dependent fashion (6–10). Many of these genes encode proteins that function in aspects of DNA repair, such as the *uvrA* gene product that acts in nucleotide excision repair (NER) (11) (see also Chapter 6). Others encode proteins that act in damage tolerance, such as the *umuDC* gene products that act in both a DNA damage checkpoint control (12,13), and in translesion DNA synthesis (TLS) (14–16) (see also Chapter 23). In addition, some encode proteins that act in regulation of cell division, such as *sulA*, whose product interacts with the cell division protein FtsZ to block cytokinesis following induction of the SOS response (17–19). Furthermore, a large number of genes are either transcriptionally repressed following DNA damage, or have their messages degraded via an unknown mechanism(s) (7). Finally, a growing number of genes whose products function in DNA replication appear to be induced following DNA damage in a *recA*-dependent but *lexA*-independent manner, including *dnaA* (DnaA protein), *dnaB* (DnaB helicase), *dnaQ* (epsilon proofreading subunit of DNA polymerase III), and *dnaN* (processivity clamp subunit of DNA polymerase III) (7,20–24). Hence, strictly speaking, these genes are regulated in a DNA damage-dependent but SOS-independent fashion.

Proper management of the SOS-regulated genes requires the products of the *recA*⁺ and *lexA*⁺ genes. LexA protein acts as a transcriptional repressor by blocking access of RNA polymerase to the promoter by binding to sites termed "SOS boxes" located nearby promoters, thereby effectively repressing transcription of the SOS-regulated genes (9). Importantly, LexA protein regulates transcription of its own gene as part of the global SOS response, ensuring that sufficient levels of LexA are present for regulation of the SOS regulon (9). Thus, under conditions in which *E. coli* has not experienced replication-blocking DNA damage, the message levels of many of the genes belonging to the SOS regulon are maintained at a very low abundance (Fig. 1). Further regulation is achieved by minor differences in the nucleotide sequence of the SOS box motif that alter its affinity for the LexA repressor. Thus, some SOS-regulated genes, such as *lexA*, *recA*, and *uvrA*, are expressed at moderate levels in the absence of DNA damage (11,25–27), while others, such as *umuDC*, are very tightly repressed under normal growth conditions and thus are present at significant levels only following SOS induction (12,28).

RecA protein, which is required for most homologous recombination (reviewed in Refs. 29–31, see also Chapter 27), as well as for the repair of double strand (ds) DNA breaks (reviewed in Refs. 29–31), the restart of stalled replication forks (reviewed in Refs. 29, 32) and most TLS (33–37) becomes activated for its role in the SOS response by binding to single stranded (ss) DNA. Following replication-blocking DNA damage, ssDNA accumulates, presumably a result of the cell's failed

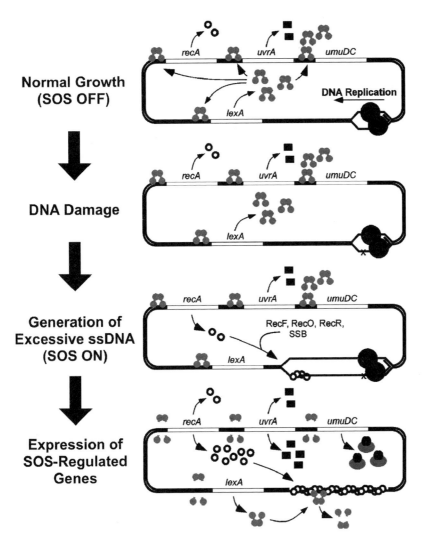

Figure 1 Drawing depicting the presumed mechanism for induction of the *E. coli* SOS response. In the absence of DNA damage, LexA acts to repress, albeit to differing extents, transcription of the different SOS-regulated genes. In response to replication-blocking DNA damage, ssDNA is generated, leading to formation of RecA–ssDNA nucleoprotein filaments. Interaction of LexA repressor with these RecA–ssDNA filaments leads to self-cleavage of LexA, which largely inactivates its repressor function, leading to transcriptional de-repression of the LexA-regulated genes. (See text for additional details.)

attempts to copy over lesions in the DNA (38). RecA forms a helical filament structure upon binding to ssDNA, and formation of these filaments seems to serve as the signal that alerts the cell to the presence of DNA damage (2,39). Consistent with its role as the cell's internal sensor that it has experienced DNA damage, RecA has been reported to bind damaged DNA more strongly than it does undamaged DNA (40,41). However, it is currently unclear whether this activity of RecA is important for induction of the SOS response.

These RecA–ssDNA filaments play an essential role in SOS induction (Fig. 1). Interaction of LexA with the RecA–ssDNA filament mediates self-cleavage of LexA, which largely inactivates its repressor function (9,10,42). Since RecA–ssDNA-mediated self-cleavage of LexA is rapid, occurring within minutes of DNA damage, it leads to a rapid decrease in the intracellular level of active LexA, which in turn leads to the rapid de-repression of the LexA-regulated genes (38). The *recA* gene is also LexA-regulated (9). LexA regulation of *recA* may act to ensure that there are sufficient levels of RecA protein following DNA damage to maintain SOS induction for a long enough period so as to repair and/or tolerate the damage, as well as to help turn off the SOS response once DNA damage has been processed by downregulating *recA* transcription.

2.2. LexA Protein–DNA Interactions

LexA protein binds specifically to a palindromic sequence termed the SOS box (9). Based on a comparison of various different functional *E. coli* SOS box sequences, the following LexA consensus binding sequence was proposed: 5'-TACTGT(AT) $_4$ACAGTA-3' (6,7,9). The SOS box is made up of two identical half sites, each of which is bound by a single LexA protomer. Dimerization of the bound LexA appears to play an important role in stabilizing the protein–DNA complex (43). The 5'-CTGT-3' inverted repeat present in each half site is particularly important for LexA binding since base substitution mutations affecting any one of these four positions severely impair LexA-dependent transcriptional repression in vivo (reviewed in Ref. 1).

Although the structure of the 84-amino acid residue LexA DNA binding domain has been determined using NMR (44), the structure of the LexA-DNA complex has not. The three-dimensional structure of the LexA DNA binding domain is made up of three alpha helices and two anti-parallel beta strands, forming a "winged" helix-turn-helix (HTH) DNA binding motif. In this motif, helix 2 and 3 comprise a variant HTH, while the beta strands that follow the HTH motif constitute a beta strand-loop-beta strand element referred to as the "wing," and is reminiscent of the DNA binding domain found in *E. coli* catabolite gene activator protein (CAP) (45).

3. STRUCTURE–FUNCTION OF THE LEXA PROTEIN FAMILY

The 230-amino acid LexA protein possesses two domains: an N-terminal DNA binding domain and a C-terminal dimerization domain. Based on the crystal structures of several mutant forms of LexA (46), the N- and C-terminal domains are connected to each other by a very short, hydrophilic and solvent-exposed linker that forms a structural component of the C-terminal dimerization domain (46). LexA dimerization properly juxtaposes the two N-terminal DNA binding domains in LexA, thereby allowing for efficient transcriptional repression of genes bearing an SOS box (9,43).

In 1984, Little (10) made the important discovery that cleavage of LexA protein results from conformational alterations in its structure caused by its specific, physical interaction with RecA–ssDNA filaments. This structural alteration presumably activates its latent auto-lytic activity (9,42). Lin and Little (10,47) went on to demonstrate that Ser119 and Lys156 of LexA, which are located within the C-terminal domain, function as a Ser–Lys dyad that catalyzes cleavage of the LexA Ala84–Gly85 bond, which is located within the linker that tethers the N- and C-terminal domains (46), to largely inactivate its repressor function. The ability of RecA–ssDNA to promote cleavage of LexA, as well as of other members of the LexA family, is referred to as its co-protease activity.

Although a variety of different mutant *lexA* alleles have been described, they can be grouped into two main classes: those defective for transcriptional repression, referred to as *lexA defective* or *lexA*(Def) (48), and those that are refractory to RecA-mediated self-cleavage, which are referred to as *lexA induction minus*, or *lexA*(Ind⁻) (49). While the *lexA*(Def) alleles are simply null mutations, *lexA*(Ind⁻) alleles bear mutations that alter the cleavage site and/or the active site dyad required for auto-digestion of LexA protein, rendering the mutant protein non-cleavable (47). As a result, the SOS response is constitutively induced in a *lexA*(Def) mutant strain, and cannot be induced in a *lexA*(Ind⁻) mutant strain. In addition to these two main classes of *lexA* mutations, another class of *lexA* alleles, referred to as *lexA*(Ind^s), encodes mutant proteins that are proficient for repression, but display a greatly increased rate of self-cleavage (50–52).

LexA has been suggested to exist in two structurally distinct conformations: a "cleavable" (C) conformation and a "non-cleavable" (NC) conformation (50). It has been estimated that at neutral pH roughly 0.01% of the total LexA protein population exists in the C conformation (50). LexA auto-digestion has been proposed to occur via a RecA–ssDNA-stabilization of the C conformation (46,50). Consistent with this model, crystal structures of several different mutant forms of LexA protein suggest that the LexA dimer may, in fact, alternate between two structurally distinct conformations that differ with respect to the positioning of the Ala84–Gly85 cleavage site relative to the Ser119–Lys156 active site dyad. In one set of crystal structures, the positioning of the cleavage site within each protomer relative to the active site dyad was compatible with cleavage (C conformation), while in another set of structures, the cleavage site was positioned approximately 20 Å away from the Ser–Lys dyad (NC conformation) (46). Alteration of the pK_a of Lys156 in the "cleavable" conformation due to its burial is presumably necessary in order for it to activate Ser119 for nucleophilic attack of the Ala84–Gly85 cleavage site.

Following its self-cleavage, the N- and C-terminal peptides of LexA are targeted to further proteolysis by the Lon and ClpXP proteases (53,54). This additional destruction of LexA appears to be biologically important because strains lacking the ClpXP protease and expressing the N-terminal domain of LexA display an increased sensitivity to UV similar to that observed for *lexA*(Ind⁻) mutants (53). Interestingly, as discussed further at the end of this chapter, the Lon and ClpXP proteases also play vital roles in managing the steady-state levels of the *umuDC* gene products, and hence play important roles in helping to regulate *umuDC*-dependent TLS, as well as in controlling the induction of the SOS response.

LexA shares homology with three other classes of proteins: (i) various bacteriophage repressors, including phage λ cI repressor, as well as others (reviewed in Ref. 1), (ii) *E. coli* UmuD, which acts in a DNA damage checkpoint control and also controls most TLS (14), and (iii) the *E. coli* type I signal peptidase (55,56). Like LexA, the phage repressors (reviewed in Ref. 1) and UmuD exist as dimers (57,58). In addition, the phage repressor proteins (59–61), as well as UmuD, each undergo a similar specific self-cleavage reaction that is mediated by their interaction with RecA–ssDNA filaments (58,62,63).

Comparison of the crystal structures of UmuD' (the cleaved form of UmuD), λ cI repressor and LexA indicates that despite their striking lack of sequence similarity, these proteins nonetheless share a remarkably high degree of structural similarity (46,64–66). Interestingly, this structural similarity translates to the signal peptidases as well (55,56). Comparison of the structure of the catalytic core of the *E. coli* type I signal peptidase (55,56) with those of *E. coli* LexA (46), *E. coli* UmuD' (the cleaved form of UmuD) (64–66), and phage λ cI repressor (67) indicates that their peptide backbones are superimposable on each other with root mean square deviations of 1.61, 1.54, and 1.54 Å, for 71, 68 and 75 common Cα atoms, respectively (68). These findings indicate that the LexA family and the type I signal peptidases collectively represent an ancient super-family of proteins.

In the case of the phage repressors, self-cleavage serves to induce the lytic life cycle of the phage. It has been suggested that certain phage may have evolved this mechanism to couple their lytic induction with their host's SOS response in order to allow for efficient "evacuation" under conditions in which their host may die (2,69).

Despite the striking structural similarity between LexA and UmuD', the *umuD* gene product lacks a detectable DNA binding activity, and hence does not act as a transcriptional repressor. Instead, the SOS-regulated *umuD* gene product partici-pates in a DNA damage checkpoint control (12,13), and acts as part of DNA pol V to promote TLS (35,36,70,71) (see also Chapter 23). By removing its N-terminal 24 residues to yield a truncated form known as UmuD', RecA–ssDNA-mediated self-cleavage of UmuD acts to regulate these two temporally separated biological roles of the different *umuDC* gene products. Intact UmuD, together with UmuC, participate in a primitive DNA damage checkpoint control which acts to arrest DNA replication, possibly by sequestering the beta sliding clamp of pol III (72–74), while UmuD', together with UmuC, enables pol V-dependent TLS (35,71). A particularly intriguing aspect of this model is that the same gene products act in two temporally distinct yet intimately related biological processes, and that post-translational modification of a single gene product (auto-digestion of UmuD to yield UmuD') plays an important role in their temporal regulation (Fig. 2).

Genetic analyses indicate that RecA–ssDNA-mediated self-cleavage of UmuD can occur inter-molecularly, with the active site dyad of one protomer cleaving between Cys24–Gly25 of its intra-dimer partner (66,75). Structural analyses of the UmuD$_2$ homodimer are consistent with this type of model (65,76). However, in contrast to UmuD, RecA–ssDNA-mediated self-cleavage of LexA is reported to occur intra-molecularly rather than inter-molecularly (10,77), consistent with the mechanism proposed based on the structures of several mutant forms of LexA protein (46). A recent determination of the K_d for the LexA$_2$ dimer indicates that it is in the picomolar range (78), consistent with LexA existing as a dimer within the cell. Taken together, these observations suggest that LexA may undergo intra-molecular self-cleavage as a dimer in vivo. Importantly, it has been reported that a rather modest conformational change in the structure of the C conformation of LexA could result in a "domain swap" such that intra-dimer LexA partners exchange their active-site domains to effect inter-molecular cleavage (46). Thus, it is currently unclear whether LexA undergoes intra- or inter-molecular cleavage. Regardless, it is possible that modest structural differences in the cleavage sites of the different LexA-like proteins determine whether or not a particular dimer undergoes inter- or intra-molecular self-cleavage.

4. RECA PROTEIN–DNA INTERACTIONS AND LEXA SELF-CLEAVAGE

The 352-amino acid RecA protein possesses both DNA-dependent ATPase activity and ATP-dependent ssDNA and dsDNA binding activities (29–31) (see also Chapter 27). Although the ADP-bound and nucleotide-free forms of RecA can also bind

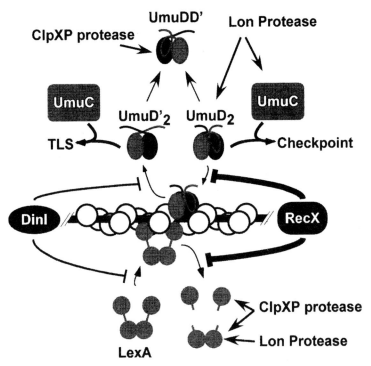

Figure 2 Drawing depicting mechanisms that act to regulate RecA-mediated self-cleavage of both UmuD and LexA. Intact UmuD, together with UmuC, acts in a DNA damage checkpoint control. Interactions of UmuD and with RecA–ssDNA filaments lead to its self-cleavage, which removes the N-terminal 24 residues of the *umuD* gene product to yield UmuD' UmuD', together with UmuC, comprise pol V that acts in TLS. RecA–ssDNA-mediated self-cleavage of UmuD and LexA is each blocked by the DinI and the RecX proteins. In vitro, RecX displays a more potent ability to inhibit the co-protease activity of the RecA filaments than does DinI. The thicker lines depicting RecX-dependent inhibition of the co-protease activity of the RecA–ssDNA filament relative to those for DinI are intended to represent this fact. See text for additional details. Susceptibilities of LexA and the different *umuDC* gene products to the Lon and ClpXP proteases are also indicated.

DNA, the filaments formed under these conditions display a more compact structure, and are inactive for catalyzing both the co-protease and the DNA strand transfer reactions (29). Interestingly, these filaments do not appear to easily inter-convert to the more extended, ATP-bound active form (29,79). Although Steitz and colleagues have described crystal structures of both the RecA monomer (80) and the ADP-bound RecA polymer (81), the form of RecA that is active for induction of the SOS response, as well as homologous recombination, is a nucleoprotein filament composed of ATP-bound RecA and ssDNA (29). Nonetheless, these "inactive" structures have provided important insights into RecA function.

RecA can form a filament on either dsDNA or ssDNA (29,30), and the structures of these filaments have been analyzed using electron microscopy combined with three-dimensional image reconstruction (82,83). Although the kinetics for filament assembly on ssDNA differs from those for dsDNA, the structures of both RecA filaments are similar except that filament formed on dsDNA appears more regular than that formed on ssDNA. Each RecA protomer within the RecA-nucleoprotein filament contacts 3-bp of DNA, as well as the adjacent RecA protomers, to form a right-handed helical structure with a diameter of approximately 100 Å (84–86). The DNA within these filaments is stretched out compared to standard B form DNA. These filaments also bear a deep helical groove in which the LexA and LexA-like proteins (i.e., λ cI and UmuD) bind (87,88). Extensive genetic and biochemical characterizations of numerous mutant forms of RecA have begun to delineate some of the sites in RecA that make critical contacts with its different substrates (i.e., Refs. 89–91 and reviewed in Refs. 1, 31). For example, four different *recA* alleles, *recA91* (bearing a G229S substitution), *recA430* (G204S), *recA1730* (S117F), and *recA1734* (R243L), exhibit differential co-protease activities towards the λ cI repressor, LexA and UmuD (reviewed in Ref. 31). Furthermore, these analyses indicate that although there appear to be some common contacts, each LexA-like protein also makes a series of unique contacts with the filament as well.

Although a discussion of how the SOS response is turned off is included at the end of this chapter, three important mechanisms that help to regulate SOS induction at the level of RecA filament function will be discussed here. One mechanism pertains to the *E. coli* single-strand DNA binding protein (SSB), which plays an important role in regulating induction of the SOS response by modulating RecA–ssDNA filament formation. SSB plays critically important roles in DNA replication, repair, and recombination (92,93). Although SSB facilitates RecA–ssDNA filament formation by removing secondary structure in ssDNA, it can also inhibit filament formation by inhibiting the initial nucleation event (29–31). The capacity of RecA protein to displace SSB from ssDNA is dramatically enhanced in mutant RecA proteins with C-terminal deletions (94,95). Thus, competition between RecA and SSB for binding to ssDNA is presumably biologically important in that it prevents induction of the SOS response in the absence of DNA damage.

A second mechanism for regulating SOS induction at the level of the RecA–ssDNA filament involves the role of accessory proteins. Genetic analyses suggest that the *recF*, *recO*, and *recR* genes act in a collaborative fashion to influence RecA function in vivo (96–98). Recent biochemical analyses support these genetic findings, and indicate that the RecF, RecO, and RecR proteins act by facilitating the loading of RecA protein onto gapped DNA that is coated with SSB in vitro (99). While catalyzing this process, the RecF–RecR complex appears to limit the extension of RecA filaments beyond ssDNA gaps in vitro (100), while the RecO–RecR complex stabilizes RecA-DNA filaments in vitro (101). Thus, the RecFOR proteins play an important role in SOS induction by influencing RecA–ssDNA filament formation and stability.

A third mechanism that affects the ability of RecA to mediate self-cleavage of LexA pertains to the relationship between the DNA strand pairing and the co-protease activities of the RecA-DNA filament. Biochemical analyses indicate that the DNA strand exchange and LexA co-protease activities of the RecA-DNA filament are competitive reactions, suggesting that homologous DNA and LexA protein bind to overlapping surfaces on the RecA-DNA filament (102,103). Consistent with this finding, electron microscopy indicates that both DNA and LexA interact within the helical groove. Furthermore, these same experiments indicated that one of the LexA contact sites on RecA mapped onto a region of the filament postulated to be a second DNA binding site important for DNA strand pairing (87,104).

5. ROLE OF DNA DAMAGE IN INDUCING THE *E. COLI* SOS RESPONSE

A variety of different types of DNA damaging agents induce the SOS response, including ultraviolet (UV) light, ionizing radiation, alkylating agents, such as methyl methanesulfonate, inter-strand cross-linking agents, such as mitomycin C, and, under certain conditions, hydrogen peroxide and ethanol, as well as a wide variety of other agents (reviewed in Ref. 1). In addition, ssDNA and dsDNA breaks, such as those induced by the type II topoisomerase inhibitor nalidixic acid, are also known to induce the SOS response (29–31). Numerous genetic studies have focused on understanding the mechanism of SOS induction. These studies demonstrated a correlation between the generation of ssDNA and SOS induction. Consistent with this conclusion, Sassanfar and Roberts (38) demonstrated that the major pathway of induction following UV irradiation required an active replication fork. This implies that gaps left in the DNA as a result of lesions that cannot be bypassed by the cell's normal replication machinery served as the source of ssDNA leading to SOS induction.

Since the chromosome is replicated in the 5′-to-3′ direction, one strand (leading strand) is replicated continuously, while the other strand (lagging strand) is replicated discontinuously as smaller pieces referred to as Okazaki fragments. Each of these Okazaki fragments is replicated independently of each other via a repetitive process of (i) priming, (ii) elongation, and (iii) maturation that is highly coordinated with the replication of the leading strand (reviewed in Ref. 105). It has been reported that the coupling between leading and lagging strand replication is disrupted after the replisome encounters a guanine-*N*-2-acetlyaminoflournene lesion (106). The coupling of leading and lagging strand synthesis is likely disrupted by other types of DNA lesions, as well, such as those discussed above. Thus, these findings suggest that replication-blocking DNA damage perturbs the way in which ssDNA at the replication fork is managed, providing a source of ssDNA for RecA protein in response to DNA damage. The increased level of ssDNA under these circumstances could allow the formation of RecA–ssDNA filaments. Likewise, processing of a dsDNA break by the RecBCD complex, which simultaneously unwinds and degrades DNA from a dsDNA end to generate a 3′-OH tail onto which RecA is loaded (29–31), could also induce the SOS response. However, it should be stressed that the precise source(s) of the ssDNA bound by RecA to induce the SOS response is currently unknown.

More recently, a reaction coupling RecA–ssDNA-mediated cleavage of LexA to de-repression of a LexA-regulated promoter has been reconstituted in vitro using purified components (107). In this reaction, de-repression of the LexA-regulated promoter was achieved by addition of ssDNA, or by generation of a dsDNA break. In the latter case, there was an absolute requirement for the RecBCD complex, which simultaneously unwinds and degrades DNA from a dsDNA end and facilitates loading of RecA onto ssDNA. Thus, this study indicates that in vitro both ssDNA and dsDNA breaks are capable of promoting activation of RecA, leading to de-repression of a LexA-regulated promoter. Interestingly, the authors noted that addition of linear dsDNA bearing a Chi site enhanced the rate of de-repression, suggesting another possible mechanism whereby the cell can regulate SOS induction (107). The Chi site is an octameric sequence involved in RecBCD-dependent recombination (29–31). The authors suggested that the effect of the Chi site on SOS induction may have evolved to enhance the efficiency by which SOS responds to damage of "self" chromosomal DNA as opposed to "non-self" viral or plasmid DNA (107). However, the biological significance of the Chi site to SOS induction remains to be determined.

5.1. Perturbation of "Normal" DNA Replication Can Induce the SOS Response

In addition to ssDNA or dsDNA breaks, and lesions in the DNA that block the cell's normal replication machinery, a variety of mutations affecting genes coding for proteins that function in various aspects of DNA metabolism have been reported to induce the SOS response in the absence of exogenous DNA damage. These genes include *polA* (which encodes DNA polymerase I) (108), *dam* (deoxyadenosine methyl transferase, or Dam) (109,110), *uvrD* (DNA helicase II) (111,112), *dnaQ* (*mutD*; epsilon proofreading subunit of pol III) (113), *rep* (Rep helicase) (114,115), and *priA* (primosome assembly factor Y, or protein n') (116,117). In addition, a point mutation in the SOS-regulated *dinD* gene, *dinD68* (also referred to as *pcsA68*), has been demonstrated to induce the SOS response when the mutant strain is grown at low temperature (118). The fact that SOS was not induced in the Δ*dinD1*::Mud1 (Ap *lac*) mutant strain suggests that *dinD68* is a gain-of-function allele. However, the biological function(s) of the *dinD* gene product is currently unknown.

DNA polymerase I (pol I), Dam protein, the epsilon proofreading subunit of pol III and PriA protein each play important roles in DNA replication. Furthermore, although the *uvrD*-encoded helicase II plays important roles in nucleotide excision repair and in methyl directed mismatch repair (reviewed in Ref. 1), genetic and biochemical evidence suggests that it may also function in DNA replication (119–121). Thus, one possible mechanism to account for SOS induction in the absence of exogenous DNA damage in the mutant strains described above is that the mutant gene products perturb DNA replication. Such a perturbation could result either from a biochemical defect of the mutant proteins that directly affects their role(s) in DNA replication, or from a structural defect of the mutants that acts to destabilize the replisome. Either of these two scenarios would likely alter leading and/or lagging strand synthesis, thus affecting the way in which ssDNA at the replication fork is managed, resulting in SOS induction. Regardless of the mechanistic basis for SOS induction in these mutant strains, these findings suggest that induction can be triggered by perturbation of DNA replication as well as by treatments that lead to DNA damage.

5.2. SOS Induction by DNA Transposition

In addition to DNA damage and perturbation of DNA replication, Roberts and Kleckner (122) found that conservative transposition by transposon Tn*10* caused a transient induction of the SOS response. They speculated that degradation of the gapped donor DNA molecule, which is a product of conservative transposition, was responsible for SOS induction, and suggested that the SOS response might play an important role in helping a cell undergoing transposition to repair the damage to the transposon donor's chromosome. Interestingly, Roberts and Kleckner (122) found that SOS induction was not required for Tn*10* transposition, and that partial induction of SOS by addition of mitomycin C did not significantly alter Tn*10* transposition frequency. In contrast, Eichenbaum and Livneh (123) found that the transposition frequency of IS*10*, which comprises the right module of Tn*10*, and can function as either an individual insertion sequence or mediate transposition of the entire Tn*10*, was stimulated more than 10-fold following exposure to UV irradiation in an SOS-regulated manner. Finally, Weinreich et al. (124) found that transposition of Tn*5* was reduced between 5- and 10-fold following the induction of the SOS response. Thus, although replication-blocking lesions in the DNA and dsDNA breaks may represent the major source of ssDNA that induces the SOS response, other types of DNA structures resulting from biologically important DNA transactions, such as transposition, can also induce SOS.

5.3. Stationary Phase and the SOS Response

In addition to replication-blocking DNA damage, and certain mutations that perturb DNA replication, as well as biologically important DNA transactions such as transposition, the SOS response is also induced in stationary phase cells (125). SOS induction in aging cultures is both *lexA*- and *recA*-dependent, but also requires camp [since a Δ*cya* mutation eliminated the stationary phase SOS response (125)]. Given that the SOS response influences biological processes in addition to just DNA repair and damage tolerance (such as transposon mobility, as discussed above) the authors of this study suggested that stationary phase induction of SOS might have adaptive value (125). This conclusion, taken together with the fact that in nature *E. coli* fluctuates between periods of feast and famine, typically spending extended periods of time in a nutrient poor stationary-like phase, suggests that in its natural environment the SOS response may play an important role in normal day-to-day *E. coli* DNA metabolism.

More recently, Finkel and colleagues (126) reported that optimal laboratory growth of *E. coli* under stationary phase conditions required the presence of functional copies of the *polB*, *dinB* and *umuDC* genes, which encode DNA pols II, IV, and V, respectively. All three of these DNA polymerases are regulated as part of the SOS response (1,127). Moreover, the *dinB*-encoded pol IV plays an important role in adaptive, or stationary phase mutagenesis (128,129). Thus, these SOS-regulated DNA polymerases clearly play important biological roles in the absence of treatments that induce DNA damage. A recent report from Rosenberg and colleagues (129) indicates that the activity of the *dinB*-encoded pol IV is influenced by other proteins in vivo, suggesting that additional SOS-regulated gene products may play important roles in generating genetic diversity under conditions of limiting nutrients.

It was reported that DNA pols II, IV, and V each interact with the *E. coli* beta sliding clamp protein, and that this interaction may be important for the biological activities of these enzymes (72,73,130–132). In light of these findings, it is interesting that transcription of *dnaN*, which encodes the beta sliding clamp, is stimulated nearly 100-fold as cells approach stationary phase (133). Although it is not yet clear whether the roles of these DNA pols during stationary phase require the presence of the beta clamp, the stationary phase-dependent stimulation of *dnaN* transcription suggests that they may.

6. UPREGULATION OF DNA REPAIR AND DNA DAMAGE TOLERANCE UNDER THE SOS RESPONSE

All DNA damage suffered by a cell must be appropriately dealt with by at least one DNA repair or damage tolerance response prior to cell division. Based on the known functions of many of the genes whose expression levels are increased as part of the SOS response (reviewed in Refs. 1, 7), TLS, RecA-mediated homologous recombination, and NER are all regulated as part of the global SOS response. Consistent with this conclusion, *lexA*(Ind⁻) mutants of *E. coli*, which are refractory to RecA–ssDNA-mediated auto-digestion, and are therefore impaired for induction of the SOS response, are severely impaired for most TLS (reviewed in Ref. 1) and are partially impaired for NER (134). Thus, SOS regulation may either tightly regulate a process, such as TLS, which is limited almost exclusively to times of SOS induction, or may allow for a temporary enhancement of an otherwise constitutive repair function for a defined time frame following an unusually high level of DNA damage, such as NER.

Recently, Crowley and Hanawalt tested the hypothesis that NER was enhanced during SOS induction. Exposure to UV irradiation results in formation of cyclobutane pyrimidine dimers (CPD) as well as the more distorting pyrimidine (6–4) pyrimidone photoproducts (6–4 photoproduct). UvrABCD-dependent NER repairs most UV-induced DNA lesions. Three of these four genes, *uvrA*, *uvrB*, and *uvrD*, are regulated as part of the SOS response. By manipulating the expression levels of the different *uvr* gene products, Crowley and Hanawalt asked whether or not elevated levels of the Uvr proteins were required for optimal repair of either UV-induced DNA lesion. Although non-SOS-induced (i.e., basal) levels of the different *uvr* gene products were able to repair the 6–4 photoproducts equally well, efficient repair of CPDs required elevated levels of the UvrA and UvrB proteins (134). Consistent with this need, they found that in wild-type *E. coli* cells, the steady-state levels of both UvrA and UvrB underwent a twofold induction within 10 min following UV irradiation (134,135). Interestingly, the basal steady-level of UvrD was sufficient for optimal NER-mediated repair of both CPDs and 6–4 photoproducts (135). Thus, the SOS response acts in part to enhance the efficiency with which NER repairs UV-induced DNA damage.

6.1. Translesion DNA Synthesis: A Major Role of the SOS Response

Despite the high degree of efficiency with which excision pathways repair DNA lesions, circumstances arise in which lesions either evade repair, or cannot be repaired. An example of this latter case includes a situation in which two or more lesions are in close proximity to each other, but reside on opposite DNA strands (i.e., clustered lesions), such as is the case following exposure to ionizing radiation (1). Attempts to repair clustered lesions via excision repair could result in a double strand DNA break, as well as a loss of genetic information, both of which represent significantly more serious types of damage than the initial lesion. Although the histone-like protein HU has recently been shown to influence repair of clustered lesions (136), it is likely that in these cases a small number of lesions persist in the DNA until such time as the replication machinery encounters them. However, because replicative DNA polymerases are by necessity high fidelity enzymes, they are unable to bypass most non-instructive or distorting DNA lesions. Therefore, when the replication fork encounters a lesion that was not repaired, the lesion must be tolerated via TLS. It is important to stress the fact that TLS is a DNA damage tolerance process in that it allows for replication past the lesion without repairing it.

The process of TLS in *E. coli* has been studied for more than 30 years (reviewed in Refs. 1, 5, 137, see also Chapter 23). Although three of the five *E. coli* DNA polymerases, namely pols II, IV, and V, are SOS-regulated, and are capable of participating in TLS, most TLS following UV irradiation or exposure to chemical carcinogens requires the *umuDC*-encoded DNA pol V (15,16). Despite the fact that TLS is inherently error prone, due in part to the relaxed fidelity of the specialized DNA polymerase that carries out TLS, as well as the fact that most lesions are either non-coding or are miscoding, it nonetheless contributes significantly to cell survival following DNA damage by allowing the completion of DNA replication. Failure to completely replicate the DNA once each cell cycle results in cell death. Consistent with TLS serving an important biological role, inactivation of the *E. coli umuDC* gene products, which eliminates most TLS, confers a modest yet significant sensitization to UV irradiation as well as to many chemical mutagens (1).

Like *E. coli*, humans also possess multiple DNA polymerases that function in TLS (reviewed in Refs. 127, 138). Importantly, mutations mapping to one of these polymerases, pol eta, which is structurally and functionally related to the *E. coli umuC* gene product (139), leads to the genetic disorder known as xeroderma pigmentosum variant (XP-V) (140,141). This disorder is characterized by a severe sensitivity to UV light due to an inability to properly tolerate UV-induced DNA lesions, resulting in a predisposition to skin cancer (142).

6.2. Coordinating DNA Replication, Recombination, and TLS Under the SOS Response

It has recently become evident that a central role of RecA-mediated homologous recombination is to underpin DNA replication by helping to restart replication forks that have stalled in the absence of SOS induction (32). Moreover, RecA, together with the RecF, RecO, and RecR, and the primosomal proteins, which act to load the replicative DNA helicase, play important roles in restarting replication forks that have stalled due to replication-blocking DNA damage that induces the SOS response (29,30,143). DNA pols II and V may also participate in this process (143,144). Thus, *E. coli* possesses both constitutive and DNA damage-inducible systems for safeguarding DNA replication. Although the mechanistic basis of both the SOS-independent and the SOS-dependent systems of replication fork maintenance are still largely unknown, the SOS-regulated system(s) is likely to be considerably more complicated due to the fact that multiple repair and damage tolerance functions are simultaneously induced as part of the SOS response. Thus, following SOS induction, the process of restarting stalled replication forks must be coordinated with ongoing repair functions. As discussed further below, this coordination appears to be achieved, at least in part, by the *umuDC* gene products.

The process of NER, which is upregulated as part of the global SOS response (134,135), involves the removal of a short patch of ssDNA containing the lesion, followed by resynthesis of the ssDNA gap (1). Thus, if the replication machinery were to encounter an NER intermediate during DNA replication, a double strand DNA break could result. Work from Walker and colleagues (12) suggests that a primitive, SOS-regulated DNA damage checkpoint control, dependent on the *umuDC* gene products, plays an important role in helping to prevent this. In their model, intact UmuD, together with UmuC, acts to block DNA replication following induction of the SOS response, thereby allowing additional time for NER to repair lesions in the genome prior to the resumption of processive DNA synthesis (12). RecA-mediated self-cleavage of UmuD to yield UmuD' serves to release the checkpoint while simultaneously enabling pol V-dependent TLS.

The model for the *umuDC*-dependent DNA damage checkpoint control grew from a series of experiments designed to investigate the ability of the *umuDC* gene products, when expressed at a higher than physiological level, to confer a cold sensitive growth phenotype (145–147). Although it had been known for more than 10 years that *umuDC*-mediated cold sensitivity, as it was referred to, correlated with a rapid cessation of DNA replication in vivo (146), it was not until the role of the *umuDC* gene products in replication restart was examined that it became clear that UmuD, together with UmuC, act to inhibit DNA replication following DNA damage (12,148).

Following UV irradiation of *E. coli*, DNA replication ceases almost immediately, and then resumes about 5–10 min later, ultimately regaining a normal rate within approximately 15 min following DNA damage (149). This inhibition and subsequent resumption of DNA replication has been termed replication restart (150), or induced replisome reactivation (IRR) (149). IRR is known to be dependent upon the *recA* gene product (149). Although the *umuDC* gene products do not appear to be required for the inhibition of DNA replication following DNA damage (149), they do influence the resumption of DNA replication in certain mutant *recA* genetic backgrounds (151). Consistent with this result, Opperman et al. (12) found that modest overexpression of the *umuDC* gene products from a plasmid significantly enhanced survival following UV irradiation. Importantly, this survival benefit was conferred by intact UmuD together with UmuC, and not by UmuD'/UmuC complex (i.e., pol V). Moreover, it correlated with a pronounced delay in the rate of replication recovery following a modest dose of UV irradiation (12). Thus, these findings suggest that intact UmuD, together with UmuC, acts to block processive DNA replication following DNA damage, thereby allowing additional time for DNA repair. In addition, physiologically relevant levels of the *umuDC* gene products also modulate the transition to rapid growth in cells that have experienced DNA damage while in stationary phase (13).

Insight into a possible mechanism for the *umuDC*-dependent checkpoint control, and the process by which the cell initiates the switch from checkpoint to TLS was offered by the finding that UmuD and UmuD' each interact physically with components of DNA pol III in reciprocal fashions: UmuD, which acts in checkpoint, interacts more strongly with the beta sliding clamp subunit of pol III than it does with the alpha catalytic subunit, while UmuD', which acts in TLS, interacts more strongly with the alpha subunit than it does the beta sliding clamp (72). Thus, intact UmuD, together with UmuC, may perturb DNA replication as part of a DNA damage checkpoint control by sequestering the beta clamp, thereby blocking processive DNA replication following induction of the SOS response (73,74). Although, it is currently unclear whether the *umuDC*-dependent DNA damage checkpoint control operates at all replication forks, or only at a subset of replication forks following DNA damage, there is a growing body of evidence suggesting that the timing of UmuD cleavage plays an important role in the restart of stalled replication forks following DNA damage (12,143,144). Taken together, these observations suggest that the auto-digestion of UmuD, which remodels the dimer interface and tertiary structure of the *umuD* gene product (65,76), serves to regulate the two temporally separate biological roles of the *umuDC* gene products, at least in part by attenuating the affinity of the different forms of the *umuD* gene product for beta (74).

Based on in vitro reconstitution of pol V-dependent TLS using purified components, beta appears to play an important role in stimulating pol V processivity by stabilizing the pol V–DNA complex (36,70,152,153). The interaction of UmuD' with the alpha catalytic subunit of pol III may also be important for TLS. Consistent with this interpretation, pol V stabilized a temperature sensitive pol III mutant in vitro, and low levels of pol V in combination with low levels of alpha were significantly more active for lesion bypass in vitro than were high levels of pol V alone (70). Taken together, these findings have been interpreted to suggest that the UmuD'-alpha interaction might act at least in part to effect a polymerase switch between pols III and V (127).

It was recently reported that human pol eta and pol iota interact with each other, and that this interaction is largely required in order for pol iota to be localized to replication foci (154). The fact that both pol eta and pol iota are structurally and functionally related to *E. coli* UmuC (pol V) suggests that polymerase–polymerase interactions may represent a general mechanism for managing the actions of the UmuC-like Y family of DNA polymerases, and may extend to other polymerase families as well.

UmuD and UmuD' each also interact with the epsilon proofreading subunit of pol III (72,148). Although it is currently unknown whether UmuD and UmuD' possess different affinities for this subunit, genetic analyses suggest that interaction of UmuD' with epsilon is dispensable for TLS (155), while the UmuD–epsilon interaction is important for the checkpoint (148).

Finally, the finding that higher than physiologically normal levels of UmuD' together with UmuC (i.e., pol V), but not intact UmuD together with UmuC (checkpoint), inhibit RecA-dependent recombination in vivo, has been interpreted to suggest that the *umuDC* gene products regulate RecA-dependent DNA homologous recombination (156). Inhibition of the DNA strand transfer activity of RecA by purified pol V has also been observed in vitro (157). Thus, RecA-mediated self-cleavage of UmuD to yield UmuD' appears to regulate as part of the SOS response both (i) a switch from the checkpoint to TLS, as well as (ii) a switch from RecA-dependent recombination-mediated repair to TLS.

7. AFTER THE DAMAGE IS REPAIRED: TURNING OFF THE SOS RESPONSE AND THE RETURN TO NORMALCY

Although the SOS response plays a vital role in both repairing as well as tolerating DNA damage, its constitutive induction confers a modest but obvious growth defect (i.e., Ref. 145), suggesting that at least some of the SOS-regulated gene products interfere with normal growth under standard laboratory conditions. An extreme example of this is the *sulA* gene product, which, when expressed constitutively, leads to lethal cell filamentation by inhibiting the action of the cell division protein FtsZ (158,159). In light of these findings, it is clear that mechanisms must exist which enable *E. coli* to turn off the SOS response after DNA damage has been dealt with and the enhanced capacity for repair is no longer needed. Although the precise molecular mechanisms by which DNA damage is monitored to ensure that the cell knows when to turn off the SOS response and resume processive DNA replication are currently unknown, it appears that at least under certain laboratory conditions, abrogation of the SOS response can be achieved within as little as 10 min from the time that the SOS-inducing signal is removed (160).

One simplified model to explain how the SOS response might be turned off assumes that the process of repairing DNA damage, or tolerating it via TLS, reduces the level of ssDNA, leading to a reduction in the amount of ssDNA and hence RecA-nucleoprotein filaments. This, in turn, allows the LexA repressor to once again accumulate to a level sufficient for effective repression of the SOS regulon (1,161,162). However, one important issue that is not addressed by this model concerns the mechanism(s) by which the co-protease activity of residual RecA–ssDNA nucleoprotein filaments is regulated so as to allow for a timely increase in the steady-state level of LexA. The fact that the steady-state level of RecA protein is induced roughly 10-fold as part of the SOS response (163,164) underscores the need to regulate the co-protease activity of these RecA–ssDNA nucleoprotein filaments.

Ohmori and colleagues (165,166) recently provided some insight into how *E. coli* might regulate the co-protease activity of RecA–ssDNA filaments to turn off the SOS response. Briefly, they found that the SOS-regulated DinI protein acts to inhibit both the co-protease and DNA strand transfer reactions catalyzed by the RecA–ssDNA filament, suggesting that DinI may play an important role in turning off the SOS response (Fig. 2). However, the precise mechanism by which this occurs is currently unclear: Ohmori and colleagues (166) reported that DinI binds to RecA–ssDNA filaments to act as a competitive inhibitor of RecA–ssDNA function, while Camerini-Otero and colleagues (167,168) reported that DinI prevents formation of the RecA–ssDNA filament by binding to free RecA protein and preventing it from binding ssDNA.

Another protein, RecX, which is encoded by the *recX* gene that is located immediately downstream form *recA*, similarly regulates the activity of the RecA–ssDNA nucleoprotein filament (169–171). Like *recA*, the *recX* gene is LexA-regulated (171). In contrast to the DinI protein, which inhibited the co-protease activity of the RecA–ssDNA nucleoprotein filaments only when present at significantly higher molar concentrations relative to RecA in vitro, RecX effectively inhibited both the strand transfer and co-protease activities of the RecA–ssDNA nucleoprotein filament at levels that were sub-stoichiometric relative to that of RecA (169,170). Thus, there appear to be at least two proteins, DinI and RecX, which act to regulate the activity of RecA following SOS induction, and may therefore play important roles in helping to turn off the SOS response following repair (Fig. 2).

Interestingly, a recent study demonstrated that although most SOS-regulated genes turn on at about the same time, presumably due to the rapid destruction of LexA following DNA damage, these same genes turn off in temporally distinct fashions, with timing differences between genes of the order of as many as 10 min (172). Moreover, the timing with which most genes examined were turned off was generally consistent with the presumed biological roles of their products (172). For example, *uvrA*, which functions in NER, and *lexA* and *recA*, which as discussed above play vitally important roles in regulation of the SOS response, are some of the earliest genes to turn off. *umuDC*, which encode DNA polymerase V that acts in most TLS, and also acts in a primitive DNA damage checkpoint control, is turned off at an intermediate time. The last genes to turn off are *polB*, which encodes DNA polymerase II. Pol II is involved in replication fork recovery following DNA damage, together with other genes involved in late stage repair processes, and, under certain conditions, plays a role in TLS. It should be stressed, however, that this study investigated the transcriptional control of these SOS-regulated genes only, and not the relative steady-state levels of the encoded proteins.

In addition to restoring LexA-dependent transcriptional regulation, the cell must also turnover the proteins expressed during SOS induction in order to fully turn off SOS. Thus, another question that remains to be fully addressed concerns the mechanisms by which the steady-state levels of the various SOS-regulated gene products return to the low, basal levels typical of non-SOS-induced conditions after SOS is turned off. Although this process is not yet understood for many of the SOS-regulated gene products, some important gains have been made with respect to understanding how the cell regulates the steady-state levels of the different *umuDC* gene products following DNA damage and SOS induction. In addition to the sophisticated series of transcriptional and post-translational controls that regulate expression of the different *umuDC* gene products, work from Woodgate and colleagues (173–176) indicates that the steady-state levels of the different *umuDC* gene products are also regulated at the level of proteolysis. UmuD and UmuC are each susceptible to degradation by the Lon protease, while UmuD', but only when present in the UmuD–UmuD' heterodimer, is susceptible to degradation by the ClpXP protease (173,175). The Lon protease is also involved in degradation of the SOS-regulated SulA protein that acts to block cell division following DNA damage (177). Thus, targeted proteolysis of the various SOS-regulated gene products may turn out to be a general mechanism by which their steady-state levels are regulated following their transcriptional repression once the SOS response has been turned off.

8. CONCLUDING REMARKS AND FUTURE PERSPECTIVES

The *E. coli* SOS response represents a highly coordinated process for managing global DNA repair and replication in response to a significant disruption of the normal replication process. Although all known mechanisms leading to SOS induction appear to channel through a common pathway involving RecA–ssDNA nucleoprotein filament formation followed by LexA self-cleavage, those involved in turning off the SOS response appear to be considerably more complex. Recent work suggests that a combination of regulatory mechanisms act together to shut off the SOS response following repair, including (i) auto-regulation of *lexA* transcription, (ii) efficient repression of the co-protease activity of the RecA–ssDNA nucleoprotein filaments by both the DinI and the RecX proteins, and (iii) efficient proteolysis of at least certain SOS-regulated gene products (i.e., the *sulA* and *umuDC* gene products). However, whether these mechanisms are necessary and sufficient for turning off the SOS response, or whether additional as yet unidentified functions are also required will require further investigation.

Other important questions concerning the SOS response that remain to be answered include the following: what is the source of the ssDNA that presumably activates RecA for induction of the SOS response? Does it come from an arrested replication fork, and if so, what is the structure of such a fork? In addition, how exactly does LexA protein undergo self-cleavage? Although significant strides have recently been made in better understanding the answer to the latter question, the structure of the LexA-DNA complex has been elusive. It will also be important to understand exactly how *E. coli* manages the actions of its five distinct DNA polymerases during SOS induction to ensure that the correct polymerase gains access to the replication fork at the appropriate time. A related question pertains to DNA repair: how many of the DNA functions that are regulated as part of the SOS response compete with each other to effect repair, and how many act together in a synergistic fashion? Finally, it will be important to understand the functions of the various SOS-regulated gene products, more than half of which have not yet been characterized. Answers to these and related questions will better define the coordination of DNA replication, recombination, repair, and damage tolerance following DNA damage in *E. coli*, and will serve as a valuable model for understanding similar control networks in other organisms.

ACKNOWLEDGMENT

This work was supported in part by Public Service Health grant GM66094.

REFERENCES

1. Friedberg EC, Walker GC, Siede W. DNA Repair and Mutagenesis. Washington, DC: ASM Press, 1995.
2. Radman M. Basic Life Sci 1975:355–367.
3. Walker GC, Smith BT, Sutton MD. In: Stroz G, Hengge-Aronis R, eds. Bacterial Stress Responses. Washington, DC: ASM Press, 2000:131–144.
4. Walker GC. Cold Spring Harb Symp Quant Biol 2000; 65:1–10.
5. Crowley DJ, Courcelle J. J Biomed Biotechnol 2002; 2:66–74.
6. Fernandez De Henestrosa AR, Ogi T, Aoyagi S, Chafin D, Hayes JJ, Ohmori H, Woodgate R. Mol Microbiol 2000; 35:1560–1572.
7. Courcelle J, Khodursky A, Peter B, Brown PO, Hanawalt PC. Genetics 2001; 158: 41–64.
8. Kenyon CJ, Walker GC. Proc Natl Acad Sci USA 1980; 77:2819–2823.
9. Little JW, Mount DW, Yanisch-Perron CR. Proc Natl Acad Sci USA 1981; 78: 4199–4203.
10. Little JW. Proc Natl Acad Sci USA 1984; 81:1375–1379.
11. Kenyon CJ, Walker GC. Nature 1981; 289:808–810.
12. Opperman T, Murli S, Smith BT, Walker GC. Proc Natl Acad Sci USA 1999; 96: 9218–9223.
13. Murli S, Opperman T, Smith BT, Walker GC. J Bacteriol 2000; 182:1127–1135.
14. Elledge SJ, Walker GC. J Mol Biol 1983; 164:175–192.
15. Kato T, Shinoura Y. Mol Gen Genet 1977; 156:121–131.
16. Steinborn G. Mol Gen Genet 1978; 165:87–93.
17. Trusca D, Bramhill D. Anal Biochem 2002; 307:322–329.
18. Bi E, Lutkenhaus J. J Bacteriol 1993; 175:1118–1125.
19. Jones C, Holland IB. Proc Natl Acad Sci USA 1985; 82:6045–6049.
20. Kleinsteuber S, Quinones A. Mol Gen Genet 1995; 248:695–702.
21. Quinones A, Juterbock WR, Messer W. Mol Gen Genet 1991; 227:9–16.
22. Quinones A, Kaasch J, Kaasch M, Messer W. EMBO J 1989; 8:587–593.
23. Kaasch M, Kaasch J, Quinones A. Mol Gen Genet 1989; 219:187–192.
24. Tadmor Y, Bergstein M, Skaliter R, Shwartz H, Livneh Z. Mutat Res 1994; 308:53–64.
25. Huisman O, D'Ari R, George J. J Bacteriol 1980; 144:185–191.
26. Miki T, Kumahara H, Nakazawa A. Mol Gen Genet 1981; 183:25–31.
27. Casaregola S, D'Ari R, Huisman O. Mol Gen Genet 1982; 185:430–439.
28. Woodgate R, Ennis DG. Mol Gen Genet 1991; 229:10–16.
29. Lusetti SL, Cox MM. Annu Rev Biochem 2002; 71:71–100.
30. Cox MM. Prog Nucl Acid Res Mol Biol 1999; 63:311–366.
31. Kowalczykowski SC, Dixon DA, Eggleston AK, Lauder SD, Rehrauer WM. Microbiol Rev 1994; 58:401–465.
32. Cox MM, Goodman MF, Kreuzer KN, Sherratt DJ, Sandler SJ, Marians KJ. Nature 2000; 404:37–41.

33. Sweasy JB, Witkin EM, Sinha N, Roegner-Maniscalco V. J Bacteriol 1990; 172: 3030–3036.
34. Dutreix M, Moreau PL, Bailone A, Galibert F, Battista JR, Walker GC, Devoret R. J Bacteriol 1989; 171:2415–2423.
35. Reuven NB, Tomer G, Livneh Z. Mol Cell 1998; 2:191–199.
36. Tang M, Bruck I, Eritja R, Turner J, Frank EG, Woodgate R, O'Donnell M, Goodman MF. Proc Natl Acad Sci USA 1998; 95:9755–9760.
37. Rajagopalan M, Lu C, Woodgate R, O'Donnell M, Goodman MF, MF, Echols H. Proc Natl Acad Sci USA 1992; 89:10777–10781.
38. Sassanfar M, Roberts JW. J Mol Biol 1990; 212:79–96.
39. Roberts JW, Phizicky EM, Burbee DG, Roberts CW, Moreau PL. Biochimie 1982; 64:805–807.
40. Lu C, Echols H. J Mol Biol 1987; 196:497–504.
41. Lu C, Scheuermann RH, Echols H. Proc Natl Acad Sci USA 1986; 83:619–623.
42. Little JW, Edmiston SH, Pacelli LZ, Mount DW. Proc Natl Acad Sci USA 1980; 77:3225–3229.
43. Kim B, Little JW. Science 1992; 255:203–206.
44. Fogh RH, Ottleben G, Ruterjans H, Schnarr M, Boelens R, Kaptein R. EMBO J 1994; 13:3936–3944.
45. Holm L, Sander C, Ruterjans H, Schnarr M, Fogh R, Boelens R, Kaptein R. Protein Eng 1994; 7:1449–1453.
46. Luo Y, Pfuetzner RA, Mosimann S, Paetzel M, Frey EA, Cherney M, Kim B, Little JW, Strynadka NC. Cell 2001; 106:585–594.
47. Lin LL, Little JW. J Bacteriol 1988; 170:2163–2173.
48. Mount DW, Low KB, Edmiston SJ. J Bacteriol 1972; 112:886–893.
49. Howard-Flanders P, Theriot L. Genetics 1966; 53:1137–1150.
50. Roland KL, Smith MH, Rupley JA, Little JW. J Mol Biol 1992; 228:395–408.
51. Smith MH, Cavenagh MM, Little JW. Proc Natl Acad Sci USA 1991; 88:7356–7360.
52. Little JW. Biochimie 1991; 73:411–421.
53. Neher SB, Flynn JM, Sauer RT, Baker TA. Genes Dev 2003; 17:1084–1089.
54. Little JW. Variations in the In Vivo Stability of LexA Repressor During the SOS Regulatory Cycle. New York: Alan Liss, 1983.
55. Paetzel M, Dalbey RE, Strynadka NC. Nature 1998; 396:186–190.
56. Paetzel M, Strynadka NC. Protein Sci 1999; 8:2533–2536.
57. Battista JR, Ohta T, Nohmi T, Sun W, Walker GC. Proc Natl Acad Sci USA 1990; 87:7190–7194.
58. Burckhardt SE, Woodgate R, Scheuermann RH, Echols H. Proc Natl Acad Sci USA 1988; 85:1811–1815.
59. Kim B, Little JW. Cell 1993; 73:1165–1173.
60. Little JW. J Bacteriol 1993; 175:4943–4950.
61. Craig NL, Roberts JW. Nature 1980; 283:26–30.
62. Shinagawa H, Iwasaki H, Kato T, Nakata A. Proc Natl Acad Sci USA 1988; 85: 1806–1810.
63. Nohmi T, Battista JR, Dodson LA, Walker GC. Proc Natl Acad Sci USA 1988; 85:1816–1820.
64. Peat TS, Frank EG, McDonald JP, Levine AS, Woodgate R, Hendrickson WA. Nature 1996; 380:727–730.
65. Ferentz AE, Walker GC, Wagner G. EMBO J 2001; 20:4287–4298.
66. Ferentz AE, Opperman T, Walker GC, Wagner G. Nat Struct Biol 1997; 4:979–983.
67. Bell CE, Frescura P, Hochschild A, Lewis M. Cell 2000; 101:801–811.
68. Paetzel M, Dalbey RE, Strynadka NC. J Biol Chem 277:9512–9519.
69. Witkin EM. Basic Life Sci 2002:369–378.
70. Tang M, Shen X, Frank EG, O'Donnell M, Woodgate R, Goodman MF. Proc Natl Acad Sci USA 1999; 96:8919–8924.

71. Reuven NB, Arad G, Maor-Shoshani A, Livneh Z. J Biol Chem 1999; 274:31763–31766.
72. Sutton MD, Opperman T, Walker GC. Proc Natl Acad Sci USA 1999; 96:12373–12378.
73. Sutton MD, Farrow MF, Burton BM, Walker GC. J Bacteriol 2001; 183:2897–2909.
74. Sutton MD, Narumi I, Walker GC. Proc Natl Acad Sci USA 99:5307–5312.
75. McDonald JP, Frank EG, Levine AS, Woodgate R. Proc Natl Acad Sci USA 2002; 95:1478–1483.
76. Sutton MD, Guzzo A, Narumi I, Costanzo M, Altenbach C, Ferentz AE, Hubbell WL, Walker GC. DNA Repair 2002; 1:77–93.
77. Slilaty SN, Rupley JA, Little JW. Biochemistry 1986; 25:6866–6875.
78. Mohana-Borges R, Pacheco AB, Sousa FJ, Foguel D, Almeida DF, Silva JL. J Biol Chem 2000; 275:4708–4712.
79. Yu X, Egelman EH. J Mol Biol 1992; 227:334–346.
80. Story RM, Steitz TA. Nature 1992; 355:374–376.
81. Story RM, Weber IT, Steitz TA. Nature 1992; 355:318–325.
82. VanLoock MS, Yu X, Yang S, Galkin VE, Huang H, Rajan SS, Anderson WF, Stohl EA, Seifert HS, Egelman EH. J Mol Biol 2003; 333:345–354.
83. Stasiak A, Egelman EH. Experientia 1994; 50:192–203.
84. Egelman EH, Stasiak A. J Mol Biol 1986; 191:677–697.
85. Egelman EH, Stasiak A. J Mol Biol 1988; 200:329–349.
86. Stasiak A, Egelman EH, Howard-Flanders P. J Mol Biol 1988; 202:659–662.
87. Yu X, Egelman EH. J Mol Biol 1993; 231:29–40.
88. Frank EG, Cheng N, Do CC, Cerritelli ME, Bruck I, Goodman MF, Egelman EH, Woodgate R, Steven AC. J Mol Biol 2000; 297:585–597.
89. Mustard JA, Little JW. J Bacteriol 2000; 182:1659–1670.
90. Konola JT, Guzzo A, Gow JB, Walker GC, Knight KL. J Mol Biol 1998; 276:405–415.
91. Sutton MD, Kim M, Walker GC. J Bacteriol 2001; 183:347–357.
92. Glassberg J, Meyer RR, Kornberg A. J Bacteriol 1979; 140:14–19.
93. Whittier RF, Chase JW. Mol Gen Genet 1981; 183:341–347.
94. Eggler AL, Lusetti SL, Cox MM. J Biol Chem 2003; 278:16389–16396.
95. Lusetti SL, Wood EA, Fleming CD, Modica MJ, Korth J, Abbott L, Dwyer DW, Roca AL, Inman RB, Cox MM. J Biol Chem 2003; 278:16372–16380.
96. Tseng YC, Hung JL, Wang TC. Mutat Res 1994; 315:1–9.
97. Clark AJ, Sandler SJ. Crit Rev Microbiol 1994; 20:125–142.
98. Liu YH, Cheng AJ, Wang TC. J Bacteriol 1998; 180:1766–1770.
99. Morimatsu K, Kowalczykowski SC. Mol Cell 2003; 11:1337–1347.
100. Webb BL, Cox MM, Inman RB. Cell 1997; 91:347–356.
101. Shan Q, Bork JM, Webb BL, Inman RB, Cox MM. J Mol Biol 1997; 265:519–540.
102. Rehrauer WM, Kowalczykowski SC. J Biol Chem 1996; 271:11996–12002.
103. Rehrauer WM, Lavery PE, Palmer EL, Singh RN, Kowalczykowski SC. J Biol Chem 1996; 271:23865–23873.
104. Egelman EH, Yu X. Science 1989; 245:404–407.
105. Kelman Z, O'Donnell M. Curr Opin Genet Dev 1994; 4:185–195.
106. Pages V, Fuchs RP. Science 2003; 300:1300–1303.
107. Anderson DG, Kowalczykowski SC. Cell 1998; 95:975–979.
108. Witkin EM. Genetics 1975; 79:199–213.
109. Goze A, Sedgwick SG. Mutat Res 1978; 52:323–331.
110. Peterson KR, Wertman KF, Mount DW, Marinus MG. Mol Gen Genet 1985; 201:14–19.
111. SaiSree L, Reddy M, Gowrishankar J. J Bacteriol 2000; 182:3151–3157.
112. Ossanna N, Mount DW. Bacteriol J 1989; 171:303–307.
113. Slater SC, Lifsics MR, O'Donnell M, Maurer R. J Bacteriol 1994; 176:815–821.
114. Lane HE, Denhardt DT. J Bacteriol 1974; 120:805–814.
115. Lane HE, Denhardt DT. J Mol Biol 1975; 97:99–112.
116. Nurse P, Zavitz KH, Marians KJ. J Bacteriol 1991; 173:6686–6693.

117. Sandler SJ, McCool JD, Do TT, Johansen RU. Mol Microbiol 2001; 41:697–704.
118. Ohmori H, Saito M, Yasuda T, Nagata T, Fujii T, Wachi M, Nagai K. J Bacteriol 1995; 177:156–165.
119. Kuhn B, Abdel-Monem M. Eur J Biochem 1982; 125:63–68.
120. Klinkert MQ, Klein A, Abdel-Monem M. J Biol Chem 1980; 255:9746–9752.
121. Moolenaar GF, Moorman C, Goosen N. J Bacteriol 2000; 182:5706–5714.
122. Roberts D, Kleckner N. Proc Natl Acad Sci USA 1988; 85:6037–6041.
123. Eichenbaum Z, Livneh Z. Genetics 1998; 149:1173–1181.
124. Weinreich MD, Makris JC, Reznikoff WS. J Bacteriol 1991; 173:6910–6918.
125. Taddei F, Matic I, Radman M. Proc Natl Acad Sci USA 1995; 92:11736–11740.
126. Yeiser B, Pepper ED, Goodman MF, Finkel SE. Proc Natl Acad Sci USA 2002; 99:8737–8741.
127. Sutton MD, Walker GC. Proc Natl Acad Sci USA 2001; 98:8342–8349.
128. McKenzie GJ, Lee PL, Lombardo MJ, Hastings PJ, Rosenberg SM. Mol Cell 2001; 7:571–579.
129. McKenzie GJ, Magner DB, Lee PL, Rosenberg SM. J Bacteriol 2003; 185:3972–3977.
130. Bonner CA, Stukenberg PT, Rajagopalan M, Eritja R, O'Donnell M, McEntee K, Echols H, Goodman MF. J Biol Chem 1992; 267:11431–11438.
131. Wagner J, Fujii S, Gruz P, Nohmi T, Fuchs RP. EMBO Rep 2000; 1:484–488.
132. Becherel OJ, Fuchs RP, Wagner J. DNA Repair 2002; 1:703–708.
133. Villarroya M, Perez-Roger I, Macian F, Armengod ME. EMBO J 1998; 17:1829–1837.
134. Crowley DJ, Hanawalt PC. J Bacteriol 1998; 180:3345–3352.
135. Crowley DJ, Hanawalt PC. Mutat Res 2001; 485:319–329.
136. Hashimoto M, Imhoff B, Ali MM, Kow YW. J Biol Chem 2003; 278:28501–28507.
137. Sutton MD, Smith BT, Godoy VG, Walker GC. Annu Rev Gen 2000; 34:479–497.
138. Friedberg EC, Feaver WJ, Gerlach VL. Proc Natl Acad Sci USA 2000; 97:5681–5683.
139. Ohmori H, Friedberg EC, Fuchs RP, Goodman MF, Hanaoka F, Hinkle D, Kunkel TA, Lawrence CW, Livneh Z, Nohmi T, Prakash L, Prakash S, Todo T, Walker GC, Wang Z, Woodgate R. Mol Cell 2001; 8:7–8.
140. Johnson RE, Kondratick CM, Prakash S, Prakash L. Science 1999; 285:263–265.
141. Masutani C, Kusumoto R, Yamada A, Dohmae N, Yokoi M, Yuasa M, Araki M, Iwai S, Takio K, Hanaoka F. Nature 1999; 399:700–704.
142. Lehmann AR, Bridges BA. Br J Dermatol 1990; 122:115–119.
143. Rangarajan S, Woodgate R, Goodman MF. Mol Microbiol 2002; 43:617–628.
144. Rangarajan S, Woodgate R, Goodman MF. Proc Natl Acad Sci USA 1999; 96:9224–9229.
145. Opperman T, Murli S, Walker GC. J Bacteriol 1996; 178:4400–4411.
146. Marsh L, Walker GC. J Bacteriol 1985; 162:155–161.
147. Sutton MD, Walker GC. J Bacteriol 2001; 183:1215–1224.
148. Sutton MD, Murli S, Opperman T, Klein C, Walker GC. J Bacteriol 2001; 183:1085–1089.
149. Khidhir MA, Casaregola S, Holland IB. Mol Gen Genet 1985; 199:133–140.
150. Echols H, Goodman MF. Annu Rev Biochem 1991; 60:477–511.
151. Witkin EM, Roegner-Maniscalco V, Sweasy JB, McCall JO. Proc Natl Acad Sci USA 1987; 84:6805–6809.
152. Maor-Shoshani A, Livneh Z. Biochemistry 2002; 41:14438–14446.
153. Pham P, Bertram JG, O'Donnell M, Woodgate R, Goodman MF. Nature 2001; 409:366–370.
154. Kannouche P, Fernandez de Henestrosa AR, Coull B, Vidal AE, Gray C, Zicha D, Woodgate R, Lehmann AR. EMBO J 2003; 22:1223–1233.
155. Slater SC, Maurer R. Proc Natl Acad Sci USA 1991; 88:1251–1255.
156. Sommer S, Bailone A, Devoret R. Mol Microbiol 1993; 10:963–971.
157. Rehrauer WM, Bruck I, Woodgate R, Goodman MF, Kowalczykowski SC. J Biol Chem 1998; 273:32384–32387.

158. Ennis DG, Little JW, Mount DW. J Bacteriol 1993; 175:7373–7382.
159. Santos D, De Almeida DF. J Bacteriol 1975; 124:1502–1507.
160. Casaregola S, D'Ari R, Huisman O. Mol Gen Genet 1982; 185:440–444.
161. Little JW, Mount DW. Cell 1982; 29:11–22.
162. Little JW. J Mol Biol 1983; 167:791–808.
163. Huisman O, D'Ari R, Casaregola S. Biochimie 1982; 64:709–712.
164. Sommer S, Boudsocq F, Devoret R, Bailone A. Mol Microbiol 1998; 28:281–291.
165. Yasuda T, Morimatsu K, Horii T, Nagata T, Ohmori H. EMBO J 1998; 17:3207–3216.
166. Yasuda T, Morimatsu K, Kato R, Usukura J, Takahashi M, Ohmori H. EMBO J 2001; 20:1192–1202.
167. Voloshin ON, Ramirez BE, Bax A, Camerini-Otero RD. Genes Dev 2001; 15:415–427.
168. Ramirez BE, Voloshin ON, Camerini-Otero RD, Bax A. Protein Sci 2000; 9:2161–2169.
169. Venkatesh R, Ganesh N, Guhan N, Reddy MS, Chandrasekhar T, Muniyappa K. Proc Natl Acad Sci USA 2002; 99:12091–12096.
170. Stohl EA, Brockman JP, Burkle KL, Morimatsu K, Kowalczykowski SC, Seifert H. J Biol Chem 2003; 278:2278–2285.
171. Pages V, Koffel-Schwartz N, Fuchs RP. DNA Repair 2003; 2:273–284.
172. Ronen M, Rosenberg R, Shraiman BI, Alon U. Proc Natl Acad Sci USA 2002; 99:10555–10560.
173. Gonzalez M, Frank EG, Levine AS, Woodgate R. Genes Dev 1998; 12:3889–3899.
174. Gonzalez M, Rasulova F, Maurizi MR, Woodgate R. EMBO J 2000; 19:5251–5258.
175. Frank EG, Ennis DG, Gonzalez M, Levine AS, Woodgate R. Proc Natl Acad Sci USA 1996; 93:10291–10296.
176. Gonzalez M, Frank EG, McDonald JP, Levine AS, Woodgate R. Acta Biochim Pol 1998; 45:163–172.
177. Mizusawa S, Gottesman S. Proc Natl Acad Sci USA 1983; 80:358–362.

36
Recognition of DNA Damage as the Initial Step of Eukaryotic Checkpoint Arrest

Wolfram Siede
Department of Cell Biology and Genetics, University of North Texas Health Science Center, Fort Worth, Texas, U.S.A.

1. INTRODUCTION

Eukaryotic cells can respond to an attack by DNA-damaging agents with arrest in various stages of the cell cycle. Such checkpoint responses are transient, actively regulated and beneficial to the organism (1,2). It is thought that transient arrest creates enhanced opportunities for DNA repair and prevents the conversion of repairable into irreparable damage, such as mutations or chromosome fragmentation, by processes, such as mitosis, that are associated with cell cycle progression. It is instructive to discuss a checkpoint process as an example of a signal-transduction pathway—with sensors that recognize DNA damage or a consequence of DNA damage, mediators that amplify and transform the signal into activation of transducers and executers that finally modify cell cycle targets. Coordinated and related responses use a largely overlapping signal-transduction network: these include decisions on cell fate (such as apoptosis or senescence), transcriptional regulation of genes involved in DNA repair and direct modification of repair proteins. Indeed, it has been argued that regulation of repair may be of more importance for resistance to DNA-damaging agents than cell cycle arrest. Failure to invoke these signal-transduction networks can give rise to genetic instability and this area of investigation has considerable ramifications for cancer research. However, a conceptually similar signal-transduction system appears to exist already in *Escherichia coli*, as mentioned in the previous chapter.

This review focuses exclusively on aspects of damage recognition that initiate checkpoint pathways in eukaryotic systems, primarily in the yeast *Saccharomyces cerevisiae* and mammalian cells. A discussion of the complex events that relate to damage signaling in replication is not included herein, as these are discussed in a separate chapter.

2. EARLY STUDIES CHARACTERIZING CHECKPOINT TRIGGERING DAMAGE AND SENSOR PROTEINS

Various DNA-damaging agents of very different nature are efficient in triggering various cell cycle checkpoints. At the outset, it was conceivable that sensing DNA damage as the initial precondition for checkpoint responses is accomplished through sensing of one or more common but indirect consequences of damage such as stalled replication or transcription. DNA-damaging agents can also create cytoplasmic damage that may serve as important stress signals, i.e., alkylating agents can damage proteins or oxidizing agents may cause membrane damage. These consequences of DNA-damaging agents may differ according to cell cycle stage and each form of cell cycle arrest may use different specific sensing mechanisms.

However, it is now quite clear that the checkpoint system employs one or more specific sensing systems that show an immediate response and function in close proximity to the initial DNA damage, possibly also in direct contact with DNA repair proteins. With a few notable exceptions, indirect consequences are not essential for triggering a signal. DNA-damage sensing during replication, however, will require special consideration (Chapter 37). Here, we are introducing several sensors which appear to function largely independently of cell cycle stage and mainly respond to structural changes within DNA.

Generally speaking, genotoxic agents induce a variety of DNA damages and it is difficult to conclude what damage type and frequency may be required for generating a checkpoint signal. But there are methods of introducing just a single type of damage, such as the introduction of double-strand breaks by cuts following restriction enzyme expression. If restriction enzymes are electroporated into human cells, stabilization of the tumor suppressor protein p53 occurs (3). This is commonly associated with checkpoint arrest and is frequently used as a downstream marker for DNA-damage sensing, signal generation, and transduction. Such a dependency on strand breakage was further confirmed by the similar effects of an agent such as camptothecin that stabilizes the cleavable DNA-topoisomerase I complex and results specifically in double-strand breakage in the vicinity of an approaching replication fork (3). Similarly, in *S. cerevisiae*, thermoconditional DNA-topoisomerase I (*top1*) mutants will trigger checkpoint arrest under restrictive conditions (4).

In *S. cerevisiae*, the HO endonuclease system allows the introduction of a defined number of targeted double-strand breaks by controlled overexpression of the endonuclease required for mating type switching (5). Double-strand breaks that are quickly repaired do not result in cell cycle arrest (6). However, if there is no donor sequence available, the double-strand break is rendered irreparable. Under these conditions, even a single double-strand break can trigger extended G2/M arrest (7). In spite of the inability to repair, this arrest following a single irreparable double-strand break is transient. However, introduction of two irreparable double-strand breaks on different chromosomes results in irreversible arrest. The phenomenon of *adaptation* is influenced by the amount of single-stranded DNA produced during double-strand break processing which appears to serve as an important signal. I will return to adaptation as a special kind of damage recognition and assessment later.

The already introduced *cdc13* mutant of *S. cerevisiae* was important in establishing single-stranded DNA as a possibly important structural signal for checkpoint activation. Cdc13 is a single-stranded DNA-binding protein that protects telomeres from uncontrolled single-strand degradation (8). Interestingly, the extent of single-stranded DNA appearing around telomeres in *cdc13* mutants under restrictive conditions (9) seems to be controlled by checkpoint proteins, with certain potential sensor proteins being required; Rad9, however, negatively regulates the process (10) and so do downstream acting kinases (11). Double-strand breaks and single-stranded DNA are also created in the context of recombination during meiosis I and again, many checkpoint proteins are involved in faithful completion of meiotic recombination; again Rad9 plays a special role and is not required (12,13).

In vertebrate cells, microinjection of damage model substrates into fibroblast nuclei proved to be highly informative (14). One can elicit p53-dependent G1 arrest by microinjection of a plasmid that had been converted into linear DNA by restriction digest at a single cleavage site, but not by injection of the covalently closed circular plasmid. Large single-stranded gaps were found to be effective, too, but not small gaps of less than 50 nucleotides. This system is highly sensitive—the required amount of DNA was estimated to be just a few plasmids or maybe even as little as a single linearized plasmid molecule. *Xenopus* egg extracts were another useful system where similar results were obtained. Addition of double-stranded DNA fragments or of single stranded-DNA (that may have been converted to double-stranded DNA by the replication machinery) results in typical phosphorylation events of downstream kinases that normally accompany checkpoint arrests (15,16).

Thus, there is overwhelming evidence that double-strand breaks can trigger cell cycle checkpoints. Most studies have therefore concentrated on agents such as ionizing radiation that cause strand breakage. If persistent, the required damage dosage may be as little as one double-strand break per genome. We may be dealing with a system that can discriminate between damaged and undamaged DNA with extremely high specificity. Or, perhaps more likely, with a network of sensors that can easily be put on alert but requires signal enforcement by positive feedback loops in order to exceed a threshold and activate a transmitted signal. There is also good evidence that single-stranded DNA of sufficient length can be an important trigger. For a double-strand break to serve as a checkpoint signal, its conversion into single-stranded DNA through exonuclease activity may indeed be required.

How does one identify candidate sensor proteins? As I will discuss in more detail, checkpoint activation depends on signal transmission through kinases that are themselves activated by phosphorylation. Obviously, a sensor protein will be needed during the earliest events. The sooner a protein gets modified after DNA-damaging treatments and the more proximal it maps to DNA damage because of independence from other phosphorylation events, the more likely it functions as a sensor.

Other types of candidates include those proteins with known or suspected DNA-binding or modifying activities whose inactivation compromises checkpoints. DNA repair proteins are certainly prime candidates. After all, some of these have evolved to recognize DNA damage highly specifically and one may wonder if specific sensors dedicated to the checkpoint system even exist. As we will see, the truth lies somewhere in between. There are indeed sensor proteins that are exclusively involved in checkpoint pathways but certain repair-related sensor proteins are also being co-opted to accelerate damage recognition or monitor ongoing repair.

3. THE ATM PROTEIN IS A KINASE AND A PUTATIVE DAMAGE SENSOR

Ataxia telangiectasia (AT) is an autosomal hereditary disease with a wide array of phenotypes (17–19). Afflicted individuals are sensitized to ionizing radiation, immundeficient, and susceptible to cancer, they show cerebellar degeneration, progressive mental retardation, and many other, seemingly unrelated phenotypes, such as dilation of blood vessels (telangiectasia) and uneven gait (ataxia). Heterozygotes in the human population can be found at a frequency of about 1% and a possible predisposition of such individuals for cancer, especially breast cancer, is of considerable concern (20). Also, the AT cells are characterized by what has long been known as "radioresistant DNA synthesis," an inability to downregulate DNA synthesis following the treatment with ionizing radiation and radiomimetic chemicals (21). More detailed studies point to multiple checkpoint defects: at the G1/S transition, in S (frequently called "intra S-phase checkpoint") and at the G2/M transition.

Cloning of the ATM (for *a*taxia *t*elangiactasia *m*utated) gene revealed a huge protein with homology to known phosphoinositol (PI-) kinases (22,23). The activity of the protein, however, is that of a protein kinase. The sequence of the ATM gene product indicates sequence similarities that extend beyond the PI-kinase domain and thus evolutionary conservation. The yeast homolog Tel1 is necessary for telomere maintenance, however, the protein on its own is not essential for DNA-damage responses but it plays a role in a redundant pathway (24). (A list of gene products involved in checkpoint control and their respective nomenclature in various organisms can be found in Chapter 37, Table 1.)

Another PI-kinase-like protein is represented by the catalytic subunit of DNA-PK (DNA-PKCS) that has already been discussed (see Chapters 28 and 29) (25). One might expect to find targeting subunits such as the Ku proteins for DNA-PKCS that link the ATM kinase to certain DNA structures. No functionally comparable protein has been identified, nevertheless, ATM-associated kinase activity is induced extremely rapidly after ionizing radiation treatment without any concomitant increase of protein abundance (26,27). There is no upstream acting protein known and thus a direct interaction with damage is conceivable. Indeed, studies on the purified protein have indicated DNA binding with a preference for double-strand DNA ends (28,29). The ATM can also be found at DNA breaks that are introduced during V(D)J recombination (30).

ATM-kinase activation by ionizing radiation was shown to be accompanied by autophosphorylation and dissociation of ATM molecules that have been rendered inactive by sequestration (31). In undamaged cells, ATM forms dimers or multimeric structures by binding of one molecule's kinase domain to the internal domain of a neighboring molecule in *trans*. Phosphorylation of the internal domain residue serine 1981 by the partner molecule releases ATM whose catalytic domain is now free to phosphorylate other substrates. This mechanism also provides a satisfying explanation for the dominant inhibitory effect of kinase-dead (kd) versions of the protein.

But what kind of event starts the reaction? The activation by ionizing radiation occurs extremely rapidly and efficiently. Within 5 min, a dose that results in only about 18 double-strand breaks per mammalian cell genome induces phosphorylation of 50% of all ATM molecules (31). The initial activation appears to take place in a distance from double-strand breaks. Immunofluorescence staining shows an initially diffuse pattern of phosphorylated ATM and a foci formation at putative sites of double-strand breaks occurs only at later stages. Based on these observations, it was speculated that an alteration of higher-order chromatin structure due to strand breakage may represent the triggering event (31). Interestingly, the diffuse pattern of activation can also be seen with agents such as hypotonic swelling or chloroquine, that do not induce strand breaks but changes in chromatin structure. Maybe one has to regard the initial effect as a mode of putting a sensor on alert in order to facilitate subsequent detection of the "real" damage that ultimately triggers downstream checkpoint responses.

4. THE ATR PROTEIN AND ITS TARGETING SUBUNIT

The ATR protein represents yet another PI-kinase related gene product with an important role in controlling various checkpoints (32,33). *ATR* stands for *ATM* and *R*ad3 *r*elated, since *Schizosaccharomyces pombe* Rad3 and its *S. cerevisiae* equivalent Mec1 are structural and, in many but not all respects, also functional homologs of vertebrate ATR. Attesting to its role in multiple areas of checkpoint responses, *S. cerevisiae MEC1* has been isolated several times—as a mitotic entry checkpoint mutant (34), as a mutant conferring extreme sensitivity to hyroxyurea (35) and, finally, as a mitotically essential gene that plays a role in repair and recombination (*esr1*) (36).

There appears to be a division of tasks between the two PI-kinase-like products that has changed during eukaryotic evolution. Whereas ATM responds mainly to strand-breaking agents such as ionizing radiation, ATR plays important roles in cellular regulation following UV-radiation treatment or the inhibition of replication: overexpression of catalytically inactive ATR results in hypersensitivity to these agents and abrogation of G2 checkpoint arrest (37). In *S. cerevisiae*, Mec1 also responds to ionizing radiation damage and plays a more general role in checkpoint regulation than the ATM-structure homolog Tel1.

The regulation of ATR protein does not resemble that of ATM. ATR does not form oligomeres and its protein associations do not change in response to DNA damage (38). Although capable in vitro, autophosphorylation does not seem to occur in vivo. Comparable to the DNA-PK paradigm, an evolutionary conserved DNA-binding protein, termed Rad26 in *Schizosaccharomyces pombe*, Lcd1 in *S. cerevisiae* or ATRIP in vertebrates, has been identified that partners with ATR/Mec1/Rad3 and targets the protein to sites of DNA damage. The *Schizosaccharomyces pombe* checkpoint protein Rad26 was already known to play a role very early in checkpoint activation and its DNA-damage-associated phosphorylation is only dependent on its interacting partner, the ATR ortholog Rad3 (39). The *S. cerevisiae* ortholog of Rad26, Lcd1 (also known as Ddc2 or Pei1) was isolated based on sequence similarity and as a Mec1-interacting protein (40–42). Whereas Mec1 kinase activity per se is not influenced by Lcd1, it seems to act as its targeting factor. As such, however, it is essential for Mec1's role in checkpoint arrest. Consequently, *lcd1* mutant cells are highly sensitive to DNA-damaging agents and inhibitors of replication, they are defective in the phosphorylation of downstream checkpoint kinases and in all cell cycle checkpoints that are affected in *mec1* mutants. Lcd1 is indeed capable of binding DNA substrates when extracts are probed and in vivo, its presence will recruit Mec1 to DNA damage (43). Extract experiments with bead-bound DNA indicate a sequence-independent binding to double-stranded and single-stranded oligonucleotides; competition experiments proved the binding to DNA ends (43). Chromatin immuno-precipitation and GFP-labeling demonstrated also in vivo binding of Lcd1 near telomeres following introduction of DNA damage in *cdc13* mutant strains or near HO-endonuclease-induced double-strand breaks (43–45). Other known or putative damage sensors such as the Ku proteins are not required for this recruiting. Although Lcd1 is subject to Mec1-dependent phosphorylation (41), the kinase activity of the latter is not required for localization of the complex (43).

The human protein termed ATRIP has been identified as an homologous *ATR*-interacting protein that is phosphorylated by ATR and colocalizes with ATR in intranuclear foci following DNA damage or inhibition of replication (46). Sequence similarity to the described yeast proteins but also to *Drosophila* Mus304, a known checkpoint protein (47), was evident. Interestingly, ATR protein stability depends on the presence of ATRIP and vice versa. Reducing ATRIP and consequently ATR levels using *si* (*s*mall *i*nterfering) RNA technology results in greatly reduced G2 arrest following γ-irradiation.

As discussed in other chapters, double-stranded DNA breaks can easily undergo exonucleolytic processing and it should come as no surprise that the heterotrimeric eukaryotic single-stranded DNA-binding complex replication factor A (RPA) (48) plays also a role in DNA-damage sensing during checkpoint arrest. This abundant protein complex will coat emerging single-stranded DNA quickly and with high affinity. In the context of cell cycle regulation, it is important to realize that RPA may be useful as an important damage *sensor* but possibly also as a *target* for cell cycle arrest pathways since it is essential for replicon initiation in S-phase.

Indeed, purified epitope-tagged ATRIP requires the presence of RPA for efficient binding to single-stranded DNA probes, *E. coli* ssDNA-binding protein cannot substitute (49). Interestingly, RPA is also required for ATRIP binding to the ends of double-stranded DNA, most likely indicating a need for single-strand degradation before recognition by ATRIP. Recruitment of the ATR–ATRIP complex to single stranded-DNA by RPA can result in the phosphorylation of other damage sensors (to be discussed below) (49), including the second-largest subunit of RPA itself (38).

Also in vivo, RPA appears to be required for the binding of human ATRIP to ionizing radiation-induced damage sites since colocalization of RPA and ATR in foci can be abolished by siRNA reducing expression of RPA70, the largest RPA subunit (49). Following UV and inhibition of replication, the same technique prevents phosphorylation of a downstream kinase. Using *S. cerevisiae*, it was demonstrated that lowered Rfa1 (= RPA70) levels or a Rfa1 mutant known to diminish checkpoint responses (50) reduces binding of the ATRIP homolog Lcd1 to HO-induced double-strand breaks (49).

Also, ATR may play a role as a damage sensor in the absence of ATRIP or RPA. ATR has RPA-independent affinity for single-stranded DNA (38) which may be concealed in the ATR–ATRIP complex (49). A possibly related provocative finding concerns the capability of ATR alone to preferentially bind to UV-irradiated DNA in vitro (51). This could indicate a role of ATR without ATRIP in direct recognition of bulky base damage without requiring conversion into strand breaks or single-stranded DNA tracts. While it has been estimated that 95% of all ATR is complexed with ATRIP in vivo, even without DNA damage (49), the likelihood of a dissociation under physiological conditions has also been discussed (38). Nevertheless, the in vivo relevance of any activities of ATR separate from ATRIP remains to be demonstrated.

5. PCNA- AND RFC-LIKE CLAMP AND CLAMP LOADER COMPLEXES FUNCTION AS DAMAGE SENSORS

Three interacting checkpoint proteins, originally described in *Schizosaccharomyces pombe* and *S. cerevisiae* and termed Rad9, Rad1, Hus1 or Ddc1, Rad17, Mec3, respectively, confer very similar defects if mutated (10,52–60). Such mutant cells are largely deficient in UV and ionizing radiation-induced cell cycle arrests. In *Schizosaccharomyces pombe*, but not in *S. cerevisiae* these proteins play also a major role in S-phase extension due to replicational stress (see Chapter 39) . Structurally and functionally homologous proteins were isolated from various higher eukaryotes (61–70), we will refer to the complex as *9-1-1 complex*. Rec1, the *Ustilago maydis* homolog of the Rad1^{Sp}/Rad17^{Sc} component has been characterized as a $3'{\to}5'$ exonuclease (71,72) but the same activity has remained controversial for the human homolog (61,62). Instead, a different protein of the same complex, hRad9, appears to have $3'{\to}5'$ exonuclease activity (73), thus suggesting that this activity is indeed somehow an important property of the complex. On the other hand, the purified *S. cerevisiae* complex lacks any exonuclease activity (74).

The components of this complex are nuclear even in unirradiated mammalian cells, however, following DNA damage a close association with chromatin is observed (75–77). The observed increased nuclear retention is correlated with phosphorylation of hRad9 and hRad1. Similarly, the *S. cerevisiae* hRad9 homolog Ddc1 interacts with Mec1 and is subject to Mec1-dependent phosphorylation in response to DNA damage (41). The significance of these phosphorylation events has not been conclusively demonstrated.

A structural analysis of the components of the complex revealed an essential fact: homology with PCNA, a homomeric trimer known as a "sliding clamp"-like tethering factor that improves processivity of replicative polymerases (58,78–80). Consequently, it was assumed that the three components form a heterotrimeric ring structure that encircles DNA. The complex was reconstituted following coexpression of the individual components in a heterologous system and its postulated structure was indeed confirmed (Fig. 1) (81–85). Although the 9-1-1 subunits are predominantly found in this complex, some flexibility in their interactions is also apparent. For instance, Rad17^{Sc} plays a role in meiotic checkpoint controls that can be dissociated from that of the other components (86). Rad17^{Sc} can also undergo self-interaction that is stimulated by certain DNA-damaging agents and an analysis of point mutations has correlated a defect in self-interaction with a checkpoint defect (87). Although not dependent on 9-1-1 complex formation, this interaction may nevertheless be used to pull several of these complexes together. Alternatively, a separate trimeric complex containing two Rad17 molecules may exist and a low level of such a complex has indeed been detected in vitro (74). Interestingly, in vivo interaction of 9-1-1 components and PCNA in response to damage has also been reported (88).

Reminiscent of the replication factor C (RFC)-PCNA paradigm, there is not only a clamp but also a clamp loader. A checkpoint phenotype that is very similar

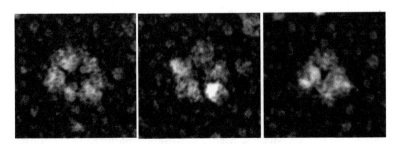

Figure 1 Transmission electron microscopy images of the human 9-1-1 complex. *Source*: Courtesy of T. Tsurimoto.

and epistatic to 9-1-1 mutants is conferred by mutations in the *S. cerevisiae* checkpoint gene *RAD24* and its *Schizosaccharomyces pombe* counterpart *RAD17*. Structural and functional orthologs were identified in higher eukaryotes (89,90). Sequence similarity with subunits of RFC was noted (91), the gene, however, is not essential for viability. Rad24^{Sc} forms an alternative RFC-like complex with 4 out of 5 RFC subunits where the checkpoint protein replaces the largest subunit, Rfc1 (92,93). This complex was reconstituted in vitro and shown to possess DNA binding and ATPase activity (85).

RFC binds to a primer–template junction, opens the PCNA ring and loads it onto DNA (94,95). This begs the obvious question: Does Rad17$^{Sp/Hs}$-RFC load the 9-1-1 complex onto DNA? The quest for interactions between both complexes in vivo yielded initially negative results (54,76), but a dependency of 9-1-1 loading on Rad17 was subsequently confirmed (44,45,96,97). In vitro supercomplex formation and loading of the 9-1-1 complex onto nicked or gapped plasmid DNA is readily detectable (74,81). The interaction is primarily established between Rad17$^{Sp/Hs}$ and Rad9$^{Sp/Hs}$ which implies that normal RFC cannot substitute for Rad17–RFC in the 9-1-1 loading reaction. Following loading, the 9-1-1 complex can slide for more than 1 kb along duplex DNA and this process may be of importance for the detection of DNA damage in a distance from the entry site which may itself constitute damage (= a double strand break) (74).

Once again, RPA functions as an important cofactor that can modulate the loading reaction. The presence of RPA stimulates the binding of Rad17–RFC to ss DNA, gapped and primed DNA structures as well as the subsequent loading of the 9-1-1 complex (98). In contrast to normal RFC, structures with both 3′ and 5′ overhangs can be accepted, an important difference that seems fitting to its role as a sensor of a broad spectrum of substrates created by DNA-damaging agents.

6. CROSSTALK BETWEEN SENSORS

We have thus described the two major types of candidate damage sensors for checkpoint responses (independent of S-phase): PI-kinase related proteins with, in case of ATR/Mec1/Rad3, a targeting subunit and the 9-1-1-Rad17-RFC clamp/clamp loader complex. Is there any relationship between these complexes? Do they act independently or are they part of a larger supercomplex? Is there a division of labor according to the specific type of DNA damage?

Concerning DNA strand breaks, recruitment and initial damage recognition occurs largely independently but there is subsequent crosstalk that critically influences downstream events. Specific binding of candidate sensor proteins to DNA damage in vivo was elegantly visualized by the use of GFP (green fluorescent protein)-fusions in *S. cerevisiae* (44). Instead of a mere demonstration of induced foci, a firm correlation with initial damage was established by introducing a defined number of damaged targets—a single double-strand break or *cdc13*-related damage at telomeres, corresponding to the number of chromosomes. Indeed, the number of foci per cell was an exact reflection of the number of damaged sites. When the 9-1-1 complex member Ddc1^{Sc} (= Rad9$^{Sp/Hs}$) is GFP-labeled, foci formation is dependent on other 9-1-1 components as well as on Rad24^{Sc} (= Rad$^{17Sp/Hs}$), as expected from the biochemical model, but not on Mec1^{Sc}. Mec1, however, is required for foci formation of Lcd1–GFP (= ATRIP) but none of the 9-1-1 or RFC-like complex components is. Such independent recruiting of the two types of complexes was also found using a chromatin immunoprecipitation approach where protein binding close to a targeted HO-endonuclease-induced double-strand break was detected (45).

Nevertheless, studies in human cells also revealed extensive crosstalk. HRad17 (= Rad24^{Sc}) can be found in a complex with ATM and ATR (99). This association is enhanced by DNA-damaging agents, in the case of ATM very selectively by ionizing radiation. HRad17 is subject to DNA-damage-induced, ATM/ATR-dependent phosphorylation of Ser 635 and Ser 645. Overexpression of a Rad17 version with alanine substitutions in both positions prevented its association with the 9-1-1 complex and abrogated G2 checkpoint arrest. In the light of the GFP-localization studies in yeast discussed above, this dependency is somewhat surprising. However, nothing is known about phosphorylation of the *S. cerevisiae* Rad17 homolog Rad24^{Sc} and it is possible that a kinase other than Mec1 (such as Tel1) can provide a back-up role for phosphorylation in the yeast system.

Another study confirmed colocalization and ATR-dependent phosphorylation of chromatin-bound hRad17 which is already chromatin-associated even without DNA damage (97). This contrasts somewhat with Rad17 in *Schizosaccharomyces pombe* which only becomes tightly associated with chromatin following DNA damage (100). ATR-dependent phosphorylation of chromatin-bound human Rad17 was not required for the recruitment of the 9-1-1 complex. The sequence of events suggested in this study (97) is the following: hRad17–RFC exists close to chromatin and loads the 9-1-1 complex onto DNA upon the occurrence of DNA damage, then ATR-dependent phosphorylation of Rad17 occurs. One has to conclude that the recruitment of the 9-1-1 complex by the Rad17 complex enables the substrate selection by ATR. This model may extend beyond Rad17 phosphorylation to many additional substrates of ATR.

7. THE MRN COMPLEX PLAYS A ROLE IN CHECKPOINT ARRESTS

In Chapter 31, this multifunctional protein complex (101) was already discussed extensively. Prompted by initial observations, that cells defective in mammalian MRN components, such as Nijmegen breakage syndrome cells (mutated in Nbs1/ nibrin), showed radioresistant DNA synthesis not unlike AT cells (102–105), a possible role as a checkpoint determinant was studied in detail in *S. cerevisiae* where viable deletion mutants of MRN components are available. Indeed, the complex was found to be essential for G1 arrest, intra S-phase arrest and to be of some importance for G2/M arrest following treatment with γ-irradiation and radiomimetic chemicals (106,107). A defect in MRN components also confers sensitivity to hydroxyurea (HU) and low doses of HU fail to slow down replication in such mutants (107). The failure to arrest was correlated with the failure to phosphorylate downstream substrates such as Rad9^{Sc}, and the downstream kinases Rad53^{Sc} or Chk1^{Sc}. However, no effect was found for UV-induced arrests (106). So, this complex appears to be specifically involved in checkpoint responses to double-strand breaks. An investigation of Mre11 point mutants suggests a critical role of its nuclease activity (107). Interestingly, Mre11 and Xrs2 (but not Rad50) are subject to phosphorylation following treatment with DNA-damaging agents (including UV) (107). This phosphorylation depends on the yeast ATM homolog Tel1 but apparently not on other checkpoint proteins. In *S. cerevisiae*, Tel1 was found to be recruited to sites of double-strand breaks by a mechanism that involves the C-terminus of Xrs2 (108).

This pathway appears to be conserved in human cells where ATM-dependent phosphorylation of Serine 343 of Nbs1 was characterized as being essential for phosphorylation of downstream substrates and for ionizing radiation but not UV-induced S-phase arrest (109–112). A role of the MRN complex for S-phase arrest specifically was also demonstrated in *Schizosaccharomyces pombe* (113).

Does the MRN complex represent a sensor for double-strand breaks that is used for both repair and checkpoint pathways? Data in *S. cerevisiae* suggest no absolute requirement for the compromised checkpoints. It is perhaps best to assign an accessory break-processing function that may assist in recognition by other sensors. Again, single-stranded DNA exposed through Mre11 exonuclease activity may be the critical feature. Since MRN components are subject to ATM/Tel1-dependent phosphorylation, one would place the complex downstream of ATM/Tel1. However, ATM activation was also found to be dependent on the MRN complex (112). Possibly, both players are connected through reinforcing feedback loops and a strictly linear sequence of events is an incorrect representation.

8. SYNOPSIS: INDEPENDENT BUT COMMUNICATING SENSORS ARE LINKED BY COMMON REQUIREMENTS

A likely scenario for the recognition of strand break damage emerges from the previous discussions (Fig. 2). The ATR–ATRIP (= Mec1^{Sc}–Lcd1^{Sc}, Rad3^{Sp}–Rad26^{Sp}) complex and Rad17–RFC are recruited to damaged sites largely independently. (It is possible that Rad17 phosphorylation is an early step and important for function, however, in the absence of ATR, other kinases may substitute.) For either recruiting event, the presence of single-stranded DNA and binding of RFA appears to be critical and this common requirement may bring both sensor complexes in close contact. At double-strand breaks without staggered ends, single-stranded DNA may be exposed through the action of the MRN complex. Thus, one has to conclude that the essential signal indicating a level of DNA damage that requires a regulatory response is single-stranded DNA and single-stranded DNA binding proteins—a remarkable unifying feature of pro- and eukaryotic "SOS-like" responses.

Subsequently, Rad17–RFC will recruit the 9-1-1 complex and its presence will enable substrate selection and phosphorylation events catalyzed by ATR. Possible substrates are 9-1-1 components themselves as well as Rad17 and, as we will discuss below, downstream acting kinases.

The role of ATM is less well defined. Initially, it may respond to signals originating from the disturbance of higher order chromatin structure. In its activated form, it may modify other sensors or helpers (such as MRN), bind to double-strand breaks and provide backup support.

9. THE GENERATION OF A TRANSDUCIBLE SIGNAL

The activation of the downstream effector kinases Chk1 and Chk2 (or Rad53 in *S. cerevisiae*, Cds1 in *Schizosaccharomyces pombe*) is the essential next step of checkpoint signaling. These events are outside of the primary scope of this review but should be discussed briefly since initial damage recognition will be without consequence if the signal is not converted into phosphorylation of downstream kinases.

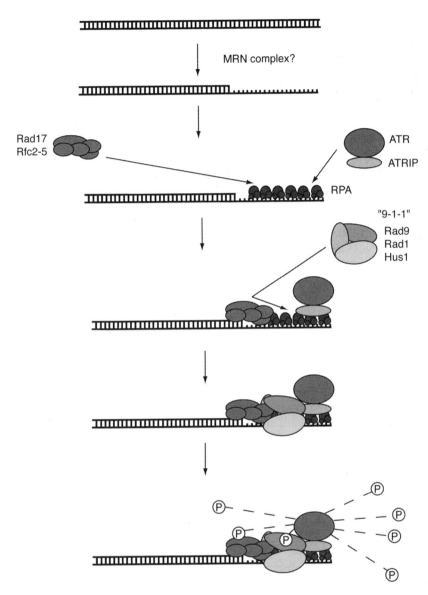

Figure 2 A likely model of strand break recognition and initial signal generation by eukaryotic checkpoint sensors. Following double-strand breakage, single-strand degradation (possibly catalyzed by the MRN complex) results in recruitment of RPA, in turn stimulating the independent binding of Rad17–RFC and ATR–ATRIP. Rad17–RFC loads the 9-1-1 complex, thus creating a scaffold that allows substrate recognition by ATR. The ensuing phosphorylation events will target Rad17, 9-1-1, and RPA components as well as other substrates such as downstream acting kinases.

Here, mediator proteins become important. These proteins serve as scaffolds that amplify the signal and allow the primary kinases to modify downstream substrates.

To exemplify the role of a mediator such as *S. cerevisiae* Rad9, we will discuss a reasonable model for this response in some detail. Very early following DNA damage (but not following inhibition of replication), Rad9[Sc] is phosphorylated at multiple SQ/TQ sites in a Mec1-dependent reaction, with Tel1 functioning as a backup kinase (114–118). As discussed, DNA-damage recognition and complex formation may be essential for this substrate selection by Mec1/ATR. Mediated by a C-terminal BRCT domain, phosphorylated Rad9 is able to form dimers or oligomers (119). Phosphorylated peptides of Rad9 can interact with the FHA2 domain of Rad53 (117) and Rad53 may thus dock onto the complex, with the result of positioning Rad53 proteins in immediate vicinity to each other enabling autophosphorylation in *trans* between two Rad53 molecules (120). Hyper-phosphorylated Rad53 looses contact to Rad9 which more or less serves here as a catalytic surface, and the Rad53 binding sites can again be occupied by unphosphorylated Rad53 (120). This mechanism ensures that, if initial Mec1-dependent phosphorylation can reach a threshold that permits Rad9 oligomer formation and initial Rad53 autophosphorylation, the signal can quickly be amplified to a level that permits downstream signaling. However, Rad53 is also subject to *trans*-phosphorylation by Mec1 and this can be considered as an initiating event that starts the autophosphorylation process (121). This pathway may also be of importance for Rad53 activation by other pathways that do not require participation of Rad9.

10. OTHER SENSOR CANDIDATES

The reader may wonder about previously discussed repair proteins with an essential role in recognizing strand break damage that have not been included in our discussion of damage sensors in the context of cell cycle arrest. Prominent examples are DNA-PK and poly (ADP-ribose) polymerase I (PARP). As clearly shown in yeast, the Ku proteins do not play a role in initiating cell cycle arrest following strand breakage (7,122). However, as we will see, they do play other roles in the checkpoint response. Similarly, the catalytic subunit of DNA-PK, the mammalian SCID protein, is not essential for most DNA-damage-induced cell cycle arrests and associated responses (123–129).

PARP has been characterized as a single-strand break sensor (see Chapter 34) . Following treatment of PARP-1 deficient or PARP-inhibited cells with ionizing radiation, an impaired accumulation of p53 has been detected (130–132). As mentioned, p53 accumulation is an important event that normally precedes G1 cell cycle arrest. The available studies, however, do not agree on an influence of PARP on G1 arrest (131,133). Interactions with the established sensor proteins have not been reported and further investigation is clearly required.

Most interesting are perhaps the connections of mismatch repair with the checkpoint pathway. Investigations with colon cancer cell lines lacking functional MutL and MutS homologs (see Chapter 21) indicate a defect in ionizing-radiation-induced G2/M arrest (134,135). But such cell lines also show radioresistant DNA synthesis and thus a S-phase checkpoint defect (136). However, ATM is activated and Nbs1 is phosphorylated in these cells. Nevertheless, the signal does not seem to reach the Chk2 downstream kinase whose activation depends on ATM. MutL interacts with ATM whereas MutS interacts with Chk2 (136): the mismatch repair complex may thus act not only as a damage sensor but also as a scaffold to enable phosphorylation of Chk2 by ATM. Since these data concern ionizing radiation, one would have to postulate that one or several of the many species of oxidative base damages is being recognized by the mismatch repair system and—in addition to strand breaks—is being used as an additional checkpoint triggering signal.

11. SENSING UV DAMAGE

The previous discussion of damage sensing was primarily focused on structural DNA damage related to strand breaks. Although UV radiation is a potent inducer of various cell cycle arrests and associated responses in lower and higher eukaryotes (e.g., 111,137,138) , our understanding of the molecular events required for sensing such bulky base damage has lagged behind that of strand breaks as a triggering signal. Of course, single-strand breaks and single-stranded DNA gaps will occur during nucleotide excision repair (NER) of pyrimidine dimers and NER may thus create a substrate for the DNA-damage sensors we already discussed. Studies have therefore concentrated on possibly defective checkpoint responses in NER-deficient cells. Again, RFA is potentially an interesting candidate mediating such damage recognition. However, such gaps as repair intermediates will be small and short-lived.

If NER-deficient *S. cerevisiae* cells synchronized in G1 are UV-irradiated, a delay of S-phase is found at much lower UV doses than are required in the wild type for a similar delay (137,139). However, unlike wild type, this delay does not occur at or upstream of START but during the early stages of replication. Only a subset of checkpoint proteins is required for this arrest that appears to be due to an interference of unrepaired damage with the replication apparatus (139). It is important to discuss such responses in S-phase separately (see Chapter 39). The absence of a true G1 arrest under these conditions can be interpreted as indicating a requirement of NER for primary damage recognition by checkpoint sensors. However, for valid conclusions, the initial UV-damage level and thus the dose needs to be comparable to that used for wild-type cells since a threshold level of primary damage may be the essential trigger, if there is something like direct UV-damage recognition by these sensors, a function suggested for ATR (51). If one increases the dose to that level, such an analysis becomes problematic due to the extreme lethality of NER-cells.

Nevertheless, a mutant hunt revealed an allele of Rad14 (the yeast XP-A homolog, see Chapter 7) that prevents checkpoint activation after UV outside of S-phase (140). In fact, all yeast mutants deficient in incision showed no rapid UV-induced phosphorylation of downstream checkpoint signal transmitters such as the Rad53 kinase. This correlates with a physical interaction of Rad14 with the 9-1-1 complex component Ddc1. Thus, a model is emerging that is different from the initial assumption of strand breaks or gaps resulting from NER as a checkpoint signal. Instead, there seems to be a close monitoring of the presence of NER protein/ DNA complexes by checkpoint sensor proteins such as the PCNA-like 9-1-1 complex that may be sequestered (following loading by the RFC-like complex at a break site?) through a Rad14–Ddc1 interaction. However, downstream acting incision proteins or events appear to be additionally required for checkpoint activation.

However, the continued presence of unrepaired UV damage in NER-deficient noncycling cells also seems to activate a possibly separate pathway of checkpoint signaling that occurs in conjunction with DNA degradation. Whether this is a severely delayed and muted response (140, Muzi-Falconi, personal communication) or, as suggested in another study, one that can virtually substitute the NER-dependent process (141) still remains to be defined more clearly. However, wild-type cells will perform NER and there is indeed evidence that the latter process does not contribute to cell cycle arrest in UV-treated cells under normal repair and optimal growth conditions (140). The described "liquid-holding" response to unrepaired damage may reflect the increasingly likely existence of active processes of eliminating irreversibly damaged cells in yeasts, resembling the apoptotic processes of higher eukaryotes (142).

The NER/checkpoint protein crosstalk in yeast may not be all that different from mammalian cells. As mentioned before, checkpoint responses in vertebrates are mediated by activation of p53 as a transcription factor. Both P53 and its antagonist MDM2 undergo a multitude of modifications that are induced by DNA damage and generally result in the stabilization of p53 levels (143). If XP-A cells are not allowed to enter replication following UV treatment, no accumulation of p53 is found (3). Again, the Rad14 ortholog XP-A appears to be required for a checkpoint response if secondary damage due to interference with replication is excluded. But there was also evidence for an NER-independent mechanism that leads to p53 accumulation in a higher dose range (3).

In cycling cells, the absence of the XP-A, XP-D, CS-A or CS-B proteins results in a sensitization of the p53 response, i.e., the minimum dose required for p53 accumulation is reduced (144,145). Remarkably, this is not the case in XP-C cells where repair of actively transcribed genes stays unaffected. This may indicate that the sensitizing events are special signals elicited by interference of UV damage with transcription (145,146).

There is indeed evidence for more than one pathway of p53 activation following UV damage (147). Serine 392 of p53 is phosphorylated in response to UV only (148,149), with a complex of casein kinase II and the chromatin transcriptional elongation factor FACT identified as potential upstream kinase (150). Details are speculative but the casein kinase II–FACT complex is certainly a promising candidate for communicating a UV-damage signal that corresponds to inhibition of transcriptional elongation. Further, P53 can also interact with several components of the transcription/repair factor TFIIH but it is unknown if this interaction has consequences for regulation (151).

Another interesting aspect of damage recognition is the activation of typical stress kinases in mammalian cells. Specifically for UV-induced G2 arrest, the relevant activated kinase is a member of the MAP-kinase pathway and was identified as p38 α/β isoforms by use of a specific inhibitor (152,153). On the other hand, inhibitors of ATM, ATR kinases involved in ionizing radiation damage signaling such as caffeine or UCN-01 had no effect on the initial arrest response following UV damage. The kinase target appears to be the Cdc25B phosphatase, thus stabilizing inhibitory phosphorylation of Cdc2, orchestrating G2/M transition. The involvement of the MAP-kinase pathway adds a novel dimension to our discussion of damage recognition during the course of checkpoint arrest (154). Many members of the MAP-kinase pathway are suspected to be primarily activated by cytoplasmic damage such as reactive oxygen species and not by DNA damage. A reasonable model suggests that a fast and transient arrest response may be triggered by cytoplasmic UV damage but further detection of concomitant DNA damage is required to reinforce and prolong checkpoint arrest.

12. ADAPTATION AND CELL CYCLE RESTART

As soon as DNA damage is repaired, the cell cycle arrest should cease but surprisingly little is known about the ways cells can turn off a checkpoint response. Furthermore, what happens if damage remains irreparable? This situation could lead to permanent withdrawal from the cell cycle or even to apoptosis. However, cells may also resume cell cycle progression in the presence of unrepaired damage, actively silence the damage signal or make signal transduction proteins more resistant to activation. As mentioned, this process of checkpoint override is called *adaptation*. It has been studied in detail in budding yeast only. Some of the employed mechanisms may play a very general role in recovery from cell cycle arrest following successful repair. Adaptation correlates with silencing of the activity of downstream effector kinases and mutants that do not adapt retain phosphorylation of Rad53 and Chk1 (155).

In *S. cerevisiae*, continuous expression of HO endonuclease or its expression in a double-strand break repair deficient background introduces irreparable damage. If a break at an HO site is introduced on a nonessential chromosome, cells arrest for a long time in G2/M but ultimately resume cell cycle progression in the presence of the unrepaired chromosome for several generations—a clear case of adaptation (156). (However, in a similar plasmid system others have observed no adaptation but cell death—immediately, or protracted by several divisions in the case of a checkpoint-deficient mutant background (157). These discrepancies have remained unexplained.)

It has already been discussed what happens to such a persistent double strand break: $3' \to 5'$ single-strand degradation, which is slowed by the presence of the Ku complex, will create substrates for RPA binding. Although the yeast Ku proteins are not required for G2/M arrest, they are required for adaptation to a single double-strand break and mutants become instead permanently arrested (7). However, even wild-type cells do not adapt following the introduction of two double-strand breaks. Interestingly, both phenotypes can be suppressed by a mutation in the largest RPA subunit (*rfa1-t11*). Interestingly, this is the same mutation we have mentioned as resulting in weakened binding by Mec1–Lcd1 sensor complexes to double-strand breaks (49). It may be assumed that cells measure the amount of persistent checkpoint-relevant damage through the amount of RPA-covered single-stranded DNA. If there is an excessive amount (as in Ku deficient cells), cells do not adapt. If the damage signal is dampened by altered RPA, this inability to adapt can be abolished (Fig. 3).

Components of the homologous recombination system were identified as additional signal emitters that define a second nonepistatic pathway involved in damage assessment (6,158,159). Cells lacking Rad51 or the Srs2 helicase fail to adapt. However, the negative effect of Rad51 is dependent on Rad52 interactions. The interaction of Rad51 with Rad52 and possibly the single-strand binding activity of Rad52 are essential. Absence of Rad51 with Rad52 present appears to enhance the damage signal (no adaptation), whereas absence of both has no effect (possibly other sensors compensate) (Fig. 3). So, the presence or absence of the conventional homologous recombination protein complex bound to DNA appears to be insufficient to explain these observations, we seem to be looking at a rather sophisticated use of the recombination components for a different purpose. Does a cell evaluate damage differently if it enters normal recombination or attracts other, unconventional types of protein interactions? Details are still very sketchy.

There is little known about adaptation in higher eukaryotes. If of significance in a multicellular organism which cannot tolerate genetic instability introduced by adaptation, modification of such a process has implications for cancer etiology and cancer therapy. But again, some of the same processes may operate during normal recovery from checkpoint arrest.

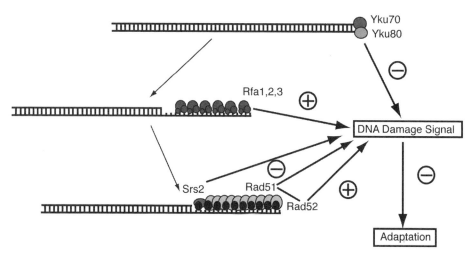

Figure 3 The signal transduction circuits of adaptation in response to persistent double-strand breaks in *S. cerevisiae*. The Ku complex prevents single-strand degradation and in its absence, the damage signal is abnormally enhanced and adaptation is prevented. Mutations in the RPA component Rfa1 can suppress this effect since RPA binding appears to signal the presence of DNA damage (a positive influence, "+") and consequently counteracts adaptation. If bound to Rad52, Rad51 and Srs2 seem to play a RPA-independent role that mutes the signal emanating from broken DNA since noninteracting mutants are adaptation defective.

ACKNOWLEDGMENTS

I thank Toshiki Tsurimoto for providing figure material. Studies in the author's laboratory were supported by grants from the National Institutes of Health (CA87381, TW01189, and ES11163).

REFERENCES

1. Zhou B-B, Elledge SJ. The DNA damage response: putting checkpoints in perspective. Nature 2000; 408:433–439.
2. Nyberg KA, Michelson RJ, Putnam CW, Weinert TA. Toward maintaining the genome: DNA damage and replication checkpoints. Annu Rev Genet 2002; 36: 617–656.
3. Nelson WG, Kastan MB. DNA strand breaks: the DNA template alterations that trigger p53-dependent DNA damage response pathways. Mol Cell Biol 1994; 14:1815–1823.
4. Levin N, Bjornsti M-A, Fink GR. A novel mutation in DNA topoisomerase I of yeast causes DNA damage and *RAD9*-dependent cell cycle arrest. Genetics 1993; 133:799–814.
5. Haber JE. Mating-type gene switching in *Saccharomyces cerevisiae*. Annu Rev Genet 1998; 32:561–599.
6. Vaze MB, Pellicioli A, Lee SE, Ira G, Liberi G, Arbel-Eden A, Foiani M, Haber JE. Recovery from checkpoint-mediated arrest after repair of a double-strand break requires Srs2 helicase. Mol Cell 2002; 10:373–385.
7. Lee SE, Moore JK, Holmes A, Umezu K, Kolodner RD, Haber JE. *Saccharomyces* Ku70, Mre11/Rad50, and RPA proteins regulate adaptation to G2/M arrest after DNA damage. Cell 1998; 94:399–409.

51. Ünsal-Kaçmaz K, Makhov AM, Griffith JD, Sancar A. Preferential binding of ATR protein to UV-damaged DNA. Proc Natl Acad Sci USA 2002; 99:6673–6678.

52. Murray JM, Carr AM, Lehmann AR, Watts FZ. Cloning and characterization of the rad9 DNA repair gene from *Schizosaccharomyces pombe*. Nucleic Acids Res 1991; 19: 3525–3531.

53. Kaur R, Kostrub CF, Enoch T. Structure-function analysis of fission yeast Hus1-Rad1-Rad9 checkpoint complex. Mol Biol Cell 2001; 12:3744–3758.

54. Kondo T, Matsumoto K, Sugimoto K. Role of a complex containing Rad17, Mec3 and Ddc1 in the yeast DNA damage checkpoint pathway. Mol Cell Biol 1999; 19:1136–1143.

55. Longhese MP, Paciotti V, Fraschini R, Zaccarini R, Plevani P, Lucchini G. The novel DNA damage checkpoint protein Ddc1p is phosphorylated periodically during the cell cycle and in response to DNA damage in budding yeast. EMBO J 1997; 16:5216–5226.

56. Paciotti V, Lucchini G, Plevani P, Longhese MP. Mec1p is essential for phosphorylation of the yeast DNA damage checkpoint protein Ddc1p, which physically interacts with Mec3p. EMBO J 1998; 17:4199–4209.

57. Siede W, Nusspaumer G, Portillo V, Rodriguez R, Friedberg EC. Cloning and characterization of *RAD17*, a gene controlling cell cycle responses to DNA damage in *Saccharomyces cerevisiae*. Nucleic Acids Res 1996; 24:1669–1675.

58. Caspari T, Dahlen M, Kanter-Smoler G, Lindsay HD, Hofmann K, Papadimitriou K, Sunnerhagen P, Carr AM. Characterization of *Schizosaccharomyces pombe* Hus1: a PCNA-related protein that associates with Rad1 and Rad9. Mol Cell Biol 2000; 74: 1254–1262.

59. Kostrub CF, Al-Khodairy F, Ghazizadeh H, Carr AM, Enoch T. Molecular analysis of *hus1*⁺, a fission yeast gene required for S-M and DNA damage checkpoints. Mol Gen Genet 1997; 254:389–399.

60. Kostrub CF, Knudsen K, Subramani S, Enoch T. Hus1p, a conserved fission yeast checkpoint protein, interacts with Rad1p and is phosphorylated in response to DNA damage. EMBO J 1998; 17:2055–2066.

61. Parker AE, Van de Weyer I, Laus MC, Oostveen I, Yon J, Verhasselt P, Luyten WHML. A human homologue of the *Schizosaccharomyces pombe rad1*⁺ checkpoint gene encodes an exonuclease. J Biol Chem 1998; 273:18332–18339.

62. Freire R, Murguía JR, Tarsounas M, Lowndes NF, Moens PB, Jackson SP. Human and mouse homologs of *Schizosaccharomyces pombe rad1*⁺ and *Saccharomyces cerevisiae RAD17*: linkage of checkpoint control and mammalian meiosis. Genes Dev 1998; 12:2560–2573.

63. Dean FB, Lian L, O'Donnell M. cDNA cloning and gene mapping of human homologs for *Schizosaccharomyces pombe rad17, rad1*, and *hus1* and cloning of homologs from mouse, *Caenorhabditis elegans*, and *Drosophila melanogaster*. Genomics 1998; 54: 424–436.

64. Ahmed S, Hodgkin J. MRT-2 checkpoint protein is required for germline immortality and telomore replication in *C. elegans*. Nature 2000; 403:159–164.

65. Weiss RS, Kostrub CF, Enoch T, Leder P. Mouse *Hus1*, a homolog of the *Schizosaccharomyces pombe hus1*⁺ cell cycle checkpoint gene. Genomics 1999; 59:32–39.

66. Hang H, Lieberman HB. Physical interactions among human checkpoint control proteins HUS1p, RAD1p, and RAD9p, and implications for the regulation of cell cycle progression. Genomics 2000; 65:24–33.

67. Hang H, Zhang Y, Dunbrack J, R.L., Wang C, Lieberman HB. Identification and characterization of a paralog of human cell cycle checkpoint gene *HUS1*. Genomics 2002; 79:487–492.

68. Weiss RS, Enoch T, Leder P. Inactivation of mouse *Hus1* results in genomic instability and impaired responses to genotoxic stress. Genes Dev 2000; 14:1886–1898.

69. Udell CM, Lee SK, Davey S. *HRAD1* and *MRAD1* encode mammalian homologues of the fission yeast *rad1*⁺ cell cycle checkpoint control gene. Nucleic Acids Res 1998; 26:3971–3976.

70. Lieberman HB, Hopkins KM, Nass M, Demetrick D, Davey S. A human homolog of the *Schizosaccharomyces pombe rad9*[+] checkpoint control gene. Proc Natl Acad Sci USA 1996; 93:13890–13895.

71. Thelen MP, Onel K, Holloman WK. The *REC1* gene of *Ustilago maydis* involved in the cellular response to DNA damage encodes an exonuclease. J Biol Chem 1994; 269: 747–754.

72. Naureckiene S, Holloman WK. DNA hydrolytic activity associated with the *Ustilago maydis REC1* gene product analyzed in hairpin oligonucleotide substrates. Biochemistry 1999; 38:14379–14386.

73. Bessho T, Sancar A. Human DNA damage checkpoint protein hRAD9 is a 3′ to 5′ exonuclease. J Biol Chem 2000; 275:7451–7454.

74. Majka J, Burgers PM. Yeast RAD17/Mec3/Ddc1: A sliding clamp for the DNA damage checkpoint. Proc Natl Acad Sci USA 2003; 100:2249–2254.

75. Burtelow MA, Kaufmann SH, Karnitz LM. Retention of the human Rad9 checkpoint complex in extraction-resistant nuclear complexes after DNA damage. J Biol Chem 2000; 275:26343–26348.

76. Volkmer E, Karnitz LM. Human homolog of *Schizosaccharomyces pombe* Rad1, Hus1, and Rad9 from a DNA damage-responsive protein complex. J Biol Chem 1999; 274:567–570.

77. St. Onge RP , Udell CM, Casselman R, Davey S. The human G2 checkpoint control protein hRAD9 is a nuclear phosphoprotein that forms complexes with hRAD1 and hHUS1. Mol Biol Cell 1999; 10:1985–1995.

78. Thelen MP, Venclovasc C, Fidelis K. A sliding clamp model for the Rad1 family of cell cycle checkpoint proteins. Cell 1999; 96:769–770.

79. Venclovas C, Thelen MP. Structure-based predictions of Rad1, Rad9, Hus1 and Rad17 participation in sliding clamp and clamp-loading complexes. Nucleic Acids Res 2000; 28:2481–2493.

80. Aravind L, Walker DR, Koonin EV. Conserved domains in DNA repair proteins and evolution of repair systems. Nucleic Acids Res 1999; 27:1223–1242.

81. Bermudez VP, Lindsey-Boltz LA, Cesare AJ, Maniwa Y, Griffith JD, Hurwitz J, Sancar A. Loading of the human 9-1-1 checkpoint complex onto DNA by the checkpoint clamp loader hRad17-replication factor C complex in vitro. Proc Natl Acad Sci USA 2003; 100:1633–1638.

82. Burtelow MA, Roos-Mattjus PMK, Rauen M, Babendure JR, Karnitz LM. Reconstitution and molecular analysis of the hRad9-hHus1-hRad1 (9-1-1) DNA damage responsive checkpoint complex. J Biol Chem 2001; 276:25903–25909.

83. Griffith JD, Lindsey-Boltz LA, Sancar A. Structures of the human Rad17-replication factor C and checkpoint Rad 9-1-1 complexes visualized by spray/low voltage microscopy. J Biol Chem 2002; 277:15233–15236.

84. Shiomi Y, Shinozaki A, Nakada D, Sugimoto K, Usukura J, Obuse C, Tsurimoto T. Clamp and clamp loader structures of the human checkpoint protein complexes, Rad9-1-1 and Rad17-RFC. Genes Cells 2002; 7:861–868.

85. Lindsey-Boltz LA, Bermudez VP, Hurwitz J, Sancar A. Purification and characterization of human DNA damage checkpoint Rad complexes. Proc Natl Acad Sci USA 2001; 98:11236–11241.

86. Hong EJ, Roeder GS. A role for Ddc1 in signaling meiotic double-strand breaks at the pachytene checkpoint. Genes Dev 2002; 16:363–376.

87. Zhang H, Zhu Z, Vidanes G, Mbangkollo D, Liu Y, Siede W. Characterization of DNA damage-stimulated self-interaction of *Saccharomyces cerevisiae* checkpoint protein Rad17p. J Biol Chem 2001; 276:26715–26723.

88. Komatsu K, Wharton W, Hang H, Wu C, Singh S, Lieberman H, Pledger WJ, Wang H-G. PCNA interacts with hHUS1/hRad9 in response to DNA damage and replication inhibition. Oncogene 2000; 19:5291–5297.

89. von Deimling F, Scharf JM, Liehr T, Rothe M, Kelter A-R, Albers P, Dietrich WF, Kunkel LM, Wernert N, Wirth B. Human and mouse RAD17 genes: identification, localization, genomic structure and histological expression from in normal testis and seminoma. Hum Genet 1995; 105:17–27.

90. Bao S, Shen X, Shen K, Liu Y, Wang X-F. The mammalian homologue Rad24 homologous to yeast *Saccharomyces cerevisiae* Rad24 and *Schizosaccharomyces pombe* Rad17 is involved in DNA damage checkpoint. Cell Growth Diff 1998; 9:961–967.

91. Griffiths DJF, Barbet NC, McCready S, Lehmann AR, Carr AM. Fission yeast *rad17*: a homologue of budding yeast *RAD24* that shares regions of sequence similarity with DNA polymerase accessory proteins. EMBO J 1995; 14:5812–5823.

92. Green CM, Erdjument-Bromage H, Tempst P, Lowndes NF. A novel RAD24 checkpoint protein complex closely related to replication factor C. Curr Biol 2000; 10:39–42.

93. Shimomura T, Ando S, Matsumoto K, Sugimoto K. Functional and physical interaction between Rad24 and Rfc5 in the yeast checkpoint pathway. Mol Cell Biol 1998; 18:5485–5491.

94. Waga S, Stillman B. The DNA replication fork in eukaryotic cells. Annu Rev Biochem 1998; 67:721–751.

95. Mossi R, Hübscher U. Clamping down on clamp loaders: the eukaryotic replication factor C. Eur J Biochem 1998; 254:209–216.

96. Rauen M, Burtelow MA, Dufault VM, Karnitz LM. The human checkpoint protein hRad17 interacts with the PCNA-like proteins hRad1, hHus1, and hRad9. J Biol Chem 2000; 275:29767–29771.

97. Zou L, Cortez D, Elledge SJ. Regulation of ATR substrate selection by Rad17-dependent loading of Rad9 complexes onto chromatin. Genes Dev 2002; 16:198–208.

98. Zou L, Liu D, Elledge SJ. Replication protein A-mediated recruitment and activation of Rad17 complexes. Proc Natl Acad Sci USA 2003; 100:13827–13832.

99. Bao S, Tibbetts RS, Brumbaugh KM, Fang Y, Richardson DA, Ali A, Chen SM, Abraham RT, Wang X-F. ATR/ATM-mediated phosphorylation of human Rad17 is required genotoxic stress responses. Nature 2001; 411:969–974.

100. Kai M, Tanaka H, Wang TS-F. Fission yeast Rad17 associates with chromatin in response to aberrant genomic structures. Mol Cell Biol 2001; 21:3289–3301.

101. D'Amours D, Jackson SP. The Mre11 complex: at the crossroads of DNA repair and checkpoint signalling. Nat Rev Mol Cell Biol 2002; 3:317–327.

102. Stewart GS, Maser RS, Stankovic T, Bressan DA, Kaplan MI, Jaspers NGJ, Raams A, Byrd PJ, Petrini JHJ, Taylor AMR. The DNA double-strand break repair gene *hMRE11* is mutated in individuals with an ataxia-telangiectasia-like disorder. Cell 1999; 99:577–587.

103. Antoccia A, Stumm M, Saar K, Ricordy R, Maraschio P, Tanzarella C. Impaired p53-mediated DNA damage response, cell-cycle disturbance and chromosome aberrations in Nijmegen breakage syndrome lymphoblastoid cell lines. Int J Radiat Biol 1999; 75:583–591.

104. Shiloh Y. Ataxia telangiectasia and the Nijmegen breakage syndrome - related disorders but genes apart. Ann Rev Genet 1997; 31:635–662.

105. Jongmans W, Vuillaume M, Chrzanowska K, Smeets D, Sperling K, Hall J. Nijmegen breakage syndrome cells fail to induce the p53-mediated DNA damage response following exposure to ionizing radiation. Mol Cell Biol 1997; 17:5016–5022.

106. Grenon M, Gilbert C, Lowndes NF. Checkpoint activation in response to doublestrand breaks requires the Mre11/Rad50/Xrs2 complex. Nature Cell Biol 2001; 3:844–847.

107. D'Amours D, Jackson SP. The yeast Xrs2 complex functions in S phase checkpoint regulation. Genes Dev 2001; 15:2238–2249.

108. Nakada D, Shimomura T, Matsumoto K, Sugimoto K. The ATM-related Tel1 protein of *Saccharomyces cerevisiae* controls a checkpoint response following phleomycin treatment. Nucleic Acids Res 2003; 31:1715–1724.

109. Lim D-S, Kim S-T, Xu B, Maser RS, Lin J, Petrini JHJ, Kastan MB. ATM phosphorylates p95/nbs1 in an S-phase checkpoint pathway. Nature 2000; 404:613–617.

110. Buscemi G, Savio C, Zannini L, Micciche F, Masnada D, Nakanishi M, Tauchi H, Komatsu K, Mizutani S, Khanna K, Chen P, Concannon P, Chessa L, Delia D. Chk2 activation dependence on Nbs1 after DNA damage. Mol Cell Biol 2001; 21:5214–5222.

111. Heffernan TP, Simpson DA, Frank AR, Heinloth AN, Paules RS, Cordeiro-Stone M, Kaufmann WK. An ATR- and Chk1-dependent S checkpoint inhibits replicon initiation following UVC-induced DNA damage. Mol Cell Biol 2002; 22:8552–8561.

112. Uziel T, Lerenthal Y, Moyal L, Andegeko Y, Mittelman L, Shiloh Y. Requirement of the MRN complex for ATM activation by DNA damage. EMBO J 2003; 22:5612–5621.

113. Chahwan C, Nakamura TM, Sivakumar S, Russell P, Rhind N. The fission yeast Rad32 (Mre11)-Rad50-Nbs1 complex is required for the S-phase DNA damage checkpoint. Mol Cell Biol 2003; 23:6564–6573.

114. Emili A. *Mec1*-dependent phosphorylation of Rad9p in response to DNA damage. Mol Cell 1998; 2:183–189.

115. Vialard JE, Gilbert CS, Green CM, Lowndes NF. The budding yeast Rad9 checkpoint protein is subjected to Mec1/Tel1-dependent hyperphosphorylation and interacts with Rad53 after DNA damage. EMBO J 1998; 17:5679–5688.

116. Sun ZX, Hsiao J, Fay DS, Stern DF. Rad53 FHA domain associated with phosphorylated Rad9 in the DNA damage checkpoint. Science 1998; 218:272–274.

117. Schwartz MF, Duong JK, Sun Z, Morrow JS, Pradhan D, Stern DF. Rad9 phosphorylation sites couple Rad53 to the *Saccharomyces cerevisiae* DNA damage checkpoint. Mol Cell 2002; 9:1055–1065.

118. Lee S-J, Schwartz MF, Duong JK, Stern DF. Rad53 phosphorylation site clusters are important for Rad53 regulation and signaling. Mol Cell Biol 2003; 23:6300–6314.

119. Soulier J, Lowndes NF. The BRCT domain of the *Saccharomyces cerevisiae* checkpoint protein Rad9 mediates a Rad9-Rad9 interaction after DNA damage. Curr Biol 1999; 9:551–554.

120. Gilbert CS, Green CM, Lowndes NF. Budding yeast Rad9 is an ATP-dependent Rad53 activating machine. Mol Cell 2001; 8:129–136.

121. Pellicioli A, Lucca C, Liberi G, Marini F, Lopes M, Plevani P, Romano A, Di Fiore PP, Foiani M. Activation of Rad53 kinase in response to DNA damage and its effect in modulating phosphorylation of the lagging strand DNA polymerase. EMBO J 1999; 18:6561–6572.

122. Siede W, Friedl AA, Dianova I, Eckardt-Schupp F, Friedberg EC. The *Saccharomyces cerevisiae* Ku autoantigen homologue affects radiosensitivity only in the absence of homologous recombination. Genetics 1996; 142:91–102.

123. Huang L, Clarkin KC, Wahl GM. p53-dependent cell cycle arrests are preserved in DNA-activated protein kinase-deficient mouse fibroblasts. Cancer Res 1996; 56: 2940–2944.

124. Wang S, Guo M, Ouyang H, Li X, Cordon-Cardo C, Kurimasa A, Chen DJ, Fuks Z, Ling CC, Li GC. The catalytic subunit of DNA-dependent protein kinase selectively regulates p53-dependent apoptosis but not cell-cycle arrest. Proc Natl Acad Sci USA 2000; 97:1584–1588.

125. Gurley KE, Kemp CJ. p53 induction, cell cycle checkpoints, and apoptosis in DNAPK-deficient scid mice. Carcinogenesis 1996; 17:2537–2542.

126. Burma S, Kurimasa A, Xie G, Taya Y, Araki R, Abe M, Crissman HA, Ouyang H, Li GC, Chen DJ. DNA-dependent protein kinase-independent activation of p53 in response to DNA damage. J Biol Chem 1999; 274:17139–17143.

127. Rathmell WK, Kaufmann WK, Hurt JC, Byrd LL, Chu G. DNA-dependent protein kinase is not required for accumulation of p53 or cell cycle arrest after DNA damage. Cancer Res 1997; 57:68–74.

128. Jhappan C, Yusufzai TM, Anderson S, Anver MR, Merlino G. The p53 response to DNA damage in vivo is independent of DNA-dependent protein kinase. Mol Cell Biol 2000; 20:4075–4083.

129. Jimenez GS, Bryntesson F, Torres-Arzayus MI, Priestley A, Beeche M, Saito S, Sakaguchi K, Appella E, Jeggo PA, Taccioli GE, Wahl GM, M Hubank. DNA-dependent protein kinase is not required for the p53-dependent response to DNA damage. Nature 1999; 400:81–85.

130. Valenzuela MT, Guerrero R, Núñez MI, de Almodóvar JMR, Sarker M, de Murcia G, Oliver FJ. PARP-1 modifies the effectiveness of p53-mediated DNA damage response. Oncogene 2002; 21:1108–1116.

131. Agarwal ML, Agarwal A, Taylor WR, Wang ZQ, Wagner EF, Stark GR. Defective induction but normal activation and function of p53 in mouse cells lacking poly-ADP-ribose polymerase. Oncogene 1997; 15:1035–1041.

132. Wang X, Ohnishi K, Takahashi A, Ohnishi T. Poly(ADP-ribosyl)ation is required for p53-dependent signal transduction induced by radiation. Oncogene 1998; 17:2819–2825.

133. Wieler S, Gagne JP, Vaziri H, Poirier GG, Benchimol S. Poly(ADP-ribose) polymerase-1 is a positive regulator of the p53-mediated G1 arrest response following ionizing radiation. J Biol Chem 2003; 278:18914–18921.

134. Davis TW, Wilson-Van Patten C, Meyers M, Kunugi KA, Cuthill S, Reznikoff C, Garces C, Boland CR, Kinsella TJ, Fishel R, Boothman DA. Defective expression of the DNA mismatch repair protein, MLH1, alters G_2-M cell cycle checkpoint arrest following ionizing radiation. Cancer Res 1998; 58:767–778.

135. Yan T, Schupp JE, Hwang H-S, Wagner MW, Berry SE, Strickfaden S, Veigl ML, Sedwick WD, Boothman DA, Kinsella TJ. Loss of DNA mismatch repair imparts defective cdc2 signaling and G_2 arrest responses without altering survival after ionizing radiation. Cancer Res 2001; 61:8290–8297.

136. Brown KD, Rathi A, Kamath R, Beardsley DI, Zhan Q, Mannino J, Baskaran R. The mismatch repair system is required for S-phase checkpoint activation. Nat Genet 2003; 33:80–84.

137. Siede W, Friedberg AS, Dianova I, Friedberg EC. Characterization of G_1 checkpoint control in the yeast *Saccharomyces cerevisiae* following exposure to DNA-damaging agents. Genetics 1994; 138:271–281.

138. Orren DK, Petersen LN, Bohr VA. A UV-responsive G_2 checkpoint in rodent cells. Mol Cell Biol 1995; 15:3722–3730.

139. Neecke H, Lucchini G, Longhese MP. Cell cycle progression in the presence of irreparable DNA damage is controlled by a Mec1- and Rad53-dependent checkpoint in budding yeast. EMBO J 1999; 18:4485–4497.

140. Giannattasio M, Lazzaro F, Longhese MP, Plevani P, Muzi-Falconi M. Physical and functional interactions between nucleotide excision repair and DNA damage checkpoint. EMBO J 2004; 23:429–438.

141. Zhang H, Taylor J, Siede W. Checkpoint arrest signaling in response to UV damage is independent of nucleotide excision repair in *Saccharomyces cerevisiae*. J Biol Chem 2003; 278:9382–9387.

142. Weinberger M, Ramachandran L, Burhans WC. Apoptosis in yeasts. IUBMB Life 2003; 55:467–472.

143. Appella E, Anderson CW. Post-translational modifications and activation of p53 by genotoxic stresses. Eur J Biochem 2001; 268:2764–2772.

144. Dumaz N, Drougard C, Quilliet X, Mezzina M, Sarasin A, Daya-Grosjean L. Recovery of the normal p53 response after UV treatment in DNA repair-deficient fibroblasts by retroviral-mediated correction with the *XPD* gene. Carcinogenesis 1998; 19:1701–1704.

145. Yamaizumi M, Sugano T. U.v.-induced nuclear accumulation of p53 is evoked through DNA damage of actively transcribed genes independent of the cell cycle. Oncogene 1994; 9:2775–2784.

146. Ljungman M, Zhang F, Chen F, Rainbow AJ, McKay BC. Inhibition of RNA polymerase II as a trigger for the p53 response. Oncogene 1999; 18:583–592.

147. Blaydes JP, Craig AL, Wallace M, Ball HM-L, Traynor NJ, Gibbs NK, Hupp TR. Synergistic activation of p53-dependent transcription by two cooperating damage recognition pathways. Oncogene 2000; 19:3829–3839.

148. Lu H, Taya Y, Ikeda M, Levine AJ. Ultraviolet radiation, but not γ radiation or etoposide-induced DNA damage, results in the phosphorylation of the murine p53 protein at serine-389. Proc Natl Acad Sci USA 1998; 95:6399–6402.

149. Kapoor M, Lozan G. Functional activation of p53 via phosphorylation following DNA damage by UV but not γ radiation. Proc Natl Acad Sci USA 1998; 95:2834–2837.

150. Keller DM, Zeng X, Wang Y, Zhang QH, Kapoor M, Shu H, Goodman R, Lozano G, Zhao Y, Lu H. A DNA damage-induced p53 serine 392 kinase complex contains CK2, hSpt16, and SSRP1. Mol Cell 2001; 7:283–292.

151. Wang XW, Yeh H, Schaeffer L, Roy R, Moncollin V, Egly J-M, Wang Z, Friedberg EC, Evans MK, Taffe BG, Bohr VA, Weeda G, Hoeijmakers JHJ, Forrester K, Harris CC. p53 modulation of TFIIH-associated nucleotide excision repair activity. Nat Genet 1995; 10:188–195.

152. Bulavin DV, Amundson SA, Fornace AJ, Jr. p38 and Chk1 kinases: different conductors for the G_2/M checkpoint symphony. Curr Opin Cell Biol 2002; 12:92–97.

153. Bulavin DV, Higashimoto Y, Popoff IJ, Gaarde WA, Basrur V, Potapova O, Appella E, Fornace AJ, Jr. Initiation of a G2/M checkpoint after ultraviolet radiation requires p38 kinase. Nature 2001; 411:102–107.

154. Pearce A, Humphrey T. Integrating stress-response and cell-cycle checkpoint pathways. Tr Cell Biol 2001; 11:426–433.

155. Pellicioli A, Lee SE, Lucca C, Foiani M, Haber JE. Regulation of *Saccharomyces* Rad53 checkpoint kinase during adaptation from DNA damage-induced G2/M arrest. Mol Cell 2001; 7:293–300.

156. Sandell LL, Zakian VA. Loss of a yeast telomere: arrest, recovery, and chromosome loss. Cell 1993; 75:729–739.

157. Bennett CB, Westmoreland TJ, Snipe JR, Resnick MA. A double-strand break within a yeast artificial chromosome (YAC) containing human DNA can result in YAC loss, deletion, or cell lethality. Mol Cell Biol 1996; 16:4414–4425.

158. Lee SE, Pellicioli A, Vaze MB, Sugawara N, Malkova A, Foiani M, Haber JE. Yeast Rad52 and Rad51 recombination proteins define a second pathway of DNA damage assessment in response to a single double-strand break. Mol Cell Biol 2003; 23: 8913–8923.

159. Lee SE, Pellicioli A, Malkova A, Foiani M, Haber JE. The *Saccharomyces* recombination protein Tid1p is required for adaptation from G2/M arrest induced by a double-strand break. Curr Biol 2001; 11:1053–1057.

37
Responses to Replication of DNA Damage

Maria Pia Longhese
Dipartimento di Biotecnologie e Bioscienze, Università di Milano-Bicocca, Milan, Italy

Marco Foiani
Istituto FIRC di Oncologia Molecolare, Dipartimento di Scienze Biomolecolari e Biotecnologie, Università di Milano, Milan, Italy

1. INTRODUCTION

The ability of cells to fully and faithfully replicate their DNA is essential for ensuring genomic integrity. Most of the genomic instability that is necessary for the development of many types of cancers can ultimately be attributed to replication problems. Although chromosome replication is a remarkably accurate process, DNA replication is potentially dangerous by itself and can be cause of DNA damage. Single- and double-strand breaks (DSBs) continuously arise due to the action of nick and closing enzymes (i.e., DNA topoisomerases) and DNA replication can be perturbed by intrinsic replication errors that can arise from nucleotide misincorporation, availability of nucleotide pools and from slippage of newly synthesized DNA on repetitive DNA sequences (1–3). Replication errors can have important consequences as they can result in the joining of sequences with very little homology, thus causing deletions or expansions of repeated sequences (4). Furthermore, the intrinsic difficulty of certain DNA regions to be replicated may cause chromosome breakage and the expression of fragile sites, specific regions in the human genome particularly prone to rearrangements and deletions (5). Finally, eukaryotic chromosomes are linear DNA molecules, which represent a dilemma for their complete replication. The ability of all known DNA polymerases to proceed only in the 5'–3' direction with each newly synthesized DNA molecule being RNA primed poses a problem for synthesizing DNA at the end of a linear replicon. In fact, while the leading strand acts as a template to synthesize a daughter strand that runs right up to the end, removal of the most distal RNA primer in the complementary strand leaves an 8- to 12-base gap at the 5' end that cannot be repaired by conventional enzymes (6,7).

The situation can become even more dramatic when cells experience DNA damage while they are replicating the genome. Environmental and endogenous DNA damaging agents can cause replication blocking lesions that interfere with the ability of cells to properly duplicate their DNA. If lesions are not repaired prior to initiation of DNA replication, a damaged template can cause replication fork stalling, leading to the accumulation of single-stranded DNA (ssDNA) regions and, under certain conditions, to the formation of DSBs (8–17).

Cells have evolved alternative options to deal with a damaged template and replication stress (Fig. 1). This review will mainly focus on the molecular mechanisms allowing cell survival in response to intra-S DNA damage and replication pausing.

2. HOW DO CELLS DEAL WITH A DAMAGED TEMPLATE DURING DNA REPLICATION?

Cells are able to achieve chromosome replication in the presence of DNA lesions by coordinating the replication process with cell cycle progression, DNA recombination, and repair. A lack of coordination between these processes inevitably results in the generation of secondary damage, accumulation of mutations, genome instability, and cancer.

A replication fork encountering a bulge in the template may stop and resume replication by re-priming DNA synthesis downstream of the lesion (Fig. 1A). This mechanism would generate highly recombinogenic daughter strand gaps, thus converting primary lesions into intrinsically unstable ssDNA regions, that are promptly channeled into homologous recombination pathways (18). This process can be achieved without coupling replication fork progression to homologous recombination and is generally known as post-replicative repair (PRR), although it is still unclear whether the recombination-mediated filling of the gaps occurs during the S phase or rather is postponed until replication is completed, in the G2 phase of the cell cycle.

When a replication fork runs into a single-strand interruption in the template DNA, it may create a double-strand end and result in a collapsed replication fork (19,20) (Fig. 1B). A collapsed fork may then trigger replication restart through a recombination-mediated step, thus re-establishing a functional replication fork. Hence, in this case, replication restart has to be coupled with recombination and therefore this process is known as replication by recombination (21) or break-induced replication (BIR) (22).

Considering that DNA recombination during the mitotic cell cycle is often associated to genome instability and both PRR and BIR engage the chromosomes in recombination events, these recombination-dependent replication processes, at least in principle, may also contribute to genomic rearrangements.

An alternative way to deal with a damaged template during replication would be to promote a template-switching event at the forks (Fig. 1C). A fork encountering a lesion on the template can transiently stall and uncouple leading and lagging strand synthesis (23). The stalled strand could then be displaced from the template and pair with the other newly synthesized chain (24). This would allow the stalled strand to copy a correct template. Although the mechanism and the genetic requirements promoting template switching are still unknown, this mechanism implies that the replication fork is assisted by specialized enzymatic activities that allow the newly synthesized chains to occasionally switch template.

Replication fork stalling could also prime fork reversal leading to the formation of a X-shaped structure, called chicken-foot, that is structurally related to the four-branched Holliday junction (Fig. 1D) (15,25–28). Fork reversal could therefore allow the re-annealing of the damaged-parental strands and eventually lesion removal

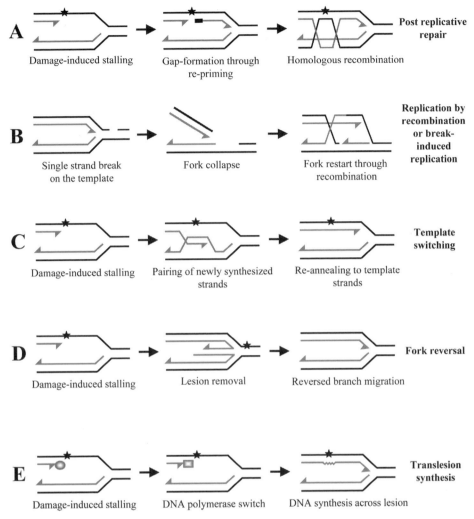

Figure 1 Mechanisms implicated in replication of a damaged template. (A) Upon encountering a lesion, DNA replication can be resumed by repriming DNA synthesis downstream of the damage, thus generating a daughter strand gap that can be channelled into homologous recombination pathways known as PRR. (B) When a replication fork runs into a SSB in the template, it generates a DSB end that may invade the chromosome and reconstitute a functional fork through recombination. (C) A stalled fork can undergo template switching by pairing the newly synthesized chains, thus bypassing the lesion on the template. (D) Upon encountering a lesion, a stalled replication fork may reverse to form a chicken-foot intermediate. Specialized repair processes can then remove the lesion on the template. Alternatively, reversed forks can be resolved by Holliday junction resolution processes (see text for details). (E) Specialized DNA polymerases (gray square) replace normal polymerases (gray circle) and synthesize DNA across the lesions. The DNA lesion is represented by a star; arrows indicate the 3′ end of the newly synthesized strands; black lines designate template DNA and gray lines newly synthesized DNA; the small black bar on newly synthesized strand in panel A represents the RNA primer.

through excision-mediated repair events. Alternatively, reversed forks could be cleaved by Holliday junction resolution enzymes leading to the formation of DSBs that could allow the resumption of DNA replication through a BIR-like mechanism (29,30). It is unclear how reversed forks generate in response to replication pausing and particularly whether recombination pathways are involved in their formation. In any case, fork reversal can engage recombination activities by creating Holliday junction-like structures.

Another way to get past a damaged template is to use specialized DNA polymerases, called translesion or bypass polymerases, that are able to specifically bypass lesions that would block normal DNA polymerases (Fig. 1E) (31). Indeed, a variety of DNA polymerases have been implicated in lesion bypass (32). This replication bypass mechanism implies that the replication machinery switches DNA polymerases when encountering certain type of lesions on the template. Indeed, this process is highly regulated and, in eukaryotes, seems to be assisted by the checkpoint response (33). Notably, certain translesion polymerases synthesize incorrect nucleotides, thus causing accumulation of mutations (31,34).

Considering the variety of options that cells can employ to cope with DNA lesions during S-phase, several mechanistic details remain still elusive. For instance, it is still unclear which replication processes represent physiological options and which, instead, are promoted selectively under pathological situation. Particularly, while reversed forks seem to accumulate in certain genetic backgrounds altered in the integrity of the replisome, any attempt to detect these structures in wild type cells has so far failed (16). Further, it is also possible that the type of DNA damage may influence the choice between the different replication mechanisms. Finally, the possible connections between the replication-related repair processes described above and specialized repair pathways that include nucleotide and base excision repair and mismatch repair need to be elucidated.

DNA damage tolerance during S-phase can be highly influenced by the establishment of sister chromatid cohesion that contributes to keep sister chromatids in close proximity until chromosome segregation takes place (35). It has been recently shown that specialized sister chromatid junctions form during an early step of DNA synthesis, at origins of replication, through mechanisms independent of Rad51 and Rad52-mediated homologous recombination (36,37). These structures migrate chasing the replication forks and resemble hemicatenanes, in which one newly synthesized strand is coiled around the other newly synthesized strand (Fig. 2A) (37). It has been suggested that the joint structures may couple chromosome replication to sister chromatid-mediated recombination and replication processes, thus assisting cells to overcome intra-S DNA damage and replication impediments (37). A tantalizing possibility is that hemicatenanes could directly mediate template switching by engaging newly synthesized strands in pairing (Fig. 2B). This is supported by the finding that, with a similar mechanism, the sister chromatid junctions contribute to the formation of reversed forks in the absence of a functional checkpoint (Fig. 2C) (37).

3. THE S-PHASE CHECKPOINT

The ability of eukaryotic cells to respond to replication interference depends on a surveillance mechanism, known as S-phase checkpoint or DNA replication checkpoint (38,39). This pathway serves two primary purposes: one is to prevent nuclear

(A) Hemicatenanes

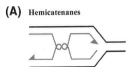

(B) Hemicatenane-mediated template switching

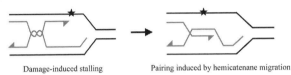

Damage-induced stalling   Pairing induced by hemicatenane migration

(C) Hemicatenane-mediated fork reversal in checkpoint defective cells

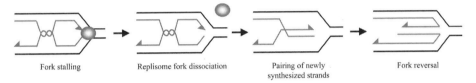

Fork stalling   Replisome fork dissociation   Pairing of newly synthesized strands   Fork reversal

Figure 2 Hemicatenanes might allow specialized replication bypass mechanisms. Specialised sister chromatid junctions resembling hemicatenanes are generated during origin firing (37). (A) Hemicatenanes at newly synthesized strands may form by coiling together the nascent chains. (B) Forks stalled at a damaged site uncouple leading and lagging strand synthesis. Hemicatenanes reach the stalled fork and promote pairing between newly synthesized strands. (C) In checkpoint defective cells, the replisome (gray circle) dissociates from stalled forks. Hemicatenanes run off in the absence of a stable replisome–fork complex causes pairing of newly synthesized strands and fork reversal.

division when DNA replication is perturbed, thus coordinating cell cycle progression with DNA repair capacity; the other one is to modulate the various repair and recombination systems that help cells to survive replication stress. This leads to the idea that the S-phase checkpoint is the principal line of defense against genomic instability.

The S-phase checkpoint can be envisaged as a signal-transduction cascade, where upstream sensors monitor and detect altered DNA molecules, while central transducers act in protein kinase cascades to regulate downstream effectors (40). Its activation occurs during S-phase and can be provoked by deoxyribonucleoside triphosphate (dNTP) depletion or by the stalling of a replication fork at the sites of DNA damage. Unless otherwise stated, we will refer to the *S. cerevisiae* checkpoint factors (see Table 1 for further details).

3.1. The Checkpoint Cascade

Central to this process is a protein kinase family related to phosphoinositide 3-kinase, among which are *S. cerevisiae* Mec1 and Tel1, *S. pombe* Rad3 and mammalian ATM and ATR (41,42) (Table 1). These protein kinases are required to activate the Chk1 and Rad53 kinases, which are responsible for the phosphorylation of downstream targets (38,39). Based on recent observations, it has been suggested that ATM in unperturbed cells is sequestered as a dimer or multimer, with its kinase domain bound

Table 1 Proteins Involved in the S-phase Checkpoint in Yeasts and Mammals

Protein function	S. cerevisiae	S. pombe	Mammals
ATM/ATR-kinases	Mec1	Rad3	ATR
	Tel1	Tel1	ATM
ATR-interacting proteins	Ddc2	Rad26	ATRIP
RFC-like proteins	Rad24	Rad17	Rad17
	Rfc2-5	Rfc2-5	Rfc2-5
PCNA-like proteins	Ddc1	Rad9	Rad9
	Rad17	Rad1	Rad1
	Mec3	Hus1	Hus1
Mediators	Rad9	Crb2	BRCA1
	Mrc1	Mrc1	Claspin
Effector kinases	Rad53	Cds1	Chk2
	Chk1	Chk1	Chk1

to a region surrounding serine 1981 of a neighboring ATM molecule (43). In this configuration, the kinase domain of each molecule is maintained inactive. DNA damage induces rapid intermolecular autophosphorylation of serine 1981, thus causing dimer dissociation and ATM activation (43). This mechanism provides a molecular explanation for the dominant inhibitory properties of catalytically -inactive ATR/ATM, Mec1, and Rad3 kinases (44–46).

One characteristic of ATR-related proteins is their need for an accessory protein: ATR, Rad3, and Mec1 have been shown to stably associate with an ancillary factor, likely involved in the regulation of the kinase. Mec1 physically interacts with the checkpoint protein Ddc2 (also called Lcd1 or Pie1) (47–49), functionally related to Rad26 and ATRIP, which bind Rad3 and ATR in S. pombe and human cells, respectively (50,51). Mec1 and Ddc2 are recruited to sites of DNA damage independently of other checkpoint proteins (52–54) and Mec1-dependent Ddc2 (as well as Rad3-dependent Rad26) phosphorylation does not require other known checkpoint factors, suggesting a pivotal role for these complexes in sensing DNA alterations (47,50). Similarly, ATR co-localizes with ATRIP in nuclear foci in response to DNA damage, indicating that the ATR–ATRIP complex may also be directly recruited at the site of DNA damage (55).

Although the Mec1–Ddc2 complex plays a key role in the S-phase checkpoint, a complete Mec1-dependent activation requires other factors, like Rad24 and the Ddc1-Rad17-Mec3 complex (56–60). This complex (referred to hereafter as PCNA-like) is structurally related to the proliferating cell nuclear antigen (PCNA) "sliding clamp" (61–63). Rad24 has instead homology with the Rfc1 large subunit of replication factor C (RFC) and forms an RFC-like complex with the four small RFC subunits, Rfc2-5 (64). Analogously to the RFC-mediated loading of PCNA during DNA replication, it has been shown that the PCNA-like complex associates with the sites of DNA damage and its recruitment depends upon the RFC-like complex but is independent of Mec1/ATR (52,53,55,65).

Once DNA perturbations are sensed, the ATM/Mec1 and ATR/Tel1 kinases propagate the signals by phosphorylating the protein kinases Chk1 and Rad53 (66,67). ATR- and ATM-dependent activation of Rad53 depends on the presence of DNA damage-specific or S-phase-specific mediators. In particular, Mrc1 mediates the response to replication blocks after treatment with the DNA synthesis inhibitor hydroxyurea (HU) and it has been found associated with moving replication forks (68–70), while Rad9 is required to activate Rad53 in response to DNA damage (71,72).

Full activation of the DNA damage checkpoint requires independent recruitment of the Mec1–Ddc2 and Rad24–RFC complexes at the sites of damage. A likely possibility is that Rad24–RFC loads the PCNA-like complex next to the Mec1–Ddc2 kinase, thus stimulating activation of the Mec1–Ddc2 complex. The independent assembly of Mec1–Ddc2 and the PCNA-like proteins at damaged DNA suggests a fail-safe mechanism for checkpoint activation (73). However, the relative contribution of the different players may change depending on the type of DNA lesion and/or the cell cycle phases. In fact, although the RFC- and PCNA-like complexes are required for checkpoint activation throughout the cell cycle, their need during S-phase (at least in *S.cerevisiae*) is highly reduced compared to G1 and G2 (74,75). Moreover, they appear dispensable for checkpoint-mediated S-phase arrest of nucleotide excision repair (NER)-defective yeast cells (76).

Although the precise roles of the RFC- and PCNA-like complexes within the checkpoint cascade have not been elucidated yet, these factors could mediate the recruitment of specialized repair activities, such as translesion DNA polymerases (33). In any case, DNA damage recognition and the activity of RFC- and PCNA-like complexes have to be coordinated with cell cycle progression, DNA replication, and repair.

3.2. Sensing DNA Damage During DNA Replication

Several types of replication blocks are able to elicit a checkpoint response during S-phase. For example, inhibition of DNA synthesis by dNTP depletion and intra-S DNA damage caused by defective replication proteins or a variety of genotoxic agents are able to cause activation of the checkpoint cascade. This has lead to the hypothesis of the existence of a common structure able to trigger the checkpoint response. Consistent with this hypothesis, several observations suggest that ssDNA, coated by the single-stranded DNA binding protein Replication Protein A (RPA), may represent the signal that triggers a checkpoint-mediated cell cycle arrest (77–84) (Fig. 3). In humans, RPA is required for the recruitment of ATR to sites of DNA damage and stimulates the binding of ATRIP to ssDNA, thus indicating that RPA–ssDNA nucleoprotein filaments are critical for checkpoint activation (85). Further, it has been shown that Ddc2 recruitment in response to DSB formation depends on RPA (85) and that RPA is required for intra-S checkpoint activation following treatment with DNA topoisomerase II inhibitors (86). Although eukaryotes have developed more complex mechanisms to respond to DNA damage, the use of ssDNA as a DNA damage signal is conserved also in bacteria, where RecA-coated ssDNA generated by DNA damage processing is a signal for the SOS response (87).

By studying the mechanisms leading to checkpoint activation in response to DSB formation in yeast, it has been shown that a threshold of ≈10 kb of ssDNA is required to promote activation of the Mec1/Rad53 pathway (88). Further, recent results indicate that yeast stalled replication forks accumulate short gaps of extra ≈100 nucleotides compared to normal forks likely because one newly synthesized strand is preferentially elongated compared to the other, leading to the asymmetric accumulation of ssDNA (16). ssDNA accumulation at stalled forks is responsible for the generation of RPA nucleofilaments that mediate the recruitment of checkpoint

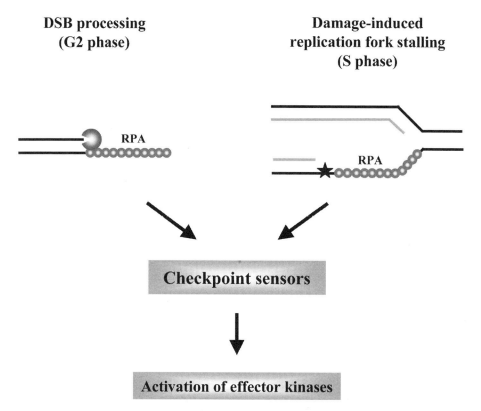

Figure 3 Signals activating the DNA damage checkpoint in response to double strand breaks or replication fork stalling. (A) DNA ends at a DSB are resected generating 3′ ssDNA overhangs. (B) Replication of a damaged template causes replication forks to stall and accumulate ssDNA regions by transiently uncoupling leading and lagging strand synthesis. In both A and B, ssDNA coated by RPA recruits the checkpoint sensors that result in the phosphorylation and activation of the effector kinases Rad53/Chk2.

factors (Fig. 3) (84,85). Thus, each fired origin in the presence of HU accumulates ≈200 nucleotides of additional ssDNA coated by RPA (16). This observation, together with the finding that a specific threshold is required for checkpoint activation, suggests that a critical number of origins (i.e., 40–60) would have to be fired to trigger a robust checkpoint activation in response to replication blocks. A logic expectation of this hypothesis is that, under unperturbed conditions, the total amount of ssDNA is kept below the threshold by the temporal control of origin firing that allows only a limited number of forks to be present at any time (89). This mechanism would therefore prevent inappropriate checkpoint activation during normal S-phase. This hypothesis is also consistent with the findings that the S-phase checkpoint is activated only when a sufficient number of replication forks is altered (90,91) and that any mutants partially defective in initiation of DNA replication due to the establishment of fewer replication forks results in the inability to efficiently activate the checkpoint following replication stress. This does not imply direct roles for the corresponding gene products in checkpoint activation, but is rather due to defects in establishing sufficient numbers of replication forks (90,91).

        ssDNA intermediates may result also by the processing of primary damage by the repair pathways. In both yeast and *E. coli*, UV-induced DNA damage and DSBs must be first processed by the DNA repair machinery prior to generate enough checkpoint signal. The presence of UV-induced DNA lesions is not sufficient to cause SOS induction in *E. coli*. Rather, the SOS inducing signal arises when cells attempt to replicate the damaged DNA, leading to the accumulation of single-stranded regions (87,92,93). In the absence of replication, the SOS response requires an intact NER pathway, since the SOS response does not take place in UV-irradiated *uvrB dnaC* double mutants (87). The situation seems to be conserved also in *Saccharomyces cerevisiae*, in which, upon UV irradiation, irreparable primary DNA lesions are not capable to activate the checkpoint outside the S-phase in NER-defective cells (76,94). This implies that active processing of UV-induced damage by DNA repair proteins is necessary to attract checkpoint proteins to the sites of damage (76,95).

In contrast to UV, the alkylating agent methyl methane sulfonate (MMS) activate Rad53 preferentially during S-phase in wild-type yeast cells, and this requires the establishment of replication (91). This implies that no efficient system is able to detect alkylated DNA and promotes checkpoint activation outside of S-phase. Similarly, the assembly of DNA replication forks is crucial for generating ssDNA in response to treatment with DNA damaging agents in *E. coli* (87,92).

In addition to the existence of an S-phase threshold for checkpoint activation, the level of Mec1 activity required to slow down replication of damaged DNA may be higher than that required to block the G2/M transition. Consistent with this hypothesis, while overproduction of Mec1 kinase-deficient variants causes dominant-negative S-phase checkpoint defects, it does not abrogate the G2/M checkpoint (44). Moreover, two hypomorphic *mec1* mutants, that were completely defective in the S-phase DNA damage checkpoint, were able to activate the G2/M checkpoint (44).

3.3. Checkpoint-Mediated Control of Replication

The S-phase checkpoint counteracts DNA replication initiation from late origins, stabilizes stalled forks by maintaining them in a replication-competent state and prevents entry into mitosis (38,39).

Chronic DNA damage during S-phase is able to slow down the rate of DNA replication even in the absence of Rad53 and Mec1; this is likely due to the presence of physical impediments represented by the template lesions (96–98). The accelerated S-phase progression that is observed when checkpoint mutants are replicating a damaged DNA is the result of inappropriate initiation events caused by unscheduled firing of late origins (96,99,100). The checkpoint-dependent suppression of late origin firing may have the dual benefit to provide a longer window of time to repair the damaged DNA and the opportunity to fire new replicons to facilitate the resumption of S-phase when the checkpoint is shut off during recovery (101).

The S-phase checkpoint plays an active role also in protecting stalled forks when the replication machinary hits a damaged template. In fact, *mec1* and *rad53* mutants fail to recover from replication blocks induced by HU treatment (102,103) and to complete DNA synthesis in the presence of MMS (96). Stalled replication forks, in the absence of a functional checkpoint, rapidly degenerate accumulating a variety of pathological structures that include long ssDNA gaps, hemireplicated DNA regions, and reversed forks (16); these abnormal structures prevent replication forks to resume replication, thus explaining the inability of checkpoint mutants to recover from replication blocks or intra-S damage. The replication fork processing observed in checkpoint mutants likely arises from the inability to maintain DNA polymerases stably associated to stalled forks (84,104), from unscheduled nucleolytic erosion of newly synthesized chains (84; C.Cotta-Ramusino and M.Foiani unpublished) and from Holliday junction resolution processes that actively engage reversed forks (103). A likely possibility is that the checkpoint controls the stability of replisome-fork complexes by modulating the phosphorylation state of key replication proteins and/or by modulating the availability of dNTPs that could influence the productive association of polymerase to the substrate (38,39). Indeed, the phosphorylation state of DNA polymerase alpha-primase complex and RPA is regulated by the S-phase checkpoint (75,105) and specific DNA primase and RPA alleles mimic the defects of checkpoint mutants (84,85,106,107).

Although it is well established that the replisome is targeted by the checkpoint pathway, based on the characterization of mutants in certain DNA replication/ recombination proteins, several fork-associated factors have been implicated in promoting activation of the checkpoint response. These include DNA polymerase epsilon (74), Tof1 (topoisomerase 1-associated factor 1) (108), Mrc1 (68–70,109), Rfc5 (110), and the two helicases Sgs1 and Srs2 (111,112). While in certain cases the role of these proteins in facilitating activation of the checkpoint is somehow expected (i.e., in the case of Mrc1), it is still unclear whether the lack of checkpoint activation observed in certain replication mutants results from the inability to generate RPA–ssDNA nucleofilaments at stalled forks, or to establish functional forks (that, consequently, would limit the amount of ssDNA gaps), or from unscheduled recombination pathways that engage the ssDNA into Rad51 filaments. In fact, since Rad51 and RPA compete for the same substrate (i.e., ssDNA), it is possible that the engagement of ssDNA at the forks into recombination intermediates might prevent the formation of extensive RPA filaments, therefore limiting the amount of checkpoint signal. Indeed, this could account for the mild checkpoint defects observed in the absence of functional Sgs1 and Srs2 helicases and is consistent with the observations that both proteins have been implicated in counteracting the accumulation of Rad51-dependent recombination structures at damaged replication forks (Liberi and Foiani, unpublished). Further this is also consistent with the finding that *rad51* mutants irreversibly accumulate checkpoint signals following DSB formation (113).

The S-phase checkpoint, besides allowing the resumption of DNA replication once the forks have stalled at damaged sites, has been implicated in mediating the recruitment of translesion DNA polymerases for mutagenic synthesis. Indeed, certain components of the S-phase checkpoint physically interact with the translesion polymerase DinB and are necessary for its loading onto chromatin (33).

It should be pointed out that even unperturbed S-phase could occasionally challenge the checkpoint response while the replication machinery engages chromosomal regions that intrinsically cause replication pausing due to interference with transcription, to chromatin context or to the presence of repetitive sequences (17,114,115). Further, forks experience a physiological collapse whenever they reach telomeres, the natural ends of eukaryotic chromosomes. This implies that replication at telomeres has to be tightly regulated in order to avoid checkpoint activation. Indeed, it has been shown that a checkpoint response in the absence of exogenous DNA damage can be elicited by unregulated telomere replication, that may uncouple telomerase-mediated extension of the TG_{1-3} strand from the synthesis of the complementary $C_{1-3}A$ strand (116). The long unprotected 3' ssDNA overhangs that may be produced by uncontrolled telomere replication may be detected as DNA damage and trigger a checkpoint response.

4. REPLICATION-RELATED GENOME INSTABILITY

It is now clear that DNA structure checkpoints are integrated into a larger DNA damage response pathway. In this view, the increased sensitivity to genotoxic agents of cells impaired in checkpoint functions is not simply caused by the failure to delay cell cycle transition but also results from the inability to mediate the efficient repair of DNA lesions. In particular, since all genotoxic agents impose stress on moving replication forks, it is now clear that cell lethality exhibited by checkpoint mutants in response to intra-S DNA damage is due to the inability to properly carry out DNA replication on a damaged template.

In budding yeast, genetic defects in checkpoint proteins acting during DNA replication significantly increases the rate of genome rearrangements, while mutations in genes specifically required for the G1 and G2 DNA damage checkpoints have little or no effects (117–119). Thus, the S-phase checkpoint seems to play a critical role in preserving genome integrity, likely by regulating cell cycle progression in response to replication errors, by modulating DNA repair functions and by stabilizing stalled replication fork. Further, since deregulation of replication origins results in increased genome instability (118–121), checkpoint-mediated regulation of origin firing may also contribute to assure genetic integrity. Notably the Mec1-mediated controls of fork stability and late origin firing can be genetically uncoupled (44,91).

The collapse of replication forks that, in checkpoint mutants, results in chromosomes breakage, does not arise at the sites of stochastic fork collapse. Instead, breaks occur in preferred locations in the genome as a consequence of replication pausing (17). Chromosome breakage in checkpoint mutants closely resembles breakage at mammalian fragile sites, which are triggered by delayed progression through normally late-replicating regions and whose stability is controlled by ATR kinase (5,122).

These events may have important implications for the enhanced genome instability observed in those human genetic syndromes caused by checkpoint defects and in cancer cells exhibiting checkpoint abnormalities (41). Due to the tight correlation between checkpoint defects, chromosomal aberrations and tumor development, future investigations aimed at understanding of the molecular mechanisms controlling the integrity of replicating chromosomes may certainly contribute new insights into the events leading to carcinogenesis.

ACKNOWLEDGMENTS

We thank G.Liberi, A.Pellicioli, G.Lucchini, and all members of M.P.L. and M.F. laboratories for helpful discussions. Work in M.F. and M.P.L. laboratories is supported by Associazione Italiana per la Ricerca sul Cancro, European Union, Telethon-Italy and Ministero della Salute.

REFERENCES

1. Echols H, Goodman MF. Annu Rev Biochem 1991; 60:477–511.
2. Kornberg A, Baker TA. DNA Replication. 2nd ed. New York: W.H. Freeman & Co., 1992.
3. Strand M, Prolla TA, Liskay RM, Petes TD. Nature 1993; 365:274–276.

4. Djian P. Cell 1998; 94:155–160.
5. Cimprich KA. Curr Biol 2003; 13:231–233.
6. Zakian VA. Annu Rev Genet 1996; 30:141–172.
7. Chakhparonian M, Wellinger RJ. Trends Gen 2003; 19:439–446.
8. Rupp WD, Howard-Flanders P. J Mol Biol 1968; 31:291–304.
9. Rupp WD, Wilde III CE, Reno DL, Howard-Flanders P. J Mol Biol 1971; 61:25–44.
10. Wolff S, Bodycote J, Painter RB. Mutat Res 1974; 25:73–81.
11. Wolff S, Rodin B, Cleaver JE. Nature 1977; 265:347–349.
12. Sarasin AR, Hanawalt PC. J Mol Biol 1980; 138:299–319.
13. Michel B, Ehrlich SD, Uzest M. EMBO J 1997; 16:430–438.
14. Rothstein R, Michel B, Gangloff S. Genes Dev 2000; 14:1–10.
15. Kuzminov A. Proc Natl Acad Sci USA 2001; 98:8461–8468.
16. Sogo JM, Lopes M, Foiani M. Science 2002; 297:599–602.
17. Cha RS, Kleckner N. Science 2002; 297:602–606.
18. Smirnova M, Klein HL. Mutat Res 2003; 532:117–135.
19. Kuzminov A. Mol Microbiol 1995; 16:373–384.
20. Kuzminov A. Microbiol Mol Biol Rev 1999; 63:751–813.
21. Kogoma T. Cell 1996; 85:625–627.
22. Kraus E, Leung W-Y, Haber JE. Proc Natl Acad Sci USA 2001; 98:8255–8262.
23. Pages V, Fuchs RP. Science 2003; 300:1300–1303.
24. Higgins NP, Kato K, Strauss B. J Mol Biol 1976; 101:417–425.
25. McGlynn P, Lloyd RG. Nat Rev Mol Cell Biol 2002; 3:859–870.
26. Cox MM. Ann Rev Genet 2001; 35:53–82.
27. Inman RB. Biochim Biophys Acta 1984; 783:205–215.
28. Gruss A, Michel B. Curr Opin Microbiol 2001; 4:595–601.
29. Helleday T. Mutat Res 2003; 532:103–115.
30. McGowan CH. Mutat Res 2003; 532:75–84.
31. Friedberg EC, Wagner R, Radman M. Science 2002; 296:1627–1630.
32. Kai MK, Wang TS-F. Mutat Res 2003; 532:59–73.
33. Kai MK, Wang TS-F. Genes Dev 2003; 17:64–76.
34. Goodman MF. Annu Rev Biochem 2002; 71:17–50.
35. Sjogren C, Nasmyth K. Curr Biol 2001; 11:991–995.
36. Benard M, Maric C, Pierron G. Mol Cell 2001; 7:971–980.
37. Lopes M, Cotta-Ramusino C, Liberi G, Foiani M. Mol Cell 2003; 12:1499–1510.
38. Longhese MP, Clerici M, Lucchini G. Mutat Res 2003; 532:41–58.
39. Muzi Falconi M, Liberi G, Lucca C, Foiani M. Cell Cycle 2003; 2:564–567.
40. Nyberg KA, Michelson RJ, Putnam CW, Weinert TA. Annu Rev Genet 2002; 36: 617–656.
41. Shiloh Y. Nat Rev Cancer 2003; 3:155–168.
42. Abraham RT. Genes Dev 2001; 15:2177–2196.
43. Bakkenist CJ, Kastan MB. Nature 2003; 421:499–506.
44. Paciotti V, Clerici M, Scotti M, Lucchini G, Longhese MP. Mol Cell Biol 2001; 21:3913–3925.
45. Bentley NJ, Holtzman DA, Flaggs G, Keegan KS, DeMaggio A, Ford JC, Hoekstra M, Carr AM. EMBO J 1996; 15:6641–6651.
46. Cliby WA, Roberts CJ, Cimprich KA, Stringer CM, Lamb JR, Schreiber SL, Friend SH. EMBO J 1998; 17:159–169.
47. Paciotti V, Clerici M, Lucchini G, Longhese MP. Genes Dev 2000; 14:2046–2059.
48. Rouse J, Jackson SP. EMBO J 2000; 19:5801–5812.
49. Wakayama T, Kondo T, Ando S, Matsumoto K, Sugimoto K. Mol Cell Biol 2001; 21:755–764.
50. Edwards RJ, Bentley NJ, Carr AM. Nat Cell Biol 1999; 1:393–398.
51. Cortez D, Guntuku S, Qin J, Elledge SJ. Science 2001; 294:1713–1716.
52. Kondo T, Wakayama T, Naiki T, Matsumoto K, Sugimoto K. Science 2001; 294:867–870.

53. Melo JA, Cohen J, Toczyski DP. Genes Dev 2001; 15:2809–2821.
54. Rouse J, Jackson SP. Mol Cell 2002; 9:857–869.
55. Zou L, Cortez D, Elledge SJ. Genes Dev 2002; 16:198–208.
56. Weinert TA, Kiser GL, Hartwell LH. Genes Dev 1994; 8:652–665.
57. Longhese MP, Fraschini R, Plevani P, Lucchini G. Mol Cell Biol 1996; 16:3235–3244.
58. Longhese MP, Paciotti V, Fraschini R, Zaccarini R, Plevani P, Lucchini G. EMBO J 1997; 16:5216–5226.
59. Paciotti V, Lucchini G, Plevani P, Longhese MP. EMBO J 1998; 17:4199–4209.
60. Longhese MP, Foiani M, Muzi-Falconi M, Lucchini G, Plevani P. EMBO J 1998; 17:5525–5528.
61. Thelen MP, Venclovas C, Fidelis K. Cell 1999; 96:769–770.
62. Majka J, Burgers PM. Proc Natl Acad Sci USA 2003; 100:2249–2254.
63. Caspari T, Dahlen M, Kanter-Smole G, Lindsay HD, Hofmann K, Papadimitriou K, Sunnerhagen P, Carr AM. Mol Cell Biol 2000; 20:1254–1262.
64. Green CM, Erdjument-Bromage H, Tempst P, Lowndes NF. Curr Biol 2000; 10:39–42.
65. Burtelow MA, Kaufmann SH, Karnitz LM. J Biol Chem 2000; 275:26343–26348.
66. Sanchez Y, Desany BA, Jones WJ, Liu Q, Wang B, Elledge SJ. Science 1996; 271:357–360.
67. Sanchez Y, Bachant J, Wang H, Hu FH, Liu D, Tetzlaff M, Elledge SJ. Science 1999; 286:1166–1171.
68. Alcasabas AA, Osborn AJ, Bachant J, Hu F, Werler PJ, Bousset K, Furuya K, Diffley JF, Carr AM, Elledge SJ. Nat Cell Biol 2001; 3:958–965.
69. Tanaka K, Russell P. Nat Cell Biol 2001; 3:966–972.
70. Katou Y, Bando M, Kanoh Y, Noguchi H, Tanaka H, Ashikari T, Sugimoto K, Shirahige K. Nature 2003; 424:1078–1083.
71. Gilbert CS, Green CM, Lowndes NF. Mol Cell 2001; 8:129–136.
72. Schwartz MF, Duong JK, Sun Z, Morrow JS, Pradhan D, Stern DF. Mol Cell 2002; 9:1055–1065.
73. Caspari T, Carr AM. Curr Biol 2002; 12:105–107.
74. Navas TA, Sanchez Y, Elledge SJ. Genes Dev 1996; 10:2632–2643.
75. Pellicioli A, Lucca C, Liberi G, Marini F, Lopes M, Plevani P, Romano A, Di PP Fiore, Foiani M. EMBO J 1999; 18:6561–6572.
76. Neecke H, Lucchini G, Longhese MP. EMBO J 1999; 18:4485–4497.
77. Witkin EM. Biochimie 1991; 73:133–141.
78. Kornbluth S, Smythe C, Newport JW. Mol Cell Biol 1992; 12:3216–3223.
79. Huang LC, Clarkin KC, Wahl GM. Proc Natl Acad Sci USA 1996; 93:4827–4832.
80. Garvik B, Carson M, Hartwell L. Mol Cell Biol 1995; 15:6128–6138.
81. Lydall D, Weinert T. Science 1995; 270:1488–1491.
82. Maringele L, Lydall D. Genes Dev 2002; 16:1919–1933.
83. Lee SE, Moore JK, Holmes A, Umezu K, Kolodner RD, Haber JE. Cell 1998; 94:399–409.
84. Lucca C, Vanoli F, Cotta-Ramusino C, Pellicioli A, Liberi G, Haber J, Foiani M. Oncogene 2004; 23:1206–1213.
85. Zou L, Elledge SJ. Science 2003; 300:1542–1548.
86. Costanzo V, Shechter D, Lupardus PJ, Cimprich KA, Gottesman M, Gautier J. Mol Cell 2003; 11:203–213.
87. Salles B, Defais M. Mutat Res 1984; 131:53–59.
88. Vaze MB, Pellicioli A, Lee SE, Ira G, Liberi G, Arbel-Eden A, Foiani M, Haber JE. Mol Cell 2002; 10:373–385.
89. Raghuraman MK, Winzeler EA, Collingwood D, Hunt S, Wodicka L, Conway A, Lockhart DJ, Davis RW, Brewer BJ, Fangman WL. Science 2001; 294:115–121.
90. Shimada K, Pasero P, Gasser SM. Genes Dev 2002; 16:3236–3252.
91. Tercero JA, Longhese MP, Diffley JFX. Mol Cell 2003; 11:1323–1336.
92. Sassanfar M, Roberts JW. J Mol Biol 1990; 212:79–96.

93. Sutton MD, Smith BT, Godoy VG, Walker GC. Annu Rev Genet 2000; 34:479–497.
94. Siede W, Friedberg AS, Dianova I, Friedberg EC. Genetics 1994; 138:271–281.
95. Giannattasio M, Lazzaro V, Longhese MP, Plevani P, Muzi-Falconi M. EMBO J 2004; 23:429–438.
96. Tercero JA, Diffley JFX. Nature 2001; 412:553–557.
97. Paulovich AG, Hartwell LH. Cell 1995; 82:841–847.
98. Paulovich AG, Margulies RU, Garvik BM, Hartwell LH. Genetics 1997; 145:45–62.
99. Santocanale C, Diffley JFX. Nature 1998; 395:615–618.
100. Shirahige K, Hori Y, Shiraishi K, Yamashita M, Takahashi K, Obuse C, Tsurimoto T, Yoshikawa H. Nature 1998; 395:618–621.
101. Pasero P, Shimada K, Duncker BP. Cell Cycle 2003; 2:568–572.
102. Desany BA, Alcasabas AA, Bachant JB, Elledge SJ. Genes Dev 1998; 12:2956–2970.
103. Lopes M, Cotta-Ramusino C, Pellicioli A, Liberi G, Plevani P, Muzi-Falconi M, Newlon C, Foiani M. Nature 2001; 412:557–561.
104. Cobb JA, Bjergbaek L, Shimada K, Frei C, Gasser SM. EMBO J 2003; 22:4325–4336.
105. Brush GS, Morrow DM, Hieter P, Kelly TJ. Proc Natl Acad Sci USA 1996; 93: 15075–15080.
106. Marini F, Pellicioli A, Paciotti V, Lucchini G, Plevani P, Stern DF, Foiani M. EMBO J 1997; 16:639–650.
107. Longhese MP, Neecke H, Paciotti V, Lucchini G, Plevani P. Nucleic Acids Res 1996; 24:3533–3537.
108. Foss EJ. Genetics 2001; 157:567–577.
109. Osborn AJ, Elledge SJ. Genes Dev 2003; 17:1755–1767.
110. Sugimoto K, Ando S, Shimomura T, Matsumoto K. Mol Cell Biol 1997; 17:5905–5914.
111. Frei C, Gasser SM. Genes Dev 2000; 14:81–96.
112. Liberi G, Chiolo I, Pellicioli A, Lopes M, Plevani P, Muzi-Falconi M, Foiani M. EMBO J 2000; 19:5027–5038.
113. Lee SE, Pellicioli A, Vaze MB, Sugawara N, Malkova A, Foiani M, Haber JE. Mol Cell Biol 2003; 23:8913–8923.
114. Deshpande AM, Newlon CS. Science 1996; 272:1030–1033.
115. Ivessa AS, Lenzmeier BA, Bessler JB, Goudsouzian LK, Schnakenberg SL, Zakian VA. Mol Cell 2003; 12:1525–1536.
116. Viscardi V, Baroni E, Romano M, Lucchini G, Longhese MP. Mol Biol Cell 2003; 14:3126–3143.
117. Myung K, Datta A, Kolodner RD. Cell 2001; 104:397–408.
118. Kolodner RD, Putnam CD, Myung K. Science 2002; 297:552–557.
119. Huang D, Koshland D. Genes Dev 2003; 17:1741–1754.
120. Tanaka S, Diffley JF. Genes Dev 2002; 16:2639–2649.
121. Lengronne A, Schwob E. Mol Cell 2002; 5:1067–1078.
122. Camper AM, Nghiem P, Arlt MF, Glover TW. Cell 2002; 111:779–789.

Index